An Introduction to Probability, Decision, and Inference

INTERNATIONAL SERIES IN DECISION PROCESSES

INGRAM OLKIN, Consulting Editor

A Basic Course in Statistics, 2d ed., T. R. Anderson and M. Zelditch, Jr.
Introduction to Statistics, R. A. Hultquist
Applied Probability, W. A. Thompson, Jr.
Elementary Statistical Methods, 3d ed., H. M. Walker and J. Lev
Reliability Handbook, B. A. Kozlov and I. A. Ushakov (edited by J. T. Rosenblatt and L. H. Koopmans)
Fundamental Research Statistics for the Behavioral Sciences, J. T. Roscoe
Statistics: Probability, Inference, and Decision, Volumes I and II, W. L. Hays and R. L. Winkler
An Introduction to Probability, Decision, and Inference, I. H. LaValle

FORTHCOMING TITLES

Elementary Probability, C. Derman, L. Gleser, and I. Olkin
Probability Theory, Y. S. Chou and H. Teicher
Statistical Inference, 2d ed., H. M. Walker and J. Lev
Statistics for Psychologists, 2d ed., W. L. Hays
Decision Theory for Business, D. Feldman and E. Seiden
Times Series Analysis, D. R. Brillinger
Statistics Handbook, C. Derman, L. Gleser, G. H. Golub, G. J. Lieberman, I. Olkin, A. Madansky, and M. Sobel
An Introduction to Statistics and Probability for Engineers, C. Derman and P. Kolesar
Statistics for Scientists, M. Zelen
Mathematics Models in Experimental Design, M. Zelen
Time Series, D. R. Brillinger
Applied Multivariate Analysis Including Bayesian Techniques, S. J. Press
Bayesian Statistics, R. Winkler
Elements of Probability and Statistics, S. Lippman
Multivariate Linear Model, J. Finn
Statistics and Computer Programming, I. Olkin and others
A First Course in the Methods of Multivariate Analysis, C. Y. Kramer

An Introduction to Probability, Decision, and Inference

Irving H. LaValle
Tulane University

HOLT, RINEHART AND WINSTON, INC.
New York • Chicago • San Francisco • Atlanta • Dallas
Montreal • Toronto • London • Sydney

Library of Congress Catalog Card Number: 77-111500
AMS 1968 Subject Classifications 6201, 6210, 6230, 6235, 9030

SBN: 03–078385–2

Printed in the United States of America
0 1 2 3   22   1 2 3 4 5 6 7 8 9

To Arnold Wood

PREFACE

This book is intended as a text for a year's introductory course in probability and statistics with a subjectivist Bayesian orientation. It evolved from my lecture notes for first-year students in the Management Science Option of the Tulane M.B.A. program. However, it is suitable for mathematically qualified upper-level undergraduate and beginning graduate students in statistics, economics, engineering, and certain mathematics curricula as well.

It is my experience that many students have some acquaintance and mechanical facility with introductory calculus and linear algebra but lack mathematical sophistication. Recognizing this, I have tailored the exposition to fit the level of difficulty of the concepts; hence, this book is somewhat nonuniform in style.

The specific prerequisites for a successful reading of this book are (1) an introductory course in the differential and integral calculus, covering integration and differentiation of polynomials and exponential functions, multiple integrals, partial derivatives, and change-of-variable theorems; (2) an introductory course in (applied) linear algebra, covering the elementary matrix and vector operations, determinants, and matrix inversion; and (3) tolerance for notational complexity. A simple and universally applicable notation for probability, decision, and inference has yet to be developed and still seems remote.

The mild calculus and linear algebra prerequisite enables one to discuss a number of topics normally restricted to advanced treatises if in book form at all; for example, prior-posterior theory for a k-variate Normal process with covariance matrix given the natural-conjugate inverted Wishart distribution. I avoid a real confrontation with advanced calculus by not proving that certain density functions of random vectors are normed. Instructors who wish to fill this gap can readily do so by standard analytic methods.

This book affords considerable latitude as to the order in which topics can be presented. The more commonly used set-theoretic material in Part I can be summarized briefly by instructors who do not wish to dwell on it for very long. Part IV on inference can be covered before Part III on Bayesian decision theory at little cost in clarity.

Part II, on basic probability for decision theory and inference, is fairly standard in content except for Chapters 9 and 14. The introduction to statistical inference in Chapter 9 includes material such as likelihood functions and sufficiency, which permits a smoother subsequent presentation of important processes. Chapter 14, on natural-conjugate analysis, develops much of the known, applied prior-posterior theory except that which is specific to regression and the analysis of variance. Here, as elsewhere, both univariate and multivariate Normal processes are discussed separately, so as to permit passage from the simpler to the more complex.

Some will no doubt be surprised at the omission of a chapter or chapters on linear models and at the inclusion of only an exercise in Chapter 12 on least-squares linear regression. My decision not to include more on linear models is based on the fact that a responsible and complete introduction is impossible within the confines of a year's course which covers the topics included in this book. Such an introduction to linear models can be acquired subsequently by consulting Fisher [**28**], Graybill [**32**], Scheffé [**71**], and Volume III of Kendall and Stuart [**41**] for orthodox (that is, non-Bayesian) treatments, and also Chapter 12 of Raiffa and Schlaifer [**66**], Chapter 24 of Pratt, Raiffa, and Schlaifer [**64**], Box and Draper [**9**], Box and Tiao [**10**]–[**12**], Hill [**37**]–[**38**], Tiao [**74**], Tiao and Box [**75**], Tiao and Tan [**76**]–[**77**], Tiao and Zellner [**78**]–[**79**], Zellner and Chetty [**86**], and Zellner and Tiao [**87**] for Bayesian treatments.

Part III covers a number of basic topics in Bayesian decision theory and begins with two chapters which develop utility and subjective probability along the Luce and Raiffa [**52**] and Anscombe and Aumann [**3**] lines. Instructors wishing to get on with applied statistical decision theory can virtually omit Chapters 15 and 16, although I personally think that these foundational topics are very important.

Part IV, on statistical inference, addresses the standard problems in both orthodox and Bayesian ways, with some comparisons and interrelations. I stress the asymptotic agreement between consistent Bayesian and consistent orthodox procedures but refrain from discussing the principle of precise measurement [**70**] or of stable estimation [**16**], since the reasoning about locally uniform priors seems hard for beginning students to grasp. Furthermore, I have taken some pains in Chapter 22 to illustrate the basic ideas of hypothesis testing, another subject which Bayesianly indoctrinated students find difficult.

Responsibility for shortcomings remains mine in all cases, but many people have contributed directly and indirectly to this undertaking. I have received the kind permission of Dr. Joseph Berkson and the Editor of the *Journal of the American Statistical Association* to reproduce the quotation from [**5**] in Chapter 22; the Editor of *Biometrika*, to publish Appendix Tables II and V; and McGraw-Hill, Inc., to publish Appendix Table III. I am indebted to the Literary Executor of the late Sir Ronald A. Fisher, F.R.S., to Dr. Frank Yates, F.R.S., and to Oliver & Boyd, Ltd., Edinburgh, for permission to reprint Table IV from their book *Statistical Tables for Biological, Agricultural and Medical Research*. The sixth edition of this work appeared in 1963.

Professor Robert M. Thrall read a preliminary draft and offered a number of valuable suggestions for its improvement. I am grateful also for the advice of Professors Joseph L. Balintfy and Martin Krakowski regarding several aspects of this work and for the helpful comments of students concerning the originally distributed lecture notes.

My wife has been a constant source of encouragement as well as an epitome of tolerance. Her intuitive appreciation of the difficulties of authorship is at times beyond belief.

The positive contribution of my secretary, Alphecca (Mrs. Abdulgadu) Muttardy, extended far beyond her transduction of my hieroglyphics into virtually error-free typing of the final draft. She also drew most of the figures in the manuscript, correctly questioned many erroneously numbered equations, sections, subscripts, and so on, and generally functioned both as secretary and research assistant.

Finally, it is a pleasure to acknowledge the interest and cooperation of Holt, Rinehart and Winston, Inc., in making this work a reality.

New Orleans, Louisiana — Irving H. LaValle

May 1970

CONTENTS

part I

Set Theory

1

ELEMENTARY SET THEORY

1.1 INTRODUCTION

This chapter concerns a mathematical abstraction of the notion of "grouping things together." It explores some properties of, and relationships among, the resulting aggregates. Some of the terms defined below will probably be familiar to you even though you may never have studied "set theory," since they are an integral part of the so-called "new mathematics." This is no reason to be concerned, however, as the new mathematics is neither very new (only newly taught at the precollege level) nor very recondite (if it were it could not be taught at that level).

1.2 ELEMENT, SET, AND BELONGING

We assume that we have in mind a class of distinguishable objects, which we shall not attempt to characterize further. Each such object is called an *element*, and an aggregate of elements is called a *set*.

If these definitions seem broader than what you think is or should be characteristic of mathematics, it is because the concepts are fundamental and of much greater generality than numerical concepts. There is a price to be paid for such generality: as one strives for greater generality, he must be content with shallower results. (In fact, as one achieves more and more generality, he says less and less about more and more until he ends up by saying nothing about everything. This is the converse of the old canard about specialists learning more and more about less and less until they end up by knowing everything about nothing.)

Examples

(1) Let x be an integer. x is an *element* of the *set* R^1 of all real numbers. x is also an element of the set of all integers. (2) Let $(x_1, x_2, \ldots, x_n)$ be an n-tuple of real numbers; that is, a list of n real numbers. Then $(x_1, x_2, \ldots, x_n)$ is an *element* of the set R^n of *all* n-tuples of real numbers. R^n is a very important set, called *real n-space*, or *n-dimensional real space*. (3) The reader is an *element* of the *set* of all readers of this book; and the author is an *element* of the *set* of all people who have written books.

Now suppose x is an element and A is a set. It is desirable to have a compact notation for the statement that x is an element of A. This statement is denoted by

$$x \in A,$$

which can be read, "x is an element of A," "x belongs to A," or simply "x is in A." Whenever x is *not* an element of A, we write

$$x \notin A.$$

It is also desirable to have a notation that defines sets in terms of the elements belonging to them. For sets that consist of only a few elements, we simply list (horizontally) the elements themselves and enclose the list in braces. Thus $\{\pi$, 0, 83,  ,  $\}$ is simply the set which consists of the numbers pi, zero, eighty-three, together with a jack-o'-lantern and a missile.

For sets which consist of many (even infinitely many) elements, a direct listing would be tedious to both writer and reader—if not altogether impossible. However, such a set is usually *defined* as the set of all elements under discussion which possess a common attribute. This attribute can be used to denote the set itself, as we now show.

Let $S(x)$ denote a statement, about elements x, which is either unambiguously true or unambiguously false for every given element x. We denote the set of all x's for which $S(x)$ is *true* by

$$\{x\colon S(x)\}.$$

Thus, if we are discussing real numbers, the set of all *nonnegative* real numbers is denoted by $\{x\colon x \geq 0\}$. If we are discussing the world population, $\{x\colon x$ is at least 35 years old$\}$ is a well-defined set.

Let $T(x)$ be another statement about elements x which is either unambiguously true or unambiguously false for each element x. The set of all x's for which *both* $S(x)$ and $T(x)$ are true is denoted by

$$\{x\colon S(x), T(x)\}.$$

Thus $\{x\colon x \in A, x \notin B\}$ is the set of all x's which are elements of A but are *not* elements of B.

1.3 EMPTY SET AND UNIVERSAL SET

We call the set of *all* elements under discussion (in a given context) the *universal set*, or *universe of discourse*. For the remainder of this chapter it will be denoted by X. Thus any *set* A under discussion will be an aggregate of some of the elements of X.

For somewhat technical reasons (which will be abundantly clear later in this chapter), we must define a set which has no elements. We call a set with no elements the *empty set*, or *void set*. It is denoted by $\emptyset$.

A set which does contain elements is called a *nonempty* set, or a *nonvoid* set.

1.4 VENN DIAGRAMS

For purely illustrative purposes, we may think of a set as a blob in the plane, or as a collection of such blobs. The whole plane (or a convenient rectangle in it) can be thought of as the universal set X. Each point in the plane (or rectangle) denotes an element. A Venn diagram depicting two sets A and B is exhibited in Figure 1.1. (Keep Figure 1.1 in mind throughout the next four sections.)

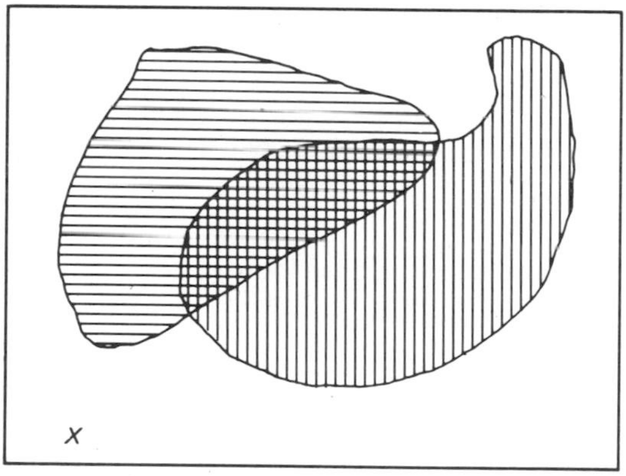

Figure 1.1 Horizontal Lines Cover A; Vertical Lines Cover B.

1.5 INCLUSION, SUBSETS, AND SET EQUALITY

If A and B are sets and if every element of A is also an element of B, then we say that *A is a subset of B*, or that *A is included in B*; and we write

$$A \subset B.$$

Examples

(1) $\{x\colon x \text{ is an even integer}\} \subset \{x\colon x \text{ is an integer}\}$. (2) $\{x\colon x \text{ is over 50 years old}\} \subset \{x\colon x \text{ is over 30 years old}\}$. (3) $\{x\colon S(x), T(x)\} \subset \{x\colon S(x)\}$, because every x for which both S and T are true is obviously an x for which S is true.

If $A \subset B$ and $B \subset A$, we say $A = B$. Clearly, $A = B$ if and only if A and B consist of precisely the same elements. Equivalently, "$A = B$" means that the statement "$x \in A$" is true when and only when the statement "$x \in B$" is true.

In Figure 1.1, A is *not* a subset of B because you can point to the horizontally but not vertically ruled portion of Figure 1.1 to show elements of A which do not belong to B. We write

$$A \not\subset B$$

to signify that A is not a subset of B. Similarly, $B \not\subset A$ in Figure 1.1.

For somewhat obvious reasons it follows from the definition of "$\subset$" that $A \subset A$ for every set A. All you need to note to prove this is that if $x \in A$, then $x \in A$, and hence $A \subset A$. QED. It may be an abuse of English to say that A is a subset of itself, but it is perfectly logical. From the definition of the universal set X, it follows that $A \subset X$ for every set A under discussion.

For ostensibly profound but really quite trivial reasons, it follows that $\varnothing \subset A$ for every set A. This is so because there is no element of $\varnothing$ which does *not* belong to A—simply because there are no elements in $\varnothing$.

1.6 UNION OF SETS

Given two sets A and B, we define their *union* $A \cup B$ as the set of all elements belonging to A or to B or to both:

$$A \cup B = \{x\colon x \in A \text{ and/or } x \in B\}.$$

In Figure 1.1, $A \cup B$ is precisely the set of those points in X within one or both of the shaded regions. Figure 1.2 exhibits $A \cup B$ explicitly for the basic Venn diagram in Figure 1.1.

As another example, let X denote the set R^1 of all real numbers. Then $\{x\colon x > 0\} \cup \{x\colon x < 0\}$ is the set of all *nonzero* real numbers.

In Exercise 5 at the end of this chapter, the reader is asked to show that *union is commutative*, in the sense that $A \cup B = B \cup A$ for any two sets A and B. In Exercise 6 the reader is asked to show that *union is associative*, in the sense that $A \cup (B \cup C) = (A \cup B) \cup C$ for any three sets A, B, and C.

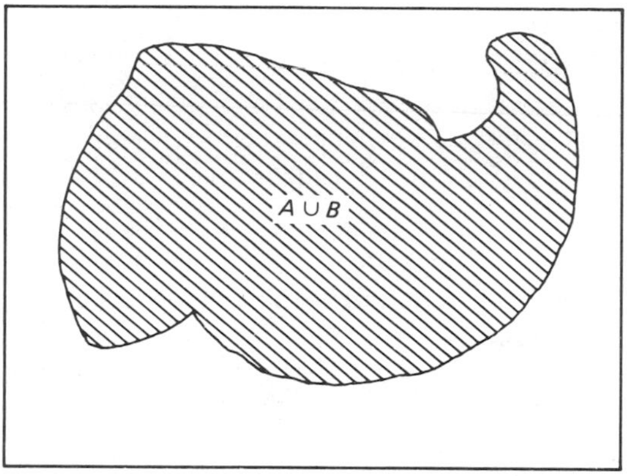

Figure 1.2

Associativity and commutativity of union are very desirable properties, as they imply that the union of several sets can be determined without regard to the order in which the pairwise (two-at-a-time) unions are taken and without regard to the order in which the given sets are listed; that is,

$$\begin{aligned}(A \cup B) \cup (C \cup D) &= A \cup (B \cup (C \cup D)) \\ &= ((D \cup C) \cup B) \cup A, \text{ and so forth.}\end{aligned}$$

Thus $\bigcup_{i=1}^{n} A_i$ is well-defined if a set A_i is given for each integer $i \in \{1, 2, \ldots, n\}$. More generally, if a set A_i is defined for every element i in some nonempty (possibly infinite) set I (called an *index set*), we may form the union of all the sets A_i:

$$\bigcup_{i \in I} A_i = \{x \colon x \in A_i, \text{ for at least one } i \in I\}.$$

Examples

(1) Let $I = R^1$ and let A_i denote the set of all points on the horizontal line of height i (a real number) above the abscissa. (A negative i means the line lies below the abscissa.) Then $\bigcup_{i \in I} A_i$ is the entire plane, because every point in the plane lies on some horizontal line. (2) Let A_i denote the set of students who achieve straight-A averages in university i this year, and let I denote the set of all universities. Then $\bigcup_{i \in I} A_i$ is the set of all university students who achieve straight-A averages this year.

The reader may show in Exercise 9 that for any set A, we have $A \subset A \cup B$ for every set B, and that $A = A \cup B$ for every B such that $B \subset A$. Moreover, for every set A we have $A \cup X = X$ and $A \cup \varnothing = A$.

If $A \cup B = X$ for two sets A and B, we say that A and B are *collectively exhaustive*. More generally, a collection $\{A_i \colon i \in I\}$ of sets is said to be collectively exhaustive if $\bigcup_{i \in I} A_i = X$.

1.7 INTERSECTION OF SETS

Given two sets A and B, we define their *intersection* $A \cap B$ as the set of all elements belonging to *both* A and B:

$$A \cap B = \{x: x \in A, x \in B\}.$$

In Figure 1.1, $A \cap B$, is precisely the cross-checked region of X; that is, the overlap of A and B.

As another example, let $X = R^1$. Then $\{x: x \geq 0\} \cap \{x: x < 3\} = \{x: 0 \leq x < 3\}$.

In Exercises 7 and 8 the reader is asked to show that intersection is *commutative* and *associative*; that is, $A \cap B = B \cap A$ for all A and B, and $A \cap (B \cap C) = (A \cap B) \cap C$ for all A, B, and C respectively. Hence, as with union, the intersection of several sets can be determined without regard to the order in which the pairwise (two-at-a-time) intersections are performed and without regard to the order in which the given sets are listed; that is,

$$\begin{aligned}(A \cap B) \cap (C \cap D) &= A \cap (B \cap (C \cap D)) \\ &= ((D \cap C) \cap B) \cap A, \text{ and so on.}\end{aligned}$$

Thus, $\bigcap_{i=1}^{n} A_i$ is well-defined if a set A_i is given for each integer $i \in \{1, 2, \ldots, n\}$. More generally, if a set A_i is defined for every element i in some nonempty index set I, we may form the intersection of all the sets A_i:

$$\bigcap_{i \in I} A_i = \{x: x \in A_i, \text{ for } \textit{every } i \in I\}.$$

Examples

(1) In Example 2 of Section 1.7, $\bigcap_{i \in I} A_i$ is the set of all students who obtain straight-A averages in *every* university. Since no student attends all universities, $\bigcap_{i \in I} A_i = \varnothing$. Similarly, in Example 1 of Section 1.7, every point in the plane belongs to only one horizontal line, and hence the intersection of any two (and hence of all) horizontal lines is empty (equals $\varnothing$). The reader may show as Exercise 11 that for any set A, we have $A \cap B \subset A$ for every set B, and that $A \cap B = A$ for every B such that $A \subset B$. Moreover, for every set A we have $A \cap X = A$ and $A \cap \varnothing = \varnothing$.

If $A \cap B = \varnothing$ for two sets A and B, we say that A and B are *disjoint*, or *mutually exclusive*. A collection $\{A_i: i \in I\}$ of sets is said to be *pairwise disjoint* if $A_i \cap A_j = \varnothing$ for every pair i and j of elements of I such that $i \neq j$.

1.8 RELATIVE COMPLEMENT AND COMPLEMENT

Given two sets A and B, the *relative complement of B with respect to A* is the set $A \setminus B$ (often written A-B) of all elements in A which are *not* in B; that is,

$$A \setminus B = \{x: x \in A, x \notin B\}.$$

Figure 1.3 depicts $A \setminus B$ and $B \setminus A$ for the sets A and B in Figure 1.1.

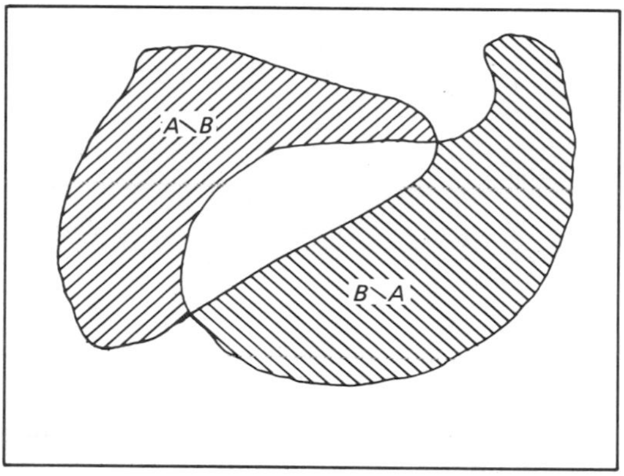

Figure 1.3

Examples

(1) Let X denote the set of all people; let $A = \{x: x \text{ is over 30 years old}\}$; and let $B = \{x: \text{you can trust } x\}$. Then $A \setminus B$ is the set of all people over 30 years old whom you cannot trust, and $B \setminus A$ is the set of all people whom you can trust and who are not over 30 years old. Hence, clearly $A \setminus B \neq B \setminus A$. (Some people, not in A, believe that $A \cap B = \varnothing$.) (2) Let X denote the set of all voters; let $A = \{x: x \text{ is worried about domestic policy}\}$; and let $B = \{x: x \text{ is worried about foreign policy}\}$. Then $A \setminus B$ is the set of all voters worried about domestic but not foreign policy, while $B \setminus A$ is the set of all voters worried about foreign but not about domestic policy. $A \cap B$ is the (usually substantial) set of all voters worried about both domestic and foreign policy; while $A \cup B$ is the set of all worried voters.

Given any set A, the *complement* of A, denoted by A^c (or by $\bar{A}$ in some works), is the set of all elements *not* in A; that is,

$$A^c = X \setminus A = \{x: x \notin A\}.$$

The following facts about complement and relative complement are frequently useful:

$$A\backslash B = A \cap B^c;$$

$$A = (A\backslash B) \cup (A \cap B);$$

$$A \cup A^c = X;$$

$$A \cap A^c = \varnothing;$$

$$X^c = \varnothing;$$

$$\varnothing^c = X;$$

and

$$(A^c)^c = A.$$

The proofs of these assertions are left as Exercise 13 for the reader.

1.9 INTERVALS OF REAL NUMBERS

We have already introduced the set R^1 of all real numbers. It can be represented as an infinite line in the plane (in fact, the abscissa, or x-axis is a natural choice). Connected pieces of the real line are very important subsets of it, and a special notation is commonly used. The following equalities define the symbol on the left. If $a < b$, then

$$[a, b] = \{x: a \leq x \leq b\};$$

$$(a, b) = \{x: a < x < b\};$$

$$[a, b) = \{x: a \leq x < b\};$$

$$(a, b] = \{x: a < x \leq b\};$$

$$(a, \infty) = \{x: a < x\};$$

$$[a, \infty) = \{x: a \leq x\};$$

$$(-\infty, a) = \{x: x < a\};$$

$$(-\infty, a] = \{x: x \leq a\};$$

and

$$(-\infty, \infty) = \{x: -\infty < x < \infty\} = R^1.$$

EXERCISES

1. Let X denote the set of all corporations whose common stocks are listed on the New York Stock Exchange. Let $A = \{x\colon x \text{ manufactures plastic}\}$; let $B = \{x\colon x \text{ had sales last year exceeding \$50 million}\}$; let $C = \{x\colon x \text{ maintains a lobbyist in Washington}\}$, and let $D = \{x\colon x \text{ has at least six vice-presidents}\}$. Describe in words the following sets:

$$A \cup B,\ A \cap B,\ A \cap B \cap C,\ A \cup (B \cap C \cap D),$$

$$A \backslash (B \cup C),\ (A \backslash B) \cup (B \backslash A), \text{ and } (A \backslash B) \cap (C \backslash D).$$

2. Let X denote the set of all housewives; let $A = \{x\colon x \text{ buys GLOP detergent}\}$; let $B = \{x\colon x \text{ employs a maid}\}$; let $C = \{x\colon x \text{ is married}\}$; and let $D = \{x\colon x \text{ is gainfully employed}\}$. Describe in words the following sets:

$$A \cup D,\ A \cap B,\ A \backslash B,\ A \cap (C \backslash B),\ B \cap D,$$

$$B \cap C \cap D,\ A \backslash (C \cap D), \text{ and } A \backslash (C \cup D).$$

3. Let $X = \{1, 2, 3, 4, 5, 6\}$ denote the set of all possible outcomes of the experiment which consisted of rolling a die (one-half of a pair of dice) and recording the number of dots on the uppermost face. Describe in words the following sets:

$$\{1, 3, 5\},\ \{2, 4, 6\},\ \{2, 4\},\ \{6\}, \text{ and } \{1\}.$$

(*Note:* the *subset* $\{6\}$ of $\{1, 2, 3, 4, 5, 6\}$ is *formally* distinct from the *element* $6 \in \{1, 2, 3, 4, 5, 6\}$, although in *practice* we usually do not care if we confuse them.)

4. Let $X = R^1$. Re-express the following sets as intervals and in the interval notation:

$$(8, 50) \cap (40, 90);$$

$$(8, 50) \backslash (0, 35);$$

$$(30, 100) \cup (-\infty, 50);$$

and

$$(-10, 10) \cap (1, 2).$$

5. Show that *union* is *commutative*; that is,

$$A \cup B = B \cup A, \text{ for all } A \text{ and all } B.$$

6. Show that *union* is *associative*; that is,

$$(A \cup B) \cup C = A \cup (B \cup C), \text{ for all } A, B, \text{ and } C.$$

7. Show that *intersection* is *commutative*; that is,

$$A \cap B = B \cap A, \text{ for all } A \text{ and all } B.$$

8. Show that *intersection* is *associative*; that is,

$$(A \cap B) \cap C = A \cap (B \cap C), \text{ for all } A, B, \text{ and } C.$$

9. Show that $A \subset A \cup B$, for every A and B, and that $A = A \cup B$ if $B \subset A$.

10. Show that $A \cup X = X$ and $A \cup \varnothing = A$, for every A.

11. Show that $A \cap B \subset A$, for every A and B, and that $A \cap B = A$ if $A \subset B$.

12. Show that $A \cap X = A$ and $A \cap \varnothing = \varnothing$, for every A.

13. Show that for every A and B we have:

$$A \backslash B = A \cap B^c;$$
$$A = (A \backslash B) \cup (A \cap B);$$
$$A \cup A^c = X;$$
$$A \cap A^c = \varnothing;$$
$$X^c = \varnothing;$$
$$\varnothing^c = X;$$

and

$$(A^c)^c = A.$$

14. Show that the following *distributive laws* hold:

$$A \cap (B \cup C) = (A \cap B) \cup (A \cap C);$$

and

$$A \cup (B \cap C) = (A \cup B) \cap (A \cup C).$$

15. Let I be a nonempty index set, and let A_i be a subset of X, for every $i \in I$. Let B be any subset of X. Show that

$$B \cap (\cup_{i \in I} A_i) = \cup_{i \in I}(B \cap A_i);$$

and

$$B \cup (\cap_{i \in I} A_i) = \cap_{i \in I}(B \cup A_i).$$

(This is an easy extension of Exercise 14.)

16. Prove De Morgan's laws:

$$(A_1 \cup A_2)^c = A_1{}^c \cap A_2{}^c;$$

and

$$(A_1 \cap A_2)^c = A_1{}^c \cup A_2{}^c.$$

17. Let I be a nonempty index set, and let A_i be a subset of X, for every $i \in I$. Let B be any subset of X. Show that

$$B \backslash (\cup_{i \in I} A_i) = \cap_{i \in I}(B \backslash A_i);$$

and

$$B \backslash (\cap_{i \in I} A_i) = \cup_{i \in I}(B \backslash A_i).$$

(This is an easy extension of Exercise 16. When $B = X$, these versions of De Morgan's laws become:

$$(\cup_{i \in I} A_i)^c = \cap_{i \in I} A_i{}^c;$$

and

$$(\cap_{i \in I} A_i)^c = \cup_{i \in I} A_i{}^c).$$

2

SOME IMPORTANT SETS OF SETS

2.1 INTRODUCTION

In Chapter 1 we discussed elements and sets without really distinguishing between them on an absolute basis. In fact, no absolute distinction exists; what the elements are and what the sets are depend upon the context of any given argument. For example, the set X of all squares in the plane is perfectly well-defined, since each square is an element and the sets A are sets of squares; but each square is itself a set of points in the plane. Hence if we adopt the more "atomic" viewpoint of letting points in the plane be the elements, we see that the set of all squares in the plane is a set of sets.

We shall adopt precisely this atomic view in Chapter 2 by assuming as given a (universal) set X and by considering various sets of subsets of X. In Section 2.2 we consider the set of *all* subsets of X, while in the ensuing sections we consider sets consisting of perhaps only some subsets of X.

2.2 THE POWER SET OF X

We define the *power set* $\mathcal{R}(X)$ as the set of *all* subsets of X; that is,

$$\mathcal{R}(X) = \{A: A \subset X\}.$$

Since $\varnothing$ is a subset of X, it follows that $\varnothing$ is an *element* of $\mathcal{R}(X)$. Since $X \subset X$, we also have $X \in \mathcal{R}(X)$.

Examples

(1) If $X = \{1, 2\}$, then $\mathcal{R}(X) = \{\varnothing, \{1\}, \{2\}, \{1, 2\}\}$. Note that $\{1\}$ is formally distinct from 1, since $\{1\}$ is a subset of $\{1, 2\}$ and 1 is an element of $\{1, 2\}$. (2) If $X = \{1, 2, 3\}$, then $\mathcal{R}(X) = \{\varnothing, \{1\}, \{2\}, \{3\}, \{1, 2\}, \{1, 3\}, \{2, 3\}, \{1, 2, 3\}\}$. (3) If X is a set of people, then $\mathcal{R}(X)$ is the set of all possible cliques (allowing for the empty clique $\varnothing$ and for all "lone wolf" cliques $\{x\}$). (4) If X is the set of all components (including connections) in a complex electronic system, then $\mathcal{R}(X)$ is the set of all possible (single or multiple) ways in which the system can fail, if we describe a given failure by the subset A (of X) of components which broke down.

$\mathcal{R}(X)$ is the "largest" set of subsets of X. In the following two sections we shall consider subsets of $\mathcal{R}(X)$; that is, sets of some (but not necessarily all) subsets of X. In order to pave the way for Section 2.3, we shall now list some properties of $\mathcal{R}(X)$ which are obvious from its definition as the set of all subsets of X and from the definitions of $\cup$, $\cap$, and ^{c} in Chapter 1. If $A \in \mathcal{R}(X)$ and $B \in \mathcal{R}(X)$, then

(1) $A \cup B \in \mathcal{R}(X)$;
(2) $A \cap B \in \mathcal{R}(X)$;
(3) $A^c \in \mathcal{R}(X)$; and
(4) $A \backslash B \in \mathcal{R}(X)$.

Moreover, let I be an arbitrary nonempty index set and let $A_i \in \mathcal{R}(X)$ for every $i \in I$. Then

(1*) $\bigcup_{i \in I} A_i \in \mathcal{R}(X)$; and
(2*) $\bigcap_{i \in I} A_i \in \mathcal{R}(X)$.

Since the left-hand side of "$\in$" is always a subset of X (by the definitions in Chapter 1), it is an element of $\mathcal{R}(X)$ [by definition of $\mathcal{R}(X)$].

2.3 FIELDS OF SUBSETS OF X

We now turn from $\mathcal{R}(X)$ and consider sets of subsets of X which do not necessarily satisfy (1*) and (2*).

A *field* $\mathfrak{F}$ of subsets of X is a nonempty set of subsets of X such that

(1) if $A \in \mathfrak{F}$ and $B \in \mathfrak{F}$, then $A \cup B \in \mathfrak{F}$; and
(3) if $A \in \mathfrak{F}$, then $A^c \in \mathfrak{F}$.

These two axioms (numbered to conform with properties of $\mathcal{R}(X)$ given in Section 2.2) imply that $\mathfrak{F}$ possesses the following additional properties:

(2) if $A \in \mathfrak{F}$ and $B \in \mathfrak{F}$, then $A \cap B \in \mathfrak{F}$;
(4) if $A \in \mathfrak{F}$ and $B \in \mathfrak{F}$, then $A \backslash B \in \mathfrak{F}$;
(1*f*) if I is a nonempty index set consisting of a *finite* number of elements i and if $A_i \in \mathfrak{F}$ for every $i \in I$, then $\bigcup_{i \in I} A_i \in \mathfrak{F}$; and
(2*f*) if I is a nonempty index set consisting of a *finite* number of elements i and if $A_i \in \mathfrak{F}$ for every $i \in I$, then $\bigcap_{i \in I} A_i \in \mathfrak{F}$.

We shall derive (2), (4), (1*f*), and (2*f*) from the axioms (1) and (3). We derive (2) first: since $A \in \mathfrak{F}$ and $B \in \mathfrak{F}$, then $A^c \in \mathfrak{F}$ and $B^c \in \mathfrak{F}$ by (3), and hence $A^c \cup B^c \in \mathfrak{F}$ by (1). Hence $(A^c \cup B^c)^c \in \mathfrak{F}$ by (3) again. But by Exercise 16 of Chapter 1, we have $A^c \cup B^c = (A \cap B)^c$, and so $(A^c \cup B^c)^c = ((A \cap B)^c)^c$. But $((A \cap B)^c)^c = A \cap B$ by Exercise 13 of Chapter 1. Hence $A \cap B = (A^c \cup B^c)^c \in \mathfrak{F}$, thus proving (2). To prove (4), we have that since $B \in \mathfrak{F}$, (3) implies that $B^c \in \mathfrak{F}$. Hence, by (2), $A \cap B^c \in \mathfrak{F}$; but by Exercise 13 of Chapter 1, we have $A \backslash B = A \cap B^c$. Hence $A \backslash B \in \mathfrak{F}$. To prove (1*f*) and (2*f*) we may assume that $I = \{1, 2, \ldots, n\}$ for some (finite) positive integer n, and we use finite induction from properties (1) and (2), the induction being permitted by associativity of $\cup$ and $\cap$.

Two points concerning fields must be stressed. The first is that $\mathfrak{F}$ need not contain *all* subsets of X. It will contain X, since $A^c \in \mathfrak{F}$ for every $A \in \mathfrak{F}$; hence $A \cup A^c = X \in \mathfrak{F}$. It will also contain $\varnothing$ $(= X^c)$. To show that some subsets of X can be omitted from $\mathfrak{F}$, we consider the set $\{\varnothing, \{1\}, \{2, 3\}, \{1, 2, 3\}\}$ of subsets of $\{1, 2, 3\}$. The reader may verify that this set of subsets is a field, although it is *not* equal to $\mathcal{R}(\{1, 2, 3\})$ (see Example 2 of Section 2.2; it omits five elements of $\mathcal{R}(\{1, 2, 3\})$.

The second point is that $\mathfrak{F}$ need *not* contain the union of an arbitrary set of subsets of X in $\mathfrak{F}$ if X is an infinite set (contains infinitely many elements).

To see this, let X denote the set of all integers and let $\mathfrak{F}$ consist of all subsets A of X which satisfy *either* (i) A is a finite set; *or* (ii) A^c is a finite set. Then contains X (because $X^c = \varnothing$, the "most finite" set possible), and hence $\mathfrak{F}$ also contains $\varnothing$. Moreover, if A satisfies (i), then A^c satisfies (ii), and vice-versa, so that $A^c \in \mathfrak{F}$ whenever $A \in \mathfrak{F}$. Thus axiom (3) is satisfied. To verify axiom (1) we start by assuming that $A \in \mathfrak{F}$ and $B \in \mathfrak{F}$. Then we have four cases to consider: (a) both A and B are finite; (b) both A^c and B^c are finite; (c) A is finite and B^c is finite; and (d) A^c is finite and B is finite. For case (a), if A and B are finite, then so is $A \cup B$ since if A contains m elements and B contains n elements, then $A \cup B$ contains *at most* $m + n$ elements. Hence $A \cup B \in \mathfrak{F}$ for case (a). For case (b), if A^c and B^c are finite, then $A^c \cap B^c$ is certainly finite. But $A^c \cap B^c = (A \cup B)^c$, and so $A \cup B$ satisfies requirement (ii) for belonging to $\mathfrak{F}$. For case (c), if B^c is finite, then $B^c \cap A^c$ must also be finite (since $A^c \cap B^c \subset B^c$). But $A^c \cap B^c = (A \cup B)^c$, so that $A \cup B$ again satisfies requirement (ii). The final case, (d), follows exactly as did (c), except that the roles of A and B are reversed. Hence $\mathfrak{F}$ is a field of subsets of the set of all integers. We shall now note that every set A_i of the form $A_i = \{1, 2, \ldots, i\}$ for i a positive integer is in $\mathfrak{F}$, since it is a finite set. But $\bigcup_{i=1}^{\infty} A_i (= \bigcup_{i \in I} A_i)$ is the set of all positive integers, and it is *not* in $\mathfrak{F}$, since it is an infinite set, the complement of which is also infinite; namely, $\{\ldots, -2, -1, 0\}$. Hence (1*) does not hold for $\mathfrak{F}$.

2.4 SIGMA-FIELDS OF SUBSETS OF X

In probability theory in particular, and in measure theory in general, we usually do not care whether a set $\mathfrak{F}$ of subsets of X contains the union and intersection of an arbitrary set of subsets of X which are all in $\mathfrak{F}$. We do care whether $\mathfrak{F}$ contains the union and the intersection of every *countable* set of subsets of X which are all in $\mathfrak{F}$.

[A set is *countable* if its elements can be placed in one-to-one correspondence with some subset of the set of integers. Every finite set is countable. The set of integers is countable. The set of all rational numbers (ratios of integers with nonzero denominators) is countable. R^1 is *not* countable.]

The example in Section 2.3 of a field $\mathfrak{F}$ which does not satisfy (1*) also showed that $\mathfrak{F}$ does not contain the union of every countable set of elements of $\mathfrak{F}$; there were only countably many A_i's.

Fields which contain the countable union (and countable intersection) of their elements (subsets of X) are called *sigma-fields*. More formally, a *sigma-field* $\mathcal{S}$ of subsets of X is a nonempty set of subsets of X such that

(3) if $A \in \mathcal{S}$, then $A^c \in \mathcal{S}$; and

(1^σ) if I is a nonempty index set consisting of a *countable* number of elements i and if $A_i \in \mathcal{S}$, for every $i \in I$, then

$$\bigcup_{i \in I} A_i \in \mathcal{S}.$$

Since (1^σ) implies (1) and (1^f) as special cases, it follows that a sigma-field is a field, and hence it possesses properties (2), (4), and (2^f) as well. It also possesses the following extension of property (2^f):

(2^σ) If I is a nonempty index set consisting of a *countable* number of elements i and if $A_i \in \mathcal{S}$, for every $i \in I$, then

$$\bigcap_{i \in I} A_i \in \mathcal{S}.$$

If X is a finite set, then $\mathcal{R}(X)$ is also a finite set and hence any field $\mathfrak{F}$ is finite. Thus any countable union must be a finite union, and hence any field of subsets of a finite set is a sigma-field. The distinction between fields and sigma-fields becomes relevant only when X is an infinite set.

Let $X = R^1$ and let $\mathcal{S}$ consist of all subsets A of X which satisfy *either* (i) A is a countable set; *or* (ii) A^c is a countable set. By using a theorem which states that the union of a countable number of sets, each of which is countable, is a countable set, we may adapt the discussion of the example in Section 2.3 of a field $\mathcal{F}$ to show that $\mathcal{S}$ is a sigma-field but does not satisfy (1*).

2.5 PARTITIONS OF X

Let I be an arbitrary nonempty index set, and let A_i be a subset of X for every $i \in I$. We say the subset $\mathcal{P} = \{A_i\colon i \in I\}$ of $\mathcal{R}(X)$ is a *partition* of X if the following conditions obtain:

(1) $\bigcup_{i \in I} A_i = X$; and
(2) $A_i \cap A_j = \emptyset$, if $i \in I, j \in I$, and $i \neq j$.

Equivalently in words (defined in Chapter 1), a set of subsets of X is a *partition* of X if the given subsets are *mutually exclusive* [condition (2)] and *collectively exhaustive* [condition (1)].

An immediate consequence of the definition of a partition $\mathcal{P}$ is the fact that every element x in X belongs to one and only one element A_i of $\mathcal{P}$. Thus a partition of X is the set of pieces which results from the "act of partitioning."

Figure 2.1 consists of a Venn diagram illustrating a partition of X. (In this diagram, you are to assume that each point on the boundary between any two A_i's has been assigned to precisely one of them.) Clearly, $A_1 \cup A_2 \cup A_3 \cup A_4 = X$, and (given the preceding assumption) the intersection of any two A_i's is empty.

Partitions of a set are very important in statistics. For example, if X denotes the set of all wage earners, we may wish to classify the wage earners

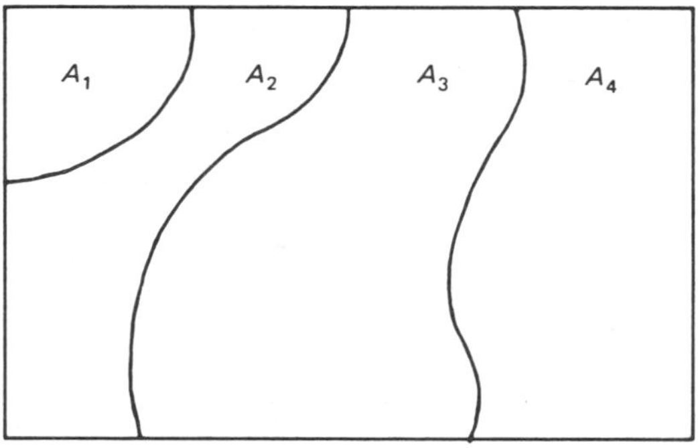

Figure 2.1

into only a few income categories. We might define $A_1 = \{x\colon x$ earns at least \$50,000/year$\}$, $A_2 = \{x\colon x$ earns at least \$10,000 but less than \$50,000/year$\}$, and $A_3 = \{x\colon x$ earns less than \$10,000/year$\}$. Then $\mathcal{P}_1 = \{A_1, A_2, A_3\}$ is a partition of X.

Now suppose that I and J are nonempty index sets, and let $\mathcal{P}_1 = \{A_i\colon i \in I\}$ and $\mathcal{P}_2 = \{B_j\colon j \in J\}$ be partitions of X. Then the set $\mathcal{P}_1\mathcal{P}_2$ of all sets of the form $A_i \cap B_j$, as i ranges through I and j ranges through J, is also a partition, called the *product* of $\mathcal{P}_1$ and $\mathcal{P}_2$. Formally,

$$\mathcal{P}_1\mathcal{P}_2 = \{A_i \cap B_j\colon i \in I, j \in J\}.$$

To continue our previous example, suppose $B_1 = \{x\colon x$ owns at least one dog$\}$ and $B_2 = \{x\colon x$ owns no dogs$\}$. Then $\mathcal{P}_2 = \{B_1, B_2\}$ is also a partition of the set X of all wage earners. The product $\mathcal{P}_1\mathcal{P}_2$ contains six elements (subsets of X): $A_1 \cap B_1$, $A_2 \cap B_1$, $A_3 \cap B_1$, $A_1 \cap B_2$, $A_2 \cap B_2$, and $A_3 \cap B_3$. Those wage earners making less than \$10,000/year and not owning a dog constitute the element $A_3 \cap B_2$ of $\mathcal{P}_1\mathcal{P}_2$ (alternatively, the subset $A_3 \cap B_2$ of X).

Thus product partitions correspond to two-way classifications of a population (many-way classifications in the case of several "factors" in the "product").

The reader may verify as an exercise that the "multiplication" of partitions of X is *commutative* and *associative*; that is, if $\mathcal{P}_1$, $\mathcal{P}_2$, and $\mathcal{P}_3$ are partitions of X, then

$$\mathcal{P}_1\mathcal{P}_2 = \mathcal{P}_2\mathcal{P}_1; \text{ and}$$

$$\mathcal{P}_1(\mathcal{P}_2\mathcal{P}_3) = (\mathcal{P}_1\mathcal{P}_2)\mathcal{P}_3.$$

Figure 2.2 exhibits a different partition (call it $\mathcal{P}^*$) of the set X depicted in Figure 2.1 with a partition $\mathcal{P} = \{A_1, A_2, A_3, A_4\}$. Figure 2.3 shows X with the product $\mathcal{P}\mathcal{P}^*$ of the partitions in Figures 2.1 and 2.2.

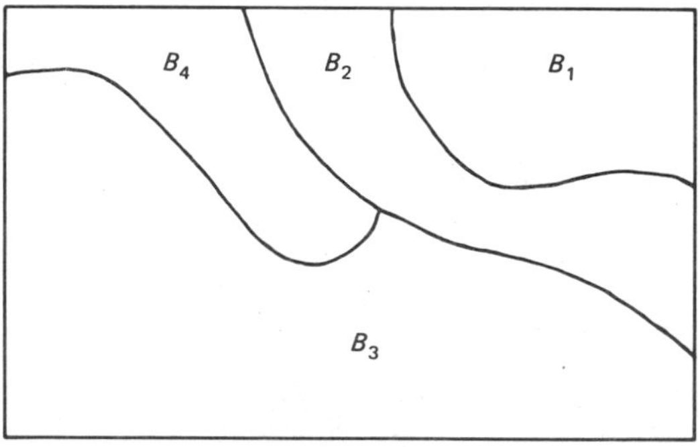

Figure 2.2

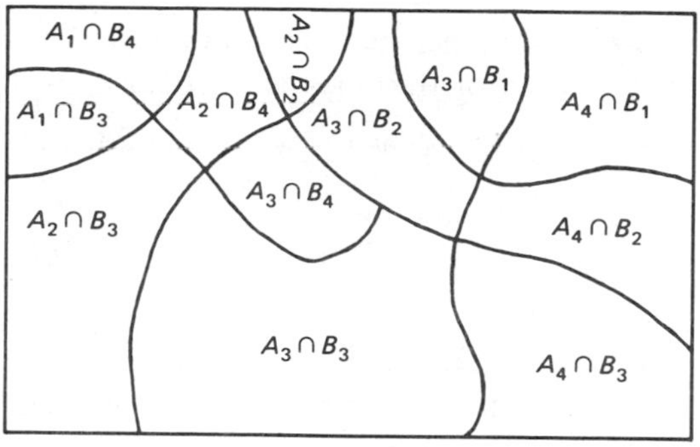

Figure 2.3

Note that $A_3 = (A_3 \cap B_1) \cup (A_3 \cap B_2) \cup (A_3 \cap B_3) \cup (A_3 \cap B_4)$, which is obvious from Figure 2.3. Also, since $A_1 \cap B_2 = \varnothing = A_1 \cap B_1$ and $S \cup \varnothing = S$ for any set S, we also have $A_1 = (A_1 \cap B_1) \cup (A_1 \cap B_2) \cup (A_1 \cap B_3) \cup (A_1 \cap B_4)$, similarly for A_2 and A_4. This phenomenon is true in a somewhat more general context, and it is important in probability theory.

Let S be any subset of X, and let $\mathcal{P} = \{A_i: i \in I\}$ be any partition of X. Then $S = \bigcup_{i \in I}(S \cap A_i)$. Moreover, if $i \neq j$, then $(S \cap A_i) \cap (S \cap A_j) \neq \varnothing$.

This assertion has a very natural interpretation: define $\mathcal{P}_S$ to be the set of all sets of the form $S \cap A_i$, where $A_i \in \mathcal{P}$. Then $\mathcal{P}_S$ is a partition of S. [To prove the assertion, we use Exercise 15 of Chapter 1 to show that $\bigcup_{i \in I}(S \cap A_i) = S \cap (\bigcup_{i \in I} A_i)$. But $\bigcup_{i \in I} A_i = X$ because $\mathcal{P}$ is a partition of X. Hence $S \cap (\bigcup_{i \in I} A_i) = S \cap X$, and $S \cap X = S$. Things equal to the same thing are equal to each other, and hence $S = \bigcup_{i \in I}(S \cap A_i)$, which is the first part of the assertion. The second is immediate, since commutativity and associativity of $\cap$ imply $(S \cap A_i) \cap (S \cap A_j) = S \cap (A_i \cap A_j) = S \cap \varnothing = \varnothing$.]

Examples

The preceding assertion sounds silly in practice. Following two cases in point. (1) A wage earner making at least \$10,000 per year (belonging to a set S) either owns at least one dog (belongs to A_1 as well as to S) or owns no dogs (belongs to A_2 as well as to S). (2) A person over 30 can either be trusted or not be trusted: let S comprise all people over 30, A_1 comprise all people who can be trusted, and $A_2 = X \setminus A_1$.

Let $\mathcal{P}_1 = \{A_i: i \in I\}$ and $\mathcal{P}_2 = \{B_j: j \in J\}$ be partitions of X. We say that $\mathcal{P}_2$ is *finer than* $\mathcal{P}_1$ if every A_i in $\mathcal{P}_1$ equals a union of sets B_j belonging to $\mathcal{P}_2$. Intuitively, this means that $\mathcal{P}_1$ can be gotten from $\mathcal{P}_2$ by erasing boundary lines, or (more generally) by joining together elements of $\mathcal{P}_2$. If

X is some population, $\mathcal{P}_2$ is finer than $\mathcal{P}_1$ if and only if $\mathcal{P}_2$ corresponds to a more precise classification of the population than does $\mathcal{P}_1$. We write

$$\mathcal{P}_2 \gtrsim \mathcal{P}_1$$

as shorthand for "$\mathcal{P}_2$ is finer than $\mathcal{P}_1$."

The following properties of "is finer than" can be verified as an exercise. Let X be fixed. Then,

(1) if $\mathcal{P}$ is any partition of X, we have $\mathcal{P} \gtrsim \mathcal{P}$;
(2) if $\mathcal{P}_1$, $\mathcal{P}_2$, and $\mathcal{P}_3$ are partitions of X, if $\mathcal{P}^1 \gtrsim \mathcal{P}_2$, and if $\mathcal{P}_2 \gtrsim \mathcal{P}_3$, then we have $\mathcal{P}_1 \gtrsim \mathcal{P}_3$;
(3) every partition $\mathcal{P}$ of X is finer than the partition $\mathcal{P}_* = \{X\}$;
(4) $\mathcal{P}^* = \{\{x\}: x \in X\}$ is finer than every partition $\mathcal{P}$ of X; and
(5) if $\mathcal{P}$ and $\mathcal{P}_1$ are partitions of X and if $\mathcal{P}_2 = \mathcal{P}\mathcal{P}_1$, then we have $\mathcal{P}_2 \gtrsim \mathcal{P}_1$.

In words: (1) every partition is finer than itself; (2) "is finer than" is a "transitive" relation (see Chapter 3); (3) there is a "coarsest," or "least fine" partition, consisting of X itself; (4) there is a "finest" partition, consisting of all one-element subsets of X; and (5) the product of any two partitions is finer than either of them. (As an aside, we might define a "pure individualist" as one who thinks the only valid way of classifying the world population is via $\mathcal{P}^*$, and a "pure collectivist" as one who thinks the only valid way of classifying the world population is via $\mathcal{P}_*$. Obviously, most individualists and collectivists are, by these definitions, impure.)

EXERCISES

1. Let $X = \{a, b\}$. Find $\mathcal{R}(X)$.
2. Let $X = \{1, 2, 3, 4\}$. Find $\mathcal{R}(X)$.
3. Is $\{\{1\}, \{1, 2\}, \{1, 2, 3\}\}$ a field of subsets of $\{1, 2, 3\}$?
4. Let X be any nonempty set. Is $\{\varnothing, X\}$ a field of subsets of X? (Is it a sigma-field?)
5. Is $\{\varnothing, \{1\}, \{2, 3\}, \{1, 2, 3\}\}$ a field of subsets of $\{1, 2, 3\}$? Is $\{\varnothing, \{1, 2\}, \{3\}, \{1, 2, 3\}\}$ a field of subsets of $\{1, 2, 3\}$?
6. Show that if $\mathfrak{F}_1$ and $\mathfrak{F}_2$ are two fields of subsets of X, then $\mathfrak{F}_1 \cap \mathfrak{F}_2$ is a field of subsets of X. ($\mathfrak{F}_1 \cap \mathfrak{F}_2 = \{A: A \in \mathfrak{F}_1, A \in \mathfrak{F}_2\}$.) (Show that $\mathcal{S}_1 \cap \mathcal{S}_2$ is a sigma-field if $\mathcal{S}_1$ and $\mathcal{S}_2$ are sigma-fields.)
7. Fields and partitions are both sets of subsets of X. Can a field be a partition? Can a partition be a field?
8. Prove assertions (1) through (5) of Section 2.5.

9. For any nonempty set X, show that $\mathcal{R}(X)$ is a field. [Show that $\mathcal{R}(X)$ is a sigma-field.]

10. Field $\mathcal{F}_1$ *is finer than* field $\mathcal{F}_2$ if every $A \in \mathcal{F}_2$ belongs to $\mathcal{F}_1$; that is, $\mathcal{F}_2 \subset \mathcal{F}_1$. (Similarly, sigma-field $\mathcal{S}_1$ *is finer than* sigma-field $\mathcal{S}_2$ if $\mathcal{S}_2 \subset \mathcal{S}_1$.) Let X be a nonempty set and let $\mathcal{A}$ be any set of subsets of X [that is, $\mathcal{A}$ is any subset of $\mathcal{R}(X)$]. Show that there exists a field $\mathcal{F}(\mathcal{A})$ such that: (1) $\mathcal{A} \subset \mathcal{F}(\mathcal{A})$; and (2) if $\mathcal{F}$ is any field such that $\mathcal{A} \subset \mathcal{F}$, then $\mathcal{F}$ is finer than $\mathcal{F}(\mathcal{A})$. We call $\mathcal{F}(\mathcal{A})$ the field *generated by* $\mathcal{A}$. [Show that there exists a sigma-field $\mathcal{S}(\mathcal{A})$ such that: (a) $\mathcal{A} \subset \mathcal{S}(\mathcal{A})$; and (b) if $\mathcal{S}$ is any sigma-field such that $\mathcal{A} \subset \mathcal{S}$, then $\mathcal{S}$ is finer than $\mathcal{S}(\mathcal{A})$. We call $\mathcal{S}(\mathcal{A})$ the sigma-field *generated by* $\mathcal{A}$.] [*Hint*: Let F be the set of all fields which satisfy (1). F is nonempty, since $\mathcal{R}(X)$ is a field and satisfies (1) (by Exercise 9). Now generalize Exercise 6 to show that $\bigcap_{\mathcal{F} \in F} \mathcal{F} = \{A: A \in \mathcal{F}, \text{ for every } \mathcal{F} \in F\}$ is a field, that it satisfies (1) and (2), and hence is $\mathcal{F}(\mathcal{A})$.]

11. In Exercise 10, show that if $\mathcal{A}$ is a field, then $\mathcal{F}(\mathcal{A}) = \mathcal{A}$. [If $\mathcal{A}$ is a sigma-field, then $\mathcal{A} = \mathcal{S}(\mathcal{A})$. However, if $\mathcal{A}$ is a field, it does not necessarily follow that $\mathcal{A} = \mathcal{S}(\mathcal{A})$. See Exercise 13.] If $\mathcal{A}$ is a sigma-field, then $\mathcal{A} = \mathcal{F}(\mathcal{A})$. [*Hint*: If $\mathcal{A}$ is a field, then $\mathcal{A} \in F$ and hence $\mathcal{F}(\mathcal{A}) \subset \mathcal{A}$. But requirement (a) implies $\mathcal{A} \subset \mathcal{F}(\mathcal{A})$.]

12. Find $\mathcal{F}(\mathcal{A})$ as defined in Exercise 10 for:

    (a) $\mathcal{A} = \{\{1\}, \{2, 3\}\}$; and
    (b) $\mathcal{A} = \{\{1, 2\}, \{3\}\}$.

13. Let X be the set of all integers and let $\mathcal{A}$ be the field of all sets which either are finite or which have finite complements, as discussed in Section 2.3. Show that $\mathcal{S}(\mathcal{A}) = \mathcal{R}(X)$ and hence $\mathcal{A} \neq \mathcal{S}(\mathcal{A})$. (*Hint*: The set of all integers is countable.)

14. Let $X = R^1$, the set of all real numbers, and let $\mathcal{A}$ denote the set of all open intervals; that is,

$$\mathcal{A} = \{(a, b): a \in R^1, b \in R^1, a < b\}.$$

By Exercise 10, there exists a sigma-field $\mathcal{S}(\mathcal{A})$ generated by $\mathcal{A}$. This sigma-field is of great importance in rigorous probability theory. It is called the *Borel field in* R^1 and is denoted by $\mathcal{B}^1$. Show that for every real number x we have $\{x\} \in \mathcal{B}^1$. [*Hint*: $\mathcal{B}^1$ satisfies (2^σ); and $\bigcap_{n=1}^{\infty}(x - 1/n,\ x + 1/n) = \{x\}$.]

3

CARTESIAN PRODUCTS (I): RELATIONS, ORDERINGS, AND FUNCTIONS

3.1 INTRODUCTION

In Chapter 1 we examined elements and subsets of a given (universal) set X, and we noted that $\cup$, $\cap$, and $\setminus$ provided ways of constructing new subsets of X from two given subsets. In this chapter we consider two (universal) sets X and Y; we form the set $X \times Y$ of all *ordered* pairs (x, y) for $x \in X$ and $y \in Y$. This set of ordered pairs is called the *Cartesian product* of X and Y (in that order). In Section 3.2 we examine properties of $X \times Y$ and some generalizations to the case of more than two "factors" X and Y.

In Section 3.3 we present a mathematical definition, based upon Cartesian products, of the intuitive idea of a "relation between things," and we examine resulting properties of relations and ways of combining them.

Section 3.4 covers certain special cases of relations, called *orderings*, which are of great importance in mathematics generally and in probability, inference, and decision theory in particular.

Section 3.5 concerns another special case of relations; namely, functions. It may come as a surprise to see that neither the domain nor the range of a function need be numerical; indeed, the concept of a function is fundamentally set-theoretic.

3.2 CARTESIAN PRODUCTS

Let X and Y be two nonempty sets. If $x \in X$ and $y \in Y$, then we can form the *ordered pair* (x, y). Now, (x, y) is neither an element of X nor an element of Y. Yet it is quite clear that if X is a set of people and Y is a set of dogs, then we are not hesitant about writing down (John Doe, Fido). Hence (x, y) must belong to *some* set.

We define the *Cartesian product* $X \times Y$ of X and Y (in that order) as the set of all ordered pairs (x, y) for $x \in X$ and $y \in Y$. More formally,

$$X \times Y = \{(x, y): x \in X, y \in Y\}.$$

Examples

(1) The Euclidean plane can be thought of as the set of all ordered pairs (x, y) of real numbers. Here, $X = R^1 = Y$; and x denotes horizontal distance while y denotes vertical distance of the given point (x, y) from the origin point (0, 0) after horizontal and vertical directions have been specified. Hence the Euclidean plane can be thought of as the Cartesian product $R^1 \times R^1$. (2) Let X denote the set of all people and let Y denote the set of all books. Then $X \times Y$ is the set of all ordered pairs of the form (person, book). Note that an ordered pair of the form (book, person) does *not* belong to $X \times Y$. Hence $X \times Y \neq Y \times X$.

For *purely intuitive* purposes, we may think of the sets X and Y as being (closed) intervals of real numbers, and we may then think of $X \times Y$ as a rectangle in the plane (see Figure 3.1).

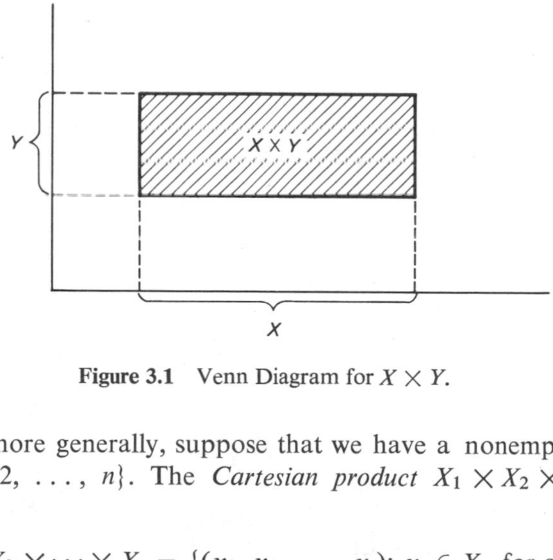

Figure 3.1 Venn Diagram for $X \times Y$.

Somewhat more generally, suppose that we have a nonempty set X_i for every $i \in \{1, 2, \ldots, n\}$. The *Cartesian product* $X_1 \times X_2 \times \cdots \times X_n$ is defined by

$$X_1 \times X_2 \times \cdots \times X_n = \{(x_1, x_2, \ldots, x_n): x_i \in X_i, \text{ for all } i\}.$$

(3) "Three-dimensional space" can be represented as the set of all triples (x, y, z) of real numbers; that is, the Cartesian product $R^3 = R^1 \times R^1 \times R^1$. We defined R^3 implicitly in Example 2 of Section 1.2 as this set of triples,

and we now see that it can be equivalently defined as a Cartesian product. (4) Similarly, real n-space R^n is the Cartesian product $R^1 \times R^1 \times \cdots \times R^1$ (a total of n "factors"). We often call an element $(x_1, x_2, \ldots, x_n)$ a *vector*, or an *n-dimensional vector*, or simply an *n-tuple*. We often abbreviate the notation by defining $\mathbf{x} = (x_1, x_2, \ldots, x_n)$; thus boldface $\mathbf{x}$ will hereafter denote a vector. (The dimensionality n will be made clear, so that no confusion will result from its omission in the abbreviated vector notation.)

3.3 RELATIONS

Given two (not necessarily distinct) sets X and Y, we say that R is a (binary) *relation from X to Y* if for every $x \in X$ and every $y \in Y$ we can, at least in principle, determine whether or not x is R-related to y; that is, whether xRy is true. xRy might be the statement, "x is a parent of y," or, "x is greater than y," or "x dislikes y," and so forth.

A very easy and natural way of representing any relation R as defined previously is to consider the subset of all (x, y) in $X \times Y$ such that x is R-related to y; that is, to identify the relation R with the set of all pairs (x, y) for which xRy is true:

$$R = \{(x, y): (x, y) \in X \times Y, xRy \text{ is true}\}.$$

In fact, relations are *defined* mathematically as subsets of $X \times Y$.

If $X = Y$, we say R is a (binary) *relation in X* rather than a relation from X to X. We write "the relation $R: X \to Y$" as shorthand for "the relation R from X to Y." Hence, "the relation R in X" can also be written "the relation $R: X \to X$." This notation is in keeping with that used for functions (see Section 3.5).

Examples

(1) Let X denote the set of all people and Y denote the set of all houses. Say that xRy is true if and only if x owns y. Then R is the set of all (person, house) pairs in which the listed person owns the listed house. (2) Let X denote the set of all employees of ABC Company, and let the relation R in X be defined by xRy if and only if x likes y. Then R can be identified with the set of all (x, y) in $X \times X$ such that x likes y. (3) Let X denote the set of prizes in some contest, and let the relation R in X be defined by xRy if and only if you would not prefer winning x to winning y.

From now on we shall think of a relation $R: X \to Y$ (or $R: X \to X$) as simply a subset of $X \times Y$ (or of $X \times X$). There are many relations $R: X \to Y$—as many, in fact, as there are subsets of $X \times Y$. Figure 3.2 depicts a

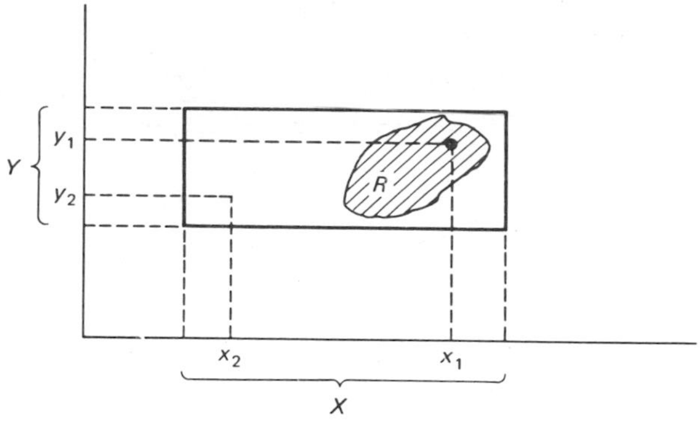

Figure 3.2

relation R: $X \to Y$; in this relation $x_1 R y_1$ is true but $x_2 R y_2$ is false, because $(x_1, y_1) \in R$ but $(x_2, y_2) \notin R$.

Figure 3.3 depicts the relation "$\leq$" between real numbers. The relation "$\leq$" consists of all points on or above the diagonal line $y = x$ in the plane.

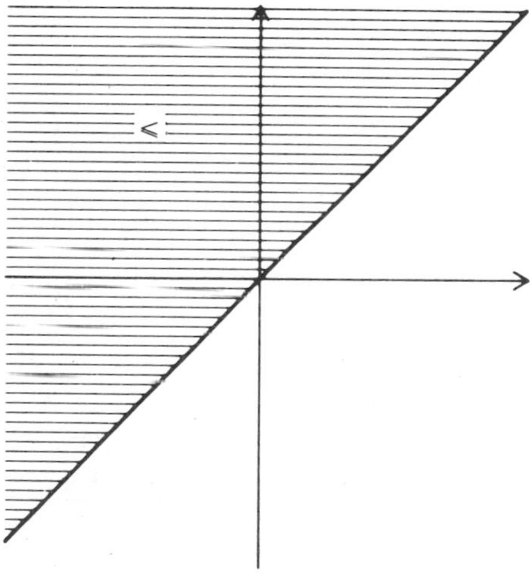

Figure 3.3

The *domain* $DOM(R)$ of a relation R: $X \to Y$ is the set of all $x \in X$ such that $(x, y) \in R$, for at least one $y \in Y$; that is,

$$DOM(R) = \{x: (x, y) \in R, \text{ for at least one } y \in Y\}.$$

Hence $DOM(R) \subset X$. Figure 3.4 depicts $DOM(R)$ for R as given in Figure 3.2. For "$\leq$," shown in Figure 3.3, $DOM(\leq) = X = R^1$.

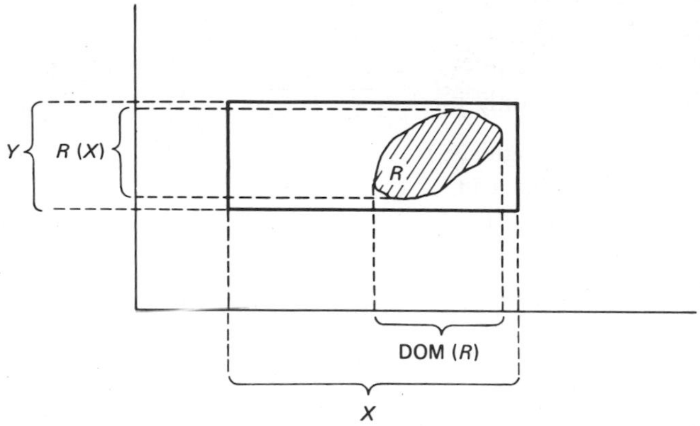

Figure 3.4

The *range* $R(X)$ of a relation $R: X \to Y$ is the set of all $y \in Y$ such that $(x, y) \in R$, for at least one $x \in X$; that is,

$$R(X) = \{y: (x, y) \in R, \text{ for at least one } x \in X\}.$$

Hence, $R(X) \subset Y$. Figure 3.4 shows $R(X)$ for R, the relation in Figure 3.2. For "$\leq$" as shown in Figure 3.3, we have $\leq(R^1) = R^1$. The range of a relation is a special case of the following idea.

Let A by any subset of X. The *R-image of A*, denoted by $R(A)$, is defined to be the set of all y such that $(x, y) \in R$ for at least one x *belonging to A*; that is,

$$R(A) = \{y: (x, y) \in R, \text{ for at least one } x \in A\}.$$

If $x \in A$ then clearly $x \in X$, because $A \subset X$; and hence for any $A \subset X$ we have $R(A) \subset R(X)$. More generally, you may verify as an exercise that if $A \subset B \subset X$, then $R(A) \subset R(B) \subset R(X)$. You may also verify that $R(\varnothing) = \varnothing$. If $A = \{x\}$, a one-element subset of X, we shall write $R(x)$ instead of $R(\{x\})$.

You may also show as an exercise that $R(x) \neq \varnothing$ if and only if $x \in DOM(R)$.

Now let $R_1: X \to Y$ and $R_2: X \to Y$ be any two relations from X to Y. Since R_1 and R_2 are subsets of $X \times Y$, we can form their union, intersection, complements, and relative complements, which are all relations (because they, too, are subsets of $X \times Y$).

To show that unions, intersections, complements, and relative complements of relations have perfectly natural interpretations, we shall use the xRy shorthand for "x is R-related to y." We have:

(1) $xR_1 \cup R_2y$ if and only if xR_1y *and/or* xR_2y;
(2) $xR_1 \cap R_2y$ if and only if xR_1y *and* xR_2y;
(3) $xR_1{}^cy$ if and only if x is *not* R_1-related to y; and
(4) $xR_1 \backslash R_2y$ if and only if xR_1y, but x is *not* R_2-related to y.

There are three distinguished relations. Since $\varnothing$ is a subset of $X \times Y$, $\varnothing$ is a relation, called the *void relation*. No x is $\varnothing$-related to any y; that is, $x\varnothing y$ is always false. The second distinguished relation is $U = X \times Y$, called the *universal* relation. Every x is U-related to every y; that is, xUy is always true.

The third distinguished relation is defined only when $Y = X$. We define the *identity relation* $I: X \rightarrow X$ by

$$I = \{(x, x): x \in X\}.$$

Let $x \in X$ and $y \in Y$. Then xIy if and only if $x = y$; that is, if and only if x and y denote the same element of X. For example, let $X = Y = R^1$; then $X \times Y = R^2$ and I is the 45-degree line constituting the boundary of "$\leq$" in Figure 3.3.

Now, suppose that $R: X \rightarrow Y$ is a given relation. We define the *inverse* R^{-1} of R as the relation $R^{-1}: Y \rightarrow X$ given by

$$R^{-1} = \{(y, x): (x, y) \in R\}.$$

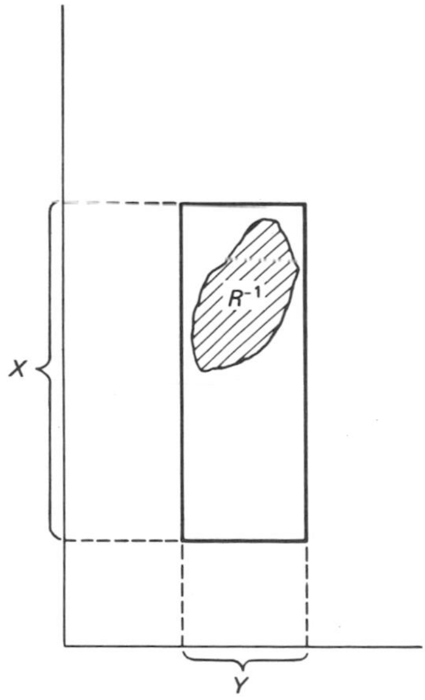

Figure 3.5

Hence $yR^{-1}x$ if and only if xRy. Figure 3.5 graphs R^{-1} for R, the relation, depicted in Figure 3.2. We emphasize the fact that R^{-1} is a subset of $Y \times X$, not of $X \times Y$.

Example

(4) Let X be all people and Y be all houses. Then say xRy if and only if x lives in y. Then R^{-1} is the relation "is lived in by"; that is, $yR^{-1}x$ if and only if y is lived in by x. Hence, roughly speaking, if R is "active voice," then R^{-1} is "passive voice," and vice versa.

By using the inverse R^{-1} of R, one can define $DOM(R)$ simply as $R^{-1}(Y)$, since (the reader may show as an exercise) $R(X) = DOM(R^{-1})$ and $R^{-1}(Y) = DOM(R)$.

The reader is asked to prove as Exercise 19 the following properties of relations $R: X \to Y$:

(1) if $A_i \subset X$ for every i in a nonempty index set I, then

$$R(\bigcup_{i \in I} A_i) = \bigcup_{i \in I} R(A_i);$$

(2) if $A_i \subset X$ for every i in a nonempty index set I, then

$$R(\bigcap_{i \in I} A_i) \subset \bigcap_{i \in I} R(A_i); \text{ and}$$

(3) if $A \subset X$ and $B \subset X$, then

$$R(B) \backslash R(A) \subset R(B \backslash A).$$

Finally, suppose $R: X \to Y$ and $S: Y \to Z$ are given relations. We define the *composition* $S \circ R$ of R and S as the relation from X to Z given by

$$S \circ R = \{(x, z): \text{for at least one } y \in Y \text{ we have } (x, y) \in R \text{ and } (y, z) \in S\}.$$

The reader may show as an exercise that $S \circ R \neq \varnothing \subset X \times Z$ if and only if $R(X) \cap DOM(S) \neq \varnothing \subset Y$.

Let $A \subset X$. We shall prove the *basic composition formula*

$$S \circ R(A) = S(R(A)).$$

[Since $A = \bigcup_{x \in A}\{x\}$, it follows from property (1) above that all we need prove is that $S \circ R(x) = S(R(x))$. (Recall that we drop the braces around one-element subsets $\{x\}$.) Now, $z \in S \circ R(x)$ if and only if there exists some $y \in Y$ such that $(x, y) \in R$ and $(y, z) \in S$, which is true if and only if $y \in R(x)$ and $z \in S(y) \subset S(R(x))$. But $S(R(x)) = S(\bigcup_{y \in R(x)}\{y\}) =$ (by property (1)) $\bigcup_{y \in R(x)} S(y)$. Hence $S \circ R(x) = S(R(x))$. For arbitrary $A \subset X$ we use property (1) repeatedly to obtain $S \circ R(A) \overset{*}{=} \bigcup_{x \in A} S \circ R(x) = \bigcup_{x \in A} S(R(x)) \overset{*}{=} S(\bigcup_{x \in A} R(x)) = S(R(\bigcup_{x \in A} x) \overset{*}{=} S(R(A))$, where "$\overset{*}{=}$" means "equals by virtue of property (1)."]

Example

(5) Let $X = Y = Z =$ the set of all people, and let R and S be defined by

xRy if and only if x is a sibling of y; and

xSy if and only if x is a child of y.

Then:

(1) $xR \cup Sy$ if and only if x is a sibling or a child of y;
(2) $xR \cap Sy$ if and only if x is both a sibling and a child of y (impossible, and hence $R \cap S = \varnothing$);
(3) xR^cy if and only if x is not a sibling of y; and
(4) $xR \circ Sz$ if and only if x is a child of some y and that y is a sibling of z; that is, x is a child of a sibling of z. That is, z is an aunt or uncle of x.

3.4 ORDERING RELATIONS

Some special cases of relations are very important in mathematics generally and in decision theory particularly. One of these cases generalizes the familiar concept of numerical equality by abstracting the following properties of "$=$": (1) $x = x$ for every x; (2) if $x = y$, then $y = x$; and (3) if $x = y$ and $y = z$, then $x = z$. Another case generalizes the following properties of "$\geq$": (1) $x \geq x$ for every x; (2) if $x \geq y$ and $y \geq x$, then $x = y$; and (3) if $x \geq y$ and $y \geq z$, then $x \geq z$.

We shall give relation-theoretic definitions of these and other properties first. Let $R: X \to X$ be a relation in X. Let I be the identity relation. Then

(1) R is *reflexive* if and only if $I \subset R$;
(2) R is *irreflexive* if and only if $I \cap R = \varnothing$;
(3) R is *symmetric* if and only if $R = R^{-1}$;
(4) R is *asymmetric* if and only if $R \cap R^{-1} = \varnothing$;
(5) R is *antisymmetric* if and only if $R \cap R^{-1} \subset I$;
(6) R is *transitive* if and only if $R \circ R \subset R$;
(7) R is *intransitive* if and only if $(R \circ R) \setminus R \neq \varnothing$; and
(8) R is *complete* if and only if $R \cup R^{-1} = X \times X$.

These eight properties are stated in a very concise manner; it will be instructive to see what they mean in xRy language:

(1′) R is *reflexive* if and only if xRx holds for every $x \in X$;
(2′) R is *irreflexive* if and only if xRx holds for no $x \in X$;
(3′) R is *symmetric* if and only if yRx whenever xRy;
(4′) R is *asymmetric* if and only if xRy and yRx cannot *both* be true;
(5′) R is *antisymmetric* if and only if $x = y$ whenever xRy and yRx;

(6′) R is *transitive* if and only if xRz whenever xRy and yRz;
(7′) R is *intransitive* if and only if there are three (not necessarily different) elements x, y, and z in X such that xRy and yRx, but xR^cz; and
(8′) R is *complete* if and only if either xRy and/or yRx must hold for every pair x and y of (not necessarily different) elements of X.

An *ordering relation* is any relation R which satisfies (6); that is, any *transitive* relation. An ordering relation is called a *weak order* if it is also *reflexive*; that is, if it satisfies (1) as well as (6). A weak order is also called a *quasi-order*.

Formally, $R: X \to X$ is a *quasi-order* if and only if it satisfies (1) and (6); that is, if and only if it is reflexive and transitive; in other words, if and only if

(1) xRx holds for every $x \in X$; and
(2) if xRy and yRz, then xRz.

A quasi-order which satisfies (5) as well as (1) and (6) is called a *partial order*, or a *proper order*. It follows that a partial order is reflexive, antisymmetric, and transitive. Hence R is a partial order if and only if

(1) xRx holds for every $x \in X$;
(2) xRy and yRx imply $x = y$; and
(3) xRy and yRz imply xRz.

A quasi-order which satisfies (8) as well as (1) and (6) is called a *total order*; that is, R is a *total order* if and only if R is reflexive, transitive, and complete.

A *partial* order which satisfies (8) as well as (1), (5), and (6) is called a *linear order*; that is, a linear order is an antisymmetric total order.

A *strong partial order* is a relation R satisfying (4) and (6). Thus R is a strong partial order if and only if R is transitive and *asymmetric*. [Note that asymmetry implies irreflexivity (property (2).]

In Figures 3.6 through 3.10, the circles represent the elements of X, and an arrow pointing from circle x to circle y denotes that xRy. All circles are assumed to represent different elements of X. A curved arrow from x to x indicates that xRx. All the following relations are assumed to be transitive (that is, to be orderings), and hence we can abbreviate

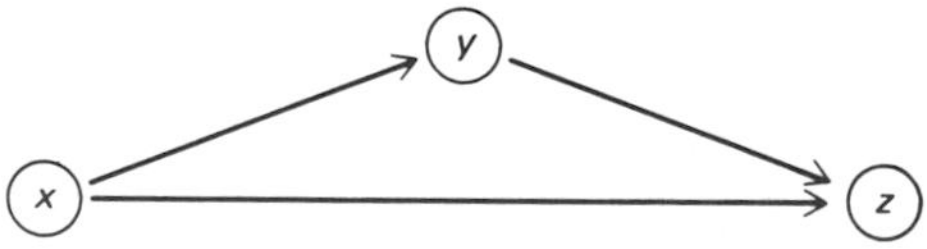

to

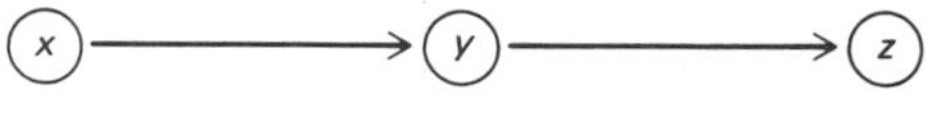

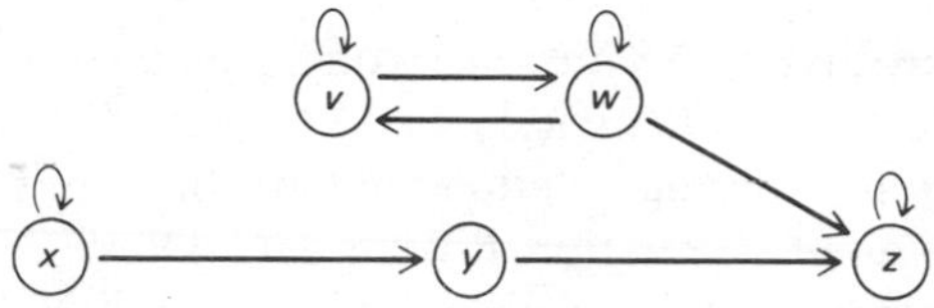

Figure 3.6 An Ordering Relation which Is Not a Quasi-order. (Here, yRy is not true and hence reflexivity fails.)

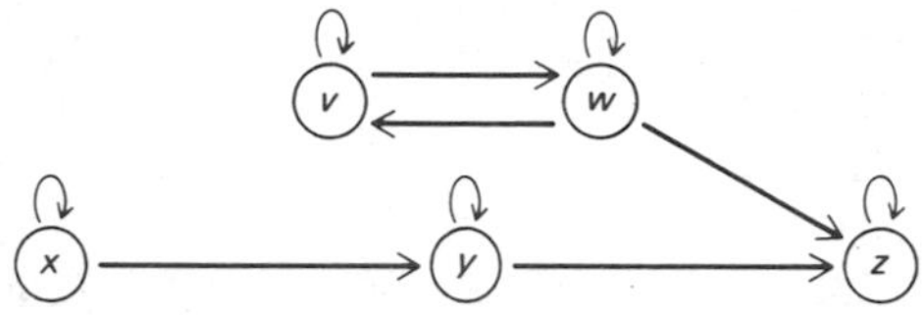

Figure 3.7 A Quasi-order which Is Not a Partial Order. (Here, yRy, so so R is reflexive. But $v \neq w$ and yet vRw and wRv; hence antisymmetry fails.)

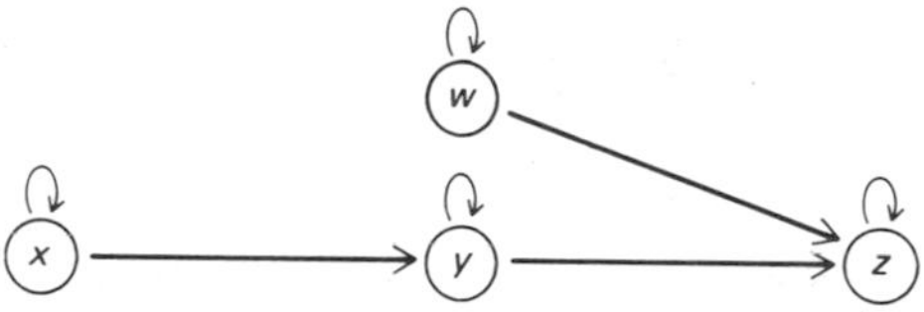

Figure 3.8 A Partial Order which Is Not a Linear Order. (Here, R is reflexive, antisymmetric, and transitive, but neither xRw nor wRx. Hence, $R \cup R^{-1} \neq X \times X$.)

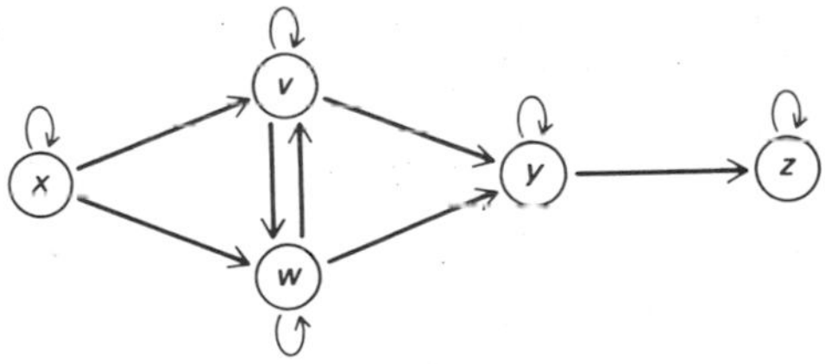

Figure 3.9 A Total Order which Is Not a Linear Order. (Here, R is reflexive, transitive, and complete, but not antisymmetric.)

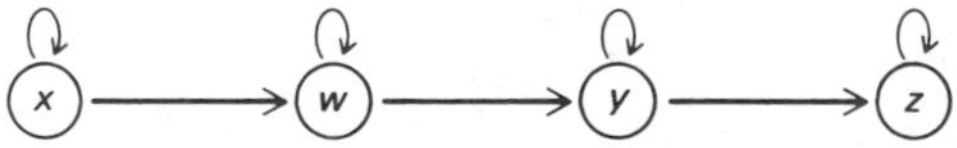

Figure 3.10 A Linear Order. (Here, we have "inserted" w as given in Figure 3.8 "between" x and y. This is tantamount to having added an arrow from x to w and an arrow from w to y and having removed the now unnecessary arrows from x to y and from w to z—unnecessary because of transitivity.)

We could make Figure 3.8 into a strong partial order by deleting the circular arrows denoting reflexivity.

A very important ordering relation that we have not yet considered is the *equivalence relation.* A relation R is an equivalence relation if and only if R is reflexive, symmetric, and transitive; that is, if and only if R possesses properties (1), (3), and (6). In other words, R is an equivalence relation if and only if

(1) xRx for every $x \in X$;
(2) yRx whenever xRy; and
(3) xRy and yRz imply xRz.

Equivalence relations possess a very important property; namely, $\{R(x): x \in X\}$ is a partition of X. This is equivalent to saying that: (1) $\bigcup_{x \in X} R(x) = X$; and (2) if $x \neq y$, then *either* $R(x) = R(y)$ *or* $R(x) \cap R(y) = \emptyset$. The second statement is a valid rendition of mutual exclusivity because we cannot distinguish between $R(x)$ and $R(y)$ if they have the same elements.

Proof

Since $x \in R(x)$, for every x, it follows that $X = \bigcup_{x \in X} R(x)$, and hence (1) holds. To prove (2), suppose *first* that xRy. By symmetry, this implies that yRx. Now let z be *any* element of $R(x)$; that is, let xRz be true. Then we have yRx and xRz, and so by transitivity we have yRz, which implies $z \in R(y)$. Hence we have shown that if xRy, then $z \in R(y)$ whenever $z \in R(x)$. Hence $R(x) \subset R(y)$. On the other hand, if xRy and zRy, then (by symmetry) yRz and so (by transitivity) xRz. Thus $z \in R(y)$ implies $z \in R(x)$. Hence $R(y) \subset R(x)$. Hence $R(x) = R(y)$ whenever xRy. Now suppose, *second,* that xR^cy. We shall show that in this case $R(x) \cap R(y) = \emptyset$ by *reductio ad absurdum.* Specifically, suppose to the contrary that $z \in R(x) \cap R(y)$. Then xRz and yRz. By symmetry, yRz implies zRy. By transitivity, xRz and zRy imply xRy, which contradicts our assumption that xR^cy. Hence no such z can exist. Therefore xR^cy implies $R(x) \cap R(y) = \emptyset$. This concludes the proof of the second assertion.

The preceding proof actually yielded more than the second assertion. We showed that if xRy, then $R(x) = R(y)$; and if xR^cy, then $R(x) \cap R(y) = \emptyset$.

Each subset $R(x) \subset X$ is called an *equivalence class.* We have just proved that the set of equivalence classes is a partition of X.

In Figure 3.11 the shaded subset of $X \times X$ represents an equivalence relation R, and $Q_1, \ldots, Q_5$ denote the equivalence classes of R.

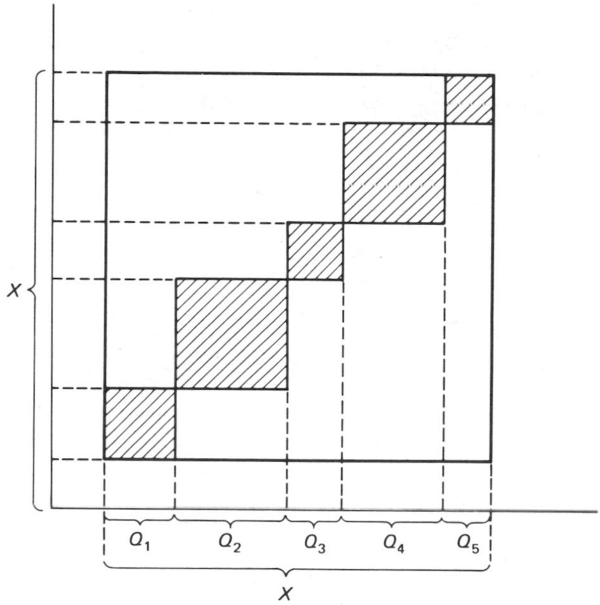

Figure 3.11 An Equivalence Relation.

The reader may show as an exercise that the identity relation I is an equivalence relation.

Let R be a quasi-order, and recall that unless R is a partial order it does not follow that xRy and yRx imply $x = y$. Hence $(R \cap R^{-1})\backslash I \neq \varnothing$ unless R is a partial order. The reader may show as an exercise that $R \cap R^{-1}$ is an equivalence relation. Roughly speaking, the set of all y's such that xRy and yRx are "lumped together" into one equivalence class of $R \cap R^{-1}$. For example, the set of equivalence classes generated by the quasi-order in Figure 3.7 is: $\{\{x\}, \{y\}, \{z\}, \{v, w\}\}$.

3.5 FUNCTIONS

Let f: $X \rightarrow Y$ be a relation. The symbol f is called a *function* if and only if

(1) $DOM(f) = X$; and
(2) for every $x \in X$, $f(x)$ consists of a single element of Y.

Thus a function is a relation whose domain is all of X and such that the image of every x is a unique $y \in Y$. Thus so-called "multiple-valued functions" are *not* functions according to our definition.

Synonyms for function are: *mapping*, *transformation*, and (in special cases) *operator* and *functional*.

A function f is, like a relation in general, identified with its *graph*, which can be written

$$f = \{(x, f(x)) \colon x \in X\}.$$

Figure 3.12 depicts a relation which is not a function, because for some x's $f(x)$ is a many-element subset of Y.

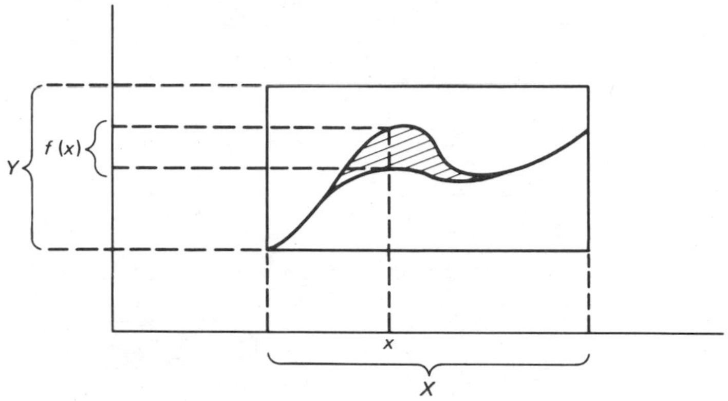

Figure 3.12 A Relation which Is Not a Function

Figure 3.13 depicts a relation which *is* a function.

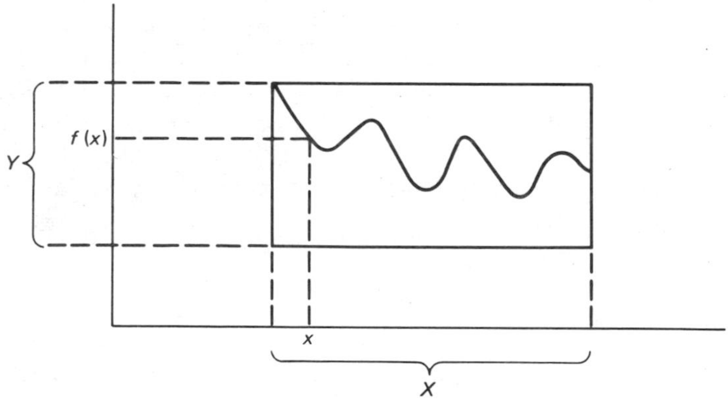

Figure 3.13 A Function

Yet another equivalent way of defining a function is to say that a function is a relation f such that for each $x \in X$ there exists precisely one $y \in Y$ such that $\{y\} = f(x)$. Again, we usually drop the braces around y.

All results in Section 3.3 on relations continue to hold for functions. However, the union, intersection, and relative complement of functions is usually no longer a function. The inverse of a function need not be a function; for example, let $X = Y = R^1$ and define f by setting $f(x) = 5$ for every $x \in R^1$. Then $f^{-1}(5) = R^1$, and so $f^{-1}(5)$ is certainly not a unique element of R^1.

The *composition* $g \circ f$ of the functions f: $X \rightarrow Y$ and g: $Y \rightarrow Z$ is a function and can be defined by specializing the definition of composition of relations. However, it is simpler to define $g \circ f$ as the function from X to Z satisfying $g \circ f(x) = g(f(x))$ for every $x \in X$.

If X and Y are finite sets and f: $X \rightarrow Y$ is a function, then f has a very simple representation: list the elements of X in a column, list the elements of Y in another column (to the right of the X column), and draw an arrow from each x to that y such that $y = f(x)$. In Figure 3.14, we assume $X = \{1, 2, 3, 4\}$, $Y = \{a, b, c\}$, and $f = \{(1, a), (2, a), (3, b), (4, b)\}$.

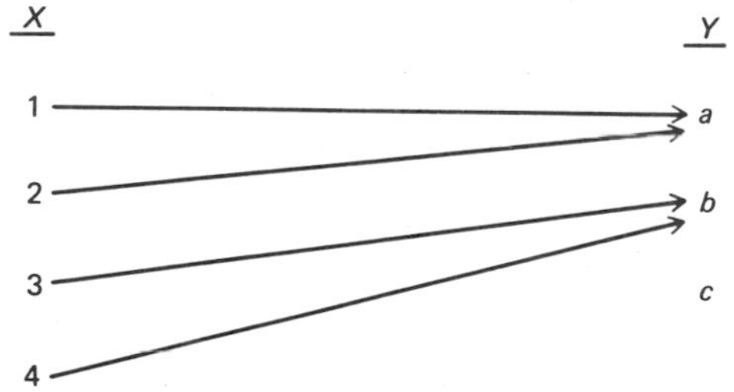

Figure 3.14 A Function f: $X \rightarrow Y$

We cannot emphasize too strongly that f is the set of *all ordered pairs* $(x, f(x))$, and that $f(x)$ is not a function but rather an element of Y. However, many authors use the phrases "the function $f(x)$," "the function $f(x) = x^2$," "$f(x)$ is an interesting function," and so forth. We shall occasionally use such phrases ourselves in later chapters and they should not cause undue confusion.

However, a preferable way of referring to functions f is to replace x in $f(x)$ by a dot. Thus a common and perfectly acceptable way of referring to f is to write $f(\cdot)$. This dot notation is especially useful when the function depends upon some parameters (that is, arbitrary constants) and you wish to assume particular values for the parameters. Thus the function f: $R^1 \rightarrow R^1$ whose value $f(x)$ at any $x \in R^1$ is $ax^2 + bx + c$, could be denoted in general by $f(\cdot \mid a, b, c)$; and, later in his discussion, if the reader wishes to specify $a = 10$, $b = 0$, and $c = -2$, he could use $f(\cdot \mid 10, 0, -2)$ in place of $f(\cdot \mid a, b, c)$.

Since a function is a relation, the inverse of f is the well-defined relation f^{-1}: $Y \rightarrow X$. As noted previously, however, f^{-1} need not be a function, even if we consider f^{-1} as defined only on its domain [so that property (1) of a function will be automatically satisfied]. The example in Figure 3.14 is a good illustration of this point: $DOM(f^{-1}) = \{a, b\} \subset Y$, and f^{-1}: $DOM(f^{-1}) \rightarrow X$ obviously satisfies (1). However, $f^{-1}(a) = \{1, 2\}$, and so property (2) fails. Hence f^{-1} is not a function in this case.

Nevertheless, f^{-1} is a very well-behaved relation. It possesses not only the numbered properties (1) through (3) of relations stated in Section 3.3, but also the following stronger versions of the second and third properties:

(2′) if $A_i \subset Y$, for every i in a nonempty index set I, then

$$f^{-1}(\bigcap_{i\in I} A_i) = \bigcap_{i\in I} f^{-1}(A_i); \text{ and}$$

(3′) if $A \subset Y$ and $B \subset Y$, then

$$f^{-1}(A)\backslash f^{-1}(B) = f^{-1}(A\backslash B).$$

The reader may prove (2′) and (3′) as an exercise.

We shall now prove an important property of the inverse f^{-1}: $Y \rightarrow X$ of a function f; namely, $\{f^{-1}(y): y \in f(X)\}$ is a *partition* $\mathcal{P}_f$ of X.

Proof

We shall proceed to demonstrate that definitional requirements (1) and (2) for partitions, stated in Section 2.5, are satisfied. First, since $DOM(f) = X$, it follows that for every $x \in X$ we have $f(x) \in f(X)$ and hence $x \in f^{-1}(f(x))$. Hence every x is in some element of $\mathcal{P}_f$, and this implies requirement (1). Second, if $y_1 \in f(X)$, $y_2 \in f(X)$ and $y_1 \neq y_2$, then we shall show that $f^{-1}(y_1) \cap f^{-1}(y_2) = \varnothing$, thus proving that requirement (2) for partitions is met. Suppose to the contrary that there is an element $z \in X$ such that $z \in f^{-1}(y_1) \cap f^{-1}(y_2)$. Since $z \in f^{-1}(y_1)$, it follows that $f(z) = y_1$. But since $z \in f^{-1}(y_2)$ it also follows that $f(z) = y_2$. But $y_1 \neq y_2$ and so f cannot be a function. This contradicts the assumption that f^{-1} is the inverse of a function f. Hence there can be no element $z \in X$ such that $z \in f^{-1}(y_1) \cap f^{-1}(y_2)$. Hence $f^{-1}(y_1) \cap f^{-1}(y_2) = \varnothing$.

As an example of how a function f induces a partition $\mathcal{P}_f$, consider the function given in Figure 3.14. Since $f(X) = \{a, b\}$, we need only note that $\{f^{-1}(a), f^{-1}(b)\} = \{\{1, 2\}, \{3, 4\}\}$ is a partition of X.

A function f from an arbitrary set X to an arbitrary set Y may possess one or both of two important properties:

(1) we say f: $X \rightarrow Y$ is *surjective*, or *a surjection*, if and only if $f(X) = Y$;
(2) we say f: $X \rightarrow Y$ is *injective*, or *an injection*, if and only if $f(x_1) \neq f(x_2)$ whenever $x_1 \neq x_2$; and
(3) we say f: $X \rightarrow Y$ is *bijective*, or *a bijection*, if and only if f is both a surjection and an injection.

Older terminology: (1) f is *onto* Y; (2) f is *one-to-one*; and (3) f is *one-to-one onto* Y, or f is a *one-to-one correspondence* from X to Y.

The function f in Figure 3.14 is neither surjective nor injective. Since $c \notin f(X)$, f is not a surjection, and since $f(1) = f(2)$ but $1 \neq 2$, f is not an injection. Figure 3.15 depicts a surjection; Figure 3.16, an injection; and Figure 3.17, a bijection.

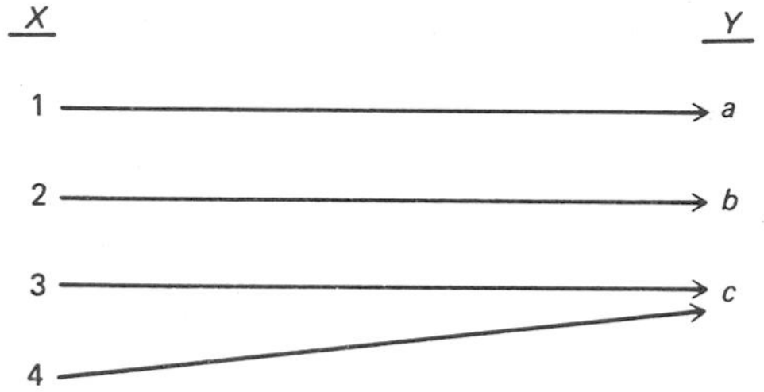

Figure 3.15 A Surjection [not an injection because $f(3) = f(4)$]

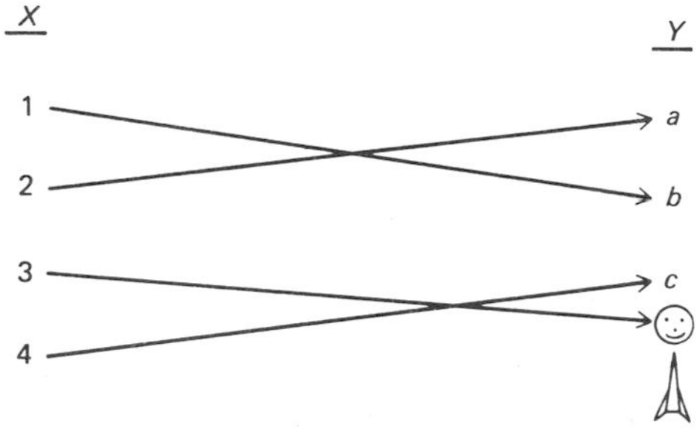

Figure 3.16 An Injection [not a surjection because  $\notin f(X)$]

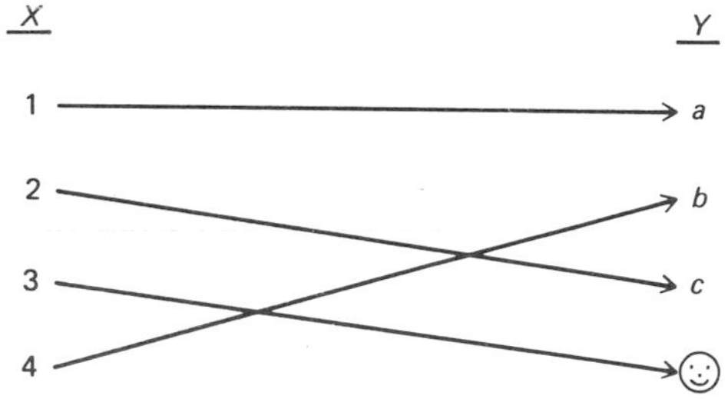

Figure 3.17 A Bijection

The reader may verify as an exercise that the function graphed in Figure 3.13 is neither a surjection nor an injection.

Next, note that nothing in the definition of function (or in the definition of relation) prevents X from being a Cartesian product $X = X_1 \times \cdots \times X_n$, or prevents Y from being a Cartesian product, $Y = Y_1 \times \cdots \times Y_m$. Indeed, one often has reason to consider functions f of the form

$$f\colon X_1 \times \cdots \times X_n \to Y_1 \times \cdots \times Y_m.$$

When all X_i's and Y_j's equal R^1, then $X_1 \times \cdots \times X_n = R^n$ and $Y_1 \times \cdots \times Y_m = R^m$, and so we would consider functions f of the form

$$f\colon R^n \to R^m.$$

When $m = 1$ we speak of a "real-valued function of n real variables," and when $n = 1$ also, we speak of a "real-valued function of a real variable."

EXERCISES

1. Let X denote the set of all positive integers, and let $Y = \{1, 2, 3, \text{go}\}$. Does (go, 15) belong to $X \times Y$? Does (go, 15) belong to $Y \times X$?
2. Let $X = \{1, 2, 3\}$, $Y = \{$$\}$, and $Z = \{a, b\}$. List all elements of $X \times Y \times Z$.
3. Consider in Figure 3.18 relation R in $X \times Y$:

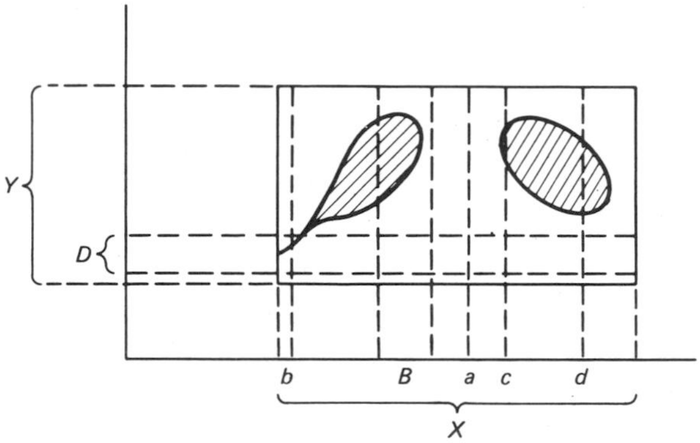

Figure 3.18

True or false:

(a) $a \in DOM(R)$
(b) $c \in R^{-1}(D)$
(c) $d \in R^{-1}(D)$
(d) $D \cap R(B) = \varnothing$
(e) $b \in R^{-1}(D)$
(f) R is reflexive
(g) R is symmetric
(h) R is a function

4. Let $X = \{a, b, c\}$ and let $Y = \{d, e, f\}$. Then $X \times Y$ may be depicted as follows:

f	·	·	·
e	·	·	·
d	·	·	·
	a	b	c

Let $R: X \to Y$ be defined by the following true statements: aRe, aRd, bRe, and cRf.

(a) Represent R as a subset of $X \times Y$ by circling those "dots" (x, y) in the preceding graph for which xRy is true.
(b) Is R a function?
(c) Specify R^{-1} as a subset of $Y \times X$.
(d) Specify R^c as a subset of $X \times Y$.
(e) Find $R \cup R^c$ as a subset of $X \times Y$.

5. Let $X = \{a, b, c\}$ and let $Y = X$. Define the relation R in X by the following graph:

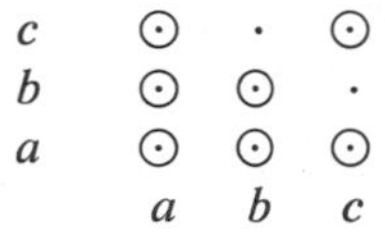

(a) Is R reflexive?
(b) Is R symmetric?
(c) Is R transitive? [*Hint*: $(b, a) \in R$ and $(a, c) \in R$.]

6. Let $X = Y = \{a, b, c\}$ as in Exercise 5. Let R be as given in the following graph:

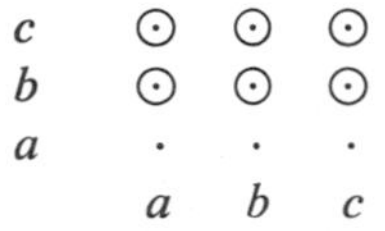

(a) Is R reflexive?
(b) Is R symmetric?
(c) Is R transitive?

7. Let $X = Y = \{a, b, c, d, e\}$, and let R be as given in the following graph:

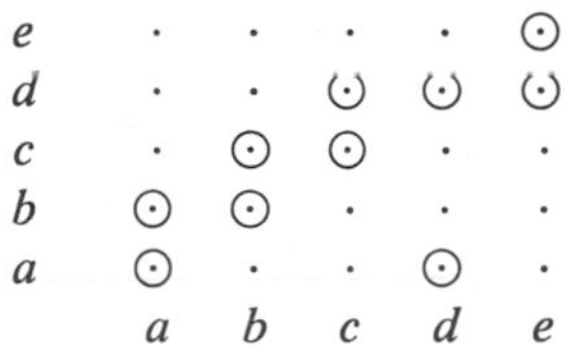

(a) What is $R \circ R$?
(b) Show that $R \subset R \circ R$.
(c) Is R transitive?
(d) What is $(R \circ R) \circ (R \circ R)$?
(e) Is $R \circ R$ transitive?

8. Show that if R is any *reflexive* relation, then $R \subset R \circ R$. (*Hint*: first show that if R, T_1, and T_2 are relations and if $T_1 \subset T_2$, then $R \circ T_1 \subset R \circ T_2$. Then show that $R \circ I = R$, where I is the identity relation. Finally, note that "R is reflexive" means "$I \subset R$" and apply your first result for $T_2 = R$.)

9. Show that composition of relations is *associative*; that is, for any relations $R_1: X \to Y$, $R_2: Y \to Z$, and $R_3: Z \to W$ we have $R_1 \circ (R_2 \circ R_3) = (R_1 \circ R_2) \circ R_3$ (a relation from X to W). Thus we may write $R_1 \circ R_2 \circ R_3$ without ambiguity.

10. Define R^m by $R^m = R \circ R \circ \ldots \circ R$ (a total of m "factors").

    (a) Show that if R is transitive, then $R^m \subset R$ for every m.
    (b) In Exercise 8 you were asked to derive R^2 and R^4. Derive R^3 and R^5.

11. Show that if $R_j: X \to X$ is a transitive relation in X, for every j in a nonempty index set J, then $\bigcap_{j \in J} R_j$ is a transitive relation in X. ($\varnothing \subset X \times X$ is transitive.)

12. Apply Exercise 11 to show that if R is *any* relation in X, then there is a "smallest" *transitive* relation $R\#$ including R; that is, there exists a unique transitive relation $R\#$ satisfying

    (1) $R \subset R\#$; and
    (2) if T is a transitive relation with $R \subset T$, then $R\# \subset T$.

    [*Hint*: Let $R\#$ be the intersection of all transitive relations T such that $R \subset T$. Such transitive relations exist, since $X \times X$ is transitive, and $R \subset X \times X$. (*Note:* Exercise 12 implies that any relation R "generates" an ordering on X.)]

13. Show that if $R_j: X \to X$ is a reflexive and transitive relation in X for every j in a nonempty index set J, then $\bigcap_{j \in J} R_j$ is a reflexive and transitive relation in X. (*Hint*: if $I \subset R_j$, for every $j \in J$, then $I \subset \bigcap_{j \in J} R_j$.)

14. Use Exercise 13 to show that if R is *any* relation in X, then there is a smallest quasi-order R^* including R; that is, there exists a unique quasi-order R^* such that:

    (1) $R \subset R^*$; and
    (2) if T is a quasi-order with $R \subset T$, then $R^* \subset T$.

    [*Hint*: Let R^* be the intersection of all quasi-orders (reflexive and transitive relations) T such that $R \subset T$. Since $X \times X$ is a quasi-order staisfying (1), the set of quasi-orders satisfying (1) is nonempty and hence the intersection is well-defined.]

15. Assume as given a set $\{c_1, \ldots, c_n\} = C$ of possible *consequences* of a decision problem. Any given c_i might be a monetary return, or it might be an n-tuple of payoffs in different commodities, or it might be a complex physical and psychological condition. All that we require is that each c_i be well-defined. Suppose that the decision-maker has determined a total order R on C with the interpretation that cRc' if and only if c is at least as desirable as c' for him.

    (a) Let $E = R \cap R^{-1}$. Show that E is an equivalence relation. [Property (8), completeness, is not needed for this proof, and hence the assertion holds for any quasi-order R.]
    (b) What is the interpretation of E; that is, what does cEc' mean to the decision-maker?

(c) Let $P = R \backslash E$. Show that P is irreflexive, asymmetric, and transitive.
(d) Show that $P \cup P^{-1} = (C \times C) \backslash E$. (This is a weak form of completeness.)
(e) What is the interpretation of P; that is, what does cPc' mean to the decision-maker?
(f) Let $C^* = \{C^*_1, \ldots C^*_k\}$ be the partition of C induced by E. Define a relation P^* in C^* by asserting that $C^*_i P^* C^*_j$ is true if and only if cPc' for all $c \in C^*_i$ and all $c' \in C^*_j$. Show that P^* is a strong partial order of C^*; and furthermore, that P^* is a "strong linear order" in the sense that P^* is transitive and asymmetric and also satisfies $P^* \cup P^{*-1} = (C^* \times C^*) \backslash I^*$, where I^* is the identity relation on C^*. [The last property, like the assertion of (d), is a weak form of completeness.]
(g) Let $R^* = P^* \cup I^*$. Show that R^* is a *linear* order of C^*. (Briefly, the assumption that R is a total order of C means that the decision-maker can *rank* all the elements of C from most to least desirable, the relative ranking of equally desirable consequences being arbitrary. Moreover, E subsumes equally desirable consequences, and R^* preserves the ranking of R after the lumping together.)

16. Show that if $A \subset B \subset X$, then $R(A) \subset R(B) \subset R(X)$ for any relation $R: X \to Y$.

17. Show that $R(x) \neq \varnothing \subset Y$, if and only if $x \in DOM(R)$.

18. Show that $R(X) = DOM(R^{-1})$ and $R^{-1}(Y) = DOM(R)$.

19. Prove the following properties of relations $R: X \to Y$:

(1) if $A_i \subset X$, for every i in a nonempty index set I, then

$$R(\cup_{i \in I} A_i) = \cup_{i \in I} R(A_i);$$

(2) if $A_i \subset X$, for every i in a nonempty index set I, then

$$R(\cap_{i \subset I} A_i) \subset \cap_{i \in I} R(A_i); \text{ and}$$

(3) if $A \subset X$ and $B \subset X$, then

$$R(B) \backslash R(A) \subset R(B \backslash A).$$

20. Show that $T \circ R \neq \varnothing \subset X \times Z$, if and only if $R(X) \cap DOM(T) \neq \varnothing \subset Y$, where $R: X \to Y$ and $T: Y \to Z$ are arbitrary relations.

21. Show that the identity relation I is an equivalence relation in X; and show that if E is any equivalence relation in X, then $E \backslash I$ is symmetric and transitive.

22. In Exercise 19, suppose that $R = f^{-1}$, where f: $Y \to X$ is a function. Show that "$\subset$" can be replaced by "$=$" in (2) and (3).

23. Let f: $X \to Y$ be a function. Define a relation E in X by saying that $x_1 E x_2$ if and only if $f(x_1) = f(x_2)$. Show that E is an equivalence relation, and that the set of E-equivalence classes is precisely $\mathcal{P}_f = \{f^{-1}(y): y \in f(X)\}$.

24. Let $X = Y = \{a, b, c, d, e\}$ and let $f = \{(a, b), (b, c), (c, d), (d, e), (e, e)\}$. Let $f^2 = f \circ f$, $f^3 = f \circ f^2$, $f^4 = f \circ f^3$, and so forth. Show that $f^4 = f^5 = f^6 =$

. . . . (Suppose a, b, c, d, and e represent possible "states" of a physical system at any point in time, and that f represents the evolution of the system in one time period. Thus, if the system is in state c now, it will be in state d one time period in the future. If the states represent "energy levels," and if $a > b > c > d > e$, what can we say about the ultimate energy level of the system? The appropriate answer to this question is a discrete and deterministic version of the second law of thermodynamics.)

4

CARTESIAN PRODUCTS (II): TOWARD DECISION THEORY AND PROBABILITY

4.1 INTRODUCTION

In this chapter we shall consider a number of issues related to Cartesian products and their application in decision theory. Section 4.2 concerns sets of subsets of $X \times Y$ (such as partitions and fields) and their relationships with given sets of subsets of X and of Y. For example, there is a smallest field of subsets of $X \times Y$ associated in a natural way with given fields $\mathfrak{F}_X$ and $\mathfrak{F}_Y$ of subsets of X and of Y respectively.

In Section 4.3 we introduce the general model of decision-making with or without "information." In this model we are concerned with several sets and functions from some to others.

In Section 4.4 we introduce the notion of *events* concerned with uncertain states or information messages, and we argue that sets of events should be fields of a certain type.

Finally, in Sections 4.5 and 4.6 we define the terms *random variable* and *random vector* as certain real-valued (or vector-valued) functions which make numerical statements about events.

Thus this chapter concludes our explicit discussion of the "new mathematics"; it introduces the general model of the central theme of this book—decision theory; and it lays the groundwork for probability theory, which will occupy us for several succeeding chapters.

4.2 SETS OF SUBSETS OF $X \times Y$

Since $X \times Y$ is a set, all results regarding union, intersection, and difference of subsets of $X \times Y$ are well-defined, exactly as in Chapter 1. Moreover, the power set $\mathcal{P}(X \times Y)$ is well-defined; and we can discuss fields, sigma-fields, and partitions of $X \times Y$ just as in Chapter 2. If the reader has doubts about these statements, he should just note that $X \times Y$ is the set Z defined by

$$Z = \{z: z = (x, y), \text{ for some } x \in X \text{ and some } y \in Y\},$$

and apply *all* definitions and results to the set Z of all elements z.

Hence it would serve no useful purpose to rehash the material in Chapters 1 and 2 just to replace notation. Of much more interest is the relationship of certain sets of subsets of $X \times Y$ (for example, partitions, fields, and sigma-fields) with the *corresponding* sets of subsets of X and of Y. We could go in two directions in investigating this relationship: (1) by considering a given set of subsets of $X \times Y$ and asking what sets of subsets of X and of Y are, in some stipulated sense, most intimately related to the given set of subsets; or (2) by considering given sets of subsets of X and of Y and asking what set of subsets of $X \times Y$ is most intimately related to the given sets of subsets. We shall pursue the latter track, as it is the more fruitful for our purposes.

We start by showing that a partition $\mathcal{P}_X$ of X and a partition $\mathcal{P}_Y$ of Y generate, in a natural way, a partition of $X \times Y$.

Let $\mathcal{P}_X = \{A_i: i \in I\}$ be a partition of X and let $\mathcal{P}_Y = \{B_j: j \in J\}$ be a partition of Y. Then $\{A_i \times B_j: i \in I, j \in J\}$ is a partition of $X \times Y$, called the *Cartesian product* of $\mathcal{P}_X$ and $\mathcal{P}_Y$ and, denoted by $\mathcal{P}_X \times \mathcal{P}_Y$. It is important not to confuse the definition of the Cartesian product of two partitions with the product of partitions as defined in Section 2.5. The Cartesian product of $\mathcal{P}_X$ and $\mathcal{P}_Y$ is a set of subsets of $X \times Y$; so even if $X = Y$ the *Cartesian* product of two partitions of X is a set of subsets of $X \times X$, whereas the product of two partitions of X is a set of subsets of X. Figure 4.1 shows the Cartesian product of partitions $\mathcal{P}_X$ and $\mathcal{P}_Y$ of the intervals X and Y in R^1, respectively.

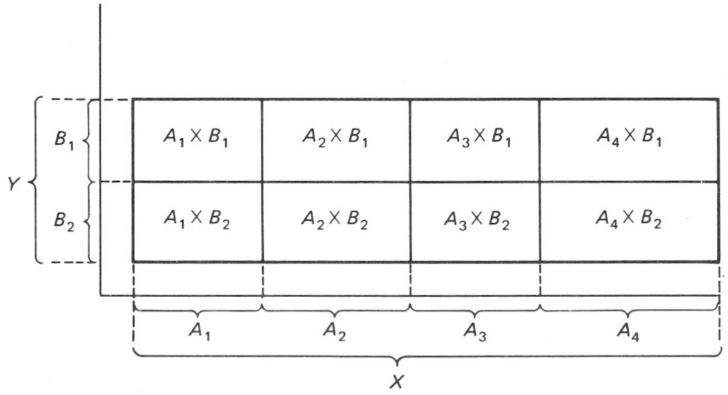

Figure 4.1

Note in our "intuition-bolstering" Figure 4.1 that every set of the form $A_i \times B_j$ is a rectangle. This is a consequence of A_i and B_j being intervals. We shall generalize the term *rectangle* by saying that *any* set $A \times B \subset X \times Y$ is a rectangle if $A \subset X$ and $B \subset Y$. Figure 4.2 depicts a set which is a rectangle according to our generalized definition.

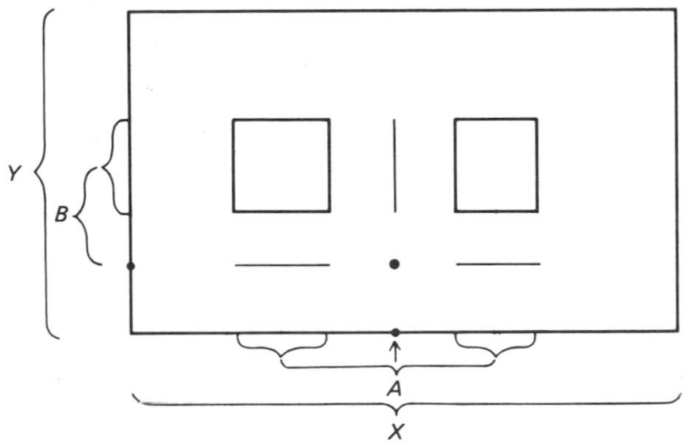

Figure 4.2

We can now generalize the notion of Cartesian product of partitions. Let $\mathfrak{A}$ be any set of subsets of X and let $\mathfrak{B}$ be any set of subsets of Y. Define the *Cartesian product* $\mathfrak{A} \times \mathfrak{B}$ of $\mathfrak{A}$ and $\mathfrak{B}$ by

$$\mathfrak{A} \times \mathfrak{B} = \{A \times B\colon A \in \mathfrak{A}, B \in \mathfrak{B}\}.$$

(When $\mathfrak{A}$ and $\mathfrak{B}$ are partitions of X and Y, respectively, this definition reduces to the preceding.) Thus $\mathfrak{A} \times \mathfrak{B}$ is a set of (generalized) rectangles in $X \times Y$.

Next, note by examining $A_3 \times B_2$ in Figure 4.1 that the complement of a rectangle need *not* be a rectangle; $(X \times Y)\backslash(A_3 \times B_2)$ is not a rectangle. Hence $\mathfrak{F}_X \times \mathfrak{F}_Y$ need not be a field of subsets of $X \times Y$ even though $\mathfrak{F}_X$ and $\mathfrak{F}_Y$ are fields of subsets of X and of Y, respectively.

Nevertheless, we can show that for any set $\mathfrak{A}$ of subsets of X and set $\mathfrak{B}$ of subsets of Y, the set $\mathfrak{A} \times \mathfrak{B}$ of rectangles in $X \times Y$ *generates*, in a natural manner, a field $\mathfrak{F}(\mathfrak{A} \times \mathfrak{B})$ of subsets of $X \times Y$. The field $\mathfrak{F}(\mathfrak{A} \times \mathfrak{B})$ will be the *smallest* field of subsets of $X \times Y$ which includes $\mathfrak{A} \times \mathfrak{B}$, and it is called the field *generated* by $\mathfrak{A} \times \mathfrak{B}$. (If the reader worked Exercise 10 of Chapter 2 he may omit the following paragraph and its proof, since he can consider $\mathfrak{A} \times \mathfrak{B}$ as the given set $\mathfrak{A}$ of subsets in that exercise.)

We assert that there exists a unique field $\mathfrak{F}(\mathfrak{A} \times \mathfrak{B})$ of subsets of $X \times Y$ with the following properties:

(1) $\mathfrak{A} \times \mathfrak{B} \subset \mathfrak{F}(\mathfrak{A} \times \mathfrak{B})$; and
(2) if $\mathfrak{A} \times \mathfrak{B} \subset \mathfrak{F}$, then $\mathfrak{F}(\mathfrak{A} \times \mathfrak{B}) \subset \mathfrak{F}$, where $\mathfrak{F}$ denotes any field of subsets of $X \times Y$.

Proof

Let F denote the set of all fields $\mathfrak{F}$ of subsets of $X \times Y$ such that $\mathcal{A} \times \mathcal{B} \subset \mathfrak{F}$. F is nonempty, since $\mathcal{R}(X \times Y)$ is a field and $\mathcal{A} \times \mathcal{B} \subset \mathcal{R}(X \times Y)$. Consider the set $\mathfrak{F}_0$ of subsets of $X \times Y$ defined by: $\mathfrak{F}_0 = \{A: A \in \mathfrak{F}$, for every $\mathfrak{F} \in F\}$. Since $A \in \mathfrak{F}$ implies $A^c \in \mathfrak{F}$, for any field $\mathfrak{F}$, we have that if $A \in \mathfrak{F}_0$, then $A^c \in \mathfrak{F}_0$ because F is a set of fields. Moreover, if $A \in \mathfrak{F}$ and $B \in \mathfrak{F}$, then $A \cup B \in \mathfrak{F}$, for every field $\mathfrak{F}$. Hence if $A \in \mathfrak{F}_0$ and $B \in \mathfrak{F}_0$, then $A \cup B \in \mathfrak{F}$ for every $\mathfrak{F} \in F$ and hence $A \cup B \in \mathfrak{F}_0$. Hence $\mathfrak{F}_0$ is a field, and clearly $\mathfrak{F}_0 \subset \mathfrak{F}$, for every $\mathfrak{F} \in F$. Finally, $\mathcal{A} \times \mathcal{B} \subset \mathfrak{F}_0$ because every $A \times B \in \mathcal{A} \times \mathcal{B}$ is also an element of every $\mathfrak{F} \in F$ (by definition of F). Hence $\mathfrak{F}_0$ is the smallest field of subsets of $X \times Y$ which includes $\mathcal{A} \times \mathcal{B}$. It is clearly unique.

We denote the field $\mathfrak{F}(\mathfrak{F}_X \times \mathfrak{F}_Y)$ of subsets of $X \times Y$ generated by fields $\mathfrak{F}_X$ and $\mathfrak{F}_Y$ of subsets of X and of Y respectively by the somewhat more compact notation $\mathfrak{F}_X \otimes \mathfrak{F}_Y$.

Examples

(1) Let $X = \{1, 2\}$, $Y = \{a, b\}$, $\mathfrak{F}_X = \{\varnothing, \{1, 2\}\}$, $\mathfrak{F}_Y = \{\varnothing, \{a\}, \{b\}, \{a, b\}\}$. Then $X \times Y = \{(1, a), (1, b), (2, a), (2, b)\}$ and $\mathfrak{F}_X \times \mathfrak{F}_Y = \{\varnothing, \{(1, a), (2, a)\}\}, \{(1, b), (2, b)\}, \{(1, a), (2, a), (1, b), (2, b)\}\}$. It is easy to verify that *in this case* $\mathfrak{F}_X \times \mathfrak{F}_Y$ is a field and hence equals $\mathfrak{F}_X \otimes \mathfrak{F}_Y$. (2) Consider $\mathcal{P}_X \times \mathcal{P}_Y$ as given in Figure 4.1. The reader may verify that $\mathfrak{F}(\mathcal{P}_X \times \mathcal{P}_Y)$ is the set of subsets of $X \times Y$ which consists of all unions of elements of $\mathcal{P}_X \times \mathcal{P}_Y$. [This result holds for any two *finite* partitions $\mathcal{P}_X$ of X and $\mathcal{P}_Y$ of Y. More generally, if $\mathcal{P}_X$ and $\mathcal{P}_Y$ are arbitrary partitions of X and Y respectively, then $\mathfrak{F}(\mathcal{P}_X \times \mathcal{P}_Y)$ is the set consisting of all unions of a finite number of elements of $\mathcal{P}_X \times \mathcal{P}_Y$ and all complements of unions of a finite number of elements of $\mathcal{P}_X \times \mathcal{P}_Y$.]

What we have done with fields can also be done with sigma-fields. From the parenthesized portion of Exercise 10 in Chapter 2, it follows that if $\mathcal{A}$ and $\mathcal{B}$ are arbitrary sets of subsets of X and Y respectively, then there exists a (unique) sigma-field $\mathcal{S}(\mathcal{A} \times \mathcal{B})$ of subsets of $X \times Y$ with the following properties:

(1) $\mathcal{A} \times \mathcal{B} \subset \mathcal{S}(\mathcal{A} \times \mathcal{B})$; and
(2) if $\mathcal{A} \times \mathcal{B} \subset \mathcal{S}$, then $\mathcal{S}(\mathcal{A} \times \mathcal{B}) \subset \mathcal{S}$, where $\mathcal{S}$ denotes any sigma-field of subsets of $X \times Y$.

We call $\mathcal{S}(\mathcal{A} \times \mathcal{B})$ the sigma-field *generated by* $\mathcal{A} \times \mathcal{B}$. If $\mathcal{S}_X$ and $\mathcal{S}_Y$ are sigma-fields of subsets of X and of Y, respectively, then we abbreviate $\mathcal{S}(\mathcal{S}_X \times \mathcal{S}_Y)$ to $\mathcal{S}_X \otimes \mathcal{S}_Y$.

Examples

(3) Let $\mathcal{P}_X$ and $\mathcal{P}_Y$ be arbitrary partitions of X and of Y, respectively. Then $\mathcal{S}(\mathcal{P}_X \times \mathcal{P}_Y)$ is the set consisting of all *countable* unions of elements of $\mathcal{P}_X \times \mathcal{P}_Y$ and all complements of countable unions of elements of $\mathcal{P}_X \times \mathcal{P}_Y$.
(4) Let $X = Y = R^1$; then $X \times Y = R^2$. If $\mathcal{S}_X = \mathcal{S}_Y = \mathcal{B}^1 =$ the Borel field generated by the sets $\mathcal{A} = \mathcal{B}$ of all open intervals (a, b) for $-\infty < a < b < +\infty$ (see Exercise 13 of Chapter 2), then $\mathcal{S}_X \otimes \mathcal{S}_Y =$ the sigma-field generated by the set of all open rectangles of the form $(a, b) \times (c, d)$ for $a < b$ and $c < d$. We denote this sigma-field by $\mathcal{B}^2$, and call it the *Borel field in* R^2. More generally, $\mathcal{B}^n$ is the *Borel field in* R^n, generated by all "open boxes" of the form $(a_1, b_1) \times (a_2, b_2) \times \cdots \times (a_n, b_n)$, where $-\infty < a_1 < b_i < +\infty$, for every $i \in \{1, 2, \ldots, n\}$.

4.3 FOUNDATIONS OF DECISION THEORY (I)

4.3.1 Introductory Example

Consider what is involved in deciding between (1) ordering dessert and coffee and, (2) ordering just coffee. Ordering dessert and coffee might cause you to be late for an important appointment, or it might not. The dessert might be too rich and cause you indigestion, or it might not. Thus if you order dessert and coffee, one of four events will occur: (a) on time with indigestion; (b) on time and feeling very satisfied, (c) late with indigestion, and (d) late but feeling very satisfied. On the other hand, if you order just coffee, then (we assume) you will surely be on time and will not have indigestion, but you will not feel as satisfied as if you had eaten a light dessert. What should you do?

We shall formalize this problem slightly in order to introduce the general decision-theoretic model. Let s_1 denote the "event" that the dessert would be served quickly enough so that you would not be late, and also that it would not be too rich. Let s_2 denote the event that the dessert would be served quickly enough but would be too rich. Let s_3 denote the event that the dessert would not be too rich but would be served too slowly for you to be on time. Let s_4 denote the event that it would be served slowly and also be too rich.

It is clear that only one of the s's can be true, and we shall assume that one of the s's must be true. (No assumption is needed if we assume that *quickly* and *slowly* are so precisely defined that one of these descriptions will surely be true.)

We define $S = \{s_1, s_2, s_3, s_4\}$. S is called the set of possible *states of the environment* for your decision problem.

Now, if you decide to order dessert and coffee, we will say that you choose *act* a_1; whereas if you decide to order just coffee, we will say that you choose act a_2. If there are no other possible courses of action, then you will therefore have to choose an element (act) of the set A of *available acts*. We define $A = \{a_1, a_2\}$.

If you choose act a_1 and state s_1 is true, then the consequence will be "feel very satisfied and productive" (because you kept the appointment). We might expand the description of this consequence by adding a few paragraphs or pages of text covering the monetary and psychological outcomes of taking act a_1 when state s_1 is true, but we shall be content here with a brief description. We shall denote this consequence by c_1. If you take act a_1 and state s_2 obtains, then you will suffer consequence c_2, which is "feel productive but ill." If you take act a_1 and state s_3 obtains, you suffer consequence c_3 which is "feel satisfied physically but unproductive professionally"; while taking act a_1 when state s_4 obtains produces consequence c_4: "feel terrible physically and professionally." On the other hand, if you take act a_2, then regardless of the true state s you will suffer consequence c_5: "feel professionally productive and moderately satisfied physically."

Hence we have defined a set $C = \{c_1, c_2, c_3, c_4, c_5\}$ of *consequences* of the decision. Moreover, since an *act* assigns an unambiguous consequence to each state s of the environment, we see that each act is a function from S to C. Figure 4.3 depicts acts a_1 and a_2 as functions.

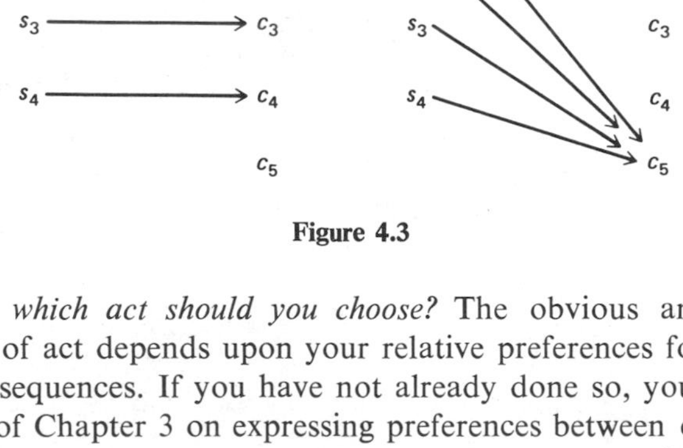

Figure 4.3

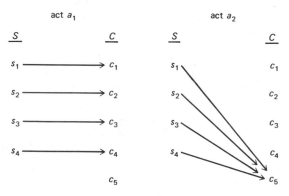

Question: *which act should you choose?* The obvious answer is that your choice of act depends upon your relative preferences for the various possible consequences. If you have not already done so, you should read Exercise 15 of Chapter 3 on expressing preferences between consequences. We shall assume that your preference-or-indifference quasi-order R is as follows:

$$c_1 R c_5 R c_2 R c_3 R c_4.$$

(Since R should be transitive, writing this sequence of preferences as a chain is admissible: it implies c_1Rc_2, c_1Rc_3, c_1Rc_4, c_5Rc_3, c_5Rc_4, c_2Rc_3, c_2Rc_4, and c_iRc_i as well as c_1Rc_5, c_5Rc_2, c_2Rc_3, and c_3Rc_4.)

Now, *if you knew* that s_1 were true, then you *should* choose a_1 in order to obtain c_1, the most preferred consequence. On the other hand, if you knew that s_2, s_3, or s_4 (that is, $s_1{}^c$) were true, then you could only experience c_2, c_3, or c_4 by choosing a_1, and so you should choose a_2 in order to obtain the more desirable c_5. The crux of your problem is that you do not know whether s_1 is true or not; that is, you must make a *decision under uncertainty*.

In real life you would try to obtain some information on whether s_1 is true; for example, you might ask the waitress to describe the ingredients of the dessert and to estimate the odds on being able to serve it within the time limit necessitated by your appointment. We shall consider information acquisition after generalizing from our example to arbitrary decision problems.

4.3.2 The General No-information Decision

The *general no-information decision problem* is specified by:

(1) a set S of possible *states of the environment* s, sometimes also called *parameter values*;
(2) a set C of possible *consequences of the decision* c; and
(3) a set A of *available acts* a, each act a being a function $a: S \to C$.

4.3.3 The General Decision Problem

Now suppose in our example that we had decided to ask the waitress the question noted at the end of Subsection 4.3.1. We call this questioning an *experiment* and denote it by e.

Examples

(1) The reader can see from the preceding how general the notion of *experiment* can be. (2) A company commander performs an experiment when he sends out a reconnaissance patrol. (3) n measurements of the length of a steel bar constitute jointly an experiment. This example is more in line with the usual connotation of the word experiment.

Given that we perform experiment e in our example problem, the waitress can make any one of a number of possible responses, called *outcomes* of the experiment, or *messages*, and denoted typically by z. We let $Z(e)$ denote the *set* of all possible outcomes z of experiment e; and we assume that one and only one outcome $z \in Z(e)$ will occur.

Now suppose that we have performed experiment e and observed outcome z. Then we must choose an act a. We shall allow for the possibility that the set A of available acts depends upon both e and z. Hence we shall denote the *set* of available acts given e and z by $A(e, z)$.

Next, note that nothing we have said implies that the waitress *knows* the true state or that she will honestly reveal it even if she does know it. Hence it might happen that she would say z_1 = "I can easily make your deadline and the dessert will certainly not cause you indigestion," and yet s_4 is true. If $a_1 \in A(e, z_1)$ and you chose a_1, then, intuitively, your *consequence* might be different than if you had *not* obtained e and observed z_1: you had to beckon the waitress and worry about her taking offense at your questioning her service and the food quality. Hence a_1 is not *really* the same function as without the (e, z_1) preface. In fact, C must be enlarged to account for the new consequence.

Finally, note that the general no-information problem can be formulated in (e, z) terms if we *define* e_0 as the *null experiment* and z_0 as the *null message* resulting from the null experiment.

We may now generalize. The *general decision problem* is specified by:

(1) a set S of possible *states of the environment*, or *parameter values* s;
(2) a set E of available *experiments* e, with the null experiment e_0 an element of E;
(3) for each $e \in E$, a set $Z(e)$ of possible *outcomes* of e, or e-messages z, with $Z(e_0) = \{z_0\}$, the null message;
(4) a set C of possible *consequences* (assumed large enough to describe all contingencies); and
(5) for each $e \in E$ and each $z \in Z(e)$, a set $A(e, z)$ of available *acts* a, with each $a \in A(e, z)$ being a function $a: S \to C$.

4.3.4 The Decision-tree Representation

Consider a general decision problem in which $C = \{c_1, c_2, c_3, c_4, c_5\}$; $S = \{s_1, s_2, s_3\}$; $E = \{e_0, e_1, e_2\}$; $Z(e_0) = \{z_0\}$; $Z(e_1) = \{z_1, z_2\}$; $Z(e_2) = \{z_3, z_4, z_5\}$; $A(e_0, z_0) = \{a_1, a_2\}$; $A(e_1, z_1) = \{a_3\}$; $A(e_1\ z_2) = \{a_4, a_5\}$; and $A(e_2, z_3) = A(e_2, z_4) = A(e_2, z_5) = \{a_6, a_7\}$. Suppose also that the functions a_i: $S \to C$ are as specified in Figure 4.4.

S	$a_1(s)$	$a_2(s)$	$a_3(s)$	$a_4(s)$	$a_5(s)$	$a_6(s)$	$a_7(s)$
s_1	c_1	c_2	c_7	c_2	c_2	c_7	c_4
s_2	c_4	c_3	c_7	c_3	c_6	c_3	c_3
s_3	c_6	c_1	c_7	c_5	c_1	c_2	c_7

Figure 4.4

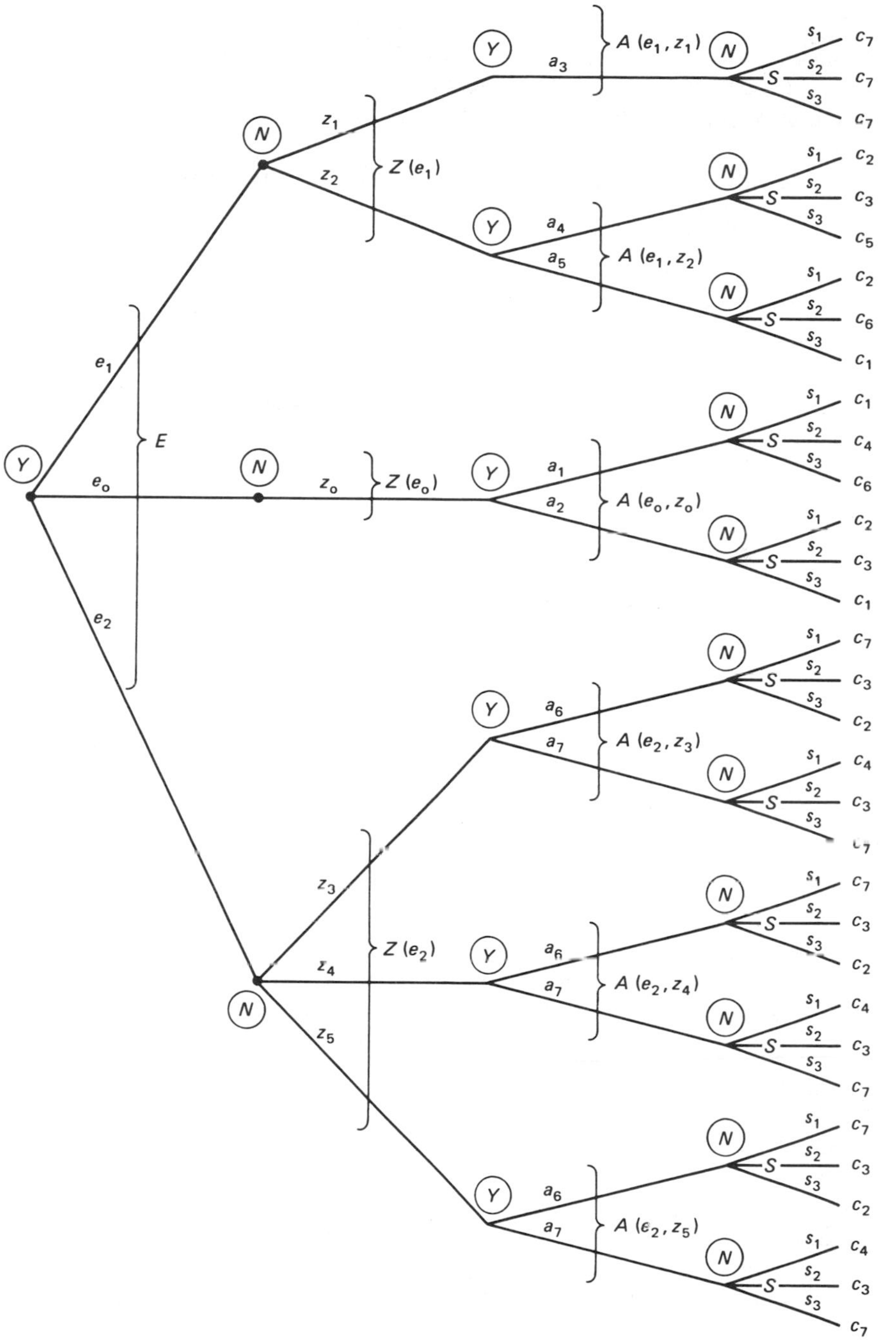

Y
e1
e0
e2
E
N
N
N
z1
z2
Z (e1)
z0
Z (e0)
z3
z4
z5
Z (e2)
Y
a3
A (e1, z1)
Y
a4
a5
A (e1, z2)
Y
a1
a2
A (e0, z0)
Y
a6
a7
A (e2, z3)
Y
a6
a7
A (e2, z4)
Y
a6
a7
A (ϵ2, z5)
N
S
s1
s2
s3
c7
c7
c7
c2
c3
c5
c2
c6
c1
c1
c4
c6
c2
c3
c1
c7
c3
c2
c4
c3
c7
c7
c3
c2
c4
c3
c7
c7
c3
c2
c4
c3
c7

Figure 4.5

Now keeping track of all this information is difficult without some systematic and reasonably concrete way of representing it all at one glance. Figure 4.5 shows an important and popular way of obtaining this representation of the decision. It exemplifies, for the preceding decision, the concept of a *decision tree*.

Here is how Figure 4.5 should be interpreted in terms of the *relative time sequencing*: (1) you choose an experiment $e \in E$; (2) you observe the outcome $z \in Z(e)$ of e; (3) you choose an act $a \in A(e, z)$; (4) you observe the true $s \in S$; and (5) you experience the consequence $c = a(s)$. Thus each of the possible complete historics of the decision consists of an (e, z, a, s)-*path* from the initial (leftmost) fork in the tree to an endpoint.

We have circled each fork in the tree and put a Y at those forks where *y*ou make a choice and an N at those forks where *n*ature, or chance determines which branch will be traveled. The tree makes it clear that you have partial, but generally incomplete, control over the path.

It is clear from the tree that the consequence attached to a given path may be considered a function of the path itself. Since each path is uniquely determined by an (e, z, a, s) quadruple, we can consider the consequences as being assigned (functionally) to the (e, z, a, s) quadruples. Thus we can consider the *consequences as being determined by a function from the set of all* (e, z, a, s) *quadruples to* C, *and we can write the consequences typically as* $c(e, z, a, s)$.

This viewpoint appears to conflict with our definition of acts as functions from S to C; that is, there appears to be circularity in the reasoning. The circularity is more apparent than actual, however, since e and z specify the set $A(e, z)$ of available acts, and hence naming e, z, and $a \in A(e, z)$ determine a unique function on S to C. This is equivalent to saying that specifying $e \in E$, $z \in Z(e)$, $a \in A(e, z)$, and $s \in S$ uniquely determines $c \in C$. The basic reason for defining acts as functions on S to C rather than as elements in an abstract set is to preclude the possibility that the choice of act might affect our opinions as to nature's choice of s.

Finally, note that the decision tree in Figure 4.5 contains no more and no less information than was given by Figure 4.4 and the paragraph preceding it. The tree does, however, indicate somewhat more *accessibly* the pertinent facts regarding *contingent* availability of acts and *contingent* possibilities of events occurring.

We shall have occasion to use trees often in the remainder of this book. The basic philosophy of tree construction is simple.

(1) A fork indicates a (relative) time point at which *either* you make a choice (of act or of experiment) *or* nature (chance) makes a choice; and
(2) branches emanating from a fork indicate the possible choices (by you or chance) at that fork.

In later chapters we shall add much more structure to the general decision problem. We shall give probabilities to the branches emanating from chance forks; and we shall give a numerical function $u: C \rightarrow R^1$ which in a real sense mirrors your preferences among elements of C. Once these additional data have been determined, we shall detail a procedure by which the general decision problem can be solved. The solution will consist of specifying: (1) for each $A(e, z)$, an act $a_{z|e}^{0}$ which you *should* adopt if you had chosen experiment e and observed outcome $z \in Z(e)$; and (2) which experiment $e^0 \in E$ you *should* choose. [To implement the solution, you will perform experiment e^0, then observe $z \in Z(e^0)$, and then adopt act $a_{z|e^0}^{0}$ corresponding to the observed z.]

Our first major task is to develop the concept of *probability*. For this, we shall need to consider sets of subsets of S and of $Z(e)$ (for each $e \in E$), and also sets of subsets of $Z(e) \times S$. These sets of subsets will be fields, and the elements of each such field will be called *events*.

4.4 EVENTS

Assume that we have chosen an experiment e, which will remain fixed throughout this section.

Let Q denote any one of S, $Z(e)$, and $Z(e) \times S$. We shall call the elements q of Q *elementary events*.

Examples

(1) Suppose an experiment e consists of flipping a coin. A space-age description of $Z(e)$ is {"heads," "tails," "on edge," "went into orbit," "went into outer space"}. Let $Q = \{q_1, q_2, q_3, q_4, q_5\} = Z(e)$, where q_i denotes the elementary event corresponding to the ith listed outcome. If you would win one dollar if the coin landed heads or went into orbit, and you would win nothing otherwise, you would be interested in whether $q \in B = \{q_1, q_4\}$: in (almost) everyday language you might claim that you would win one dollar if the event that the coin landed heads or went into orbit occurs. (2) In the dessert example of Section 4.3, your choice between a_1 and a_2 ideally depends upon whether the *event* that $s \in \{s_2, s_3, s_4\}$ is true or not. Here we would set $Q = S$.

From these two examples it is clear that we would like to use the term *event* to denote some subsets of Q.

There are two reasonable requirements that the subsets of Q called *events* should satisfy:

(1) if B is an event, then B^c is an event; and
(2) if B_1 and B_2 are events, then $B_1 \cup B_2$ is an event.

The first requirement says that if it is meaningful to discuss whether B will occur, then it is meaningful to discuss whether B *will not* occur (that is, if B^c will occur). The second requirement says that if it is meaningful to discuss whether B_1 will occur and also meaningful to discuss whether B_2 will occur, then it is meaningful to discuss whether either B_1 and /or B_2 will occur (that is, whether $B_1 \cup B_2$ will occur).

These requirements look suspiciously familiar. They are simply reworded versions of the axioms of a field. Thus *the set* $\mathcal{E}_Q$ *of all events is a field.* Hence $\mathcal{E}_Q$ possesses all properties, (1) through (4) and (1^f) and (2^f), of fields.

[In rigorous probability theory, it is usually assumed that the set $\mathcal{E}_Q$ of all events is a *sigma*-field. We shall not develop any motivation for events satisfying axiom (1^σ). In fact, some probabilists consider axiom (1^σ) in the nature of a regularity condition, imposed for technical reasons, and not necessarily corresponding to anything operationally important that will concern us.]

A third reasonable requirement is that if q is an elementary event, then $\{q\}$ is an event. Now, in *practice* we do not usually distinguish between the objects q and $\{q\}$, even though they are logically distinct. Hence the third requirement can be nicely "jargonized": *elementary events are events.*

Example

Suppose the elementary events are 1, 2, and 3. Then the field $\mathfrak{F} = \{\varnothing, \{1\}, \{2, 3\}, \{1, 2, 3\}\}$ does *not* qualify as a field of events, since the elementary events 2 and 3 are not elements of $\mathfrak{F}$. The only field of events is $\mathcal{R}(\{1, 2, 3\})$, as given in Section 4.3.

This example generalizes. The reader may show as an exercise that if the set of elementary events is *finite*, then the set of all events must be the power set of the set of elementary events. In other words, every set of elementary events is an event if there are only finitely many elementary events. (If there are only *countably* many elementary events, and if $\mathcal{E}_Q$ is required to be a sigma-field, then $\mathcal{E}_Q = \mathcal{R}(Q)$; that is, every set of elementary events is an event.)

We shall usually lapse in rigor in Chapters 5 and 6 on basic probability theory and assume that the set of events is always $\mathcal{R}(Q)$, even when Q is an infinite set—in which case one can run into trouble in rigorous probability theory. (The trouble is that when there are too many events, there are too few useful ways of assigning probabilities to events.)

It is worth commenting at this point that the reasons for the apparently wasteful duplication of terminology, in *element versus elementary event, universal set versus sure event*, and *empty set versus impossible event* lie partly in historical usage and partly in the fact that probability theory and the analysis of events is an application of pure set theory; and applications of mathematics generally grow their own intuitively meaningful synonyms for mathematical terms.

Finally, suppose that we have specified a field $\mathcal{E}_S$ of s-events and a field $\mathcal{E}_{Z(e)}$ of z-events. Let $Q = Z(e) \times S$. We ask if there is a natural field $\mathcal{E}_Q$ of (z, s)-events given the fields $\mathcal{E}_S$ and $\mathcal{E}_{Z(e)}$. If (as probabilists usually do) we interpret *natural* to mean "smallest field including all the rectangles $\mathcal{E}_{Z(e)} \times \mathcal{E}_S$," the answer is "yes," by Section 4.2. There is a smallest field $\mathcal{F}(\mathcal{E}_{Z(e)} \times \mathcal{E}_S)$ of subsets of $Z(e) \times S$ which contains every rectangle in $\mathcal{E}_{Z(e)} \times \mathcal{E}_S$.

But is $\mathcal{F}(\mathcal{E}_{Z(e)} \times \mathcal{E}_S)$ a field of *events?* That is, is every $\{(z, s)\}$ an element of $\mathcal{F}(\mathcal{E}_{Z(e)} \times \mathcal{E}_S)$? The answer is yes: since $\{s\} \in \mathcal{E}_S$ and $\{z\} \in \mathcal{E}_{Z(e)}$ for every $s \in S$ and $z \in Z(e)$, it follows that every $\{(z, s)\}$ is a rectangle and hence belongs to $\mathcal{F}(\mathcal{E}_{Z(e)} \times \mathcal{E}_S)$. We thus see that $\mathcal{F}(\mathcal{E}_{Z(e)} \times \mathcal{E}_S)$ is a field of events; we shall denote it hereafter by $\mathcal{E}_{Z(e)} \otimes \mathcal{E}_S$.

We shall henceforth assume that when $Z(e)$ is associated with a field $\mathcal{E}_{Z(e)}$ of z-events and S is associated with a field $\mathcal{E}_S$ of s-events, then $Z(e) \times S$ is associated with the field $\mathcal{E}_{Z(e)} \otimes \mathcal{E}_S$ of (z, s)-events.

4.5 RANDOM VARIABLES

Often Q does not possess a natural numerical description, and yet we are interested in the properties of a numerical function $\tilde{x}: Q \to R^1$. For example, Q might consist of all elements q of the form, "q is a person who might be selected for your sample." If you are interested in q's weight, you might define it as $\tilde{x}(q)$. Now, $\tilde{x}(q)$ obviously varies as q ranges through Q. Hence $\tilde{x}: Q \to R^1$ = set of all possible weights is a perfectly good example of a numerical function defined on Q and of (potentially) operational interest (for example you are interested in the necessary strength of a floor in your house).

A *random variable* $\tilde{x}$ is simply a function $\tilde{x}: Q \to R^1$. The special terminology is widely used and of historical importance.

[In rigorous probability theory, one is given the sigma-field $\mathcal{E}_Q$ of events, and the term *random variable* is reserved for functions $\tilde{x}: Q \to R^1$ satisfying the following "measurability property":

$$\{q: \tilde{x}(q) \leq x\} \in \mathcal{E}_Q, \text{ for every } x \in R^1.$$

A less simple, but equivalent stipulation is:

$$\{q: \tilde{x}(q) \in B\} \in \mathcal{E}_Q, \text{ for every } B \in \mathcal{B}^1 = \text{the Borel field in } R^1.]$$

Random variables are of basic importance in probability theory, as their values constitute the basic data in most cases. Later in this book we shall be almost exclusively concerned with random variables.

The purpose of the tilde ($\smile$) superscript is to enable us to distinguish succinctly between a random variable $\tilde{x}$ (a function from Q to R^1) and the typical values $x \in R^1$ assumed by $\tilde{x}$. The need for such a distinction is acute in later chapters.

4.6 RANDOM VECTORS

The reader will recall that a *vector* is simply an (ordered) n-tuple of real numbers.

Suppose that instead of a single random variable $\tilde{x}: Q \to R^1$ we are given n random variables $\tilde{x}_1, \tilde{x}_2, \ldots, \tilde{x}_n$. We can consider the n random variables as constituting a *single* function $\tilde{\mathbf{x}}: Q \to R^n$, defined by

$$\tilde{\mathbf{x}}(q) = (\tilde{x}_1(q), \tilde{x}_2(q), \ldots, \tilde{x}_n(q))' \text{ for every } q \in Q.$$

(The superscript "$'$" denotes *transpose*, since we assume that all vectors will be of the column type unless otherwise noted. Also note the boldfacing of $\tilde{\mathbf{x}}(q)$ and of $\tilde{\mathbf{x}}$. Boldfacing a symbol will denote that it is a vector.)

The n-tuples of random variables are called *random vectors*. In applications, n is of course known.

For example, let Q consist of all elements q of the form "q could be selected for observation." If you record the height, weight, and intelligence of any subject q, then you might define $\tilde{x}_1(q)$ to be q's height, $\tilde{x}_2(q)$ to be q's weight, and $\tilde{x}_3(q)$ to be q's intelligence quotient. Then $\tilde{\mathbf{x}}$ is a perfectly respectable random vector, where

$$\tilde{\mathbf{x}}(q) = (\tilde{x}_1(q), \tilde{x}_2(q), \tilde{x}_3(q))' \text{ for every } q \in Q.$$

[In rigorous probability theory the term *random vector* is reserved for those functions $\tilde{\mathbf{x}}: Q \to R^1$ which satisfy

$$\{q: \tilde{x}_i(q) \leq y\} \in \mathcal{E}_Q, \text{ for every } y \in R^1 \text{ and every } i \in \{1, 2, \ldots, n\};$$

or equivalently,

$$\{q: \tilde{x}_i(q) \in B\} \in \mathcal{E}_Q, \text{ for every } B \in \mathcal{B}^1 \text{ and every } i \in \{1, 2, \ldots, n\};$$

or equivalently,

$$\{q: \tilde{\mathbf{x}}(q) \in B\} \in \mathcal{E}_Q, \text{ for every } B \in \mathcal{B}^n.$$

These conditions are equivalent for any (finite) positive integer n. All random variables and random vectors that we shall consider in this book will automatically satisfy this and the preceding conditions.]

EXERCISES

1. Let $X = \{1, 2, 3\}$, $Y = \{a, b, c\}$, $\mathcal{P}_X = \{\{1, 2\}, \{3\}\}$, and $\mathcal{P}_Y = \{\{a\}, \{b, c\}\}$. Find $\mathcal{P}_X \times \mathcal{P}_Y$.

2. Let $X \times Y = \{(1, a), (1, b), (2, a), (2, b), (3, a), (3, b)\}$. Then $\{\{(1, a), (1, b), (2, a), (3, a)\}, (2, b), (3, b)\}\}$ is a partition of $X \times Y$. Is it the Cartesian product of a partition $\mathcal{P}_X$ and a partition $\mathcal{P}_Y$? [*Hint*: $\{(2, b), (3, b)\} = \{2, 3\} \times \{b\}$ is a rectangle.]

3. Let X and Y each be a finite set.

   (a) Argue that $X \times Y$ is also a finite set.
   (b) Let $\mathcal{A}$ be a set of subsets of X and $\mathcal{B}$ be a set of subsets of Y. Then $\mathcal{A}$ and $\mathcal{B}$ are necessarily finite sets. Show that $\mathcal{A} \times \mathcal{B}$ is a finite set.
   (c) Define an equivalence relation $E^{\mathcal{A}}$ in X by saying $x_1 E^{\mathcal{A}} x_2$ if and only if for every $A \in \mathcal{A}$ we have *either* $x_1 \in A$ and $x_2 \in A$ *or* $x_1 \notin A$ and $x_2 \notin A$. Show that $E^{\mathcal{A}}$ is an equivalence relation. The partition consisting of the $E^{\mathcal{A}}$-equivalence classes is denoted by $\mathcal{P}_X{}^{\mathcal{A}}$ and called the *partition* (of X) *generated by* $\mathcal{A} \subset \mathcal{R}(X)$.
   (d) Show that $\mathcal{F}(\mathcal{A})$ is the set of all unions of elements of $\mathcal{P}_X{}^{\mathcal{A}}$, together with $\varnothing$. (For purposes of taking unions, each element of $\mathcal{P}_X{}^{\mathcal{A}}$ is regarded as a subset of X.) (*Note*: Finiteness of $\mathcal{A}$ is essential for this result.)
   (e) Define an equivalence relation $E^{\mathcal{B}}$ in Y in a manner analogous to the definition of $E^{\mathcal{A}}$; it generates the partition $\mathcal{P}_Y{}^{\mathcal{B}}$ of Y. Argue that $\mathcal{P}_{X \times Y}{}^{\mathcal{A} \times \mathcal{B}} = \mathcal{P}_X{}^{\mathcal{A}} \times \mathcal{P}_Y{}^{\mathcal{B}}$.
   (f) Argue that $\mathcal{F}(\mathcal{P}_{X \times Y}{}^{\mathcal{A} \times \mathcal{B}}) = \mathcal{F}(\mathcal{P}_X{}^{\mathcal{A}}) \otimes \mathcal{F}(\mathcal{P}_Y{}^{\mathcal{B}})$.
   [*Note*: Assertions (c), (e), and (f) hold regardless of finiteness of X, Y, $\mathcal{A}$, and $\mathcal{B}$.]

4. Let $\mathcal{P}_X$ be a partition of X. Define $\mathcal{P}'_X \subset \mathcal{R}(X \times Y)$ by

$$\mathcal{P}'_X = \{A \times Y : A \in \mathcal{P}_X\}.$$

   (a) Show that $\mathcal{P}'_X$ is a partition of $X \times Y$.
   (b) Show that $f: \mathcal{P}_X \to \mathcal{P}'_X$, defined by

$$f(A) = A \times Y, \text{ for every } A \in \mathcal{P}_X,$$

   is a bijection.
   (c) Define $\mathcal{P}'_X \subset \mathcal{R}(X \times Y)$ by

$$\mathcal{P}'_X = \{X \times B : B \in \mathcal{P}_Y\}$$

   for some given partition $\mathcal{P}_Y$ of Y. Then $\mathcal{P}'_Y$ corresponds to $\mathcal{P}_Y$ in exactly the same way as $\mathcal{P}'_X$ corresponds to $\mathcal{P}_X$. Show that the *Cartesian* product $\mathcal{P}_X \times \mathcal{P}_Y$ of $\mathcal{P}_X$ and $\mathcal{P}_Y$ coincides with the (usual) product $\mathcal{P}'_X \mathcal{P}'_Y$ of $\mathcal{P}'_X$ and $\mathcal{P}'_Y$; that is, show that if $A \times B \in \mathcal{P}_X \times \mathcal{P}_Y$, then there is a unique $C \in \mathcal{P}'_X$ and a unique $D \in \mathcal{P}'_Y$ such that $A \times B = C \cap D$—and conversely. [*Hint*: Set $C = A \times Y$.]

5. Let $X = \{\text{good, bad, indifferent}\}$, $Y = \{\text{yes, no, maybe}\}$, $\mathcal{A} = \{\{\text{good, indifferent}\}, \{\text{bad, indifferent}\}, \{\text{bad}\}\}$, and $\mathcal{B} = \{\{\text{yes}\}, \{\text{yes, no}\}, \{\text{maybe}\}\}$. What is $\mathcal{A} \times \mathcal{B}$?

6. You will win a bet with me if the Democratic candidate wins the next presidential election, and I will win otherwise. Define S in a relevant fashion.

7. You are to roll a die (one half of a pair of dice) once. Define S relevantly.

8. You are to roll five dice simultaneously. You will win one dollar if the sum of the dice exceeds ten and you will lose ten dollars if the sum of the dice does not exceed ten. Give as many natural definitions of S as you can think of.

9. Let a set $C = \{c_1, c_2, \ldots, c_6\}$ of consequences be given; say cPc' if and only if you would prefer experiencing c to experiencing c'; and say cEc' if and only if c and c' are equally desirable to you. Suppose c_1Pc_2, c_2Pc_3, c_3Pc_4, c_4Pc_5, c_5Ec_6, and $P \cup E$ is transitive (see Exercise 15 of Chapter 3). Consider the decision tree in Figure 4.6.

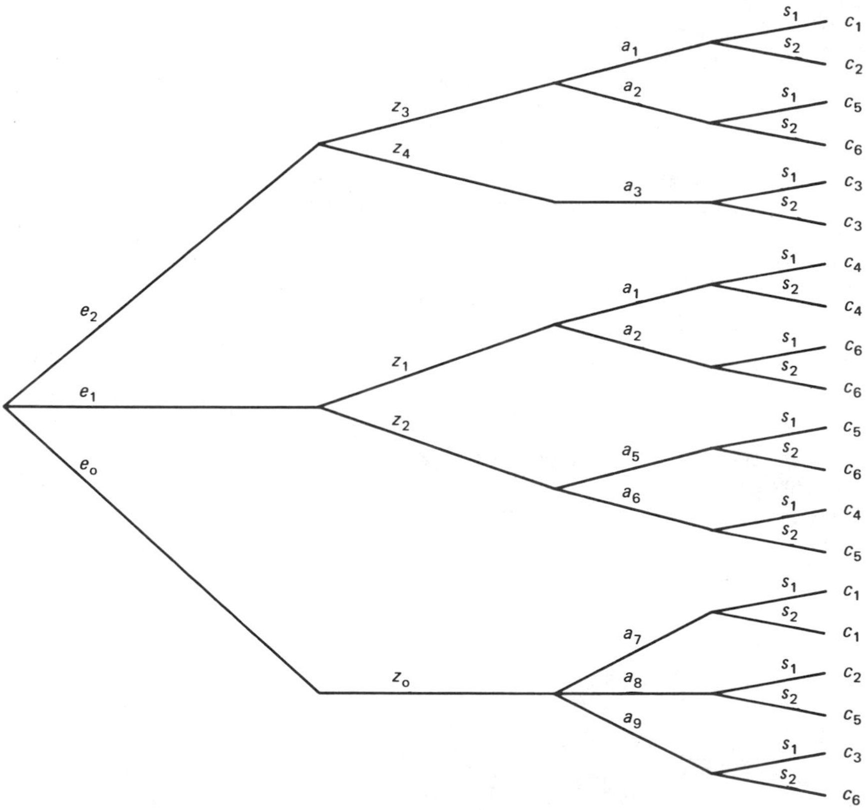

Figure 4.6

(a) Suppose you have chosen e_0 and observed, naturally, z_0. Show that you *should* choose act a_7 in order to guarantee yourself c_1.
(b) Suppose you have chosen e_1 and observed z_1. Show that you should choose act a_1 in order to guarantee yourself c_4. Suppose that you have chosen e_1 and observed z_2. Show that a_6 is a better choice than a_5 regardless of which state s is true, and that by choosing a_6 you guarantee yourself at least c_5 and perhaps c_4. Thus show that you can obtain at least c_5 by choosing e_1.
(c) Suppose you have chosen e_2 and observed z_3. Show that you should choose a_1 in order to obtain c_2 or better. If you have observed z_4, you will obtain c_3. Hence show that e_2 leads to obtaining c_3 or better.
(d) Which of e_0, e_1, and e_2 should you choose?
(e) Suppose the branches emanating from z_0 were changed to those shown in Figure 4.7. Can you *now* make a choice of act without a more precise statement regarding your preferences and without some quantification of your judgments regarding the chances of s_1 versus s_2? Assuming the rest of the tree remains as originally given, can you still reach an unambiguous choice of experiment? [The correct answers to these questions, and an understanding of our *modus operandi* in (a) through (d) suggests the way we shall analyze decision problems. Roughly, the first step is the evaluation of each act and the choice of an optimal act given each (e, z). This evaluation and choice then lead to an evaluation of each e, from which choosing an optimal e will be easy.]

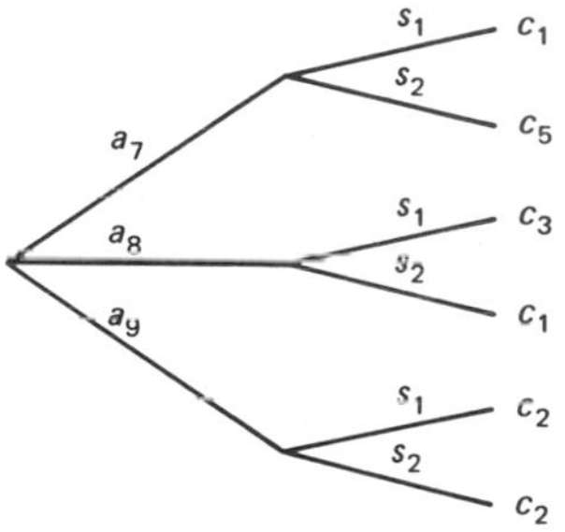

Figure 4.7

10. Let $\mathcal{E}_S$ and $\mathcal{E}_{Z(e)}$ be given fields of s-events and z-events, respectively.

(a) Show that every rectangle of the form $Z(e) \times B$ belongs to $\mathcal{E}_{Z(e)} \otimes \mathcal{E}_S$; that is, is a (z, s)-event.
(b) Let $\mathcal{E}'_S = \{Z(e) \times B: B \in \mathcal{E}_S\}$. Show that $f: \mathcal{E}_S \to \mathcal{E}'_S$, defined by

$$f(B) = Z(e) \times B, \text{ for every } B \in \mathcal{E}_S,$$

is a bijection.

(c) Argue that B and $Z(e) \times B$ describe the same contingency in the general decision problem.

(d) Let $\mathcal{E}'_{Z(e)} = \{A \times S\colon A \in \mathcal{E}_{Z(e)}\}$. Then $\mathcal{E}'_{Z(e)}$ corresponds to $\mathcal{E}_{Z(e)}$ in exactly the same way as $\mathcal{E}'_S$ corresponds to $\mathcal{E}_S$. Show that $\mathcal{E}_{Z(e)} \otimes \mathcal{E}_S$ is the field generated by $\mathcal{C}$, where

$$\mathcal{C} = \{C\colon C \in \mathcal{E}'_S \text{ and/or } C \in \mathcal{E}'_{Z(e)}\}.$$

part II

Probability

5

PROBABILITY (I): BASIC CONCEPTS

5.1 INTRODUCTION

In this chapter we begin our study of probability theory, a central building block for decision theory and inference. *Probability* is a concept with which everyone has some familiarity. We hear about the "probability of rain" on the weather report; we hear that someone's success in the stock market is due to his being a "shrewd player of the probabilities"; and, finally, we hear that something is "a possibility but not a probability."

The tasks of this chapter are to introduce the *mathematical* notion of probability and to discuss philosophical points of view regarding *admissible applications* of mathematical probability to practical problems.

Probability, as mathematically defined, possesses all properties of *proportion*. Hence we shall corroborate all statements about probability by discussing relevant facts about proportions. Section 5.2, on proportions, is a prelude to Section 5.3, on the mathematical definition of probability.

With the mathematical groundwork laid in Section 5.3, Section 5.4 discusses the different viewpoints on the "true nature" of probability and the sorts of events upon which the machinery of probability can be brought to bear.

Section 5.5 returns to the discussion of proportions in order to introduce Section 5.6, a continuation of the mathematics of probability.

5.2 PROPORTIONS (I)

We shall state all proportions as numbers between 0 and 1 inclusive; that is, if p is a proportion, then $p \in [0, 1]$. We shall not, therefore, be discussing *percents*, which are in [0, 100]. [A percent (%) can always be converted to a proportion (p) by dividing it by 100; that is, $\%/100 = p$. Conversely, to obtain the percent corresponding to a proportion p, set $\% = 100p$.]

Suppose that we have a population of 10,000 people, and that we are interested in their level of education. Let A_1 denote the set of all people in X who have college degrees; let A_2 denote the set of all people in X who have had some college but have not obtained degrees, and let A_3 denote all people in X who belong to neither A_1 nor A_2. Then $\{A_1, A_2, A_3\}$ is a partition of X.

Let $\#(A)$ denote the *number* of people in subset A of X; and suppose that we have found $\#(A_1) = 3000$, $\#(A_2) = 2000$, and $\#(A_3) = 5000$. Since $\{A_1, A_2, A_3\}$ is a partition of X, the A_i's are mutually exclusive, and hence $\#(A_1 \cap A_2) = \#(A_1 \cap A_3) = \#(A_2 \cap A_3) = 0$. Moreover, it clearly follows that:

$$\#(A_1 \cup A_2) = \#(A_1) + \#(A_2) = 5000;$$

$$\#(A_1 \cup A_3) = \#(A_1) + \#(A_3) = 8000; \text{ and}$$

$$\#(A_2 \cup A_3) = \#(A_2) + \#(A_3) = 7000.$$

Obviously, $\#(A_1 \cup A_2 \cup A_3) = 10{,}000$.

Now let $p(A)$ denote the *proportion* of people in subset A or X, Obviously,

$$p(A) = \frac{\#(A)}{\#(X)}.$$

We summarize the proportions for our example in Figure 5.1.

A	$\#(A)$	$p(A)$
A_1	3000	.30
A_2	2000	.20
A_3	5000	.50
$A_1 \cup A_2$	5000	.50
$A_1 \cup A_3$	8000	.80
$A_2 \cup A_3$	7000	.70
X	10,000	1.00

Figure 5.1

Note that disjointness of A_1 and A_2 implies that $\#(A_1 \cup A_2) = \#(A_1) + \#(A_2)$, which in turn implies that

$$\begin{aligned} p(A_1 \cup A_2) &= \frac{\#(A_1) + \#(A_2)}{\#(X)} \\ &= \frac{\#(A_1)}{\#(X)} + \frac{\#(A_2)}{\#(X)} \\ &= p(A_1) + p(A_2). \end{aligned}$$

Also note that $p(X) = 1$, which should be obvious, and which is an equivalent way of saying that 100 percent of the people in X are in X. Also note that if a subset $A \subset X$, defined as the set of all x's possessing a given attribute, has no people in it, then $\#(A) = 0$ and $A = \varnothing$, and we see that $p(A) = p(\varnothing) = 0$.

We may now abstract from our example and state three properties of proportions as axioms from which *all* properties of proportions can be derived.

Let X be a nonempty set with $\#(X)$ elements. For any subset $A \subset X$, we define $\#(A)$ as the number of elements in A; and we define $p(A)$ as the proportion of elements in X which belong to A. Then $p(A) = \#(A)/\#(X)$. Our three basic properties are:

($p1$) $0 \leq p(A) \leq 1$, for every $A \subset X$;
($p2$) $p(X) = 1$; and
($p3$) if $A_1 \cap A_2 = \varnothing$, then $p(A_1 \cup A_2) = p(A_1) + p(A_2)$.

We have derived ($p1$) through ($p3$) for our introductory example. Those derivations are stated in terms which apply directly to the general case here.

Five other properties of proportions can be either proved directly from the definition of $p(A)$ as $\#(A)/\#(X)$ and basic arithmetic properties of counting, or they can be proved as consequences of "axioms" ($p1$) through ($p3$). We shall prove them as consequences of ($p1$) through ($p3$), viewed as axioms for a proportion function $p\colon \mathcal{R}(X) \to R^1$.

Other properties of proportions

($p4$) $p(A^c) = 1 - p(A)$, for every $A \in \mathcal{R}(X)$;
($p5$) $p(\varnothing) = 0$;
($p6$) $p(A_1 \cup A_2) = p(A_1) + p(A_2) - p(A_1 \cap A_2)$, for all A_1 and A_2 in $\mathcal{R}(X)$;
($p7$) if $A_1 \subset A_2$, then $p(A_1) \leq p(A_2)$; and
($p3'$) if $A_1, \ldots, A_m$ are mutually exclusive subsets of X, then

$$p(\cup_{i=1}^{m} A_i) = \sum_{i=1}^{m} p(A_i),$$

[where $\sum_{i=1}^{m} p(A_1)$ is simply a useful and popular shorthand for $p(A_1) + p(A_2) + \cdots + p(A_m)$].

Proof

To prove ($p4$), note that $A \cap A^c = \varnothing$, and so by ($p3$) we have $p(X) = p(A \cup A^c) = p(A) + p(A^c)$. But by ($p2$) we have $p(X) = 1$, from which ($p4$) follows. To prove ($p5$) we note that $X^c = \varnothing$ and we use ($p4$) to write $p(\varnothing) = 1 - p(X) = 1 - 1 = 0$. To prove ($p6$), note that $A_1 \cup A_2 = A_1 \cup (A_2 \backslash A_1)$, and that A_1 and $A_2 \backslash A_1$ are disjoint. Hence $p(A_1 \cup A_2) =$

$p(A_1) + p(A_2 \backslash A_1)$, by $(p3)$. But $A_2 = (A_2 \backslash A_1) \cup (A_1 \cap A_2)$, and $A_2 \backslash A_1$ and $A_1 \cap A_2$ are disjoint; and so by $(p3)$ we have $p(A_2) = p(A_2 \backslash A_1) + p(A_1 \cap A_2)$. Hence $p(A_2 \backslash A_1) = p(A_2) - p(A_1 \cap A_2)$; and when we substitute this equation for $p(A_2 \backslash A_1)$ into the preceding equation for $p(A_1 \cup A_2)$, we obtain $(p6)$. To prove $(p7)$, note that if $A_1 \subset A_2$, then $A_2 = A_1 \cup (A_2 \backslash A_1)$, and A_1 and $A_2 \backslash A_1$ are disjoint. Thus $(p3)$ implies that $p(A_2) = p(A_1) + p(A_2 \backslash A_1)$ By $(p1)$, $p(A_2 \backslash A_1) \geq 0$, and so $p(A_2) = p(A_1) + p(A_2 \backslash A_1) \geq p(A_1) + 0 = p(A_1)$, which is $(p7)$. The proof of $(p3^f)$ is a simple induction on m.

5.3 THE AXIOMS OF PROBABILITY

Let Q be a set of elementary events q, as defined in Section 4.4; and let $\mathcal{E}_Q$ be a given field of q-events. [Recall that Q can be any one of S, $Z(e)$, and $Z(e) \times S$.]

A probability on $\mathcal{E}_Q$ is a function $P: \mathcal{E}_Q \rightarrow R^1$ such that:

$(P1)$ $0 \leq P(A) \leq 1$, for every $A \in \mathcal{E}_Q$;
$(P2)$ $P(Q) = 1$; and
$(P3)$ if $A_1 \in \mathcal{E}_Q$, $A_2 \in \mathcal{E}_Q$, and $A_1 \cap A_2 = \varnothing$, then

$$P(A_1 \cup A_2) = P(A_1) + P(A_2).$$

Thus a probability P on a field $\mathcal{E}_Q$ of events assigns a number in $[0, 1]$ to each event in $\mathcal{E}_Q$. When Q is a finite set, you may think of $\mathcal{E}_Q$ as the set $\mathcal{R}(Q)$ of all subsets of Q (which it is in this case). However, when Q is an infinite set, $\mathcal{E}_Q$ may not contain all subsets of Q. In fact, when Q is uncountable, there are very good reasons for *not* calling every subset of Q an event: in this case there would be too few useful ways of assigning probabilities to the events.

Note that axioms $(P1)$ through $(P3)$ of probability are identical in form to axioms $(p1)$ through $(p3)$ of proportion. This identity of form implies that a probability P on $\mathcal{E}_Q$ possesses the following additional properties:

$(P4)$ $P(A^c) = 1 - P(A)$;
$(P5)$ $P(\varnothing) = 0$;
$(P6)$ $P(A_1 \cup A_2) = P(A_1) + P(A_2) - P(A_1 \cap A_2)$;
$(P7)$ if $A_1 \subset A_2$, then $P(A_1) \leq P(A_2)$; and
$(P3^f)$ if $A_1, \ldots, A_m$ are mutually exclusive, then

$$P(\cup_{i=1}^{m} A_i) = \sum_{i=1}^{m} P(A_i).$$

In all cases, it is assumed that the events in question are elements of $\mathcal{E}_Q$, since their probabilities would otherwise not be defined.

[In rigorous probability theory, $\mathcal{E}_Q$ is assumed to be a *sigma*-field, and ($P3$) is replaced by

($P3^\sigma$) if $\{A_1, A_2 \cdots\} \subset \mathcal{E}_Q$ and $A_i \cap A_j = \varnothing$ whenever $i \neq j$, then

$$P(\cup_{i=1}^{\infty} A_i) = \sum_{i=1}^{\infty} P(A_i).$$

This axiom implies both ($P3$) and ($P3^f$). It ensures the validity of certain equations involving taking limits.]

Take any event $A \in \mathcal{E}_Q$ such that $P(A) < 1$. The ratio $P(A)/P(A^c)$ is often called the *odds in favor of A*, and one often hears people say, "the odds in favor of A are $P(A)/P(A^c)$ to one." [Naturally, they replace $P(A)/P(A^c)$ by some number.] Slightly more generally, if $A \in \mathcal{E}_Q$, $B \in \mathcal{E}_Q$, and $P(B) > 0$, we say that "A is $P(A)/P(B)$ times as likely as B." Thus a probability on $\mathcal{E}_Q$ enables one to state the odds in favor of events in $\mathcal{E}_Q$ and also to give relative likelihoods of these events.

In that vein, suppose that $\mathcal{A} = \{A_1, \ldots, A_m\}$ is a partition of Q and that $\mathcal{A} \subset \mathcal{E}_Q$. Since $\mathcal{A} \subset \mathcal{E}_Q$, every A_i is an event and hence has a probability, $P(A_i)$. Since $\mathcal{A}$ is a partition, $\cup_{i=1}^{m} A_i = X$ and hence $P(\cup_{i=1}^{m} A_i) = P(X) = 1$. But by property ($P3^f$) we have that $P(\cup_{i=1}^{m} A_i) = \sum_{i=1}^{m} P(A_i)$. Hence $\sum_{i=1}^{m} P(A_i) = 1$. We have thus shown that if $\mathcal{A} \subset \mathcal{E}_Q$ and $\mathcal{A}$ is a partition, then the set $\{P(A_1), \ldots, P(A_m)\}$ of nonnegative [by($P1$)] numbers adds up to one. This set of numbers can then be used to state relative odds of the elements of $\mathcal{A}$.

Examples

(1) Suppose $Q = \{\text{head, tail}\}$ is the set of elementary events associated with one flip of a coin. If you have no reason for thinking heads any more or less likely than tails, then you should set $P(\text{heads}) = P(\text{tails})$. Since $\{\{\text{heads}\}\ \{\text{tails}\}\}$ is a partition of Q, it follows that $P(\text{heads}) + P(\text{tails}) = 1$, and so $P(\text{heads}) = P(\text{tails}) = \frac{1}{2}$. (2) If an urn contains 100 balls, $\#(R)$ of which are red and $\#(W)$ of which are white, and $\#(B) = 100 - \#(R) - \#(W)$ of which are black; and if one ball is to be drawn from the urn by a blindfolded person after a thorough mixing of the balls, then you might believe that each one of the 100 balls is no more or less likely to be drawn from the urn than any other of the balls. Hence you would assess the odds in favor of the drawn ball being red equal to $\#(R)/(100 - \#(R))$. Similarly for white and black. These assessments lead directly to assigning the event, "ball drawn will be of color C" $= \mathbf{C}$ the *probability equal to the proportion of balls of color* C; that is, $P(\mathbf{R}) = p(R)$, $P(\mathbf{W}) = p(W)$, and $P(\mathbf{B}) = p(B)$. Since only red, white, and black balls are in the urn, it follows that $p(R) +$

$p(W) + p(B) = 1$, which implies $P(\mathbf{R}) + P(\mathbf{W}) + P(\mathbf{B}) = 1$. These probability assignments then imply the probabilities of all other events in $\mathcal{E}_Q = \mathcal{R}(Q)$; for example, $P(\mathbf{R} \cup \mathbf{B}) = P(\mathbf{R}) + P(\mathbf{B})$. We stress the fact that this assignment of probabilities depends crucially upon the assumption that you believe that each of the 100 balls is equally as likely to be drawn as is any other ball. (3) Suppose the aforementioned urn containg only red and black balls, with $\#(B) = 100 - \#(R)$. Also suppose: (a) that you would rather bet me ten dollars that a red ball will be drawn (by a blindfolded person after a thorough mixing of the balls) than bet me ten dollars that it will rain tomorrow, providing $\#(R) \geq 61$; (b) that you would rather bet on rain tomorrow if $\#(R) \leq 59$; and (c) you find the bets equally attractive if $\#(R) = 60$. We then say that *your* probability of rain tomorrow is .60; that is, we find that proportion $p(R)$ of reds for which you are indifferent between the bets, and we equate your $P(\text{rain})$ to it. (In this case we are clearly assuming that $Q = \{\text{rain, no rain}\}$; and if we were serious about the betting or some other decision in which "rain tomorrow" was a factor, we would be very specific about what constitutes "rain tomorrow": how much, where, and so forth. (4) By the comparison-of-bets method used in (3), you could assess your $P(s_1)$ for state s_1 as defined in Section 4.3 for the dessert problem. Similarly, you could assess $P(s_2)$, $P(s_3)$, and $P(s_4)$ for that problem. On your first try you may find that $\sum_{i=1}^{4} P(s_i) \neq 1$. If so, you are violating some desirable decision criteria, and you will want to either revise your assessments or use a different (but related) assessment method, as described later.

Not everyone would agree that all of the preceding probability assignments are valid, or even meaningful. The objections lie with Examples (3) and (4) and stem fundamentally from philosophical differences of opinion. We shall now sketch extremes of these viewpoints.

5.4 THE MEANING AND SCOPE OF PROBABILITY

This section furnishes *only* a sketch of some *extreme* philosophical viewpoints on the foundations of probability. In fact, it might not be too much of an exaggeration to say that there are as many viewpoints on this subject as there are statisticians and probabilists.

5.4.1 Classical "Principle of Insufficient Reason"

This viewpoint dates back to the seventeenth century, and is expressible roughly as follows: if you do not know *precisely* how to assign unequal probabilities to the elements $A_1, \ldots, A_m$ of a partition $\mathcal{A}$ of Q, then you should assign $P(A_1) = P(A_2) = \cdots = P(A_m) = 1/m$. We tacitly used this principle in Examples (1) and (2) of the previous section.

However, the principle leads to ambiguities. Suppose that we modify partition $\mathcal{A}$ as given above by defining $B_1 = A_1, B_2 = A_2. \ldots, B_{m-2} = A_{m-2}$, and $B_{m-1} = A_{m-1} \cup A_m$. Then $\{B_1, \ldots, B_{m-1}\}$ is a perfectly good partition of Q, but the principle of insufficient reason dictates $P(B_1) = P(B_2) = \cdots = P(B_{m-1}) = 1/(m-1)$. But $A_1 = B_1$, and hence in one case the principle dictates $P(A_1) = 1/m$ but in the other it requires $P(A_1) = 1/(m-1)$.

Proponents would retort that if you had started with $\mathcal{A}$, then the axiom ($P3$) would tell you exactly how to assign unequal probabilities to the B's. However, the theory does not tell you with which partition to start.

Hence we are left with a serious ambiguity, an example of which is *D'Alembert's paradox*. D'Alembert said that if we flip a coin twice, then there are three possible outcomes: one head and one tail, two heads, and two tails. By the principle of insufficient reason, P(two heads) = P(two tails) = P(one head and one tail) = $\frac{1}{3}$. Now, the assignment of P(two heads) = $\frac{1}{3}$ is contrary to experience, which suggests the following definition of Q: $Q = \{HH, HT, TH, TT\}$, where the first letter in each pair refers to the first flip and the second letter refers to the second flip. Applying the principle of insufficient reason to Q directly, instead of to D'Alembert's partition $\{\{HH\}, \{HT, TH\}, \{TT\}\}$, yields P(two heads) = P(two tails = $\frac{1}{4}$ and P(one head and one tail) = $P(HT) + P(TH) = \frac{1}{2}$.

Somewhat more sophisticated reformulations of the principle of insufficient reason rescue it from D'Alembert's paradox, but they introduce other possible ambiguities by requiring, for example, notions of "relevance of evidence."

Nevertheless, the class of events to which probabilities (regardless of their meaningfulness) can be attached is unrestricted by the principle of insufficient reason. The same cannot be said for the frequentist viewpoint.

5.4.2 Limiting Relative Frequency Viewpoint

Perhaps the most extreme of several closely related frequency viewpoints maintains that probability is a concept meaningful only for events whose occurrence or nonoccurrence can be observed on each of an indefinitely long sequence of experimental trials conducted under identical conditions.

To define the probability $P(A)$ that the outcome q of any trial will belong to a subset A of the set Q of all possible trial outcomes, we start by assuming that our experiment will consist of an indefinitely long sequence of trials, and we start performing trials. After each trial i we record the total number $t_i(A)$ of trials *to date* in which q belonged to A, and we calculate the *relative frequency* $P_i(A)$, defined as equal to $t_i(A)/i$, of trials on which event A occurred.

If we stop the experiment after n trials, then (provided n is reasonably large) it seems reasonable to use the observed relative frequency $P_n(A) = t_n(A)/n$ as an "estimate" of the probability that $q \in A$ will occur on any given trial, because the trials are supposedly performed under identical conditions.

Furthermore, it is also apparent that if n is large, then $P_{n+1}(A)$ will differ very little from $P_n(A)$ regardless of whether $q \in A$ or $q \in A^c$ on trial $n + 1$. To see this, note that either $t_{n+1}(A) = t_n(A)$ (if $q \in A^c$ on trial $n + 1$) or $t_{n+1}(A) = t_n(A) + 1$ (if $q \in A$ on trial $n + 1$). Hence either

$$P_{n+1}(A) = \frac{t_n(A)}{(n+1)} = \frac{n}{(n+1)} \cdot \frac{t_n(A)}{n}$$
$$= \frac{n}{(n+1)} P_n(A)$$

or

$$P_{n+1}(A) = \frac{t_n(A) + 1}{n+1} = \frac{n}{n+1} \cdot \frac{t_n(A)}{n} + \frac{1}{n+1}$$
$$= \frac{n}{n+1} \cdot P_n(A) + \frac{1}{n+1}.$$

Now, for large n, we have $n/(n + 1) \doteq 1$ and $1/(n + 1) \doteq 0$, and hence $P_{n+1}(A) \doteq P_n(A)$.

Thus when n becomes large the succeeding estimates $P_n(A)$ must become closer to their neighbors $P_{n-1}(A)$ and $P_{n+1}(A)$. This suggests (but does not imply) that $\lim_{n \to \infty} P_n(A)$ exists. This limit may fail to exist, but we shall henceforth *assume* that it does.

We then *define the probability* $P(A)$ *of* A *to be that limit*:

$$P(A) = \lim_{n \to \infty} P_n(A).$$

The limiting relative frequency definition of probability has the property of being based solely upon directly observed evidence. Two experimenters obtaining the same outcomes on each trial of the experiment will obtain identical estimates $P_n(A)$ of $P(A)$ regardless of their prior feelings about the "odds in favor of A," provided that A is a well-defined set of outcomes q.

Additional support for the frequentist approach is furnished by the following "random sampling" argument. Suppose each trial consists of selecting "at random" an element q in some population Q and observing whether q belongs to the subset A which consists of all q possessing a given attribute or characteristic. Selecting at random means that on each trial every element $q \in Q$ is no more or no less likely to be selected than any other element $q' \in Q$. We say that event A occurred on trial i if the selected q possessed the attribute in question. Let $p(A)$ denote the proportion of elements of Q which are in A.

It is intuitively appealing to state that the limit $P(A)$ of the relative frequencies $P_n(A)$ in this experiment coincides with $p(A)$. The stated equality $p(A) = P(A)$ $(= \lim_{n \to \infty} P_n(A))$ will in fact be almost always true, in the following sense: if we repeat the *experiment* (consisting of an infinite number of trials) an infinite number of times and denote by $P^m(A)$ the limiting relative frequency in the mth experiment, then

$$\{m: P^m(A) \neq p(A)\}$$

is a *finite* set. In other words, the experiments in which $P(A) \neq p(A)$ are so rare as to be worth ignoring for all practical purposes.

This result implies that if n is very large, then $P_n(A)$ is: (1) a good estimate of $P(A)$; and (2) a good estimate of the proportion $p(A)$ of elements in the population possessing the attribute in question. Hence it makes good sense to equate $P(A)$ to $p(A)$.

In the preceding subsection we seemed to settle on the probability $P(\text{two heads}) = \frac{1}{4}$ of obtaining two heads in two flips of a coin. Suppose, however, that the coin was biased in such a way that it usually landed tails. If we did not *know precisely how* this supposition should be taken into account in assigning probabilities, the principle of insufficient reason would lead us back to $P(\text{two heads}) = \frac{1}{4}$, which clearly seems too high in view of our assumption that the coin is tail-biased. The frequentist definition, however, admits values for $P(\text{two heads})$ different from $\frac{1}{4}$: let each trial consist of two flips of the coin, let $t_n(\text{two heads})$ denote the total number of trials up through n on which two heads occurred, define $P_n(\text{two heads}) = t_n(\text{two heads})/n$, and estimate $P(\text{two heads})$ by P_n (two heads) if n is large. Tail-biasedness should be reflected by $P_n(\text{two heads}) < \frac{1}{4}$.

Despite its merits, the frequentist position suffers from two serious defects. The *first* is that it is not as completely objective as it appears. Subjectivity enters in two respects: first, the number of trials which is "large enough" to make an estimate $P_n(A)$ of $P(A)$ "good" is subjectively determined; and second, "randomness" of selecting q (or the performance of all trials under identical conditions) can never be verified with certainty.

The *second* defect is that many events of importance in decision-making are simply not capable of being tested on an indefinitely large number of trials; hence they possess no probabilities according to the limiting relative frequency viewpoint.

For example, we cannot order dessert and coffee in the example of Subsection 4.3.1 an infinite number of times, eat the dessert, and find out if it is digestible and served sufficiently promptly. Hence elementary event

s_1 is simply without a probability according to the limiting relative frequency definition. Yet it seems reasonable to base our decision in this case at least in part upon our odds $P(s_1)/P(s_1^c)$-to-one in favor of s_1.

5.4.3 Subjective Probability

We can widen the class of events to which probabilities can be meaningfully assigned if we are willing to retreat from the degree of objectivity embodied in the limiting relative frequency approach. What is involved in the subjective probability viewpoint is essentially the judgmental approach of the classical school without the confining principle of insufficient reason.

The following extreme subjectivist approach is also the most applicable in decision theory. In the following approach the decision-maker is required to make hypothetical choices between hypothetical bets. The choice situation is hypothetical and related (in this formulation) to the given decision problem only via the event $A \in \mathcal{E}_Q$, which is the same in both.

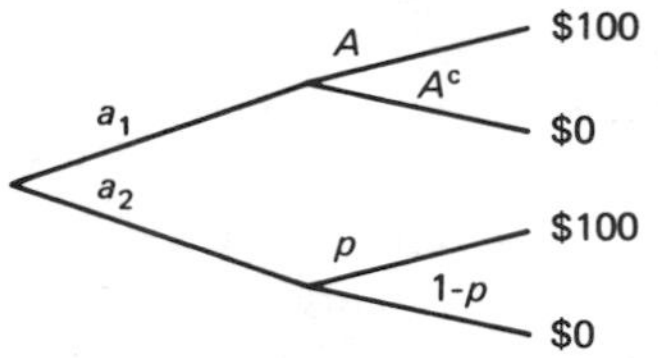

Figure 5.2

The decision-maker is asked to consider the tree in Figure 5.2, where p is the proportion of red balls in a given urn. Now, providing the decision-maker can imagine that the selection of a ball from the urn is truly random, we can proceed as follows: (1) determine those values of p for which the decision-maker prefers a_1 to a_2—this will be an interval, say $[0, p_1]$; (2) determine those values of p for which the decision-maker prefers a_2 to a_1—this will be an interval, say $[p_2, 1]$; and (3) *define* $P(A)$ as some number in $[p_1, p_2]$. Ideally, the decision-maker will be able to specify $p_1 = p_2$, in which case $P(A)$ is uniquely determined. We shall hereafter assume that $p_1 = p_2$ and use p_A to denote their common value.

Certain reasonable axioms of consistency imply that if the events $A_1, \ldots, A_m$ are mutually exclusive, then $P(\bigcup_{i=1}^m A_i) = \sum_{i=1}^{m} P(A_i)$ and hence $(P3f)$ [and thus also $(P3)$] hold for subjective probability. Moreover, $P(Q) = 1$. (Why? Examine the assessment procedure.) Furthermore, the probability P on $\mathcal{E}_Q$ is not affected if we replace \$100 and \$0 by any two consequences c and c' such that c and c' are not equally desirable. We shall present one such set of axioms in Chapter 16.

By defining $P(A)$ as the "indifference proportion" p_A in the urn bet, we are defining probability essentially in terms of indifference betting odds, or equivalent betting odds: the decision-maker's (subjective) odds $P(A)/P(A^c)$ to one in favor of A are defined as equal to $p_A/(1 - p_A)$.

Examples (3) and (4) of Section 5.3 involve assessment of subjective probabilities. We shall return to assessment techniques later.

The subjectivist approach is not without its drawbacks. The chief objection to subjective probability is that people may disagree on the probability of any given event $A \neq Q$ or $\varnothing$. *You*, for example, might assess P(ABC Company's year-end common stock price $\geq$ \$60/share) = .85, while *I* might assess the probability of that event to be .05. The theory furnishes no means by which this drastic difference of opinion can be reconciled or compromised without further evidence.

One occasionally aired criticism of subjective probability is that it may ignore available objective evidence in the form of frequency data. Most subjectivists would reply that *if* the relative frequency $P_n(A)$ were determined from a reasonably large number n of trials, and *if* the decision-maker believes that the next trial will be performed under conditions identical to those obtaining on the previous n trials, *then he should assess $P(A)$ very close to $P_n(A)$*. In summary, responsible probability assessors do not ignore *pertinent* evidence.

Nevertheless, the subjectivist may still ignore the available objective evidence if he believes that the experimental conditions governing the next trial might differ from those governing the previous n trials. But this possibility is equally present in the frequentest approach, which dictates that under such conditions one can usually say *nothing* about the probability of A on the next trial.

5.5 MARGINAL PROPORTIONS

Suppose again that we are considering a given population X of, say, people. Suppose that $\mathcal{A} = \{A_1, \ldots, A_m\}$ and $\mathcal{B} = \{B_1, \ldots, B_n\}$ are two partitions of X. Then consider the product partition $\mathcal{A}\mathcal{B} = \{A_i \cap B_j: i \in \{1, \ldots, m\}, j \in \{1, \ldots, n\}\}$.

Let $\#(A_i, B_j) = \#(A_i \cap B_j)$ denote the number of elements of X which belong to $A_i \cap B_j$. If $A_i \cap B_j = \varnothing$, then $\#(A_i, B_j) = 0$. The *proportion* of elements in $A_i \cap B_j$ is denoted by $p(A_i, B_j)$ or by $p(A_i \cap B_j)$. Since $A_i \cap B_j$ is a subset of X, it is clear from Section 5.2 that

$$p(A_i, B_j) = p(A_i \cap B_j) = \frac{\#(A_i, B_j)}{\#(X)}$$

for every $i \in \{1, \ldots, m\}$ and every $j \in \{1, \ldots, n\}$. If $\#(A_i, B_j) = 0$, then obviously $p(A_i, B_j) = 0$.

Example

Let $\mathcal{A} = \{A_1, A_2\}$, $\mathcal{B} = \{B_1, B_2, B_3\}$, $\#(A_1, B_1) = 10$, $\#(A_2, B_1,) = 15$, $\#(A_1, B_2) = 20$, $\#(A_2, B_2) = 5$, $\#(A_1, B_3) = 0$, and $\#(A_2, B_3) = 50$. Adding all the $\#(A_i, B_j)$'s yields $\#(X) = 100$. We can arrange the $\#(A_i, B_j)$'s in a table.

	B_1	B_2	B_3
A_1	10	20	0
A_2	15	5	50

By dividing all the $\#(A_i, B_j)$'s by 100 ($= \#(X)$) we obtain the corresponding table of *joint proportions*.

	B_1	B_2	B_3
A_1	.10	.20	.00
A_2	.15	.05	.50

Obviously, the double sum $\sum_{i=1}^{2} \sum_{j=1}^{3} p(A_i, B_j)$ equals 1.

We call the number $p(A_i, B_j)$ the *joint proportion* of elements in X which are in both A_i and B_j.

Now suppose that we are given a table of joint proportions, as in the preceding example, but also suppose that we are interested solely in the proportions in each of B_1, B_2, and B_3; that is, we want $p(B_1)$, $p(B_2)$, and $p(B_3)$.

The proportion of elements of X in each B_j can be derived easily from the joint proportions. First, note that

$$\begin{aligned} B_j &= X \cap B_j && \text{(because } B_j \subset X\text{)} \\ &= (\cup_{i=1}^{m} A_i) \cap B_j && \text{(because } \cup_{i=1}^{m} A_i = X\text{)} \\ &= \cup_{i=1}^{m} (A_i \cap B_j) && \text{(by distributivity of } \cap\text{).} \end{aligned}$$

Note also that if $i_1 \neq i_2$, then

$$\begin{aligned} (A_{i_1} \cap B_j) \cap (A_{i_2} \cap B_j) &= (A_{i_1} \cap A_{i_2}) \cap B_j \\ &= \varnothing \cap B_j \qquad (\mathcal{A} \text{ a partition}) \\ &= \varnothing. \end{aligned}$$

Hence $\{A_i \cap B_j: i \in \{1, \ldots, m\}\}$ is a family of mutually exclusive subsets of B_j whose union equals B_j. Thus by property ($p3'$) of proportions, we have

$$\begin{aligned} p(B_j) &= p(\cup_{i=1}^{m} (A_i \cap B_j)) \\ &= \sum_{i=1}^{m} p(A_i \cap B_j) \\ &= \sum_{i=1}^{m} p(A_i, B_j). \end{aligned}$$

We now have our desired result:

$$p(B_j) = \sum_{i=1}^{m} p(A_i, B_j).$$

Therefore, to find the proportion $p(B_j)$ of elements of X in B_j, simply add up all the m joint proportions $p(A_i, B_j)$ as i ranges through $\{1, \ldots, m\}$.

Note that we also have that

$$p(B_j) = \sum_{i=1}^{m} p(A_i, B_j)$$

implies

$$\sum_{j=1}^{n} p(B_j) = \sum_{j=1}^{n} \sum_{i=1}^{m} p(A_i, B_j).$$

But the double summation on the right-hand side of this equation is simply the sum of all the numbers $p(A_i, B_j)$ as i ranges through $\{1, \ldots, m\}$ and j ranges through $\{1, \ldots, n\}$. This sum equals $p(X)$, which is 1. Hence we have shown that

$$\sum_{j=1}^{n} p(B_j) = 1,$$

which is only to be expected in view of the fact that $\mathcal{B}$ is a partition of X.

The numbers $p(B_1), \ldots, p(B_n)$ are called the *marginal proportions* of the B_j's. The reason for the adjective *marginal* lies in the tabular format given in the preceding example: $p(B_1) = .10 + .15 = .25$; $p(B_2) = .20 + .05 = .25$; and $p(B_3) = .00 + .50 = .50$. Also, $p(B_1) + p(B_2) + p(B_3) = .25 + .25 + .50 = 1.00$ for our example. The numbers $p(B_j)$ can be written as "column totals" of the table of joint proportions, as in the following table:

	B_1	B_2	B_3	$p(A_i)$
A_1	.10	.20	.00	.30
A_2	.15	.05	.50	.70
$p(B_j)$	.25	.25	.50	1.00

This table also shows the marginal proportions $p(A_i)$. You can see that the numbers $p(B_j)$ are written in the "lower margin" of the table, while the numbers $p(A_i)$ are written in the "right margin" of the table.

The qualifier *marginal* is always understood as being relative to some other data, as is the qualifier *joint*. For example, without being given any information about a partition $\mathcal{A}$ of X we would have no reason to call $p(B_1)$, $p(B_2)$, and $p(B_3)$ anything but proportions (without qualification).

Joint proportions can also be marginal. Suppose there were a third partition $\mathcal{C} = \{C_1, \ldots, C_r\}$ of X and that we had been given a three-way (three-dimensional) table of joint proportions $p(A_i, B_j, C_k)$. It is easy enough to show that

$$p(A_i, B_j) = \sum_{k=1}^{r} p(A_i, B_j, C_k),$$

for every $i \in \{1, \ldots, m\}$ and every $j \in \{1, \ldots, n\}$.
The proportions $p(A_i, B_j)$ are now considered *joint* as far as $\mathcal{A}$ and $\mathcal{B}$ are concerned, but *marginal* with respect to $\mathcal{C}$.

This ambiguity, or relativity, will not be a cause for confusion, because appropriate definitions of the product partition will be given.

5.6 MARGINAL PROBABILITY

Let Q be a set of elementary events q; let $\mathcal{E}_Q$ be a field of events; and let $P: \mathcal{E}_Q \to R^1$ be a probability on $\mathcal{E}_Q$. Suppose $\mathcal{A} = \{A_1, \ldots, A_m\} \subset \mathcal{E}_Q$ and $\mathcal{B} = \{B_1, \ldots, B_n\} \subset \mathcal{E}_Q$ are partitions of Q. Then every $A_1 \cap B_j \in \mathcal{E}_Q$, and hence the product partition $\mathcal{A}\mathcal{B}$ is a subset of $\mathcal{E}_Q$. Therefore all the $P(A_i \cap B_j)$'s are defined.

We define $P(A_i \cap B_j)$ to be the *joint probability* of the events A_i and B_j and will often denote it by $P(A_i, B_j)$.

Given the joint probabilities $P(A_i, B_j)$ of all events in the product $\mathcal{A}\mathcal{B}$ of two partitions $\mathcal{A}$ and $\mathcal{B}$ of Q, we call $P(B_j)$ the *marginal probability* of $B_j \in \mathcal{B}$ and $P(A_i)$ the *marginal probability* of $A_i \in \mathcal{A}$, for every i and every j.

The derivations of the preceding section translate verbatim (replacing all p's by P's) to show that

$$P(A_i) = \sum_{j=1}^{n} P(A_i, B_j), \text{ for all } i \in \{1, \ldots, m\};$$

$$\sum_{i=1}^{m} P(A_i) = 1;$$

$$P(B_j) = \sum_{i=1}^{m} P(A_i, B_j), \text{ for all } j \in \{1, \ldots, n\}; \text{ and}$$

$$\sum_{j=1}^{n} P(B_j) = 1.$$

Example

(1) In the general decision problem, suppose that you have selected an experiment $e \in E$; that $Z(e) = \{z_1, z_2, z_3\}$; that $S = \{s_1, s_2, s_3, s_4\}$; that $Q = Z(e) \times S$; and hence (because Q is a finite set) $\mathcal{E}_Q = \mathcal{R}(Z(e) \times S)$.

Suppose a probability $P_{z,s}: \mathcal{E}_Q \to R^1$ is given by the following table, where the number in the ith row and jth column is $P_{z,s}(z_i, s_j)$:

	s_1	s_2	s_3	s_4
z_1	.10	.13	.05	.02
z_2	.05	.12	.20	.03
z_3	.02	.01	.02	.25

From this table we can compute the probability of any subset A of $Z(e) \times S$ simply by adding the probabilities of the elements of A. Hence the table furnishes sufficient information to specify $P_{z,s}$ on all of $\mathcal{E}_Q$ unambiguously. We may compute the following marginal probabilities, where P_s denotes marginal probabilities of the s's and P_z denotes marginal probabilities of the z's:

$$
\begin{aligned}
P_s(s_1) &= .10 + .05 + .02 = .17 \\
P_s(s_2) &= .13 + .12 + .01 = .26 \\
P_s(s_3) &= .05 + .20 + .02 = .27 \\
P_s(s_4) &= .02 + .03 + .25 = .30
\end{aligned}
$$

$$P_s(s_1) + P_s(s_2) + P_s(s_3) + P_s(s_4) = 1.00$$

$$
\begin{aligned}
P_z(z_1) &= .10 + .13 + .05 + .02 = .30 \\
P_z(z_2) &= .05 + .12 + .20 + .03 = .40 \\
P_z(z_3) &= .02 + .01 + .02 + .25 = .30
\end{aligned}
$$

$$P_z(z_1) + P_z(z_2) + P_z(z_3) = 1.00$$

Examples of this sort [with different numbers of states s and outcomes $z \in Z(e)$] are very common in decision theory.

EXERCISES

1. Let $\#(X) = 10{,}000$; $\#(A) = 4500$; $\#(B) = 6300$; and $\#(A \cap B) = 3200$. Compute $p(A)$, $p(B)$, $p(A \cap B)$, $p(A \cup B)$, $p(A \backslash B)$, $p(B \backslash A)$, $p(A^c)$, $p(B^c)$, and $p(A^c \cup B^c)$.

2. In a town of 2000 people, it was determined that 1200 people read the local newspaper; 1800 people watch the evening television news program; 1500 people listen to the noon radio news; 1000 people listen to both evening television and noon radio news, and 500 of these also read the local paper. Four hundred people read the paper and watch only the television news, and

only 100 people read the paper but listen to neither television nor radio news.

(a) Find:

(i) the number of people who read the paper and listen to the noon radio news;

(ii) the number of people who listen only to the noon radio news and do not read the paper;

(iii) the number of people who either read the paper or watch the evening television news, or both; and

(iv) the number of people who neither read the paper nor listen to either of the news programs.

(b) Find the proportions of people in each of the categories mentioned in the problem statement and in part (a).
[*Hint*: Fill in the following Venn diagram properly.]

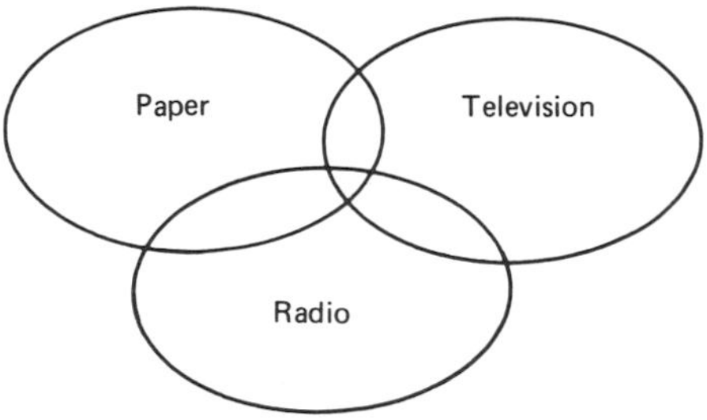

(c) If an individual is selected from this town at random, what is the probability that he watches the evening television news but does not read the paper?

(d) Using the same assumptions as (c), what is the probability that he does not listen to the noon radio news?

3. Derive properties $(p1)$ through $(p7)$ and $(p3')$ from the definition of proportion.

4. Using the principle of insufficient reason for the roll of one die:

(a) what is the probability of i, for $i = 1, \ldots, 6$?
(b) what is the probability of obtaining an even outcome?
(c) what is the probability of obtaining an odd outcome?
(d) what is the probability of obtaining "5 or 6"?

5. Assume the die in Exercise 4 yields outcomes i with the following probabilities:

i	$P(i)$
1	$\frac{1}{6}$
2	$\frac{1}{12}$
3	$\frac{1}{12}$
4	$\frac{1}{3}$
5	$\frac{1}{12}$
6	$\frac{1}{4}$

Answer questions (b) through (d) of Exercise 4 for this loaded die.

6. Assess *your* probability that some rain will fall tomorrow on the building in which you are now sitting. Use the method introduced in Subsection 5.4.3.

7. Suppose X is a population of $\#(X)$ balls in an urn. Suppose $\#(R, 0)$ of these balls are red and marked with a zero; that $\#(R, 1)$ of them are red and marked with a one; that $\#(W, 0)$ are white and marked with a zero; and that $\#(W, 1)$ are white and marked with a one. Suppose that $\#(X) = \#(R, 0) + \#(R, 1) + \#(W, 0) + \#(W, 1)$. Give formulas for $p(R)$, $p(W)$, $p(0)$, and $p(1)$.

8. In Example (1) of Section 5.6, suppose that you have now observed outcome z_1 of e. Are your judgments regarding relative likelihoods of the s's still given by the marginal probabilities $P_s(s)$? Explain why they are or are not; and if not, then state what probabilities on S do express your judgments at this point in the decision.

9. Let A and B be any events. Show that

$$P(A \cap B) \geq P(A) + P(B) - 1.$$

6

PROBABILITY (II): CONDITIONAL PROBABILITY AND INFORMATION

6.1 INTRODUCTION

This chapter continues our study of probability theory by attacking the problem of how *information* should be taken into account in revising originally given probability judgments. Exercise 8 of Chapter 5 introduced you to the subject of this chapter.

We shall follow our previous strategy by discussing conditional proportions in Section 6.2 as a prelude to conditional probability, the basics of which are established in Section 6.3. The most important part of Section 6.3 is Bayes' Theorem, since it plays a central role in decision theory.

Section 6.4 introduces the role of Bayes' Theorem in decision theory It shows precisely how probability judgments are to be revised in light of experimental evidence.

Section 6.5, on "independence," indicates conditions under which experimental evidence will induce no revision in probability judgments.

6.2 CONDITIONAL PROPORTIONS

Suppose that you are given an m(rows) by n(columns) table of joint proportions $p(A_i, B_j)$ corresponding to two partitions $\mathcal{A}$ and $\mathcal{B}$ of the population X. Consider the subpopulation consisting of all $x \in X$ who belong to $B_j \in \mathcal{B}$. We ask what proportion of the elements in B_j belong to A_1? To A_2? . . . to A_m?

The answer should be intuitively clear: since $\#(B_j)$ x's belong to B_j and $\#(A_i, B_j)$ x's belong to both A_i and B_j, the proportion of x's in B_j that belong to A_i is $\#(A_i, B_j)/\#(B_j)$. We define $\#(A_i, B_j)/\#(B_j)$ to be the *conditional proportion in A_i given B_j* and denote it by $p(A_i \mid B_j)$. Thus

$$p(A_i \mid B_j) = \frac{\#(A_i, B_j)}{\#(B_j)}.$$

Since $0/0$ is undefined, $p(A_i \mid B_j)$ is defined only when $\#(B_j) > 0$.

Since

$$\frac{\#(A_i, B_j)}{\#(B_j)} = \frac{\#(A_i, B_j)/\#(X)}{\#(B_j)/\#(X)} = \frac{p(A_i, B_j)}{p(B_j)},$$

it follows that

$$p(A_i \mid B_j) = \frac{p(A_i, B_j)}{p(B_j)}$$

whenever $p(B_j) > 0$. Even when $p(B_j) = 0$, we may write

$$p(A_i, B_j) = p(A_i \mid B_j) \cdot p(B_j).$$

Moreover, since $\sum_{i=1}^{m} p(A_i, B_j) = p(B_j)$, it follows (when $p(B_j) > 0$) that

$$\sum_{i=1}^{m} p(A_i \mid B_j) = 1.$$

Finally, note that whenever $p(A_{i_1} \mid B_j) > 0$ we have

$$\frac{p(A_{i_2} \mid B_j)}{p(A_{i_1} \mid B_j)} = \frac{p(A_{i_2} \mid B_j) \cdot p(B_j)}{p(A_{i_1} \mid B_j) \cdot p(B_j)}$$

$$= \frac{p(A_{i_2}, B_j)}{p(A_{i_1}, B_j)}.$$

Hence the ratios of the joint proportions of (A_{i_1}, B_j) and (A_{i_2}, B_j) are preserved by the conditional proportions in A_{i_1} and in A_{i_2} given B_j.

Example

(1) In the example of Section 5.6, we may compute

$$p(A_1 \mid B_1) = \frac{p(A_1, B_1)}{p(B_1)} = \frac{.10}{.25} = .40,$$

$$p(A_2 \mid B_1) = \frac{p(A_2, B_1)}{p(B_1)} = \frac{.15}{.25} = .60,$$

and note that $p(A_1 \mid B_1) + p(A_2 \mid B_1) = .40 + .60 = 1.00$.

Also note that $p(A_1 \mid B_1)/p(A_2 \mid B_1) = .40/.60 = \frac{2}{3} = p(A_1, B_1)/p(A_2, B_1)$. Similarly, we may compute

$$p(A_1 \mid B_2) = \frac{p(A_1, B_2)}{p(B_2)} = \frac{.20}{.25} = .80,$$

$$p(A_2 \mid B_2) = \frac{p(A_2, B_2)}{p(B_2)} = \frac{.05}{.25} = .20.$$

Note that $p(A_1 \mid B_2) + p(A_2 \mid B_2) = .80 + .20 = 1.00$, and also that $p(A_1 \mid B_2)/p(A_2 \mid B_2) = .80/.20 = 4.00 = .20/.05 = p(A_1, B_2)/p(A_2, B_2)$. Finally, we see that

$$p(A_1 \mid B_3) = \frac{p(A_1, B_3)}{p(B_3)} = \frac{.00}{.50} = .00,$$

$$p(A_2 \mid B_3) = \frac{p(A_2, B_3)}{p(B_3)} = \frac{.50}{.50} = 1.00,$$

that $p(A_1 \mid B_3) + p(A_2 \mid B_3) = .00 + 1.00 = 1.00$, and also that $p(A_1 \mid B_3)/p(A_2 \mid B_3) = .00/1.00 = 0 = 0/.50 = p(A_1, B_3)/p(A_2, B_3)$. You may compute the conditional proportions $p(B_1 \mid A_i)$, $p(B_2 \mid A_i)$, and $p(B_3 \mid A_i)$ for each i as an exercise.

Another way of looking at conditional proportions given, say, B_j, is to assume that $X \backslash B_j$ can be forgotten about—assume that the elements of $X \backslash B_j$ have been discarded—and to then *re*compute the proportions in the various A_i's under this assumption. The recomputation will yield the conditional proportions $p(A_i \mid B_j)$, for $i \in \{1, \ldots, m\}$.

From this viewpoint we can re-express our original proportions in the form $p(A \mid X)$, since these original proportions were naturally conditional upon our definition of the original population X.

Example

(2) Recall that in the random sampling context of the frequency definition of probability, we set $P(A) = p(A)$, for every $A \in \mathcal{E}_Q$. Suppose an element q has been drawn at random from Q and you observe that it belongs to B_j. It now makes sense to use the conditional proportions $p(A \mid B_j)$ to express your probability judgments that q also belongs to any given event $A \in \mathcal{E}_Q$.

We shall generalize this example in the next section.

Finally, suppose we have a third partition $\mathcal{C} = \{C_1, \ldots, C_r\}$ of X and that we are originally given the joint proportions $\{p(A_i, B_j, C_k)$:

$i \in \{1, \ldots, m\}, j \in \{1, \ldots, n\}, k \in \{1, \ldots, r\}\}$. We may consider $\mathcal{BC}$ as a single partition in its own right, and hence use the preceding discussion to write

$$p(A_i \mid B_j, C_k) = \frac{p(A_i, B_j, C_k)}{p(B_j, C_k)},$$

which implies

$$p(A_i, B_j, C_k) = p(A_i \mid B_j, C_k) \cdot p(B_j, C_k).$$

But

$$p(B_j, C_k) = p(B_j \mid C_k) \cdot p(C_k),$$

and by substituting this into the preceding equation implies that

$$p(A_i, B_j, C_k) = p(A_i \mid B_j, C_k) \cdot p(B_j \mid C_k) \cdot p(C_k).$$

Now, nothing in our definition of conditional proportions necessarily required that the events A_i, B_j, and so forth, be elements in partitions. Thus we can generalize the preceding multiplication formula for conditional proportions to assert that if $E_1, E_2, \ldots, E_n$ are any n events for any finite n, then

$$p(E_1, \ldots, E_n) = p(E_1 \mid E_2, \ldots, E_n) \cdot p(E_2 \mid E_3, \ldots, E_n) \cdot \ldots \cdot p(E_n).$$

6.3 CONDITIONAL PROBABILITY

Let $\mathcal{A} \subset \mathcal{E}_Q$ and $\mathcal{B} \subset \mathcal{E}_Q$ be two partitions of Q. We denote the *conditional probability of event A_i given event B_j* by $P(A_i \mid B_j)$ and define it when $P(B_j) > 0$ by

$$P(A_i \mid B_j) = \frac{P(A_i, B_j)}{P(B_j)}.$$

When $P(B_j) = 0$, we use a limiting process to define $P(A_i \mid B_j)$. We shall not go into the matter here.

By arguments similar to those used with conditional proportions, we find

$$\sum_{i=1}^{m} P(A_i \mid B_j) = 1, \tag{6.3.1}$$

for every $j \in \{1, \ldots, n\}$ such that $P(B_j) > 0$; and

$$\frac{P(A_{i_2} \mid B_j)}{P(A_{i_1} \mid B_j)} = \frac{P(A_{i_2}, B_j)}{P(A_{i_1}, B_j)} \tag{6.3.2}$$

whenever either ratio is well-defined.

Since $P(A_i \mid B_j) = \dfrac{P(A_i, B_j)}{P(B_j)}$, it follows by elementary algebra that

$$P(A_i \mid B_j) \cdot P(B_j) = P(A_i, B_j), \tag{6.3.3}$$

which holds even when $P(B_j) = 0$. But we could have started the discussion by assuming that A_i was given and that we wanted the conditional probability of B_j given A_i. For this case, (6.3.3) becomes

$$P(B_j \mid A_i) \cdot P(A_i) = P(B_j, A_i) = P(A_i, B_j). \tag{6.3.3$'$}$$

Things equal to the same thing [namely, $P(A_i, B_j)$] are equal to each other, and hence we obtain

$$P(B_j \mid A_i) \cdot P(A_i) = P(A_i \mid B_j) \cdot P(B_j). \tag{6.3.4}$$

We divide both sides of Equation (6.3.4) by $P(A_i)$ to obtain

$$P(B_j \mid A_i) = \frac{P(A_i \mid B_j) \cdot P(B_j)}{P(A_i)}. \tag{6.3.5}$$

Now,

$$\begin{aligned} P(A_i) &= \sum_{k=1}^{n} P(A_i, B_k) \\ &= \sum_{k=1}^{n} P(A_i \mid B_k) \cdot P(B_k), \end{aligned}$$

which we substitute into (6.3.5) to obtain

$$P(B_j \mid A_i) = \frac{P(A_i \mid B_j) \cdot P(B_j)}{\sum_{k=1}^{n} P(A_i \mid B_k) \cdot P(B_k)}, \tag{6.3.6}$$

all $j \in \{1, \ldots, n\}$.

Equation (6.3.6) is of fundamental importance in decision theory; it is called *Bayes' Theorem*, after its discoverer, the Reverend Thomas Bayes (1702–1761).

Our partition-based definition of conditional probability is well-suited for applications in decision theory. However, it is simpler to first define $P(E_1, E_2) = P(E_1 \cap E_2)$ for any two events E_1 and E_2, and then to define

$$P(E_1 \mid E_2) = \frac{P(E_1, E_2)}{P(E_2)},$$

whenever $P(E_2) > 0$. We then obtain

$$P(E_1, E_2) = P(E_1 \mid E_2) \cdot P(E_2)$$

[even when $P(E_2) = 0$]. For any n events $E_1, \ldots, E_n$ we have the multiplication formula

$$P(E_1, E_2, \ldots, E_n) = P(E_1 \mid E_2, \ldots, E_n) \cdot P(E_2 \mid E_3, \ldots, E_n) \cdot \ldots \cdot P(E_n), \tag{6.3.7}$$

just as with proportions. Equation (6.3.7) is often called the *chain rule* of conditional probability.

6.4 BAYES' THEOREM IN DECISION THEORY

In decision theory the events A_i correspond to experimental outcomes z and the events B_j to states s of the environment. The decision-maker is usually given the probabilities $P(A_i \mid B_j)$ of the experimental outcomes A_i given the states B_j and assesses his probabilities $P(B_j)$ of the states B_j. Then Bayes' Theorem shows how his probabilities on the states B_j should be revised in light of the message A_i, so that his judgments regarding the states B_j are now expressed by the conditional probabilities $P(B_j \mid A_i)$.

We shall now make these ideas more precise. In the general decision model, let $Q = Z(e) \times S$ and assume, for the sake of simplicity, that *both* Z(e) *and* S *are finite sets.* (We shall relax this assumption at the end of the section.) Then $\mathcal{E}_{Z(e)}$, $\mathcal{E}_S$, and $\mathcal{E}_{Z(e) \times S}$ are the sets of all subsets of their respective subscript sets.

Next, assume that the decision-maker has specified a probability P_s: $\mathcal{E}_S \rightarrow [0, 1]$. This probability is completely determined by the set $\{P_s(s): s \in S\}$ of probabilities of all elementary s-events [because S is finite and hence any $A \in \mathcal{E}_S$ is the union of a finite number of (mutually exclusive) elementary events s; use ($P3'$) to prove this].

The probability P_s is called the decision-maker's *prior* probability on $\mathcal{E}_S$. The adjective *prior* is intended to suggest that P_s expresses the decision-maker's judgments regarding the state-events (in $\mathcal{E}_S$) *prior* to observing any message-event (in $\mathcal{E}_{Z(e)}$). The prior probability P_s is often subjective, as the decision-maker may lack relative frequency data on the events in $\mathcal{E}_S$.

Finally, we assume that the decision-maker has specified a conditional probability $P^e_{z|s}(\cdot \mid s)$: $\mathcal{E}_{Z(e)} \rightarrow [0, 1]$ on $\mathcal{E}_{Z(e)}$ given $s \in S$ for every s such that $P_s(s) > 0$. [Since S is finite, we clearly may, and shall, ignore any state s such that $P_s(s) = 0$.] In many applications, $P^e_{z|s}$ can be objectively determined from relative frequency data for each $s \in S$.

From the specifications P_s and $P^e_{z|s}$ (all s) we can deduce the joint probability $P^e_{z,s}$ on $\mathcal{E}_{Z(e) \times S}$ by first deriving $P^e_{z,s}(z, s)$ for every $(z, s) \in Z(e) \times S$. This derivation is very easy, since the relevant analogue of Equation (6.3.3) is

$$P^e_{z,s}(z, s) = P^e_{z|s}(z \mid s) \cdot P_s(s).$$

The deduction is completed by observing that any event D in $\mathcal{E}_{Z(e)\times S} = \mathcal{E}_{Z(e)} \otimes \mathcal{E}_S$ is a finite union of the one-element events $\{(z, s)\}$ belonging to D. Hence

$$P^e_{z,s}(D) = \sum\{P^e_{z,s}(z, s): (z, s) \in D\}.$$

If the decision-maker has derived the joint probabilities $\{P^e_{z,s}(z, s): (z, s) \in Z(e) \times S\}$, he can compute the marginal probability

$$P^e_z(z) = \sum_{s\in S} P^e_{z,s}(z, s) = \sum_{s\in S} P^e_{z|s}(z \mid s)\cdot P_s(s),$$

for every $z \in Z(e)$.

Finally, the definition of conditional probability implies

$$P^e_{s|z}(s \mid z) = \frac{P^e_{z|s}(z \mid s)\cdot P_s(s)}{\sum_{s\in S} P^e_{z|s}(z \mid s)\cdot P_s(s)}, \tag{6.4.1}$$

for every $s \in S$ and every $z \in Z(e)$. This is Bayes' Theorem for the general decision problem with both S and $Z(e)$ finite sets.

Bayes' Theorem shows how the decision-maker should revise his judgments regarding relative likelihoods of state-events given the experimental outcome z. His revised judgments are given by the conditional probability $P^e_{s|z}(\cdot \mid z)$, defined on $\mathcal{E}_S$ by setting

$$P^e_{s|z}(B \mid z) = \sum_{s\in B} P^e_{s|z}(s \mid z)$$

for any $B \in \mathcal{E}_S$. We are usually satisfied (for S finite) in reporting $\{P^e_{s|z}(s \mid z): s \in S\}$ for each $z \in Z(e)$.

The number $P^e_{s|z}(B \mid z)$ is called the decision-maker's *posterior* probability of B given z. The adjective *posterior* is intended to connote that $P^e_{s|z}$ expresses the decision-maker's judgments regarding the states *posterior* to observing the experimental outcome $z \in Z(e)$.

Examples

(1) Let $S = \{s_1, s_2, s_3, s_4\}$, and let $Z(e) = \{z_1, z_2, z_3\}$. Suppose the decision-maker assessed $P_s(s_i) = .25$ for $i = 1, 2, 3, 4$. Suppose also that he has assessed the following conditional probabilities given s:

z	$p^e_{z\mid s}(z \mid s_1)$	$p^e_{z\mid s}(z \mid s_2)$	$p^e_{z\mid s}(z \mid s_3)$	$p^e_{z\mid s}(z \mid s_4)$
z_1	.30	$\frac{1}{3}$	.70	.00
z_2	.10	$\frac{1}{3}$	.20	1.00
z_3	.60	$\frac{1}{3}$	.10	.00
	1.00	1.00	1.00	1.00

We may then compute the joint probabilities [Example: $P^e_{z|s}(z_2, s_3) = P^e_{z|s}(z_2,| s_3) \cdot P_s(s_3) = (.20) \cdot (.25) = .05$].

	s_1	s_2	s_3	s_4	$p^e_z(z)$
z_1	.075	$\frac{1}{12}$	.175	.000	$\frac{40}{120}$
z_2	.025	$\frac{1}{12}$	.050	.250	$\frac{49}{120}$
z_3	.150	$\frac{1}{12}$	.025	.000	$\frac{31}{120}$
$P_s(s)$	.250	.250	.250	.250	1.000

From these data we may compute the posterior probabilities by dividing each row of the joint-probability table by its total [$= P^e_z(z)$].

		s_1	s_2	s_3	s_4	*total*
	z_1	.225	.250	.525	.000	1.000
given	z_2	$\frac{3}{49}$	$\frac{10}{49}$	$\frac{6}{49}$	$\frac{30}{49}$	1.000
	z_3	$\frac{10}{31}$	$\frac{10}{39}$	$\frac{3}{31}$	$\frac{0}{31}$	1.000
	prior	.250	.250	.250	.250	1.000

Note that the decision-maker's prior probability of s_4 is .250 while his posterior probability of s_4 given z_2 is 30/49, and that it is zero given either z_1 or z_2. Thus his judgments regarding the likelihood of s_4 are rather drastically affected by the experimental outcome z. Now recall the extreme generality of our definition of experiment to see how widely Bayes' Theorem is applicable to real life. (2) In Example (1) of Section 5.6 we were given the joint probabilities $P^e_{z,s}(z, s)$ and computed the marginal probabilities P_s and P^e_z. Those data are reproduced here for convenience.

	s_1	s_2	s_3	s_4	$p^e_z(z)$
z_1	.10	.13	.05	.02	.30
z_2	.05	.12	.20	.03	.40
z_3	.02	.01	.02	.25	.30
$p_s(s)$	.17	.26	.27	.30	1.00

The prior probabilities of the s's are given in the fourth row of the table. We can find the posterior probabilities just as in Example (1) by dividing each row by its (right-hand) total, $P^e_z(z)$.

	s_1	s_2	s_3	s_4	*total*
z_1	$\frac{10}{30}$	$\frac{13}{30}$	$\frac{5}{30}$	$\frac{2}{30}$	1.00
z_2	$\frac{5}{40}$	$\frac{12}{40}$	$\frac{20}{40}$	$\frac{3}{40}$	1.00
z_3	$\frac{2}{30}$	$\frac{1}{30}$	$\frac{2}{30}$	$\frac{25}{40}$	1.00
$P_s(s)$	.17	.26	.27	.30	1.00

For purposes of making comparisons, it helps to express all posterior probabilities in decimals.

	s_1	s_2	s_3	s_4	*total*
z_1	.333	.433	.167	.067	1.000
z_2	.125	.300	.500	.075	1.000
z_3	.067	.033	.067	.833	1.000
$P_s(s)$	.170	.260	.270	.300	1.000

Now, suppose that for decision-making purposes you are solely interested in whether s_2 is true. Your prior probability $P_s(s_2)$ is .260. If you receive "message" z_1, your posterior probability of s_2 given z_1 rises to .433, whereas if you receive "message" z_3, your posterior probability is .033, a drastic decline from the prior probability. Thus your judgments regarding the "chances" of s_2 can be significantly influenced by the experimental outcome.

The decision-maker's prior probability P_s is irrelevant once he has ascertained the experimental outcome z. After observing z his judgments regarding s are expressed by his posterior probability $P^e_{s|z}(\cdot \mid z)$. Figure 6.1 renders this statement graphically.

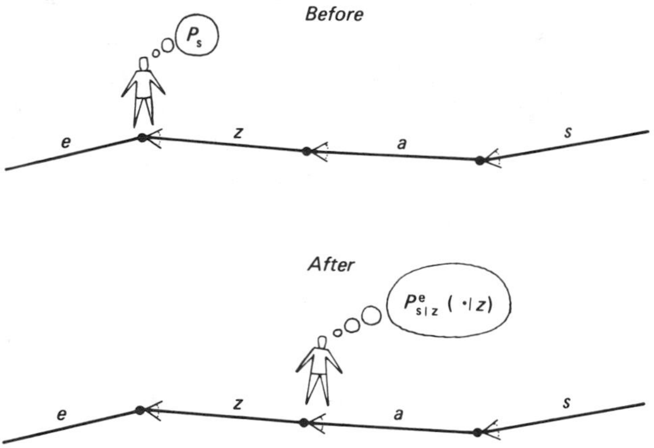

Figure 6.1

Another way of expressing this fact is to state that probabilities are always conditional upon the information possessed by the decision-maker, and that his prior probability is based upon all the information at his disposal before he ascertains z.

Once z has been ascertained, the decision-maker will base his choice of act $a \in A(e, z)$ in part upon his posterior probabilities $P^e_{s|z}(\cdot \mid z)$ of the events in $\mathcal{E}_S$.

6.4.1 Technical Note

[Strictly speaking, our prior probability P_s is not defined on $\mathcal{E}_S$ but rather on $\mathcal{E}_S' = \{Z(e) \times B\colon B \in \mathcal{E}_S\}$, as given in Exercise (10) of Chapter 4. Moreover, $P^e_{z|s}(\cdot \mid s)$ is not defined on $\mathcal{E}_{Z(e)}$ but rather on $\mathcal{E}^s_{Z(e)} = \{A \times \{s\}\colon A \in \mathcal{E}_{Z(e)}\}$. These comments show that the marginal probability P_s and the conditional probabilities $P^e_{z|s}$ are in fact derivable from the joint probability $P^e_{z,s}$, which is the basic probability function P used in Section 6.3.

In the infinite S and/or $Z(e)$ case, we assume that all event fields are sigma-fields and that all probabilities satisfy property ($P3^\sigma$) (see Chaper 5). Moreover, for every $A \in \mathcal{E}_{Z(e)}$ we assume that $P^e_{z|s}(A \mid \cdot)\colon S \to [0, 1]$ is well-behaved (that is, measurable). Then we define

$$P^e_{z,s}(A, B) = \int_B P^e_{z|s}(A \mid s) dP_s(s),$$

for every $B \in \mathcal{E}_S$, where the integral is defined in the sense of Lebesgue. This definition specifies $P^e_{z,s}$ on the set $\mathcal{E}_{Z(e)} \times \mathcal{E}_S$ of all rectangles in $\mathcal{E}_{Z(e)} \otimes \mathcal{E}_S$. It can be shown that there is one and only one way of defining $P^e_{z,s}$ on $\mathcal{E}_{Z(e)} \otimes \mathcal{E}_S$ so that the result coincides with the preceding definition on $\mathcal{E}_{Z(e)} \times \mathcal{E}_S$ and so that $P^e_{z,s}$ possesses property ($P3^\sigma$).]

6.5 INDEPENDENCE, OR THE IRRELEVANCE OF INFORMATION

In each of the examples of prior and posterior probability to date, the posterior probability $P^e_{s|z}$ has not coincided with the prior probability. Assume, however, that we have two partitions $\mathcal{A}$ and $\mathcal{B}$ of X and the following table of joint pobabilities:

	B_1	B_2	$P(A)$
A_1	.10	.15	.25
A_2	.10	.15	.25
A_3	.20	.30	.50
$P(B)$	.40	.60	1.00

We again find the conditional probability $P(\cdot \mid A_i)$ for each i by dividing each row by its total.

	B_1	B_2	*total*
A_1	.40	.60	1.000
A_2	.40	.60	1.000
A_3	.40	.60	1.000
prior	.40	.60	1.000

In this example, $P(B_j \mid A_i) = P(B_j)$ for every B_j and every A_i. Hence if the A_i's represent outcomes z and the B_j's represent states s, we see that the posterior probabilities are unaffected by the experimental outcome z, and hence the outcome is irrelevant in an operational sense. It follows that experiment e is an exercise in make-work.

We now define the circumstance exemplified previously. Let $\mathcal{A} \subset \mathcal{E}_Q$ and $\mathcal{B} \subset \mathcal{E}_Q$ be two partitions of Q, and let P: $\mathcal{E}_Q \to [0, 1]$ be a given probability on $\mathcal{E}_Q$. We say that the partitions $\mathcal{A}$ and $\mathcal{B}$ are *pairwise independent* if and only if

$$P(A_i, B_j) = P(A_i) \cdot P(B_j), \text{ for every } A_i \in \mathcal{A} \text{ and every } B_j \in \mathcal{B}.$$

If $P(A_i) > 0$, for every i, the preceding definition of pairwise independence is equivalent to:

$$P(B_j \mid A_i) = P(B_j), \text{ for every } B_j \in \mathcal{B} \text{ and every } A_i \in \mathcal{A}.$$

[In general, $P(B_j \mid A_i) = P(A_i, B_j)/P(A_i)$ if $P(A_i) > 0$; but if $\mathcal{A}$ and $\mathcal{B}$ are independent, then $P(A_i, B_j) = P(A_i) \cdot P(B_j)$. Thus $P(B_j \mid A_i) = P(A_i) \cdot P(B_j)/P(A_i)$. Cancel $P(A_i)$.]

Similarly, if $P(B_j) > 0$, for every j, the preceding definition of pairwise independence is equivalent to:

$$P(A_i \mid B_j) = P(A_i), \text{ for every } A_i \in \mathcal{A} \text{ and every } B_j \in \mathcal{B}.$$

Partitions $\mathcal{A}^1, \mathcal{A}^2, \ldots, \mathcal{A}^n \subseteq \mathcal{E}_Q$ with $\mathcal{A}^k = \{A_i{}^k: i \in \{1, \ldots, m_i\}\}$ are called *mutually independent* if and only if

$$P(A_{i_1}{}^1, A_{i_2}{}^2, \ldots, A_{i_n}{}^n) = P(A_{i_1}{}^1) \cdot P(A_{i_2}{}^2) \cdot \ldots \cdot P(A_{i_n}{}^n),$$

for every $A_{i_1}{}^1 \in \mathcal{A}^1$, every $A_{i_2}{}^2 \in \mathcal{A}^2, \ldots,$ and every $A_{i_n}{}^n \in \mathcal{A}^n$.

Pairwise independence does not imply mutual independence. However, we shall be interested solely in mutual independence in this book. We shall use the simpler term *independence* in place of *mutual independence*.

Independence of partitions is less basic than independence of events. We say that the events $E_1, \ldots, E_n$ are *mutually independent* if and only if

$$P(E_1, \ldots, E_n) = P(E_1)\cdot P(E_2)\cdot \ldots \cdot P(E_n).$$

An important application of the notion of independent events is in formulating the concept of *random sampling*, as introduced in Chapter 5. Let Q be a population of elements q each of which is as likely to be selected for observation on any trial of the experiment as is any other. Let E_i denote the event that the element selected in the ith trial belongs to the given subset E of Q with proportion $p(E)$.

Now, the assumption of randomness on trial 1 implies that $P(E_1) = p(E)$; the assumption of randomness on trial 2 implies that $P(E_2 \mid E_1) = p(E)$ (since the outcome of trial 1 does not affect the outcome of trial 2); and so on: $P(E_n \mid E_1, \ldots, E_{n-1}) = p(E)$. But to say that every element is equally likely to be selected at every trial implies that for every $i \geq 2$ we have event E_i occurring with probability $P(E_i) = p(E)$. Hence it follows that

$$P(E_i \mid E_1, \ldots, E_{i-1}) = P(E_i) = p(E),$$

for every $i \geq 2$. Thus, from the "chain rule,"

$$P(E_1, \ldots, E_n) = P(E_1)\cdot P(E_2)\cdot \ldots \cdot P(E_n).$$

Thus we see the following:

If $p(E)$ denotes the proportion of elements $q \in Q$ which belong to $E \subset Q$ and if E_i denotes the event that the observation on trial i belongs to E, then for any n the events $E_1, \ldots, E_n$ are mutually independent if the sampling is random on every trial.

The converse of this assertion is also true: the sampling is not random on some trial if the events $E_1, \ldots, E_n$ are not mutually independent.

Examples

(1) Many people consider different flips of a coin to be different trials performed under identical conditions. Let $P(H)$ denote the probability of heads ($\in Q$) on any given flip. Then the probability of heads on each of two flips is $P(H_1, H_2) = P(H)\cdot P(H)$. Thus if $P(H) = \frac{1}{2}$, then $P(H_1, H_2) = \frac{1}{4}$.
(2) People who say, "This isn't my night," or "My luck is bound to change," in games *of chance* are really finding fault with the independence assumption. If pressed, they would have to admit that their statements imply the agency of some supernatural entity in creating a dependence of future trials upon the past.

EXERCISES

1. Consider the following table of joint probabilities:

	B_1	B_2	B_3	B_4
A_1	.10	.05	.15	.05
A_2	.05	.10	.15	.05
A_3	.15	.05	.10	.00

   (a) Compute the marginal probabilities $P(A_1)$, $P(A_2)$, and $P(A_3)$ and show that their sum is 1.00.
   (b) Compute the marginal probabilities $P(B_1), \ldots, P(B_4)$ and show that their sum is 1.00.
   (c) Compute the conditional probabilities $P(B_1 \mid A_1), \ldots, P(B_4 \mid A_1)$ of the B's given A_1 and show that their sum is 1.00. Do the same for the conditional probabilities of the B's given A_2, and given A_3.
   (d) Compute the conditional probabilities of the A's given each B_j, and show that $\sum_{i=1}^{3} P(A_i \mid B_j) = 1.00$, for $j = 1, 2, 3, 4$.

2. Consider the following table of joint probabilities:

	B_1	B_2	B_3	B_4
A_1	.12	.08	.16	.04
A_2	.12	.08	.16	.04
A_3	.06	.04	.08	.02

   Show that $\mathcal{A} = \{A_1, A_2, A_3\}$ and $\mathcal{B} = \{B_1, B_2, B_3, B_4\}$ are independent partitions.

3. If $\frac{1}{2}$ of all unmarried housewives use Brand X, $\frac{3}{4}$ of all married housewives use Brand X, and $\frac{3}{4}$ of all housewives are married, then:

   (a) what proportion of all housewives use Brand X; and
   (b) what proportion of the housewives who use Brand X are married?

4. If $\frac{2}{3}$ of all suburban males read *Bon Vivant* magazine, what is the probability that *all* members of a random sample consisting of ten suburban males read this publication? (*Hint*: You must assume that in each of the ten trials every suburban male had an equal chance of being selected.)

5. There are two urns on a table. Urn 1 contains nine red balls and one black ball, and urn 2 contains eight black and two red balls. You are blindfolded, an urn is selected at random, and you draw one ball out of the selected urn. It is red.

(a) Without making any computations, what was your *prior* probability that urn 1 was selected (before you knew you had drawn a red ball from the selected urn)? *Also without computations*, what is your posterior probability that urn 1 was selected (after knowing that you had drawn a red ball from the selected urn)?
(b) Argue that $P(\text{red} \mid \text{urn 1}) = .9$, $P(\text{red} \mid \text{urn 2}) = .2$, and $P(\text{urn 1}) = .5$.
(c) Compute $P(\text{red})$.
(d) Compute $P(\text{urn 1} \mid \text{red})$. How far off was your intuitive estimate in part (a)?

6. Let $S = \{s_1, \ldots, s_n\}$ and consider an experiment e^* such that:

$$Z(e^*) = \{z_1, \ldots, z_n\}; \text{ and}$$

$$P_{z|s}^{e^*}(z_i \mid s_j) = \begin{cases} 0, & i \neq j \\ 1, & i \neq j, \end{cases}$$

for $i = 1, 2, \ldots, n$ and $j = 1, 2, \ldots, n$. These assumptions imply the following table of conditional probabilities $P_{z|s}^{e^*}(z \mid s)$ when $n = 4$.

	s_1	s_2	s_3	s_4
z_1	1	0	0	0
z_2	0	1	0	0
z_3	0	0	1	0
z_4	0	0	0	1

Let $P_s(s_i)$ denote the prior probability of s_i given experiment e^*.

(a) Use Bayes' Theorem to show that

$$P_{s|z}^{e^*}(s_j \mid z_i) = \begin{cases} 1, & i = j \\ 0, & i \neq j; \end{cases}$$

that is, once you have observed z_i, then you are *certain* that $s = s_i$. The symbol e^* is called *perfect information*, or the *ultimate experiment*.
(b) Show that $P_z^{e^*}(z_i) = P_s(s_i)$, for every $i \in \{1, \ldots, n\}$.

7. Since the "null message" z_0 is the only possible outcome of the "null experiment" e_0 the table of conditional probabilities $P_{z|s}^{e_0}$ has one row:

	s_1	$s_2 \cdots s_n$
z_0	1	$1 \cdots 1$

(a) Use Bayes' Theorem to show that

$$P_{s|z}^{e_0}(s_i \mid z_0) = P_s(s_i), \text{ for every } i \in \{1, \ldots, n\}.$$

(b) Show that $P_z^{e_0}(z_0) = 1$.

8. *A Problem in Criminology*: Suppose that 90 percent of all robberies committed in town A are committed by residents of town A, the remainder being committed by residents of town B. Suppose that only one percent of the residents in town A smoke pipes, but that 40 percent of the residents of town B smoke pipes. A robbery was committed in town A by a pipe smoker. What is the probability that he is a resident of town A?

9. *A Problem in Sales Management*: Ninety percent of the sales of egg beaters by a door-to-door salesman are made to women, and eighty percent of the "no-sales" were on calls to men. His contact log shows that twenty percent of his calls resulted in sales.

   (a) What percentage of his calls were to men?
   (b) What percentage of his calls to men resulted in a sale?
   (c) What percentage of his calls to women resulted in a sale?

10. *A Problem in Investments*: Let $\mathcal{A} = \{A_1, A_2\}$, where A_1 = "no recession" and A_2 = "recession." Let $\mathcal{B} = \{B_1, B_2, B_3\}$, where

B_1 = "*XYZ* raises its dividend";
B_2 = "*XYZ*'s dividend is unchanged"; and
B_3 = "*XYZ* reduces or omits its dividend."

Suppose you have assessed the following conditional probabilities on $\mathcal{B}$ given the A_i's:

	B_1	B_2	B_3
A_1	.30	.70	.00
A_2	.00	.70	.30

and further suppose that you have assessed $P(A_1) = .60$.

   (a) Find and interpret $P(B_1)$, $P(B_2)$, and $P(B_3)$.
   (b) Now suppose that $\mathcal{C} = \{C_1, C_2\}$, where

C_1 = "*F*. Orcaster says there will be no recession";
C_2 = "*F*. Orcaster says there will be a recession";

and further suppose that you have assessed the following conditional probabilities $P(C_j \mid A_i)$ on Orcaster's forecasting acumen:

	C_1	C_2
A_1	.90	.10
A_2	.60	.40

Finally, assume that the forecast has no influence upon the $P(B_k \mid A_i)$'s, in the sense that $P(B_k \mid A_i, C_j) = P(B_k \mid A_i)$, for every i, j, and k. Find $P(B_k \mid C_j)$, for every j and k.

(c) By comparing the $P(B_k \mid C_j)$'s with the $P(B_k)$'s, discuss the effect of Orcaster's forecast upon your judgments regarding the behavior of XYZ's dividend.
[*Hint for part* (*b*): First find $P(A_i \mid C_j)$, for every i and j. Then verify that

$$P(B_k \mid C_j) = \sum_{i=1}^{2} P(B_k, A_i \mid C_j)$$

$$= \sum_{i=1}^{2} P(B_k \mid A_i, C_j) \cdot P(A_i \mid C_j)$$

$$= \sum_{i=1}^{2} P(B_k \mid A_i) \cdot P(A_i \mid C_j), \text{ (by our final assumption).]}$$

11. *Conditional Independence*: Partitions $\mathcal{A}$ and $\mathcal{B}$ are *conditionally independent given partition* $\mathcal{C}$ if and only if

$$P(A, B \mid C) = P(A \mid C) \cdot P(B \mid C), \text{ for all } A \in \mathcal{A}, B \in \mathcal{B}, \text{ and } C \in \mathcal{C}$$

such that all terms are defined.

(a) Show that if $\mathcal{A}$ and $\mathcal{B}$ are conditionally independent given $\mathcal{C}$, then

$$P(A \mid B, C) = P(A \mid C)$$

and

$$P(B \mid A, C) = P(B \mid C)$$

whenever defined.

(b) Let $\mathcal{A} = \{A_1, A_2\}$, $\mathcal{B} = \{B_1, B_2\}$, and $\mathcal{C} = \{C_1, C_2\}$. Assume the following joint probabilities:

$$\begin{aligned}
P(A_1, B_1, C_1) &= .294 \\
P(A_1, B_1, C_2) &= .006 \\
P(A_1, B_2, C_1) &= .126 \\
P(A_1, B_2, C_2) &= .024 \\
P(A_2, B_1, C_1) &= .196 \\
P(A_2, B_1, C_2) &= .054 \\
P(A_2, B_2, C_1) &= .084 \\
P(A_2, B_2, C_2) &= .216
\end{aligned}$$

(i) Compute $P(C_1)$ and $P(C_2)$.
(ii) Compute all eight $P(A_i, B_j \mid C_k)$'s.
(iii) Compute all four $P(A_i \mid C_k)$'s.
(iv) Compute all four $P(B_j \mid C_k)$'s.
(v) Show that $P(A_i, B_j \mid C_k) = P(A_i \mid C_k) \cdot P(B_j \mid C_k)$, for all i, j, and k, and hence that $\mathcal{A}$ and $\mathcal{B}$ are conditionally independent given $\mathcal{C}$.
(vi) Compute all four $P(A_i, B_j)$'s.
(vii) Compute both $P(A_i)$'s and both $P(B_j)$'s.
(viii) Show that $P(A_i, B_j) \neq P(A_i) \cdot P(B_j)$, for all i and j and hence that $\mathcal{A}$ and $\mathcal{B}$ are not independent. Thus conditional independence does not imply independence.

(c) Let $\mathcal{A}$, $\mathcal{B}$, and $\mathcal{C}$ be as given in part (b), but now assume the following joint probabilities:

$$\begin{aligned}
P(A_1, B_1, C_1) &= .10\\
P(A_1, B_1, C_2) &= .14\\
P(A_1, B_2, C_1) &= .26\\
P(A_1, B_2, C_2) &= .10\\
P(A_2, B_1, C_1) &= .15\\
P(A_2, B_1, C_2) &= .01\\
P(A_2, B_2, C_1) &= .04\\
P(A_2, B_2, C_2) &= .20
\end{aligned}$$

(i) Compute all four $P(A_i, B_j)$'s.
(ii) Compute both $P(A_i)$'s and both $P(B_j)$'s.
(iii) Show that $\mathcal{A}$ and $\mathcal{B}$ are independent.
(iv) Compute all eight $P(A_i, B_j \mid C_k)$'s.

(v) Compute all four $P(A_i \mid C_k)$'s and all four $P(B_j \mid C_k)$'s.
(vi) Show that $\mathcal{A}$ and $\mathcal{B}$ are not conditionally independent given $\mathcal{C}$.

12. *Prior and Posterior Odds*: Show that if A and B are events, then

$$\frac{P(B \mid A)}{P(B^c \mid A)} = \frac{P(B)}{P(B^c)} \cdot \frac{P(A \mid B)}{P(A \mid B^c)},$$

provided that $0 < \{\min P(B^c), P(A \mid B^c)\}$.

[*Note*: This result shows that when B is an s-event and A is a z-event, then the decision-maker's "posterior odds $P(B \mid A)/P(B^c \mid A)$ in favor of B" equal his "prior odds $P(B)/P(B^c)$ in favor of B," times a factor $P(A \mid B)P(A \mid B^c)$ measuring the relative likelihood of A given B vis-à-vis B^c.]

7

THE DISTRIBUTION OF A RANDOM VARIABLE

7.1 INTRODUCTION

Recall from Section 4.5 that a *random variable* $\tilde{x}$ is a function $\tilde{x}: Q \rightarrow R^1$ with the "measurability property" that $Q(x) \equiv \{q: \tilde{x}(q) \leq x\} \in \mathcal{E}_Q$ for every $x \in R^1$; that is, every $Q(x)$ is an event and therefore has a probability $P(Q(x))$.

In this chapter we shall discuss how the given probability $P(\cdot)$ on $\mathcal{E}_Q$ induces, or generates, probabilities $P[\cdot]$ of certain events in R^1 (actually in $\mathcal{B}^1$) defined by statements about the value assumed by $\tilde{x}$. Any event that will be of interest to us will be a union of events of one of the following forms: $\tilde{x} = x$; $\tilde{x} \in (a, b]$; $\tilde{x} \in [a, b]$; $\tilde{x} \in (a, b)$; $\tilde{x} \in [a, b)$; $\tilde{x} \in (x, \infty)$; $\tilde{x} \in [x, \infty)$; $\tilde{x} \in (-\infty, x)$; or $\tilde{x} \in (-\infty, x]$, where $-\infty < a \leq b < \infty$ and $x \in R^1$. For example, $\tilde{x} \in (a, b]$ is "the event that the random variable $\tilde{x}$ takes on a value in the interval $(a, b]$"; $\tilde{x} \in (a, b]$ is also often written as $a < \tilde{x} \leq b$.

Now, the event $\tilde{x} \in (a, b]$ will occur if and only if $q \in \tilde{x}^{-1}((a, b])$; that is, if and only if q is such that $\tilde{x}(q) \in (a, b]$. As noted in Chapter 4, if x is a "legitimate" random variable, then $\tilde{x}^{-1}((a, b]) \in \mathcal{E}_Q$ for every interval $(a, b] \subset R^1$. In particular, $\tilde{x}^{-1}((-\infty, x]) \equiv Q(x)$ belongs to $\mathcal{E}_Q$ for every $x \in R^1$.

We shall now show that to every event listed previously there corresponds a q-event definable in terms of events of the form $Q(X)$:

$\tilde{x}$-event	-corresponds to-	q-event	
$\tilde{x} = x$		$Q(x)\backslash\bigcup_{n=1}^{\infty}Q(x - 1/n)$	(*)
$\tilde{x} \leq x$		$Q(x)$	
$\tilde{x} < x$		$\bigcup_{n=1}^{\infty}Q(x - 1/n)$	(*)
$\tilde{x} > x$		$Q\backslash Q(x)$	(*)
$\tilde{x} \geq x$		$Q\backslash\bigcup_{n=1}^{\infty}Q(x - 1/n)$	
$a < \tilde{x} \leq b$		$Q(b)\backslash Q(a)$	
$a \leq \tilde{x} \leq b$		$Q(b)\backslash\bigcup_{n=1}^{\infty}Q(a - 1/n)$	(*)
$a \leq \tilde{x} < b$		$[\bigcup_{n=1}^{\infty}Q(b - 1/n)]\bigcup_{n=1}^{\infty}Q(a - 1/n)$	(*)
$a < \tilde{x} < b$		$[\bigcup_{n=1}^{\infty}Q(b - 1/n)]\backslash Q(a)$	(*)

The reader may verify these assertions as Exercise 7; the starred ones require $\mathcal{E}_Q$ to be a sigma-field.

To obtain the probability of any $\tilde{x}$-event, it is clear that we must equate the probability of the $\tilde{x}$-event to the probability of its corresponding $\mathcal{E}_Q$-event. So doing for every $\tilde{x}$-event completely specifies the probability $P[\cdot]$ on the sigma-field $\mathcal{B}^1$ of *all* $\tilde{x}$-events.

But the preceding table shows that all we need do is specify the P-probabilities of the $\tilde{x}$-events of the form $\tilde{x} \leq x$, as all other events have probabilities obtainable from them by applying axioms ($P1$), ($P2$), and ($P3^\sigma$) of probability to $P[\cdot]$. [If $\mathcal{E}_Q$ is a sigma-field and $P(\cdot)$ satisfies ($P3^\sigma$), then $P[\cdot]$ automatically satisfies ($P3^\sigma$) on the Borel (sigma-)field $\mathcal{B}^1$ of $\tilde{x}$-events.] The resulting probability $P[\cdot]$ on $\mathcal{B}^1$ is called the *probability distribution, distribution, probability law,* or *law* of the random variable $\tilde{x}$.

A very significant fact for applications is that *once we have specified the distribution of $\tilde{x}$, all probabilities of $\tilde{x}$-events can be obtained without appealing to $\mathcal{E}_Q$ and/or $P(\cdot)$ on $\mathcal{E}_Q$.* Hence ($\mathcal{E}_Q$, $P(\cdot)$) can be replaced by ($\mathcal{B}^1$, $P[\cdot]$) if the possible values of $\tilde{x}$ are all that concern us. The significance of this assertion lies in the facts (1) that our basic, observable data in decision and inference problems often constitute values of random variables; and (2) that it may be completely *impossible* to determine q. We have stated that determining q is also completely *unnecessary*.

In Section 7.2 we introduce the *cumulative distribution function* F: $R^1 \to R^1$ of $\tilde{x}$, defined by

$$F(x) \equiv P[\tilde{x} \leq x] = P(Q(x)).$$

The cumulative distribution function F of $\tilde{x}$ should not be confused with the distribution of $\tilde{x}$, although some authors abbreviate *cumulative distribution function* to *distribution function.* F completely specifies the distribution $P[\cdot]$ of $\tilde{x}$, and (obviously) vice versa. Moreover, defining F is much simpler than defining $P[\cdot]$ mathematically. Section 7.2 shows how the probability of any interval can be obtained directly from F. Also, the mathematical properties of F are used to define three types of random variables.

F is a very *useful* way of specifying P. Section 7.3 defines the general concept of a *measure of location* of any $\tilde{x}$; defines fractiles, or quantiles, or percentiles of $\tilde{x}$ directly from F; and shows that fractiles are measures of location.

Sections 7.4, 7.5, and 7.6 discuss some other means of specifying the distribution of $\tilde{x}$—means which are useful in many contexts and which are directly related to F. These three sections correspond to the three types of random variables defined in Section 7.2.

7.2 THE CUMULATIVE DISTRIBUTION FUNCTION OF A RANDOM VARIABLE

Let $\tilde{x}$: $Q \to R^1$ be a random variable. We define the *cumulative distribution function* $F(\cdot)$ of $\tilde{x}$ by

$$F(x) = P(Q(x)), \text{ every } x \in R^1.$$

Example 7.2.1

Let $Q = \{a, b, c\}$; $P(a) = .10$, $P(b) = .30$, $P(c) = .60$; $\tilde{x}(a) = 0$, $\tilde{x}(b) = 3$, and $\tilde{x}(c) = 10$. Then

$$Q(x) = \begin{cases} \varnothing, & x < 0 \\ \{a\}, & 0 \le x < 3 \\ \{a, b\}, & 3 \le x < 10 \\ \{a, b, c\}, & 10 \le x. \end{cases}$$

Hence

$$F(x) = \begin{cases} P(\varnothing), & x < 0 \\ P(a), & 0 \le x < 3 \\ P(\{a, b\}), & 3 \le x < 10 \\ P(\{a, b, c\}), & 10 \le x \end{cases}$$

$$= \begin{cases} 0, & x < 0 \\ .10, & 0 \le x < 3 \\ .40, & 3 \le x < 10 \\ 1.00, & 10 \le x. \end{cases}$$

$F(\cdot)$ is graphed in Figure 7.1.

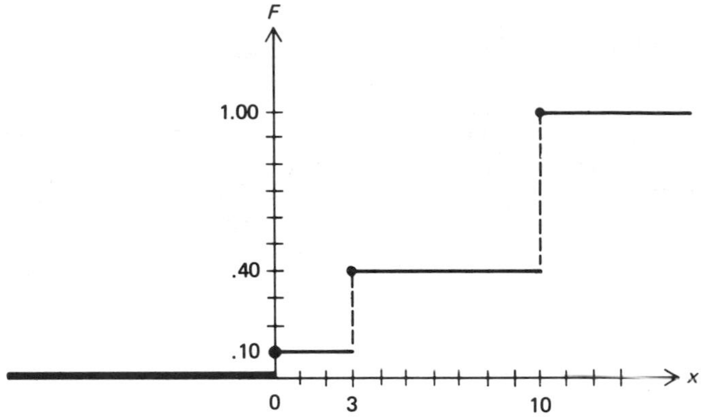

Figure 7.1

Clearly, $F(\cdot)$ is a nondecreasing step function and continuous from the right. That is: (1) if $x \in R^1$, $x' \in R^1$, and $x' > x$, then $F(x') \geq Fx)$ (nondecreasingness); (2) $F(\cdot)$ is constant except at a countable (here, finite) number of "jumps," "steps," or "saltuses" (that is, F is a step function); and (3) if $\{x^i: i = 1, 2, \ldots\}$ is a decreasing sequence of numbers (that is, if $j \geq i$ then $x^j \leq x^i$) such that $\lim_{i \to \infty} x^i = x$, then $\lim_{i \to \infty} F(x^i) = F(x)$ (continuous from the right).

In the preceding example we have seen a number of properties which *any* cumulative distribution function (hereafter abbreviated *cdf*) $F(\cdot)$ should possess. We shall state these, and other properties formally, and prove them from the definition of a cdf.

Properties of Cumulative Distribution Functions
Let $F(\cdot)$ be the cdf of $\tilde{x}$. Then:
(cdf 1) $0 \leq F(x) \leq 1$, for every $x \in R^1$;
(cdf 2) if $x' > x$, then $F(x') \geq F(x)$;
(cdf 3) if $\{x^i: i = 1, 2, \ldots\}$ is a decreasing sequence of numbers with limit x, then $\lim_{i \to \infty} F(x^i) = F(x)$;
(cdf 4) $\lim_{x \to -\infty} F(x) = 0$; and
(cdf 5) $\lim_{x \to \infty} F(x) = 1$.

Proof

Since $F(x) = P(Q(x))$, for every $x \in R^1$, $0 \leq F(x) \leq 1$ follows from axiom ($P1$) of probability. For (cdf 2), if $x < x'$ then $Q(x) \subset Q(x')$. Hence, by property ($P7$) of probability, $P(Q(x)) \leq P(Q(x'))$, from which (2) follows by definition of F. Properties (cdf 3) through (cdf 5) require

$\mathcal{E}_Q$ to be a sigma-field and P to satisfy property ($P3^\sigma$), countable additivity. Property (cdf 4) follows from (cdf 3) by noting that $F(-\infty) = P(\{q: \tilde{x}(q) \leq -\infty\}) = P(\varnothing) = 0$, because $-\infty < \tilde{x}(q) < \infty$ for every q. Hence also $P(\{q: \tilde{x}(q) \leq +\infty\}) = P(Q) = 1 = F(+\infty) = \lim_{x\to\infty} F(x)$. To prove (cdf 3), note first of all that if $\{x^i: i = 1, 2, \ldots\}$ is a decreasing sequence of numbers with limit x, then $\ldots \subset Q(x^3) \subset Q(x^2) \subset Q(x^1)$ and $\bigcap_{i=1}^{\infty} Q(x^i) = Q(x)$. Now, the limit assertion of (cdf 3) is equivalent to

$$\lim_{i\to\infty} (1 - F(x^i)) = 1 - F(x); \text{ and clearly we have}$$

$$1 - F(x^i) = P(Q \backslash Q(x^i)),$$

for every i. Since $\{Q(x^i): i = 1, 2, \ldots\}$ is a decreasing sequence with intersection $Q(x)$, $\{Q\backslash Q(x^i): i = 1, 2, \ldots\}$ is an increasing sequence with union $Q\backslash Q(x) = Q\backslash \bigcap_{i=1}^{\infty} Q(x^i)$. Hence we must show that $\lim_{i\to\infty} P(Q\backslash Q(x^i)) = P(Q\backslash Q(x))$. This will follow from the more general assertion:

($p7^\sigma$) if $\{E_i: i = 1, 2, \ldots\} \subset \mathcal{E}_Q$ and
$E_1 \subset E_2 \subset E_3 \subset \cdots$, then $\lim_{i\to\infty} P(E_i) = P(\bigcup_{i=1}^{\infty} E_i)$.

The reader is asked to prove ($P7^\sigma$) as Exercise 8.

Note that if $F(x) = 0$ for all sufficiently small values of x (as in our example), then (cdf 4) is satisfied. Similarly, if $F(x) = 1$ for all sufficiently large values of x (as in our example), then (cdf 5) is satisfied.

It is also true that if $F(\cdot)$ is *any* real-valued function of a real variable x and possesses properties (cdf 1) through (cdf 5), then there is a set Q of elementary events, a field $\mathcal{E}_Q$ of events, and a random variable $\tilde{x}$ on Q such that $F(\cdot)$ is the cdf of $\tilde{x}$.

To see this, let $Q = R^1$, $\mathcal{E}_Q = \mathcal{B}^1$ = the Borel field in R^1, and use a result that any probability P on $\mathcal{B}^1$ is determined by the probabilities of all half-open intervals of the form $(-\infty, x]$. Let $\tilde{x}: R^1 \to R^1$ be the identity function; that is, $\tilde{x}(x) = x$ for every x. So if $Q = R^1$, $\mathcal{E}_Q = \mathcal{B}^1$, and P is the unique probability determined by the probabilities $P((-\infty, x]) = F(x)$, then $F(\cdot)$ is the cdf of the identity random variable.

Example 7.2.2

Define $F(\cdot)$ on R^1 by

$$F(x) = \begin{cases} 0, & x \leq 0 \\ x, & 0 < x < 1 \\ 1, & 1 \leq x. \end{cases}$$

Then $F(\cdot)$ possesses properties (cdf 1) through (cdf 5) and hence is the cdf of a random variable $\tilde{x}$.

Example 7.2.3

Define $F(\cdot)$ on R^1 by

$$F(x) = \begin{cases} 0, & x \leq 0 \\ 1 - e^{-\lambda x}, & x > 0, \end{cases}$$

where $\lambda > 0$. Then $F(\cdot)$ possesses properties (cdf 1) through (cdf 5) and hence is the cdf of a random variable $\tilde{x}$.

Example 7.2.4

Define $F(\cdot)$ on R^1 by

$$F(x) = \begin{cases} 0, & x \leq 0 \\ x^\rho, & 0 < x < 1 \\ 1, & 1 \leq x, \end{cases}$$

where $\rho > 0$. Then $F(\cdot)$ is the cdf of a random variable $\tilde{x}$.

Example 7.2.5

Let Q be a countable set. Then P on $\mathcal{E}_Q$ is completely determined by the set $\{P(q): q \in Q\}$ of probabilities of the elementary events, and $\tilde{x}: Q \to R^1$ suffices to determine a countable set $\{x^i: i = 1, 2, \ldots\}$ (which may be finite) such that $Q \backslash \{q: \tilde{x}(q) \in \{x^i: i = 1, 2, \ldots\}\} = \varnothing$. Now, $Q(x) = \bigcup_{x^i \leq x} \{q: \tilde{x}(q) = x^i\}$, where the union on the right is of all events $\{q: \tilde{x}(q) = x^i\}$ such that $x^i \leq x$. Since for $i \neq j$ the events $\{q: \tilde{x}(q) = x^i\}$ and $\{q: \tilde{x}(q) = x^j\}$ are mutually exclusive, we have

$$P(Q(x)) = \sum_{x^i \leq x} P(\{q: \tilde{x}(q) = x^i\});$$

that is,

$$F(x) = \sum \{P[\tilde{x} = x^i]: x^i \leq x\},$$

where $P[\tilde{x} = x^i] = P(\{q: \tilde{x}(q) = x^i\})$. Hence $F(\cdot)$ is a step function, as in Example 7.2.1, with a jump of distance $P[x^i]$ at x^i for every i, and continuous on the right. The only novelty in this example vis-à-vis Example 7.2.1 is that here there may be (countably) many jumps instead of only three. You are asked to show as an exercise that any random variable $\tilde{x}$ on a *countable* set Q of elementary events has a step function for its cdf.

Figure 7.1 depicted a typical step cumulative distribution function. Figure 7.2 graphs the continuous cdf of Example 7.2.3 for $\lambda = 1$. It is easy to show that the cdf in Example 7.2.3 is differentiable at every x except for $x = 0$.

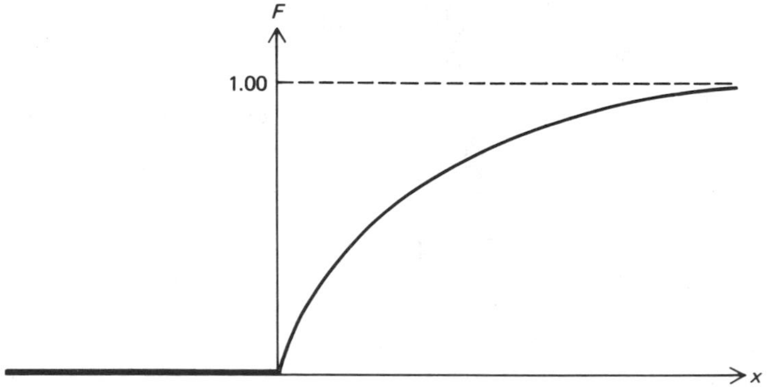

Figure 7.2

It is a fact of advanced analysis that continuous functions can be poorly differentiable. Some very ill-behaved cdf's still possessing properties (cdf 1) through cdf (5) do exist, but we shall not consider them. Instead, we shall assume that *every cdf* $F(\cdot)$ *possesses a derivative at all points where it is continuous, with at most a countable number of exceptions.* Thus we shall assume in this book that every cdf $F(\cdot)$ possesses only a countable number of points of continuity but nondifferentiability (that is, "corners").

With this restriction, there are precisely three classes of cdf's. These classes are then used to classify random variables.

(1) $\tilde{x}$ is called a *discrete random variable* if its cdf $F(\cdot)$ is a pure step function, continuous on the right everywhere, with a countable (finite or denumerable) number of jumps.

(2) $\tilde{x}$ is called a *continuous random variable* if its cdf $F(\cdot)$ is everywhere continuous, and also differentiable at all but a countable number of values of x. [$F(\cdot)$ is certainly allowed to be everywhere differentiable, for in this case the number of nondifferentiability points is zero, a finite number.]

(3) $\tilde{x}$ is called a *mixed random variable* if its cdf is:
  (a) continuous except at a countable number of jumps, and everywhere continuous on the right;
  (b) not everywhere constant between adjacent jumps; and
  (c) differentiable at all except a countable number of its points of continuity.

Examples 7.2.1 and 7.2.3, and the related Figures 7.1 and 7.2 respectively, constitute cdf's of a discrete and of a continuous random variable respectively. Examples 7.2.2 and 7.2.4 are cdf's of continuous random variables, while Example 7.2.5 is the cdf of a discrete random variable. Figure 7.3 depicts the cdf of a mixed random variable. Jumps occur at x^1, x^2, and x^3, while x^4 is a point of continuity but nondifferentiability.

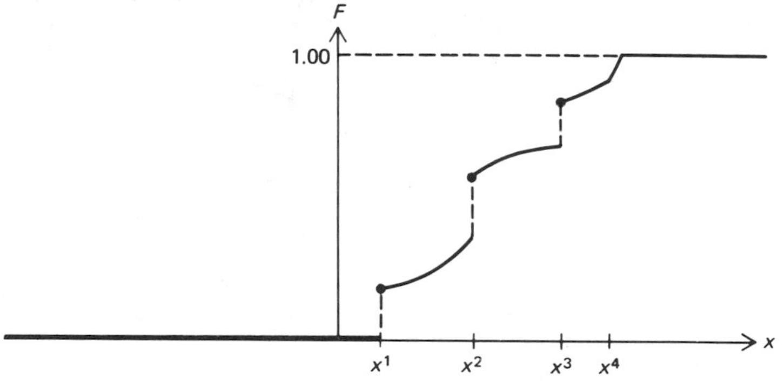

Figure 7.3

We shall have little occasion to use mixed random variables in this book. Moreover, we shall have no occasion to consider random variables $\tilde{x}$ whose cdf's possess more than two or three points of continuity but nondifferentiability. Hence we shall be concerned primarily with discrete random variables and with very well-behaved continuous random variables.

We close this section by showing how probabilities of x-intervals can be ascertained from the cdf F of $\tilde{x}$. The reader may verify (cdf 6) through (cdf 14) as an exercise.

(cdf 6) $P[\tilde{x} \leq x] = F(x)$;
(cdf 7) $P[\tilde{x} = x] = F(x) - F(x^-)$;
(cdf 8) $P[\tilde{x} < x] = F(x^-)$;
(cdf 9) $P[\tilde{x} > x] = 1 - F(x)$;
(cdf 10) $P[\tilde{x} \geq x] = 1 - F(x^-)$;
(cdf 11) $P[a < \tilde{x} \leq b] = F(b) - F(a)$;
(cdf 12) $P[a \leq \tilde{x} \leq b] = F(b) - F(a^-)$;
(cdf 13) $P[a \leq \tilde{x} < b] = F(b^-) - F(a^-)$; and
(cdf 14) $P[a < \tilde{x} < b] = F(b^-] - F(a)$.

Probabilities of unions of disjoint intervals can be found by applying property ($P3^f$) (or ($P3^\sigma$)) to $P[\cdot]$.

Note that $P[\tilde{x} = x]$ equals the height of the jump on the cdf at the point x, if any. Thus if F is continuous at x, then $P[\tilde{x} = x] = 0$, which may appear startling to the reader in view of the fact that x may be a possible value of

$\tilde{x}$; that is, $\{q: \tilde{x}(q) = x\} \neq \varnothing$. Exercise 10 should convince the reader that we have not fallen prey to a logical contradiction.

Example 7.2.6

In Example 7.2.1, we seek the probability $P[2 < \tilde{x} \leq 10]$. We have $P[2 < \tilde{x} \leq 10] = F(10) - F(2) = 1.00 - .10 = .90$.
However,

$$\begin{aligned} P[2 < \tilde{x} < 10] &= F(10) - F(2) - P[\tilde{x} = 10] \\ &= 1.00 - .10 - .60 = .30, \end{aligned}$$

because

$$P[\tilde{x} = 10] = F(10) - F(10^-) = 1.00 - .40 = .60.$$

Example 7.2.7

In Example 7.2.4, F is continuous and hence $P[\tilde{x} = x] = 0$ for every $x \in R^1$. Therefore, $P[a \leq \tilde{x} \leq b] = P[a \leq \tilde{x} < b] = P[a < \tilde{x} < b] = P[a < \tilde{x} \leq b]$ for all a and b such that $a < b$. We have

$$P[a < \tilde{x} \leq b] = F(b) - F(a) = \begin{cases} 0; & b \leq 0 \\ b^p; & a \leq 0,\ 0 < b < 1 \\ b^p - a^p; & 0 < a < b < 1 \\ 1; & a \leq 0,\ b \geq 1 \\ 1 - a^p; & 0 < a < 1,\ b \geq 1 \\ 0; & a \geq 1. \end{cases}$$

7.3 FRACTILES, OR QUANTILES OF A RANDOM VARIABLE

In this brief section we shall consider a direct application of the definition of cdf in giving *measures of location* of a random variable $\tilde{x}$.

Let $p \in (0, 1)$ be a given number. We define a pth *fractile* ${}_px$ *of* $\tilde{x}$ as

(1) any number y such that $F(y) = p$; *or*, if no y satisfies (1), as
(2) the smallest number y such that $F(y) > p$.

This definition of ${}_px$ is equivalent to the geometric procedure of drawing the horizontal line through $F = p$ on the graph of $F(\cdot)$, noting the intersection of this line with $F(\cdot)$, and reading down to the x-axis below this intersection, provided that the intersection exists. If the intersection fails to exist, then the horizontal line intersects the dotted line at a jump, in which case we record the value ${}_px$ at the jump.

Figure 7.4 indicates a unique fractile ${}_px$ found by criterion (1).

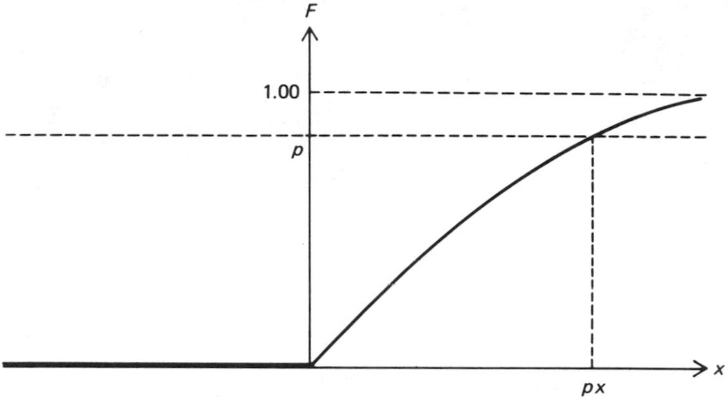

Figure 7.4

Figure 7.5 indicates a nonunique fractile determined by criterion (1).

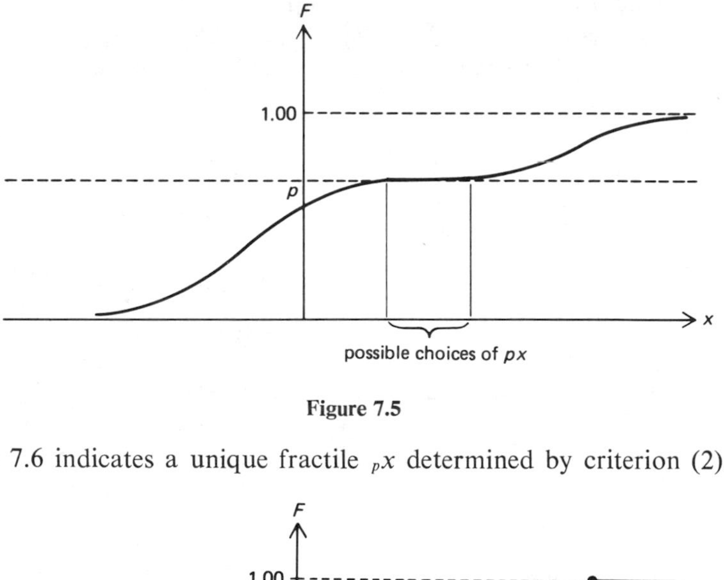

Figure 7.5

Figure 7.6 indicates a unique fractile ${}_px$ determined by criterion (2).

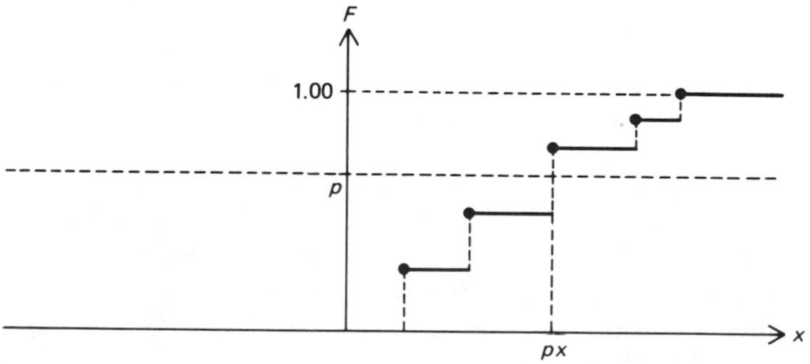

Figure 7.6

A pth fractile ${}_p x$ of $\tilde{x}$ is also called a pth *quantile* or a 100 pth *percentile.* The symbol ${}_{0.50}x$ is also called a *median* of $\tilde{x}$; and ${}_{0.25}x$ and ${}_{0.75}x$ are respectively called a *first quartile* and a *third quartile* of $\tilde{x}$.

We have referred to fractiles as measures of location. To show that this is so, we first define precisely what we mean by a measure of location.

A *measure of location* is a real-valued function $L(\cdot)$ on the set of all random variables such that if $\tilde{x}_2(q) = \tilde{x}_1(q) + \alpha$ for all $q \in Q$ and some real number α, then we have $L(\tilde{x}_2) = L(\tilde{x}_1) + \alpha$. In other words, adding a constant to all values of a random variable causes the measure of location to also change by that constant.

To see that pth fractiles are measures of location, first observe that if $\tilde{x}_2(q) = \tilde{x}_1(q) + \alpha$ for all q and some α, then

$$\begin{aligned}\{q\colon \tilde{x}_2(q) \leq x_1 + \alpha\} &= \{q\colon \tilde{x}_1(q) + \alpha \leq x_1 + \alpha\}\\ &= \{q\colon \tilde{x}_1(q) \leq x_1\}\end{aligned}$$

for every $x_1 \in R^1$, and hence that

$$P(\{q\colon \tilde{x}_2(q) \leq x_1 + \alpha\}) = P(\{q\colon \tilde{x}_1(q) \leq x_1\})$$

for every $x_1 \in R^1$, where P is the originally given probability on $\mathcal{E}_Q$. Now let $F_i(\cdot)$ denote the cdf of $\tilde{x}_i$ for $i = 1, 2$, so that we have

$$F_2(x_1 + \alpha) = F_1(x_1) \qquad \text{for every } x_1 \in R^1.$$

Finally, note that if ${}_p x_1$ satisfies

$$p = F_1({}_p x_1) = F_2({}_p x_1 + \alpha).$$

then it follows that ${}_p x_1$ is a pth fractile of $\tilde{x}_1$ and ${}_p x_1 + \alpha$ is a pth fractile of $\tilde{x}_2$. Thus one adds α to the fractiles of $\tilde{x}_1$ to obtain the corresponding fractiles of $\tilde{x}_2 = \tilde{x}_1 + \alpha$. Hence the pth fractiles are measures of location, as asserted.

The previous discussion can be approximated on an intuitive level somewhat more succinctly. First, if one adds a constant to all values of a random variable $\tilde{x}_1$, the result is a random variable whose cdf is obtained by sliding the cdf of $\tilde{x}_1$ (through a distance equal to that constant) along the x_1-axis without distorting its shape. Hence all points on the cdf of $\tilde{x}_1$ at height p are also slid through a distance equal to the constant added to $\tilde{x}_1$. A somewhat more important measure of location is given in Chapter 10.

7.4 MASS FUNCTIONS OF DISCRETE RANDOM VARIABLES

Section 7.2 introduced the cumulative distribution function $F(\cdot)$ as a convenient way of describing the probability distribution of a random variable $\tilde{x}$. This section furnishes an alternative for the class of *discrete* random variables.

Let $\{x^i: i = 1, 2, \ldots\}$ denote the (countable) set of possible values of a discrete random variable $\tilde{x}$. We define the *probability mass function*, or *mass function* $g(\cdot)$ of $\tilde{x}$ by

$$g(x) = \begin{cases} 0, & x \notin \{x^i: i = 1, 2, \ldots\} \\ P[\tilde{x} = x^i], & x = x^i \text{ for some } i. \end{cases}$$

It will be convenient to abbreviate *mass function* to *mf* in the remainder of this book.

The mf $g(\cdot)$ of a random variable $\tilde{x}$ can be obtained very easily from the cdf $F(\cdot)$ of $\tilde{x}$. For example, the cdf in Figure 7.1 induces the mf $g(\cdot)$ defined by

$$g(x) = \begin{cases} 0, & x \notin \{0, 3, 10\} \\ .10, & x = 0 \\ .30, & x = 3 \\ .60, & x = 10. \end{cases}$$

This mass function is graphed in Figure 7.7, which shows that $g(\cdot)$ is a simple "line graph."

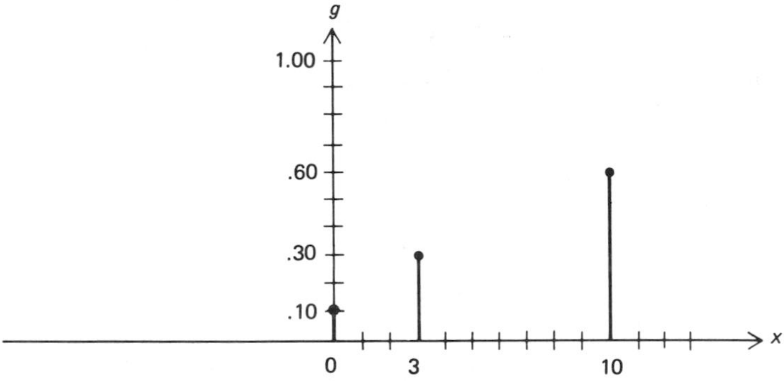

Figure 7.7

It is evident that the number $g(x)$ expresses the probability that $\tilde{x}$ will equal precisely x. Hence

(mf 1) $g(x) \geq 0$, for every $x \in R^1$,

because probabilities are nonnegative. Also,

(mf 2) $g(x) = 0$, if $x \notin \{x^i: i = 1, 2, \ldots\}$

(where $\{x^i: i = 1, 2, \ldots\} = \{0, 3, 10\}$ in our example). Moreover, since $P(Q) = 1$, we have

(mf 3) $\sum \{g(x^i): i = 1, 2, \ldots\} = 1.$

Properties (mf 1) through (mf 3) characterize mf's of discrete random variables, in the following sense:

If $g(\cdot)$ is the mf of some discrete random variable $\tilde{x}$, then there is a countable subset $\{x^i: i = 1, 2, \ldots\}$ of R^1 such that $g(\cdot)$ possesses properties (mf 1) through (mf 3).

The converse of this assertion is also true: interchange *if* and *then* and rearrange the sentence.

The cdf of a discrete random variable $\tilde{x}$ can be obtained from its mf just as in Example 7.2.5:

(mf 4) $F(x) = \sum \{g(t): t \leq x\}$.

Hence the mf and the cdf of a *discrete* random variable $\tilde{x}$ are equally good representatives of its distribution.

A note of caution is in order: a mass function is not considered at all for a *continuous* random variable $\tilde{x}$, since it is *not* a useful description of the distribution of $\tilde{x}$, In fact, if $\tilde{x}$ is a continuous random variable, then $g(x) = 0$, for every $x \in R^1$. (The reader may verify as Exercise 11.)

Example 7.4.1

Let $p \in (0, 1)$, where p is a parameter. Define g by

$$g(x) = \begin{cases} 0, & x \notin \{0, 1, 2, 3, \ldots\} \\ p(1-p)^x, & x \in \{0, 1, 2, 3, \ldots\}. \end{cases}$$

Then $g(x) \geq 0$ for every $x \in R^1$ so g possesses properties (mf 1) and (mf 2). Moreover,

$$\begin{aligned} \sum \{g(x)&: x = 0, 1, 2, \ldots\} \\ &= \sum_{i=0}^{\infty} p(1-p)^i \\ &= p[1 + (1-p) + (1-p)^2 + \cdots] \\ &= p\left[\frac{1}{1-(1-p)}\right] \\ &= \frac{p}{p} \\ &= 1, \end{aligned}$$

and so (mf 3) is also an attribute of g. Hence g is the mf of a discrete random variable $\tilde{x}$.

We close this section by showing how probabilities of x-intervals can be ascertained from the mf g of the discrete random variable $\tilde{x}$. You may verify (mf 4) through (mf 12) as Exercise 13.

(mf 4) $F(x) = \sum \{g(t): t \leq x\}$;
(mf 5) $P[\tilde{x} = x] = g(x)$;
(mf 6) $P[\tilde{x} < x] = \sum \{g(t): t < x\}$;
(mf 7) $P[\tilde{x} > x] = \sum \{g(t): t > x\}$;
(mf 8) $P[\tilde{x} \geq x] = \sum \{g(t): t \geq x\}$;
(mf 9) $P[a < \tilde{x} \leq b] = \sum \{g(t): a < t \leq b\}$;
(mf 10) $P[a \leq \tilde{x} \leq b] = \sum \{g(t): a \leq t \leq b\}$;
(mf 11) $P[a \leq \tilde{x} < b] = \sum \{g(t): a \leq t < b\}$; and
(mf 12) $P[a < \tilde{x} < b] = \sum \{g(t): a < t < b\}$.

7.5 DENSITY FUNCTIONS OF CONTINUOUS RANDOM VARIABLES

Let $\tilde{x}$ be a continuous random variable with cdf $F(\cdot)$. Then $F(\cdot)$ is differentiable at all except for a countable number of points x, and we have noted that in future applications the number of points of nondifferentiability will be very small.

We define the *probability density function*, or *density function* $f(\cdot)$ of $\tilde{x}$ by

$$f(x) = \begin{cases} 0, & F \text{ is not differentiable at } x \\ \dfrac{d}{dx}F(x), & F \text{ is differentiable at } x. \end{cases}$$

Thus $f(x)$ is the derivative of F at x where that derivative exists, and is zero otherwise. We shall abbreviate *density function* to *df*.

To motivate the concept of the df $f(\cdot)$ of $\tilde{x}$, we note that

$$P\left(x - \frac{dx}{2} < \tilde{x} < x + \frac{dx}{2}\right) = F\left(x + \frac{dx}{2}\right) - F\left(x - \frac{dx}{2}\right),$$
$$= \frac{\left[F\left(x + \frac{dx}{2}\right) - F\left(x - \frac{dx}{2}\right)\right] \cdot dx}{dx},$$

for all increments $dx > 0$ to x. When dx is sufficiently small and when F is differentiable at x, we have

$$\frac{F\left(x + \frac{dx}{2}\right) - F\left(x - \frac{dx}{2}\right)}{dx} \doteq f(x).$$

Hence

$$P\left(x - \frac{dx}{2} < \tilde{x} < x + \frac{dx}{2}\right) \doteq f(x)dx.$$

Thus we see that the so-called *probability element* $f(x)\,dx$ is an approximation to the probability that $\tilde{x}$ will fall in the interval of length dx centered at x. The approximation gets better as dx gets smaller.

Example 7.5.1

In Example 7.2.4, we differentiate F to obtain

$$f(x) = \begin{cases} 0, & x \leq 0 \\ \rho x^{\rho-1}, & 0 < x < 1 \\ 0, & x \geq 1. \end{cases}$$

This function is graphed for $\rho = \frac{1}{2}$, 1, and 2 in Figure 7.8.

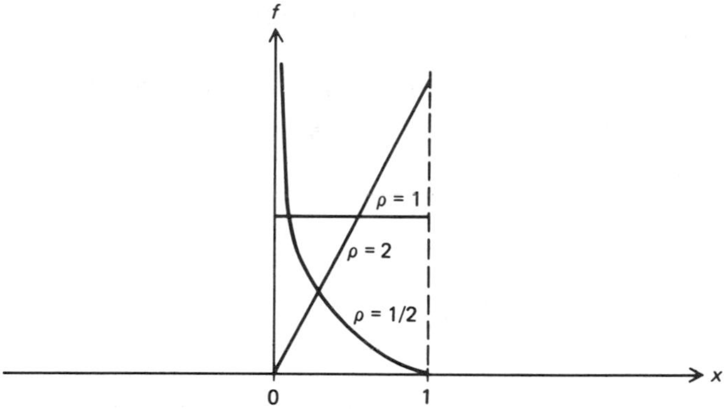

Figure 7.8

Example 7.5.2

In Example 7.2.3 we differentiate F to obtain

$$f(x) = \begin{cases} 0, & x \leq 0 \\ \lambda e^{-\lambda x}, & x > 0. \end{cases}$$

This function is graphed, for $\lambda = \frac{1}{2}$, 1, and 2 in Figure 7.9.

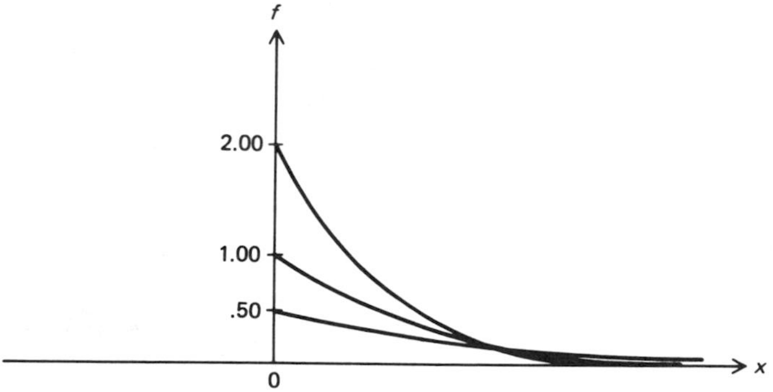

Figure 7.9

Note in these figures that the *density* function of $\tilde{x}$ can exceed one. In fact, $\lim_{x\to 0^+} \rho x^{\rho-1} = +\infty$ for $\rho = \frac{1}{2}$.

Note also that the property (cdf 2) in Section 7.2 implies that the derivative of F is never negative when it exists. Hence $f(x) \geq 0$ whenever $f(x) = (d/dx)F(x)$. Otherwise, by our definition, $f(x) = 0$. Hence we have

(df 1) $f(x) \geq 0$, for every x.

Moreover, since (except at points of nondifferentiability of F)

$$f(t) = \frac{d}{dt}F(t),$$

it follows from the fundamental theorem of integral calculus that

(df 3) $F(x) = \int_{-\infty}^{x} f(t)dt.$

Since $F(+\infty) = 1$, we have

(df 2) $\int_{-\infty}^{\infty} f(t)dt = 1.$

[In (df 2) and (df 3) we use t as a *dummy variable* of integration; we could just as easily have used another letter. The main point is that it is undesirable to use x both as a limit of integration and also as the variable of integration.]

Properties (df 1) and (df 2) completely characterize df's of continuous random variables $\tilde{x}$, in the following sense:

If $f(\cdot)$ is the df of some continuous random variable $\tilde{x}$, then $f(\cdot)$ possesses properties (df 1) and (df 2).

Conversely, *any* function f: $R^1 \to R^1$ which possesses properties (df 1) and (df 2) is the df of some continuous random variable $\tilde{x}$; namely, the random variable whose cdf is given by (df 3).

Example 7.5.3

Let f be defined by

$$f(x) = \begin{cases} 0, & x \leq \Lambda \\ \dfrac{r\Lambda^r}{x^{r+1}}, & x > \Lambda, \end{cases}$$

where $r > 0$ and $\Lambda > 0$. Then $f(x) \geq 0$ for every $x \in R^1$, and so (df 1) is satisfied. Moreover,

$$\begin{aligned}\int_{-\infty}^{\infty} f(t)dt &= \int_{-\infty}^{\Lambda} 0 \cdot dt + \int_{\Lambda}^{\infty} \frac{r\Lambda^r}{t^{r+1}}\,dt \\ &= 0\Big|_{-\infty}^{\Lambda} - \left(\frac{\Lambda}{t}\right)^r\Big|_{\Lambda}^{\infty} \\ &= 0 - 0 + \left(\frac{\Lambda}{\Lambda}\right)^r \\ &= 1,\end{aligned}$$

and hence (df 2) is satisfied. We use (df 3) to find $F(\cdot)$:

$$\begin{aligned}F(x) &= \int_{-\infty}^{x} f(t)dt \\ &= \begin{cases} 0, & x \leq \Lambda \\ \int_{\Lambda}^{x} f(t)dt, & x > \Lambda. \end{cases}\end{aligned}$$

Now,

$$\int_{\Lambda}^{x} f(t)dt = \int_{\Lambda}^{x} \frac{r\Lambda^r}{t^{r+1}}\,dt = -\left(\frac{\Lambda}{x}\right)^r\Big|_{\Lambda}^{x} = 1 - \left(\frac{\Lambda}{x}\right)^r.$$

Hence

$$F(x) = \begin{cases} 0, & x \leq \Lambda \\ 1 - \left(\frac{\Lambda}{x}\right)^r, & x > \Lambda. \end{cases}$$

The reader may easily verify that F possesses properties (cdf 1) through (cdf 5) and is, in fact, the cdf of a continuous random variable x.

What about density functions of *discrete* random variables? The reader can verify as Exercise 12 that the definition of density function implies that $f(x) = 0$, for every $x \in R^1$ if $\tilde{x}$ is a discrete random variable with step cdf $F(\cdot)$. Hence the density function of a discrete random variable tells us nothing about its distribution.

We close this section by showing how probabilities of $\tilde{x}$-events can be obtained directly from the df $f(\cdot)$ of the continuous random variable $\tilde{x}$. The reader may prove (df 4) through (df 11) as Exercise 14.

(df 3) $F(x) = \int_{-\infty}^{x} f(t)dt;$

(df 4) $P[\tilde{x} = x] = 0;$

(df 5) $$P[\tilde{x} < x] = \int_{-\infty}^{x} f(t)dt;$$

(df 6) $$P[x > x] = \int_{x}^{\infty} f(t)dt;$$

(df 7) $$P[x \geq x] = \int_{x}^{\infty} f(t)dt;$$

(df 8) $$P[a < x \leq b] = \int_{a}^{b} f(t)dt;$$

(df 9) $$P[a < x \leq b] = \int_{a}^{b} f(t)dt;$$

(df 10) $$P[a \leq x < b] = \int_{a}^{b} f(t)dt; \text{ and}$$

(df 11) $$P[a < x < b] = \int_{a}^{b} f(t)dt.$$

7.6 DENSITY AND MASS FUNCTIONS OF MIXED RANDOM VARIABLES

Let $\tilde{x}$ be a mixed random variable with cdf $F(\cdot)$. Then the mass function $g(\cdot)$ of $\tilde{x}$ is not identically zero, as is the case with continuous random variables, because at the jumps x^i of $F(\cdot)$ we have $g(x^i) = P[\tilde{x} = x^i] = F(x^i) - F(x^{i-}) > 0$. Clearly, $g(x) \geq 0$, for every $x \in R^1$, but since $\tilde{x}$ is a mixed rather than discrete random variable, $\sum \{g(x^i): i = 1, 2, \ldots\} < 1$. But since $F(\cdot)$ is not constant between every pair of adjacent jumps, it follows that the *density* function $f(\cdot)$ of $\tilde{x}$ is also not identically zero. Hence neither the mf nor the df of $\tilde{x}$ is superfluous if $\tilde{x}$ is a mixed random variable, and both are needed to specify the distribution of $\tilde{x}$ completely.

We conclude this section by showing how probabilities of $\tilde{x}$-events can be obtained from the df f and the mf g of a mixed random variable $\tilde{x}$.

(df and mf 1) $$F(x) = \int_{-\infty}^{x} f(t)dt + \sum \{g(t): t \leq x\};$$

(df and mf 2) $$P(\tilde{x} = x) = g(x);$$

(df and mf 3) $$P(\tilde{x} < x) = \int_{-\infty}^{x} f(t)dt + \sum \{g(t): t < x\};$$

(df and mf 4) $$P(\tilde{x} > x) = \int_{x}^{\infty} f(t)dt + \sum \{g(t): t > x\};$$

(df and mf 5) $$P(\tilde{x} \geq x) = \int_{x}^{\infty} f(t)dt + \sum \{g(t): t \geq x\};$$

(df and mf 6) $$P(a < \tilde{x} \leq b) = \int_{a}^{b} f(t)dt + \sum \{g(t): a < t \leq b\};$$

(df and mf 7) $P(a \leq \tilde{x} \leq b) = \int_a^b f(t)dt + \sum \{g(t): a \leq t \leq b\};$

(df and mf 8) $P(a \leq \tilde{x} < b) = \int_a^b f(t)dt + \sum \{g(t): a \leq t < b\};$

(df and mf 9) $P(a < \tilde{x} < b) = \int_a^b f(t)dt + \sum \{g(t): a < t < b\}.$

You may verify (df and mf 1) through (df and mf 9) as Exercise 15.

EXERCISES

1. Let $Q = \{a, b, c, d\}$; let $\tilde{x}(a) = 0$, $\tilde{x}(b) = 1$, $\tilde{x}(c) = 2$, $\tilde{x}(d) = 3$; and let $P(a) = .20$, $P(b) = .25$, $P(c) = .40$, and $P(d) = .15$. Find the cdf F of $\tilde{x}$.

2. Let $H: R^1 \to R^1$ be defined by

$$H(x) = \begin{cases} 0, & x \leq 0 \\ 1 - e^{-\lambda x^\rho}, & x > 0, \end{cases}$$

where $\lambda > 0$ and $\rho > 0$ are parameters.

   (a) Is H a cdf?
   (b) If H is a cdf, find the df and the mf of $\tilde{x}$.
   (c) If H is a cdf, find $P(-10 < \tilde{x} \leq 0)$, $P(-2 < \tilde{x} \leq 5)$, $P(\tilde{x} = 20)$, and $P(\tilde{x} \geq 10)$.
   (d) If $0 < p < 1$, express a pth fractile ${}_px$ of $\tilde{x}$ as a function of p.

3. If $0 < p < 1$, express a pth fractile of $\tilde{x}$ as a function of p, where $\tilde{x}$ is as defined in:

   (a) Example 7.2.1;
   (b) Example 7.2.3;
   (c) Example 7.2.4;
   (d) Exercise 1.

4. Find the mf g of $\tilde{x}$ as defined in Exercise 1.

5. Let $H: R^1 \to R^1$ be defined by

$$H(x) = \begin{cases} 0, & x \notin (0, 1) \\ 12x^2(1 - x), & x \in (0, 1). \end{cases}$$

   (a) Is H the df of a continuous random variable $\tilde{x}$?
   (b) If so, find the cdf F of $\tilde{x}$.
   (c) If so, and if $p \in (0, 1)$, express a pth fractile ${}_px$ of $\tilde{x}$ as a function of p.
   (d) If so, find $P(-10 < x < -2)$, $P(-1 < x < \frac{1}{2})$, $P(\frac{1}{4} < x < \frac{1}{2})$, and $P(\frac{1}{2} < x \leq 4)$.

6. Let $H: R^1 \to R^1$ be defined by

$$H(x) = \begin{cases} 0; & x \leq 0 \\ 1 - e^{-x}; & 0 < x < 1 \\ 1 - e^{-2x}; & 1 \leq x < 2 \\ 1 - e^{-3x}; & 2 \leq x < \infty. \end{cases}$$

(a) Is H the cdf of some random variable $\tilde{x}$?
(b) If so, find the df and the mf of $\tilde{x}$.
(c) If so, and if $p \in (0, 1)$, express a pth fractile ${}_px$ of $\tilde{x}$ as a function of p.
(d) If so, find $P(-10 < x \leq 0)$, $P(-2 < x \leq \frac{1}{2})$, $P(\frac{1}{2} < x < 1)$, $P(\frac{1}{2} < x \leq 1)$, and $P(x > 1)$.

7. Verify the correspondences between $\tilde{x}$-events and q-events asserted in Section 7.1.

8. Prove property ($p7^\sigma$) of probability, used in the proof of (cdf 3).

[*Hints*: (1) Define $D_1 = E_1$, $D_2 = E_2 \backslash E_1, \ldots, D_n = E_n \backslash E_{n-1}, \ldots$.
(2) Show that $\{D_n: n = 1, 2, \ldots\}$ is a set of mutually exclusive events and that $\bigcup_{n=1}^{i} D_n = E_i$, for every i.
(3) Show that $\bigcup_{n=1}^{\infty} D_n = \bigcup_{i=1}^{\infty} E_i$.
(4) Apply ($P3^\sigma$) to show that

$$P(\bigcup_{n=1}^{\infty} E_i) = P(\bigcup_{n=1}^{\infty} D_n) = \sum_{n=1}^{\infty} P(D_n)$$

$$= \lim_{i \to \infty} \sum_{n=1}^{i} P(D_n).$$

(5) From (2), $\sum_{n=1}^{i} P(D_n) = P(E_i)$.]

9. Verify (cdf 6) through (cdf 14).

10. Suppose we have an urn containing only the poker chips numbered from one to $n(\geq 1)$, and suppose that you are blindfolded and draw a chip out at random.

(a) If $n = 100$, what is the probability of drawing chip number 25?
(b) If $n = 1{,}000{,}000$, what is the probability of drawing chip number 25?
(c) If $n = +\infty$, what is the probability of drawing chip number 25? Is drawing chip number 25 impossible?

11. Show that if $\tilde{x}$ is a continuous random variable, then $g(x) = 0$, for every $x \in R^1$.

12. Show that if $\tilde{x}$ is a discrete random variable, then $f(x) = 0$, for every $x \in R^1$.

13. Verify (mf 4) through (mf 12).

14. Verify (df 4) through (df 11).

15. Verify (df and mf 1) through (df and mf 9).

16. *Degenerate Random Variables*: A random variable $\tilde{x}$ is said to be *degenerate* if, for every $q \in Q$, $\tilde{x}(q) =$ a given $x^* \in R^1$. Thus $\tilde{x}$ is not really "random." Assume $\tilde{x}$ is a degenerate random variable.

 (a) Show that the cdf F of $\tilde{x}$ is given by

$$F(x) = \begin{cases} 0, & x < x^* \\ 1, & x \geq x^*. \end{cases}$$

 (b) Show that the mf g of $\tilde{x}$ is given by

$$g(x) = \begin{cases} 0, & x \neq x^* \\ 1, & x = x^*. \end{cases}$$

17. *Distributions of Functions of Random Variables*: Let $h\colon \tilde{x}_1(Q) \to R^1$ be a continuous and strictly increasing function on the range of $\tilde{x}_1$. Then the composition $h \circ \tilde{x}_1\colon Q \to R^1$ is a random variable; call it $\tilde{x}_2$.

 (a) Show that the cdf's F_1 and F_2 of $\tilde{x}_1$ and $\tilde{x}_2$ are related to each other by

$$F_2(x_2) = \begin{cases} 0, & x_2 < \text{all } x \in h(\tilde{x}_1(Q)) \\ F_1(h^{-1}(x_2)), & x_2 \in h(\tilde{x}_1(Q)) \\ 1, & x_2 > \text{all } x \in h(\tilde{x}_1(Q)) \end{cases}$$

 and

$$F_1(x_1) = \begin{cases} 0, & x_1 < \text{all } x \in h^{-1}(h(x_1(Q))) \\ F_2(h(x_1)), & x_1 \in h^{-1}(h(x_1(Q))) \\ 1, & x_1 > \text{all } x \in h^{-1}(h(x_1(Q))). \end{cases}$$

 [*Hint*: $\{q\colon h \cdot x_1(q) \leq x_2\} = \{q\colon x_1(q) \leq h^{-1}(x_2)\}$.]

 (b) Show that if h is also differentiable and $\tilde{x}_1$ is a continuous random variable, then the df's f_1 and f_2 of $\tilde{x}_1$ and $\tilde{x}_2$ are related by

$$f_2(x_2) = f_1(h^{-1}(x_2)) \cdot \frac{dh^{-1}(x_2)}{dx_2}$$

 and

$$f_1(x_1) = f_2(h(x_1)) \cdot \frac{dh(x_1)}{dx_1}.$$

 [*Hint*: Use part (a), the definition of df, and the chain rule for derivatives. These are change-of-variable formulae in integration.]

 (c) Suppose h is a strictly decreasing function on $\{x_1\colon f_1(x_1) > 0\}$. Show that (b) remains true if the absolute value of each derivative is used in place of that derivative itself.

18. Let $\tilde{x}_1$ be defined as $\tilde{x}$ was in Example 7.2.3. Find F_2 and f_2, where $\tilde{x}_2 \equiv h(\tilde{x}_1)$ and h is defined by

 (a) $h(x) = x^{-y}$, for all $x > 0$ and some $y > 0$
 (b) $h(x) = x^{y}$, for all $x > 0$ and some $y > 0$.
 [*Hint*: Use the results of Exercise 17.]

8

THE DISTRIBUTION OF SEVERAL RANDOM VARIABLES

8.1 INTRODUCTION

In this chapter we shall extend the discussion of Chapter 7 to the case of several random variables, say, $\tilde{x}_1, \tilde{x}_2, \ldots, \tilde{x}_w$. We shall derive their joint distribution, joint cdf, joint density (if their joint cdf is continuous), and so forth. Our basic program will be quite similar to that of Chapter 7; namely, show how the given probability P on $\mathcal{E}_Q$ induces probabilities of events in R^w defined by statements about the values $x_1, \ldots, x_w$ assumed by $\tilde{x}_1, \ldots, \tilde{x}_w$. The resulting probability $P[\cdot]$ on the (sigma)-field $\mathcal{B}^w$ of events in R^w is called the *joint probability distribution*, or *joint distribution*, of $\tilde{x}_1, \ldots, \tilde{x}_w$.

Moreover, once we know the joint distribution of $\tilde{x}_1, \ldots, \tilde{x}_w$ we may dispense with $\mathcal{E}_Q$ and P, since all probability statements about $\tilde{x}_1, \ldots, \tilde{x}_w$ can be made from their joint distribution or from one of its representatives [for example, a cdf, a mf, a df, or a density-mass function (dmf)]. Section 8.2 discusses alternative representations of the joint distribution of $\tilde{x}_1, \ldots, \tilde{x}_w$.

Section 8.3 introduces a topic which could not have been considered in Chapter 7; namely, the *marginal* joint distribution of *some* of $\tilde{x}_1, \ldots, \tilde{x}_w$—say, of $\tilde{x}_1, \ldots, \tilde{x}_m$—and the representations of that marginal joint distribution. We shall examine the way in which these representations are related to the analogous representations of the distribution of the entire set $\tilde{x}_1, \ldots, \tilde{x}_w$ of random variables.

Section 8.4 defines the notion of the *conditional* joint distribution of $\tilde{x}_{m+1}, \ldots, \tilde{x}_w$ given observed values $x_1, \ldots, x_m$ of $\tilde{x}_1, \ldots, \tilde{x}_m$. Representations of this conditional distribution are discussed and related to their analogues in Section 8.2 and 8.3.

The reader will note that in the preceding paragraphs we could have enclosed $\tilde{x}_1, \ldots, \tilde{x}_w$ in parentheses and called it a random *vector*. We shall feel free to do so whenever desirable. Our conventions are that $(x_1, \ldots, x_w)'$ always denotes a *column* vector; that a prime (′) denotes *transpose;* and hence that "$(x_1, \ldots, x_w)$" denotes a *row* vector.

Section 8.5 briefly presents the generalization obtained from Sections 8.2 through 8.4, when each $\tilde{x}_i$ is replaced by a random vector $\tilde{\mathbf{x}}_i \equiv (\tilde{x}_{1i}, \ldots, \tilde{x}_{ni})'$. No new theory is involved, since $(\tilde{\mathbf{x}}_1', \ldots, \tilde{\mathbf{x}}_w')'$ is simply the nw-dimensional random vector $(\tilde{x}_{11}, \ldots, \tilde{x}_{n1}, \tilde{x}_{12}, \ldots, \tilde{x}_{n2}, \ldots\ldots, \tilde{x}_{1w}, \ldots, \tilde{x}_{nw})'$.

Section 8.6 presents Bayes' Theorem for random variables and random vectors, together with a popular alternative notation for the decision problems. This alternative notation is used extensively in Chapter 9.

We shall adopt three conventions that will greatly simplify the subsequent presentation. *First,* we shall denote density functions *and* mass functions (and also density-mass functions defined later) by the letter f, and refer to them generically as *distributing functions* (abbreviated *dgf*). (The term *distributing function* is not standard. It is believed to be due to Pratt, Raiffa and Schlaifer [64].)

Second, we shall denote all sums and integrals by the integral sign; thus

$$\int_A f(x)\, dx = \sum \{f(x): x \in A\}$$

if f is the mf of a discrete random variable, while the meaning of $\int_A f(x)\, dx$ is evident when f is the df of a continuous random variable. Moreover, for multiple integrals we shall use

$$\int \cdots \int_A f(x_1, \ldots, x_w)\, dx_1 \ldots dx_w$$

even when we mean summation with respect to some x_i's.

The *third* convention is the useful heresy of using $f(x_1, \ldots, x_w)$ to denote the function whose value at $(x_1, \ldots, x_w)$ is $f(x_1, \ldots, x_w)$. This convention avoids a very cumbersome subscripting system which would not be very useful except in rare instances, and which would conflict with the subscripting conventions developed later.

8.2 REPRESENTATIONS OF THE JOINT DISTRIBUTION OF $\tilde{x}_1, \ldots, \tilde{x}_w$.

We define the *cumulative distribution function* (cdf) $F: R^w \to R^1$ of $\tilde{x}_1, \ldots, \tilde{x}_w$ by

$$\begin{aligned} F(x_1, \ldots, x_w) &= P(\{q: \tilde{x}_i(q) \leq x_i, \text{ for } i = 1, \ldots, w\}) \\ &= P(\cap_{i=1}^{w} Q_i(x_i)), \end{aligned}$$

for every $(x_1, \ldots, x_w) \in R^w$, where

$$Q_i(x_i) \equiv \{q: \tilde{x}_i(q) \leq x_i\},$$

for every $x_i \in R^1$ (as in Chapter 7).

Example 8.2.1

Let $Q = \{a, b, c\}$; $P(a) = .10$, $P(b) = .30$, $P(c) = .60$; $\tilde{\mathbf{x}} = (\tilde{x}_1, \tilde{x}_2)'$; $\tilde{\mathbf{x}}(a) = (1, 2)'$, $\tilde{\mathbf{x}}(b) = (3, 1)'$, and $\tilde{\mathbf{x}}(c) = (10, 4)'$. Then

$$Q_1(x_1) = \begin{cases} \varnothing, & x_1 < 1 \\ \{a\}, & 1 \leq x_1 < 3 \\ \{a, b\}, & 3 \leq x_1 < 10 \\ Q, & x_1 \geq 10 \end{cases}$$

$$Q_2(x_2) = \begin{cases} \varnothing, & x_2 < 1 \\ \{b\}, & 1 \leq x_2 < 2 \\ \{a, b\}, & 2 \leq x_2 < 4 \\ Q, & x_2 \geq 4 \end{cases}$$

and hence

$$Q_1(x_1) \cap Q_2(x_2) = \begin{cases} \varnothing, & \mathbf{x} \ngeq (1, 1)' \text{ or } \mathbf{x} < (3,2)' \\ \{a\}, & 1 \leq x_1 < 3,\ x_2 \geq 2 \\ \{b\}, & x_1 \geq 3,\ 1 \leq x_2 < 2 \\ \{a, b\}, & \mathbf{x} \geq (3, 2)' \text{ but } \mathbf{x} \ngeq (10, 4)' \\ Q, & \mathbf{x} \geq (10, 4)'. \end{cases}$$

Hence

$$F(x_1, x_2) = \begin{cases} 0; & \mathbf{x} \ngeq (1, 1)' \text{ or } \mathbf{x} < (3, 2)' \\ .10; & 1 \leq x_1 < 3,\ x_2 \geq 2 \\ .30; & x_1 \geq 3,\ 1 \leq x_2 < 2 \\ .40; & \mathbf{x} \geq (3, 2)' \text{ but } \mathbf{x} \ngeq (10, 4)' \\ 1.00; & \mathbf{x} \geq (10, 4)'. \end{cases}$$

Figure 8.1 graphs the sets of constant F-value.

Note that $F(\mathbf{b}) \geq F(\mathbf{a})$ whenever $\mathbf{b} \geq \mathbf{a}$. Clearly, $\lim_{\mathbf{x}\to\infty} F(\mathbf{x}) = 1$ and $\lim_{x_i\to-\infty} F(\mathbf{x}) = 0$ for each component x_i of $\mathbf{x}$. (The term $\lim_{\mathbf{x}\to\infty}$ means the limit as *both* x_1 and x_2 increase without bound.)

We abstract from this example to establish a number of properties which any joint cdf $F(x_1, \ldots, x_w)$ of w random variables should possess.

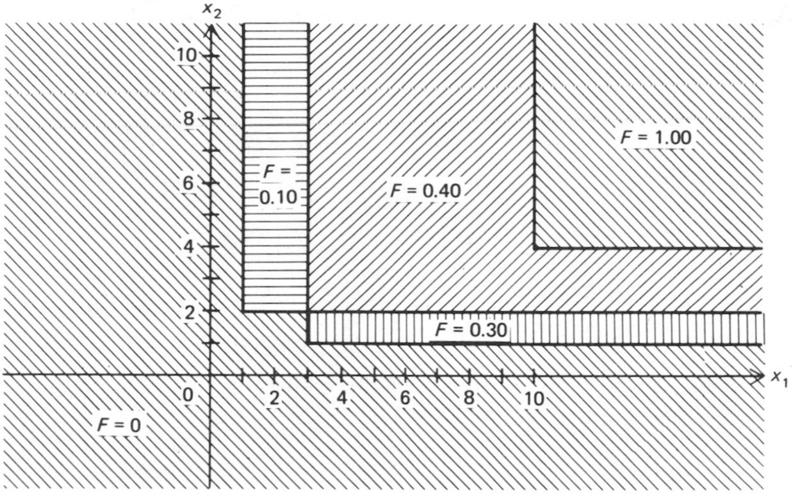

Figure 8.1

Properties of the cdf of $\tilde{\mathbf{x}}$:
Let $F(\cdot)$ be the cdf of $\tilde{\mathbf{x}}$. Then

(cdf 1) $0 \leq F(\mathbf{x}) \leq 1$, for every $\mathbf{x} \in R^w$;

(cdf 2) if $\mathbf{b} \geq \mathbf{a}$, then $F(\mathbf{b}) \geq F(\mathbf{a})$;

(cdf 3) if $\{\mathbf{x}^i\colon i = 1, 2, \ldots\}$ is a *decreasing* sequence of w-tuples with limit $\mathbf{x}$, then
$\lim_{i \to \infty} F(\mathbf{x}^i) = F(\mathbf{x})$;

(cdf 4) $\lim_{x_i \to \infty} F(\mathbf{x}) = 0$, for every $\in \{1, \ldots, w\}$; and

(cdf 5) $\lim_{x \to \infty} F(\mathbf{x}) = 1$.

(The term $\lim_{\mathbf{x} \to \infty}$ denotes the limit as *all* x_i increase without bound.)

Proof

From the definition of F, (cdf 1) follows immediately. For (cdf 2), if $\mathbf{b} \geq \mathbf{a}$, then $b_i \geq a_i$, for every $i \in \{1, \ldots, w\}$, and hence $Q_i(a_i) \subset Q_i(b_i)$ for every i. Hence $\bigcap_{i=1}^{w} Q_i(a_i) \subset \bigcap_{i=1}^{w} Q_i(b_i)$, and (cdf 2) now follows from property ($P7$) of probability and the definition of F. As in the one-dimensional case, (cdf 4) and (cdf 5) follow from (cdf 3), and (cdf 3) is established by an argument similar to that used for (cdf 3) in Chapter 7.

Note that (cdf 4) says that $F(\mathbf{x})$ tends to zero as *any* component x_i of $\mathbf{x}$ tends to $-\infty$, while (cdf 5) says that $F(\mathbf{x})$ tends to one as *all* components x_i of $\mathbf{x}$ tend to $+\infty$.

Example 8.2.2

Let $\{\mathbf{t}^i: i = 1, 2, \ldots\}$ be a countable subset of R^w and let $f: R^w \to R^1$ satisfy

$$f(\mathbf{x}) \geq 0, \text{ for every } \mathbf{x} \in R^w;$$
$$f(\mathbf{x}) = 0, \text{ if } \mathbf{x} \notin \{\mathbf{t}^i: i = 1, 2, \ldots\}; \text{ and}$$
$$\sum \{f(\mathbf{t}^i): i = 1, 2, \ldots\} = 1.$$

Define $F: R^w \to R^1$ by

$$F(\mathbf{x}) = \sum \{f(\mathbf{t}^i): \mathbf{t}^i \leq \mathbf{x}\}, \text{ for every } \mathbf{x} \in R^w.$$

Then F possesses properties (cdf 1) through (cdf 5) and is therefore the joint cdf of w random variables $\tilde{x}_1, \ldots, \tilde{x}_w$.

Example 8.2.3

Let $f: R^w \to R^1$ satisfy

(dgf 1) $f(\mathbf{x}) \geq 0$, for every $\mathbf{x} \in R^w$; and

(dgf 2) $\int_{-\infty}^{\infty} \cdots \int_{-\infty}^{\infty} f(t_1, \ldots, t_w)dt_1 \ldots dt_w.$

Define $F: R^w \to R^1$ by

(dgf 3) $F(x_1, \ldots, x_w) = \int_{-\infty}^{x_w} \cdots \int_{-\infty}^{x_1} f(t_1. \ldots, t_w)dt, \ldots dt_w,$

for every $(x_1, \ldots, x_w) \in R^w$.

Then F possesses properties (cdf 1) through (cdf 5) and is therefore the joint cdf of w random variables $\tilde{x}_1, \ldots, \tilde{x}_w$—or, alternatively, the cdf of a w-dimensional random vector $\tilde{\mathbf{x}} \equiv (\tilde{x}_1, \ldots, \tilde{x}_w)$. In (dgf 2) and (dgf 3) the integrals are to be understood as sums with respect to any x_i such that $\tilde{x}_i$ is a discrete random variable. This example generalizes Example 8.2.2.

We say that $f: R^w \to R^1$ is the *joint distributing function* of w random variables $\tilde{x}_1, \ldots, \tilde{x}_w$ if f possesses the aforementioned properties (dgf 1) through (dgf 3). In that event we also refer to f as the distributing function of the w-dimensional random vector $\tilde{\mathbf{x}} \equiv (\tilde{x}_1, \ldots, \tilde{x}_w)$. In the applications it will always be clear which components $\tilde{x}_i$ are discrete random variables (and hence which integrals $\int[\cdot]\, dx_i$ are really sums).

We could show how probabilities of "rectangles" of the form $(a_1, b_1] \times \cdots \times (a_w, b_w)$ can be obtained from the joint cdf $F(x_1, \ldots, x_w)$ of $\tilde{x}_1, \ldots, \tilde{x}_w$; but the requisite notation is somewhat involved, and we shall not need the results in later work. [Finding probabilities of rectangles $(a_1, b_1] \times (a_2, b_2\}$ in R^2 from $F(x_1, x_2)$ constitutes Exercise 11.]

The more usual, and more *useful* way of describing the joint distribution of $\tilde{x}_1, \ldots, \tilde{x}_w$ is via the distributing function $f(x_1, \ldots, x_w)$. It can be shown that if $\tilde{x}_1, \ldots, \tilde{x}_w$ has the joint distributing function (dgf) $f(x_1, \ldots, x_w)$, then

$$\text{(dgf 4)} \quad P[(\tilde{x}_1, \ldots, \tilde{x}_w) \in A \subset R^w] = \int \cdots \int_A f(t_1, \ldots, t_w) dt_1 \ldots dt_w,$$

where, again, integrals are to be interpreted as sums wherever necessary.

The following three examples will illustrate the application of the preceding notions.

Example 8.2.4

Let $w = 2$ and define $f\colon R^2 \to R^1$ by

$$f(x_1, x_2) = \begin{cases} 12e^{-4x_1-3x_2}; & x_1 > 0, x_2 > 0 \\ 0; & \text{elsewhere.} \end{cases}$$

It is plain that $f(x_1, x_2) \geq 0$ for every $(x_1, x_2) \in R^2$ and hence (dgf 1) obtains. Moreover,

$$\int_{-\infty}^{\infty} \int_{-\infty}^{\infty} f(t_1, t_2)\, dt_1 dt_2 = \int_0^{\infty} \int_0^{\infty} 12e^{-4t_1} \cdot e^{-3t_2}\, dt_1\, dt_2,$$

where

$$\int_0^{\infty} 12e^{-4t_1} \cdot e^{-3t_2}\, dt_1 = 3e^{-3t_2}.$$

Hence

$$\int_{-\infty}^{\infty} \int_{-\infty}^{\infty} f(t_1, t_2)\, dt_1\, dt_2 = \int_0^{\infty} 3e^{-3t_2}\, dt_2 = 1,$$

and therefore (dgf 2) obtains, which concludes the proof that f is the distributing function of a two-dimensional random vector $\tilde{\mathbf{x}}$. We find the cdf F of $\tilde{\mathbf{x}}$ by applying (dgf 3):

$$F(x_1, x_2) = \int_{-\infty}^{x_2} \int_{-\infty}^{x_1} f(t_1, t_2) dt_1 dt_2.$$

Now,

$$\int_{-\infty}^{x_1} f(t_1, t_2) dt_1 = \begin{cases} 0; & t_2 \leq 0 \\ 0; & t_2 > 0, x_1 \leq 0 \\ \int_0^{x_1} 12e^{-4t_1} e^{-3t_2} dt_1; & t_2 > 0, x_1 > 0 \end{cases}$$

$$= \begin{cases} 0; & t_2 \leq 0 \\ 0; & t_2 > 0, x_1 \leq 0 \\ e^{-3t_2}(1 - e^{-4x_1})\ ; & t_2 > 0, x_1 > 0. \end{cases}$$

Hence, by again considering special cases of (x_1, x_2), we obtain

$$F(x_1, x_2) = \begin{cases} (1 - e^{-4x_1})(1 - e^{-3x_2}); & x_1 > 0,\ x_2 > 0 \\ 0; & \text{elsewhere.} \end{cases}$$

The reader may verify as Exercise 12 that F possesses properties (cdf 1) through (cdf 5).

In Example 8.2.4 it is clear that $F(x_1, x_2)$ is a continuous function. Whenever $\tilde{x}_1, \ldots, \tilde{x}_w$ are such that $F(x_1, \ldots, x_w)$ is a continuous function, we shall say that $\tilde{x}_1, \ldots, \tilde{x}_w$ are *jointly continuous random variables;* or that $\tilde{\mathbf{x}} \equiv (\tilde{x}_1, \ldots, \tilde{x}_w)'$ is a *continuous random vector.* If $\tilde{x}_1, \ldots, \tilde{x}_w$ are jointly continuous random variables, we say that $f(x_1, \ldots, x_w)$ is a (joint) *density function.* In Figure 8.2 we graph the joint density function of $(\tilde{x}_1, \tilde{x}_2)$ as given in Example 8.2.4.

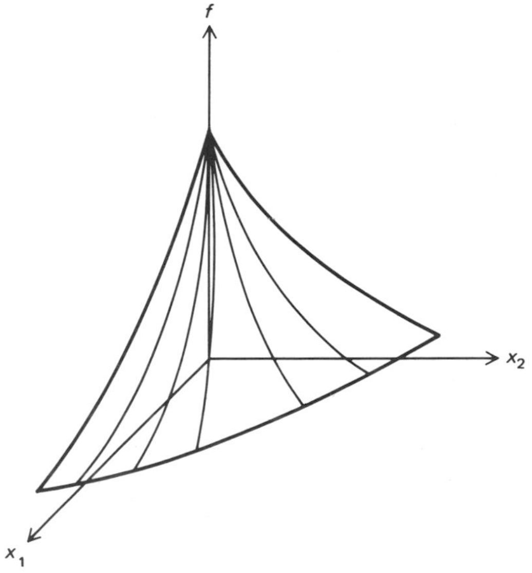

Figure 8.2

If $\tilde{x}_1, \ldots, \tilde{x}_w$ are jointly continuous random variables, the joint dgf of $\tilde{x}_1, \ldots, \tilde{x}_w$ will be a density function, given by

$$f(x_1, \ldots, x_w) = \frac{\partial^w}{\partial x_1 \partial x_2 \ldots \partial x_w} F(x_1, \ldots, x_w),$$

where this mixed partial derivative exists, and by $f(x_1, \ldots, x_w) = 0$ otherwise.

When $f(x_1, \ldots, x_w)$ is a density function, (dgf 4) asserts that $P[(\tilde{x}_1, \ldots, \tilde{x}_w) \in A]$ is simply the "volume" under f over "base" A. (For $w = 2$, the quotes are removable.)

Example 8.2.5

Let $w = 2$ and define $f\colon R^2 \rightarrow R^1$ by

$$f(x_1, x_2) = \begin{cases} p_1 p_2 (1 - p_1)^{x_1} (1 - p_2)^{x_2}, & x_i \in \{0, 1, 2, \ldots\}, \text{ for } i = 1, 2 \\ 0, & \text{elsewhere.} \end{cases}$$

Here, $f(x_1, x_2)$ is nonzero only at a countable number of points (x_1, x_2) and hence the integrals in (dgf 2) and (dgf 3) are sums. Hence

$$\begin{aligned} \int_{-\infty}^{\infty} \int_{-\infty}^{\infty} f(t_1, t_2)\,dt_1\,dt_2 &= \sum_{j=0}^{\infty} \sum_{i=0}^{\infty} p_1 p_2 (1 - p_1)^i (1 - p_2)^j \\ &= \sum_{j=0}^{\infty} p_2 (1 - p_2)^j \left\{ \frac{p_1}{1 - (1 - p_1)} \right\} \\ &= \sum_{j=0}^{\infty} p_2 (1 - p_2)^j \\ &= \frac{p_2}{1 - (1 - p_2)} \\ &= 1, \end{aligned}$$

and hence (dgf 2) is satisfied. Moreover, (dgf 3) implies that whenever $x_i \in \{0, 1, 2, \ldots\}$ for $i = 1, 2$ we have

$$\begin{aligned} F(x_1, x_2) &= \sum_{j=0}^{x_1} \sum_{i=0}^{x_1} p_1 p_2 (1 - p_1)^i (1 - p_2)^j \\ &= [1 - (1 - p_1)^{x_2+1}][1 - (1 - p_2)^{x_2+1}]. \end{aligned}$$

Hence, in general,

$$F(x_1, x_2) = \begin{cases} 0, \text{ unless } x_i \geq 0, \text{ for } i = 1, 2 \\ [1 - (1 - p_1)^{y_1+1}][1 - (1 - p_2)^{y_2+1}], \\ \qquad y_i \in \{0, 1, 2, \ldots\} \text{ and} \\ \qquad y_i \leq x_i < y_i + 1, \text{ for } i = 1, 2. \end{cases}$$

In Example 8.4.5, the range of $\tilde{\mathbf{x}} \equiv (\tilde{x}_1, \tilde{x}_2)$ is a countable subset (the pairs of nonnegative integers) of R^2 and integration was always summation. Whenever $F(x_1, \ldots, x_w)$ is obtained solely by summations, we say that $\tilde{x}_1, \ldots, \tilde{x}_w$ are *jointly discrete* random variables, or that $\tilde{\mathbf{x}} \equiv (\tilde{x}_1, \ldots, \tilde{x}_w)'$ is a *discrete random vector.* If $\tilde{x}_1, \ldots, \tilde{x}_w$ are jointly discrete random variables, we say that $f(x_1, \ldots, x_w)$ is a (joint) *mass function.*

Example 8.2.6

Let $w = 2$ and define f: $R^2 \to R^1$ by

$$f(x_1, x_2) = \begin{cases} \rho x_1{}^{\rho}(1 - x_1)^{x_2}; & 0 < x_1 < 1,\ x_2 \in \{0, 1, 2, \ldots\} \\ 0; & \text{elsewhere,} \end{cases}$$

where $\rho > 0$ is a parameter. Then $f(x_1, x_2) \geq 0$ for all $(x_1, x_2) \in R^2$ and hence (dgf 1) obtains. Moreover, we have

$$\begin{aligned} \int_{-\infty}^{\infty} \left[\sum \{f(x_1, x_2)\colon x_2 = 0, 1, 2, \ldots\} \right] dx_1 \\ &= \int_0^1 \left[\sum_{x_2=0}^{\infty} \rho x_1{}^{\rho}(1 - x_1)^{x_2} \right] dx_1 \\ &= \int_0^1 \rho x_1{}^{\rho-1} dx_1 \\ &= 1, \end{aligned}$$

and so (dgf 2) obtains as well.

The cdf $F(x_1, x_2)$ is rather difficult to evaluate; we shall not do so.

Roughly speaking, in Example 8.2.6, $\tilde{x}_2$ is discrete and $\tilde{x}_1$ continuous. In such cases, we say that $(\tilde{x}_1, \tilde{x}_2)$ is a *mixed* random vector. More generally, we say that $\tilde{x} \equiv (\tilde{x}_1, \ldots, \tilde{x}_w)'$ is a *mixed* random vector if $\tilde{\mathbf{x}}$ is neither a continuous nor a discrete random vector.

Continuous random vectors occur most frequently in decision theory and inference. We conclude the section with another example of them.

Example 8.2.7

Let $w = 2$ and define f: $R^2 \to R^1$ by

$$f(x_1, x_2) = \begin{cases} 12(x_1 - x_2)^2, & 0 < x_2 < x_1 < 1 \\ 0, & \text{elsewhere.} \end{cases}$$

Since $12(x_1 - x_2)^2 \geq 0$, for every (x_1, x_2), it follows that $f(x_1, x_2) \geq 0$, for every $(x_1, x_2) \in R^2$ and hence (dgf 1) obtains. Moreover,

$$\begin{aligned} \int_{-\infty}^{\infty} \int_{-\infty}^{\infty} f(t_1, t_2)\, dt_1 dt_2 &= \int_0^1 \int_0^{t_1} 12(t_1 - t_2)^2\, dt_2 dt_1 \\ &= \int_0^1 \int_{-t_1}^{0} 12y^2\, dy dt_1, \qquad (y \equiv t_2 - t_1) \\ &= \int_0^1 4t_1{}^3\, dt_1 \\ &= t_1{}^4 \Big|_0^1 \\ &= 1, \end{aligned}$$

and hence (dgf 2) obtains. Therefore f is the density function of a two-dimensional random vector $\tilde{\mathbf{x}}$. We find the cdf F of $\tilde{\mathbf{x}}$ by applying (dgf 3):

$$F(x_1, x_2) = \int_{-\infty}^{x_1} \int_{-\infty}^{x_2} f(t_1, t_2)\, dt_2 dt_1,$$

which reduces, after a tedious consideration of special cases, to

$$F(x_1, x_2) = \begin{cases} 0; & x_1 \leq 0 \text{ and/or } x_2 \leq 0 \\ x_1{}^4; & 0 < x_1 < \min\{x_2, 1\} \\ x_1{}^4 - (x_1 - x_2)^4; & 0 < x_2 < x_1 < 1 \\ 1 - (1 - x_2)^4; & 0 < x_2 \leq 1 \leq x_1 \\ 1; & x_1 \geq 1,\ x_2 \geq 1. \end{cases}$$

8.3 MARGINAL DISTRIBUTIONS AND THEIR REPRESENTATIONS

Let $\tilde{x}_1, \ldots, \tilde{x}_w$ be w random variables with joint cdf $F(x_1, \ldots, x_w)$ and joint dgf $f(x_1, \ldots, x_w)$, each of which represents the joint distribution of $\tilde{x}_1, \ldots, \tilde{x}_w$.

Take the first m $(<w)$ random variables $\tilde{x}_1, \ldots, \tilde{x}_m$ which together comprise the subset $\{\tilde{x}_1, \ldots, \tilde{x}_m\}$ of $\{\tilde{x}_1, \ldots, \tilde{x}_w\}$. Clearly, $\tilde{x}_1, \ldots, \tilde{x}_m$ has a joint distribution in its own right, and this joint distribution is represented by a cdf $F(x_1, \ldots, x_m)$ and a dgf $f(x_1, \ldots, x_m)$.

We shall show in this section how $F(x_1, \ldots x_m)$ is related to $F(x_1, \ldots, x_w)$ $[= F(x_1, \ldots, x_m, x_{m+1}, \ldots, x_w)]$. These relations are such that $F(x_1, \ldots, x_m)$ and $f(x_1, \ldots x_m)$ can be ascertained without having to appeal to the basic probability P on $\mathcal{E}_Q$. We prove, *first*,

$$F(x_1, \ldots, x_m) = \lim_{\substack{x_{m+1} \to \infty \\ \vdots \\ x_w \to \infty}} F(x_1, \ldots, x_m, x_{m+1}, \ldots, x_w). \tag{8.3.1}$$

Proof

From the definition of cdf it follows that

$$F(x_1, \ldots, x_m) = P(\cap_{i=1}^m Q_i(x_i)).$$

But

$$\cap_{i=1}^m Q_i(x_i) = [\cap_{i=1}^m Q_i(x_i)] \cap Q,$$

and

$$Q = \lim_{\substack{x_{m+1} \to \infty \\ \vdots \\ x_w \to \infty}} \cap_{i=m+1}^w Q_i(x_i).$$

Hence

$$\bigcap_{i=1}^{m} Q_i(x_i) = \lim_{\substack{x_{m+1}\to\infty \\ \vdots \\ x_w\to\infty}} [\bigcap_{i=1}^{m} Q_i(x_i)] \cap [\bigcap_{i=m+1}^{w} Q_i(x_i)]$$

$$= \lim_{\substack{x_{m+1}\to\infty \\ \vdots \\ x_w\to\infty}} \bigcap_{i=1}^{w} Q_i(x_i).$$

Now, by property ($P7^{\sigma}$) (see Chapter 7) we have

$$P(\bigcap_{i=1}^{m} Q_i(x_i)) = \lim_{\substack{x_{m+1}\to\infty \\ \vdots \\ x_w\to\infty}} P(\bigcap_{i=1}^{w} Q_i(x_i));$$

that is,

$$F(x_1, \ldots, x_m) = \lim_{\substack{x_{m+1}\to\infty \\ \vdots \\ x_w\to\infty}} F(x_1, \ldots, x_w),$$

as was to be shown.

It is clear that Equation (8.3.1) continues to hold for *any* subset of $\{\tilde{x}_1, \ldots, \tilde{x}_w\}$: to find the cdf $F(x_{i_1}, \ldots, x_{i_m})$ of *any* subset of $\{\tilde{x}_1, \ldots, \tilde{x}_w\}$, let

$$\{j_1, \ldots, j_{w-m}\} = \{1, \ldots, w\} \backslash \{i_1, \ldots, i_m\}$$

and take the limit of $F(x_1, \ldots, x_w)$ as every x_{j_h} tends to $+\infty$.

The cdf $F(x_1, \ldots, x_m)$: $R^m \to R^1$ is called the *marginal* cdf of $\tilde{x}_1, \ldots, \tilde{x}_m$; the (joint) cdf of $\tilde{x}_1, \ldots, \tilde{x}_m$; *marginal upon* $\tilde{x}_{m+1}, \ldots, \tilde{x}_w$; or the *unconditional* cdf of $\tilde{x}_1, \ldots, \tilde{x}_m$.

Example 8.3.1

In Example 8.2.4, we find

$$F(x_1) = \lim_{x_2\to\infty} F(x_1, x_2)$$

$$= \begin{cases} \lim_{x_2\to\infty} (1 - e^{-4x_1})(1 - e^{-3x_2}), & x_1 > 0 \\ 0, & x_1 \le 0 \end{cases}$$

$$= \begin{cases} 1 - e^{-4x_1}, & x_1 \ge 0 \\ 0, & x_1 \le 0. \end{cases}$$

Similarly,

$$F(x_2) = \lim_{x_1\to\infty} F(x_1, x_2)$$
$$= \begin{cases} \lim_{x_1\to\infty} (1 - e^{-4x_1})(1 - e^{-3x_2}), & x_2 > 0 \\ 0, & x_2 \leq 0 \end{cases}$$
$$= \begin{cases} 1 - e^{-3x_2}, & x_2 > 0 \\ 0, & x_2 \leq 0. \end{cases}$$

Notice that in this example we have $F(x_1, x_2) = F(x_1)\cdot F(x_2)$ for all $(x_1, x_2) \in R^2$. This condition is one way of stating that $\tilde{x}_1$ and $\tilde{x}_2$ are *independent* random variables.

Example 8.3.2

In Example 8.2.7 we find

$$F(x_1) = \lim_{x_2\to\infty} F(x_1, x_2)$$
$$= \begin{cases} \lim_{x_2\to\infty} 0; & x_1 \leq 0 \\ \lim_{x_2\to\infty} x_1^4; & 0 < x_1 < 1 \\ \lim_{x_2\to\infty} 1; & x_1 \geq 1 \end{cases}$$
$$= \begin{cases} 0, & x_1 \leq 0 \\ x_1^4; & 0 < x_1 < 1 \\ 1; & x_1 \geq 1. \end{cases}$$

Similarly,

$$F(x_2) = \lim_{x_1\to\infty} F(x_1, x_2)$$
$$= \begin{cases} \lim_{x_1\to\infty} 0; & x_2 \leq 0 \\ \lim_{x_1\to\infty} 1 - (1 - x_2)^4; & 0 < x_2 < 1 \\ 1; & x_2 \geq 1 \end{cases}$$
$$= \begin{cases} 0, & x_2 \leq 0 \\ 1 - (1 - x_2)^4, & 0 < x_2 < 1 \\ 1, & x_2 \geq 1. \end{cases}$$

Here, $F(x_1, x_2) \neq F(x_1)\cdot F(x_2)$ for some pairs (x_1, x_2). Hence (anticipating future developments) $\tilde{x}_1$ and $\tilde{x}_2$ are *not* independent random variables.

The second task of this section is to relate the *marginal* dgf $f(x_1, \ldots, x_m)$ of $\tilde{x}_1, \ldots, \tilde{x}_m$ to the dgf $f(x_1, \ldots, x_w)$ of $\tilde{x}_1, \ldots, \tilde{x}_w$. We assert that

$$f(x_1, \ldots, x_m) = \int_{-\infty}^{\infty} \cdots \int_{-\infty}^{\infty} f(x_1, \ldots, x_m, t_{m+1}, \ldots, t_w)\, dt_{m+1} \ldots dt_w. \quad (8.3.2)$$

Proof

$(\tilde{x}_1, \ldots, \tilde{x}_m) \in A \subset R^m$ if and only if $(\tilde{x}_1, \ldots, \tilde{x}_m) \in A$ *and* $(\tilde{x}_{m+1}, \ldots, \tilde{x}_w) \in R^{m-w}$ [since $(\tilde{x}_{m+1}, \ldots, \tilde{x}_w) \in R^{m-w}$ is sure to occur]. Hence $(\tilde{x}_1, \ldots, \tilde{x}_m) \in A$ if and only if $(\tilde{x}_1, \ldots, \tilde{x}_w) \in A \times R^{m-w}$. Therefore

$$P[\tilde{x}_1, \ldots, \tilde{x}_m) \in A] = P[(\tilde{x}_1, \ldots, \tilde{x}_w) \in A \times R^{m-w}].$$

But by (dgf 4),

$$P[(\tilde{x}_1, \ldots, \tilde{x}_w) \in A \times R^{m-w}]$$

$$= \int_{-\infty}^{\infty} \cdots \int_{-\infty}^{\infty} \int_A \cdots \int f(x_1, \ldots, x_m, t_{m+1}, \ldots, t_w) dx_1 \ldots dx_m dt_{m+1} \ldots dt_w$$

$$= \int_A \cdots \int \int_{-\infty}^{\infty} \cdots \int_{-\infty}^{\infty} f(x_1, \ldots, x_m, t_{m+1}, \ldots, t_w) dt_{m+1} \ldots dt_w dx_1 \ldots dx_m$$

(reversing the order of integration, which is valid also when the integral represents a summation). But the preceding equalities are valid for *every* $A \subset R^m$. Now, the truth of (dgf 4) for $\tilde{x}_1, \ldots \tilde{x}_m$ implies

$$P[(\tilde{x}_1, \ldots, \tilde{x}_m) \in A] = \int_A \cdots \int f(x_1, \ldots, x_m) dx_1 \ldots dx_m.$$

Hence $f(x_1, \ldots, x_m)$ may be taken as equal to the term in the brackets above; and a theorem in advanced calculus asserts that (essentially) it *must* be so taken. You may verify as Exercise 13 that $f(x_1, \ldots, x_m)$ as *defined* by Equation (8.3.2) satisfies (dgf 1) through (dgf 3), where in (dgf 3) $F(x_1, \ldots, x_m)$ is defined by Equation (8.3.1).

Example 8.3.3

In Example 8.2.4 we apply Equation (8.3.2) to obtain

$$f(x_1) = \begin{cases} \int_{-\infty}^{\infty} 0 \cdot dt_2, & x_1 \leq 0 \\ \int_{-\infty}^{0} 0 \cdot dt_2 + \int_0^{\infty} 12e^{-4x_1 - 3t_2}\, dt_2, & x_1 > 0 \end{cases}$$

$$= \begin{cases} 0, & x_1 \leq 0 \\ 4e^{-4x_1}, & x_1 > 0. \end{cases}$$

Note that $f(x_1) = (d/dx_1)F(x_1)$, where $F(x_1)$ is as given in Example 8.3.1. Similarly,

$$f(x_2) = \begin{cases} 0, & x_2 \leq 0 \\ \int_0^\infty 12e^{-4t_1-3x_2}dt_1, & x_2 > 0 \end{cases}$$

$$= \begin{cases} 0, & x_2 \leq 0 \\ 3e^{-3x_2}, & x_2 > 0, \end{cases}$$

and that $f(x_2) = (d/dx_2)F(x_2)$.
Also note that

$$f(x_1, x_2) = f(x_1)\cdot f(x_2)$$

for all $(x_1, x_2) \in R^2$; this condition is another criterion that $\tilde{x}_1$ and $\tilde{x}_2$ be independent random variables.

Example 8.3.4

For Example 8.2.6, we use Equation (8.3.2) to obtain

$$f(x_1) = \begin{cases} 0, & x_1 \notin (0, 1) \\ \sum_{x_2=0}^{\infty} \rho x_1{}^\rho(1 - x_1)^{x_2}, & x_1 \in (0, 1) \end{cases}$$

$$= \begin{cases} 0, & x_1 \notin (0, 1) \\ \rho x_1{}^{\rho-1}, & x_1 \in (0, 1). \end{cases}$$

We shall not derive $f(x_2)$ for this example.

8.4 CONDITIONAL DISTRIBUTIONS AND THEIR REPRESENTATIVES

Suppose that you have ascertained that $(\tilde{x}_1, \ldots, \tilde{x}_m)$ assumes a value in a given (measurable) set $A' \subset R^m$; and also suppose that you now wish to determine the probability that $(\tilde{x}_{m+1}, \ldots, \tilde{x}_w)$ belongs to any given (measurable) subset $B' \subset R^{w-m}$. At this point *the marginal distribution* of $(\tilde{x}_{m+1}, \ldots, \tilde{x}_w)$ *is irrelevant* (see Figure 6.11, which expresses the same idea in the context of marginal probabilities on an abstract field of events). Instead of the marginal distribution, we want the *conditional* probabilities

$$P[(\tilde{x}_{m+1}, \ldots, \tilde{x}_w) \in B' \mid (\tilde{x}_1, \ldots, \tilde{x}_m) \in A'] \equiv P[B' \mid A'].$$

Now, the event $A' \in \mathcal{B}^m$ occurs if and only if the event $A \equiv \{q\colon (\tilde{x}_1(q), \ldots, \tilde{x}_m(q)) \in A'\} \in \mathcal{E}_Q$ occurs; and $B' \in \mathcal{B}^{w-m}$ occurs if and only if $B \equiv \{q\colon (\tilde{x}_{m+1}(q), \ldots, \tilde{x}_w(q)) \in B'\} \in \mathcal{E}_Q$ occurs. Hence it follows that

$$P[B' \mid A'] = P(B \mid A),$$

provided that $P(B \mid A)$ is defined, which will be true if $P(A) > 0$ and may be true in certain cases where $P(A) = 0$.

The reader will recall from Chapter 6 that

$$P(B \mid A) = \frac{P(A, B)}{P(A)}$$

whenever $P(A) > 0$. Hence also

$$P[B' \mid A'] = \frac{P[A', B']}{P[A'],} \tag{8.4.1}$$

provided that $P[A'] > 0$. In view of (dgf 4), (8.4.1) implies that

$$P[B' \mid A']$$

$$= \frac{\underbrace{\int \cdots \int}_{B'} \underbrace{\int \cdots \int}_{A'} f(x_1, \ldots, x_m, x_{m+1}, \ldots, x_w) dx_1 \ldots dx_m dx_{m+1} \ldots dx_w}{\underbrace{\int \cdots \int}_{A'} f(x_1, \ldots, x_m) dx_1 \ldots dx_m}$$

$$= \underbrace{\iint}_{B'} \left[\frac{\underbrace{\int \cdots \int}_{A'} f(x_1, \ldots, x_m, x_{m+1}, \ldots, x_w) dx_1 \ldots dx_m}{\underbrace{\int \cdots \int}_{A'} f(x_1, \ldots, x_m) dx_1 \ldots dx_m} \right] dx_{m+1} \ldots dx_w$$

$$= \underbrace{\iint}_{B'} f(x_{m+1}, \ldots, x_w \mid A') dx_{m+1} \ldots dx_w, \tag{8.4.2}$$

where $f(x_{m+1}, \ldots, x_w \mid A')$ is defined as the ratio of integrals in the brackets. Now, (8.4.2) simply asserts that conditional probabilities of $(\tilde{x}_{m+1}, \ldots, \tilde{x}_w)$-events B' given the event $(\tilde{x}_1, \ldots, \tilde{x}_m) \in A'$ can be found by integrating, via (dgf 4), a conditional dgf $f(x_{m+1}, \ldots, x_w \mid A')$; that is:

If $P[A'] > 0$, then the conditional probability that $(\tilde{x}_{m+1}, \ldots, \tilde{x}_w) \in B'$ given that $(\tilde{x}_1, \ldots, \tilde{x}_m) \in A'$ is found by applying (dgf 4) to (8.4.3)

$$f(x_{m+1}, \ldots, x_w \mid A')$$

$$\equiv \frac{\underbrace{\int \cdots \int}_{A'} f(x_1, \ldots, x_m, x_{m+1}, \ldots, x_w) dx_1 \ldots dx_m}{\underbrace{\int \cdots \int}_{A'} f(x_1, \ldots, x_m) dx_1, \ldots dx_m}; \tag{8.4.4}$$

that is,

$$P[(\tilde{x}_{m+1}, \ldots, \tilde{x}_w) \in B' \mid (\tilde{x}_1, \ldots, \tilde{x}_m) \in A'] = \underbrace{\int \cdots \int}_{B'} f(x_{m+1}, \ldots, x_w \mid A') dx_{m+1} \ldots dx_w, \tag{8.4.5}$$

for every $B' \in \mathcal{B}^{w-m}$.

What happens if the denominator of (8.4.4) is zero; in particular, if $A' = \{(x_1, \ldots, x_m)\}$, a single point in R^m? This question may seem trivial, but it is really of fundamental importance. The answer is essentially as follows:

If $f(x_1, \ldots, x_m) > 0$, then the conditional distribution

$$P[\cdot \mid x_1, \ldots, x_m]: \mathcal{B}^{w-m} \to R^1 \tag{8.4.6}$$

of $(\tilde{x}_{m+1}, \ldots, \tilde{x}_w)$ given $(\tilde{x}_1, \ldots, \tilde{x}_m) = (x_1, \ldots, x_m)$

is represented by a conditional dgf, given by

$$f(x_{m+1}, \ldots, x_w \mid x_1, \ldots, x_m) = \frac{f(x_1, \ldots, x_m, x_{m+1}, \ldots, x_w)}{f(x_1, \ldots, x_m)}, \tag{8.4.7}$$

and such that

$$P[B' \mid x_1, \ldots, x_m] = \underbrace{\int \cdots \int}_{B'} f(x_{m+1}, \ldots, x_w \mid x_1, \ldots, x_m)\, dx_{m+1} \ldots dx_w.$$

Proof

The following argument is not quite rigorous, but suggests the rigorous proof attainable by advanced analysis. We know that result (8.4.4) holds for every A' such that $P[A'] > 0$. When $A' = \{(x_1, \ldots, x_m)\}$, a single point in R^m with $f(x_1, \ldots, x_m) > 0$, choose a subset $A^* \subset R^m$ such that $(x_1, \ldots, x_m) \in A^*$ and such that both numerator and denominator of (8.4.4) (with A^* replacing A') are small but positive. By the mean value theorem, there exist points $(\zeta'_1, \ldots, \zeta'_m)$ and $(\zeta''_1, \ldots, \zeta''_m)$ in A^* such that the numerator equals $f(\zeta'_1, \ldots, \zeta'_m, x_{m+1}, \ldots, x_w)V(A^*)$ and the denominator equals $f(\zeta''_1, \ldots, \zeta''_m)V(A^*)$, where $V(A^*)$ denotes the volume of A^*. Hence the volumes $V(A^*)$ cancel in (8.4.4):

$$f(x_{m+1}, \ldots, x_w \mid A^*) = \frac{f(\xi'_1, \ldots, \xi'_m, \ldots, x_w)}{f(\xi''_1, \ldots, \xi''_m)}.$$

This equation must hold for all A^* of arbitrarily small probability; and if we let A^* shrink to $A' \equiv \{(x_1, \ldots, x_m)\}$, then $(\zeta'_1, \ldots, \zeta'_m)$ and $(\zeta''_1, \ldots, \zeta''_m)$ are both forced to $(x_1, \ldots, x_m)$.

It is worth stressing that the hypothesis in (8.4.6) which states $f(x_1, \ldots x_m) > 0$ is crucial. It also makes good sense on an intuitive level:

Why condition probabilities [of $(\tilde{x}_{m+1}, \ldots, \tilde{x}_w)$-events] upon impossibilities [points $(x_1, \ldots, x_m)$ such that $f(x_1, \ldots, x_m) = 0$]?

In the future we shall say that $(x_1, \ldots, x_m)$ is *possible* if and only if $f(x_1, \ldots, x_m) > 0$.

Equation (8.4.7) implies the following *chain rule*: if $(x_1, \ldots, x_w)$ is possible, then

$$\begin{aligned} f(x_1, \ldots, x_w) &= f(x_1)\cdot f(x_2 \mid x_1)\cdot \ldots \cdot f(x_w \mid x_1, \ldots, x_{w-1}) \\ &= f(x_1) \cdot \prod_{i=1}^{w-1} f(x_{i+1} \mid x_1, \ldots, x_i). \end{aligned} \tag{8.4.8}$$

[Recall Equation (6.3.7).]

Now, our approach thus far in this chapter has been to show how the joint distribution of $\tilde{x}_1, \ldots, \tilde{x}_w$ induces the marginal distribution of $\tilde{x}_1, \ldots, \tilde{x}_m$ and, for certain $(\tilde{x}_1, \ldots, \tilde{x}_m)$-events A', the conditional distribution of $(\tilde{x}_{m+1}, \ldots, \tilde{x}_w)$ given A'. In practice, however, you are usually given $f(x_1, \ldots, x_m)$ and also $f(x_{m+1}, \ldots, x_w \mid x_1, \ldots, x_m)$ for every $(x_1, \ldots, x_m)$ such that $f(x_1, \ldots, x_m) > 0$; and you are required to find, first, $f(x_1, \ldots, x_w)$, and then both $f(x_{m+1}, \ldots, x_w)$ and $f(x_1, \ldots, x_m \mid x_{m+1}, \ldots, x_w)$ [for every possible $(x_{m+1}, \ldots, x_w)$]. It is obvious that you should define

$$f(x_1, \ldots, x_w) = \begin{cases} 0, & f(x_1, \ldots, x_m) = 0 \\ f(x_{m+1}, \ldots, x_w \mid x_1, \ldots, x_m)\cdot f(x_1, \ldots, x_m), & \text{otherwise.} \end{cases} \tag{8.4.9}$$

Then use (8.3.2) to compute

$$f(x_{m+1}, \ldots, x_w) = \int_{-\infty}^{\infty} \cdots \int_{-\infty}^{\infty} f(x_1, \ldots, x_w)dx_1 \ldots dx_m;$$

and use (8.4.7) to obtain

$$f(x_1, \ldots, x_m \mid x_{m+1}, \ldots, x_w) = \frac{f(x_1, \ldots, x_w)}{f(x_1, \ldots, x_m)},$$

whenever $f(x_1, \ldots, x_m) > 0$. We shall return to this topic in Section 8.7, on Bayes' Theorem.

Example 8.4.1

Suppose $\tilde{x}_1, \tilde{x}_2$ have joint dgf given by

$$f(x_1, x_2) = \begin{cases} .10, & (x_1, x_2) = (0, 1) \\ .20, & (x_1, x_2) = (1, 1) \\ .25, & (x_1, x_2) = (2, 1) \\ .30, & (x_1, x_2) = (1, 0) \\ .15, & (x_1, x_2) = (2, 0) \\ 0, & \text{elsewhere.} \end{cases}$$

The marginal dgf of $\tilde{x}_1$ is a mass function, given by

$$f(x_1) = \begin{cases} .10, & x_1 = 0 \\ .50, & x_1 = 1 \\ .40, & x_1 = 2 \\ 0, & \text{elsewhere.} \end{cases}$$

Hence

$$f(x_2 \mid x_1 = 0) = \frac{f(x_1 = 0, x_2)}{f(x_1 = 0)}$$

$$= \begin{cases} \dfrac{.10}{.10} = 1, & x_2 = 1 \\ 0, & x_2 \neq 1; \end{cases}$$

$$f(x_2 \mid x_1 = 1) = \frac{f(x_1 = 1, x_2)}{f(x_1 = 1)}$$

$$= \begin{cases} \dfrac{.30}{.50} = .60, & x_2 = 0 \\ \dfrac{.20}{.50} = .40, & x_2 = 1 \\ 0, & \text{elsewhere;} \end{cases}$$

and

$$f(x_2 \mid x_1 = 2) = \frac{f(x_1 = 2, x_2)}{f(x_1 = 2)}$$

$$= \begin{cases} \dfrac{.15}{.40} = \dfrac{3}{8}, & x_2 = 0 \\ \dfrac{.25}{.40} = \dfrac{5}{8}, & x_2 = 1 \\ 0, & \text{elsewhere.} \end{cases}$$

The reader may derive $f(x_2)$, $f(x_1 \mid x_2 = 0)$, and $f(x_1 \mid x_2 = 1)$ as Exercise 14.

Example 8.4.2

In Example 8.3.4 we solved for $f(x_1)$ when $\tilde{x}_1$, $\tilde{x}_2$, had the joint dgf defined in Example 8.2.6. Using (8.4.7), we find, for $x_1 \in (0, 1)$,

$$f(x_2 \mid x_1) = \begin{cases} \dfrac{\rho x_1{}^{\rho}(1 - x_1)^{x_2}}{\rho x_1{}^{\rho-1}}, & x_2 \in \{0, 1, 2, \ldots\} \\ 0, & \text{elsewhere} \end{cases}$$

$$= \begin{cases} x_1(1 - x_1)^{x_2}, & x_2 \in \{0, 1, 2, \ldots\} \\ 0, & \text{elsewhere.} \end{cases}$$

Example 8.4.3

From Examples 8.3.3 and 8.2.4 we obtain, for $x_1 > 0$

$$f(x_2 \mid x_1) = \begin{cases} 3e^{-3x_2}, & x_2 > 0 \\ 0, & x_2 \leq 0. \end{cases}$$

Note that $f(x_2 \mid x_1) = f(x_2)$, for every x_2 and every $x_1 > 0$. This is an independence condition for the random variables $\tilde{x}_1$ and $\tilde{x}_2$.

Example 8.4.4

From Examples 8.3.2 and 8.2.7 we obtain

$$F(x_1) = \begin{cases} 0, & x_1 \leq 0 \\ x_1{}^4, & 0 < x_1 < 1 \\ 1, & x_1 \geq 1. \end{cases}$$

$F(x_1)$ is continuous, and hence $f(x_1) = (d/dx_1)F(x_1)$ where $F(x_1)$ is differentiable, and it equals zero elsewhere. Hence

$$f(x_1) = \begin{cases} 4x_1{}^3, & 0 < x_2 < 1 \\ 0, & \text{elsewhere.} \end{cases}$$

Therefore,

$$f(x_2 \mid x_1) = \begin{cases} \dfrac{12(x_1 - x_2)^2}{4x_1{}^3}, & 0 < x_2 < 1 \\ 0, & x_2 \notin (0, 1), \end{cases}$$

provided that $0 < x_2 < 1$. [Naturally, $f(x_2 \mid x_1)$ is undefined for $x_1 \notin (0, 1)$.] Also from Example 8.3.2, we obtain

$$f(x_2) = \begin{cases} 4(1 - x_2)^3, & 0 < x_2 < 1 \\ 0, & x_2 \notin (0, 1), \end{cases}$$

and hence

$$f(x_1 \mid x_2) = \begin{cases} \dfrac{12(x_1 - x_2)^2}{4(1 - x_2)^3}, & x_1 \in (x_2, 1) \\ 0, & x_2 \notin (x_2, 1), \end{cases}$$

provided that $0 < x_2 < 1$.

8.5 JOINT DISTRIBUTIONS OF RANDOM VECTORS

So far in this chapter we have considered joint distributions of several random variables, $\tilde{x}_1, \ldots, \tilde{x}_w$, the marginal distribution of a subset $\tilde{x}_1, \ldots \tilde{x}_m$ of them, and conditional distributions of some, $(\tilde{x}_{m+1}, \ldots, \tilde{x}_w)$, given the others, $(\tilde{x}_1, \ldots, \tilde{x}_m)$. We have occasionally aggregated these random variables into random vectors $(\tilde{x}_1, \ldots, \tilde{x}_w)$, $(\tilde{x}_1, \ldots, \tilde{x}_m)$, and so forth. Using boldface type to denote vectors, we define

$$\begin{aligned}\tilde{\mathbf{x}} &\equiv (\tilde{x}_1, \quad \ldots, \tilde{x}_w)',\\ \tilde{\mathbf{x}}_1 &\equiv (\tilde{x}_1, \quad \ldots, \tilde{x}_m)', \text{ and}\\ \tilde{\mathbf{x}}_2 &\equiv (\tilde{x}_{m+1}, \ldots, \tilde{x}_w)'.\end{aligned}$$

Then the pair $\tilde{\mathbf{x}}_1$, $\tilde{\mathbf{x}}_2$ of random vectors is easily seen to correspond to $\tilde{\mathbf{x}}$. Hence the theme of Sections 8.2 through 8.4 was the interrelationships among $f(\mathbf{x})$, $f(\mathbf{x}_1)$, $f(\mathbf{x}_2 \mid \mathbf{x}_1)$, and so forth.

More generally, we might wish to consider w random *vectors* $\tilde{\mathbf{x}}_1, \ldots, \tilde{\mathbf{x}}_w$, where $\tilde{\mathbf{x}}_i \equiv (\tilde{x}_{1i}, \ldots, \tilde{x}_{k_i i})'$ is a k_i-dimensional random vector. It is easy to verify that $(\tilde{\mathbf{x}}_1', \ldots, \tilde{\mathbf{x}}_w')'$ corresponds to a partitioning of the $\sum_{i=1}^{w} k_i$-dimensional random vector $(\tilde{x}_{11}, \ldots, \tilde{x}_{k_1 1}, \tilde{x}_{12}, \ldots, \tilde{x}_{k_2 2}, \ldots\ldots, \tilde{x}_{w1}, \ldots, \tilde{x}_{k_w w})'$. Hence the marginal dgf $f(\mathbf{x}_1, \ldots, \mathbf{x}_m)$ of $(\tilde{\mathbf{x}}_1', \ldots, \tilde{\mathbf{x}}_m')'$ is just a rewriting of the marginal dgf of $(\tilde{x}_{11}, \ldots, \tilde{x}_{k_1 1}, \ldots, \tilde{x}_{1m}, \ldots, \tilde{x}_{k_m m})'$; and so all of the results in the previous sections continue to obtain when the x's are all written in boldface (our vector notation), and when w is replaced by $\sum_{i=1}^{w} k_i$, m is replaced by $\sum_{i=1}^{m} k_i$, and w-m is replaced by $\sum_{i=m+1}^{w} k_i$. For example, (8.3.2) becomes

$$f(\mathbf{x}_1, \ldots, \mathbf{x}_m) = \underbrace{\int \cdots \int}_{R^{k*}} f(\mathbf{x}_1, \ldots, \mathbf{x}_m, \mathbf{t}_{m+1}, \ldots, \mathbf{t}_w) d\mathbf{t}_{m+1} \ldots d\mathbf{t}_w, \quad (8.3.2')$$

where $d\mathbf{t}_i \equiv \prod_{j=1}^{k_i} dt_{ji}$ and $k^* \equiv \sum_{i=m+1}^{w} k_i$; and (8.4.7) becomes

$$f(\mathbf{x}_{m+1}, \ldots, \mathbf{x}_w \mid \mathbf{x}_1, \ldots, \mathbf{x}_m) = \frac{f(\mathbf{x}_1, \ldots, \mathbf{x}_m, \mathbf{x}_{m+1}, \ldots, \mathbf{x}_w)}{f(\mathbf{x}_1, \ldots, \mathbf{x}_m)}, \quad (8.4.7')$$

whenever $f(\mathbf{x}_1, \ldots, \mathbf{x}_m) > 0$.

8.6 INDEPENDENCE AND CONDITIONAL INDEPENDENCE

We say that $\tilde{x}_1, \ldots, \tilde{x}_w$ are (mutually) *independent* random variables if for every $\tilde{x}_1$-event A_1', every $\tilde{x}_2$-event A_2', $\ldots$, and every $\tilde{x}_w$-event A_w', we have

$$P[(\tilde{x}_1, \ldots, \tilde{x}_w) \in A_1' \times A_2' \times \cdots \times A_w'] = \prod_{i=1}^{w} P[A_i']. \quad (8.6.1)$$

Let $A_i \equiv \{q: \tilde{x}_i(q) \in A_i'\}$. Then (8.6.1) is equivalent to the definition of event-independence in Section 6.5. It can be shown, by an argument involving shrinking each A_i' to a point x_i, that

$\tilde{x}_1, \ldots, \tilde{x}_w$ are mutually independent if and only if

$$f(x_1, \ldots, x_w) = f(x_1)\cdot\ldots\cdot f(x_w), \text{ for every } (x_1, \ldots, x_w)' \in R^w; \qquad (8.6.2)$$

that is, if and only if the joint distributing function of $\tilde{x}_1, \ldots, \tilde{x}_w$ factors into the product of the w marginal distributing functions $f(x_i)$.

The reader may show as Exercise 15 that (8.6.1) implies

$\tilde{x}_1, \ldots, \tilde{x}_w$ are mutually independent if and only if

$$F(x_1, \ldots, x_w) = F(x_1)\cdot\ldots\cdot F(x_w), \text{ for every } (x_1, \ldots, x_w) \in R^w. \qquad (8.6.3)$$

We say that $\tilde{\mathbf{x}}_1, \ldots, \tilde{\mathbf{x}}_n$ are (mutually) *independent* random vectors if, for every $\tilde{\mathbf{x}}_1$-event $A_1' \subset R^{k_1}, \ldots,$ and every $\tilde{\mathbf{x}}_n$-event $A_n' \subset R^{k_n}$, we have

$$P[(\tilde{\mathbf{x}}_1, \ldots, \tilde{\mathbf{x}}_n) \in A_1' \times \cdots \times A_n'] = \prod_{i=1}^{n} P[A_i']; \qquad (8.6.1')$$

that is, (8.6.1) is essentially unchanged.

$\tilde{\mathbf{x}}_1, \ldots, \tilde{\mathbf{x}}_n$ are mutually independent if and only if (8.6.2′)

$$f(\mathbf{x}_1, \ldots, \mathbf{x}_n) = f(\mathbf{x}_1)\cdot\ldots\cdot f(\mathbf{x}_n), \text{ for every } (\mathbf{x}_1, \ldots, \mathbf{x}_n)' \in R^{\Sigma^n{}_{i=1} k_i};$$

that is, if and only if

$$f(x_{11}, \ldots, x_{k_1 1}, x_{12}, \ldots, x_{k_2 2}, \ldots, x_{1n}, \ldots, x_{k_n n}) = \prod_{i=1}^{n} f(x_{1i}, \ldots, x_{k_i i}), \text{ for}$$

every $(x_{11}, \ldots, x_{k_n n})' \in R^{\Sigma^n{}_{i=1} k_i}$.

Now, (8.6.1′) and (8.6.2′) do *not* imply that two components of $\tilde{\mathbf{x}}_i$, say $\tilde{x}_{ji}$ and $\tilde{x}_{ki}$, need be independent. In fact, let $\tilde{y}_1$ and $\tilde{y}_2$ be two independent random variables, and define

$$\tilde{x}_{j1}(q) = \tilde{y}_1(q), \text{ for every } q \text{ and } j = 1, \ldots, 4,$$

and

$$\tilde{x}_{j2}(q) = y_2(q), \text{ for every } q \text{ and for } j = 1, \ldots, 8.$$

Also define $\tilde{\mathbf{x}}_1 \equiv (\tilde{x}_{11}, \ldots, \tilde{x}_{41})'$ and $\tilde{\mathbf{x}}_2 \equiv (\tilde{x}_{12}, \ldots, \tilde{x}_{82})'$. Then $\tilde{\mathbf{x}}_1$ and $\tilde{\mathbf{x}}_2$ are independent, but the components of $\tilde{\mathbf{x}}_1$ are certainly as mutually dependent as they can be, and the same is true of the components of $\tilde{\mathbf{x}}_2$. Nonetheless, each component of $\tilde{\mathbf{x}}_1$ is independent of every component of $\tilde{\mathbf{x}}_2$. More generally, the reader may prove as Exercise 16 the following:

If $\tilde{\mathbf{x}}_1, \ldots, \tilde{\mathbf{x}}_n$ are mutually independent random vectors, then $\tilde{x}_{j_1 1}, \ldots, \tilde{x}_{j_n n}$ are mutually independent random variables, where each $\tilde{x}_{j_i i}$ is an arbitrary component of $\tilde{\mathbf{x}}_i$ for $i = 1, \ldots, n$. (8.6.4)

Example 8.6.1

Suppose $\tilde{x}_1, \tilde{x}_2, \tilde{x}_3$ have the joint density function

$$f(x_1, x_2, x_3) = \begin{cases} \left[\prod_{i=1}^{3} (b_i - a_i)\right]^{-1}; & a_i < x_i < b_i, \text{ for } i = 1, 2, 3 \\ 0; & \text{elsewhere,} \end{cases}$$

where $a_i < b_i$ for $i = 1, 2, 3$. It is easily verified that

$$f(x_i) = \begin{cases} (b_i - a_i)^{-1}, & a_i < x_i < b_i \\ 0, & \text{elsewhere,} \end{cases}$$

for $i = 1, 2, 3$. Hence we have $f(x_1, x_2, x_3) = f(x_1)\cdot f(x_2)\cdot f(x_3)$ for all $(x_1, x_2, x_3) \in R^3$. Hence $\tilde{x}_1, \tilde{x}_2, \tilde{x}_3$ are mutually independent.

We say that $\tilde{\mathbf{x}}_1, \ldots, \tilde{\mathbf{x}}_n$ are *conditionally independent* given $\tilde{\mathbf{x}}_0$ if, for every $\mathbf{x}_0$ such that $f(\mathbf{x}_0) > 0$,

$$P[(\tilde{\mathbf{x}}_1, \ldots, \tilde{\mathbf{x}}_n) \in A'_1 \times \cdots \times A'_n \mid \mathbf{x}_0] = \prod_{i=1}^{n} P[A'_i \mid \mathbf{x}_0], \qquad (8.6.5)$$

for every $A'_1 \subset R^{k_1}, \ldots$, and every $A'_n \subset R^{k_n}$. By a "shrinking" argument, we have

$\tilde{\mathbf{x}}_1, \ldots, \tilde{\mathbf{x}}_n$ are conditionally independent given $\tilde{\mathbf{x}}_0$ if

$$f(\mathbf{x}_1, \ldots, \mathbf{x}_n \mid \mathbf{x}_0) = \prod_{i=1}^{n} f(\mathbf{x}_i \mid \mathbf{x}_0), \qquad (8.6.6)$$

for every $\mathbf{x}_0$ such that $f(\mathbf{x}_0) > 0$ and every

$$(\mathbf{x}_1, \ldots, \mathbf{x}_n) \in R^{\Sigma^n{}_{i=1} k_i}.$$

Equations (8.6.5) and (8.6.6) specialize to random variables by putting $k_1 = k_2 = \cdots = k_n = 1$.

Example 8.6.2

Let $\tilde{x}_1, \ldots, \tilde{x}_n$ have the conditional df

$$f(x_1, \ldots, x_n \mid a, b) = \begin{cases} (b - a)^{-n}, & a < x_i < b, \text{ for all } i \\ 0, & \text{elsewhere,} \end{cases}$$

for all $(a, b) \equiv \mathbf{x}_0$ such that $f(\mathbf{x}_0) > 0$, where $f(\mathbf{x}_0) = 0$ when $a \geq b$. It is easy to show that

$$f(x_i \mid a, b) = \begin{cases} (b - a)^{-1}, & a < x_i < b \\ 0, & \text{elsewhere,} \end{cases}$$

for every i and every (a, b) with $f(a, b) > 0$. Clearly,

$$f(x_1, \ldots, x_n \mid a, b) = \prod_{i=1}^{n} f(x_i \mid a, b),$$

and hence $\tilde{x}_1, \ldots, \tilde{x}_n$ are conditionally independent given (a, b).

If $\tilde{\mathbf{x}}_1, \ldots \tilde{\mathbf{x}}_n$ are not only mutually independent but also have the *same* distribution (that is, the same cdf F, the same dgf f, and so forth) then we say that $\tilde{\mathbf{x}}_1, \ldots \tilde{\mathbf{x}}_n$ are *independent and identically distributed,* which phrase is often abbreviated to *iid.* If $\tilde{\mathbf{x}}_1, \ldots, \tilde{\mathbf{x}}_n$ are not only conditionally independent given $\tilde{\mathbf{x}}_0$ but also have the same conditional distribution given every possible $\tilde{\mathbf{x}}_0$-event, then we say that $\tilde{\mathbf{x}}_1, \ldots, \tilde{\mathbf{x}}_n$ are "iid given $\tilde{\mathbf{x}}_0$." We often say, "$\tilde{\mathbf{x}}_1, \ldots, \tilde{\mathbf{x}}_n$ are iid given $\tilde{\mathbf{x}}_0$ with common dgf $f(\mathbf{x} \mid \mathbf{x}_0)$ defined by. . ." (and a formula for the common dgf follows).

Example 8.6.3

If $\tilde{x}_1, \ldots, \tilde{x}_n$ are iid given $(\tilde{a}, \tilde{b})$ with common dgf $f(x \mid a, b)$ given by

$$f(x \mid a, b) = \begin{cases} (b - a)^{-1}, & a < x < b \\ 0, & \text{elsewhere,} \end{cases}$$

then

$$f(x_1, \ldots, x_n \mid a, b) = \prod_{i=1}^{n} f(x_i \mid a, b).$$

(See Example 8.6.2.)

8.7 BAYES' THEOREM FOR RANDOM VARIABLES AND RANDOM VECTORS

Recall from Section 6.4 that Bayes' Theorem is an *easy* consequence of the definitions of conditional and marginal probability, but also a very *useful* consequence in showing how prior judgments become modified in the light of evidence. This section formulates Bayes' Theorem for dgf's of random variables and random vectors.

First, we shall introduce some notation of considerable currency in decision theory. We let the random vector $\tilde{\boldsymbol{\theta}} \equiv (\tilde{\theta}_1, \ldots, \tilde{\theta}_y)$ denote the vector of quantities of interest to the decision-maker. It plays the role of s in Chapter 6.

Next, we shall suppose that the decision-maker may perform an experiment e, the outcome of which is a random vector $\tilde{\mathbf{x}}$. In all applications in this book, we assume that the decision-maker knows the conditional dgf $f^e(\mathbf{x} \mid \boldsymbol{\theta})$ of $\tilde{\mathbf{x}}$ given $\boldsymbol{\theta}$ for every possible value $\boldsymbol{\theta}$ of $\tilde{\boldsymbol{\theta}}$.

Finally, we suppose that, given experiment e, the decision-maker has quantified his judgments about $\tilde{\boldsymbol{\theta}}$ in the form of a dgf $f(\boldsymbol{\theta})$, which will be called the *prior* dgf.

The preceding assumptions enable the computation of:

(1) the joint dgf $f^e(\mathbf{x}, \boldsymbol{\theta})$ of $\tilde{\mathbf{x}}$ and $\tilde{\boldsymbol{\theta}}$;
(2) the marginal dgf $f^e(\mathbf{x})$ of $\tilde{\mathbf{x}}$; and
(3) the conditional dgf $f^e(\boldsymbol{\theta} \mid \mathbf{x})$ of θ given $\mathbf{x}$ for every possible value $\mathbf{x}$ of $\tilde{\mathbf{x}}$.

We have:

$$f^e(\mathbf{x}, \boldsymbol{\theta}) = \begin{cases} f^e(\mathbf{x} \mid \boldsymbol{\theta}) f(\boldsymbol{\theta}), & f(\boldsymbol{\theta}) > 0 \\ 0, & f(\boldsymbol{\theta}) = 0 \end{cases}$$

by Equation (8.4.9);

$$f^e(\mathbf{x}) = \int_{-\infty}^{\infty} \cdots \int_{-\infty}^{\infty} f^e(\mathbf{x}, \boldsymbol{\theta}) d\theta_1 \ldots d\theta_y$$
$$= \int_{-\infty}^{\infty} \cdots \int_{-\infty}^{\infty} f^e(\mathbf{x} \mid \boldsymbol{\theta}) f(\boldsymbol{\theta}) d\theta_1 \ldots d\theta_y$$

by (8.3.2′); and $f^e(\boldsymbol{\theta} \mid \mathbf{x}) = f^e(\mathbf{x}, \boldsymbol{\theta})/f^e(\mathbf{x})$, for all $\mathbf{x}$ with $f^e(\mathbf{x}) > 0$, by (8.4.7′). Simple substitutions now yield

$$f^e(\boldsymbol{\theta} \mid \mathbf{x}) = \frac{f^e(\mathbf{x} \mid \boldsymbol{\theta}) f(\boldsymbol{\theta})}{\int_{-\infty}^{\infty} \cdots \int_{-\infty}^{\infty} f^e(\mathbf{x} \mid \boldsymbol{\theta}) f(\boldsymbol{\theta}) d\theta_1 \ldots d\theta_y}, \tag{8.7.1}$$

$$\text{all } \boldsymbol{\theta}, f^e(\mathbf{x}) > 0.$$

Equation (8.7.1) is Bayes' Theorem for dgf's. It shows how the *prior* dgf (representing the decision-maker's prior judgments about $\tilde{\boldsymbol{\theta}}$) interacts with $f^e(\mathbf{x} \mid \boldsymbol{\theta})$ (representing the distribution of the experimental outcome $\tilde{\mathbf{x}}$) so as to produce the so-called *posterior* dgf $f^e(\boldsymbol{\theta} \mid \mathbf{x})$, which represents the decision-maker's judgments about $\tilde{\boldsymbol{\theta}}$ *after* observing the value $\mathbf{x}$ of $\tilde{\mathbf{x}}$; that is, his judgments *posterior* to the outcome $\mathbf{x}$ of e.

Many actual experiments e consist in performing a predetermined number n of independent trials under identical conditions (see Section 5.4) and are such that each trial i will have as outcome a k-tuple $\mathbf{x}_i \equiv (x_{1i}, \ldots, x_{ki})'$ which, before the experiment is performed, is a random vector, $\tilde{\mathbf{x}}_i$. Before the experiment, the as yet unobserved outcome is the ($w =$) kn-dimensional random vector $\tilde{\mathbf{x}} \equiv (\tilde{\mathbf{x}}_1', \ldots, \tilde{\mathbf{x}}_n')'$. Moreover, "independence of trials" and "identical conditions" imply that $\tilde{\mathbf{x}}_1, \ldots, \tilde{\mathbf{x}}_n$ are iid given the true state $\tilde{\boldsymbol{\theta}}$ of affairs, with some common dgf $f^e(\mathbf{x}_i \mid \boldsymbol{\theta})$. Hence (8.6.5) implies

$$f^e(\mathbf{x}_1, \ldots, \mathbf{x}_n \mid \boldsymbol{\theta}) = \prod_{i=1}^{n} f^e(\mathbf{x}_i \mid \boldsymbol{\theta}),$$

for every possible $\boldsymbol{\theta}$. In such cases, adding the decision-maker's prior dgf $f(\boldsymbol{\theta})$ yields the following specialization of Bayes' Theorem:

$$f^e(\boldsymbol{\theta} \mid \mathbf{x}_1, \ldots, \mathbf{x}_n) = \frac{\prod_{i=1}^{n} f^e(\mathbf{x}_i \mid \boldsymbol{\theta}) f(\boldsymbol{\theta})}{\int_{-\infty}^{\infty} \cdots \int_{-\infty}^{\infty} \prod_{i=1}^{n} f^e(\mathbf{x}_i \mid \boldsymbol{\theta}) f(\boldsymbol{\theta}) d\theta_1 \ldots d\theta_y}. \tag{8.7.2}$$

Example 8.7.1

Suppose $\tilde{x}_1, \ldots, \tilde{x}_n$ are iid given $\tilde{\boldsymbol{\theta}} = (\tilde{\theta}_1, \tilde{\theta}_2)'$ with common dgf

$$f^e(x_i \mid \theta_1, \theta_2) = \begin{cases} (\theta_2 - \theta_1)^{-1}, & \theta_1 < x_i < \theta_2 \\ 0, & x_i \notin (\theta_1, \theta_2), \end{cases}$$

for every possible (θ_1, θ_2), where the set of possible (θ_1, θ_2)'s is defined as $\{(\theta_1, \theta_2)\colon \theta_2 > \theta_1\}$. Example 8.6.2 and 8.6.3 imply

$$\begin{aligned} f(x_1, \ldots, x_n \mid \theta_1, \theta_2) &= \begin{cases} (\theta_2 - \theta_1)^{-n}, & \theta_1 < \text{every } x_i < \theta_2 \\ 0, & \text{some } x_i \notin (\theta_1, \theta_2) \end{cases} \\ &= \begin{cases} (\theta_2 - \theta_1)^{-n}, & \theta_1 < m < M < \theta_2 \\ 0, & \text{elsewhere,} \end{cases} \end{aligned} \tag{8.7.3}$$

where $M \equiv \max\{x_1, \ldots, x_n\}$ and $m \equiv \min\{x_1, \ldots x_n\}$. Now suppose that the prior dgf is

$$f(\theta_1, \theta_2) \equiv \begin{cases} \dfrac{(\nu_0 - 1)(\nu_0 - 2)}{(M_0 - m_0)^{2-\nu_0}} (\theta_2 - \theta_1)^{-\nu_0}, & -\infty < \theta_1 < m_0 < M_0 < \theta_2 < \infty \\ 0, & \text{elsewhere,} \end{cases} \tag{8.7.4}$$

where m_0, M_0, and ν_0 are parameters such that $M_0 > m_0$ and $\nu_0 > 2$. The reader may verify as Exercise 17 that $f(\theta_1, \theta_2)$ is a density function. Now, by (8.4.9) we find

$$f^e(\mathbf{x}, \boldsymbol{\theta}) = \frac{(\nu_0 - 1)(\nu_0 - 2)}{(M_0 - m_0)^{2-\nu_0}} (\theta_2 - \theta_1)^{-(\nu_0+n)},$$

for all $\theta_1, \theta_2, x_1, \ldots, x_n$ such that $-\infty < \theta_1 < \min\{m, m_0\} \equiv m_1 < M_1 \equiv \max\{M, M_0\} < \theta_2 < \infty$ [why?], and $f^e(\mathbf{x}, \boldsymbol{\theta}) = 0$ otherwise. Hence (8.3.2′) implies

$$\begin{aligned} f^e(\mathbf{x}) &= \int_{M_1}^{\infty} \int_{-\infty}^{m_1} \frac{(\nu_0 - 1)(\nu_0 - 2)}{(M_0 - m_0)^{2-\nu_0}} (\theta_2 - \theta_1)^{-(\nu_0+n)} d\theta_1 d\theta_2 \qquad (8.7.5) \\ &= \frac{(\nu_0 - 1)(\nu_0 - 2)}{(\nu_1 - 1)(\nu_1 - 2)} \frac{(M_1 - m_1)^{2-\nu_1}}{(M_0 - m_0)^{2-\nu_0}}, \end{aligned}$$

where $m_1 = \min \{m_0, x_1, \ldots, x_n\}$, $\nu_1 \equiv \nu + n$, and $M_1 = \max \{M_0, x_1, \ldots, x_n\}$. Note that this formula for $f^e(\mathbf{x})$ is valid for all $\mathbf{x} \in R^n$. Now, knowing $f^e(\mathbf{x}, \boldsymbol{\theta})$ and $f^e(\mathbf{x})$, we may use (8.4.7′) to deduce that

$$f^e(\boldsymbol{\theta} \mid x_1, \ldots, x_n) = \begin{cases} \dfrac{(\nu_1 - 1)(\nu_1 - 2)}{(M_1 - m_1)^{2-\nu_1}}(\theta_2 - \theta_1)^{-\nu_1}, & -\infty < \theta_1 < m_1 < M_1 < \theta_2 \\ 0, & \text{elsewhere.} \end{cases} \tag{8.7.6}$$

This example will be referred to at several points in the sequel, as it illustrates several important features of the so-called Bayesian approach to inference. *One* feature, discussed in Chapter 9, is the fact that $f^e(\boldsymbol{\theta} \mid x_1, \ldots x_n)$ depends upon $x_1, \ldots, x_n$ only through three numbers, $m \equiv \min \{x_1, \ldots, x_n\}$, $M \equiv \max \{x_1, \ldots, x_n\}$, and n. *Another* feature is that the posterior dgf $f^e(\boldsymbol{\theta} \mid x_1, \ldots, x_n)$ is of exactly the same mathematical form as the prior dgf $f(\boldsymbol{\theta})$, but with parameters ν_1, m_1, and M_1 instead of ν_0, m_0 and M_0. This feature will be discussed in Chapter 14.

EXERCISES

1. Let $Q = \{a, b\}$, $P(a) = .25$, and $P(b) = .75$. Let $w = 2$.

   (*a*) Let $\tilde{\mathbf{x}}(a) = (1, 1)'$ and $\tilde{\mathbf{x}}(b) = (2, 3)'$. Find the cdf $F(x_1, x_2)$ of $\tilde{\mathbf{x}} = (\tilde{\mathbf{x}}_1, \tilde{\mathbf{x}}_2)'$ and draw the figure analogous to Figure 8.1.
   (b) Let $\tilde{\mathbf{x}}(a) = (1, 1)'$ and $\tilde{\mathbf{x}}(b) = (-1, 2).'$ Find the cdf $F(x_1, x_2)$ of $\tilde{\mathbf{x}} = (\tilde{\mathbf{x}}_1, \tilde{x}_2)'$ and draw the figure analogous to Figure 8.1.

2. Find the mf $f(x_1, x_2)$ of $(\tilde{x}_1, \tilde{x}_2),'$ where $(\tilde{x}_1, \tilde{x}_2)' \equiv \tilde{\mathbf{x}}$ is as defined in

   (a) part (a) of Exercise 1; and
   (b) part (b) of Exercise 1.

3. Let $(\tilde{x}_1, \tilde{x}_2)'$ have joint mf $f(x_1, x_2)$ as defined in Example 8.2.5.

   (a) Find the marginal mf's $f(x_1)$ and $f(x_2)$.
   (b) Find the conditional mf's $f(x_1 \mid x_2)$ and $f(x_2 \mid x_1)$. Be sure to state precisely the domains of these functions.
   (c) Are $\tilde{x}_1$ and $\tilde{x}_2$ independent?

4. Let $f(x_1, x_2, x_3)$ be defined by

$$f(x_1, x_2, x_3) = \begin{cases} 0, & \text{some } x_i \leq 0 \\ (2\lambda t)^3 x_1 x_2 x_3 e^{-\lambda t(x_1^2 + x_2^2 + x_3^2)}, & \\ & \text{all } x_i > 0. \end{cases}$$

   Is f the density function of some three-dimensional random vector? If so, find $F(x_1, x_2, x_3)$, $f(x_1)$, $f(x_2)$, $f(x_3)$, and $f(x_2, x_3)$.

5. Define $f(x_1, x_2)$ by

$$f(x_1, x_2) \equiv \begin{cases} \lambda_1\lambda_2 e^{-(\lambda_1-\lambda_2)x_1}e^{-\lambda_2 x_2}, & 0 < x_1 < x_2 \\ 0, & \text{elsewhere.} \end{cases}$$

Is f the density function of some two-dimensional random vector? If so, find $F(x_1, x_2)$, $f(x_1)$, $f(x_2)$, $f(x_1, x_2)$, $f(x_2 \mid x_1)$, and $F(x_1 \mid x_2)$.

6. Let $\tilde{x}_1$, $\tilde{x}_2$ be jointly continuous with density function $f(x_1, x_2)$. Show that

$$F(x_1, x_2) = \int_{-\infty}^{x_2} F(x_1 \mid t_2) f(t_2) dt_2. \tag{8.8.1}$$

[*Hint*: Replace the range $(-\infty, x_2]$ of integration by $\{t_2 : f(t_2) > 0, t_2 \le x_2\} \equiv A$. The value of the integral is unchanged, and

$$\int_A \int_{-\infty}^{x_1} f(t_1, t_2) dt_1 dt_2 = \int_A \int_{-\infty}^{x_1} f(t_2) \cdot \frac{f(t_1, t_2)}{f(t_2)} dt_1 dt_2.]$$

7. Let $f(\theta_1, \theta_2)$ be the prior density function defined in Example 8.7.1.

(a) Find the marginal prior density functions $f(\theta_1)$ of $\tilde{\theta}_1$ and $f(\theta_2)$ of $\tilde{\theta}_2$.
(b) Find the marginal *posterior* density functions $f^e(\theta_1 \mid x_1, \ldots, x_n)$ and $f^e(\theta_2 \mid x_1, \ldots, x_n)$ of $\tilde{\theta}_1$ and $\tilde{\theta}_2$, respectively.
(c) Does $f^e(\theta_1 \mid x_1, \ldots, x_n)$ depend upon $x_1, \ldots, x_n$ only through m? Through M? Through both m and M? Same question for $f^e(\theta_2 \mid x_1, \ldots, x_n)$.

8. (Continuation of Exercise 7.) Suppose $\nu_0 = 3$, $m_0 = 1$, and $M_0 = 4$. Further, suppose e resulted in $x_1 = 2$, $x_2 = 1$, $x_3 = 6$, $x_4 = 4$, $x_5 = \frac{1}{2}$, $x_6 = 7$, and $x_7 = 5$.

(a) Find $P[\tilde{\theta}_2 \in (0, 8)]$; this is the decision-maker's *prior* probability that $\tilde{\theta}_2 \in (0, 8)$.
(b) Find $p^e[\tilde{\theta}_2 \in (0, 8) \mid x_1, \ldots, x_7]$; this is the decision-maker's *posterior* probability that $\tilde{\theta}_2 \in (0, 8)$.
(c) Graph $f(\theta_2)$ and $f^e(\theta_2 \mid x_1, \ldots, x_7)$ for the given values of $x_1, \ldots, x_7$. How has the result of the experiment altered the decision-maker's judgments?
(d) Repeat parts (a) through (c) for $\tilde{\theta}_1$ replacing $\tilde{\theta}_2$, and (0, 3) replacing (0, 8).

9. *Fuctions of Random Vectors*: Let $\tilde{x}_1, \ldots, \tilde{x}_w$ be jointly continuous random variables with joint cdf F_1 and joint dgf f_1. Let $h: R^w \to R^w$ be a differentiable injection, and let $\tilde{y}_1 \equiv h_i(\tilde{x}_1, \ldots, \tilde{x}_w)$. Then $\tilde{y}_1, \ldots, \tilde{y}_w$ are jointly continuous random variables with a joint cdf F_2 and joint dgf f_2. Show that

$$f_2(\mathbf{y}) = f_1(h^{-1}(\mathbf{y})) \mid J_{h^{-1}}(\mathbf{y}) \mid; \text{ and} \tag{8.8.2}$$

$$f_1(\mathbf{x}) = f_2(h(\mathbf{x})) \mid J_h(\mathbf{x}) \mid, \tag{8.8.3}$$

where $|J_{h^{-1}}(\mathbf{y})|$ is the absolute value of the Jacobian of h^{-1} evaluated at $\mathbf{y}$ and $|J_h(\mathbf{x})|$ is the absolute value of the Jacobian of h evaluated at $\mathbf{x}$.
[*Hint*: These are standard change-of-variable results.]

10. *Sums of Random Variables:* Let $\tilde{x}_1$ and $\tilde{x}_2$ be random variables; let $F_{1|2}(\cdot \mid \cdot)$ denote the conditional cdf of $\tilde{x}_1$ given $\tilde{x}_2$, and let f_2 denote the dgf of $\tilde{x}_2$. Let $\tilde{x}_3 \equiv \tilde{x}_1 + \tilde{x}_2$; that is, $\tilde{x}_3(q) = \tilde{x}_1(q) + \tilde{x}_2(q)$ for every $q \in Q$; and let $F_3(\cdot)$ and $f_3(\cdot)$ denote respectively the cdf and dgf of $\tilde{x}_3$.

(a) Show that if $\tilde{x}_2$ is discrete, then

$$F_3(x_3) = \sum \{F_{1|2}(x_3 - x_2 \mid x_2)\cdot f_2(x_2): \quad \text{all } x_2 \in (-\infty, \infty)\};$$

and if $\tilde{x}_2$ is continuous, by

$$F_3(x_3) = \int_{-\infty}^{\infty} F_{1|2}(x_3 - x_2 \mid x_2)\cdot f_2(x_2)dx_2.$$

[*Hint*: In the discrete case, what is $F_{1|2}(x_3 - x_2 \mid x_2)\cdot f_2(x_2)$in probability terms?]

(b) Show that the dgf f_3 of $\tilde{x}_3$ is given by

$$f_3(x_3) = \sum \{f_{1|2}(x_3 - x_2 \mid x_2)\cdot f_2(x_2): \quad \text{all } x_2\}$$

if $\tilde{x}_2$ is discrete, and by

$$f_3(x_3) = \int_{-\infty}^{\infty} f_{1|2}(x_3 - x_2 \mid x_2)\cdot f_2(x_2)dx_2$$

if $\tilde{x}_2$ is continuous.

11. Let $\tilde{x}_1$, $\tilde{x}_2$ have joint cdf $F(x_1, x_2)$. Show that

$$P[(\tilde{x}_1, \tilde{x}_2) \in (a_1, b_1] \times (a_2, b_2]] = F(b_1, b_2) - F(a_1, b_2) - F(b_1, a_2) + F(a_1, a_2).$$

Hint:

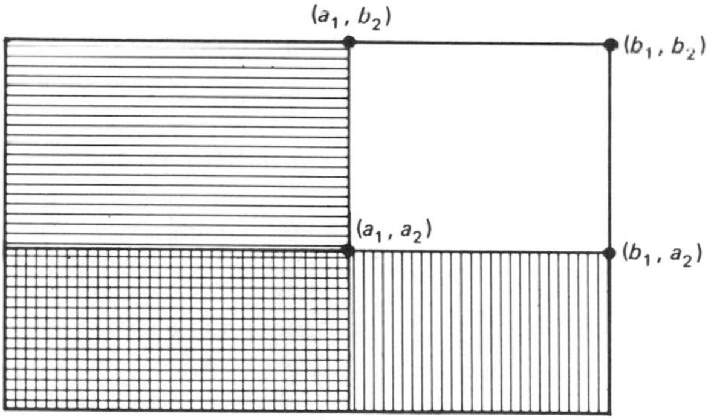

12. Verify that $F(x_1, x_2)$ as derived in Example 8.2.4 possesses properties (cdf 1) through (cdf 5).

13. Verify that $f(x_1, \ldots, x_m)$ as defined by (8.3.2) possesses properties (dgf 1) through (dgf 3).

14. Derive $f(x_2)$, $f(x_1 \mid x_2 = 0)$, and $f(x_1 \mid x_2 = 1)$ in Example 8.4.1.

15. Show that (8.6.1) implies (8.6.3).

16. Derive (8.6.4).

17. Show that $f(\theta_1, \theta_2)$ as defined by (8.7.4) is a density function [that is, satisfies (dgf 1) through (dgf 3)].

18. Derive the *chain rule* for random vectors:

$$f(\mathbf{x}_1, \ldots, \mathbf{x}_w) = f(\mathbf{x}_1)\cdot\prod_{i=1}^{w-1} f(\mathbf{x}_{i+1} \mid \mathbf{x}_1, \ldots, \mathbf{x}_i), \tag{8.4.8'}$$

whenever $f(\mathbf{x}_1, \ldots, \mathbf{x}_w) > 0$.

19. *Distribution of the Maximum*: Let $\tilde{x}_1, \ldots, \tilde{x}_n$ be iid with common cdf $F(x)$, and define $\tilde{M} \equiv \max\{\tilde{x}_1, \ldots, \tilde{x}_n\}$; that is, $\tilde{M}(q) \equiv \max\{\tilde{x}_1(q), \ldots, \tilde{x}_n(q)\}$ for every $q \in Q$. Show that the cdf $F_M(M)$ of $\tilde{M}$ is given by

$$F_M(M) = [F(M)]^n, \quad -\infty < M < \infty; \tag{8.8.4}$$

that is,

$$P[\tilde{M} \leq M] = \{P[\tilde{x} \leq M\}^n.$$

20. *Distribution of the Minimum*: Let $\tilde{x}_1, \ldots, \tilde{x}_n$ be iid with common cdf $F(x)$, and define $\tilde{m} \equiv \min\{\tilde{x}_1, \ldots, \tilde{x}_n\}$; that is $\tilde{m}(q) \equiv \min\{\tilde{x}_1(q), \ldots, \tilde{x}_n(q)\}$, for every $q \in Q$. Show that the cdf $F_m(m)$ of $\tilde{m}$ given by

$$F_m(m) = 1 - [1 - F(m)]^n; \tag{8.8.5}$$

that is,

$$P[\tilde{m} \leq m] = 1 - [1 - F(m)]^n, \text{ or}$$
$$P[\tilde{m} > m] = \{P[\tilde{x} > m\}]^n.$$

21. *Generalization of Exercises 19 and 20*: Let $\tilde{x}_1, \ldots, \tilde{x}_n$ have joint cdf $F(x_1, \ldots, x_n)$, and define $\tilde{m}$ as in Exercise 20 and $\tilde{M}$ as in 19.

(a) Show that the cdf $F_M(M)$ of $\tilde{M}$ is given by

$$F_M(M) = F(M, M, \ldots, M) \tag{8.8.6}$$

and that

$$G_m(m) = G(m, m, \ldots, m), \tag{8.8.6'}$$

where $G_m(m) \equiv P[\tilde{m} > m]$ and $G(x_1, \ldots, x_n) \equiv P[\tilde{x}_1 > x_1, \ldots, \tilde{x}_n > x_n]$.

22. *A Discrete Example*: Let $\tilde{x}_1$ have mf $f(x_1)$ defined by

$$f(x_1) = \begin{cases} .10, & x_1 = 0 \\ .15, & x_1 = 1 \\ .20, & x_1 = 2 \\ .55, & x_1 = 3 \\ 0, & x_1 \notin \{0, 1, 2, 3\}. \end{cases}$$

Given $x_1 \in \{0, \ldots, 3\}$ let $\tilde{x}_2$ have mf $f(x_2 \mid x_1)$ (defined by the following table):

x_2 \ x_1	0	1	2	3
0	.25	.75	.10	.10
1	.25	.10	.10	.10
2	.25	.10	.05	.75
3	.25	.05	.75	.05
sum	1.00	1.00	1.00	1.00

(a) Find $f(x_1 \mid x_2 = 0)$, $f(x_1 \mid x_2 = 1)$, $f(x_1 \mid x_2 = 2)$, and $f(x_1 \mid x_2 = 3)$ by Bayes' Theorem for $x_1 \in \{0, 1, 2, 3\}$.

(b) Find $f(x_2)$ for all x_2.

23. *Summary on Joint Distributions of Random Variables*: (Also exemplifying a fairly common subscripting convention). Let $\tilde{x}_1, \ldots, \tilde{x}_8$ be continuous random variables and suppose that you are given the dgf $f_{123}\colon R^3 \to R^1$ of $(\tilde{x}_1, \tilde{x}_2, \tilde{x}_3)$ and also the conditional dgf $f_{45678|123}(\cdot, \ldots, \cdot \mid x_1, x_2, x_3)\colon R^5 \to R^1$ of $(\tilde{x}_4, \tilde{x}_5, \tilde{x}_6, \tilde{x}_7, \tilde{x}_8)$ given every (x_1, x_2, x_3) such that $f_{123}(x_1, x_2, x_3) > 0$

(a) Show that

$$f_{1\ldots8}(x_1, \ldots, x_8) = \begin{cases} 0, & f_{123}(x_1, x_2, x_3) = 0 \\ f_{4\ldots8|123}(x_4, \ldots, x_8 \mid x_1, x_2, x_3) f_{123}(x_1, x_2, x_3), & \\ & \text{otherwise.} \end{cases}$$

Show that

$$f_{78}(x_7, x_8) = \int_{-\infty}^{\infty} \cdots \int_{-\infty}^{\infty} f_{1\ldots8}(t_1, \ldots, t_6, x_7, x_8)\,dt_1 \ldots dt_6.$$

(c) Show that

$$f_{1278}(x_1, x_2, x_7, x_8\} = \int_{-\infty}^{\infty} \cdots \int_{-\infty}^{\infty} f_{1\ldots8}(x_1, x_2, t_3, \ldots . t_6, x_7, x_8)\,dt_3 \ldots dt_6.$$

(d) Show that

$$f_{12|78}(x_1, x_2 \mid x_7, x_8) = f_{1278}(x_1, x_2, x_7, x_8)/f_{78}(x_7, x_8),$$

wherever $f_{78}(x_7, x_8) > 0$.

(e) Show that

$$f_{1\ldots6|78}(x_1, \ldots, x_6 \mid x_7, x_8) = f_{1\ldots8}(x_1, \ldots, x_8)/f_{78}(x_7, x_8),$$

wherever $f_{78}(x_7, x_8) > 0$.

(f) Show that

$$f_{12|78}(x_1, x_2 \mid x_7, x_8)$$
$$= \int_{-\infty}^{\infty} \cdots \int_{-\infty}^{\infty} f_{1\ldots6|78}(x_1, x_2, t_3, \ldots, t_6 \mid x_7, x_8)\,dt_3 \ldots dt_6,$$

wherever $f_{78}(x_7, x_8\} > 0$.

9

AN INTRODUCTION TO STATISTICAL INFERENCE

9.1 INTRODUCTION

We now have at our disposal sufficient fundamentals to introduce some ingredients of statistical inference. The purposes of this chapter are, first, to furnish a useful and precise definition of *statistical experiment;* and second, to sketch two approaches to statistical inference.

Section 9.2 begins by abstracting the notion of "random sampling of a population." One assumption as to the means of selecting individuals for observation leads to independent trials, while another assumption does not.

Not all experimental situations consist of random sampling from populations, at least not from "obvious" populations. Hence we generalize in Section 9.3 by defining a statistical experiment as consisting of two main ingredients; first, a *data-generating process,* which characterizes the observations and their joint distributions; and second, a *stopping process,* which determines the number of observations in the experiment.

Section 9.4 then presents a brief sketch of two principal approaches to statistical inference; namely, Bayesian inference and orthodox inference, which correspond to the incorporation and nonincorporation respectively of prior judgments about the unknowns of interest.

Sections 9.5 and 9.6 deal with two avenues of simplification in the description and analysis of a statistical experiment. Specifically, Section 9.5, on sufficient statistics, concerns ways of distilling the germane information in the experiment without losing any; while Section 9.6, on *noninformative stopping,* gives conditions under which an experiment can be regarded for inference purposes *as if* the number n of observations had been predetermined, even when such is not the case.

9.2 RANDOM SAMPLING FROM A POPULATION

Suppose that you wish to obtain information on some attribute of a population Q; that you can measure k numerical characteristics $x_1(q)$, . . . , $x_k(q)$ of any individual member $q \in Q$; and that you can perform an experiment, each trial of which consists in selecting an individual "at random" (see Section 5.4) from among those eligible for selection. Each trial, or observation, i then consists of the k-tuple $\mathbf{x}_i \equiv (x_{1i}, \ldots, x_{ki})$ of characteristics of the individual selected on trial i. Before performing trial i, $\mathbf{x}_i$ is unknown (else why perform trial i?) and is thus a random vector, $\tilde{\mathbf{x}}_i$, with a distribution that is determined by the method of selecting individuals for observation.

We shall consider two methods. The first is called *random sampling with replacement*: on any trial i each individual q is *equally as likely* to be selected for observation as is any other individual q' *regardless* of whether q and/or q' have been selected on a previous trial. Random sampling with replacement implies clearly that $\tilde{\mathbf{x}}_1, \tilde{\mathbf{x}}_2, \ldots$ are iid random vectors with a common distribution represented by the dgf (a mass function)

$$P(\tilde{\mathbf{x}}_i = \mathbf{x}) \equiv f(\mathbf{x}) = \frac{\#(\{q: \tilde{\mathbf{x}}(q) = \mathbf{x}\})}{\#(Q)},$$

where (as in Chapter 5) $\#(A)$ denotes the number of individuals in A. The dgf $f(\cdot)$ defined by (9.2.1) is called the *population* dgf. We may think of the sample as being taken in order to obtain information about $f(\cdot)$; for example, about that $\mathbf{x}$ which maximizes $f(\cdot)$—the "most likely $\mathbf{x}$"—or about the .99 fractile of the first component $\tilde{x}_1$ of $\tilde{x}$.

Random sampling with replacement is such a common experimental situation that it has given rise to a synonym for "$\tilde{\mathbf{x}}_1, \ldots, \tilde{\mathbf{x}}_n$ are iid with common dgf $f(\mathbf{x})$"; namely, "$\tilde{\mathbf{x}}_1, \ldots, \tilde{\mathbf{x}}_n$ is a random sample from $f(\mathbf{x})$." (The words "with replacement" are to be understood.)

The second method of selection that we shall consider is *random sampling without replacement*: an individual q is eligible for selection on trial i if and only if he has not been selected on any previous trial; and on trial i each *eligible* individual is equally as likely to be selected as is any other *eligible* individual. Hence on trial i there are $\#(Q)-(i-1)$ eligibles, and each eligible individual has probability $1/[\#(Q)-(i-1)]$ of being selected. Since the probability that q will be selected on trial i depends both upon the trial number i and upon the outcomes of trials of the first $i-1$ trials, it is clear that $\tilde{\mathbf{x}}_1, \tilde{\mathbf{x}}_2$ are *not* independent. Hence we shall wish to consider experiments the trials 1, 2, . . . which yield observations $\tilde{\mathbf{x}}_1, \tilde{\mathbf{x}}_2, \ldots$ which are *not* iid.

Thus we shall not want to rule out dependence of the observations $\tilde{\mathbf{x}}_1, \tilde{\mathbf{x}}_2, \ldots$ in our general definition of an experiment.

Sometimes the set Q of elementary events is not at all obvious. For example, your sample might consist of making repeated weighings of this book on a rusty scale. Clearly, $k = 1$ and the observation x_i are the readings on the weighings i. The state, about which you desire information, is the true weight θ of this book. Moreover, $\tilde{x}_i$ will have a dgf $f(x_i \mid x_1 \ldots, x_{i-1}, \theta)$ given the previous weighings, the trial number, and the true weight. Now, the elementary events could be defined in various ways, but the results of Chapters 7 and 8 imply that Q *need not* be specified once $f(x_i \mid x_1, \ldots, x_{i-1}, \theta)$ has been specified for all possible values of $x_1, \ldots, x_{i-1}, \theta$.

Hence an experiment is described in large part by the sequence (finite or infinite) of dgf's $f(\mathbf{x}_i \mid \mathbf{x}_1, \ldots, \mathbf{x}_{i-1}, \theta)$, for i = 1, 2, ... of $\tilde{\mathbf{x}}_i$g iven all possible $\mathbf{x}_1, \ldots, \mathbf{x}_{i-1}, \boldsymbol{\theta}$. [When $i = 1$, the dgf is $f(\mathbf{x}_1 \mid \boldsymbol{\theta})$.]

The remaining part of the description of an experiment consists in the specification of the total number n of observations. In most experiments the total number of observations is predetermined, but in others it is not. For example, the experimentation may continue until nightfall, until Congress cuts the appropriation, or until luncheon is announced. In each of these cases, the total number of observations is not known beforehand for certain. Hence we shall want to describe for each i the probability that observation $i + 1$ will be made, given the state vector $\boldsymbol{\theta}$ and the previous observations, and given some (additional) vector $\boldsymbol{\theta}_\alpha$.

Sometimes the state vector $\boldsymbol{\theta}$ is not specified, and experimentation is performed in order to obtain information about the arbitrary common dgf $f(\mathbf{x})$ of n iid observations $\tilde{\mathbf{x}}_1, \ldots, \tilde{\mathbf{x}}_n$. These inference problems are called *nonparametric* and will be introduced in Chapter 22.

9.3 DATA-GENERATING AND STOPPING PROCESSES

We refer to the *data-generating process* as the producer of a sequence (finite or infinite) of random vectors $\tilde{\mathbf{x}}_1, \tilde{\mathbf{x}}_2, \ldots$, denoting possible observations of some statistical experiment. The data-generating process is characterized by the sequence $f(\mathbf{x}_1 \mid \boldsymbol{\theta}), f(\mathbf{x}_2 \mid \mathbf{x}_1, \boldsymbol{\theta}), \ldots$ of conditional dgf's given the outcomes of previous trials and given the state vector $\boldsymbol{\theta}$. We interpret $f(\mathbf{x}_i \mid \mathbf{x}_1, \ldots, \mathbf{x}_{i-1}, \boldsymbol{\theta})$ as the dgf of $\tilde{\mathbf{x}}_i$ given $\mathbf{x}_1, \ldots, \mathbf{x}_{i-1}, \boldsymbol{\theta}$, and given that observation i is made in the experiment.

Example 9.3.1

We say that the data-generating process yields iid observations $\tilde{\mathbf{x}}_1, \tilde{\mathbf{x}}_2, \ldots$ with common dgf $f(x \mid \boldsymbol{\theta}) \equiv f(x \mid \theta_1, \theta_2)$ defined by

$$f(x \mid \theta_1, \theta_2) = \begin{cases} (\theta_2 - \theta_1)^{-1}, & \theta_1 < x < \theta_2 \\ 0, & x \notin (\theta_1, \theta_2), \end{cases}$$

for some (θ_1, θ_2) such that $\theta_1 < \theta_2$. If we now predetermine n and take n observations, we have Example 8.7.1.

From the sequence $\{f(\mathbf{x}_i \mid \mathbf{x}_1, \ldots, \mathbf{x}_{i-1}, \boldsymbol{\theta}): i = 1, 2, \ldots\}$ we may compute from the chain rule (8.4.8′) the joint dgf of the first i observations:

$$f(\mathbf{x}_1, \ldots, \mathbf{x}_i \mid \boldsymbol{\theta}) = f(\mathbf{x}_1 \mid \boldsymbol{\theta}) \cdot \prod_{j=1}^{i-1} f(\mathbf{x}_{j+1} \mid \mathbf{x}_1, \ldots, \mathbf{x}_j, \boldsymbol{\theta}). \tag{9.3.1}$$

Now, a statistical experiment is described only in part by the data-generating process. The remainder of its description is called the *stopping process*: and it may be thought of as determining the total number n of observations. The stopping process is characterized by a sequence (finite or infinite) of *stopping probabilities* $\pi(0 \mid \boldsymbol{\theta}, \boldsymbol{\theta}_\alpha)$,

$$\begin{aligned}&\pi(1 \mid \mathbf{x}_1, \boldsymbol{\theta}, \boldsymbol{\theta}_\alpha), \pi(2 \mid \mathbf{x}_1, \mathbf{x}_2, \boldsymbol{\theta}, \boldsymbol{\theta}_\alpha), \ldots,\\ &\pi(i \mid \mathbf{x}_1, \ldots, \mathbf{x}_i, \boldsymbol{\theta}, \boldsymbol{\theta}_\alpha), \ldots, \text{ where}\\ &\pi(i \mid \mathbf{x}_1, \ldots, \mathbf{x}_i, \boldsymbol{\theta}, \boldsymbol{\theta}_\alpha)\end{aligned}$$

is the probability that the experiment terminates immediately after observation i; given that i observations were made, given their values $\mathbf{x}_1, \ldots, \mathbf{x}_i$, given $\boldsymbol{\theta}$ (the state vector of the data-generating process), and given a vector $\boldsymbol{\theta}_\alpha$ of additional parameters. The term $\pi(0 \mid \boldsymbol{\theta}, \boldsymbol{\theta}_\alpha)$ gives the probability that no observations will be made.

Example 9.3.2

Let $\pi(0 \mid \boldsymbol{\theta}, \boldsymbol{\theta}_\alpha) = 0$ and

$$\pi(i \mid \mathbf{x}_1, \ldots, \mathbf{x}_i, \boldsymbol{\theta}, \boldsymbol{\theta}_\alpha) = \begin{cases} 0, & 1 \leqq i < n \\ 1, & i \geqq n, \end{cases}$$

for every $(\mathbf{x}_1, \ldots, \mathbf{x}_i, \boldsymbol{\theta}, \boldsymbol{\theta}_\alpha)$. This stopping process is tantamount to predetermining that the experiment will consist of precisely n observations.

Example 9.3.3

Suppose that you have a deformed coin which lands "heads" with probability $\theta_\alpha \in (0, 1)$ (not necessarily $= \frac{1}{2}$) on any given flip. Also suppose that you will not take the first observation from a given data-generating process if you flip the coin and it lands heads; and that you will stop after the ith observation if you flip the coin and it lands heads. Then the stopping probabilities are given by

$$\pi(0 \mid \boldsymbol{\theta}, \theta_\alpha) = \theta_\alpha = \pi(i \mid \mathbf{x}_1, \ldots, \mathbf{x}_i, \boldsymbol{\theta}, \theta_\alpha),$$

for all i and all $\mathbf{x}_1, \ldots, \mathbf{x}_i, \boldsymbol{\theta}$.

Once we have the sequence of stopping probabilities, we may compute the probability $S(i \mid \mathbf{x}_1, \ldots, \mathbf{x}_i, \boldsymbol{\theta}, \boldsymbol{\theta}_\alpha)$ that an experiment with that stopping process will consist of *precisely* i observations, given $\mathbf{x}_1, \ldots, \mathbf{x}_i$, $\boldsymbol{\theta}$ and $\boldsymbol{\theta}_\alpha$; we have

$$S(0 \mid \boldsymbol{\theta}, \boldsymbol{\theta}_\alpha) = \pi(0 \mid \boldsymbol{\theta}, \boldsymbol{\theta}_\alpha);$$
$$S(1 \mid \mathbf{x}_1, \boldsymbol{\theta}, \boldsymbol{\theta}_\alpha) = \pi(1 \mid \mathbf{x}_1, \boldsymbol{\theta}, \boldsymbol{\theta}_\alpha)\cdot[1 - \pi(0 \mid \boldsymbol{\theta}, \boldsymbol{\theta}_\alpha)];$$
$$S(2 \mid \mathbf{x}_1, \mathbf{x}_2, \boldsymbol{\theta}, \boldsymbol{\theta}_\alpha) = \pi(2 \mid \mathbf{x}_1, \mathbf{x}_2, \boldsymbol{\theta}, \boldsymbol{\theta}_\alpha)\cdot[1 - \pi(0 \mid \boldsymbol{\theta}, \boldsymbol{\theta}_\alpha)][1 - \pi(1 \mid \mathbf{x}_1, \boldsymbol{\theta}, \boldsymbol{\theta}_\alpha)];$$

and, in general,

$$S(i \mid \mathbf{x}_1, \ldots, \mathbf{x}_i, \boldsymbol{\theta}, \boldsymbol{\theta}_\alpha) = \pi(i \mid \mathbf{x}_1, \ldots, \mathbf{x}_i, \boldsymbol{\theta}, \boldsymbol{\theta}_\alpha)\cdot \prod_{j=0}^{i-1}[1 - \pi(j \mid \mathbf{x}_1, \ldots, \mathbf{x}_j, \boldsymbol{\theta}, \boldsymbol{\theta}_\alpha)], \tag{9.3.2}$$

where $\pi(0 \mid \mathbf{x}_1, \mathbf{x}_0, \boldsymbol{\theta}, \boldsymbol{\theta}_\alpha) \equiv \pi(0 \mid \boldsymbol{\theta}, \boldsymbol{\theta}_\alpha)$.

Example 9.3.4

For the stopping process defined in Example 9.3.3, Equation (9.3.2) implies

$$S(i \mid \mathbf{x}_1, \ldots, \mathbf{x}_i, \boldsymbol{\theta}, \boldsymbol{\theta}_\alpha) = \theta_\alpha(1 - \theta_\alpha)^i,$$

for all $i \in \{0, 1, 2, \ldots\}$, every $\mathbf{x}_1, \mathbf{x}_2, \ldots, \mathbf{x}_i, \boldsymbol{\theta}$, and every possible $\theta_\alpha \in (0, 1)$.

Now suppose that we have completely specified a statistical experiment by giving its data-generating process and its stopping process. We may compute the joint dgf of $(\tilde{n}, \tilde{\mathbf{x}}_1, \ldots, \tilde{\mathbf{x}}_n)$ given $\boldsymbol{\theta}$ and $\boldsymbol{\theta}_\alpha$ readily, from (9.3.1) and (9.3.2):

$$f(n, \mathbf{x}_1, \ldots, \mathbf{x}_n \mid \boldsymbol{\theta}, \boldsymbol{\theta}_\alpha) = S(n \mid \mathbf{x}_1, \ldots, \mathbf{x}_n, \boldsymbol{\theta}, \boldsymbol{\theta}_\alpha)\cdot f(\mathbf{x}_1, \ldots, \mathbf{x}_n \mid \boldsymbol{\theta}). \tag{9.3.3}$$

Proof

$f(\mathbf{x}_1, \ldots, \mathbf{x}_n \mid \boldsymbol{\theta}) = f(\mathbf{x}_1, \ldots, \mathbf{x}_n \mid \boldsymbol{\theta}, \boldsymbol{\theta}_\alpha)$, for all $\boldsymbol{\theta}_\alpha$ and all $\mathbf{x}_1, \ldots, \mathbf{x}_n, \boldsymbol{\theta}$ because $\boldsymbol{\theta}_\alpha$ is a vector of additional parameters of the stopping process. Hence (9.3.3) is of the form $f(n, \mathbf{x} \mid \boldsymbol{\theta}, \boldsymbol{\theta}_\alpha) = f(n \mid \mathbf{x}, \boldsymbol{\theta}, \boldsymbol{\theta}_\alpha)\cdot f(\mathbf{x} \mid \boldsymbol{\theta}, \boldsymbol{\theta}_\alpha)$. QED

Example 9.3.5

We combine Examples 9.3.1 and 9.3.4 to obtain, via (8.7.3),

$$f(n, x_1, \ldots, x_n \mid \theta_1, \theta_2, \theta_\alpha) = \theta_\alpha(1 - \theta_\alpha)^n f(x_1, \ldots, x_n \mid \theta_1, \theta_2)$$
$$= \begin{cases} \theta_\alpha(1 - \theta_\alpha)^n(\theta_2 - \theta_1)^{-n}, & \begin{cases}\theta_1 < m < M < \theta_2 \\ n \in \{0, 1, 2, \ldots\}\end{cases} \\ 0, & \text{elsewhere,} \end{cases}$$

where $m \equiv \min\{x_1, \ldots, x_n\}$ and $M \equiv \max\{x_1, \ldots, x_n\}$.

Example 9.3.6

For the data-generating process in Example 9.3.1, we define a stopping process by $\pi(0 \mid \theta_1, \theta_2, \theta_\alpha) = 0$ and

$$\pi(i \mid x_1, \ldots, x_i, \theta_1, \theta_2, \theta_\alpha) = \begin{cases} \left.\begin{matrix} 0, & i < 5 \\ 1, & i \geqq 5 \end{matrix}\right\} & \theta_1 > 1 \\ \left.\begin{matrix} 0, & i < 10 \\ 1, & i \geqq 10 \end{matrix}\right\} & \theta_1 \leqq 1, \end{cases}$$

for all $x_1, \ldots, x_i, \theta_2, \theta_\alpha$ (an "empty vector") and all θ_1. Intuitively, this stopping process yields five observations if $\theta_1 > 1$ and ten if $\theta_1 \leqq 1$; and intuition is corroborated by (9.3.2), which yields

$$S(i \mid x_1, \ldots, x_i, \theta_1, \theta_2, \theta_\alpha) = \begin{cases} \left.\begin{matrix} 1, & i = 5 \\ 0, & i \neq 5 \end{matrix}\right\} & \theta_1 > 1 \\ \left.\begin{matrix} 1, & i = 10 \\ 0, & i \neq 10 \end{matrix}\right\} & \theta_1 \leqq 1. \end{cases}$$

For the data-generating process in Example 9.3.1 we therefore obtain

$$f(n, x_1, \ldots, x_n \mid \theta_1, \theta_2, \theta_\alpha) = \begin{cases} (\theta_2 - \theta_1)^{-5}, & \begin{cases} 1 < \theta_1 < m < M < \theta_2 \\ n = 5 \end{cases} \\ (\theta_2 - \theta_1)^{-10}, & \begin{cases} \theta_1 \geqq 1, \theta_1 < m < M < \theta_2 \\ n = 10 \end{cases} \\ 0, & \text{elsewhere.} \end{cases}$$

Now suppose that you perform a statistical experiment e and that the outcome is $(n, \mathbf{x}_1, \ldots, \mathbf{x}_n)$. A crucial building block for making inferences about, and decisions affected by $(\boldsymbol{\theta}, \boldsymbol{\theta}_\alpha)$ is $f(n, \mathbf{x}_1, \ldots, \mathbf{x}_n \mid \boldsymbol{\theta}, \boldsymbol{\theta}_\alpha)$, *considered as a function of the possible values* $(\boldsymbol{\theta}, \boldsymbol{\theta}_\alpha)$, with parameters $(n, \mathbf{x}_1, \ldots, \mathbf{x}_n)$, instead of as a function of $(n, \mathbf{x}_1, \ldots, \mathbf{x}_n)$ with parameters $(\boldsymbol{\theta}, \boldsymbol{\theta}_\alpha)$. When $f(n, \mathbf{x}_1, \ldots, \mathbf{x}_n \mid \boldsymbol{\theta}, \boldsymbol{\theta}_\alpha)$ is considered a function of $(\boldsymbol{\theta}, \boldsymbol{\theta}_\alpha)$ with parameters $(n, \mathbf{x}_1, \ldots, \mathbf{x}_n)$, we call it the *likelihood function;* and we define

$$L(\boldsymbol{\theta}, \boldsymbol{\theta}_\alpha \mid n, \mathbf{x}_1, \ldots, \mathbf{x}_n) = f(n, \mathbf{x}_1, \ldots, \mathbf{x}_n \mid \boldsymbol{\theta}, \boldsymbol{\theta}_\alpha) \qquad (9.3.4)$$

whenever the right-hand side is defined. The only reason for defining the likelihood function is to emphasize the change of focus.

Example 9.3.7

Suppose $\boldsymbol{\theta}$ is one-dimensional, denoted λ; that $\boldsymbol{\theta}_\alpha$ is empty (zero-dimensional); that $n = 5$ is predetermined; and that the data-generating process yields iid observations $\tilde{\mathbf{x}}_1, \tilde{\mathbf{x}}_2, \ldots$ with common density function

$$f(x \mid \lambda) = \begin{cases} \lambda e^{-\lambda x}, & x > 0 \\ 0, & x \leqq 0, \end{cases}$$

for all "possible" $\lambda > 0$. Then, because $n = 5$ is predetermined, we have

$$f(n, x_1, \ldots, x_n \mid \lambda) = \begin{cases} \lambda^n e^{-\lambda r}; & \text{all } x_i > 0,\ n = 5 \\ 0; & \text{elsewhere}, \end{cases}$$
$$\equiv L(\lambda \mid n, x_1, \ldots, x_n),$$

where $r \equiv \sum_{i=1}^{n} x_i$. If you now perform experiment e and obtain the five observations 3, 4, 2, 6, 12, then $r = 27$. You may graph $L(\lambda \mid 5, 3, 4, 2, 6, 12)$ as an exercise.

9.4 INTRODUCTION TO STATISTICAL INFERENCE

Statistical inference is concerned with drawing conclusions about the parameter $(\boldsymbol{\theta}, \boldsymbol{\theta}_\alpha)$ of a statistical experiment based upon the outcome $(n, \mathbf{x}_1, \ldots, \mathbf{x}_n)$ of that experiment. In fact, one performs the experiment solely to obtain information on this parameter. After he has obtained the outcome $(n, \mathbf{x}_1, \ldots, \mathbf{x}_n)$ the experimenter is faced with the decision as to what to infer about $(\boldsymbol{\theta}, \boldsymbol{\theta}_\alpha)$. This decision will yield a consequence which depends upon his inference, upon the experimental outcome $(n, \mathbf{x}_1, \ldots, \mathbf{x}_n)$, and upon the true value $(\boldsymbol{\theta}, \boldsymbol{\theta}_\alpha)$. Hence inference is a subtopic of decision theory, and the general decision model in Chapter 4 can be invoked.

However, there are good reasons for considering inference separately from decision theory. First, many people do not wish to specify the set C of consequences very precisely or, what is more difficult, to specify a total "preference" order in C. Second, many interesting and useful statements about $(\boldsymbol{\theta}, \boldsymbol{\theta}_\alpha)$ can be made without explicit consideration of the potential consequences. Finally, many interesting and useful things can be said about $(\boldsymbol{\theta}, \boldsymbol{\theta}_\alpha)$ without taking into account a prior dgf $f(\boldsymbol{\theta}, \boldsymbol{\theta}_\alpha)$ on $(\tilde{\boldsymbol{\theta}}, \tilde{\boldsymbol{\theta}}_\alpha)$, considered as random, *and* without taking into account the potential consequences.

A suitable provisional definition of *statistical inference* for our purposes is: "the subject of drawing conclusions about $(\boldsymbol{\theta}, \boldsymbol{\theta}_\alpha)$ without explicit consideration of the consequences of those conclusions."

With this definition, we may now summarize the two principal approaches to inference. One approach, *Bayesian inference*, holds that all conclusions about $(\boldsymbol{\theta}, \boldsymbol{\theta}_\alpha)$ should be based upon the *posterior dgf* $f(\boldsymbol{\theta}, \boldsymbol{\theta}_\alpha \mid n, \mathbf{x}_1, \ldots, \mathbf{x}_n)$ given the experimental outcome $(n, \mathbf{x}_1, \ldots, \mathbf{x}_n)$. In this approach, it is necessary to consider $(\boldsymbol{\theta}, \boldsymbol{\theta}_\alpha)$ as random and to specify its prior dgf $f(\boldsymbol{\theta}, \boldsymbol{\theta}_\alpha)$. Herein lie objections, for in many statistical problems there is no *objective* way of specifying $f(\boldsymbol{\theta}, \boldsymbol{\theta}_\alpha)$ even though the data-generating and stopping processes may be objectively known for every possible $(\boldsymbol{\theta}, \boldsymbol{\theta}_\alpha)$. Therefore, those who reject the subjectivist philosophy of probability also reject

Bayesian inference for those problems in which $f(\boldsymbol{\theta}, \boldsymbol{\theta}_\alpha)$ is not objectively determinable.

The alternative is to base all conclusions about $(\boldsymbol{\theta}, \boldsymbol{\theta}_\alpha)$ upon the *likelihood function* $L(\boldsymbol{\theta}, \boldsymbol{\theta}_\alpha \mid n, \mathbf{x}_1, \ldots, \mathbf{x}_n)$ of $(\boldsymbol{\theta}, \boldsymbol{\theta}_\alpha)$ given the experimental outcome $(n, \mathbf{x}_1, \ldots, \mathbf{x}_n)$ as defined by (9.3.4). Note that the arguments of the likelihood function coincide in location with those of posterior dgf (when it exists). Often *all possible* values of the "parameters" $(n, \mathbf{x}_1, \ldots, \mathbf{x}_n)$ are relevant to inferences.

This approach to statistical inference, in which conclusions about $(\boldsymbol{\theta}, \boldsymbol{\theta}_\alpha)$ are based upon the likelihood function, considered as a function of *all* of its arguments, will be called *orthodox inference*, rather than "non-Bayesian inference." The latter term may seem unintentionally pejorative and only serves to underscore the bias in favor of Bayesian inference to which we have already admitted in Chapter 5.

Some statisticians are concerned with the reconciliation of Bayesian and likelihood inference, in the following sense. They are concerned with ascertaining which *prior* dgf's $f(\boldsymbol{\theta}, \boldsymbol{\theta}_\alpha)$ produce, for every $(n, \mathbf{x}_1, \ldots, \mathbf{x}_n)$, a posterior dgf which coincides with the likelihood function. When such prior dgf's exist, any estimates of $(\boldsymbol{\theta}. \boldsymbol{\theta}_\alpha)$ based upon a given way of dealing with the *likelihood* function will coincide with estimates of $(\boldsymbol{\theta}, \boldsymbol{\theta}_\alpha)$ based upon the *same* way of dealing with the *posterior* dgf. Hence in such cases there is justification for saying that the Bayesian approach *includes* the orthodox approach as special cases (that is, for the reconciling prior dgf's). It turns out in several important cases, however, that there is no *proper* prior dgf which equates the posterior dgf and the likelihood function; those functions which do so fail to satisfy (dgf 2). We call such functions $f(\boldsymbol{\theta}, \boldsymbol{\theta}_\alpha)$ *improper prior* dgf's.

We shall have much more to say about Bayesian and orthodox inference in later chapters, in the context of specific data-generating processes. In some instances we shall comment briefly upon reconciling prior dgf's, proper (that is, valid) or improper.

It is worth noting that up until Section 9.3 all prior dgf's have been on $\tilde{\boldsymbol{\theta}}$ alone, but that in this section we have been discussing *joint* prior dgf's $f(\boldsymbol{\theta}, \boldsymbol{\theta}_\alpha)$ on $(\tilde{\boldsymbol{\theta}}, \tilde{\boldsymbol{\theta}}_\alpha)$. The reason for this complication is that the "additional parameter" $\boldsymbol{\theta}_\alpha$ of the stopping process may be codependent with $\boldsymbol{\theta}$ and that information conveyed by $(n, \mathbf{x}_1, \ldots, \mathbf{x}_n)$ about $\boldsymbol{\theta}_\alpha$ may be relevant in making inferences about $\boldsymbol{\theta}$, the vector of primary interest.

9.5 SUFFICIENT STATISTICS

We have already noted that the outcome $(n, \mathbf{x}_1, \ldots, \mathbf{x}_n)$ of an experiment e may be extremely cumbersome to process, transmit, or otherwise handle. We shall therefore seek to condense this information via a (measurable) function $T(\cdot, \cdot, \ldots, \cdot \mid n)$: $R^{k_1+\cdots+k_n} \longrightarrow R^{K_n}$, for some $K_n \in \{1, 2, \ldots\}$, where k_i is the dimension of $\mathbf{x}_i$. (Note: One function T is given for each $n \in \{1, 2, \ldots\}$.)

A *statistic* is a set $\{T(\ldots \mid n): n = 1, 2, \ldots\}$ of functions such that, for each n, $T(\cdot, \ldots, \cdot \mid n)$ is a function of the values $\mathbf{x}_1, \ldots, \mathbf{x}_n$ of (the first) n observations from the data-generating process, but neither a function of $\boldsymbol{\theta}$ and/or $\boldsymbol{\theta}_\alpha$ nor having $\boldsymbol{\theta}$ and/or $\boldsymbol{\theta}_\alpha$ as parameters.

Note that before the experiment is conducted its outcome $\tilde{n}, \tilde{\mathbf{x}}_1, \ldots, \tilde{\mathbf{x}}_n$ is random, and hence $T(\tilde{\mathbf{x}}_1, \ldots, \tilde{\mathbf{x}}_{\tilde{n}} \mid \tilde{n})$ is a random vector which we shall denote by $\tilde{\mathbf{x}}_T$. We shall use the term *statistic* in referring either to $\{T(\ldots \mid n): n = 1, 2, \ldots\}$, *or* to $\tilde{\mathbf{x}}_T$ *or* to a value $\mathbf{x}_T$. In common parlance, *statistic* is used solely to denote $\mathbf{x}_T$'s. In many typical experiments n is predetermined and hence possesses a degenerate mf; however, it can still be considered a random variable (see Exercise 16 of Chapter 7).

Example 9.5.1

Let $T(\tilde{\mathbf{x}}_1, \ldots, \tilde{\mathbf{x}}_n \mid \tilde{n}) \equiv \tilde{\mathbf{x}}_T = (\tilde{\mathbf{x}}_1, \tilde{n})$, for every n. Then $\tilde{\mathbf{x}}_T$ is a statistic.

Example 9.5.2

$T(\tilde{\mathbf{x}}_1, \ldots, \tilde{\mathbf{x}}_n \mid \tilde{n}) \equiv (\tilde{n}, \tilde{\mathbf{x}}_1, \ldots, \tilde{\mathbf{x}}_n)$ is a statistic.

There is an important qualitative difference between the statistics in Examples 9.5.1 and 9.5.2. In Example 9.5.1, the dimensionality K_n of $\tilde{\mathbf{x}}_T$ equals $k_1 + 1$ regardless of the number n of observations, whereas in Example 9.5.2 the dimensionality K_n of $\tilde{\mathbf{x}}_T$ equals $1 + \sum_{i=1}^{n} k_i$ and hence depends upon the number n of observations. Statistics T for which K_n is independent of n are called statistics of *fixed dimensionality*. Most statistics in this book will be of fixed dimensionality.

Many statistics, such as those shown in Example 9.5.1, discard information about $(\boldsymbol{\theta}, \boldsymbol{\theta}_\alpha)$ which the experiment furnishes. On the other hand, the statistic in Example 9.5.2 discards no information but also fails to condense it. We now define statistics which discard no information, and will then indicate how "economical" statistics of this sort can be ascertained.

A statistic $\tilde{\mathbf{x}}_T$ is *sufficient* for $(\boldsymbol{\theta}, \boldsymbol{\theta}_\alpha)$ if there are nonnegative real-valued functions $g(\mathbf{x}_T \mid \boldsymbol{\theta}, \boldsymbol{\theta}_\alpha)$ of $\mathbf{x}$ given $\boldsymbol{\theta}$ and $\boldsymbol{\theta}_\alpha$ and $g(n, \mathbf{x}_1, \ldots, \mathbf{x}_n)$ of $n, \mathbf{x}_1, \ldots, \mathbf{x}_n$ such that $g(n, \mathbf{x}_1, \ldots, \mathbf{x}_n)$ is independent of $\boldsymbol{\theta}$ and $\boldsymbol{\theta}_\alpha$; $g(\mathbf{x}_T \mid \boldsymbol{\theta}, \boldsymbol{\theta}_\alpha)$ depends upon $(n, \mathbf{x}_1, \ldots, \mathbf{x}_n)$ only via $\mathbf{x}_T$; and

$$f(n, \mathbf{x}_1, \ldots, \mathbf{x}_n \mid \boldsymbol{\theta}, \boldsymbol{\theta}_\alpha) = g(\mathbf{x}_T \mid \boldsymbol{\theta}, \boldsymbol{\theta}_\alpha) \cdot g(n, \mathbf{x}_1, \ldots, \mathbf{x}_n), \qquad (9.5.1)$$

for every possible $(\boldsymbol{\theta}, \boldsymbol{\theta}_\alpha)$ and every $(n, \mathbf{x}_1, \ldots, \mathbf{x}_n)$.

Example 9.5.3

In Example 9.3.5, define $T(x_1, \ldots, x_n \mid n) = (n, m, M)$, for every n, $x_1, \ldots, x_n$, where m and M are respectively the smallest and largest observations. Let $g(n, x_1, \ldots, x_n) = 1$, for all $n, x_1, \ldots, x_n$ and let

$$g(\mathbf{x}_T \mid \theta_1, \theta_2, \theta_\alpha) = \begin{cases} \theta_\alpha(1 - \theta_\alpha)^n(\theta_2 - \theta_1)^{-n}, & \begin{cases} \theta_1 < m < M < \theta_2 \\ n \in \{0, 1, 2, \ldots\} \end{cases} \\ 0, & \text{elsewhere.} \end{cases}$$

Then (9.5.1) obviously holds and hence (n, m, M) is a sufficient statistic for $(\theta_1, \theta_2, \theta_\alpha)$. It is easy to verify that $g(n, x_1, \ldots, x_n)$ is independent of $(\theta_1, \theta_2, \theta_\alpha)$ and that $g(\mathbf{x}_T \mid \theta_1, \theta_2, \theta_\alpha)$ depends upon $(n, x_1, \ldots, x_n)$ only through $\mathbf{x}_T$.

Example 9.5.4

For Example 9.3.7, define $\mathbf{x}_T = (n, r) \equiv (n, \sum_{i=1}^{n} x_i)$; $g(n, x_1, \ldots, x_n) \equiv 1$, for all $n, x_1, \ldots, x_n$; and

$$g(\mathbf{x}_T \mid \lambda) = \begin{cases} \lambda^n e^{-\lambda r}; & \text{all } x_i > 0,\ n = 5 \\ 0; & \text{elsewhere.} \end{cases}$$

Then (9.5.1) obviously holds and $\mathbf{x}_T$ is a sufficient statistic for λ.

It is difficult to exaggerate the practical importance of sufficient statistics of fixed dimensionality. Suppose in Example 9.5.3 that n had been 100,000. Then $n, x_1, \ldots, x_n$ would be 100,001 numbers—difficult to record, store, and transmit, and virtually impossible to render intuitively meaningful en masse. However, $\mathbf{x}_T \equiv (n, m, M)$ consists of only three numbers, each of which *is* meaningful.

It can be shown that when $\mathbf{x}_T$ is a sufficient statistic for $(\boldsymbol{\theta}, \boldsymbol{\theta}_\alpha)$, then there is a positive function $g(\mathbf{x}_T) > 0$ such that the dgf $f(\mathbf{x}_T \mid \boldsymbol{\theta}, \boldsymbol{\theta}_\alpha)$ of $\tilde{\mathbf{x}}_T$ given $\boldsymbol{\theta}$ and $\boldsymbol{\theta}_\alpha$ is given by

$$f(\mathbf{x}_T \mid \boldsymbol{\theta}, \boldsymbol{\theta}_\alpha) = g(\mathbf{x}_T \mid \boldsymbol{\theta}, \boldsymbol{\theta}_\alpha) \cdot g(\mathbf{x}_T) \tag{9.5.2}$$

and the conditional dgf $f(n, \mathbf{x}_1, \ldots, \mathbf{x}_n \mid \mathbf{x}_T, \boldsymbol{\theta}, \boldsymbol{\theta}_\alpha)$ of $(\tilde{n}, \tilde{\mathbf{x}}_1, \ldots, \tilde{\mathbf{x}}_n)$ given $\mathbf{x}_T$, $\boldsymbol{\theta}$, and $\boldsymbol{\theta}_\alpha$ is given by

$$f(n, \mathbf{x}_1, \ldots, \mathbf{x}_n \mid \mathbf{x}_T, \boldsymbol{\theta}, \boldsymbol{\theta}_\alpha) = \frac{g(n, \mathbf{x}_1, \ldots, \mathbf{x}_n)}{g(\mathbf{x}_T)} \tag{9.5.3}$$

whenever all terms make sense. Thus (9.5.3) asserts that the conditional distribution of $(\tilde{n}, \tilde{\mathbf{x}}_1, \ldots, \tilde{\mathbf{x}}_n)$ given $\tilde{\mathbf{x}}_T$ is independent of $\boldsymbol{\theta}$ and $\boldsymbol{\theta}_\alpha$. In fact, (9.5.3) is often taken as the *definition* of a sufficient statistic $\tilde{\mathbf{x}}_T$, as it implies that the codependence of $(\boldsymbol{\theta}, \boldsymbol{\theta}_\alpha)$ and $(\tilde{n}, \tilde{\mathbf{x}}_1, \ldots, \tilde{\mathbf{x}}_n)$ is fully represented by

$f(\mathbf{x}_T \mid \boldsymbol{\theta}, \boldsymbol{\theta}_\alpha)$, and that once $\mathbf{x}_T$ has been recorded there is not the slightest reason for bothering about the basic data $n, \mathbf{x}_1, \ldots, \mathbf{x}_n$.

Given (9.5.2) and (9.5.3), we may rewrite (9.5.1) as

$$f(n, \mathbf{x}_1, \ldots, \mathbf{x}_n \mid \boldsymbol{\theta}, \boldsymbol{\theta}_\alpha) = f(\mathbf{x}_T \mid \boldsymbol{\theta}, \boldsymbol{\theta}_\alpha) \cdot f(n, \mathbf{x}_1, \ldots, \mathbf{x}_n \mid \mathbf{x}_T). \qquad (9.5.4)$$

The reason for defining sufficiency of $\mathbf{x}_T$ via (9.5.1) rather than via (9.5.4) lies in the fact that (9.5.1) is much easier to verify than (9.5.4).

There are two fundamental results concerning sufficient statistics. The first asserts, roughly, that the *posterior* distributions, given $\tilde{\mathbf{x}}_T$ and given $(\tilde{n}, \tilde{\mathbf{x}}_1, \ldots, \tilde{\mathbf{x}}_n)$, are identical for any common prior distribution on $(\tilde{\boldsymbol{\theta}}, \tilde{\boldsymbol{\theta}}_\alpha)$:

If $\tilde{\mathbf{x}}_T$ is a sufficient statistic for $(\boldsymbol{\theta}, \boldsymbol{\theta}_\alpha)$. then

$$f(\boldsymbol{\theta}, \boldsymbol{\theta}_\alpha) \mid \mathbf{x}_T) = f(\boldsymbol{\theta}, \boldsymbol{\theta}_\alpha \mid n, \mathbf{x}_1, \ldots, \mathbf{x}_n), \qquad (9.5.5)$$

for every $(\boldsymbol{\theta}, \boldsymbol{\theta}_\alpha)$ and for every possible $(n, \mathbf{x}_1, \ldots, \mathbf{x}_n)$, provided that each of these posterior dgf's was obtained by Bayes' Theorem from the same, arbitrary prior dgf $f(\boldsymbol{\theta}, \boldsymbol{\theta}_\alpha)$ of $(\tilde{\boldsymbol{\theta}}, \tilde{\boldsymbol{\theta}}_\alpha)$.

Proof

The following chain of equalities, valid for every possible $(n, \mathbf{x}_1, \ldots, \mathbf{x}_n)$, suffices:

$f(\boldsymbol{\theta}, \boldsymbol{\theta}_\alpha \mid n, \mathbf{x}_1, \ldots, \mathbf{x}_n)$

$$(1) \qquad = \frac{f(n, \mathbf{x}_1, \ldots, \mathbf{x}_n \mid \boldsymbol{\theta}, \boldsymbol{\theta}_\alpha) \cdot f(\boldsymbol{\theta}, \boldsymbol{\theta}_\alpha)}{\int\int f(n, \mathbf{x}_1, \ldots, \mathbf{x}_n \mid \boldsymbol{\theta}, \boldsymbol{\theta}_\alpha) f(\boldsymbol{\theta}, \boldsymbol{\theta}_\alpha) d\boldsymbol{\theta} d\boldsymbol{\theta}_\alpha}$$

$$(2) \qquad = \frac{f(n, \mathbf{x}_1, \ldots, \mathbf{x}_n \mid \boldsymbol{\theta}, \boldsymbol{\theta}_\alpha) f(\boldsymbol{\theta}, \boldsymbol{\theta}_\alpha)}{d(n, \mathbf{x}_1, \ldots, \mathbf{x}_n) \cdot \int\int f(\mathbf{x}_T \mid \boldsymbol{\theta}, \boldsymbol{\theta}_\alpha) f(\boldsymbol{\theta}, \boldsymbol{\theta}_\alpha) d\boldsymbol{\theta} d\boldsymbol{\theta}_\alpha}$$

$$(3) \qquad = \frac{d(n, \mathbf{x}_1, \ldots, \mathbf{x}_n) \cdot f(\mathbf{x}_T \mid \boldsymbol{\theta}, \boldsymbol{\theta}_\alpha) f(\boldsymbol{\theta}, \boldsymbol{\theta}_\alpha)}{d(n, \mathbf{x}_1, \ldots, \mathbf{x}_n) \cdot \int\int f(\mathbf{x}_T \mid \boldsymbol{\theta}, \boldsymbol{\theta}_\alpha) f(\boldsymbol{\theta}, \boldsymbol{\theta}_\alpha) d\boldsymbol{\theta} d\boldsymbol{\theta}_\alpha}$$

$$(4) \qquad = \frac{f(\mathbf{x}_T \mid \boldsymbol{\theta}, \boldsymbol{\theta}_\alpha) \cdot f(\boldsymbol{\theta}, \boldsymbol{\theta}_\alpha)}{\int\int f(\mathbf{x}_T \mid \boldsymbol{\theta}, \boldsymbol{\theta}_\alpha) f(\boldsymbol{\theta}, \boldsymbol{\theta}_\alpha) d\boldsymbol{\theta} d\boldsymbol{\theta}_\alpha}$$

$$(5) \qquad = f(\boldsymbol{\theta}, \boldsymbol{\theta}_\alpha \mid \mathbf{x}_T).$$

Equalities (1) and (5) are obvious from Bayes' Theorem, and the integrals are to be understood as being over-all possible values of $(\boldsymbol{\theta}, \boldsymbol{\theta}_\alpha)$. In (2) and (3), $d(n, \mathbf{x}_1, \ldots, \mathbf{x}_n) \equiv f(n, \mathbf{x}_1, \ldots, \mathbf{x}_n \mid \mathbf{x}_T, \boldsymbol{\theta}, \boldsymbol{\theta}_\alpha)$, and hence

$$f(n, \mathbf{x}_1, \ldots, \mathbf{x}_n \mid \boldsymbol{\theta}, \boldsymbol{\theta}_\alpha) = d(n, \mathbf{x}_1, \ldots, \mathbf{x}_n) \cdot f(\mathbf{x}_T \mid \theta, \theta_\alpha),$$

which implies for (2) that $d(n, \mathbf{x}_1, \ldots, \mathbf{x}_n)$ is a constant factor and can be moved outside the integral. Canceling, (3) implies (4). (Recall that integrals are sums where appropriate.) QED

The *second* fundamental result is that the *converse* of the first assertion is also true; namely,

If $\tilde{\mathbf{x}}_T$ is a statistic and if (9.5.6)

$$f(\boldsymbol{\theta}, \boldsymbol{\theta}_\alpha \mid \mathbf{x}_T) = f(\boldsymbol{\theta}, \boldsymbol{\theta}_\alpha \mid n, \mathbf{x}_1, \ldots, \mathbf{x}_n),$$

for every $(\boldsymbol{\theta}, \boldsymbol{\theta}_\alpha)$ and for every possible $(n, \mathbf{x}_1, \ldots, \mathbf{x}_n)$ and every possible *common* prior dgf $f(\boldsymbol{\theta}, \boldsymbol{\theta}_\alpha)$, then $\tilde{\mathbf{x}}_T$ is a sufficient statistic for $(\tilde{\boldsymbol{\theta}}, \tilde{\boldsymbol{\theta}}_\alpha)$.

A proof of this assertion may be found in Raiffa and Schlaifer [**66**], 34–35, who *define* sufficiency by (9.5.6).

The implications for inference of this section can be stated succinctly: in either Bayesian or orthodox inference, there are sufficient statistics. However, we can go further in *Bayesian* inference and consider statistics which are sufficient for $\tilde{\boldsymbol{\theta}}$ but not necessarily for $\tilde{\boldsymbol{\theta}}_\alpha$ as well.

Recall that our primary interest is in the "parameter" $\boldsymbol{\theta}$ of the *data-generating process*, but that the additional parameter $\boldsymbol{\theta}_\alpha$ had to be carried along because of the possible codependence of $\boldsymbol{\theta}$ and $\boldsymbol{\theta}_\alpha$ and the significance of the stopping process for inferences about $\boldsymbol{\theta}$. In fact, for decision-making purposes we may be interested only in a *sub*vector of the parameter vector $\boldsymbol{\theta}$ of the data-generating process. Moreover, it is conceivable, though rare, that a decision-maker is solely interested in a subvector of $\boldsymbol{\theta}$ *and* in a subvector $\boldsymbol{\theta}_1$ of $(\boldsymbol{\theta}, \boldsymbol{\theta}_\alpha)$. Let $\boldsymbol{\theta}_2$ denote the components of $(\boldsymbol{\theta}, \boldsymbol{\theta}_\alpha)$ which are not in $\boldsymbol{\theta}_1$. Thus $(\boldsymbol{\theta}_1, \boldsymbol{\theta}_2)$ is a rearrangement of the components of $(\boldsymbol{\theta}, \boldsymbol{\theta}_\alpha)$.

You may verify, by integrating all terms in the equality chain in the proof of (9.5.5), that if $\tilde{\mathbf{x}}_T$ is sufficient for $(\boldsymbol{\theta}, \boldsymbol{\theta}_\alpha)$, then the *marginal* posterior dgf's of $\tilde{\boldsymbol{\theta}}_1$ given $(n, \mathbf{x}_1, \ldots, \mathbf{x}_n)$ and given $\mathbf{x}_T$ are the same; that is,

$$f(\boldsymbol{\theta}_1 \mid \mathbf{x}_T) = f(\boldsymbol{\theta}_1 \mid n, \mathbf{x}_1, \ldots, \mathbf{x}_n), \tag{9.5.7}$$

for all $\boldsymbol{\theta}_1$ and every possible $(n, \mathbf{x}_1, \ldots, \mathbf{x}_n)$. This result can be re-expressed: If $\tilde{\mathbf{x}}_T$ is sufficient for $(\boldsymbol{\theta}, \boldsymbol{\theta}_\alpha)$, then $\tilde{\mathbf{x}}_T$ is *marginally sufficient* for any subvector $\tilde{\boldsymbol{\theta}}_1$ of $(\tilde{\boldsymbol{\theta}}, \tilde{\boldsymbol{\theta}}_\alpha)$.

More generally, let $J \equiv \{j\colon j \in J\}$ be a *set of prior dgf's* $f(\boldsymbol{\theta}_1, \boldsymbol{\theta}_2)$ of $(\tilde{\boldsymbol{\theta}}_1, \tilde{\boldsymbol{\theta}}_2)$. [Clearly, each $f(\boldsymbol{\theta}_1, \boldsymbol{\theta}_2)$ corresponds uniquely to some $f(\boldsymbol{\theta}, \boldsymbol{\theta}_\alpha)$] Let $f^j(\boldsymbol{\theta}_1 \mid n, \mathbf{x}_1, \ldots, \mathbf{x}_n)$ denote the posterior dgf of $\tilde{\boldsymbol{\theta}}_1$ given $(n, \mathbf{x}_1, \ldots, \mathbf{x}_n)$ determined by the prior dgf $j \in J$, and define $f^j(\boldsymbol{\theta}_1 \mid \mathbf{x}_T)$ similarly. We say that $\tilde{\mathbf{x}}_T$ is *marginally sufficient for* $\tilde{\boldsymbol{\theta}}_1$ *relative to* J if

$$f^j(\boldsymbol{\theta}_1 \mid \mathbf{x}_T) = f^j(\boldsymbol{\theta}_1 \mid n, \mathbf{x}_1, \ldots, \mathbf{x}_n), \tag{9.5.8}$$

for every $\boldsymbol{\theta}_1$, every $(n, \mathbf{x}_1, \ldots, \mathbf{x}_n)$ such that $f^j(n, \mathbf{x}_1, \ldots, \mathbf{x}_n) > 0$ *and* every $j \in J$.

We have shown that if $\tilde{\mathbf{x}}_T$ is sufficient for $(\tilde{\boldsymbol{\theta}}, \tilde{\boldsymbol{\theta}}_\alpha)$, then $\tilde{\mathbf{x}}_T$ is marginally sufficient for any subvector $\tilde{\boldsymbol{\theta}}_1$ of $(\tilde{\boldsymbol{\theta}}, \tilde{\boldsymbol{\theta}}_\alpha)$ relative to the set of *all* prior dgf's $f(\boldsymbol{\theta}_1, \boldsymbol{\theta}_2)$.

The most common set J of prior dgf's is the set J_I of *independent* prior dgf's: $J_I = \{f(\boldsymbol{\theta}_1, \boldsymbol{\theta}_2): f(\boldsymbol{\theta}_1, \boldsymbol{\theta}_2) = f(\boldsymbol{\theta}_1)f(\boldsymbol{\theta}_2)$ for all $(\boldsymbol{\theta}_1, \boldsymbol{\theta}_2)\}$. For this set there is a result similar to the equivalence of (9.5.1) and (9.5.5) for (complete) sufficiency:

> $\tilde{\mathbf{x}}_T$ is marginally sufficient for $\tilde{\boldsymbol{\theta}}_1$ relative to the set J_I of all independent prior dgf's $f(\boldsymbol{\theta}_1, \boldsymbol{\theta}_2)$ if and only if there are nonnegative functions $g(\mathbf{x}_T \mid \boldsymbol{\theta}_1)$ and $g(n, \mathbf{x}_1, \ldots, \mathbf{x}_n \mid \boldsymbol{\theta}_2)$ such that (9.5.9)
>
> $$f(n, \mathbf{x}_1, \ldots, \mathbf{x}_n \mid \boldsymbol{\theta}_1, \boldsymbol{\theta}_2) = g(\mathbf{x}_T \mid \boldsymbol{\theta}_1)\cdot g(n, \mathbf{x}_1, \ldots, \mathbf{x}_n \mid \boldsymbol{\theta}_2) \qquad (9.5.10)$$
>
> whenever the left-hand side is defined. [Note that $f(n, \mathbf{x}_1, \ldots, \mathbf{x}_n \mid \boldsymbol{\theta}_1, \boldsymbol{\theta}_2)$ is easily obtainable from $f(x, \mathbf{x}_1, \ldots, \mathbf{x}_n \mid \boldsymbol{\theta}, \boldsymbol{\theta}_\alpha)$ by rearrangement of components, and vice versa.]

The importance of marginally sufficient statistics lies in the fact that there may be a very *convenient* statistic $\tilde{\mathbf{x}}_T$ marginally sufficient for $\tilde{\boldsymbol{\theta}}_1$ relative to some class J of prior dgf's, this class contains the decision-maker's prior dgf, but $\tilde{\mathbf{x}}_T$ may *not* be sufficient for $(\tilde{\boldsymbol{\theta}}_1, \tilde{\boldsymbol{\theta}}_2)$. If all that concerns him is $\tilde{\boldsymbol{\theta}}_1$, then he need record no more than $\tilde{\mathbf{x}}_T$.

The components of $\boldsymbol{\theta}_2$ are often called *nuisance parameters*, since they usually have to be carried through the analysis but have no direct relevance for the decision-maker.

9.6 NONINFORMATIVE STOPPING PROCESSES

Equation (9.3.3) expressed the likelihood function $L(\boldsymbol{\theta}, \boldsymbol{\theta}_\alpha \mid n, \mathbf{x}_1, \ldots, \mathbf{x}_n)$ as the product of: (1) the conditional likelihood function $L(\boldsymbol{\theta} \mid \mathbf{x}_1, \ldots, \mathbf{x}_n)$ given the first n observations $\mathbf{x}_1, \ldots, \mathbf{x}_n$; and (2) the probability $S(n \mid \mathbf{x}_1, \ldots, \mathbf{x}_n, \boldsymbol{\theta}, \boldsymbol{\theta}_\alpha)$ that observation $n + 1$ will *not* be made given $\boldsymbol{\theta}, \boldsymbol{\theta}_\alpha$, and the first n observations $\mathbf{x}_1, \ldots, \mathbf{x}_n$. From this equation it is clear that $\boldsymbol{\theta}_\alpha$ is a nuisance parameter when we are solely interested in making inferences about $\boldsymbol{\theta}$. Moreover, if we let $\tilde{\mathbf{x}}_T \equiv (n, \tilde{\mathbf{x}}_1, \ldots, \tilde{\mathbf{x}}_n)$ and define

$$g(\mathbf{x}_T \mid \boldsymbol{\theta}) \equiv L(\boldsymbol{\theta} \mid \mathbf{x}_1, \ldots, \mathbf{x}_n)$$

and

$$g(n, \mathbf{x}_1, \ldots, \mathbf{x}_n \mid \boldsymbol{\theta}, \boldsymbol{\theta}_\alpha) \equiv S(n \mid \mathbf{x}_1, \ldots, \mathbf{x}_n, \boldsymbol{\theta}, \boldsymbol{\theta}_\alpha),$$

then Equation (9.3.3) *almost* satisfies the marginal-sufficiency factorization criterion (9.5.10) [almost, in that (9.5.10) requires $g(n, \mathbf{x}_1, \ldots, \mathbf{x}_n \mid \boldsymbol{\theta}, \boldsymbol{\theta}_\alpha)$ to be independent of $\boldsymbol{\theta}$].

Now, if possible, we would like to avoid the potentially vexatious calculation of the stopping probabilities $S(n \mid \mathbf{x}_1, \ldots, \mathbf{x}_n, \boldsymbol{\theta}, \boldsymbol{\theta}_\alpha)$ in our derivation of the marginal posterior dgf $f(\boldsymbol{\theta} \mid n, \mathbf{x}_1, \ldots, \mathbf{x}_n)$. The following completely satisfactory sufficient condition is due to Raiffa and Schlaifer [**66**], 37–38:

If:

(1) $\pi(n \mid \mathbf{x}_1, \ldots, \mathbf{x}_n, \boldsymbol{\theta}, \boldsymbol{\theta}_\alpha)$ does not depend upon $\boldsymbol{\theta}$ for any n; and (9.6.1)

(2) the decision-maker deems $\tilde{\boldsymbol{\theta}}$ and $\tilde{\boldsymbol{\theta}}_\alpha$ independent in the sense that $f(\boldsymbol{\theta}, \boldsymbol{\theta}_\alpha) = f(\boldsymbol{\theta})\cdot f(\boldsymbol{\theta}_\alpha)$ for all $\boldsymbol{\theta}$ and $\boldsymbol{\theta}_\alpha$, then,

$$f(\boldsymbol{\theta} \mid n, \mathbf{x}_1, \ldots, \mathbf{x}_n) = \frac{f(\mathbf{x}_1, \ldots, \mathbf{x}_n \mid \boldsymbol{\theta}) f(\boldsymbol{\theta})}{\int f(\mathbf{x}_1, \ldots, \mathbf{x}_n \mid \boldsymbol{\theta}) f(\boldsymbol{\theta}) d\boldsymbol{\theta}} \tag{9.6.2}$$

wherever defined.

Proof

$f(\boldsymbol{\theta} \mid n, \mathbf{x}_1, \ldots, \mathbf{x}_n)$

(1) $= \int f(\boldsymbol{\theta}, \boldsymbol{\theta}_\alpha \mid n, \mathbf{x}_1, \ldots, \mathbf{x}_n) d\boldsymbol{\theta}_\alpha$

(2) $= \int \left\{ \frac{f(n, \mathbf{x}_1, \ldots, \mathbf{x}_n \mid \boldsymbol{\theta}, \boldsymbol{\theta}_\alpha) f(\boldsymbol{\theta}, \boldsymbol{\theta}_\alpha)}{\int\int f(n, \mathbf{x}_1, \ldots, \mathbf{x}_n \mid \boldsymbol{\theta}, \boldsymbol{\theta}_\alpha) f(\boldsymbol{\theta}, \boldsymbol{\theta}_\alpha) d\boldsymbol{\theta} d\boldsymbol{\theta}_\alpha} \right\} d\boldsymbol{\theta}_\alpha$

(3) $= \int \left\{ \frac{f(\mathbf{x}_1, \ldots, \mathbf{x}_n \mid \boldsymbol{\theta}) S(n \mid \mathbf{x}_1, \ldots, \mathbf{x}_n, \boldsymbol{\theta}_\alpha) f(\boldsymbol{\theta}) f(\boldsymbol{\theta}_\alpha)}{\int\int f(\mathbf{x}_1, \ldots, \mathbf{x}_n \mid \boldsymbol{\theta}) S(n \mid \mathbf{x}_1, \ldots, \mathbf{x}_n, \boldsymbol{\theta}_\alpha) f(\boldsymbol{\theta}) f(\boldsymbol{\theta}_\alpha) d\boldsymbol{\theta} d\boldsymbol{\theta}_\alpha} \right\} d\boldsymbol{\theta}_\alpha$

(4) $= \int \left\{ \frac{[f(\mathbf{x}_1, \ldots, \mathbf{x}_n \mid \boldsymbol{\theta}) \cdot f(\boldsymbol{\theta})][S(n \mid \mathbf{x}_1, \ldots, \mathbf{x}_n, \boldsymbol{\theta}_\alpha) f(\boldsymbol{\theta}_\alpha)]}{[\int S(n \mid \mathbf{x}_1, \ldots, \mathbf{x}_n, \boldsymbol{\theta}_\alpha) f(\boldsymbol{\theta}_\alpha) d\boldsymbol{\theta}_\alpha][\int f(\mathbf{x}_1, \ldots, \mathbf{x}_n \mid \boldsymbol{\theta}) f(\boldsymbol{\theta}) d\boldsymbol{\theta}]} \right\} d\boldsymbol{\theta}_\alpha$

(5) $= \frac{f(\mathbf{x}_1, \ldots, \boldsymbol{\theta}_n \mid \boldsymbol{\theta}) f(\boldsymbol{\theta}) \cdot \int S(n \mid \mathbf{x}_1, \ldots, \mathbf{x}_n, \boldsymbol{\theta}_\alpha) f(\boldsymbol{\theta}_\alpha) d\boldsymbol{\theta}_\alpha}{[\int f(\mathbf{x}_1, \ldots, \boldsymbol{\theta}_n \mid \boldsymbol{\theta}) f(\boldsymbol{\theta}) d\boldsymbol{\theta}] \int S(n \mid \mathbf{x}_1, \ldots, \mathbf{x}_n, \boldsymbol{\theta}_\alpha)\, d(\boldsymbol{\theta}_\alpha) d\boldsymbol{\theta}_\alpha}$

(6) $= \frac{f(\mathbf{x}_1, \ldots, \boldsymbol{\theta}_n \mid \boldsymbol{\theta}) f(\boldsymbol{\theta})}{\int f(\mathbf{x}_1, \ldots, \mathbf{x}_n \mid \boldsymbol{\theta}) f(\boldsymbol{\theta}) d\boldsymbol{\theta}}$,

as asserted. Equality (3) is where hypotheses (1) and (2) are used; you may easily verify that hypothesis (1) implies that all $S(n \mid \mathbf{x}_1, \ldots, \mathbf{x}_n, \boldsymbol{\theta}, \boldsymbol{\theta}_\alpha)$ are independent of $\boldsymbol{\theta}$. Equality (1) follows from (8.3.2), and equality (2) from (8.7.1). Equalities (4) and (5) follow from standard facts about iterated integration and the moving of constant factors outside integrals. Equality (6) is a simple cancellation. QED

What result (9.6.1) really says is that if hypotheses (1) and (2) hold, then for purposes of Bayesian inference the experiment can be treated *as if* the number n of observations had been predetermined, and the computation of stopping probabilities is completely avoidable.

When hypotheses (1) and (2) of (9.6.1) obtain, we say that the stopping process is noninformative about $\tilde{\boldsymbol{\theta}}$; we also refer to "obtaining n observations $\tilde{\mathbf{x}}_1, \ldots, \tilde{\mathbf{x}}_n$ from a data-generating process with noninformative stopping."

Example 9.6.1

The stopping process in Example 9.3.2 is noninformative because it has no additional parameters $\boldsymbol{\theta}_\alpha$ for the decision-maker to consider [hence hypothesis (2) obtains], while all π's are independent of $\boldsymbol{\theta}$. Thus, experiments with predetermined numbers of observations have noninformative stopping processes.

Example 9.6.2

Hypothesis (1) of (9.6.1) is satisfied by the stopping process in Example 9.3.3. Hence it is a noninformative stopping process if the decision-maker judges $\tilde{\theta}_\alpha$ and $\tilde{\boldsymbol{\theta}}$ to be independent [that is, $f(\theta_\alpha, \boldsymbol{\theta}) = f(\theta_\alpha)f(\boldsymbol{\theta})$ for all $\theta_\alpha, \boldsymbol{\theta}$.].

Example 9.6.3

The stopping process in Example 9.3.6 clearly *is* informative, as the π's do depend upon $\boldsymbol{\theta}$ through θ_1.

We shall sometimes use "$\tilde{\mathbf{x}}_1, \ldots, \tilde{\mathbf{x}}_n$ is a random sample of size n from $f(\mathbf{x} \mid \boldsymbol{\theta})$" to mean, "$\tilde{\mathbf{x}}_1, \ldots, \tilde{\mathbf{x}}_n$ are n iid observations with common dgf $f(\mathbf{x} \mid \boldsymbol{\theta})$, obtained with noninformative stopping."

9.7 THREE IMPORTANT PROBLEMS IN INFERENCE

We conclude the chapter by stating three classes of problems that, in one form or another, dominate the subject of statistical inference. We shall state them in terms of $\boldsymbol{\theta}$ rather than $(\boldsymbol{\theta}, \boldsymbol{\theta}_\alpha)$, without loss of generality, since $\boldsymbol{\theta}$ can be partitioned as $(\boldsymbol{\theta}_1, \boldsymbol{\theta}_2)$, where $\boldsymbol{\theta}_1$ denotes the parameters of the data-generating process and $\boldsymbol{\theta}_2$ denotes the additional parameters (if any) of the stopping process.

Problem 1: Furnish a "close" *estimate* $\hat{\boldsymbol{\theta}}$ of $\boldsymbol{\theta}$ based upon the outcome $(n, \mathbf{x}_1, \ldots, \mathbf{x}_n)$ of an experiment e.

It is clear that if the experiment is at all relevant, then $\boldsymbol{\theta}$ will depend functionally upon $(n, \mathbf{x}_1, \ldots, \mathbf{x}_n)$; accordingly, it is often denoted by $\hat{\boldsymbol{\theta}}(n, \mathbf{x}_1, \ldots, \mathbf{x}_n)$. The word *close* needs to be made much more precise; there are several ways of making it more precise, and Bayesian decision theory shows how *close should* be interpreted within the context of any given decision problem.

Problem 2: Furnish a "fairly small" subset $I(n, \mathbf{x}_1, \ldots, \mathbf{x}_n)$ of $\boldsymbol{\theta}$ to which "it is reasonable to suppose" that $\boldsymbol{\theta}$ belongs.

Problem 2 is a more modest version of Problem 1, in that it recognizes that a *point estimate* of $\boldsymbol{\theta}$ via Problem 1, is almost certain to be wrong if there are a great many possible values of $\boldsymbol{\theta}$. If θ is one-dimensional, $I(n, \mathbf{x}_1, \ldots, \mathbf{x}_n)$ is called a *confidence interval* for θ. The phrase, "it is reasonable to suppose," has one interpretation in Bayesian inference and another in orthodox inference.

Problem 3: Let $\boldsymbol{\Theta}_0$ be a subset of the set $\boldsymbol{\Theta}$ of possible values of $\boldsymbol{\theta}$, and let H_0 denote the hypothesis (assertion; statement) that $\boldsymbol{\theta} \in \boldsymbol{\Theta}_0$. Accept or reject H_0 on the basis of the outcome $(n, \mathbf{x}_1, \ldots, \mathbf{x}_n)$ of e.

Problem 3 is more closely related to Problem 2 than may be apparent at first glance. It concerns the credibility of assertions, or hypotheses about $\boldsymbol{\theta}$; for example, the hypothesis that average demand θ for a product exceeds 50 [here, $\theta_0 = (50, \infty)$].

These three problems are introduced in greater depth in Chapters 20 through 22 successively.

EXERCISES

1. Let the dgf $f(x_1, \ldots, x_n \mid r, \Lambda)$ be given by

$$f(x_1, \ldots, x_n \mid r, \Lambda) = \begin{cases} r^n \Lambda^{nr} / \prod_{i=1}^{n} x_i^{r+1}, & \text{all } x_i > \Lambda \\ 0, & \text{some } x_i \leq \Lambda, \end{cases}$$

where $r > 0$ and $\Lambda > 0$ are parameters.

   (a) Are $\tilde{x}_1, \ldots, \tilde{x}_n$ iid given (r, Λ)?
   (b) Is $\tilde{\mathbf{x}}_T \equiv (n, \tilde{m}, \prod_{i=1}^{n} \tilde{x}_i)$ sufficient for (r, Λ), where $\tilde{m} \equiv \min \{\tilde{x}_1, \ldots, \tilde{x}_n\}$?
   (c) Is $(n, \tilde{m})$ marginally sufficient for Λ relative to the set J_I of all independent prior dgf's, of the form $f(r, \Lambda) = f(r)f(\Lambda)$?
   (d) Is $(n, \prod_{i=1}^{n} \tilde{x}_i)$ marginally sufficient for r relative to the set J_I of all independent prior dgf's?
   (e) Find the dgf of $\tilde{m}$ given (n, r, Λ). [See Exercise 20 in Chapter 8, Equation (8.8.7).]

2. (a) Show that if $\tilde{\mathbf{x}}_T$ is sufficient for $(\tilde{\boldsymbol{\theta}}, \tilde{\boldsymbol{\theta}})$, and of fixed dimensionality, then so is $k \cdot k_T$, where k is any nonzero number.
   (b) Show that if $\tilde{\mathbf{x}}_T$ is a sufficient statistic for $(\boldsymbol{\theta}, \boldsymbol{\theta}_\alpha)$ of fixed dimensionality k, and if $h: R^k \rightarrow R^k$ is a (measurable) bijection, then $\tilde{\mathbf{x}}_{hT} \equiv h(\tilde{\mathbf{x}}_T)$ is also a sufficient statistic for $(\boldsymbol{\theta}, \boldsymbol{\theta}_\alpha)$.

3. We say that the k-dimensional random vector $\tilde{\mathbf{x}}$ has a density function *of exponential type* if

$$f(\mathbf{x} \mid \boldsymbol{\theta}) = \begin{cases} g_1(\boldsymbol{\theta}) g_2(\mathbf{x}) \exp\left(\sum_{j=1}^{\ell} g_{3j}(\boldsymbol{\theta}) g_{4j}(\mathbf{x})\right), & \mathbf{x} \in D(\boldsymbol{\theta}) \\ 0, & \mathbf{x} \notin D(\boldsymbol{\theta}), \end{cases} \tag{9.8.1}$$

for every $\boldsymbol{\theta}$ in some given subset $\boldsymbol{\Theta} \subset R^m$, where $g_2(\mathbf{x}) > 0$, for every $\mathbf{x} \in \bigcup_{\theta \in \Theta} D(\boldsymbol{\theta})$ and $g_1(\boldsymbol{\theta}) > 0$, for every $\boldsymbol{\theta} \in \boldsymbol{\Theta}$.

(a) Let $\tilde{\mathbf{x}}_1, \ldots, \tilde{\mathbf{x}}_n$ be a random sample of size n from $f(\mathbf{x} \mid \boldsymbol{\theta})$. Show that

$$f(\mathbf{x}_1, \ldots, \mathbf{x}_n \mid \boldsymbol{\theta}) \equiv \prod_{i=1}^{n} f(\mathbf{x}_i \mid \boldsymbol{\theta})$$

$$= \begin{cases} [g_1(\boldsymbol{\theta})]^n \cdot [\prod_{i=1}^{n} g_2(\mathbf{x}_i)] \exp\left(\sum_{j=1}^{\ell} \sum_{i=1}^{n} g_{3j}(\boldsymbol{\theta}) g_{4j}(\mathbf{x}_i)\right), & \text{all } \mathbf{x}_i \in D(\boldsymbol{\theta}), \\ 0, & \text{some } \mathbf{x}_i \notin D(\boldsymbol{\theta}). \end{cases} \tag{9.8.2}$$

(b) Show that $\tilde{\mathbf{x}}_T \equiv (n, \tilde{T}_1, \ldots, \tilde{T}_\ell)$ is a sufficient statistic for $\boldsymbol{\theta}$, where $\tilde{T}_j \equiv \sum_{i=1}^{n} g_{4j}(\tilde{\mathbf{x}}_i)$ for $j = 1, \ldots, \ell$, provided that $D(\boldsymbol{\theta})$ is independent of $\boldsymbol{\theta}$; that is, $D(\boldsymbol{\theta}) \equiv D$ for every $\boldsymbol{\theta} \in \boldsymbol{\Theta}$.

(c) Show that if $D(\boldsymbol{\theta})$ is independent of $\boldsymbol{\theta}$ and if $\tilde{\boldsymbol{\theta}}$ has a prior dgf of the form

$$f(\boldsymbol{\theta}) \equiv \begin{cases} K(n_0, T_{10}, \ldots, T_{\ell 0})[g_1(\boldsymbol{\theta})]^{n_0} \exp\left(\sum_{j=1}^{\ell} g_{3j}(\boldsymbol{\theta}) \cdot T_{j0}\right), & \boldsymbol{\theta} \in \boldsymbol{\Theta} \\ 0, & \boldsymbol{\theta} \notin \boldsymbol{\Theta}, \end{cases} \tag{9.8.3}$$

where $K(n_0, T_{10}, \ldots, T_{\ell 0})$ is a positive factor chosen such that (dgf 2) obtains, $n_0 > 0$, and $T_{10}, \ldots, T_{\ell 0}$ are parameters; *then* the *posterior* dgf is of the form (9.7.3) with $n_0, T_{10}, \ldots, T_{\ell 0}$ replaced by $n_1 \equiv n_0 + n$, $T_{11} \equiv T_{10} + T_1, \ldots$, and $T_{\ell 1} \equiv T_{\ell 0} + T_\ell$.

(d) Under the assumptions of (c), show that (9.8.2) and (9.8.3) imply that the unconditional joint dgf of $\tilde{\mathbf{x}}_1, \ldots, \tilde{\mathbf{x}}_n$ is given by

$$f(\mathbf{x}_1, \ldots, \mathbf{x}_n) \equiv \begin{cases} \dfrac{K(n_0, T_{10}, \ldots, T_{\ell 0})}{K(n_1, T_{11}, \ldots, T_{\ell 1})} \prod_{i=1}^{n} g_2(\mathbf{x}_i), & \text{all } \mathbf{x}_i \in D \\ 0, & \text{some } \mathbf{x}_i \notin D. \end{cases} \tag{9.8.4}$$

4. Let $\tilde{x}_1, \ldots, \tilde{x}_n$ be a random sample of size n from

$$f(x \mid \mu, \sigma^2) \equiv (2\pi\sigma^2)^{-1/2} \exp\left(-\frac{1}{2\sigma^2}[x - \mu]^2\right), \quad -\infty < x < \infty,$$

where $-\infty < \mu < \infty$ and $\sigma^2 > 0$ are parameters.

(a) Show that

$$L(\mu, \sigma^2 \mid x_1, \ldots, x_n)$$

$$= \begin{cases} (2\pi\sigma^2)^{-n/2} \exp\left(-\dfrac{n}{2\sigma^2}\left[(\bar{x}_n - \mu)^2 + \dfrac{n-1}{n} s_n^2\right]\right), & \begin{cases} \mu \in R^1, \\ \sigma^2 > 0 \end{cases} \\ 0 & \text{elsewhere,} \end{cases}$$

for any $(x_1, \ldots, x_n) \in R^n$, where

$$\bar{x}_n \equiv \frac{1}{n}\sum_{i=1}^{n} x_i$$

and

$$s_n^2 \equiv \frac{1}{n-1}\sum_{i=1}^{n} (x_i - \bar{x}_n)^2.$$

[*Hint:* (1) $\sum_{i=1}^{n} (x_i - \mu)^2 = \sum_{i=1}^{n} [(x_i - \bar{x}_n) + (\bar{x}_n - \mu)]^2$; and

(2) $\sum_{i=1}^{n} (x_i - \bar{x}_n) = 0$.]

(b) Show that $(n, \tilde{\bar{x}}_n, \tilde{s}_n^2) \equiv \tilde{\mathbf{x}}_T$ is a sufficient statistic for $\boldsymbol{\theta} \equiv (\mu, \sigma^2)$.

5. *Conditional Sufficiency*: Let $(\tilde{\boldsymbol{\theta}}, \tilde{\boldsymbol{\theta}}_\alpha) = (\tilde{\boldsymbol{\theta}}_1, \tilde{\boldsymbol{\theta}}_2)$ and define $\tilde{\mathbf{x}}_T(\boldsymbol{\theta}_2) \equiv T(n, \tilde{\mathbf{x}}_1, \ldots, \tilde{\mathbf{x}}_n, \boldsymbol{\theta}_2)$ for every possible $\boldsymbol{\theta}_2$. We say that $\tilde{\mathbf{x}}_T(\boldsymbol{\theta}_2)$ is *conditionally sufficient for* $\tilde{\boldsymbol{\theta}}_1$ *given* $\boldsymbol{\theta}_2$, or that $\tilde{\mathbf{x}}_T(\boldsymbol{\theta}_2)$ *is sufficient* for $\tilde{\boldsymbol{\theta}}_1$ *when* $\boldsymbol{\theta}_2$ *is known*, if there exist functions $g(\mathbf{x}_T(\boldsymbol{\theta}_2) \mid \boldsymbol{\theta}_1, \boldsymbol{\theta}_2)$ and $g(n, \mathbf{x}_1, \ldots, \mathbf{x}_n \mid \boldsymbol{\theta}_2)$ such that

$$f(n, \mathbf{x}_1, \ldots, \mathbf{x}_n \mid \boldsymbol{\theta}_1, \boldsymbol{\theta}_2) = g(\mathbf{x}_T(\boldsymbol{\theta}_2) \mid \boldsymbol{\theta}_1, \boldsymbol{\theta}_2) \cdot g(n, \mathbf{x}_1, \ldots, \mathbf{x}_n \mid \boldsymbol{\theta}_2) \quad (9.8.5)$$

whenever the left-hand side is defined.

(a) Prove the following:

If $\tilde{\mathbf{x}}_T(\boldsymbol{\theta}_2)$ is sufficient for $\tilde{\boldsymbol{\theta}}_1$ when $\boldsymbol{\theta}_2$ is known, then

$$f(\boldsymbol{\theta}_1 \mid \boldsymbol{\theta}_2, n, \mathbf{x}_1, \ldots, \mathbf{x}_n) = f(\boldsymbol{\theta}_1 \mid \boldsymbol{\theta}_2, \mathbf{x}_T(\boldsymbol{\theta}_2)) \quad (9.8.6)$$

whenever either (and hence each) side is defined, and whenever each side is derived from the same *arbitrary* joint prior dgf $f(\boldsymbol{\theta}_1, \boldsymbol{\theta}_2)$.

[*Hint:* See proof of (9.5.5); and, since $\tilde{\boldsymbol{\theta}}_2 = \boldsymbol{\theta}_2$ will always be assumed known, you might as well pass from any $f(\boldsymbol{\theta}_1, \boldsymbol{\theta}_2)$ to $f(\boldsymbol{\theta}_1 \mid \boldsymbol{\theta}_2)f(\boldsymbol{\theta}_2)$ and perform the entire analysis starting from an arbitrary conditional dgf $f(\boldsymbol{\theta}_1 \mid \boldsymbol{\theta}_2)$.]

(b) Show that $\tilde{\mathbf{x}}_T(\sigma^2) \equiv (n, \tilde{\bar{x}}_n)$ is sufficient for μ in Exercise 4 when σ^2 is known.

(c) Show that $\tilde{\mathbf{x}}_T(\mu) \equiv (n, \tilde{\hat{\sigma}}_n^2)$ is sufficient for σ^2 when μ is known, where

$$\tilde{\hat{\sigma}}_n^2 \equiv \frac{1}{n}\sum_{i=1}^{n} (\tilde{x}_i - \mu)^2.$$

(d) In Exercise 1, show that $\tilde{\mathbf{x}}_T \equiv (n, \tilde{\bar{m}})$ is sufficient for Λ when r is known, and that $(n, \prod_{i=1}^{n} \tilde{x}_i^{r+1})$ is sufficient for $\tilde{r}$ when Λ is known.

6. You will draw bolts, with replacement, from a bin and weigh them in order to obtain information about their average weight θ. You will continue the weighings until quitting time. Discuss whether the stopping is noninformative.

7. *The Likelihood Principle*: Many statisticians, of both objectivist and subjectivist persuasions, accept the following assertion as a sensible axiom for inference.

> If the outcomes $(\tilde{n}, \tilde{\mathbf{x}}_1, \ldots, \tilde{\mathbf{x}}_n) \equiv \tilde{\mathbf{x}}$ and $(\tilde{n}^*, \tilde{\mathbf{x}}_1^*, \ldots, \tilde{\mathbf{x}}^*_{n^*}) \equiv \tilde{\mathbf{x}}^*$ of experiments e and e^* have dgf's $f(\mathbf{x} \mid \boldsymbol{\theta}, \boldsymbol{\theta}_\alpha)$ and $f(\mathbf{x}^* \mid \boldsymbol{\theta}, \boldsymbol{\theta}_\alpha)$ with the same parameters $(\boldsymbol{\theta}, \boldsymbol{\theta}_\alpha)$; and if the actual outcomes $\mathbf{x}$ and $\mathbf{x}^*$ are such that $L(\boldsymbol{\theta}, \boldsymbol{\theta}_\alpha \mid \mathbf{x}) = kL(\boldsymbol{\theta}, \boldsymbol{\theta}_\alpha \mid \mathbf{x}^*)$ for every possible $(\boldsymbol{\theta}, \boldsymbol{\theta}_\alpha)$ and for some constant $k > 0$, then any inference about $(\boldsymbol{\theta}, \boldsymbol{\theta}_\alpha)$ based upon e and $\mathbf{x}$ must be the same as the inference about $(\boldsymbol{\theta}, \boldsymbol{\theta}_\alpha)$ based upon e^* and $\mathbf{x}^*$. (9.8.7)

In summary the likelihood principle requires that all inferences be based solely upon the likelihood function of $(\boldsymbol{\theta}, \boldsymbol{\theta}_\alpha)$ given the *actual* outcome of the experiment.

(a) Show that Bayesian inference satisfies the likelihood principle.

(b) Experiments are often conducted for the purpose of obtaining some *estimate* (a statistic $(\hat{\boldsymbol{\theta}}(\mathbf{x}), \hat{\boldsymbol{\theta}}_\alpha(\mathbf{x}))$ of $(\boldsymbol{\theta}, \boldsymbol{\theta}_\alpha)$. A *maximum likelihood estimate* $(\hat{\boldsymbol{\theta}}(\mathbf{x}), \hat{\boldsymbol{\theta}}_\alpha(\mathbf{x}))$ is any possible value of $(\boldsymbol{\theta}, \boldsymbol{\theta}_\alpha)$ such that

$$L(\hat{\boldsymbol{\theta}}(\mathbf{x}), \hat{\boldsymbol{\theta}}_\alpha(\mathbf{x}) \mid \mathbf{x}) \geq L(\boldsymbol{\theta}, \boldsymbol{\theta}_\alpha \mid \mathbf{x}), \text{ for every } (\boldsymbol{\theta}, \boldsymbol{\theta}_\alpha). \tag{9.8.8}$$

Show that the maximum likelihood estimation rule satisfies the likelihood principle.

8. Prove that if $\tilde{\mathbf{x}}_T$ is sufficient for $(\tilde{\boldsymbol{\theta}}, \tilde{\boldsymbol{\theta}}_\alpha)$, then $\tilde{\mathbf{x}}_T$ is marginally sufficient for any subvector $\tilde{\boldsymbol{\theta}}_1$ of $(\tilde{\boldsymbol{\theta}}, \tilde{\boldsymbol{\theta}}_\alpha)$.

9. Prove the "if" part of assertion (9.5.9).

10

EXPECTATION AND MOMENTS

10.1 INTRODUCTION

Chapters 7, 8, and 9 provided a reasonably complete introduction to random variables, random vectors, their distributions, and the rudiments of statistical inference. The central focus of those chapters was the distribution of a random variable or vector, usually as represented by a dgf f.

Dgf's being functions, are usually too complicated to convey good qualitative information about the distribution of a random variable (or random vector). In Section 10.3 we shall consider some summary measures, called *moments*, of the distribution of a random variable or of the joint distribution of n random variables. The most important such measures are: *expectation*, or *mean* of a random variable, which is a measure of location; *variance* of a random variable, which is a measure of uncertainty about the value it will assume; and *covariance* of two random variables, which measures their tendency to be simultaneously large and sumultaneously small.

Moments of random variables and random vectors are defined in terms of the expectation operator E, which sends random variables to constants. E is studied in Section 10.2; roughly speaking, the expectation $E(\tilde{x})$ of a random variable $\tilde{x}$ is the weighted average of the possible values x of $\tilde{x}$ with their probabilities as weights. The expectation $E(h(\tilde{x}))$ of a function $h(\tilde{x}) \equiv h \circ \tilde{x}\colon Q \to R^1$ of $\tilde{x}$ is easily found to be the weighted average of the $h(x)$'s with the probabilities of the x's as weights.

In Section 10.4 we deduce some facts about means and variances of components of random vectors. Great economy of notation is attained via the use of linear algebra.

Section 10.5 contains a discussion of the probabilistic relationship between the common mean μ of iid random variables $\tilde{x}_1, \ldots, \tilde{x}_n$ and their simple average $\bar{\tilde{x}}_n \equiv (1/n) \sum_{i=1}^{n} \tilde{x}_i$, called the *sample mean*. This discussion is in terms of Chebychev's inequality and culminates in (a version of) the "weak law of large numbers."

Section 10.6 introduces moment-generating functions of random variables, which provide a useful method for finding moments and also a very good way of finding the distribution of a linear function of several independent random variables. A mathematically more acceptable concept, the characteristic function, is introduced in Exercise 30 and is related to the moment-generating function.

10.2 EXPECTATION

Let $\tilde{x}$ be a discrete random variable with mf f. We define the *expectation* $E(\tilde{x})$ of $\tilde{x}$ by

$$E(\tilde{x}) = \sum \{xf(x): f(x) > 0\};$$

if $\tilde{x}$ is a continuous random variable with *df* f, its expectation is defined as

$$E(\tilde{x}) = \int_{-\infty}^{\infty} xf(x)dx;$$

while if $\tilde{x}$ is a mixed random variable with df f and mf f^*, its expectation is defined as

$$E(\tilde{x}) = \int_{-\infty}^{\infty} xf(x)dx + \sum \{xf^*(x): f^*(x) > 0\}.$$

In all three cases, $E(\tilde{x})$ is a weighted average of the possible values of $\tilde{x}$, whose distribution provides the relevant weights via a mf, a df, or both. [Readers familiar with Stieltjes integrals will recognize that all three cases are expressed by the Stieltjes integral

$$E(\tilde{x}) = \int_{-\infty}^{\infty} xdF(x)$$

of x as integrand and F as integrator. Thus the expectation of a random variable is the Stieltjes integral of x with respect to the cdf F of $\tilde{x}$.]

Example 10.2.1

Let $\tilde{x}$ be discrete with mf f given by

$$f(x) = \begin{cases} .10, & x = 1 \\ .30, & x = 2 \\ .60, & x = 3 \\ 0, & x \notin \{1, 2, 3\}. \end{cases}$$

Then $E(\tilde{x}) = (1)(.10) + (2)(.30) + (3)(.60) = 2.50.$

Example 10.2.2

Let $\tilde{x}$ be continuous with df f given, for $\lambda > 0$, by

$$f(x) = \begin{cases} 0, & x \leqq 0 \\ \lambda e^{-\lambda x}, & x > 0, \end{cases}$$

then

$$\begin{aligned} E(\tilde{x}) &= \int_{-\infty}^{0} x \cdot 0 \cdot dx + \int_{0}^{\infty} \lambda x e^{-\lambda x} dx \\ &= 0 + \int_{0}^{\infty} x d(-e^{-\lambda x}) \\ &= -xe^{-\lambda x}\Big|_{0}^{\infty} + \int_{0}^{\infty} e^{-\lambda x} dx \\ &= 0 + \int_{0}^{\infty} d\left(\frac{-e^{-\lambda x}}{\lambda}\right) \\ &= \frac{1}{\lambda}. \end{aligned}$$

Example 10.2.3

Let $\tilde{x}$ be continuous with df f given, for $\rho > 0$, by

$$f(x) = \begin{cases} 0, & x \notin (0, 1) \\ \rho x^{\rho - 1}, & x \in (0, 1), \end{cases}$$

then

$$\begin{aligned} E(\tilde{x}) &= \int_{0}^{1} x \cdot \rho x^{\rho - 1} dx \\ &= \int_{0}^{1} \rho x^{\rho} dx \\ &= \frac{\rho}{\rho + 1}. \end{aligned}$$

Note from Example 10.2.1 that $E(\tilde{x})$ need not be a possible value of $\tilde{x}$.

More generally, let g: $R^1 \rightarrow R^1$ be a continuous function. Then $g \circ x \equiv g(\tilde{x})$ is a random variable, and its expectation is given by

$$E(g(\tilde{x})) = \begin{cases} \sum\{g(x)\cdot f(x) : f(x) > 0\}, & \tilde{x} \text{ discrete} \\ \int_{-\infty}^{\infty} g(x)f(x)dx, & \tilde{x} \text{ continuous.} \end{cases} \quad (10.2.1)$$

We shall no longer formally consider expectations of functions of mixed random variables. [Equation (10.2.1) requires proof in view of the fact that since $\tilde{g} \equiv g(\tilde{x})$ is a random variable, with a dgf, it has an expectation $E(\tilde{g}) \equiv \int_{-\infty}^{\infty} gf(g)dg$. It is not hard to show that $E(\tilde{g}) = E(g(\tilde{x}))$ as given by (10.2.1).]

In (10.2.1), as in the definition of expectation, we assume that the expectation *exists;* that is, it is a *finite* number and also that the defining sum or integral converges *absolutely*. Hence $E(g(\tilde{x}))$ exists if and only if $E(|g(\tilde{x})|) < \infty$.

Let g: $R^w \rightarrow R^1$ be a continuous function, and let $\tilde{\mathbf{x}}$ be a w-dimensional random vector. We may compute the expectation $E(g(\tilde{\mathbf{x}}))$ of the random variable $g(\tilde{\mathbf{x}})$ by

$$E(g(\tilde{\mathbf{x}})) = \int_{R^w} g(\mathbf{x})f(\mathbf{x})d\mathbf{x} \quad (10.2.2)$$

if $\tilde{\mathbf{x}}$ is continuous. If $\tilde{\mathbf{x}}$ is discrete, or mixed in the sense of Chapter 8, then the integration is replaced by a summation, or by a combination of summation and integration. We shall henceforth use the convention in Chapter 8 that an integral can be used to denote a sum where appropriate, and that $d\mathbf{x} \equiv dx_1 \cdot dx_2 \cdot \ldots \cdot dx_w$.

Let $\tilde{\mathbf{x}} \equiv (\tilde{x}_1, \ldots, \tilde{x}_w)$ be partitioned into $\tilde{\mathbf{x}}_1 \equiv (\tilde{x}_1, \ldots, \tilde{x}_m)$ and $\tilde{\mathbf{x}}_2 \equiv (\tilde{x}_{m+1}, \ldots, \tilde{x}_w)$. Then (10.2.2) can be rewritten as

$$E(g(\tilde{\mathbf{x}}_1, \tilde{\mathbf{x}}_2)) = \int_{R^m}\int_{R^{w-m}} g(\mathbf{x}_1, \mathbf{x}_2)f(\mathbf{x}_1, \mathbf{x}_2)dx_2dx_1. \quad (10.2.3)$$

Now

$$f(\mathbf{x}_1, \mathbf{x}_2) = f(\mathbf{x}_2 \mid \mathbf{x}_1)\cdot f(\mathbf{x}_1)$$

implies

$$\int_{R^m}\int_{R^{w-m}} g(\mathbf{x}_1, \mathbf{x}_2)f(\mathbf{x}_1, \mathbf{x}_2)d\mathbf{x}_2d\mathbf{x}_1$$

$$= \int_{R^m}\left[\int_{R^{w-m}} g(\mathbf{x}_1, \mathbf{x}_2)f(\mathbf{x}_2 \mid \mathbf{x}_1)d\mathbf{x}_2\right]f(\mathbf{x}_1)d\mathbf{x}_1,$$

where the outer integral is interpreted as being taken over those $\mathbf{x}_1$'s for which $f(\mathbf{x}_1) > 0$. But the term in brackets is simply the expectation of the function $g(\mathbf{x}_1, \cdot)$ of $\tilde{\mathbf{x}}_2$ with respect to the *conditional* dgf $f(\mathbf{x}_2 \mid \mathbf{x}_1)$ of $\tilde{\mathbf{x}}_2$. We call it the *conditional expectation of* $g(\tilde{\mathbf{x}}_1, \tilde{\mathbf{x}}_2)$ *with respect to* $\tilde{\mathbf{x}}_2$ given $\mathbf{x}_1$. Explicitly:

$$E(g(\mathbf{x}_1, \tilde{\mathbf{x}}_2) \mid \mathbf{x}_1) \equiv \int_{R^{w-m}} g(\mathbf{x}_1, \mathbf{x}_2) f(\mathbf{x}_2 \mid \mathbf{x}_1) d\mathbf{x}_2. \qquad (10.2.4)$$

Hence (10.2.3) can be rewritten as

$$E(g(\tilde{\mathbf{x}}_1, \tilde{\mathbf{x}}_2)) = \int_{R^m} E(g(\mathbf{x}_1, \tilde{\mathbf{x}}_2) \mid \mathbf{x}_1) f(\mathbf{x}_1) d\mathbf{x}_1.$$

But $E(g(\cdot, \tilde{\mathbf{x}}_2) \mid \cdot): R^m \to R^1$ is a function of $\tilde{\mathbf{x}}_1$, and $\int_{R^m} [\cdot] d\mathbf{x}_1$ is simply expectation $E(\cdot)$ with respect to $f(x_1)$. Hence (10.2.3) reduces ultimately to

$$E(g(\tilde{\mathbf{x}}_1, \tilde{\mathbf{x}}_2)) = E[E(g(\tilde{\mathbf{x}}_1, \tilde{\mathbf{x}}_2 \mid \tilde{\mathbf{x}}_1)]. \qquad (10.2.5)$$

In many important cases in the remainder of this book, $g(\cdot, \cdot)$ is independent of $\mathbf{x}_1$ and hence we may define $h(\mathbf{x}_2) = g(\mathbf{x}_1, \mathbf{x}_2)$, for all $(\mathbf{x}_1, \mathbf{x}_2) \in R^m \times R^{w-m}$, whereupon (10.2.5) becomes

$$E(h(\tilde{\mathbf{x}}_2)) = E[E(h(\tilde{\mathbf{x}}_2) \mid \tilde{\mathbf{x}}_1)]. \qquad (10.2.5')$$

Equation (10.2.5) generalizes easily to cases in which $\tilde{g} \equiv g(\tilde{\mathbf{x}}_1, \ldots, \tilde{\mathbf{x}}_n)$. For example, when $n = 3$ we may write

$$E(g(\tilde{\mathbf{x}}_1, \tilde{\mathbf{x}}_2, \tilde{\mathbf{x}}_3)) = E\{E(E[g(\tilde{\mathbf{x}}_1, \tilde{\mathbf{x}}_2, \tilde{\mathbf{x}}_3) \mid \tilde{\mathbf{x}}_1, \tilde{\mathbf{x}}_2] \mid \tilde{\mathbf{x}}_1)\}, \qquad (10.2.6)$$

the innermost (and first performed) expectation being taken with respect to $f(\mathbf{x}_3 \mid \mathbf{x}_1, \mathbf{x}_2)$; the middle expectation being taken with respect to $f(\mathbf{x}_2 \mid \mathbf{x}_1)$, and the outer expectation being taken with respect to $f(\mathbf{x}_1)$.

It should be clear that the *order* in which $f(\mathbf{x}_1, \ldots, \mathbf{x}_n)$ is factored via the chain rule (8.4.8), and the subsequent order in which the integrations are performed, are immaterial whenever $E(g(\tilde{\mathbf{x}}_1, \ldots, \tilde{\mathbf{x}}_n))$ exists, since that condition suffices for interchanging the order of integration and hence the composition of conditional expectations.

Example 10.2.4

Let $(\tilde{x}_1, \tilde{x}_2)$ be discrete with

$$f(x_1, x_2) = \begin{cases} .30, & (x_1, x_2) = (1, 5) \\ .10, & (x_1, x_2) = (1, 3) \\ .40, & (x_1, x_2) = (2, 3) \\ .20, & (x_1, x_2) = (2, 5) \\ 0, & \text{elsewhere;} \end{cases}$$

and let

$$g(x_1, x_2) = \begin{cases} 10, & (x_1, x_2) = (1, 5) \\ 15, & (x_1, x_2) = (1, 3) \\ 30, & (x_1, x_2) = (2, 3) \\ 20, & (x_1, x_2) = (2, 5). \end{cases}$$

To compute $E(g(\tilde{x}_1, \tilde{x}_2) \mid x_1)$ we need $f(x_2 \mid x_1)$; it is given by the following table:

x_2 \ x_1	1	2
3	$\frac{1}{4}$	$\frac{2}{3}$
5	$\frac{3}{4}$	$\frac{1}{3}$
sum	1	1

Now

$$E(g(1, \tilde{x}_2) \mid 1) = g(1, 3)(\tfrac{1}{4}) + g(1, 5)(\tfrac{3}{4}) = \frac{45}{4},$$

while

$$E(g(2, \tilde{x}_2) \mid 2) = g(2, 3)(\tfrac{2}{3}) + g(2, 5)(\tfrac{1}{3}) = \frac{80}{3}.$$

Now,

$$f(x_1) = \begin{cases} .40, & x_1 = 1 \\ .60, & x_1 = 2 \\ 0, & x_1 \notin \{1, 2\}, \end{cases}$$

and hence

$$E[E(g(\tilde{x}_1, \tilde{x}_2) \mid \tilde{x}_1)] = \left(\frac{45}{4}\right)(.40) + \left(\frac{80}{3}\right)(.60) = 20.5.$$

Now,

$$E(g(\tilde{x}_1, \tilde{x}_2)) = (15)(.10) + (10)(.30) + (30)(.40) + (20)(.20) = 20.5,$$

which obviously equals $E[E(g(\tilde{x}_1, \tilde{x}_2) \mid \tilde{x}_1)]$.

Example 10.2.5

In Examples 8.2.7, 8.3.2, and 8.4.4, we had $(\tilde{x}_1, \tilde{x}_2)$ continuous with

$$f(x_2 \mid x_1) = \begin{cases} 3(x_1 - x_2)^2 x_1^{-3}, & 0 < x_2 < x_1 \\ 0, & x_2 \notin (0, x_1), \end{cases}$$

provided that $0 < x_1 < 1$, and

$$f(x_1) = \begin{cases} 4x_1^3, & 0 < x_1 < 1 \\ 0, & x_1 \notin (0, 1). \end{cases}$$

Let $g(x_1, x_2) \equiv h(x_2) \equiv x_2$ for all (x_1, x_2). Then

$$E(\tilde{x}_2 \mid x_1) = \int_0^{x_1} x_2 \cdot 3(x_1 - x_2)^2 x_1^{-3} dx_2$$

$$= \frac{x_1}{4}$$

provided that $x_1 \in (0, 1)$. Hence

$$E[E(\tilde{x}_2 \mid \tilde{x}_1)] = \int_0^1 \left(\frac{x_1}{4}\right) \cdot 4x_1^3 dx_1$$

$$= \tfrac{1}{5}.$$

Now, this result could have been obtained by using the original definition of expectation and recalling that

$$f(x_2) = \begin{cases} 4(1 - x_2)^3, & 0 < x_2 < 1 \\ 0, & x_2 \notin (0, 1): \end{cases}$$

$$E(\tilde{x}_2) = \int_0^1 x_2 \cdot 4(1 - x_2)^3 dx_2$$

$$= \int_0^1 (1 - y) 4y^3 dy, \quad (\text{for } y = 1 - x_2)$$

$$= \tfrac{1}{5}.$$

Example 10.2.6

In Example 10.2.3, define $g(x) = x^3$, for every $x_1 \in R$. Then

$$E(g(\tilde{x})) = \int_0^1 x^3 \rho x^{\rho - 1} dx$$

$$= \int_0^1 \rho x^{\rho + 2} dx$$

$$= \frac{\rho}{\rho + 3}.$$

Now, $g(\tilde{x}) \equiv \tilde{g}$ is a random variable, and

$$P[\tilde{g} \leqq g] = P[\tilde{x}^3 \leqq g]$$
$$= \begin{cases} 0, & g \leqq 0 \\ P[\tilde{x} \leqq g^{1/3}], & 0 < g < 1 \\ 1, & g \geqq 1; \end{cases}$$

and hence

$$F(g) = \begin{cases} 0, & g \leqq 0 \\ g^{\rho/3}, & 0 < g < 1 \\ 1, & g \geqq 1. \end{cases}$$

Hence

$$f(g) = \begin{cases} 0, & g \notin (0, 1) \\ \left(\frac{\rho}{3}\right) g^{(\rho/3)-1}, & g \in (0, 1). \end{cases}$$

Therefore

$$E(\tilde{g}) = \int_0^1 g\left(\frac{\rho}{3}\right) g^{(\rho/3)-1} dg$$
$$= \int_0^1 \left(\frac{\rho}{3}\right) g^{\rho/3} dg$$
$$= \frac{\rho/3}{(\rho/3) + 1}$$
$$= \frac{\rho}{\rho + 3},$$

which agrees with $E(g(\tilde{x}))$ but requires much more tedious effort to obtain.

The following properties of expectation are of fundamental importance. All are consequences of elementary properties of integrals (or sums, or sums-integrals).

Properties of Expectation

Let k, k_1, and k_2 be any real numbers, and let $\tilde{x}$, $\tilde{x}_1$, $\tilde{x}_2$ be any random variables. Then:

$$E(k_1\tilde{x}_1 + k_2\tilde{x}_2) = k_1E(\tilde{x}_1) + k_2E(\tilde{x}_2); \tag{10.2.7}$$

$$E(k\tilde{x}) = kE(\tilde{x}); \tag{10.2.7'}$$

$$E(\tilde{x} + k) = k + E(\tilde{x}); \text{ and} \tag{10.2.7''}$$

$$E(\tilde{x}_1) = E[E(\tilde{x}_1 \mid \tilde{x}_2)]. \tag{10.2.8}$$

Proof

For (10.2.7) we have

$E(k_1\tilde{x}_1 + k_2\tilde{x}_2)$

(1) $$= \int_{-\infty}^{\infty}\int_{-\infty}^{\infty} (k_1x_1 + k_2x_2)f(x_1, x_2)dx_1dx_2$$

(2) $$= k_1\int_{-\infty}^{\infty}\int_{-\infty}^{\infty} x_1f(x_1, x_2)dx_1dx_2 + k_2\int_{-\infty}^{\infty}\int_{-\infty}^{\infty} x_2f(x_1, x_2)dx_1dx_2$$

(3) $$= k_1\int_{-\infty}^{\infty} x_1f(x_1)\left[\int_{-\infty}^{\infty} f(x_2 \mid x_1)dx_2\right]dx_1$$
$$+k_2\int_{-\infty}^{\infty} x_2f(x_2)\left[\int_{-\infty}^{\infty} f(x_1 \mid x_2)dx_1\right]dx_2$$

(4) $$= k_1\int_{-\infty}^{\infty} x_1f(x_1)dx_1 + k_2\int_{-\infty}^{\infty} x_2f(x_2)dx_2$$

(5) $$= k_1E(\tilde{x}_1) + kE_2(\tilde{x}_2),$$

equalities (1) and (5) being the definition of expectation, (2) being a consequence of linearity of integrals with respect to integrands, (3) being rearrangement and invocation of the multiplication rule for df's, and (4) following from the fact that each of the inner integrals in brackets equals 1. Property (10.2.7′) follows from (10.2.7) by letting $k_1 = k$ and $k_2 = 0$ in (10.2.7). Property (10.2.7″) follows from (10.2.7) by letting $\tilde{x} = \tilde{x}_1$, $k_1 = 1$, $k_2 = k$, and $\tilde{x}_2(q) = 1$, for every $q \in Q$. Property (10.2.8) has already been derived as (10.2.5). QED

10.3 MOMENTS OF RANDOM VARIABLES

We have introduced expectations and conditional expectations of random variables and of functions of several random variables. Moreover, you may verify by inspection that (10.2.7″) shows that expectation is a measure of location, in the sense discussed in Section 7.3. In this section we shall consider other summary measures of the distribution of $\tilde{x}$, of the joint distribution of $\tilde{x}_1$, $\tilde{x}_2$, and (more generally) of the joint distribution of $\tilde{x}_1, \ldots, \tilde{x}_w$.

Let r be any real number. We define the rth *moment* of $\tilde{x}$ *about zero* as $E(\tilde{x}^r)$ [$= E(g(\tilde{x}))$ for $g(x) \equiv x^r$] when it exists, and we write

$$\mu^{(r)} \equiv E(\tilde{x}^r). \tag{10.3.1}$$

You may verify as Exercise 31 that

$$\mu^{(0)} = 1, \tag{10.3.2}$$

for any random variable $\tilde{x}$. The most useful moments $\mu^{(r)}$ about zero are those in which r is a positive integer, usually $r = 1$ or $r = 2$.

Example 10.3.1

Let $f(x)$ be as defined in Example 10.2.3. Then the rth moments exist for all $r > -\rho$, and we have

$$\begin{aligned}\mu^{(r)} = E(\tilde{x}^r) &= \int_0^1 x^r \rho x^{\rho-1} dx \\ &= \int_0^1 \rho x^{\rho+r-1} dx \\ &= \frac{\rho}{\rho + r}.\end{aligned}$$

As with expectations in general, the rth moment $\mu^{(r)}$ is said to *exist* if and only if $E(|x^r|)$ is finite. It can be shown that if $\mu^{(r)}$ exists for some $r > 0$, then $\mu^{(r')}$ also exists for every $r' \in [0, r]$.

The first moment $\mu^{(1)}$ of $\tilde{x}$ about zero is by far the most important. It is, from (10.3.1), simply the expectation $E(\tilde{x})$ of $\tilde{x}$. Where no confusion can result, we also denote $\mu^{(1)}$ by μ.

There is an important frequency interpretation of μ as a measure of location. Suppose you perform an experiment with a predetermined, large number n of independent trials whose outcomes are iid random variables $\tilde{x}_1, \tilde{x}_2, \ldots, \tilde{x}_n$ with common mf f defined by

$$f(x) = \begin{cases} .10, & x = 1 \\ .60, & x = 2 \\ .30, & x = 3 \\ \ \ 0, & x \notin \{1, 2, 3\}. \end{cases}$$

It stands to reason that *if* n is large and *if* $\tilde{x}_1, \ldots, \tilde{x}_n$ are iid with common mf f, *then* $x_i = 1$ on roughly 10 percent of the trials, $x_i = 2$ on roughly 60 percent of the trials, and $x_i = 3$ on roughly 30 percent of the trials. Hence the *sample mean* $\bar{\tilde{x}}_n$, defined by

$$\bar{x}_n \equiv \left(\frac{1}{n}\right) \sum_{i=1}^{n} x_i,$$

should equal roughly

$$\left(\frac{1}{n}\right)[(.10n)(1) + (.60n)(2) + (.30n)(3)]$$

$$= (.10)(1) + (.60)(2) + (.30)(3) = 2.20$$

$$= E(\tilde{x}) \equiv \mu,$$

the expectation of each one of the $\tilde{x}_i$'s (because they are iid).

The approximate equality between μ, also called the *population mean,* and the sample mean $\tilde{x}_n$ is probabilistic in nature. In Section 10.5 we shall state the approximate equality in precise form as the "weak law of large numbers."

Let r be any real number. The rth *central moment* of $\tilde{x}$ is defined as

$$\mu_{(r)} \equiv E([\tilde{x} - \mu]^r). \tag{10.3.8}$$

If $E(|\tilde{x} - \mu|^r)$ is finite for some positive number r, then $\mu_{(r')}$ exists for every $r' \in [0, r]$.

By far the most important rth central moment is the second; that is, $r = 2$. It is called the *variance* of $\tilde{x}$ and is denoted variously by v, $\sigma^2(\tilde{x})$, var $(\tilde{x})$, V, σ^2, and $V(\tilde{x})$. We shall most frequently use the last two of these symbols.

Example 10.3.2

The variance $V(\tilde{x})$ for $\tilde{x}$ as given in Example 10.2.3 is

$$V(\tilde{x}) = \int_0^1 \left[x - \frac{\rho}{\rho + 1}\right]^2 \rho x^{\rho-1} dx$$

$$= \frac{\rho}{\rho + 2} - \left[\frac{\rho}{\rho + 1}\right]^2$$

$$= \frac{\rho}{(\rho + 1)^2(\rho + 2)}.$$

By change-of-variable theorems it can be shown that if $\tilde{\mathbf{x}}$ is any w-dimensional random vector and $g: R^w \rightarrow R^1$ is continuous, then

$$V(g(\tilde{\mathbf{x}})) = E([g(\tilde{\mathbf{x}}) - E(g(\tilde{\mathbf{x}}))]^2)$$

$$= \int_{R^w} [g(\mathbf{x}) - E(g(\tilde{\mathbf{x}}))]^2 f(\mathbf{x}) d\mathbf{x}. \tag{10.3.9}$$

Example 10.3.3

Let $\tilde{x}$ be as in Example 10.3.2 and define $g\colon R^1 \to R^1$ by $g(x) = x^3$, for every $x \in R^1$. Then

$$E(g(\tilde{x})) = E(\tilde{x}^3) = \frac{\rho}{\rho + 3},$$

as shown in Example 10.2.6. Hence

$$\begin{aligned} V(g(\tilde{x})) &= E\left(\left[\tilde{x}^3 - \frac{\rho}{\rho+3}\right]^2\right) \\ &= \int_0^1 \left[x^3 - \frac{\rho}{\rho+3}\right]^2 \rho x^{\rho-1} dx \\ &= \frac{\rho}{\rho+6} - \left[\frac{\rho}{\rho+3}\right]^2 \\ &= \frac{9\rho}{(\rho+3)^2(\rho+6)}. \end{aligned}$$

Alternatively, recall that in Example 10.2.6 we derived the df $f(g)$ of $\tilde{g} \equiv g(\tilde{x})$; from it we find

$$\begin{aligned} V(g) &= E\left(\left[\tilde{g} - \frac{\rho}{\rho+3}\right]^2\right) \\ &= \int_0^6 \left[g - \frac{\rho}{\rho+3}\right]^2 (\rho/3) g^{\rho/3-6} dg \\ &= \frac{\rho}{\rho+6} - \left[\frac{\rho}{\rho+3}\right]^2, \end{aligned}$$

which agrees with $V(g(\tilde{x}))$—as it must in view of its nature as a change-of-variable manipulation. In most applications it is easier to compute $E(g(\tilde{x}))$ and $V(g(\tilde{x}))$ than $E(\tilde{g})$ and $V(\tilde{g})$.

The variance of a random variable is an *index of uncertainty* as to the value it will assume. To see this, consider the random variable $\tilde{x}_i$ which can assume only the values i and $-i$ (for $i > 0$), each with probability $\frac{1}{2}$. Then $E(\tilde{x}_i) = (i)(\frac{1}{2}) + (-i)(\frac{1}{2}) = 0$, for every i. However, $V(\tilde{x}_i) = (i - 0)^2 (\frac{1}{2}) + (-i - 0)^2(\frac{1}{2}) = (i^2)(\frac{1}{2}) + (i^2)(\frac{1}{2}) = i^2$, which is an *increasing* function of i. Thus $V(\tilde{x}_1) = 1$ and $V(\tilde{x}_5) = 25$. Now note that there is much more uncertainty about $\tilde{x}_5$ than there is about $\tilde{x}_1$, since $\tilde{x}_1$ will deviate from its expectation only by one unit while $\tilde{x}_5$ will deviate from its expectation by five units. Alternatively, the possible values of $\tilde{x}_1$ are only two units apart while the possible values of $\tilde{x}_5$ are ten units apart.

Note that $V(\tilde{x})$ is measured in "x-units squared"; for example, if $\tilde{x}$ measures height in inches, $V(\tilde{x})$ is inches squared. For purposes of interpretation and explanation it is desirable to have a measure of uncertainty of $\tilde{x}$ that is stated in x-units themselves. To obtain such a measure we simply take the positive square root of $V(\tilde{x})$, define it as the *standard deviation* of $\tilde{x}$, and denote it by $\upsilon^{1/2}$, $\sigma(\tilde{x})$, $\text{var}^{1/2}(\tilde{x})$, $V^{1/2}$, σ, or $V^{1/2}(\tilde{x})$.

There are a number of important properties of variance which are of both conceptual and computational significance:

Properties of Variance

Let k, k_1, and k_2 be any real numbers, and let $\tilde{x}$, $\tilde{x}_1$, . . . , $\tilde{x}_n$ be random variables. Then

$$V(k_1\tilde{x}_1 + k_2\tilde{x}_2) = k_1{}^2V(\tilde{x}_1) + k_2{}^2V(\tilde{x}_2), \tag{10.3.10}$$

provided that $\tilde{x}_1$ and $\tilde{x}_2$ are *independent*;

$$V\left(\sum_{i=1}^{n} \tilde{x}_1\right) = \sum_{i=1}^{n} V(\tilde{x}_i) \tag{10.3.10$'$}$$

provided that $\tilde{x}_1, \ldots, \tilde{x}_n$ are *independent*;

$$V(\tilde{x}) = E(\tilde{x}^2) - [E(\tilde{x})]^2; \tag{10.3.11}$$

$$V(\tilde{x}) \geqq 0; \tag{10.3.12}$$

$$V(k\tilde{x}) = k^2V(\tilde{x}); \tag{10.3.13}$$

$$V(\tilde{x}) = 0 \tag{10.3.14}$$

if $\tilde{x}$ is a degenerate random variable (that is, assumes some given $x \in R^1$ with probability 1).

Proof

Exercise 19.

Let $\tilde{\mathbf{x}} = (\tilde{x}_1, \ldots, \tilde{x}_w)$ and let $(r_1, \ldots, r_w) \in R^w$. The $(r_1, \ldots, r_w)$th *mixed moment of* $\tilde{\mathbf{x}}$ *about zero* is defined as

$$\mu^{(r_1,\ldots,r_w)} \equiv E(\tilde{x}_1{}^{r_1}\cdot\tilde{x}_2{}^{r_2}\cdot \ldots \cdot\tilde{x}_w{}^{r_w}) \tag{10.3.15}$$

(provided that it exists). The most important such moments are those in which i and j are different integers in $\{1, \ldots, w\}$, $r_i = r_j = 1$, and $r_k = 0$ whenever $i \neq k \neq j$; that is, moments of the form $E(\tilde{x}_i \cdot \tilde{x}_j)$.

Let $\mu_i \equiv E(\tilde{x}_i)$, for $i = 1, \ldots, w$. The $(r_1, \ldots, r_w)$th *mixed central moment* of $\tilde{\mathbf{x}}$ is defined as

$$\mu_{(r_1,\ldots,r_w)} \equiv E\left(\prod_{i=1}^{n} [\tilde{x}_i - \mu_i]^{r_i}\right) \tag{10.3.16}$$

(provided that it exists). By far the most important mixed central moments are those of the form $E([\tilde{x}_i - \mu_i][\tilde{x}_j - \mu_j])$; they are called *covariances* and are denoted by v_{ij}, cov $(\tilde{x}_i, \tilde{x}_j)$, V_{ij}, σ_{ij}^2, or $V(\tilde{x}_i, \tilde{x}_j)$. [Note that $V(\tilde{x}_i, \tilde{x}_j)$ has *two* arguments while $V(\tilde{x}_i)$ has only one. Hence no confusion will result.]

The covariance of $\tilde{x}_i$ and $\tilde{x}_j$ is a measure of the degree to which $\tilde{x}_i$ and $\tilde{x}_j$ tend to *simultaneously* exceed or fail to exceed their expectations. If $V(\tilde{x}_i, \tilde{x}_j) > 0$, then $\tilde{x}_i$ and $\tilde{x}_j$ do tend to be sumultaneously large and simultaneously small; and if $V(\tilde{x}_i, \tilde{x}_j) < 0$, then $\tilde{x}_i$ tends to be large when $\tilde{x}_j$ is small and vice versa. You may show that if $\tilde{x}_i$ and $\tilde{x}_j$ are independent, then $V(\tilde{x}_i, \tilde{x}_j) = 0$ in Exercise 20. However, the converse is false in general: $V(\tilde{x}_i, \tilde{x}_j) = 0$ does not imply that $\tilde{x}_i$ and $\tilde{x}_j$ are independent (see Exercise 21). Note also that if $i = j$, then $V(\tilde{x}_i, \tilde{x}_j) = V(\tilde{x}_i)$, a fact that the v_{ij}, V_{ij}, and σ_{ij}^2 notions accommodate quite nicely.

Example 10.3.4

Let $(\tilde{x}_1, \tilde{x}_2)$ be as given in Example 10.2.4. Then

$$\begin{aligned} E(\tilde{x}_1) &= (1)(.40) + (2)(.60) = 1.60 \text{ and} \\ E(\tilde{x}_2) &= (3)(.50) + (5)(.50) = 4.00; \text{ and} \\ V(\tilde{x}_1, \tilde{x}_2) &= (1 - 1.60)(5 - 4.00)(.30) \\ &\quad + (1 - 1.60)(3 - 4.00)(.10) \\ &\quad + (2 - 1.60)(3 - 4.00)(.40) \\ &\quad + (2 - 1.60)(5 - 4.00)(.20) = -.20, \end{aligned}$$

which indicates that $\tilde{x}_2$ tends to be large when x_1 is small; this qualitative statement is corroborated by a comparison of $f(x_2 \mid x_1 = 1)$ with $f(x_2 \mid x_1 = 1)$ (see Example 10.2.4).

The covariance of $\tilde{x}_i$ and $\tilde{x}_j$ is stated in dimensions "x_1-units times x_2-units." It is useful to have a dimensionless measure of the tendency of $\tilde{x}_i$ and $\tilde{x}_j$ to vary together. We define the *correlation* between $\tilde{x}_i$ and $\tilde{x}_j$, denoted ρ_{ij} or $\rho(\tilde{x}_i, \tilde{x}_j)$, by

$$\rho_{ij} \equiv \frac{\sigma_{ij}^2}{\sigma(\tilde{x}_i)\cdot\sigma(\tilde{x}_j)} \equiv \frac{V(\tilde{x}_i, \tilde{x}_j)}{\sqrt{V(\tilde{x}_i)\cdot V(\tilde{x}_j)}}. \tag{10.3.17}$$

Example 10.3.5

For Example 10.3.4 we may compute

$$V_{11}^{1/2} = [(1 - 1.60)^2(.40) + (2 - 1.60)^2(.60)]^{1/2} = (.24)^{1/2} = .49,$$

and

$$V_{22}^{1/2} = [(3 - 4)^2(.50) + (5 - 4)^2(.50)]^{1/2} = (1.00)^{1/2} = 1.00,$$

and hence

$$\rho_{12} = \frac{-.20}{[(0.49)(1.00)]} = -.41.$$

Let $\tilde{\mathbf{x}} = (\tilde{\mathbf{x}}_1, \tilde{\mathbf{x}}_2)$, where $\tilde{\mathbf{x}}_1 = (\tilde{x}_1, \ldots, \tilde{x}_m)$. Let $(r_1, \ldots, r_m) \in R^m$. Then the $(r_1, \ldots, r_m)$th *conditional mixed moment of* $\tilde{\mathbf{x}}_1$ *about zero, given* $\tilde{\mathbf{x}}_2$, is defined as

$$\mu^{(r_1,\ldots,r_m)}(\mathbf{x}_2) \equiv E\left(\prod_{i=1}^{m} x_i^{r_i} \mid \mathbf{x}_2\right), \tag{10.3.18}$$

for every $\mathbf{x}_2$ such that $f(\mathbf{x}_2) > 0$; and the $(r_1, \ldots, r_m)$th *conditional mixed central moment of* $\tilde{\mathbf{x}}_1$ *given* $\tilde{\mathbf{x}}_2$ is defined as

$$\mu_{(r_1,\ldots,r_m)}(\mathbf{x}_2) \equiv E\left(\prod_{i=1}^{m} [\tilde{x}_i - E(\tilde{x}_i \mid \mathbf{x}_2)]^{r_i} \mid \mathbf{x}_2\right), \tag{10.3.19}$$

for every $\mathbf{x}_2$ with $f(\mathbf{x}_2) > 0$. The most important special cases of these conditional moments are:

(1) *conditional expectation*, or *conditional mean*, $E(\tilde{x}_i \mid \mathbf{x}_2)$, denoted by $\mu_i(\mathbf{x}_2)$;
(2) *conditional variance*, $E([\tilde{x}_i - \mu_i(\mathbf{x}_2)]^2 \mid \mathbf{x}_2]$, denoted by $V(\tilde{x}_i \mid \mathbf{x}_2)$, $V_{ii}(\mathbf{x}_2)$, or $\sigma_{ii}{}^2(\mathbf{x}_2)$; and
(3) *conditional covariance*, $E([\tilde{x}_i - \mu_i(\mathbf{x}_2)][\tilde{x}_j - \mu_j(\mathbf{x}_2)] \mid \mathbf{x}_2)$, denoted by $V(\tilde{x}_i, \tilde{x}_j \mid \mathbf{x}_2)$, $V_{ij}(\mathbf{x}_2)$, or $\sigma_{ij}{}^2(\mathbf{x}_2)$.

We also define the *conditional correlation between* $\tilde{x}_i$ *and* $\tilde{x}_j$ *given* $\mathbf{x}_2$, denoted by $\rho(\tilde{x}_i, \tilde{x}_j \mid \mathbf{x}_2)$ or $\rho_{ij}(\mathbf{x}_2)$:

$$\rho_{ij}(\mathbf{x}_2) \equiv \frac{\sigma_{ij}{}^2(\mathbf{x}_2)}{\sqrt{\sigma_{ii}{}^2(\mathbf{x}_2)\cdot\sigma_{jj}{}^2(\mathbf{x}_2)}} \equiv \frac{V(\tilde{x}_i, \tilde{x}_j \mid \mathbf{x}_2)}{\sqrt{V(\tilde{x}_i \mid \mathbf{x}_2)\cdot V(\tilde{x}_j \mid \mathbf{x}_2)}}. \tag{10.3.20}$$

The simplest way of recalling the definitions of these conditional analogues of variance, covariance, and correlation is to suppose that you have learned that $\tilde{\mathbf{x}}_2 = \mathbf{x}_2$ and to note that your judgments regarding $\tilde{\mathbf{x}}_1$ are *now* expressed fully by $f(\mathbf{x}_1 \mid \mathbf{x}_2)$.

Since conditional variance, covariance, and correlation (and, indeed, all conditional moments) depend upon $\mathbf{x}_2$, they are functions of $\tilde{\mathbf{x}}_2$ and hence random variables.

More Properties of Variance, Covariance, and Correlation

Let $\tilde{x}_i$ and $\tilde{x}_j$ be any random variables, neither a component of $\tilde{\mathbf{x}}_2$; and let k_i, k_j, m_i, and m_j be real numbers. Then:

$$V(\tilde{x}_i, \tilde{x}_j) = V(\tilde{x}_j, \tilde{x}_i); \tag{10.3.21}$$

$$V(k_i\tilde{x}_i + m_i, k_j\tilde{x}_j + m_j) = k_ik_jV(\tilde{x}_i, \tilde{x}_j); \tag{10.3.22}$$

$$V(k_i\tilde{x}_i + k_j\tilde{x}_j) = k_i^2V(\tilde{x}_i) + k_j^2V(\tilde{x}_j) + 2k_ik_jV(\tilde{x}_i, \tilde{x}_j); \tag{10.3.23}$$

$$V(\tilde{x}_i, \tilde{x}_j) = E(\tilde{x}_i \cdot \tilde{x}_j) - \mu_i \cdot \mu_j = E[\tilde{x}_i \cdot \mu_j(\tilde{x}_i)] - \mu_i \cdot \mu_j; \tag{10.3.24}$$

$$\rho(k_i\tilde{x}_i + m_i, k_j\tilde{x}_j + m_j) = \rho(\tilde{x}_i, \tilde{x}_j) \cdot \left[\frac{k_ik_j}{|k_ik_j|}\right], \text{ provided that } k_i \neq 0 \neq k_j; \tag{10.3.25}$$

$$-1 \leqq \rho(\tilde{x}_i, \tilde{x}_j) \leqq 1; \tag{10.3.26}$$

$$V(\tilde{x}_i) = E[V(\tilde{x}_i \mid \tilde{\mathbf{x}}_2)] + V[\mu_i(\tilde{\mathbf{x}}_2)]; \text{ and} \tag{10.3.27}$$

$$V(\tilde{x}_i) \geqq E[V(\tilde{x}_i \mid \tilde{\mathbf{x}}_2)]. \tag{10.3.28}$$

Proof

The property (10.3.26) follows from the Cauchy-Schwarz inequality in analysis; (10.3.21) is obvious from the definition of covariance; (10.3.28) follows from (10.3.27) via (10.3.12) applied to $\tilde{x} \equiv \mu_i(\tilde{\mathbf{x}}_2)$. Properties (10.3.22) through (10.3.25) are left to the reader as Exercise 22. To prove (10.3.27), use the definition of $V(\tilde{x}_i)$ to obtain

$$\begin{aligned} V(\tilde{x}_i) \equiv E([\tilde{x}_i - \mu_i]^2) &= E([\{\tilde{x}_i - \mu_i(\tilde{\mathbf{x}}_2)\} + \{\mu_i(\tilde{\mathbf{x}}_2) - \mu_i\}]^2) \\ (*) \qquad &= E[E([\{\tilde{x}_i - \mu_i(\tilde{\mathbf{x}}_2)\} + \{\mu_i(\tilde{\mathbf{x}}_2) - \mu_i\}]^2 \mid \mathbf{x}_2)]. \end{aligned}$$

Now, by expanding the square and evaluating $E(\cdot \mid \mathbf{x}_2)$, we obtain

$$E([\{\tilde{x}_i - \mu_i(\mathbf{x}_2)\} + \{\mu_i(\mathbf{x}_2) - \mu_i\}]^2 \mid \mathbf{x}_2) = V(\tilde{x}_i \mid \mathbf{x}_2) + [\mu_i(\mathbf{x}_2) - \mu_i]^2,$$

for all $\mathbf{x}_2$ where the conditional expectations are defined; namely, where $f(\mathbf{x}_2) > 0$. Now, $\mu_i = E(\mu_i(\tilde{\mathbf{x}}_2))$ by (10.2.8); and hence (*) becomes

$$\begin{aligned} &E[V(\tilde{x}_i \mid \tilde{\mathbf{x}}_2) + E([\mu_i(\tilde{\mathbf{x}}_2) - E(\mu_i(\tilde{\mathbf{x}}_2))]^2)] \\ &\qquad = E[V(\tilde{x}_i \mid \tilde{\mathbf{x}}_2)] + E([\mu_i(\tilde{\mathbf{x}}_2) - E(\mu_i(\tilde{\mathbf{x}}_2))]^2] \\ &\qquad = E[V(\tilde{x}_i \mid \tilde{\mathbf{x}}_2)] + V[\mu_i(\tilde{\mathbf{x}}_2)], \text{ as asserted.} \quad \text{QED} \end{aligned}$$

Note that (10.3.23) is the generalization of (10.3.10) to the case in which $\tilde{x}_i$ and $\tilde{x}_j$ are possibly *not* independent. Intuitively, if $V(\tilde{x}_i, \tilde{x}_j) > 0$, the variance $V(\tilde{x}_1 + \tilde{x}_2)$ should be large because fluctuations of $\tilde{x}_i$ and $\tilde{x}_j$ tend to reinforce each other rather than to be mutually canceling, as they would be if $V(\tilde{x}_i, \tilde{x}_j)$ were negative.

Property (10.3.28) implies that if you use variance to measure your uncertainty about random variables and if you are to observe the value $\mathbf{x}_2$ of $\tilde{\mathbf{x}}_2$, then the expectation $E[V(\tilde{\mathbf{x}}_i \mid \tilde{\mathbf{x}}_2)]$ of the remaining uncertainty $V(\tilde{\mathbf{x}}_i \mid \tilde{\mathbf{x}}_2)$ about $\tilde{x}_i$ after observing $\mathbf{x}_2$ is never more than your current uncertainty $V(\tilde{x}_i)$ about $\tilde{x}_i$. In summary, you "expect" that your uncertainty about $\tilde{x}_i$ will not be increased.

Example 10.3.6

In Example 10.2.5, let $i = 2$ and $\mathbf{x}_2 = \tilde{x}_1$ to obtain

$$V(\mu_2(\tilde{x}_1)) = V\left(\frac{\tilde{x}_1}{4}\right) = \frac{V(\tilde{x}_1)}{16}.$$

To find $V(\tilde{x}_1)$, we first compute

$$E(\tilde{x}_1{}^2) = \int_0^1 x_1{}^2 \cdot 4x_1{}^3 dx_1 = \tfrac{2}{3}$$

and

$$E(\tilde{x}_1) = \int_0^1 x_1 \cdot 4x_1{}^3 dx_1 = \tfrac{4}{5}.$$

Hence, by (10.3.11)

$$V(\tilde{x}_1) = \frac{2}{3} - \left(\frac{4}{5}\right)^2 = \frac{2}{75}.$$

Therefore

$$V(\mu_2(\tilde{x}_1)) = \frac{\left(\frac{2}{75}\right)}{16} = \frac{1}{600}.$$

Now, by (10.3.11),

$$\begin{aligned} V(\tilde{x}_2) &= E(\tilde{x}_2{}^2) - [E(\tilde{x}_1)]^2 \\ &= \int_0^1 x_2{}^2 . 4(1 - x_2)^3 dx_2 - (\tfrac{1}{5})^2 \\ &= \frac{1}{15} - \frac{1}{25} = \frac{2}{75}. \end{aligned}$$

Hence

$$E[V(\tilde{x}_2 \mid \tilde{x}_1)] = \frac{2}{75} - \frac{1}{600} = \frac{1}{40}.$$

Now,

$$V(\tilde{x}_2 \mid x_1) = E(\tilde{x}_2{}^2 \mid x_1) - [E(\tilde{x}_2 \mid x_1)]^2$$
$$= \int_0^{x_1} x_2{}^2 \cdot 3(x_1 - x_2)^2 \cdot x_1{}^{-3} dx_2 - \left(\frac{x_1}{4}\right)^2$$
$$= \left(\frac{3}{80}\right) x_1{}^2.$$

Hence

$$E[V(\tilde{x}_2 \mid \tilde{x}_1)] = \int_0^1 \left(\frac{3}{80}\right) x_1{}^2 \cdot 4x_1{}^3 dx_1 = \frac{1}{40},$$

as obtained previously. But note that $V(\tilde{x}_2 \mid x_1) > V(\tilde{x}_2)$ if $x_1 \in [(4\sqrt{10}/15), 1]$, which is possible. Hence if $\tilde{x}_1$ is "too big," your uncertainty about $\tilde{x}_2$ will be *increased* by learning x_1. However, (10.3.27) asserts, roughly, that you do not expect $\tilde{x}_1$ to be too big.

Example 10.3.7

For the same $f(x_1, x_2)$ as in Example 10.3.6, we shall compute $V(\tilde{x}_1, \tilde{x}_2)$ by using the computationally convenient result (10.3.24) to write

$$V(\tilde{x}_1, \tilde{x}_2) = E(\tilde{x}_1 \cdot E(\tilde{x}_2 \mid \tilde{x}_1)] - E(\tilde{x}_1)E(\tilde{x}_2)$$
$$= E\left[\tilde{x}_1 \cdot \left(\frac{\tilde{x}_1}{4}\right)\right] - \left(\frac{4}{5}\right)\left(\frac{1}{5}\right)$$
$$= \frac{E(\tilde{x}_1{}^2)}{4} - \frac{4}{25}$$
$$= \frac{(2/3)}{4} - \frac{4}{25}$$
$$= \frac{1}{150}.$$

The reader may compute $E(\tilde{x}_1 \mid \tilde{x}_2)$, $V(\tilde{x}_1 \mid \tilde{x}_2)$, and so forth, and $\rho(\tilde{x}_1, \tilde{x}_2)$ as Exercise 24.

Occasionally we shall consider the *precision* $\sigma^{-2}(\tilde{x})$ of a random variable $\tilde{x}$. It is defined as the reciprocal of the variance:

$$\sigma^{-2}(\tilde{x}) \equiv [\sigma^2(\tilde{x})]^{-1} \equiv [V(\tilde{x})]^{-1}. \qquad (10.3.29)$$

Since $V(\tilde{x})$ measures one's *uncertainty* about $\tilde{x}$, it follows that $\sigma^{-2}(\tilde{x})$ measures one's *certainty*, or *information* about $\tilde{x}$. The term *precision*, apparently due to Raiffa and Schlaifer [**66**], is intended to connote "precision *of information*," although here (as elsewhere) it can be dangerous to carry intuition too far into analysis.

10.4 EXPECTATIONS AND VARIANCES OF RANDOM VECTORS

Let $\tilde{\mathbf{x}} = (\tilde{x}_1, \ldots, \tilde{x}_w)'$ be a random vector. We define the expectation $\mu \equiv E(\tilde{\mathbf{x}})$ of $\tilde{\mathbf{x}}$ as the vector of expectations of the components of $\tilde{x}$; that is,

$$\mu \equiv (E(\tilde{x}_1). \ldots, E(\tilde{x}_w)).' \tag{10.4.1}$$

More generally, if X is the random matrix

$$\begin{bmatrix} \tilde{x}_{11} & \cdots & \tilde{x}_{1w_2} \\ \tilde{x}_{21} & \cdots & \tilde{x}_{2w_2} \\ \cdot & & \\ \cdot & & \\ \cdot & & \\ \tilde{x}_{w_1 1} & \cdots & \tilde{x}_{w_1 w_2} \end{bmatrix},$$

then $u \equiv E(\tilde{X})$ is the corresponding matrix of expectations.

$$\boldsymbol{\mu} \equiv \begin{bmatrix} E(\tilde{x}_{11}) & \cdots & E(\tilde{x}_{1w_2}) \\ \cdot & & \cdot \\ \cdot & & \cdot \\ \cdot & & \cdot \\ E(\tilde{x}_{w_1 1}) & \cdots & E(\tilde{x}_{w_1 w_2}) \end{bmatrix}. \tag{10.4.2}$$

We define the *covariance matrix* $V(\tilde{\mathbf{x}}) \equiv \boldsymbol{\sigma}^2$ of $\tilde{\mathbf{x}}$ as the matrix of covariances $V(\tilde{x}_i, \tilde{x}_j) \equiv \sigma_{ij}{}^2$, for all $i \in \{1, \ldots, w\}$ and all $j \in \{1, \ldots, w\}$:

$$\boldsymbol{\sigma}^2 \equiv \begin{bmatrix} \sigma_{11}{}^2 & & \sigma_{1w}{}^2 \\ \sigma_{21}{}^2 & \cdots & \sigma_{2w}{}^2 \\ \cdot & & \cdot \\ \cdot & & \cdot \\ \sigma_{w1}{}^2 & \cdots & \sigma_{ww}{}^2 \end{bmatrix}. \tag{10.4.3}$$

The reader may show as Exercise 29 that

$$\boldsymbol{\sigma}^2 = E[(\tilde{\mathbf{x}} - \boldsymbol{\mu})(\tilde{\mathbf{x}} - \boldsymbol{\mu})']. \tag{10.4.4}$$

[Recall that all vectors are column vectors unless transposed; hence the argument of $E[\cdot]$ in (10.4.4) is a $w \times w$ matrix.]

From (10.3.21) we deduce that $\boldsymbol{\sigma}^2$ is *symmetric*: $\sigma_{ij}{}^2 = \sigma_{ji}{}^2$ for all i and j. The reader may show as Exercise 25 that $\boldsymbol{\sigma}^2$ is *nonnegative definite*: $\mathbf{d}'\boldsymbol{\sigma}^2\mathbf{d} \geqq 0$ for every $\mathbf{d} \in R^w$. We say that $\tilde{\mathbf{x}}$ is *nondegenerate*, or *nonsingular* if and only if $\boldsymbol{\sigma}^2$ is *positive definite*; that is, $\boldsymbol{\sigma}^2$ is nonnegative definite and $\mathbf{d}'\boldsymbol{\sigma}^2\mathbf{d} = 0$ *only* if $\mathbf{d} = \mathbf{0} \in R^w$.

We digress to give a useful sufficient condition for $\boldsymbol{\sigma}^2$ to be positive definite:

$\boldsymbol{\sigma}^2$ is positive definite if and only if det (**A**) > 0 for every leading principal submatrix **A** of $\boldsymbol{\sigma}^2$. (A leading principal submatrix **A** is obtained by deleting the last j rows and columns of $\boldsymbol{\sigma}^2$, for $j = 0, 1, \ldots, w - 1$. See Hadley [**33**], p. 260.) (10.4.5)

Let $\tilde{\mathbf{x}}$ be a random vector and let **B** be an $m \times w$ matrix. Define $\tilde{\mathbf{y}} \equiv \mathbf{B} \cdot \tilde{\mathbf{x}}$. The reader may show as Exercise 26 that

$$E(\tilde{\mathbf{y}}) = \mathbf{B} \cdot E(\tilde{\mathbf{x}}) = \mathbf{B}\boldsymbol{\mu} \tag{10.4.6}$$

and

$$V(\tilde{\mathbf{y}}) = \mathbf{B} \cdot V(\tilde{\mathbf{x}}) \cdot \mathbf{B}' = \mathbf{B}\boldsymbol{\sigma}^2\mathbf{B}'. \tag{10.4.7}$$

Equations (10.4.6) and (10.4.7) show that finding the mean vector and covariance matrix of a set of *linear combinations* of the components of $\tilde{\mathbf{x}}$ is a simple task, at least conceptually if not always computationally.

The reader may show as Exercise 27 that $\tilde{\mathbf{y}} = \mathbf{B}\tilde{\mathbf{x}}$ is nondegenerate if and only if $\tilde{\mathbf{x}}$ is nondegenerate, $m \leqq w$, and rank (**B**) = m.

We shall have occasion to consider the *precision matrix* $\boldsymbol{\sigma}^{-2}$ of $\tilde{\mathbf{x}}$; it is defined as the inverse of $\boldsymbol{\sigma}^2$:

$$\boldsymbol{\sigma}^{-2} \equiv (\boldsymbol{\sigma}^2)^{-1}. \tag{10.4.8}$$

[Compare (10.4.8) with (10.3.29) for single random variables.] Naturally, $\boldsymbol{\sigma}^{-2}$ is defined only when $\boldsymbol{\sigma}^2$ is nonsingular.

Now suppose that $\tilde{\mathbf{x}}$ is partitioned into $\tilde{\mathbf{x}}_1 \equiv (\tilde{\mathbf{x}}_1, \ldots, \tilde{x}_m)'$ and $\tilde{\mathbf{x}}_2 = (\tilde{x}_{m+1}, \ldots, \tilde{x}_w).'$ Partition $\boldsymbol{\mu} = (\boldsymbol{\mu}'_1, \boldsymbol{\mu}'_2)'$ similarly; and partition $\boldsymbol{\sigma}^2$ into four submatrices:

$$\boldsymbol{\sigma}^2 = \begin{bmatrix} \boldsymbol{\sigma}_{11}{}^2 & \boldsymbol{\sigma}_{12}{}^2 \\ \boldsymbol{\sigma}_{21}{}^2 & \boldsymbol{\sigma}_{22}{}^2 \end{bmatrix},$$

where $\boldsymbol{\sigma}_{11}{}^2$ is the $m \times m$ covariance matrix of $\tilde{\mathbf{x}}_1$ (and is the upper left-hand corner of $\boldsymbol{\sigma}^2$); $\boldsymbol{\sigma}_{22}{}^2$ is the $(w - m) \times (w - m)$ covariance matrix of $\tilde{\mathbf{x}}_2$ (and is the lower right-hand corner of $\boldsymbol{\sigma}^2$); $\boldsymbol{\sigma}_{12}{}^2$ is the $m \times (w - m)$ matrix of covariances of the elements of $\tilde{\mathbf{x}}_1$ with the elements of $\mathbf{x}_2$, and $\boldsymbol{\sigma}_{21}{}^2$ is the $(w - m) \times m$ matrix of covariances of the elements of $\tilde{\mathbf{x}}_2$ with the elements of $\tilde{x}_1$. Symmetry of $\boldsymbol{\sigma}^2$ implies that $\boldsymbol{\sigma}_{12}{}^2 = (\boldsymbol{\sigma}_{21}{}^2)'$.

Example 10.4.1

Let $(\tilde{x}_1, \tilde{x}_2, \tilde{x}_3)' \equiv \tilde{\mathbf{x}}$ have some dgf $f(x_1, x_2, x_3)$ such that $E(\tilde{x}_1) = 2$, $E(\tilde{x}_2) = 4$, $E(\tilde{x}_3) = -5$, $V(\tilde{x}_1) = V(\tilde{x}_3) = 16$, $V(\tilde{x}_2) = 49$, $V(\tilde{x}_1, \tilde{x}_2) = V(\tilde{x}_2, \tilde{x}_3) = 0$, and $V(\tilde{x}_1, \tilde{x}_3) = -12$. Then

$$\boldsymbol{\mu} = E(\tilde{\mathbf{x}}) = (2, 4, -5)'$$

and

$$\tilde{\sigma}^2 = \begin{bmatrix} 16 & 0 & -12 \\ 0 & 16 & 0 \\ -12 & 0 & 49 \end{bmatrix}.$$

Let $\tilde{y}_1 = \tilde{x}_1 + \tilde{x}_2 - 2\tilde{x}_3$ and $y_2 = \tilde{x}_1 + \tilde{x}_3$; and define $\tilde{\mathbf{y}} = (\tilde{y}_1, \tilde{y}_2)'$. Clearly,

$$\mathbf{B} = \begin{bmatrix} 1 & 1 & -2 \\ 1 & 0 & 1 \end{bmatrix}.$$

Hence

$$E(\tilde{\mathbf{y}}) = \begin{bmatrix} 1 & 1 & -2 \\ 1 & 0 & 1 \end{bmatrix} \begin{pmatrix} 2 \\ 4 \\ -5 \end{pmatrix} = \begin{pmatrix} 16 \\ -3 \end{pmatrix}$$

and

$$V(\tilde{\mathbf{y}}) = \begin{bmatrix} 1 & 1 & -2 \\ 1 & 0 & 1 \end{bmatrix} \begin{bmatrix} 16 & 0 & -12 \\ 0 & 16 & 0 \\ -12 & 0 & 49 \end{bmatrix} \begin{bmatrix} 1 & 1 \\ 1 & 0 \\ -2 & 1 \end{bmatrix}$$

$$= \begin{bmatrix} 276 & -70 \\ -70 & 41 \end{bmatrix}.$$

Example 10.4.2

Suppose $\tilde{\mathbf{x}}$ is as in Example 10.4.1, but that $V(\tilde{x}_1, \tilde{x}_2) = 0$. (The reader may suppose, even further, that $\tilde{x}_1$, $\tilde{x}_2$, and $\tilde{x}_3$ are mutually independent.) Let $\tilde{\mathbf{y}} = \mathbf{B}\tilde{\mathbf{x}}$ where $\mathbf{B}$ is as in Example 10.4.1. Then $E(\tilde{\mathbf{y}}) = (16, -3)$ as before, but

$$V(\tilde{\mathbf{y}}) = \begin{bmatrix} 1 & 1 & -2 \\ 1 & 0 & 1 \end{bmatrix} \begin{bmatrix} 16 & 0 & 0 \\ 0 & 16 & 0 \\ 0 & 0 & 49 \end{bmatrix} \begin{bmatrix} 1 & 1 \\ 1 & 0 \\ -2 & 1 \end{bmatrix}$$

$$= \begin{bmatrix} 288 & -82 \\ -82 & 65 \end{bmatrix}.$$

Hence $\tilde{y}_1$ and $\tilde{y}_2$ are *not* independent.

10.5 CHEBYCHEV'S INEQUALITY AND THE WEAK LAW OF LARGE NUMBERS

In Section 10.3 we promised to make precise the *approximate equality* between the *sample mean* $\bar{x}_n$ and the *population mean* μ. To do this we first prove a general result, which follows.

Chebychev's Inequality

Let $\tilde{x}$ be any random variable with finite variance σ^2 (hence finite standard deviation σ and finite mean μ), and let k be a positive real number. Then

$$P[\mu - k\sigma < \tilde{x} < \mu + k\sigma] \geqq 1 - \frac{1}{k^2}. \qquad (10.5.1)$$

Proof

Assume $\tilde{x}$ is continuous and write

$$\begin{aligned}\sigma^2 &= \int_{-\infty}^{\infty} [x - \mu]^2 f(x)dx \\ &= \int_{-\infty}^{\mu-k\sigma} [x - \mu]^2 f(x)dx + \int_{\mu-k\sigma}^{\mu+k\sigma} [x - \mu]^2 f(x)dx + \int_{\mu+k\sigma}^{\infty} [x - \mu]^2 f(x)dx.\end{aligned}$$

Now, the middle integral is nonnegative because its integrand is, and hence we cannot increase the right-hand side of this equation if we discard it:

$$\sigma^2 \geqq \int_{-\infty}^{\mu-k\sigma} [x - \mu]^2 f(x)dx + \int_{\mu+k\sigma}^{\infty} [x - \mu]^2 f(x)dx.$$

Note now that if $x \leq \mu - k\sigma$, then $(-k\sigma)^2 = [\mu - k\sigma - \mu]^2 \leqq [x - \mu]^2$, and hence

$$\int_{-\infty}^{\mu-k\sigma} [x - \mu]^2 f(x)dx \geqq \int_{-\infty}^{\mu-k\sigma} k^2\sigma^2 f(x)dx = k^2\sigma^2 F(\mu - k\sigma);$$

and if $x \geq \mu + k\sigma$, then

$$(+k\sigma)^2 = [\mu + k\sigma - \mu]^2 \leqq [x - \mu]^2,$$

and hence

$$\int_{\mu+k\sigma}^{\infty} [x - \mu]^2 f(x)dx \geqq \int_{\mu+k\sigma}^{\infty} k^2\sigma^2 f(x)dx = k^2\sigma^2[1 - F(\mu + k\sigma)].$$

Substituting these expressions into the first inequality (which obviously will remain valid) yields $\sigma^2 \geqq k^2\sigma^2[1 - \{F(\mu + k\sigma) - F(\mu - k\sigma)\}]$; that is,

$$\sigma^2 \geqq k^2\sigma^2(1 - P[\mu - k\sigma < \tilde{\mathbf{x}} < \mu + k\sigma]),$$

from which the stated inequality follows readily. It is clear how this proof is to be modified if $\tilde{x}$ is discrete or mixed. QED

Next, assume that $\tilde{\mathbf{x}}_1, \ldots, \tilde{x}_n$ constitute a *random sample of size n from* $f(x)$[that is, $\tilde{x}_1, \ldots, \tilde{x}_n$ are iid with common dgf $f(x)$]. Before the sample is taken, or before the observations have been obtained, the sample mean is a random variable (it is a statistic in the sense of Chapter 9; in fact, the most important statistic in applications):

$$\bar{\tilde{x}}_n \equiv \left(\frac{1}{n}\right) \sum_{i=1}^{n} \tilde{x}_i.$$

The following result is basic:

If $\tilde{x}_1, \ldots, \tilde{x}_n$ are iid with common dgf $f(x)$, common mean μ, and common variance σ^2, then

$$E(\bar{\tilde{x}}_n) = \mu \tag{10.5.2}$$

and

$$V(\bar{\tilde{x}}_n) = \frac{\sigma^2}{n};$$

that is, the expectation of the sample mean equals the population mean, and the variance of the sample mean equals the population variance divided by the number n of observations.

Proof

$$\begin{aligned}
E(\bar{\tilde{x}}_n) &= E\left[\left(\frac{1}{n}\right) \sum_{i=1}^{n} \tilde{x}_i\right] && \text{(definition of } \bar{\tilde{x}}_n\text{)} \\
&= \left(\frac{1}{n}\right) E\left(\sum_{i=1}^{n} \tilde{x}_i\right) && \text{[by (10.2.7')]} \\
&= \left(\frac{1}{n}\right) \sum_{i=1}^{n} E(\tilde{x}_i) && \text{[by (10.2.7)]} \\
&= \left(\frac{1}{n}\right) \cdot n\mu && [E(\tilde{x}_i) = \mu, \text{ for all } i] \\
&= \mu && \text{[cancellation]}
\end{aligned}$$

proving (10.5.2). For (10.5.3),

$$\begin{aligned}
V(\bar{\tilde{x}}_n) &= V\left[\left(\frac{1}{n}\right) \sum_{i=1}^{n} \tilde{x}_i\right] && \text{[definition of } \bar{\tilde{x}}_n\text{]} \\
&= \left(\frac{1}{n}\right)^2 V\left[\sum_{i=1}^{n} \tilde{x}_i\right] && \text{[by (10.3.13)]} \\
&= \left(\frac{1}{n}\right)^2 \sum_{i=1}^{n} V(\tilde{x}_i) && \text{[by (10.3.10)]} \\
&= \left(\frac{1}{n}\right)^2 \cdot n\sigma^2 && [V(\tilde{x}_i) = \sigma, \text{ for all } i] \\
&= \frac{\sigma^2}{n}. && \text{[cancellation] QED}
\end{aligned}$$

This result goes part of the way to making the *approximate equality* precise in a probabilistic sense: no matter how large or small the sample is, the *expectation of* $\tilde{x}_n$ equals (exactly the population mean μ; and as the sample size n becomes larger, the variance of (that is, your uncertainty about) $\tilde{x}_n$ gets smaller and the probability that $\tilde{x}_n$ lies in any given interval containing μ increases.

We may make a precise statment by combining this result with Chebychev's inequality: substitute $\sigma/\sqrt{n}$ for σ to obtain

$$P\left[\tilde{x}_n \in \left(\mu - \frac{k\sigma}{\sqrt{n}}, \mu + \frac{k\sigma}{\sqrt{n}}\right)\right] \geq 1 - \frac{1}{k^2}. \tag{10.5.4}$$

Thus if k is predetermined, the interval $(\mu - k\sigma/\sqrt{n}, \mu + k\sigma/\sqrt{n})$ centered at μ diminishes as n increases; while if $k/\sqrt{n}$ is predetermined, k decreases with n, $-1/k^2$ decreases with n, and the right-hand side of (10.5.4) increases with n up to 1; this says that the probability that $\tilde{x}_n$ falls within any interval around μ tends to one as n becomes very large. This, together with $E(\tilde{x}_n) = \mu$, is the precise sense of the approximate equality.

The *weak law of large numbers* expresses (10.5.4) somewhat more succinctly:

If $\tilde{x}_1, \tilde{x}_2, \ldots$ are iid with common dgf $f(x)$ common (finite) expectation μ and common (finite) variance σ^2 then for every $\delta > 0$,

$$\lim_{n\to\infty} P[\mu - \delta < \tilde{x}_n < \mu + \delta] = 1; \tag{10.5.5}$$

equivalently, for every $\epsilon > 0$ and every $\delta > 0$, there is an $n(\epsilon, \delta)$ such that

$$P[\mu - \delta < \tilde{x}_n < \mu + \delta] \geqq 1 - \epsilon,$$

for every $n \geqq n(\epsilon, \delta)$.

The weak law of large numbers, and (10.5.4), are useful when one wants to ascertain how large a sample one must take in order to be sure that the sample mean $\tilde{x}_n$ will fall within δ of the population mean μ with probability at least $1 - \epsilon$; one chooses ϵ and δ and solves for $n(\epsilon, \delta)$. The reader may show as Exercise 28 that

$$n(\epsilon, \delta) = \frac{\sigma^2}{(\delta^2 \cdot \epsilon)}. \tag{10.5.6}$$

However, $n(\epsilon, \delta)$ is usually extremely large. This is so because it works for *any* dgf f such that $\sigma^2 < \infty$, and hence in many cases of "reasonably well-behaved" f's, using (10.5.6) is rather like driving a thumbtack with a pile driver. For most purposes, the Central Limit Theorem (Chapter 13) provides much smaller, approximate, and more practical results.

The weak law of large numbers remains true even when the assumption that $\sigma^2 < \infty$ fails.

10.6 MOMENT-GENERATING FUNCTIONS

Occasionally one finds the computation of moments $E(\tilde{x}^r)$ of $\tilde{x}$ about zero very difficult if performed directly from their definition. It is sometimes easier to follow a circuitous route if by so doing one can travel much faster and less painfully.

Given a random variable $\tilde{x}$, we define the *moment-generating function* $M(\cdot)$ of $\tilde{x}$ by

$$M(u) = E(e^{u\tilde{x}}), \tag{10.6.1}$$

for all u in some interval $(-\delta, \delta)$ about 0 if this expectation exists. We show that if $r \epsilon \{1, 2, \ldots\}$ and $E(\tilde{x}^r)$ exists, then

$$E(\tilde{x}^r) = \left[\frac{d^r}{du^r}M(u)\right]\Bigg|_{u=0}. \tag{10.6.2}$$

Proof

Assume $\tilde{x}$ is continuous with df $f(x)$, and recall that

$$\frac{d^r}{du^r}e^{ux} = x^r e^{ux}.$$

Then

$$M(u) \equiv \int_{-\infty}^{\infty} e^{ux}f(x)dx \qquad \text{(definition of } M\text{);}$$

hence

$$\begin{aligned}\frac{d^r}{du^r}M(u) &= \frac{d^r}{du^r}\int_{-\infty}^{\infty} e^{ux}f(x)dx \\ &= \int_{-\infty}^{\infty}\left[\frac{d^r}{du^r}e^{ux}f(x)\right]dx \\ &= \int_{-\infty}^{\infty} x^r e^{ux}f(x)dx.\end{aligned}$$

Now put $u = 0$ to obtain

$$\begin{aligned}\frac{d^r}{du^r}M(u)\Bigg|_{u=0} &= \int_{-\infty}^{\infty} x^r e^{0\cdot x}f(x)dx \\ &= \int_{-\infty}^{\infty} x^r\cdot 1\cdot f(x)dx \\ &= E(\tilde{x}^r),\end{aligned}$$

as asserted. The interchange of differentiation and integration will be permissible in all applications in this book. The preceding proof continues to obtain when $\tilde{x}$ is discrete or mixed.

Example 10.6.1

Let $\tilde{x}$ have df $f(x)$ given by

$$f(x) = \begin{cases} 0, & x \leqq 0 \\ \lambda e^{-\lambda x}, & x > 0, \end{cases}$$

where $\lambda > 0$. Then

$$\begin{aligned} M(u) &= \int_0^\infty e^{ux}\lambda e^{-\lambda x} dx \\ &= \frac{\lambda}{\lambda - \mu} \int_0^\infty (\lambda - u)e^{-(\lambda-u)x} dx \\ &= \frac{\lambda}{\lambda - u}, \end{aligned}$$

provided that $u < \lambda$ (otherwise the integral will not converge). Hence

$$\begin{aligned} E(\tilde{x}) &= \frac{d}{du}\left[\frac{\lambda}{\lambda - u}\right]\Bigg|_{u=0} \\ &= \frac{\lambda}{(\lambda - u)^2}\Bigg|_{u=0} \\ &= \frac{\lambda}{\lambda^2} = \frac{1}{\lambda}, \end{aligned}$$

as given in Example 10.2.2.

More generally, let $\tilde{\mathbf{x}}$ be a random vector with dgf $f(\mathbf{x})$. We define the moment-generating function $M(\mathbf{u})$ of $\tilde{\mathbf{x}}$ by

$$M(\mathbf{u}) = E(e^{\mathbf{u}'\tilde{\mathbf{x}}}), \tag{10.6.3}$$

for all $\mathbf{u}$ in some neighborhood of $\mathbf{0} \in R^w$ if this expectation exists. By a proof similar to that of (10.6.2), it can be shown that if $r_i \epsilon \{0, 1, 2, \ldots\}$, for all $i \epsilon \{1, \ldots, w\}$ then,

$$E(\tilde{x}_1^{r_1}, \ldots, \tilde{x}_w^{r_w}) = \left[\frac{\partial^{r_1+\cdots+r_w}}{\partial u_1^{r_1} \cdots \partial u_w^{r_w})} M(u_1, \ldots, u_w)\right]\Bigg|_{\mathbf{u}=\mathbf{0}}, \tag{10.6.4}$$

provided that this moment exists.

Moment-generating functions are useful in finding moments, but they are even more useful in finding the distributions of the sum and sample mean of independent random variables:

(a) Let $\tilde{x}_1, \ldots, \tilde{x}_n$ be independent random variables, and let $M_i(\cdot)$ be the moment-generating function of $\tilde{x}_i$. Let $y = \sum_{i=1}^{n} \tilde{x}_i$ and $\tilde{\bar{x}}_n = \tilde{y}/n$ with moment-generating functions $M_y(\cdot)$ and $M_{\bar{x}_n}(\cdot)$, respectively. Then

$$M_y(u) = \prod_{i=1}^{n} M_i(u) \tag{10.6.5}$$

and

$$M_{\bar{x}_n}(u) = \prod_{i=1}^{n} M_i\left(\frac{u}{n}\right). \tag{10.6.6}$$

(b) In particular, if $\tilde{x}_1, \ldots, \tilde{x}_n$ are iid with common moment-generating function $M(\cdot)$, then

$$M_y(u) = [M(u)]^n \tag{10.6.7}$$

and

$$M_{\bar{x}_n}(u) = \left[M\left(\frac{u}{n}\right)\right]^n. \tag{10.6.8}$$

(c) If $\tilde{x}_1, \ldots, \tilde{x}_n$ are independent random variables and $\tilde{x}_L \equiv k_1\tilde{x}_1 + \cdots + k_n\tilde{x}_n$, then the moment-generating function $M_L(\cdot)$ of $\tilde{x}_L$ is given by

$$M_L(u) = \prod_{i=1}^{n} M_i(k_i u). \tag{10.6.9}$$

Proof

We prove (c) first, as it implies (a) and (b).

$$\begin{aligned} M_L(u) &\equiv E(e^{u\tilde{x}_L}) \\ &= \{E(\exp[u(k_1\tilde{x}_1 + \cdots + k_n\tilde{x}_n)])\} \\ &= E[\exp(k_1u\tilde{x}_1)\cdot \ldots \cdot \exp(k_nu\tilde{x}_n)]. \end{aligned}$$

We evaluate this expectation by using (10.2.8) in the form $E(e^{\tilde{y}_1}\cdot \ldots \cdot e^{\tilde{y}_n}) = E[E(e^{\tilde{y}_1}\cdot \ldots \cdot e^{\tilde{y}_n} \mid \tilde{x}_1, \ldots, \tilde{x}_{n-1})]$. Now,

$$\begin{aligned} E(e^{k_1u\tilde{x}_1}\cdot \ldots \cdot e^{k_nu\tilde{x}_n} \mid x_1, \ldots, x_{n-1}) &= e^{k_1ux_1}\cdot \ldots \cdot e^{k_{n-1}ux_{n-1}}E(e^{k_nu\tilde{x}_n} \mid x_1, \ldots, x_{n-1}) \\ &= e^{k_1ux_1}\cdot \ldots \cdot e^{k_{n-1}ux_{n-1}}E(e^{k_nu\tilde{x}_n}) \end{aligned}$$

[because $\tilde{x}_1, \ldots, \tilde{x}_n$ are independent and hence

$$E(e^{\tilde{y}} \mid x_1, \ldots, x_{n-1}) = E(e^{\tilde{y}})]$$

$$= e^{k_1ux_1}\cdot \ldots \cdot e^{k_{n-1}ux_{n-1}}\cdot M_n(k_nu).$$

Hence we have shown that

$$M_L(u) = M_n(k_nu)\cdot E(e^{k_1u\tilde{x}_1}\cdot \ldots \cdot e^{k_{n-1}u\tilde{x}_{n-1}}).$$

Applying the same procedure, using $E(e^{\tilde{y}_1}\cdot \ldots \cdot e^{\tilde{y}_{n-1}}) = E[E(e^{\tilde{y}_1}\cdot \ldots \cdot e^{\tilde{y}_{n-1}} \mid \tilde{x}_1, \ldots, \tilde{x}_{n-2})]$ yields

$$M_L(u) = M_n(k_n u)\cdot M_{n-1}(k_{n-1}u)\cdot E(e^{k_1 u \tilde{x}_1}\cdot \ldots \cdot e^{k_{n-2}u\tilde{x}_{n-2}});$$

and so forth, yielding (c).

Part (a) follows from (c) by putting $k_1 = \cdots = k_n = 1$ to obtain $\tilde{x}_L = \tilde{y}$ and putting $k_1 = \cdots = k_n = 1/n$ to obtain $\tilde{x}_L = \bar{x}_n$. Part (b) follows from (a) by putting $M_1 = \cdots = M_n = M$. QED

There are ways of obtaining the dgf of a random variable from its moment-generating function, and hence the obvious thing to do to obtain the dgf of $\tilde{x}_L$ as defined in (c) is to use one of these ways. However, they involve some subtleties which are unnecessary for our later applications. Our procedure will be to use (a), (b), or (c) to obtain the moment-generating function M_L of $\tilde{x}_L$, and then to *recognize* M_L as the moment-generating function of a random variable $\tilde{x}_L$ with *known* dgf. It can be shown that the moment-generating function $M_L(\cdot)$ of $\tilde{x}_L$—*if* it exists—determines the distribution of $\tilde{x}_L$ uniquely.

Example 10.6.2

Let $\tilde{x}_1, \ldots, \tilde{x}_n$ be iid with common mass function $f(x)$ defined by

$$f(x) = \begin{cases} 0, & x \notin \{0, 1, 2, \ldots\} \\ e^{-\lambda t}\dfrac{(\lambda t)^x}{x!}, & x \in \{0, 1, 2, \ldots\}, \end{cases}$$

where $\lambda t > 0$.

Then

$$\begin{aligned} M(u) = E(e^{u\tilde{x}}) &= \sum_{x=0}^{\infty} e^{ux}e^{-\lambda t}\frac{(\lambda t)^x}{x!} \\ &= e^{-\lambda t}\sum_{x=0}^{\infty}\frac{(\lambda t e^u)^x}{x!} \\ &= e^{-\lambda t}\cdot e^{\lambda t e^u} \\ &= e^{\lambda t(e^u - 1)}. \end{aligned}$$

The moment-generating function $M_y(\cdot)$ of $\tilde{y} = \tilde{x}_1 + \cdots + \tilde{x}_n$ is given by

$$\begin{aligned} M_y(u) &= [e^{\lambda t(e^u-1)}]^n \\ &= e^{n\lambda t(e^u-1)}, \end{aligned}$$

which we recognize as the moment-generating function of the random variable $\tilde{y}$ with mass function $f(y)$ given by

$$f(y) = \begin{cases} 0, & y \notin \{0, 1, 2, \ldots\} \\ e^{-n\lambda t}\dfrac{(n\lambda t)^y}{y!}, & y \in \{0, 1, 2, \ldots\}. \end{cases}$$

Two other results on moment-generating functions are easy consequences of their definition:

(d) Let $(\tilde{\mathbf{x}}'_1, \tilde{\mathbf{x}}'_2)'$ be a partitioning of $\tilde{\mathbf{x}}$ and let $\mathbf{u} = (\mathbf{u}'_1, \mathbf{u}'_2)' \in R^m \times R^{w-m}$ be the corresponding partitioning of $\mathbf{u} \in R^w$. Then the moment-generating function $M_1(\cdot)$ of $\tilde{\mathbf{x}}_1$ is given by

$$M_1(\mathbf{u}_1) = M(\mathbf{u}_1, \mathbf{0}), \tag{10.6.10}$$

where $M(\cdot,\cdot)$ is the moment-generating function of $\tilde{\mathbf{x}}$. (e) Let $M(\mathbf{u}_1 \mid \mathbf{x}_2)$ denote the conditional moment-generating function of $\tilde{\mathbf{x}}_1$ given $\tilde{\mathbf{x}}_2 = \mathbf{x}_2$. Then the moment-generating function $M_1(\cdot)$ of $\tilde{\mathbf{x}}_1$ is given by

$$M_1(\mathbf{u}_1) = E[M(\mathbf{u}_1 \mid \tilde{\mathbf{x}}_2)]. \tag{10.6.11}$$

In closing, it is well to point out that the defining expectation (10.6.1) [or (10.6.3)] of the moment-generating function sometimes fails to exist for *any* nonzero u(or $\mathbf{u}$), and hence the moment-generating function itself fails to exist. For this reason, there is not a one-to-one correspondence between moment-generating functions and (distributions of) random variables (or random vectors). Exercise 30 introduces a more satisfactory relative of the moment-generating function, *the characteristic function*, which always exists, and is such that if $\tilde{\mathbf{x}}_1$ and $\tilde{\mathbf{x}}_2$ have different cdf's, then they have different characteristic functions. Hence the correspondence between characteristic functions and distributions is one-to-one.

EXERCISES

1. Let $\tilde{x}_i$ be a discrete random variable with mf f defined by

$$f(x_i) = \begin{cases} \dfrac{1}{i}, & x_i = -1 \\ \dfrac{(i-2)}{i}, & x_i = 0 \\ \dfrac{1}{i}, & x_i = 1 \\ 0, & x_i \notin \{-1, 0, 1\}, \end{cases}$$

where $i \geqq 2$. Find $E(\tilde{x}_i)$, $V(\tilde{x}_i)$, and $E(\tilde{x}_i^r)$ for $r = 2, 3, 4, \ldots$.

2. Let $\tilde{x}$ be a continuous random variable with df f defined by

$$f(x) = \begin{cases} 4(x-5)^3, & x \in (5, 6) \\ 0, & x \notin (5, 6). \end{cases}$$

Find $E(\tilde{x})$ and $V(\tilde{x})$.

3. Let $\tilde{x}_i$ have expectation $\mu_i = i^6$, for $i = 1, 2, \ldots$. Find $E(\tilde{x}_1 + 2\tilde{x}_2 + \cdots + 10\tilde{x}_{10})$.

4. Suppose that $\tilde{x}_2$ has conditional density, given x_1, defined for $x_1 > 0$ by

$$f(x_2 \mid x_1) = \begin{cases} x_1^{-1}e^{-x_2/x_1}, & x_2 > 0 \\ 0, & x_2 \leqq 0, \end{cases}$$

and suppose that $\tilde{x}_1$ has density f_1 defined by

$$f(x_1) = \begin{cases} \lambda e^{-\lambda x_1}, & x_1 > 0 \\ 0, & x_1 \leqq 0, \end{cases}$$

where λ is a positive parameter.

(a) Find $E(\tilde{x}_2 + 4 \mid x_1)$.
(b) Find $E(\tilde{x}_2 + 4)$.
(c) Find $V(\tilde{x}_2 + 4 \mid x_1)$.
(d) Find $V(\tilde{x}_2 + 4)$.
(e) Find $V(\tilde{x}_1, \tilde{x}_2)$.

5. Suppose $\tilde{x}_1, \ldots, \tilde{x}_n$ represent the production costs of n items and that they have identical expectations $\mu_i = \mu$ and variances $\sigma_{ii}^2 = \sigma^2$.

(a) Show that if $\tilde{x}_1, \ldots, \tilde{x}_n$ are independent, then $V(\tilde{x}_1 + \cdots + \tilde{x}_n) = n\sigma^2$.
(b) Show that if $\tilde{x}_1(q) = \tilde{x}_2(q) = \cdots = \tilde{x}_n(q)$, for every q, then

$$(\tilde{x}_1 + \cdots + \tilde{x}_n)(q) = n \cdot \tilde{x}_i(q),$$

for any i and that $V(\tilde{x}_1 + \cdots + \tilde{x}_n) = n^2\sigma^2$.
(c) Explain why the variance of total cost as given in (b) exceeds that of total cost as given in (a).

6. Find $V(\tilde{x})$, where $\tilde{x}$ is as given

(a) in Example 10.2.1; and
(b) by

$$f(x) = \begin{cases} .10, & x = 1 \\ .60, & x = 2 \\ .30, & x = 3 \\ 0, & x \notin \{1, 2, 3\}. \end{cases}$$

(c) Explain the difference. (*Hint*: Graph the two mf's.)

7. What is the expectation of the outcome $\tilde{x}$ of the roll of one die? What is $V(\tilde{x})$?

8. The moment generating function $M(\cdot)$ of a random variable $\tilde{x}$ is given by

$$M(u) = e^{u\mu + \frac{1}{2}u^2\sigma^2},$$

where $\mu \in (-\infty, \infty)$ and $\sigma^2 \in (0, \infty)$ are parameters.

(a) Find $E(\tilde{x})$ and $V(\tilde{x})$.
(b) Suppose $\tilde{x}_1, \ldots, \tilde{x}_n$ are iid with $e^{u\mu+\frac{1}{2}u^2\sigma^2}$ as common moment generating functions. Find the moment generating function $M_y(\cdot)$ of $\tilde{y} \equiv \sum_{i=1}^{n} \tilde{x}_i$ and the moment generating function $M_{\bar{x}_n}(\cdot)$ of $\tilde{\bar{x}}_n \equiv \tilde{y}/n$.

9. Let $\tilde{x}$ be a random variable with moment generating function $M(\cdot)$.

(a) Show that the moment generating function M^* of $\tilde{x} + k$. for k, a real number, is given by

$$M^*(u) = e^{uk}M(u), \text{ for all } u.$$

(b) Show that the moment generating function M^* of $k\tilde{x}$, for k, a real number, is given by

$$M^*(u) = M(ku).$$

10. Let $M(u_1, \ldots, u_n)$ be the moment generating function of $(\tilde{x}_1, \ldots, \tilde{x}_n)'$. Show that the moment generating function $M_y(\cdot)$ of $\tilde{y} \equiv \sum_{i=1}^{n} \tilde{x}_i$ is given by

$$M_y(u) = M(u, u, \ldots u), \text{ for all } u,$$

and that the moment generating function $M_{\bar{x}_n}(\cdot)$ of $\tilde{\bar{x}}_n \equiv \tilde{y}/n$ is given by

$$M_{\bar{x}_n}(u) = M\left(\frac{u}{n}, \frac{u}{n}, \ldots, \frac{u}{n}\right), \text{ for all } u.$$

(*Note*: These results hold regardless of independence or dependence of $\tilde{x}_1, \ldots, \tilde{x}_n$).

11. Let $(\tilde{\mathbf{x}}'_1, \ldots, \tilde{\mathbf{x}}'_n)'$ be a partitioning of $\tilde{\mathbf{x}}$ and suppose that each $\tilde{\mathbf{x}}_i$ is a k-dimensional random vector. Let $M(\cdot)$ be the moment-generating function of $\tilde{\mathbf{x}}$, and let $\tilde{\mathbf{y}} \equiv \sum_{i=1}^{n} \tilde{\mathbf{x}}_i$ and $\tilde{\bar{\mathbf{x}}}_n \equiv (1/n)\tilde{\mathbf{y}}$.

(a) Show that

$$M_{\mathbf{y}}(\mathbf{u}) = M(\mathbf{u}, \mathbf{u}, \ldots, \mathbf{u})$$

and that

$$M_{\bar{\mathbf{x}}_n}(\mathbf{u}) = M\left(\frac{1}{n}\mathbf{u}, \frac{1}{n}\mathbf{u}, \ldots, \frac{1}{n}\mathbf{u}\right).$$

(b) Show that

$$M_{\tilde{y}}(\mathbf{u}) = \prod_{i=1}^{n} M_i(\mathbf{u})$$

and

$$M_{\bar{\tilde{x}}_n}(\mathbf{u}) = \prod_{i=1}^{n} M_i\left(\frac{1}{n}\mathbf{u}\right)$$

if $\tilde{\mathbf{x}}_1, \ldots, \tilde{\mathbf{x}}_n$ are mutually independent with moment generating functions $M_1(\cdot), \ldots, M_n(\cdot)$.

(c) What do the results in (b) become if $\tilde{\mathbf{x}}_1, \ldots, \tilde{\mathbf{x}}_n$ are iid?

12. Suppose $\tilde{\mathbf{x}}_1, \ldots, \tilde{\mathbf{x}}_n$ are iid with common moment-generating function

$$M(\mathbf{u}) = \exp\,(\mathbf{u}'\boldsymbol{\mu} + \tfrac{1}{2}\mathbf{u}'\boldsymbol{\sigma}^2\mathbf{u}),$$

where $\boldsymbol{\mu} \in R^k$ and $\boldsymbol{\sigma}^2$ is a $k \times k$, symmetric, and positive-definite matrix ($\boldsymbol{\mu}$ and $\boldsymbol{\sigma}^2$ are parameters). Using Exercise 12, find the moment generating functions of $\tilde{\mathbf{x}}_1 + \cdots + \tilde{\mathbf{x}}_n$ and $(1/n)\,(\tilde{\mathbf{x}}_1 + \cdots + \tilde{\mathbf{x}}_n)$.

13. In Figure 6-1, suppose the outcome z of experiment e is expressed by a random vector $\tilde{\mathbf{x}}$ and the state s is expressed by a random variable $\tilde{\theta}$.

(a) Fill in the bubble below:

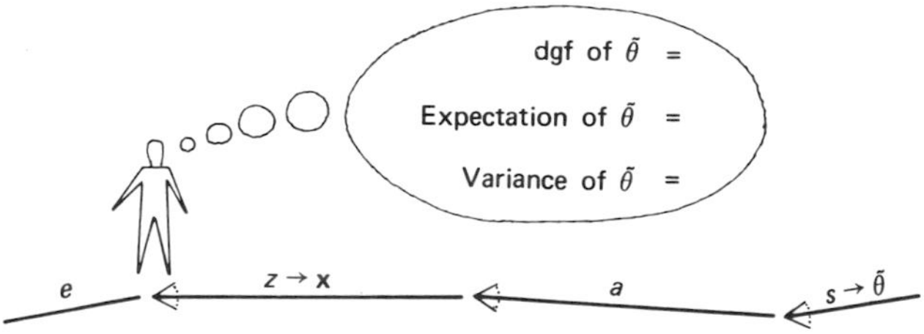

(b) Fill in the following bubble:

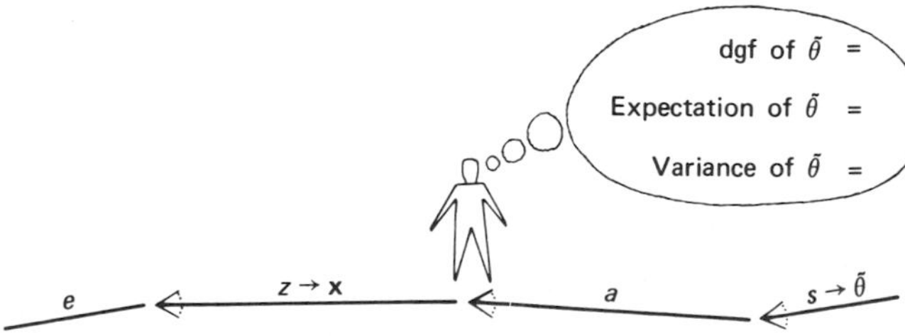

14. Let $\tilde{\mathbf{X}}$ be a random $w_1 \times w_2$ matrix. Then $\tilde{\mathbf{X}}$ may be represented as $(\tilde{\mathbf{x}}'_1, \ldots, \tilde{\mathbf{x}}'_{w_2})' \in R^{w_1 w_2}$, where $\tilde{\mathbf{x}}_j$ denotes the jth *column* of $\tilde{\mathbf{X}}$.

(a) Let $v_{jk}{}^{c}$ denote the matrix of covariances of the components of column j with those of column k:

$$v_{jk}{}^{c} \equiv E[(\tilde{\mathbf{x}}_j - \boldsymbol{\mu}_j)(\tilde{\mathbf{x}}_k - \boldsymbol{\mu}_k)'].$$

Then ${}_{j}v_{k}{}^{c}$ is a $w_1 \times w_1$ matrix, for every $\{j, k\} \subset \{1, \ldots, w_2\}$.

$$\text{Let } v_{\text{col}} \equiv \begin{bmatrix} v_{11}{}^{c} & \cdots & v_{1w_2}{}^{c} \\ v_{w_2 1}{}^{c} & \cdots & v_{w_2 w_2}{}^{c} \end{bmatrix}.$$

Show that $v_{jk}{}^{c} = v_{kj}{}^{c\prime}$ and that v_{col} is symmetric.

(b) Let $\tilde{\mathbf{x}}^i$ denote the ith *row* (vector) of $\tilde{\mathbf{X}}$, let $\boldsymbol{\mu}^i$ denote the *row* vector expectation of $\tilde{\mathbf{x}}^i$, and let $v_{ih}{}^{r}$ denote the matrix of covariances of the components of row i with those of row h:

$$v_{ih}{}^{r} \equiv E[(\tilde{\mathbf{x}}^i - \boldsymbol{\mu}^i)'(\tilde{\mathbf{x}}^h - \boldsymbol{\mu}^h)].$$

Then $v_{ih}{}^{r}$ is a $w_2 \times w_2$ matrix, for every $\{i, h\} \subset \{1, \ldots, w_1\}$. Let

$$v_{\text{row}} \equiv \begin{bmatrix} v_{11}{}^{r} & \cdots & v_{1w_1}{}^{r} \\ v_{w_1 1}{}^{r} & \cdots & v_{w_1 w_1}{}^{r} \end{bmatrix}.$$

Show that $v_{ih}{}^{r} = v_{hi}{}^{r\prime}$ and that v_{row} is symmetric.

(c) Show that v_{col} is the covariance matrix of the random vector

$$(\tilde{x}_{11}, \ldots, \tilde{x}_{w_1 1}, \tilde{x}_{12}, \ldots, \tilde{x}_{w_1 2}, \ldots, \tilde{x}_{1w_2}, \ldots, \tilde{x}_{w_1 w_2})' \equiv (\tilde{\mathbf{x}}'_1, \ldots, \tilde{\mathbf{x}}'_{w_2})',$$

while v_{row} is the covariance matrix of the random vector $(\tilde{x}_{11}, \ldots, \tilde{x}_{1w_2}, \tilde{x}_{21}, \ldots, \tilde{x}_{2w_2}, \ldots, \tilde{x}_{w_1 1}, \ldots, \tilde{x}_{w_1 w_2})' \equiv (\tilde{\mathbf{x}}^1, \ldots, \tilde{\mathbf{x}}^{w_1})'$.

(d) Let

$$\tilde{\mathbf{X}} \equiv \begin{bmatrix} \tilde{x}_{11} & \tilde{x}_{12} \\ \tilde{x}_{21} & \tilde{x}_{22} \\ \tilde{x}_{31} & \tilde{x}_{32} \end{bmatrix}$$

and let $v_{ij,kl} \equiv V(\tilde{x}_{ij}, \tilde{x}_{kl})$.

Express v_{col} and v_{row} in terms of the $v_{ij,kl}$'s.

(e) Show that if the columns of $\tilde{\mathbf{X}}$ are mutually independent, then

$$v_{\text{col}} = \begin{bmatrix} v_{11}{}^{c} & \mathbf{0} & \cdots & \mathbf{0} & \mathbf{0} \\ \mathbf{0} & v_{22}{}^{c} & \cdots & \mathbf{0} & \mathbf{0} \\ \cdot & \cdot & & \cdot & \cdot \\ \cdot & \cdot & & \cdot & \cdot \\ \cdot & \cdot & & \cdot & \cdot \\ \mathbf{0} & \mathbf{0} & \cdots & \mathbf{0} & v_{w_1 w_1}{}^{c} \end{bmatrix}$$

(that is, $\upsilon_{ij}{}^c = \mathbf{0}$ if $i \neq j$ and υ_{col} is block diagonal); and also show that every $\upsilon_{ij}{}^r$ is diagonal; that is, of the form

$$\upsilon_{ij}{}^{(r)} = \begin{bmatrix} \alpha_1 & & & & 0 \\ & \alpha_2 & & & \\ & & \cdot & & \\ & & & \cdot & \\ & & & & \cdot \\ & 0 & & & \\ & & & & \alpha_{w_2} \end{bmatrix}.$$

(f) Show that if the rows of X are mutually independent, then

$$\upsilon_{\text{row}} = \begin{bmatrix} \upsilon_{11}{}^r & \mathbf{0} & \cdots & \mathbf{0} & \mathbf{0} \\ \mathbf{0} & \upsilon_{12}{}^r & \cdots & \mathbf{0} & \mathbf{0} \\ \cdot & \cdot & & \cdot & \cdot \\ \cdot & \cdot & & \cdot & \cdot \\ \cdot & \cdot & & \cdot & \cdot \\ \mathbf{0} & \mathbf{0} & \cdots & \mathbf{0} & \upsilon_{w_1 w_1}{}^r \end{bmatrix}$$

and also show that every $\upsilon_{ij}{}^c$ is diagonal.

15. Let $\tilde{\mathbf{x}} = (\tilde{x}_1, \tilde{x}_2, \tilde{x}_3, \tilde{x}_4),'$ $\boldsymbol{\mu} = (3, 1, -5, 10)'$, and

$$\boldsymbol{\sigma}^2 = \begin{bmatrix} 8 & 0 & 0 & 0 \\ 0 & 9 & 0 & 0 \\ 0 & 0 & 4 & 0 \\ 0 & 0 & 0 & 16 \end{bmatrix}.$$

Let

$$\mathbf{B} = \begin{bmatrix} 10 & 0 & 0 & -10 \\ 1 & 1 & 1 & 1 \\ 0 & 8 & 6 & 4 \end{bmatrix},$$

and let $\tilde{\mathbf{y}} \equiv \mathbf{B}\tilde{\mathbf{x}}$. Find $E(\tilde{\mathbf{y}})$ and $V(\tilde{\mathbf{y}})$.

16. Show that if $\tilde{x}_1, \ldots, \tilde{x}_n$ are independent, then

$$E(\tilde{x}_1 \cdot \tilde{x}_2 \cdot \ldots \cdot \tilde{x}_n) = \mu_1 \cdot \mu_2 \cdot \ldots \cdot \mu_n,$$

where $\mu_i \equiv E(\tilde{x}_i)$ for every i.

17. Let $Q = \{a, b, c, d\}$ and let $(\tilde{x}_1, \tilde{x}_2, \tilde{x}_3) \equiv \tilde{\mathbf{x}}'$ be defined by

$$\tilde{\mathbf{x}}(q) = \begin{cases} (10, 5, 2)', & q = a \\ (0, 1, 2)', & q = b \\ (-2, 4, 0)', & q = c \\ (5, -1, 0)', & q = d. \end{cases}$$

Let $P(a) = \frac{1}{2}$, $P(b) = \frac{1}{4}$, and $P(c) = P(d) = \frac{1}{8}$.

(a) Find the joint mf $f(x_1, x_2, x_3)$ of $(\tilde{x}_1, \tilde{x}_2, \tilde{x}_3)'$.
(b) Find $(\mu_1, \mu_2, \mu_3)' = \boldsymbol{\mu}$.
(c) Find $\boldsymbol{\sigma}^2$.
(d) Find ρ_{12}, ρ_{13}, and ρ_{23}.
(e) Find $f(x_1, x_2 \mid x_3)$, $\mu_1(x_3)$, $\mu_2(x_3)$,

$$\begin{bmatrix} \sigma_{11}^2(x_3) & \sigma_{12}^2(x_3) \\ \sigma_{21}^2(x_3) & \sigma_{22}^2(x_3) \end{bmatrix},$$

and $\rho_{12}(x_3)$ for $x_3 = 2$.
(f) Same as (e), but for $x_3 = 0$.

18. Derive (10.3.10) through (10.3.14), assuming all random variables are continuous (so that only integrals are required; the discrete and mixed cases are derived similarly).

19. Show that if $\tilde{x}_i$ and $\tilde{x}_j$ are independent, then $\sigma_{ij}^2 = 0$.

20. Let $\tilde{x}_1$ and $\tilde{x}_2$ be discrete random variables with joint mf $f(x_1, x_2)$ as given in the following table:

$\tilde{x}_1$ \ $\tilde{x}_2$	-2	-1	1	2
1	$\frac{1}{4}$	0	0	$\frac{1}{4}$
-1	0	$\frac{1}{4}$	$\frac{1}{4}$	0

Find μ_1, μ_2, σ_{11}^2, σ_{22}^2, and show that $\sigma_{12}^2 = 0$ even though $\tilde{x}_1$ and $\tilde{x}_2$ are obviously not independent.

21. Derive (10.3.22) through (10.3.25), assuming all random variables are (jointly) continuous.

22. Show that if $\tilde{x}_i$ and $\tilde{\mathbf{x}}_2$ are independent, then $\sigma_{ii}^2(\mathbf{x}_2) = \sigma_{ii}^2$ for all possible $\mathbf{x}_2$.

23. Find $\mu_1(x_2)$, $\sigma_1^2(x_2)$, and ρ_{12} for $(\tilde{x}_1, \tilde{x}_2)$ as given in Example 10.3.7.

24. Prove that if $\boldsymbol{\sigma}^2$ is a covariance matrix, then $\mathbf{d}'\boldsymbol{\sigma}^2\mathbf{d} \geqq 0$ for every $\mathbf{d} \in R^w$. [*Hint*: Let $\tilde{y} \equiv \mathbf{d}'\tilde{\mathbf{x}}$, where $\boldsymbol{\sigma}^2$ is the covariance matrix of $\tilde{\mathbf{x}}$. What is the variance of $\tilde{y}$ as given by Equation (10.4.6)? Apply (10.3.12).]

25. Derive Equations (10.4.6) and (10.4.7).

26. Prove that $\tilde{\mathbf{y}} \equiv \mathbf{B}\tilde{\mathbf{x}}$ is nondegenerate if $\tilde{\mathbf{x}}$ is nondegenerate, $m < w$, and rank $(\mathbf{B}) = m$, where $\mathbf{B}$ is $m \times w$.

27. Derive Equation (10.5.6).

28. Derive Equation (10.4.4).

29. Let $\tilde{x}$ be any random variable, with dgf $f(x)$. Define $\phi(u) \equiv E(e^{iux})$, where $u \in R^1$ and $i \equiv \sqrt{-1}$. The function $\phi(u)$ is called the *characteristic function* of $\tilde{x}$, and is therefore a complex-valued function of a real variable. Since $e^{iux} = \cos(ux) + i\sin(ux)$, for any u and any x, it follows that $\phi(u) = E[\cos(u\tilde{x})] + i\,E[\sin(u\tilde{x})]$. Each summand exists, and hence $\phi(u)$ *always* exists.

(a) Show that $E(\tilde{x}^r) = \left(\frac{1}{i^r}\right)\frac{d^r}{du^r}\,\phi(u)\Big|_{u=0}$, for $r = 1, 2, \ldots,$ if $E(\tilde{x}^r)$ exists.

(b) Show that if $\tilde{x}_1$ and $\tilde{x}_2$ are independent random variables with respective characteristic functions $\phi_1(u)$ and $\phi_2(u)$, then $\tilde{y} \equiv \tilde{x}_1 + \tilde{x}_2$ has characteristic function $\phi_1(u)\cdot\phi_2(u)$.

Note: You may show that the characteristic function $\phi(u)$ possesses all properties (10.6.5) through (10.6.9) of the moment generating function and has the added advantage of always existing. This fact will be used in Chapter 13. Moreover, it can be shown that the characteristic function of a random variable determines its distribution uniquely, and that

$$P[a \leq x < b] = \lim_{y\to\infty}\frac{1}{2\pi}\int_{-y}^{y}\frac{e^{-iua} - e^{-iub}}{iu}\,\phi(u)du, \tag{10.7.1}$$

whenever $-\infty < a < b < \infty$. These facts rescue weak arguments based upon the moment generating function, since $\phi(u)$ exists whenever $M(u)$ does, and they are related by

$$\phi(u) = M(iu) \tag{10.7.2}$$

and

$$M(u) = \phi\left(\frac{u}{i}\right); \tag{10.7.3}$$

that is, to obtain the characteristic function of $\tilde{x}$, replace u in $M(u)$ by iu wherever it appears.

(c) Show that if $\tilde{x}$ is degenerate, with $P[\tilde{x} = x] = 1$, for a given $x \in R^1$, then $\phi(u) = e^{iux}$.

(d) Show that if $\tilde{x}$ has the moment generating function defined in Exercise 9, then

$$\phi(u) = e^{iu\mu - \frac{1}{2}u^2\sigma^2}. \tag{10.7.4}$$

If $\tilde{\mathbf{x}} \equiv (\tilde{x}_1, \ldots, \tilde{x}_k)'$, the characteristic function $\phi(\mathbf{u})$ of $\tilde{\mathbf{x}}$ is defined, for all $\mathbf{u} \in R^k$, by

$$\phi(\mathbf{u}) \equiv E(e^{i\mathbf{u}'\tilde{\mathbf{x}}}). \tag{10.7.5}$$

(e) Show that

$$E(\tilde{x}_1^{r_1}\cdot\tilde{x}_2^{r_2}\cdot\ldots\cdot\tilde{\mathbf{x}}_k^{r_k}) = \left(\frac{1}{i^{r_1+\ldots+r_k}}\right)\frac{\partial^{r_1+\ldots+r_k}}{\partial u_1^{r_1}\ldots\partial u_k^{r_k}}\,\phi(\mathbf{u})\Big|_{\mathbf{u}=0} \tag{10.7.6}$$

30. Show that $\mu^{(0)} = 1$ for any random variable $\tilde{x}$.

31. Let $\tilde{\mathbf{x}}_1, \ldots, \tilde{\mathbf{x}}_n$ be iid k-dimensional random vectors with common mean vector $\boldsymbol{\mu}$ and common covariance matrix $\boldsymbol{\sigma}^2$. Define $\bar{\tilde{\mathbf{x}}}_n \equiv \frac{1}{n}\sum_{i=1}^{n} \tilde{\mathbf{x}}_i$. Show that

$$E(\bar{\tilde{\mathbf{x}}}_n) = \boldsymbol{\mu} \tag{10.7.7}$$

and

$$V(\bar{\tilde{\mathbf{x}}}_n) = \frac{1}{n}\boldsymbol{\sigma}^2, \tag{10.7.8}$$

for every n.

32. Generalization of (10.3.27)—show that

$$V(\tilde{x}_i, \tilde{x}_j) = E[V(\tilde{x}_i, \tilde{x}_j \mid \tilde{\mathbf{x}}_2)] + V[\mu_i(\tilde{\mathbf{x}}_2), \mu_j(\tilde{\mathbf{x}}_2)], \tag{10.7.9}$$

for any random variables $\tilde{x}_1$ and $\tilde{x}_2$ and any random vector $\tilde{\mathbf{x}}_2$. [*Hint*: Duplicate as far as possible the proof of (10.3.27).]

11

SOME IMPORTANT PROBABILITY DISTRIBUTIONS AND DATA-GENERATING PROCESSES (I)

11.1 INTRODUCTION

In Chapters 7, 8, and 10 we have developed in large part the so-called "calculus of random variables (and random vectors)." We have discussed representations of the distribution of a random variable or random vector in terms of its cdf and, more importantly, in terms of its distributing function. In Chapter 10 we considered some summary measures of distributions in terms of *location* (means) and of *uncertainty* (variance—or its inverse, precision). We have shown the influence of partial information in the form of conditional dgf's, means, and variances, and their relationships with their marginal (or unconditional) counterparts.

In Chapter 9, and in Section 10.5, we have sketchily considered some topics in the statistical applications of probability theory by examining the constituents of a statistical experiment, sufficiency, noninformative stopping, random sampling, two basic approaches to statistical inference and the probabilistic relationship between the mean $\tilde{\bar{x}}_n$ of a random sample of size n from a dgf $f(x)$ and the population mean μ $(= E(\tilde{x}))$.

The time has come to consider some data-generating processes that are very important in applications. In this chapter we shall consider five data-generating processes, their sufficient statistics, the distribution of the (random part of the) sufficient statistic, and typical applications of these data-generating processes.

Section 11.2 is preliminary in nature and concerns permutations and combinations. It may be omitted by readers who are familiar with elementary combinatorial analysis.

Section 11.3 concerns the *Bernoulli process*, which generates iid observations $\tilde{x}_1, \tilde{x}_2, \ldots, \tilde{x}_i, \ldots$ with common mf defined by

$$f(x \mid p) = \begin{cases} p, & x = 1 \\ 1 - p, & x = 0 \\ 0, & x \notin \{0, 1\}, \end{cases}$$

where $p \in (0, 1)$ is a parameter. This data-generating process is frequently used in random sampling with replacement to make inferences about the proportion p of defective items in a production lot, the proportion p of voters who support a given candidate, the proportion p of housewives who buy a given product, and so forth.

Section 11.4 generalizes Section 11.3 and concerns the "k-variate Bernoulli process," which generates iid k-tuples $\tilde{\mathbf{x}}_i \equiv (\tilde{x}_{1i}, \ldots, \tilde{x}_{ki})$ with common mf defined by

$$f_{\mathbf{x}|p}(x_1, \ldots, x_k \mid p_1, \ldots, p_k) = \begin{cases} p_1; & x_1 = 1, x_2 = x_3 = \cdots = x_k = 0 \\ p_2; & x_2 = 1, x_1 = x_3 = \cdots = x_k = 0 \\ \cdot & \\ \cdot & \\ \cdot & \\ p_k; & x_1 = x_2 = \cdots = x_{k-1} = 0, x_k = 1 \\ 1 - \sum_{i=1}^{k} p_i; & \mathbf{x} = \mathbf{0} \\ 0\ ; & \text{elsewhere,} \end{cases}$$

where $p_j \in (0, 1)$ for $j = 1, \ldots, k$ and $\sum_{j=1}^{k} p_j < 1$. This data-generating process is frequently used in more complicated situations similar to those for the (simple) Bernoulli process; namely random sampling with replacement to make inferences about the proportions $p_1, \ldots, p_k$ of the population which fall into categories $k_1, \ldots, k_k$ (for example, opinion polling with "yes," "no," "maybe," "undecided," and "get out" responses).

Section 11.5 concerns the *hypergeometric process*, which is the analogue of the Bernoulli process for the case of random sampling *without* replacement. Here the observations are *not* iid. Section 11.6 concerns the similar analogue of Section 11.4; namely, the "k-variate hypergeometric process."

Section 11.7 concerns the *Poisson* process, in which the number $\tilde{x}$ of "events" occurring in any given period t (>0) of time has mf

$$f(x \mid \lambda t) = \begin{cases} e^{-\lambda t}\dfrac{(\lambda t)^x}{x!}, & x \in \{0, 1, 2, \ldots\} \\ 0, & x \notin \{0, 1, 2, \ldots\}, \end{cases}$$

where $\lambda \in (0, \infty)$ is a parameter. The Poisson process crops up in considering problems of estimating mean rate λ of demand for a product per time period; mean rate λ of occurrences of accidents per time period; mean rate λ of incoming calls to a switchboard; mean rate λ of arrival of cars at an inspection station; and so forth.

Section 11.8 furnishes means, variances, and (where applicable) covariances of the random variables (and random vectors) whose distributions are derived in Sections 11.3 through 11.7.

11.2 ELEMENTARY COMBINATORIAL ANALYSIS

There are two fundamental principles of elementary combinatorial analysis, a *compounding law* and an *addition law*. Each can be stated in set-theoretic terminology. Let A, B, B_1, B_2, and so forth all be finite sets with $\#(A)$, $\#(B)$, $\#(B_1)$, $\#(B_2)$, and so forth, distinct elements respectively.

Addition Law: If $A \cap B = \varnothing$, then $\#(A \cup B) = \#(A) + \#(B)$; (11.2.1)

and

Compounding Law: If a set B_i is associated with each $i \in A$, then (11.2.2)

$$\#(\{(i, j): j \in B_i, i \in A\}) = \sum_{i=1}^{\#(A)} \#(B_i).$$

Proof

The addition law is obvious and was used tacitly in Chapter 5. The compounding law is proved by an appeal to a tree of the form in Figure 11.1,

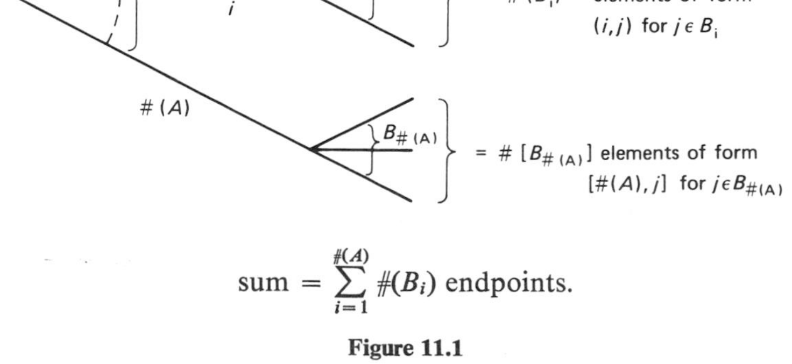

$$\text{sum} = \sum_{i=1}^{\#(A)} \#(B_i) \text{ endpoints.}$$

Figure 11.1

Note that all pairs (i, j) are distinct from one another by fiat; namely, the assumption that elements in a given set are distinct and an assumption that elements in different sets can be distinguished from each other. QED

The compounding law has a very important corollary; namely, if $B_1 = \cdots = B_{\#(A)} \equiv B$, then $\{(i, j): j \in B, i \in A\} = A \times B$, and hence

$$\#(A \times B) = \#(A)\#(B). \tag{11.2.3}$$

Example 11.2.1

Given n distinguishable objects (say, books with different titles), how many different lists of precisely two different objects are there, if two lists are considered different when the *order* of listing differs? Clearly, there are $n - 1$ ways of specifying the second item on the list given each specification of the first item. Let B_i denote the $n - 1$ possible second choices given item i was the first choice, and let A denote the set of all $n(= \#(A))$ objects. Then $\#(\{(i,j): j \in B_i, i \in A\}) = \sum_{i=1}^{n} (n - 1) = n(n - 1)$.

Example 11.2.1 generalizes easily. You may show that there are precisely $n(n - 1)(n - 2)\cdot\ldots\cdot(n - r + 1)$ different lists of $r(\leqq n)$ different elements of a set $X(\#(X) \equiv n)$, where two lists are considered different even if they contain precisely the same elements but in different orderings. We call this number of different lists the *number of r-permutations of n*, and denote it by $(n)_r$; thus

$$(n)_r \equiv n\cdot(n - 1)\cdot(n - 2)\cdot\ldots\cdot(n - r + 1), \tag{11.2.4}$$

provided that $r \leq n$. If $r = n$, the number of n-permutations of n is simply the total number of ways of arranging the given set X of n things; The designation $(n)_n$ is given a special name, *n factorial*, and a special symbol, $n!$. The reader is cautioned against supposing that $n!$ denotes enthusiasm or histrionics; in mathematics, $n!$ is *defined* by

$$n! = 1\cdot 2\cdot\ldots\cdot n, \tag{11.2.5}$$

for n a positive integer. By convention (and to make certain formulae of wider applicability), we define

$$0! \equiv 1. \tag{11.2.6}$$

The reader may show as Exercise 18 that

$$(n)_r = \frac{n!}{(n - r)!}. \tag{11.2.7}$$

It follows from (11.2.6) and (11.2.7) that

$$(n)_0 = 1. \tag{11.2.8}$$

Now, we have seen that the number of r-item lists of distinct elements in X is $(n)_r$, where $n \equiv \#(X)$. For each r-element subset of X, there are $r!$ ways of arranging this subset, and each such arrangement produces a different r-element list. Therefore, each different (unordered) r-element subset of X is counted exactly $r!$ times in the number $(n)_r$ of r-item lists. *Hence, there are $(n)_r/r!$ distinct r-element subsets of X.* We define $\binom{n}{r}$ by

$$\binom{n}{r} \equiv (n)_r/r! \equiv n!/[r!(n-r)!] \tag{11.2.9}$$

for any integers r and n such that $0 \leqq r \leqq n$. We read $\binom{n}{r}$ as "the *number of r-combinations* of n," or a "binomial coefficient."

Example 11.2.2

Let $X \equiv \{a, b, c\}$. There are six two-item lists of distinct elements of X (because $6 = 3!/(3-2)! = (3 \cdot 2 \cdot 1/1)$; they are:

#1	#2	#3	#4	#5	#6
a	b	a	c	b	c
b	a	c	a	c	b

However, there are three ($= 6/2! = 6/2$) two-element subsets of X; namely

$\{a, b\}$ (corresponding to lists #1 and #2),
$\{a, c\}$ (corresponding to lists #3 and #4),
and $\{b, c\}$ (corresponding to lists #5 and #6).

Example 11.2.3

There is precisely one empty ($r = 0$) subset of a set X of n elements, since $\binom{n}{0} = n!/[0!(n-0)!] = n!/(1 \cdot n!) = 1$. Similarly, there is precisely one n-element subset of X if $\#(X) = n$, since $\binom{n}{n} = n!/[n!(n-n)!] = n!/(n! \cdot 0!) = 1$.

The reader may show as Exercise 19 that

$$\binom{n}{r} = \binom{n}{n-r} \tag{11.2.10}$$

if r and n are nonnegative integers such that $0 \leq r \leq n$.

Example 11.2.4

The *binomial theorem* of high school algebra asserts that if a and b are any real numbers and n is any positive integer, then

$$(a+b)^n = \sum_{r=0}^{n} \binom{n}{r} a^r b^{n-r}. \tag{11.2.11}$$

The binomial theorem is the motivation for calling $\binom{n}{r}$ a binomial coefficient.

In Example 11.2.4, if we let $a = b = 1$ in the binomial theorem, we obtain

$$2^n = \sum_{r=0}^{n} \binom{n}{r}. \tag{11.2.12}$$

This result shows that *there are precisely* 2^n *subsets* (including $\varnothing$) of *any n-element set*, because $\binom{n}{r}$ is the number of r-element subsets and we sum from $r = 0$ ($\varnothing$ is the unique 0-element subset of X) to $r = n$ (X itself is the unique n-element subset of X).

There is a second interpretation of $\binom{n}{r}$ which follows readily from the first: $\binom{n}{r}$ gives the number of ways of dividing a set X of n elements into two disjoint subsets, one containing r elements and the other containing $n - r$.

But this second interpretation suggests a generalization: what is the number of ways of dividing X (with $\#(X) \equiv n$) into *three* disjoint subsets, the first containing n_1 elements, the second containing n_2, and the third containing the rest, $n_3 = n - n_1 - n_2$? We reason that this division can be performed into two stages. In the *first* stage we divide X into two subsets of n_1 and $n - n_1$ elements, respectively. There are $\binom{n}{n_1}$ ways of doing this. In the *second* stage we divide the $n - n_1$ element subset into two subsets consisting of n_2 and $(n - n_1) - n_2$ elements respectively. It is clear that for each of the $\binom{n}{n_1}$ ways of performing the first stage, there are $\binom{n-n_1}{n_2}$ ways of performing the second stage. Hence by the compounding law (with A denoting the set of ways of accomplishing stage one and B_i denoting the number of ways of accomplishing stage two given "way i" of accomplishing stage one, and with $\#(A) = \binom{n}{n_1}$ and $\#(B_i) = \binom{n-n_1}{n_2}$ for every $i \in \{1, \ldots, \binom{n}{n_1}\}$) we obtain the result that *there are*

$$\binom{n}{n_1} \cdot \binom{n - n_1}{n_2} = \frac{n!}{n_1!(n - n_1)!} \cdot \frac{(n - n_1)!}{n_2!(n - n_1 - n_2)!}$$
$$= \frac{n!}{n_1!n_2!(n - n_1 - n_2)!}$$

ways of dividing a set of n elements into three disjoint subsets of n_1, n_2, *and* $n - n_1 - n_2$ *elements.*

This result generalizes. The number of ways of dividing a set of n elements into $k + 1$ disjoint subsets of $n_1, n_2, \ldots, n_k$, and $n_{k+1} \equiv n - n_1 - \cdots - n_k$ elements is denoted by $\binom{n}{n_1 \ldots n_k}$, which is defined by

$$\binom{n}{n_1 \ldots n_k} = \frac{n!}{n_1!n_2! \ldots n_k!n_{k+1}!}, \tag{11.2.13}$$

where $n, n_1, \ldots, n_k, n_{k+1}$ are nonnegative integers and $\sum_{i=1}^{k+1} n_i = n$. We call $\binom{n}{n_1 \ldots n_k}$ a *multinomial coefficient*, or the *number of* $(n_1, \ldots, n_k)$*-combinations of n*.

Example 11.2.5

Let $X = \{a, b, c, d\}$ and hence $n = 4$. Let $k = 2, n_1 = 2, n_2 = 1$ (and hence $n_3 = 4 - 2 - 1 = 1$). There are 12 $(= 4!/(2!1!1!))$ ways of dividing X into three sets of two, one, and one elements; namely,

$$\begin{aligned}
&(\{a, b\}, \{c\}, \{d\});\ (\{b, c\}, \{a\}, \{d\});\\
&(\{a, b\}, \{d\}, \{c\});\ (\{b, c\}, \{d\}, \{a\});\\
&(\{a, c\}, \{b\}, \{d\});\ (\{b, d\}, \{a\}, \{c\});\\
&(\{a, c\}, \{d\}, \{b\});\ (\{b, d\}, \{c\}, \{a\});\\
&(\{a, d\}, \{b\}, \{c\});\ (\{c, d\}, \{a\}, \{b\});\\
\text{and}\quad &(\{a, d\}, \{c\}, \{b\});\ (\{c, d\}, \{b\}, \{a\}).
\end{aligned}$$

Note that the order *in which the subsets are listed* counts, but not the order in which the elements are listed *within a subset.*

Example 11.2.6

The *multinomial theorem* is: if $a_1, \ldots, a_{k+1}$ are real numbers and n is a positive integer, then

$$(a_1 + \cdots + a_{k+1})^n = \sum_A \binom{n}{n_1 \ldots n_k} a_1^{n_1} \cdot a_2^{n_2} \cdot \ldots \cdot a_{k+1}^{n_{k+1}}, \tag{11.2.14}$$

where $A \equiv \{(n_1, \ldots, n_k): \sum_{i=1}^{k} n_i \leqq n,\ n_i \in \{0, 1, \ldots, n\}, \text{ for every } i\}$, and $n_{k+1} \equiv n - \sum_{i=1}^{k} n_i$.

If in (11.2.14) we put $a_1 = a_2 = \ldots = a_{k+1} = 1$, we obtain

$$(k + 1)^n = \sum_A \binom{n}{n_1 \ldots n_k}; \tag{11.2.15}$$

which, of course, reduces to (11.2.11) if $k = 1$.

11.3 THE BERNOULLI PROCESS

Consider an opinion poll in which the individuals are randomly selected with replacement and answer either "yes" or "no" to a question regarding, say, their advocacy of a political candidacy. Suppose the population consists of N people and exactly R of them would say "yes" if asked. You wish to make inferences about the proportion $p = R/N$ of "yesses."

Alternatively, consider taking a random sample with replacement from a production lot of N items in order to make inferences about the proportion $R/N = p$ of *defective* items in that lot.

In each of these cases, the assumption of sampling *with* replacement implies that the ith observation will be "yes" (or defective) with probability $p(= R/N)$. We define the random variable $\tilde{x}_i$ characterizing the ith observation by

$$\tilde{x}_i = \begin{cases} 1, & \text{"yes" (or defective)} \\ 0, & \text{"no" (or not defective).} \end{cases}$$

Sampling *with* replacement implies that $\tilde{x}_1, \tilde{x}_2, \ldots$ are iid with common conditional mf given p defined by

$$f(x \mid p) = \begin{cases} p^x(1-p)^{1-x}, & x \in \{0, 1\} \\ 0, & x \notin \{0, 1\}. \end{cases} \qquad (11.3.1)$$

Note that (11.3.1) is simply a more succinct way of writing

$$f(x \mid p) = \begin{cases} p, & x = 1 \\ 1-p, & x = 0 \\ 0, & x \notin \{0, 1\}. \end{cases}$$

We shall find it convenient to think in terms of an "urn model," in which an urn contains N balls, R of which are red and the remaining $N - R$ of which are black. Then the preceding examples correspond to drawing balls from the urn at random and with replacement, and recording the colors of the balls drawn. Here, $\tilde{x}_i$ is one if a red ball is drawn and zero if a black ball is drawn.

The abstraction of the urn model, and the examples, is the *Bernoulli (data-generating) process*, which produces iid observations $\tilde{x}_1, \tilde{x}_2, \ldots$ given p with common mf defined by (11.3.1).

Suppose the stopping process is noninformative (in the sense of Section 9.6), and that the experiment terminated after the nth observation. Then the joint mf of $(\tilde{x}_1, \ldots, \tilde{x}_n)$ given n and p is, from (11.3.1),

$$f(x_1, \ldots, x_n \mid n, p) = \begin{cases} \prod_{i=1}^{n} p^{x_i}(1-p)^{1-x_i}, & \text{all } x_i \in \{0, 1\} \\ 0, & \text{some } x_i \notin \{0, 1\} \end{cases} \qquad (11.3.2)$$

$$= \begin{cases} p^r(1-p)^{n-r}, & r \in \{0, 1, \ldots, n\} \\ 0, & r \notin \{0, 1, \ldots, n\}, \end{cases}$$

where $r \equiv \sum_{i=1}^{n} x_i$.

It is clear from (9.5.1) that $\mathbf{x}_T \equiv (r, n)$ is a sufficient statistic for p: let $p \equiv (\boldsymbol{\theta}, \boldsymbol{\theta}_\alpha)$, $g(n, x_1, \ldots, x_n) = 1$, and $g(\mathbf{x}_T \mid p) = f(x_1, \ldots, x_n \mid n, p)$ as defined by (11.3.2).

We shall now consider two special noninformative stopping processes. In the *first*, n is predetermined and hence the random part of (r, n) is r. That is, before the sample was taken, $\tilde{r}$ was a random variable (hence tilded). We shall now derive the mf of $\tilde{r}$ given n and p.

Note, first, that all $(x_1, \ldots, x_n)$'s with precisely r ones have the same probability; namely, $p^r(1-p)^{n-r}$. Now,

$$P(\tilde{r} = r \mid n, p) = \sum \{p^r(1-p)^{n-r}\text{: all } (x_1, \ldots, x_n\} \text{ with precisely } r \text{ ones}\}.$$

But the number of $(x_1, \ldots, x_n)$'s with precisely r ones is simply the number of ways of selecting r components of $(x_1, \ldots, x_n)$ to assume the value one, and this number is $\binom{n}{r}$. Hence

$$P(\tilde{r} = r \mid n, p) = \begin{cases} \binom{n}{r} p^r(1-p)^{n-r}, & r \in \{0, 1, \ldots, n\} \\ 0, & r \notin \{0, 1, \ldots, n\} \end{cases}$$

which defines a perfectly good mf for any values of its parameters $n \in \{1, 2, \ldots\}$ and $p \in (0, 1)$; it is nonnegative, and $\sum_{r=0}^{n} \binom{n}{r} p^r(1-p)^{n-r} = [p + (1-p)]^n = 1^n = 1$ by the binomial theorem (Example 11.2.4). It is called the *binomial mass function.*

The binomial mf is our first big "name" distributing function. For it, and for other "name" dgf's we shall use a suggestive subscripting convention. We define the *binomial mf* $f_b(\cdot \mid n, p)$ by

$$f_b(r \mid n, p) = \begin{cases} \binom{n}{r} p^r(1-p)^{n-r}, & r \in \{0, 1, \ldots, n\} \\ 0, & r \notin \{0, 1, \ldots, n\}, \end{cases} \tag{11.3.3}$$

where $n \in \{1, 2, \ldots\}$ and $p \in (0, 1)$ are paramenters.

We have thus derived the following important result:

If an experiment is to consist of a predetermined number n of observations from a Bernoulli process with parameter p, then the mf of $\tilde{r} \equiv \sum_{i=1}^{n} \tilde{x}_i$ is the binomial mf given by (11.3.3). (11.3.4)

Figures 11.2, 11.3, and 11.4 depict $f_b(\cdot \mid 4, \frac{1}{4})$, $f_b(\cdot \mid 4, \frac{1}{2})$, and $f_b(\cdot \mid 4, \frac{3}{4},)$ respectively.

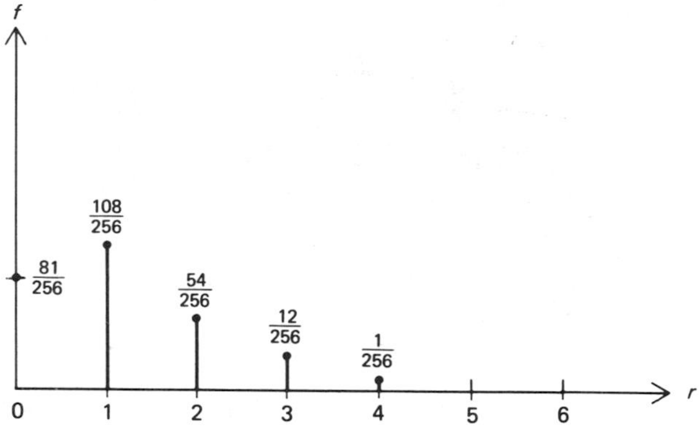

Figure 11.2 $f_b(\cdot \mid 4, \frac{1}{4})$

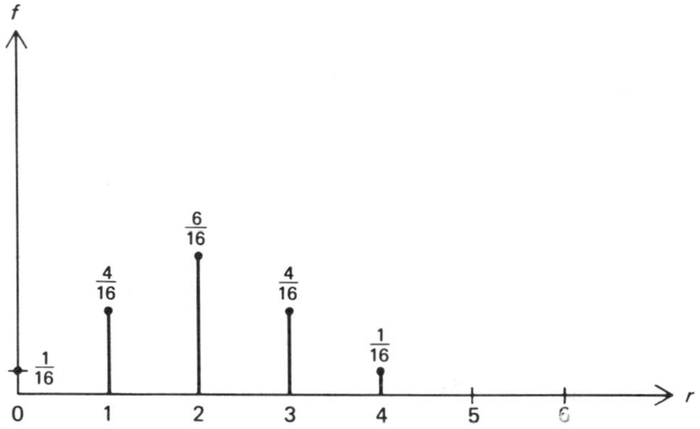

Figure 11.3 $f_b(\cdot \mid 4, \frac{1}{2})$

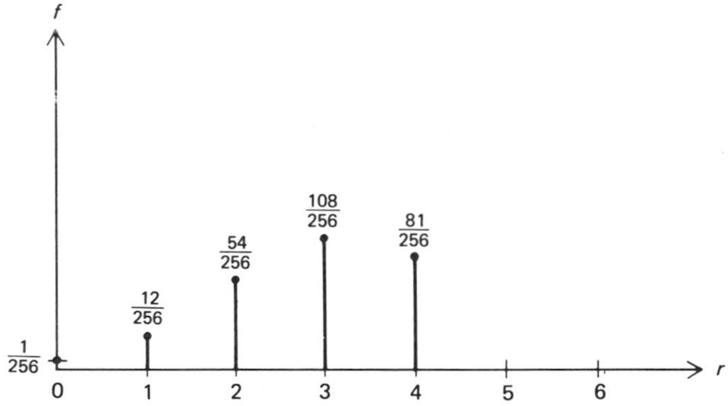

Figure 11.4 $f_b(\cdot \mid 4, \frac{3}{4})$

Example 11.3.1

If a production lot (urn) consists of $N = 1000$ parts (balls and) $R = 250$ of them are defective (red), and if you take a random sample of $n = 4$ parts with replacement, then the probability of obtaining $\tilde{r} = 4$ defectives (red balls) is only 1/256. Thus if someone had told you that $p = \frac{1}{4}$ and then you found $\tilde{r} = 4$ is a sample of four items, you might suspect the veracity of your informant.

Example 11.3.2

If you randomly selected four voters with replacement and found $\tilde{r} = 0$ supporting Candidate X, then you might doubt the correspondent who asserts that 75 percent ($= 100p$) of the voters support Candidate X. (He might be trying to create a "bandwagon effect.")

The binomial mf describes the distribution of the number $\tilde{r}$ of observations out of n which possess a certain characteristic, given that each observation has probability p of possessing that characteristic and the observations are independent. It is difficult to overestimate the importance of the binomial distribution in applications. The reader should reflect upon sampling situations which give rise to it.

We now turn to our *second* (and final) stopping process for Bernoulli experiments. Suppose observations are continued until precisely r ones are obtained: for example, keep randomly sampling voters (with replacement) until you obtain four supporting candidate X. Alternatively, in the urn model, keep sampling balls with replacement until you have obtained precisely r ones.

This stopping process is easy to describe. In the notation of Section 9.3, we have $\pi(0 \mid p) = 0$ and

$$\pi(i \mid x_1, \ldots, x_i, p) = \begin{cases} 0, & \sum_{j=1}^{i} x_j < r \\ 1, & \sum_{j=1}^{i} x_j \geqq r, \end{cases}$$

from which it is clear that the stopping process is noninformative about p. Experiments with this stopping process are said to be *inverse sampling procedures*.

For the fixed-r stopping process, the number $\tilde{n}$ of observations is random, and we shall derive the mf of $\tilde{n}$ given r and p. First, consider the (fully described) final outcome $(x_1, \ldots, x_n)$ of the experiment if it terminates after the nth observation; and note that in this case x_n *will have to be a one*, as the experiment terminates with the rth one.

Now, any $(x_1, \ldots, x_{n-1}, 1)$ which contains precisely $r - 1$ ones in $(x_1, \ldots, x_{n-1})$ will yield the sufficient statistic (r, n), and all such $(x_1, \ldots, x_{n-1}, 1)$'s have the same probability; namely, $p^r(1 - p)^{n-r}$. Hence

$$P(\tilde{n} = n \mid r, p) = \begin{cases} 0, & n \notin \{r, r+1, \ldots\} \\ \sum \{p^r(1-p)^{n-r}: & \text{all } (x_1, \ldots, x_{n-1}, 1) \text{ with precisely} \\ & r - 1 \text{ ones in } (x_1, \ldots, x_{n-1})\}, \\ & n \in \{r, r+1, \ldots\}. \end{cases} \tag{11.3.5}$$

But the number of $(x_1, \ldots, x_{n-1}, 1)$'s with $r - 1$ ones in $(x_1, \ldots, x_{n-1})$ is just the number of ways of selecting a subset of $r - 1$ components from among the first $n - 1$ components of $(x_1, \ldots, x_{n-1})$ to put ones in, and is therefore $\binom{n-1}{r-1}$. Hence (11.3.5) becomes

$$P(\tilde{n} = n \mid r, p) = \begin{cases} \binom{n-1}{r-1} p^r(1-p)^{n-r}, & n \in \{r, r+1, \ldots\} \\ 0, & n \notin \{r, r+1, \ldots\}. \end{cases}$$

We define the *Pascal* mf $f_{Pa}(\cdot \mid r, p)$ by

$$f_{Pa}(n \mid r, p) = \begin{cases} \binom{n-1}{r-1} p^r(1-p)^{n-r}, & n \in \{r, r+1, \ldots\} \\ 0, & n \notin \{r, r+1, \ldots\}, \end{cases} \tag{11.3.6}$$

where $r \in \{1, 2, \ldots\}$ and $p \in (0, 1)$ are parameters. It characterizes the distribution of the total number $\tilde{n}$ of Bernoulli observations required to produce precisely r one's (or red balls, or yesses, or defectives).

A distribution closely related to the Pascal is the *negative binomial.* It is the distribution of the total number $\tilde{\zeta} \equiv \tilde{n} - r$ of zeros occurring before the rth one; from (11.3.6) we easily obtain the negative binomial mf:

$$f_{nb}(\zeta \mid r, p) \equiv \begin{cases} \binom{\zeta + r - 1}{r - 1} p^r(1-p)^{\zeta}, & \zeta \in \{0, 1, 2, \ldots\} \\ 0, & \zeta \notin \{0, 1, 2, \ldots\}, \end{cases} \tag{11.3.7}$$

where $r \in \{1, 2, \ldots\}$ and $p \in (0, 1)$ are parameters. Another name for the negative binomial distribution is the *binomial waiting time* distribution. Moreover, be cautioned that some authors also call (11.3.7) the *Pascal distribution*, and/or call (11.3.6) the *negative binomial distribution.*

It can be shown that

$$1 = \sum_{n=r}^{\infty} f_{Pa}(n \mid r, p) = \sum_{\zeta=0}^{\infty} f_{nb}(\zeta \mid r, p),$$

for every $r \in \{1, 2, \ldots\}$ and $p \in (0, 1)$. Hence there is probability zero (that is, virtual impossibility) of having to keep on observing forever to obtain r ones.

The special case of (11.3.7) in which $r = 1$ is called the *geometric* mf, and is given by

$$f_g(\zeta \mid p) = \begin{cases} p(1-p)^{\zeta}, & \zeta \in \{0, 1, 2, \ldots\} \\ 0, & \zeta \notin \{0, 1, 2, \ldots\}. \end{cases} \tag{11.3.8}$$

The reader may verify as Exercise 20 that the cdf $F_g(\cdot \mid p)$ of ζ is given by

$$F_g(\zeta \mid p) = \begin{cases} 0; & \zeta < 0 \\ 1 - (1-p)^{i+1}; & i \leq \zeta < i+1, \ i \in \{0, 1, 2, \ldots\}. \end{cases} \tag{11.3.9}$$

We have now derived the following important result for inverse sampling from a Bernoulli process:

> If an experiment is to consist of a predetermined number r of observations equal to one from a Bernoulli process with parameter p, then the distribution of the number $\tilde{n}$ of observations is characterized by the Pascal mf (11.3.6), and the distribution of the number ζ of zeros is characterized by the negative binomial mf (11.3.7). (11.3.10)

Example 11.3.2

How many transistors are required in order to obtain precisely four nondefectives if each is defective with probability .10 and the quality of each is independent of the qualities of the others? The number $\tilde{n}$ of transistors required to produce $r = 4$ with P(nondefective) $= .9$ has the Pascal mf $f_{Pa}(\cdot \mid 4, .9)$.

Example 11.3.3

If 60 percent of all suburban males read *Bon Vivant* magazine, what is the probability that exactly six observations (with replacement) are required to produce four readers? Clearly, $f_{Pa}(6 \mid 4, .6) = \binom{5}{3}(.6)^4(.4)^2 = .20736$.

Example 11.3.4

If 50 percent of all housewives use GLOP detergent, what is the probability that a random sample with replacement of 100 housewives (predetermined n) found only five ($= r$) users of GLOP? Clearly $f_b(5 \mid 100, \frac{1}{2})$, which is a *very* small number.

Example 11.3.5

Three missiles are aimed at a target point. The distances $\tilde{\delta}_1$, $\tilde{\delta}_2$, $\tilde{\delta}_3$ of their points of impact from the target point are assumed to be iid random variables with common df $f(\delta \mid \alpha, \beta)$ defined by

$$f(\delta \mid \alpha, \beta) = \begin{cases} \dfrac{\alpha(1 - \delta/\beta)^{\alpha-1}}{\beta}, & \delta \in (0, \beta) \\ 0, & \delta \notin (0, \beta), \end{cases}$$

where $\alpha > 0$ and $\beta > 0$ are parameters. A "kill" is scored if at least two missiles fall within a distance γ $(< \beta)$ of the target point. What is the probability of a "kill"? First, note that the probability of *any one* missile falling within γ of the target point is

$$\int_0^\gamma \alpha\beta^{-1}\left(1 - \frac{\delta}{\beta}\right)^{\alpha-1} d\delta = 1 - \left(1 - \frac{\gamma}{\beta}\right)^\alpha.$$

Next, define $\tilde{x}_i$ by

$$\tilde{x}_i = \begin{cases} 1, & \tilde{\delta}_i \in (0, \gamma) \\ 0, & \tilde{\delta}_i \notin (0, \gamma). \end{cases}$$

Then $\tilde{x}_1$, $\tilde{x}_2$, $\tilde{x}_3$ are observations from the Bernoulli process with parameter $p \equiv 1 - (1 - \gamma/\beta)^\alpha$, and the probability of a "kill" is $f_b(2 \mid 3, p) + f_b(3 \mid 3, p)$.

Example 11.3.6

This is the same problem as in Example 11.3.5, except that we now wish to ascertain how many missiles to fire in order that the "kill probability" be at least $1 - \epsilon$ for a predetermined $\epsilon \in (0, 1)$. Again let $\tilde{x}_i$ be one if a "hit" $(\tilde{\delta}_i < \gamma)$ is scored. We have

$$P(\text{"kill"}) = \sum_{r=2}^{n} \binom{n}{r}\left(1 - \left[1 - \frac{\gamma}{\beta}\right]^\alpha\right)^r \left(\left[1 - \frac{\gamma}{\beta}\right]^\alpha\right)^{n-r}.$$

Given α, β, and γ, you can compute $P(\text{"kill"})$ for various values of n in order to find the smallest n for which $P(\text{"kill"}) \geq 1 - \epsilon$.

11.4 THE *k*-VARIATE BERNOULLI PROCESS

Consider an urn which contains N balls of $k + 1$ colors, $c_1, \ldots, c_{k+1}$, and suppose that n balls are drawn at random and with replacement from this urn. Record the outcome of the ith draw by a k-tuple $\tilde{\mathbf{x}}_i$, defined by

$$\tilde{\mathbf{x}}_i \equiv \begin{cases} (1, 0, 0, \ldots, 0), & \text{ball is of color } c_1 \\ (0, 1, 0, \ldots, 0), & \text{ball is of color } c_2 \\ \cdot & \\ \cdot & \\ \cdot & \\ (0, 0, \ldots, 0, 1), & \text{ball is of color } c_k \\ (0, 0, 0, \ldots, 0), & \text{ball is of color } c_{k+1}. \end{cases} \tag{11.4.1}$$

Suppose there are R_j balls of color c_j in the urn, and define $p_j \equiv R_j/N$.

Random drawing with replacement implies that $\tilde{\mathbf{x}}_1, \ldots, \tilde{\mathbf{x}}_n$ are iid random vectors with common mf $f(\mathbf{x} \mid p)$ defined for any $\tilde{\mathbf{x}}_i \equiv (x_{1i}, \ldots, x_{ki})$ by

$$f(\mathbf{x}_i \mid \mathbf{p}) = \begin{cases} p_1, & \mathbf{x}_i = (1, 0, 0, \ldots, 0, 0) \\ p_2, & \mathbf{x}_i = (0, 1, 0, \ldots, 0, 0) \\ \quad\cdot & \\ \quad\cdot & \\ \quad\cdot & \\ p_k, & \mathbf{x}_i = (0, 0, \ldots, 0, 0, 1) \\ 1 - \sum_{j=1}^{k} p_j, & \mathbf{x}_i = (0, 0, 0, \ldots, 0, 0) \\ 0, & \text{elsewhere.} \end{cases}$$

The preceding can be written more succinctly as

$$f(\mathbf{x}_i \mid \mathbf{p}) = \begin{cases} p_1{}^{x_{1i}} \cdot \ldots \cdot p_k{}^{x_{ki}} p_{k+1}{}^{x_{(k+1)i}}, & \mathbf{x}_i \text{ possible} \\ 0, & \mathbf{x}_i \text{ impossible,} \end{cases} \tag{11.4.2}$$

where $x_{(k+1)i} \equiv 1 - \sum_{j=1}^{k} x_{ji}$, $p_{k+1} \equiv 1 - \sum_{j=1}^{k} p_j$, $0 < p_j < 1$, for every $j \in \{1, \ldots, k + 1\}$, and $\mathbf{p} \equiv (p_1, \ldots, p_k)$.

The *k-variate Bernoulli process* generates iid k-tuples $\tilde{\mathbf{x}}_i$ with common mf given by (11.4.2).

Since $\tilde{\mathbf{x}}_1, \ldots, \tilde{\mathbf{x}}_n$ are iid, their joint mf is easily found from (11.4.2) to be

$$f(\mathbf{x}_1, \ldots, \mathbf{x}_n \mid n, \mathbf{p})$$

$$= \begin{cases} p_1{}^{r_1} \cdot \ldots \cdot p_k{}^{r_k} p_{k+1}{}^{r_{k+1}}, & \begin{cases} \text{all } r_j \in \{0, 1, \ldots, n\} \\ \text{and } \sum_{j=1}^{k+1} r_j = n \end{cases} \\ 0, & \text{elsewhere,} \end{cases} \tag{11.4.3}$$

where $\sum_{j=1}^{k+1} p_j = 1$, every $p_j > 0$, and $r_j \equiv \sum_{i=1}^{n} x_{ji}$ (which is simply the total number of balls of color c_j in the sample). Let $\mathbf{r} \equiv (r_1, \ldots, r_k)$.

From (11.4.3) it is apparent that $(\mathbf{r}, n)$ is a sufficient statistic for $\mathbf{p}$. Moreover, when n is predetermined, $\tilde{\mathbf{r}}$ is a random vector whose mf is determined in a manner similar to that for the binomial mf. Since $\tilde{\mathbf{r}} = \sum_{i=1}^{n} \tilde{x}_i$, it follows that $\tilde{\mathbf{r}} = \mathbf{r}$ if and only if $\sum_{i=1}^{n} \tilde{\mathbf{x}}_i = \mathbf{r}$, which is true if and only if $(\tilde{\mathbf{x}}_1, \ldots, \tilde{\mathbf{x}}_n) \in \{(\mathbf{x}_1, \ldots, \mathbf{x}_n)\colon \sum_{i=1}^{n} \mathbf{x}_i = \mathbf{r}\} \equiv A$. Hence

$$P(\tilde{\mathbf{r}} = \mathbf{r} \mid n, \mathbf{p}) = \sum \{f(\mathbf{x}_1, \ldots, \mathbf{x}_n \mid n, \mathbf{p})\colon (\mathbf{x}_1, \ldots, \mathbf{x}_n) \in A\}.$$

But $f(\mathbf{x}_1, \ldots, \mathbf{x}_n \mid n, \mathbf{p})$ is given by (11.4.3) for every $(\mathbf{x}_1, \ldots, \mathbf{x}_n) \in A$ and hence constant on A. Moreover $(\mathbf{x}_1, \ldots, \mathbf{x}_n) \in A$ if and only if precisely r_1 balls are of color c_1, r_2 of color c_2, . . ., and r_k of color c_k. There are $\binom{n}{r_1 \ldots r_k}$ ways of selecting r_1 trials for the outcome to be a ball of color c_1, selecting r_2 other trials for the outcome to be a ball of color c_2, . . . , and selecting r_k trials for the outcome to be a ball of color c_k. Hence A contains precisely $\binom{n}{r_1 \ldots r_k}$ elements, and we conclude that

$$P(\tilde{\mathbf{r}} = \mathbf{r} \mid n, \mathbf{p}) = \binom{n}{r_1 \ldots r_k} \prod_{j=1}^{k+1} p_j^{r_j},$$

for all possible $\mathbf{r}$.

We have shown that the distribution of $\tilde{\mathbf{r}}$ is characterized by the k-variate *multinomial* mass function $f_{mu}^{(k)}(\cdot \mid n, \mathbf{p})$ given by

$$f_{mu}^{(k)}(\mathbf{r} \mid n, \mathbf{p}) = \begin{cases} \binom{n}{r_1 \ldots r_k} \prod_{j=1}^{k+1} p_j^{r_j}, & \begin{cases} \text{all } r_j \in \{0, 1, \ldots, n\} \\ \text{and } \sum_{j=1}^{k+1} r_j = n \end{cases} \\ 0, & \text{elsewhere,} \end{cases} \tag{11.4.4}$$

where $p_{k+1} \equiv 1 - \sum_{j=1}^{k} p_k$ and $r_{k+1} \equiv n - \sum_{j=1}^{k} r_j$.

A multinomial distribution is also called a $(k + 1)$-*nomial* distribution. The reader may show that (11.4.4) reduces to the binomial mass function when $k = 1$.

Example 11.4.1

For $k = 2$ we speak of the *trinomial* distribution with parameters n and (p_1, p_2). Let $n = 4$, $p_1 = \frac{1}{4}$, and $p_2 = \frac{1}{2}$. For any (r_1, r_2) such that $r_i \in \{0, 1, \ldots, 4\}$ for $i = 1, 2$ and $r_1 + r_2 \leqq 4$, the definition (11.4.4) becomes

$$\begin{aligned} &f_{mu}^{(2)}(r_1, r_2 \mid 4, (\tfrac{1}{4}, \tfrac{1}{2})) \\ &\quad = \frac{4!}{r_1! r_2! (4 - r_1 - r_2)!} \left(\frac{1}{4}\right)^{r_1} \left(\frac{1}{2}\right)^{r_2} \left(\frac{1}{4}\right)^{4-r_1-r_2} \\ &\quad = \frac{n!}{r_1! r_2! (4 - r_1 - r_2)!} \left(\frac{1}{4}\right)^{4-r_2} \left(\frac{1}{2}\right)^{r_2}. \end{aligned}$$

From this formula we may easily compute the following table of joint probabilities:

r_1 \ r_2	0	1	2	3	4	*row sum*
0	$\frac{1}{256}$	$\frac{8}{256}$	$\frac{24}{256}$	$\frac{32}{256}$	$\frac{16}{256}$	$\frac{81}{256}$
1	$\frac{4}{256}$	$\frac{24}{256}$	$\frac{48}{256}$	$\frac{32}{256}$	0	$\frac{108}{256}$
2	$\frac{6}{256}$	$\frac{24}{256}$	$\frac{24}{256}$	0	0	$\frac{54}{256}$
3	$\frac{4}{256}$	$\frac{8}{256}$	0	0	0	$\frac{12}{356}$
4	$\frac{1}{256}$	0	0	0	0	$\frac{1}{256}$
column sum	$\frac{1}{16}$	$\frac{4}{16}$	$\frac{6}{16}$	$\frac{4}{16}$	$\frac{1}{16}$	1

The marginal mf's of $\tilde{r}_1$ and of $\tilde{r}_2$ look suspiciously regular. The reader may check that the marginal mf of $\tilde{r}_1$ is $f_b(\cdot \mid 4, \frac{1}{4})$ and the marginal mf of $\tilde{r}_2$ is $f_b(\cdot \mid 4, \frac{1}{2})$.

That the marginal distributions of $\tilde{r}_1$ and $\tilde{r}_2$ in Example 11.4.1 are binomial is no mere happenstance. The following result is true in general:

If the k-dimensional random vector $\tilde{\mathbf{r}}$ has the multinomial mf $fmu^{(k)}(\cdot|n, \mathbf{p})$; if $\tilde{\mathbf{r}}_1 \equiv (\tilde{r}_1, \ldots, \tilde{r}_j)$ is a $j(<k)$-dimensional subvector of $\tilde{\mathbf{r}}$; and if $\mathbf{p}_1$ is the corresponding subvector of $\mathbf{p}$, then the marginal mf of $\tilde{\mathbf{r}}_1$ is $f_{mu}{}^{(j)}(\cdot \mid n, \mathbf{p}_1)$. (11.4.5)

A rigorous proof is, of course, possible, but we shall be content with the following argument. If we adopt the cruder description of the observations in which a ball of color $c_{j+1}, \ldots, c_{k+1}$ is called of color "not j" (denoted c^*_{j+1}) then the general derivation of the multinomial mf proceeds just as before, with j replacing k everywhere and c^*_{j+1} replacing c_{k+1}.

Once we have the distribution of $\tilde{\mathbf{r}}$ and the marginal distribution of $\tilde{\mathbf{r}}_1$, we can derive the conditional distribution of the $(k-j)$-dimensional random vector $\tilde{\mathbf{r}}_2 \equiv (\tilde{r}_{j+1}, \ldots, \tilde{r}_k)$ given $\tilde{\mathbf{r}}_1 = \mathbf{r}_1$. The reader may prove as Exercise 21 the following result:

If $\tilde{\mathbf{r}}$ has mf $f_{mu}{}^{(k)}(\cdot|n, \mathbf{p})$; $0 < j < k$; $\tilde{\mathbf{r}} = (\tilde{\mathbf{r}}_1, \tilde{\mathbf{r}}_2)$ and $\mathbf{r}_1 \equiv (r_1, \ldots, r_j)$; $n^*(\mathbf{r}_1) \equiv n - \sum_{i=1}^{j} r_i$; and $\mathbf{p}^*_2 \equiv [1/(1 - \sum_{i=1}^{j} p_i)]\mathbf{p}_2$; then the conditional mf of $\tilde{\mathbf{r}}_2$ given $\tilde{\mathbf{r}}_1 = \mathbf{r}_1$ is $f_{mu}{}^{(k-j)}(\cdot|n^*(\mathbf{r}_1), \mathbf{p}^*_2)$. (11.4.6)

Example 11.4.2

Let $j = 1$ in Example 11.4.1. Then

$$f_{mu}{}^{(k-j)}(\cdot \mid n^*(r_1),\ p^*_2) = f_b(\cdot \mid n^*(r_1),\ p^*_2),$$

where $n^*(r_1) = 4 - r_1$ and $p^*_2 = p_2/(1 - p_1) = (\frac{1}{2})/(\frac{3}{4}) = \frac{2}{3}$. If $r_1 = 1$, then $n^* = 3$ and $\tilde{r}_2$ has the conditional mf $f_b(\cdot \mid 3, \frac{2}{3})$ given by

$$f_b(r_2 \mid 3, \tfrac{2}{3}) = \begin{cases} \frac{1}{27}, & r_2 = 0 \\ \frac{6}{27}, & r_2 = 1 \\ \frac{12}{27}, & r_2 = 2 \\ \frac{8}{27}, & r_2 = 3 \\ 0, & r_2 \notin \{0, 1, 2, 3\}. \end{cases}$$

The reader may check that the same result obtained by dividing the $r_1 = 1$ row of the table in Example 11.4.1 by its total.

It should be apparent that (11.4.6) continues to obtain when $\mathbf{r}_1$ is *any* j-dimensional subvector of $\mathbf{r}$, since the components of $\tilde{\mathbf{r}}$ (colors, classes) can be renumbered so that (11.4.6) applies as stated. Without the re-numbering, define $\mathbf{r}_1 \equiv (r_{i_1}, \ldots, r_{i_j})$, $n^*(\mathbf{r}_1) = n - \sum_{h=1}^{j} r_i$, and $p^*_2 = [1/(1 - \sum_{h=1}^{j} p_{i_h})]p_2$; and then conclude that the subector $\tilde{\mathbf{r}}_2$ of components of $\tilde{\mathbf{r}}$ with subscripts *not* in $\{i_1, \ldots, i_j\}$ has conditional mf

$$f_{mu}{}^{(k-j)}(\cdot \mid n^*(\mathbf{r}_1), \mathbf{p}^*_2)$$

as before.

11.5 THE HYPERGEOMETRIC PROCESS

Consider the urn model of Section 11.3 in which the urn contains R red and $N - R$ black balls, but now assume that sampling from the urn is without replacement. Hence $\tilde{x}_1, \tilde{x}_2, \ldots$ are *not* independent. Furthermore, the largest possible number of observations is N.

The first observation $\tilde{x}_1$ obviously has the mass function

$$f(x_1 \mid R, N) = \begin{cases} \frac{R}{N}, & x_1 = 1 \\ \frac{(N - R)}{N}, & x_1 = 0 \\ 0, & x_1 \notin \{0, 1\}, \end{cases}$$

which (with purpose aforethought) we re-express as

$$f(x_1 \mid R, N) = \begin{cases} \frac{(R)_{x_1}(N - R)_{1-x_1}}{(N)_1}, & x_1 \in \{0, 1\} \\ 0, & x_1 \notin \{0, 1\}. \end{cases} \quad (11.5.1)$$

The reader may verify, by invoking the relation $(a)_b = a!/(a-b)!$, that the two characterizations of $f(x_1 \mid R, N)$ coincide.

After the first observation, the urn will contain $N-1$ balls, $R - x_1$ of which are red and $N - R - (1 - x_1)$ of which are black. More generally, if the first i observations produced $\sum_{j=1}^{i} x_j$ red and $i - \sum_{j=1}^{i} x_j$ black balls, then the assumption that sampling is without replacement implies that after the ith observation the urn will contain $N - i$ balls, of which $R - \sum_{j=1}^{i} x_j$ are red and $N - R - \left(i - \sum_{j=1}^{i} x_j\right)$ are black. Obviously, $R - \sum_{j=1}^{i} x_j \geqq 0$ and $N - R - \left(i - \sum_{j=1}^{i} x_j\right) \geqq 0$, which means that $\sum_{j=1}^{i} x_j$ must satisfy

$$\max\{0, i - (N-R)\} \leq \sum_{j=1}^{i} x_j \leqq \min\{i, R\}. \tag{11.5.2}$$

Hence the mass function of $\tilde{x}_{i+1}$, given $(x_1, \ldots, x_i)$, is

$$f(x_{i+1} \mid x_1, \ldots, x_i, R, N)$$
$$= \begin{cases} \dfrac{\left(R - \sum_{j=1}^{i} x_j\right)_{x_{i+1}} \left(N - R - i + \sum_{j=1}^{i} x_j\right)_{1-x_{i+1}}}{(N-i)_1}, & x_{i+1} \in \{0, 1\} \\ 0, & x_{i+1} \notin \{0, 1\}, \end{cases} \tag{11.5.3}$$

for every possible $(x_1, \ldots, x_i)$; that is, satisfying (11.5.2).

The *hypergeometric* (data-generating) *process* with parameters R and N (positive integers with $R < N$) is characterized by the unconditional mf (11.5.1) of $\tilde{x}_1$ and the family (11.5.3) of conditional mf's of $\tilde{x}_{i+1}$ given all possible $(x_1, \ldots, x_i)$.

Now take (11.5.3) for $i = 1$:

$$f(x_2 \mid x_1, R, N) = \begin{cases} \dfrac{(R - x_1)_{x_2}(N - R - 1 + x_1)_{1-x_2}}{(N-1)_1}, & x_2 \in \{0, 1\} \\ 0, & x_2 \notin \{0, 1\}, \end{cases} \tag{11.5.4}$$

for $\max\{0, 1 - (N-R)\} \leqq x_1 \leqq \min\{(1, R\}$.

Multiply (11.5.4) by (11.5.1) to obtain the joint mf of $\tilde{x}_1$ and $\tilde{x}_2$:

$$f(x_1, x_2 \mid R, N)$$
$$= \begin{cases} \dfrac{(R)_{x_1}(R - x_1)_{x_2}(N-R)_{1-x_1}(N - R - 1 + x_1)_{1-x_2}}{(N)_1(N-1)_1}; & \begin{cases} x_1 \in \{0, 1\} \\ x_2 \in \{0, 1\} \end{cases} \\ 0; & \text{elsewhere} \end{cases}$$
$$= \begin{cases} \dfrac{(R)_{x_1+x_2}(N-R)_{2-(x_1+x_2)}}{(N)_2}; & \begin{cases} x_1 \in \{0, 1\} \\ x_2 \notin \{0, 1\} \end{cases} \\ 0; & \text{elsewhere,} \end{cases} \tag{11.5.5}$$

where the second equality follows from the first from the definition of $(a)_b$, in view of $x_2 \in \{0, 1\}$. The reader may fill in the details as Exercise 22.

It is now clear how we should proceed for the general case; in Exercise 23 the reader may prove by induction that:

If $r \equiv \sum_{j=1}^{n} x_j$, then

$$f(x_1, \ldots, x_n \mid R, N) \tag{11.5.6}$$
$$= \begin{cases} \dfrac{(R)_r(N-R)_{n-r}}{(N)_n}; & \max\{0, n-(N-R) \le r \le \min\{n, R\}, r \text{ integer} \\ 0; & \text{otherwise.} \end{cases}$$

From this it follows that (r, n) is a sufficient statistic for (R, N).

As in Section 11.3, we now consider two stopping processes. In the *first* stopping process the number n of observations is predetermined and $\tilde{r}$ is a random variable (before the experiment is conducted). Again as in Section 11.3, there are precisely $\binom{n}{r}$ n-tuples $(x_1, \ldots, x_n)$ such that $\sum_{j=1}^{n} x_j = r$ for each possible value of r, and hence

$$P(\tilde{r} = r \mid n, R, N)$$
$$= \begin{cases} \dbinom{n}{r}\dfrac{(R)_r(N-R)_{n-r}}{(N)_n}; & \max\{0, n-(N-R)\} \leqq r \leqq \min\{n, R\}, \ r \text{ an integer} \\ 0; & \text{elsewhere.} \end{cases}$$

Now, by writing $\binom{n}{r} = n!/[r!(n-r)!]$, $(R)_r = R!/(R-r)!$, and so forth, it follows upon rearrangement and association of terms that

$$\binom{n}{r}\frac{(R)_r(N-R)_{n-r}}{(N)_n} = \frac{\dbinom{R}{r}\dbinom{N-R}{n-r}}{\dbinom{N}{n}}.$$

The reader may fill in the details as Exercise 24.

We have thus obtained the following result:

If an experiment consists of a predetermined number n of observations from the hypergeometric process, then $\tilde{r} \equiv \sum_{i=1}^{n} \tilde{x}_i$ has the *hypergeometric* distribution, with mass function given by

$$f_h(r \mid n, R, N) = \begin{cases} \dfrac{\dbinom{R}{r}\dbinom{N-R}{n-r}}{\dbinom{N}{n}}, & \begin{cases} r \geqq \max\{0, n-(N-R)\} \\ r \leqq \min\{n, R\} \\ r \text{ an integer} \end{cases} \\ 0, & \text{elsewhere,} \end{cases} \tag{11.5.7}$$

where R and N are positive integers and $N \geqq \max\{n, R\}$.

Example 11.5.1

Examples 11.3.1 and 11.3.3 may be altered so as to serve as samples of hypergeometric processes, if we now assume that sampling is *without* replacement and in Example 11.3.3 we replace 60 percent with a (R, N) pair such that $R = .60N$. Note that when the sampling is without replacement, R and N are needed individually, and not solely in the aggregated form $R/N \equiv p$ (as in the Bernoulli process). For Example 11.3.1, we find the following hypergeometric analogues of the binomial probabilities, graphed in Figure 11.2. All are accurate to four decimal places.

r	$f_h(r \mid 4250{,}1000)$	$f_b(r \mid 4,\frac{1}{4})$
0	.3158	.3164
1	.4227	.4219
2	.2111	.2109
3	.0466	.0469
4	.0038	.0039

The reader will notice how very slight is the difference between the binomial and hypergeometric probabilities. This means, of course, that the assumption of sampling *with* versus *without* replacement is not very crucial to useful conclusions in this problem. However, when n is large relative to R and N-R, there may be marked differences between binomial and hypergeometric results, as in the following example:

Example 11.5.2

An urn contains $N = 2$ balls, $R = 1$ of which is red. A sample of predetermined size 2 is drawn without replacement. Few calculations are necessary to derive the following table:

r	$f_h(r \mid 2,2,1)$	$f_b(r \mid 2,\frac{1}{2})$
0	0	$\frac{1}{4}$
1	1	$\frac{1}{2}$
2	0	$\frac{1}{4}$

In this table, the effect of the replacement versus nonreplacement assumption is striking.

There are hypergeometric-process analogues to the Pascal and negative binomial mass functions, which arose from "inverse sampling" (that is, predetermination of r). We have already shown that this stopping process is noninformative. Moreover, if the experiment ends after the nth observation, then $x_n = 1$ and the appropriate factor is found, as in Section 11.3, to be $\binom{n-1}{r-1}$. After rearrangement, we obtain the following:

If an experiment is to continue until a predetermined number r of ones have been obtained from a hypergeometric process with parameters R and N, then the total number $\tilde{n}$ of observations has the *Pascal-hypergeometric* distribution, with mass function

$$f_{Pah}(n \mid r, R, N) = \begin{cases} \dfrac{\dbinom{n-1}{r-1}\dbinom{N-n}{R-r}}{\dbinom{N}{R}}; & n \in \{r, r+1, \ldots, N-R+r\} \\ 0; & \text{elsewhere,} \end{cases} \quad (11.5.8)$$

provided that $r \leqq R$. If $r > R$, then all N items are drawn successively and $P(\tilde{n} = N \mid r, R, N) = 1$. Moreover, the number $\tilde{\zeta} \equiv \tilde{n} - r$ of zeros (black balls) until the rth one (red ball) has the *negative hypergeometric*, or *hypergeometric waiting-time* distribution, with mass function

$$f_{nh}(\zeta \mid r, R, N) = \begin{cases} \dfrac{\dbinom{\zeta+r-1}{r-1}\dbinom{N-\zeta-r}{R-r}}{\dbinom{N}{R}}; & \zeta \in \{0, 1, \ldots, N-R\} \\ 0; & \text{elsewhere,} \end{cases} \quad (11.5.9)$$

provided that $r \leqq R$.

We have seen in Example 11.5.1 how close the binomial and hypergeometric mass functions $f_b(\cdot \mid n, R/N)$ and $f_h(\cdot \mid n, R, N)$ tend to be when n is small relative to R and N. The binomial mf $f_b(\cdot \mid n, R/N)$ is a good approximation to $f_h(\cdot \mid n, R, N)$ when n is small relative to *both* R and $N - R$. And the mathematical manipulation of the binomial is much easier than that of the hypergeometric!

11.6 THE k-VARIATE HYPERGEOMETRIC PROCESS

As in Section 11.4, we consider an urn containing N balls, R_1 of which are of color $c_1, \ldots,$ and R_{k+1} of which are of color c_{k+1}. Here, however, we assume that balls are to be drawn from the urn *without* replacement. If we define $R \equiv (\mathbf{R}_1, \ldots, R_k)$, $R_{k+1} \equiv N - \sum_{j=1}^{k} R_j$, the k-tuple observations $\tilde{\mathbf{x}}_i \equiv (\tilde{x}_{1i}, \ldots, \tilde{x}_{ki})$ as in Section 11.4, and $\tilde{x}_{(k+1)i} \equiv 1 - \sum_{j=1}^{k} \tilde{x}_{ji}$, then we deduce

$$f(\mathbf{x}_1 \mid \mathbf{R}, N) = \frac{(R_1)_{x_{11}} \cdot (R_2)_{x_{21}} \cdot \ldots \cdot (R_{k+1})_{x_{(k1+1)}}}{(N)_1}, \quad (11.6.1)$$

wherever $\mathbf{x}_1$ is possible; that is, belongs to the set

$$\{(1, 0, 0, \ldots, 0, 0), (0, 1, 0, \ldots, 0, 0), \ldots, (0, 0, 0, \ldots, 0, 1), (0, \ldots, 0, 0)\}.$$

After i observations, the urn contains $N - i$ balls, of which $R_j - \sum_{h=1}^{i} x_{jh}$ are of color c_j, for every $j \in \{1, \ldots, k + 1\}$. Hence,

$$f(\mathbf{x}_{i+1} \mid \mathbf{x}_1, \ldots, \mathbf{x}_i, \mathbf{R}, N) = \frac{\prod_{j=1}^{k+1} \left(R_j - \sum_{h=1}^{i} x_{jh}\right)_{x_{j(i+1)}}}{(N - i)_1}, \qquad (11.6.2)$$

wherever $\mathbf{x}_i$ is possible.

Equations (11.6.1) and (11.6.2) characterize the *k-variate hypergeometric process.*

The joint mf of $(\tilde{\mathbf{x}}_1, \ldots, \tilde{\mathbf{x}}_n)$ follows from (11.6.1) and (11.6.2) just as (11.5.6) followed from (11.5.1) and (11.5.3):

$$f(\mathbf{x}_1, \ldots, \mathbf{x}_n \mid n, \mathbf{R}, N) = \begin{cases} \dfrac{\prod_{j=1}^{k+1} (R_j)_{r_j}}{(N)_n}; & \begin{cases} r_j \text{ an integer, } 0 \leqq r_j \leqq \min\{n, R_j\} \text{ for} \\ \text{every } j, \text{ and } \sum_{j=1}^{k} r_j \geqq n - \left(N - \sum_{j=1}^{k} R_j\right) \end{cases} \\ 0; & \text{elsewhere,} \end{cases} \qquad (11.6.3)$$

where $r_j \equiv \sum_{i=1}^{n} x_{ji}$ for $j \in \{1, \ldots, k + 1\}$; $N, R_1, \ldots, R_{k+1}$ are positive integers, $N \geqq \max\{n, R_1, \ldots, R_{k+1}\}$, and $\sum_{j=1}^{k+1} R_j = N$.

From (11.6.3) it is clear that $(\mathbf{r}, n)$ is a sufficient statistic for $\mathbf{R}, N)$, where $\mathbf{r} \equiv (r_1, \ldots, r_k)$. Moreover, when n is predetermined, $\tilde{\mathbf{r}}$ is a random k-tuple; the factor $\binom{n}{r_1 \ldots r_k}$ derived in Section 11.4 applies, and hence we obtain the distribution of $\tilde{\mathbf{r}}$:

If an experiment consists of a predetermined number n of observations from the k-variate hypergeometric process, then $\tilde{\mathbf{r}} \equiv \sum_{i=1}^{n} \tilde{\mathbf{x}}_i$ has the *k-variate hypergeometric* distribution, with mass function given by

$$f_h^{(k)}(\mathbf{r} \mid n, \mathbf{R}, N) = \begin{cases} \dfrac{\prod_{j=1}^{k+1} \binom{R_j}{r_j}}{\binom{N}{n}}; & \begin{cases} r_j \text{ integer and } 0 \leqq r_j \leqq \min\{n, R_j\} \text{ for} \\ \text{every } j, \text{ and } \sum_{j=1}^{k} r_j \geqq n - \left(N - \sum_{j=1}^{k} R_j\right) \end{cases} \\ 0; & \text{elsewhere,} \end{cases} \qquad (11.6.4)$$

where $N, R_1, \ldots, R_{k+1}$ are positive integers, $N \geqq \max\{n, R_1, \ldots, R_{k+1}\}$, and $\sum_{j=1}^{k+1} R_j = N$.

By reasoning as in Section 11.4 we deduce that the marginal distributions of subvectors of $\tilde{\mathbf{r}}$ are also hypergeometric:

If $\tilde{\mathbf{r}}$ has mf $f_h^{(k)}(\cdot \mid n, \mathbf{R}, N)$, $0 \leqq j \leqq k$, $\tilde{\mathbf{r}}_1 \equiv (\tilde{r}_1, \ldots, \tilde{r}_j)$, $\tilde{\mathbf{r}} = (\tilde{\mathbf{r}}_1, \tilde{\mathbf{r}}_2)$, and $\mathbf{R}_1$ is the corresponding subvector of $\mathbf{R}$, then $\tilde{\mathbf{r}}_1$ has mf $f_h^{(j)}(\cdot \mid n, \mathbf{R}_1, N)$. (11.6.5)

These two results imply that $\tilde{\mathbf{r}}_2$ has a hypergeometric distribution given any possible value of $\tilde{\mathbf{r}}_1$:

If $\tilde{\mathbf{r}}$ has mf $f_h^{(k)}(\cdot \mid n, \mathbf{R}, N)$, $j \in \{1, \ldots, k-1\}$, $\tilde{\mathbf{r}}_1 = (\tilde{r}_1, \ldots, \tilde{r}_j)$, and $\tilde{\mathbf{r}} = (\tilde{\mathbf{r}}_1, \tilde{\mathbf{r}}_2)$, then the distribution of $\tilde{\mathbf{r}}_2$ conditional upon $\tilde{\mathbf{r}}_1 = \mathbf{r}_1$ has mass function $f_h^{(k-j)}(\cdot \mid n^*(\mathbf{r}_1), \mathbf{R}_2, N^*)$, where $n^*(\mathbf{r}_1) \equiv n - \sum_{i=1}^{j} r_i$ and $N^* \equiv N - \sum_{i=1}^{j} R_j$.

(11.6.6)

The concluding remarks of Section 11.4 apply to (11.6.6) with the obvious modifications.

Example 11.6.1

Let $k = 3$, $N = 8$, $R_1 = 2$, $R_2 = 4$, and $n = 4$. Note that $R_1/N = \frac{1}{4}$ and $R_2/N = \frac{1}{2}$, for comparability with Example 11.4.1. The following table specifies $f_h^{(2)}(\cdot \mid 4, (2, 4), 8)$.

r_1 \ r_2	0	1	2	3	4	*row sum*
0	0	0	$\frac{6}{70}$	$\frac{8}{70}$	$\frac{1}{70}$	$\frac{15}{70}$
1	0	$\frac{8}{70}$	$\frac{24}{70}$	$\frac{8}{70}$	0	$\frac{40}{70}$
2	$\frac{1}{70}$	$\frac{8}{70}$	$\frac{6}{70}$	0	0	$\frac{15}{70}$
3	0	0	0	0	0	0
4	0	0	0	0	0	0
column sum	$\frac{1}{70}$	$\frac{16}{70}$	$\frac{36}{70}$	$\frac{16}{70}$	$\frac{1}{70}$	1

The reader may convert this table and that of Example 11.4.1 to decimals as Exercise 25 in order to facilitate numerical comparisons. Note in this example the effect of the constraints $r_j \leqq R_j$ and $\sum_{j=1}^{k} r_j \geqq n - \left(N - \sum_{j=1}^{k} R_j\right)$.

11.7 THE POISSON PROCESS

In the preceding sections we have been concerned with data-generating processes which arise naturally in sampling from populations (of people, of items in a production lot, of bridge hands, and so forth). In this section we shall be concerned with a data-generating process that describes the repeated occurrences of some physical (or social) phenomenon at random points of time; for example, occurrences of auto accidents, of burnouts of lightbulbs in a building, of earthquakes at a given spot, and of arrivals of customers at a checkout counter.

We shall assume that each such phenomenon occurs at an instant in (point of) time, and that the "interarrival time" $\tilde{t}$ between any two successive occurrences has the *exponential* distribution, with density function

$$f_e(t \mid \lambda) = \begin{cases} 0, & t \leqq 0 \\ \lambda e^{-\lambda t}, & t > 0, \end{cases} \tag{11.7.1}$$

for some positive parameter λ. (See Figure 7.9.)

We define the *Poisson process* with ("intensity") parameter λ as the generator of iid random variables (interarrival times) $\tilde{t}_1$, $\tilde{t}_2$, . . . with common df (11.7.1).

In Figure 7.9 we graphed $f_e(\cdot \mid \lambda)$ for three values of λ, and it was clear that the interarrival times tend to be short when λ is large. Hence the term *intensity* for λ is quite appropriate: the phenomena in question tend to occur at an intense pace when λ is large, which implies that the interarrival times tend to be small.

The Poisson process is by far the most important data-generating process used to describe repeated occurrences of a phenomenon at random points in time. Its importance stems both from the fact that interarrival times often seem to be iid with common df $f_e(\cdot \mid \lambda)$ for some λ, and the fact that this assumption leads to results of great mathematical tractability.

As in previous sections, our interest will be in observing a Poisson process in order to make inferences about its intensity λ. Unlike previous sections, however, our observation will not be of a discrete number n of outcomes x_i (or $\mathbf{x}_i$) resulting in the statistic r (or $\mathbf{r}$), but rather of some period t of time, in which r phenomena of the specified type occurred.

Now, in a general data-generating process describing repeated occurrences of some phenomenon, it is necessary to specify *when* you start observing the process relative to the time of the preceding occurrence. To see this, let τ_0 denote the time of the occurrence immediately preceding the time T at which observation of the process starts; and for $i = 1, 2, \ldots$ let $\tilde{\tau}_i$ denote the time of the ith occurrence following the start of observation. Let $\tilde{t}_1 \equiv \tilde{\tau}_1 - \tilde{\tau}_0$ and $\tilde{t}_i \equiv \tilde{\tau}_i - \tilde{\tau}_{i-1}$, for $i = 2, 3, \ldots$ denote the interrarrival times, and suppose that $\tilde{t}_1, \tilde{t}_2, \ldots$ are iid with some common density function f_t such that $F_t(0) = 0$. Now, note that $\tau_0 < T < \tilde{\tau}_1$, but that you will not be able to observe the value of $\tilde{t}_1$ unless you know τ_0. But if you knew τ_0, then *in effect* you started observation at τ_0. Hence you might have to wait until $\tilde{\tau}_1$ to start obtaining *usable* data $t_2, t_3, \ldots$. This waiting is unrealistic for processes with very large typical interarrival times.

Hence you would like to be able to start observation at an *arbitrary* point T between $\tilde{\tau}_0$ and $\tilde{\tau}_1$, and to know the distribution of the time $\tilde{y} \equiv \tilde{\tau}_1 - T$ from the starting point T to the first occurrence after T.

The df $f(y)$ of $\tilde{y}$ is obtainable from $f(t)$ by noting:

(a) $\tilde{y} = \tilde{t}_1 - (T - \tau_0)$;
(b) $\tilde{y} > 0$ (because $\tau_0 < T < \tau_1$);

hence

(c) $\tilde{t}_1 > T - \tau_0$;

and therefore

(d) $P[\tilde{y} > y] = P[\tilde{y} > y \mid \tilde{y} > 0] = P[\tilde{t}_1 > y + T - \tau_0 \mid \tilde{t}_1 > T - \tau_0]$.

But

(e) $$P[\tilde{t}_1 > y + (T - \tau_0) \mid \tilde{t}_1 > T - \tau_0]$$
$$= \frac{P[\tilde{t}_1 > y + (T - \tau_0), \tilde{t}_1 > T - \tau_0]}{P[\tilde{t}_1 > T - \tau_0]}$$
$$= \frac{P[\tilde{t}_1 > y + (T - \tau_0)]}{P[\tilde{t}_1 > T - \tau_0]},$$

and this depends in general upon $T - \tau_0$, which is undesirable.

A very important characteristic of the *Poisson* process is

$$P[\tilde{t}_1 > y + T - \tau_0 \mid \tilde{t}_1 > T - \tau_0] = P[\tilde{t}_1 > y]; \tag{11.7.2}$$

that is, the event that you have to wait at least y time units until the first occurrence after the start of observation is independent of the occurrences prior to the start of observation.

Proof

In view of (e) it is sufficient to prove that

(*) $$\frac{[1 - F_e(y + T - \tau^0 \mid \lambda)]}{[1 - F_e(T - \tau_0 \mid \lambda)]} = 1 - F_e(y \mid \lambda).$$

Now,

$$1 - F_e(x \mid \lambda) = \int_x^\infty \lambda e^{-\lambda t} dt = e^{-\lambda x},$$

for any $x > 0$ and $\lambda > 0$. Hence the left-hand side of (*) equals

$$\frac{e^{-\lambda(y+T-\tau_0)}}{e^{-\lambda(T-\tau_0)}},$$

which simplifies to $e^{-\lambda y}$, the right-hand side of (*). QED

It can be shown, conversely, that the Poisson process is the *only* inter-arrival-time process with $\tilde{t}_1, \tilde{t}_2, \ldots$ iid and continuous, with common cdf $F(t)$ such that $F(0) = 0$, and (11.7.2) holds. (See Raiffa and Schlaifer [**66**], pp. 275–276 for proof of this assertion.)

Thus we may begin observing a Poisson process at any point in time and let $\tilde{t}_1$ denote the time from the start of observation until the first occurrence.

Now suppose you observe a Poisson process for t time periods, during which there were r occurrences following interarrival times $t_1, \ldots,$ and t_r. Clearly, $\sum_{i=1}^{r} t_i \leqq t$. If $\sum_{i=1}^{r} t_i < t$, then you have obtained more information about λ than is embodied in $(t_1, \ldots, t_r)$ alone, because you have waited for $t - \sum_{i=1}^{r} t_i$ time units for occurrence $r + 1$. Let $\tilde{t}_{r+1}$ denote the interarrival time between the rth and $(r + 1)$th occurrence. The joint density of $(\tilde{t}_1, \ldots, \tilde{t}_{r+1})$ is, from (11.7.1),

$$f(t_1, \ldots, t_{r+1} \mid t, \lambda) = \begin{cases} \prod_{i=1}^{r+1} \lambda e^{-\lambda t_i}; & \begin{cases} \text{every } t_i > 0 \\ \sum_{i=1}^{r} t_i \leqq t < \sum_{i=1}^{r+1} t_i \end{cases} \\ 0; & \text{elsewhere} \end{cases}$$

$$= \begin{cases} \lambda^{r+1} \exp\left(-\lambda \sum_{i=1}^{r+1} t_i\right); & \text{every } t_i > 0 \\ & \sum_{i=1}^{r} t_i \leqq t < \sum_{i=1}^{r+1} t_i \\ 0; & \text{elsewhere.} \end{cases} \tag{11.7.3}$$

However, we will not know the value of $\tilde{t}_{r+1}$ for sure, because we cease observing before the $(r + 1)$th occurrence. All we know at the end of the observation is that $\tilde{t}_{r+1} \in (t - \sum_{i=1}^{r} t_i, \infty)$. Hence we must integrate (11.7.3) with respect to t_{r+1} from $t - \sum_{i=1}^{r} t_i$ to ∞. For notational convenience let $a \equiv \sum_{i=1}^{r} t_i$. Then

$$\int_{t-a}^{\infty} \lambda^{r+1} \cdot e^{-\lambda(a+t_{r+1})} dt_{r+1}$$

$$= \lambda^r e^{-\lambda a} \int_{t-a}^{\infty} \lambda \cdot e^{-\lambda t_{r+1}} dt_{r+1}$$

$$= \lambda^r e^{-\lambda a} \cdot e^{-\lambda(t-a)}$$

$$= \lambda^r e^{-\lambda t}.$$

Hence we obtain

$$f(t_1, \ldots, t_r \mid t, \lambda) = \begin{cases} \lambda^r e^{-\lambda t}, & t \geqq \sum_{i=1}^{r} t_i \\ 0, & \text{elsewhere,} \end{cases} \tag{11.7.4}$$

from which it is clear that (r, t) is a sufficient statistic for λ when the stopping is noninformative.

The analogue of a fixed sample size n for the Poisson process is a fixed time period t of observation. If t is fixed, then the number $\tilde{r}$ of occurrences is a random variable, with range $= \{0, 1, 2, \ldots\}$.

We shall now derive the mf of $\tilde{r}$ given λ and t. Let $f(r \mid t)$ denote the probability of finding precisely r occurrences in a time period $t > 0$. From (11.7.1) it is clear that

$$f(0 \mid t) = e^{-\lambda t}. \tag{11.7.5}$$

Next suppose that there are r occurrences during $(0, t]$, and suppose that the *first* occurrence falls in the infinitesimal subinterval $(\tau, \tau + d\tau)$ of $(0, t]$. Then $r - 1$ occurrences must fall in the (approximate) subinterval $(\tau, t]$, and the probability of this is (by definition) $f(r - 1 \mid t - \tau)$. But the probability that the first occurrence falls in $(\tau, \tau + d\tau)$ is (approximately) $\lambda e^{-\lambda\tau} d\tau$, and hence the joint probability of

(1) first occurrence in $(\tau, \tau + d\tau)$

and

(2) last $r - 1$ occurrences in $(\tau, t]$

is (approximately)

$$f(r - 1 \mid t - \tau) \cdot \lambda e^{-\lambda\tau} d\tau;$$

and hence

$$f(r \mid t) = \int_0^t f(r - 1 \mid t - \tau) \lambda e^{-\lambda\tau} d\tau. \tag{11.7.6}$$

The above "derivation" of (11.7.6) is heuristic but can be made completely rigorous.

Equations (11.7.5) and (11.7.6) can be solved recursively; for $r = 1$ from $r = 0$; for $r = 2$ from $r = 1$; and so forth. You may prove by induction as Exercise 26 that

$$f(r \mid t) = \frac{e^{-\lambda t}(\lambda t)^r}{r!}, \quad r \in \{0, 1, 2, \ldots\}.$$

This is called the *Poisson* mf, and hence we have obtained:

If an experiment consists in observing a Poisson process with intensity parameter λ for a predetermined time period of length t, then $\tilde{r} \equiv$ number of occurrences in that period) has the Poisson distribution with parameter λt, with mass function given by

$$f_{Po}(r \mid \lambda t) \equiv \begin{cases} \dfrac{e^{-\lambda t}(\lambda t)^r}{r!}, & r \in \{0, 1, 2, \ldots\} \\ 0, & r \notin \{0, 1, 2, \ldots\}, \end{cases} \tag{11.7.7}$$

where $\lambda t > 0$.

Example 11.7.1

Suppose $t = 1$ minute and $\lambda = 2$ customers/minute is the intensity of arrivals at a checkout counter. Then the first few terms of $f_{Po}(\cdot \mid 2)$ are (to three places) as follows:

r	$f_{Po}(r \mid 2)$
0	.135
1	.271
2	.271
3	.180
4	.090
5	.036
6	.012
7	.003
8	.001
9	.000

Example 11.7.2

Bortkiewicz [8] found that the number $\tilde{r}$ of deaths per corps per year in the (mid-to-late-nineteenth century) Prussian army due to horse kicks appears to have the Poisson mf $f_{Po}(\cdot \mid \lambda t = .61)$. Since $t = 1$ year, the intensity λ is seen to be .61 deaths per corps due to a horse kick/year. [Since the nineteenth-century Prussian army corps totaled approximately 42,500 men, each man therefore had probability approximately .61/42,500 (= .0000143) of being killed by a horse kick in any given year, assuming equal exposure of all corps soldiers to horse kicks.]

Example 11.7.2 is rather bizarre, but it illustrates the use of the Poisson distribution for a random variable $\tilde{r}$ whose only possible values are in $\{0, 1, 2, \ldots\}$ and which (for all practical purposes) can assume arbitrarily large integer values. We may show, in fact, that the Poisson mass function can be derived as a limit of certain *binomial* mass functions:

Let n tend to $+\infty$ and define $p(n) \equiv \lambda t/n$ for a fixed $\lambda t > 0$. Then for every $r \in \{0, 1, 2, \ldots\}$, we have

$$\lim_{n\to\infty} f_b(r \mid n, p(n)) = f_{Po}(r \mid \lambda t). \tag{11.7.8}$$

Proof

Write

$$\begin{aligned} f_b(r \mid n, p(n)) &\equiv \frac{n!}{r!(n-r)!}\left(\frac{\lambda t}{n}\right)^r\left(1-\frac{\lambda t}{n}\right)^{n-r} \\ &= \frac{(\lambda t)^r}{r!}\left(1-\frac{\lambda t}{n}\right)^n \cdot \frac{(n)_r}{n^r} \cdot \left(1-\frac{\lambda t}{n}\right)^{-r} \\ &= \frac{(\lambda t)^r}{r!}\left(1-\frac{\lambda t}{n}\right)^n \cdot \left[\left(1-\frac{\lambda t}{n}\right)^{-r} \cdot \prod_{i=1}^{r-1}\left(1-\frac{i}{n}\right)\right] \end{aligned}$$

[because $(n)_r/n^r = \prod_{i=1}^{r-1}(1 - i/n)$]. Now, $\lim f_b(r \mid n, p(n))$ exists and equals the product of the terms

$$\lim_{n\to\infty}\left\{\frac{(\lambda t)^r}{r!}\right\} = \frac{(\lambda t)^r}{r!},$$

$$\lim_{n\to\infty}\left\{\left(1-\frac{\lambda t}{n}\right)^n\right\} = e^{-\lambda t},$$

$$\lim_{n\to\infty}\left\{\left(1-\frac{\lambda t}{n}\right)^{-r}\right\} = 1,$$

and

$$\lim_{n\to\infty}\left\{\prod_{i=1}^{r-1}\left(1-\frac{i}{n}\right)\right\} = 1.$$

Multiplying these four limits together yields $f_{Po}(r \mid \lambda t)$, as asserted. QED

There is an "inverse sampling" procedure for the Poisson process. Suppose that you continue observing the process until the rth occurrence and then *immediately* cease observation. Thus our random component of the sufficient statistic is $\tilde{t} \equiv \sum_{i=1}^{r} \tilde{t}_i$. Now, $\tilde{t}_1, \ldots, \tilde{t}_r$ are iid with common df $f_e(\cdot \mid \lambda)$ and common moment-generating function $M_e(u) \equiv \lambda/(\lambda - u)$ (by Example 10.6.1). Hence $\tilde{t} = \sum_{i=1}^{r} \tilde{t}_i$ has moment-generating function

$$M(u) \equiv \frac{\lambda^r}{(\lambda - u)^r}. \tag{11.7.9}$$

We shall show that (11.7.9) is the moment-generating function corresponding to the *gamma* density $f_\gamma(\cdot \mid r, \lambda)$, defined by

$$f_\gamma(t \mid r, \lambda) \equiv \begin{cases} \dfrac{\lambda e^{-\lambda t}(\lambda t)^{r-1}}{\Gamma(r)}, & t > 0 \\ 0, & t \leqq 0, \end{cases} \tag{11.7.10}$$

where $r > 0$ and $\lambda > 0$ are real parameters, and $\Gamma(r)$ is defined for all $r > 0$ by

$$\Gamma(r) \equiv \int_0^\infty e^{-x}x^{r-1}dx. \tag{11.7.11}$$

The reader may show as Exercise 27 that if r is a positive integer, then $\Gamma(r) = (r-1)!$ (Clearly, r *will* be a positive integer when r represents a total number of Poisson-process events.)

Now we derive the moment-generating function $M_\gamma(u \mid r, \lambda)$ of the gamma random variable:

$$\begin{aligned} E(e^{u\tilde{t}}) &= \int_0^\infty \frac{e^{ut}\lambda e^{-\lambda t}(\lambda t)^{r-1}dt}{\Gamma(r)} \\ &= \int_0^\infty \frac{\lambda^r e^{-(\lambda-u)t}t^{r-1}dt}{\Gamma(r)} \\ &= \frac{\lambda^r}{(\lambda-u)^r}\int_0^\infty \frac{(\lambda-u)e^{-(\lambda-u)t}[(\lambda-u)t]^{r-1}dt}{\Gamma(r)}. \end{aligned}$$

But the integral in the last term is simply $F_\gamma(+\infty \mid r, \lambda - u)$ and hence equals one provided that $u < \lambda$. Hence $M_\gamma(u \mid r, \lambda) = E(e^{u\tilde{t}}) = \lambda^r/(\lambda - u)^r$, in agreement with (11.7.9). We have thus shown the following:

If a Poisson process with intensity parameter λ is observed until precisely r "events" have occurred following inception of observation, then the total observation time $\tilde{t}$ has the *gamma* distribution with density function $f_\gamma(\cdot \mid r, \lambda)$ as defined by (11.7.10). (11.7.12)

It is clear from (11.7.10) that $f_e(t \mid \lambda) = f_\gamma(t \mid 1, \lambda)$ for all real t and $\lambda > 0$. Hence the exponential distribution is simply the special case $r = 1$ of the gamma distribution. When $0 < r < 1$, the densities $f_\gamma(\cdot \mid r, \lambda)$ become infinite as t tends to zero, while for $r = 1$, $f_\gamma(0+ \mid 1, \lambda) = \lambda$; and for $r > 1$, $f_\gamma(0+ \mid r, \lambda) = 0$. Thus for $0 < r < 1$ the densities $f_\gamma(\cdot \mid r, \lambda)$ are even more drastically *J*-shaped than in Figure 7.9.

Figure 12.2 indicates the various available "shapes" of $f_\gamma(\cdot \mid r, \lambda)$. It depicts $f_\gamma(\cdot \mid r, r)$ for various values of $r > 0$ and provides sufficient generality in view of the role of λ as a "scale parameter"; all it does is to change the scale of the abscissa.

Note that the role of f_γ as the density of the random part $\tilde{t}$ of the sufficient statistic of the Poisson process with inverse sampling makes the gamma distribution analogous to the negative binomial distribution in the Bernoulli process. Since for $r = 1$ the gamma distribution is the exponential and the negative binomial is the geometric, it is worth asking whether the geometric distribution satisfies Equation (11.7.2). The answer is "yes," and if $\tilde{t}_1$ satisfies (11.7.2) and is a discrete random variable with possible values being the nonnegative integers, then the distribution of $\tilde{t}_1$ *must* be the geometric for some $p \in (0, 1)$.

11.8 SOME MOMENTS AND MOMENT-GENERATING FUNCTIONS

11.8.1 Binomial Distribution

If $\tilde{r}$ has mf $f_b(\cdot \mid n, p)$, then $\tilde{r}$ can be considered as defined by the sum $\sum_{i=1}^{n} \tilde{x}_i$, where $\tilde{x}_1, \ldots, \tilde{x}_n$ are iid with common mf $f_b(\cdot \mid 1, p)$. Now,

$$f_b(x \mid 1, p) = \begin{cases} p, & x = 1 \\ 1 - p, & x = 0 \\ 0, & x \notin \{0, 1\}, \end{cases}$$

and hence the moment-generating function $M_b(u \mid 1, p)$ corresponding to $f_b(\cdot \mid 1, p)$ is given by

$$\begin{aligned} M_b(u \mid 1, p) &= E_b(e^{u\tilde{x}} \mid 1, p) \\ &= e^{u\cdot 1}(p) + e^{u\cdot 0}(1 - p) \\ &= 1 + p(e^u - 1). \end{aligned}$$

Hence we apply Equation (10.6.6) to obtain

$$\begin{aligned} M_b(u \mid n, p) &= [M_b(u \mid 1, p)]^n \\ &= [1 + p(e^u - 1)]^n, \end{aligned} \tag{11.8.1}$$

from which the reader may deduce [via (10.6.2) and (10.3.11)] that

$$E_b(\tilde{r} \mid n, p) = np \tag{11.8.2}$$

and

$$V_b(\tilde{r} \mid n, p) = np(1 - p). \tag{11.8.3}$$

Now suppose $\tilde{r}_i$ has mf $f_b(\cdot \mid n_i, p)$ for $i = 1, \ldots, m$ and that $\tilde{r}_1, \ldots, \tilde{r}_m$ are independent. Then $\sum_{i=1}^{m} \tilde{r}_i \equiv \tilde{r}$ has moment-generating function

$$\prod_{i=1}^{m} M_b\,(u \mid n_i, p),$$

which (from 11.8.1) is easily found to be

$$[1 + p(e^u - 1)]^{\sum_{i=1}^{m} n_i},$$

thus implying that the mf of $\tilde{r} \equiv \sum_{i=1}^{m} \tilde{r}_i$ is $f_b(\cdot \mid \sum_{i=1}^{m} n_i, p)$.

11.8.2 Negative Binomial Distribution

If $\tilde{\zeta}$ has mf $f_g(\cdot \mid 1, p)$, then $\tilde{\zeta}$ has moment-generating function

$$\begin{aligned} M_g(u \mid 1, p) &= \sum_{\zeta=0}^{\infty} e^{u\zeta} p(1-p)^{\zeta} \\ &= p \sum_{\zeta=0}^{\infty} [e^u(1-p)]^{\zeta} \qquad (11.8.4) \\ &= \frac{p}{[1-(1-p)e^u]}, \end{aligned}$$

provided that $(1 - p)e^u < 1$.

More generally, if $\tilde{\zeta}$ has mf $f_{nb}(\cdot \mid r, p)$, then $\tilde{\zeta}$ has moment-generating function

$$\begin{aligned} M_{nb}(u \mid r, p) &= \sum_{\zeta=0}^{\infty} e^{u\zeta} \binom{\zeta + r - 1}{r - 1} p^r (1-p)^{\zeta} \\ &= \frac{p^r}{[1-(1-p)e^u]^r} \sum_{\zeta=0}^{\infty} \binom{\zeta + r - 1}{r - 1} [1-(1-p)e^u]^r [(1-p)e^u]^{\zeta} \\ &= \left[\frac{p}{1-(1-p)e^u}\right]^r \sum_{\zeta=0}^{\infty} f_{nb}(\zeta \mid r, 1-(1-p)e^u) \qquad (11.8.5) \\ &= \left[\frac{p}{1-(1-p)e^u}\right]^r, \end{aligned}$$

provided that $(1 - p)e^u < 1$.

We have thus shown that if $\tilde{\zeta}$ has mf $f_{nb}(\cdot \mid r, p)$, then $\tilde{\zeta}$ can be represented as the sum of r iid random variables $\tilde{\zeta}_i$ with common mf $f_g(\cdot \mid 1, p)$. More generally, the reader may show as Exercise 28 [via (11.8.5) and (10.6.5)] that if $\tilde{\zeta}_i$ has mf $f_{nb}(\cdot \mid r_i, p)$ for $i = 1, \ldots, m$ and $\tilde{\zeta}_1, \ldots, \tilde{\zeta}_m$ are independent, then $\tilde{\zeta} \equiv \sum_{i=1}^{m} \tilde{\zeta}_i$ has mf $f_{nb}(\cdot \mid \sum_{i=1}^{m} r_i, p)$. From (11.8.5) we easily obtain

$$E_{nb}(\tilde{\zeta} \mid r, p) = \frac{r(1-p)}{p} \qquad (11.8.6)$$

and

$$V_{nb}(\tilde{\zeta} \mid r, p) = \frac{r(1-p)}{p^2}. \qquad (11.8.7)$$

11.8.3 Pascal Distribution

If $\tilde{\xi}_i$ has the negative binomial mf $f_{nb}(\cdot \mid r_i, p)$, then $\tilde{n}_i \equiv \tilde{\xi}_i + r_i$ has the Pascal mf $f_{Pa}(\cdot \mid r_i, p)$. Therefore, applying Exercise 10.10(a) to (11.8.5) yields

$$M_{Pa}(u \mid r, p) = \left[\frac{pe^u}{1 - (1 - p)e^u}\right]^r, \tag{11.8.8}$$

which implies that if $\tilde{n}_i$ has mf $f_{Pa}(\cdot \mid r_i, p)$, for $i = 1, \ldots, m$, and if $\tilde{n}_1 \ldots, \tilde{n}_m$ are independent, then $\tilde{n} \equiv \sum_{i=1}^{m} \tilde{n}_i$ has mf $f_{Pa}(\cdot \mid \sum_{i=1}^{m} r_i, p)$; and if $\tilde{n}$ has mf $f_{Pa}(\cdot \mid r, p)$, then $\tilde{n}$ can be represented as the sum of r iid random variables $\tilde{n}_i$ with common mf $f_{Pa}(\cdot \mid 1, p)$. From (10.2.7″), (10.3.10), and (10.3.14) it follows that

$$E_{Pa}(\tilde{n} \mid r, p) = r + E_{nb}(\tilde{n} - r \mid r, p) = \frac{r}{p} \tag{11.8.9}$$

and

$$V_{Pa}(\tilde{n} \mid r, p) = V_{nb}(\tilde{n} - r \mid r, p) = \frac{r(1 - p)}{p^2}. \tag{11.8.10}$$

11.8.4 Poisson Distribution

In Example 10.6.2 we have shown that if $\tilde{r}$ has the Poisson mf $f_{Po}(\cdot \mid \lambda t)$, then

$$M_{Po}(u \mid \lambda t) = e^{\lambda t(e^u - 1)}. \tag{11.8.11}$$

Example 10.6.2 also showed that if $\tilde{r}_i$ has the Poisson mf $f_{Po}(\cdot \mid (\lambda t)_i)$ for $i = 1, \ldots, m$, and if $\tilde{r}_1, \ldots, \tilde{r}_m$ are independent, then $\tilde{r} \equiv \sum_{i=1}^{m} \tilde{r}_i$ has mf $f_{Po}(\cdot \mid \sum_{i=1}^{m} (\lambda t)_i)$. In particular, if one observes a Poisson process with fixed intensity λ for a time interval of length t_1 and then for a second (and non-overlapping) time interval of length t_2, then the total number $\tilde{r}_1 + \tilde{r}_2$ of occurrences in both intervals has mf $f_{Po}(\cdot \mid \lambda(t_1 + t_2))$.

From (11.8.11) the reader may verify that

$$E_{Po}(\tilde{r} \mid \lambda t) = \lambda t \tag{11.8.12}$$

and

$$V_{Po}(\tilde{r} \mid \lambda t) = \lambda t. \tag{11.8.13}$$

11.8.5 Gamma Distribution

We have already shown that if $\tilde{t}$ has df $f_\gamma(\cdot \mid r, \lambda)$, then $\tilde{t}$ has moment-generating function

$$M_\gamma(u \mid r, \lambda) = \left[\frac{\lambda}{(\lambda - u)}\right]^r \tag{11.8.14}$$

(and conversely), where r and λ are positive parameters. It follows readily that

$$E_\gamma(\tilde{t} \mid r, \lambda) = \frac{r}{\lambda} \tag{11.8.15}$$

and

$$V_\gamma(\tilde{t} \mid r, \lambda) = \frac{r}{\lambda^2}. \tag{11.8.16}$$

From (11.8.14) it is also clear that if $\tilde{t}_i$ has df $f_\gamma(\cdot \mid r_i, \lambda)$ for $i = 1, \ldots, m$ and if $\tilde{t}_1, \ldots, \tilde{t}_m$ are independent, then $\tilde{t} \equiv \sum_{i=1}^{m} \tilde{t}_i$ has df $f_\gamma(\cdot \mid \sum_{i=1}^{m} r_i, \lambda)$; and if $\tilde{t}$ has df $f_\gamma(\cdot \mid r, \lambda)$ for r a positive integer, then $\tilde{t}$ can be represented as $\tilde{t} \equiv \sum_{i=1}^{r} \tilde{t}_i$, where $\tilde{t}_i$ are iid with common df $f_e(\cdot \mid \lambda)$ $[\equiv f_\gamma(\cdot \mid 1, \lambda)]$.

11.8.6 Multinomial Distribution

The reader may imitate the argument of Subsection 11.8.1 to show that the moment-generating function $M_{mu^{(k)}}(\mathbf{u} \mid 1, \mathbf{p})$ of the k-tuple $\tilde{\mathbf{x}}$ having mf $f_{mu^{(k)}}(\cdot \mid 1, \mathbf{p})$ is

$$M_{mu^{(k)}}(\mathbf{u} \mid 1, \mathbf{p}) = \sum_{i=1}^{k} p_i e^{u_i} + p_{k+1}. \tag{11.8.17}$$

Moreover, if $\tilde{\mathbf{r}} = \sum_{i=1}^{n} \tilde{\mathbf{x}}_i$, where $\tilde{\mathbf{x}}_1, \ldots, \tilde{\mathbf{x}}_n$ are iid with common mf $f_{mu^{(k)}}$ $(\cdot \mid 1, \mathbf{p})$, then (as we have shown, $\tilde{\mathbf{r}}$ has mf $f_{mu^{(k)}}(\cdot \mid n, \mathbf{p})$. Hence the moment-generating function $M_{mu^{(k)}}(\mathbf{u} \mid n, \mathbf{p})$ of a multinomial random variable $\tilde{\mathbf{r}}$ with parameters n and $\mathbf{p}$ is

$$\begin{aligned} M_{mu^{(k)}}(\mathbf{u} \mid n, \mathbf{p}) &= \left[\sum_{i=1}^{k} p_i e^{u_i} + p_{k+1}\right]^n \\ &= \left[\sum_{i=1}^{k} p_i (e^{u_i} - 1) + 1\right]^n. \end{aligned} \tag{11.8.18}$$

From (11.8.18) we easily obtain that if $\tilde{\mathbf{r}}_i$ has mf $f_{mu^{(k)}}(\cdot \mid n_i, \mathbf{p})$, for $i = 1, \ldots, m$ and if $\tilde{\mathbf{r}}_1, \ldots, \tilde{\mathbf{r}}_m$ are mutually independent, then $\tilde{\mathbf{r}} \equiv \sum_{i=1}^{m} \tilde{\mathbf{r}}_i$ has mf $f_{mu^{(k)}}(\cdot \mid \sum_{i=1}^{m} n_i, \mathbf{p})$.

From (10.6.2) and (10.3.23), we infer from (11.8.18) that

$$E_{mu}{}^{(k)}(\tilde{r}_i \mid n, \mathbf{p}) = np_i, \text{ for every } i \in \{1, \ldots, k+1\}, \tag{11.8.19}$$

$$V_{mu}{}^{(k)}(\tilde{r}_i \mid n, \mathbf{p}) = np_i(1 - p_i), \tag{11.8.20}$$

and

$$V_{mu}{}^{(k)}(\tilde{r}_i, \tilde{r}_j \mid n, \mathbf{p}) = -np_ip_j. \tag{11.8.21}$$

Hence if $\tilde{\mathbf{r}}$ has mf $f_{mu}{}^{(k)}(\cdot \mid n, \mathbf{p})$, then (in the notation of Section 10.4),

$$\mathbf{u} = n\mathbf{p} \tag{11.8.22}$$

and

$$\boldsymbol{\sigma}^2 = n\cdot\begin{bmatrix} p_1(1-p_1) & -p_1p_2 & \cdots & -p_1p_k \\ -p_2p_1 & p_2(1-p_2) & \cdots & -p_2p_k \\ \cdot & & & \\ \cdot & & & \\ \cdot & & & \\ -p_kp_1 & -p_kp_2 & \cdots & p_k(1-p_k) \end{bmatrix}. \tag{11.8.23}$$

11.8.7 Hypergeometric Distribution

Moment-generating functions are of no utility whatever with the hypergeometric family. We shall have to proceed directly, to find

$$\begin{aligned} E_h(\tilde{r} \mid n, R, N) &= \sum_{r=0}^{n} \frac{r\binom{R}{r}\binom{N-R}{n-r}}{\binom{N}{n}} \\ &= \sum_{r=1}^{n} \frac{r\binom{R}{r}\binom{N-R}{n-r}}{\binom{N}{n}} \\ &= n\frac{R}{N}\sum_{r=1}^{n} \frac{\binom{R-1}{r-1}\binom{(N-1)-(R-1)}{(n-1)-(r-1)}}{\binom{N-1}{n-1}} \\ &= n\frac{R}{N}\sum_{\rho=0}^{n-1} f_h(\rho \mid n-1, R-1, N-1) \\ &= n\frac{R}{N}. \end{aligned}$$

Hence

$$E_h(\tilde{r} \mid n, R, N) = n\frac{R}{N}. \tag{11.8.24}$$

[Compare with (11.8.2) for $p \equiv R/N$.]

To find $V_h(\tilde{r} \mid n, R, N)$, we shall first find $E_h(\tilde{r}(\tilde{r} - 1) \mid n, R, N)$, which will equal $E_h(\tilde{r}^2 \mid n, R, N) - E_h(\tilde{r} \mid n, R, N)$; then find $E_h(\tilde{r}^2 \mid n, R, N)$, and then $V_h(\tilde{r} \mid n, R, N)$:

$$\begin{aligned}
&E_h(\tilde{r}(\tilde{r} - 1) \mid n, R, N) \\
&\qquad = \sum_{r=0}^{n} \frac{r(r-1)\binom{R}{r}\binom{N-R}{n-r}}{\binom{N}{n}} \\
&\qquad = n(n-1)\frac{R(R-1)}{N(N-1)} \sum_{\rho=0}^{n-2} f_h(\rho \mid n-2, R-2, N-2) \\
&\qquad = n(n-1)\frac{R(R-1)}{N(N-1)}.
\end{aligned}$$

Hence

$$E_h(\tilde{r}^2 \mid n, R, N) = n(n-1)\frac{R(R-1)}{N(N-1)} + n\frac{R}{N}.$$

Hence (10.3.11) implies

$$V_h(\tilde{r} \mid n, R, N) = \frac{n(n-1)R(R-1)}{N(N-1)} + n\frac{R}{N} - \frac{n^2R^2}{N^2},$$

which, after rearrangement, becomes

$$V_h(\tilde{r} \mid n, R, N) = \left[\frac{N-n}{N-1}\right] n\frac{R}{N}\left(1 - \frac{R}{N}\right), \tag{11.8.25}$$

which is written so as to facilitate comparison with (11.8.3) (in which p corresponds to R/N). Note that the ratio of the hypergeometric variance to the corresponding binomial variance is $(N - n)/(N - 1)$, which is less than one for all $n \in \{2, \ldots, N\}$ and hence implies that the variance of $\tilde{r}$ is *smaller* in sampling without replacement than in sampling with replacement. This is entirely reasonable, as one can convince oneself by assuming $n = N$ and comparing (11.8.25) and (11.8.3).

11.8.8 Pascal-Hypergeometric Distribution

Again working directly, we compute

$$E_{Pah}(\tilde{n} \mid r, R, N)$$

$$= \sum_{n=r}^{N-R+r} \frac{n\binom{n-1}{r-1}\binom{N-n}{R-r}}{\binom{N}{R}}$$

$$= \frac{r(N+1)}{R+1} \sum_{n'=r'}^{N'-R'+r'} \frac{\binom{n'-1}{r'-1}\binom{N'-n'}{R'-r'}}{\binom{N'}{R'}}$$

$$= \frac{r(N+1)}{(R+1)} \sum_{n'=r'}^{N'-R'+r'} f_{nh}(n' \mid r', N', R')$$

$$= \frac{r}{(R+1)/(N+1)},$$

where $N' \equiv N + 1$, $R' \equiv R + 1$, $r' \equiv r + 1$, and $n' \equiv n + 1$. Thus

$$E_{Pah}(\tilde{n} \mid r, R, N) = \frac{r}{(R+1)/(N+1)}. \tag{11.8.26}$$

This is not the same as (11.8.9) for the Pascal; for $p = R/N$, (11.8.9) becomes $E_{Pa}(\tilde{n} \mid r, p) = r/[R/N]$, which exceeds $r/[(R + 1)/(N + 1)]$. This means that $\tilde{n}$ is expected to be greater in sampling with replacement than in sampling without. (Why is this reasonable?)

By the same sort of substitution as in the derivation of (11.8.26), we find

$$E_{Pah}(\tilde{n}(\tilde{n}+1) \mid r, R, N) = \frac{r(r+1)(N+1)(N+2)}{[(R+1)(R+2)]},$$

which implies

$$V_{Pah}(\tilde{n} \mid r, R, N) = \frac{r(N+1)(N-R)(R-r+1)}{(R+1)^2(R+2)}$$

$$= \left[\frac{R+1-r}{R+2}\right]\frac{r(1-p^*)}{p^{*2}}, \tag{11.8.27}$$

where $p^* \equiv (R + 1)/(N + 1)$. The second formula is furnished for comparability with (11.8.10).

11.8.9 Negative Hypergeometric Distribution

If $\tilde{n}$ has mf $f_{Pah}(\cdot \mid r, R, N)$, then $\tilde{\xi} \equiv \tilde{n} - r$ has mf $f_{nh}(\cdot \mid r, R, N)$ and conversely. Hence (11.8.26) and (11.8.27) imply, via (10.2.7″), (10.3.10), and (10.3.14), that

$$E_{nh}(\tilde{\xi} \mid r, R, N) = \frac{r(N - R)}{(R + 1)} \tag{11.8.28}$$

and

$$V_{nh}(\tilde{\xi} \mid r, R, N) = \frac{r(N + 1)(N - R)(R - r + 1)}{(R + 1)^2(R + 2)}. \tag{11.8.29}$$

11.8.10 The k-Variate Hypergeometric Distribution

Since subvectors $\tilde{\mathbf{r}}_1$ of a hypergeometrically distributed random vector $\tilde{\mathbf{r}}$ have hypergeometric distributions, we have

$$E_h^{(k)}(\tilde{r}_i \mid n, \mathbf{R}, N) = \frac{nR_i}{N}, \; i = 1, \ldots, k \tag{11.8.30}$$

and

$$V_h^{(k)}(\tilde{r}_i \mid n, \mathbf{R}, N) = \left(\frac{N - n}{N - 1}\right)\frac{nR_i}{N}\left(1 - \frac{R_i}{N}\right), \tag{11.8.31}$$

for $i \in \{1, \ldots, k\}$. Moreover, the reader may show as Exercise 29, by using the general approach leading up to (11.8.24), that

$$V_h^{(k)}(\tilde{r}_i, \tilde{r}_j \mid n, R, N) = -\left(\frac{N - n}{N - 1}\right)\frac{nR_iR_j}{N^2}, \tag{11.8.32}$$

for $\{i, j\} \subset \{1, \ldots, k\}$ and $i \neq j$.

EXERCISES

1. How many different bridge hands are there? (A bridge hand consists of 13 different cards.)

2. What is the probability of being dealt a hand consisting of four spades and three each in the other suits? [*Hint:* $R_i = 13$, for $i \in$ {spades, hearts, diamonds}, $N = 52$, and $n = 13$. Why do we not need to mention clubs?]

3. If 75 percent of all voters favor Candidate X, and if a random sample of size 100 is to be drawn, then what is the expectation of the number $\tilde{r}$ of voters sampled who favor Candidate X,

   (a) assuming random sampling with replacement,

   and

   (b) assuming random sampling without replacement?

4. Show that $V_b(\tilde{r} \mid n, p)$ is maximized at $p = \frac{1}{2}$, for every $n > 0$, and explain why this should be true from the interpretation of variance as an index of uncertainty.

5. Argue that $P_b(\tilde{r} \leqq r \mid n,p) = P_{Pa}(\tilde{n} \geqq n \mid r, p)$ and hence

$$F_b(r \mid n, p) = 1 - F_{Pa}(n - 1 \mid r, p), \tag{11.9.1}$$

all integers $r > 0$ and $n \geqq r$, and all $p \in (0, 1)$. (Apply to Example 11.3.2.) [*Note:* A rigorous proof of Exercise 5 is possible, but you are required to furnish only an intuitive argument. Appeal to the Bernoulli process.]

6. Argue that $P_{Po}(\tilde{r} \leqq r \mid \lambda t) = P_\gamma(\tilde{t} \geqq t \mid r + 1, \lambda)$ whenever r is a nonnegative integer, and hence that

$$F_{Po}(r \mid \lambda t) = 1 - F_\gamma(t \mid r + 1, \lambda). \tag{11.9.2}$$

[*Note:* Again, an intuitive argument is desirable, although you may show that integration by parts implies

$$F_\gamma(t \mid r + 1, \lambda) = 1 - \sum_{j=0}^{r} \frac{e^{-\lambda t}(\lambda t)^j}{j!}, \tag{11.9.3}$$

in view of which the assertion is mathematically obvious.]

7. The Bernoulli process may be considered the *discrete-time* analogue of the Poisson process, since the Bernoulli process may be *defined* as the generator of iid random variables $\tilde{\zeta}_1, \tilde{\zeta}_2, \ldots$ with common mf $f_g(\cdot \mid p)$, where $\tilde{\zeta}_i$ is the "interarrival time" (in integer units) between the $(i - 1)$th and ith one. Here, "interarrival time" is measured as number of zeroes.

   (a) Show that $f_g(\cdot \mid p)$ satisfies (11.7.2), with ζ and $\tilde{\zeta}_1$ replacing t and $\tilde{t}_1$ respectively.

   (b) Derive the binomial mf by supposing that you predetermine the number n of observations ("time units") and let number $\tilde{r}$ of occurrences ("ones") be random.

8. The iid interarrival-times model for repeated occurrences of some phenomenon is fundamental in queueing theory (to which we are indebted for the word *interarrival*). In the general case [that is, not satisfying (11.7.2)] suppose observation starts immediately after an occurrence and hence $\tilde{t}_1 = \tilde{y}$.

   (a) Suppose $\tilde{t}_1, \tilde{t}_2, \ldots$ are iid with common df $f_\gamma(\cdot \mid r^*, \lambda)$. Show that the time $\tilde{t} \equiv \sum_{i=1}^{r} \tilde{t}_i$ required to produce exactly r occurrences has df $f_\gamma(\cdot \mid r \cdot r^*, \lambda)$.

   (b) Suppose $\tilde{t}_1, \tilde{t}_2, \ldots$ are iid with common df $f_\gamma(\cdot \mid r^*, \lambda)$, as in (a); and suppose that you will observe the process for a predetermined period t of time, which commences immediately after an occurrence (which is not to be counted). By using the appropriate analogues of (11.7.5) and (11.7.6), show that if r^* is a positive integer, then

$$P[\tilde{r} = r \mid t, r^*, \lambda] = e^{-\lambda t} \sum_{i=0}^{r^*-1} \frac{(\lambda t)^{i+rr^*}}{(i + rr^*)!} \tag{11.9.4}$$

is the probability of precisely r occurrences in the period of length t, given r^* and λ.

(c) The process described in (a) and (b) is called the *Erlang process* when r^* is a positive integer. The *Palm process* is more general still: $\tilde{t}_1, \tilde{t}_2, \ldots$ are iid with common dgf f_t such that $F_t(0) = 0$. What are the appropriate analogues of (11.7.5) and (11.7.6) for the Palm process?

9. *The Stuttering Poisson Distribution:* Suppose the number $\tilde{m}$ of orders (for example, invoices) arriving during any time period t has mf $f_{Po}(\cdot \mid \lambda t)$ for some intensity λ. Suppose that the numbers $\tilde{n}_1, \ldots, \tilde{n}_m$ of items demanded in the first, . . . , mth orders are iid random variables with common mf $f_{Pa}(\cdot \mid 1, p)$.

(a) If $\tilde{m} = m$ orders are received in a time period of length t, what is the distribution of the total number $\tilde{n} \equiv \sum_{i=1}^{m} \tilde{n}_i$ of items demanded? That is, what is the conditional mf $f(n \mid m)$?

(b) Find the joint mf $f(n, m)$ of the total number $\tilde{n}$ of items demanded and the total number $\tilde{m}$ of orders.

(c) Find the marginal mf f_n of the total number $\tilde{n}$ of items demanded (this mf is called the *stuttering Poisson*).

[*Hint:* $f_n(0) = f_{Po}(0) = e^{-\lambda t}$: no items can result only when there are no orders, as each order demands at least one item (by definition of f_{Pa} $(\cdot \mid 1, p)$).]

10. *Poisson Generator of Orders, arbitrary iid demands per order:* Suppose the number $\tilde{m}$ of orders received in any time period of length t has mf $f_{Po}(\cdot \mid \lambda t)$, and that the numbers $\tilde{n}_1, \ldots, \tilde{n}_m$ of items demanded in the first, . . . , mth orders are iid with some common mf f^* and common moment-generating function M^*. Then $\tilde{n} \equiv \sum_{i=1}^{\tilde{m}} \tilde{n}_i$ has conditional moment-generating function

$$M(u \mid m) \equiv [M^*(u)]^m,$$

given m.

(a) Argue that the *un*conditional moment-generating function of $\tilde{n}$ is given by

$$\begin{aligned} M(u) &= E[M(u \mid \tilde{m})] \\ &= E[\{M^*(u)\}^{\tilde{m}}] \\ &= \sum_{m=0}^{\infty} e^{-\lambda t} \frac{[\lambda t M^*(u)]^m}{m!}, \end{aligned}$$

and show that this reduces to

$$M(u) = e^{\lambda t[M^*(u)-1]}. \tag{11.9.5}$$

(b) Conclude from (11.9.5) that

$$E(\tilde{n}) = \lambda t \mu \tag{11.9.6}$$

and

$$V(\tilde{n}) = \lambda t(\sigma^2 + \mu^2), \tag{11.9.7}$$

where $\mu \equiv E(\tilde{n}_i)$, for every i and $\sigma^2 + \mu^2 \equiv E(\tilde{n}_i^2)$, for every i.

[*Hint*: If $M_x(\cdot)$ is any moment-generating function, then $M_x(0) = E(e^{0 \cdot \tilde{x}}) = E(1) = 1$.]

(c) Suppose the number $\tilde{m}$ of people who pass a hot-dog stand during any time period of length t has mf $f_{Po}(\cdot \mid \lambda t)$, and that we define $\tilde{x}_i = 1$ if individual i buys a hot dog, and $\tilde{x}_i = 0$ otherwise. Assume $\tilde{x}_1, \tilde{x}_2, \ldots$ are iid with common mf $f_b(\cdot \mid 1, p)$. Show that the mf of the total sales $\tilde{n}$ of hot dogs is $f_{Po}(\cdot \mid p\lambda t)$.

[*Note:* Exercises 8 through 10 illustrate important ideas in queueing and inventory theory. We have as yet said nothing about the problems of: (1) estimating the parameters of a given model; and, more fundamentally, (2) ascertaining whether a given model is an adequate approximation to reality. Later chapters deal with these problems.]

11. Show that if the number $\tilde{\zeta}$ of occurrences of a phenomenon has mf $f_{Po}(\cdot \mid (1-p)t)$ given t, and if $\tilde{t}$ is a random variable with df $f_\gamma(\cdot \mid r, p)$, then the marginal mf of $\tilde{\zeta}$ is the following continuous-r generalization of the *negative binomial:*

$$f_{nb}(\zeta \mid r, p) \equiv \frac{\Gamma(r+\zeta)}{\Gamma(r)\Gamma(\zeta+1)} p^r (1-p)^\zeta, \tag{11.9.8}$$

for $\zeta \in \{0, 1, 2, \ldots\}$ (and $= 0$ otherwise), where r is any positive real number and $0 < p < 1$.

12. Show that if $\tilde{y} \equiv 1/\tilde{t}$ and $\tilde{t}$ has df $f_\gamma(\cdot|r, \lambda)$, then $\tilde{y}$ has the *inverted gamma* distribution, with df $f_{i\gamma}(\cdot \mid r/\lambda, 2r)$, where $f_{i\gamma}(\cdot \mid \xi, \nu)$ is defined by

$$f_{i\gamma}(y \mid \xi, \nu) \equiv \begin{cases} 0, & y \leqq 0 \\ \dfrac{e^{-\frac{1}{2}\xi\nu/y}\left(\dfrac{\frac{1}{2}\nu\xi}{y}\right)^{\frac{1}{2}(\nu+1)}}{[\frac{1}{2}\nu\xi\Gamma(\frac{1}{2}\nu)]}, & y > 0. \end{cases}$$

13. (a) If $\tilde{t}$ has df $f_\gamma(\cdot \mid r, \lambda)$, then show that $\tilde{z} \equiv \lambda\tilde{t}$ has df $f_\gamma(\cdot \mid r, 1)$. We define the *standardized gamma* distribution, and mf $f_{\gamma^*}(\cdot \mid r)$, by

$$f_{\gamma^*}(z \mid r) \equiv f_\gamma(z \mid r, 1), \tag{11.9.10}$$

for every $z \in R^1$ and every $r > 0$.

(b) Show that

$$F_\gamma(t \mid r, \lambda) = F_{\gamma^*}(\lambda t \mid r), \tag{11.9.11}$$

for all positive r, t, and λ.

(c) Show that

$$F_{i\gamma}(y \mid \xi, \nu) = 1 - F_{\gamma^*}(\tfrac{1}{2}\nu\xi/y \mid \tfrac{1}{2}\nu). \tag{11.9.12}$$

14. A useful form of the gamma density in applications concerning distributions of uncertain variances is the so-called *gamma-2* density. We say that $\tilde{h}$ has the *gamma-2* density $f_{\gamma 2}(\cdot \mid \xi, \nu)$ with parameters $\xi > 0$ and $\nu > 0$ if $\tilde{z} \equiv \frac{1}{2}\nu\xi\tilde{h}$ has the standardized gamma density $f_{\gamma^*}(\cdot \mid \frac{1}{2}\nu)$. Show that

$$f_{\gamma 2}(h \mid \xi, \nu) = f_{\gamma}(h \mid \tfrac{1}{2}\nu, \tfrac{1}{2}\nu\xi), \tag{11.9.13}$$

and that

$$F_{\gamma 2}(h \mid \xi, \nu) = F_{\gamma^*}(\tfrac{1}{2}\nu\xi h \mid \tfrac{1}{2}\nu). \tag{11.9.14}$$

15. We say that $\tilde{\sigma}$ has the *inverted gamma-2* distribution if the df of $\tilde{\sigma}$ is

$$f_{i\gamma 2}(\sigma \mid \xi^{1/2}, \nu) \equiv \begin{cases} 0, & \sigma \leq 0 \\ \dfrac{2e^{-\frac{1}{2}\nu\xi/\sigma^2}\left(\dfrac{\frac{1}{2}\nu\xi}{\sigma^2}\right)^{\frac{1}{2}(\nu+1)}}{\Gamma(\frac{1}{2}\nu)(\frac{1}{2}\nu\xi)}, & \sigma > 0. \end{cases} \tag{11.9.15}$$

Show that $\tilde{\sigma}$ has the inverted gamma-2 distribution if and only if $\tilde{h} \equiv 1/\tilde{\sigma}^2$ has the gamma 2 distribution with parameters ξ and ν.

[*Note:* The choice of $\tilde{\sigma}$ as the symbol for an inverted gamma-2 random variable is *intended* to connote standard deviation. Using the results of Exercise 14, it follows that if $\tilde{\sigma}$ has the inverted gamma-2 density $f_{i\gamma 2}(\cdot \mid \xi^{\frac{1}{2}}, \nu)$, then $\tilde{V} \equiv \tilde{\sigma}^2$ has the inverted gamma density $f_{i\gamma}(\cdot \mid \xi, \nu)$. The symbol $\tilde{V}$ is supposed to connote variance.]

16. *Moments of Gammas:* Derive the following moments:

(a) $$E_{\gamma}(\tilde{l}^q \mid r, \lambda) = \frac{\lambda^{-q}\Gamma(r+q)}{\Gamma(r)}, \quad \text{for all } q > -r. \tag{11.9.16}$$

(b) $$E_{\gamma 2}(\tilde{h}^q \mid \xi, \nu) = \frac{(\frac{1}{2}\nu\xi)^{-q}\Gamma(\frac{1}{2}\nu+q)}{\Gamma(\frac{1}{2}\nu)}, \quad \text{for all } q > -\tfrac{1}{2}\nu. \tag{11.9.17}$$

(c) $$E_{i\gamma}(\tilde{y}^q \mid \xi, \nu) = \frac{(\frac{1}{2}\nu\xi)^{q}\Gamma(\frac{1}{2}\nu-q)}{\Gamma(\frac{1}{2}\nu)}, \quad \text{for all } q < \tfrac{1}{2}\nu. \tag{11.9.18}$$

(d) $$E_{i\gamma 2}(\tilde{\sigma}^q \mid \xi^{1/2}, \nu) = \frac{(\frac{1}{2}\nu\xi)^{\frac{1}{2}q}\Gamma(\frac{1}{2}(\nu-q))}{\Gamma(\frac{1}{2}\nu)}, \quad \text{for all } q < \nu. \tag{11.9.19}$$

17. *Generating Functions, or "z-transforms"*: For discrete random variable $\tilde{x}$ with range a subset of $\{0, 1, 2, \ldots\}$, a useful alternative to the moment-generating function $M(\cdot)$. [or to the characteristic function $\phi(\cdot)$] of $\tilde{x}$ is the *generating function* $\psi(\cdot)\colon [-1, 1] \to R^1$ by

$$\psi(s) = \sum_{x=0}^{\infty} s^x f(x). \tag{11.9.20}$$

If $\tilde{x}$ is any nonnegative, discrete random variable, then $\psi(s) \in [-1, 1]$ [because $|s^x| \leq 1$ for all $s \in [-1, 1]$ and hence

$$\begin{aligned} \left|\sum_{x=0}^{\infty} s^x f(x)\right| &\leq \sum_{x=0}^{\infty} |s^x| f(x) \\ &\leq \sum_{x=0}^{\infty} 1 \cdot f(x) \\ &= 1]. \end{aligned}$$

(a) Show that $\psi(\cdot)$ determines $f(\cdot)$ uniquely, in that we may recapture $f(\cdot)$ from $\psi(\cdot)$ via

$$f(x) = \left(\frac{1}{x!}\right) \frac{d^x \psi(s)}{ds^x}\bigg|_{s=0}. \tag{11.9.21}$$

(b) Show that if $\tilde{x}_1$ and $\tilde{x}_2$ are independent, nonnegative-integer-valued random variables with respective generating functions $\psi_1(\cdot)$ and $\psi_2(\cdot)$, then the moment-generating function $\psi_y(\cdot)$ of $\tilde{y} \equiv \tilde{x}_1 + \tilde{x}_2$ is given by

$$\psi_y(s) = \psi_1(s) \cdot \psi_2(s). \tag{11.9.22}$$

[*Hints:* (1) $f_y(i) = \sum_{j=0}^{i} f_2(i - j) \cdot f_1(j)$,

(2) $\psi_k(s) = \sum_{j=0}^{\infty} s^j f_k(j)$ for $k = 1, 2$.

(3) $\left[\sum_{j=0}^{\infty} f_1(j) s^j\right]\left[\sum_{j=0}^{\infty} f_2(j) s^j\right] = \sum_{i=0}^{\infty} \left\{\sum_{j=0}^{i} f_1(j) f_2(i - j)\right\} s^i$

(by rearrangement of terms in the left-hand product of two sums).]

(c) Let $\tilde{n}$ be a nonnegative-integer-valued random variable with mf f_n and generating function $\psi_n(\cdot)$. Let $\tilde{x}_1, \tilde{x}_2, \ldots$ be iid nonnegative integer valued random variables with common mf $f_x(\cdot)$ and generating function $\psi_x(\cdot)$. Let $\tilde{y} \equiv \sum_{i=1}^{\tilde{n}} \tilde{x}_i$ denote the sum of the first (random number) $\tilde{n}$ of the $\tilde{x}_i$'s. Show that

$$\psi_y(s) = \psi_n[\psi_x(s)]. \tag{11.9.23}$$

[*Hint:* Given n, $\psi_{y\,|\,n}(s) = [\psi_x(s)]^n$ by (b). Hence,

$$\psi_y(s) = \sum_{n=0}^{\infty} [\psi_x(s)]^n \cdot f_n(n).]$$

(d) Show that $E(\tilde{x}_{(r)}) \equiv E(\tilde{x}(\tilde{x} - 1) \ldots (\tilde{x} - r + 1))$ is given by

$$E(\tilde{x}_{(r)}) = \frac{d^r \psi_x(s)}{ds^r}\bigg|_{s=1}. \tag{11.9.24}$$

(e) Show that the moment generating function $M(\cdot)$ of $\tilde{x}$ is related to $\psi(\cdot)$ by

$$M(u) = \psi(e^u)$$

and, conversely,

$$\psi(s) = M(\log_e s).$$

18. Show that $(n)_r = n!/(n-r)!$ whenever n and r are nonnegative integers such that $r \leq n$. (Recall that $0! = 1$.)

19. Show that $\binom{n}{r} = \binom{n}{n-r}$ whenever n and r are nonnegative integers such that $r \leq n$.

20. Derive (11.3.8) from (11.3.7).

21. Prove that if $\tilde{\mathbf{r}}$ has mf $f_{mu}^{(k)}(\cdot \mid n, \mathbf{p})$ then the $(k-j)$-dimensional subvector $\tilde{\mathbf{r}}_2$ has conditional mf $f_{mu}^{(k-j)}(\cdot \mid n^*(\mathbf{r}_1), \mathbf{p}^*{}_2)$ given $\tilde{\mathbf{r}}_1 = \mathbf{r}_1$ where

$$n^*(\mathbf{r}_1) \equiv n - \sum_{i=1}^{j} r_j$$

and

$$\mathbf{p}^*{}_2 \equiv \left[\frac{1}{\left(1 - \sum_{i=1}^{j} p_j\right)}\right] \mathbf{p}_2.$$

22. Derive (11.5.5).

23. Derive (11.5.6).

24. Show that

$$\binom{n}{r}\frac{(R)_r(N-R)_{n-r}}{(N)_n} = \frac{\binom{R}{r}\binom{N-R}{n-r}}{\binom{N}{n}}.$$

25. Convert the tables in Examples 11.4.1 and 11.6.1 to decimals and compare them. For what (r_1, r_2) is agreement good? Poor?

26. Derive the Poisson mf $f_{Po}(r \mid \lambda t)$ from (11.7.5) and (11.7.6).

27. Prove, by induction and integration by parts, that (11.7.10) implies $\Gamma(r) = (r-1)\Gamma(r-1)$ whenever $r \geqq 1$, and hence $\Gamma(r) = (r-1)!$ whenever r is a positive integer.

28. Show that if $\tilde{\xi}_i$ has mf $f_{nb}(\cdot \mid r_i, p)$ for $i = 1, \ldots, m$, and if $\tilde{\xi}_1, \ldots, \tilde{\xi}_m$ are independent, then $\tilde{\xi} \equiv \sum_{i=1}^{m} \tilde{\xi}_i$ has mf $f_{nb}(\cdot \mid \sum_{i=1}^{m} r_i, p)$.

29. Prove (11.8.32).

12

SOME IMPORTANT PROBABILITY DISTRIBUTIONS AND DATA-GENERATING PROCESSES (II)

12.1 INTRODUCTION

This chapter continues the exposition of important families of probability distributions, but with a focus somewhat different from that of Chapter 11. Here, we will not be concerned with data-generating processes as such, but rather only with distributions of continuous random variables. We shall develop almost all of the distribution theory essential for the remainder of this book, and more immediately for Chapter 13.

Section 12.2 consists of a brief recapitulation of the families of gamma densities (gamma, gamma-2, inverted gamma, inverted gamma-2, and chi-square) introduced in Section 11.7 and Exercises 12 through 15 in Chapter 11. All are related to the standardized gamma and chi-square distributions. Some chi-square fractiles are furnished in Table II of the Appendix.

Section 12.3 introduces the families of beta and inverted beta-2 distributions. Section 12.4 concerns the family of univariate Normal distributions—by far the most important family in all of statistics and decision theory. Some reasons for its importance will be explained in Chapter 13.

In Section 12.5 we define the univariate Student family and thereby conclude the exposition of families of univariate distributions (that is, distributions of random variables, rather than of random vectors).

Certain distributions of random *vectors* are discussed in Sections 12.6 through 12.8. Readers who have been deferring a review of linear algebra are advised that that subject is utilized extensively in Sections 12.6 through 12.10.

Section 12.6 concerns the k-variate *beta* distribution (also called the k-variate *Dirichlet* distribution). Section 12.7 introduces k-variate (nonsingular) *Normal* distributions, while Section 12.8 introduces k-variate (nonsingular) *Student* distributions.

Section 12.9 concerns useful families of distributions of random *matrices*; namely, the *Wishart* and *inverted Wishart* distributions of kth order positive-definite, symmetric, random matrices. By the symmetry assumption, only $k(k + 1)/2$ elements of such a matrix are functionally independent, and therefore the Wishart and inverted Wishart distribution can be thought of as distributions of $k(k + 1)/2$-dimensional random vectors.

Section 12.10 introduces another family of distributions of random matrices; namely, the kth order inverted beta-2 distributions. This family is not well known but has proved useful in the Bayesian study of k-variate Normal sampling theory.

The reader should not attempt to absorb all of this chapter before continuing with its sequels. However, the basic facts about each family of distributions, such as available shapes of df's in the family, should be noted.

The reader should pay special attention to the Normal and k-variate Normal families, since they are of central importance throughout the remainder of this book.

12.2 THE GAMMA FAMILIES

We say that $\tilde{z}$ has the *standardized gamma distribution* with parameter $r > 0$ if the df of $\tilde{z}$ is

$$f_{\gamma^*}(z \mid r) \equiv \begin{cases} 0, & z \leqq 0 \\ \dfrac{\exp(-z)z^{r-1}}{\Gamma(r)}, & z > 0, \end{cases} \tag{12.2.1}$$

where

$$\Gamma(r) \equiv \int_0^\infty e^{-t}t^{r-1}dt, \text{ for every } r > 0. \tag{12.2.2}$$

Fractiles ${}_p\gamma^*{}_{(r)}$ of $\tilde{z}$ are, naturally, defined implicitly by

$$p = F_{\gamma^*}({}_p\gamma^*{}_{(r)} \mid r). \tag{12.2.3}$$

We say that $\tilde{t}$ has the *gamma distribution with parameters* r *and* λ if the df of $\tilde{t}$ is

$$f_{\gamma}(t \mid r, \lambda) \equiv \begin{cases} 0, & t \leqq 0 \\ \dfrac{\lambda \exp(-\lambda t)(\lambda t)^{r-1}}{\Gamma(r)}, & t > 0, \end{cases} \tag{12.2.4}$$

where $r > 0$ and $\lambda > 0$. Clearly, $\tilde{t}$ has the gamma distribution with parameters r and λ if and only if $\tilde{z} \equiv \lambda\tilde{t}$ has the standardized gamma distribution with parameter r. Hence

$$F_\gamma(t \mid r, \lambda) = F_{\gamma^*}(\lambda t \mid r), \tag{12.2.5}$$

for every $r > 0$ and $\lambda > 0$, and therefore the pth fractile ${}_p\gamma_{(r,\lambda)}$ of $\tilde{t}$ is given by

$${}_p\gamma_{(r,\lambda)} = \frac{{}_p\gamma^*_{(r)}}{\lambda}, \tag{12.2.6}$$

for all $p \in (0, 1)$, $r > 0$, and $\lambda > 0$. Hence for any $(p, q) \subset (0, 1)$ the fractile ratio

$$\frac{{}_p\gamma_{(r,\lambda)}}{{}_q\gamma_{(r,\lambda)}} = \frac{{}_p\gamma^*_{(r)}}{{}_q\gamma^*_{(r)}} \tag{12.2.7}$$

and is independent of λ.

We say that $\tilde{x}$ has the *gamma*-2 distribution with parameters $\psi > 0$ *and* $\nu > 0$ if the df of $\tilde{x}$ is

$$f_{\gamma 2}(x \mid \psi, \nu) \equiv \begin{cases} 0, & x \leqq 0 \\ \dfrac{\frac{1}{2}\nu\psi \exp(-\frac{1}{2}\nu\psi x)(\frac{1}{2}\nu\psi x)^{\frac{1}{2}\nu-1}}{\Gamma(\frac{1}{2}\nu)}, & x > 0. \end{cases} \tag{12.2.8}$$

Clearly, $\tilde{x}$ has the gamma-2 distribution with parameters ψ and ν if and only if $\tilde{z} \equiv \frac{1}{2}\nu\psi\tilde{x}$ has the standardized gamma distribution with parameter $\frac{1}{2}\nu$.

We say that $\tilde{Q}$ has the *chi-square distribution with* ν *degrees of freedom* if the df of $\tilde{Q}$ is

$$f_{\chi^2}(Q \mid \nu) = \begin{cases} 0, & Q \leqq 0 \\ \dfrac{\frac{1}{2} \exp(-\frac{1}{2}Q)(\frac{1}{2}Q)^{\frac{1}{2}\nu-1}}{\Gamma(\frac{1}{2}\nu)}, & Q > 0. \end{cases} \tag{12.2.9}$$

Clearly, the chi-square distribution with ν degrees of freedom is the gamma distribution with parameters $r = \frac{1}{2}\nu$ and $\lambda = \frac{1}{2}$. Moreover, the chi-square distribution with ν degrees of freedom is the gamma-2 distribution with parameters $\psi = 1/\nu$ and ν. The justification for its separate definition is based upon its importance in Normal-process theory, in Chapter 13.

Fractiles ${}_p\chi^2_{(\nu)}$ of the chi-square distributions with ν degrees of freedom are given in Table II of the Appendix. We shall indicate later in the section how these fractiles can be used to obtain fractiles of distributions belonging to other gamma families.

We say that $\tilde{y}$ has the *inverted gamma distribution with parameters* $\psi > 0$ *and* $\nu > 0$ if the df of $\tilde{y}$ is

$$f_{i\gamma}(y \mid \psi, \nu) \equiv \begin{cases} 0, & y \leqq 0 \\ \dfrac{\exp(-\frac{1}{2}\nu\psi/y)(\frac{1}{2}\nu\psi/y)^{\frac{1}{2}\nu+1}}{[\frac{1}{2}\nu\psi\Gamma(\frac{1}{2}\nu)]}, & y > 0; \end{cases} \tag{12.2.10}$$

that is, if $\tilde{x} \equiv 1/\tilde{y}$ has the gamma-2 distribution with parameters ψ and ν.

Finally, we say that $\tilde{w}$ has the *inverted gamma-2 distribution with parameters* $\psi > 0$ *and* $\nu > 0$ if the df of $\tilde{w}$ is

$$f_{i\gamma 2}(w \mid \psi, \nu) \equiv \begin{cases} 0, & w \leqq 0 \\ \dfrac{2 \exp\left(-\frac{1}{2}\nu\psi/w^2\right)\left(\frac{1}{2}\nu\psi/w^2\right)^{\frac{1}{2}(\nu+1)}}{\left[\left(\frac{1}{2}\nu\psi\right)^{1/2}\Gamma\left(\frac{1}{2}\nu\right)\right]}, & w > 0; \end{cases} \tag{12.2.11}$$

that is, if $\tilde{y} \equiv w^2$ has the inverted gamma distribution with parameters ψ and ν.

Table II of the Appendix can be used to obtain some fractiles of any of the preceding gamma families. We have:

$$_p\gamma^*{}_{(r)} = \frac{_p\chi^2{}_{(2r)}}{2}; \tag{12.2.12}$$

$$_p\gamma_{(r,\lambda)} = \frac{_p\chi^2{}_{(2r)}}{2\lambda}; \tag{12.2.13}$$

$$_p(\gamma 2)_{(\psi,\nu)} = \frac{_p\chi^2{}_{(\nu)}}{\nu\psi}; \tag{12.2.14}$$

$$_p(i\gamma)_{(\psi,\nu)} = \frac{\nu\psi}{_{(1-p)}\chi^2{}_{(\nu)}}; \tag{12.2.15}$$

and

$$_p(i\gamma 2)_{(\psi,\nu)} = \left[\frac{\nu\psi}{_{(1-p)}\chi^2{}_{(\nu)}}\right]^{1/2}. \tag{12.2.16}$$

Proof

For (12.2.12), $\tilde{z}$ has df $f_{\gamma^*}(\cdot \mid r)$ if and only if $\tilde{Q} = 2\tilde{z}$ has df $f_{\chi^2}(\cdot \mid 2r)$ (Exercise 1), so that $P[\tilde{z} \leqq {}_p\gamma^*{}_{(r)}] = p = P[\tilde{Q} \leqq 2 \cdot {}_p\gamma^*{}_{(r)}] = P[\tilde{Q} \leqq {}_p\chi^2{}_{(2r)}]$. The cdf $F_{\chi^2}(\cdot \mid 2r)$ is strictly increasing on $(0, \infty)$, and hence $2 \cdot {}_p\gamma^*{}_{(r)} = {}_p\chi^2{}_{(2r)}$. Equation (12.2.13) is immediate from (12.2.12) via (12.2.6) Since $\tilde{x}$ has df $f_{\gamma 2}(\cdot \mid \psi, \nu)$ if and only if $\tilde{z} \equiv \frac{1}{2}\nu\psi\tilde{x}$ has df $f_{\gamma^*}(\cdot \mid \frac{1}{2}\nu)$ if and only if $\tilde{Q} \equiv \nu\psi\tilde{x}$ has df $f_{\chi^2}(\cdot \mid \nu)$, it follows that $P[\tilde{x} \leqq {}_p(\gamma 2)_{(\psi,\nu)}] = p = P[\tilde{Q} \leqq \nu\psi \cdot {}_p(\gamma 2)_{(\psi,\nu)}] = P[\tilde{Q} \leqq {}_p\chi^2{}_{(\nu)}]$, so that $\nu\psi \cdot {}_p(\gamma 2)_{(\psi,\nu)} = {}_p\chi^2{}_{(\nu)}$, which yields (12.2.14). For (12.2.15), $\tilde{y}$ has df $f_{i\gamma}(\cdot \mid \psi, \nu)$ if and only if $\tilde{x} \equiv 1/\tilde{y}$ has df $f_{\gamma 2}(\cdot \mid \psi, \nu)$, so that $P[\tilde{y} \leqq {}_p(i\gamma)_{(\psi,\nu)}] = p = P[\tilde{x} \geqq 1/{}_p(i\gamma)_{(\psi,\nu)}] = 1 - P[\tilde{x} \leqq 1/{}_p(i\gamma)_{(\psi,\nu)}]$. Hence $P[\tilde{x} \leqq 1/{}_p(i\gamma)_{(\psi,\nu)}] = 1 - p = P[\tilde{x} \leqq {}_{(1-p)}(\gamma 2)_{(\psi,\nu)}] = P[\tilde{x} \leqq {}_{(1-p)}\chi^2{}_{(\nu)}/(\nu\psi)]$, by (12.2.14), so that $[{}_p(i\gamma)_{(\psi,\nu)}]^{-1} = {}_{(1-p)}\chi^2{}_{(\nu)}/(\nu\psi)$, from which (12.2.15) follows. For (12.2.16), $P[\tilde{w} \leqq {}_p(i\gamma 2)_{(\psi,\nu)}] = P[\tilde{w}^2 \leqq \{{}_p(i\gamma 2)_{(\psi,\nu)}\}^2] = P[\tilde{w}^2 \leqq {}_p(i\gamma)_{(\psi,\nu)}]$ because $\tilde{w}$ has df $f_{i\gamma}(\cdot \mid \psi, \nu)$; if and only if $\tilde{y} \equiv \tilde{w}^2$ has df $f_{i\gamma}(\cdot \mid \psi, \nu)$; and hence ${}_p(i\gamma 2)_{(\psi,\nu)} = [{}_p(i\gamma)_{(\psi,\nu)}]^{1/2}$, from which (12.2.16) follows via (12.2.15). QED

Table 12.1 exhibits the expectation and variance of each of the preceding distributions. The reader may derive all entries as Exercise 2.

Table 12.1

random variable $\tilde{x}$	*df*	$E(\tilde{x})$	$V(\tilde{x})$
$\tilde{z}$	$f_{\gamma*}(\cdot \mid r)$	r	r
$\tilde{t}$	$f_{\gamma}(\cdot \mid r,\lambda)$	r/λ	r/λ^2
$\tilde{x}$	$f_{\gamma 2}(\cdot \mid \psi,\nu)$	$1/\psi$	$1/(\frac{1}{2}\nu\psi^2)$
$\tilde{Q}$	$f_{\chi^2}(\cdot \mid \nu)$	ν	2ν
$\tilde{y}$	$f_{i\gamma}(\cdot \mid \psi,\nu)$	$\nu\psi/(\nu - 2)$	$2\nu^2\psi^2/[(\nu-2)^2(\nu-4)]$
$\tilde{w}$	$f_{i\gamma 2}(\cdot \mid \psi,\nu)$	$\psi^{1/2}(\frac{1}{2}\nu)^{1/2}\Gamma(\frac{1}{2}[\nu-1])/\Gamma(\frac{1}{2}\nu)$	$\nu\psi/[\nu-2] - \{E(\tilde{w})\}^2$

Figure 12.1 depicts $f_\gamma(\cdot \mid r, r)$ for several values of r to indicate the various available shapes of gamma df's having expectation one. Figures 12.2 and 12.3 show $f_{i\gamma}(\cdot \mid 1, \nu)$ and $f_{i\gamma 2}(\cdot \mid 1, \nu)$, respectively. Note that as r or ν increases, each df becomes "taller and thinner," concentrating more and more probability in any given interval about its expectation. Then check Table 12.1 to verify that the variance of each is a decreasing function of r or ν.

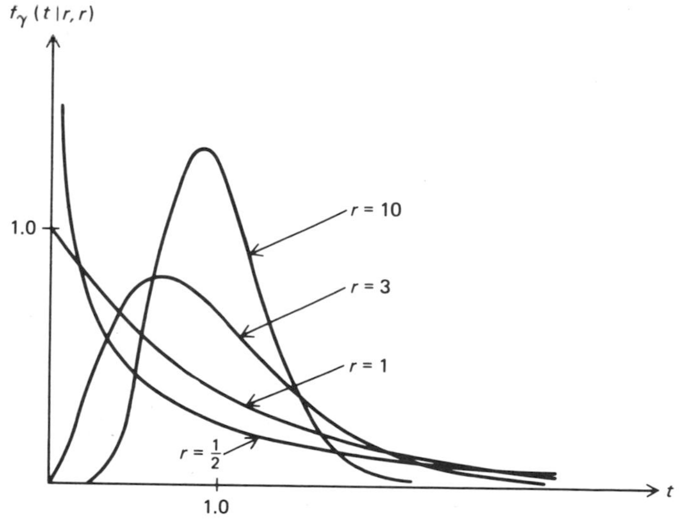

Figure 12.1

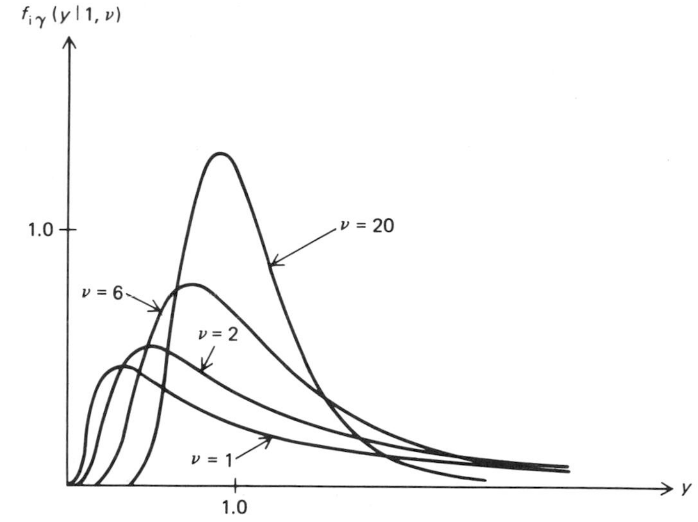

Figure 12.2

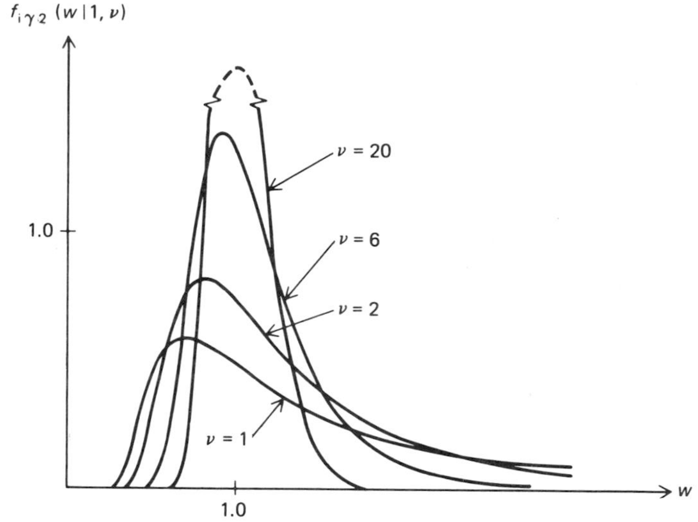

Figure 12.3

The following results on sums of gamma-distributed, independent random variables are of considerable importance in our later applications:

If $\tilde{Q}_1, \ldots, \tilde{Q}_n$ are independent and the df of $\tilde{Q}_i$ is $f_{\chi^2}(\cdot \mid \nu_i)$, for every i, then the df of $\tilde{Q} \equiv \sum_{i=1}^{n} \tilde{Q}_i$ is $f_{\chi^2}(\cdot \mid \sum_{i=1}^{n} \nu_i)$; and (12.2.17)

If $\tilde{t}_1, \ldots, \tilde{t}_n$ are independent and the df of $\tilde{t}_i$ is $f_\gamma(\cdot \mid r_i, \lambda)$, for every i, then the df of $\tilde{t} \equiv \sum_{i=1}^{n} \tilde{t}_i$ is $f_\gamma(\cdot \mid \sum_{i=1}^{n} r_i, \lambda)$. (12.2.18)

Proof

Statement (12.2.17) follows from (12.2.18) as the special case $\tilde{Q}_i = t_i$ with df $f_\gamma(\cdot \mid \frac{1}{2}\nu_i, \frac{1}{2})$. Statement (12.2.18) was proved in Chapter 11. QED

12.3 THE BETA FAMILIES

We say that $\tilde{z}$ has the *beta distribution with parameters* $\rho > 0$ *and* $\nu > \rho$ if the df of $\tilde{z}$ is

$$f_\beta(z \mid \rho, \nu) \equiv \begin{cases} 0, & z \notin (0, 1) \\ \dfrac{z^{\rho-1}(1-z)^{\nu-\rho-1}}{B(\rho, \nu)}, & z \in (0, 1), \end{cases} \tag{12.3.1}$$

where $B(\rho, \nu)$ is defined, for all $\rho > 0$ and $\nu > \rho$ by

$$B(\rho, \nu) = \int_0^1 z^{\rho-1}(1-z)^{\nu-\rho-1}dz. \tag{12.3.2}$$

It can be shown that

$$B(\rho, \nu) = \frac{\Gamma(\rho)\Gamma(\nu-\rho)}{\Gamma(\nu)}, \tag{12.3.3}$$

from which it is obvious that $B(\rho, \nu) = B(\nu - \rho, \nu)$, for all $\nu > \rho > 0$. Moreover, the reader may verify as Exercise 3 that $\tilde{z}$ has df $f_\beta(\cdot \mid \rho, \nu)$ if and only if $\tilde{w} \equiv 1 - \tilde{z}$ has df $f_\beta(\cdot \mid \nu - \rho, \nu)$.

Table III of the Appendix provides fractiles of beta distributions for integral values of ν up to 29. The preceding comment enables us to record the fractiles only for ρ's less than or equal to $\nu/2$, since

$$_p\beta_{(\rho,\nu)} = {}_{(1-p)}\beta_{(\nu-\rho,\nu)}, \tag{12.3.4}$$

which the reader may verify as Exercise 4.

As Exercise 5, the reader may also show that

$$E_\beta(\tilde{z}^q \mid \rho, \nu) = \frac{\Gamma(\nu)\,\Gamma(\rho+q)}{\Gamma(\rho)\,\Gamma(\nu+q)}, \tag{12.3.5}$$

which by Chapter 11, Exercise 27 implies that

$$E_\beta(\tilde{z} \mid \rho, \nu) = \rho/\nu \tag{12.3.6}$$

and together with (10.3.11), that

$$\begin{aligned} V_\beta(\tilde{z} \mid \rho, \nu) &= \frac{\rho(\nu-\rho)}{\nu^2(\nu+1)} \\ &= \frac{E_\beta(\tilde{z} \mid \rho, \nu)[1 - E_\beta(\tilde{z} \mid \rho, \nu)]}{(\nu+1)}. \end{aligned} \tag{12.3.7}$$

Figure 12.4 indicates the rich variety of shapes that the beta df may assume. Since the range of possible values of $\tilde{z}$ is (0, 1), the beta df appears to be a natural vehicle for expressing judgments about an unknown proportion or probability $\tilde{p}$. That the beta df is indeed well-suited for this role will be made clear in Chapter 14.

Actually, (12.3.1) is a *standardized* beta df. Occasionally one encounters "nonstandardized" beta distributions. We say that $\tilde{x}$ has the nonstandard-

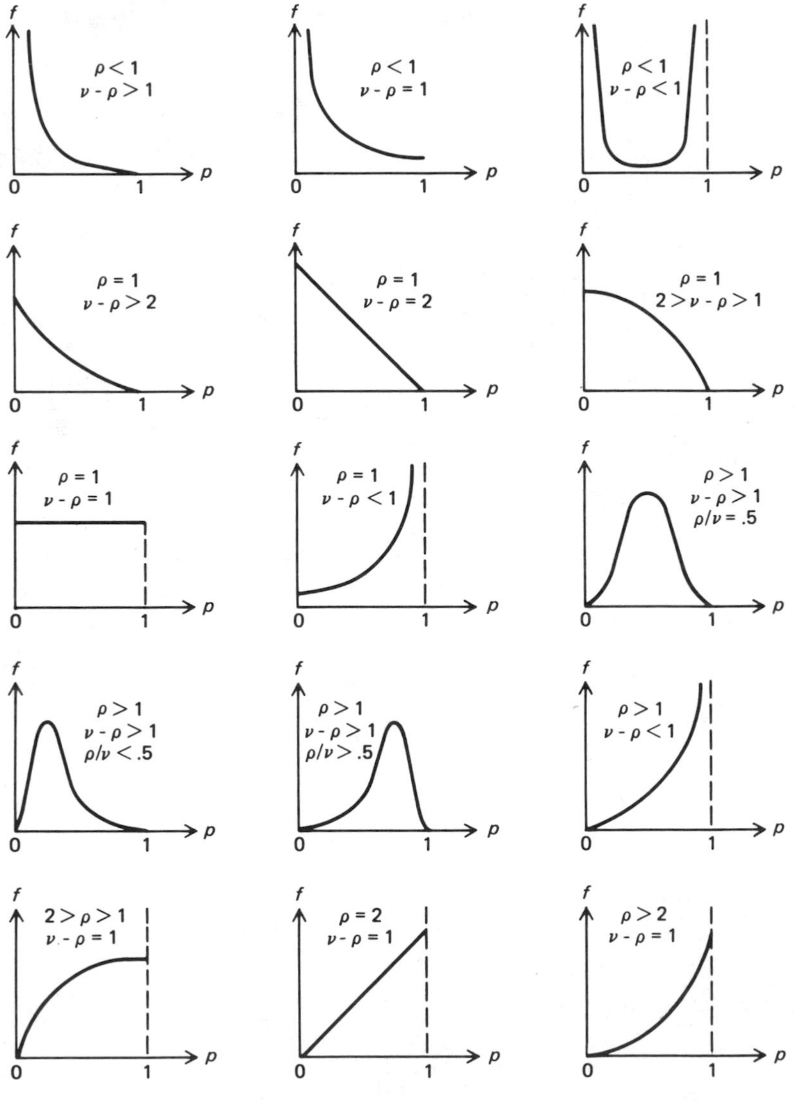

Figure 12.4 Shapes of Beta Densities

ized beta distribution with parameters $\rho > 0$, $\nu > \rho$, $a \in (-\infty, \infty)$, and $b > a$ if and only if $\tilde{z} \equiv (\tilde{x} - a)/(b - a)$ has the (standardized) beta distribution with parameters ρ and ν. Using elementary change-of-variable facts in Exercise 8.9, the reader may verify as Exercise 6 that the df $f_{\beta-}(\cdot \mid \rho, \nu, a, b)$ of $\tilde{x}$ is given by

$$f_{\beta-}(x \mid \rho, \nu, a, b) \equiv \begin{cases} 0, & x \notin (a, b) \\ \dfrac{(x - a)^{\rho-1}(b - x)^{\nu-\rho-1}}{(b - a)^{\nu-1}B(\rho, \nu)}, & x \in (a, b), \end{cases} \tag{12.3.8}$$

that

$$E_{\beta-}(\tilde{x} \mid \rho, \nu, a, b) = \frac{b\rho + a(\nu - \rho)}{\nu}, \tag{12.3.9}$$

and that

$$V_{\beta-}(\tilde{x} \mid \rho, \nu, a, b) = (b - a)^2 V_\beta(\tilde{z} \mid \rho, \nu). \tag{12.3.10}$$

The nonstandardized beta distribution is sometimes used to express judgments about the uncertain (but surely finite) times required to perform each of an interdependent network of tasks in a large project. The scheduling and control of such projects has been formalized as "Pert" (Program Evaluation Review Technique), and is discussed in many modern books on operations research and production management; for example, Hillier and Lieberman [**39**].

Since Exercise 9 in Chapter 8 implies that

$$F_{\beta-}(x \mid \rho. \nu, a, b) = F_\beta\left(\frac{x - a}{b - a} \mid \rho, \nu\right), \tag{12.3.11}$$

it follows that

$${}_p\beta_{(\rho,\nu,a,b)} = a + (b - a)\,{}_p\beta_{(\rho,\nu)}, \tag{12.3.12}$$

for every $p \in (0, 1)$.

We say that $\tilde{y}$ has the *inverted beta-2 distribution with parameters* $\rho > 0$, $\nu > \rho$, *and* $b > 0$ if the df of $\tilde{y}$ is given by

$$f_{i\beta2}(y \mid \rho, \nu, b) \equiv \begin{cases} 0, & y \geqq 0 \\ \dfrac{b^{\nu-\rho}y^{\rho-1}}{B(\rho, \nu)(b + y)^{\nu}}, & y > 0. \end{cases} \tag{12.3.13}$$

Alternatively, $\tilde{y}$ has df $f_{i\beta2}(\cdot \mid \rho, \nu, b)$ if and only if $\tilde{z} \equiv \tilde{y}/(\tilde{y} + b)$ has df $f_\beta(\cdot \mid \rho, \nu)$. The reader may verify as Exercise 7 that

$$F_{i\beta2}(y \mid \rho, \nu, b) = F_\beta\left(\frac{y}{b + y} \mid \rho, \nu\right), \tag{12.3.14}$$

for all real y, and hence that

$${}_p(i\beta2)_{(\rho,\nu,b)} = \frac{b \cdot {}_p\beta_{(\rho,\nu)}}{1 - {}_p\beta_{(\rho,\nu)}}, \tag{12.3.15}$$

which implies that inverted beta-2 fractiles can be readily computed from entries in Table III (see Appendix).

The reader may also verify, by direct application of the definition, that

$$E_{i\beta 2}(\tilde{y}^q \mid \rho, \nu, b) = \frac{b^q \Gamma(\rho + q)\Gamma(\nu - \rho - q)}{\Gamma(\rho)\Gamma(\nu - \rho)}, \tag{12.3.16}$$

provided that $q \in (-\rho, \nu - \rho)$. From (12.3.16) we obtain

$$E_{i\beta 2}(\tilde{y} \mid \rho, \nu, b) = \frac{b\rho}{(\nu - \rho - 1)}, \tag{12.3.17}$$

provided that $\nu - \rho > 1$, and

$$V_{i\beta 2}(\tilde{y} \mid \rho, \nu, b) = \frac{b^2 \rho(\nu - 1)}{(\nu - \rho - 1)^2(\nu - \rho - 2)}, \tag{12.3.18}$$

provided that $\nu - \rho > 2$. If the respective hypotheses $\nu - \rho > 1$ and $\nu - \rho > 2$ in (12.3.17) and (12.3.18) do not obtain, then the respective moments do not exist; that is, the defining integrals diverge to $+\infty$.

Figure 12.5 depicts some representative inverted beta-2 df's. It is worth remarking that there is an inverted beta df, but we shall not need it in this book.

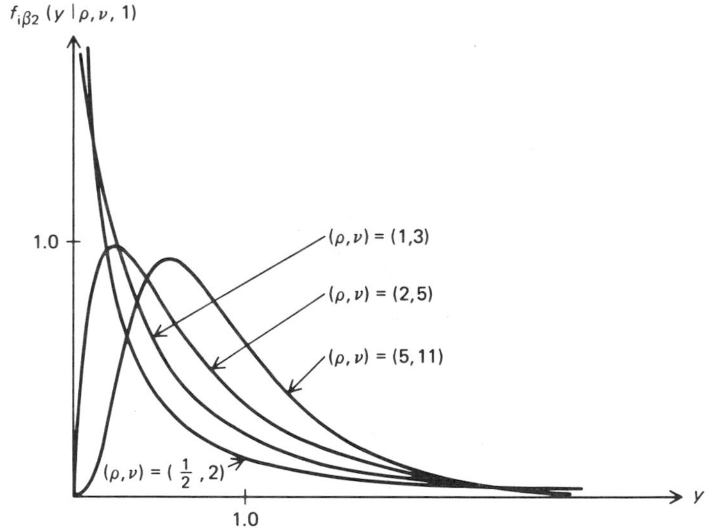

Figure 12.5

We say that $\tilde{\phi}$ has the *F-distribution with* ν_1 *and* ν_2 *degrees of freedom* if $\tilde{\phi}$ has the inverted beta-2 distribution with parameters $\rho = \frac{1}{2}\nu_1$, $\nu = \frac{1}{2}(\nu_1 + \nu_2)$, and $b = \nu_2/\nu_1$; that is,

$$f_F(\phi \mid \nu_1, \nu_2) = f_{i\beta 2}\left(\phi \mid \tfrac{1}{2}\nu_1, \tfrac{1}{2}(\nu_1 + \nu_2), \frac{\nu_2}{\nu_1}\right). \tag{12.3.19}$$

Table V in the Appendix furnishes selected fractiles of the F-distribution for selected values of ν_1 and ν_2, despite the fact that Table III and (12.3.15) would suffice; it is convenient to have a direct table of the F-fractiles for applications in the later chapters.

12.4 THE UNIVARIATE NORMAL DISTRIBUTIONS

We say that $\tilde{z}$ has the *standardized Normal distribution* if the df of $\tilde{z}$ is given by

$$f_{N^*}(z) \equiv (2\pi)^{-1/2}e^{-\frac{1}{2}z^2}, \qquad \text{all real } z. \tag{12.4.1}$$

This density is tabulated for z going in steps of .01 from 0.00 to 4.00 in Table VI of the Appendix. From (12.4.1) it is clear that $f_{N^*}(\cdot)$ is symmetric about zero and hence that

$$f_{N^*}(-z) = f_{N^*}(z), \qquad \text{all real } z. \tag{12.4.2}$$

The cumulative distribution function $F_{N^*}(\cdot)$ of $\tilde{z}$ cannot be expressed in "closed form"; that is, as an algebraic formula. It is tabulated in Table I of the Appendix for the same $\tilde{z}$'s as Table VI; and symmetry of $f_{N^*}(\cdot)$ implies that

$$F_{N^*}(-z) = 1 - F_{N^*}(z), \qquad \text{all real } z, \tag{12.4.3}$$

the verification of which is left to the reader as Exercise 8.

We say that $\tilde{x}$ has the *Normal distribution with parameters* $\mu \in R^1$ and $\sigma^2 > 0$ if the df of $\tilde{x}$ is given by

$$f_N(x \mid \mu, \sigma^2) = (2\pi\sigma^2)^{-1/2}e^{-\frac{1}{2}(x-\mu)^2/\sigma^2}, \qquad \text{all real } x. \tag{12.4.4}$$

The reader may verify as Exercise 9 that $\tilde{x}$ has df $f_N(\cdot \mid \mu, \sigma^2)$ if and only if $\tilde{z} \equiv (\tilde{x} - \mu)/\sigma$ has df $f_{N^*}(\cdot)$, where $\sigma \equiv +\sqrt{\sigma^2}$. Hence

$$f_N(x \mid \mu, \sigma^2) = (1/\sigma)f_{N^*}([x - \mu]/\sigma), \tag{12.4.5}$$

and fractiles can be found from

$${}_pN_{(\mu,\sigma^2)} = \mu + \sigma \cdot {}_pN^*, \tag{12.4.6}$$

which results are left to the reader as Exercise 10.

The linguistically perceptive reader will note that by capitalizing the word *normal* we have violated our previous practice of capitalizing only proper-name distributions. Our reasons are twofold: first, *normal* is an adjective of greater currency than *binomial* or *hypergeometric*, and we wish to avoid confusion; second, capitalization helps to emphasize the profound respect which the Normal df's are due. A synonym for normal distribution is *Gaussian distribution*, after the mathematician Gauss.

Figure 12.6 depicts three Normal df's with identical μ's and different σ^2's. It shows that all Normal df's are "bell curves," and that small σ^2's correspond to "tall and skinny" density functions which concentrate high probability in any given interval containing μ, the axis of symmetry of $f_N(\cdot \mid \mu, \sigma^2)$. Conversely, a high value of σ^2 corresponds to a "short and broad" density function which diffuses considerable probability far from μ. We shall see in the following that $E(\tilde{x}) = \mu$ and $V(\tilde{x}) = \sigma^2$, so that these comments are consistent with our policy of calling variance an index of uncertainty.

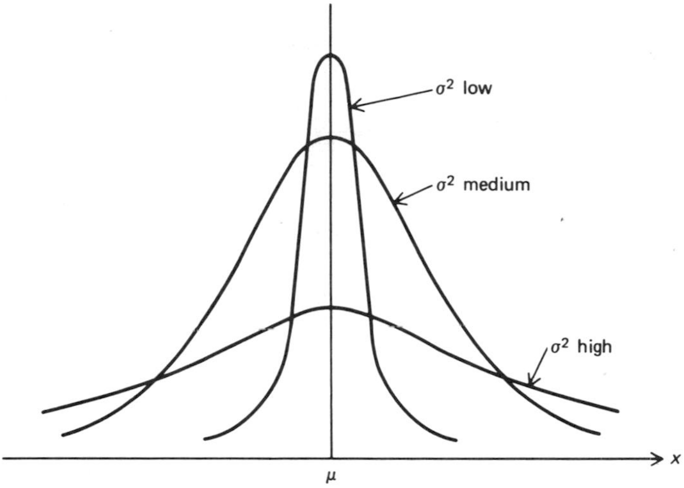

Figure 12.6

When laymen speak of a "bell curve," they are almost invariably referring to a univariate Normal density function. Many real-life phenomena exhibit bell-curve-shaped relative frequencies that are approximated fairly accurately by some Normal df. An explanation as to why relative frequencies are often bell-shaped is given in Section 13.1.

To derive moments and other facts about Normal distributions, we shall first derive the moment-generating function.

If $\tilde{x}$ has df $f_N(x \mid \mu, \sigma^2)$, then the moment-generating function $M_N(u \mid \mu, \sigma^2)$ of $\tilde{x}$ is given by (12.4.7)

$$M_N(u \mid \mu, \sigma^2) = \exp(u\mu + \tfrac{1}{2}u^2\sigma^2),$$

for every $u \in R^1$.

Proof

By completing a square in the exponent to obtain (*) below, we find

$$M_N(u \mid \mu, \sigma^2) = E_N(e^{u\tilde{x}} \mid \mu, \sigma^2)$$

$$= \int_{-\infty}^{\infty} (2\pi\sigma^2)^{-1/2} e^{ux} \exp\left[-\tfrac{1}{2}(x-\mu)^2/\sigma^2\right] dx$$

$$(*) = \exp\{u\mu + \tfrac{1}{2}u^2\sigma^2\} \int_{-\infty}^{\infty} (2\pi\sigma^2)^{-1/2} \exp\{-\tfrac{1}{2}[x-(\mu+u\sigma^2)]^2/\sigma^2\} dx$$

$$= \exp\{u\mu + \tfrac{1}{2}u^2\sigma^2\} \int_{-\infty}^{\infty} f_N(x \mid \mu + u\sigma^2, \sigma^2)\, dx$$

$$= \exp\{u\mu + \tfrac{1}{2}u^2\sigma^2\}. \qquad \text{QED}$$

The reader may use (12.4.7) to obtain

$$E_N(\tilde{x} \mid \mu, \sigma^2) = \mu \tag{12.4.8}$$

and

$$V_N(\tilde{x} \mid \mu, \sigma^2) = \sigma^2 \tag{12.4.9}$$

as Exercise 11 (a), thus showing that our usage of μ and σ^2 to denote the parameters is indeed consistent with our use of these symbols in Chapter 10 to denote expectation and variance. The generalization (12.4.7) readily yields the following result about linear functions of independent, normally distributed random variables:

If $\tilde{x}_1, \ldots, \tilde{x}_n$ are independent, $\tilde{x}_i$ has df $f_N(\cdot \mid \mu_i, \sigma_{ii}^2)$ for $i = 1, \ldots, n$, $c \in R^1$, and $b \equiv (b_1, \ldots, b_n)' \in R^n$ with $b \neq 0$, then (12.4.10)

(a) the df of $\tilde{y} \equiv c + \sum_{i=1}^{n} b_i\tilde{x}_i$ is

$$f_N(\cdot \mid c + \sum_{i=1}^{n} b_i\mu_i, \sum_{i=1}^{n} b_i^2\sigma_{ii}^2);$$

and if $\tilde{x}_1, , \ldots, \tilde{x}_n$ are iid with common df $f_N(\cdot \mid \mu, \sigma^2)$, then

(b) the df of $\tilde{y} \equiv \sum_{i=1}^{n} \tilde{x}_i$ is $f_N(\cdot \mid n\mu, n\sigma^2)$

and

(c) the df of $\bar{\tilde{x}} \equiv (1/n) \sum_{i=1}^{n} \tilde{x}_i$ is $f_N(\cdot \mid \mu, \sigma^2/n)$.

Proof

Exercise 12.

In Section 12.7 we shall derive the generalization of (12.4.10) to the case in which the $\tilde{x}_i$'s need not be independent.

12.5 THE UNIVARIATE STUDENT DISTRIBUTIONS

We say that $\tilde{x}$ *has the univariate Student distribution* with parameters $\mu \in R^1$, $\sigma^2 > 0$, and $\nu > 0$ if:

(1) there is a random variable $\tilde{h}$ with df $f_{\gamma 2}(h \mid 1, \nu)$; and
(2) conditional upon $\tilde{h}$, the df of $\tilde{x}$ is $f_N(x \mid \mu, \sigma^2/h)$.

If (1) and (2) obtain, the marginal df of $\tilde{x}$ is by definition the univariate Student df with parameters $\mu \in R^1$, $\sigma^2 > 0$, and $\nu > 0$. The univariate Student df $f_S(\cdot \mid \mu, \sigma^2, \nu)$ is given by

$$f_S(x \mid \mu, \sigma^2, \nu) \equiv \frac{S(\nu)(\sigma^2)^{-1/2}}{[\nu + (x - \mu)^2/\sigma^2]^{\frac{1}{2}(\nu+1)}}, \tag{12.5.1}$$

for every $x \in R^1$, where

$$S(\nu) \equiv \frac{(\frac{1}{2}\nu)^{\frac{1}{2}\nu}\Gamma(\frac{1}{2}\nu + \frac{1}{2})}{(2\pi)^{1/2}\Gamma(\frac{1}{2}\nu)}. \tag{12.5.2}$$

Proof

The joint df of $\tilde{x}$ is, by assumptions (1) and (2), $f_N(x \mid \mu, \sigma^2/h) \cdot f_{\gamma 2}(h \mid 1, \nu)$, and hence the marginal df of $\tilde{x}$ is

$$\begin{aligned}
&f_S(x \mid \mu, \sigma^2, \nu)\\
&\equiv \int_0^\infty f_N\left(x \mid \mu, \frac{\sigma^2}{h}\right) f_{\gamma 2}(h \mid 1, \nu)\,dh\\
&= \int_0^\infty \left(\frac{h}{2\pi\sigma^2}\right)^{1/2} \exp\left[-\tfrac{1}{2}h\left[\left(\frac{x-\mu}{\sigma}\right)^2\right]\right](\tfrac{1}{2}\nu) \exp\left[-\tfrac{1}{2}\nu h\right](\tfrac{1}{2}\nu h)^{\frac{1}{2}\nu - 1}\left(\frac{dh}{\Gamma(\frac{1}{2}\nu)}\right)\\
&= \left\{\frac{(2\pi\sigma^2)^{-1/2}(\frac{1}{2}\nu)^{\frac{1}{2}\nu}}{\Gamma(\frac{1}{2}\nu)}\right\} \int_0^\infty \exp\left\{-\tfrac{1}{2}h\left[\nu + \left(\frac{x-\mu}{\sigma}\right)^2\right]\right\} h^{\frac{1}{2}\nu + \frac{1}{2} - 1}\,dh.
\end{aligned}$$

Change variable in the final integral from h to $z \equiv \frac{1}{2}h[\nu + (x - \mu)^2/\sigma^2]$ to obtain, after a little algebra,

$$\begin{aligned}
f_S(x \mid \mu, \sigma^2, \nu) &= \left\{\frac{S(\nu)(\sigma^2)^{-1/2}}{[\nu + (x - \mu)^2/\sigma^2]^{\frac{1}{2}(\nu+1)}}\right\} \int_0^\infty f_{\gamma *}(z \mid \tfrac{1}{2}\nu + \tfrac{1}{2})\,dz\\
&= \frac{S(\nu)(\sigma^2)^{-1/2}}{[\nu + (x - \mu)^2/\sigma^2]^{\frac{1}{2}(\nu+1)}}. \qquad \text{QED}
\end{aligned}$$

We say that $\tilde{z}$ has the *standardized Student distribution with* $\nu(>0)$ *degrees of freedom*, or the *Student-t distribution with* ν *degrees of freedom* if the df of $\tilde{z}$ is $f_S(\cdot \mid 0, 1, \nu)$; and we define the standardized Student df $f_{S^*}(\cdot \mid \nu)$ by

$$f_{S^*}(z \mid \nu) \equiv f_S(z \mid 0, 1, \nu), \qquad \text{all real } z. \tag{12.5.3}$$

The reader may show as Exercise 13 that $\tilde{x}$ has df $f_S(\cdot \mid \mu, \sigma^2, \nu)$ if and only if $\tilde{z} \equiv (\tilde{x} - \mu)/\sigma$ has df $f_{S^*}(\cdot \mid \nu)$. Table IV in the Appendix contains fractiles ${}_pS^*_{(\nu)}$ of the standardized Student distribution. Symmetry of $f_{S^*}(\cdot \mid \nu)$ about zero implies that

$$f_{S^*}(-z \mid \nu) = f_{S^*}(z \mid \nu), \qquad \text{all real } z, \tag{12.5.4}$$

and

$$F_{S^*}(-z \mid \nu) = 1 - F_{S^*}(z \mid \nu), \qquad \text{all real } z, \tag{12.5.5}$$

both facts constituting Exercise 14 for verfication. Moreover (Exercise 14),

$$F_S(x \mid \mu, \sigma^2, \nu) = F_{S^*}\left(\frac{x - \mu}{\sigma} \mid \nu\right), \tag{12.5.6}$$

$$f_S(x \mid \mu, \sigma^2, \nu) = \left(\frac{1}{\sigma}\right) f_{S^*}\left(\frac{x - \mu}{\sigma} \mid \nu\right), \tag{12.5.7}$$

and nonstandardized Student fractiles are obtained from Table IV via

$${}_pS_{(\mu,\sigma^2,\nu)} = \mu + \sigma \cdot {}_pS^*_{(\nu)} \tag{12.5.8}$$

[compare (12.4.6)].

It is instructive to compare $f_{S^*}(\cdot \mid \nu)$ with $f_{N^*}(\cdot)$. In Figure 12.7 are graphed $f_{S^*}(\cdot \mid 1)$, $f_{S^*}(\cdot \mid 5)$ and $f_{N^*}(\cdot)$. Note that $f_{S^*}(\cdot \mid 5)$ concentrates probability around zero better than does $f_{S^*}(\cdot \mid 1)$, but not as well as does $f_{N^*}(\cdot)$. Moreover, it can be shown that

$$\lim_{\nu \to \infty} f_{S^*}(z \mid \nu) = f_{N^*}(z), \qquad \text{for every real } z, \tag{12.5.9}$$

so that $f_{N^*}(\cdot)$ could be considered the special case $f_{S^*}(\cdot \mid \infty)$.

Table VII in the Appendix furnishes values of $f_{S^*}({}_pS^*_{(\nu)} \mid \nu)$; that is, the ordinates of the standardized Student df at the fractile abscissa values, for the fractiles given in Table IV. We shall need these values in Chapters 18 and 19.

The expectation and variance of the Student random variable $\tilde{x}$ with parameters μ, σ^2, and ν are

$$E_S(\tilde{x} \mid \mu, \sigma^2, \nu) = \mu, \tag{12.5.10}$$

provided that $\nu > 1$, and

$$V_S(\tilde{x} \mid \mu, \sigma^2, \nu) = \sigma^2 \left[\frac{\nu}{\nu - 2}\right], \tag{12.5.11}$$

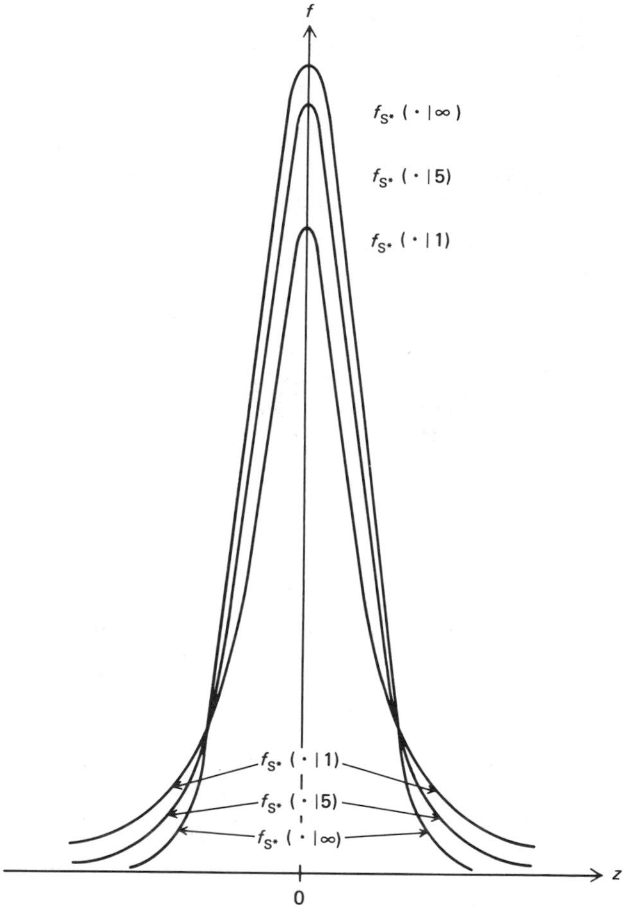

Figure 12.7

provided that $\nu > 2$. If $\nu \leqq 1$, the expectation does not exist, while if $\nu \leqq 2$, the variance does not exist. Appearances can be deceptive; $f_{S^*}(\cdot \mid 1)$ in Figure 12.7 seems nice enough—viewed casually, it can be confused with $f_{N^*}(\cdot)$—but it is in fact the pathological "Cauchy density." Equations (12.5.10) and (12.5.11) are left to the reader as Exercise 15.

The following result is analogous to (12.4.10), which is used in its proof:

If $\tilde{h}$ has df $f_{\gamma_2}(\cdot \mid 1, \nu)$, $\tilde{x}_1, \ldots, \tilde{x}_n$ are conditionally independent given $\tilde{h}$ with $\tilde{x}_i$ having conditional df $f_N(\cdot \mid \mu_i, \sigma_{ii}^2/h)$, $c \in R^1$, and $b \in R^n$ with $b \neq 0$, then (12.5.12)

(a) the unconditional df $\tilde{y} \equiv c + \sum_{i=1}^{n} b_i \tilde{x}_i$ is

$$f_S(\cdot \mid c + \sum_{i=1}^{n} b_i \mu_i, \sum_{i=1}^{n} b_i^2 \sigma_{ii}^2, \nu);$$

and if $\mu_i = \mu$ and $\sigma_{ii}^2 = \sigma^2$, for all i, then

(b) the df of $\tilde{y} \equiv \sum_{i=1}^{n} \tilde{x}_i$ is $f_S(\cdot \mid n\mu, n\sigma^2, \nu)$

and

(c) the df of $\bar{\tilde{x}} \equiv (1/n) \sum_{i=1}^{n} \tilde{x}_i$ is $f_S(\cdot \mid \mu, \sigma^2/n, \nu)$.

Proof

Exercise 16.

Finally, note that in the definition of $f_S(\cdot \mid \mu, \sigma^2, \nu)$ we used the special case of $f_{\gamma 2}(\cdot \mid \psi, \nu)$ in which $\psi = 1$. We do not gain any significant generality by replacing $f_{\gamma 2}(\cdot \mid 1, \nu)$ with $f_{\gamma 2}(\cdot \mid \psi, \nu)$ in (12.5.1), because

$$\int_0^\infty f_N\left(x \mid \mu, \frac{\sigma^2}{h}\right) \cdot f_{\gamma 2}(h \mid \psi, \nu) dh = f_S(\cdot \mid \mu, \psi\sigma^2, \nu). \qquad (12.5.13)$$

Proof

From (12.2.8) it is clear that $\tilde{h}$ has df $f_{\gamma 2}(\cdot \mid \psi, \nu)$ if and only if $\tilde{y} \equiv \psi \tilde{h}$ has df $f_{\gamma 2}(\cdot \mid 1, \nu)$. By changing variable from h to y, we replace σ^2/h by $\psi\sigma^2/y$ and obtain

$$\int_0^\infty f_N\left(x \mid \mu, \frac{\sigma^2}{h}\right) f_{\gamma 2}(h \mid \psi, \nu) dh = \int_0^\infty f_N\left(x \mid \mu, \frac{\psi\sigma^2}{y}\right) f_{\gamma 2}(y \mid 1, \nu) dy$$
$$\equiv f_S(x \mid \psi\sigma^2, \nu). \qquad \text{QED}$$

12.6 THE k-VARIATE BETA DISTRIBUTIONS

We say that the k-dimensional random vector $\tilde{\mathbf{z}}$ has the *k-variate beta distribution with parameters* $\boldsymbol{\rho} \in R^k$ ($\boldsymbol{\rho} > \mathbf{0}$) *and* $\nu > \sum_{i=1}^{k} \rho_i$ if the df of $\tilde{\mathbf{z}}$ is

$$f_{\beta^{(k)}}(\mathbf{z} \mid \boldsymbol{\rho}, \nu) \equiv \begin{cases} \dfrac{\prod_{i=1}^{k+1} z_i^{\rho_i - 1}}{B_k(\boldsymbol{\rho}, \nu)}, & \begin{cases} z_i > 0, \text{ for every } i \\ \sum_{i=1}^{k+1} z_i = 1 \end{cases} \\ 0, & \text{elsewhere,} \end{cases} \qquad (12.6.1)$$

where $z_{k+1} \equiv 1 - \sum_{i=1}^{k} z_i$, $\rho_{k+1} \equiv \nu - \sum_{i=1}^{k} \rho_i$, and

$$B_k(\boldsymbol{\rho}, \nu) \equiv \frac{\prod_{i=1}^{k+1} \Gamma(\rho_i)}{\Gamma(\nu)} \tag{12.6.2}$$

for $\Gamma(\cdot)$ as defined by (12.2.2). It is clear that $B_1(\rho, \nu) = B(\rho, \nu)$ as defined by (12.3.3).

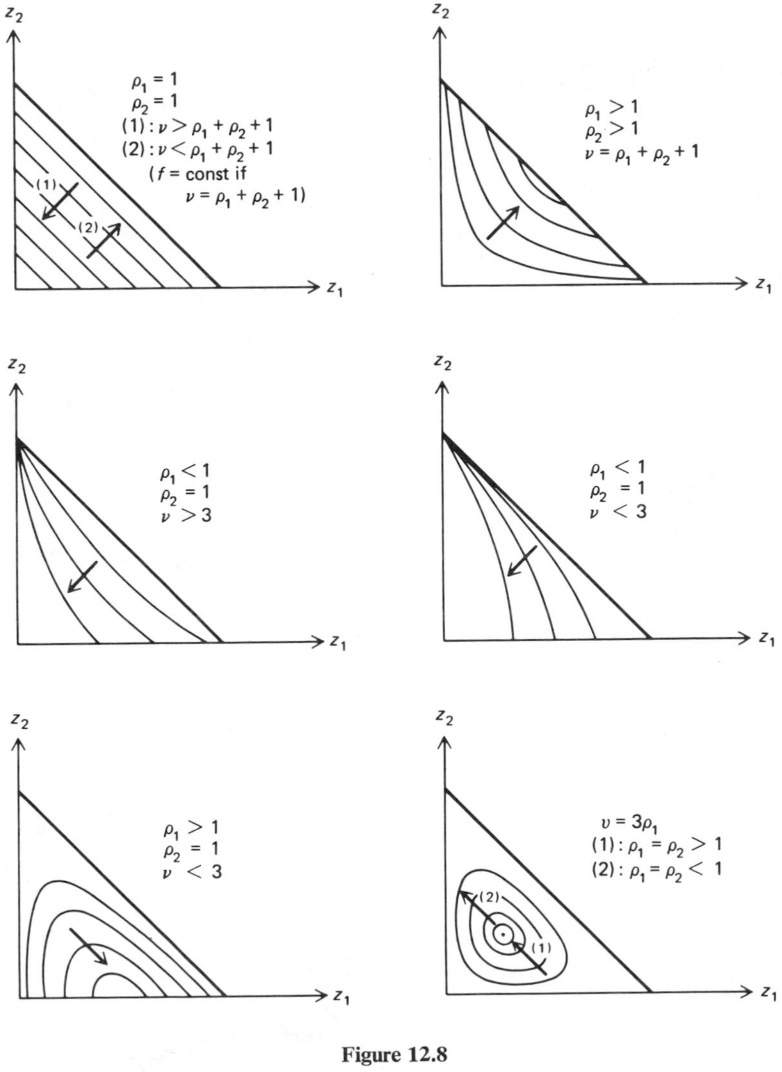

Figure 12.8

The k-variate beta distribution is also called the (k-variate) *Dirichlet distribution*. When $k = 1$, (12.6.1) reduces to (12.3.1) because in that case $z_2 \equiv 1 - z_1 \equiv 1 - z$.

Figure 12.8 depicts some level curves $\{\mathbf{z}: f_\beta^{(2)}(\mathbf{z} \mid \boldsymbol{\varrho}, \nu) = \text{const}\}$ for various bivariate beta df's. The arrow in each graph of the figure points in the direction of increasing f.

When considering distributions of $k > 1$-dimensional random vectors, we shall be concerned with marginal and conditional df's of j-dimensional subvectors. Our first result roughly states that marginal df's of k-variate betas are j-variate betas.

If $\tilde{\mathbf{z}}$ has df $f_\beta^{(k)}(\cdot \mid \boldsymbol{\varrho}, \nu)$, $\tilde{\mathbf{z}}_1$ is a subvector consisting of j components of $\tilde{\mathbf{z}}$, and $\boldsymbol{\varrho}_1$ is the corresponding subvector of $\boldsymbol{\varrho}$; then the df of $\tilde{\mathbf{z}}_1$ is $f_\beta^{(j)}(\cdot \mid \boldsymbol{\varrho}_1, \nu)$. (12.6.3)

Proof

Suppose, without loss of generality, that $\tilde{\mathbf{z}}_1 \equiv (\tilde{z}_1, \ldots, \tilde{z}_j)'$, since the components of $\tilde{\mathbf{z}}$ and $\boldsymbol{\varrho}$ can be rearranged as desired. We derive the joint density $f(z_1, \ldots, z_j)$ for a fixed $(z_1, \ldots, z_j)$. Let $z^* \equiv 1 - \sum_{i=1}^{j} z_i$, $\nu^* \equiv \nu - \sum_{i=1}^{j} \rho_i$, and

$$A \equiv \{(z_{j+1}, \ldots, z_k): \sum_{i=j+1}^{k} z_i < z^*,\ z_i > 0, \text{ for } i = j+1, \ldots, k\}.$$

Then

$$f(z_1, \ldots, z_j) = \left[\frac{1}{B_k(\boldsymbol{\varrho}, \nu)}\right] \times \underbrace{\int \cdots \int}_{A} \left\{\prod_{i=1}^{k} z_i^{\rho_i - 1}\right\}\left(1 - \sum_{i=1}^{k} z_i\right)^{\nu - \Sigma_{i=1}^{k}\rho_i - 1} dz_{j+1} \ldots dz_k$$

$$= \left\{\frac{\prod_{i=1}^{j} z_i^{\rho_i - 1}}{B_k(\rho, \nu)}\right\} \times \underbrace{\int \cdots \int}_{A} \left\{\prod_{i=j+1}^{k} z_i^{\rho_i - 1}\right\}\left(z^* - \sum_{i=j+1}^{k} z_i\right)^{\nu^* - \Sigma_{i=j+1}^{k}\rho_i - 1} dz_{j+1} \ldots dz_k$$

$$= \left\{\prod_{i=1}^{j} z_i^{\rho_i - 1}\right\} \frac{(z^*)^{\nu^* - 1}}{B_k(\boldsymbol{\varrho}, \nu)} \times \underbrace{\int \cdots \int}_{A^*} \left\{\prod_{i=j+1}^{j} y_i^{\rho_i - 1}\right\}\left(1 - \sum_{i=j+1}^{k} y_i\right)^{\nu^* - \Sigma_{i=j+1}^{k}\rho_i - 1} dy_{j+1} \ldots dy_k,$$

where $y_i \equiv z_i/z^*$ for $i = j+1, \ldots, k$ and

$$A^* = \{(y_{j+1}, \ldots, y_k): \sum_{i=j+1}^{k} y_i < 1, y_i > 0, \text{ for } i = j+1, \ldots, k\} = A.$$

Hence the integral in the last equality can be expressed as

$$\int_{A^*}\!\!\cdots\!\int B_{k-j}(\boldsymbol{\varrho}_2, \nu^*) f_\beta^{(k-j)}(\tilde{\mathbf{y}} \mid \boldsymbol{\varrho}_2, \nu^*) dy_{j+1} \ldots dy_k;$$

and, since A^* is the set of all possible y's $(= (z^*)^{-1}\tilde{\mathbf{z}}_2$'s), its value is $B_{k-j}(\boldsymbol{\varrho}_2, \nu^*)$. Hence

$$f(z_1, \ldots, z_j) = \left\{\prod_{i=1}^{j} z_i^{\rho_i - 1}\right\}\left(1 - \sum_{i=1}^{j} z_i\right)^{\nu - \Sigma_{i=1}^{j}\rho_i - 1} \times \frac{B_{k-j}(\boldsymbol{\varrho}_2, \nu_*)}{B_k(\boldsymbol{\varrho}, \nu)},$$

which coincides with $f_\beta^{(j)}(\mathbf{z}_1 \mid \boldsymbol{\varrho}_1, \nu)$ because $B_{k-j}(\boldsymbol{\varrho}_2, \nu^*)/B_k(\boldsymbol{\varrho}, \nu) = 1/B_j(\boldsymbol{\varrho}_1, \upsilon)$, which follows readily from (12.6.2). QED

Result (12.6.3) implies, in particular, that the marginal df of $\tilde{z}_i$ is $f_\beta(\cdot \mid \rho_i, \nu)$ for every i (including $i = k+1$).

Now that the marginal df of $\tilde{\mathbf{z}}_1$ has been determined, it is simple to find the conditional df $f(\mathbf{z}_2 \mid \mathbf{z}_1) = f(\mathbf{z})/f(\mathbf{z}_1)$. Actually, we shall determine the conditional df of a positive multiple of $\tilde{\mathbf{z}}_2$.

Let $\tilde{\mathbf{z}}$ have df $f_\beta^{(k)}(\cdot \mid \boldsymbol{\varrho}, \nu)$ and let $\tilde{\mathbf{z}}_1$ be a j-dimensional subvector of $\tilde{\mathbf{z}}$ for $0 < j < k$. Let

$$\begin{aligned} J &\equiv \{i: \tilde{z}_i \text{ is a component of } \tilde{\mathbf{z}}_1\}, \\ z^*(\mathbf{z}_1) &\equiv 1 - \sum_{i \in J} z_i, \\ \nu^* &\equiv \nu - \sum_{i \in J} \rho_i, \qquad (12.6.4) \\ \tilde{\mathbf{z}}^*_2 &\equiv \left[\frac{1}{z^*(\mathbf{z}_1)}\right]\tilde{\mathbf{z}}_2, \end{aligned}$$

and $\boldsymbol{\varrho}_2$ denote the subvector of $\boldsymbol{\varrho}$ corresponding to $\tilde{\mathbf{z}}_2$. Then the conditional df of $\tilde{\mathbf{z}}^*_2$ given $\mathbf{z}_1$ is $f_\beta^{(k-j)}(\cdot \mid \boldsymbol{\varrho}_2, \nu^*)$.

Proof

Exercise 17.

Notice that (12.6.4) gives the df of a rescaled version of $\tilde{\mathbf{z}}_2$, and not the df of $\tilde{\mathbf{z}}_2$ itself. The conditional df of $\tilde{\mathbf{z}}_2$ given $\mathbf{z}_1$ is a "nonstandardized" $(k-j)$-variate beta df.

Example 12.6.1

If $\tilde{\mathbf{z}}$ has df $f_\beta^{(5)}(\cdot \mid (8, 9, 4, 1, 1)', 28)$, then $\tilde{\mathbf{z}}_1 \equiv (\tilde{z}_1, \tilde{z}_2, \tilde{z}_3)'$ has marginal df $f_\beta^{(3)}(\cdot \mid (8, 9, 4)', 28)$, and $\tilde{\mathbf{z}}_2^* \equiv [1/(1 - \sum_{i=1}^{3} z_i)]\tilde{\mathbf{z}}_2$ has conditional df $f_\beta^{(2)}(\cdot \mid (1, 1)', 7)$ given $\tilde{\mathbf{z}}_1 = (z_1, z_2, z_3)'$. Note that the conditional distribution of $\tilde{\mathbf{z}}_2$ given $\mathbf{z}_1$ depends upon the value of $\mathbf{z}_1$ only through $\sum_{i=1}^{3} z_i$.

We next obtain

$$E_\beta^{(k)}\left(\prod_{i=1}^{k+1} \tilde{z}_i^{q_i} \mid \boldsymbol{\varrho}, \nu\right) = \frac{B_k\left(\boldsymbol{\varrho} + \mathbf{q}, \nu + \sum_{i=1}^{k+1} q_i\right)}{B_k(\boldsymbol{\varrho}, \nu)}, \tag{12.6.5}$$

which exists whenever $(q_1, \ldots, q_k)' \equiv \mathbf{q} > -\boldsymbol{\varrho}$ and $q_{k+1} > -\rho_{k+1} = \sum_{i=1}^{k} \rho_i - \nu$.

Proof

$$E_\beta^{(k)}\left(\prod_{i=1}^{k+1} \tilde{z}_i^{q_i} \mid \boldsymbol{\varrho}, \nu\right)$$
$$= \int_A \left\{\prod_{i=1}^{k+1} z_i^{q_i}\right\}\left\{\prod_{i=1}^{k+1} z_i^{\rho_i - 1}\right\}\frac{d\mathbf{z}}{B_k(\boldsymbol{\varrho}, \nu)}$$
$$= \left\{\frac{B_k(\boldsymbol{\varrho} + \mathbf{q}, \nu + \sum_{i=1}^{k+1} q_i)}{B_k(\boldsymbol{\varrho}, \nu)}\right\} \int_A \prod_{i=1}^{k+1} z_i^{\rho_i + q_i - 1} \frac{d\mathbf{z}}{B_k(\boldsymbol{\varrho} + \mathbf{q}, \nu + \sum_{i=1}^{k+1} q_i)},$$

where $A = \{\mathbf{z}: z_i > 0, \text{ for every } i, \sum_{i=1}^{k} z_i < 1\}$. But the last integral is simply

$$\int_A f_\beta^{(k)}(\mathbf{z} \mid \boldsymbol{\varrho} + q, \nu + \sum_{i=1}^{k+1} q_i)\, d\mathbf{z} = 1. \qquad \text{QED}$$

Equation (12.6.5) readily yields

$$E_\beta^{(k)}(\tilde{z}_i \mid \boldsymbol{\varrho}, \nu) = \frac{\rho_i}{\nu}, \qquad \text{(all } i\text{)}, \tag{12.6.6}$$

$$V_\beta^{(h)}(\tilde{z}_i \mid \boldsymbol{\varrho}, \nu) = \frac{\rho_i(\nu - \rho_i)}{\nu^2(\nu + 1)}, \qquad \text{(all } i\text{)}, \tag{12.6.7}$$

and

$$V_\beta^{(k)}(\tilde{z}_i, \tilde{z}_j \mid \boldsymbol{\varrho}, \nu) = \frac{-\rho_i\rho_j}{\nu^2(\nu + 1)}, \qquad (i \neq j); \tag{12.6.8}$$

details are left to the reader as Exercise 18.

We noted in Section 12.3 that the univariate beta df is useful in representing judgments about an unknown proportion or probability $\tilde{p}$. The k-variate beta df is useful in similar but more involved contexts; namely, representing judgments about the unknown proportions $\tilde{p}_1, \ldots, \tilde{p}_k$ of a population which belong to (mutually exclusive but nonexhaustive) classes $c_1, \ldots, c_k$, where c_{k+1} denotes the remainder class containing proportion $\tilde{p}_{k+1} \equiv 1 - \sum_{i=1}^{k} \tilde{p}_i$ of the population.

12.7 THE k-VARIATE NORMAL DISTRIBUTIONS

We say that the k-dimensional random vector $\tilde{\mathbf{z}}$ has the *unit spherical k-variate Normal distribution* if the components $\tilde{z}_1, \ldots, \tilde{z}_k$ of $\tilde{\mathbf{z}}$ are iid with common df the standardized Normal df $f_{N^*}(\cdot)$; equivalently, if the df of $\tilde{\mathbf{z}}$ is given by

$$f_N{}^{(k)}(\mathbf{z} \mid \mathbf{0}, \mathbf{I}) \equiv (2\pi)^{-\frac{1}{2}k} e^{-\frac{1}{2}\mathbf{z}'\mathbf{z}}, \ \mathbf{z} \in R^k \tag{12.7.1}$$

where $\mathbf{0} \equiv (0, 0. \ldots, 0)' \in R^k$ and $\mathbf{I}$ is the identity matrix of order k.

We say that $\tilde{\mathbf{x}}$ has the (nonsingular) *k-variate Normal distribution* with parameters $\boldsymbol{\mu} \in R^k$ and $\boldsymbol{\sigma}^2$ (a kth order, positive definite, symmetric matrix) if the df of $\tilde{\mathbf{x}}$ is

$$f_N{}^{(k)}(\mathbf{x} \mid \boldsymbol{\mu}, \boldsymbol{\sigma}^2) \equiv (2\pi)^{-\frac{1}{2}k} |\boldsymbol{\sigma}^2|^{-1/2} \exp\left[-\tfrac{1}{2}(\mathbf{x} - \boldsymbol{\mu})'\boldsymbol{\sigma}^{-2}(\mathbf{x} - \boldsymbol{\mu})\right], \tag{12.7.2}$$

for all $\mathbf{x} \in R^k$, where $\boldsymbol{\sigma}^{-2} \equiv (\boldsymbol{\sigma}^2)^{-1}$.

If $\tilde{\mathbf{x}}$ has df $f_N{}^{(k)}(\mathbf{x} \mid \boldsymbol{\mu}, \boldsymbol{\sigma}^2)$, then there is a nonsingular kth order matrix $\mathbf{A}$ such that $\tilde{\mathbf{z}} \equiv \mathbf{A}^{-1}(\tilde{\mathbf{x}} - \boldsymbol{\mu})$ has df $f_N{}^{(k)}(\cdot \mid \mathbf{0}, \mathbf{I})$. (12.7.3)

Proof

A theorem in linear algebra asserts that there exists a nonsingular kth order matrix $\mathbf{A}$ such that $\mathbf{AA}' = \boldsymbol{\sigma}^2$ and therefore $\boldsymbol{\sigma}^{-2} \equiv (\boldsymbol{\sigma}^2)^{-1} = (\mathbf{AA}')^{-1} = (\mathbf{A}')^{-1}\mathbf{A}^{-1} = (\mathbf{A}^{-1})'\mathbf{A}^{-1}$. Hence

$$\begin{aligned}(\mathbf{x} - \boldsymbol{\mu})'\boldsymbol{\sigma}^{-2}(\mathbf{x} - \boldsymbol{\mu}) &= (\mathbf{x} - \boldsymbol{\mu})'(\mathbf{A}^{-1})'\mathbf{A}^{-1}(\mathbf{x} - \boldsymbol{\mu}) \\ &= [\mathbf{A}^{-1}(\mathbf{x} - \boldsymbol{\mu})]'[\mathbf{A}^{-1}(\mathbf{x} - \boldsymbol{\mu})] \\ &\equiv \mathbf{z}'\mathbf{z}.\end{aligned}$$

Furthermore, $|\boldsymbol{\sigma}^{-2}| = |\mathbf{A}^{-1}|^2 = |\mathbf{A}|^{-2}$ and hence $|\mathbf{A}| = |\boldsymbol{\sigma}^{-2}|^{-1/2}$. We change variable in (12.7.1) from $\mathbf{x}$ to $\mathbf{z}$, obtaining

$$f(\mathbf{z}) = (2\pi)^{-\frac{1}{2}k} |\boldsymbol{\sigma}^{-2}|^{\frac{1}{2}} e^{-\frac{1}{2}\mathbf{z}'\mathbf{z}} |J|,$$

where

$$|J| \equiv |\partial(x_1, \ldots, x_k)/\partial(z_1, \ldots, z_k)|.$$

But $|J| = |\mathbf{A}| = |\boldsymbol{\sigma}^{-2}|^{-1/2}$, which cancels $|\boldsymbol{\sigma}^2|^{-1/2}$ to show that $f(\mathbf{z}) = f_N{}^{(k)}(\mathbf{z} \mid \mathbf{0}, \mathbf{I})$, as asserted. QED

Result (12.7.3) shows that any nonsingular k-variate Normal random vector is obtainable as a nonsingular linear function (technically, a nonsingular affine transformation) of the k-variate unit spherical Normal random vector; namely, $\tilde{\mathbf{x}} = \mathbf{A}\tilde{\mathbf{z}} + \boldsymbol{\mu}$.

If $\tilde{\mathbf{x}} = \mathbf{A}\tilde{\mathbf{z}} + \boldsymbol{\mu}$ where $\tilde{\mathbf{z}}$ has the k-variate unit spherical Normal distribution but $\mathbf{A}$ is singular, then at least one component of $\tilde{\mathbf{x}}$ is functionally determined by the others, and $\boldsymbol{\sigma}^2 \equiv \mathbf{AA}'$ is symmetric but only positive semidefinite. In this case we say that $\tilde{\mathbf{x}}$ has a *singular* k-variate Normal distribution with parameters $\boldsymbol{\mu}$ and $\boldsymbol{\sigma}^2$. Singular Normal distributions do not have density functions given by (12.7.2), since $\boldsymbol{\sigma}^{-2}$ does not exist if rank $(\boldsymbol{\sigma}^2) < k$.

We now show that any singular k-variate Normal random vector $\tilde{\mathbf{x}}$ is obtainable as a linear function of a w-variate unit spherical Normal random vector, where $w < k$.

Let $\tilde{\mathbf{x}}$ be a singular k-variate Normal random vector with parameters $\boldsymbol{\mu}$ and $\boldsymbol{\sigma}^2$, where rank $(\boldsymbol{\sigma}^2) \equiv w \in \{1, \ldots, k-1\}$. Then there exists a $k \times w$ matrix $\mathbf{A}^*$ such that $\tilde{\mathbf{x}} = \mathbf{A}^*\tilde{\mathbf{z}}^* + \boldsymbol{\mu}$, where $\tilde{\mathbf{z}}^*$ is the w-variate unit spherical Normal random vector. The definition of $\boldsymbol{\sigma}^2$ is unambiguous: $\boldsymbol{\sigma}^2 = \mathbf{AA}' = \mathbf{A}^*\mathbf{A}^{*\prime}$.

Proof

Suppose that $\tilde{\mathbf{x}} - \boldsymbol{\mu} = \mathbf{A}\tilde{\mathbf{z}}$, where $\mathbf{A}$ is $k \times k$ and $\tilde{\mathbf{z}}$ is k-variate unit spherical Normal. From linear algebra, rank $(\boldsymbol{\sigma}^2) \equiv$ rank $(\mathbf{AA}') = w$ if and only if rank $(\mathbf{A}) = w$. We may suppose, without loss of generality (because components of $\tilde{\mathbf{x}}$, $\tilde{\mathbf{z}}$ and $\boldsymbol{\mu}$ and rows of $\mathbf{A}$ can be renumbered), that the first w rows of $\mathbf{A}$ are linearly independent. Partition $\mathbf{A}$ as

$$\mathbf{A} = \begin{bmatrix} \mathbf{A}_1 \\ \mathbf{A}_2 \end{bmatrix}$$

and $\tilde{\mathbf{x}} - \boldsymbol{\mu}$ as

$$\tilde{\mathbf{x}} - \boldsymbol{\mu} = \begin{bmatrix} \tilde{\mathbf{x}}_1 - \boldsymbol{\mu}_1 \\ \tilde{\mathbf{x}}_2 - \boldsymbol{\mu}_1 \end{bmatrix},$$

where (by assumption) $\mathbf{A}_1$ is $w \times k$ of rank w. Then there exists a $(w - k) \times w$ matrix $\mathbf{D}$ such that

$$\mathbf{A}_2 = \mathbf{DA}_1$$

because the last $w - k$ rows of $\mathbf{A}$ are linearly dependent upon the first w rows. Now, from $\tilde{\mathbf{x}} - \boldsymbol{\mu} = \mathbf{A}\tilde{\mathbf{z}}$ we obtain

$$\begin{bmatrix} \tilde{\mathbf{x}}_1 - \boldsymbol{\mu}_1 \\ \tilde{\mathbf{x}}_2 - \boldsymbol{\mu}_2 \end{bmatrix} = \begin{bmatrix} \mathbf{A}_1 \\ \mathbf{DA}_1 \end{bmatrix} \tilde{\mathbf{z}} = \begin{bmatrix} \mathbf{A}_1\tilde{\mathbf{z}} \\ \mathbf{DA}_1\tilde{\mathbf{z}} \end{bmatrix},$$

from which $\tilde{\mathbf{x}}_2 - \boldsymbol{\mu}_2 = \mathbf{D}\mathbf{A}_1\tilde{\mathbf{z}} = \mathbf{D}(\tilde{\mathbf{x}}_1 - \boldsymbol{\mu}_1)$ follows, showing the functional dependence of $\tilde{\mathbf{x}}_2$ upon $\tilde{\mathbf{x}}_1$, where [because rank $(\mathbf{A}_1) = w$] $\tilde{\mathbf{x}}_1$ is nonsingular w-variate Normal. Hence by (12.7.3) there exists a wth order nonsingular matrix $\mathbf{A}^*_1$ such that $\tilde{\mathbf{x}}_1 - \boldsymbol{\mu}_1 = \mathbf{A}^*_1\tilde{\mathbf{z}}^*$ for $\tilde{\mathbf{z}}^*$ the w-variate unit spherical Normal random vector. But $\tilde{\mathbf{x}}_2 - \boldsymbol{\mu}_2 = \mathbf{D}(\tilde{\mathbf{x}}_1 - \boldsymbol{\mu}_1) = \mathbf{D}\mathbf{A}^*_1\tilde{\mathbf{z}}^*$, and hence by defining $\mathbf{A}^*_2 \equiv \mathbf{D}\mathbf{A}^*_1$ and

$$\mathbf{A}^* \equiv \begin{bmatrix} \mathbf{A}^*_1 \\ \mathbf{A}^*_2 \end{bmatrix},$$

we obtain $\tilde{\mathbf{x}} - \boldsymbol{\mu} = \mathbf{A}^*\tilde{\mathbf{z}}^*$, which completes the proof of the first assertion. To show that $\mathbf{A}\mathbf{A}' = \mathbf{A}^*\mathbf{A}^{*\prime}$, note that $\mathbf{A}_1\mathbf{A}'_1 = \mathbf{A}^*_1\mathbf{A}^{*\prime}_1$ by definition of $\mathbf{A}^*_1$, and hence

$$\begin{aligned}
\mathbf{A}\mathbf{A}' &= \begin{bmatrix} \mathbf{A}_1 \\ \mathbf{A}_2 \end{bmatrix}[\mathbf{A}'_1 \quad \mathbf{A}'_2] \\
&= \begin{bmatrix} \mathbf{A}_1 \\ \mathbf{D}\mathbf{A}_1 \end{bmatrix}[\mathbf{A}'_1 \quad \mathbf{A}'_1\mathbf{D}'] \\
&= \begin{bmatrix} \mathbf{A}_1\mathbf{A}'_1 & \mathbf{A}_1\mathbf{A}'_1\mathbf{D}' \\ \mathbf{D}\mathbf{A}_1\mathbf{A}'_1 & \mathbf{D}\mathbf{A}_1\mathbf{A}'_1\mathbf{D}' \end{bmatrix} \\
&= \begin{bmatrix} \mathbf{A}^*_1\mathbf{A}^{*\prime}_1 & \mathbf{A}^*_1\mathbf{A}^{*\prime}_1\mathbf{D}' \\ \mathbf{D}\mathbf{A}^*_1\mathbf{A}^{*\prime}_1 & \mathbf{D}'\mathbf{A}^*_1\mathbf{A}^{*\prime}_1\mathbf{D}' \end{bmatrix} \\
&= \begin{bmatrix} \mathbf{A}^*_1 \\ \mathbf{D}\mathbf{A}^*_1 \end{bmatrix}[\mathbf{A}^{*\prime}_1 \quad \mathbf{A}^{*\prime}_1\mathbf{D}'] \\
&= \mathbf{A}^*\mathbf{A}^{*\prime},
\end{aligned}$$

and the proof is complete. QED

Result (12.7.4) does not address the case in which rank $(\mathbf{A}) \equiv w = 0$ because it is trivial: if $w = 0$, then $\mathbf{A}$ must be the zero matrix, and $P[\tilde{\mathbf{x}} = \boldsymbol{\mu}] = 1$. The triviality of this case explains why we did not bother to mention in Section 12.4 the possibility of singular univariate Normal distributions, in which $\sigma^2 = 0$ and $P[\tilde{x} = \mu] = 1$.

The main purpose of (12.7.4) is to prepare the way for (12.7.5), which essentially states that all linear functions of Normal random vectors are Normal random vectors.

Let $\tilde{\mathbf{x}}$ have the k-variate Normal distribution with parameters $\boldsymbol{\mu}$ and $\boldsymbol{\sigma}^2$, where rank $(\boldsymbol{\sigma}^2) \equiv w \leqq k$. Let $\mathbf{c} \in R^m$ and $\mathbf{B}$ be an $m \times k$ matrix of rank r. Define $\tilde{\mathbf{y}} \equiv \mathbf{B}\tilde{\mathbf{x}} + c$. Then $\tilde{\mathbf{y}}$ has the m-variate Normal distribution with parameters $\mathbf{B}\boldsymbol{\mu} + \mathbf{c}$ and $\mathbf{B}\boldsymbol{\sigma}^2\mathbf{B}'$. If $m > k$, then $\tilde{\mathbf{y}}$ is singular. If $w = k$, $m \leqq k$, and $r = m$, then $\tilde{\mathbf{y}}$ is nonsingular.

Proof

Define $\mathbf{A}^*$ and $\tilde{\mathbf{z}}^*$ as in (12.7.4), noting that $\mathbf{A}^* = \mathbf{A}$ and $\tilde{\mathbf{z}}^* = \tilde{\mathbf{z}}$ if $w = k$; and write

$$\tilde{\mathbf{y}} = \mathbf{B}\tilde{\mathbf{x}} + \mathbf{c} = \mathbf{B}[\mathbf{A}^*\tilde{\mathbf{z}}^* + \boldsymbol{\mu}] + \mathbf{c} = (\mathbf{B}\mathbf{A}^*)\tilde{\mathbf{z}}^* + (\mathbf{B}\boldsymbol{\mu} + \mathbf{c}),$$

where rank $(\mathbf{BA}^*) \equiv s \leqq \min\{m, w\}$. By the procedures of (12.7.4) and its proof, there is an $m \times s$ matrix $\mathbf{G}$ such that

$$\tilde{\mathbf{y}} = \mathbf{G}\tilde{\mathbf{z}}^{**} + (\mathbf{B}\boldsymbol{\mu} + \mathbf{c}),$$

where $\tilde{\mathbf{z}}^{**}$ is the s-variate unit spherical Normal random vector, and hence $\tilde{\mathbf{y}}$ has the m-variate Normal distribution with parameters $\boldsymbol{\mu}_y \equiv \mathbf{B}\boldsymbol{\mu} + \mathbf{c}$ and $\boldsymbol{\sigma}_y^2 = \mathbf{G}\mathbf{G}'$. But by (12.7.4), $\mathbf{GG}' = (\mathbf{BA}^*)(\mathbf{BA}^*)' = (\mathbf{BA}^*)(\mathbf{A}^{*\prime}\mathbf{B}') = \mathbf{B}(\mathbf{A}^*\mathbf{A}^{*\prime})\mathbf{B}' \equiv \mathbf{B}\boldsymbol{\sigma}^2\mathbf{B}'$, and the first assertion is proved. The second and third are immediate consequences of the elementary result

$$\text{rank}\,(\mathbf{M}_1\mathbf{M}_2) \leqq \min\{\text{rank}\,(\mathbf{M}_1), \text{rank}\,(\mathbf{M}_2)\}$$

in linear algebra. QED

From (12.7.5) we readily obtain the marginal distributions of subvectors of a k-variate Normal random vector.

If $\tilde{\mathbf{x}}$ has the k-variate Normal distribution with parameters $\boldsymbol{\mu}$ and $\boldsymbol{\sigma}^2$, $\tilde{\mathbf{x}}_1$ is a j-dimensional subvector of $\tilde{\mathbf{x}}$, $\boldsymbol{\mu}_1$ is the corresponding subvector of $\boldsymbol{\mu}$, and $\boldsymbol{\sigma}_{11}^2$ is the submatrix of $\boldsymbol{\sigma}^2$ obtained by deleting all rows i and columns i of $\boldsymbol{\sigma}^2$ for which $\tilde{x}_i$ is not in $\tilde{\mathbf{x}}_1$; then $\tilde{\mathbf{x}}_1$ has the j-variate Normal distribution with parameters $\boldsymbol{\mu}_1$ and $\boldsymbol{\sigma}_{11}^2$. (12.7.6)

Proof

In (12.7.5), let $\mathbf{B}$ be the matrix formed from the kth order identity matrix $\mathbf{I}$ by deleting all rows i for which $\tilde{x}_i$ is not a component of $\tilde{\mathbf{x}}_1$. Assertion (12.7.6) now follows by direct computations which are left to the reader as Exercise 19. QED

In (12.7.6) it is clear that nonsingularity of $\tilde{\mathbf{x}}$ is sufficient for nonsingularity of $\tilde{\mathbf{x}}_1$, but it is not necessary. Let $k = 2$, $\tilde{\mathbf{x}} = \mathbf{A}\tilde{\mathbf{z}}$, and

$$\mathbf{A} = \begin{bmatrix} 1 & 0 \\ 1 & 0 \end{bmatrix},$$

so that $\tilde{x}_1 = \tilde{x}_2$; then the marginal distribution of each $\tilde{x}_i$ is (nonsingular, univariate) standardized Normal with df $f_{N*}(\cdot)$.

Result (12.7.6) has the corollary that if $\tilde{\mathbf{x}}$ has df $f_N^{(k)}(\cdot \mid \boldsymbol{\mu}, \boldsymbol{\sigma}^2)$ (and hence is nonsingular), then $\tilde{\mathbf{x}}_1$ has df $f_N^{(j)}(\cdot \mid \boldsymbol{\mu}_1, \boldsymbol{\sigma}_{11}^2)$ (because it too is nonsingular).

We now obtain the moment-generating function of the k-variate Normal random vector $\tilde{\mathbf{x}}$ with parameters $\boldsymbol{\mu}$ and $\boldsymbol{\sigma}^2$, from which it will follow that $E(\tilde{\mathbf{x}}) = \boldsymbol{\mu}$ and $V(\tilde{\mathbf{x}}) = \boldsymbol{\sigma}^2$, showing that our usage of $\boldsymbol{\mu}$ and $\boldsymbol{\sigma}^2$ to denote parameters is consistent with the moment notation of Chapter 10.

If $\tilde{\mathbf{x}}$ has the k-variate Normal distribution with parameters $\boldsymbol{\mu}$ and $\boldsymbol{\sigma}^2$, then the moment-generating function of $\tilde{\mathbf{x}}$ is (12.7.7)

$$M_{N^{(k)}}(\mathbf{u} \mid \boldsymbol{\mu}, \boldsymbol{\sigma}^2) = \exp\,[(\mathbf{u}'\boldsymbol{\mu} + \tfrac{1}{2}\mathbf{u}'\boldsymbol{\sigma}^2\mathbf{u})].$$

Proof

$$\begin{aligned} M_{N^{(k)}}(\mathbf{u} \mid \boldsymbol{\mu}, \boldsymbol{\sigma}^2) &\equiv [E(\exp \mathbf{u}'\tilde{\mathbf{x}})] \\ &= \int_{R^w} \exp\,[\mathbf{u}'(\mathbf{A}\mathbf{z} + \boldsymbol{\mu})](2\pi)^{-\frac{1}{2}w} \exp\,[-\tfrac{1}{2}\mathbf{z}'\mathbf{z}]\,dz, \\ &= \exp\,[\mathbf{u}'\boldsymbol{\mu}] \int_{R^w} (2\pi)^{-\frac{1}{2}w} \exp\,[-\tfrac{1}{2}(\mathbf{z}'\mathbf{z} - 2\mathbf{u}'\mathbf{A}\mathbf{z})]\,dz, \end{aligned}$$

where the second equality follows by (12.7.4); we have dropped asterisks from $\mathbf{A}^*$ and $\mathbf{z}^*$ to simplify notation. The task now is to complete the square in the exponent of the integrand, obtaining (verify as Exercise 20)

$$-\tfrac{1}{2}(\mathbf{z}'\mathbf{z} - 2\mathbf{u}'\mathbf{A}\mathbf{z}) = -\tfrac{1}{2}(\mathbf{z} - \mathbf{A}'\mathbf{u})'(\mathbf{z} - \mathbf{A}'\mathbf{u}) + \tfrac{1}{2}\mathbf{u}'\mathbf{A}\mathbf{A}'\mathbf{u},$$

so that

$$\begin{aligned} &M_{N^{(k)}}(\boldsymbol{\mu} \mid \mathbf{u}, \boldsymbol{\sigma}^2) \\ &\quad= \exp\,[\mathbf{u}'\boldsymbol{\mu}] \int_{R^w} (2\pi)^{-\frac{1}{2}w} \exp\,[-\tfrac{1}{2}(\mathbf{z} - \mathbf{A}'\mathbf{u})'(\mathbf{z} - \mathbf{A}'\mathbf{u}) \exp\,[(\tfrac{1}{2}\mathbf{u}'\mathbf{A}\mathbf{A}'\mathbf{u})]\,dz \\ &\quad= \exp\,[(\mathbf{u}'\boldsymbol{\mu} + \tfrac{1}{2}\mathbf{u}'\mathbf{A}\mathbf{A}'\mathbf{u}] \int_{R^w} f_{N^{(w)}}(\mathbf{z} \mid \mathbf{A}'\mathbf{u}, \mathbf{I})\,dz \\ &\quad= \exp\,[(\mathbf{u}'\boldsymbol{\mu} + \tfrac{1}{2}\mathbf{u}'\mathbf{A}\mathbf{A}'\mathbf{u})] \end{aligned}$$

because the last integral is one. But $\mathbf{A}\mathbf{A}' = \boldsymbol{\sigma}^2$, by (12.7.4). QED

As an easy consequence of (12.7.7), we have

If $\tilde{\mathbf{x}}$ has the k-variate Normal distribution with parameters $\boldsymbol{\mu}$ and $\boldsymbol{\sigma}^2$, then (12.7.8)

$$E(\tilde{\mathbf{x}}) = \boldsymbol{\mu}$$

and

$$V(\tilde{\mathbf{x}}) = \boldsymbol{\sigma}^2.$$

Proof

Exercise 21.

The moment-generating function $M_N^{(k)}(\boldsymbol{\mu} \mid \boldsymbol{\mu}, \boldsymbol{\sigma}^2)$ constitutes the basis of an alternative proof of (12.7.6); the reader may carry it out as Exercise 22.

One more involved argument remains; namely, the derivation of the conditional distribution of $\tilde{\mathbf{x}}_2$ given $\mathbf{x}_1$ when $(\tilde{\mathbf{x}}'_1, \tilde{\mathbf{x}}'_2)' \equiv \tilde{\mathbf{x}}$ has a *nonsingular* k-variate Normal distribution, with df $f_N^{(k)}(\mathbf{x} \mid \boldsymbol{\mu}, \boldsymbol{\sigma}^2)$. We shall suppose that the components of $\tilde{\mathbf{x}}$ have already been relabeled (if necessary, along with the corresponding components of $\boldsymbol{\mu}$ and elements of $\boldsymbol{\sigma}^2$) so that $\tilde{\mathbf{x}}_1 \equiv (\tilde{x}_1, \ldots, \tilde{x}_j)'$ and $\tilde{x}_2 \equiv (\tilde{x}_{j+1}, \ldots, \tilde{x}_k)'$. We partition

$$\boldsymbol{\mu} \equiv (\boldsymbol{\mu}'_1, \boldsymbol{\mu}'_2)'$$

and

$$\boldsymbol{\sigma}^2 \equiv \begin{bmatrix} \boldsymbol{\sigma}_{11}^2 & \boldsymbol{\sigma}_{12}^2 \\ \boldsymbol{\sigma}_{21}^2 & \boldsymbol{\sigma}_{22}^2 \end{bmatrix}$$

conformably with the partition $(\tilde{\mathbf{x}}'_1, \tilde{\mathbf{x}}'_2)'$ of $\tilde{\mathbf{x}}$.

If $\tilde{\mathbf{x}} \equiv (\tilde{\mathbf{x}}'_1, \tilde{\mathbf{x}}'_2)'$ has df $f_N^{(k)}(\cdot \mid \boldsymbol{\mu}, \boldsymbol{\sigma}^2)$, then the conditional distribution of $\tilde{\mathbf{x}}_2$ given that $\tilde{\mathbf{x}}_1 = \mathbf{x}_1$ is nonsingular $(k-j)$-variate Normal with df (12.7.9)

$$f_N^{(k-j)}(\cdot \mid \boldsymbol{\mu}_{2|1}(\mathbf{x}_1), \boldsymbol{\sigma}_{22|1}^2),$$

where

$$\boldsymbol{\mu}_{2|1}(\mathbf{x}_1) \equiv \boldsymbol{\mu}_2 + \boldsymbol{\sigma}_{21}^2(\boldsymbol{\sigma}_{11}^2)^{-1}(\mathbf{x}_1 - \boldsymbol{\mu}_1)$$

and

$$\boldsymbol{\sigma}_{22|1}^2 \equiv \boldsymbol{\sigma}_{22}^2 - \boldsymbol{\sigma}_{21}^2(\boldsymbol{\sigma}_{11}^2)^{-1}\boldsymbol{\sigma}_{12}^2.$$

Proof

The proof requires a great deal of linear algebra, and hence we list some facts before turning to the proof proper. Let $\mathbf{H} \equiv \boldsymbol{\sigma}^{-2} \equiv (\boldsymbol{\sigma}^2)^{-1}$, which exists due to the nonsingularity assumption. Partition $\mathbf{H}$ into

$$\mathbf{H} = \begin{bmatrix} \mathbf{H}_{11} & \mathbf{H}_{12} \\ \mathbf{H}_{21} & \mathbf{H}_{22} \end{bmatrix}$$

conformably with the partition of $\boldsymbol{\sigma}^2$; then

(a) $(\boldsymbol{\sigma}_{11}^2)^{-1} = \mathbf{H}_{11} - \mathbf{H}_{12}\mathbf{H}_{22}^{-1}\mathbf{H}_{21}$,
(b) $\mathbf{H}_{22} = (\boldsymbol{\sigma}_{22}^2 - \boldsymbol{\sigma}_{21}^2(\boldsymbol{\sigma}_{11}^2)^{-1}\boldsymbol{\sigma}_{12}^2)^{-1}$,

and

(c) $\mathbf{H}_{22}^{-1}\mathbf{H}_{21} = -\boldsymbol{\sigma}_{21}^2(\boldsymbol{\sigma}_{11}^2)^{-1}$.

[The reader may verify (a) through (c) as Exercise 23.] Now to the proof itself: Let $\tilde{\mathbf{y}} \equiv \tilde{\mathbf{x}} - \boldsymbol{\mu}$, and partition $\tilde{\mathbf{y}} \equiv (\tilde{\mathbf{y}}'_1, \tilde{\mathbf{y}}'_2)'$ conformably with the partition of $\tilde{\mathbf{x}}$. Then the exponential term of $f_N^{(k)}(\mathbf{x} \mid \boldsymbol{\mu}, \boldsymbol{\sigma}^2)$ is $\exp(-\frac{1}{2}Q)$, where

$$
\begin{aligned}
Q &= (\mathbf{x} - \boldsymbol{\mu})'\boldsymbol{\sigma}^{-2}(\mathbf{x} - \boldsymbol{\mu}) \\
&= (\mathbf{y}'_1, \mathbf{y}'_2)\begin{bmatrix} \mathbf{H}_{11} & \mathbf{H}_{12} \\ \mathbf{H}_{21} & \mathbf{H}_{22} \end{bmatrix}\begin{pmatrix} \mathbf{y}_1 \\ \mathbf{y}_2 \end{pmatrix} \\
(*) \qquad &= \mathbf{y}'_1\mathbf{H}_{11}\mathbf{y}_1 + \mathbf{y}'_1\mathbf{H}_{12}\mathbf{y}_2 + \mathbf{y}'_2\mathbf{H}_{21}\mathbf{y}_1 + \mathbf{y}'_2\mathbf{H}_{22}\mathbf{y}_2.
\end{aligned}
$$

Now, the exponential term of $f_N^{(j)}(\mathbf{x}_1 \mid \boldsymbol{\mu}_1, \boldsymbol{\sigma}_{11}^2)$ is $\exp(-\frac{1}{2}Q_1,)$ where

$$
\begin{aligned}
Q_1 &= (\mathbf{x}_1 - \boldsymbol{\mu}_1)'(\boldsymbol{\sigma}_{11}^2)^{-1}(\mathbf{x}_1 - \boldsymbol{\mu}_1) \\
(**) \qquad &= \mathbf{y}'_1(\mathbf{H}_{11} - \mathbf{H}_{12}\mathbf{H}_{22}^{-1}\mathbf{H}_{21})\mathbf{y}_1,
\end{aligned}
$$

and hence

$$
\begin{aligned}
f(\mathbf{x}_2 \mid \mathbf{x}_1) &\equiv \frac{f_N^{(k)}(\mathbf{x} \mid \boldsymbol{\mu}, \boldsymbol{\sigma}^2)}{f_N^{(j)}(\mathbf{x}_1 \mid \boldsymbol{\mu}_1, \boldsymbol{\sigma}_{11}^2)} \\
&= K \exp[-\tfrac{1}{2}(Q - Q_1)
\end{aligned}
$$

for some positive K which does not depend upon $\mathbf{y}$. The rest of the proof amounts to showing that $Q - Q_1 \equiv Q_{2|1}$, where $\exp(-\frac{1}{2}Q_{2|1})$ is the exponential term in $f_N^{(k-j)}(\mathbf{x}_2 \mid \boldsymbol{\mu}_{2|1}(\mathbf{x}_1), \boldsymbol{\sigma}_{22|1}^2)$. From (*) and (**),

$$
Q - Q_1 = \mathbf{y}'_1(\mathbf{H}_{12}\mathbf{H}_{22}^{-1}\mathbf{H}_{21})\mathbf{y}_1 + \mathbf{y}'_1\mathbf{H}_{12}\mathbf{y}_2 + \mathbf{y}'_2\mathbf{H}_{21}\mathbf{y}_1 + \mathbf{y}'_2\mathbf{H}_{22}\mathbf{y}_2.
$$

But $\mathbf{H}_{22} = \mathbf{H}'_{22}$, $\mathbf{H}_{22}^{-1} = \mathbf{H}'^{-1}_{22} = \mathbf{H}^{-1}{}_{22}'$, and $\mathbf{H}_{12} = \mathbf{H}'_{21}$, by symmetry of $\boldsymbol{\sigma}^2$ and thus of $\mathbf{H}$, and $\mathbf{H}_{22}^{-1}\mathbf{H}_{22} = I$, so that

$$
\begin{aligned}
Q - Q_1 &= \mathbf{y}'_1(\mathbf{H}'_{21}\mathbf{H}'^{-1}_{22}\mathbf{H}'_{22}\mathbf{H}'^{-1}_{22}\mathbf{H}_{21})\mathbf{y}_1 + \mathbf{y}'_1(\mathbf{H}'_{21}\mathbf{H}'^{-1}_{22}\mathbf{H}_{22})\mathbf{y}_2 \\
&\quad + \mathbf{y}'_2\mathbf{H}_{22}\mathbf{H}_{22}^{-1}\mathbf{H}_{21}\mathbf{y}_1 + \mathbf{y}'_2\mathbf{H}_{22}\mathbf{y}_2 \\
(***) \qquad &= (\mathbf{y}_2 + \mathbf{H}_{22}^{-1}\mathbf{H}_{21}\mathbf{y}_1)'\mathbf{H}_{22}(\mathbf{y}_2 + \mathbf{H}_{22}^{-1}\mathbf{H}_{21}\mathbf{y}_1).
\end{aligned}
$$

The reader may verify these equalities as Exercise 24. Now, (***) shows that the conditional df of $\tilde{\mathbf{y}}_2$ given $\mathbf{y}_1$ is $f_N^{(k-j)}(\cdot \mid -\mathbf{H}_{22}^{-1}\mathbf{H}_{21}\mathbf{y}_1, \mathbf{H}_{22}^{-1})$, so the conditional df of $\tilde{\mathbf{x}}_2$ given $\mathbf{x}_1$ is

$$
f_N^{(k-j)}(\cdot \mid \boldsymbol{\mu}_2 - \mathbf{H}_{22}^{-1}\mathbf{H}_{21}(\mathbf{x}_1 - \boldsymbol{\mu}_1), \mathbf{H}_{22}^{-1}).
$$

The desired result now following upon applying algebraic results (b) and (c). QED

Two important conclusions follow at once from (12.7.9). First, the conditional covariance matrix $\boldsymbol{\sigma}_{22|1}^2$ of $\tilde{\mathbf{x}}_2$ given $\mathbf{x}_1$ does not depend upon $\mathbf{x}_1$, only upon $\boldsymbol{\sigma}_2$. This fact is of major importance in applications, especially in Chapters 14, 18, and 19. And second, the vector $\boldsymbol{\mu}_{2|1}(\mathbf{x}_1)$ of conditional expectations is a linear function (an affine transformation) of $\mathbf{x}_1$.

The next result is a direct generalization of (12.4.10).

If $\tilde{\mathbf{x}}_1, \ldots, \tilde{\mathbf{x}}_n$ are independent, $\tilde{\mathbf{x}}_i$ has df $f_N^{(k)}(\cdot \mid \boldsymbol{\mu}_i, \boldsymbol{\sigma}_{ii}^2)$, for $i = 1, \ldots, n$, $\mathbf{c} \in R^k$, and $\mathbf{b} = (b_1, \ldots, b_n)' \in R^n$ with $\mathbf{b} \neq \mathbf{0}$, then (12.7.10)

(a) the df of $\tilde{\mathbf{y}} \equiv \mathbf{c} + \sum_{i=1}^{n} b_i \tilde{\mathbf{x}}_i$ is

$$f_N^{(k)}(\cdot \mid \mathbf{c} + \sum_{i=1}^{n} b_i \boldsymbol{\mu}_i, \sum_{i=1}^{n} b_i^2 \boldsymbol{\sigma}_{ii}^2);$$

and if $\tilde{\mathbf{x}}_1, \ldots, \tilde{\mathbf{x}}_n$ are iid with common df $f_N^{(k)}(\cdot \mid \boldsymbol{\mu}, \boldsymbol{\sigma}^2)$, then

(b) the df of $\tilde{\mathbf{y}} \equiv \sum_{i=1}^{n} \tilde{\mathbf{x}}_i$ is $f_N^{(k)}(\cdot \mid n\boldsymbol{\mu}, n\boldsymbol{\sigma}^2)$, and

(c) the df of $\bar{\tilde{\mathbf{x}}} \equiv (1/n) \sum_{i=1}^{n} \tilde{\mathbf{x}}_i$ is $f_N^{(k)}(\cdot \mid \boldsymbol{\mu}, (1/n)\boldsymbol{\sigma}^2)$.

Proof

Exercise 25.

Result (12.7.10) also obtains when the distributions of the $\tilde{\mathbf{x}}_i$'s may be singular k-variate Normal; in this case, (a) becomes "the distribution of $\tilde{\mathbf{y}} \equiv \mathbf{c} + \sum_{i=1}^{n} b_i \tilde{\mathbf{x}}_i$ is k-variate Normal with parameters $\mathbf{c} + \sum_{i=1}^{n} b_i \boldsymbol{\mu}_i$ and $\sum_{i=1}^{n} b_i^2 \boldsymbol{\sigma}_{ii}^2$."

There is a common generalization of both (12.7.10) and (12.7.5), which gives the joint distribution of several linear functions of several k-variate Normal random vectors.

Let $\tilde{\mathbf{x}}$ have the kn-variate Normal distribution with parameters $\boldsymbol{\mu}$ and $\boldsymbol{\sigma}^2$, and partition $\tilde{\mathbf{x}} \equiv (\tilde{\mathbf{x}}'_1, \ldots, \tilde{\mathbf{x}}'_n)'$, where $\tilde{\mathbf{x}}'_i \equiv (\tilde{x}_{k(i-1)+1}, \ldots, \tilde{x}_{ki})$. Partition $\boldsymbol{\mu} \equiv (\boldsymbol{\mu}'_1, \ldots, \boldsymbol{\mu}'_n)'$ and (12.7.11)

$$\boldsymbol{\sigma}^2 = \begin{bmatrix} \boldsymbol{\sigma}_{11}^2 & \cdots & \boldsymbol{\sigma}_{1n}^2 \\ \cdot & & \cdot \\ \cdot & & \cdot \\ \cdot & & \cdot \\ \boldsymbol{\sigma}_{n1}^2 & \cdots & \boldsymbol{\sigma}_{nn}^2 \end{bmatrix}$$

conformably with the partition of $\tilde{\mathbf{x}}$. Let $\mathbf{B}_{ji}$ denote an $r \times k$ matrix and $\mathbf{c}_j$ a point in R^r, for $i = 1, \ldots, n$ and $j = 1, \ldots, m$. Define $\tilde{\mathbf{y}}_j \equiv \mathbf{c}_j + \sum_{i=1}^{n} \mathbf{B}_{ji} \tilde{\mathbf{x}}_i$, for $j = 1, \ldots, m$, and $\tilde{\mathbf{y}} \equiv (\tilde{\mathbf{y}}'_1, \ldots, \tilde{\mathbf{y}}'_m)'$. Then the distribution of $\tilde{\mathbf{y}}$ is rm-variate Normal, with

$$E(\tilde{\mathbf{y}}_j) = \mathbf{c}_j + \sum_{i=1}^{n} \mathbf{B}_{ji}\boldsymbol{\mu}_i$$

and

$$V(\tilde{\mathbf{y}}_j, \tilde{\mathbf{y}}_h) = \sum_{g=1}^{n}\sum_{i=1}^{n} \mathbf{B}_{ji}\boldsymbol{\sigma}_{ig}{}^2\mathbf{B}'_{hg}.$$

Proof

Although (12.7.11) is a generalization of (12.7.5), its proof is based squarely upon (12.7.5). Define $\mathbf{c} \equiv (\mathbf{c}'_1, \ldots, \mathbf{c}'_m)'$ and

$$\mathbf{B} \equiv \begin{bmatrix} \mathbf{B}_{11} & \cdots & \mathbf{B}_{1n} \\ \cdot & & \cdot \\ \cdot & & \cdot \\ \cdot & & \cdot \\ \mathbf{B}_{m1} & \cdots & \mathbf{B}_{mn} \end{bmatrix};$$

it is clear from the definitions of $\tilde{\mathbf{y}}_j$ and $\tilde{\mathbf{y}}$ that

$$\tilde{\mathbf{y}} = \mathbf{B}\tilde{\mathbf{x}} + \mathbf{c}.$$

Hence by (12.7.5) the distribution of $\tilde{\mathbf{y}}$ is rm-variate Normal with parameters $\mathbf{B}\boldsymbol{\mu} + \mathbf{c}$ and $\mathbf{B}\boldsymbol{\sigma}^2\mathbf{B}'$. By appropriate partitioning, we obtain

$$E(\tilde{\mathbf{y}}) = \begin{bmatrix} \mathbf{B}_{11} & \cdots & \mathbf{B}_{1n} \\ \cdot & & \cdot \\ \cdot & & \cdot \\ \cdot & & \cdot \\ \mathbf{B}_{m1} & \cdots & \mathbf{B}_{mn} \end{bmatrix} \begin{pmatrix} \boldsymbol{\mu}_1 \\ \cdot \\ \cdot \\ \cdot \\ \boldsymbol{\mu}_n \end{pmatrix} + \begin{pmatrix} \mathbf{c}_1 \\ \cdot \\ \cdot \\ \cdot \\ \mathbf{c}_m \end{pmatrix}$$

$$= \begin{bmatrix} \mathbf{c}_1 + \sum_{i=1}^{n} \mathbf{B}_{1i}\boldsymbol{\mu}_i \\ \cdot \\ \cdot \\ \cdot \\ \mathbf{c}_m + \sum_{i=1}^{n} \mathbf{B}_{mi}\boldsymbol{\mu}_i \end{bmatrix}$$

and

$$\mathbf{B}\boldsymbol{\sigma}^2\mathbf{B}' = \begin{bmatrix} \mathbf{B}_{11} & \cdots & \mathbf{B}_{1n} \\ \cdot & & \\ \cdot & & \\ \cdot & & \\ \mathbf{B}_{m1} & \cdots & \mathbf{B}_{mn} \end{bmatrix} \begin{bmatrix} \boldsymbol{\sigma}_{11}{}^2 & \cdots & \boldsymbol{\sigma}_{1n}{}^2 \\ \cdot & & \\ \cdot & & \\ \cdot & & \\ \boldsymbol{\sigma}_{n1}{}^2 & \cdots & \boldsymbol{\sigma}_{nn}{}^2 \end{bmatrix} \begin{bmatrix} \mathbf{B}'_{11} & \cdots & \mathbf{B}'_{m1} \\ \cdot & & \\ \cdot & & \\ \cdot & & \\ \mathbf{B}'_{1n} & \cdots & \mathbf{B}'_{mn} \end{bmatrix},$$

the (j, h)th submatrix of which is given by

$$\sum_{g=1}^{n}\left(\sum_{i=1}^{n}\mathbf{B}_{ji}\boldsymbol{\sigma}_{ig}^{2}\right)\mathbf{B}'_{hg} = \sum_{g=1}^{n}\sum_{i=1}^{n}\mathbf{B}_{ji}\boldsymbol{\sigma}_{ig}^{2}\mathbf{B}'_{hg},$$

as asserted. QED

When $m = k = r = 1$ in (12.7.11), we obtain the generalization of [12.4.10(a)] for the case in which $\tilde{x}_1, \ldots, \tilde{x}_n$ are jointly Normally distributed but not necessarily independent. Notice also that (12.7.11) imposes no restrictions of nonsingularity.

Our final general result is a weak converse of (12.7.6) that will prove useful in Chapter 13.

If $\tilde{\mathbf{x}}$ is a k-variate random vector with expectation vector $\boldsymbol{\mu}$ and covariance matrix $\boldsymbol{\sigma}^2$, and if for every $\mathbf{d} \in R^k$ the distribution of $\mathbf{d}'\tilde{\mathbf{x}}$ is univariate Normal with parameters $\mathbf{d}'\boldsymbol{\mu}$ and $\mathbf{d}'\boldsymbol{\sigma}^2 d$, then $\tilde{\mathbf{x}}$ has the k-variate Normal distribution with parameters $\boldsymbol{\mu}$ and $\boldsymbol{\sigma}^2$. (12.7.12)

Proof

For any random vector $\tilde{\mathbf{x}}$, the moment-generating function $M^{(1)}(u \mid \mathbf{d})$ of $\mathbf{d}'\tilde{\mathbf{x}}$ is $E(e^{u[\mathbf{d}'\mathbf{x}]})$, and hence is related to the moment-generating function $M^{(k)}(\mathbf{u})$ of $\tilde{\mathbf{x}}$ by

(*) $$M^{(1)}(u \mid \mathbf{d}) = M^{(k)}(u\mathbf{d}),$$

for those $u \in R^1$ and $\mathbf{d} \in R^k$ such that one (and hence each) side of (*) exists. By assumption, $\mathbf{d}'\tilde{\mathbf{x}}$ here has the Normal distribution with moment-generating function $M_N(u \mid \mathbf{d}'\boldsymbol{\mu}, \mathbf{d}'\boldsymbol{\sigma}^2\mathbf{d})$, for every $\mathbf{d} \in R^k$, and hence also for every $u \in R^1$. Let $\mathbf{u} \equiv u\mathbf{d}$. Then

$$\begin{aligned} M_N(u \mid \mathbf{d}'\boldsymbol{\mu}, \mathbf{d}'\boldsymbol{\sigma}^2\mathbf{d}) & \\ &= \exp[(u\mathbf{d}'\boldsymbol{\mu} + \tfrac{1}{2}u^2\mathbf{d}'\boldsymbol{\sigma}^2\mathbf{d})] && \text{[by (12.7.7)]} \\ &\equiv \exp[(\mathbf{u}'\boldsymbol{\mu} + \tfrac{1}{2}\mathbf{u}'\boldsymbol{\sigma}^2\mathbf{u})] && [\mathbf{u} \equiv u\mathbf{d}] \\ &= M_N^{(k)}(\mathbf{u} \mid \boldsymbol{\mu}, \boldsymbol{\sigma}^2) && \text{[by (*)].} \end{aligned}$$

Hence the distribution of $\tilde{\mathbf{x}}$ must be k-variate Normal with parameters $\boldsymbol{\mu}$ and $\boldsymbol{\sigma}^2$. QED

It is instructive to consider the geometry of the nonsingular k-variate Normal distributions. For this purpose, we note from (12.7.2) that the density function is constant on the level sets

$$\lambda(c) \equiv \{\mathbf{x}: \mathbf{x} \in R^k, (\mathbf{x} - \boldsymbol{\mu})'\boldsymbol{\sigma}^{-2}(\mathbf{x} - \boldsymbol{\mu}) = c\}, \tag{12.7.13}$$

which are nonempty for every $c \geqq 0$ and consist of infinitely many points for $c > 0$. Since $\boldsymbol{\sigma}^2$ is positive definite and symmetric, so too is $\boldsymbol{\sigma}^{-2}$, and hence $\lambda(c)$ is a [$(k - 1)$-dimensional] ellipsoidal surface in R^k.

Geometric intuition is best fostered by the case $k = 2$, for which graphs are possible. The graph of $f_N^{(2)}(\cdot \mid \boldsymbol{\mu}, \boldsymbol{\sigma}^2)$ really does suggest a "bell" as indicated in Figure 12.9 for $\boldsymbol{\mu} = (3, 3)'$ and

$$\boldsymbol{\sigma}^2 = \begin{bmatrix} 1 & \frac{1}{2} \\ \frac{1}{2} & 1 \end{bmatrix}.$$

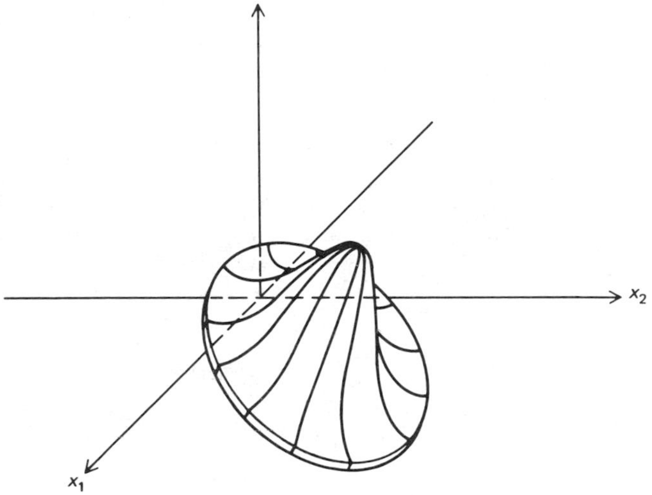

Figure 12.9

The level curves for the nonsingular bivariate Normal distribution are confocal ellipses with center at $(\mu_1, \mu_2)'$. The major axis of these ellipses has positive slope if $\sigma_{12}^2 > 0$ and negative slope if $\sigma_{12}^2 < 0$. If $\sigma_{12}^2 = 0$, then the ellipses are circles. Typical cases are depicted in Figure 12.10.

It is interesting to specialize (12.7.9) to the case $k = 2$, for which it becomes

If $\tilde{\mathbf{x}}$ has df $f_N^{(2)}(\cdot \mid \boldsymbol{\mu}, \boldsymbol{\sigma}^2)$, then the conditional df of $\tilde{x}_2$ given $\tilde{x}_1 = x_1$ is

$$f_N(\cdot \mid \mu_2 + (\sigma_{21}^2/\sigma_{11}^2)(x_1 - \mu_1), \sigma_{22}^2 - \sigma_{22}^2\sigma_{12}^2/\sigma_{11}^2)$$
$$= f_N(\cdot \mid \mu_2 + \rho_{12}(\sigma_{22}/\sigma_{11})(x_1 - \mu_1), \sigma_{22}^2(1 - \rho_{12}^2)),$$

where $\sigma_{ii} \equiv \sqrt{\sigma_{ii}^2}$ for $i = 1, 2$ and ρ_{12} denotes the correlation coefficient $\rho(\tilde{x}_1, \tilde{x}_2)$. (12.7.9′)

Proof

Exercise 26.

The most important insight furnished by (12.7.9′) is the fact that $\sigma_{22|1}^2 = \sigma_{22}^2(1 - \rho_{12}^2)$. which implies that ρ_{12}^2 represents the proportionate

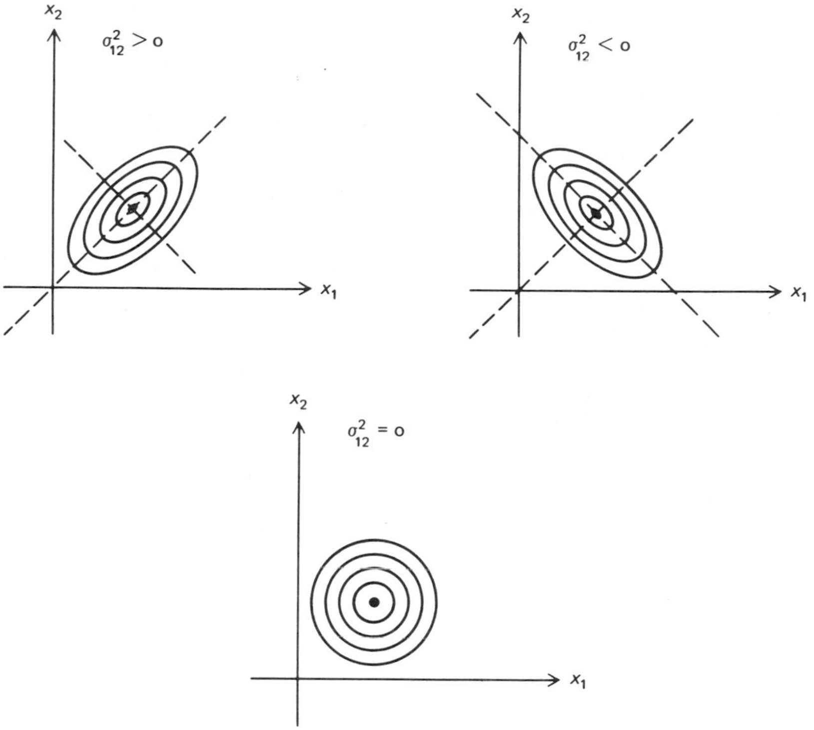

Figure 12.10

reduction in $\sigma_{22}{}^2$ caused by learning the value x_1 of $\tilde{x}_1$. Exercises 47 and 48 develop some deeper aspects of correlation theory for both Normal and more general distributions.

In the remainder of this book we shall adopt a few simple phraseological conventions to distinguish between singular and nonsingular Normal distributions. We shall mean

"$\tilde{\mathbf{x}}$ has df $f_N^{(k)}(\cdot \mid \boldsymbol{\mu}, \boldsymbol{\sigma}^2)$,"
"$\tilde{\mathbf{x}}$ has a k-variate Normal df (or density),"

or similar references to a density, as *implying* that $\boldsymbol{\sigma}^2$ is positive definite and hence that the distribution of $\tilde{\mathbf{x}}$ is *non*singular. Singular normal distributions rarely occur in subsequent chapters and will be clearly specified as such when they do.

12.8 THE k-VARIATE STUDENT DISTRIBUTION

We say that the k-dimensional random vector $\tilde{\mathbf{x}}$ has the *k-variate Student distribution with parameters* $\boldsymbol{\mu} \in R^k$, $\boldsymbol{\sigma}^2$ a positive semidefinite symmetric matrix, and $\nu > 0$ if:

(1) there is a random variable $\tilde{h}$ with df $f_{\gamma 2}(h \mid 1, \nu)$;

and

(2) conditional upon $\tilde{h}$, $\tilde{\mathbf{x}}$ has the k-variate Normal distribution with parameters $\mathbf{u}$ and $(1/h)\boldsymbol{\sigma}^2$.

If (1) and (2) obtain, the marginal distribution of $\tilde{\mathbf{x}}$ is by definition k-variate Student with parameters $\boldsymbol{\mu}$, $\boldsymbol{\sigma}^2$, and ν. If in addition the conditional distribution of $\tilde{\mathbf{x}}$ given $\tilde{h}$ is *nonsingular* Normal (that is, $\boldsymbol{\sigma}^2$ is positive definite), then the marginal distribution of $\tilde{\mathbf{x}}$ is nonsingular k-variate Student and has a density function.

If $\tilde{\mathbf{x}}$ has the nonsingular k-variate Student distribution with parameters $\boldsymbol{\mu}$, $\boldsymbol{\sigma}^2$, and ν, then $\boldsymbol{\sigma}^2$ is positive definite and the df of $\tilde{\mathbf{x}}$ is given by (12.8.1)

$$f_{S^{(k)}}(\mathbf{x} \mid \boldsymbol{\mu}, \boldsymbol{\sigma}^2, \nu) = \frac{S_k(\nu)|\boldsymbol{\sigma}^2|^{-1/2}}{[\nu + (\mathbf{x} - \boldsymbol{\mu})'\boldsymbol{\sigma}^{-2}(\mathbf{x} - \boldsymbol{\mu})]^{\frac{1}{2}(\nu+k)}},$$

for every $\mathbf{x} \in R^k$, where

$$S_k(\nu) \equiv \frac{(\frac{1}{2}\nu)^{\frac{1}{2}\nu}\Gamma(\frac{1}{2}\nu + \frac{1}{2}k)}{[(2\pi)^{\frac{1}{2}k}\Gamma(\frac{1}{2}\nu)]}.$$

Proof

Exercise 27; imitate the proof of (12.5.1).

The k-variate Student df $f_{S^{(k)}}(\mathbf{x} \mid \boldsymbol{\mu}, \boldsymbol{\sigma}^2, \nu)$ clearly generalizes (12.5.1). Moreover, we can show that

$$\lim_{\nu \to \infty} f_{S^{(k)}}(\mathbf{x} \mid \boldsymbol{\mu}, \boldsymbol{\sigma}^2, \nu) = f_{N^{(k)}}(\mathbf{x} \mid \boldsymbol{\mu}, \boldsymbol{\sigma}^2), \quad \text{all } \mathbf{x} \in R^k. \tag{12.8.2}$$

There are two arguments establishing (12.8.2). The first is purely analytic and based upon the formula in (12.8.1). The second is more intuitive (but rigorous) and based upon the fact that, in the limit as $\nu \to \infty$, $f_{\gamma 2}(\cdot \mid 1, \nu)$ concentrates probability one on the event $\tilde{h} = 1$. Hence, interchanging passage to the limit and integration yields the result, which can be stated in somewhat greater generality; namely, the (singular or nonsingular) k-variate Student distribution with parameters $\boldsymbol{\mu}$, $\boldsymbol{\sigma}^2$, and $\nu = \infty$ is the k-variate Normal distribution with parameters $\boldsymbol{\mu}$ and $\boldsymbol{\sigma}^2$.

The moment-generating function of the k-variate Student distribution does not exist, but it is not needed. Most of the basic results about k-variate Normal distributions have almost literal translations for k-variate Student distributions, because by (10.2.8) we are justified in interchanging orders of $\tilde{\mathbf{x}}$- and $\tilde{h}$-expectation so as to obtain the following:

If $\tilde{\mathbf{x}}$ has the k-variate Student distribution with parameters $\boldsymbol{\mu}$, $\boldsymbol{\sigma}^2$, and ν, and if $E_{S^{(k)}}[\mathbf{g}(\tilde{\mathbf{x}}) \mid \boldsymbol{\mu}, \boldsymbol{\sigma}^2, \nu]$ exists for a function $\mathbf{g}$: $R^k \to R^m$, then (12.8.3)

$$E_{S^{(k)}}[\mathbf{g}(\tilde{x}) \mid \boldsymbol{\mu}, \boldsymbol{\sigma}^2, \nu] = E_{\gamma 2}\left\{E_{N^{(k)}}\left[\mathbf{g}(\tilde{\mathbf{x}}) \mid \boldsymbol{\mu}, \left(\frac{1}{\tilde{h}}\right)\boldsymbol{\sigma}^2\right] \mid 1, \nu\right\}.$$

With (12.8.3) in hand, we may now proceed briskly with the Student analogues of useful results in 12.7.

Let $\tilde{\mathbf{x}}$ have the k-variate Student distribution with parameters $\boldsymbol{\mu}$, $\boldsymbol{\sigma}^2$, and ν, where rank $(\boldsymbol{\sigma}^2) \equiv w \leqq k$). Let $\mathbf{c} \in R^m$ and $\mathbf{B}$ be an $m \times k$ matrix of rank r. Define $\tilde{\mathbf{y}} \equiv \mathbf{B}\tilde{\mathbf{x}} + \mathbf{c}$. Then $\tilde{\mathbf{y}}$ has the m-variate Student distribution with parameters $\mathbf{B}\boldsymbol{\mu} + \mathbf{c}$, $\mathbf{B}\boldsymbol{\sigma}^2\mathbf{B}'$, and ν. If $m > k$, then $\tilde{\mathbf{y}}$ is singular. If $w = k$, $m \leqq k$, and $r = m$, then $\tilde{\mathbf{y}}$ is nonsingular, with df $f_{S^{(k)}}(\cdot \mid \mathbf{B}\boldsymbol{\mu} + \mathbf{c}, \mathbf{B}\boldsymbol{\sigma}^2\mathbf{B}', \nu)$. (12.8.4)

Proof

Exercise 28; compare (12.7.5).

If $\tilde{\mathbf{x}}$ has the k-variate Student distribution with parameters $\boldsymbol{\mu}$, $\boldsymbol{\sigma}^2$, and ν, $\tilde{\mathbf{x}}_1$ is a j-dimensional subvector of $\tilde{\mathbf{x}}$, $\boldsymbol{\mu}_1$ is the corresponding subvector of $\boldsymbol{\mu}$, and $\boldsymbol{\sigma}_{11}{}^2$ is the submatrix of $\boldsymbol{\sigma}^2$ obtained by deleting all rows i and columns i of $\boldsymbol{\sigma}^2$ for which $\tilde{\mathbf{x}}_i$ is not in $\tilde{\mathbf{x}}_1$; then $\tilde{\mathbf{x}}_1$ has the j-variate Student distribution with parameters $\boldsymbol{\mu}_1$, $\boldsymbol{\sigma}_{11}{}^2$, and ν. (12.8.5)

Proof

Exercise 29; compare (12.7.6).

If $\tilde{\mathbf{x}} \equiv (\tilde{\mathbf{x}}'_1, \tilde{\mathbf{x}}'_2)'$ has df $f_{S^{(k)}}(\cdot \mid \boldsymbol{\mu}, \boldsymbol{\sigma}^2, \nu)$, then the conditional distribution of $\tilde{\mathbf{x}}_2$ given that $\tilde{\mathbf{x}}_1 = \mathbf{x}_1$ is nonsingular $(k - j)$-variate Student, with df (12.8.6)

$$f_{S^{(k-j)}}(\cdot \mid \boldsymbol{\mu}_{2|1}(\mathbf{x}_1), \boldsymbol{\sigma}_{22}{}^2{}_{|1}, \nu),$$

where

$$\boldsymbol{\mu}_{2|1}(\mathbf{x}_1) \equiv \boldsymbol{\mu}_2 + \boldsymbol{\sigma}_{21}{}^2(\boldsymbol{\sigma}_{11}{}^2)^{-1}(\mathbf{x}_1 - \boldsymbol{\mu}_1)$$

and

$$\boldsymbol{\sigma}_{22}{}^2{}_{|1} \equiv \boldsymbol{\sigma}_{22}{}^2 - \boldsymbol{\sigma}_{21}{}^2(\boldsymbol{\sigma}_{11}{}^2)^{-1}\boldsymbol{\sigma}_{12}{}^2.$$

Proof

Exercise 30; compare (12.7.9).

If $\tilde{\mathbf{x}}_1, \ldots, \tilde{\mathbf{x}}_n$ are conditionally independent given $\tilde{h}$, $\tilde{h}$ has df $f_{\gamma 2}(\cdot \mid 1, \nu)$, $\tilde{\mathbf{x}}_i$ has df $f_{N^{(k)}}(\cdot \mid \boldsymbol{\mu}_i, (1/h)\boldsymbol{\sigma}_{ii}{}^2)$, for $i = 1, \ldots, n$, $\mathbf{c} \in R^k$, and $\mathbf{b} = (b_1, \ldots, b_n)' \in R^n$ with $\mathbf{b} \neq \mathbf{0}$, then (12.8.7)

(a) the marginal (upon $\tilde{h}$) df of $\tilde{\mathbf{y}} \equiv \mathbf{c} + \sum_{i=1}^{n} b_i \tilde{\mathbf{x}}_i$ is

$$f_S^{(k)}(\cdot \mid \mathbf{c} + \sum_{i=1}^{n} b_i \boldsymbol{\mu}_i, \sum_{i=1}^{n} b_i^2 \boldsymbol{\sigma}_{ii}^2, \nu);$$

and if $\tilde{\mathbf{x}}_1, \ldots, \tilde{\mathbf{x}}_n$ are conditionally (upon $\tilde{h}$) iid with common df

$$f_N^{(k)}(\cdot \mid \boldsymbol{\mu}, (1/h)\boldsymbol{\sigma}^2),$$

then

(b) the marginal (upon $\tilde{h}$) df of $\tilde{\mathbf{y}} \equiv \sum_{i=1}^{n} \tilde{\mathbf{x}}_i$ is

$$f_S^{(k)}(\cdot \mid n\boldsymbol{\mu}, n\boldsymbol{\sigma}^2, \nu),$$

and

(c) the marginal (upon $\tilde{h}$) df of $\bar{\mathbf{x}} \equiv (1/n) \sum_{i=1}^{n} \tilde{\mathbf{x}}_i$ is

$$f_S^{(k)}(\cdot \mid \boldsymbol{\mu}, (1/n)\boldsymbol{\sigma}^2, \nu).$$

Proof

Exercise 31; compare (12.7.10).

The most general result along these lines is the Student analogue of (12.7.11):

Let $\tilde{\mathbf{x}}$ have the kn-variate Student distribution with parameters $\boldsymbol{\mu}$, $\boldsymbol{\sigma}^2$, and ν, and partition $\tilde{\mathbf{x}} \equiv (\tilde{\mathbf{x}}'_1, \ldots, \tilde{\mathbf{x}}'_n)'$, where $\tilde{\mathbf{x}}'_i \equiv (\tilde{x}_{k(i-1)+1}, \ldots, \tilde{x}_{ki})$. Partition $\boldsymbol{\mu} \equiv (\boldsymbol{\mu}'_1, \ldots, \boldsymbol{\mu}'_n)'$ and (12.8.8)

$$\boldsymbol{\sigma}^2 \equiv \begin{bmatrix} \boldsymbol{\sigma}_{11}^2 & \cdots & \boldsymbol{\sigma}_{1n}^2 \\ \cdot & & \cdot \\ \cdot & & \cdot \\ \cdot & & \cdot \\ \boldsymbol{\sigma}_{n1}^2 & \cdots & \boldsymbol{\sigma}_{nn}^2 \end{bmatrix}$$

conformably with the partition of $\tilde{\mathbf{x}}$. Let $\mathbf{B}_{ji}$ denote an $r \times k$ matrix and $\mathbf{c}_j$ a point in R^r, for $i = 1, \ldots, n$ and $j = 1, \ldots, m$. Define $\tilde{\mathbf{y}}_j \equiv \mathbf{c}_j + \sum_{i=1}^{n} \mathbf{B}_{ji} \tilde{\mathbf{x}}_i$ for $j = 1, \ldots, m$, and $\tilde{\mathbf{y}} \equiv (\tilde{\mathbf{y}}'_1, \ldots, \tilde{\mathbf{y}}'_m)'$. Then the distribution of $\tilde{\mathbf{y}}$ is rm-variate Student, with parameters $\boldsymbol{\mu}_y$, $\boldsymbol{\sigma}_y^2$, and ν, where

$$\boldsymbol{\mu}_y = (E(\tilde{\mathbf{y}}_1)', \ldots, E(\tilde{\mathbf{y}}_m)')'$$

for

$$E(\tilde{\mathbf{y}}_j) = \mathbf{c}_j + \sum_{i=1}^{n} \mathbf{B}_{ji}\boldsymbol{\mu}_i,$$

and

$$\boldsymbol{\sigma}_y^2 = \begin{bmatrix} \mathbf{V}_{11} & \cdots & \mathbf{V}_{1m} \\ \cdot & & \\ \cdot & & \\ \cdot & & \\ \mathbf{V}_{m1} & \cdots & \mathbf{V}_{mm} \end{bmatrix}$$

for

$$\mathbf{V}_{jh} \equiv V(\tilde{\mathbf{y}}_j, \tilde{\mathbf{y}}_h) = \sum_{g=1}^{n}\sum_{i=1}^{n} \mathbf{B}_{ji}\boldsymbol{\sigma}_{ig}{}^2\mathbf{B}'_{hg}.$$

Proof

Exercise 32.

The reader should note carefully in (12.8.9) that the covariance matrix of a k-variate Student random vector is not $\boldsymbol{\sigma}^2$ but rather $[\nu/(\nu - 2)]\boldsymbol{\sigma}^2$. The larger Student covariances reflect the greater uncertainty introduced by h, and correspond to the flatter shape of $f_{S*}(\cdot \mid 1)$ in Figure 12.7.

If $\tilde{\mathbf{x}}$ has the k-variate Student distribution with parameters $\boldsymbol{\mu}$, $\boldsymbol{\sigma}^2$, and ν, then (12.8.9)

$$E(\tilde{\mathbf{x}}) = \boldsymbol{\mu}, \qquad \text{provided that } \nu > 1$$

and

$$V(\tilde{\mathbf{x}}) = [\nu/(\nu - 2)]\boldsymbol{\sigma}^2, \qquad \text{provided that } \nu > 2.$$

Proof

By (12.8.3),

$$\begin{aligned} E(\tilde{\mathbf{x}}) &= E_{\gamma 2}[E_N{}^{(k)}(\tilde{\mathbf{x}} \mid \boldsymbol{\mu}, (1/\tilde{h})\boldsymbol{\sigma}^2) \mid 1, \nu] \\ &= E_{\gamma 2}[\boldsymbol{\mu} \mid 1, \nu] \qquad \text{(by 12.7.8)} \\ &= \boldsymbol{\mu}, \end{aligned}$$

proving the first assertion. Again by (12.8.3),

$$\begin{aligned} V(\tilde{\mathbf{x}}) &= E_{\gamma 2}[E_N{}^{(k)}\{(\tilde{\mathbf{x}} - \boldsymbol{\mu})(\tilde{\mathbf{x}} - \boldsymbol{\mu})' \mid \boldsymbol{\mu}, (1/\tilde{h})\boldsymbol{\sigma}^2\} \mid 1, \nu] \\ &= E_{\gamma 2}[1/\tilde{h})\boldsymbol{\sigma}^2 \mid 1, \nu] \\ &= \{E_{\gamma 2}(\tilde{h}^{-1} \mid 1, \nu)\}\boldsymbol{\sigma}^2. \end{aligned}$$

But $\tilde{h}$ has df $f_{\gamma 2}(\cdot \mid 1, \nu)$ if and only if $\tilde{y} \equiv \tilde{h}^{-1}$ has df $f_{i\gamma}(\cdot \mid 1, \nu)$, so that

$$E_{\gamma 2}(\tilde{h}^{-1} \mid 1, \nu) = E_{i\gamma}(\tilde{y} \mid 1, \nu)$$
$$= \frac{\nu}{\nu - 2}$$

from Table 12.1. The stipulations that $\nu > 1$ for $E(\tilde{\mathbf{x}})$ to exist and $\nu > 2$ for $V(\tilde{\mathbf{x}})$ to exist will not be proven here; they require advanced reasoning. See Raiffa and Schlaifer [**66**], 257–258. QED

We noted in (12.5.13) that no great generality is gained by replacing $f_{\gamma 2}(h \mid 1, \nu)$ with $f_{\gamma 2}(h \mid \psi, \nu)$ for arbitrary $\psi > 0$. The same holds true for the k-variate Student distribution.

If $\tilde{h}$ has df $f_{\gamma 2}(\cdot \mid \psi, \nu)$ and the conditional distribution of $\tilde{\mathbf{x}}$ given $\tilde{h}$ is k-variate Normal with parameters $\boldsymbol{\mu}$ and $(1/h)\boldsymbol{\sigma}^2$, then the unconditional distribution of $\tilde{\mathbf{x}}$ is k-variate Student with parameters $\boldsymbol{\mu}$, $\psi\boldsymbol{\sigma}^2$, and ν. (12.8.10)

Proof

Exercise 33.

Assertion (12.8.10) obtains even when $\boldsymbol{\sigma}^2$ is only positive semidefinite and both the conditional and unconditional distributions of $\tilde{\mathbf{x}}$ are singular.

The geometry of the k-variate Student distributions is qualitatively identical to that of the k-variate Normal distributions. In particular, level sets are ellipsoidal surfaces.

We shall use the same phraseological conventions with k-variate Student distributions as with k-variate Normal distributions; that is,

"$\tilde{x}$ has df $f_{S^{(k)}}(\cdot \mid \boldsymbol{\mu}, \boldsymbol{\sigma}^2, \nu)$"

and

"$\tilde{\mathbf{x}}$ has a k-variate Student df (or density)"

will imply that $\boldsymbol{\sigma}^2$ is positive definite. Singular Student distributions (with $\boldsymbol{\sigma}^2$ only positive semidefinite) will always be so specified, on the rare occasions when they are invoked.

12.9 THE WISHART FAMILIES

We say that the kth order symmetric random matrix $\tilde{\mathbf{X}}$ has the kth *order Wishart distribution with parameters* $\boldsymbol{\psi}$ *and* ν, where $\boldsymbol{\psi}$ is a positive definite and symmetric matrix and $\nu > 0$, if the df of $\tilde{\mathbf{X}}$ is given by

$$f_{W^{(k)}}(\mathbf{X} \mid \boldsymbol{\psi}, \nu) \equiv \begin{cases} W_k(\nu)|\boldsymbol{\psi}|^{\frac{1}{2}(\nu+k-1)}|\mathbf{X}|^{\frac{1}{2}(\nu-2)} \exp\left[-\tfrac{1}{2}\nu \operatorname{tr}\{\boldsymbol{\psi}\mathbf{X}\}\right], & \mathbf{X} \text{ is PDS} \\ 0, & \mathbf{X} \text{ not PDS}, \end{cases} \tag{12.9.1}$$

where PDS stands for *positive definite and symmetric*,

$$W_k(\nu) \equiv \frac{1}{\{(2/\nu)^{k(\nu+k-1)/2}\pi^{k(k-1)/4} \prod_{i=1}^{k} \Gamma(\frac{1}{2}[\nu + k - i])\}}, \tag{12.9.2}$$

and *tr* denotes *trace*.

The trace of a square matrix $\mathbf{A}$ is the sum of its diagonal elements; that is, $\text{tr}\,\{\mathbf{A}\} \equiv \sum_{i=1}^{k} A_{ii}$. If $\mathbf{A}$, $\mathbf{B}$, and $\mathbf{C}_1, \ldots, \mathbf{C}_r$ are symmetric kth order matrices, $m \in R^1$, and $\mathbf{y} \in R^k$, then the reader may show as Exercise 34 that (12.9.3)

$$\text{tr}\,\{\mathbf{AB}\} = \text{tr}\,\{\mathbf{BA}\}; \tag{12.9.3a}$$

$$\text{tr}\,\{\mathbf{A}\cdot \sum_{i=1}^{r} \mathbf{C}_i\} = \sum_{i=1}^{r} \text{tr}\,\{\mathbf{AC}_i\}; \tag{12.9.3b}$$

$$\text{tr}\,\{\mathbf{A}\cdot\mathbf{yy}'\} = \mathbf{y}'\mathbf{Ay}; \text{ and} \tag{12.9.3c}$$

$$m\cdot\text{tr}\,\{\mathbf{AB}\} = \text{tr}\,\{m\mathbf{A}\cdot\mathbf{B}\} = \text{tr}\,\{\mathbf{A}\cdot m\mathbf{B}\}. \tag{12.9.3d}$$

Note that $f_W{}^{(k)}(\cdot \mid \boldsymbol{\psi}, \nu)$ is positive only at *symmetric* matrices $\tilde{\mathbf{X}}$, which have $k(k+1)/2$ functionally independent elements. By writing $\tilde{\mathbf{X}}$ as $(\tilde{X}_{11}, \ldots, \tilde{X}_{1k}, \tilde{X}_{22}, \ldots, \tilde{X}_{2k}, \ldots, \tilde{X}_{ii}, \ldots, \tilde{X}_{ik}, \ldots, \tilde{X}_{kk})'$, the Wishart distribution can be thought of as a distribution of $[k(k+1)/2]$-dimensional random vectors.

So far we have encountered positive-definite and symmetric matrices only as covariance matrices of (nonsingular) random vectors. Thus the fact that $f_W{}^{(k)}(\mathbf{X} \mid \boldsymbol{\psi}, \nu) > 0$ only if $\mathbf{X}$ is positive-definite and symmetric, suggests that the Wishart distribution has something to do with random covariance matrices; and so it does, as Chapter 13 indicates.

From (12.9.1) it might be suspected that obtaining useful results about Wishart distributions involves some complicated, if not advanced, mathematics. This too is true, and hence we shall merely state the relevant results, referring the reader to Anderson [**1**] for their derivations. (Anderson's notation differs from ours.) The following table furnishes the necessary translations:

Ours	*Anderson's*
$\mathbf{X}$	$\mathbf{A}$
k	p
ν	$n - p + 1$
$\boldsymbol{\psi}$	$(1/[n-p+1)]\boldsymbol{\Sigma}^{-1}$
$\nu + k - 1$	n
$(1/\nu)\boldsymbol{\psi}^{-1}$	$\boldsymbol{\Sigma}$

If $\tilde{\mathbf{X}}$ has df $f_W^{(k)}(\cdot \mid \boldsymbol{\psi}, \nu)$, then (12.9.4)

$$E_W^{(k)}(\tilde{\mathbf{X}} \mid \boldsymbol{\psi}, \nu) = \left[\frac{(\nu + k - 1)}{\nu}\right]\boldsymbol{\psi}^{-1}$$

and

$$V_W^{(k)}(\tilde{X}_{ij}, \tilde{X}_{k\ell} \mid \boldsymbol{\psi}, \nu) = [(\nu + k - 1)/\nu^2][\psi^{-1}{}_{ik}\psi^{-1}{}_{j\ell} + \psi^{-1}{}_{i\ell}\psi^{-1}{}_{jk}],$$

where $\psi^{-1}{}_{\alpha\beta}$ denotes the (α, β)th element of $\boldsymbol{\psi}^{-1}$.

Proof

Anderson [**1**], p. 161.

We cannot characterize in an easy fashion the marginal distribution of an *arbitrary* subset of the components of $\tilde{\mathbf{X}}$ when $\tilde{\mathbf{X}}$ has the Wishart distribution, but we can do so for subsets that constitute principal submatrices of $\tilde{\mathbf{X}}$. By rearranging rows and corresponding columns of $\tilde{\mathbf{X}}$ and $\boldsymbol{\psi}$, the following becomes a general result in this direction.

Let $\tilde{\mathbf{X}}$ have df $f_W^{(k)}(\cdot \mid \boldsymbol{\psi}, \nu)$, and partition $\tilde{\mathbf{X}}$, $\boldsymbol{\psi}$, and $\boldsymbol{\psi}^{-1}$ into (12.9.5)

$$\mathbf{X} \equiv \begin{bmatrix} \tilde{\mathbf{X}}_{11} & \tilde{\mathbf{X}}_{12} \\ \tilde{\mathbf{X}}_{21} & \tilde{\mathbf{X}}_{22} \end{bmatrix},$$

$$\boldsymbol{\psi} \equiv \begin{bmatrix} \boldsymbol{\psi}_{11} & \boldsymbol{\psi}_{12} \\ \boldsymbol{\psi}_{21} & \boldsymbol{\psi}_{22} \end{bmatrix},$$

and

$$\boldsymbol{\psi}^{-1} \equiv \begin{bmatrix} (\boldsymbol{\psi}^{-1})_{11} & (\boldsymbol{\psi}^{-1})_{12} \\ (\boldsymbol{\psi}^{-1})_{21} & (\boldsymbol{\psi}^{-1})_{22} \end{bmatrix},$$

where $\mathbf{A}_{11}$ is of jth order $(0 < j < k)$ in each case. Then the marginal df of $\tilde{\mathbf{X}}_{11}$ is $f_W^{(j)}(\cdot \mid [(\boldsymbol{\psi}^{-1})_{11}]^{-1}, \nu)$, where

$$[(\boldsymbol{\psi}^{-1})_{11}]^{-1} = \boldsymbol{\psi}_{11} - \boldsymbol{\psi}_{12}(\boldsymbol{\psi}_{22})^{-1}\boldsymbol{\psi}_{21}.$$

Proof

Anderson [**1**], p. 163. The formula for $[(\boldsymbol{\psi}^{-1})_{11}]^{-1}$ is from (b) in the proof of (12.7.9).

Of occasional interest is the fact that the sum of certain Wishart-distributed, independent random matrices has a Whishart distribution.

If $\tilde{\mathbf{X}}_1, \ldots, \tilde{\mathbf{X}}_r$ are independent random kth order random matrices and $\tilde{\mathbf{X}}_i$ has df $f_W^{(k)}(\cdot \mid \nu_i^{-1}\boldsymbol{\zeta}, \nu_i)$, then $\tilde{\mathbf{X}} \equiv \sum_{i=1}^{r} \tilde{\mathbf{X}}_i$ has df $f_W^{(k)}(\cdot \mid \nu_*^{-1}\boldsymbol{\zeta}, \nu_*)$, where (12.9.6)

$$\nu_* \equiv (r - 1)(k - 1) + \sum_{i=1}^{r} \nu_i.$$

Proof

Anderson [**1**], p. 162.

Positive scalar multiples of Wishart-distributed matrices are Wishart distributed:

If $\tilde{\mathbf{X}}$ has df $f_{W^{(k)}}(\cdot \mid \boldsymbol{\psi}, \nu)$ and $m > 0$, then $m\tilde{\mathbf{X}}$ has df $f_{W^{(k)}}(\cdot \mid m^{-1}\boldsymbol{\psi}, \nu)$. (12.9.7)

Proof

Exercise 35.

The Wishart distribution is a generalization of the gamma-2 distribution:

$$f_{W^{(1)}}(X \mid \psi, \nu) = f_{\gamma 2}(X \mid \psi, \nu), \qquad \text{all } X \in R^1. \tag{12.9.8}$$

Proof

Exercise 36.

We say that the kth order symmetric random matrix $\tilde{\mathbf{Y}}$ has the kth order *inverted Wishart distribution with parameters* $\boldsymbol{\psi}$ *and* ν, where $\boldsymbol{\psi}$ is a positive definite and symmetric matrix and $\nu > 0$, if the df of $\tilde{\mathbf{Y}}$ is

$$f_{iW^{(k)}}(\mathbf{Y} \mid \boldsymbol{\psi}, \nu) \equiv \begin{cases} W_k(\nu) \mid \boldsymbol{\psi}|^{\frac{1}{2}(\nu+k-1)}|\mathbf{Y}^{-1}|^{\frac{1}{2}(\nu+2k)}, & \exp\left(-\tfrac{1}{2}\nu \operatorname{tr}\{\mathbf{Y}^{-1}\boldsymbol{\psi}\}\right), \\ & \mathbf{Y} \text{ is PDS} \\ 0, & \mathbf{Y} \text{ not PDS}, \end{cases} \tag{12.9.9}$$

where PDS stands for *positive definite and symmetric*, $W_k(\nu)$ is given by (12.9.2), and tr again denotes *trace*.

$\tilde{\mathbf{Y}}$ has df $f_{iW^{(k)}}(\cdot \mid \boldsymbol{\psi}, \nu)$ if and only if $\tilde{\mathbf{X}} \equiv (\tilde{\mathbf{Y}})^{-1}$ has df $f_{W^{(k)}}(\cdot \mid \boldsymbol{\psi}, \nu)$. (12.9.10)

Proof

By Exercise 8.9 in matrix notation,

$$\begin{aligned} f_{\mathbf{Y}}(\mathbf{Y})d(\mathbf{Y}) &= f_{\mathbf{Y}}(\mathbf{g}[\mathbf{X}]) \left| \frac{d(\mathbf{g}[\mathbf{X}])}{d(\mathbf{X})} \right| d(\mathbf{X}) \\ &= f_{\mathbf{X}}(\mathbf{X})d(\mathbf{X}). \end{aligned}$$

where $|d(g[\mathbf{X}])/d(\mathbf{X})|$ is the Jacobian. Hence here, $g[\mathbf{X}] = \mathbf{X}^{-1}$ and $(\mathbf{X}^{-1})^{-1} = \mathbf{X}$ implies

$$\begin{aligned} f_{\mathbf{Y}}(\mathbf{Y})d(\mathbf{Y}) &\equiv f_{iW^{(k)}}(\mathbf{Y} \mid \boldsymbol{\psi}, \nu)d(\mathbf{Y}) \\ &= K|\mathbf{Y}^{-1}|^{\frac{1}{2}(\nu+2k)} \cdot \exp\left(-\tfrac{1}{2}\nu \operatorname{tr}\{\mathbf{Y}^{-1}\boldsymbol{\psi}\}\right)d(\mathbf{Y}) \\ &= K|\mathbf{X}|^{\frac{1}{2}(\nu+2k)} \cdot \exp\left(-\tfrac{1}{2}\nu \operatorname{tr}\{\mathbf{X}\boldsymbol{\psi}\}\right)d(\mathbf{Y}) \\ (*) \qquad &= K|\mathbf{X}|^{\frac{1}{2}(\nu+2k)} \cdot \exp\left(-\tfrac{1}{2}\nu \operatorname{tr}\{\boldsymbol{\psi}\mathbf{X}\}\right)d(\mathbf{Y}) \end{aligned}$$

by (12.9.3), where $K \equiv W_k(\nu)|\boldsymbol{\psi}|^{\frac{1}{2}(\nu+k-1)}$. But

$$|d(\mathbf{X}^{-1})/d(\mathbf{X})| = |\mathbf{X}|^{-(k+1)}$$

(compare Tiao and Zellner, [**79**] 279), so that $d(\mathbf{Y}) = |\mathbf{X}|^{-(k+1)}d(\mathbf{X})$ and (*) becomes

$$\begin{aligned} W_k(\nu)|\boldsymbol{\psi}|^{\frac{1}{2}(\nu+k-1)}|\mathbf{X}|^{\frac{1}{2}(\nu-2)} \exp\left(-\tfrac{1}{2}\nu \text{ tr } \{\boldsymbol{\psi}\mathbf{X}\}\right)d(\mathbf{X}) \\ = f_{W^{(k)}}(\mathbf{X} \mid \boldsymbol{\psi}, \nu)d(\mathbf{X}) \\ = f_{\mathbf{X}}(\mathbf{X})d(\mathbf{X}). \end{aligned}$$

The reasoning is reversible, which completes the proof. QED

From (12.9.4) and (12.9.10), we know that if $\tilde{\mathbf{Y}}$ has df $f_{iW^{(k)}}(\cdot \mid \boldsymbol{\psi}, \nu)$, then $E(\tilde{\mathbf{Y}}^{-1}) = [(\nu + k - 1)/\nu]\boldsymbol{\psi}^{-1}$, but that is of no particular interest to us.

If $\tilde{\mathbf{Y}}$ has df $f_{iW^{(k)}}(\cdot \mid \boldsymbol{\psi}, \nu)$, then (12.9.11)

$$E(\tilde{\mathbf{Y}}) = [\nu/(\nu - 2)]\boldsymbol{\psi}, \qquad \text{provided that } \nu > 2.$$

Proof

See Evans [**20**], p. 282. Our $\tilde{\mathbf{Y}}$ is his $\sum$; our ν is his $N - q + 1$; and our $\nu\boldsymbol{\psi}$ is his $\boldsymbol{\Upsilon}$.

Not all of the variances $V_{iW^{(k)}}(\tilde{Y}_{ij} \mid \boldsymbol{\psi}, \nu)$ can be determined by elementary methods, but the $V_{iW^{(k)}}(\tilde{Y}_{ij} \mid \boldsymbol{\psi}, \nu)$'s can be determined from (12.9.15) and (12.9.12).

$$f_{iW^{(1)}}(Y \mid \psi, \nu) = f_{i\gamma}(Y \mid \psi, \nu). \qquad (12.9.12)$$

Proof

Exercise 37.

If $\tilde{\mathbf{Y}}$ has df $f_{iW^{(k)}}(\cdot \mid \boldsymbol{\psi}, \nu)$ and $m > 0$, then $m\tilde{\mathbf{Y}}$ has df $f_{iW^{(k)}}(\cdot \mid m\boldsymbol{\psi}, \nu)$. (12.9.13)

Proof

Exercise 38.

If $\tilde{\mathbf{Y}}$ has df $f_{iW^{(k)}}(\cdot \mid \boldsymbol{\psi}, \nu)$ and $\tilde{\mathbf{Y}}$ and $\boldsymbol{\psi}$ are partitioned into (12.9.14)

$$\tilde{\mathbf{Y}} \equiv \begin{bmatrix} \tilde{\mathbf{Y}}_{11} & \tilde{\mathbf{Y}}_{12} \\ \tilde{\mathbf{Y}}_{21} & \tilde{\mathbf{Y}}_{22} \end{bmatrix}$$

and

$$\boldsymbol{\psi} \equiv \begin{bmatrix} \boldsymbol{\psi}_{11} & \boldsymbol{\psi}_{12} \\ \boldsymbol{\psi}_{21} & \boldsymbol{\psi}_{22} \end{bmatrix}$$

respectively, where $\mathbf{A}_{11}$ is of jth order $(0 < j < k)$ in each case, then $\tilde{\mathbf{Y}}_{11} - \tilde{\mathbf{Y}}_{12}(\tilde{\mathbf{Y}}_{22})^{-1}\tilde{\mathbf{Y}}_{21} \equiv \tilde{\mathbf{Y}}_{11|2}$ has df $f_{iW}^{(j)}(\cdot \mid \boldsymbol{\psi}_{11} - \boldsymbol{\psi}_{12}(\boldsymbol{\psi}_{22})^{-1}\boldsymbol{\psi}_{21}, \nu)$.

Proof

By (12.9.10), $\tilde{\mathbf{X}} \equiv \tilde{\mathbf{Y}}^{-1}$ has df $f_W^{(k)}(\cdot \mid \boldsymbol{\psi}, \nu)$. Partition $\tilde{\mathbf{X}}$ as $\tilde{\mathbf{Y}}$ and $\boldsymbol{\psi}$; by (12.9.5), $\tilde{\mathbf{X}}_{11}$ has df $f_W^{(j)}(\cdot \mid \boldsymbol{\psi}_{11} - \boldsymbol{\psi}_{12}(\boldsymbol{\psi}_{22})^{-1}\boldsymbol{\psi}_{21}, \nu)$ and hence $(\tilde{\mathbf{X}}_{11})^{-1}$ has df $f_{iW}^{(j)}(\cdot \mid \boldsymbol{\psi}_{11} - \boldsymbol{\psi}_{12}(\boldsymbol{\psi}_{22})^{-1}\boldsymbol{\psi}_{21}, \nu)$. But by (b) in the proof of (12.7.9), $(\tilde{\mathbf{X}}_{11})^{-1} = \tilde{\mathbf{Y}}_{11} - \tilde{\mathbf{Y}}_{12}(\tilde{\mathbf{Y}}_{22})^{-1}\tilde{\mathbf{Y}}_{21}$. QED

Tiao and Zellner have derived marginal distributions of jth order principal submatrices of $\tilde{\mathbf{Y}}$.

If $\tilde{\mathbf{Y}}$ has df $f_{iW}^{(k)}(\cdot \mid \boldsymbol{\psi}, \nu)$, $\tilde{\mathbf{Y}}_{11}$ is a jth order principal submatrix of $\tilde{\mathbf{Y}}$, and $\boldsymbol{\psi}_{11}$ is the corresponding submatrix of $\boldsymbol{\psi}$, then the marginal df of $\tilde{\mathbf{Y}}_{11}$ is $f_{iW}^{(j)}(\cdot \mid \boldsymbol{\psi}_{11}, \nu)$. (12.9.15)

Proof

See Tiao and Zellner [**79**], p. 280. Their ν is our $\nu + 2k$; their p, our j; their m, our k; and their $\mathbf{S}$, our $\nu\boldsymbol{\psi}$.

From (12.9.15) and (12.9.12) it follows readily that the marginal df of every diagonal element $\tilde{Y}_{ii}$ is $f_{i\gamma}(\cdot \mid \psi_{ii}, \nu)$, and hence that $V_{iW}^{(k)}(\tilde{\mathbf{Y}}_{ii} \mid \boldsymbol{\psi}, \nu) = 2\nu^2\psi_{ii}^2/[\nu - 2)^2(\nu - 4)]$ provided that $\nu > 4$, by Table 12.1.

12.10 kth ORDER INVERTED BETA-2 DISTRIBUTIONS

We say that the kth order symmetric matrix $\tilde{\mathbf{X}}$ has the kth *order inverted beta-2 distribution with parameters* $\boldsymbol{\Phi}$, ρ, ν, and τ, where $\boldsymbol{\Phi}$ is a positive definite and symmetric kth order matrix, and ρ, ν, and τ are positive real numbers, if:

(1) there is a kth order matrix $\tilde{\boldsymbol{\psi}}$ such that $[(\nu/\tau)\tilde{\boldsymbol{\psi}}]^{-1}$ has df $f_{iW}^{(k)}(\cdot \mid \boldsymbol{\Phi}, \rho)$;

and

(2) conditional upon $\tilde{\boldsymbol{\psi}}$, $\tilde{\mathbf{X}}$ has df $f_W^{(k)}(\cdot \mid \boldsymbol{\psi}, \nu)$. If (1) and (2) obtain, the marginal distribution of $\tilde{\mathbf{X}}$ is by definition the kth order inverted beta-2 distribution with parameters $\boldsymbol{\Phi}$, ρ, ν, and τ, where df is denoted by $f_{i\beta 2}^{(k)}(\cdot \mid \boldsymbol{\Phi}, \rho, \nu, \tau)$.

$$f_{i\beta 2}^{(k)}(\mathbf{X} \mid \boldsymbol{\Phi}, \rho, \nu, \tau) = \begin{cases} B_k(\rho, \nu, \tau) \,|\, (\rho/\tau)\boldsymbol{\Phi}|^{\frac{1}{2}(\rho+k-1)}|\mathbf{X}|^{\frac{1}{2}(\nu-2)}|\mathbf{X} + (\rho/\tau)\boldsymbol{\Phi}|^{-\frac{1}{2}(\nu+\rho+2k-2)}, & \mathbf{X} \text{ is PDS} \\ 0, & \mathbf{X} \text{ not PDS,} \end{cases}$$

where

$$B_k(\rho, \nu, \tau) \equiv \frac{W_k(\nu)W_k(\rho)}{W_k(\nu + \rho + k - 1)} \cdot \frac{(\nu + \rho + k - 1)^{k(\nu+\rho+2k-2)/2}}{\nu^{k(\nu+k-1)/2}\rho^{k(\rho+k-1)/2}}.$$

Proof

We follow the same procedure as in the proof of (12.5.1). By assumption, the joint df of $\tilde{\mathbf{X}}$ and $\tilde{\mathbf{Y}} \equiv [(\nu/\tau)\tilde{\boldsymbol{\psi}}]^{-1}$ is

$$\begin{aligned} f(\mathbf{X}, \mathbf{Y}) &= f(\mathbf{X} \mid \mathbf{Y})f(\mathbf{Y}) \\ &= f_W{}^{(k)}(\mathbf{X} \mid (\tau/\nu)\mathbf{Y}^{-1}, \nu) f_{iW}{}^{(k)}(\mathbf{Y} \mid \boldsymbol{\Phi}, \rho) \\ &= \{ W_k(\nu)|(\tau/\nu)Y^{-1}|^{\frac{1}{2}(\nu+k-1)}|\mathbf{X}|^{\frac{1}{2}(\nu-2)} \exp[-\tfrac{1}{2}\nu \operatorname{tr}\{(\tau/\nu)\mathbf{Y}^{-1}\mathbf{X}\}]\} \\ &\quad \times \{ W_k(\rho)|\boldsymbol{\Phi}|^{\frac{1}{2}(\rho+k-1)}|\mathbf{Y}^{-1}|^{\frac{1}{2}(\rho+2k)} \cdot \exp[-\tfrac{1}{2}\rho \operatorname{tr}\{\mathbf{Y}^{-1}\boldsymbol{\Phi}\}]\} \\ &= \{ W_k(\nu)W_k(\rho)(\tau/\nu)^{k(\nu+k-1)/2}|\boldsymbol{\Phi}|^{\frac{1}{2}(\rho+k-1)}|\mathbf{X}|^{\frac{1}{2}(\nu-2)}\} \\ &\quad \times \{|\mathbf{Y}^{-1}|^{\frac{1}{2}(\nu+\rho+k-1\ +2k)} \cdot \exp[-\tfrac{1}{2} \operatorname{tr}\{\mathbf{Y}^{-1}[\tau\mathbf{X} + \rho\boldsymbol{\Phi}]\}]\} \\ &= \{ W_k(\nu)W_k(\rho)(\tau/\nu)^{k(\nu+k-1)/2}|\boldsymbol{\Phi}|^{\frac{1}{2}(\rho+k-1)}|\mathbf{X}|^{\frac{1}{2}(\nu-2)} \\ &\quad \times \{ W_k(\nu^*)|\mathbf{v}|^{\frac{1}{2}(\nu^*+k-1)}\}^{-1} \\ &\qquad \times \{ W_k(\nu^*)|\mathbf{v}|^{\frac{1}{2}(\nu^*+k-1)}|\mathbf{Y}^{-1}|^{\frac{1}{2}(\nu^*+2k)} \cdot \exp[-\tfrac{1}{2}\nu^* \operatorname{tr}\{\mathbf{Y}^{-1}\mathbf{v}\}]\}, \end{aligned}$$

where $\nu^* \equiv \nu + \rho + k - 1$ and $\mathbf{v} \equiv (\tau/\nu^*)\mathbf{X} + (\rho/\nu^*)\boldsymbol{\Phi}$. But the third bracketed factor in the last member of the equality chain is simply

$$f_{iW}{}^{(k)}(\mathbf{Y} \mid \mathbf{v}, \nu^*).$$

Denoting the other two factors (both independent of $\mathbf{Y}$) by K, we obtain

$$\int f(\mathbf{X} \mid \mathbf{Y})f(\mathbf{Y})d(\mathbf{Y}) = \int K f_{iW}{}^{(k)}(\mathbf{Y} \mid \mathbf{v}, \nu^*)d(\mathbf{Y}) = K.$$

Thus $f_{i\beta 2}{}^{(k)}(\mathbf{X} \mid \boldsymbol{\Phi}, \rho, \nu, \tau)$ is the product of the first two factors when $\mathbf{X}$ is positive definite and symmetric, and the assertion now follows by elementary algebra (recalling that $|b\mathbf{c}|^t = b^{kt}|\mathbf{c}|^t$ for $b > 0$ and $\mathbf{c}$ a kth order matrix). QED

For $k = 1$, the reader may verify as Exercise 39 that

$$f_{i\beta 2}{}^{(1)}(X \mid \phi, \rho, \nu, \tau) = f_{i\beta 2}(X \mid \tfrac{1}{2}\nu, \tfrac{1}{2}(\nu + \rho), \rho\phi/\tau), \qquad (12.10.2)$$

where the right-hand side is as defined by (12.3.13).

Equation (12.10.2) suggests that the parameter τ is redundant and could be subsumed into $\boldsymbol{\Phi}$. That is true; we have

$$f_{i\beta 2}{}^{(k)}(\mathbf{X} \mid \boldsymbol{\Phi}, \rho, \nu, \tau) = f_{i\beta 2}{}^{(k)}(\mathbf{X} \mid \tau^{-1}\boldsymbol{\Phi}, \rho, \nu, 1), \qquad \text{all } \mathbf{X}. \quad (12.10.3)$$

However, it is convenient to keep τ around for the same later purposes that motivated the statement of (12.8.10).

Not very much is known regarding the kth order inverted beta-2 distribution. The following three results are easily obtained by interchanging order of expectations.

If $\tilde{\mathbf{X}}$ has df $f_{i\beta 2}{}^{(k)}(\cdot \mid \boldsymbol{\Phi}, \rho, \nu, \tau)$, then (12.10.4)

$$E(\tilde{\mathbf{X}}) = \left[\frac{\rho}{\rho - 2}\right]\left[\frac{\nu + k - 1}{\tau}\right]\boldsymbol{\Phi},$$

provided that $\rho > 2$.

Proof

Exercise 41.

If $\tilde{\mathbf{X}}_1, \ldots, \tilde{\mathbf{X}}_r$ are conditionally independent random kth order matrices given $\tilde{\boldsymbol{\zeta}}$, $\tilde{\boldsymbol{\zeta}}^{-1}$ has df $f_{iW}{}^{(k)}(\cdot \mid \boldsymbol{\Phi}, \rho)$, and the conditional (upon $\tilde{\boldsymbol{\zeta}}$) df of $\tilde{\mathbf{X}}_i$ is $f_W{}^{(k)}(\cdot \mid \nu_i^{-1} \boldsymbol{\zeta}, \nu_i)$, then the unconditional df of $\tilde{\mathbf{X}} \equiv \sum_{i=1}^{r} \tilde{\mathbf{X}}_i$ is $f_{i\beta 2}{}^{(k)}(\cdot \mid \boldsymbol{\Phi}, \rho, \nu^*, 1)$, where ν^* is as defined in (12.9.6). (12.10.5)

Proof

Exercise 42.

If $\tilde{\mathbf{X}}$ has df $f_{i\beta 2}{}^{(k)}(\cdot \mid \boldsymbol{\Phi}, \rho, \nu, \tau)$ and $m > 0$, then $m\tilde{\mathbf{X}}$ has df (12.10.6)

$$f_{i\beta 2}{}^{(k)}(\cdot \mid \boldsymbol{\Phi}, \rho, \nu, \tau/m).$$

Proof

Exercise 43.

Moreover, principal submatrices have jth order inverted beta-2 distributions.

If $\tilde{\mathbf{X}}$ has df $f_{i\beta 2}{}^{(k)}(\cdot \mid \boldsymbol{\Phi}, \rho, \nu, \tau)$, $\tilde{\mathbf{X}}_{11}$ is a jth order principal submatrix of $\tilde{\mathbf{X}}$, and $\boldsymbol{\Phi}_{11}$ is the corresponding jth order submatrix of $\boldsymbol{\Phi}$, then the marginal df of $\tilde{\mathbf{X}}_{11}$ is $f_{i\beta 2}{}^{(j)}(\cdot \mid \boldsymbol{\Phi}_{11}, \rho, \nu, \tau)$. (12.10.7)

Proof

Assume, without loss of generality, that $\tilde{\mathbf{X}}_{11}$ is the upper left-hand corner of $\tilde{\mathbf{X}}$. By (12.9.5), the conditional df of $\tilde{\mathbf{X}}_{11}$ given $\boldsymbol{\psi}$ is $f_W{}^{(j)}(\cdot \mid [(\boldsymbol{\psi}^{-1})_{11}]^{-1}, \nu)$, which $\boldsymbol{\psi}$ is partitioned conformably with $\tilde{\mathbf{X}}$. But if $[(\nu/\tau)\tilde{\boldsymbol{\psi}}]^{-1}$ has df

$$f_{iW}{}^{(k)}(\cdot \mid \boldsymbol{\Phi}, \rho),$$

then by (12.9.15) the df of $\{(\nu/\tau)[(\tilde{\boldsymbol{\psi}}^{-1})_{11}]^{-1}\}^{-1} = (\tau/\nu)(\tilde{\boldsymbol{\psi}}^{-1})_{11}$ is

$$f_{iW}{}^{(j)}(\cdot \mid \boldsymbol{\Phi}_{11}, \rho).$$

Hence the assertion follows by interchanging order of integrations. QED

EXERCISES

1. Show that $\tilde{z}$ had df $f_{\gamma*}(\cdot \mid r)$ if and only if $\tilde{Q} \equiv 2\tilde{z}$ has df $f_{\gamma^2}(\cdot \mid 2r)$.
2. Derive each element of columns 3 and 4 of Table 12.1.
3. Show that $\tilde{z}$ has df $f_\beta(\cdot \mid \rho, \nu)$ if and only if $\tilde{w} \equiv 1 - \tilde{z}$ has df $f_\beta(\cdot \mid \nu - \rho, \nu)$.
4. Derive (12.3.4).
5. Derive (12.3.5).
6. Derive (12.3.8).
7. Derive (12.3.14) and (12.3.15).
8. Derive (12.4.3).
9. Show that $\tilde{x}$ has df $f_N(\cdot \mid \mu, \sigma^2)$ if and only if $\tilde{z} \equiv (\tilde{x} - \mu)/\sigma$ has df $f_{N*}(\cdot)$.
10. Verify (12.4.6).
11. Derive (12.4.8) and (12.4.9):
    (a) from (12.4.7), according to (10.6.2); and
    (b) by direct integration, according to the definition of expectation and variance.
12. Prove (12.4.10).
13. Show that $\tilde{x}$ has df $f_S(\cdot \mid \mu, \sigma^2, \nu)$ if and only if $\tilde{z} \equiv (\tilde{x} - \mu)/\sigma$ has df $f_{S*}(\cdot \mid \nu)$.
14. Derive (12.5.4) through (12.5.8).
15. Derive (12.5.10) and (12.5.11).
16. Prove (12.5.12).
17. Prove (12.6.4).
18. Verify (12.6.6) through (12.6.8).
19. Complete the proof of (12.7.6).
20. Show that if $\{\mathbf{z}, \mathbf{u}\} \subset R^k$ and $\mathbf{A}$ is a kth order matrix, then
$$\mathbf{z}'\mathbf{z} - 2\mathbf{u}'\mathbf{A}\mathbf{z} = (\mathbf{z} - \mathbf{A}'\mathbf{u})'(\mathbf{z} - \mathbf{A}'\mathbf{u}) - \mathbf{u}'\mathbf{A}\mathbf{A}'\mathbf{u}.$$
21. Prove (12.7.8).
22. Use (12.7.8) to derive (12.7.6) via (10.6.4).
23. Prove equations (a) through (c) in the proof of (12.7.9).
24. Verify equality (***) in the proof of (12.7.9).

25. Prove (12.7.10).

26. Prove (12.7.9′).

27. Prove (12.8.1).

28. Prove (12.8.4).

29. Prove (12.8.5).

30. Prove (12.8.6).

31. Prove (12.8.7).

32. Prove (12.8.8).

33. Prove (12.8.10).

34. Derive Equations (12.9.3).

35. Prove (12.9.7).

36. Prove (12.9.8).

37. Prove (12.9.12).

38. Prove (12.9.13).

39. Prove (12.10.2).

40. Prove (12.10.3).

41. Provc (12.10.4).

42. Prove (12.10.5).

43. Prove (12.10.6).

44. Anderson ([**1**], 162) proves the following result:

If $\tilde{\mathbf{X}}$ has df $f_W^{(k)}(\cdot \mid \boldsymbol{\psi}, \nu)$, $\mathbf{B}$ is a nonsingular kth order matrix, and $\tilde{\mathbf{Y}} \equiv \mathbf{B}\tilde{\mathbf{X}}\mathbf{B}'$, then $\tilde{\mathbf{Y}}$ has df $f_W^{(k)}(\cdot \mid (\mathbf{B}^{-1})'\boldsymbol{\psi}(\mathbf{B}^{-1}), \nu)$. (12.11.1)

(a) Prove the following:

If $\tilde{\mathbf{Y}}$ has df $f_{iW}^{(k)}(\cdot \mid \boldsymbol{\psi}, \nu)$, $\mathbf{B}$ is a nonsingular kth order matrix, and $\tilde{\mathbf{Z}} \equiv \mathbf{B}\tilde{\mathbf{Y}}\mathbf{B}'$, then $\tilde{\mathbf{Z}}$ has df $f_{iW}^{(k)}(\cdot \mid \mathbf{B}\boldsymbol{\psi}\mathbf{B}', \nu)$. (12.11.2)

(b) Prove the following:

If $\tilde{\mathbf{X}}$ has df $f_{i\beta 2}^{(k)}(\cdot \mid \boldsymbol{\Phi}, \rho, \nu, \tau)$, $\mathbf{B}$ is a nonsingular kth order matrix, and $\tilde{\mathbf{Y}} \equiv \mathbf{B}\tilde{\mathbf{X}}\mathbf{B}'$, then $\tilde{\mathbf{Y}}$ has df $f_{i\beta 2}^{(k)}(\cdot \mid \mathbf{B}\boldsymbol{\Phi}\mathbf{B}', \rho, \nu, \tau)$. (12.11.3)

45. *The (Univariate) Lognormal Family*: We say that $\tilde{x}$ has the *lognormal distribution with parameters* μ *and* σ^2 if $\log_e \tilde{x}$ has the Normal distribution with parameters μ and σ^2. The df of $\tilde{x}$ is denoted by $f_L(\cdot \mid \mu, \sigma^2)$.

(a) Show that

$$f_L(x \mid \mu, \sigma^2) = \begin{cases} (2\pi\sigma^2)^{-1/2}x^{-1} \exp(-\frac{1}{2}(\log_e x - \mu)^2/\sigma^2), & x > 0; \\ 0, & x \leqq 0; \end{cases} \tag{12.11.4a}$$

$$F_L(x \mid \mu, \sigma^2) = \begin{cases} 0, & x \leqq 0 \\ F_{N*}([\log_e x - \mu]/\sigma), & x > 0; \end{cases} \tag{12.11.4b}$$

and that the pth fractile ${}_pL_{(\mu,\sigma^2)}$ of $\tilde{x}$ is given by

$${}_pL_{(\mu,\sigma^2)} = \exp(\mu + \sigma \cdot {}_pN^*). \tag{12.11.5}$$

(b) Show that

$$E_L(\tilde{x}^r \mid \mu, \sigma^2) = \exp(r\mu + \tfrac{1}{2}r^2\sigma^2), \tag{12.11.6}$$

and hence that

$$E_L(\tilde{x} \mid \mu. \pi^2) = \exp(\mu + \tfrac{1}{2}\sigma^2) \tag{12.11.7}$$

and

$$V_L(\tilde{x} \mid \mu, \sigma^2) = \exp(2\mu + \sigma^2)(\exp(\sigma^2) - 1). \tag{12.11.8}$$

[*Hint*: $E_L(\tilde{x}^r \mid \mu, \sigma^2) = E_L(\exp[r \cdot \log_e \tilde{x}] \mid \mu, \sigma^2)$.]

(c) Prove the following analogue of (12.4.10):

If $\tilde{x}_1, \ldots, \tilde{x}_n$ are independent, $\tilde{x}_i$ has df $f_L(\cdot \mid \mu_i, \sigma_{ii}{}^2)$, $c > 0$, and $\mathbf{b} \in R^n$ with $\mathbf{b} \neq \mathbf{0}$, then (12.11.9)

(i) $c \prod_{i=1}^{n} \tilde{x}_i{}^{b_i}$ has df $f_L(\cdot \mid \sum_{i=1}^{n} b_i\mu_i, \sum_{i=1}^{n} b_i{}^2\sigma_{ii}{}^2)$; and if $\tilde{x}_1, \ldots, \tilde{x}_n$ are iid with common df $f_L(\cdot \mid \mu, \sigma^2)$, then

(ii) $\prod_{i=1}^{n} \tilde{x}_i$ has df $f_L(\cdot \mid n\mu, n\sigma^2)$ and

(iii) the geometric mean $\prod_{i=1}^{n} \tilde{x}_i{}^{1/n}$ has df $f_L(\cdot \mid \mu, \sigma^2/n)$.

46. *Generalization of Exercise 45. The Three-Parameter Lognormal Family*: We say that $\tilde{x}$ has the (three-parameter) lognormal distribution with parameters μ, σ^2, and Λ if $\tilde{y} \equiv \tilde{x} - \Lambda$ has df $f_L(\cdot \mid \mu, \sigma^2)$.

(a) Show that the three-parameter lognormal df $f_L(\cdot \mid \mu, \sigma^2, \Lambda)$ is given by

$$f_L(y \mid \mu, \sigma^2, \Lambda) = \begin{cases} (2\pi\sigma^2)^{-1/2} \exp(-\frac{1}{2}[\log_e(x - \Lambda) - \mu]^2/\sigma^2), & x > \Lambda \\ 0, & x \leqq \Lambda; \end{cases} \tag{12.11.4a'}$$

that

$$F_L(x \mid \mu, \sigma^2, \Lambda) = \begin{cases} 0, & y \leqq \Lambda \\ F_{N*}([\log_e(x - \Lambda) - \mu]/\sigma), & y > \Lambda \end{cases} \tag{12.11.4b'}$$

and hence that the pth fractile ${}_pL_{(\mu,\sigma^2,\Lambda)}$ satisfies

$${}_pL_{(\mu,\sigma,{}^2\Lambda)} = \Lambda + \exp(\mu + \sigma \cdot {}_pN^*). \tag{12.11.5'}$$

(b) Show that

$$E_L(\tilde{x} \mid \mu, \sigma^2, \Lambda) = \Lambda + \exp(\mu + \tfrac{1}{2}\sigma^2) \qquad (12.11.7')$$

and

$$V_L(\tilde{x} \mid \mu, \sigma^2, \Lambda) = \exp(2\mu + \sigma^2)(\exp(\sigma^2) - 1). \qquad (12.11.8')$$

47. *Generalization of* (12.7.9′)

(a) Verify:

If $\tilde{\mathbf{x}}$ has df $f_N^{(k)}(\cdot \mid \boldsymbol{\mu}, \boldsymbol{\sigma}^2)$, $\tilde{\mathbf{x}}_1 \equiv (\tilde{x}_1, \ldots, \tilde{x}_{k-1})'$, and $\tilde{\mathbf{x}}$, $\boldsymbol{\mu}$, and $\boldsymbol{\sigma}^2$ are conformably partitioned into $= (\tilde{\mathbf{x}}'_1, \tilde{x}_k)'$, $\boldsymbol{\mu} = (\boldsymbol{\mu}'_1, \mu_k)'$, and

$$\boldsymbol{\sigma}^2 = \begin{bmatrix} \boldsymbol{\sigma}_{11}^2 & \boldsymbol{\sigma}_{1k}^2 \\ \boldsymbol{\sigma}_{k1}^2 & \sigma_{kk}^2 \end{bmatrix}, \qquad (12.7.9'')$$

then the conditional df of $\tilde{x}_k$ given that $\tilde{\mathbf{x}}_1 = \mathbf{x}_1$ is

$$f_N(\cdot \mid \mu_k + \boldsymbol{\sigma}_{k1}^2(\boldsymbol{\sigma}_{11}^2)^{-1}(\mathbf{x}_1 - \boldsymbol{\mu}_1), \sigma_{kk}^2 - \boldsymbol{\sigma}_{k1}^2(\boldsymbol{\sigma}_{11}^2)^{-1}\boldsymbol{\sigma}_{1k}^2),$$

where $\mu_k + \boldsymbol{\sigma}_{k1}^2(\boldsymbol{\sigma}_{11}^2)^{-1}(\mathbf{x}_1 - \boldsymbol{\mu}_1)$ is a linear function of $\mathbf{x}_1$, and

$$\sigma_{kk}^2 - \boldsymbol{\sigma}_{k1}^2(\boldsymbol{\sigma}_{11}^2)^{-1}\boldsymbol{\sigma}_{1k}^2 > 0.$$

(b) Let

$$\rho_{kk.12\ldots(k-1)} \equiv \boldsymbol{\sigma}_{k1}^2(\boldsymbol{\sigma}_{11}^2)^{-1}\boldsymbol{\sigma}_{1k}^2/[\sigma_{kk}^2\boldsymbol{\sigma}_{k1}^2(\boldsymbol{\sigma}_{11}^2)^{-1}\boldsymbol{\sigma}_{1k}^2]^{1/2}.$$

Show that the conditional variance of $\tilde{x}_k$ given $\tilde{\mathbf{x}}_1 = \mathbf{x}_1$ can be expressed as $\sigma_{kk}^2(1 - \rho^2_{kk\cdot 12\ldots(k-1)})$. We call $\rho_{kk\cdot 12\ldots(k-1)}$ the *multiple correlation coefficient* between $\tilde{x}_k$ and $\tilde{x}_1, \ldots, \tilde{x}_{k-1}$; and $(1 - \rho^2_{kk\cdot 12\ldots(k-1)})\sigma_{kk}^2$ the least squares residual variance of $\tilde{x}_k$ on $\tilde{x}_1, \ldots, \tilde{x}_{k-1}$ (for reasons which become clear in the next exercise).

(c) Show that $\rho_{kk\cdot 12\ldots(k-1)} \geqq 0$, with equality obtaining only if $\boldsymbol{\sigma}_{1k}^2$ $(= \boldsymbol{\sigma}_{k1}^{2\prime}) = 0$.

48. *Introduction to Least Squares Linear Regression Theory*: Let $\tilde{\mathbf{x}}$ be a k-dimensional random vector with expectation vector $\boldsymbol{\mu}$ and positive-definite covariance matrix $\boldsymbol{\sigma}^2$. The distribution of $\tilde{\mathbf{x}}$ need not be Normal. Our purpose is to find a linear function $\tilde{\hat{x}}_k \equiv \beta_0 + \sum_{i=1}^{k-1} \beta_i \tilde{x}_i \equiv \beta_0 + \boldsymbol{\beta}'\tilde{\mathbf{x}}_1$ which best predicts $\tilde{x}_k$ in the sense of minimizing the penalty (or loss) function

$$l(\beta_0, \boldsymbol{\beta}) \equiv E([\tilde{x}_k - \tilde{\hat{x}}_k]^2) = E([\tilde{x}_k - \beta_0 - \boldsymbol{\beta}'\tilde{\mathbf{x}}_1]^2). \qquad (12.11.10)$$

(The reason for not incorporating β_0 into the definition of $\boldsymbol{\beta}$ will be made clear shortly.)

(a) Let $\tilde{\mathbf{y}} \equiv (\tilde{\hat{x}}_k, \tilde{x}_k)'$. Show that

$$E(\tilde{\mathbf{y}}) = (\beta_0 + \boldsymbol{\beta}'\boldsymbol{\mu}_1, \mu_k)'$$

and

$$V(\tilde{\mathbf{y}}) = \begin{bmatrix} \boldsymbol{\beta}'\boldsymbol{\sigma}_{11}^2 & \boldsymbol{\beta}'\boldsymbol{\sigma}_{1k}^2 \\ \boldsymbol{\sigma}_{k1}^2\boldsymbol{\beta} & \sigma_{kk}^2 \end{bmatrix}.$$

[*Hint*: $\tilde{\mathbf{y}} = \mathbf{B}\tilde{\mathbf{x}} + \mathbf{c}$ for appropriate definitions of $\mathbf{B}$ and $\mathbf{c}$; (10.4.6) and (10.4.7) are revelant.]

(b) Let $\tilde{w} \equiv \tilde{x}_k - \hat{\tilde{x}}_k \equiv \tilde{y}_2 - \tilde{y}_1$. Then $(\beta_0, \boldsymbol{\beta})$ minimizes $\ell(\beta_0, \boldsymbol{\beta})$ if and only if $(\beta_0, \boldsymbol{\beta})$ minimizes $E(\tilde{w}^2) = V(\tilde{w}) + [E(\tilde{w})]^2$. Show that

$$E(\tilde{w}^2) = \boldsymbol{\beta}'\boldsymbol{\sigma}_{11}{}^2\boldsymbol{\beta} - 2\boldsymbol{\beta}'\boldsymbol{\sigma}_{1k}{}^2 + \sigma_{kk}{}^2 + [\mu_k - \beta_0 - \boldsymbol{\beta}'\boldsymbol{\mu}_1]^2,$$

and argue that the optimal choice $\beta_0{}^0$ of β_0 must satisfy $\beta_0{}^0 = \mu_k - \boldsymbol{\beta}^{0\prime}\boldsymbol{\mu}_1$, where $\boldsymbol{\beta}^0$ denotes the optimal choice of $\boldsymbol{\beta}$.

(c) Show that $E(\tilde{w}^2)$ [and hence $\ell(\beta_0, \boldsymbol{\beta})$] is minimized at

$$(\beta_0{}^0, \boldsymbol{\beta}^{0\prime})' = (\mu_k - \boldsymbol{\sigma}_{k1}{}^2(\boldsymbol{\sigma}_{11}{}^{-1})\boldsymbol{\mu}_1, \{(\boldsymbol{\sigma}_{11}{}^2)^{-1}\boldsymbol{\sigma}_{1k}{}^2\}')'.$$

[*Hint*: It is easy to verify the following (column) gradient vectors: $\partial(\boldsymbol{\beta}'\boldsymbol{\sigma}_{11}{}^2\boldsymbol{\beta})/\partial\boldsymbol{\beta} = 2\boldsymbol{\sigma}_{11}{}^2\boldsymbol{\beta}$; and $\partial(\boldsymbol{\beta}'\boldsymbol{\sigma}_{1k}{}^2)/\partial\boldsymbol{\beta} = \boldsymbol{\sigma}_{1k}{}^2$. Solve for $\boldsymbol{\beta}^0$ first, and then for $\beta_0{}^0$.]

(d) Show that the best linear function $\hat{\tilde{x}}_k$ of $\tilde{x}_1, \ldots, \tilde{x}_{k-1}$ for predicting $\tilde{x}_k$ is

$$\hat{\tilde{x}}_k = \mu_k + \boldsymbol{\sigma}_{k1}{}^2(\boldsymbol{\sigma}_{11}{}^2)^{-1}(\tilde{\mathbf{x}}_1 - \boldsymbol{\mu}_1), \tag{12.11.11}$$

and that

$$\ell(\beta_0{}^0, \boldsymbol{\beta}^0) = \sigma_{kk}{}^2 - \boldsymbol{\sigma}_{k1}{}^2(\boldsymbol{\sigma}_{11}{}^2)^{-1}\boldsymbol{\sigma}_{1k}{}^2 = V(\tilde{w}(\beta_0{}^0, \boldsymbol{\beta}^0)). \tag{12.11.12}$$

The random variable $\tilde{w}$ is called the *residual*; it represents the discrepancy between $\tilde{x}_k$ and its linear prediction or estimate $\hat{\tilde{x}}_k$. $\ell(\beta^0, \boldsymbol{\beta}^0)$ is frequently denoted by $\sigma_{k\cdot 12\ldots k}{}^2$ or $\sigma_{kk\cdot 12\ldots k}{}^2$, which is called the *least squares residual variance* of $\tilde{x}_k$ *on* $\tilde{x}_1, \ldots, \tilde{x}_{k-1}$.

(e) Show that $V(\hat{\tilde{x}}_k(\beta_0{}^0, \boldsymbol{\beta}^0), \tilde{x}_k) = \boldsymbol{\sigma}_{k1}{}^2(\boldsymbol{\sigma}_{11}{}^2)^{-1}\boldsymbol{\sigma}_{1k}{}^2$, and hence that the correlation between $\tilde{x}_k$ and $\hat{\tilde{x}}_k(\beta_0{}^0, \boldsymbol{\beta}^0)$ is given by

$$\rho_{kk\cdot 12\ldots(k-1)} \equiv \frac{\boldsymbol{\sigma}_{k1}{}^2(\boldsymbol{\sigma}_{11}{}^2)^{-1}\boldsymbol{\sigma}_{1k}{}^2}{[\sigma_{kk}{}^2\boldsymbol{\sigma}_{k1}{}^2(\boldsymbol{\sigma}_{11}{}^2)^{-1}\boldsymbol{\sigma}_{1k}{}^2]^{1/2}} = \left[\frac{\boldsymbol{\sigma}_{k1}{}^2(\boldsymbol{\sigma}_{11}{}^2)^{-1}\boldsymbol{\sigma}_{1k}{}^2}{\sigma_{kk}{}^2}\right]^{1/2}.$$

The quantity $\rho_{kk\cdot 12\ldots(k-1)}$, also denoted by $\rho_{k\cdot 12\ldots(k-1)}$, is called the *multiple correlation coefficient between* $\tilde{x}_k$ and $\tilde{x}_1, \ldots, \tilde{x}_{k-1}$.

(f) Show that $\sigma_{kk\cdot 12\ldots(k-1)}{}^2 = \sigma_{kk}{}^2(1 - \rho_{kk\cdot 12\ldots(k-1)}{}^2)$, and thus $1 - \rho_{kk\cdot 12\ldots(k-1)}{}^2$ represents that fraction of the variance of $\tilde{x}_k$ "left unexplained" by $\hat{\tilde{x}}_k(\beta_0{}^0, \boldsymbol{\beta}^0)$.

(g) Show that $\hat{x}_k(\beta_0{}^0, \boldsymbol{\beta}^0)$ and $\mu_{k|12\ldots(k-1)}$ coincide only when $(\mu_{k|12\ldots(k-1)} - \mu_k)$ is a linear transformation of $\mathbf{x}_1 - \boldsymbol{\mu}_1$.

49. *An Investment Portfolio Problem*: Suppose that $\tilde{x}_i$ represents the market value at year-end of one dollar invested now in the common stock of company i, for $i = 1, \ldots, k$.

(a) Is it reasonable to believe that $\tilde{x}_1, \ldots, \tilde{x}_k$ are independent random variables?

(b) Suppose that an investor has expressed his judgments about $\tilde{\mathbf{x}} \equiv (\tilde{x}_1, \ldots, \tilde{x}_k)'$ in the form of a k-variate Normal df $f_N^{(k)}(\cdot \mid \boldsymbol{\mu}, \boldsymbol{\sigma}^2)$; that he has T dollars to invest; and that he plans to invest T_i dollars in security i ($i = 1, \ldots, k$), where $\sum_{i=j}^{k} T_i = T$ and every $T_i \geqq 0$. What is the df of his total return $\sum_{i=1}^{k} T_i\tilde{x}_i$? (Assume that any positive amount can be invested in any security i without affecting the return rate $\tilde{x}_i$.)

(c) The correct answer to (b) will imply that if $k = 2$, $T = 1$, $T_1 \equiv \pi \in [0, 1]$ (and hence $T_2 \equiv 1 - \pi$), $\boldsymbol{\mu} = (\mu_1, \mu_2)'$, and

$$\boldsymbol{\sigma}^2 \equiv \begin{bmatrix} \sigma_{11}{}^2 & \sigma_{12}{}^2 \\ \sigma_{21}{}^2 & \sigma_{22}{}^2 \end{bmatrix},$$

then $\pi\tilde{x}_1 + (1 - \pi)\tilde{x}_2$ has df $f_N(\cdot \mid \mu(\pi), \sigma^2(\pi))$, where

$$\mu(\pi) \equiv \mu_2 + \pi(\mu_1 - \mu_2)$$

and

$$\sigma^2(\pi) = \pi^2\sigma_{11}{}^2 + (1 - \pi)^2\sigma_{22}{}^2 + 2\pi(1 - \pi)\sigma_{12}{}^2.$$

Depict the codependence of $\mu(\pi)$ and $\sigma^2(\pi)$ upon π graphically in the (μ, σ^2)-plane for:

(i) $\mu_1 = 1.2$, $\mu_2 = 1.3$, $\sigma_{11}{}^2 = .36$, $\sigma_{22}{}^2 = 1$, $\sigma_{12}{}^2 = .3$;
(ii) $\mu_1 = 1.2$, $\mu_2 = 1.3$, $\sigma_{11}{}^2 = .36$, $\sigma_{22}{}^2 = 1$, $\sigma_{12}{}^2 = 0$;
(iii) $\mu_1 = 1.2$, $\mu_2 = 1.3$, $\sigma_{11}{}^2 = .36$, $\sigma_{22}{}^2 = 1$, $\sigma_{12}{}^2 = -.3$; and
(iv) $\mu_1 = 1.2$, $\mu_2 = 1.3$, $\sigma_{11}{}^2 = 1$, $\sigma_{22}{}^2 = .36$, $\sigma_{12}{}^2 = .3$.

50. *Skewness and Excess*: The *coefficient of skewness* γ_1 of a random variable $\tilde{x}$ is defined by

$$\gamma_1 \equiv \frac{\mu_{(3)}}{[\mu_{(2)}]^{3/2}} = \frac{[\tilde{x} - E(\tilde{x})]^3}{[V(\tilde{x})]^{3/2}},$$

while the *coefficient of excess* γ_2 of $\tilde{x}$ is defined by

$$\gamma_2 \equiv \frac{\mu_{(4)}}{[\mu_{(2)}]^2} - 3 \equiv \frac{E[\tilde{x} - E(\tilde{x})]^4}{[V(\tilde{x})]^2} - 3.$$

(a) Suppose the dgf $f(x)$ of $\tilde{x}$ is symmetric about a point μ; that is, $f(\mu - a) = f(\mu + a)$ for all $a \in R^1$. Show that $E(\tilde{x}) = \mu$ and $\mu_{(i)} = 0$ for all odd integers i such that $\mu_{(i)}$ exists. Hence $\gamma_1 = 0$ for symmetric dgf's.

(b) Show that if $\tilde{x}$ has df $f_N(x \mid \mu, \sigma^2)$, then $\gamma_1 = 0$ and $\gamma_2 = 0$. [The subtraction of 3 in (12.11.15) is intended to make $\gamma_2 = 0$ for Normal random variables. If $\gamma_2 > 0$ for a (continuous) random variable $\tilde{x}$, its df tends to be more "peaked" about μ than the Normal df with the same mean and variance. Conversely, $\gamma_2 < 0$ implies that the df of $\tilde{x}$ is less than Normally peaked about μ.]

(c) Let $\tilde{x}$ have df $f_S(x \mid \mu, \sigma^2, \nu)$. Show that $\gamma_1 = 0$, provided that $\nu > 3$, and $\gamma_2 = 6/(\nu - 4)$, provided that $\nu > 4$.

[*Hints*: (1) $f_S(x \mid \mu, \sigma^2, \nu)$ is symmetric about μ.
(2) $E_S([\tilde{x} - \mu]^4 \mid \mu, \sigma^2, \nu) = E_{\gamma_2}\{E_N([\tilde{x} - \mu]^4 \mid \mu, \sigma^2/\tilde{h}) \mid 1, \nu\}$.
(3) $\Gamma(\frac{1}{2}\nu) = (\frac{1}{4})(\nu - 2)(\nu - 4)\Gamma(\frac{1}{2}\nu - 2)$.]

(d) Let $\tilde{t}$ have df $f_\gamma(t \mid r, \lambda)$. Show that $\gamma_1 = 2/\sqrt{r}$ and $\gamma_2 = 6/r$.

51. *The Generalized Laplace Distribution*: The continuous random variable $\tilde{x}$ is said to have the *generalized Laplace distribution* with parameters $\mu \in R^1$, $\phi > 0$, and $\zeta > 0$ if the df of $\tilde{x}$ is

$$f_{gLa}(x \mid \mu, \phi, \zeta) = G(\phi, \zeta) \exp\left[-\frac{1}{\zeta}\left|\frac{x - \mu}{\phi}\right|^\zeta\right], \quad -\infty < x < \infty, \tag{12.11.16}$$

where $|a|$ denotes the absolute value of a and

$$G(\phi,\zeta) \equiv [2\phi\Gamma(\zeta^{-1})\zeta^{(1-\zeta)/\zeta}]^{-1}. \tag{12.11.17}$$

(a) Show that $\int_{-\infty}^{\infty} f_{gLa}(x \mid \mu, \phi, \zeta)dx = 1$.

[*Hints*: (1) $f_{gLa}(x \mid \mu, \phi, \zeta)$ is symmetric about μ;

(2) To find $\int_a^{\infty} f_{gLa}(x \mid u, \phi, \zeta)dx$, change variable from x to

$$z \equiv \zeta^{-1}[(x - \mu)/\phi]^\zeta.]$$

(b) Show that if $\tilde{x}$ has df $f_{gLa}(x \mid \mu, \sigma^2, \nu)$, then $\mu_{(i)} = 0$, for all odd integers $i > 0$.

(c) Sketch $f_{gLa}(\cdot \mid 0, 1, 1)$. [The function $f_{gLa}(\cdot \mid \mu, \phi, 1)$ is the "nongeneralized" Laplace df.]

(d) Show that $f_{gLa}(x \mid \mu, \phi, 2) = f_N(x \mid \mu, \phi^2)$, for all real x.

(e) Show that $\mu_{(i)} = (\phi\zeta^{1/\zeta})^i\Gamma(\zeta^{-1}[1 + i])/\Gamma(\zeta^{-1})$, for all even integers $i > 0$.

[*Hints*: (1) $\int_{-\infty}^{\infty} (x - \mu)^i f_{gLa}(x \mid \mu, \phi, \zeta)dx$

$$= 2\int_{\mu}^{\infty} (x - \mu)^i f_{gLa}(x \mid \mu, \phi, \zeta)dx, \qquad \text{for } i \text{ even.}$$

(2) See Hint (2) of (a).]

(f) Show that if $\tilde{x}$ has df $f_{gLa}(x \mid \mu, \phi, \zeta)$, then $\gamma_1 = 0$ and

$$\gamma_2 = \Gamma(5\zeta^{-1})(\zeta^{-1})/[\Gamma(3\zeta^{-1})]^2 - 3.$$

(g) Show from (f) that if $\tilde{x}$ has the Laplace df $f_{gLa}(x \mid \mu, \phi, 1)$, then $\gamma_2 = 3$.

(h) Show that γ_2 in (f) can be re-expressed as

$$\gamma_2 = -3 + (9/5)\{\Gamma(1 + 5\zeta^{-1})\Gamma(1 + \zeta^{-1})/[\Gamma(1 + 3\zeta^{-1})]^2\},$$

and hence that $\lim_{\zeta \to \infty} \gamma_2 = -6/5$.

52. (a) Prove the following:

If the k-dimensional random vector $\tilde{\mathbf{x}}$ has df $f_N^{(k)}(\mathbf{x} \mid \boldsymbol{\mu}, \boldsymbol{\sigma}^2)$ and $\boldsymbol{\sigma}^2$ is a diagonal matrix [that is, $\sigma_{ij}^2 = 0$ whenever $i \neq j$], then $\tilde{x}_1, \ldots, \tilde{x}_k$ are independent random variables. (12.11.18)

(The hypothesis that the components $\tilde{x}_1, \ldots, \tilde{x}_k$ of $\tilde{\mathbf{x}}$ have a joint normal distribution is not superfluous; recall from Exercise 10.20 that $\sigma_{ij}^2 = 0$ does not imply in general that $\tilde{x}_i$ and $\tilde{x}_j$ are independent.)

(b) Does (12.11.18) remain true if $f_S^{(k)}(\mathbf{x} \mid \boldsymbol{\mu}, \boldsymbol{\sigma}^2, \nu)$ is substituted for $f_N^{(k)}(\mathbf{x} \mid \boldsymbol{\mu}, \boldsymbol{\sigma}^2)$?

53. *Modes*: A *mode* of a discrete or continuous random vector $\tilde{\mathbf{x}}$ is any value $\hat{\mathbf{x}}$ of $\tilde{x}$ which satisfies

$$f(\hat{\mathbf{x}}) \geqq f(\mathbf{x}), \qquad \text{all } \mathbf{x} \in R^k, \tag{12.11.19}$$

where $f(\cdot)$ is the df [or mf] of $\tilde{\mathbf{x}}$ if $\tilde{\mathbf{x}}$ is a continuous [or discrete] random vector. [If $\tilde{x}$ is a mixed random variable, a mode of $\tilde{x}$ is any number $\hat{x}$ which satisfies $P[\tilde{x} = \hat{x}] \geqq P[\tilde{x} = x]$, all $x \in R^1$.] Prove the following:

If $\tilde{t}$ has df $f_\gamma(t \mid r, \lambda)$, then $\hat{t} = (r - 1)/\lambda$, provided that $r > 1$. If $0 < r \leqq 1$, then $\tilde{t}$ does not possess a mode. (12.11.20)

If $\tilde{x}$ has df $f_{\gamma 2}(x \mid \psi, \nu)$, then $\hat{x} = \psi^{-1}(\nu - 2)/\nu$, provided that $\nu > 2$. If $0 < \nu \leqq 2$, then $\tilde{x}$ does not possess a mode. (12.11.21)

If $\tilde{y}$ has df $f_{i\gamma}(y \mid \psi, \nu)$, then $\hat{y} = \psi\nu/(\nu + 2)$. (12.11.22)

If $\tilde{w}$ has df $f_{i\gamma 2}(w \mid \psi, \nu)$, then $\hat{w} = [\nu\psi/(\nu + 1)]^{1/2}$. (12.11.23)

If $\tilde{z}$ has df $f_\beta(z \mid \rho, \nu)$, then $\hat{z} = (\rho - 1)/(\nu - 2)$, provided that $\rho > 1$ and $\nu - \rho > 1$. If $\rho \leqq 1$ and/or $\nu - \rho \leqq 1$, then $\tilde{z}$ does not possess a mode. (12.11.24)

If $\tilde{y}$ has df $f_{i\beta 2}(y \mid \rho, \nu, b)$, then $\hat{y} = b(\rho - 1)/(\nu - \rho + 1)$, provided that $\rho > 1$. If $\rho \leqq 1$, then $\tilde{y}$ does not possess a mode. (12.11.25)

If $\tilde{x}$ has df $f_N(x \mid \mu, \sigma^2)$, then $\hat{x} = \mu$. (12.11.26)

If $\tilde{x}$ has df $f_S(x \mid \mu, \sigma^2, \nu)$, then $\hat{x} = \mu$. (12.11.27)

If $\tilde{\mathbf{z}}$ has df $f_\beta^{(k)}(\mathbf{z} \mid \boldsymbol{\rho}, \nu)$, then $\hat{\mathbf{z}} = (\nu - k - 1)^{-1}(\boldsymbol{\rho} - \mathbf{1})$, where $\mathbf{1} \equiv (1, 1, \ldots, 1)'$, provided that $\nu > \sum_{i=1}^{k} \rho_i + 1$ and $\rho_i > 1$, for $i = 1, \ldots, k$; otherwise, $\tilde{\mathbf{z}}$ does not possess a mode. (12.11.28)

If $\tilde{\mathbf{x}}$ has df $f_N^{(k)}(\mathbf{x} \mid \mu, \sigma^2)$, then $\hat{\mathbf{x}} = \boldsymbol{\mu}$. (12.11.29)

If $\tilde{\mathbf{x}}$ has df $f_S^{(k)}(\mathbf{x} \mid \boldsymbol{\mu}, \boldsymbol{\sigma}^2, \nu)$, then $\hat{\mathbf{x}} = \boldsymbol{\mu}$. (12.11.30)

For the information of the reader, it can be shown that the following two results obtain:

If $\tilde{\mathbf{X}}$ has df $f_{W}^{(k)}(\mathbf{X} \mid \boldsymbol{\psi}, \nu)$, then $\hat{\mathbf{X}} = [(\nu - 2)/\nu]\boldsymbol{\psi}^{-1}$ provided that $\nu > 2$. If $\nu \leqq 2$, then $\tilde{\mathbf{X}}$ does not possess a mode. [compare (12.11.21)]. (12.11.31)

If $\tilde{\mathbf{Y}}$ has df $f_{iW}^{(k)}(\mathbf{Y} \mid \boldsymbol{\psi}, \nu)$, then $\hat{\mathbf{Y}} = [\nu/(\nu + 2k)]\boldsymbol{\psi}$ [compare (12.11.22)]. (12.11.32)

54. *Stratified Sampling (I)*: In stratified sampling the population is partitioned into r classes, or strata $C_1, \ldots, C_r$ and a random sample of size $n_i \geqq 0$ is drawn from stratum C_i for $i = 1, \ldots, r$. Suppose that observations $\tilde{\mathbf{x}}_i$ from stratum C_i are iid with common df $f_N^{(k)}(\mathbf{x}_i \mid \boldsymbol{\mu}_i, \boldsymbol{\sigma}_{ii}^2)$ and that π_i denotes the (known) proportion of the total population contained in C_i, for $i = 1, \ldots, r$. Hence $\pi_i \geqq 0$, for every i and $\sum_{i=1}^{r} \pi_i = 1$.

(a) Let $\boldsymbol{\mu}$ and $\boldsymbol{\sigma}^2$ denote respectively the expectation vector and covariance matrix of an observation $\tilde{\mathbf{x}}$ drawn at random from the total population. Show that

$$\boldsymbol{\mu} = \sum_{i=1}^{r} \pi_i \boldsymbol{\mu}_i \tag{12.11.33}$$

and

$$\boldsymbol{\sigma}^2 = \sum_{i=1}^{r} \pi_i \boldsymbol{\sigma}_{ii}^2 + \sum_{i=1}^{r} \pi_i (\boldsymbol{\mu}_i - \boldsymbol{\mu})(\boldsymbol{\mu}_i - \boldsymbol{\mu})'. \tag{12.11.34}$$

[*Hint*: Consider the selection of $\tilde{\mathbf{x}}$ as proceeding in two stages: (1) a randomization to select the stratum, in which C_i is selected with probability π_i; and (2) a randomization within the selected stratum i, in which an observation can be considered as being generated from $f_N^{(k)}(\cdot \mid \boldsymbol{\mu}_i, \boldsymbol{\sigma}_{ii}^2)$. Then use (10.2.5) and (10.7.9).]

(b) Define

$$\boldsymbol{\sigma}_w^2 \equiv \sum_{i=1}^{r} \pi_i \boldsymbol{\sigma}_{ii}^2 \tag{12.11.35}$$

and

$$\boldsymbol{\sigma}_b^2 \equiv \sum_{i=1}^{r} \pi_i (\boldsymbol{\mu}_i - \boldsymbol{\mu})(\boldsymbol{\mu}_i - \boldsymbol{\mu})'. \tag{12.11.36}$$

Then (12.11.34) can be rewritten as

$$\boldsymbol{\sigma}^2 = \boldsymbol{\sigma}_w^2 + \boldsymbol{\sigma}_b^2. \tag{12.11.37}$$

The subscript of $\boldsymbol{\sigma}_w^2$ connotes that $\boldsymbol{\sigma}_w^2$ represents the average covariance matrix of an observation, measured with respect to the mean $\boldsymbol{\mu}_i$ of its stratum and averaged with respect to the stratum proportions: $\boldsymbol{\sigma}_w^2$ is called the (average) *within*-stratum covariance matrix. On the other hand, $\boldsymbol{\sigma}_b^2$ is called the *between*-stratum covariance matrix; it is a measure of the degree to which the stratum mean vectors $\boldsymbol{\mu}_i$ differ from each other. Show that $\boldsymbol{\sigma}_b^2$ is the covariance matrix of a discrete random vector $\tilde{\mathbf{y}}$ which equals $\boldsymbol{\mu}_i$ with probability π_i, for $i = 1, \ldots, r$.

(c) Suppose that n observations are to be taken at random from the total population and that the sample mean $\tilde{\bar{x}}$ is to be recorded. Show that

$$E(\tilde{\bar{x}}) = \boldsymbol{\mu} \tag{12.11.38}$$

and

$$V(\tilde{\bar{x}}) = n^{-1}(\boldsymbol{\sigma}_w^2 + \boldsymbol{\sigma}_b^2). \tag{12.11.39}$$

(d) Suppose that $n_i > 0$ observations are to be taken at random from stratum C_i, for $i = 1, \ldots, r$, and that the sample means $\tilde{\bar{x}}_1, \ldots, \tilde{\bar{x}}_r$ are to be recorded. Let $\tilde{\bar{\mathbf{x}}}^* \equiv \sum_{i=1}^{r} \pi_i \tilde{\bar{\mathbf{x}}}_i$. Show that

$$E(\tilde{\bar{x}}^*) = \boldsymbol{\mu}. \tag{12.11.40}$$

$$V(\tilde{\bar{\mathbf{x}}}^*) = \sum_{i=1}^{r} \pi_i^2 n_i^{-1} \boldsymbol{\sigma}_{ii}^2; \tag{12.11.41}$$

and, if $n_i = \pi_i n$, for every i, then

$$V(\tilde{\bar{x}}^*) = n^{-1} \boldsymbol{\sigma}_w^2. \tag{12.11.42}$$

(e) Let $k = 1$. By comparing (12.11.39) and (12.11.42), show that $\tilde{\bar{x}}^*$ provides more information about μ than does $\tilde{\bar{x}}$.

(f) Show that the df of $\tilde{\bar{x}}^*$ is $f_N^{(k)}(\cdot \mid \boldsymbol{\mu}, \sum_{i=1}^{r} \pi_i^2 n_i^{-1} \boldsymbol{\sigma}_{ii}^2)$.

(g) Show that, conditional upon $n_1, \ldots, n_r$ observations coming from strata $C_1, \ldots, C_r$, the df of

$$\tilde{\bar{\mathbf{x}}} \left(= (1/n) \sum_{i=1}^{r} n_i \tilde{\bar{\mathbf{x}}}_i\right) \text{ is } f_N^{(k)}\left(\cdot \mid \sum_{i=1}^{r} (n_i/n)\boldsymbol{\mu}_i, \sum_{i=1}^{r} (n_i/n^2)\boldsymbol{\sigma}_{ii}^2\right).$$

[The primary advantage of stratified sampling over random sampling as previously considered—called *simple random sampling*—is that stratified sampling can exploit within-stratum homogeneity (small $\boldsymbol{\sigma}_w^2$) vis-à-vis between strata heterogeneity (large $\boldsymbol{\sigma}_b^2$) by eliminating $\boldsymbol{\sigma}_b^2$ from the variance of the statistic $\tilde{\bar{\mathbf{x}}}^*$ (used to estimate $\boldsymbol{\mu}$); compare (12.11.39) and (12.11.42).]

13

NORMAL PROCESSES AND CENTRAL LIMIT THEORY

13.1 INTRODUCTION

In Chapter 12 we asserted that the family of Normal distributions is of utmost importance in probability and statistics. A *superficial* reason for its importance is that many natural and behavioral phenomena (for example, individuals' heights, aptitude test scores, annual rainfall) have relative frequencies which *seem* normal, in that they are well approximated by bell curves. A somewhat more profound reason explains the apparent normality of these phenomena in terms of the *Central Limit Theorem*, the subject of Sections 13.8 and 13.9.

Sections 13.2 through 13.4 concern the univariate Normal data-generating process, in which the observations $\tilde{x}_1, \tilde{x}_2, \ldots$ are assumed to be iid with common df $f_N(x \mid \mu, \sigma^2)$. In Section 13.2 we derive the likelihood function for n observations with noninformative stopping, and determine sufficient statistics under each of three assumptions concerning prior knowledge of μ and σ^2. In Section 13.3 we determine the conditional distributions of the sufficient statistics given μ and/or σ^2. Section 13.4 concerns situations in which observations are obtained from each of two univariate Normal df's for the purpose of comparing the parameters of these processes; we derive distributions of functions of the individual-process sufficient statistics.

Sections 13.5 through 13.7 concern the k-variate Normal data-generating process, in which the observations $\tilde{\mathbf{x}}_1, \tilde{\mathbf{x}}_2, \ldots$ are assumed to be iid k-tuples with common df $f_N^{(k)}(\mathbf{x} \mid \boldsymbol{\mu}, \boldsymbol{\sigma}^2)$. There is an exact topical correspondence of Sections 13.5, 13.6, and 13.7 to Sections 13.2, 13.3, and 13.4, respectively.

Sections 13.8 through 13.9 concern the Central Limit Theorem and some of its ramifications. We conclude this introduction by discussing the (univariate) Central Limit Theorem, as its consequences constitute motivation for the detailed examination of Normal processes to follow.

The Central Limit Theorem proved in 13.8 asserts that the sample mean $\tilde{\bar{x}} \equiv 1/n \sum_{i=1}^{n} \tilde{x}_i$ of n iid random variables $\tilde{x}_1, \ldots, \tilde{x}_n$ with common dgf $f(x)$ [and hence common expectation, or "population mean" μ and common, finite variance $\sigma^2 > 0$] has, for large n, a distribution which is approximately Normal with mean μ and variance σ^2/n. That is, if n is large, then

$$F(\bar{x}) \doteq F_N\left(\bar{x} \mid \mu, \frac{\sigma^2}{n}\right), \text{ for every } \bar{x} \in R^1. \tag{13.1.1}$$

Equivalently, the sample sum $\tilde{y} \equiv n\tilde{\bar{x}}$ has approximate distribution given by

$$F(y) \doteq F_N(y \mid n\mu, n\sigma^2). \tag{13.1.2}$$

From (13.1.2) we can explain the apparent normality of many real-life phenomena. In many circumstances, the over-all *deviation* of an individual observation x_i from its population mean μ is explainable as the sum of many small, independent, specific disturbances with individual expectations approximately zero. The Central Limit Theorem, and generalizations of it, imply that if the specific disturbances are small enough and numerous enough, then the over-all deviation will be approximately Normal with mean zero, and hence the individual observation will be approximately Normal with mean μ [by (12.4.10) applied to $\tilde{x}_i = \tilde{d}_i + \mu$ for $\tilde{d}_i \equiv$ deviation], thus the apparent normality of individual observations in many cases.

We hasten to caution the reader that many real phenomena do *not* have approximately Normal df's; their individual deviations *cannot* be explained as sums of small, independent disturbances from μ. Yet even in such cases the sample mean $\tilde{\bar{x}}$ of a large sample $\tilde{x}_1, \ldots, \tilde{x}_n$ will be approximately normally distributed about the population mean μ with variance σ^2/n, by virtue of (13.1.1).

Now, many experiments consist in obtaining a predetermined number n of iid observations $\tilde{x}_1, \ldots, \tilde{x}_n$ in order to make inferences about (their common $=$) the population mean $\mu(= E(\tilde{x} \mid \boldsymbol{\theta})$, where $\boldsymbol{\theta}$ denotes the unknown parameter(s) of the common dgf $f(x \mid \boldsymbol{\theta})$ of $\tilde{x}_1, \ldots, \tilde{x}_n)$. The Central Limit Theorem implies that if n is large, then we can make inferences about μ just as we would if the common dgf of $\tilde{x}_1, \ldots, \tilde{x}_n$ were $f_N(x \mid \mu, \sigma^2)$ instead of the more arbitrary $f(x \mid \boldsymbol{\theta})$. How large n must be in order for the approximation to be sufficiently accurate depends both upon the true nature of $f(x \mid \boldsymbol{\theta})$ and upon the precise rendition of "sufficiently accurate." Furthermore, an additional restriction is required when the common variance σ^2 of $\tilde{x}_1, \ldots, \tilde{x}_n$ is unknown.

13.2 THE UNIVARIATE NORMAL PROCESS (I): LIKELIHOOD FUNCTIONS

In this and the following section we assume that an experiment will result in n iid observations $\tilde{x}_1, \ldots, \tilde{x}_n$ with common df $f_N(x \mid \mu, \sigma^2)$, obtained with noninformative stopping. Define the random variables

$$\tilde{\bar{x}} \equiv \left(\frac{1}{n}\right) \sum_{i=1}^{n} \tilde{x}_i, \tag{13.2.1}$$

$$\tilde{s}^2 \equiv \left(\frac{1}{n-1}\right) \sum_{i=1}^{n} (\tilde{x}_i - \tilde{\bar{x}})^2. \tag{13.2.2}$$

and

$$\tilde{\xi} \equiv \left(\frac{1}{n}\right) \sum_{i=1}^{n} (\tilde{x}_i - \mu)^2. \tag{13.2.3}$$

Note that $\tilde{\xi}$ is not a statistic unless μ is known.

There are four possible states of prior knowledge about μ and σ^2 which may obtain: (1) σ^2 is known but μ is unknown; (2) μ is known but σ^2 is unknown; (3) neither μ nor σ^2 is known; and (4) both μ and σ^2 are known. The fourth state is uninteresting, as nothing remains to be inferred about the process.

Our first result is for case (2):

If $\tilde{x}_1, \ldots, \tilde{x}_n$ are iid with common df $f_N(x \mid \mu, \sigma^2)$ and μ is known, then $(n, \tilde{\xi})$ is conditionally sufficient for σ^2 given μ, and (13.2.4)

$$L(\sigma^2 \mid n, x_1, \ldots, x_n, \mu) = K \cdot (\sigma^2)^{-n/2} \exp\left(-\tfrac{1}{2} n\xi/\sigma^2\right),$$

where $K \equiv (2\pi)^{-n/2}$, all provided that the stopping process is noninformative.

Proof

The joint df of $\tilde{x}_1, \ldots, \tilde{x}_n$ is

$$\begin{aligned} f(x_1, \ldots, x_n \mid \mu, \sigma^2) &= \prod_{i=1}^{n} f_N(x_i \mid \mu, \sigma^2) \\ &= \prod_{i=1}^{n} (2\pi\sigma^2)^{-1/2} \exp\left[-\tfrac{1}{2}(x_i - \mu)/\sigma^2\right] \\ &= (2\pi\sigma^2)^{-n/2} \exp\left\{-[\tfrac{1}{2}/\sigma^2] \sum_{i=1}^{n} (x_i - \mu)^2\right\} \\ &= (2\pi\sigma^2)^{-n/2} \exp\left[-\tfrac{1}{2} n\xi/\sigma^2\right], \end{aligned}$$

all equalities being elementary to verify. This proves the second assertion, by virtue of the definition (9.3.4) of likelihood function. Conditional sufficiency of ξ is immediate from the second assertion and (9.7.5), in which $\boldsymbol{\theta}_2 = \mu$, $g(n, x_1, \ldots, x_n) = K$, $\boldsymbol{\theta}_1 = \sigma^2$, $\mathbf{x}_T(\mu) = \xi$, and $g(\xi \mid \mu, \sigma^2) = (\sigma^2)^{-n/2} \exp[-\frac{1}{2}n\xi/\sigma^2]$. QED

If $\tilde{x}_1, \ldots, \tilde{x}_n$ are iid with common df $f_N(x \mid \mu, \sigma^2)$ and σ^2 is known, then $(n, \tilde{\bar{x}})$ is conditionally sufficient for μ given σ^2, and (13.2.5)

$$L(\mu \mid n, x_1, \ldots, x_n, \sigma^2) = K \cdot \exp[-\tfrac{1}{2}n(\bar{x} - \mu)^2/\sigma^2],$$

where $K \equiv (2\pi\sigma^2)^{-n/2} \exp[-\frac{1}{2}(n-1)s^2/\sigma^2]$, all provided that the stopping process is noninformative.

Proof

As in the proof of (13.2.4).

$$\begin{aligned} L(\mu \mid n, x_1, \ldots, x_n, \sigma^2) &= f(x_1, \ldots, x_n \mid \mu, \sigma^2) \\ &= (2\pi\sigma^2)^{-n/2} \exp[-\tfrac{1}{2}n\xi/\sigma^2]. \end{aligned}$$

To prove the second assertion, it clearly suffices to prove that

$$\sum_{i=1}^{n} (x_i - \mu)^2 = n(\bar{x} - \mu)^2 + \sum_{i=1}^{n} (x_i - \bar{x})^2.$$

For this, we have

$$\begin{aligned} &\sum_{i=1}^{n} (x_i - \mu)^2 \\ &= \sum_{i=1}^{n} [(x_i - \bar{x}) + (\bar{x} - \mu)]^2 \\ &= \sum_{i=1}^{n} \{(x_i - \bar{x})^2 + 2(\bar{x} - \mu)(x_i - \bar{x}) + (\bar{x} - \mu)^2\} \\ &= \sum_{i=1}^{n} (x_i - \bar{x})^2 + 2(\bar{x} - \mu)\sum_{i=1}^{n} (x_i - \bar{x}) + n(\bar{x} - \mu)^2, \end{aligned}$$

which is the desired result in view of the fact that $\sum_{i=1}^{n} (x_i - \bar{x}) = 0$. Hence

$$(*) \quad \begin{aligned} f(x_1, \ldots, x_n \mid \mu, \sigma^2) &= (2\pi)^{-n/2}(\sigma^2)^{-n/2} \exp[-\tfrac{1}{2}(n-1)s^2/\sigma^2] \\ &\quad \times \exp[-\tfrac{1}{2}n(\bar{x} - \mu)^2/\sigma^2], \end{aligned}$$

from which the second assertion is obvious. (We star this equation for later use.) The first assertion now follows from (9.7.5). QED

If $\tilde{x}_1, \ldots, \tilde{x}_n$ are iid with common df $f_N(x \mid \mu, \sigma^2)$ and neither μ nor σ^2 is known, then $(n, \tilde{\bar{x}}, \tilde{s}^2)$ is sufficient for (μ, σ^2), and (13.2.6)

$$L(\mu, \sigma^2 \mid n, x_1, \ldots, x_n) = K(\sigma^2)^{-n/2} \exp[-\tfrac{1}{2}(n-1)s^2/\sigma^2] \exp[-\tfrac{1}{2}n(\bar{x} - \mu^2)/\sigma^2],$$

where $K \equiv (2\pi)^{-n/2}$, all provided that the stopping process is noninformative.

Proof

The second assertion is immediate from (*) in the proof of (13.2.5), and the first assertion follows from it by (9.5.1), in which $g(n, x_1, \ldots, x_n) \equiv (2\pi)^{-n/2}$. QED

13.3 THE UNIVARIATE NORMAL PROCESS (II): DISTRIBUTIONS OF STATISTICS

We now build upon the results of 13.2 by deriving the distributions of sufficient statistics and of functions of them, all conditional upon (μ, σ^2). First, a preliminary result follows:

If $\tilde{z}_1, \ldots, \tilde{z}_n$ are iid with common df $f_{N*}(z)$, then $\tilde{Q} \equiv \sum_{i=1}^{n} \tilde{z}_i^2$ has df $f_{\chi^2}(Q \mid n)$. (13.3.1)

Proof

Independence of the $\tilde{z}_i$'s implies independence of the $\tilde{z}_i^2$'s, so that if the assertion is true for $n = 1$ [that is, if $\tilde{z}^2$ had df $f_{\chi^2}(z^2 \mid 1)$ when $\tilde{z}$ has df $f_{N*}(z)$], then it is also true for every n by virtue of (12.2.17). We shall prove the assertion for $n = 1$ by finding the df of $\tilde{Q} \equiv \tilde{z}^2$. First,

$$F(Q) \equiv P(\tilde{Q} \leqq Q) = P(\tilde{z}^2 \leqq Q) = P(\tilde{z} \in [-\sqrt{Q}, \sqrt{Q}])$$

and hence

$$F(Q) = 1 - 2F_{N*}(-\sqrt{Q})$$

(by symmetry of the standardized Normal distribution about zero). Therefore

$$\begin{aligned} f(Q) = \frac{dF(Q)}{dQ} &= -2\frac{dF_{N*}(-\sqrt{Q})}{dQ} \\ &= -f_{N*}(-\sqrt{Q})\frac{d(-\sqrt{Q})}{dQ} \\ &= Q^{-1/2} f_{N*}(-\sqrt{Q}). \end{aligned}$$

But

$$\begin{aligned} Q^{-1/2} f_{N*}(-\sqrt{Q}) &= (2\pi)^{-1/2} Q^{-1/2} \exp\left[-\tfrac{1}{2}Q\right] \\ &= f_\gamma(Q \mid \tfrac{1}{2}, \tfrac{1}{2}) \\ &= f_{\chi^2}(Q \mid 1), \end{aligned}$$

by virtue of the fact that $\Gamma(\frac{1}{2}) = \sqrt{\pi}$. QED

We shall also require

If $\tilde{\chi}^2$ has df $f_{\chi^2}(\chi^2 \mid \nu)$ and $a > 0$, then $\tilde{Q} \equiv a\tilde{\chi}^2$ has df $f_{\gamma 2}(Q \mid (a\nu)^{-1}, \nu)$. (13.3.2)

Proof

Exercise 1.

We can now state some useful results:

If $\tilde{x}_1, \ldots, \tilde{x}_n$ are iid with common df $f_N(x \mid \mu, \sigma^2)$, then $\tilde{\bar{x}}$ has df $f_N(\bar{x} \mid \mu, \sigma^2/n)$ given (μ, σ^2). (13.3.3)

Proof

This is simply a restatement of (12.4.10c).

If $\tilde{x}_1, \ldots, \tilde{x}_n$ are iid with common df $f_N(x \mid \mu, \sigma^2)$, then $\tilde{Q} \equiv n\tilde{\xi}/\sigma^2$ has df $f_{\chi^2}(Q \mid n)$ and $\tilde{\xi}$ has df $f_{\gamma 2}(\xi \mid 1/\sigma^2, n)$ given σ^2. (13.3.4)

Proof

$n\tilde{\xi}/\sigma^2 \equiv \sum_{i=1}^{n} [(\tilde{x}_i - \mu)/\sigma]^2 = \sum_{i=1}^{n} \tilde{z}_i^2$, where the $\tilde{z}_i$'s are iid with common df $f_{N^*}(z)$. Hence the first assertion follows by (13.3.1), and the second, by (13.3.2). QED

If $\tilde{x}_1, \ldots, \tilde{x}_n$ are iid with common df $f_N(x \mid \mu, \sigma^2)$, then $\tilde{Q} \equiv n(\tilde{\bar{x}} - \mu)^2/\sigma^2$ has df $f_{\chi^2}(Q \mid 1)$. (13.3.5)

Proof

By (13.3.3), $(\tilde{\bar{x}} - \mu)/(\sigma/n^{1/2}) \equiv n^{1/2}(\tilde{\bar{x}} - \mu)/\sigma \equiv \tilde{z}$ has df $f_{N^*}(z)$, and hence $n(\tilde{\bar{x}} - \mu)^2/\sigma^2 = \tilde{z}^2$ has the asserted df, by (13.3.2). QED

If $\tilde{x}_1, \ldots, \tilde{x}_n$ are iid with common df $f_N(x \mid \mu, \sigma^2)$, then: $\tilde{Q} \equiv (n-1)\tilde{s}^2/\sigma^2$ has df $f_{\chi^2}(Q \mid n-1)$; $\tilde{s}^2$ has df $f_{\gamma 2}(s^2 \mid \sigma^{-2}, n-1)$; and $\tilde{\bar{x}}$ and $\tilde{s}^2$ are conditionally independent given (μ, σ^2), provided that $n \geqq 2$. (13.3.6)

Proof

A proof of the first and third assertions may be found in Wilks [**83**], pp. 208–210. The second follows from the first by (13.3.2). QED

The converse of the third part of (13.3.6) is true: if $\tilde{\bar{x}}$ and $\tilde{s}^2$ as defined by (13.2.1) and (13.2.2) are conditionally independent given (μ, σ^2), then the common dgf $f(x)$ of $\tilde{x}_1, \ldots, \tilde{x}_n$ must be $f_N(x \mid \mu, \sigma^2)$ for some mean μ and variance σ^2.

The following celebrated result was first proved by "Student," the pseudonym of W. S. Gossett, who was employed by an English firm which apparently discouraged publication.

If $\tilde{z}$ has df $f_{N*}(z)$, $\tilde{Q}$ has df $f_{\chi^2}(Q \mid \nu)$, and $\tilde{z}$ and $\tilde{Q}$ are independent, then (13.3.7)

$$\tilde{t} \equiv \frac{\tilde{z}}{(\tilde{Q}/\nu)^{1/2}}$$

has df $f_{S*}(t \mid \nu)$ [see (12.5.2)].

Proof

See Wilks [**83**], pp. 184–185.

As a corollary, the reader may show (Exercise 2) that

If $\tilde{x}_1, \ldots, \tilde{x}_n$ are iid with common df $f_N(x \mid \mu, \sigma^2)$ and $n \geqq 2$, then (13.3.8)

$$\tilde{t} \equiv (\bar{\tilde{x}} - \mu)/\sqrt{\tilde{s}^2/n}$$

has df $f_{S*}((t \mid n - 1)$.

It is in the form of (13.3.8) that Student's result is of great value in statistical inference, as it enables the statistician to make conclusions about μ when σ^2 is unknown.

Another result of great usefulness in inference follows:

If $\tilde{Q}_i$ has df $f_{\chi^2}(Q_i \mid \nu_i)$ for $i = 1, 2$ and $\tilde{Q}_1$ and $\tilde{Q}_2$ are independent, then $\tilde{\phi} \equiv (\tilde{Q}_1/\nu_1)/(\tilde{Q}_2/\nu_2)$ has the F-distribution $f_F(\phi \mid \nu_1, \nu_2)$ [see (12.3.19)]. (13.3.9)

Proof

See Wilks [**83**], pp. 186–187.

The F-distribution was discovered by Snedecor, who named it in honor of Sir Ronald A. Fisher, who had discovered the distribution of a related random variable. It is useful in making inferences about the relationship between the variances of two univariate Normal processes, as will be clear in the next section; it appears again in Section 13.6.

13.4 STATISTICS FOR COMPARING TWO UNIVARIATE NORMAL PROCESSES

This short section derives the distributions of statistics useful in comparing the means or the variances of two univariate Normal processes. Inferences based upon these statistics are introduced in Chapters 21 and 22.

We assume that $\tilde{x}_{11}, \ldots, \tilde{x}_{n_11}$ are iid with common df $f_N(x_1 \mid \mu_1, \sigma_1^2)$, that $\tilde{x}_{12}, \ldots, \tilde{x}_{n_22}$ are iid with common df $f_N(x_2 \mid \mu_2, \sigma_2^2)$, and that all $\tilde{x}_{j1}$'s are independent of the $\tilde{x}_{k2}$'s. These assumptions amount to obtaining, independently, n_1 observations from $f_N(x_1 \mid \mu_1, \sigma_1^2)$ and n_2 observations from $f_N(x_2 \mid \mu_2, \sigma_2^2)$.

Define

$$\bar{\tilde{x}}_i \equiv \left(\frac{1}{n_i}\right) \sum_{j=1}^{n_i} \tilde{x}_{ji}, \text{ for } i = 1, 2; \tag{13.4.1}$$

$$\tilde{s}_i^2 \equiv \left(\frac{1}{n_i - 1}\right) \sum_{j=1}^{n_i} (\tilde{x}_{ji} - \bar{\tilde{x}}_i)^2, \text{ for } i = 1, 2; \tag{13.4.2}$$

$$\tilde{\xi}_i \equiv \left(\frac{1}{n_i}\right) \sum_{j=1}^{n_i} (\tilde{x}_{ji} - \mu_i)^2, \text{ for } i = 1, 2; \tag{13.4.3}$$

$$\tilde{s}^2 \equiv \frac{(n_1 - 1)\tilde{s}_1^2 + (n_2 - 1)\tilde{s}_2^2}{n_1 + n_2 - 2}; \tag{13.4.4}$$

$$\Delta\bar{\tilde{x}} \equiv \bar{\tilde{x}}_1 - \bar{\tilde{x}}_2; \tag{13.4.5}$$

$$\Delta\mu = \mu_1 - \mu_2; \tag{13.4.6}$$

$$\tilde{\zeta}_i \equiv \begin{cases} \tilde{s}_i^2, & \text{if } \mu_i \text{ is unknown} \\ \tilde{\xi}_i, & \text{if } \mu_i \text{ is known;} \end{cases} \tag{13.4.7}$$

and

$$\nu_i \equiv \begin{cases} n_i - 1, & \text{if } \mu_i \text{ is unknown} \\ n_i, & \text{if } \mu_i \text{ is known.} \end{cases} \tag{13.4.8}$$

Our basic results can now be stated succinctly.

With notation and assumptions as in the preceding, the df of (13.4.9)

$(\Delta\bar{\tilde{x}} - \Delta\mu)/[(1/n_1)\sigma_1^2 + (1/n_2)\sigma_2^2]^{1/2} = \tilde{z}$ is $f_{N*}(z)$.

Proof

$\bar{\tilde{x}}_i$ has df $f_N(\bar{x}_i \mid \mu_i, \sigma_i^2, n_i)$, by (13.3.3), for $i = 1, 2$; and $\bar{\tilde{x}}_1$ and $\bar{\tilde{x}}_2$ are independent. Hence $\Delta\bar{\tilde{x}} \equiv \bar{\tilde{x}}_1 - \bar{\tilde{x}}_2$ has df $f_N(\Delta\bar{x} \mid \Delta\mu, \sigma_1^2/n_1 + \sigma_2^2/n_2)$, by (12.4.8a′). The rest is obvious. QED.

With notation and assumptions as in the preceding, the df of $(\tilde{\zeta}_1/\sigma_1^2)/(\tilde{\zeta}_2/\sigma_2^2) \equiv \tilde{\phi}$ is $f_F(\phi \mid \nu_1, \nu_2)$. (13.4.10)

Proof

By (13.3.4) and (13.3.6), the df of $\tilde{Q}_i \equiv \nu_i\tilde{\zeta}_i/\sigma_i^2$ is $f_{\chi^2}(Q_i \mid \nu_i)$; and $\tilde{Q}_1$ and $\tilde{Q}_2$ are independent. Hence by (13.3.9), $\tilde{\phi} \equiv (\tilde{Q}_1/\nu_1)/(\tilde{Q}_2/\nu_2) = (\tilde{\zeta}_1/\sigma_1^2)/(\tilde{\zeta}_2/\sigma_2^2)$ has the asserted df. QED

Finally, we have the following:

> With notation and assumptions as in the preceding, and with the additional assumption that $\sigma_1^2 = \sigma_2^2 \equiv \sigma^2$, the df of $(\Delta\tilde{\bar{x}} - \Delta\mu)/[(n_1^{-1} + n_2^{-1})\tilde{\bar{s}}^2]^{1/2} \equiv \tilde{t}$ is $f_{S*}(t \mid n_1 + n_2 - 2)$. (13.4.11)

Proof

Exercise 3.

It is instructive to compare (13.4.9) and (13.4.11), both of which are used in making inferences about the difference $\Delta\mu$. If $\sigma_1^2 = \sigma_2^2$ in (13.4.9), that result becomes identical to (13.4.11) except that $\tilde{\bar{s}}^2$ in (13.4.11) replaces σ^2 in (13.4.9). Roughly speaking, in (13.4.11) we use $\tilde{\bar{s}}^2$ as an estimate of the common variance σ^2 when it is unknown.

13.5 THE k-VARIATE NORMAL PROCESS (I): LIKELIHOOD FUNCTIONS

This section imitates as closely as possible the development of 13.2. We shall assume that an experiment will result in n iid observations $\tilde{\mathbf{x}}_1, \ldots, \tilde{\mathbf{x}}_n$ with common df $f_N{}^{(k)}(\mathbf{x} \mid \boldsymbol{\mu}, \boldsymbol{\sigma}^2)$, obtained with noninformative stopping.

There are four possible sets of knowledge assumptions, obtained by replacing μ and σ^2 in 13.2 with their vector and matrix counterparts; but there are two others which correspond to an intermediate knowledge assumption about $\boldsymbol{\sigma}^2$. Specifically, we shall occasionally assume that $\boldsymbol{\sigma}^2$ is known up to a positive multiple, or equivalently, that the ratios $\sigma_{ij}{}^2/\sigma_{k\ell}{}^2$ of all pairs of elements of $\boldsymbol{\sigma}^2$ are known. For this intermediate knowledge assumption, we define

$$\sigma_*{}^2 \equiv |\boldsymbol{\sigma}^2|^{1/k}; \tag{13.5.1}$$

that is, $\sigma_*{}^2$ is the kth root of the determinant of $\boldsymbol{\sigma}^2$. We also define

$$\boldsymbol{\sigma}_*{}^2 \equiv (1/\sigma_*{}^2)\boldsymbol{\sigma}^2, \tag{13.5.2}$$

so that $|\boldsymbol{\sigma}_*{}^2| = 1$ and $\boldsymbol{\sigma}^2 = \sigma_*{}^2\boldsymbol{\sigma}_*{}^2$.

The result is that there are five knowledge assumptions of interest: (1) $\boldsymbol{\mu}$ unknown and $\boldsymbol{\sigma}^2$ known; (2) $\boldsymbol{\mu}$ and $\sigma_*{}^2$ unknown but $\boldsymbol{\sigma}_*{}^2$ known; (3) $\boldsymbol{\mu}$ and $\boldsymbol{\sigma}^2$ unknown: (4) $\boldsymbol{\mu}$ and $\boldsymbol{\sigma}_*{}^2$ known but $\sigma_*{}^2$ unknown; and (5) $\boldsymbol{\mu}$ known but $\boldsymbol{\sigma}^2$ unknown (that is, both $\sigma_*{}^2$ and $\boldsymbol{\sigma}_*{}^2$ unknown). There are three other logically possible cases: both $\boldsymbol{\mu}$ and $\boldsymbol{\sigma}^2$ known (and nothing left to estimate); $\boldsymbol{\mu}$ and $\sigma_*{}^2$ known but $\boldsymbol{\sigma}_*{}^2$ unknown; and $\sigma_*{}^2$ known but $\boldsymbol{\mu}$ and $\boldsymbol{\sigma}_*{}^2$ unknown. We shall not consider these and the still other cases obtained by stipulating that some components of $\boldsymbol{\mu}$ and/or elements of $\boldsymbol{\sigma}^2$ are known.

We define the following random variables:

$$\tilde{\bar{\mathbf{x}}} \equiv \left(\frac{1}{n}\right) \sum_{i=1}^{n} \tilde{\mathbf{x}}_i; \tag{13.5.3}$$

$$\tilde{\mathbf{s}}^2 \equiv \left(\frac{1}{n-1}\right) \sum_{i=1}^{n} (\tilde{\mathbf{x}}_i - \tilde{\bar{\mathbf{x}}})(\tilde{\mathbf{x}}_i - \tilde{\bar{\mathbf{x}}})'; \tag{13.5.4}$$

$$\tilde{\boldsymbol{\xi}} \equiv \left(\frac{1}{n}\right) \sum_{i=1}^{n} (\tilde{\mathbf{x}}_i - \boldsymbol{\mu})(\tilde{\mathbf{x}}_i - \boldsymbol{\mu})'; \tag{13.5.5}$$

$$\tilde{s}_*^2 \equiv \left(\frac{1}{k(n-1)}\right) \sum_{i=1}^{n} (\tilde{\mathbf{x}}_i - \tilde{\bar{\mathbf{x}}})' \boldsymbol{\sigma}_*^{-2} (\tilde{\mathbf{x}}_i - \tilde{\bar{\mathbf{x}}}); \tag{13.5.6}$$

and

$$\tilde{\xi}_* \equiv \left(\frac{1}{kn}\right) \sum_{i=1}^{n} (\tilde{\mathbf{x}}_i - \boldsymbol{\mu})' \boldsymbol{\sigma}_*^{-2} (\tilde{\mathbf{x}}_i - \boldsymbol{\mu}). \tag{13.5.7}$$

Note that $\tilde{\boldsymbol{\xi}}$, $\tilde{s}_*^2$, and $\tilde{\xi}_*$ are not statistics unless $\boldsymbol{\mu}$, $\boldsymbol{\sigma}_*^2$, and $(\boldsymbol{\mu}, \boldsymbol{\sigma}_*^2)$, respectively, are known.

We may now furnish the k-variate analogues of (13.2.4) through (13.2.6).

If $\tilde{\mathbf{x}}_1, \ldots, \tilde{\mathbf{x}}_n$ are iid with common df $f_N^{(k)}(\mathbf{x} \mid \boldsymbol{\mu}, \boldsymbol{\sigma}^2)$ and $\boldsymbol{\mu}$ (but neither σ_*^2 nor $\boldsymbol{\sigma}_*^2$) is known, then $(n, \tilde{\boldsymbol{\xi}})$ is conditionally sufficient for $\boldsymbol{\sigma}^2$ given $\boldsymbol{\mu}$, and (13.5.8)

$$L(\boldsymbol{\sigma}^2 \mid n, \mathbf{x}_1, \ldots, \mathbf{x}_n, \boldsymbol{\mu}) = K \mid \boldsymbol{\sigma}^2 \mid^{-1/2n} \exp[-\tfrac{1}{2}n \cdot \operatorname{tr}\{\boldsymbol{\sigma}^{-2}\boldsymbol{\xi}\}],$$

where $K \equiv (2\pi)^{-nk/2}$, all provided that the stopping process is noninformative.

Proof

For the second assertion,

$$L(\boldsymbol{\sigma}^2 | n, \mathbf{x}_1, \ldots, \mathbf{x}_n, \boldsymbol{\mu})$$

$$= f(\mathbf{x}_1, \ldots, \mathbf{x}_n \mid \boldsymbol{\mu}, \boldsymbol{\sigma}^2)$$

$$= \prod_{i=1}^{n} f_N^{(k)}(\mathbf{x}_i \mid \boldsymbol{\mu}, \boldsymbol{\sigma}^2)$$

$$= \prod_{i=1}^{n} (2\pi)^{-k/2} |\boldsymbol{\sigma}^2|^{-1/2} \exp[-\tfrac{1}{2}(\mathbf{x}_i - \boldsymbol{\mu})' \boldsymbol{\sigma}^{-2} (\mathbf{x}_i - \boldsymbol{\mu})]$$

(1) $$= \prod_{i=1}^{n} (2\pi)^{-k/2} |\boldsymbol{\sigma}^2|^{-1/2} \exp[-\tfrac{1}{2} \operatorname{tr}\{\boldsymbol{\sigma}^{-2}[(\mathbf{x}_i - \boldsymbol{\mu})(\mathbf{x}_i - \boldsymbol{\mu})']\}]$$

$$= (2\pi)^{-nk/2} |\boldsymbol{\sigma}^2|^{-n/2} \exp[-\tfrac{1}{2} \sum_{i=1}^{n} \operatorname{tr}\{\boldsymbol{\sigma}^{-2}[(\mathbf{x}_i - \boldsymbol{\mu})(\mathbf{x}_i - \boldsymbol{\mu})']\}]$$

(2) $$= (2\pi)^{-nk/2} |\boldsymbol{\sigma}^2|^{-n/2} \exp[-\tfrac{1}{2} \operatorname{tr}\{\boldsymbol{\sigma}^{-2} \sum_{i=1}^{n} (\mathbf{x}_i - \boldsymbol{\mu})(\mathbf{x}_i - \boldsymbol{\mu})'\}]$$

$$= (2\pi)^{-nk/2} |\boldsymbol{\sigma}^2|^{-n/2} \exp[-\tfrac{1}{2} \operatorname{tr}\{\boldsymbol{\sigma}^{-2} \cdot n\boldsymbol{\xi}\}]$$

(3) $$= (2\pi)^{-nk/2} |\boldsymbol{\sigma}^2|^{-n/2} \exp[-\tfrac{1}{2}n \cdot \operatorname{tr}\{\boldsymbol{\sigma}^{-2}\boldsymbol{\xi}\}],$$

where equalities (1) through (3) follow from elementary properties (12.9.3) of "tr" which were developed as Exercise 12.34. QED

Before proceeding further with the sufficient statistics and likelihood functions, we shall digress to prove a result from which the remainder of the section will follow readily.

For any set of values of n, $\mathbf{x}_1, \ldots, \mathbf{x}_n$, $\boldsymbol{\mu}$ and $\boldsymbol{\sigma}^2$, the following are equal to each other: (13.5.9)

(1) $n \cdot \text{tr}\,\{\boldsymbol{\sigma}^{-2}\boldsymbol{\xi}\}$;
(2) $(n-1)\,\text{tr}\,\{\boldsymbol{\sigma}^{-2}\mathbf{s}^2\} + n(\bar{\mathbf{x}} - \boldsymbol{\mu})'\boldsymbol{\sigma}^{-2}(\bar{\mathbf{x}} - \boldsymbol{\mu})'$;
(3) $k(n-1)s_*^2/\sigma_*^2 + n[(\bar{\mathbf{x}} - \boldsymbol{\mu})'\boldsymbol{\sigma}_*^{-2}(\bar{\mathbf{x}} - \boldsymbol{\mu})]/\sigma_*^2$; and
(4) $kn\xi_*/\sigma_*^2$.

Proof

We shall prove: (a) (1) = (4); (b) (2) = (3); and (c) (1) = (2), since these three equalities imply the remaining ones by transitivity of "=." In these proofs we make use of the fact that if $\mathbf{A}$ is a symmetric kth order matrix and $\mathbf{y} \in R^k$, then $\text{tr}\,\{\mathbf{Ayy}'\} = \mathbf{y}'\mathbf{Ay}$, which the reader has been asked to prove in Exercise 12.34.

Proof of (a)

$$\begin{aligned}
n \cdot \text{tr}\,\{\boldsymbol{\sigma}^{-2}\boldsymbol{\xi}\}
&= \text{tr}\,\{\boldsymbol{\sigma}^{-2}n\boldsymbol{\xi}\} \\
&= \text{tr}\left\{\boldsymbol{\sigma}^{-2}\sum_{i=1}^{n}(\mathbf{x}_i - \boldsymbol{\mu})(\mathbf{x}_i - \boldsymbol{\mu})'\right\} \\
&= \sum_{i=1}^{n}\text{tr}\,\{\boldsymbol{\sigma}^{-2}(\mathbf{x}_i - \boldsymbol{\mu})(\mathbf{x}_i - \boldsymbol{\mu})'\} \\
&= \sum_{i=1}^{n}(\mathbf{x}_i - \boldsymbol{\mu})'\boldsymbol{\sigma}^{-2}(\mathbf{x}_i - \boldsymbol{\mu}) \\
&= \sum_{i=1}^{n}(\mathbf{x}_i - \boldsymbol{\mu})'[\sigma_*^2\boldsymbol{\sigma}_*^2]^{-1}(\mathbf{x}_i - \boldsymbol{\mu}) \\
&= \frac{\left[\sum_{i=1}^{n}(\mathbf{x}_i - \boldsymbol{\mu})'\boldsymbol{\sigma}_*^{-2}(\mathbf{x}_i - \boldsymbol{\mu})\right]}{\sigma_*^2} \\
&= \frac{kn\xi_*}{\sigma_*^2},
\end{aligned}$$

in which all equalities are easy to verify.

Proof of (b)

The second summand of (2) clearly equals the second summand of (3) because $\boldsymbol{\sigma}^{-2} = (1/\sigma_*{}^2)\boldsymbol{\sigma}_*{}^{-2}$. Thus we must show that the first summands are equal:

$$
\begin{aligned}
(n-1)\,\mathrm{tr}\,\{\boldsymbol{\sigma}^{-2}\mathbf{s}\}
&= \mathrm{tr}\,\{\boldsymbol{\sigma}^{-2}(n-1)\mathbf{s}\}\\
&= \mathrm{tr}\,\left\{\boldsymbol{\sigma}^{-2}\left\{\sum_{i=1}^{n}(\mathbf{x}_i-\bar{\mathbf{x}})(\mathbf{x}_i-\bar{\mathbf{x}})'\right\}\right.\\
&= \sum_{i=1}^{n}\mathrm{tr}\,\{\boldsymbol{\sigma}^{-2}(\mathbf{x}_i-\bar{\mathbf{x}})(\mathbf{x}_i-\bar{\mathbf{x}})'\}\\
&= \sum_{i=1}^{n}(\mathbf{x}_i-\bar{\mathbf{x}})'\boldsymbol{\sigma}^{-2}(\mathbf{x}_i-\bar{\mathbf{x}})\\
&= \sum_{i=1}^{n}\frac{(\mathbf{x}_i-\bar{\mathbf{x}})'\boldsymbol{\sigma}_*{}^{-2}(\mathbf{x}_i-\bar{\mathbf{x}})}{\sigma_*{}^2}\\
&= \frac{k(n-1)s_*{}^2}{\sigma_*{}^2}.
\end{aligned}
$$

Proof of (c)

$n\cdot\mathrm{tr}\,\{\boldsymbol{\sigma}^{-2}\boldsymbol{\xi}\}$

$$
\begin{aligned}
&= \mathrm{tr}\,\{\boldsymbol{\sigma}^{-2}\sum_{i=1}^{n}(\mathbf{x}_i-\boldsymbol{\mu})(\mathbf{x}_i-\boldsymbol{\mu})'\}\\
&= \mathrm{tr}\,\{\boldsymbol{\sigma}^{-2}\sum_{i=1}^{n}[\{(\mathbf{x}_i-\bar{\mathbf{x}})+(\bar{\mathbf{x}}-\boldsymbol{\mu})\}\{(\mathbf{x}_i-\bar{\mathbf{x}})+(\bar{\mathbf{x}}-\boldsymbol{\mu})\}']\}\\
&= \mathrm{tr}\,\{\boldsymbol{\sigma}^{-2}\sum_{i=1}^{n}[(\mathbf{x}_i-\bar{\mathbf{x}})(\mathbf{x}_i-\bar{\mathbf{x}})'+(\bar{\mathbf{x}}-\boldsymbol{\mu})(\mathbf{x}_i-\bar{\mathbf{x}})'\\
&\qquad +(\mathbf{x}_i-\bar{\mathbf{x}})(\bar{\mathbf{x}}-\boldsymbol{\mu})'+(\bar{\mathbf{x}}-\boldsymbol{\mu})(\bar{\mathbf{x}}-\boldsymbol{\mu})']\}.\\
&= \mathrm{tr}\,\left\{\boldsymbol{\sigma}^{-2}\left[\sum_{i=1}^{n}(\mathbf{x}_i-\bar{\mathbf{x}})(\mathbf{x}_i-\bar{\mathbf{x}})'+(\bar{\mathbf{x}}-\boldsymbol{\mu})\sum_{i=1}^{n}(\mathbf{x}_i-\bar{\mathbf{x}})'\right.\right.\\
&\qquad \left.\left.+\left\{\sum_{i=1}^{n}(\mathbf{x}_i-\bar{\mathbf{x}})\right\}(\bar{\mathbf{x}}-\boldsymbol{\mu})'+n(\bar{\mathbf{x}}-\boldsymbol{\mu})(\bar{\mathbf{x}}-\boldsymbol{\mu})'\right]\right\}.
\end{aligned}
$$

But

$$
\sum_{i=1}^{n}(\mathbf{x}_i-\bar{\mathbf{x}})' = \sum_{i=1}^{n}(\mathbf{x}_i-\bar{\mathbf{x}}) = \mathbf{0}
$$

and hence

$$(\bar{\mathbf{x}} - \boldsymbol{\mu}) \sum_{i=1}^{n} (\mathbf{x}_i - \bar{\mathbf{x}})' = \mathbf{0} = \left\{ \sum_{i=1}^{n} (\mathbf{x}_i - \bar{\mathbf{x}}) \right\} (\bar{\mathbf{x}} - \boldsymbol{\mu})'.$$

Therefore the last term in the preceding equality chain equals

$$\begin{aligned}
&\operatorname{tr} \left\{ \boldsymbol{\sigma}^{-2} \left[\sum_{i=1}^{n} (\mathbf{x}_i - \bar{\mathbf{x}})(\mathbf{x}_i - \bar{\mathbf{x}})' + n(\bar{\mathbf{x}} - \boldsymbol{\mu})(\bar{\mathbf{x}} - \boldsymbol{\mu})' \right] \right\} \\
&\quad = \operatorname{tr} \left\{ \boldsymbol{\sigma}^{-2} \sum_{i=1}^{n} (\mathbf{x}_i - \bar{\mathbf{x}})(\mathbf{x}_i - \bar{\mathbf{x}})' \right\} + \operatorname{tr} \{ \boldsymbol{\sigma}^{-2} \cdot n(\bar{\mathbf{x}} - \boldsymbol{\mu})(\bar{\mathbf{x}} - \boldsymbol{\mu})' \} \\
&\quad = \operatorname{tr} \{ \boldsymbol{\sigma}^{-2}(n-1)\mathbf{s}^2 \} + n \cdot \operatorname{tr} \{ \boldsymbol{\sigma}^{-2}(\bar{\mathbf{x}} - \boldsymbol{\mu})(\bar{\mathbf{x}} - \boldsymbol{\mu})' \} \\
&\quad = (n-1) \operatorname{tr} \{ \boldsymbol{\sigma}^{-2}\mathbf{s}^2 \} + n(\bar{\mathbf{x}} - \boldsymbol{\mu})' \boldsymbol{\sigma}^{-2} (\bar{\mathbf{x}} - \boldsymbol{\mu}),
\end{aligned}$$

which concludes the proof of (c). QED

We are now ready to handle easily the remaining likelihood functions and sufficient statistics.

If $\tilde{\mathbf{x}}_1, \ldots, \tilde{\mathbf{x}}_n$ are iid with common df $f_N^{(k)}(\mathbf{x} \mid \boldsymbol{\mu}, \boldsymbol{\sigma}^2)$ and $\boldsymbol{\sigma}^2$ (but not $\boldsymbol{\mu}$) is known, then $(n, \bar{\tilde{\mathbf{x}}})$ is conditionally sufficient for $\boldsymbol{\mu}$ given $\boldsymbol{\sigma}^2$, and (13.5.10)

$$L(\boldsymbol{\mu} \mid n, \mathbf{x}_1, \ldots, \mathbf{x}_n, \boldsymbol{\sigma}^2) = K \exp \left[-\tfrac{1}{2} n(\bar{\mathbf{x}} - \boldsymbol{\mu})' \boldsymbol{\sigma}^{-2} (\bar{\mathbf{x}} - \boldsymbol{\mu}) \right],$$

where $K \equiv (2\pi)^{-nk/2} |\boldsymbol{\sigma}^2|^{-n/2} \exp \left[-\tfrac{1}{2}(n-1) \operatorname{tr} \{ \boldsymbol{\sigma}^{-2}\mathbf{s}^2 \} \right]$, all provided that the stopping process is noninformative.

Proof

The likelihood function follows immediately from (13.5.8) via "(1) = (2)" of (13.5.9). Since K does not depend upon $\boldsymbol{\mu}$, sufficiency of $\bar{\mathbf{x}}$ for $\boldsymbol{\mu}$ is evident from the likelihood function. QED

If $\tilde{\mathbf{x}}_1, \ldots, \tilde{\mathbf{x}}_n$ are iid with common df $f_N^{(k)}(\mathbf{x} \mid \boldsymbol{\mu}, \boldsymbol{\sigma}^2)$ and neither $\boldsymbol{\mu}$ nor $\boldsymbol{\sigma}^2$ is known, then $(n, \bar{\tilde{\mathbf{x}}}, \tilde{\mathbf{s}}^2)$ is sufficient for $(\boldsymbol{\mu}, \boldsymbol{\sigma}^2)$, and (13.5.11)

$$\begin{aligned}
&L(\boldsymbol{\mu}, \boldsymbol{\sigma}^2 \mid n, \mathbf{x}_1, \ldots, \mathbf{x}_n) \\
&\quad = K \cdot |\boldsymbol{\sigma}^2|^{-n/2} \exp \left[-\tfrac{1}{2}(n-1) \operatorname{tr} \{ \boldsymbol{\sigma}^{-2}\mathbf{s}^2 \} \right] \exp \left[-\tfrac{1}{2} n(\bar{\mathbf{x}} - \boldsymbol{\mu})' \boldsymbol{\sigma}^{-2} (\bar{\mathbf{x}} - \boldsymbol{\mu}) \right],
\end{aligned}$$

where $K \equiv (2\pi)^{-nk/2}$, all provided that the stopping process is noninformative.

Proof

Simply write the likelihood function in (13.5.10) with its K explicit, thus obtaining the likelihood function for this case. The sufficiency assertion is immediate from the likelihood function. QED

If $\tilde{\mathbf{x}}_1, \ldots, \tilde{\mathbf{x}}_n$ are iid with common df $f_N^{(k)}(\mathbf{x} \mid \boldsymbol{\mu}, \boldsymbol{\sigma}^2)$, and $\boldsymbol{\mu}$ and $\boldsymbol{\sigma}_*^2$ (but not σ_*^2) are known, then (n, ξ_*) is conditionally sufficient for σ_*^2 given $(\boldsymbol{\mu}, \boldsymbol{\sigma}_*^2)$, and (13.5.12)

$$L(\sigma_*^2 \mid n, \mathbf{x}_1, \ldots, \mathbf{x}_n, \boldsymbol{\mu}, \boldsymbol{\sigma}_*^2) = K\cdot(\sigma_*^2)^{-nk/2} \exp\left[-\tfrac{1}{2}kn\xi_*/\sigma_*^2\right],$$

where $K \equiv (2\pi)^{-nk/2}$.

Proof

A comparison of the asserted likelihood function with the likelihood function (13.5.8) shows that the exponents of e agree by virtue of "(1) = (4)" of (13.5.9). It remains to show that $|\boldsymbol{\sigma}^2|^{-n/2} = (\sigma_*^2)^{-nk/2}$, which is obvious from (13.5.1). The sufficiency assertion is immediate from the likelihood function. QED

If $\tilde{\mathbf{x}}_1, \ldots, \tilde{\mathbf{x}}_n$ are iid with common df $f_N^{(k)}(\mathbf{x} \mid \boldsymbol{\mu}, \boldsymbol{\sigma}^2)$ and $\boldsymbol{\sigma}_*^2$ (but neither $\boldsymbol{\mu}$ nor σ_*^2) is known, then $(n, \bar{\tilde{\mathbf{x}}}, \tilde{s}_*^2)$ is conditionally sufficient for $(\boldsymbol{\mu}, \sigma_*^2)$ given $\boldsymbol{\sigma}_*^2$, and (13.5.13)

$$L(\boldsymbol{\mu}, \sigma_*^2 \mid n, \mathbf{x}_1, \ldots, \mathbf{x}_n, \boldsymbol{\sigma}_*^2)$$
$$= K\cdot(\sigma_*^2)^{-nk/2} \exp\left[-\tfrac{1}{2}(n-1)s_*^2/\sigma_*^2\right] \exp\left[-\tfrac{1}{2}n[(\bar{\mathbf{x}} - \boldsymbol{\mu})'\boldsymbol{\sigma}^{-2}(\bar{\mathbf{x}} - \boldsymbol{\mu})]/\sigma_*^2\right]$$

where $K \equiv (2\pi)^{-nk/2}$, all provided that the stopping process is noninformative

Proof

Exercise 4.

13.6 THE k-VARIATE NORMAL PROCESS (II): DISTRIBUTION OF STATISTICS

We now imitate the development of 13.3 by characterizing the distributions of some statistics given in 13.5 and functions of those statistics. Our first result is the k-variate counterpart of (13.3.1).

If $\tilde{\mathbf{z}}_1, \ldots, \tilde{\mathbf{z}}_n$ are iid with common df $f_N^{(k)}(\mathbf{z} \mid \mathbf{0}, \boldsymbol{\sigma}^2)$, then $\tilde{\mathbf{V}} \equiv \sum_{i=1}^{n} \tilde{\mathbf{z}}_i\tilde{\mathbf{z}}'_i$ has df $f_W^{(k)}(\mathbf{V} \mid (n-k+1)^{-1}\boldsymbol{\sigma}^{-2}, n-k+1)$, provided that $n \geqq k$. (13.6.1)

Proof

See Anderson [1], pp. 154–157, and our notational translation table preceding (12.9.4).

Statement (12.9.7) is the counterpart of (13.3.2).

If $\tilde{\mathbf{x}}_1, \ldots, \tilde{\mathbf{x}}_n$ are iid with common df $f_N^{(k)}(\mathbf{x} \mid \underset{\sim}{\boldsymbol{\mu}}, \tilde{\boldsymbol{\sigma}}^2)$, then $\bar{\tilde{\mathbf{x}}}$ has df (13.6.2)

$$f_N^{(k)}(\bar{\mathbf{x}} \mid \underset{\sim}{\boldsymbol{\mu}}, n^{-1}\tilde{\boldsymbol{\sigma}}^2),$$

given $(\underset{\sim}{\boldsymbol{\mu}}, \tilde{\boldsymbol{\sigma}}^2)$.

Proof

This is simply a restatement of (12.7.10c).

If $\tilde{\mathbf{x}}_1, \ldots, \tilde{\mathbf{x}}_n$ are iid with common df $f_N^{(k)}(\mathbf{x} \mid \underset{\sim}{\boldsymbol{\mu}}, \tilde{\boldsymbol{\sigma}}^2)$, then $\tilde{\mathbf{V}} \equiv n\tilde{\boldsymbol{\xi}}$ has df (13.6.3)

$$f_W^{(k)}(\mathbf{V} \mid (n-k+1)^{-1}\tilde{\boldsymbol{\sigma}}^{-2},\ n-k+1),$$

and $\tilde{\boldsymbol{\xi}}$ has df

$$f_W^{(k)}(\boldsymbol{\xi} \mid n(n-k+1)^{-1}\tilde{\boldsymbol{\sigma}}^{-2},\ n-k+1),$$

given $(\underset{\sim}{\boldsymbol{\mu}}, \tilde{\boldsymbol{\sigma}}^2)$, provided that $n \geqq k$.

Proof

The first assertion is immediate from (13.6.1) in view of the facts that $n\tilde{\boldsymbol{\xi}} = \tilde{\mathbf{V}} = \sum_{i=1}^{n} (\tilde{\mathbf{x}}_i - \underset{\sim}{\boldsymbol{\mu}})(\tilde{\mathbf{x}}_i - \underset{\sim}{\boldsymbol{\mu}})'$ and that if $\tilde{\mathbf{x}}_1, \ldots, \tilde{\mathbf{x}}_n$ are iid with common df $f_N^{(k)}(\mathbf{x} \mid \underset{\sim}{\boldsymbol{\mu}}, \tilde{\boldsymbol{\sigma}}^2)$, then $\tilde{\mathbf{z}}_1 \equiv (\tilde{\mathbf{x}}_1 - \underset{\sim}{\boldsymbol{\mu}}), \ldots, \tilde{\mathbf{z}}_n \equiv (\tilde{\mathbf{x}}_n - \underset{\sim}{\boldsymbol{\mu}})$ are iid with common df $f_N^{(k)}(\mathbf{z} \mid \mathbf{0}, \tilde{\boldsymbol{\sigma}}^2)$. The second assertion follows from the first via (12.9.7). QED

If $\tilde{\mathbf{x}}_1, \ldots, \tilde{\mathbf{x}}_n$ are iid with common df $f_N^{(k)}(\mathbf{x} \mid \underset{\sim}{\boldsymbol{\mu}}, \tilde{\boldsymbol{\sigma}}^2)$ and $n \geqq k+1$, then $\tilde{\mathbf{V}} \equiv (n-1)\tilde{\mathbf{s}}^2$ has df $f_W^{(k)}(\mathbf{V} \mid (n-k)^{-1}\tilde{\boldsymbol{\sigma}}^{-2}, n-k)$; $\tilde{\mathbf{s}}^2$ has df (13.6.4)

$$f_W^{(k)}(\mathbf{s}^2 \mid (n-1)(n-k)^{-1}\tilde{\boldsymbol{\sigma}}^{-2},\ n-k);$$

and $\bar{\tilde{\mathbf{x}}}$ and $\tilde{\mathbf{s}}^2$ are conditionally independent given $(\underset{\sim}{\boldsymbol{\mu}}, \tilde{\boldsymbol{\sigma}}^2)$.

Proof

A proof of the first and third assertions may be found in Anderson [**1**] pp. 53 and 157. The second assertion follows from the first via (12.9.7). QED

As with (13.3.6), the converse of the third assertion is true: if $\bar{\tilde{\mathbf{x}}}$ and $\tilde{\mathbf{s}}^2$ as defined by (13.5.3) and (13.5.4) are conditionally independent given $(\underset{\sim}{\boldsymbol{\mu}}, \tilde{\boldsymbol{\sigma}}^2)$, then the common dgf $f(\mathbf{x})$ of $\tilde{\mathbf{x}}_1, \ldots, \tilde{\mathbf{x}}_n$ must be $f_N^{(k)}(\mathbf{x} \mid \underset{\sim}{\boldsymbol{\mu}}, \tilde{\boldsymbol{\sigma}}^2)$ for some $\underset{\sim}{\boldsymbol{\mu}}$ and $\tilde{\boldsymbol{\sigma}}^2$.

Now (13.6.3) and (13.6.4) are the k-variate generalizations of (13.3.4) and (13.3.6), respectively for the cases where neither σ_*^2 nor $\tilde{\boldsymbol{\sigma}}_*^2$ is known.

In order to obtain the corresponding generalizations for the case where $\boldsymbol{\sigma}_*^2$ is known, we need another k-variate counterpart of (13.3.1):

If $\tilde{\mathbf{y}}$ has df $f_N^{(k)}(\mathbf{y} \mid \boldsymbol{\mu}, \boldsymbol{\sigma}^2)$, then $\tilde{Q} \equiv (\tilde{\mathbf{y}} - \boldsymbol{\mu})'\boldsymbol{\sigma}^{-2}(\tilde{\mathbf{y}} - \boldsymbol{\mu})$ has df $f_{\chi^2}(Q \mid k)$. (13.6.5)

Proof

From linear algebra, there exists a nonsingular matrix $\mathbf{C}$ such that $\mathbf{C}'\mathbf{C} = \boldsymbol{\sigma}^{-2}$. Define $\tilde{\mathbf{z}} \equiv \mathbf{C}(\tilde{\mathbf{y}} - \boldsymbol{\mu})$. Then

$$\begin{aligned}\tilde{Q} &\equiv (\tilde{\mathbf{y}} - \boldsymbol{\mu})'\mathbf{C}'\mathbf{C}(\tilde{\mathbf{y}} - \boldsymbol{\mu}) \\ &= [\mathbf{C}(\tilde{\mathbf{y}} - \boldsymbol{\mu})]'(\mathbf{C}(\tilde{\mathbf{y}} - \boldsymbol{\mu})] \\ &= \tilde{\mathbf{z}}'\tilde{\mathbf{z}} \\ &\equiv \sum_{i=1}^{k} \tilde{\mathbf{z}}_i^2.\end{aligned}$$

But from Section 12.7, $\tilde{\mathbf{y}}$ has df $f_N^{(k)}(\mathbf{y} \mid \boldsymbol{\mu}, \boldsymbol{\sigma}^2)$ if and only if $\tilde{\mathbf{z}}$ has df $f_N^{(k)}(\mathbf{z} \mid \mathbf{0}, \mathbf{I})$; that is, if and only if the $\tilde{\mathbf{z}}_i$'s are iid with common df $f_{N*}(z)$. Hence by (13.3.1), the df of $\sum_{i=1}^{k} \tilde{\mathbf{z}}_i^2 \equiv \tilde{Q}$ is $f_{\chi^2}(Q \mid k)$. QED

A number of important results follow from (13.6.5). First is an analog of (13.3.4) and (13.6.3).

If $\tilde{\mathbf{x}}_1, \ldots, \tilde{\mathbf{x}}_n$ are iid with common df $f_N^{(k)}(\mathbf{x} \mid \boldsymbol{\mu}, \boldsymbol{\sigma}^2)$, then $\tilde{Q} \equiv kn\tilde{\xi}_*/\sigma_*^2$ has df $f_{\chi^2}(Q \mid kn)$, and $\tilde{\xi}_*$ has df $f_{\gamma 2}(\xi_* \mid 1/\delta_*^2, kn)$, provided that $n \geqq k$. (13.6.6)

Proof

$\tilde{Q} \equiv kn\tilde{\xi}/\sigma_*^2 - \sum_{i=1}^{n} (\tilde{\mathbf{x}}_i - \boldsymbol{\mu})\boldsymbol{\sigma}^{-2}(\tilde{\mathbf{x}}_i - \boldsymbol{\mu})$ because $\boldsymbol{\sigma}_*^{-2}(1/\sigma_*^2) = \boldsymbol{\sigma}^{-2}$. By (13.6.5), the summands in $\sum_{i=1}^{n} (\tilde{\mathbf{x}}_i - \boldsymbol{\mu})'\boldsymbol{\sigma}^{-2}(\tilde{\mathbf{x}}_i - \boldsymbol{\mu})$ are iid with common df $f_{\chi^2}(\cdot \mid k)$, and hence (12.2.17) implies that the df of the sum $\tilde{Q}$ is $f_{\chi^2}(\cdot \mid kn)$. The second assertion follows from the first via (13.3.2). QED

Next is an analog of (13.3.6) and (13.6.4).

If $\tilde{\mathbf{x}}_1, \ldots, \tilde{\mathbf{x}}_n$ are iid with common df $f_N^{(k)}(\mathbf{x} \mid \boldsymbol{\mu}, \boldsymbol{\sigma}^2)$, then $\tilde{Q} \equiv k(n-1)\tilde{s}_*^2/\sigma_*^2$ has df $f_{\chi^2}(Q \mid k(n-1))$; $\tilde{s}^2_*$ has df $f_{\gamma 2}(s_*^2 \mid 1/\sigma_*^2, k(n-1))$; and $\bar{\tilde{\mathbf{x}}}$ and $\tilde{s}_*^2$ are conditionally independent given $(\boldsymbol{\mu}, \sigma_*^2)$. (13.6.7)

Proof

As in the proof of (13.6.6), it follows easily that

$$\tilde{Q} = \sum_{i=1}^{n} (\tilde{\mathbf{x}}_i - \bar{\tilde{\mathbf{x}}})'\boldsymbol{\sigma}^{-2}(\tilde{\mathbf{x}}_i - \bar{\tilde{\mathbf{x}}}).$$

Again define $\tilde{\mathbf{z}}_i \equiv \mathbf{C}(\tilde{\mathbf{x}}_i - \boldsymbol{\mu})$, where $\mathbf{C}'\mathbf{C} = \boldsymbol{\sigma}^{-2}$. Then we readily obtain

$$\bar{\tilde{\mathbf{z}}} = \mathbf{C}(\bar{\tilde{\mathbf{x}}} - \boldsymbol{\mu}),$$
$$\tilde{\mathbf{z}}_i - \bar{\tilde{\mathbf{z}}} = \mathbf{C}(\tilde{\mathbf{x}}_i - \boldsymbol{\mu}),$$
$$(\tilde{\mathbf{z}}_i - \bar{\tilde{\mathbf{z}}})'(\tilde{\mathbf{z}}_i - \bar{\tilde{\mathbf{z}}}) = (\tilde{\mathbf{x}}_i - \bar{\tilde{\mathbf{x}}})'\boldsymbol{\sigma}^{-2}(\tilde{\mathbf{x}}_i - \bar{\tilde{\mathbf{x}}}),$$

and

$$\begin{aligned}\tilde{Q} &= \sum_{i=1}^{n} (\tilde{\mathbf{z}}_i - \bar{\tilde{\mathbf{z}}})'(\tilde{\mathbf{z}}_i - \bar{\tilde{\mathbf{z}}}) \\ &= \sum_{i=1}^{n}\sum_{j=1}^{k} (\tilde{z}_{ij} - \bar{\tilde{\mathbf{z}}}_j)^2 \\ &= \sum_{j=1}^{k}\sum_{i=1}^{n} (\tilde{z}_{ij} - \bar{\tilde{\mathbf{z}}}_j)^2,\end{aligned}$$

where $\tilde{z}_{ij}$ and $\bar{\tilde{\mathbf{z}}}_j$ are the jth components of $\tilde{\mathbf{z}}_i$ and $\bar{\tilde{\mathbf{z}}}$, respectively. Now, the $\tilde{\mathbf{z}}_i$'s are independent because the $\tilde{\mathbf{x}}_i$'s are, and (by definition of $\mathbf{C}$) the $\tilde{z}_{ij}$'s are independent for each i. Hence $\{\tilde{z}_{11}, \ldots, \tilde{z}_{n1}, \tilde{z}_{12}, \ldots, \tilde{z}_{n2}, \ldots, \tilde{z}_{1k}, \ldots, \tilde{z}_{nk}\}$ is a set of mutually independent and identically distributed random variables with common df $f_{N*}(\cdot)$, and therefore:

(1) $$\tilde{Q}_j \equiv \sum_{i=1}^{n} (\tilde{z}_{ij} - \bar{\tilde{\mathbf{z}}}_j)^2$$

has df $f_{\chi^2}(Q_j \mid n - 1)$, by (13.3.6); and

(2) $$\tilde{Q} \equiv \sum_{j=1}^{k} \tilde{Q}_j$$

has df $f_{\chi^2}(Q \mid k(n - 1))$, because the $\tilde{Q}_j$'s are iid with common df $f_{\chi^2}(\cdot \mid n - 1)$ and thus by (12.2.17) their sum has df $f_{\chi^2}(\cdot \mid k(n - 1))$. This proves the first assertion. The second is immediate from the first via (13.3.2). A proof of the third assertion may be found in Raiffa and Schlaifer [**66**], 317. QED

In Chapter 21, we shall need the following:

If $\tilde{\mathbf{x}}_1, \ldots, \tilde{\mathbf{x}}_n$ are iid with common df $f_N^{(k)}(\mathbf{x} \mid \boldsymbol{\mu}, \boldsymbol{\sigma}^2)$, then the df of (13.6.8)

$$\tilde{Q} \equiv n(\bar{\tilde{\mathbf{x}}} - \boldsymbol{\mu})'\boldsymbol{\sigma}^{-2}(\bar{\tilde{\mathbf{x}}} - \boldsymbol{\mu})$$

is $f_{\chi^2}(Q \mid k)$.

Proof

Exercise 5.

Assertion (13.6.8) is required for region estimates of $\boldsymbol{\mu}$ when $\boldsymbol{\sigma}^2$ is known. When $\boldsymbol{\sigma}^2$ is unknown, we shall need the corollary (13.6.10) of the following:

If $\tilde{\mathbf{z}}$ has df $f_N^{(k)}(\mathbf{z} \mid \mathbf{0}, \mathbf{\sigma}^2)$, $(n - i)\tilde{\mathbf{V}}$ has df $f_W^{(k)}((n - i)\mathbf{V} \mid (n - i - k + 1)^{-1}\mathbf{\sigma}^{-2}, n - i - k + 1)$, and $\tilde{\mathbf{z}}$ and $\tilde{\mathbf{V}}$ are independent given $\mathbf{\sigma}^2$, then (13.6.9)

$$\tilde{\phi} \equiv [(n - i - k + 1)/(k[n - i])]\tilde{\mathbf{z}}'\tilde{\mathbf{V}}^{-1}\tilde{\mathbf{z}}$$

has df $f_F(\phi \mid k, n - i - k + 1)$.

Proof

See Anderson [**1**], pp. 105–107.

If $\tilde{\mathbf{x}}_1, \ldots, \tilde{\mathbf{x}}_n$ are iid with common df $f_N^{(k)}(\mathbf{x} \mid \boldsymbol{\mu}, \mathbf{\sigma}^2)$, then (13.6.10)

$$\tilde{\phi} \equiv (n/k)([n - k]/[n - 1])(\tilde{\bar{\mathbf{x}}} - \boldsymbol{\mu})'\tilde{\mathbf{s}}^{-2}(\tilde{\bar{\mathbf{x}}} - \boldsymbol{\mu})$$

has df $f_F(\phi \mid k, n - k)$, provided that $n \geqq k + 1$, where (as always) $\tilde{\mathbf{s}}^{-2} \equiv (\tilde{\mathbf{s}}^2)^{-1}$.

Proof

See Anderson [**1**], pp. 105–107.

The analogue of (13.6.10) for the case in which $\mathbf{\sigma}_*^2$ is known is:

If $\tilde{\mathbf{x}}_1, \ldots, \tilde{\mathbf{x}}_n$ are iid with common df $f_N^{(k)}(\mathbf{x} \mid \boldsymbol{\mu}, \mathbf{\sigma}^2)$, then (13.6.11)

$$\tilde{\phi} \equiv \left(\frac{n}{k}\right)(\tilde{\bar{\mathbf{x}}} - \boldsymbol{\mu})'(\tilde{s}_*^2)^{-1}\mathbf{\sigma}_*^{-2}(\tilde{\bar{\mathbf{x}}} - \boldsymbol{\mu})$$

has df $f_F(\phi \mid k, k(n - 1))$, provided that $n \geqq 2$.

Proof

By (13.6.8), $\tilde{Q}_1 = [n(\tilde{\bar{\mathbf{x}}} - \boldsymbol{\mu})'\mathbf{\sigma}_*^{-2}(\tilde{\bar{\mathbf{x}}} - \boldsymbol{\mu})]/\sigma_*^2$ has df $f_{\chi^2}(Q_1 \mid k)$; and by (13.6.7), $\tilde{Q}_2 = k(n - 1)\tilde{s}_*^2/\sigma_*^2$ has df $f_{\chi^2}(Q_2 \mid k(n - 1))$. Furthermore, the independence of $\tilde{s}_*^2$ and $\tilde{\bar{\mathbf{x}}}$ asserted in (13.6.7) implies the independence of $\tilde{Q}_1$ and $\tilde{Q}_2$. Hence we apply (13.3.9) to conclude that

$$\tilde{\phi} \equiv \left\{\frac{\tilde{Q}_1}{k}\right\} \Big/ \left\{\frac{\tilde{Q}_2}{k(n - 1)}\right\}$$

$$= \frac{\left\{\left(\frac{n}{k}\right)\frac{(\tilde{\bar{\mathbf{x}}} - \boldsymbol{\mu})'\mathbf{\sigma}_*^{-2}(\tilde{\bar{\mathbf{x}}} - \boldsymbol{\mu})}{\sigma_*^2}\right\}}{\{s_*^2/\sigma_*^2\}}$$

$$= \left(\frac{n}{k}\right)(\tilde{\bar{\mathbf{x}}} - \boldsymbol{\mu})'(\tilde{s}_*^2)^{-1}\mathbf{\sigma}_*^{-2}(\tilde{\bar{\mathbf{x}}} - \boldsymbol{\mu})$$

has the F distribution with k and $k(n - 1)$ degrees of freedom. QED

Chapter 21 also requires an analogue of (13.6.5) for the k-variate Student distribution.

If $\tilde{\mathbf{y}}$ has df $f_{S}^{(k)}(\mathbf{y} \mid \boldsymbol{\mu}, \boldsymbol{\sigma}^2, \nu)$, then $\tilde{Q} \equiv (\tilde{\mathbf{y}} - \boldsymbol{\mu})'\boldsymbol{\sigma}^{-2}(\tilde{\mathbf{y}} - \boldsymbol{\mu})$ has df (13.6.12)

$$f_{i\beta 2}(Q \mid \tfrac{1}{2}k, \tfrac{1}{2}(\nu + k), \nu).$$

Proof

By the definition (12.8.1) of the k-variate Student distribution, $\tilde{\mathbf{y}}$ has *df* $f_{S}^{(k)}(\mathbf{y} \mid \boldsymbol{\mu}, \boldsymbol{\sigma}^2, \nu)$ if and only if there is some random variable $\tilde{h}$ such that the conditional df of $\tilde{\mathbf{y}}$ given h is $f_{N}^{(k)}(\mathbf{y} \mid \boldsymbol{\mu}, h^{-1}\boldsymbol{\sigma}^2)$ and the df of $\tilde{h}$ is $f_{\gamma 2}(h \mid 1, \nu)$. Hence, conditional upon h, the df of $\tilde{Q}' = h\tilde{Q}$ is $f_{\chi^2}(Q' \mid k)$, by (13.6.5), and thus the conditional df of $\tilde{Q}$ is $f_{\gamma 2}(Q \mid h/k, k)$ given h, by (13.3.2). Thus the unconditional df of $\tilde{Q}$ is

$$\int_0^\infty f_{\gamma 2}\left(Q \mid \frac{h}{k}, k\right) \cdot f_{\gamma 2}(h \mid 1, \nu)\,dh$$

$$= \int_0^\infty \frac{\frac{1}{2}h \exp[-\frac{1}{2}hQ](\frac{1}{2}hQ)^{\frac{1}{2}k-1}}{\Gamma(\frac{1}{2}k)} \cdot \frac{\frac{1}{2}\nu \exp[-\frac{1}{2}\nu h](\frac{1}{2}\nu h)^{\frac{1}{2}\nu-1}}{\Gamma(\frac{1}{2}\nu)}\,dh$$

$$= \frac{\nu^{\nu/2} Q^{\frac{1}{2}k-1}}{\Gamma(\frac{1}{2}\nu)\Gamma(\frac{1}{2}k)} \int_0^\infty \tfrac{1}{2} \exp[-\tfrac{1}{2}(\nu + Q)h](\tfrac{1}{2}h)^{\frac{1}{2}(\nu+k)-1}\,dh$$

$$= \frac{\Gamma(\frac{1}{2}[\nu + k])}{\Gamma(\frac{1}{2}\nu)\Gamma(\frac{1}{2}k)} \frac{\nu^{\frac{1}{2}\nu} Q^{\frac{1}{2}k-1}}{(\nu + Q)^{\frac{1}{2}(\nu+k)}}$$

$$\times \int_0^\infty \frac{\frac{1}{2}(\nu + Q) \exp[-\frac{1}{2}(\nu + Q)h][\frac{1}{2}(\nu + Q)h]^{\frac{1}{2}(\nu+k)-1}}{\Gamma(\frac{1}{2}[\nu + k])}\,dh$$

$$= f_{i\beta 2}(Q \mid \tfrac{1}{2}k, \tfrac{1}{2}(\nu + k), \nu) \cdot \int_0^\infty f_\gamma(h \mid \tfrac{1}{2}(\nu + k), \nu + Q)\,dh$$

$$= f_{i\beta 2}(Q \mid \tfrac{1}{2}k, \tfrac{1}{2}(\nu + k), \nu). \quad \text{QED}$$

13.7 STATISTICS FOR COMPARING TWO k-VARIATE NORMAL PROCESSES

Throughout this section we assume that $\tilde{\mathbf{x}}_{11}, \ldots, \tilde{\mathbf{x}}_{n_1 1}$ are iid with common df $f_{N}^{(k)}(\mathbf{x}_1 \mid \boldsymbol{\mu}_1, \boldsymbol{\sigma}_1^2)$, that $\tilde{\mathbf{x}}_{12}, \ldots, \tilde{\mathbf{x}}_{n_2 2}$ are iid with common df

$$f_{N}^{(k)}(\mathbf{x}_2 \mid \boldsymbol{\mu}_2, \boldsymbol{\sigma}_2^2),$$

and that all $\tilde{\mathbf{x}}_{j1}$'s are independent of the $\tilde{\mathbf{x}}_{k2}$'s. Thus, as in 13.4, the assumptions are tantamount to obtaining, independently, n_i observations from each of two k-variate Normal processes $i = 1, 2$.

Define:

$$\tilde{\bar{\mathbf{x}}}_i \equiv \left(\frac{1}{n_i}\right)\sum_{j=1}^{n_i} \tilde{\mathbf{x}}_{ji}, \text{ for } i = 1, 2; \tag{13.7.1}$$

$$\tilde{\mathbf{s}}_i^2 \equiv \left(\frac{1}{n_i - 1}\right)\sum_{j=1}^{n_i} (\tilde{\mathbf{x}}_{ji} - \tilde{\bar{\mathbf{x}}}_i)(\tilde{\mathbf{x}}_{ji} - \tilde{\bar{\mathbf{x}}}_i)', \text{ for } i = 1, 2; \tag{13.7.2}$$

$$\tilde{\xi} \equiv \left(\frac{1}{n_i}\right)\sum_{j=1}^{n_i} (\tilde{\mathbf{x}}_{ji} - \boldsymbol{\mu}_i)(\tilde{\mathbf{x}}_{ji} - \boldsymbol{\mu}_i)', \text{ for } i = 1, 2; \tag{13.7.3}$$

$$\tilde{s}_{*i}^2 \equiv \left(\frac{1}{k(n_i - 1)}\right)\sum_{j=1}^{n_i} (\tilde{\mathbf{x}}_{ji} - \tilde{\bar{\mathbf{x}}}_i)'\boldsymbol{\sigma}_{*i}^{-2}(\tilde{\mathbf{x}}_{ji} - \tilde{\bar{\mathbf{x}}}_i), \text{ for } i = 1, 2; \tag{13.7.4}$$

$$\tilde{\xi}_{*i} \equiv \left(\frac{1}{kn_i}\right)\sum_{j=1}^{n_i} (\tilde{\mathbf{x}}_{ji} - \boldsymbol{\mu}_i)'\boldsymbol{\sigma}_{*i}^{-2}(\tilde{\mathbf{x}}_{ji} - \boldsymbol{\mu}_i), \text{ for } i = 1, 2; \tag{13.7.5}$$

$$\sigma_{*i}^2 \equiv |\boldsymbol{\sigma}_i^2|^{1/k}, \text{ for } i = 1, 2; \tag{17.7.6}$$

$$\boldsymbol{\sigma}_{*i}^2 \equiv \left(\frac{1}{\sigma_{*i}^2}\right)\boldsymbol{\sigma}_i^2, \text{ for } i = 1, 2; \tag{13.7.7}$$

$$\tilde{\mathbf{s}}^2 \equiv \left(\frac{1}{n_1 + n_2 - 2}\right)[(n_1 - 1)\tilde{\mathbf{s}}_1^2 + (n_2 - 1)\tilde{\mathbf{s}}_2^2]; \tag{13.7.8}$$

$$\tilde{s}_*^2 \equiv \left(\frac{1}{k(n_1 + n_2 - 2)}\right)[(n_1 - 1)\tilde{s}_{*1}^2 + (n_2 - 1)\tilde{s}_{*2}^2]; \tag{13.7.9}$$

$$\Delta\tilde{\bar{\mathbf{x}}} \equiv \tilde{\bar{\mathbf{x}}}_1 - \tilde{\bar{\mathbf{x}}}_2; \tag{13.7.10}$$

and

$$\Delta\boldsymbol{\mu} \equiv \boldsymbol{\mu}_1 - \boldsymbol{\mu}_2. \tag{13.7.11}$$

The results which follow furnish no analogue to (13.4.10), for comparing the covariance matrices $\boldsymbol{\sigma}_1^2$ and $\boldsymbol{\sigma}_2^2$. Such results are more complicated to derive, and we shall omit them. The applications of this section in Chapters 21 and 22 are primarily concerned with comparing the mean vectors $\boldsymbol{\mu}_1$ and $\boldsymbol{\mu}_2$; for this purpose the following results will suffice.

With definitions and assumptions as in the preceding, the df of (13.7.12)

$$\tilde{Q} \equiv (\Delta\tilde{\bar{\mathbf{x}}} - \Delta\boldsymbol{\mu})'(n_1^{-1}\boldsymbol{\sigma}_1^2 + n_2^{-1}\boldsymbol{\sigma}_2^2)^{-1}(\Delta\tilde{\bar{\mathbf{x}}} - \Delta\boldsymbol{\mu}) \text{ is } f_{\chi^2}(Q \mid k).$$

Proof

By (13.6.2), the df of $\tilde{\bar{\mathbf{x}}}_i$ is $f_N^{(k)}(\bar{\mathbf{x}}_i \mid \boldsymbol{\mu}_i, n_i^{-1}\boldsymbol{\sigma}_i^2)$, for $i = 1, 2$; and by assumption of independence of the $\tilde{\mathbf{x}}_{j1}$'s of the $\tilde{\mathbf{x}}_{k2}$'s, the $\tilde{\bar{\mathbf{x}}}_i$'s are independent. Hence by (12.7.10a), $\Delta\tilde{\bar{\mathbf{x}}}$ has df $f_N^{(k)}(\Delta\bar{\mathbf{x}} \mid \Delta\boldsymbol{\mu}, n_1^{-1}\boldsymbol{\sigma}_1^2 + n_2^{-1}\boldsymbol{\sigma}_2^2)$, from which the assertion follows via (13.6.5). QED

With definitions and assumptions as in the preceding, and with the additional assumption that $\tilde{\sigma}_1{}^2 = \tilde{\sigma}_2{}^2 \equiv \tilde{\sigma}^2$, it follows that the df of (13.7.13)

$$\tilde{\phi} \equiv K(\Delta\tilde{\bar{\mathbf{x}}} - \Delta\boldsymbol{\mu})'\tilde{\mathbf{s}}^{-2}(\Delta\tilde{\bar{\mathbf{x}}} - \Delta\boldsymbol{\mu})$$

is $f_F(\phi \mid k, n_1 + n_2 - k - 1)$, where

$$\tilde{\mathbf{s}}^{-2} \equiv (\tilde{\mathbf{s}}^2)^{-1}$$

and

$$K \equiv (n_1 + n_2 - k - 1)\frac{(n_1{}^{-1} + n_2{}^{-1})^{-1}}{k(n_1 + n_2 - 2)}.$$

Proof

By the proof of (13.7.12), the df of $\Delta\tilde{\bar{\mathbf{x}}}$ is $f_{N^{(k)}}(\Delta\bar{\mathbf{x}} \mid \Delta\boldsymbol{\mu}, (n_1{}^{-1} + n_2{}^{-1})\tilde{\sigma}^2)$ when $\tilde{\sigma}_1{}^2 = \tilde{\sigma}_2{}^2 \equiv \tilde{\sigma}^2$, and hence the df of $\tilde{\mathbf{z}} \equiv (n_1{}^{-1} + n_2{}^{-1})^{-1/2}(\Delta\tilde{\bar{\mathbf{x}}} - \Delta\boldsymbol{\mu})$ is $f_{N^{(k)}}(\mathbf{z} \mid \mathbf{0}, \tilde{\sigma}^2)$. Next, note that the independence assumption implies that the $(n_i - 1)\tilde{\mathbf{s}}_i{}^2$'s are independent with df's $f_{W^{(k)}}(\cdot \mid (n_i - k)^{-1}\tilde{\sigma}^{-2}, n_i - k)$ and hence, by (12.9.6), $\tilde{\mathbf{V}} \equiv (n_1 + n_2 - 2)\tilde{\mathbf{s}}^2 = (n_1 - 1)\tilde{\mathbf{s}}_1{}^2 + (n_2 - 1)\tilde{\mathbf{s}}_2{}^2$ has df $f_{W^{(k)}}(\mathbf{V} \mid (n_1 + n_2 - k - 1)^{-1}\tilde{\sigma}^{-2}, n_1 + n_2 - k - 1)$. Moreover, by (13.6.4), $\tilde{\mathbf{s}}_i{}^2$ is independent of $\tilde{\bar{\mathbf{x}}}_i$, for each i, and this, together with the initial independence assumption, implies that $\tilde{\mathbf{V}}$ and $\tilde{\mathbf{z}}$ are independent. The assertion now follows readily from (13.6.9) by direct substitutions. QED

With definitions and assumptions as in the preceding, and with the additional assumption that $\sigma_{*1}{}^2 = \sigma_{*2}{}^2 \equiv \sigma_*{}^2$, the df of (13.7.14)

$$\tilde{\phi} \equiv \left(\frac{1}{k}\right)(\Delta\tilde{\bar{\mathbf{x}}} - \Delta\boldsymbol{\mu})'[\tilde{s}_*{}^2(n_1{}^{-1}\tilde{\sigma}_{*1}^2 + n_2{}^{-1}\tilde{\sigma}_{*2}^2)]^{-1}(\Delta\tilde{\bar{\mathbf{x}}} - \Delta\boldsymbol{\mu})$$

is $f_F(\phi \mid k, k(n_1 + n_2 - 2))$.

Proof

By (13.7.12), the df of

$$\tilde{Q}_1 \equiv \frac{(\Delta\tilde{\bar{\mathbf{x}}} - \Delta\boldsymbol{\mu})'(n_1{}^{-1}\tilde{\sigma}_{*1}^2 + n_2{}^{-1}\tilde{\sigma}_{*2}^2)^{-1}(\Delta\tilde{\bar{\mathbf{x}}} - \Delta\boldsymbol{\mu})}{\sigma_*{}^2}$$

is $f_{\chi^2}(Q_1 \mid k)$ given the hypothesis that $\sigma_{*1}{}^2 = \sigma_{*2}{}^2 \equiv \sigma_*{}^2$. By (13.6.7) and the initial independence assumption, the $k(n_i - 1)\tilde{s}_{*i}{}^2/\sigma_*{}^2$'s are independent with df's $f_{\chi^2}(\cdot \mid k(n_i - 1))$ and hence their sum, $k(n_1 + n_2 - 2)\tilde{s}_*{}^2/\sigma_*{}^2 \equiv \tilde{Q}_2$, has df $f_{\chi^2}(\cdot \mid k(n_1 + n_2 - 2))$. Thus by (13.3.9), $\tilde{\phi} \equiv (\tilde{Q}_1/k)/(\tilde{Q}_2/[k(n_1 + n_2 - 2)])$ has df $f_F(\phi \mid k, k(n_1 + n_2 - 2))$. But $\tilde{\phi}$ is of the asserted form after direct substitution for $\tilde{Q}_1$ and $\tilde{Q}_2$. QED

In the course of proving (13.7.14), we have established a special case of the following:

With definitions and assumptions as in the preceding, with the assumption that $\sigma_{*1}^2 = \sigma_{*2}^2 \equiv \sigma_*^2$, and with the definitions (13.7.15)

$$\nu_i \equiv \begin{cases} n_i - 1, & \text{if } \mathbf{\mu}_i \text{ is unknown} \\ n_i, & \text{if } \mathbf{\mu}_i \text{ is known,} \end{cases}$$

$$\tilde{\zeta}_{*i} \equiv \begin{cases} \tilde{s}_{*i}^2, & \text{if } \mathbf{\mu}_i \text{ is unknown} \\ \tilde{\xi}_{*i}, & \text{if } \mathbf{\mu}_i \text{ is known,} \end{cases}$$

$$\nu \equiv \nu_1 + \nu_2,$$

and

$$\tilde{\zeta}_* \equiv \nu^{-1}(\nu_1 \tilde{\zeta}_{*1} + \nu_2 \tilde{\zeta}_{*2}),$$

it follows that the df of $\tilde{Q} \equiv k\nu\tilde{\zeta}_*/\sigma_*^2$ is $f_{\chi^2}(Q \mid k\nu)$, and the df of $\tilde{\zeta}_*$ is

$$f_{\gamma 2}(\zeta_* \mid 1/\sigma_*^2, k\nu).$$

Proof

The second assertion is immediate from the first via (13.3.2). To prove the first, it follows from the definitions and (13.6.6) and (13.6.7) that $k\nu_i\tilde{\zeta}_i/\sigma_*^2$ has df $f_{\chi^2}(\cdot \mid k\nu_i)$, and independence implies that $\tilde{Q} \equiv k\nu\tilde{\zeta}_*/\sigma_*^2$ has df $f_{\chi^2}(\cdot \mid k(\nu_1 + \nu_2))$. QED

Note that if the $\tilde{\zeta}_{*i}$'s are interpreted as estimates of the σ_{*i}^2's, then their weighted average $\tilde{\zeta}_{*i}$ is an estimate of the assumed common value σ_*^2 of σ_{*1}^2 and σ_{*2}^2. A similar result obtains for the assumption that $\boldsymbol{\sigma}_1^2 = \boldsymbol{\sigma}_2^2 = \boldsymbol{\sigma}^2$:

With definitions and assumptions as in the preceding, with the assumption that $\boldsymbol{\sigma}_1^2 = \boldsymbol{\sigma}_2^2 \equiv \boldsymbol{\sigma}^2$, and with the definitions (13.7.16)

$$\nu_i \equiv \begin{cases} n_i - 1, & \text{if } \mathbf{\mu}_i \text{ is unknown} \\ n_i, & \text{if } \mathbf{\mu}_i \text{ is known,} \end{cases}$$

$$\tilde{\boldsymbol{\zeta}}_i \equiv \begin{cases} \mathbf{s}_i^2, & \text{if } \mathbf{\mu}_i \text{ is unknown} \\ \tilde{\boldsymbol{\xi}}_i, & \text{if } \mathbf{\mu}_i \text{ is known,} \end{cases}$$

$$\nu \equiv \nu_1 + \nu_2,$$

and

$$\tilde{\boldsymbol{\zeta}} \equiv \nu^{-1}(\nu_1 \tilde{\boldsymbol{\zeta}}_1 + \nu_2 \tilde{\boldsymbol{\zeta}}_2),$$

it follows that the df of $\nu\tilde{\boldsymbol{\zeta}} \equiv \tilde{\mathbf{V}}$ is $f_{W^{(k)}}(\mathbf{V} \mid (\nu - k + 1)^{-1}\boldsymbol{\sigma}^{-2}, \nu - k + 1)$, and the df of $\tilde{\boldsymbol{\zeta}}$ is $f_{W^{(k)}}(\boldsymbol{\zeta} \mid \nu(\nu - k + 1)^{-1}\boldsymbol{\sigma}^{-2}, \nu - k + 1)$.

Proof

The second assertion is immediate from the first via (12.9.7). To prove the first, it follows from the definitions and (13.6.3) and (13.6.4) that

$\nu_i\tilde{\zeta}_i$ has df $f_{W^{(k)}}(\cdot \mid (\nu_i - k + 1)^{-1}\boldsymbol{\sigma}^{-2}, \nu_i - k + 1)$. The $\nu_i\tilde{\zeta}_i$'s are clearly independent, and hence $\nu\tilde{\zeta} = \nu_1\tilde{\zeta}_1 + \nu_2\tilde{\zeta}_2$ has df $f_{W^{(k)}}(\cdot \mid (\nu - k + 1)^{-1}\boldsymbol{\sigma}^{-2}, \nu - k + 1)$, by (12.9.6). QED

13.8 THE UNIVARIATE CENTRAL LIMIT THEOREM

We shall prove that version of the Central Limit Theorem which is of greatest use in statistics. It is due, in the form stated, to Lindeberg [**47**] and Lévy [**46**]. Two more general versions, the Lyapunov Theorem and the Lindeberg-Feller Theorem, are of considerable interest but are not needed here. They, as well as many other similar results about limiting distributions, can be found in Gnedenko and Kolmogorov [**30**] or Loéve [**51**].

The remainder of this section consists of two subsections, the first of which contains the Central Limit Theorem and some of its more immediate consequences. The proof is deferred to the second subsection and may be omitted by readers impatient with the necessarily mathematical detail.

13.8.1 The Central Limit Theorem and Some of Its Consequences

(Lindeberg-Lèvy) Central Limit Theorem: Let $\tilde{x}_1, \tilde{x}_2, \ldots$ be iid random variables with common dgf $f(x)$. Assume that their common variance σ^2 is *positive* and *finite*. (Hence their common expectation μ is also finite and they are not degenerate random variables.) Define (13.8.1)

$$\bar{\tilde{x}}_n \equiv (1/n) \sum_{i=1}^{n} \tilde{x}_i$$

and

$$\tilde{z}_n \equiv (\bar{\tilde{x}}_n - \mu)/(\sigma^2/n)^{1/2} = \left(\sum_{i=1}^{n} \tilde{x}_i - n\mu\right)/(n\sigma^2)^{1/2},$$

for $n = 1, 2, \ldots$. Then

$$\lim_{n\to\infty} P[\tilde{z}_n \leqq z] = F_{N^*}(z), \text{ for all } z \in R^1.$$

Note from its definition in (13.8.1) that $E(\tilde{z}_n) = 0$ and $V(\tilde{z}_n) = 1$ for every n (Exercise 6), and hence $\tilde{z}_n$ has the same expectation and variance as the standardized Normal random variable $\tilde{z}$. The importance of (13.8.1) lies in its assertion that the sequence $F_1(\cdot), F_2(\cdot), \ldots$ of *cdf's* of $\tilde{z}_1, \tilde{z}_2, \ldots$, respectively approaches the *cdf* $F_{N^*}(\cdot)$ of $\tilde{z}$ as a limit,

Therefore, (13.8.1) enables us to assert that if n is "large enough," the cdf $F_n(\cdot)$ of $\tilde{z}_n$ is "approximately" $F_{n^*}(\cdot)$ It can be shown that the approximation $F_n(z_n) \doteq F_{N^*}(z_n)$ is better for values of z_n near zero than for values far from zero.

Some relatively immediate consequences of (13.8.1) constitute the following:

Let $\tilde{x}_1, \tilde{x}_2, \ldots$ satisfy the assumptions of (13.8.1), and let $-\infty \leqq a < b \leqq \infty$. Then for large n, (13.8.2)

(a) $P[a < \tilde{\bar{x}}_n \leqq b] \doteq F_{N*}\left(\frac{b-\mu}{\sigma/\sqrt{n}}\right) - F_{N*}\left(\frac{a-\mu}{\sigma/\sqrt{n}}\right)$,

(b) $P\left[a < \sum_{i=1}^{n} \tilde{x}_i \leqq b\right] \doteq F_{N*}\left(\frac{b-n\mu}{\sigma\sqrt{n}}\right) - F_{N*}\left(\frac{a-n\mu}{\sigma\sqrt{n}}\right)$,

(c) $P[a < \tilde{\bar{x}}_n \leqq b] \doteq F_N\left(b \mid \mu, \frac{\sigma^2}{n}\right) - F_N\left(a \mid \mu, \frac{\sigma^2}{n}\right)$, and

(d) $P\left[a < \sum_{i=1}^{n} \tilde{x}_i \leqq b\right] \doteq F_N(b \mid n\mu, n\sigma^2) - F_N(a \mid n\mu, n\sigma^2)$,

where "$\doteq$" denotes approximate equality.

Proof

Exercise 7.

As an application of (13.8.2), we obtain the historically earliest version of the Central Limit Theorem:

de Moivre-Laplace Theorem; or a Normal Approximation to the Binomial cdf: If n is large, then (13.8.3)

$$F_b(r \mid n, p) \doteq F_{N*}\left(\frac{r-np}{\sqrt{np(1-p)}}\right),$$

for $r \in \{0, 1, \ldots, n\}$.

Proof

If $\tilde{r}$ has mf $f_b(r \mid n, p)$, then $\tilde{r} = \sum_{i=1}^{n} \tilde{x}_i$, where $\tilde{x}_1, \ldots, \tilde{x}_n$ are iid with common mf $f_b(\cdot \mid 1, p)$, common expectation $\mu = p$, and common variance $\sigma^2 = p(1-p)$. Hence (13.8.3) follows from (13.8.2b) for $a \equiv -\infty$ and $b \equiv r$. QED

A somewhat better approximation than that given by (13.8.3) is obtained by replacing r with $r + \frac{1}{2}$ in the argument of F_{N*}. Slightly more generally we have

$$P_b[a < \tilde{r} \leqq b \mid n, p] = F_{N*}\left(\frac{b+\frac{1}{2}-np}{\sqrt{np(1-p)}}\right) - F_{N*}\left(\frac{a+\frac{1}{2}-np}{\sqrt{np(1-p)}}\right), \tag{13.8.4}$$

for $\{a, b\} \subset \{0, 1, \ldots, n\}$ and $b > a$.

Example 13.8.1

If $100p$ percent = 50 percent of the voters in a given state support Candidate X, what is the probability that at least 66 voters in a random sample (with replacement) of 100 voters support X? Let $\tilde{r}$ denote the number who support X out of $n = 100$. Then $\tilde{r}$ has mf $f_b(r \mid 100, \frac{1}{2})$, and

$$\begin{aligned} P_b[r \geqq 66 \mid 100, \tfrac{1}{2}] &= 1 - P_b[r \leqq 65 \mid 100, \tfrac{1}{2}] \\ &\doteq 1 - F_{N*}\left(\frac{65.5 - 50}{\sqrt{100 \cdot \frac{1}{2} \cdot \frac{1}{2}}}\right) \\ &= 1 - F_{N*}(3.1) \\ &= .0010 \text{ (from Table I).} \end{aligned}$$

The exact value of $P_b[\tilde{r} \geqq 66 \mid 100, \frac{1}{2}]$ is .0009, which shows that the Normal approximation is quite good even for $z = 3.1$ standard deviations. (Thus some outcome $r \geqq 66$ of the poll would be almost a miracle if in fact only 50 percent of the voters support X, and if $r \geqq 66$, then X might be led to doubt claims by his opponent that X failed to have the support of a majority of the electorate. These ideas will be more fully developed in Chapters 21 and 22.)

A Normal Approximation to the Gamma cdf: If r is large, then (13.8.5)

$$F_{\gamma*}(z \mid r) \doteq F_{N*}\left(\frac{z - r}{\sqrt{r}}\right),$$

for $z \in R^1$.

Proof

Assume r is a large positive integer, and let $\tilde{z} = \sum_{i=1}^{n} \tilde{z}_i$, where $\tilde{z}_1, \ldots, \tilde{z}_r$ are iid with common df $f_e(z_i \mid 1)$, defined by (11.7.1). In Section 11.7 we showed that $\tilde{z}$ has df $f_\gamma(z \mid r, 1) \equiv f_{\gamma*}(z \mid r)$. Since the common expectation and variance of the z_i's are both one, the assertion follows readily from (13.8.2b). If r is large but nonintegral, the assertion follows by a continuity argument. QED

A somewhat better approximation is available as an alternative to (13.8.5):

The Wilson-Hilferty Approximation: If r is large, then (13.8.6)

$$F_{\gamma*}(z \mid r) \doteq F_{N*}\left(3\sqrt{r}\left[\sqrt{\frac{z}{r}} + \frac{1}{9r} - 1\right]\right).$$

Proof

See Kendall and Stuart [**41**], Vol. I.

A Normal Approximation to the Poisson cdf: If α is large, then

$$F_{Po}(r \mid \alpha) \doteq F_{N*}\left(\frac{r - \alpha}{\sqrt{\alpha}}\right). \tag{13.8.7}$$

Proof

For large α, choose a large positive integer n and define $\lambda = \alpha/n$, so that $\alpha = \lambda n$. From Example 10.6.2 it follows that $\tilde{r} = \sum_{i=1}^{n} \tilde{r}_i$, where $\tilde{r}_1, \ldots, \tilde{r}_n$ are iid with common mf $f_{Po}(r_i \mid \lambda)$, and hence common expectation and variance both equal to λ. The assertion now follows from (13.8.2b). QED

A Normal Approximation to the Beta df: If ν is large, then

$$F_\beta(p \mid \rho, \nu) \doteq F_{N*}\left(\frac{(\nu - 1)p - \rho + \frac{1}{2}}{\sqrt{(\nu - 1)p(1 - p)}}\right) \tag{13.8.8}$$

Proof

In Chapter 7 of Raiffa and Schlaifer [**66**] it is shown that

$$F_\beta(p \mid \rho, \nu) = 1 - F_b(\rho - 1 \mid \nu - 1, p),$$

whenever ρ and ν are positive integers such that $\nu > \rho \geqq 1$. By (13.8.4),

$$1 - F_b(\rho - 1 \mid \nu - 1) \doteq 1 - F_{N*}\left(\frac{\rho - 1 + \frac{1}{2} - (\nu - 1)p}{\sqrt{(\nu - 1)p(1 - p)}}\right)$$

$$= F_{N*}\left(\frac{(\nu - 1)p - \rho + \frac{1}{2}}{\sqrt{(\nu - 1)p(1 - p)}}\right). \quad \text{QED}$$

An alternative Normal approximation to the beta cdf for large ν is given in Chapter 18 of Pratt, Raiffa, and Schlaifer [**64**]; we lack the distribution theory at this point to derive it.

If ν is large and $\rho > 1$, then (13.8.9)

$$F_\beta(p \mid \rho, \nu) \doteq F_{N*}\left(\frac{p - \pi}{\sqrt{\pi(1 - \pi)/n}}\right),$$

where $\pi \equiv (\rho - 1)/(\nu - 2)$.

13.8.2 Proof of the Central Limit Theorem

We shall not be able to give a completely self-contained proof of (13.8.1) because such a proof requires results about characteristic functions which in turn require powerful methods of analysis to prove. We shall state the necessary results about characteristic functions without proof, referring the interested reader to Loéve [**51**], Gnedenko and Kolmogorov [**30**], or Lukacs [**53**] for a complete treatment.

Let $\tilde{x}$ be a random variable. The *characteristic function* $\phi(u)$ of $\tilde{x}$ is defined by

$$\phi(u) \equiv E(\exp[iu\tilde{x}]), \tag{13.8.10}$$

where $i \equiv \sqrt{-1}$. It is an elementary fact of complex analysis that

$$\exp[iux] = \cos(ux) + i \cdot \sin(ux), \tag{13.8.11}$$

for every real x and u. Since $\cos(\cdot)$ and $\sin(\cdot)$ are bounded above and below by $+1$ and -1, respectively, $E[\cos(u\tilde{x})]$ and $E[\sin(u\tilde{x})]$ both belong to $[-1, 1]$. Since $E(\exp[iu\tilde{x}]) = E[\cos(u\tilde{x})] + iE[\sin(u\tilde{x})]$, it follows that $\phi(u)$ exists for every u and *every* random variable $\tilde{x}$, unlike the moment-generating function $M(u)$ of $\tilde{x}$, which is given by

$$M(u) = \phi\left(\frac{u}{i}\right) \tag{13.8.12}$$

when it exists. From (12.4.5) and (13.8.12), it follows that the characteristic function $\phi_{N^*}(u)$ of the standardized Normal random variable $\tilde{z}$ is given by

$$\phi_{N^*}(u) = \exp[-\tfrac{1}{2}u^2]. \tag{13.8.13}$$

Moments $E(\tilde{x}^k)$ of $\tilde{x}$ are easily obtained by differentiation of the characteristic function. More precisely, we have the following:

If $E(\tilde{x}^m)$ exists for m a positive integer, then $E(\tilde{x}^j)$ exists for every $j \in \{1, \ldots, m\}$, and (13.8.14)

$$E(\tilde{x}^j) = \left(\frac{1}{i^j}\right)\left[\frac{d^j\phi(u)}{du^j}\right]\Bigg|_{u=0}.$$

(Partially) conversely, if m is an *even* positive integer, existence of

$$\frac{d^m\phi(u)}{du^m}\Bigg|_{u=0}$$

implies existence of $E(\tilde{x}^m)$, the validity of the preceding characterization of $E(\tilde{x}^m)$, and the Maclaurin expansion (Taylor, about $u = 0$)

$$\phi(u) = 1 + \sum_{j=1}^{m} \frac{E(\tilde{x}^j)}{j!}(iu)^j + R_m(u^m),$$

where $R_m(\cdot)$ is such that

$$\lim_{y\to 0}\left\{\frac{R_m(y)}{y}\right\} = 0.$$

The following result is very similar to (10.6.8):

If $\tilde{y}_1, \ldots, \tilde{y}_n$ are iid with common characteristic function $\phi(u)$, then the characteristic function of (13.8.15)

$$\left(\sum_{i=1}^{n}\tilde{y}_i\right)/(\sigma\sqrt{n}) \quad \text{is} \quad [\phi(u/[\sigma\sqrt{n}])]^n.$$

Proof

Exercise 8.

A final needed result concerns the correspondence between cdf's and characteristic functions, and between limits of sequences of these functions.

Let $\tilde{x}_1, \tilde{x}_2, \ldots$ have cdf's $F_1(\cdot), F_2(\cdot), \ldots$ and characteristic functions $\phi_1(\cdot), \phi_2(\cdot), \ldots$, and let $\tilde{x}$ be a continuous random variable with cdf $F(\cdot)$ and characteristic function $\phi(\cdot)$. Then (13.8.16)

$$\lim_{n\to\infty} F_n(x) = F(x), \text{ for every } x \in R^1$$

if and only if

$$\lim_{n\to\infty} \phi_n(u) = \phi(u), \text{ for every } u \in R^1.$$

Assertion (13.8.16) is a special case of Theorem 3.6.1 in Lukacs [**53**].

Proof of (13.8.1)

Let $\phi^*(\cdot)$ denote the common characteristic function of $\tilde{y}_1, \tilde{y}_2, \ldots$, where $\tilde{y}_i \equiv \tilde{x}_i - \mu$, and let $\phi_n(\cdot)$ denote the characteristic function of $\tilde{z}_n$. Since $\tilde{z}_n = \left(\sum_{i=1}^{n}\tilde{y}_i\right)/(\sigma\sqrt{n})$, follows from (13.8.15) that

(a) $$\phi_n(u) = \left[\phi^*\left(\frac{u}{\sigma\sqrt{n}}\right)\right]^n,$$

for every $n = 1, 2, \ldots$ and every $u \in R^1$. Moreover, it is clear that $E(\tilde{y}_i) = 0$ and $E(\tilde{y}_i^2) = \sigma^2$, for every i, and hence by (13.8.14) for $m = 2$ and (a),

(b) $$\begin{aligned}\phi_n(u) &= \left\{1 + \left[\frac{\sigma^2}{2}\right]\left(\frac{iu}{\sigma\sqrt{n}}\right)^2 + R_2\left(\left\{\frac{u}{\sigma\sqrt{n}}\right\}^2\right)\right\}^n \\ &= \left[1 - \frac{u^2}{2n} + R_2\left(\frac{u^2}{n\sigma^2}\right)\right]^n \\ &= \left[1 + n^{-1}\left\{-\tfrac{1}{2}u^2 + nR_2\left(\frac{u^2}{n\sigma^2}\right)\right\}\right]^n.\end{aligned}$$

Hence

(c)
$$\lim_{n\to\infty} \phi_n(u) = \lim_{n\to\infty} (1 + n^{-1}[-\tfrac{1}{2}u^2])^n$$
$$= \exp[-\tfrac{1}{2}u^2].$$

But

$$\lim_{n\to\infty} \phi_n(u) = \exp[-\tfrac{1}{2}u^2]$$

implies (13.8.1) by virtue of (13.8.16) and (13.8.13). QED

For use in the next section, we give the following easy consequence of (13.8.1):

Let $\tilde{w}_n \equiv \sigma\tilde{\bar{z}}_n$, where $\tilde{\bar{z}}_n$ is as defined in (13.8.1). Then (13.8.17)

$$\lim_{n\to\infty} P[\tilde{w}_n \leqq w] = F_N(w \mid 0, \sigma^2), \text{ for every } w \in R^1.$$

Proof

Exercise 9.

13.9 A k-VARIATE CENTRAL LIMIT THEOREM

It is easy to obtain the following k-dimensional version of the Lindebey-Lévy Theorem.

Let $\tilde{\mathbf{x}}_1, \tilde{\mathbf{x}}_2, \ldots$ be iid k-dimensional random vectors with common dgf $f(\mathbf{x})$, common expectation vector $\boldsymbol{\mu}$, and common positive definite and symmetric covariance matrix $\boldsymbol{\sigma}^2$. Define $\tilde{\bar{\mathbf{x}}}_n \equiv n^{-1}\sum_{i=1}^{n} \tilde{\mathbf{x}}_i$ and $\tilde{\mathbf{w}}_n \equiv n^{1/2}(\tilde{\bar{\mathbf{x}}}_n - \boldsymbol{\mu})$. Then

$$\lim_{n\to\infty} P[\tilde{\mathbf{w}}_n \leqq \mathbf{w}] = F_{N^{(k)}}(\mathbf{w} \mid \mathbf{0}, \boldsymbol{\sigma}^2), \tag{13.9.1}$$

for every $w \in R^k$.

Proof

It is clear from (10.7.7) and (10.7.8) that $E(\tilde{\mathbf{w}}_n) = \mathbf{0}$ and $V(\tilde{\mathbf{w}}_n) = \boldsymbol{\sigma}^2$ for every n. Hence if $\mathbf{d} \in R^k$, then $E(\mathbf{d}'\tilde{\mathbf{w}}_n) = \mathbf{0}$ and $V(\mathbf{d}'\tilde{\mathbf{w}}_n) = \mathbf{d}'\boldsymbol{\sigma}^2\mathbf{d}$, and if $\mathbf{d} \neq \mathbf{0}$, then $\mathbf{0} < \mathbf{d}'\boldsymbol{\sigma}^2\mathbf{d} < \infty$, by positive definiteness of $\boldsymbol{\sigma}^2$. Define $\tilde{x}^*_i \equiv \mathbf{d}'\tilde{\mathbf{x}}_i$, $\mu^* \equiv \mathbf{d}'\boldsymbol{\mu}$, $\tilde{\bar{x}}^*_n \equiv \mathbf{d}'\tilde{\bar{\mathbf{x}}}_n$, and $\tilde{w}^*_n \equiv n^{1/2}(\tilde{\bar{x}}^*_n - \mu^*)$, so that $\tilde{w}^*_n = \mathbf{d}'\tilde{\mathbf{w}}_n$. Now $\tilde{w}^*_1, \tilde{w}^*_2, \ldots$ satisfy the hypotheses of (13.8.1) and (13.8.17) and therefore (13.8.17) implies that

$$\lim_{n\to\infty} P[\tilde{w}^*_n \leqq \mathbf{d}'\mathbf{w}] = F_N(\mathbf{d}'\mathbf{w} \mid 0, \mathbf{d}'\boldsymbol{\sigma}^2\mathbf{d}),$$

for *every* $\mathbf{d} \neq \mathbf{0}$. Hence by (12.7.12) the distribution of $\tilde{\mathbf{w}}_n$ must tend to the k-variate Normal with parameters $\mathbf{0}$ and $\boldsymbol{\sigma}^2$. QED

If n is large and $\tilde{\mathbf{x}}_1, \ldots, \tilde{\mathbf{x}}_n$ satisfy the hypotheses of (13.9.1), then (13.9.2)

(a) $P[\tilde{\bar{\mathbf{x}}}_n \leqq \bar{\mathbf{x}}_n] \doteq F_N(\bar{\mathbf{x}}_n \mid \boldsymbol{\mu}, n^{-1}\boldsymbol{\sigma}^2)$; and

(b) $P\left[\sum_{i=1}^{n} \tilde{\mathbf{x}}_i \leqq \mathbf{y}\right] \doteq F_N(\mathbf{y} \mid n\boldsymbol{\mu}, n\boldsymbol{\sigma}^2)$.

Proof

Exercise 10.

As one might suspect, there is a k-variate generalization of the de Moivre-Laplace Theorem, as follows:

If n is large, then (13.9.3)

$$F_{mu}{}^{(k)}(\mathbf{r} \mid n, \mathbf{p}) \doteq F_N{}^{(k)}(\mathbf{r} \mid n\mathbf{p}, n\boldsymbol{\sigma}^2),$$

where the (i, j)th element $\sigma_{ij}{}^2$ of $\boldsymbol{\sigma}^2$ is given by

$$\sigma_{ij}{}^2 = \begin{cases} p_i(1 - p_i), & i = j \\ -p_i p_j, & i \neq j. \end{cases}$$

Proof

If $\tilde{\mathbf{r}}$ has mf $f_{mu}{}^{(k)}(\cdot \mid n, \mathbf{p})$, then $\tilde{\mathbf{r}} = \sum_{i=1}^{n} \tilde{x}_i$, where $\tilde{\mathbf{x}}_1, \ldots, \tilde{\mathbf{x}}_n$ are iid with common mf $f_{mu}{}^{(k)}(\cdot \mid 1, \mathbf{p})$, common mean vector $\mathbf{p}$, and covariance $\boldsymbol{\sigma}^2$ as asserted. (13.9.3) then follows via (13.9.2b). QED

We also state without proof the k-variate generalization of (13.8.9):

If ν is large and $\boldsymbol{\rho} > \mathbf{1}$, then (13.9.4)

$$F_\beta{}^{(k)}(\mathbf{p} \mid \boldsymbol{\rho}, \nu) \doteq F_N{}^{(k)}(\mathbf{p} \mid \boldsymbol{\pi}, \nu^{-1}\boldsymbol{\sigma}^2),$$

where $\mathbf{1} \equiv (1, 1, \ldots, 1)'$,

$$\boldsymbol{\pi} \equiv (\nu - k - 1)^{-1}(\boldsymbol{\rho} - \mathbf{1}),$$

and the (i, j)th element $\sigma_{ij}{}^2$ of $\boldsymbol{\sigma}^2$ is given by

$$\sigma_{ij}{}^2 \equiv \begin{cases} \pi_i(1 - \pi_i), & i = j \\ -\pi_i \pi_j, & i \neq j. \end{cases}$$

13.10 CHI-SQUARE LIMIT THEOREMS

In Sections 13.8 and 13.9 we have shown that, under the mild variance (or covariance matrix) restriction, the sample mean (or sample mean vector) has for large n approximately the corresponding distribution under the assumption that the observations themselves were Normally distributed; compare (13.8.2) with (13.3.3), and (13.9.1) with (13.6.2), for example.

It is tempting, but erroneous, to conclude that if n is large and $\tilde{\xi}$, say, is computed from $\tilde{x}_1, \ldots, \tilde{x}_n$ via its definition (13.2.3), then the gamma-2 distribution with parameters $1/\sigma^2$ and n is a good approximation to the actual distribution of $\tilde{\xi}$, even though $f(x)$ is not Normal.

The difficulty in this line of reasoning is essentially that the good behavior of $\tilde{\xi}$ requires that its variance exist (be finite), and existence of $V(\tilde{\xi})$ requires existence of the common *fourth* moment $E(\tilde{x}^4)$ of $\tilde{x}_1, \ldots, \tilde{x}_n$; the reader may develop some of the details as Exercise 11.

In some cases, however, we may obtain results, the validity of which requires only the same restrictions as do (13.8.1) and (13.9.1). First, a general result follows:

Let $\mathbf{g}$: $R^k \to R^m$ be continuous, let $\tilde{\mathbf{w}}_1, \tilde{\mathbf{w}}_2, \ldots$ be a sequence of k-dimensional random vectors, and let $\tilde{\mathbf{g}}_i \equiv \mathbf{g}(\tilde{\mathbf{w}}_i)$, for every i. If there is a random k-tuple $\tilde{\mathbf{w}}$ such that (13.10.1)

$$\lim_{n \to \infty} P[\tilde{\mathbf{w}}_n \leqq \mathbf{w}] = P[\tilde{\mathbf{w}} \leqq \mathbf{w}], \text{ for every } \mathbf{w} \in R^k,$$

then

$$\lim_{n \to \infty} P[\tilde{\mathbf{g}}_n \leqq \mathbf{g}] = P[\mathbf{g}(\tilde{\mathbf{w}}) \leqq \mathbf{g}], \text{ for every } \mathbf{g} \in R^m.$$

Proof

See Wilks [**83**], Chapter 4.

We then obtain

If $\tilde{\mathbf{x}}_1, \ldots, \tilde{\mathbf{x}}_n$ are iid k-tuples with common dgf $f(\mathbf{x})$ satisfying the hypotheses of (13.9.1), and if (13.10.2)

$$\tilde{g}_n \equiv n(\bar{\tilde{\mathbf{x}}}_n - \boldsymbol{\mu})'\boldsymbol{\sigma}^{-2}(\bar{\tilde{\mathbf{x}}}_n - \boldsymbol{\mu}),$$

then

$$\lim_{n \to \infty} P[\tilde{g}_n \leqq g] = F_{\chi^2}(g \mid k), \text{ for every } g \in R^1.$$

Proof

$g: R^k \to R^1$ is clearly continuous, the limiting distribution of $\tilde{\bar{\mathbf{x}}}_n$ is $f_N(\bar{\mathbf{x}}_n \mid \boldsymbol{\mu}, n^{-1}\boldsymbol{\sigma}^2)$ by (13.9.2), and hence (13.10.1) implies the limiting distribution of $\tilde{g}_n$ is given by (13.6.8). QED

Note that $\tilde{g}_n$ in (13.10.2) may be re-expressed as

$$\tilde{g}_n = (\tilde{\mathbf{r}} - n\boldsymbol{\mu})'(n\boldsymbol{\sigma}^2)^{-1}(\tilde{\mathbf{r}} - n\boldsymbol{\mu}), \tag{13.10.3}$$

where $\tilde{\mathbf{r}} \equiv \sum_{i=1}^{n} \tilde{\mathbf{x}}_i$. This suggests the following consequence of (13.10.2).

If n is large and $\tilde{\mathbf{r}}$ has mf $f_{mu}^{(k)}(\cdot \mid n, \mathbf{p})$, then (13.10.4)

$$\tilde{g}_n \equiv \sum_{i=1}^{k+1} \frac{(\tilde{r}_i - np_i)^2}{np_i}$$

has approximate cdf $F_{\chi^2}(\cdot \mid k)$.

Proof

From (13.9.3), the limiting distribution of $\tilde{\mathbf{r}}$ is $F_N(\mathbf{r} \mid n\mathbf{p}, n\boldsymbol{\sigma}^2)$ for $\boldsymbol{\sigma}^2$ as defined there. By (13.10.2), $F_{\chi^2}(\cdot \mid k)$ is the limiting distribution of

$$(\tilde{\mathbf{r}} - n\mathbf{p})'(n\boldsymbol{\sigma}^2)^{-1}(\tilde{\mathbf{r}} - n\mathbf{p}).$$

It is not hard to verify the (i, j)th element σ_{ij}^{-2} of $\boldsymbol{\sigma}^{-2}$ is given by

$$\sigma_{ij}^{-2} = \begin{cases} p_i^{-1} + p_{k+1}^{-1}, & j = i \\ p_{k+1}^{-1}, & j \neq i, \end{cases}$$

when $\boldsymbol{\sigma}^2$ is as defined in (13.9.3). As Exercise 12, the reader may complete the proof by substituting these values of σ_{ij}^{-2} into $(\mathbf{r} - n\mathbf{p})'(n\boldsymbol{\sigma}^2)^{-1}(\mathbf{r} - n\mathbf{p})$ and simplifying to obtain the asserted form of $\tilde{g}_n$. QED

The random variable $\tilde{g}_n$, as defined in (13.10.4), is called *Pearson's Chi-square statistic*. It is particularly useful in solving problems in inference, such as deciding whether it is reasonable to assume that iid observations $\mathbf{x}_1, \ldots, \mathbf{x}_n$ have some hypothesized common dgf $f(\mathbf{x})$.

Notice that $\tilde{g}_n$ in (13.10.4) is also given by

$$\tilde{g}_n = \sum_{i=1}^{k+1} \frac{(\tilde{r}_i/n - p_i)^2}{p_i} \tag{13.10.5}$$

by factoring an n out of the terms $(\tilde{r}_i - np_i)^2$ and simplifying.

It is logical to suspect that a result similar to (13.10.5) obtains for the k-variate beta df:

If $\tilde{\mathbf{p}}$ has df $f_\beta^{(k)}(\mathbf{p} \mid \boldsymbol{\varrho}, \nu)$, ν is large, and $\boldsymbol{\varrho} > \mathbf{1}$, then (13.10.6)

$$\tilde{g}_\nu \equiv \nu \sum_{i=1}^{k+1} \frac{(\tilde{p}_i - \pi_i)^2}{\pi_i}$$

has approximate cdf $F_{\chi^2}(\cdot \mid k)$, where $\pi_i \equiv (\rho_i - 1)/(\nu - k - 1)$, for $i = 1, \ldots, k + 1$.

Proof

The proof follows from (13.9.4) just as (13.10.5) followed from (13.9.3). In particular, the (i, j)th element σ_{ij}^{-2} of the inverse $\boldsymbol{\sigma}^{-2}$ of the matrix $\boldsymbol{\sigma}^2$ defined in (13.9.4) is

$$\sigma_{ij}^{-2} = \begin{cases} \pi_i^{-1} + \pi_{k+1}^{-1}, & j = i \\ \pi_{k+1}^{-1}, & j \neq i. \end{cases}$$

EXERCISES

1. Derive (13.3.2).
2. Derive (13.3.8).
3. Derive (13.4.11).
4. Derive (13.5.13).
5. Derive (13.6.8).
6. Show that $E(\tilde{z}_n) = 0$ and $V(\tilde{z}_n) = 1$ for $\tilde{z}_n$ as defined in (13.8.1).
7. Derive (13.8.2).
8. Derive (13.8.15).
9. Derive (13.8.17).
10. Derive (13.9.2).
11. Let $\tilde{x}_1, \ldots, \tilde{x}_n$ be iid with common dgf $f(x)$, and define $\tilde{s}^2$ and $\tilde{\xi}$ by (13.2.2) and (13.2.3), respectively even though $f(x)$ need not be a Normal df. Show that
    (a) $E(\tilde{\xi}) = \sigma^2 \equiv V(\tilde{x})$;
    (b) $V(\tilde{\xi}) = n^{-1}[\mu_{[4]} - (\sigma^2)^2]$;
    (c) $E(\tilde{s}^2) = \sigma^2$; and
    (d) $V(\tilde{s}^2) = \dfrac{\mu_{[4]} - (n-3)(\sigma^2)^2}{n(n-1)}$,

    where $\mu_{[4]} \equiv E([\tilde{x} - \mu]^4)$.

12. Complete the proof of (13.10.4).

13. It is instructive to see that gross violations of the hypotheses (independence, identical distribution, finite positive variance) of (13.8.1) destroy the Central Limit Property. This exercise develops some of the pathologies which may occur.

    (a) Let $\tilde{x}_i(q) = \tilde{x}_0(q)$ for every q, and let the cdf of $\tilde{x}_0$ be non-Normal. Show that $\bar{\tilde{x}}_n(q) = \tilde{x}_0(q)$ for every n, and hence that the Central Limit Property fails. (Here, "independence" fails decisively.)
    (b) Let $\tilde{x}_1, \tilde{x}_2, \ldots$ be iid with common df $f_S(\cdot \mid 0, 1, 1)$. Show that the df of $\tilde{z}_n$ is $f_S(\cdot \mid 0, 1, 1)$ and hence (13.8.1) fails. (Here "$\sigma^2 < \infty$" fails.)

14. *The Lognormal Central Limit Theorem*: The Lognormal distribution was defined in Exercise 45 of Chapter 12. Prove the following multiplicative Central Limit Theorem:

    Let $\tilde{x}_1, \tilde{x}_2, \ldots$ be iid random variables with common dgf $f(x)$ such that $F(0) = 0$ and $V(\log_e \tilde{x}) \equiv \sigma^2 \in (0, \infty)$ [and thus $E(\log_e \tilde{x}) \equiv \mu \in (-\infty, \infty)$]. Then for large n, (13.11.1)

    (a) the product $\prod_{i=1}^{n} \tilde{x}_i$ has approximate cdf $F_L(\cdot \mid n\mu, n\sigma^2)$;

    (b) the geometric mean $\left[\prod_{i=1}^{n} \tilde{x}_i\right]^{1/n}$ has approximate cdf $F_L(\cdot \mid \mu, \sigma^2/n)$; and

    (c) $\left[\exp(-n\mu) \prod_{i=1}^{n} \tilde{x}_i\right]^{1/(\sigma\sqrt{n})}$ has approximate cdf $F_L(\cdot \mid 0, 1)$.

    [*Hint*: Let $\exp(\tilde{y}_i) = \tilde{x}_i$ and work in terms of the $\tilde{y}_i$'s.]

14

NATURAL CONJUGATE DISTRIBUTION FAMILIES

14.1 INTRODUCTION

This chapter concludes Part II on basic probability theory, with some facts useful in Bayesian decision theory and inference. It introduces the concept of a natural conjugate prior df $f(\theta)$ of the parameter θ (or vector parameter $\boldsymbol{\theta}$) of a data-generating process.

A family of df's $f(\theta)$ is *natural conjugate* for θ if every member of that family, used as a prior df, produces a posterior df via (8.7.1) which also belongs to that family. Thus, as we shall see, if the decision-maker uses a beta df $f_\beta(p \mid \rho, \nu)$ to express his judgments about the parameter $\tilde{p}$ of a Bernoulli process, then (with noninformative stopping) his posterior df will also be a beta df, but with different parameters reflecting the sufficient statistic (r, n) of the experiment.

The desirability of using natural conjugate prior df's stems from their mathematical tractability; it will be very *simple* to express the posterior df which characterizes one's judgments as revised by the outcome of the experiment. Moreover, most of the natural conjugate families considered subsequently are reasonably *flexible*, in that a wide variety of judgments can be represented, to a reasonable degree of accuracy, by some member of the natural conjugate family.

We refer the interested reader to Chapter 3 of Raiffa and Schlaifer [**66**] for a full discussion of the general theory of natural conjugacy. In this chapter we shall be content with presenting the basic results for the Bernoulli, k-variate Bernoulli, and Poisson processes discussed in Chapter 11, and for random samples of size n from univariate and k-variate normal df's.

For each case, our plan of attack will be to: (1) recall the general form of the likelihood function and the sufficient statistics; (2) state the general form of a natural conjugate prior df; (3) derive the posterior df; and (4) derive the *unconditional* distribution of the random part of the sufficient statistic.

It is only in step (4) that the specific form of the noninformative stopping process becomes relevant. For step (3) we shall use a convenient trick; namely, we inspect the joint dgf

$$f(\theta, \mathbf{x}_T) = f(\theta)f(\mathbf{x}_T \mid \theta) = K(x_1, \ldots, x_n)f(\theta)L(\theta \mid \mathbf{x}_T)$$

to see how it depends functionally upon θ, and [because $f(\theta)$ will be a natural conjugate prior df] we shall recognize that $f(\cdot, \mathbf{x}_T)$ is a function proportional to some member $f(\cdot \mid \mathbf{x}_T)$ of the natural conjugate family. Since (dgf 2) must obtain, we easily conclude that that member $f(\cdot \mid \mathbf{x}_T)$ is indeed the posterior df, since for every θ

$$\begin{aligned} f(\theta \mid \mathbf{x}_T) &= \frac{f(\theta, \mathbf{x}_T)}{f(\mathbf{x}_T)} \\ &= \frac{K(x_1, \ldots, x_n)L(\theta \mid \mathbf{x}_T)f(\theta)}{f(\mathbf{x}_T),} \end{aligned}$$

in which the factor $K(x_1, \ldots, x_n)/f(\mathbf{x}_T)$ is independent of θ.

Finally, knowing the posterior df $f(\theta \mid \mathbf{x}_T)$ enables us to write down the dgf $f(\mathbf{x}_T)$ of the random part of the sufficient statistic as $f(\mathbf{x}_T) = f(\theta, \mathbf{x}_T)/f(\theta \mid \mathbf{x}_T)$.

This approach, by which we may avoid some potentially rather complicated integration in determining $f(\mathbf{x}_T)$, is sometimes called the *recognition method* (Roberts [**68**], for example).

The preceding, basic ideas should not appear too novel, since Example 8.7.1 carried out steps (1) through (4) essentially *in toto*.

Sections 14.2 and 14.3 concern the Bernoulli process and the k-variate Bernoulli process, respectively, while 14.4 concerns the Poisson process. The Normal and k-variate Normal processes are covered in two long sections, 14.5 and 14.6. Their length is due to the number of different assumptions which concern the degree of prior knowledge regarding the parameters μ and σ^2 or $\boldsymbol{\mu}$ and $\boldsymbol{\sigma}^2$.

14.2 THE BERNOULLI PROCESS

Recall from (11.3.2) that k iid observations from a Bernoulli process with parameter p produce the likelihood function

$$L(p \mid n, x_1, \ldots, x_n) = \begin{cases} p^r(1-p)^{n-r}, & r \in \{0, 1, 2, \ldots, n\} \\ 0, & \text{elsewhere,} \end{cases} \tag{14.2.1}$$

for all $p \in (0, 1)$, provided that the stopping process is noninformative.

A natural conjugate prior dgf for $\tilde{p}$ is a beta df $f_\beta(p \mid \rho, \nu)$ as defined by (12.3.1), for any ρ and ν such that $\nu > \rho > 0$. Figure 12.4 suggests that a decision-maker enjoys considerable flexibility of choice in expressing his judgments about $\tilde{p}$ even when restricted to a natural conjugate prior df.

From (14.2.1) and (12.3.1), we conclude that the posterior dgf of $\tilde{p}$ depends upon p only through the product

$$f(p, r, n) \equiv L(p \mid r, n) \cdot f_\beta(p \mid \rho, \nu)$$
$$= \frac{1}{B(\rho, \nu)} p^{\rho+r-1}(1 - p)^{\nu+n-(\rho+r)-1}$$

(where positive), and hence that the posterior dgf of $\tilde{p}$ *must* be

$$f_\beta(p \mid \rho + r, \nu + n)$$
$$\equiv \begin{cases} \dfrac{1}{B(\rho + r, \nu + n)} p^{\rho+r-1}(1 - p)^{\nu+n-(\rho+r)-1}, & 0 < p < 1 \\ 0, & p \notin (0, 1), \end{cases} \quad (14.2.2)$$

regardless of whether r was predetermined, n was predetermined, or neither. (Possibly, observations were stopped because of additional funds.) Hence we have shown the following:

If an experiment consists in observing a Bernoulli process with noninformative stopping and yields the sufficient statistic (r, n), and if the Bernoulli parameter p has the natural conjugate prior df $f_\beta(p \mid \rho, \nu)$, then the posterior df is $f_\beta(p \mid \rho + r, \nu + n)$. (14.2.3)

Note how simple a task it is to characterize the decision-maker's posterior df of $\tilde{p}$ given the sufficient statistic (r, n): simply add r to ρ and n to ν. Moreover, (12.3.6) implies that the decision-maker's expectation of $\tilde{p}$ is revised from ρ/ν to $(\rho + r)/(\nu + n)$.

Suppose now that n is predetermined, as is the case in many random samples from populations. Then $\tilde{r}$ has the binomial mf $f_b(r \mid n, p)$ given p, and if $\tilde{p}$ has the prior df $f_\beta(p \mid \rho, \nu)$, then $(\tilde{r}, \tilde{p})$ has the joint dgf

$$f(r, p) = f_b(r \mid n, p) \cdot f_\beta(p \mid \rho, \nu)$$
$$= \frac{n!}{r!(n - r)!B(\rho, \nu)} p^{\rho+r-1}(1 - p)^{(\nu+n)-(\rho+r)-1}$$
$$= \frac{\Gamma(n + 1)\Gamma(\nu)}{\Gamma(r + 1)\Gamma(n - r + 1)\Gamma(\rho)\Gamma(\nu - \rho)} p^{\rho+r-1}(1 - p)^{(\nu+n)-(\rho+r)-1}$$

(where positive), where the last equality follows from (12.3.3) and the fact that $\Gamma(i) = (i - 1)!$ for any positive integer i. From $f(r, p)$ and $f(p \mid r) = f_\beta(p \mid \rho + r, (\nu + n)$, we find

$$f(r) = \frac{f(r, p)}{f(p \mid r)}$$
$$= \frac{\Gamma(n + 1)\Gamma(\nu)\Gamma(\rho + r)\Gamma(\nu + n - (\rho + r))}{\Gamma(r + 1)\Gamma(n - r + 1)\Gamma(\rho)\Gamma(\nu - \rho)\Gamma(\nu + n)}, \tag{14.2.4}$$

for $r \in \{0, 1, \ldots, n\}$ and $f(r) = 0$ elsewhere.

The mass function defined by (14.2.4) is called variously the *beta-binomial* [**66**] or the *hyper-binomial* [**64**]. We shall use the latter term and denote it by $f_{hb}(r \mid \rho, \nu, n)$. The reader may prove as Exercise 1 that if $\tilde{r}$ has mf $f_{hb}(r \mid \rho, \nu, n)$, then

$$E_{hb}(\tilde{r} \mid \rho, \nu, n) = \frac{n\rho}{\nu} \tag{14.2.5}$$

and

$$V_{hb}(\tilde{r} \mid \rho, \nu, n) = n(n + \nu)\frac{\rho(\nu - \rho)}{\nu^2(\nu + 1)}. \tag{14.2.6}$$

Example 14.2.1

Suppose that a politician has expressed his prior judgments about the proportion $\tilde{p}$ of voters who support his candidacy in the form of the beta df $f_\beta(\cdot \mid 112, 200)$, and that a random sample of $n = 9800$ voters (with replacement) was taken, with noninformative stopping. If $r = 4900$ supported him, his revised judgments about p are expressed by the df $f_\beta(\cdot \mid 5012, 10{,}000)$. By comparing $E_\beta(\tilde{p} \mid 112, 200) = .56$ with $E_\beta(\tilde{p} \mid 5012, 10{,}000) = .5012$, we note the effect of the sample on his expectation of $\tilde{p}$. By comparing $V_\beta^{1/2}(\tilde{p} \mid 112, 200) \doteq .035$ with $V_\beta^{1/2}(\tilde{p} \mid 5012, 10{,}000) \doteq .005$, we obtain some indication of the effect of the sample in reducing his uncertainty. We shall now suppose that $n = 9800$ was predetermined. Before $\tilde{r} = 4900$ was observed, the politician's judgments about r were expressed by $f_{hb}(\cdot \mid 112, 200, 9800)$, with $E_{hb}(\tilde{r} \mid 112, 200, 9800) = 5488$, Note that $\tilde{r} = 4900$ fell far below the expectation.

Now suppose that r rather than n is predetermined. Then $\tilde{n}$ has the Pascal mf $f_{Pa}(n \mid r, p)$ given p, and if $\tilde{p}$ has the prior df $f_\beta(p \mid \rho, \nu)$, then $(\tilde{n}, \tilde{p})$ has the joint dgf

$$f(n, p) = f_{Pa}(n \mid r, p) f_\beta(p \mid \rho, \nu)$$
$$= \frac{(n - 1)!}{(r - 1)!(n - r)!B(\rho, \nu)} p^{\rho+r-1}(1 - p)^{\nu+n-(\rho+r)-1}$$
$$= \frac{\Gamma(n)\Gamma(\nu)}{\Gamma(r)\Gamma(n - r + 1)\Gamma(\rho)\Gamma(\nu - \rho)} p^{\rho+r-1}(1 - p)^{\nu+n-(\rho+r)-1}$$

(where positive); and since the posterior df $f(p \mid n) = f_\beta(p \mid \rho + r, \nu + n)$, we obtain

$$f(n) = \frac{f(n, p)}{f(p \mid n)}$$
$$= \frac{\Gamma(n)\Gamma(\nu)\Gamma(\rho + r)\Gamma(\nu + n - (\rho + r))}{\Gamma(r)\Gamma(n - r + 1)\Gamma(\rho)\Gamma(\nu - \rho)\Gamma(\nu + n)}, \tag{14.2.7}$$

for $n \in \{r, r + 1, \ldots\}$, and $f(n) = 0$ elsewhere.

The mass function defined by (14.2.7) is called variously the *beta-Pascal* [**66**] or the *hyper-Pascal* [**64**]. We shall use the latter term and denote it by $f_{hPa}(n \mid \rho, \nu, r)$. The reader may prove as Exercise 2 that if n has mf $f_{hPa}(n \mid \rho, \nu, r)$, then

$$E_{hPa}(\tilde{n} \mid \rho, \nu, r) = r\frac{\nu - 1}{\rho - 1}, \tag{14.2.8}$$

which exists only if $\rho > 1$, and

$$V_{hPa}(\tilde{n} \mid \rho, \nu, r) = r(r + \rho - 1)\frac{(\nu - 1)(\nu - \rho)}{(\rho - 1)^2(\rho - 2)}, \tag{14.2.9}$$

which exists only if $\rho > 2$.

The reason for the qualified existence of (14.2.8) and (14.2.9) is suggested by Figure 12.4. For instance, when $\rho \leqq 1$ and $\nu > 1$, very small values of $\tilde{p}$ are most likely, and if $\tilde{p}$ is very small, then the number $\tilde{n}$ of observations required to produce precisely r ones (successes, supporting voters, and so on) is likely to be very large.

Example 14.2.2

In Example 14.2.1, suppose that $r = 4900$ has been predetermined. Then before the sample, $\tilde{n}$ has mf $f_{hPa}(n \mid 112, 200, 4900)$, with $E_{hPa}(\tilde{n} \mid 112, 200, 4900) \doteq 8785$. Hence $\tilde{n} = 9800$ exceeded $E(\tilde{n})$ by a considerable amount, and thus (roughly speaking) furnished evidence that $\tilde{p}$ was lower than expected.

When ρ and ν are integers, the hyper-binomial and hyper-Pascal mass functions can be re-expressed in terms of the hyper-geometric mf. Recursion formulas and a computer program exist for evaluating these mf's. (See Pratt, Raiffa and Schlaifer, [**64**] for further references.)

14.3 THE k-VARIATE BERNOULLI PROCESS

In (11.4.3) we showed that the likelihood function of the (vector) parameter $\mathbf{p} \equiv (p_1, \ldots, p_k)'$ of the k-variate Bernoulli process is

$$L(\mathbf{p} \mid n, x_1, \ldots, x_n) = \prod_{i=1}^{k+1} p_i^{r_i} \tag{14.3.1}$$

(where positive), provided that the stopping is noninformative, where $p_{k+1} \equiv 1 - \sum_{i=1}^{k} p_i$ and $r_{k+1} \equiv n - \sum_{i=1}^{k} r_i$.

A natural conjugate prior dgf for $\tilde{\mathbf{p}}$ is a k-variate beta df $f_{\beta^{(k)}}(\mathbf{p} \mid \boldsymbol{\rho}, \nu)$, as defined by (12.6.1).

From (14.3.1) and (12.6.1) we conclude that the posterior dgf of $\tilde{\mathbf{p}}$ depends upon $\mathbf{p}$ only through the product

$$f(\mathbf{p}, \mathbf{r}, n) \equiv L(\mathbf{p} \mid \mathbf{r}, n) \cdot f_{\beta^{(k)}}(\mathbf{p} \mid \boldsymbol{\rho}, \nu)$$
$$= \frac{1}{B_k(\rho, \nu)} \prod_{i=1}^{k+1} p_i^{\rho_i + r_i - 1}.$$

Hence the posterior dgf of $\tilde{\mathbf{p}}$ must be $f_{\beta^{(k)}}(\mathbf{p} \mid \boldsymbol{\rho} + \mathbf{r}, \nu + n)$, regardless of the stopping process, provided only that it be noninformative. Hence we have the following:

If an experiment consists in observing a k-variate Bernoulli process with noninformative stopping and yields the sufficient statistic $(\mathbf{r}, n)$, and if the parameter $\tilde{\mathbf{p}}$ has the natural-conjugate prior df $f_{\beta^{(k)}}(\mathbf{p} \mid \boldsymbol{\rho}, \nu)$, then the posterior df is

$$f_{\beta^{(k)}}(\mathbf{p} \mid \boldsymbol{\rho} + \mathbf{r}, \nu + n). \tag{14.3.2}$$

If n is predetermined, as is most often the case, then $\tilde{\mathbf{r}}$ has the multinomial mf $f_{mu^{(k)}}(\mathbf{r} \mid n, \mathbf{p})$ given $\mathbf{p}$, and if $\tilde{\mathbf{p}}$ has the prior df $f_{\beta^{(k)}}(\mathbf{p} \mid \boldsymbol{\rho}, \nu)$, then

$$\begin{aligned} f(\mathbf{r}) &= \frac{f(\mathbf{r}, \mathbf{p})}{f(\mathbf{p} \mid \mathbf{r})} \\ &= f_{mu^{(k)}}(\mathbf{r} \mid n, \mathbf{p}) \cdot \frac{f_{\beta^{(k)}}(\mathbf{p} \mid \boldsymbol{\rho}, \nu)}{f_{\beta^{(k)}}(\mathbf{p} \mid \boldsymbol{\rho} + \mathbf{r}, \nu + n)} \\ &= \frac{B_k(\boldsymbol{\rho} + \mathbf{r}, \nu + n)}{B_k(\boldsymbol{\rho}, \nu)} \cdot \frac{\Gamma(n+1)}{\prod_{i=1}^{k+1} \Gamma(r_i + 1)}, \end{aligned} \tag{14.3.3}$$

provided that every $r_i \geqq 0$ and $\sum_{i=1}^{k} r_i \leqq n$; $f(\mathbf{r}) = 0$ otherwise.

The mass function defined by (14.3.3) will be called the *hyper-multinomial* and denoted by $f_{hmu}^{(k)}(\mathbf{r} \mid \boldsymbol{\rho}, \nu, n)$.

$$E_{muh}^{(k)}(\tilde{r}_i \mid \boldsymbol{\rho}, \nu, n) = \frac{n\rho_i}{\nu}, \tag{14.3.4}$$

$$V_{hmu}^{(k)}(\tilde{r}_i \mid \boldsymbol{\rho}, \nu, n) = n(n+\nu)\frac{\rho_i(\nu - \rho_i)}{\nu^2(\nu+1)}, \tag{14.3.5}$$

and

$$V_{hmu}^{(k)}(\tilde{r}_i, \tilde{r}_j \mid \boldsymbol{\rho}, \nu) = -n(n+\nu)\frac{\rho_i \rho_j}{\nu^2(\nu+1)}, \quad i \neq j. \tag{14.3.6}$$

Proof

Equations (14.3.4) and (14.3.5) follow from (14.2.5) and (14.2.6) and the fact that the marginal mf of each $\tilde{r}_i$ is $f_b(\cdot \mid \rho_i, \nu)$ [compare (11.4.5)]. To prove (14.3.6), we make use of (10.7.9) in the following form:

(*) $$V(\tilde{r}_i, \tilde{r}_j) = E[V(\tilde{r}_i, r_j \mid \tilde{\mathbf{p}})] + V[E(\tilde{r}_i \mid \tilde{\mathbf{p}}), E(\tilde{r}_j \mid \tilde{\mathbf{p}})].$$

Now, (11.8.21) implies

$$V(\tilde{r}_i, \tilde{r}_j \mid \tilde{\mathbf{p}}) = -n\tilde{p}_i\tilde{p}_j$$

and hence (10.3.27), (12.6.6), and (12.6.7) imply

(a) $$\begin{aligned} E[V(\tilde{r}_i, \tilde{r}_i \mid \tilde{\mathbf{p}})] &= -nE(\tilde{p}_i\tilde{p}_j) \\ &= -n[V(\tilde{p}_i, \tilde{p}_j) + E(\tilde{p}_i)E(\tilde{p}_j)] \\ &= -n\left[-\frac{\rho_i\rho_j}{\nu^2(\nu+1)} + \frac{\rho_i\rho_j(\nu+1)}{\nu^2(\nu+1)}\right] \\ &= -n\nu\frac{\rho_i\rho_j}{\nu^2(\nu+1)}. \end{aligned}$$

Moreover,

(b) $$\begin{aligned} V[E(\tilde{r}_i \mid \tilde{\mathbf{p}}), E(\tilde{r}_j \mid \tilde{\mathbf{p}})] &= V[n\tilde{p}_i, n\tilde{p}_j] && \text{[by (11.8.19)]} \\ &= n^2V(\tilde{p}_i, \tilde{p}_j) && \text{[by (10.3.22)]} \\ &= -n^2\frac{\rho_i\rho_j}{\nu^2(\nu+1)}. && \text{[by (12.6.8)]} \end{aligned}$$

Adding (a) and (b) yields (14.3.6). QED

Example 14.3.1

Suppose the politician of Example 14.2.1 wanted to classify the voters as either (1) supporting his candidacy; (2) supporting his opponent's candidacy; or (3) undecided. Let p_i denote the proportion in class (i) for $i = 1, 2, 3$. Then any two of the p_i's—say (p_1, p_3)—can be assessed a bivariate (that is, $k = 2$) beta df. Suppose the politician assesses $f_\beta^{(2)}(p_1, p_3 \mid \rho_1, \rho_3, \nu) = f_\beta^{(2)}(p_1, p_3 \mid 112, 10, 200)$ as his prior df of $(\tilde{p}_1, \tilde{p}_3)$, and that a random sample of 9800 voters is then taken and yields $r_1 = 4900$, $r_2 = 1900$, and $r_3 = 3000$. Then his posterior df of $(\tilde{p}_1, \tilde{p}_3)$ is

$$f_\beta^{(2)}(p_1, p_3 \mid 112 + 4900, 10 + 3000, 10{,}000) = f_\beta^{(2)}(p_1, p_3 \mid 5012, 3010, 10{,}000).$$

Note that his *marginal* prior df of $\tilde{p}_1$ and his *marginal* posterior df of $\tilde{p}_1$ agree with the result in Example 14.2.1. However, his marginal prior and posterior df's of $\tilde{p}_3$ are $f_\beta(p_3 \mid 10, 200)$ and $f_\beta(p_3 \mid 3010, 10{,}000)$, respectively. The respective expectations are $10/200 = .0500$ and $3010/10{,}000 = .3010$, which suggest that the undecided vote is much larger than he had foreseen.

14.4 THE POISSON PROCESS

Recall from (11.7.4) that if a Poisson process is observed with noninformative stopping for t time units during which there were r occurrences of the given phenomenon, then the likelihood function of the intensity λ is

$$L(\lambda \mid r, t) = \lambda^r e^{-\lambda t}, \tag{14.4.1}$$

where positive; that is, for all $\lambda > 0$.

A natural conjugate prior dgf for $\tilde{\lambda}$ is a gamma density $f_\gamma(\lambda \mid \rho, \tau)$, as defined by (12.2.5) for ρ in place of r and τ in place of λ. From (14.4.1) and (12.2.5) we see that the posterior dgf of $\tilde{\lambda}$ depends upon λ only through the product

$$\begin{aligned} f(\lambda, r, t) &\equiv L(\lambda \mid r, t) \cdot f_\gamma(\lambda \mid \rho, \tau) \\ &= \frac{\tau^\rho}{\Gamma(\rho)} \exp\left[-(\tau + t)\lambda\right]\lambda^{\rho+r-1}, \end{aligned}$$

which shows that the posterior df of $\tilde{\lambda}$ must be $f_\gamma(\lambda \mid \rho + r, \tau + t)$ regardless of the stopping process, provided that it is noninformative. Hence, we have the following:

If an experiment consists in observing a Poisson process with noninformative stopping and yields the sufficient statistic (r, t), and if the process intensity parameter $\tilde{\lambda}$ has the natural conjugate prior df $f_\gamma(\lambda \mid \rho, \tau)$, then the posterior df of $\tilde{\lambda}$ is $f_\gamma(\lambda \mid \rho + r, \tau + t)$. (14.4.2)

If t is predetermined, as in many experiments on a Poisson process, then $\tilde{r}$ has the Poisson mf $f_{Po}(r \mid \lambda t)$ given λ, and if $\tilde{\lambda}$ has the prior df $f_\gamma(\lambda \mid \rho, \tau)$, then

$$\begin{aligned} f(r) &= \frac{f(r, \lambda)}{f(\lambda \mid r)} \\ &= f_{PO}(r \mid \lambda t) \cdot \frac{f_\gamma(\lambda \mid \rho, \tau)}{f_\gamma(\lambda \mid \rho + r, \tau + t)} \\ &= \frac{\Gamma(\rho + r)\tau}{\Gamma(r + 1)\Gamma(\rho)(t + \tau)} \cdot \frac{\exp\left[-\lambda(\tau + t)\right](\lambda t)^r(\lambda\tau)^{\rho-1}}{\exp\left[-\lambda(\tau + t)\right]\left[\lambda(\tau + t)\right]^{\rho+r-1}} \qquad (14.4.3) \\ &= \frac{\Gamma(\rho + r)}{\Gamma(r + 1)\Gamma(\rho)}\left(\frac{\tau}{\tau + t}\right)^\rho\left(1 - \frac{\tau}{\tau + t}\right)^r \\ &= f_{nb}\left(r \mid \rho, \frac{\tau}{\tau + t}\right) \end{aligned}$$

as defined for ρ not necessarily an integer in Exercise 11.11. As an easy generalization and application of (11.8.6) and (11.8.7), we obtain

$$E_{nb}\left(\tilde{r} \mid \rho, \frac{\tau}{\tau + t}\right) = \frac{\rho t}{\tau} \tag{14.4.4}$$

and

$$V_{nb}\left(\tilde{r} \mid \rho, \frac{\tau}{\tau + t}\right) = \frac{\rho t(\tau + t)}{\tau^2}. \tag{14.4.5}$$

We could have derived

$$f(r) = f_{nb}(r \mid \rho, \tau/[\tau + t])$$

directly via Exercise 11.11 rather than by using the *recognition method* as in (14.4.3).

Suppose that the Poisson process is observed until exactly r occurrences have transpired. Then $\tilde{t}$ is the random part of (r, t), and we know from (11.7.12) that the df of $\tilde{t}$, given λ, is $f_\gamma(t \mid r, \lambda)$. If $\tilde{\lambda}$ is given the conjugate prior df $f_\gamma(\lambda \mid \rho, \tau)$, then the marginal df of $\tilde{t}$ is inverted beta-2:

$$\begin{aligned} f(t) &= \frac{f(t, \lambda)}{f(\lambda \mid t)} \\ &= f_\gamma(t \mid r, \lambda) \cdot f_\gamma(\lambda \mid \rho, \tau) / f_\gamma(\lambda \mid \rho + r, \tau + t) \\ &= \frac{\Gamma(\rho + r)\tau^\rho \lambda^r \exp[-\lambda(\tau + t)] t^{r-1} \lambda^{\rho - 1}}{\Gamma(\rho)\Gamma(r)(\tau + t)^{\rho + r} \exp[-\lambda(\tau + t)] \lambda^{\rho + r - 1}} \\ &= \frac{\Gamma(\rho + r)}{\Gamma(\rho)\Gamma(r)} \left(\frac{\tau}{\tau + t}\right)^\rho \left(\frac{t}{\tau + t}\right)^r \cdot \frac{1}{t} \\ &= \frac{t^{r-1}\tau^\rho}{B(r, \rho + r)(t + \tau)^{\rho + r}} \\ &= f_{i\beta 2}(t \mid r, \rho + r, \tau), \end{aligned} \tag{14.4.6}$$

for all $t > 0$ [compare (12.3.13)]. It follows from (12.3.17) and (12.3.18) that

$$E_{i\beta 2}(\tilde{t} \mid r, \rho + r, \tau) = \frac{r\tau}{(\rho - 1)} \tag{14.4.7}$$

provided that $\rho > 1$, and

$$V_{i\beta 2}(\tilde{t} \mid r, \rho + r, \tau) = \frac{r\tau^2(\rho + r - 1)}{(\rho - 1)^2(\rho - 2)} \tag{14.4.8}$$

provided that $\rho > 2$. The interpretation of (14.4.7) is that if the odds of small values of $\tilde{\lambda}$ are too large [that is, $\rho \leqq 1$], then the odds of very large values of the time $\tilde{t}$ required to produce r occurrences are also large.

Example 14.4.1

Assume that the times between successive cars crossing a checkpoint are iid random variables with common df $f_e(t \mid \lambda)$. Hence the data-generating process is Poisson. If a policeman has assessed $f_\gamma(\lambda \mid \rho, \tau) = f(\lambda \mid 60, 6)$ as his prior df of $\tilde{\lambda}$, and if 100 cars were counted crossing the checkpoint in two time units (with noninformative stopping), then his posterior df of $\tilde{\lambda}$ is $f_\gamma(\lambda \mid 160, 8)$. Thus his prior and posterior expectations of $\tilde{\lambda}$ are $60/6 = 10$ and $160/8 = 20$, respectively. Moreover, if $t = 2$ had been predetermined, then before the counting commenced his mf of $\tilde{r}$ was $f_{nb}(r \mid 60, \frac{3}{4})$; whereas if $r = 100$ had been predetermined, then before the counting commenced his df of $\tilde{t}$ was $f_{i\beta 2}(t \mid 100, 160, 6)$.

14.5 THE UNIVARIATE NORMAL PROCESS

In this section we consider experiments consisting of n iid observations $\tilde{x}_1, \ldots, \tilde{x}_n$ obtained from $f_N(\cdot \mid \mu, \sigma^2)$ with noninformative stopping. The purpose of such an experiment is to obtain information about μ and/or σ^2, depending upon which (if either) is known beforehand.

We shall follow the pattern of Chapter 13 by considering the following cases:

Case A: σ^2 is known but μ is unknown;
Case B: μ is known but σ^2 is unknown, and
Case C: neither μ nor σ^2 is known.

Since the derivations of results for the Normal process are more involved than for the Poisson and Bernoulli processes, we shall follow the more formal assertion-proof format in this section.

For *Case A* (σ^2 known): if an experiment consists in obtaining n observations from $f_N(x \mid \mu, \sigma^2)$ with noninformative stopping and yields the sufficient statistic $(n, \bar{x})$, and if the process mean μ has the natural conjugate prior df $f_N(\mu \mid \mu_0, \sigma^2/n_0)$, then the posterior df of $\tilde{\mu}$ is $f_N(\mu \mid \mu_1, \sigma^2/n_1)$, where (14.5.1)

$$\mu_1 \equiv (n_0 + n)^{-1}(n_0\mu_0 + n\bar{x})$$

and

$$n_1 \equiv n_0 + n.$$

Proof

By (13.2.5), the likelihood function may be written as

$$L(\mu \mid n, \bar{x}, \sigma^2) = K \exp\left[-\tfrac{1}{2}n(\bar{x} - \mu)^2/\sigma^2\right].$$

If the prior df of $\tilde{\mu}$ is

$$f_N(\mu \mid \mu_0, \sigma^2/n_0) = K' \exp\,[-\tfrac{1}{2}n_0(\mu - \mu_0)^2/\sigma^2],$$

then the posterior df is proportional to

$$(*) \qquad KK' \exp\,[-\tfrac{1}{2}[n_0(\mu - \mu_0)^2 + n(\bar{x} - \mu)^2]/\sigma^2].$$

Now,

$$\begin{aligned}
n_0(\mu - \mu_0)^2 &+ n(\bar{x} - \mu)^2 \\
&= n_0\mu^2 - 2n_0\mu\mu_0 + n_0\mu_0^2 + n\bar{x}^2 - 2n\bar{x}\mu + n\mu^2 \\
&= (n_0 + n)\mu^2 - 2(n_0\mu_0 + n\bar{x})\mu + n_0\mu_0^2 + n\bar{x}^2 \\
&\equiv (n_0 + n)\mu^2 - 2(n_0 + n)\mu\mu_1 + (n_0 + n)\mu_1^2 - (n_0 + n)\mu_1^2 + n_0\mu_0^2 \\
&\quad + n\bar{x}^2 \\
&= n_1(\mu - \mu_1)^2 + y,
\end{aligned}$$

where n_1 and μ_1 are as defined in the assertion and

$$y \equiv n_0\mu_0^2 + n\bar{x}^2 - (n_0 + n)\mu_1^2,$$

which does not involve μ. Thus (*) becomes

$$KK'Y \exp\,[-\tfrac{1}{2}n_1(\mu - \mu_1)^2/\sigma^2],$$

where $Y \equiv e^{-\frac{1}{2}n_1 y/\sigma^2}$, which is proportional (with respect to μ) to $f_N(\mu \mid \mu_1, \sigma^2/n_1)$. QED

Moreover, the unconditional (upon $\tilde{\mu}$) df of $\tilde{x}$ is also Normal.

> *For Case A*: Under the assumptions of (14.5.1) the unconditional df of $\tilde{\bar{x}}$ given n is $f_N(\bar{x} \mid \mu_0, (n_0^{-1} + n^{-1})\sigma^2)$. (14.5.2)

Proof

Given n, the conditional df of $\tilde{\bar{x}}$ given μ is $f_N(\bar{x} \mid \mu, \sigma^2/n)$ [by (13.3.3)] so that

$$\begin{aligned}
f(\bar{x}) &= \frac{f(\bar{x} \mid \mu) f(\mu)}{f(\mu \mid \bar{x})} \\
&= f_N\!\left(\bar{x} \mid \mu, \frac{\sigma^2}{n}\right) \frac{f_N(\mu \mid \mu_0, \sigma^2/n_0)}{f_N\!\left(\bar{x} \mid \mu_1, \dfrac{\sigma^2}{n_1}\right)},
\end{aligned}$$

by (14.5.1). But, by the proof of (14.5.1), this is given by

$$K''' \exp\,(-\tfrac{1}{2}y/\sigma^2),$$

where

$$y \equiv n_0\mu_0^2 + n\bar{x}^2 - (n_0 + n)\mu_1^2$$
$$= n_0\mu_0^2 + n\bar{x}^2 - \frac{(n_0\mu_0 + n\bar{x})^2}{(n_0 + n)},$$

so that

$$(n_0 + n)y = (n_0 + n)n_0\mu_0^2 + (n_0 + n)n\bar{x}^2 - n_0^2\mu_0^2 - 2n_0n\mu_0\bar{x} - n^2\bar{x}^2$$
$$= nn_0(\bar{x} - \mu_0)^2;$$

that is, the marginal df of $\bar{x}$ is

$$K^{iv} \exp\left[-\tfrac{1}{2}[nn_0/(n_0 + n)](\bar{x} - \mu_0)^2/\sigma^2\right]$$

which is proportional to $f_N(\bar{x} \mid \mu_0, [(n_0 + n)/nn_0)]\sigma^2)$, from which (14.5.2) follows upon noting that $(n_0 + n)/(nn_0) = n_0^{-1} + n^{-1}$. QED

In (14.5.1) it is clear that the decision-maker is called upon to assess a Normal prior df for $\tilde{\mu}$, and that there are two parameters for him to fit. Usually, he *assesses* $f_N(\mu \mid \mu_0, \sigma_0^2)$ for some positive variance $\sigma_0,^2$ and then *defines* $n_0 \equiv \sigma^2/\sigma_0^2$ in order to apply (14.5.1) and (14.5.2). We could have stated these results with σ^2/σ_0^2 replacing n_0, but the replacement makes them much more cumbersome.

It is instructive to note in (14.5.1) that the posterior expectation μ_1 of $\tilde{\mu}$ is a weighted average of the prior expectation μ_0 and of the sample mean $\bar{x}$; namely,

$$\mu_1 = \left(\frac{n_0}{n_0 + n}\right)\mu_0 + \left(\frac{n}{n_0 + n}\right)\bar{x}.$$

It is apparent that as n increases, proportionally more weight is placed upon the "objective evidence" $\bar{x}$, and less upon the "subjective prior judgment" μ_0.

Furthermore, as n increases the variance $\sigma^2/(n_0 + n)$ of the posterior df of $\tilde{\mu}$ decreases, thus indicating that the larger the number of observations, the smaller the posterior uncertainty about $\tilde{\mu}$ (as measured by its variance).

Example 14.5.1

Suppose that the hourly outputs $\tilde{x}_1, \tilde{x}_2, \ldots$ of a chemical process are iid random variables with common df $f_N(x \mid \mu, \sigma^2)$, and that $\sigma^2 = 100$ is known. Suppose also that the production manager has assessed the prior df $f_N(\mu \mid \mu_0, \sigma_0^2) = f_N(\mu \mid 50, 25)$. Hence $n_0 = \sigma^2/\sigma_0^2 = 4$. If the process is observed for six hours $(= n)$ and $\bar{x}$ was computed to be 63, then

$$\frac{n_0\mu_0 + n\bar{x}}{n_0 + n} = \frac{(4)(50) + (6)(63)}{10} = 57.8$$

and

$$\frac{\sigma^2}{n_0 + n} = \frac{100}{10} = 10,$$

thus implying that his posterior df of $\tilde{\mu}$ is $f_N(\mu \mid 57.8, 10)$. Before observing $\tilde{\bar{x}}$, but having determined $n = 6$, his marginal df of $\tilde{\bar{x}}$ was

$$f_N(\bar{x} \mid 50, [\tfrac{1}{4}, + \tfrac{1}{6}]\,100) = f_N(\bar{x} \mid 50, 1000/24).$$

Next, we turn to case B.

For Case B (μ known): if an experiment consists in obtaining n observations from $f_N(x \mid \mu, \sigma^2)$ with noninformative stopping and yields the sufficient statistic (n, ξ), and if the process variance $\tilde{\sigma}^2$ has the natural conjugate prior df $f_{i\gamma}(\sigma^2 \mid \psi_0, \nu_0)$, then the posterior df of $\tilde{\sigma}^2$ is $f_{i\gamma}(\sigma^2 \mid \psi_1, \nu_1)$, where (14.5.3)

$$\psi_1 \equiv (\nu_0 + n)^{-1}[\nu_0\psi_0 + n\xi]$$

and

$$\nu_1 \equiv \nu_0 + n.$$

Proof

By (13.2.4), the likelihood function may be written as

$$L(\sigma^2 \mid n, \xi, \mu) = K(\sigma^2)^{-\frac{1}{2}n} \exp(-\tfrac{1}{2}n\xi/\sigma^2).$$

If the prior df of $\tilde{\sigma}^2$ is [by (12.2.10)]

$$f_{i\gamma}(\sigma^2 \mid \psi_0, \nu_0) = K'(\sigma^2)^{-\frac{1}{2}(\nu_0+2)} \exp[-\tfrac{1}{2}\nu_0\psi_0/\sigma^2],$$

then the posterior df of $\tilde{\sigma}^2$ is proportional to the product

$$KK'(\sigma^2)^{-\frac{1}{2}(\nu_0+n+2)} \exp[-\tfrac{1}{2}[\nu_0\psi_0 + n\xi]/\sigma^2] \equiv KK'(\sigma^2)^{-\frac{1}{2}(\nu_1+2)} \exp(-\tfrac{1}{2}\nu_1\psi_1/\sigma^2),$$

and hence the posterior df of $\tilde{\sigma}^2$ must be $f_{i\gamma}(\sigma^2 \mid \psi_1, \nu_1)$. QED

For Case B (μ known): under the assumptions of (14.5.3), the unconditional df of $\tilde{\xi}$ given n is $f_{i\beta 2}(\xi \mid \tfrac{1}{2}n, \tfrac{1}{2}\nu_1, \nu_0\,\psi_0/n)$. (14.5.4)

Proof

Given n, the conditional df of ξ given σ^2 is $f_{\gamma 2}(\xi \mid 1/\sigma^2, n)$ [by (13.3.4)] so that

$$\begin{aligned} f(\xi) &= f(\xi \mid \sigma^2)\frac{f(\sigma^2)}{f(\sigma^2 \mid \xi)} \\ &= f_{\gamma 2}(\xi \mid 1/\sigma^2, n)\frac{f_{i\gamma}(\sigma^2 \mid \psi_0, \nu_0)}{f_{i\gamma}(\sigma^2 \mid \psi_1, \nu_1)} \\ &= \frac{f_{\gamma 2}(\xi \mid 1/\sigma^2, n)\, f_{i\gamma}(\sigma^2 \mid \psi_0, \nu_0)}{f_{i\gamma}(\sigma^2 \mid \psi_1, \nu_1)} \\ &= f_{i\beta 2}(\xi \mid \tfrac{1}{2}n, \tfrac{1}{2}\nu_1, \nu_0\psi_0/n), \end{aligned}$$

the third equality following from the second upon cancellation and rearrangement of terms which appear in the fraction after each density has been written explicitly. QED

Note that ψ_1 is a weighted average of ψ_0 and ξ which places increasing weight upon the objective evidence ξ as n increases.

Example 14.5.2

Suppose that a balance scale is known to produce readings $\tilde{x}_1, \tilde{x}_2, \ldots$ which are iid with common df $f_N(x \mid \nu, \sigma^2)$, where μ denotes the true weight of the object. If ten readings on an object of known weight 10 ($= \mu$) grams resulted in $\xi = 0.20$, and if the prior df of $\tilde{\sigma}^2$ was $f_{i\gamma}(\sigma^2 \mid 1.50, 3)$, then the posterior df is again inverted gamma, with parameters $\nu_1 = 3 + 10 = 13$ and $\psi_1 = (13)^{-1}[(3)(1.50) + (10)(.20))] = .50$. The prior expectation of $\tilde{\sigma}^2$ was (by Table 12.1), $[3/(3-2)](1.50) = 4.50$, in contrast to the posterior expectation $[13/(13-2)](.50) \doteq .59$.

The most involved of the univariate Normal cases is, quite naturally, that in which neither μ nor σ^2 is known. It is clear that we shall have to assign $(\tilde{\mu}, \tilde{\sigma}^2)$ a bivariate prior df. The natural conjugate prior df of $(\tilde{\mu}, \tilde{\sigma}^2)$ is the *Normal-inverted gamma* df $f_{Ni\gamma}(\mu, \sigma^2 \mid \mu_0, \psi_0, n_0, \nu_0)$, defined by

$$f_{Ni\gamma}(\mu, \sigma^2 \mid \mu_0, \psi_0, n_0, \nu_0) \equiv \begin{cases} 0, & \sigma^2 \leqq 0 \\ f_N(\mu \mid \mu_0, \sigma^2/n_0) f_{i\gamma}(\sigma^2 \mid \psi_0, \nu_0), & \sigma^2 > 0, \end{cases}$$

for all $(\mu, \sigma^2) \in R^2$, where $\mu_0 \in R^1$, $n_0 > 0$, $\psi_0 > 0$, and $\nu_0 > 0$ are parameters.

Before giving the result on the parameters of the posterior df, we pause to deduce an important fact about the Normal-inverted gamma df (14.5.5).

If $(\tilde{\mu}, \tilde{\sigma}^2)$ has df $f_{Ni\gamma}(\mu, \sigma^2 \mid \mu_0, \psi_0, n_0, \nu_0)$, then the unconditional df of $\tilde{\mu}$ is $f_S(\mu \mid \mu_0, \psi_0/n_0, \nu_0)$. (14.5.6)

Proof

From the definition (14.5.5) it follows that

$$f(\mu) = \int_0^\infty f_N(\mu \mid \mu_0, \sigma^2/n_0) f_{i\gamma}(\sigma^2 \mid \psi_0, \nu_0) d\sigma^2.$$

But $\tilde{\sigma}^2$ has df $f_{i\gamma}(\sigma^2 \mid \psi_0, \nu_0)$ if and only if $\tilde{h} \equiv (\tilde{\sigma}^2)^{-1}$ has df $f_{\gamma 2}(h \mid \psi_0, \nu_0)$, and hence

$$f(\mu) = \int_0^\infty f_N(\mu \mid \mu_0, (n_0 h)^{-1}) f_{\gamma 2}(h \mid \psi_0, \nu_0) dh$$
$$= f_S(\mu \mid \mu_0, \psi_0/n_0, \nu_0),$$

by (12.5.1). QED

Raiffa and Schlaifer [**66**] define the Normal and Student distributions in terms of precision h instead of variance σ^2. From Exercise 8.9, it is clear that the same probabilistic conclusions will obtain if everything is done in terms of h, provided that variables are changed appropriately. We have chosen to do everything in terms of the more popular concept of variance, σ^2.

For Case C (neither μ nor σ^2 known): if an experiment consists in obtaining $n \geqq 2$ observations from $f_N(x \mid \mu, \sigma^2)$ with noninformative stopping and yields the sufficient statistic $(n, \bar{x}, s^2)$, and if $(\tilde{\mu}, \tilde{\sigma}^2)$ has the natural conjugate prior df $f_{Ni\gamma}(\mu, \sigma^2 \mid \mu_0, \psi_0, n_0, \nu_0)$, then the posterior df of $(\tilde{\mu}, \tilde{\sigma}^2)$ is $f_{Ni\gamma}(\mu, \sigma^2 \mid \mu_1, \psi_1, n_1, \nu_1)$, where (14.5.7)

$$\mu_1 \equiv (n_0 + n)^{-1}(n_0\mu_0 + n\bar{x}),$$
$$n_1 \equiv n_0 + n;$$
$$\nu_1 \equiv \nu_0 + n;$$

and

$$\psi_1 \equiv \nu_1^{-1}[\nu_0\psi_0 + (n - 1)s^2 + (n_0^{-1} + n^{-1})^{-1}(\bar{x} - \mu_0)^2].$$

The marginal prior and posterior df's of $\tilde{\mu}$ are $f_S(\mu \mid \mu_0, \psi_0/n_0, \nu_0)$ and $f_S(\mu \mid \mu_1, \psi_1/n_1, \nu_1)$, respectively. The marginal prior and posterior df's of $\tilde{\sigma}^2$ are $f_{i\gamma}(\sigma^2 \mid \psi_0, \nu_0)$ and $f_{i\gamma}(\sigma^2 \mid \psi_1, \nu_1)$, respectively.

Proof

The second assertion is immediate from the first via (14.5.6). The third also follows from the first in view of the fact, obvious from (14.5.5), that the marginal df of $\tilde{\sigma}^2$ is the second factor in (14.5.5). To prove the first assertion, note that (13.2.6) implies that the likelihood function can be written as

$$L(\mu, \sigma^2 \mid n, \bar{x}, s^2) = K(\sigma^2)^{\frac{1}{2}n} \exp [\tfrac{1}{2}(n - 1)s^2/\sigma^2] \exp [-\tfrac{1}{2}n(\bar{x} - \mu)^2/\sigma^2].$$

Similarly, the prior df $f_{Ni\gamma}(\mu, \sigma^2 \mid \mu_0, \psi_0, n_0, \nu_0)$ is proportional to

$$K'(\sigma^2)^{-1/2} \exp [-\tfrac{1}{2}n_0(\mu - \mu_0)^2/\sigma^2](\sigma^2)^{-\frac{1}{2}(\nu_0+2)} \exp [-\tfrac{1}{2}\nu_0\psi_0/\sigma^2].$$

Hence the posterior df of $(\tilde{\mu}, \tilde{\sigma}^2)$ is proportional to

$$(*) \quad KK'\{(\sigma^2)^{-1/2} \exp [-\tfrac{1}{2}n_0(\mu - \mu_0)^2/\sigma^2] \exp [-\tfrac{1}{2}n(\bar{x} - \mu)^2/\sigma^2]\}$$
$$\times \{(\sigma^2)^{-\frac{1}{2}(\nu_0+n+2)} \exp [-\tfrac{1}{2}[\nu_0\psi_0 + (n - 1)s^2]/\sigma^2]\}.$$

By the proof of (14.5.1) and (14.5.2), the first factor in braces is

$$(\sigma^2)^{-1/2} \exp [-\tfrac{1}{2}n_1(\mu - \mu_1)^2/\sigma^2] \exp [-\tfrac{1}{2}[n_0^{-1} + n^{-1}]^{-1}(\bar{x} - \mu_0)^2/\sigma^2],$$

and hence (*) can be rewritten as

$$(**) \quad KK'\{(\sigma^2)^{-\frac{1}{2}} \exp [-\tfrac{1}{2}n_1(\mu - \mu_1)^2/\sigma^2]\}$$
$$\times \{(\sigma^2)^{-(\nu_0+n+2)/2} \exp [-\tfrac{1}{2}[\nu_0\psi_0 + (n - 1)s^2 + (n_0^{-1} + n^{-1})^{-1}(\bar{x} - \mu_0)^2]/\sigma^2]\},$$

in which the first factor in braces is proportional to $f_N(\mu_1 \mid \mu_1, \sigma^2/n_1)$, while the second factor is proportional to $f_{i\gamma}(\sigma^2 \mid \psi_1, \nu_1)$. Hence the posterior df of $(\tilde{\mu}, \tilde{\sigma}^2)$ is $f_{Ni\gamma}(\mu, \sigma^2 \mid \mu_1, \psi_1, n_1, \nu_1)$. QED

As a fairly tedious but otherwise straightforward exercise in the recognition method and the calculus, the reader may prove the following:

For Case C: under the assumptions and notation of (14.5.7) and given n, (14.5.8)

(a) the df $f(\bar{x} \mid s^2)$ of $\tilde{\bar{x}}$ conditional upon s^2 but unconditional as regards $(\tilde{\mu}, \tilde{\sigma}^2)$, is given by

$$f(\bar{x} \mid s^2) = f_S(\bar{x} \mid \mu_0, (n_0^{-1} + n^{-1})\psi^*, \nu_0 + n - 1),$$

where $\psi^* \equiv [\nu_0/(\nu_0 + n - 1)]\psi_0 + [(n-1)/(\nu_0 + n - 1)]s^2$;

(b) the unconditional df $f(\bar{x})$ of $\tilde{\bar{x}}$ is given by

$$f(\bar{x}) = f_S(\bar{x} \mid \mu_0, (n_0^{-1} + n^{-1})\psi_0, \nu_0);$$

(c) the df $f(s^2 \mid \bar{x})$ of $\tilde{s}^2$ conditional upon $\tilde{\bar{x}}$ but unconditional as regards $(\tilde{\mu}, \tilde{\sigma}^2)$ is given by

$$f(s^2 \mid \bar{x}) = f_{i\beta 2}(s^2 \mid \tfrac{1}{2}(n-1), \tfrac{1}{2}(\nu_0 + n), b)$$

for $b \equiv (n-1)^{-1}[\nu_0\psi_0 + (n_0^{-1} + n^{-1})^{-1}(\bar{x} - \mu_0)^2]$; and

(d) the unconditional df $f(s^2)$ of $\tilde{s}^2$ is given by

$$f(s^2) = f_{i\beta 2}(s^2 \mid \tfrac{1}{2}(n-1), \tfrac{1}{2}(\nu_0 + n - 1), (n-1)^{-1}\nu_0\psi_0).$$

Proof

Exercise 3.

Example 14.5.3

Suppose that the scores on a certain test are known, on the basis of general theory, to be normally distributed about some mean μ with some variance σ^2, neither of which is known. Suppose further that the decision-maker has expressed his judgments about $(\tilde{\mu}, \tilde{\sigma}^2)$ by $f_{Ni\gamma}(\mu, \sigma^2 \mid 80, 100, 1, 3)$. If now 99$(= n)$ test scores are obtained from the population at random and $\bar{x} = 86$, $s^2 = 25$, then the posterior df of $(\tilde{\mu}, \tilde{\sigma}^2)$ is $f_{Ni\gamma}(\mu, \sigma^2 \mid \mu_1, \psi_1, n_1, \nu_1)$, where $\nu_1 = 3 + 99 = 102$; $n_1 = 1 + 99 = 100$; $\mu_1 = (1/100)(80) + (99/100)86 = 85.94$; and $\psi_1 = (102)^{-1}[(3)(100) + (98)(25) + (1^{-1} + 99^{-1})^{-1}(86 - 80^2)] = (102)^{-1}(2785.64) \doteq 27.3$.

The marginal prior and posterior df's of $\tilde{\mu}$ are $f_S(\mu \mid 80, 100, 3)$ and $f_S(\mu \mid 85.94, .273, 102)$. It is apparent that the posterior df of $\tilde{\mu}$ is much more tightly concentrated about $\mu_1 = 85.94$ than the prior df is about $\mu_0 = 80$. In fact, *before* obtaining the information, the decision-maker assessed probability $\frac{1}{2}$ to $\tilde{\mu}$ belonging to the interval $\mu_0 \pm (\psi_0/n_0)^{1/2}{}_{.75}S^*_{(3)} =$

$80 \pm (10)(.765) = (72.35, 87.65)$, whereas *after* obtaining the information, he assessed probability $\frac{1}{2}$ to $\tilde{\mu}$ belonging to the interval $\mu_1 \pm (\psi_1/n_1)^{1/2}{}_{.75}S^*_{(102)} = 85.95 \pm (.273)^{1/2}(.678) \doteq (85.60, 86.30)$. [Here we have used $.678 \doteq {}_{.75}S^*_{(102)}$ and $.765 = {}_{.75}S^*_{(3)}$ from Table IV in the Appendix, together with (12.5.8) for obtaining Student fractiles.]

Example 14.5.4

A regional manager for a supermarket chain is considering the introduction of a new brand of bread, and more specifically, wishes to obtain information about the average demand $\tilde{\mu}$ on nonholiday-weekend Saturdays for this product in his region. He has reason to believe that the total demands $\tilde{x}_1$ on nonholiday-weekend Saturdays are Normally distributed about μ with some common variance σ^2 and are independent. He has expressed his prior judgments about $(\tilde{\mu}, \tilde{\sigma}^2)$ in the form of $f_{Ni\gamma}(\mu, \sigma^2 \mid 10{,}000, 4 \times 10^6, 1, 5)$. Accordingly, he assesses probability $\frac{1}{2}$ to the interval $\mu_0 \pm (\psi_0/n_0)^{1/2}{}_{.75}S^*_{(5)} = 10{,}000 \pm (2000)(.727) = (8546, 11{,}454)$. (Why?) Now suppose that he test-markets the product under usual conditions on three nonholiday weekends, obtaining the statistics $n = 3$, $\bar{x} = 3000$, and $s^2 = 160{,}000$. Then his posterior df of $(\tilde{\mu}, \tilde{\sigma}^2)$ is $f_{Ni\gamma}(\mu, \sigma^2 \mid \mu_1, \psi_1, n_1, \nu_1)$ for $n_1 = 1 + 3 = 4$; $\nu_1 = 5 + 3 = 8$; $\mu_1 = (\frac{1}{4})(10{,}000) + (\frac{3}{4})(3000) = 19{,}000/4 = 4750$; and $\psi_1 = (\frac{1}{8})[(5)(4 \times 10^6) + (2)(160{,}000) + (1^{-1} + 3^{-1})^{-1}(3000 - 10{,}000)^2] = 7{,}133{,}750$.

Thus his marginal posterior df of $\tilde{\mu}$ is $f_S(\mu \mid 4750, 1{,}783{,}437, 8)$, and he assesses probability $\frac{1}{2}$ to the interval $\mu_1 \pm (\psi_1/n_1)^{1/2}{}_{.75}S^*_{(8)} \doteq 4750 \pm (1330)(.706) \doteq (3811, 5689)$.

We remark that (14.5.7) and (14.5.8b) remain true for $n = 1$.

14.6 THE k-VARIATE NORMAL PROCESS

In this section we generalize Section 14.5 by considering experiments which consist of n iid observations with common df $f_{N^{(k)}}(\mathbf{x} \mid \boldsymbol{\mu}, \boldsymbol{\sigma}^2)$.

We shall consider the same five prior-knowledge cases that were introduced in Section 13.5; namely:

Case A: $\boldsymbol{\sigma}^2$ is known but $\boldsymbol{\mu}$ is unknown;
Case B: $\boldsymbol{\mu}$ is known but $\boldsymbol{\sigma}^2$ is unknown;
Case C: neither $\boldsymbol{\mu}$ nor $\boldsymbol{\sigma}^2$ is known;
Case D: $\boldsymbol{\mu}$ and $\boldsymbol{\sigma}_*^2$ are known but σ_*^2 is unknown; and
Case E: neither $\boldsymbol{\mu}$ nor σ_*^2 is known, but $\boldsymbol{\sigma}_*^2$ is known.

Cases A through C correspond directly to cases A through C of Section 14.5, while Cases A, D, and E correspond to A through C of 14.5 respectively under the assumption that $\boldsymbol{\sigma}_*^2$ is always known.

Given the rudimentary knowledge of linear algebra which sufficed for Sections 12.9, 12.10, 13.5, and 13.6, the k-variate Normal cases are no more difficult than their univariate counterparts in 14.5.

As suggested by (14.5.1), the natural conjugate prior df for $\tilde{\boldsymbol{\mu}}$ in Case A is k-variate Normal:

For Case A ($\boldsymbol{\sigma}^2$ known): if an experiment consists in obtaining $n \geqq 1$ observations from $f_N^{(k)}(\mathbf{x} \mid \boldsymbol{\mu}, \boldsymbol{\sigma}^2)$ with noninformative stopping and yields the sufficient statistic $(n, \bar{\mathbf{x}})$, and if the process mean vector $\tilde{\boldsymbol{\mu}}$ has the natural conjugate prior df $f_N^{(k)}(\boldsymbol{\mu} \mid \boldsymbol{\mu}_0, \boldsymbol{\sigma}_0^2)$, then the posterior df of $\tilde{\boldsymbol{\mu}}$ is $f_N^{(k)}(\boldsymbol{\mu} \mid \boldsymbol{\mu}_1, \boldsymbol{\sigma}_1^2)$, where (14.6.1)

$$\boldsymbol{\sigma}_1^2 \equiv (\boldsymbol{\sigma}_0^{-2} + n\boldsymbol{\sigma}^{-2})^{-1}$$

and

$$\boldsymbol{\mu}_1 \equiv \boldsymbol{\sigma}_1^2(\boldsymbol{\sigma}_0^{-2}\boldsymbol{\mu}_0 + n\boldsymbol{\sigma}^{-2}\bar{\mathbf{x}}).$$

Proof

By (13.5.10), $(n, \tilde{\bar{\mathbf{x}}})$ is sufficient for $\boldsymbol{\mu}$, given $\boldsymbol{\sigma}^2$ and the likelihood function may be written as

$$L(\boldsymbol{\mu} \mid n, \bar{\mathbf{x}}) = K \exp\left[-\tfrac{1}{2}n(\bar{\mathbf{x}} - \boldsymbol{\mu})'\boldsymbol{\sigma}^{-2}(\bar{\mathbf{x}} - \boldsymbol{\mu})\right].$$

If the prior df of $\tilde{\boldsymbol{\mu}}$ is [see (12.7.2)]

$$f_N^{(k)}(\boldsymbol{\mu} \mid \boldsymbol{\mu}_0, \boldsymbol{\sigma}_0^2) = K' \exp\left[-\tfrac{1}{2}(\boldsymbol{\mu} - \boldsymbol{\mu}_0)'\boldsymbol{\sigma}_0^{-2}(\boldsymbol{\mu} - \boldsymbol{\mu}_0)\right],$$

then the posterior df is proportional to

$$(*) \qquad KK' \exp\left\{-\tfrac{1}{2}[n(\bar{\mathbf{x}} - \boldsymbol{\mu})'\boldsymbol{\sigma}^{-2}(\bar{\mathbf{x}} - \boldsymbol{\mu}) + (\boldsymbol{\mu} - \boldsymbol{\mu}_0)'\boldsymbol{\sigma}_0^{-2}(\boldsymbol{\mu} - \boldsymbol{\mu}_0)]\right\}.$$

The bracketed term in the exponent of (*) is re-expressible as

$$\begin{aligned}
&(\boldsymbol{\mu} - \bar{\mathbf{x}})'(n\boldsymbol{\sigma}^{-2})(\boldsymbol{\mu} - \bar{\mathbf{x}}) + (\boldsymbol{\mu} - \boldsymbol{\mu}_0)'\boldsymbol{\sigma}_0^{-2}(\boldsymbol{\mu} - \boldsymbol{\mu}_0)\\
&\quad = \boldsymbol{\mu}'(n\boldsymbol{\sigma}^{-2})\boldsymbol{\mu} - \bar{\mathbf{x}}'(n\boldsymbol{\sigma}^{-2})\boldsymbol{\mu} - \boldsymbol{\mu}'(n\boldsymbol{\sigma}^{-2})\bar{\mathbf{x}} + \bar{\mathbf{x}}'(n\boldsymbol{\sigma}^{-2})\bar{\mathbf{x}}\\
&\qquad + \boldsymbol{\mu}'\boldsymbol{\sigma}_0^{-2}\boldsymbol{\mu} - \boldsymbol{\mu}'_0\boldsymbol{\sigma}_0^{-2}\boldsymbol{\mu} - \boldsymbol{\mu}'\boldsymbol{\sigma}_0^{-2}\boldsymbol{\mu}_0 + \boldsymbol{\mu}'_0\boldsymbol{\sigma}_0^{-2}\boldsymbol{\mu}_0\\
&\quad = \boldsymbol{\mu}'(\boldsymbol{\sigma}_0^{-2} + n\boldsymbol{\sigma}^{-2})\boldsymbol{\mu} - [\bar{\mathbf{x}}'(n\boldsymbol{\sigma}^{-2}) + \boldsymbol{\mu}'_0\boldsymbol{\sigma}_0^{-2}]\boldsymbol{\mu} - \boldsymbol{\mu}'[\boldsymbol{\sigma}_0^{-2}\boldsymbol{\mu}_0 + n\boldsymbol{\sigma}^{-2}\bar{\mathbf{x}}]\\
&\qquad + \bar{\mathbf{x}}'(n\boldsymbol{\sigma}^{-2})\bar{\mathbf{x}} + \boldsymbol{\mu}'_0\boldsymbol{\sigma}_0^{-2}\boldsymbol{\mu}_0\\
&\quad = \boldsymbol{\mu}'(\boldsymbol{\sigma}_0^{-2} + n\boldsymbol{\sigma}^{-2})\boldsymbol{\mu} - [\boldsymbol{\sigma}_0^{-2}\boldsymbol{\mu}_0 + n\boldsymbol{\sigma}^{-2}\bar{\mathbf{x}}]'\boldsymbol{\mu} - \boldsymbol{\mu}'[\boldsymbol{\sigma}_0^{-2}\boldsymbol{\mu}_0 + n\boldsymbol{\sigma}^{-2}\bar{\mathbf{x}}]\\
&\qquad + \bar{\mathbf{x}}'(n\boldsymbol{\sigma}^{-2})\bar{\mathbf{x}} + \boldsymbol{\mu}'_0\boldsymbol{\sigma}_0^{-2}\boldsymbol{\mu}_0\\
&\quad = \boldsymbol{\mu}'\boldsymbol{\sigma}_1^{-2}\boldsymbol{\mu} - (\boldsymbol{\sigma}_1^{-2}\boldsymbol{\mu}_1)'\boldsymbol{\mu} - \boldsymbol{\mu}'\boldsymbol{\sigma}_1^{-2}\boldsymbol{\mu}_1\\
&\qquad + \bar{\mathbf{x}}'(n\boldsymbol{\sigma}^{-2})\bar{\mathbf{x}} + \boldsymbol{\mu}'_0\boldsymbol{\sigma}_0^{-2}\boldsymbol{\mu}_0\\
(**)&\quad = \{\boldsymbol{\mu}'\boldsymbol{\sigma}_1^{-2}\boldsymbol{\mu} - \boldsymbol{\mu}'_1\boldsymbol{\sigma}_1^{-2}\boldsymbol{\mu} - \boldsymbol{\mu}'\boldsymbol{\sigma}_1^{-2}\boldsymbol{\mu}_1 + \boldsymbol{\mu}_1\boldsymbol{\sigma}_1^{-2}\boldsymbol{\mu}_1\}\\
&\qquad + \{\bar{\mathbf{x}}'(n\boldsymbol{\sigma}^{-2})\bar{\mathbf{x}} + \boldsymbol{\mu}'_0\boldsymbol{\sigma}_0^{-2}\boldsymbol{\mu}_0 - \boldsymbol{\mu}'_1\boldsymbol{\sigma}_1^{-2}\boldsymbol{\mu}_1\}\\
&\quad = (\boldsymbol{\mu} - \boldsymbol{\mu}_1)'\boldsymbol{\sigma}_1^{-2}(\boldsymbol{\mu} - \boldsymbol{\mu}_1) + y,
\end{aligned}$$

where y is the second term in braces in (**), it does not involve $\boldsymbol{\mu}$. We have thus shown that the posterior df of $\boldsymbol{\mu}$ is proportional to

$$\{KK' \exp(-\tfrac{1}{2}y)\} \exp[-\tfrac{1}{2}(\boldsymbol{\mu} - \boldsymbol{\mu}_1)'\boldsymbol{\sigma}_1^{-2}(\boldsymbol{\mu} - \boldsymbol{\mu}_1)]$$

and hence must be $f_N{}^{(k)}(\boldsymbol{\mu} \mid \boldsymbol{\mu}_1, \boldsymbol{\sigma}_1{}^2)$. QED

For Case A: under the assumptions of (14.6.1), the unconditional df of $\tilde{\bar{\mathbf{x}}}$ given n is $f_N{}^{(k)}(\bar{\mathbf{x}} \mid \boldsymbol{\mu}_0, \boldsymbol{\sigma}_0{}^2 + n^{-1}\boldsymbol{\sigma}^2)$. (14.6.2)

Proof

By (13.6.2), the conditional df of $\tilde{\bar{\mathbf{x}}}$ given $\boldsymbol{\mu}$ is $f_N{}^{(k)}(\bar{\mathbf{x}} \mid \boldsymbol{\mu}, n^{-1}\boldsymbol{\sigma}^2)$ given n, so that

$$\begin{aligned} f(\bar{\mathbf{x}}) &= \frac{f(\bar{\mathbf{x}} \mid \boldsymbol{\mu})f(\boldsymbol{\mu})}{f(\boldsymbol{\mu} \mid \bar{\mathbf{x}})} \\ &= f_N{}^{(k)}(\bar{\mathbf{x}} \mid \boldsymbol{\mu}, n^{-1}\boldsymbol{\sigma}^2)\frac{f_N{}^{(k)}(\boldsymbol{\mu} \mid \boldsymbol{\mu}_0, \boldsymbol{\sigma}_0{}^2)}{f_N{}^{(k)}(\boldsymbol{\mu} \mid \boldsymbol{\mu}_1, \boldsymbol{\sigma}_1{}^2)} \\ &= K^* \exp(-\tfrac{1}{2}y) \exp[-\tfrac{1}{2}(\boldsymbol{\mu} - \boldsymbol{\mu}_1)'\boldsymbol{\sigma}_1^{-2}(\boldsymbol{\mu} - \boldsymbol{\mu}_1)]/ \\ &\qquad\qquad \exp[-\tfrac{1}{2}(\boldsymbol{\mu} - \boldsymbol{\mu}_1)'\boldsymbol{\sigma}_1^{-2}(\boldsymbol{\mu} - \boldsymbol{\mu}_1)] \\ &= K^* \exp(-\tfrac{1}{2}y), \end{aligned}$$

where y is as defined in the proof of (14.6.1) and K^* involves neither $\boldsymbol{\mu}$ nor $\bar{\mathbf{x}}$. To prove (14.6.2) it therefore suffices to show that

$$\begin{aligned} y &\equiv \bar{\mathbf{x}}'(n\boldsymbol{\sigma}^{-2})\bar{\mathbf{x}} + \boldsymbol{\mu}'_0\boldsymbol{\sigma}_0{}^{-2}\boldsymbol{\mu}_0 - \boldsymbol{\mu}'_1\boldsymbol{\sigma}_1{}^{-2}\boldsymbol{\mu}_1 \\ (*) \qquad &= (\bar{\mathbf{x}} - \boldsymbol{\mu}_0)'(\boldsymbol{\sigma}_0{}^2 + n^{-1}\boldsymbol{\sigma}^2)^{-1}(\bar{\mathbf{x}} - \boldsymbol{\mu}_0). \end{aligned}$$

This follows as the result of some involved linear algebra. A much simpler proof is based upon the fact (10.6.11) that the unconditional moment-generating function $M^{(k)}(\mathbf{u})$ of $\tilde{\bar{\mathbf{x}}}$ is the expectation of its conditional moment-generating function given $\boldsymbol{\mu}$:

$$\begin{aligned} M^{(k)}(\mathbf{u}) &= E_N{}^{(k)}[M_N{}^{(k)}(\mathbf{u} \mid \tilde{\boldsymbol{\mu}}, n^{-1}\boldsymbol{\sigma}^2) \mid \boldsymbol{\mu}_0, \boldsymbol{\sigma}_0{}^2] \\ &= E_N{}^{(k)}[\exp[\mathbf{u}'\tilde{\boldsymbol{\mu}} + \tfrac{1}{2}\mathbf{u}'n^{-1}\boldsymbol{\sigma}^2\mathbf{u}] \mid \boldsymbol{\mu}_0, \boldsymbol{\sigma}_0{}^2] \\ &= \exp[\tfrac{1}{2}\mathbf{u}'n^{-1}\boldsymbol{\sigma}^2\mathbf{u}]E_N{}^{(k)}[\exp(\mathbf{u}'\tilde{\boldsymbol{\mu}}) \mid \boldsymbol{\mu}_0, \boldsymbol{\sigma}_0{}^2] \\ &= \exp[\tfrac{1}{2}\mathbf{u}'n^{-1}\boldsymbol{\sigma}^2\mathbf{u}]M_N{}^{(k)}(\mathbf{u} \mid \boldsymbol{\mu}_0, \boldsymbol{\sigma}_0{}^2) \\ &= \exp[\tfrac{1}{2}\mathbf{u}'n^{-1}\boldsymbol{\sigma}^2\mathbf{u}] \exp[\mathbf{u}'\boldsymbol{\mu}_0 + \tfrac{1}{2}\mathbf{u}'\boldsymbol{\sigma}_0{}^2\mathbf{u}] \\ &= \exp[\mathbf{u}'\boldsymbol{\mu}_0 + \tfrac{1}{2}\mathbf{u}'(\boldsymbol{\sigma}_0{}^2 + n^{-1}\boldsymbol{\sigma}^2)\mathbf{u}] \\ &= M_N{}^{(k)}(\boldsymbol{\mu} \mid \boldsymbol{\mu}_0, \boldsymbol{\sigma}_0{}^2 + n^{-1}\boldsymbol{\sigma}^2). \end{aligned}$$

which implies that the unconditional df of $\tilde{\bar{\mathbf{x}}}$ is as asserted. QED

Example 14.6.1

Let $\tilde{\mathbf{x}}_1, \ldots, \tilde{\mathbf{x}}_n$ be iid with common df $f_N{}^{(2)}(\mathbf{x} \mid \boldsymbol{\mu}, \boldsymbol{\sigma}^2)$, where the first component of each $\tilde{\mathbf{x}}$ measures income [in thousands of dollars] and the second component measures level of educational attainment [in years of formal education] of the observed individual. Suppose it is known that

$$\boldsymbol{\sigma}^2 = \begin{bmatrix} 10 & 6 \\ 6 & 10 \end{bmatrix};$$

and suppose that $\tilde{\boldsymbol{\mu}}$ is given a prior df $f_N^{(2)}(\boldsymbol{\mu} \mid \boldsymbol{\mu}_0, \boldsymbol{\sigma}_0^2)$ for $\boldsymbol{\mu}_0 = (8, 10)'$ and

$$\boldsymbol{\sigma}_0^2 = \begin{bmatrix} 16 & 0 \\ 0 & 8 \end{bmatrix}.$$

Finally, suppose that $n = 5$ individuals are randomly selected and $\bar{\mathbf{x}}$ is found to be $(12, 16)'$. We first find $\boldsymbol{\sigma}_1^2$. Recall the formula

$$\begin{bmatrix} a & b \\ b & c \end{bmatrix}^{-1} = (ac - b^2)^{-1} \begin{bmatrix} c & -b \\ -b & a \end{bmatrix} \tag{14.6.3}$$

from linear algebra. Then

$$\boldsymbol{\sigma}_0^{-2} = \left(\frac{1}{128}\right) \begin{bmatrix} 8 & 0 \\ 0 & 16 \end{bmatrix} = \begin{bmatrix} \frac{2}{32} & 0 \\ 0 & \frac{4}{32} \end{bmatrix},$$

$$5\boldsymbol{\sigma}^{-2} = \left(\frac{5}{64}\right) \begin{bmatrix} 10 & -6 \\ -6 & 10 \end{bmatrix} = \begin{bmatrix} \frac{25}{32} & -\frac{15}{32} \\ -\frac{15}{32} & \frac{25}{32} \end{bmatrix},$$

and

$$\boldsymbol{\sigma}_1^{-2} = \boldsymbol{\sigma}_0^{-2} + 5\boldsymbol{\sigma}^{-2} = \begin{bmatrix} \frac{27}{32} & -\frac{15}{32} \\ -\frac{15}{32} & \frac{29}{32} \end{bmatrix},$$

which implies via (14.6.3) that

$$\boldsymbol{\sigma}_1^2 = \left(\frac{1024}{558}\right) \begin{bmatrix} \frac{29}{32} & \frac{15}{32} \\ \frac{15}{32} & \frac{27}{32} \end{bmatrix} = \begin{bmatrix} \frac{464}{279} & \frac{240}{279} \\ \frac{240}{279} & \frac{432}{279} \end{bmatrix}.$$

Note that although the prior covariance of $\tilde{\mu}_1$ and $\tilde{\mu}_2$ is zero, the posterior covariance is positive due to the positive covariance of $\tilde{x}_1$ and $\tilde{x}_2$. Next, we compute $\boldsymbol{\mu}_1$:

$$\tilde{\boldsymbol{\sigma}}_0^{-2}\boldsymbol{\mu}_0 = \begin{bmatrix} \frac{2}{32} & 0 \\ 0 & \frac{4}{32} \end{bmatrix} \begin{pmatrix} 8 \\ 10 \end{pmatrix} = \left(\frac{16}{32}, \frac{40}{32}\right)',$$

$$5\tilde{\boldsymbol{\sigma}}^{-2}\bar{\mathbf{x}} = \begin{bmatrix} \frac{25}{32} & -\frac{15}{32} \\ -\frac{15}{32} & \frac{25}{32} \end{bmatrix} \begin{pmatrix} 12 \\ 16 \end{pmatrix} = \left(\frac{60}{32}, \frac{220}{32}\right)',$$

$$\tilde{\boldsymbol{\sigma}}_0^{-2}\boldsymbol{\mu}_0 + 5\tilde{\boldsymbol{\sigma}}^{-2}\bar{\mathbf{x}} = \left(\frac{76}{32}, \frac{260}{32}\right)',$$

and hence

$$\begin{aligned} \boldsymbol{\mu}_1 &= \begin{bmatrix} \frac{464}{279} & \frac{240}{279} \\ \frac{240}{279} & \frac{432}{279} \end{bmatrix} \begin{pmatrix} \frac{76}{32} \\ \frac{260}{32} \end{pmatrix} \\ &= \left(\frac{3052}{279}, \frac{4080}{279}\right)' \\ &\doteq (10.94, 14.62)'. \end{aligned}$$

Although $\boldsymbol{\mu}_0 \leqq \boldsymbol{\mu}_1 \leqq \bar{\mathbf{x}}$ in this example, the reader should not expect each component of $\boldsymbol{\mu}_1$ to belong to the line segment joining the corresponding components of $\boldsymbol{\mu}_0$ and $\bar{\mathbf{x}}$; Exercise 4 furnishes a counterexample.

In Case B, $\boldsymbol{\mu}$ is known but $\tilde{\boldsymbol{\sigma}}^2$ is unknown. The natural conjugate prior df of $\tilde{\boldsymbol{\sigma}}^2$ is inverted Wishart.

For Case B ($\boldsymbol{\mu}$ known): if an experiment consists in obtaining $n \geqq k$ observations from $f_N^{(k)}(\mathbf{x} \mid \boldsymbol{\mu}, \boldsymbol{\sigma}^2)$ with noninformative stopping and yields the sufficient statistic $(n, \boldsymbol{\xi})$, and if the process covariance matrix $\tilde{\boldsymbol{\sigma}}^2$ has the natural conjugate prior df $f_{iW}^{(k)}(\boldsymbol{\sigma}^2 \mid \boldsymbol{\psi}_0, \nu_0)$, then the posterior df of $\tilde{\boldsymbol{\sigma}}^2$ is $f_{iW}^{(k)}(\boldsymbol{\sigma}^2 \mid \boldsymbol{\psi}_1, \nu_1)$, where (14.6.4)

$$\boldsymbol{\psi}_1 \equiv (\nu_0 + n)^{-1}[\nu_0\boldsymbol{\psi}_0 + n\boldsymbol{\xi}]$$

and

$$\nu_1 \equiv \nu_0 + n.$$

Proof

By (13.5.8), $(n, \tilde{\boldsymbol{\xi}})$ is sufficient for $\boldsymbol{\sigma}^2$ given $\boldsymbol{\mu}$ and the likelihood function may be written as

$$L(\boldsymbol{\sigma}^2 \mid n, \boldsymbol{\xi}) = K \mid \boldsymbol{\sigma}^2|^{-\frac{1}{2}n} \exp [-\tfrac{1}{2}n \cdot \text{tr } \{\boldsymbol{\sigma}^{-2}\boldsymbol{\xi}\}].$$

If the prior df of $\tilde{\boldsymbol{\sigma}}^2$ is [compare (12.9.9)]

$$f_{iW}{}^{(k)}(\boldsymbol{\sigma}^2 \mid \boldsymbol{\psi}_0, \nu_0) = K'|\boldsymbol{\sigma}^{-2}|^{\frac{1}{2}(\nu+2k)} \exp [-\tfrac{1}{2}\nu_0 \cdot \text{tr } \{\boldsymbol{\sigma}^{-2}\boldsymbol{\psi}_0\}],$$

then (because $|\boldsymbol{\sigma}^2|^{-\frac{1}{2}n} = |\boldsymbol{\sigma}^{-2}|^{\frac{1}{2}n}$) the posterior df of $\tilde{\boldsymbol{\sigma}}^2$ is proportional to the product

$$KK'|\boldsymbol{\sigma}^2|^{\frac{1}{2}(\nu_0+n+2k)} \exp [-\tfrac{1}{2} \text{ tr } \{\boldsymbol{\sigma}^{-2}[\nu_0\psi_0 + n\boldsymbol{\xi}]\}]$$

(*) $$= KK' \, |\boldsymbol{\sigma}^2|^{\frac{1}{2}(\nu_1+2k)} \exp [-\tfrac{1}{2}\nu_1 \cdot \text{tr } \{\boldsymbol{\sigma}^{-2}\boldsymbol{\psi}_1\}]$$

[where we have used (12.9.3) several times to re-express the exponent of e.] But (*) is proportional to $f_{iW}{}^{(k)}(\boldsymbol{\sigma}^2 \mid \boldsymbol{\psi}_1, \nu_1)$, which is therefore the required posterior df of $\tilde{\boldsymbol{\sigma}}^2$. QED

For Case B: under the assumptions of (14.6.4), the unconditional df of $\tilde{\boldsymbol{\xi}}$ given n is $f_{i\beta 2}{}^{(k)}(\boldsymbol{\xi} \mid \boldsymbol{\psi}_0, \nu_0, n - k + 1, n)$. (14.6.5)

Proof

By (13.6.3), the conditional df of $\tilde{\boldsymbol{\xi}}$ given $\boldsymbol{\sigma}^2$ is $f_W{}^{(k)}(\boldsymbol{\xi} \mid n(n - k + 1)^{-1}\boldsymbol{\sigma}^{-2}$, $n - k + 1)$, and $\tilde{\boldsymbol{\sigma}}^2 = \{[(n - k + 1)/n]n(n - k + 1)^{-1}\boldsymbol{\sigma}^{-2}\}^{-1}$ has df $f_{iW}{}^{(k)}(\boldsymbol{\sigma}^2 \mid \boldsymbol{\psi}_0, \nu_0)$, by assumption in (14.6.4). Hence (12.10.1) applies for $\nu \equiv n - k + 1$, $\tau = n$, $\boldsymbol{\psi} \equiv n(n - k + 1)^{-1}\boldsymbol{\sigma}^{-2}$, $\boldsymbol{\Phi} \equiv \boldsymbol{\psi}_0$, and $\rho \equiv \nu_0$. QED

It is worth noting that (14.6.4) continues to obtain when $n < k$, although such cases are rarely of interest.

For Case C, in which $\boldsymbol{\mu}$ nor $\boldsymbol{\sigma}^2$ is known beforehand, a conjugate prior df for $(\tilde{\boldsymbol{\mu}}, \tilde{\boldsymbol{\sigma}}^2)$ is the *Normal-inverted Wishart* df $f_{NiW}{}^{(k)}(\boldsymbol{\mu}, \boldsymbol{\sigma}^2 \mid \boldsymbol{\mu}_0, \boldsymbol{\psi}_0, n_0, \nu_0)$ defined by

$$f_{NiW}{}^{(k)}(\boldsymbol{\mu}, \boldsymbol{\sigma}^2 \mid \boldsymbol{\mu}_0, \boldsymbol{\psi}_0, n_0, \nu_0) = \begin{cases} 0, & \boldsymbol{\sigma}^2 \text{ is not PDS} \\ f_N{}^{(k)}(\boldsymbol{\mu} \mid \boldsymbol{\mu}_0, n_0{}^{-1}\boldsymbol{\sigma}^2)\, f_{iW}{}^{(k)}(\boldsymbol{\sigma}^2 \mid \boldsymbol{\psi}_0, \nu_0), & \boldsymbol{\sigma}^2 \text{ is PDS,} \end{cases} \quad (14.6.6)$$

where PDS stands for *positive definite and symmetric*. Before stating our main results, we shall derive the marginal prior distribution of $\tilde{\boldsymbol{\mu}}$. It is helpful that the parallel of (14.6.6) with (14.5.5) continues to obtain, with $\tilde{\boldsymbol{\mu}}$ having the k-variate Student marginal prior df $f_S{}^{(k)}(\boldsymbol{\mu} \mid \boldsymbol{\mu}_0, n_0{}^{-1}\boldsymbol{\psi}_0, \nu_0)$.

If $(\tilde{\boldsymbol{\mu}}, \tilde{\boldsymbol{\sigma}}^2)$ has df $f_{NiW}{}^{(k)}(\boldsymbol{\mu}, \boldsymbol{\sigma}^2 \mid \boldsymbol{\mu}_0, \boldsymbol{\psi}_0, n_0, \nu_0)$, then the unconditional df of $\tilde{\boldsymbol{\mu}}$ is $f_S{}^{(k)}(\boldsymbol{\mu} \mid \boldsymbol{\mu}_0, n_0{}^{-1}\boldsymbol{\psi}_0, \nu_0)$. (14.6.7)

Proof

By (14.6.6) and the definitions (12.7.2) of $f_N^{(k)}$ and (12.9.9) of $f_{iW}^{(k)}$,

$$
\begin{aligned}
f_{NiW}^{(k)}(\boldsymbol{\mu}, \boldsymbol{\sigma}^2 \mid \boldsymbol{\mu}_0, \boldsymbol{\psi}_0, n_0, \nu_0) \\
&= K|\boldsymbol{\sigma}^{-2}|^{\frac{1}{2}} \exp\left[-\tfrac{1}{2}n_0(\boldsymbol{\mu} - \boldsymbol{\mu}_0)'\boldsymbol{\sigma}^{-2}(\boldsymbol{\mu} - \boldsymbol{\mu}_0)\right]|\boldsymbol{\sigma}^{-2}|^{\frac{1}{2}(\nu_0+2k)} \\
&\quad \times \exp\left[-\tfrac{1}{2}\nu_0 \cdot \operatorname{tr}\{\boldsymbol{\sigma}^{-2}\boldsymbol{\psi}_0\}\right] \\
&= K|\boldsymbol{\sigma}^{-2}|^{\frac{1}{2}[(\nu_0+1)+2k]} \exp\left[-\tfrac{1}{2}n_0 \cdot \operatorname{tr}\{\boldsymbol{\sigma}^{-2}(\boldsymbol{\mu} - \boldsymbol{\mu}_0)(\boldsymbol{\mu} - \boldsymbol{\mu}_0)'\right] \\
&\quad \times \exp\left[-\tfrac{1}{2}\nu_0 \cdot \operatorname{tr}\{\boldsymbol{\sigma}^{-2}\boldsymbol{\psi}_0\}\right] \\
&= K|\boldsymbol{\sigma}^{-2}|^{\frac{1}{2}[(\nu_0+1)+2k]} \exp\left[-\tfrac{1}{2}\operatorname{tr}\{\boldsymbol{\sigma}^{-2}[n_0(\boldsymbol{\mu} - \boldsymbol{\mu}_0)(\boldsymbol{\mu} - \boldsymbol{\mu}_0)' + \nu_0\boldsymbol{\psi}_0]\}\right] \\
&= K|\boldsymbol{\sigma}^{-2}|^{\frac{1}{2}[\rho+2k]} \exp\left[-\tfrac{1}{2}\rho \cdot \operatorname{tr}\{\boldsymbol{\sigma}^{-2}\boldsymbol{\Phi}\}\right] \\
&= KW_k(\nu_0 + 1)|\Phi|^{-\frac{1}{2}(\rho+k-1)} f_{iW}^{(k)}(\boldsymbol{\sigma}^2 \mid \boldsymbol{\Phi}, \rho),
\end{aligned}
$$

where $\rho \equiv \nu_0 + 1$, $K \equiv (2\pi)^{-\frac{1}{2}k} W_k(\nu_0) \mid \boldsymbol{\psi}_0|^{\frac{1}{2}(\nu_0+k-1)}$, and

$$\boldsymbol{\Phi} \equiv \rho^{-1}[n_0(\boldsymbol{\mu} - \boldsymbol{\mu}_0)(\boldsymbol{\mu} - \boldsymbol{\mu}_0)' + \nu_0\boldsymbol{\psi}_0].$$

This last term is of the form $f(\boldsymbol{\mu})f(\boldsymbol{\sigma}^2 \mid \boldsymbol{\mu})$, and hence the factors preceding $f_{iW}^{(k)}(\boldsymbol{\sigma}^2 \mid \boldsymbol{\Phi}, \rho)$ constitute the df of $\tilde{\boldsymbol{\mu}}$. We shall show that these factors reduce to

$$(*) \qquad K''' \mid \nu_0 + (\boldsymbol{\mu} - \boldsymbol{\mu}_0)'(n_0\boldsymbol{\psi}_0^{-1})(\boldsymbol{\mu} - \boldsymbol{\mu}_0) \mid^{-\frac{1}{2}(\nu_0+k)},$$

where K''' is a constant not involving $\boldsymbol{\mu}$. We have:

$$
\begin{aligned}
KW_k(\nu_0 + 1)|\boldsymbol{\Phi}|^{-\frac{1}{2}(\nu_0+1+k-1)} \\
&= K'|\rho^{-1}n_0(\boldsymbol{\mu} - \boldsymbol{\mu}_0)(\boldsymbol{\mu} - \boldsymbol{\mu}_0)' + \rho^{-1}\nu_0\boldsymbol{\psi}_0|^{-\frac{1}{2}(\nu_0+k)} \\
&= K''|(\boldsymbol{\mu} - \boldsymbol{\mu}_0)(\boldsymbol{\mu} - \boldsymbol{\mu}_0)' + n_0^{-1}\nu_0\boldsymbol{\psi}_0|^{-\frac{1}{2}(\nu_0+k)},
\end{aligned}
$$

where $K'' = K'/(\rho^{-1}n_0)^{\frac{1}{2}k(\nu_0+k)}$. Now, in linear algebra there is a theorem which asserts that if $\mathbf{y} \in R^k$ and $\mathbf{M}$ is kth order nonsingular, then $|\mathbf{yy}' + \mathbf{M}| = |\mathbf{M}|\ |1 + \mathbf{y}\mathbf{M}^{-1}\mathbf{y}|$. Hence

$$
\begin{aligned}
K'' \mid (\boldsymbol{\mu} - \boldsymbol{\mu}_0)(\boldsymbol{\mu} - \boldsymbol{\mu}_0)' + n_0^{-1}\nu_0\boldsymbol{\psi}_0 \mid^{-\frac{1}{2}(\nu_0+k)} \\
&= K''|n_0^{-1}\nu_0\boldsymbol{\psi}_0|^{-\frac{1}{2}(\nu_0+k)}|1 + (\boldsymbol{\mu} - \boldsymbol{\mu}_0)'\nu_0^{-1}n_0\boldsymbol{\psi}_0^{-1}(\boldsymbol{\mu} - \boldsymbol{\mu}_0) \mid^{-\frac{1}{2}(\nu_0+k)} \\
&= K''|n_0^{-1}\nu_0\boldsymbol{\psi}_0|^{-\frac{1}{2}(\nu_0+k)}\nu_0^{\frac{1}{2}k(\nu_0+k)}|\nu_0 + (\boldsymbol{\mu} - \boldsymbol{\mu}_0)'n_0\boldsymbol{\psi}_0^{-1}(\boldsymbol{\mu} - \boldsymbol{\mu}_0) \mid^{-\frac{1}{2}(\nu_0+k)},
\end{aligned}
$$

which is of the asserted form (*). QED

The main results now follow.

For Case C (neither $\boldsymbol{\mu}$ nor $\boldsymbol{\sigma}^2$ known): if an experiment consists in obtaining $n \geqq k + 1$ observations from $f_N^{(k)}(\mathbf{x} \mid \boldsymbol{\mu}, \boldsymbol{\sigma}^2)$ with noninformative stopping and yields the sufficient statistic $(n, \bar{\mathbf{x}}, \mathbf{s}^2)$, and if $(\tilde{\boldsymbol{\mu}}, \tilde{\boldsymbol{\sigma}}^2)$ has the natural conjugate prior df $f_{NiW}^{(k)}(\boldsymbol{\mu}, \boldsymbol{\sigma}^2 \mid \boldsymbol{\mu}_0, \boldsymbol{\psi}_0, n_0, \nu_0)$, then the posterior df of $(\tilde{\boldsymbol{\mu}}, \tilde{\boldsymbol{\sigma}}^2)$ is (14.6.8)

$$f_{NiW}^{(k)}(\boldsymbol{\mu}, \boldsymbol{\sigma}^2 \mid \boldsymbol{\mu}_1, \boldsymbol{\psi}_1, n_1, \nu_1),$$

where

$$\boldsymbol{\mu}_1 \equiv \left(\frac{n_0}{n_0 + n}\right)\boldsymbol{\mu}_0 + \left(\frac{n}{n_0 + n}\right)\bar{\mathbf{x}},$$

$$n_1 \equiv n_0 + n,$$

$$\nu_1 \equiv \nu_0 + n,$$

and

$$\boldsymbol{\psi}_1 \equiv \nu_1^{-1}[\nu_0\boldsymbol{\psi}_0 + (n-1)\mathbf{s}^2 + (n_0^{-1} + n^{-1})^{-1}(\bar{\mathbf{x}} - \boldsymbol{\mu}_0)(\bar{\mathbf{x}} - \boldsymbol{\mu}_0)'].$$

The marginal prior and posterior df's of $\tilde{\boldsymbol{\mu}}$ are $f_S^{(k)}(\boldsymbol{\mu} \mid \boldsymbol{\mu}_0, n_0^{-1}\boldsymbol{\psi}_0, \nu_0)$ and $f_S^{(k)}(\boldsymbol{\mu} \mid \boldsymbol{\mu}_1, n_1^{-1}\boldsymbol{\psi}_1, \nu_1)$ respectively. The marginal prior and posterior df's of $\tilde{\boldsymbol{\sigma}}^2$ are $f_{iW}^{(k)}(\boldsymbol{\sigma}^2 \mid \boldsymbol{\psi}_0, \nu_0)$ and $f_{iW}^{(k)}(\boldsymbol{\sigma}^2 \mid \boldsymbol{\psi}_1, \nu_1)$, respectively.

Proof

The second assertion is immediate from the first via (14.6.7). The third also follows from the first in view of the fact, obvious from (14.6.6), that the marginal df of $\tilde{\boldsymbol{\sigma}}^2$ is the second factor in (14.6.6). To prove the first, we note by (13.5.11) that $(n, \tilde{\bar{\mathbf{x}}}, \tilde{\mathbf{s}}^2)$ is sufficient for $(\boldsymbol{\mu}, \boldsymbol{\sigma}^2)$, and the likelihood function may be written as

$$L(\boldsymbol{\mu}, \boldsymbol{\sigma}^2 | n, \bar{\mathbf{x}}, \mathbf{s}^2) = K|\boldsymbol{\sigma}^2|^{-\frac{1}{2}n} \exp[-\tfrac{1}{2}(n-1)\operatorname{tr}\{\boldsymbol{\sigma}^{-2}\mathbf{s}^2\}] \cdot \exp[-\tfrac{1}{2}n(\bar{\mathbf{x}} - \boldsymbol{\mu})'\boldsymbol{\sigma}^{-2}(\bar{\mathbf{x}} - \boldsymbol{\mu})].$$

But the prior df $f_{NiW}^{(k)}(\boldsymbol{\mu}, \boldsymbol{\sigma}^2 \mid \boldsymbol{\mu}_0, \boldsymbol{\psi}_0, n_0, \nu_0)$ may be written as

$$K'|\boldsymbol{\sigma}^2|^{-1/2} \exp[-\tfrac{1}{2}n_0(\boldsymbol{\mu} - \boldsymbol{\mu}_0)'\boldsymbol{\sigma}^{-2}(\boldsymbol{\mu} - \boldsymbol{\mu}_0)] \cdot |\boldsymbol{\sigma}^2|^{-\frac{1}{2}(\nu_0+2k)} \exp[-\tfrac{1}{2}\nu_0 \cdot \operatorname{tr}\{\boldsymbol{\sigma}^{-2}\boldsymbol{\psi}_0\},$$

and hence the posterior df is proportional to the product

$$(*) \quad KK'|\boldsymbol{\sigma}^2|^{-1/2} \exp[-\tfrac{1}{2}[n_0(\boldsymbol{\mu} - \boldsymbol{\mu}_0)'\boldsymbol{\sigma}^{-2}(\boldsymbol{\mu} - \boldsymbol{\mu}_0) + n(\bar{\mathbf{x}} - \boldsymbol{\mu})'\boldsymbol{\sigma}^{-2}(\bar{\mathbf{x}} - \boldsymbol{\mu})]] \cdot |\boldsymbol{\sigma}^2|^{-\frac{1}{2}(\nu_0+n+2k)} \exp[-\tfrac{1}{2}\operatorname{tr}\{\boldsymbol{\sigma}^{-2}[\nu_0\boldsymbol{\psi}_0 + (n-1)\mathbf{s}^2]\}].$$

The reader may show as Exercise 5 that

$$(**) \quad n_0(\boldsymbol{\mu} - \boldsymbol{\mu}_0)'\boldsymbol{\sigma}^{-2}(\boldsymbol{\mu} - \boldsymbol{\mu}_0) + n(\bar{\mathbf{x}} - \boldsymbol{\mu})'\boldsymbol{\sigma}^{-2}(\bar{\mathbf{x}} - \boldsymbol{\mu}) = n_1(\boldsymbol{\mu} - \boldsymbol{\mu}_1)'\boldsymbol{\sigma}^{-2}(\boldsymbol{\mu} - \boldsymbol{\mu}_1) + \operatorname{tr}\{\boldsymbol{\sigma}^{-2}[(n_0^{-1} + n^{-1})^{-1}(\bar{\mathbf{x}} - \boldsymbol{\mu}_0)(\bar{\mathbf{x}} - \boldsymbol{\mu}_0)']\}.$$

By (**) it follows that the product (*) to which the posterior df of $(\tilde{\boldsymbol{\mu}}, \tilde{\boldsymbol{\sigma}}^2)$ is proportional can be written as

$$K''|\boldsymbol{\sigma}^2|^{-1/2} \exp[-\tfrac{1}{2}n_1(\boldsymbol{\mu} - \boldsymbol{\mu}_1)'\boldsymbol{\sigma}^{-2}(\boldsymbol{\mu} - \boldsymbol{\mu}_1)] \cdot |\boldsymbol{\sigma}^2|^{-\frac{1}{2}(\nu_1+2k)} \exp[-\tfrac{1}{2}\nu_1 \cdot \operatorname{tr}\{\boldsymbol{\sigma}^{-2}\boldsymbol{\psi}_1\}],$$

which implies that the posterior df of $(\tilde{\boldsymbol{\mu}}, \tilde{\boldsymbol{\sigma}}^2)$ must be as asserted. QED

For Case C: under the assumptions of (14.6.8) and given n, (14.6.9)

(a) the df $f(\bar{\mathbf{x}} \mid \mathbf{s}^2)$ of $\tilde{\bar{\mathbf{x}}}$ conditional upon $\mathbf{s}^2$ but unconditional as regards $(\tilde{\boldsymbol{\mu}}, \tilde{\boldsymbol{\sigma}}^2)$ is given by

$$f(\bar{\mathbf{x}} \mid \mathbf{s}^2) = f_{S^{(k)}}(\bar{\mathbf{x}} \mid \boldsymbol{\mu}_0, (n_0^{-1} + n^{-1})\boldsymbol{\psi}^*, \nu_0 + n - 1),$$

where

$$\boldsymbol{\psi}^* \equiv \left(\frac{\nu_0}{\nu_0 + n - 1}\right)\boldsymbol{\psi}_0 + \left(\frac{n-1}{\nu_0 + n - 1}\right)\mathbf{s}^2;$$

(b) the unconditional df $f(\bar{\mathbf{x}})$ of $\tilde{\bar{\mathbf{x}}}$ is given by

$$f(\bar{\mathbf{x}}) = f_{S^{(k)}}(\bar{\mathbf{x}} \mid \boldsymbol{\mu}_0, (n_0^{-1} + n^{-1})\boldsymbol{\psi}_0, \nu_0);$$

(c) the df $f(\mathbf{s}^2 \mid \bar{\mathbf{x}})$ of $\tilde{\mathbf{s}}^2$ conditional upon $\bar{\mathbf{x}}$ but unconditional as regards $(\tilde{\boldsymbol{\mu}}, \tilde{\boldsymbol{\sigma}}^2)$ is given by

$$f(\mathbf{s}^2 \mid \bar{\mathbf{x}}) = f_{i\beta_2}{}^{(k)}(\mathbf{s}^2 \mid \boldsymbol{\Phi}^*, \nu_0 + 1, n - k, n - 1),$$

where

$$\boldsymbol{\Phi}^* \equiv \left(\frac{\nu_0}{\nu_0 + 1}\right)\boldsymbol{\psi}_0 + \left(\frac{1}{\nu_0 + 1}\right)(n_0^{-1} + n^{-1})^{-1}(\bar{\mathbf{x}} - \boldsymbol{\mu}_0)(\bar{\mathbf{x}} - \boldsymbol{\mu}_0)';$$

and

(d) the unconditional df $f(\mathbf{s}^2)$ of $\tilde{\mathbf{s}}^2$ is given by

$$f(\mathbf{s}^2) = f_{i\beta_2}{}^{(k)}(\mathbf{s}^2 \mid \boldsymbol{\psi}_0, \nu_0, n - k, n - 1).$$

Proof

By (13.6.2) and (13.6.4), $\tilde{\bar{\mathbf{x}}}$ and $\tilde{\mathbf{s}}^2$ are conditionally independent given $(\boldsymbol{\mu}, \boldsymbol{\sigma}^2)$ with conditional joint df

$$(*) \quad f(\bar{\mathbf{x}}, \mathbf{s}^2 \mid \boldsymbol{\mu}, \boldsymbol{\sigma}^2) = f_{N^{(k)}}(\bar{\mathbf{x}} \mid \boldsymbol{\mu}, n^{-1}\boldsymbol{\sigma}^2) f_{W^{(k)}}(\mathbf{s}^2 \mid (n-1)(n-k)^{-1}\boldsymbol{\sigma}^{-2}, n-k).$$

Since the prior and posterior df's of $(\tilde{\boldsymbol{\mu}}, \tilde{\boldsymbol{\sigma}}^2)$ are of the form

$$f(\boldsymbol{\mu}, \boldsymbol{\sigma}^2 \mid \bar{\mathbf{x}}, s^2) = f_{NiW}{}^{(k)}(\boldsymbol{\mu}, \boldsymbol{\sigma}^2 \mid \boldsymbol{\mu}_i, \boldsymbol{\psi}_i, n_i, \nu_i),$$

we may apply the recognition method easily to obtain (after a lot of cancellation, rearrangement, and so forth

$$\begin{aligned} f(\bar{\mathbf{x}}, \mathbf{s}^2) &= f(\bar{\mathbf{x}}, \mathbf{s}^2 \mid \boldsymbol{\mu}, \boldsymbol{\sigma}^2) f(\boldsymbol{\mu}, \boldsymbol{\sigma}^2) / f(\boldsymbol{\mu}, \boldsymbol{\sigma}^2 \mid \bar{\mathbf{x}}, \mathbf{s}^2) \\ (*) \qquad &= K|\boldsymbol{\psi}_0|^{\frac{1}{2}(\nu_0+k-1)}|\mathbf{s}^2|^{\frac{1}{2}(n-k-2)}|\boldsymbol{\psi}_1|^{-\frac{1}{2}(\nu_1+k-1)}, \end{aligned}$$

where

$$K \equiv [2\pi(n_0^{-1} + n^{-1})]^{-\frac{1}{2}k} \frac{W_k(\nu_0) W_k(n-k)}{W_k(\nu_0 + n)}.$$

Now, $f(\bar{\mathbf{x}}, \mathbf{s}^2)$ is cumbersome to integrate, and hence we shall use it only to obtain the conditional df's $f(\bar{\mathbf{x}} \mid \mathbf{s}^2)$ and $f(\mathbf{s}^2 \mid \bar{\mathbf{x}})$, while employing other methods for deriving $f(\bar{\mathbf{x}})$ and $f(\mathbf{s}^2)$.

Proof of (b)

$$f(\bar{\mathbf{x}}) = \iint f(\bar{\mathbf{x}} \mid \tilde{\boldsymbol{\mu}}, \tilde{\boldsymbol{\sigma}}^2) f(\tilde{\boldsymbol{\mu}}, \tilde{\boldsymbol{\sigma}}^2) d\tilde{\boldsymbol{\mu}} d\tilde{\boldsymbol{\sigma}}^2$$

$$= \int \left\{ \int f_N{}^{(k)}(\bar{\mathbf{x}} \mid \tilde{\boldsymbol{\mu}}, n^{-1}\tilde{\boldsymbol{\sigma}}^2) f_N{}^{(k)}(\tilde{\boldsymbol{\mu}} \mid \boldsymbol{\mu}_0, n_0{}^{-1}\tilde{\boldsymbol{\sigma}}^2) d\tilde{\boldsymbol{\mu}} \right\} f_{iW}{}^{(k)}(\tilde{\boldsymbol{\sigma}}^2 \mid \boldsymbol{\psi}_0, \nu_0) d\tilde{\boldsymbol{\sigma}}^2.$$

The inner integral is $f_N{}^{(k)}(\bar{\mathbf{x}} \mid \boldsymbol{\mu}_0, (n_0{}^{-1} + n^{-1})\tilde{\boldsymbol{\sigma}}^2)$, by (14.6.2) for $\boldsymbol{\sigma}_0{}^2 = n_0{}^{-1}\tilde{\boldsymbol{\sigma}}^2$. Hence

$$f(\bar{\mathbf{x}}) = \int f_N{}^{(k)}(\bar{\mathbf{x}} \mid \boldsymbol{\mu}_0, (n_0{}^{-1} + n^{-1})\tilde{\boldsymbol{\sigma}}^2) f_{iW}{}^{(k)}(\tilde{\boldsymbol{\sigma}}^2 \mid \boldsymbol{\psi}_0, \nu_0) d\tilde{\boldsymbol{\sigma}}^2$$

$$= \int f_{NiW}{}^{(k)}(\bar{\mathbf{x}}, \tilde{\boldsymbol{\sigma}}^2 \mid \boldsymbol{\mu}_0, \boldsymbol{\psi}_0, (n_0{}^{-1} + n^{-1})^{-1}, \nu_0) d\tilde{\boldsymbol{\sigma}}^2$$

$$= f_S{}^{(k)}(\bar{\mathbf{x}} \mid \boldsymbol{\mu}_0, (n_0{}^{-1} + n^{-1})\boldsymbol{\psi}_0, \nu_0),$$

by (14.6.6) and (14.6.7).

Proof of (c)

By (a) and (*),

$$f(\mathbf{s}^2 \mid \bar{\mathbf{x}}) = \frac{f(\bar{\mathbf{x}}, \mathbf{s}^2)}{f(\bar{\mathbf{x}})}$$

(**) $$= K'|\mathbf{s}^2|^{\frac{1}{2}(n-k-2)}$$

$$\times |\mathbf{s}^2 + \{(n-1)^{-1}[\nu_0\boldsymbol{\psi}_0 + (n_0{}^{-1} + n^{-1})^{-1}(\bar{\mathbf{x}} - \boldsymbol{\mu}_0)\cdot(\bar{\mathbf{x}} - \boldsymbol{\mu}_0)']\}|^{-\frac{1}{2}(\nu_1 - 1 - k)}$$

in which K' includes all factors not involving $\mathbf{s}^2$; but by (12.10.1), (**) is proportional to $f_{i\beta 2}{}^{(k)}(\mathbf{s}^2 \mid \boldsymbol{\Phi}^*, \nu_0 + 1, n - k, n - 1)$.

Proof of (d)

From (13.6.4), $\tilde{\mathbf{s}}^2$ has conditional df $f_W{}^{(k)}(\mathbf{s}^2 \mid (n-1)(n-k)^{-1}\tilde{\boldsymbol{\sigma}}^{-2}, n-k)$ given $(\tilde{\boldsymbol{\mu}}, \tilde{\boldsymbol{\sigma}}^2)$ and is thus independent of $\tilde{\boldsymbol{\mu}}$ given $\tilde{\boldsymbol{\sigma}}^2$. Hence

$$f(\mathbf{s}^2) = \int f_W{}^{(k)}(\mathbf{s}^2 \mid (n-1)(n-k)^{-1}\tilde{\boldsymbol{\sigma}}^{-2}, n-k) f_{iW}{}^{(k)}(\tilde{\boldsymbol{\sigma}}^2 \mid \boldsymbol{\psi}_0, \nu_0) d\tilde{\boldsymbol{\sigma}}^2$$

$$= f_{i\beta 2}{}^{(k)}(\mathbf{s}^2 \mid \boldsymbol{\psi}_0, \nu_0, n-k, n-1)$$

by (12.10.1) (with $\nu = n - k$, $\tau = n - 1$).

Proof of (a)

By (d) and (*),

$$f(\bar{\mathbf{x}} \mid \mathbf{s}^2) = \frac{f(\bar{\mathbf{x}}, \mathbf{s}^2)}{f(\mathbf{s}^2)}$$

$$(***) \qquad = K''|\nu_0\boldsymbol{\psi}_0 + (n-1)\mathbf{s}^2 + (n_0^{-1} + n^{-1})^{-1}(\bar{\mathbf{x}} - \boldsymbol{\mu}_0)(\bar{\mathbf{x}} - \boldsymbol{\mu}_0)'|^{-\frac{1}{2}(\nu_1-1+k)},$$

in which K'' includes all factors not involving $\bar{\mathbf{x}}$. But if $\boldsymbol{\Upsilon} \equiv (n_0^{-1} + n^{-1})$ $[\nu_0\boldsymbol{\psi}_0 + (n-1)\,\mathbf{s}^2]$, then

$$\begin{aligned} &= K'''|\boldsymbol{\upsilon} + (\bar{\mathbf{x}} - \boldsymbol{\mu}_0)(\bar{\mathbf{x}} - \boldsymbol{\mu}_0)'|^{-\frac{1}{2}(\nu_1-1+k)} \\ &= K'''|\boldsymbol{\upsilon}|^{-\frac{1}{2}(\nu_1-1+k)}|1 + (\bar{\mathbf{x}} - \boldsymbol{\mu}_0)'\boldsymbol{\upsilon}^{-1}(\bar{\mathbf{x}} - \boldsymbol{\mu}_0)|^{-\frac{1}{2}(\nu_1-1+k)} \\ &= K^{iv}|(\nu_1 - 1) + (\bar{\mathbf{x}} - \boldsymbol{\mu}_0)'(\nu_1 - 1)\boldsymbol{\upsilon}^{-1}(\bar{\mathbf{x}} - \boldsymbol{\mu}_0)|^{-\frac{1}{2}[(\nu_1-1)+k]}, \end{aligned}$$

which implies that the df of $\boldsymbol{\upsilon}$ is $f_S^{(k)}(\bar{\mathbf{x}} \mid \boldsymbol{\mu}_0, (\nu_1 - 1)^{-1}\boldsymbol{\upsilon}, \nu_1 - 1)$. But $(\nu_1 - 1)^{-1}\tilde{\bar{\mathbf{x}}} = (\nu_0 + n - 1)^{-1}(n_0^{-1} + n^{-1})[\nu_0\boldsymbol{\psi}_0 + (n-1)\mathbf{s}^2] \equiv (n_0^{-1} + n^{-1})\boldsymbol{\psi}^*$ and $\nu_1 - 1 = \nu_0 + n - 1$. In the next-to-last equality above we have used the fact that $|\mathbf{y}\mathbf{y}' + \mathbf{M}| = |\mathbf{M}||1 + \mathbf{y}'\mathbf{M}^{-1}\mathbf{y}|$. QED

Assertions (14.6.8) and (14.6.9b) continue to obtain when $n < k + 1$. However, in that case $\tilde{\mathbf{s}}^2$ has a so-called *singular* Wishart distribution which does not have a df given by (12.9.1). A complete discussion of the cases $n \leqq k + 1$ can be found in Ando and Kaufman [2].

Example 14.6.2

Suppose in Example 14.6.1 that $\boldsymbol{\sigma}^2$ was not known beforehand and that the decision-maker had assessed the prior df $f_{iW}^{(2)}(\boldsymbol{\sigma}^2 \mid \boldsymbol{\psi}_0, \nu_0)$ for $\nu_0 = 4$ and

$$\boldsymbol{\psi}_0 = \begin{bmatrix} 5 & 3 \\ 3 & 5 \end{bmatrix}.$$

Note that for these parameters, $E(\tilde{\boldsymbol{\sigma}}^2) = 2\boldsymbol{\psi}_0 =$ the assumed value of $\boldsymbol{\sigma}^2$ in Example 14.6.1. Suppose further that the prior df of $\tilde{\boldsymbol{\mu}}$ given $\tilde{\boldsymbol{\sigma}}^2$ was assessed to be $f_N^{(2)}(\boldsymbol{\mu} \mid \boldsymbol{\mu}_0, n_0^{-1}\boldsymbol{\sigma}^2)$ for $\boldsymbol{\mu}_0 = (8, 10)'$ (as in Example 14.6.1) and $n_0 = .4$. Finally, suppose that the $n = 5$ observations yielded the statistics $\bar{\mathbf{x}} = (12, 16)'$ (as before) and

$$\mathbf{s}^2 = \begin{bmatrix} 10 & 7 \\ 7 & 5 \end{bmatrix}.$$

Then the posterior df of $(\tilde{\boldsymbol{\mu}}, \tilde{\boldsymbol{\sigma}}^2)$ will also be Normal-inverted Wishart, with parameters $n_1 = 0.4 + 5 = 5.4$, $\nu_1 = 4 + 5 = 9$,

$$\begin{aligned} \boldsymbol{\mu}_1 &= \left(\frac{.4}{5.4}\right)(8, 10)' + \left(\frac{5}{5.4}\right)(12, 16)' \\ &= \left(\frac{316}{27}, \frac{420}{27}\right)', \end{aligned}$$

and

$$\boldsymbol{\psi}_1 = \left(\frac{4}{9}\right)\begin{bmatrix}5 & 3\\ 3 & 5\end{bmatrix} + \left(\frac{4}{9}\right)\begin{bmatrix}10 & 7\\ 7 & 5\end{bmatrix} + \left(\frac{1}{9}\right)\left(\frac{10}{27}\right)\binom{4}{6}(4, 6)$$

$$= \begin{bmatrix}\frac{1780}{243} & \frac{1320}{243}\\ \frac{1320}{243} & \frac{1440}{243}\end{bmatrix}.$$

The bivariate Student marginal posterior df of $\tilde{\boldsymbol{\mu}}$ can be found easily via (14.6.8) and used for purposes of making inferences about or decisions affected by $\tilde{\boldsymbol{\mu}}$; for example, to answer the question, "What is the probability that $\tilde{\mu}_1$ exceeds 13 given that $\tilde{\mu}_2 = 10$?"

Example 14.6.3

We shall modify the preceding question to, "What is the probability that the income $\tilde{x}_1$ of another individual, picked at random, exceeds 13, given that $\tilde{x}_2 = 10$?" To answer the question, we note that *prior* to selecting this individual, our judgments about $(\tilde{\boldsymbol{\mu}}, \tilde{\sigma}^2)$ are expressed by $f_{NiW}^{(2)}(\boldsymbol{\mu}, \sigma^2 \mid \boldsymbol{\mu}_1, \boldsymbol{\psi}_1, n_1, \nu_1)$, which serves as a *prior* df for the second "sample" size of $n - 1$. By (14.6.9b), the joint df of the (income, education) pair $\tilde{\mathbf{x}}$ of one additional individual is $f_S^{(2)}(\mathbf{x} \mid \boldsymbol{\mu}_1, (n_1^{-1} + 1)^{-1}\boldsymbol{\psi}_1, \nu_1)$, where $\boldsymbol{\mu}_1$, $\boldsymbol{\psi}_1$, n_1, and ν_1 are as computed in Example 14.6.2. By (12.8.6), the conditional df of $\tilde{x}_1$ given x_2 is $f_S(x_1 \mid \mu_1 + V_{12}V_{22}^{-1}(x_2 - \mu_2),\ V_{11} - V_{12}V_{22}^{-1}V_{21},\ \nu_1)$, where

$$\mathbf{V} \equiv (n_1^{-1} + 1)^{-1}\boldsymbol{\psi}_1$$

$$= \left(\frac{5.4}{6.4}\right)\begin{bmatrix}\frac{1780}{243} & \frac{1320}{243}\\ \frac{1320}{243} & \frac{1440}{243}\end{bmatrix}$$

$$= \begin{bmatrix}\frac{445}{72} & \frac{330}{72}\\ \frac{330}{72} & \frac{369}{72}\end{bmatrix}$$

from which we calculate that $\tilde{x}_1$ has df $f_S(x_1 \mid \mu_{1|2}, \sigma_{11|2}{}^2, 9)$ given $\tilde{x}_2 = 10$, where

$$\mu_{1|2} = \frac{316}{27} + \left(\frac{11}{12}\right)\left(\frac{-150}{27}\right) = \frac{1071}{162} \doteq 6.611$$

and

$$\sigma_{11|2}{}^2 = \frac{445}{72} - \left(\frac{330}{72}\right)\left(\frac{72}{360}\right)\left(\frac{330}{72}\right) = \frac{1710}{864} \doteq 1.979.$$

The desired quantity is

$$\begin{aligned}P[\tilde{x}_1 > 13] &= 1 - F_S(13 \mid 6.611, 1.979, 9)\\ &= 1 - F_{S*}\left(\frac{13 - 6.611}{1.41} \mid 9\right)\\ &\doteq 1 - F_{S*}(4.55 \mid 9)\\ &< .005.\end{aligned}$$

In Case D there is only a single, one-dimensional unknown parameter, σ_*^2, the natural conjugate prior df of which is $f_{i\gamma}(\sigma_*^2 \mid \psi_0, \nu_0)$. (14.6.10)

For Case D (σ_*^2 unknown, $\boldsymbol{\mu}$ and $\boldsymbol{\sigma}_*^2$ known): if an experiment consists in obtaining n observations from $f_N^{(k)}(\mathbf{x} \mid \boldsymbol{\mu}, \boldsymbol{\sigma}^2)$ with noninformative stopping and yields the sufficient statistic (n, ξ_*), and if $\tilde{\sigma}_*^2$ has the natural conjugate prior df $f_{i\gamma}(\sigma_*^2 \mid \psi_0, \nu_0)$, then the posterior df of $\tilde{\sigma}_*^2$ is $f_{i\gamma}(\sigma_*^2 \mid \psi_1, \nu_1)$, where

$$\psi_1 \equiv \left(\frac{\nu_0}{\nu_0 + kn}\right)\psi_0 + \left(\frac{kn}{\nu_0 + kn}\right)\xi_*$$

and

$$\nu_1 \equiv \nu_0 + kn.$$

Proof

By (13.5.12), the likelihood function may be written as

$$L(\sigma_*^2 \mid n, \xi_*) = K(\sigma_*^2)^{-\frac{1}{2}kn} \exp\left[-\tfrac{1}{2}kn\xi_*/\sigma_*^2\right].$$

But by (12.2.10),

$$f_{i\gamma}(\sigma_*^2 \mid \psi_0, \nu_0) = K'(\sigma_*^2)^{-\frac{1}{2}\nu_0 - 1} \exp\left[-\tfrac{1}{2}\nu_0\psi_0/\sigma_*^2\right],$$

and hence the posterior df of σ_*^2 is proportional to the product

$$\begin{aligned}&KK'(\sigma_*^2)^{-\frac{1}{2}(\nu_0 + kn) - 1} \exp\left[-\tfrac{1}{2}(\nu_0\psi_0 + kn\xi_*)/\sigma_*^2\right]\\ &\qquad = KK'(\sigma_*^2)^{-\frac{1}{2}\nu_1 - 1} \exp\left[-\tfrac{1}{2}\nu_1\psi_1/\sigma_*^2\right],\end{aligned}$$

which implies that it must be $f_{i\gamma}(\sigma_*^2 \mid \psi_1, \nu_1)$. QED

For Case D: under the assumptions and definitions of (14.6.10) and given n, the unconditional df of $\tilde{\xi}_*$ is (14.6.11)

$$f_{i\beta 2}(\xi_* \mid \tfrac{1}{2}kn, \tfrac{1}{2}\nu_1, \nu_0\psi_0/(kn)).$$

Proof

By (13.6.6) the conditional df of $\tilde{\xi}_*$ given σ_*^2 is $f_{\gamma 2}(\xi_* \mid 1/\sigma_*^2, kn)$, and hence

$$f(\xi_*) = f_{\gamma 2}(\xi_* \mid 1/\sigma_*^2, kn) f_{i\gamma}(\sigma_*^2 \mid \psi_0, \nu_0)/f_{i\gamma}(\sigma_*^2 \mid \psi_1, \nu_1).$$

The remainder of the proof is identical to the proof of (14.5.4), but with kn replacing n throughout. QED

Complexity returns with Case E, in which the natural conjugate prior df of $(\tilde{\boldsymbol{\mu}}, \tilde{\sigma}_*^2)$ is the k-variate *normal-inverted gamma* df

$$f_{Ni\gamma}^{(k)}(\boldsymbol{\mu}, \sigma_*^2 \mid \boldsymbol{\mu}_0, \psi_0, \boldsymbol{\sigma}_{0*}^2, \nu_0),$$

defined by

$$f_{Ni\gamma}^{(k)}(\boldsymbol{\mu}, \sigma_*^2 \mid \boldsymbol{\mu}_0, \psi_0, \boldsymbol{\sigma}_{0*}^2, \nu_0) \equiv \begin{cases} 0, & \sigma_*^2 \leqq 0 \\ f_N^{(k)}(\boldsymbol{\mu} \mid \boldsymbol{\mu}_0, \sigma_*^2 \boldsymbol{\sigma}_{0*}^2) f_{i\gamma}(\sigma_*^2 \mid \psi_0, \nu_0), & \\ & \sigma_*^2 > 0. \end{cases} \quad (14.6.12)$$

If $(\tilde{\boldsymbol{\mu}}, \tilde{\sigma}_*^2)$ has df $f_{Ni\gamma}^{(k)}(\boldsymbol{\mu}_0, \sigma_*^2\ \psi_0, \boldsymbol{\sigma}_{0*}^2, \nu_0)$, then the unconditional df of $\tilde{\boldsymbol{\mu}}$ is $f_S^{(k)}(\boldsymbol{\mu} \mid \boldsymbol{\mu}_0, \psi_0 \boldsymbol{\sigma}_{0*}^2, \nu_0)$. (14.6.13)

Proof

The quantity $\tilde{\sigma}_*^2$ has df $f_{i\gamma}(\sigma_*^2 \mid \psi_0, \nu_0)$ if and only if $\tilde{h}$ has df $f_{\gamma 2}(h \mid \psi_0, \nu_0)$, and hence

$$\begin{aligned} f(\boldsymbol{\mu}) &= \int_0^\infty f_N^{(k)}(\boldsymbol{\mu} \mid \boldsymbol{\mu}_0, h^{-1} \boldsymbol{\sigma}_{0*}^2) f_{\gamma 2}(h \mid \psi_0, \nu_0)\, dh \\ &= f_S^{(k)}(\boldsymbol{\mu} \mid \boldsymbol{\mu}_0, \psi_0 \boldsymbol{\sigma}_{0*}^2, \nu_0), \end{aligned}$$

by (12.8.1). QED

For Case E ($\boldsymbol{\mu}$ and σ_*^2 unknown, $\boldsymbol{\sigma}_*^2$ known): if an experiment consists in obtaining $n \geqq 2$ observations from $f_N^{(k)}(\mathbf{x} \mid \boldsymbol{\mu}, \boldsymbol{\sigma}^2)$ with noninformative stopping and yields the sufficient statistic $(n, \bar{\mathbf{x}}. s_*^2)$, and if $(\tilde{\boldsymbol{\mu}}, \tilde{\sigma}_*^2)$ has the natural conjugate prior df $f_{Ni\gamma}^{(k)}(\boldsymbol{\mu}, \sigma_*^2, \mid \boldsymbol{\mu}_0, \psi_0, \boldsymbol{\sigma}_0{}^2{}_*, \nu_0)$, then the posterior df of $(\tilde{\boldsymbol{\mu}}, \tilde{\boldsymbol{\sigma}}_*^2)$ is (14.6.14)

$$f_{Ni\gamma}^{(k)}(\boldsymbol{\mu}, \sigma_*^2 \mid \boldsymbol{\mu}_1, \psi_1, \boldsymbol{\sigma}_{1*}^2, \nu_1),$$

where

$$\begin{aligned} \nu_1 &\equiv \nu_0 + kn, \\ \psi_1 &\equiv \nu_1^{-1}[\nu_0\psi_0 + k(n-1)s_*^2 + (\bar{\mathbf{x}} - \boldsymbol{\mu}_0)'(\boldsymbol{\sigma}_{0*}^2 + n^{-1}\boldsymbol{\sigma}_*^2)^{-1}(\bar{\mathbf{x}} - \boldsymbol{\mu}_0)], \\ \boldsymbol{\sigma}_{1*}^2 &\equiv (\boldsymbol{\sigma}_{0*}^{-2} + n\boldsymbol{\sigma}_*^{-2})^{-1}, \end{aligned}$$

and

$$\boldsymbol{\mu}_1 \equiv \boldsymbol{\sigma}_{1*}^2(\boldsymbol{\sigma}_{0*}^{-2}\boldsymbol{\mu}_0 + n\boldsymbol{\sigma}_*^{-2}\bar{\mathbf{x}}).$$

The marginal prior and posterior df's of $\tilde{\boldsymbol{\mu}}$ are $f_S^{(k)}(\boldsymbol{\mu} \mid \boldsymbol{\mu}_0, \psi_0 \boldsymbol{\sigma}_{0*}^2, \nu_0)$ and $f_S^{(k)}(\boldsymbol{\mu} \mid \boldsymbol{\mu}_1\ \psi_1 \boldsymbol{\sigma}_{1*}^2, \nu_1)$, respectively. The marginal prior and posterior df's of $\tilde{\sigma}_*^2$ are $f_{i\gamma}(\sigma_*^2 \mid \psi_0, \nu_0)$ and $f_{i\gamma}(\sigma_*^2 \mid \psi_1, \nu_1)$, respectively.

Proof

The second assertion is immediate from the assumed Normal-inverted gamma prior df and (14.6.13), while the third is immediate from (14.6.12). By (13.5.13), the likelihood function may be written as

$$L(\boldsymbol{\mu}, \sigma_*^2 \mid n, \bar{\mathbf{x}}, s_*^2) = K(\sigma_*^2)^{-\frac{1}{2}kn} \exp[-\tfrac{1}{2}k(n-1)s_*^2/\sigma_*^2] \cdot \exp[-\tfrac{1}{2}n[(\bar{\mathbf{x}}-\boldsymbol{\mu})'\boldsymbol{\delta}_*^{-2}(\bar{\mathbf{x}}-\boldsymbol{\mu})]/\sigma_*^2].$$

By (14.6.12), (12.7.2), and (12.2.10), the prior df $f_{Ni\gamma}^{(k)}(\boldsymbol{\mu}, \sigma_*^2 \mid \boldsymbol{\mu}_0, \psi_0, \boldsymbol{\delta}_{0*}^2, \nu_0)$ can be written as

$$K'\{(\sigma_*^2)^{-\frac{1}{2}k}|\boldsymbol{\delta}_{0*}^2|^{-\frac{1}{2}} \exp[-\tfrac{1}{2}[(\boldsymbol{\mu}-\boldsymbol{\mu}_0)'\boldsymbol{\delta}_{0*}^{-2}(\boldsymbol{\mu}-\boldsymbol{\mu}_0)]/\sigma_*^2]\} \{(\sigma_*^2)^{-\frac{1}{2}\nu_0-1} \exp[-\tfrac{1}{2}\nu_0\psi_0/\sigma_*^2]\}.$$

Hence the posterior df is proportional to

$$K''(\sigma_*^2)^{-\frac{1}{2}k} \exp\{-\tfrac{1}{2}[(\boldsymbol{\mu}-\boldsymbol{\mu}_0)'\boldsymbol{\delta}_{0*}^{-2}(\boldsymbol{\mu}-\boldsymbol{\mu}_0) + n(\bar{\mathbf{x}}-\boldsymbol{\mu})'\boldsymbol{\delta}_*^{-2}(\bar{\mathbf{x}}-\boldsymbol{\mu})]/\sigma_*^2\} \cdot(\sigma_*^2)^{-\frac{1}{2}(\nu_0+nk)-1} \exp[-\tfrac{1}{2}[\nu_0\psi_0 + k(n-1)s_*^2]/\sigma_*^2].$$

But by the proofs of (14.6.1) and (14.6.2) (with asterisked subscripts),

$$(\boldsymbol{\mu}-\boldsymbol{\mu}_0)'\boldsymbol{\delta}_{0*}^{-2}(\boldsymbol{\mu}-\boldsymbol{\mu}_0) + n(\bar{\mathbf{x}}-\boldsymbol{\mu})'\boldsymbol{\delta}_*^{-2}(\bar{\mathbf{x}}-\boldsymbol{\mu}) = (\boldsymbol{\mu}-\boldsymbol{\mu}_1)'\boldsymbol{\delta}_{1*}^{-2}(\boldsymbol{\mu}-\boldsymbol{\mu}_1) + (\bar{\mathbf{x}}-\boldsymbol{\mu}_0)'(\boldsymbol{\delta}_{0*}^2 + n^{-1}\boldsymbol{\delta}_*^2)^{-1}(\bar{\mathbf{x}}-\boldsymbol{\mu}_0),$$

and hence the preceding product $f(\bar{\mathbf{x}}, s_*^2 \mid \boldsymbol{\mu}, \sigma_*^2)f(\boldsymbol{\mu}, \sigma_*^2)$ can be rearranged as

$$K''\{(\sigma_*^2)^{-\frac{1}{2}k} \exp[-\tfrac{1}{2}[(\boldsymbol{\mu}-\boldsymbol{\mu}_1)'\boldsymbol{\delta}_1^{-2}(\boldsymbol{\mu}-\boldsymbol{\mu}_1)]/\sigma_*^2]\} \{(\sigma_*^2)^{-\frac{1}{2}\nu_1-1} \exp[-\tfrac{1}{2}\nu_1\psi_1/\sigma_*^2]\},$$

which is proportional to the asserted posterior df. QED

For Case E: under the assumptions of (14.6.14) and given n, (14.6.15)

(a) the df $f(\bar{\mathbf{x}} \mid s_*^2)$ of $\tilde{\bar{\mathbf{x}}}$ conditional upon s_*^2 but unconditional as regards $(\tilde{\boldsymbol{\mu}}, \tilde{\sigma}_*^2)$ is given by

$$f(\bar{\mathbf{x}} \mid s_*^2) = f_S{}^{(k)}(\bar{\mathbf{x}} \mid \boldsymbol{\mu}_0, \boldsymbol{\psi}_*, \nu_0 + k(n-1)),$$

where

$$\boldsymbol{\psi}_* \equiv \left\{\frac{\nu_0\psi_0 + k(n-1)s_*^2}{\nu_0 + k(n-1)}\right\}(\boldsymbol{\delta}_{0*}^2 + n^{-1}\boldsymbol{\delta}_*^2);$$

(b) the unconditional df $f(\bar{\mathbf{x}})$ of $\tilde{\bar{\mathbf{x}}}$ is given by

$$f(\bar{\mathbf{x}}) = f_S{}^{(k)}(\bar{\mathbf{x}} \mid \boldsymbol{\mu}_0, \psi_0(\boldsymbol{\delta}_{0*}^2 + n^{-1}\boldsymbol{\delta}_*^2), \nu_0);$$

(c) the df $f(s_*^2 \mid \bar{\mathbf{x}})$ of $\tilde{s}_*^2$ conditional upon $\bar{\mathbf{x}}$ but unconditional as regards $(\tilde{\boldsymbol{\mu}}, \tilde{\sigma}_*^2)$ is given by

$$f(s_*^2 \mid \bar{\mathbf{x}}) = f_{i\beta 2}(s_*^2 \mid \tfrac{1}{2}k(n-1), \tfrac{1}{2}(\nu_0 + kn), b),$$

where

$$b \equiv [k(n-1)]^{-1}[\nu_0\psi_0 + (\bar{\mathbf{x}} - \boldsymbol{\mu}_0)'(\boldsymbol{\sigma}_{0*}^2 + n^{-1}\boldsymbol{\sigma}_*^2)^{-1}(\bar{\mathbf{x}} - \boldsymbol{\mu}_0)]; \text{ and}$$

(d) the unconditional df $f(s_*^2)$ of $\tilde{s}_*^2$ is given by

$$f(s_*^2) = f_{i\beta 2}(s_*^2 \mid \tfrac{1}{2}k(n-1), \tfrac{1}{2}[\nu_0 + k(n-1)], [k(n-1)]^{-1}\nu_0\psi_0).$$

Proof

Exercise 6.

Example 14.6.4

A stamping machine produces supposedly square figures, with longrun average length μ_1 and width μ_2, but unknown beforehand. Let $\boldsymbol{\mu} \equiv (\mu_1, \mu_2)'$. Each stamping has measurements $(\tilde{x}_{1i}, \tilde{x}_{2i})' \equiv \tilde{\mathbf{x}}_i$, which are iid with common df $f_N{}^{(2)}(\mathbf{x} \mid \boldsymbol{\mu}, \boldsymbol{\sigma}^2)$, where

$$\boldsymbol{\sigma}^2 = \sigma_*^2 \begin{bmatrix} 1 & 0 \\ 0 & 1 \end{bmatrix}.$$

That is, the length and width are independent and have equal variances, conditional upon $\boldsymbol{\mu}$. Suppose that the production manager's judgments about $(\tilde{\boldsymbol{\mu}}, \tilde{\sigma}_*^2)$ are expressed via the prior df $f_{Ni\gamma}{}^{(2)}(\boldsymbol{\mu}, \sigma_*^2 \mid \boldsymbol{\mu}_0, \psi_0, \boldsymbol{\sigma}_{0*}^2, \nu_0)$ for $\boldsymbol{\mu}_0 \equiv (30, 30)'$, $\psi_0 = .05$, $\nu_0 = 7$, and

$$\boldsymbol{\sigma}_{0*}^2 = \begin{bmatrix} 1.00 & .90 \\ .90 & 1.10 \end{bmatrix}.$$

If ten stampings are produced and measured, yielding the sufficient statistic $(n, \bar{\mathbf{x}}, s_*^2) = (10, (24, 26)', .15)$, then the production manager's posterior df of $(\tilde{\boldsymbol{\mu}}, \tilde{\sigma}_*^2)$ is $f_{Ni\gamma}{}^{(2)}(\boldsymbol{\mu}, \sigma_*^2 \mid \boldsymbol{\mu}_1, \psi_1, \boldsymbol{\sigma}_{1*}^2, \nu_1)$, where

$$\nu_1 = 7 + 2(10) = 27;$$

$$\begin{aligned}
\boldsymbol{\sigma}_{1*}^{-2} &= \boldsymbol{\sigma}_{0*}^{-2} + n\boldsymbol{\sigma}_*^{-2} \\
&= \begin{bmatrix} \dfrac{1.10}{.29} & \dfrac{-.90}{.29} \\[2ex] \dfrac{-.90}{.29} & \dfrac{1.00}{.29} \end{bmatrix} + \begin{bmatrix} n & 0 \\ 0 & n \end{bmatrix} \\
&= \begin{bmatrix} \dfrac{4.00}{.29} & \dfrac{-.90}{.29} \\[2ex] \dfrac{-.90}{.29} & \dfrac{3.90}{.29} \end{bmatrix}
\end{aligned}$$

implying that

$$\tilde{\sigma}_{1*}{}^{2} = \begin{bmatrix} .0765 & .0176 \\ .0176 & .0784 \end{bmatrix};$$

$$\begin{aligned}
\boldsymbol{\mu}_1 &= \tilde{\sigma}_{1*}{}^{2}\left\{\begin{bmatrix} \dfrac{1.10}{.29} & \dfrac{-.90}{.29} \\ \dfrac{-.90}{.29} & \dfrac{1.00}{.29} \end{bmatrix}\begin{pmatrix} 30 \\ 30 \end{pmatrix} + \begin{bmatrix} 10 & 0 \\ 0 & 10 \end{bmatrix}\begin{pmatrix} 24 \\ 26 \end{pmatrix}\right\} \\
&= \begin{bmatrix} .0765 & .0176 \\ .0176 & .0784 \end{bmatrix}\begin{pmatrix} \dfrac{75.60}{.29} \\ \dfrac{78.40}{.29} \end{pmatrix} \\
&= \begin{pmatrix} \dfrac{5.78340}{.29} + \dfrac{1.37984}{.29} \\ \dfrac{1.33056}{.29} + \dfrac{6.14656}{.29} \end{pmatrix} \\
&= \begin{pmatrix} \dfrac{7.16324}{.29} \\ \dfrac{7.47712}{.29} \end{pmatrix} \\
&\doteq \begin{pmatrix} 24.70 \\ 25.78 \end{pmatrix};
\end{aligned}$$

and

$$\begin{aligned}
\psi_1 &= (27)^{-1}\left\{(7)(.05) + (2)(9)(.15) + (-6, -4)\begin{bmatrix} 1.10 & .90 \\ .90 & 1.20 \end{bmatrix}^{-1}\begin{pmatrix} -6 \\ -4 \end{pmatrix}\right\} \\
&= (27)^{-1}\{(7)(.05) + (2)(9)(.15) + 34.51\} \\
&= 1.39.
\end{aligned}$$

14.7 INDEPENDENT UNIVARIATE NORMAL PROCESSES RELATED BY THE PRIOR df

In a number of important inference and decision problems, the decision-maker is interested in the parameter $\boldsymbol{\theta}_i$ of each of k data-generating processes i, and he can obtain n_i observations $\tilde{\mathbf{x}}_{ij}$ from process i. For example, in 13.7 we introduced the problem of making inferences about the relationship between Normal-process mean vectors $\boldsymbol{\mu}_1$ and $\boldsymbol{\mu}_2$ based upon n_i observations from process $i (i = 1, 2)$.

In what follows we shall assume that the observations $\tilde{\mathbf{x}}_{1i}, \ldots, \tilde{\mathbf{x}}_{n_i i}$ are iid with common dgf $f_i(\mathbf{x}_i \mid \boldsymbol{\theta}_i)$, and that the experimentation on different processes is such that the experiments are independent, in the sense that $\tilde{\mathbf{x}}_{11}, \ldots, \tilde{\mathbf{x}}_{n_1 i}, \tilde{\mathbf{x}}_{12}, \ldots, \tilde{\mathbf{x}}_{n_2 2}, \ldots, \tilde{\mathbf{x}}_{1k}, \ldots, \tilde{\mathbf{x}}_{n_k k}$ are mutually independent.

Let $\tilde{\mathbf{x}}_{Ti}$ be a sufficient statistic for the n_i observations on process i. From the preceding paragraph it is clear that $\tilde{\mathbf{x}}_{T1}, \ldots, \tilde{\mathbf{x}}_{Tk}$ are independent given $\boldsymbol{\theta} \equiv (\boldsymbol{\theta}'_1, \ldots, \boldsymbol{\theta}'_k)'$, and hence that the joint likelihood function of $\boldsymbol{\theta}_1, \ldots, \boldsymbol{\theta}_k$ is the product of the individual likelihood functions $L(\boldsymbol{\theta}_i \mid \mathbf{x}_{Ti})$:

$$L(\boldsymbol{\theta}_1, \ldots, \boldsymbol{\theta}_k \mid \mathbf{x}_{T1}, \ldots, \mathbf{x}_{Tk}) = \prod_{i=1}^{k} L(\boldsymbol{\theta}_i \mid \mathbf{x}_{Ti}). \tag{14.7.1}$$

Now, (14.7.1) furnishes a good foundation for natural conjugacy analysis, since by virtue of Bayes' Theorem any joint prior dgf $f(\boldsymbol{\theta}_1, \ldots, \boldsymbol{\theta}_k)$ of $\tilde{\boldsymbol{\theta}}_1, \ldots, \tilde{\boldsymbol{\theta}}_k$ induces a posterior dgf

$$f(\boldsymbol{\theta}_1, \ldots, \boldsymbol{\theta}_k \mid \mathbf{x}_{T1}, \ldots, \mathbf{x}_{Tk}) = f(\boldsymbol{\theta}_1, \ldots, \boldsymbol{\theta}_k) \prod_{i=1}^{k} K_i(\mathbf{x}_{Ti}) L(\boldsymbol{\theta}_i \mid \mathbf{x}_{Ti}), \tag{14.7.2}$$

where $K_i(\mathbf{x}_{Ti}) > 0$ for all $\mathbf{x}_{Ti}$ and all i.

Furthermore, it is clear from (14.7.1) and (14.7.2) that if the decision-maker deems $\tilde{\boldsymbol{\theta}}_1, \ldots, \tilde{\boldsymbol{\theta}}_k$ independent a priori, so that

$$f(\boldsymbol{\theta}_1, \ldots, \boldsymbol{\theta}_k) = \prod_{i=1}^{k} f(\boldsymbol{\theta}_i),$$

then his posterior dgf will preserve this judgment, because

$$\begin{aligned} f(\boldsymbol{\theta}_1, \ldots, \boldsymbol{\theta}_k \mid \mathbf{x}_{T1}, \ldots, \mathbf{x}_{Tk}) &= \left[\prod_{i=1}^{k} f(\boldsymbol{\theta}_i)\right]\left[\prod_{i=1}^{k} K_i(\mathbf{x}_{Ti}) L(\boldsymbol{\theta}_i \mid \mathbf{x}_{Ti})\right] \\ &= \prod_{i=1}^{k} [f(\boldsymbol{\theta}_i) K_i(\mathbf{x}_{Ti}) L(\boldsymbol{\theta}_i \mid \mathbf{x}_{Ti})] \\ &= \prod_{i=1}^{k} f(\boldsymbol{\theta}_i \mid \mathbf{x}_{Ti}). \end{aligned}$$

On the other hand, if $\tilde{\boldsymbol{\theta}}_1, \ldots, \tilde{\boldsymbol{\theta}}_k$ are a priori dependent, then they will be a posteriori dependent, given $\mathbf{x}_{T1}, \ldots, \mathbf{x}_{Tk}$ despite the independence of $\tilde{\mathbf{x}}_{T1}, \ldots, \tilde{\mathbf{x}}_{Tk}$.

We have shown that the *general* Bayesian analysis of observations from different processes whose parameters are related by a joint prior dgf is perfectly straightforward. We stress the word *general*, because it is not at all obvious that *flexible* and *tractable* natural conjugate prior dgf's for $\tilde{\boldsymbol{\theta}} \equiv (\tilde{\boldsymbol{\theta}}'_1, \ldots, \tilde{\boldsymbol{\theta}}'_k)'$ can be found.

However, when each $\tilde{\boldsymbol{\theta}}_i$ is the pair $(\mu_i, \sigma_i^2)'$ of parameters of a univariate normal process, natural conjugate prior df's are easy to find under some restrictions involving the degree of prior knowledge about $\{\sigma_1^2, \ldots, \sigma_k^2\}$. They are of precisely the same form as for Cases A and E of Section 14.6.

First, suppose that every process variance σ_i^2 is known. Then $(n_i, \tilde{\bar{x}}_i)$ is conditionally sufficient for μ_i, for each i; and, by (13.2.5), (14.7.1), and (12.7.2), the likelihood function of $\boldsymbol{\mu} \equiv (\mu_1, \ldots \mu_k)'$ may be written as

$$L(\boldsymbol{\mu} \mid \mathbf{n}, \bar{\mathbf{x}}, \boldsymbol{\sigma}^2(\mathbf{n})) = f_N^{(k)}(\bar{\mathbf{x}} \mid \boldsymbol{\mu}, \boldsymbol{\sigma}^2(\mathbf{n})), \tag{14.7.3}$$

where

$$\bar{\mathbf{x}} \equiv (\bar{\mathbf{x}}_1, \ldots, \bar{\mathbf{x}}_k)' \tag{14.7.4}$$

$$\mathbf{n} \equiv (n_1, \ldots, n_k)', \tag{14.7.5}$$

and

$$\boldsymbol{\sigma}^2(\mathbf{n}) \equiv \begin{bmatrix} n_1^{-1}\sigma_1^2 & & & 0 \\ & \cdot & & \\ & & \cdot & \\ & & & \cdot \\ 0 & & & n_k^{-1}\sigma_k^2 \end{bmatrix}. \tag{14.7.6}$$

The reader may verify this assertion as Exercise 7. Because all σ_i^2's are known beforehand, so is $\boldsymbol{\sigma}^2(\mathbf{n})$, and by (14.7.3) and (14.6.1), it should follow that a natural conjugate prior df for $\tilde{\boldsymbol{\mu}}$ is k-variate Normal.

If an experiment consists in obtaining $n_i \geqq 1$ iid observations with common df $f_N(x_i \mid \mu_i, \sigma_i^2)$ for $i = 1, \ldots, k$, if all σ_i^2's are known beforehand, and if $\tilde{\boldsymbol{\mu}} \equiv (\tilde{\mu}_1, \ldots, \tilde{\mu}_k)'$ has the natural conjugate prior df $f_N^{(k)}(\boldsymbol{\mu} \mid \boldsymbol{\mu}_0, \boldsymbol{\sigma}_0^2)$, then the posterior df of $\tilde{\boldsymbol{\mu}}$ is $f_N^{(k)}(\boldsymbol{\mu} \mid \boldsymbol{\mu}_1, \boldsymbol{\sigma}_1^2)$, where (14.7.7)

$$\boldsymbol{\sigma}_1^2 \equiv [\boldsymbol{\sigma}_0^{-2} + \boldsymbol{\sigma}^{-2}(\mathbf{n})]^{-1},$$

$$\boldsymbol{\mu}_1 \equiv \boldsymbol{\sigma}_1^2[\boldsymbol{\sigma}_0^{-2}\boldsymbol{\mu}_0 + \boldsymbol{\sigma}^{-2}(\mathbf{n})\bar{\mathbf{x}}],$$

and $\boldsymbol{\sigma}^{-2}(\mathbf{n}) = [\boldsymbol{\sigma}^2(\mathbf{n})]^{-1}$ for $\boldsymbol{\sigma}^2(\mathbf{n})$ as defined by (14.7.6).

Proof

Exercise 8.

Under the assumptions of (14.7.7) and given $\mathbf{n}$, the unconditional df of $\tilde{\bar{\mathbf{x}}}$ is $f_N^{(k)}(\bar{\mathbf{x}} \mid \boldsymbol{\mu}_0, \boldsymbol{\sigma}_0^2 + \boldsymbol{\sigma}^2(\mathbf{n}))$. (14.7.8)

Proof

Exercise 9.

Second, suppose that the ratios σ_i^2/σ_j^2 of all pairs of process variances are known. This supposition is tantamount to assuming that

$$\boldsymbol{\sigma}^2(\mathbf{1}) \equiv \begin{bmatrix} \sigma_1^2 & & & 0 \\ & \cdot & & \\ & & \cdot & \\ 0 & & & \cdot \\ & & & \sigma_k^2 \end{bmatrix}$$

is known up to a positive multiple. If we define σ_*^2 by

$$\sigma_*^2 \equiv \left(\prod_{i=1}^{k} \sigma_i^2 \right)^{1/k}, \qquad (14.7.9)$$

and $\boldsymbol{\sigma}_*^2(\mathbf{1})$ by

$$\boldsymbol{\sigma}_*^2(\mathbf{1}) \equiv \sigma_*^{-2}\boldsymbol{\sigma}^2(\mathbf{1}), \qquad (14.7.10a)$$

it follows easily that $|\boldsymbol{\sigma}_*^2(\mathbf{1})| = 1$. Similarly, we define $\boldsymbol{\sigma}_*^2(\mathbf{n})$ by

$$\boldsymbol{\sigma}_*^2(\mathbf{n}) \equiv \sigma_*^{-2}\boldsymbol{\sigma}^2(\mathbf{n}), \qquad (14.7.10b)$$

for $\boldsymbol{\sigma}^2(\mathbf{n})$ as defined by (14.7.6). Given $\mathbf{n}$, $\boldsymbol{\sigma}_*^2(\mathbf{n})$ (but not σ_*^2) is known under our current assumptions.

Reasoning as for the case in which σ_*^2 is known, we deduce (Exercise 10) that if every $n_i \geqq 1$, the likelihood function of $(\boldsymbol{\mu}, \sigma_*^2)$ may be written as

$$L(\boldsymbol{\mu}, \sigma_*^2 \mid \mathbf{n}, \bar{\mathbf{x}}, s_*^2) = f_N^{(k)}(\bar{\mathbf{x}} \mid \boldsymbol{\mu}, \sigma_*^2\boldsymbol{\sigma}_*^2(\mathbf{n}))\, f_{\gamma 2}(s_*^2 \mid 1/\sigma_*^2, \nu), \qquad (14.7.11)$$

where

$$\nu \equiv \sum_{i=1}^{k} (n_i - 1) \qquad (14.7.12)$$

and

$$s_*^2 \equiv \nu^{-1} \sum_{i=1}^{k} \frac{(n_i - 1)s_i^2}{\sigma_{*i}^2}, \qquad (14.7.13)$$

for s_i^2 as defined for process i by (13.2.2) and σ_{*i}^2 the ith diagonal element of $\boldsymbol{\sigma}_*^2(\mathbf{1})$.

If an experiment consists in obtaining n_i iid observations with common df $f_N(x_i \mid \mu_i, \sigma_i^2)$, for $i = 1, \ldots, k$, if all ratios σ_i^2/σ_j^2 are known beforehand, and if $(\tilde{\mathbf{u}}, \tilde{\sigma}_*^2)$ has the natural conjugate prior df $f_{Ni\gamma}^{(k)}(\mathbf{u}, \sigma_*^2 \mid \mathbf{u}_0, \psi_0, \boldsymbol{\sigma}_{0*}^2, \nu_0)$, then the posterior df of $(\tilde{\mathbf{u}}, \tilde{\sigma}_*^2)$ is $f_{Ni\gamma}^{(k)}(\mathbf{u}, \sigma_*^2 \mid \mathbf{u}_1, \psi_1, \boldsymbol{\sigma}_{1*}^2, \nu_1)$, where (14.7.14)

$$\boldsymbol{\sigma}_{1*}^2 \equiv [\boldsymbol{\sigma}_{0*}^{-2} + \boldsymbol{\sigma}_*^{-2}(\mathbf{n})]^{-1},$$

$$\mathbf{u}_1 \equiv \boldsymbol{\sigma}_{1*}^2[\boldsymbol{\sigma}_{0*}^{-2}\mathbf{u}_0 + \boldsymbol{\sigma}_*^{-2}(\mathbf{n})\bar{\mathbf{x}}].$$

$$\nu_1 \equiv \nu_0 + \sum_{i=1}^{k} n_i,$$

and

$$\psi_1 \equiv \nu_1^{-1}[\nu_0\psi_0 + \nu s_*^2 + (\bar{\mathbf{x}} - \mathbf{u}_0)'[\boldsymbol{\sigma}_{0*}^2 + \boldsymbol{\sigma}_*^2(\mathbf{n}]^{-1}(\bar{\mathbf{x}} - \mathbf{u}_0)].$$

The marginal prior and posterior df's of $\tilde{\mathbf{u}}$ are $f_{S^{(k)}}(\mathbf{u}_0 \mid \mathbf{u}_0, \psi_0\boldsymbol{\sigma}_{0*}^2, \nu_0)$ and $f_{S^{(k)}}(\mathbf{u} \mid \mathbf{u}_1, \psi_1\boldsymbol{\sigma}_{1*}^2(\mathbf{n}), \nu_1)$, respectively. The marginal prior and posterior df's of $\tilde{\sigma}_*^2$ are $f_{i\gamma}(\sigma_*^2 \mid \psi_0, \nu_0)$ and $f_{i\gamma}(\sigma_*^2 \mid \psi_1, \nu_1)$, respectively.

Proof

Exercise 11.

The reader may verify that (14.6.14) continues to obtain, but with $\nu \equiv \sum_{i=1}^{k} n_i - k$ replacing $k(n-1)$, ν_1 replacing $\nu_0 + kn$, and $\boldsymbol{\sigma}_*^2(\mathbf{n})$ replacing $n^{-1}\boldsymbol{\sigma}_*^2$ everywhere.

EXERCISES

1. Derive (14.2.5) and (14.2.6).
2. Derive (14.2.8) and (14.2.9).
3. Derive (14.5.8).
4. One observation $\mathbf{x} = (2, 0)'$ is obtained from a bivariate Normal process with known covariance matrix

$$\boldsymbol{\sigma}^2 = \begin{bmatrix} .01 & .90 \\ .90 & 100 \end{bmatrix}.$$

The process mean vector $\tilde{\mathbf{u}}$ is assigned a bivariate Normal prior df with parameters $\mathbf{u}_0 = (1, 1)'$ and

$$\boldsymbol{\sigma}_0^2 = \begin{bmatrix} 1 & .90 \\ .90 & 1 \end{bmatrix}.$$

Derive the parameters of the posterior df, and show that the second component of $\mathbf{u}_1$ fails to lie on the line segment joining the second components of $\mathbf{u}_0$ and $\mathbf{x}$.

5. Prove (**) in the proof of (14.6.8).

6. Prove (14.6.15).

7. Verify (14.7.3).

8. Prove (14.7.7).

9. Prove (14.7.8).

10. Verify (14.7.11).

11. Prove (14.7.14).

12. *Obtaining Information about* $\tilde{\sigma}^2$ *without Considering* μ: Suppose in the context of Case C in Section 14.5, that interest centers solely upon $\tilde{\sigma}^2$. Then assessing a joint prior df for $\tilde{\mu}$ and $\tilde{\sigma}^2$ is somewhat annoying, especially if the decision-maker's opinions about $\tilde{\mu}$ are very vague in the sense of being very easily changed by a few observations. There is a result which shows that opinions about $\tilde{\sigma}^2$ can be revised without any consideration of μ at all. Prove the following:

If an experiment consists in obtaining $n \geqq 2$ iid observations from $f_N(x \mid \mu, \sigma^2)$ with noninformative stopping and yields the statistic (n, s^2), and if the process variance $\tilde{\sigma}^2$ has the natural conjugate prior df $f_{i\gamma}(\sigma^2 \mid \psi_0, \nu_0)$, then the posterior df of $\tilde{\sigma}^2$ is $f_{i\gamma}(\sigma^2 \mid \psi_1, \nu_1)$, where (14.8.1)

$$\nu_1 \equiv \nu_0 + n - 1$$

and

$$\psi_1 \equiv \nu_1^{-1}[\nu_0\psi_0 + (n-1)s^2];$$

and

under the assumptions and definitions of (14.8.1) and given n, the unconditional df of $\tilde{s}^2$ is $f_{i\beta 2}(s^2 \mid \tfrac{1}{2}(n-1), \tfrac{1}{2}\nu_1, \nu_0\psi_0/(n-1))$. (14.8.2)

13. *k-variate Normal Generalizations of Exercise* 12: This exercise develops generalizations of (14.8.1) and (14.8.2) appropriate to Cases C and E of Section 14.6.

(a) Prove the following:

If an experiment consists in obtaining $n \geqq k+1$ iid observations from $f_N^{(k)}(\mathbf{x} \mid \boldsymbol{\mu}, \boldsymbol{\sigma}^2)$ with noninformative stopping and yields the statistic $(n, \tilde{\boldsymbol{\sigma}}^2)$, and if the process covariance matrix $\tilde{\boldsymbol{\sigma}}^2$ has the natural conjugate prior df $f_{iW}^{(k)}(\boldsymbol{\sigma}^2 \mid \boldsymbol{\psi}_0, \nu_0)$, then the posterior df of $\tilde{\boldsymbol{\sigma}}^2$ is $f_{iW}^{(k)}(\boldsymbol{\sigma}^2 \mid \boldsymbol{\psi}_1, \nu_1)$, where (14.8.3)

$$\nu_1 \equiv \nu_0 + n - 1$$

and

$$\boldsymbol{\psi}_1 \equiv \nu_1^{-1}[\nu_0\boldsymbol{\psi}_0 + (n-1)\mathbf{s}^2];$$

and

under the assumptions and definitions of (14.8.3) and given n, the unconditional df of $\tilde{\mathbf{s}}^2$ is (14.8.4)

$$f_{i\beta 2}^{(k)}(\mathbf{s}^2 \mid \boldsymbol{\psi}_0, \nu_0, n-k, n-1).$$

(b) Prove the following:

If an experiment consists in obtaining $n \geqq 2$ iid observations from $f_N{}^{(k)}(\mathbf{x} \mid \boldsymbol{\mu}, \boldsymbol{\sigma}^2)$ with noninformative stopping and yields the statistic $(n, s_*{}^2)$, if $\boldsymbol{\sigma}_*{}^2$ is known, and if $\tilde{\sigma}_*{}^2$ has the natural conjugate prior df $f_{i\gamma}(\sigma_*{}^2 \mid \psi_0, \nu_0)$, then the posterior df of $\tilde{\sigma}_*{}^2$ is $f_{i\gamma}(\sigma_*{}^2 \mid \psi_1, \nu_1)$, where (14.8.5)

$$\nu_1 \equiv \nu_0 + k(n-1)$$

and

$$\psi_1 \equiv \nu_1^{-1}[\nu_0\psi_0 + k(n-1)s_*{}^2];$$

and

under the assumptions and definitions of (14.8.5) and given n, the unconditional df of $\tilde{s}_*{}^2$ is (14.8.6)

$$f_{i\beta 2}\left(s_*{}^2 \mid \tfrac{1}{2}k(n-1), \tfrac{1}{2}\nu_1, \frac{\nu_0\psi_0}{k(n-1)}\right).$$

(c) Why are (14.8.6), (14.8.4), and (14.8.2) identical to part (d) of (14.6.15), (14.6.9), and (14.5.8), respectively?

14. *Inverse Observation on the Erlang Process*: The Erlang process was introduced in Exercise 11.8. Prove the following:

If observation on an Erlang process is commenced immediately after an occurrence and terminated immediately after the rth subsequent occurrence, and if $\tilde{\lambda}$ has the natural conjugate prior df $f_\gamma(\lambda \mid \rho, \tau)$, then the posterior df of $\tilde{\lambda}$ is $f_\gamma(\lambda \mid \rho + r^*r, \tau + t)$, where t denotes total observation time (and $r^* = 1$ if the process is Poisson); (14.8.7)

and

under the assumptions of (14.8.7), the unconditional df of $\tilde{t}$ is (14.8.8)

$$f_{i\beta 2}(t \mid r^*r, \rho + r^*r, \tau).$$

15. *Review and Extension of Example* (8.7.1): We say that a continuous random variable $\tilde{x}$ has the *uniform distribution* with parameters $\upsilon_1 \in R^1$ and $\upsilon_2 > \upsilon_1$ if the df of $\tilde{x}$ is $f_u(x \mid \upsilon_1, \upsilon_2)$, defined by (14.8.9)

$$f_u(x \mid \upsilon_1, \upsilon_2) = \begin{cases} (\upsilon_2 - \upsilon_1)^{-1}, & x \in (\upsilon_1, \upsilon_2) \\ 0, & x \notin (\upsilon_1, \upsilon_2). \end{cases}$$

$$= (\upsilon_2 - \upsilon_1)^{-1}\delta(x - \upsilon_1)\delta(\upsilon_2 - x), \; x \in R^1,$$

where

$$\delta(z) \equiv \begin{cases} 1, & z > 0 \\ 0, & z \leqq 0. \end{cases} \qquad (14.8.10)$$

(The delta function is a convenient means of expressing discontinuous functions by a one-line formula.) The *uniform* (data-generating) *process* produces iid observations $\tilde{x}_1, \tilde{x}_2, \ldots$ with common df $f_u(x \mid \upsilon_1, \upsilon_2)$.

(a) Prove the following:

If an experiment consists in obtaining n iid observations from $f_u(x \mid \upsilon_1, \upsilon_2)$ with noninformative stopping, then the likelihood function is (14.8.11)

$$L(\upsilon_1, \upsilon_2 \mid n, x_1, \ldots, x_n) = f(x_1, \ldots, x_n \mid n, \upsilon_1, \upsilon_2)$$
$$= (\upsilon_2 - \upsilon_1)^{-n}\delta(m - \upsilon_1)\delta(\upsilon_2 - M),$$

where

$$M \equiv \max\{x_1, \ldots, x_n\}$$

and

$$m \equiv \min\{x_1, \ldots, x_n\}.$$

The triple $(n, \tilde{m}, \tilde{M})$ is a sufficient statistic for (υ_1, υ_2); $(n, \tilde{M})$ is conditionally sufficient for υ_2 given υ_1; and $(n, \tilde{m})$ is conditionally sufficient for υ_1 given υ_2.

(b) Show that the joint df of $\tilde{m}$ and $\tilde{M}$ given $(n, \upsilon_1, \upsilon_2)$ is

$$f(m, M \mid n, \upsilon_1, \upsilon_2)$$
$$= n(n-1)(M-m)^{n-2}(\upsilon_2 - \upsilon_1)^{-n}\delta(m - \upsilon_1)\delta(\upsilon_2 - M)\delta(M - m). \tag{14.8.12}$$

[*Hint*: *First* show that if the observations $x_1, \ldots, x_n$ are to be rearranged so that $m \equiv x_{(1)} < x_{(2)} < \cdots < x_{(n-1)} < x_{(n)} \equiv M$ (ties have probability zero), then the joint df of $\tilde{x}_{(1)}, \ldots, \tilde{x}_{(n)}$ given $(n, \upsilon_1, \upsilon_2)$ is

$$f(x_{(1)}, \ldots, x_{(n)} \mid n, \upsilon_1, \upsilon_2) = n!\, f(x_1, \ldots, x_n \mid n, \upsilon_1, \upsilon_2).$$

Second, evaluate

$$f(x_{(1)}, x_{(n)} \mid n, \upsilon_1, \upsilon_2) = f(m, M \mid n, \upsilon_1, \upsilon_2)$$

$$= \int_{x_{(1)}}^{x_{(n)}} \int_{x_{(1)}}^{x_{(n-1)}} \cdots \int_{x_{(1)}}^{x_{(4)}} \int_{x_{(1)}}^{x_{(3)}} f(x_{(1)}, \ldots, x_{(n)} \mid n, \upsilon_1, \upsilon_2)dx_{(2)} \ldots dx_{(n-1)}.$$

Further hints: (1) There are $n!$ permutations of the components of $(x_1, \ldots, x_n)$. (2) Evaluate the multiple integral inductively.]

(c) Show that

$$f(m \mid n, \upsilon_1, \upsilon_2) = n(\upsilon_2 - m)^{n-1}(\upsilon_2 - \upsilon_1)^{-n}\delta(m - \upsilon_1)\delta(\upsilon_2 - m)$$
$$= f_{\beta-}(m \mid 1, n+1, \upsilon_1, \upsilon_2) \tag{14.8.13}$$

[compare (12.3.8)]; and

$$f(M \mid n, \upsilon_1, \upsilon_2) = n(M - \upsilon_1)^{n-1}(\upsilon_2 - \upsilon_1)^{-n}\delta(M - \upsilon_1)\delta(\upsilon_2 - M)$$
$$= f_{\beta-}(M \mid n, n+1, \upsilon_1, \upsilon_2). \tag{14.8.14}$$

(d) We define the *uniform complete conjugate* df $f_{ucc}(\upsilon_1, \upsilon_2 \mid m_0, M_0, \nu_0)$ by

$$f_{ucc}(\upsilon_1, \upsilon_2 \mid m_0, M_0, \nu_0) \equiv (\nu_0 - 1)(\nu_0 - 2)(M_0 - m_0)^{\nu_0 - 2}(\upsilon_2 - \upsilon_1)^{-\nu_0} \cdot \delta(\upsilon_2 - M_0)\delta(m_0 - \upsilon_1), \quad (14.8.15)$$

where $-\infty < m_0 < M_0 < \infty$ and $\nu_0 > 2$. Show that

$$f_{ucc}(\upsilon_1, \upsilon_2 \mid m_0, M_0, \nu_0)$$

satisfies (dfg 2), and prove the following:

If an experiment consists in obtaining $n \geqq 2$ iid observations from $f_u(x \mid \upsilon_1, \upsilon_2)$ with noninformative stopping and yields the statistic (n, m, M), and if $(\tilde{\upsilon}_1, \tilde{\upsilon}_2)$ has the natural conjugate prior df $f_{ucc}(\upsilon_1, \upsilon_2 \mid m_0, M_0, \nu_0)$, then the posterior df of $(\tilde{\upsilon}_1, \tilde{\upsilon}_2)$ is $f_{ucc}(\upsilon_1, \upsilon_2 \mid m_1, M_1, \nu_1)$, where (14.8.16)

$$m_1 \equiv \min\{m_0, m\},$$

$$M_1 \equiv \max\{M_0, M\},$$

and

$$\nu_1 \equiv \nu_0 + n;$$

and

under the assumptions and definitions of (14.8.16) and given n, the df $f(m, M \mid n)$ of $(\tilde{m}, \tilde{M})$ [unconditional as regards $(\tilde{\upsilon}_1, \tilde{\upsilon}_2)$] is given by (14.8.17)

$$f(m, M) = K(M - m)^{n-2} \frac{\delta(M - m)}{(M_1 - m_1)^{\nu_1 - 2}},$$

where $K \equiv (\nu_0 - 1)(\nu_0 - 2)n(n - 1)\dfrac{(M_0 - m_0)^{\nu_0 - 2}}{(\nu_1 - 1)(\nu_1 - 2)}$.

(e) We define the *uniform right conjugate* df $f_{urc}(\upsilon_2 \mid m_0, M_0, \nu_0)$ by

$$f_{urc}(\upsilon_2 \mid m_0, M_0, \nu_0) \equiv (\nu_0 - 1)(M_0 - m_0)^{\nu_0 - 1}(\upsilon_2 - m_0)^{-\nu_0}\delta(\upsilon_2 - M_0), \quad (14.8.18)$$

where $-\infty < m_0 < M_0 < \infty$ and $\nu_0 > 1$. Show that

$$F_{urc}(\upsilon_2 \mid m_0, M_0, \nu_0) = \begin{cases} 0, & \upsilon_2 \leqq M_0 \\ 1 - [(M_0 - m_0)/(\upsilon_2 - m_0)]^{\nu_0 - 1}, & \upsilon_2 > M_0 \end{cases}$$

and thus that (dgf 2) obtàins. Then prove the following:

If an experiment consists in obtaining $n \geqq 1$ iid observations from $f_u(x \mid \upsilon_1, \upsilon_2)$ with noninformative stopping, if υ_1 is known beforehand, if the statistic (n, M) is obtained, and if $\tilde{\upsilon}_2$ has the natural conjugate prior df $f_{urc}(\upsilon_2 \mid \upsilon_1, M_0, \nu_0)$, then the posterior df of $\tilde{\upsilon}_2$ is $f_{urc}(\upsilon_2 \mid \upsilon_1, M_1, \nu_1)$, where (14.8.19)

$$\nu_1 \equiv \nu_0 + n$$

and

$$M_1 \equiv \max\{M_0, M\};$$

and

under the assumptions and definitions of (14.8.19) and given n, the df $f(M \mid n)$ of $\tilde{M}$ (unconditional as regards $\tilde{v}_2$) is given by (14.8.20)

$$f(M \mid n) = K(M - v_1)^{n-1}\frac{\delta(M - v_1)}{(M_1 - v_1)^{\nu_1 - 1}},$$

where $K \equiv (\nu_0 - 1)\dfrac{n(M_0 - v_1)^{\nu_0 - 1}}{(\nu_1 - 1)}$.

(f) We define the *uniform left conjugate* df $f_{ulc}(v_1 \mid m_0, M_0, \nu_0)$ by

$$f_{ulc}(v_1 \mid m_0, M_0, \nu_0) = (\nu_0 - 1)(M_0 - m_0)^{\nu_0 - 1}(M_0 - v_1)^{-\nu_0}\delta(m_0 - v_1), \tag{14.8.21}$$

where $-\infty < m_0 < M_0 < \infty$ and $\nu_0 > 1$. Show that

$$F_{ulc}(v_1 \mid m_0, M_0, \nu_0) = \begin{cases} [(M_0 - m_0)/(M_0 - v_1)]^{\nu_0 - 1}, & v_1 < m_0 \\ 1, & v_1 \geqq m_0, \end{cases}$$

and thus that (dgf 2) obtains. Then prove the following:

If an experiment consists in obtaining $n \geqq 1$ iid observations from $f_u(x \mid v_1, v_2)$ with noninformative stopping, if v_2 is known beforehand, if the statistic (n, m) is obtained, and if $\tilde{v}_1$ has the natural conjugate prior df (14.8.22)

$$f_{ulc}(v_1 \mid m_0, v_2, \nu_0),$$

then the posterior df of $\tilde{v}_1$ is $f_{ulc}(v_1 \mid m_1, v_2, \nu_1)$, where

$$\nu_1 = \nu_0 + n$$

and

$$m_1 \equiv \min\{m_0, m\};$$

and

under the assumptions and definitions of (14.8.22) and given n, the df $f(m \mid n)$ of $\tilde{m}$ (unconditional as regards $\tilde{v}_1$) is given by (14.8.23)

$$f(m \mid n) = K(v_2 - m)^{n-1}\frac{\delta(v_2 - m)}{(v_2 - m_1)^{\nu_1 - 1}},$$

where $K \equiv (\nu_0 - 1)\dfrac{n(v_2 - m_0)^{\nu_0 - 1}}{(\nu_1 - 1)}$.

16. *Continuation of 15 and Preparation for 17*: Assume that a random variable $\tilde{y}$ can have such a thing as the *widget density* $f_{wi}(y \mid \theta_1, \ldots, \theta_k)$ with parameters $\theta_1, \ldots, \theta_k$. (Aside from humor, we seek generality: f_{wi} is intended to represent

any "named" univariate df.) We say that $\tilde{z}$ has the (linearly) *transformed widget density* $f_{twi}(z \mid \theta_1, \ldots, \theta_k; b, c)$ with parameters $\theta_1, \ldots, \theta_k$, b, and c if and only if $\tilde{y} \equiv c^{-1}(\tilde{z} - b)$ has the widget density $f_{wi}(y \mid \theta_1, \ldots, \theta_k)$.

(a) Show that

$$E_{twi}(\tilde{z} \mid \theta_1, \ldots, \theta_k; b, c) = b + cE_{wi}(\tilde{y} \mid \theta_1, \ldots, \theta_k) \qquad (14.8.24)$$

and

$$V_{twi}(\tilde{z} \mid \theta_1, \ldots, \theta_k; b, c) = c^2 V_{wi}(\tilde{y} \mid \theta_1, \ldots, \theta_k) \qquad (14.8.25)$$

(b) Apropos of Exercise 15, prove the following:

If $(\tilde{v}_1, \tilde{v}_2)$ has df $f_{ucc}(v_1, v_2 \mid m_0, M_0, \nu_0)$, then: (14.8.26)

(1) the marginal df of $\tilde{v}_1$ is $f_{ulc}(v_1 \mid m_0, M_0, \nu_0 - 1)$;
(2) the conditional df of $\tilde{v}_2$ given v_1 is

$$f_{ti\beta 2}(v_2 \mid 1, \nu_0, M_0 - v_1; M_0, 1);$$

(3) the marginal df of $\tilde{v}_2$ is $f_{urc}(v_2 \mid m_0, M_0, \nu_0 - 1)$;

and

(4) the conditional df of $\tilde{v}_1$ given v_2 is

$$f_{ti\beta 2}(v_1 \mid 1, \nu_0, v_2 - m_0; m_0, -1).$$

17. *The Translated Exponential Process*: The *translated exponential process* produces iid observations $\tilde{x}_1, \tilde{x}_2, \ldots$ with common df (14.8.27)

$$f_{te}(x \mid \lambda; \Omega, 1) \equiv \lambda \exp[-\lambda(x - \Omega)]\delta(x - \Omega),$$

where $\lambda > 0$, $\Omega \in R^1$, and $\delta(\cdot)$ is defined by (14.8.10).

(a) Prove the following:

If an experiment consists in obtaining n iid observations from $f_{te}(x \mid \lambda; \Omega, 1)$ with noninformative stopping, then the likelihood function is (14.8.28)

$$L(\lambda, \Omega \mid n, x_1, \ldots, x_n) = \lambda^n \exp[-\lambda(t - n\Omega)]\delta(m - \Omega),$$

where

$$t \equiv \sum_{i=1}^{n} x_i$$

and

$$m \equiv \min\{x_1, \ldots, x_n\}.$$

The triple $(n, \tilde{m}, \tilde{t})$ is a sufficient statistic for (λ, Ω); $(n, \tilde{m})$ is conditionally sufficient for Ω given λ, and $(n, \tilde{t})$ is conditionally sufficient for λ given Ω.

(b) Prove the following:

Under assumptions and definitions as in (14.8.28), the conditional joint df $f(m, t \mid n, \lambda, \Omega)$ of $\tilde{m}$ and $\tilde{t}$ given n, λ, and Ω is given by

$$f(m, t \mid m, \lambda, \Omega) \tag{14.8.29}$$

$$= n\lambda^n \exp\,[-\lambda(t - n\Omega)]\frac{(t - nm)^{n-2}\delta(t - nm)\delta(m - \Omega)}{\Gamma(n - 1)}.$$

(i) The df $f(m \mid n, \lambda, \Omega)$ of $\tilde{m}$ conditional on (n, λ, Ω) but unconditional as regards $\tilde{t}$ is given by

$$f(m \mid n, \lambda, \Omega) = f_{te}(m \mid n\lambda; \Omega, 1);$$

(ii) the df $f(t \mid m, n, \lambda, \Omega)$ of $\tilde{t}$ conditional upon m, n, λ and Ω is given by

$$f(t \mid m, n, \lambda, \Omega) = f_{t\gamma}(t \mid n - 1, \lambda; nm, 1);$$

(iii) the df $f(t \mid n, \lambda, \Omega)$ of $\tilde{t}$ conditional on (n, λ, Ω) but unconditional as regards $\tilde{m}$ is given by

$$f(t \mid n, \lambda, \Omega) = f_{t\gamma}(t \mid n, \lambda; n\Omega, 1);$$

and

(iv) the df $f(m \mid t, n, \lambda, \Omega)$ of $\tilde{m}$ conditional on t, n, λ and Ω is given by

$$f(m \mid t, n, \lambda, \mho) = f_{t\beta}\left(m \mid n - 1, n, \frac{t}{n}, -n^{-1}[t - n\Omega]\right).$$

[*Hints*: (1) First prove (i) using (8.8.7).
(2) Then prove (ii) by first showing that, conditional upon m, $\tilde{y} \equiv \tilde{t} - nm$ is the sum of $n - 1$ iid random variables with common df $f_e(\cdot \mid \lambda)$.
(3) Using (i) and (ii), find $f(m, t \mid n, \lambda, \Omega)$.
(4) Find

$$f(t \mid n, \lambda, \Omega) = \int_{\Omega}^{t/n} f(m, t \mid n, \lambda, \Omega)dm.$$

(5) Then prove (iv).]

(c) We say that $(\tilde{\lambda}, \tilde{\Omega})$ has the *translated exponential complete conjugate df* $f_{tecc}(\lambda, \Omega \mid m_0, n_0, \rho_0, \tau_0)$ if the df of $(\tilde{\lambda}, \tilde{\Omega})$ is

$$f_{tecc}(\lambda, \Omega \mid m_0, n_0, \rho_0, \tau_0) \equiv f_{te}(\Omega \mid n_0\lambda; m_0, -1)f_\gamma(\lambda \mid \rho_0, \tau_0). \tag{14.8.30}$$

Prove the following:

If $(\tilde{\lambda}, \tilde{\Omega})$ has df $f_{tecc}(\lambda, \Omega \mid m_0, n_0, \rho_0, \tau_0)$, then: (14.8.31)

(1) the marginal df of $\tilde{\lambda}$ is $f_\gamma(\lambda \mid \rho_0, \tau_0)$;
(2) the conditional df of $\tilde{\Omega}$ given λ is $f_{te}(\Omega \mid n_0\lambda; m_0, -1)$ [that is, the df of $m_0 - \tilde{\Omega}$ is $f_e(\cdot \mid n_0\,\lambda)$ given λ];

(3) the marginal df of $\tilde{\Omega}$ is $f_{ti\beta 2}(\Omega \mid 1, \rho_0 + 1, \tau_0/n_0; m_0, -1)$ [that is, the df of $m_0 - \tilde{\Omega}$ is $f_{i\beta 2}(\cdot \mid 1, \rho_0 + 1, \tau_0/n_0)$];

and

(4) the conditional df of $\tilde{\lambda}$ given Ω is $f_\gamma(\lambda \mid \rho_0 + 1, \tau_0 + n_0(m_0 - \Omega))$.

(d) Prove the following:

If an experiment consists in obtaining n iid observations from $f_{te}(x \mid \lambda; \Omega, 1)$ with noninformative stopping, and yields the statistic (n, m, t) and if $(\tilde{\lambda}, \tilde{\Omega})$ has the natural conjugate prior df $f_{tecc}(\lambda, \Omega \mid m_0, n_0, \rho_0, \tau_0)$, then the posterior df of $(\tilde{\lambda}, \tilde{\Omega})$ is $f_{tecc}(\lambda, \Omega \mid m_1, n_1, \rho_1, \tau_1)$, where (14.8.32)

$$n_1 \equiv n_0 + n,$$
$$m_1 \equiv \min\{m_0, m\},$$
$$\rho_1 \equiv \rho_0 + n,$$

and

$$\tau_1 \equiv \tau_0 + t + n_0 m_0 - n_1 m_1;$$

and

under the assumptions and definitions of (14.8.32) and given n, the df $f(m, t \mid n)$ of $(\tilde{m}, \tilde{t})$ [unconditional as regards $(\tilde{\lambda}, \tilde{\Omega})$] is given by (14.8.33)

$$f(m, t \mid n) = K(t - nm)^{n-2} \frac{\delta(t - nm)}{[\tau_0 + t + n_0 m_0 - n_1 m_1]^{\rho_0 + n}},$$

where $K \equiv \dfrac{\tau_0^{\rho_0}(n_0^{-1} + n^{-1})^{-1}}{\Gamma(\rho_0)\Gamma(n - 1)/\Gamma(\rho_0 + n)}$;

(1) the unconditional df $f(m \mid n)$ of $\tilde{m}$ is given by

$$f(m \mid n) = (n_0^{-1} + n^{-1})^{-1} \rho_0 \tau_0^{\rho_0} / [\tau_0 + n_0 m_0 + nm - n_1 m_1]^{\rho_0 + 1}$$
$$= \left(\frac{n_0}{n_0 + n}\right) f_{ti\beta 2}\left(m \mid 1, \rho_0 + 1, \frac{\tau_0}{n}; m_0, 1\right)$$
$$+ \left(\frac{n}{n_0 + n}\right) f_{ti\beta 2}\left(m \mid 1, \rho_0 + 1, \frac{\tau_0}{n}; m_0, -1\right);$$

and

(2) the df $f(t \mid m, n)$ of $\tilde{t}$ conditional on m but unconditional as regards $(\tilde{\lambda}, \tilde{\Omega})$ is given by

$$f(t \mid m, n) = f_{ti\beta 2}(t \mid n - 1, \rho_0 + n, \tau_0 + n_0 m_0 + nm - n_1 m_1; nm, 1).$$

[The df's $f(t \mid n)$ and $f(m \mid t, n)$ are rather cumbersome.]

(e) Prove the following:

If an experiment consists in obtaining n iid observations from $f_{te}(x \mid \lambda; \Omega, 1)$ with noninformative stopping and yields the statistic (n, m), *if* λ *is known beforehand*, and if $\tilde{\Omega}$ has the natural conjugate prior df $f_{te}(\Omega \mid n_0\lambda;\ m_0,\ -1)$, then the posterior df of $\tilde{\Omega}$ is $f_{te}(\Omega \mid n_1\lambda;\ m_1,\ -1)$, where (14.8.34)

$$n_1 \equiv n_0 + n$$

and

$$m_1 \equiv \min\{m_0, m\};$$

and

under the assumptions and definitions of (14.8.34) and given n, the unconditional df $f(m \mid n)$ of $\tilde{m}$ is given by (14.8.35)

$$\begin{aligned} f(m \mid n) &= (n_0^{-1} + n^{-1})\lambda \exp\left[-\lambda(n_0 m_0 + nm - n_1 m_1)\right] \\ &= \left(\frac{n}{n_0 + n}\right) f_{te}(m \mid n_0\lambda;\ m_0,\ -1) \\ &\quad + \left(\frac{n_0}{n_0 + n}\right) f_{te}(m \mid n\lambda;\ m_0,\ 1). \end{aligned}$$

(f) Prove the following:

If an experiment consists in obtaining n iid observations from $f_{te}(x \mid \lambda,\ \Omega,\ 1)$ with noninformative stopping and yields the statistic (n, t), *if Ω is known beforehand*, and if $\tilde{\lambda}$ has the natural conjugate prior df $f_\gamma(\lambda \mid \rho_0,\ \tau_0)$, then the posterior df of $\tilde{\lambda}$ is $f_\gamma(\lambda \mid \rho_1,\ \tau_1)$, where (14.8.36)

$$\rho_1 \equiv \rho_0 + n$$

and

$$\tau_1 \equiv \tau_0 + t - n\Omega;$$

and

under the assumptions and definitions of (14.8.36) and given n, the unconditional df $f(t \mid n)$ of $\tilde{t}$ is given by (14.8.37)

$$f(t \mid n) = f_{ti\beta 2}(t \mid n,\ \rho_0 + n,\ \tau_0;\ n\Omega,\ 1).$$

(g) *The Pareto Process*: We say that $\tilde{y}$ has the *Pareto distribution* with parameters $\lambda > 0$ and $\omega > 0$ if the df of $\tilde{y}$ is

$$f_{Par}(y \mid \lambda,\ \omega) = \frac{\lambda\omega^\lambda \delta(y - \omega)}{y^{\lambda+1}}.$$

Show that $\tilde{y}$ has df $f_{Par}(y \mid \lambda,\ \omega)$ if and only if $\tilde{x} \equiv \log_e \tilde{y}$ has df $f_{te}(x \mid \lambda;\ \Omega, 1)$, where $\Omega \equiv \log_e \omega$, and hence that all previous results can be transformed to apply to the Pareto process, which generates iid observations $\tilde{y}_1,\ \tilde{y}_2,\ \ldots$ with common df $f_{Par}(y \mid \lambda,\ \omega)$.

18. *Introduction to Monte Carlo Simulation*: An actual sequence $\tilde{x}_1,\ \tilde{x}_2,\ \ldots$ of iid random variables with common df $f_u(x \mid 0,\ 1)$ can be generated to any practicably desired number of decimal places on a digital computer, or by

repeated rolling of a twenty-sided die on exactly two sides of which each digit from zero through nine appears, or by reading from a table of random digits.

(a) An actual sequence $\tilde{y}_1, \tilde{y}_2, \ldots$ of iid observations with common *continuous and strictly increasing* cdf $F_y(\cdot)$ can be generated by first generating a sequence $\tilde{x}_1, \tilde{x}_2, \ldots$ of observations from $f_u(\cdot \mid 0, 1)$ and then setting $y(x_i) \equiv y_i = F_y^{-1}(x_i)$, for every i, where F_y^{-1}: $(0, 1) \to R^1$ is the inverse of F_y: $R^1 \to (0, 1)$. For this procedure to be valid, it suffices to prove that $P(\tilde{y}_i \leqq y) = p$ if and only if $P(\tilde{x}_i \leqq p) = p$, which is immediate from the following result, and which the reader is required to prove.

If $\tilde{y}$ has continuous and strictly increasing cdf $F_y(\cdot)$, then $\tilde{x} \equiv F_y(\tilde{y})$ has cdf $F_u(x \mid 0, 1)$ [and df $f_u(x \mid 0, 1)$]. (14.8.38)

(b) To generalize (a), suppose that $\tilde{y}_1, \tilde{y}_2, \ldots$ are continuous random variables, but with common cdf $F_y(\cdot)$ not necessarily strictly increasing. Show that (14.8.38) continues to obtain provided that $y(x)$ is defined by

$$y(x) \equiv \min \{z\colon F_y(z) = x\}.$$

(c) Generalizing (b), suppose that $\tilde{y}_1, \tilde{y}_2, \ldots$ are random variables with common cdf $F_y(\cdot)$, which need not be continuous. (Perhaps $\tilde{y}_1, \tilde{y}_2, \ldots$ are mixed random variables.) Show that (14.8.38) continues to obtain provided that $y(x)$ is defined by $y(x) \equiv \min \{z\colon F_y(z) \geqq x\}$.

(d) A principal use of Monte Carlo simulation is to study the overall behavior of a system, the subsystems of which have known probabilistic behavior; see Hammersley and Handscomb [**35**], Naylor, Balintfy, Burdick, and Chu [**57**], and Tocher [**80**], for example; also see the RAND Corporation table [**65**] of one million random digits. Simulation is essential when the interaction of the subsystems is too complicated for analytic methods. Consider the system in a grocery store comprised of T checkout counters and a special check-cashing authorizer; suppose the checkout clerks are authorized to cash checks for up to twenty-five dollars without further authorization; that $\tilde{n}_i$ denotes the number of checks coming to checkout counter i during one day; that the amounts of checks at counter i are iid random variables $\tilde{y}_{i1}, \tilde{y}_{i2}, \ldots,$ that the special check-cashing authorizer can handle only N checks per day, and that $\tilde{n}_1, \ldots, \tilde{n}_T$ are independent random variables. What is the distribution of the number $\tilde{m}$ of checks authorized by the special authorizer?

(1) Define

$$\begin{aligned} \tilde{x}_{ij} &\equiv \delta(\tilde{y}_{ij} - 25) \\ &= \begin{cases} 1, & \tilde{y}_{ij} > 25 \\ 0, & \tilde{y}_{ij} \leqq 25. \end{cases} \end{aligned}$$

Show that

$$\tilde{m} \equiv \min \{N, \sum_{i=1}^{T} \sum_{j=1}^{\tilde{n}_i} \tilde{x}_{ij}\}.$$

(2) Suppose that N and T are specified and that $F(n_1), \ldots, F(n_T)$ are known, as well as the common cdf $F_i(y_i)$ of the check amounts at each counter i. Show that the following procedure simulates *one* observation from $F_m(m)$:

(i) Simulate n_i from $F(n_i)$ for each $i = 1, \ldots, T$;
(ii) given n_i, simulate n_i observations y_{ij} from $F_i(y_i)$ and record x_{ij} for each y_{ij}, for $i = 1, \ldots, T$;
(iii) compute $\sum_{i=1}^{T} \sum_{j=1}^{n_i} x_{ij} \equiv X$;
(iv) set $m \equiv \min\{N, X\}$.

Repetition of steps (i) through (iv) many times enables us to estimate features of the distribution of $\tilde{m}$; the "observation" from $F_m(m)$ produced by steps (i) through (iv) really *is* an observation in the statistical sense. Here, we perform steps (i) through (iv) repeatedly to obtain independent observations for making inferences about the distribution of m—just as in random sampling from populations (compare Chapter 9). Two such inference procedures are introduced in (3) and (4).

(3) Suppose interest centers on the expected number $\mu \equiv E(\tilde{m})$ of specially authorized checks per day. Since $\tilde{m} \in \{0, 1, \ldots, N\}$, it follows that $\sigma^2 \equiv V(\tilde{m})$ and $\mu_{[4]}$ are finite. Let m_h denote the observation produced by the hth performance of (i) through (iv) in (2), and define $\overline{m}_\nu \equiv \nu^{-1} \sum_{h=1}^{\nu} m_h$ and $s_\nu^2 \equiv (\nu - 1)^{-1} \sum_{h=1}^{\nu} (m_h - \overline{m}_\nu)^2$. Show that:

(i) If $\epsilon > 0$ and $\delta > 0$ are predetermined, then there exists a ν large enough so that $P(\mu - \delta < \overline{m}_\nu < \mu + \delta) \geqq 1 - \epsilon$;
(ii) $E(\tilde{\overline{m}}_\nu) = \mu$ and $E(\tilde{s}_\nu^2) = \sigma^2$;
(iii) for large ν, the cdf of $\tilde{\overline{m}}_\nu$ is approximately $F_N(\overline{m}_\nu \mid \mu, \sigma^2/\nu)$;
(iv) for large ν, the cdf of $(\tilde{\overline{m}}_\nu - \mu)/\sqrt{\tilde{s}_\nu^2/\nu} \equiv \tilde{t}$ is approximately $F_{S^*}(t \mid \nu)$;

and

(v) if the decision-maker has assessed a Normal-inverted gamma prior df $f_{Ni\gamma}(\mu, \sigma^2 \mid \mu_0, \psi_0, n_0, \nu_0)$ for $(\tilde{\mu}, \tilde{\sigma}^2)$, then his posterior df is, for large ν, approximately $f_{Ni\gamma}(\mu, \sigma^2 \mid \mu_1, \psi_1, n_1, \nu_1)$ for

$$n_1 \equiv n_0 + \nu,$$
$$\nu_1 \equiv \nu_0 + \nu,$$
$$\psi_1 \equiv \nu_1^{-1}[\nu_0\psi_0 + (\nu - 1)s_\nu^2 + (n_0^{-1} + \nu^{-1})^{-1}(\overline{m}_\nu - \mu_0)^2],$$

and

$$\mu_1 \equiv (n_0 + \nu)^{-1}(n_0\mu_0 + \nu\overline{m}_\nu).$$

His marginal posterior df of $\tilde{\mu}$ is approximately

$$f_S\left(\mu \mid \mu_1, \frac{\psi_1}{n_1}, \nu_1\right).$$

[*Hints*: For (i) and (ii), Chapter 10 is relevant; for (iii) and (iv), Chapter 13; and for (v), this chapter.]

(4) Suppose interest centers on the probability $p \equiv P(\tilde{m} > M)$, where M is the nervous breakdown threshold of the special check authorizer. Define

$$\tilde{z}_h \equiv \delta(\tilde{m}_h - M) = \begin{cases} 1, & \tilde{m}_h > M \\ 0, & \tilde{m}_h \leqq M, \end{cases}$$

for $h = 1, \ldots, \nu$ and $\nu = 1, 2, \ldots$. Show that:

(i) For every ν, $\tilde{r}_\nu \equiv \sum_{h=1}^{\nu} \tilde{z}_h$ has mf $f_b(r_\nu \mid \nu, p)$;

(ii) $E(\tilde{r}_\nu/\nu) = p$;

and

(iii) if the decision-maker had assessed a beta prior df $f_\beta(p \mid \rho_0, \nu_0)$ for $\tilde{p}$, then his posterior df of $\tilde{p}$ is $f_\beta(p \mid \rho_0 + r_\nu, \nu_0 + \nu)$.

19. *Monte Carlo Simulation* (*continued*); *Estimating the Area of a Blob*: This exercise furnishes an idea of how Monte Carlo simulation can be used to approximate the solution of otherwise very difficult equations. Suppose one wishes to determine the area of a very irregular blob B, as in Figure 14.1

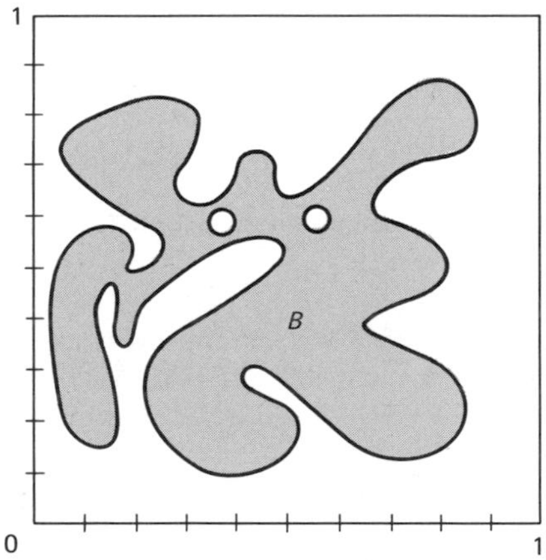

Figure 14.1

Several solution procedures are available. *One* hammer-and-tongs approach is to superimpose a fine-grid graph paper over B, add the areas of all squares included in B (an underestimate of the area of B), and also add the areas of all squares *intersecting* (included in or overlapping) B (an overestimate of the area of B). *Another* procedure is to superimpose the graph paper with axes (linearly) rescaled such that B is included in the unit square $(0, 1) \times (0, 1)$. Let R denote the ratio of the actual (unrescaled) area covered by $(0, 1) \times (0, 1)$ to the area 1 of the unit square, and let p denote the area of the blob in the unit-square coordinates. Then area $(B) = pR$. Show that:

(a) If $\tilde{x}_{11}, \tilde{x}_{12}, \tilde{x}_{21}, \tilde{x}_{22}, \ldots, \tilde{x}_{i1}, \tilde{x}_{i2}, \ldots$ are iid with common df $f_u(\cdot \mid 0, 1)$, then $(\tilde{x}_{11}, \tilde{x}_{12}), (\tilde{x}_{21}, \tilde{x}_{22}), \ldots$ are iid with common df

$$f_u^{(2)}(x_1, x_2 \mid (0, 1) \times (0, 1)) \equiv \begin{cases} 1, & (x_1, x_2) \in (0, 1) \times (0, 1) \\ 0, & \text{elsewhere.} \end{cases}$$

(b) Let

$$\tilde{z}_i \equiv \begin{cases} 1, & (\tilde{x}_{i1}, \tilde{x}_{i2}) \in B \\ 0, & \text{elsewhere.} \end{cases}$$

Then $\tilde{z}_1, \tilde{z}_2, \ldots,$ are iid with common mf $f_b(\cdot \mid 1, p)$.

(c) Let $\tilde{r}_n \equiv \sum_{i=1}^{n} \tilde{z}_i$. Then $\tilde{r}_n$ has mf $f_b(\cdot \mid n, p)$.

(d) $E(\tilde{r}_n/n) = p$ and $V(\tilde{r}_n/n) = p(1 - p)/n \leqq 1/(4n)$.

(e) For large n, the cdf of $(\tilde{r}_n - np)/\sqrt{np(1 - p)}$ is approximately $F_{N*}(\cdot)$.

(f) Let $\pi \in (0, 1)$. Then $P(-{}_\zeta N^* \leqq (\tilde{r}_n - np)/\sqrt{np(1 - p)} \leqq {}_\zeta N^*) \doteq \pi$ for large n, where $\zeta \equiv (1 + \pi)/2$; and

$$P(\tilde{r}_n/n - {}_\zeta N^*[\tfrac{1}{2}n^{-1/2}] \leqq p \leqq \tilde{r}_n/n + {}_\zeta N^*[\tfrac{1}{2}n^{-1/2}]) \dot{\geqq} \pi.$$

(g) For large n,

$$P(A \in R[\tilde{r}_n/n - {}_\zeta N^*[\tfrac{1}{2}n^{-1/2}], \tilde{r}/n + {}_\zeta N^*[\tfrac{1}{2}n^{-1/2}]]) \dot{\geqq} \pi,$$

where $A \equiv$ area (B).

(h) $E(R\tilde{r}_n/n) = A$ and $V(R\tilde{r}/n) = R^2p(1 - p)/n \leqq R^2/(4n)$. [$R\tilde{r}_n/n$ is used to estimate A; Chapter 20 concerns the general theory of *point estimation*. In (h), $E(R\tilde{r}_n/n) = A$ means that the expectation of error $R\tilde{r}/n - A$ is zero, while

$$\lim_{n\to\infty} V\left(\frac{Rr}{n}\right) \leqq \lim_{n\to\infty} \left(\frac{R^2}{4n}\right) = 0$$

means that as n grows, the distribution of the error becomes more and more concentrated about zero. Part (g) furnishes an *interval* estimate of A; the interpretation is given in Chapter 21.]

(i) If p is considered a random variable and given a prior df $f_\beta(p \mid \rho_0, \nu_0)$, then (for any n) the posterior df of $\tilde{p}$ is $f_\beta(p \mid \rho_0 + r_n, \nu_0 + n)$; and the posterior df of $\tilde{A} \equiv R\tilde{p}$ is $f_{\beta-}(A \mid \rho_0 + r_n, \nu_0 + n, 0, R)$.

[The simulation can be carried out quite interestingly if the blob is superimposed on a dart board, the simulator stands far back, has a few cocktails, and only counts as observations the darts which hit the dart board. The cocktails and standing far back are to ensure a uniform distribution of darts on the dart board; for readers who may not legally consume alcohol, we advise overexercise beforehand.]

20. *Nonnatural Conjugacy*: The definition of a natural conjugate family of distributions in Section 14.1 is too weak, in that it technically should include the condition that the family be the *smallest* one to afford reasonable flexibility. We say that a family of distributions is *conjugate* for a given process if the posterior distribution belongs to the family whenever the prior distribution does.

(a) Show that the family of *all* (proper or improper) k-variate probability distributions is conjugate for any process with a k-variate parameter $\tilde{\boldsymbol{\theta}}$.

(b) Let $\{\boldsymbol{\theta}^1, \ldots, \boldsymbol{\theta}^m] \subset \boldsymbol{\Theta} \subset R^k$, and let $\mathcal{C} \equiv \{\mathbf{p}: \mathbf{p} \in R^m, \mathbf{p} \geqq \mathbf{0}, \sum_{i=1}^{m} p_i = 1\}$. We interpret $\mathbf{p} \equiv (p_1, \ldots, p_m)'$ as assigning probability p_i to $\boldsymbol{\theta}^i$ for $i = 1, \ldots, m$. Show that $\mathcal{C}$ is conjugate for any process with k-variate parameter $\boldsymbol{\theta}$. (This result, with the arbitrariness of m and of $\boldsymbol{\theta}^1, \ldots, \boldsymbol{\theta}^m$, shows the existence of infinitely many conjugate families.)

(c) This part lays groundwork for (d) through (f). Let $\tilde{\boldsymbol{\theta}}$ be a k-dimensional random vector. Show that:

(i) If $m \in \{1, 2, \ldots, \infty\}$; if $F_i(\cdot)$ is the cdf of a k-dimensional random vector and $w_i \geqq 0$, for $i = 1, 2, \ldots, m$; and if $\sum_{i=1}^{m} w_i = 1$; then $F(\cdot)$, defined by

$$F(\boldsymbol{\theta}) \equiv \sum_{i=1}^{m} w_i F_i(\boldsymbol{\theta}),\ \boldsymbol{\theta} \in R^k$$

is a cdf.

(ii) Let $F(\cdot)$ be the cdf of a mixed random variable. There exist unique cdf's $F_c(\cdot)$ of a continuous random variable and $F_d(\cdot)$ of a discrete random variable such that

$$F(\theta) = w_c F_c(\theta) + w_d F_d(\theta), \quad \theta \in R^1,$$

for some $w_c \in (0, 1)$ and $w_d \equiv 1 - w_c$, also unique.

[*Hint*: Suppose the discontinuities of $F(\cdot)$ occur at $\theta^1, \theta^2, \ldots$ and the respective jumps are of length $d_1, d_2, \ldots$ respectively. Define

$$\Phi_d(\theta) = \begin{cases} 0, & \theta < \theta^1 \\ d_1, & \theta^1 \leqq \theta < \theta^2 \\ \vdots & \vdots \\ \sum_{j=1}^{i} d_j, & \theta^i \leqq \theta < \theta^{i+1} \\ \vdots & \vdots, \end{cases}$$

$$w_d \equiv \sum_{j=1}^{\infty} d_j,$$

$$F_d(\theta) = \Phi_d(\theta)/w_d,$$

$$\Phi_c(\theta) \equiv F(\theta) - \Phi_d(\theta),$$

$$w_c \equiv 1 - w_d,$$

and

$$F_c(\theta) \equiv \Phi_c(\theta)/w_c.]$$

(d) The family of beta distributions is natural conjugate for the Bernoulli-process parameter p; it is the smallest conjugate family. Let $\mathcal{C}(m)$ denote the set of *all* "convex combinations" $\sum_{i=1}^{m} w_i f_\beta(\cdot \mid \rho_i, \nu_i)$ of beta df's, for which $\sum_{i=1}^{m} w_i = 1$, every $w_i \geqq 0$, and every $\nu_i > \rho_i > 0$, but are otherwise arbitrary. Show that $\mathcal{C}(m)$ is conjugate for the Bernoulli-process parameter p; more specifically, prove the following:

If an experiment consists in obtaining n Bernoulli observations with noninformative stopping and yields the sufficient statistic (r, n) and if the prior df of $\tilde{p}$ is $\sum_{i=1}^{m} w_i f_\beta(p \mid \rho_i, \nu_i) \in \mathcal{C}(m)$, then the posterior df of $\tilde{p}$ is (14.8.39)

$$\sum_{i=1}^{m} w'_i f_\beta(p \mid \rho_i + r, \quad \nu_i + n) \in \mathcal{C}(m),$$

where

$$w'_i \equiv \frac{w_i B(\rho_i + r, \nu_i + n)/B(\rho_i, \nu_i)}{\sum_{j=1}^{m} \{w_j B(\rho_j + r, \nu_j + n)/B(\rho_j, \nu_j)\}},$$

for $B(\cdot, \cdot)$ as defined by (12.3.2); the posterior relative weights' ratios w'_h/w'_i satisfy

$$\frac{w'_h}{w'_i} = \left(\frac{w_h}{w_i}\right)\left[\frac{B(\rho_h + r, \nu_h + n)/B(\rho_h, \nu_h)}{B(\rho_i + r, \nu_i + n)/B(\rho_i, \nu_i)}\right].$$

Notice how a posterior relative weights' ratio is a multiple of the corresponding relative weight ratio, the mutiple depending upon the sufficient statistic (r, n) and all prior parameters—but not upon the prior relative weights.

(e) Let $\mathcal{C}(m)$ denote the set of all convex combinations $\sum_{i=1}^{m} w_i f_\gamma(\lambda \mid \rho_i, \tau_i)$ of gamma df's, for which $\sum_{i=1}^{m} w_i = 1$, every $w_i \geqq 0$, and every $\rho_i > 0$ and $\tau_i > 0$ but otherwise arbitrary. Prove the following:

If an experiment consists in observing a Poisson process with noninformative stopping, and yields the sufficient statistic (r, t), and if the prior df of the process intensity $\tilde{\lambda}$ is $\sum_{i=1}^{m} w_i f_\gamma(\lambda \mid \rho_i, \tau_i) \in \mathcal{C}(m)$, then the posterior df of $\tilde{\lambda}$ is $\sum_{i=1}^{m} w'_i f_\gamma(\lambda \mid \rho_i + r, \quad \tau_i + t) \in \mathcal{C}(m)$, where (14.8.40)

$$w'_i \equiv \frac{w_i \tau_i^{\rho_i} \Gamma(\rho_i + r)/[(\tau_i + t)^{\rho + r} \Gamma(\rho_i)]}{\sum_{j=1}^{m} \{\tau_j^{\rho_j} \Gamma(\rho_j + r)/[(\tau_j + t)^{\rho_j + r} \Gamma(\rho_j)]},$$

so that

$$\frac{w'_h}{w'_i} = \left(\frac{w_h}{w_i}\right)\left[\frac{\tau_h{}^{\rho_h}\Gamma(\rho_h + r)/[(\tau_h + t)^{\rho_h+r}\Gamma(\rho_h)]}{\tau_i{}^{\rho_i}\Gamma(\rho_i + r)/[(\tau_i + t)^{\rho_i+r}\Gamma(\rho_i)]}\right],$$

for all i and h in $\{1, \ldots, m\}$.

(f) Let $\mathcal{C}(m)$ denote the set of all convex combinations

$$\sum_{i=1}^{m} w_i f_{N i\gamma}(\mu, \sigma^2 \mid \mu_{oi}, \psi_{oi}, n_{oi}, \nu_{oi})$$

of Normal-inverted gamma df's, for which $\sum_{i=1}^{m} w_i = 1$, every $w_i \geqq 0$, and all ψ_{oi}, n_{oi}, and ν_{oi} are positive. Prove the following:

If an experiment consists in obtaining $n \geqq 2$ iid observations from $f_N(x \mid \mu, \sigma^2)$ with noninformative stopping and yields the sufficient statistic $(n, \bar{x}, s^2)$, and if the prior df of $(\tilde{\mu}, \tilde{\sigma}^2)$ is (14.8.41)

$\sum_{i=1}^{m} w_i f_{N i\gamma}(\mu, \sigma^2 \mid \mu_{oi}, \psi_{oi}, n_{oi}, \nu_{oi}) \in \mathcal{C}(m)$, then the posterior df of $(\tilde{\mu}, \tilde{\sigma}^2)$ is

$\sum_{i=1}^{m} w'_i f_{N i\gamma}(\mu, \sigma^2 \mid \mu_{1i}, \psi_{1i}, n_{1i}, \nu_{1i}) \in \mathcal{C}(m)$, where

$$\begin{aligned} n_{1i} &\equiv n_{oi} + n, \\ \nu_{1i} &\equiv \nu_{oi} + n, \\ \mu_{1i} &\equiv n_{1i}{}^{-1}(n_{oi}\mu_{oi} + n\bar{x}), \\ \psi_{1i} &\equiv \nu_{1i}{}^{-1}[\nu_{oi}\psi_{oi} + (n-1)s^2 + (n_{oi}{}^{-1} + n^{-1})^{-1}(\bar{x} - \mu_{oi})^2], \end{aligned}$$

and

$$w'_i \equiv \frac{w_i K(\mu_{oi}, \psi_{oi}, n_{oi}, \nu_{oi}, n, \bar{x}, s^2)}{\sum_{j=1}^{m} w_j K(\mu_{oj}, \psi_{oj}, n_{oj}, \nu_{oj}, n, \bar{x}, s^2)},$$

for

$$K(\mu_{oi}, \psi_{oi}, n_{oi}, \nu_{oi}, n, \bar{x}, s^2) \equiv \frac{n_{0i}{}^{1/2}(\frac{1}{2}\nu_{oi}\psi_{oi})^{\frac{1}{2}\nu_{oi}}\Gamma(\frac{1}{2}\nu_{1i})}{n_{1i}{}^{1/2}(\frac{1}{2}\nu_{1i}\psi_{1i})^{\frac{1}{2}\nu_{1i}}\Gamma(\frac{1}{2}\nu_{oi})}.$$

The marginal prior and posterior df's of $\tilde{\mu}$ are

$$\sum_{i=1}^{m} w_i f_S(\mu \mid \mu_{oi}, \psi_{oi}/n_{oi}, \nu_{oi}) \text{ and } \sum_{i=1}^{m} w'_i f_S(\mu \mid \mu_{1i}, \psi_{1i}/n_{1i}, \nu_{1i}),$$

respectively.

(g) Let $\mathcal{C}_m(p_1, \ldots, p_q)$ for $0 \leqq p_1 < p_2 < \cdots < p_q \leqq 1$ denote the family of all probability distributions with cdf's of the form

$$F(p) = \sum_{i=1}^{m} w_i F_\beta(p \mid \rho_i, \nu_i) + w_{m+1} F_d(p),$$

for all $w_i \geqq 0$, $\sum_{i=1}^{m+1} w_i = 1$, and $F_d(\cdot)$ the cdf of some discrete random variable with possible values a subset of $\{p_1, \ldots, p_q\}$. Let $L(p \mid r, n)$ be the likelihood function of the Bernoulli-process parameter p, extended from $(0, 1)$ to $[0, 1]$:

$$L(p \mid r, n) = \begin{cases} (1-p)^n, & r = 0 \\ p^r(1-p)^{n-r}, & 0 < r < n \\ p^n, & r = n, \end{cases} \tag{14.8.42}$$

for every $p \in [0, 1]$. If the prior distribution of $\tilde{p}$ belongs to $\mathcal{C}_m(p_1, \ldots, p_q)$, then the posterior cdf of $\tilde{p}$ is given for every $z \in R^1$ by

$$\begin{aligned} F(z \mid r, n) & \\ &\equiv P[\tilde{p} \leqq z \mid r, n] \\ &= \frac{P[\tilde{p} \leqq z, r, n]}{P[r, n]} \\ &= \int_{-\infty}^{z} \frac{P[r, n \mid p]dF(p)}{P[r, n]} \quad \text{(a Stieltjes integral)} \\ &= K(r, n) \int_0^z L(p \mid r, n)dF(p) \quad \text{(a Stieltjes integral)}, \end{aligned}$$

where $K(r, n) = 1 \big/ \int_0^1 L(p \mid r, n)dF(p)$, and the Stieltjes integral is evaluated by

$$\int_{-\infty}^{z} h(x)dG(x) = \zeta_1 \int_{-\infty}^{z} h(x)g_c(x)dx + \zeta_2 \sum \{h(x_i)\Delta_i\colon x_i \leqq z\} \tag{14.8.43}$$

whenever $h\colon R^1 \to (0, \infty)$ is continuous, $\zeta_1 \geqq 0$, $\zeta_2 \geqq 0$, $G(x) \equiv \zeta_1 G_c(x) + \zeta_2 G_d(x)$, $G_c(\cdot)$ is nondecreasing with derivative $g_c(\cdot)$, and $G_d(\cdot)$ is a nondecreasing step function with jumps Δ_i at x_i $(i = 1, 2, \ldots)$.

(i) Prove the following:

If an experiment consists in obtaining n Bernoulli observations with noninformative stopping and yields the sufficient statistic (r, n), and if the prior distribution of $\tilde{p}$ belongs to $\mathcal{C}_m(p_1, \ldots, p_q)$ with cdf $\sum_{i=1}^{m} w_i F_\beta(p \mid \rho_i, \nu_i) + w_{m+1}F_d(p)$ [where $F_d(\cdot)$ implies the occurrence of $p_h \in [0, 1]$ with probability π_h, for $h = 1, \ldots, q$], then the posterior distribution of $\tilde{p}$ belongs to $\mathcal{C}_m(p_1, \ldots, p_q)$ and has cdf $\sum_{i=1}^{m} w'_i F_\beta(p \mid \rho_i + r, \nu_i + n) + w'_{m+1}F_d(p \mid r, n)$, where (14.8.44)

$$w'_i \equiv \frac{w_i B(\rho_i + r, \nu_i + n)/B(\rho_i, \nu_i)}{\sum_{j=1}^{m} [w_j B(\rho_j + r, \nu_j + n)/B(\rho_j, \nu_j)] + w_{m+1} \sum_{k=1}^{q} L(p_k \mid r, n)\pi_k},$$

for $i = 1, \ldots, m$,

$$w'_{m+1} \equiv 1 - \sum_{i=1}^{m} w'_i,$$

$L(p_k \mid r, n)$ is given by (14.8.42), and $F_d(\cdot \mid r, n)$ is undefined (and also irrelevant) when $w'_{m+1} = 0$ but is otherwise the cdf of a discrete random variable which equals $p_h \in [0, 1]$ with probability $L(p_h \mid r, n)\pi_h / \sum_{k=1}^{q} L(p_k \mid r, n)\pi_k$ for $h = 1, \ldots, q$.

Statement (14.8.44) shows that the posterior relative weights' ratios coincide with those given in (14.8.39) [which is the special case $w_{m+1} = 0$] for $\{i, h\} \subset \{1, \ldots, m\}$; otherwise,

$$w'_{m+1}/w'_i = (w_{m+1}/w_i)\left[\frac{\sum_{k=1}^{q} L(p_k \mid r, n)\pi_k}{B(\rho_i + r, \nu_i + n)/B(\rho_i, \nu_i)}\right].$$

(ii) Prove the following special case of (14.8.35) for $m = 1$, $q = 2$, $p_1 = 0$, and $p_2 = 1$:

Under the assumptions of (14.8.44), if the prior distribution of $\tilde{p}$ belongs to $\mathcal{C}_1(0, 1)$, so does the posterior distribution, which has cdf $F(p \mid r, n) = w'_1 F_\beta(p \mid \rho + r, \nu + n) + w'_2 F_d(p \mid r, n)$, (14.8.44′)

where

$$w'_1 \equiv \begin{cases} 1, & r \notin \{0, n\} \\ w_1 B(\rho, \nu + n)/[w_1 B(\rho, \nu + n) + w_2 \pi B(\rho, \nu)], & r = 0 \\ w_1 B(\rho + n, \nu + n)/[w_1 B(\rho + n, \nu + n) + w_2 \pi B(\rho, \nu)], & r = n, \end{cases}$$

$w_2 = 1 - w_1$ and $w'_2 = 1 - w'_1$,

and

$$F_d(p \mid r, n) = \begin{cases} \text{indeterminate and irrelevant}, \; r \notin \{0, n\} \\ \begin{cases} 0, & p < 0 \\ \pi', & 0 \leqq p < 1, \; r \in \{0, n\} \\ 1, & p \geqq 1 \end{cases} \end{cases}$$

for

$$\pi' = \begin{cases} 1, & r = 0 \\ 0, & r = n. \end{cases}$$

Statement (14.8.44′) indicates how w'_{m+1} can vanish in (14.8.44): if $m = 2$, $(p_1, p_2) = (0, 1)$ and $0 < r < n$; if $m = 1$, $p_1 = 0$, and $r > 0$, and if $m = 1$, $p_1 = 1$, and $r < n$.

(iii) Show in (14.8.44′) that $w'_1 < w_1$ if $r \in \{0, n\}$.

21. Observations are to be obtained on a Poisson process with unknown intensity $\tilde{\lambda}$, which has been given the natural conjugate prior df $f_\gamma(\lambda \mid \frac{1}{2}, 5)$.

(a) If the process describes the occurrence of accidents in a given plant, what is the probability that there will be no accidents during the first month of observation?

(b) Suppose that one accident occurred during the first month of observation. What is the posterior df of $\tilde{\lambda}$ at the end of the first month? What is the probability, at that time, of no accidents during the *second* month?

[*Hint*: The posterior df of $\tilde{\lambda}$ for the first month is the prior df of $\tilde{\lambda}$ for the second month.]

22. A production controller has assessed a prior df $f_\beta(p \mid 1, 4)$ for the proportion $\tilde{p}$ of defective parts in an outgoing production lot and plans to take a random sample of five parts with replacement.

(a) Sketch his posterior df of $\tilde{p}$ given that $\tilde{r} = 0$, $\tilde{r} = 1$, $\tilde{r} = 2$, and $\tilde{r} = 5$ defectives are found?
(b) Suppose that he had decided not to ship the parts out unless his posterior probability that $\tilde{p} \geqq .25$ were less than .30. For what values of $\tilde{r}$ would he ship the parts out?
(c) Now suppose that the sample is taken and $\tilde{r} = 1$ is observed, but that the production controller does not feel that he knows enough about $\tilde{p}$ to make a good decision and hence decides to take five more observations with replacement. What is his prior df of $\tilde{p}$ *now*? What will his posterior df be if no defectives are found? Before obtaining the second five observations, what is the probability that no defectives are observed?

23. An economist has assessed a joint prior df $f_{Ni\gamma}(\mu, \sigma^2 \mid 3000, 10^6, 7, 4)$ for the average family income $\tilde{\mu}$ in Paraurbia and the variance of $\tilde{\sigma}^2$ of individual family incomes in that region about μ. Suppose that he obtains a random sample of 100 families, who accurately and truthfully report their incomes, and that the statistics $\bar{x} = 2600$ and $s^2 = 250{,}000$ were computed.

(a) What is his posterior df of $(\tilde{\mu}, \tilde{\sigma}^2)$?
(b) What is his marginal posterior df of $\tilde{\sigma}^2$?
(c) What is his marginal posterior df of $\tilde{\mu}$?
(d) What was his prior probability that $\tilde{\sigma} \leqq 600$? What is his posterior probability of this event? (*Hint*: $\tilde{\sigma} \leqq 600$ if and only if $\tilde{\sigma}^2 \leqq 360{,}000$.)
(e) What was his prior probability that $\tilde{\mu} \leqq 3300$? What is his posterior probability of this event?

24. The owner of a fleet of cars wishes to convince his insurance company that his accident rate is less than the known nationwide average σ_0 for fleets. He produces complete and undistorted evidence that during the past five years his cars have traveled 5,000,000 miles and been involved in 50 accidents. Let t denote miles measured in multiples of 10,000, and suppose that the number r of accidents has mf $f_{Po}(r \mid \lambda t)$ given the accident rate λ, measured in accidents per 10,000 miles.

(a) Suppose the insurer assesses a prior df $f_\gamma(\lambda \mid \rho_0, \tau_0)$ for the accident rate $\tilde{\lambda}$ of the fleet owner in question, where $\rho_0 \equiv \tau_0\lambda_0$ so that $E_\gamma(\tilde{\lambda} \mid \rho_0, \tau_0) = \lambda_0$ by Table 12.1. Describe the behavior of $f_\gamma(\lambda \mid \tau_0\lambda_0, \tau_0)$ as τ_0 varies. Describe the behavior of his posterior df $f_\gamma(\lambda \mid \tau_0\lambda_0 + 50, \tau_0 + 500)$ as τ_0 varies.

(b) Continuation of (a). Show that

(i) $\lim_{\tau_0 \to \infty} E_\gamma(\tilde{\lambda} \mid \tau_0\lambda_0 + 50, \tau_0 + 500) = \lambda_0$;

(ii) $\lim_{\tau_0 \to 0} E_\gamma(\tilde{\lambda} \mid \tau_0\lambda_0 + 50, \tau_0 + 500) = 1/10$;

(iii) $E_\gamma(\tilde{\lambda} \mid \tau_0\lambda_0 + 50, \tau_0 + 500)$ is a decreasing function of τ_0 if $\lambda_0 \leqq 1/10$ and an increasing function of τ_0 if $\lambda_0 \geqq 1/10$.

Interpret these results in the context of this exercise.

(c) Show that if the insurer's prior cdf of $\tilde{\lambda}$ is given by

(iv) $$F(\lambda) = \begin{cases} 0, & \lambda < \lambda_0 \\ 1, & \lambda \geqq \lambda_0, \end{cases}$$

then his posterior cdf of $\tilde{\lambda}$ will also be given by (iv), and hence the information (r, t) will have no effect on his judgments regarding $\tilde{\lambda}$. Is this reasonable?

25. *Security Price Changes*: An investment analyst believes that the daily closing-price *changes* $\tilde{x}_1$ and $\tilde{x}_2$ of common stocks number 1 and number 2 are jointly Normally distributed with some mean vector $\boldsymbol{\mu} \equiv (\mu_1, \mu_2)$ and some covariance matrix

$$\boldsymbol{\sigma}^2 \equiv \begin{bmatrix} \sigma_{11}{}^2 & \sigma_{12}{}^2 \\ \sigma_{21}{}^2 & \sigma_{22}{}^2 \end{bmatrix}.$$

He assesses a joint prior df $f^{(2)}_{NiW}(\boldsymbol{\mu}, \boldsymbol{\sigma}^2 \mid \boldsymbol{\mu}_0, \boldsymbol{\psi}_0, n_0, \nu_0)$ with $\boldsymbol{\mu}_0 = (\frac{1}{4}, \frac{1}{2})'$, $n_0 = 1$, $\nu_0 = 2$, and

$$\boldsymbol{\psi}_0 = \begin{bmatrix} \frac{1}{4} & -\frac{1}{16} \\ -\frac{1}{16} & \frac{1}{9} \end{bmatrix}.$$

(a) Six days' observations have produced five days' price changes, which resulted in $\bar{\mathbf{x}} = (1, 0)'$ and

$$\mathbf{s}^2 = \begin{bmatrix} 1 & \frac{1}{2} \\ \frac{1}{2} & \frac{3}{4} \end{bmatrix}.$$

What is the analyst's posterior distribution of $(\tilde{\boldsymbol{\mu}}, \tilde{\boldsymbol{\sigma}}^2)$? What is his marginal posterior df of $\tilde{\boldsymbol{\mu}}$? of $\tilde{\boldsymbol{\sigma}}^2$?

(b) Given the results in (a), what is the analyst's marginal df of the *next* price change vector $\tilde{\mathbf{x}}$? [*Hint*: The df of $(\tilde{\boldsymbol{\mu}}, \tilde{\boldsymbol{\sigma}}^2)$ *posterior* to the fifth price change is the df of $(\tilde{\boldsymbol{\mu}}, \tilde{\boldsymbol{\sigma}}^2)$ *prior* to the sixth.]

26. *Stratified Sampling (II)*: Stratified sampling was introduced in Exercise 54 of Chapter 12. Suppose in this exercise that $k = 1$; n_i denotes the number of

observations from stratum i; and $\bar{x}_i$ and s_i^2 the statistics of the sample from stratum i as defined (without subscript i) by (13.2.1) and (13.2.2), respectively. Let $\mathbf{n} \equiv (n_1, \ldots, n_r)'$, $\bar{\bar{\mathbf{x}}} \equiv (\bar{x}_1, \ldots, \bar{x}_r)'$, and $\boldsymbol{\mu} \equiv (\mu_1, \ldots, \mu_r)'$. (*Caution*: $\boldsymbol{\mu}$ was defined differently in Exercise 54 of Chapter 12.)

(a) Prove the following:

Under the preceding assumptions, if the stratum variances $\sigma_1^2, \ldots, \sigma_r^2$ are all known beforehand, and if $\tilde{\boldsymbol{\mu}}$ has the natural conjugate prior df $f_N^{(r)}(\boldsymbol{\mu} \mid \boldsymbol{\mu}_0, \boldsymbol{\sigma}_0^2)$, then the prior df of the over-all population mean $\tilde{\mu} \equiv \sum_{i=1}^{r} \pi_i \tilde{\mu}_i = \boldsymbol{\pi}'\tilde{\boldsymbol{\mu}}$ is $f_N(\mu \mid \boldsymbol{\pi}'\boldsymbol{\mu}_0, \boldsymbol{\pi}'\boldsymbol{\sigma}_0^2\boldsymbol{\pi})$; and the posterior df's of $\tilde{\boldsymbol{\mu}}$ and $\tilde{\mu}$ are $f_N^{(r)}(\boldsymbol{\mu} \mid \boldsymbol{\mu}_1, \boldsymbol{\sigma}_1^2(\mathbf{n}))$ and $f_N(\mu \mid \boldsymbol{\pi}'\boldsymbol{\mu}_1, \boldsymbol{\pi}'\boldsymbol{\sigma}_1^2(\mathbf{n}))\boldsymbol{\pi})$, where $\boldsymbol{\mu}_1$ and $\boldsymbol{\sigma}_1^2(\mathbf{n})$ are as defined in (14.7.7). (14.8.54)

(b) Prove the following:

Under the preceding assumptions, if the ratios $\sigma_i^2/\boldsymbol{\sigma}_j^2$ of all pairs of stratum variances are known beforehand, and if $(\tilde{\boldsymbol{\mu}}, \tilde{\sigma}_*^2)$ has the natural conjugate prior df $f_{Ni\gamma}^{(r)}(\boldsymbol{\mu}, \sigma_*^2 \mid \boldsymbol{\mu}_0, \psi_0, \boldsymbol{\sigma}_{0*}^2, \nu_0)$, then the prior and posterior df's of the over-all population mean $\tilde{\mu} \equiv \boldsymbol{\pi}'\tilde{\boldsymbol{\mu}}$ are $f_S(\mu \mid \boldsymbol{\pi}'\boldsymbol{\mu}_0, \psi_0\boldsymbol{\pi}'\boldsymbol{\sigma}_{0*}^2\boldsymbol{\pi}, \nu_0)$ and $f_S(\mu \mid \boldsymbol{\pi}'\boldsymbol{\mu}_1, \psi_1\boldsymbol{\pi}'\boldsymbol{\sigma}_{1*}^2(\mathbf{n})\boldsymbol{\pi}, \nu_1)$, respectively, where $\boldsymbol{\mu}_1, \psi_1, \boldsymbol{\sigma}_{1*}^2(\mathbf{n})$, and ν_1 are as defined in (14.7.14). (14.8.46)

27. *Generalizations of* (14.6.1) *and* (14.6.2): In some applications, n iid observations $\tilde{\mathbf{x}}_1, \ldots, \tilde{\mathbf{x}}_n$ are obtained from $f_N^{(k)}(\mathbf{x} \mid \mathbf{A}\boldsymbol{\theta}, \boldsymbol{\sigma}^2)$, where $\mathbf{A}$ is $k \times r$ of rank at least one, and r can be less than, equal to, or greater than k. Prove the following:

If an experiment consists in obtaining $n \geqq 1$ observations from $f_N^{(k)}(\mathbf{x} \mid \mathbf{A}\boldsymbol{\theta}, \boldsymbol{\sigma}^2)$ with $\boldsymbol{\sigma}^2$ known and noninformative stopping, if the sufficient statistic $(n, \bar{\mathbf{x}})$ is observed, and if $\tilde{\boldsymbol{\theta}}$ has the natural conjugate prior df $f_N^{(r)}(\boldsymbol{\theta} \mid \boldsymbol{\theta}_0, \boldsymbol{\sigma}_0^2)$, then the posterior df of $\tilde{\boldsymbol{\theta}}$ is $f_N^{(r)}(\boldsymbol{\theta} \mid \boldsymbol{\theta}_1, \boldsymbol{\sigma}_1^2)$, where (14.8.47)

$$\boldsymbol{\sigma}_1^2 \equiv (\boldsymbol{\sigma}_0^{-2} + n\mathbf{A}'\boldsymbol{\sigma}^{-2}\mathbf{A})^{-1}$$

and

$$\boldsymbol{\theta}_1 \equiv \boldsymbol{\sigma}_1^2(\boldsymbol{\sigma}_0^{-2}\boldsymbol{\theta}_0 + n\mathbf{A}'\boldsymbol{\sigma}^{-2}\bar{\mathbf{x}});$$

and

under the assumptions of (14.8.47), the unconditional df of $\bar{\tilde{\mathbf{x}}}$ given n is $f_N^{(k)}(\bar{\mathbf{x}} \mid \mathbf{A}\boldsymbol{\theta}_0, \mathbf{A}\boldsymbol{\sigma}_0^2\mathbf{A}' + n^{-1}\boldsymbol{\sigma}^2)$. (14.8.48)

28. *Generalizations of* (14.6.14) *and* (14.6.15): As in Exercise 27, let $\boldsymbol{\mu} \equiv \mathbf{A}\boldsymbol{\theta}$. Prove the following:

If an experiment consists in obtaining $n \geqq 2$ observations from $f_N^{(k)}(\mathbf{x} \mid \mathbf{A}\boldsymbol{\theta}, \boldsymbol{\sigma}^2)$ with $\boldsymbol{\sigma}_*^2$ known and noninformative stopping, if the sufficient statistic $(\mathrm{n}, \bar{\mathbf{x}}, s_*^2)$ is observed, and if $(\tilde{\boldsymbol{\theta}}, \tilde{\sigma}_*^2)$ has the natural conjugate prior df

$$f_{Ni\gamma}^{(r)}(\boldsymbol{\theta}, \sigma_*^2 \mid \boldsymbol{\theta}_0, \psi_0, \boldsymbol{\sigma}_*^2, \nu_0),$$

then the posterior df of $(\tilde{\boldsymbol{\theta}}, \tilde{\sigma}_*^2)$ is $f_{Ni\gamma}^{(r)}(\boldsymbol{\theta}, \sigma_*^2 \mid \boldsymbol{\theta}_1, \psi_1, \boldsymbol{\sigma}_{1*}^2, \nu_1)$, where (14.8.49)

$$\begin{aligned}
\nu_1 &\equiv \nu_0 + kn,\\
\psi_1 &\equiv \nu_1^{-1}[\nu_0\psi_0 + k(n-1)s_*^2 + (\bar{\mathbf{x}} - \mathbf{A}\boldsymbol{\theta}_0)'(\mathbf{A}\boldsymbol{\sigma}_{0*}^2\mathbf{A}' + n^{-1}\boldsymbol{\sigma}_*^2)^{-1}(\bar{\mathbf{x}} - \mathbf{A}\boldsymbol{\theta}_0)],\\
\boldsymbol{\sigma}_{1*}^2 &\equiv (\boldsymbol{\sigma}_{0*}^{-2} + \mathbf{A}'\boldsymbol{\sigma}_*^{-2}n\mathbf{A})^{-1}, \text{ and}\\
\boldsymbol{\theta}_1 &\equiv \boldsymbol{\sigma}_{1*}^2(\boldsymbol{\sigma}_{0*}^{-2}\boldsymbol{\theta}_0 + n\mathbf{A}'\boldsymbol{\sigma}_*^{-2}\bar{\mathbf{x}}).
\end{aligned}$$

The marginal prior and posterior df's of $\tilde{\boldsymbol{\theta}}$ are $f_S^{(r)}(\boldsymbol{\theta} \mid \boldsymbol{\theta}_0, \psi_0\boldsymbol{\sigma}_{0*}^2, \nu_0)$ and $f_S^{(r)}(\boldsymbol{\theta} \mid \boldsymbol{\theta}_1, \psi_1, \boldsymbol{\sigma}_{1*}^2, \nu_1)$, respectively. The marginal prior and posterior df's of $\tilde{\sigma}_*^2$ are $f_{i\gamma}(\sigma_*^2 \mid \psi_0, \nu_0)$ and $f_{i\gamma}(\sigma_*^2 \mid \psi_1, \nu_1)$, respectively. Under the assumptions of (14.8.49) and given n, (14.8.50)

(a) the df $f(\bar{\mathbf{x}} \mid s_*^2)$ of $\tilde{\bar{\mathbf{x}}}$ conditional upon s_*^2 but unconditional as regards $(\tilde{\boldsymbol{\theta}}, \tilde{\sigma}_*^2)$ is given by $f(\bar{\mathbf{x}} \mid s_*^2) = f_S^{(k)}(\bar{\mathbf{x}} \mid \mathbf{A}\boldsymbol{\theta}_0, \boldsymbol{\psi}_*, \nu_0 + k(n-1))$, where $\boldsymbol{\psi}_* \equiv \{[\nu_0\psi_0 + k(n-1)s_*^2]/[\nu_0 + k(n-1)]\}(\mathbf{A}\boldsymbol{\sigma}_{0*}^2\mathbf{A}' + n^{-1}\boldsymbol{\sigma}_*^2)$;

(b) the unconditional df $f(\bar{\mathbf{x}})$ of $\tilde{\bar{\mathbf{x}}}$ is given by

$$f(\bar{\mathbf{x}}) = f_S^{(k)}(\bar{\mathbf{x}} \mid \mathbf{A}\boldsymbol{\theta}_0, \psi_0(\mathbf{A}\boldsymbol{\sigma}_{0*}^2\mathbf{A}' + n^{-1}\boldsymbol{\sigma}_*^2), \nu_0);$$

(c) the df $f(s_*^2 \mid \bar{\mathbf{x}})$ of $\tilde{s}_*^2$ conditional upon $\bar{\mathbf{x}}$ but unconditional as regards $(\tilde{\boldsymbol{\theta}}, \tilde{\sigma}_*^2)$ is given by $f(s_*^2 \mid \bar{\mathbf{x}}) = f_{i\beta 2}(s_*^2 \mid \tfrac{1}{2}k(n-1), \tfrac{1}{2}(\nu_0 + kn), b)$, where $b \equiv \nu_1\psi_1/[k(n-1)] - s_*^2$; and

(d) the unconditional df $f(s_*^2)$ of $\tilde{s}_*^2$ is given by

$$f(s_*^2) = f_{i\beta 2}(s_*^2 \mid \tfrac{1}{2}k(n-1), \tfrac{1}{2}[\nu_0 + k(n-1)], [k(n-1)]^{-1}\nu_0\psi_0).$$

part III

Bayesian Decision Theory

The third part of this book consists of five chapters which together furnish an introduction to Bayesian (statistical) decision theory.

The reader will recall from Chapter 4 the structure of the general decision problem under uncertainty, specified by: (1) a set E of available *experiments* e; (2) for each $e \in E$, a set $Z(e)$ of possible experimental *outcomes* z; (3) for each $e \in E$ and each $z \in Z(e)$, a set $A(e, z)$ of available *acts* a; (4) a set Θ of possible *states* of θ of the environment; (5) a set C of potential *consequences* c; and (6) a function assigning a consequence $c(e, z, a, \theta)$ to each (e, z, a, θ) quadruple.

In Chapter 4 we promised, essentially, to complete the specification of the general decision problem under uncertainty by assigning probabilities to (z, θ)-events and by quantifying preferences for the c's. With Part II as background, the reader may now appreciate that when probability functions have been assigned to all events in each $Z(e) \times \Theta$, the choice of an $a \in A(e, z)$ for each $z \in Z(e)$, uniquely specifies a probability function on an appropriate set of events in C, because with e and an a for each $z \in Z(e)$ determined, only (z, θ) remains to be determined in order to specify $c(e, z, a, \theta)$. Hence the decision-maker's problem is equivalent to choosing one of the possible probability functions on (appropriate subsets of) C.

The decision-maker's choice of a probability function on C depends in an essential fashion upon his relative preferences for the elements c of C. The task of Chapter 15 is to show that a set of (we believe) reasonable assumptions about his preferences for c's and for certain, simple probability functions on C imply the existence of a real-valued function $u\colon C \rightarrow R^1$, called a *utility function*, with the properties that $u(c) > u(c')$ whenever he prefers c to c', and that the choice between probability functions on C can be made by comparing expectations of the $u(c)$'s.

Chapter 16 shows that if the decision-maker has assessed a utility function on C, then his acceptance of two (reasonable) axioms implies that he can assess subjective probabilities to all θ-events [or (z, θ)-events, if need be]. We show how this assessment can be performed when Θ is a finite set, and also how the parameters of a subjective, conjugate prior df (as discussed in Chapter 14) can be fitted.

The net result of Chapters 15 and 16 is therefore the completion of the specification of the general decision problem. Chapter 17 begins by stating the complete specification and goes on to discuss its solution. By increasingly specializing the problem, we show how information can be evaluated.

In Chapters 18 and 19, we specialize still further to consider some important classes of decision problems, with long histories in statistical inference testifying to their relevance. These chapters draw heavily upon the results of Chapter 14.

We caution the reader that these chapters constitute *only* an introduction. Another introduction, much more complete than this in many respects, is in Pratt, Raiffa, and Schlaifer [**64**]. Additional references are furnished when relevant.

15

ELEMENTS OF BAYESIAN DECISION THEORY (I): UTILITY

15.1 INTRODUCTION

This chapter presents and discusses conditions under which a decision-maker's preferences between the elements c of a finite set $C \equiv \{c_1, \ldots, c_r\}$ of possible consequences of some decision problem can (1) be represented by a real-valued function u: $C \rightarrow R^1$; and (2) extended to the set of all probability functions on C in a useful manner.

A suggestive synonym for "probability function on C" is "lottery with prizes in C." Finiteness of L implies that any lottery L with prizes in C can be represented as a finite-branched tree of the form given in Figure 15.1.

In addition, we shall wish to consider *compound lotteries*, which are lotteries with prizes being lotteries with prizes in C. Any such lottery can be

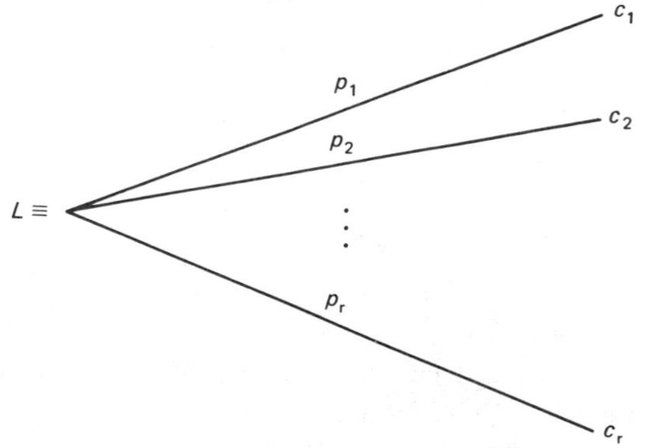

Figure 15.1

depicted as in Figure 15.2, where $L^{(i)}$ denotes the (simple) lottery which yields c_j as prize with probability $p_j^{(i)}$, for $j = 1, \ldots, r$. It is clear that $p_j^{(i)}$ is the conditional probability of receiving prize c_j given that the first-stage lottery yielded $L^{(i)}$ as prize. Hence the *marginal* probability of receiving c_j in the compound lottery is $q_j \equiv \sum_{i=1}^{m} p_j^{(i)} p_i$.

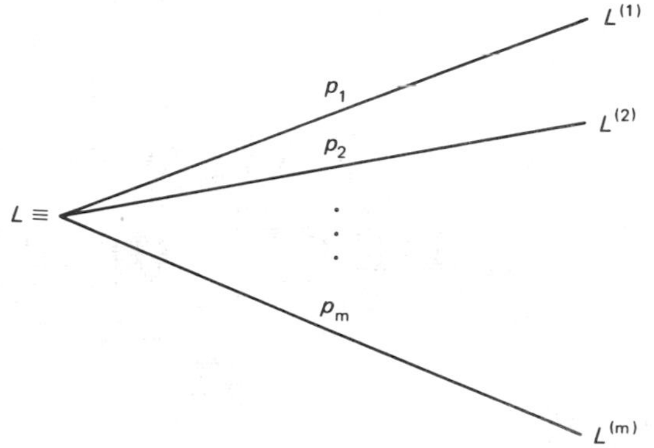

Figure 15.2

It is also clear that if the decision-maker is able to express his preferences between *any* pair of simple or compound lotteries with (ultimate) prizes in C, then he will be able to express his preferences between those pairs of lotteries which result from choosing an $e \in E$ and an $a \in A(z, e)$ for each $z \in Z(e)$ in a decision problem.

Section 15.2 presents a set of axioms which, if accepted by the decision-maker, imply that some simple preference assessments on his part imply the existence of utility function u: $C \rightarrow R^1$ with the following properties:

(1) He prefers c to c' if and only if $u(c) > u(c')$;

and

(2) He prefers $L \equiv (p_1, \ldots, p_r)$ to $L' \equiv (p_1', \ldots, p_r')$ if and only if $\sum_{j=1}^{r} u(c_j)p_j > \sum_{j=1}^{r} u(c_j)p'_j$.

The summations in (2) are in fact expectations of the discrete random variable $\tilde{u}$ which in lottery L assumes the value $u_j \equiv u(c_j)$ with probability p_j and in lottery L' assumes value u_j with probability p'_j. Hence the utility function is, in a sense, the decision-maker's subjective currency for C, and decisions (choices between C-valued lotteries) should be made so as to maximize the expectation of utility (the expected "value" in the utility currency).

Section 15.3 consists of a discussion of a procedure for assessing a utility function when C is a finite set. This procedure is implicitly suggested by the axiomatic development in Section 15.2.

In Section 15.4 we consider the question of assessing a utility function when C is an infinite set. The suggested procedure is applicable only under certain assumptions about the form of C, such as when $C = R^1 =$ money and the decision-maker always prefers more money to less.

Section 15.5 consists of remarks about the history of *utility* and variants of the concept discussed here.

All probabilities in this chapter may be considered as objective. Hence all lotteries can be thought of as abstractions of roulette games, drawing balls from urns, and so forth. In Chapter 16 we shall furnish additional axioms which imply the existence of subjective probabilities which can be used anywhere in place of objective probabilities. Hence it suffices for this chapter to assume that all probabilities are objective.

15.2 AN INFORMAL AXIOMATIC DEVELOPMENT OF UTILITY

We shall suppose that the decision-maker is confronted with a set $C \equiv \{c_1, \ldots, c_r\}$ of possible consequences of a decision problem and wishes to assess his preferences between any two lotteries L and L' with ultimate prizes in C. He will be able to do so in a very simple fashion if he accepts the following six axioms and performs the requisite assessments.

We write $c \gtrsim c'$ if and only if c is at least as desirable to the decision-maker as is c'. Thus "$\gtrsim$" is a relation in the sense of Chapter 3; it is, in fact, the relation R of Exercise 15 in that chapter. We shall call "$\gtrsim$" the *preference-or-difference relation.* We also define the indifference relation "$\sim$" by

$$c \sim c' \text{ if and only if } c \gtrsim c' \text{ and } c' \gtrsim c;$$

and the strict preference relation "$\succ$" by

$$c \succ c' \text{ if and only if } c \gtrsim c' \text{ but } \textit{not } c' \gtrsim c.$$

The first axiom requires that "$\gtrsim$" be a total order in C; that is, (a) the decision-maker can compare any two consequences c and c' preference-wise, and (b) that "$\gtrsim$" is transitive.

Axiom 1: Ordering of Consequences

(a) For every pair c, c' of consequences, either $c \gtrsim c'$ or $c' \gtrsim c$ or both;

and

(b) if $c \gtrsim c'$ and $c' \gtrsim c''$, then $c \gtrsim c''$.

Axiom 1 thus requires that the decision-maker be able to *rank* the set of consequences from most to least desirable, with the relative placement of indifferent consequences being arbitrary. It is clear that $c \sim c$ for every $c \in C$.

Part (b), transitivity, has often been questioned as to its applicability. Certainly, many people (even decision theorists) frequently behave intransitively in real life, and hence (b) is poor from a *descriptive* viewpoint. However, the following example can be used to show a decision-maker that he would *want* his preferences to be transitive.

Example 15.2.1

Suppose that $c \succsim c'$, $c' \succsim c''$, and $c'' \succ c$. Then Axiom 1(b) is violated, since $c'' \succ c$ means, by definition of "$\succ$," that c is *not* preferred or indifferent to $c.''$ Now suppose that you have c and c' in the "bank"and that the decision-maker possesses $c.''$ Since $c' \succsim c''$, he should be willing to trade in c'' for your c'. Now you have c and c'' in the bank and the decision-maker owns c'. Now, since $c \succsim c'$, he should be willing to trade in his c' for your c, so that he will have c and you will have $c'$and c''. But now, since $c'' \succ c$, he should be willing to pay you some small amount ϵ—say, 10^{-9} mil—for the privilege of trading in his c for your c''. But now he owns c'' and you own c and c'; and this is just where you both started, with one exception: *he is* 10^{-9} *mil poorer*. Moreover, unless he revises his preferences he should be willing to continue playing this game indefinitely, until bankruptcy. A persistently intransitive decision-maker has been aptly termed a "money pump." Most decision-makers are not persistent in their intransitivity; faced with the logic of the preceding example, they think harder and revise their preference assertions.

There are objections to Axiom 1 (a) as well. Some preference comparisons are so repugnant that many people refuse to make them. For example, c might be "lose your wife" and c' might be "lose your child." In drastic examples such as this, many individuals react strongly against making a choice; but, if the decision problem *must* be solved, then such an awful comparison *must* be made, at least tacitly. The difficulty of the comparison is not really removed by denying its necessity.

For further ramifications of Axiom 1, see Exercise 15 in Chapter 3 on general total orders. Those results, with E corresponding to "$\sim$" and P to "$\succ$," obviously apply to "$\succsim$."

We shall hereafter let c_b denote a "best" consequence; that is, c_b satisfies

$$c_b \succsim c_j, \text{ for all } j \in \{1, \ldots, r\}.$$

Similarly, we define a "worst" consequence c_w by

$$c_j \succsim c_w, \text{ for all } j \in \{1, \ldots, r\}.$$

Since C is assumed to be a finite set, c_b and c_w exist. (This need not be so when C is infinite: let $C = R^1 = \$$ and let $c \succsim c'$ if and only if $\$c \geqq \$c'$. There are no largest and smallest amounts of money, because $+\infty$ and $-\infty$ are not real numbers.)

Furthermore, to avoid triviality, we shall assume that $c_b \succ c_w$: otherwise $c_i \sim c_j$, for all i and j. (Why? Exercise 1.)

Axiom 2 essentially states that the decision-maker is indifferent between a compound lottery and the simple lottery which offers the same *marginal* probability at each consequence c_j as does the compound lottery.

Axiom 2: Reduction of Compound Lotteries

If L is a compound lottery as depicted in Figure 15.2, and if $Ł$ is the simple lottery which yields c_j with probability $q_j \equiv \sum_{i=1}^{m} p_j^{(i)} p_i$, then

$$L \sim Ł.$$

Note that Axiom 2 begins the process of extending "$\succsim$" from being defined on C to being defined on the set of all lotteries with ultimate prizes in C.

There is a serious objection to Axiom 2 for descriptive work. Namely, it abstracts away the "joy of gambling," in the sense that the two-stage, compound lottery L would furnish more suspense and excitement than would $Ł$, and hence $L \succ Ł$. By the same token, people who fear gambling assess $Ł \succ L$.

But this objection lacks relevance for *prescriptive* purposes of actually making decisions, because *if* a decision-maker does enjoy gambling, then the *description of the consequences* in L is different from that in $Ł$, and hence Axiom 2 does not apply to the preference-wise comparison of L and $Ł$.

Axiom 3 asserts, essentially, that every consequence c is equally as desirable as some "very simple" lottery offering at most both c_b and c_w with positive probability.

Axiom 3: Continuity

For every consequence c there is a number $u(c) \in [0, 1]$ such that $c \sim c_L$, where c_L is the lottery in Figure 15.3. We digress briefly at this point to remark that Axiom 3, in conjunction with Axiom 5 and 6, implies that if $c_b \succ c$ and $c \succ c_w$, then $u(c) \in (0, 1)$. It is this implication that seems to fail in extreme situations, as when $c_b \equiv$ "receive \$0.02," $c \equiv$ "receive \$0.01," and $c_w \equiv$ "be hanged at sundown"; for you would probably state that there is no $u(c) \in (0, 1)$—not even $1-10^{-100}$—sufficiently close but not equal to one which would make you indifferent between having c and having the lottery c_L. In other words, you would accept no probability whatsoever

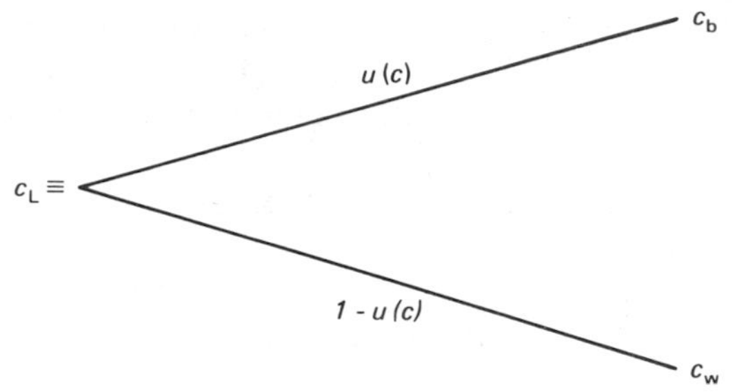

Figure 15.3

of being hanged in order to have the complementary chance of receiving the extra cent.

This line of reasoning seems valid enough until we note that we accept lotteries of the form c_L every day (for example, in running yellow lights or jaywalking).

Axiom 4 is a weak converse of Axiom 2, which stated that every compound lottery is equally as desirable as some simple lottery. Axiom 4 states that c_L can be substituted for c wherever it appears in a (simple or compound) lottery L without changing the desirability of L.

Axiom 4

Let L^* be obtained from L by replacing c in L with c_L wherever it appears, as defined in Axiom 3. Then $L \sim L^*$.

Alternatively, the decision-maker is not only indifferent between c and c_L per se (as Axiom 3 stipulates), but he is also indifferent between any two lotteries that are identical except that c in one is replaced everywhere by c_L in the other. For example, if $c_b \equiv$ "filet mignon," $c_w \equiv$ "pork chop," $c \equiv$ "eggs Benedict," and $c' \equiv$ "trout Marguery," and if I am indifferent between c and c_L, as shown in Figure 15.4, then I must also be indifferent between the Figures 15.5 and 15.6.

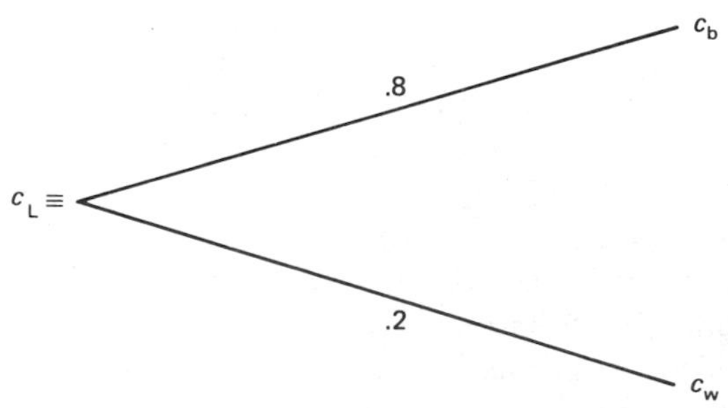

Figure 15.4

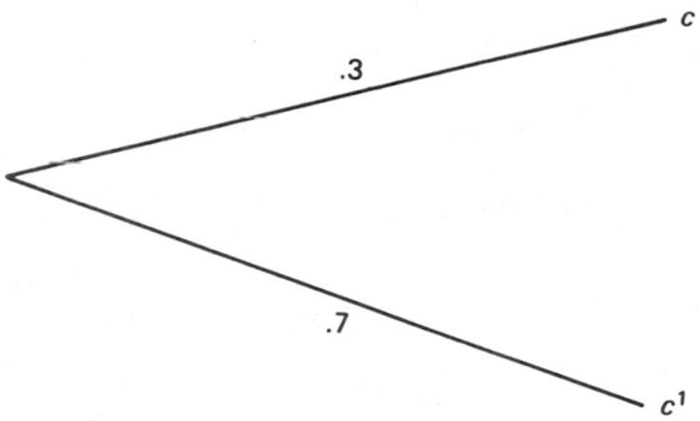

Figure 15.5

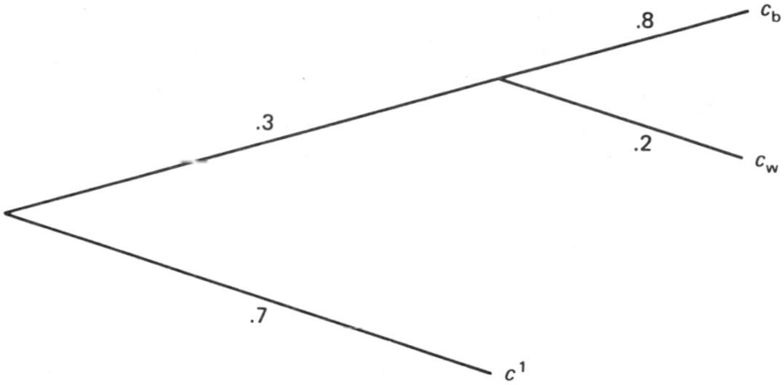

Figure 15.6

We may now show that any simple lottery L, of the form shown in Figure 15.1, is equally as desirable as a simple lottery involving only c_b and c_w. To see that this is so, we first apply Axioms 3 and 4 each r times so as to obtain a compound lottery L^* of the form shown in Figure 15.7, which, by Axioms 3 and 4, satisfies

$$L \sim L^*.$$

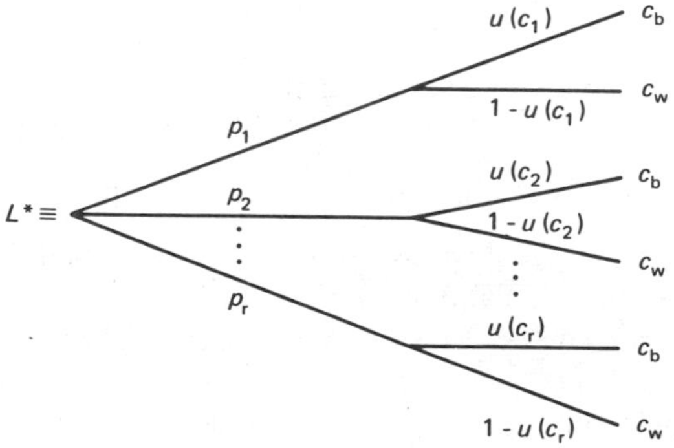

Figure 15.7

Now, by Axiom 2,

$$L^* \sim Ł^*,$$

where $Ł^*$ is as defined in Figure 15.8

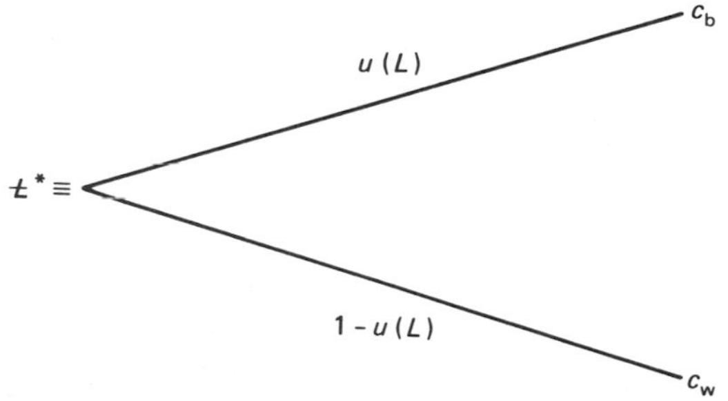

Figure 15.8

in which

$$u(L) \equiv \sum_{j=1}^{r} u(c_j)p_j, \tag{15.2.1}$$

Can we conclude that since $L \sim L^*$ and $L^* \sim Ł^*$, then $L \sim Ł^*$? We cannot do so on the basis of the preceding axioms. We shall need to assume that the relations $\succ$ and $\sim$ of preference and indifference, respectively, are transitive when applied to lotteries with prizes in C as well as when applied to C.

Axiom 5

For any lotteries L, L', and L'', all with ultimate prizes in C, if $L \succ L'$ and $L' \succ L''$, then $L \succ L''$; and if $L \sim L'$ and $L' \sim L''$, then $L \sim L''$.

Example 15.2.1 remains pertinent in the case of a decision-maker who is persistently intransitive. Hence, with Axiom 5, we may conclude that $L \sim Ł^*$. Note that $Ł^*$ is fully specified by the number $u(L) \equiv \sum_{j=1}^{r} u(c_j)p_j$.

Finally, suppose that we have two simple lotteries L and L', that $L \sim Ł^*$ and $L' \sim Ł'^*$, and that $u(L)$ and $u(L')$ have been determined. It is intuitively appealing to assert that the decision-maker prefers $Ł^*$ to $Ł'^*$ if and only if $u(L) > u(L')$, and hence, by transitivity, $L \succ L'$ if and only if $u(L) > u(L')$; but the preceding axioms are not sufficient for this assertion. We need the following axiom:

Axiom 6

Let L_i be defined as in Figure 15.9

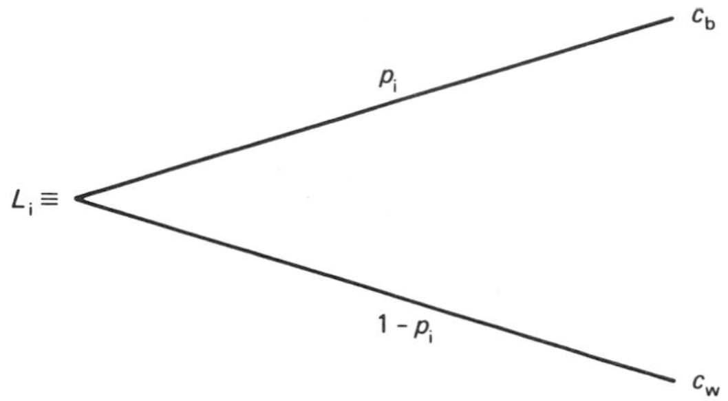

Figure 15.9

for $i = 1, 2$. Then $L_1 \gtrsim L_2$ if and only if $p_1 \geqq p_2$.

It is now clear how the decision-maker may evaluate and make choices among lotteries. Since $L \sim Ł^*$ and $L' \sim Ł'^*$, and $Ł^* \gtrsim Ł'^*$ if and only if $u(L) \geqq u(L')$ (by Axiom 6), we conclude (from Axiom 5) that $L \gtrsim L'$ if and only if $u(L) \geqq u(L')$.

Moreover, every $c \in C$ can be identified with the lottery $L(c)$ in which $p(c) = 1$ and $p(c') = 0$ for every $c' \neq c$. This paragraph, together with the preceding, can be summarized as follows:

Let C be a finite set of consequences, and suppose the decision-maker's preference-or-difference order satisfies Axioms 1 through 6. Then there exists a real-valued function u: $C \to R^1$ such that (15.2.2)

(a) $c \gtrsim c'$ if and only if $u(c) \geqq u(c')$; and
(b) if L and L' are lotteries with prizes in C, then $L \gtrsim L'$ if and only if $u(L) \geqq u(L')$.

Note that $u(L)$ can be considered as the expectation of the utilities of the c_j's with their probabilities as given by L [similarly for $u(L')$]. That is, if we replace each c by $u(c)$ in L, then L becomes a lottery $L_\#$ with real numbers as "prizes," which can be considered as a discrete [C finite] random variable. Two such lotteries $L_\#$ and $L'_\#$ are comparable preference-wise strictly on the basis of their expectations. Hence $u(L)$ is often called the expected utility of L.

The reader may show as Exercise 2 that $u(c_b) = 1$ and $u(c_w) = 0$; it is clear from Axiom 3 that $u(c) \in [0, 1]$, for every $c \in C$.

Assertion (15.2.2) showed that a function u: $C \to R^1$ satisfying (a) and (b) *exists*, and now it is natural to ask if u is unique. The answer is no, but any two such functions are related in a simple fashion as follows:

If $u: C \to R^1$ satisfies (a) and (b) of (15.2.2), then $u_0: C \to R^1$ also satisfies these conditions if and only if there is a $b \in R^1$ and an $a > 0$ such that (15.2.3)

$$u_0(c) = au(c) + b,$$

for every $c \in C$.

Proof

For the "if" assertion, we suppose that $u_0(c) = au(c) + b$, for every $c \in C$, some $b \in R^1$, and some $a > 0$; and we must show that u_0 satisfies (a) and (b) of (15.2.2). Since (a) is a special case of (b), in which c and c' are replaced by $L(c)$ and $L(c')$ respectively, it suffices to prove (b):

$$L \gtrsim L'$$

if and only if $u(L) \geqq u(L')$ [by (15.2.2b)]

if and only if $\sum_{j=1}^{r} u(c_j)p_j \geqq \sum_{j=1}^{r} u(c_j)p'_j$ [by (15.2.1)]

if and only if $a\sum_{j=1}^{r} u(c_j)p_j + b \geqq a\sum_{j=1}^{r} u(c_j)p'_j + b$ (since $a > 0$)

if and only if $\sum_{j=1}^{r} [au(c_j) + b]p_j \geqq \sum_{j=1}^{r} [au(c_j) + b]p'_j$

(since $\sum_{j=1}^{r} bp_j = \sum_{j=1}^{r} bp'_j = b$)

if and only if $\sum_{j=1}^{r} u_0(c_j)p_j \geqq \sum_{j=1}^{r} u_0(c_j)p'_j$ (definition of u_0)

if and only if $u_0(L) \geqq u_0(L')$ [by (15.2.1)].

Hence $L \gtrsim L'$ if and only if $u_0(L) \geqq u_0(L')$ and, u_0 satisfies (b) [and hence also (a)] of (15.2.2). This concludes the proof of the "if" assertion. To prove the "only if" assertion, we shall show that if $u_0(c)$ is *not* always equal to $au(c) + b$ for some b and positive a, then u_0 fails to satisfy (b) of (15.2.2). We assume that there is a consequence c such that $c_b \succ c$ and $c \succ c_w$, since otherwise there would be only two values of u and of u_0. Hence a and b could be found such that $u_0(c) = au(c) + b$ for every $c \in C$. Now choose a and b such that $u_0(c_b) = au(c_b) + b$ and $u_0(c_w) = au(c_w) + b$, and suppose that $u_0(c) \neq au(c) + b$. Let c_L be the lottery depicted in Figure 15.10

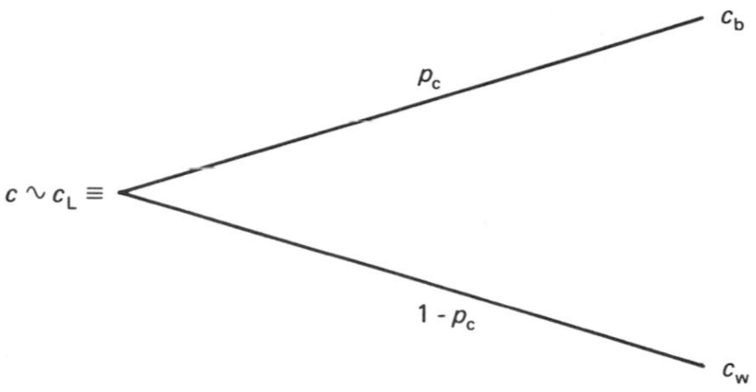

Figure 15.10

By (15.2.2), $u(c) = u(c_L)$. But by (15.2.1),

$$\begin{aligned} u_0(c_L) &= u_0(c_b)p_c + u_0(c_w)(1 - p_c) \\ &= [au(c_b) + b]p_c + [au(c_w) + b](1 - p_c) \\ &= a[u(c_b)p_c + u(c_w)(1 - p_c)] + b \\ &= au(c_L) + b. \\ &= au(c) + b \quad \text{(because } c_L \sim c\text{).} \end{aligned}$$

But by the contrary assumption,

$$u_0(c) \neq au(c) + b,$$

and so

$$u_0(c) \neq u_0(c_L).$$

Hence u_0 gainsays $c \sim c_L$ and consequently violates (15.2.2). We have thus shown that if u and u_0 both satisfy (15.2.2), then there is some $b \in R^1$ and some $a > 0$ such that, for every $c \in C$.

$$u_0(c) = au(c) + b. \quad \text{QED}$$

We conclude from (15.2.3) that there are many utility functions u: $C \to R^1$ which represent the decision-maker's preferences between consequences and between lotteries with ultimate prizes in C; but that any one of these functions can be obtained from any other by a change of *scale* (multiplying by $a > 0$) and a change of *origin* (adding b). Utility is similar to temperature in this respect: there are several popular temperature scales (for example, Fahrenheit, centigrade, and Kelvin), but all are related in the same form,

$$T_1 = aT_2 + b,$$

as are u and u_0.

Statement (15.2.3) is occasionally summarized by saying that any utility function u_0 is a *positive linear transformation* of any other, u; or that utility is determined only up to a positive linear transformation. The adjective *positive* means that $a > 0$. Another phraseological remark is that part (b) of (15.2.2) is often referred to as the expected utility "hypothesis," and that a utility function satisfying (b) is a *linear utility function*. We shall not use either of these terms in this book; in fact, *linear utility* will be used to denote something quite different in Section 15.4.

It is worth noting that this axiomatic development can be extended without essential change to the case of an infinite C. The only such case of interest in this book is that in which $C = R^1$ and the probabilities $p_1, \ldots, p_r$ in L are replaced by a dgf $f_L(c)$. When c is considered a random variable, it is easy to see that (b) of (15.2.2) becomes the following:

If L and L' are lotteries with prizes in C, then (15.2.2b′)

$$L \succsim L' \text{ if and only if } u(L) \geqq u(L'),$$

$$\text{where } u(L) \equiv E[u(\tilde{c}) \mid L] \equiv \int_{-\infty}^{\infty} u(c) f_L(c)\,dc, \text{ and similarly for } u(L').$$

Statements (15.2.2a) and (15.2.3) continue to obtain as stated.

15.3 ASSESSING UTILITY WHEN C IS A FINITE SET

The axiomatic development in Section 15.2 suggests a procedure by which a decision-maker who accepts Axioms 1 through 6 in principle can assess a utility function $u\colon C \to R^1$ when C is a finite set. We shall continue to assume that $c_b \succ c_w$, since if $c_b \sim c_w$ then he is indifferent between any two consequences and hence his decision problem is trivial, in that he is not concerned with which experiment and acts are chosen.

As *step one*, the decision-maker determines a c_b and a c_w: he asks himself which of the consequences he would be at least as happy to have as any other, and which he would be at least as reluctant to have as any other. If there is a "tie" for c_b (or for c_w) he chooses any one of the tied c's as c_b(or c_w). He then assigns $u(c_b) = 1$ and $u(c_w) = 0$.

Step two consists in assessing the $u(c)$'s for c's differing from c_b and c_w. First, he sets $u(c) = 1$, for every c such that $c \sim c_b$(if any), and he sets $u(c) = 0$, for every c such that $c \sim c_w$ (if any). Next, he assesses the utility of each c such that $c_b \succ c \succ c_w$. This task is based squarely upon Axiom 3, and involves the comparison of c with simple lotteries of the form shown in Figure 15.11.

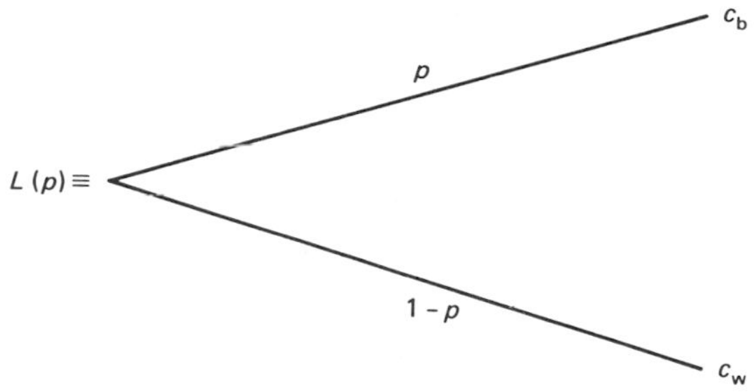

Figure 15.11

This comparison requires that he be able to imagine a randomization device (such as a fair roulette wheel) that will generate an event with probability p, conditional upon which he would receive c_b, and otherwise recieve c_w. Now he asks himself questions of the form, "If $p = p_0$, would I rather have $L(p_0)$ or c for sure?" Presumably, for p_0 very close to one he would prefer $L(p_0)$ to c; and for p_0 very close to zero he would prefer c to $L(p_0)$. Axiom 3 requires the existence of a single number $p = u(c) \in (0, 1)$ such that:

(1) he prefers $L(p)$ to c whenever $p > u(c)$;
(2) he prefers c to $L(p)$ whenever $p < u(c)$; and
(3) he is indifferent between c and $L(p)$ when $p = u(c)$.

Viewed in the assessment context, Axiom 3 appears to be a counsel of perfection. Almost every decision-maker would like to name two numbers, p_1 and p_2, such that $0 < p_1 < p_2 < 1$, with the properties that:

(1′) he prefers $L(p)$ to c whenever $p > p_2$;
(2′) he prefers c to $L(p)$ whenever $p < p_1$; and
(3′) he is indifferent between c and $L(p)$ whenever $p \in [p_1, p_2]$.

It is clear that (1′) through (3′) are less specific than (1) through (3); but all that we can say on the basis of (1′) through (3′) is that $u(c) \in [p_1, p_2]$. Fishburn [**23**] has investigated the extent to which such assessments can be used in solving the general decision problem.

We shall, however, assume that the decision-maker is able to assess $p_1 = p_2 = u(c)$ precisely.

Example 15.3.1

In the first example following Axiom 4 in Section 15.2, we might have arrived at $u(c) \equiv u(\text{eggs Benedict})$ by imagining an urn containing 100 balls, exactly x of which are red, and assuming that in $L(x/100)$ we would receive filet mignon if a red ball is drawn and a pork chop otherwise. By an act of will we can imagine that we really have a choice between $L(x/100)$ and $c =$ eggs Benedict; and we "converge" on $u(c) = .8$ as indicated in the following table:

$p = x/100$	*Preference*
1	$L(1)$
.0	c
.5	c
.95	$L(.95)$
.7	c
.9	$L(.9)$
.75	c
.85	$L(.85)$
.8	indifferent

Now, step two is performed for every c such that $c_b \succ c \succ c_w$. Once this has been accomplished, the decision-maker has determined $u(c)$ for every $c \in C$ and so appears to have finished. He should be cautioned, however, to test the implications of his assessments for consistency with his basic preferences between elements of C. This can be done by noting that, from (a) of (15.2.2), if $u(c) > u(c')$, then he prefers c to c'. He can ask himself for all such pairs, "Would I really prefer c to c'?" If the answer is yes, then he can check all pairs such that $u(c) = u(c')$ (and hence presumably $c \sim c'$) to see if he is really indifferent between c and c'. If so, then u has passed the simple consistency test.

A more stringent consistency test is to take every subset of three consequences such that $c \succ c' \succ c''$, determine that p for which $u(c') = pu(c) + (1 - p)u(c')$, and ask if he is really indifferent between: (1) c' for certain; and (2) the lottery shown in Figure 15.12.

If u fails any of the consistency tests, then it does not represent the decision-maker's preferences and must be reassessed. No specific rules can be given for the reassessment, except that the activity of assessment, checking,

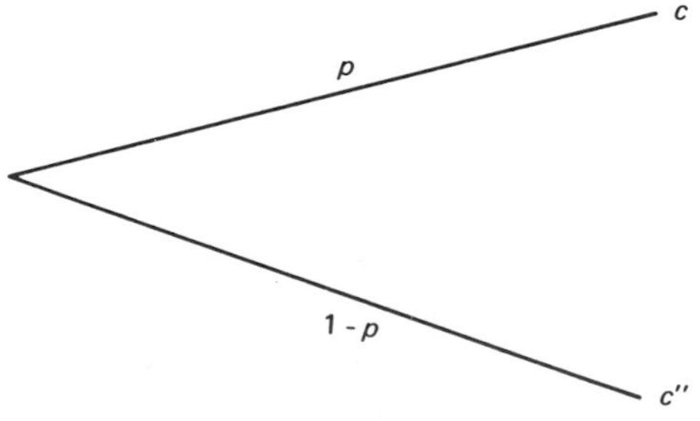

Figure 15.12

and reassessment often help the decision-maker clarify in his own mind his preferences, and hence one or two reassessments usually suffice.

Example 15.3.2

In a simple decision problem under uncertainty involving military strategy, posed by the author to a class in decision theory, there were four consequences: c_1, c_2, c_3, c_4. A student who is also a career Army officer assessed $u(c_1) = 1$, $u(c_2) = .85$ $u(c_3) = .8$, and $u(c_4) = 0$, but then noted that $c_1 \succ c_3 \succ c_2 \succ c_4$, and recognized the inconsistency of $.85 = u(c_2) > u(c_3) = .8$. His more careful reassessment yielded $u(c_2) = .85$ and $u(c_3) = .95$. He then found that he was indifferent between c_2 for certain and the lottery shown in Figure 15.13, as is required by the condition that

$$.85 = u(c_2) = \frac{17}{19}(.95) + \frac{2}{19}(0) = \frac{17}{19}u(c_3) + \frac{2}{19}u(c_4).$$

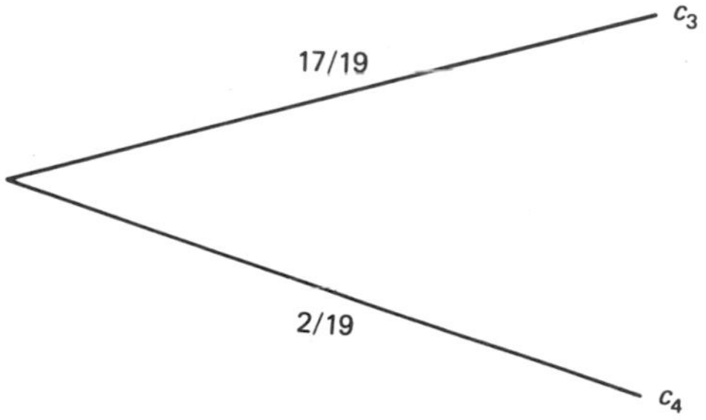

Figure 15.13

He was also indifferent between c_3 for certain and the lottery shown in Figure 15.14, which (the reader may verify as Exercise 4) is required by

$$u(c_3) = \frac{2}{3}u(c_1) + \frac{1}{3}u(c_2)$$

and the reassessed utility function.

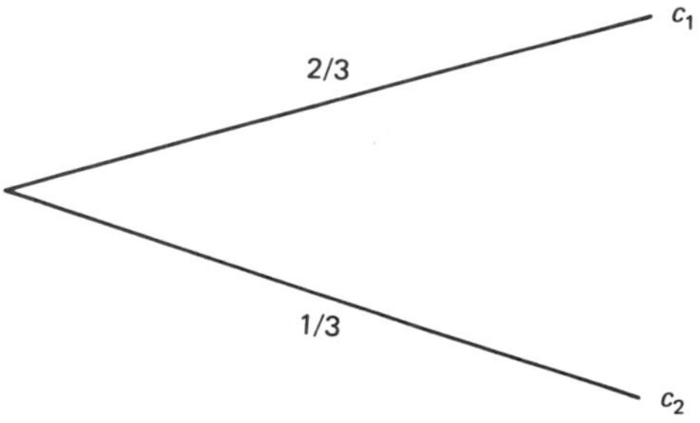

Figure 15.14

15.4 ASSESSING UTILITY WHEN C IS AN INFINITE SET

We have already noted that essentially the same axioms as in Section 15.2 imply the existence of a utility function $u: C \to R^1$ when C is an infinite set, but it should be clear that the assessment procedure of Section 15.3 breaks down when C is infinite and, in fact, becomes increasingly cumbersome as the still-infinite number of elements of C increases.

In this section we shall first discuss the assessment of a utility function $u: C \to R^1$ when $C = R^1$ and represents the set of all conceivable changes in the decision-maker's monetary assets from some fixed level—say, his assets at the time of the assessment. Second, we shall discuss assessments of utility when c is not entirely described as a monetary increment.

A natural prefatory question at this point is why utility is needed when C is monetary increments; after all, c is an entirely numerical description and admits the taking of expectations, and so forth. But the essential point is that *money need not represent the decision-maker's utility for money*—even when $\$c \succ \$c'$ if and only if $\$c > \$c'$, in which case setting $u(\$c) \equiv \$c$ for every $\$c \in R^1$ satisfies (a) of (15.2.2). For if $u(c) = c$ satisfied (b) of (15.2.2) as well, he would have to be completely indifferent between the lottery in Figure 15.15, and \$0 for sure.

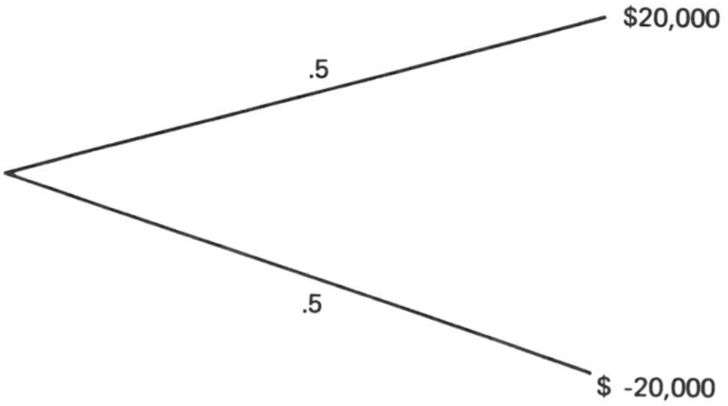

Figure 15.15

Few people are indifferent between flipping a (fair) coin for \$20,000 and not playing.

We shall describe a procedure for assessing $u(c)$ for $c \in [c_0, c_1]$, where $c_0 > -\infty$ and $c_1 < \infty$. This procedure exploits the connectedness of an interval of real numbers, together with two assumptions that most decision-makers would wish to accept:

(a) if $c > c'$, then $u(c) > u(c')$; that is, more money is always more desirable than less; and

(b) $u: [c_0, c_1] \to R^1$ is continuous; that is, small changes in money create small changes in utility.

To start the procedure, *first* arbitrarily set $u(c_0) = 0$ and $u(c_1) = 1$. *Second*, find a value c such that he is indifferent between: (1) receiving c for certain; and (2) playing the lottery which offers c_0 and c_1 as prizes, each with probability $\frac{1}{2}$. This c will be unique by virtue of (a) and (b); call it $c_{1/2}$. *Third*, find the (necessarily unique) value $c_{1/4}$ of c such that he is indifferent between: (1) receiving $c_{1/4}$ for certain; and (2) playing the lottery which offers c_0 and $c_{1/2}$ as prizes, each with probability $\frac{1}{2}$. *Fourth*, find the (necessarily unique) value $c_{3/4}$ of c such that he is indifferent between: (1) receiving $c_{3/4}$ for certain; and (2) playing the lottery which offers $c_{1/2}$ and c_1 as prizes, each with probability $\frac{1}{2}$.

We shall now show that steps two through four amount to assessing the $u = \frac{1}{2}$, $u = \frac{1}{4}$, and $u = \frac{3}{4}$ points, respectively. By (b) of (15.2.2) and the definition of $c_{1/2}$, it follows that

$$\begin{aligned} u(c_{1/2}) &= (\tfrac{1}{2})u(c_0) + (\tfrac{1}{2})u(c_1) \\ &= (\tfrac{1}{2})(0) + (\tfrac{1}{2})(1) \\ &= \tfrac{1}{2}; \end{aligned}$$

by the definition of $c_{1/4}$,

$$\begin{aligned} u(c_{1/4}) &= (\tfrac{1}{2})u(c_0) + (\tfrac{1}{2})u(c_{1/2}) \\ &= (\tfrac{1}{2})(0) + (\tfrac{1}{2})(\tfrac{1}{2}) \\ &= \tfrac{1}{4}; \end{aligned}$$

and by the definition of $c_{3/4}$,

$$\begin{aligned} u(c_{3/4}) &= (\tfrac{1}{2})u(c_{1/2}) + (\tfrac{1}{2})u(c_1) \\ &= (\tfrac{1}{2})(\tfrac{1}{2}) + (\tfrac{1}{2})(1) \\ &= \tfrac{3}{4}. \end{aligned}$$

The reader may now graph these assessed points and "fit" an increasing, continuous curve sucht hat $u(c_i) = i$, for $i = 0, \frac{1}{4}, \frac{1}{2}, \frac{3}{4}, 1$; or the reader may imitate steps two through four to determine c_i, for $i = \frac{1}{8}, \frac{3}{8}, \frac{5}{8}, \frac{7}{8}; \frac{1}{16}, \frac{3}{16}, \frac{5}{16}, \frac{7}{16}, \frac{9}{16}, \frac{11}{16}, \frac{13}{16}, \frac{15}{16}$, and so forth before fitting the curve.

Example 15.4.1

An individual was asked to assess his utility function for increments c to his present assets belonging to [0, 5000]. He assessed $u(0) = 0$, $u(5000) = 1$, $u(1000) = \frac{1}{2}$, $u(2000) = \frac{3}{4}$, $u(300) = \frac{1}{4}$, and $u(2800) = \frac{7}{8}$. A curve was fitted through these assessed values (indicated by heavy dots) as shown in Figure 15.16. Note that the expected *value* of the lottery offering \$0 and \$5000 with equal probability is $(\frac{1}{2})(\$0) + (\frac{1}{2})(\$5000) = \$2500$, but the decision-maker would *much* rather have \$2500 for certain than this lottery. Moreover, if we define $u_0(c) = 5000u(c)$, then $u_0(\$0) = 0$ and $u_0(\$5000) = 5000$, but $u_0(\$1000) = 2500$ and $u_0(\$2500) = 4100$. If money itself had been used as utility, we would of course have had 1000 and 2500 as the utility of \$1000 and \$2500, respectively.

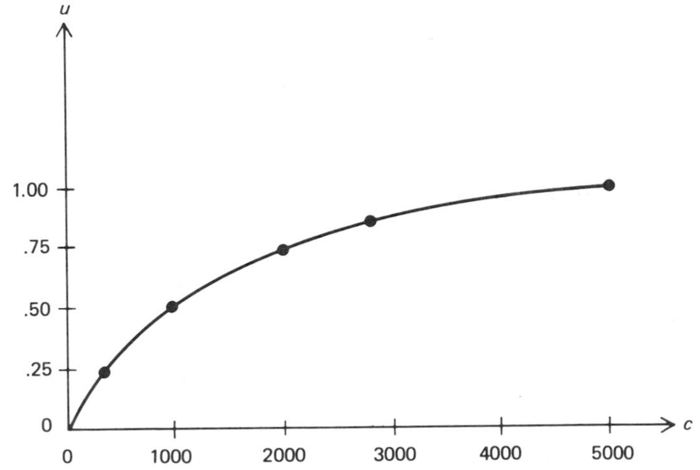

Figure 15.16

Now, once the decision-maker has assessed u: $[c_0, c_1] \to R^1$, he may *extend* this function to the right (above c_1) and/or to the left (below c_0). To extend u to the right of c_1, he may assess the (necessarily unique) value c_2 of c such that he is indifferent between: (1) receiving c_1 for certain, and (2) playing the lottery which offers c_0 and c_2 as prizes, each with probability $\frac{1}{2}$. That $u(c_2) = 2$ follows from

$$u(c_1) = (\tfrac{1}{2})u(c_0) + (\tfrac{1}{2})u(c_2),$$

which implies that

$$u(c_2) = 2u(c_1) - u(c_0) = 2 - 0 = 2.$$

To extend u to the left of c_0, he may assess the (necessarily unique) value c_{-1} of c such that he is indifferent between (1) receiving c_0 for certain, and (2) playing the lottery which offers c_{-1} and c_1 as prizes, each with probability $\frac{1}{2}$. That $u(c_{-1}) = -1$ follows from

$$u(c_0) = (\tfrac{1}{2})u(c_{-1}) + (\tfrac{1}{2})u(c_1);$$

that is,

$$u(c_{-1}) = 2u(c_0) - u(c_{-1}) = 0 - 1 = -1.$$

These extension procedures may be repeated indefinitely.

To motivate the need for utility even when C is an interval of incremental monetary returns, we gave an example of a substantial lottery offering equal chances at winning and losing \$20,000. For small amounts of money —say, one dollar—instead of 20,000, a decision-maker may be indifferent between the lottery and its expected monetary value. In many cases, decision-makers do have such preferences when the amounts of money involved do not constitute potentially substantial changes in their assets.

We shall now give a condition which, if satisfied, will enable the decision-maker to use the monetary consequences themselves as utility. In such cases we say the decision-maker's utility function is *linear in money*.

The condition is as follows:

Let $C = [c_0, c_1]$ be an interval of monetary returns, and let $L(p)$ denote the lottery which offers c_1 and c_0 as prizes with respective probabilities p and $1 - p$. If for every $p \in (0, 1)$ the decision-maker is indifferent between (1) receiving $pc_1 + (1 - p)c_0$ for certain, and (2) playing $L(p)$, then his utility function may be taken as $u(c) = c$. (15.4.1)

Proof

If the decision-maker is indifferent between alternatives (1) and (2), it follows from (b) of (15.2.2) that

$$u(pc_1 + (1 - p)c_0) = pu(c_1) + (1 - p)u(c_0).$$

Let $u(c_1)$ and $u(c_0)$ be originally defined as equal to one and zero, respectively. Then

$$u(pc_1 + (1 - p)c_0) = p(1) + (1 - p)(0) = p.$$

Now define $a = c_1 - c_0$ and $b = c_0$, and thus obtain

$$\begin{aligned} u_0(pc_1 + (1 - p)c_0) &= (c_1 - c_0)u(pc_1 + (1 - p)c_0) + c_0 \\ &= (c_1 - c_0)p + c_0 \\ &= pc_1 + (1 - p)c_0. \end{aligned}$$

We have shown that if $c \equiv pc_1 + (1 - p)c_0$, then $u_0(c) = c$. But every $c \in [c_0, c_1]$ is equal to $pc_1 + (1 - p)c_0$ for some $p \in [0, 1]$. The condition now follows by dropping the subscript of u_0. QED

A distinction must be made between a decision-maker's utility for his *own* money, and his utility for the money of *someone else*, say his employer corporation, for whom he acts as agent. Many decision-makers justifiably consider their corporation's utility to be linear in money.

Finally, it is necessary to consider situations in which some c's involve both an incremental monetary return and other, physical or psychological factors. In such cases it may be possible for the decision-maker to state an amount of money $x(c)$ such that he would be indifferent between: (1) $x(c)$, together with the *status quo* regarding the nonmonetary factors, and

(2) c itself. He may then replace the originally given consequences c of the decision problem with their monetary equivalents $x(c)$ and use his utility function for monetary returns. We say that he uses money as a *numeraire* for the consequences as originally defined.

15.5 RESULTS OF WEAKENING THE AXIOMS

The term *utility* is venerable in economics, dating at least as far back as the early nineteenth century to Jeremy Bentham, James Mill, and their colleagues, who were called *utilitarians* because of their axiom that man always acts so as to maximize his own pleasure and minimize his own pain. The ramifications of this axiom constitute what was called the *felicific calculus* (Latin *felix* ≡ "happy"). Its drawback was a lack of sufficient concreteness to explain or recommend any but very trivial and obvious types of behavior or decisions.

More specifically, numerical versions of utility were subsequently introduced in which Axioms 2 through 6 were abandoned and hence all that was required was Axiom 1, that "$\gtrsim$" be a total order in C.

Axiom 1, together with an axiom which essentially requires that there be at least as many real numbers as equivalence classes $\{c'\colon c' \sim c\}$ of "$\sim$," suffices for the existence of a real-valued function $u\colon C \to R^1$ satisfying (a) of (15.2.2).

But since abandonment of Axioms 2 through 6 means that nothing at all is said regarding the decision-maker's preferences between *lotteries*, it follows that (b) of (15.2.2) is meaningless, and that the utility theory so developed is irrelevant to decision-making under uncertainty.

A utility function satisfying (a) of (15.2.2) is called *ordinal*. Its sole task is to represent the "preference" and "indifference" relations via the "greater than" and "equality" relations in R^1. Moreover, it is determinate only up to a strictly increasing (not necessarily linear) transformation. To see this, note that if $g\colon R^1 \to R^1$ is a strictly increasing function [that is, $y > x$ if and only if $g(y) > g(x)$], and if $u\colon C \to R^1$ is an ordinal utility function, then so is u_0, defined by

$$u_0(c) = g[u(c)],$$

for every $c \in C$.

Proof

$c \succ c'$

if and only if $u(c) > u(c')$	[by (a) of (15.2.2)]
if and only if $g[u(c)] > g[u(c')]$	[g strictly increasing)
if and only if $u_0(c) > u_0(c')$	(definition of u_0).

Hence $c \succ c'$ if and only if $u_0(c) > u_0(c')$ and so u_0 also satisfies (a) of (15.2.2). Therefore u_0 is also an ordinal utility function. QED

Ordinal utility is sufficient for decisions under *certainty*: decisions in which $E = \{e_0\}$, $Z(e_0) = \{z_0\}$, and every act $a \in A(e_0, z_0)$ is associated with one and only one consequence $c(a)$. In such cases, it is easy to see that choosing an act is tantamount to choosing a consequence, and hence (a) of (15.2.2) is sufficient.

But decisions under certainty are of primary importance as analytical idealizations of decisions under uncertainty. The task of providing the decision-maker with a procedure for assigning to every act a number reflecting the relative desirability of that act when the decision is under uncertainty was begun (even before ordinal utility) by Daniel Bernoulli, but it awaited the efforts of John von Neumann and Oskar Morgenstern [**58**] for practical completion.

What we have called utility in previous sections is frequently called *von Neumann-Morgenstern utility*, or *cardinal utility*, to distinguish it from ordinal utility.

Cardinal utility has been axiomatized in several different (but essentially similar) ways since the appearance of von Neumann and Morgenstern [**58**] in 1944. Savage [**69**], Blackwell and Girshick [7], Raiffa and Schlaifer [**66**], and Pratt, Raiffa, and Schlaifer [**64**] are important references.

The remainder of this section is concerned with the consequences of dropping or modifying one or both of the more controversial Axioms 1 and 3.

J. von Neumann and O. Morgenstern [**58**] showed that if Axiom 3 is dropped, then $u(c)$ is no longer a real number but rather a *vector* of real numbers; that is, $u\colon C \to R^n$, for some $n > 0$. Moreover, one can show that $L \succsim L'$ if and only if $u(L) \geqq u(L')$, where "$\geqq$" is the "lexicographic order" of R^n, in which $(x_1, \ldots, x_n) \geqq (y_1, \ldots, y_n)$ if and only if: (1) either $x_1 > y_1$; (2) or $x_1 = y_1$ and $x_2 > y_2$; (3) or $x_1 = y_1$ and $x_2 = y_2$ but $x_3 > y_3$; . . . ; (n) or $x_i = y_i$, for $i = 1, \ldots, n-1$ but $x_n \geqq y_n$. Note that, roughly speaking, each coordinate i is "infinitely more important" than its successor coordinate $i+1$; for example, $(1, 1) \geqq (1 - 10^{-90}, 1 + 10^{90})$. Hausner [**36**] has provided a self-contained development of multidimensional utility theory without the continuity axiom, and Thrall [**72**] has shown that much of decision theory requires little modification with multidimensional utility theory.

Axiom 1 requires that "$\succsim$" be a *total* order in C, and hence in particular any two consequences must be comparable. Aumann [**4**] investigates the ramifications of replacing Axiom 1 by an axiom requiring that "$\succsim$" be only a *partial* order in C. Then a real-valued utility function on C is still possible (when C is infinite), and that function continues to *represent* preferences in the sense that if $c \succ c'$, then $u(c) > u(c')$, but the converse is no longer valid: $u(c) > u(c')$ does not necessarily mean that $c \succ c'$. The extent to which u is determinate, and the implications of such a utility function for decision theory, are given in [**4**].

The axiomatic development in Section 15.2 is that given in Chapter 2 of Luce and Raiffa [**52**]. Fishburn [**25**] has contributed a comprehensive survey of utility theory, together with references to other relevant books and papers.

EXERCISES

1. Show that if $c_b \sim c_w$, then $c_1 \sim c_2 \sim \cdots \sim c_r$, provided that Axiom 1 is satisfied.

2. Show that if $u: C \rightarrow R^1$ is as defined by Axiom 3, then $u(c_b) = 1$ and $u(c_w) = 0$. Also show that if $c \sim c_b$, then $u(c) = 1$, and if $c \sim c_w$, then $u(c) = 0$.

3. Show that in (15.2.2), (a) is a special case of (b) by showing that (b) implies (a). [*Hint*: Let $L = L(c)$ and $L' = L(c')$.]

4. Show in Example 15.3.2 that $u(c_1) = 1$, $u(c_2) = .85$ and $u(c_3) = .95$ imply that the assessor is indifferent between c_3 for certain and the lottery offering c_1 and c_2 as prizes with probabilities $\frac{2}{3}$ and $\frac{1}{3}$, respectively.

5. Assess *your* utility function for incremental monetary returns, given your current asset position y.

6. Let y denote the current asset position, x denote an incremental monetary return, and $u_y(x)$ a utility function given y. Show that if u_y has been assessed, for all $x \in R^1$, then the function $u^*: R^1 \rightarrow R^1$ defined by

$$u^*[x + y] = u_y(x), \text{ all } x \in R^1$$

is a utility function for *asset positions.*

7. Let $\tilde{x}$ denote a random incremental return—alternatively, a lottery with monetary prizes—and let u^* be as defined in Exercise 6.

   (a) Show that $E(u^*[y + \tilde{x}])$ is the decision-maker's (expected) utility of his as yet unknown asset position *after* playing the lottery.

   (b) Define $S(\tilde{x} \mid y)$ by

   $$u^*[y + S(\tilde{x} \mid y)] \equiv E(u^*[y + \tilde{x}]).$$

   Show that $S(\tilde{x} \mid y)$ exists for every initial asset position y such that $E(u^*[y + \tilde{x}])$ exists.

   (c) Argue that the decision-maker should sell his right to play the lottery for $\$\Delta$ if $\Delta > S(\tilde{x} \mid y)$, and should not sell if $\Delta < S(\tilde{x} \mid y)$. Thus $S(\tilde{x} \mid y)$ represents his minimum selling price.

(d) Show that if utility u^* is linear in money, then $S(\tilde{x} \mid y) = E(\tilde{x})$, for every $y \in R^1$.

(e) Argue that if the decision-maker does not own but wishes to purchase the right to play the lottery, then he should pay no more than $B(\tilde{x} \mid y)$, defined implicitly by

$$u^*(y) \equiv E(u^*[y + \tilde{x} - B(\tilde{x} \mid y)]).$$

[*Note*: Only in special cases is it true that $B(\tilde{x} \mid y) = S(\tilde{x} \mid y)$. Pfanzagl [**61**] and Pratt [**62**] have shown that if $S(\tilde{x} \mid y) = B(\tilde{x} \mid y)$ and both are independent of y for all $\tilde{x}$, then u^* must be representable as (must be a positive linear transformation of) $u^*[y] = y$, $u^*[y] = -\exp(-my)$ for some $m > 0$, or $u^*[y] = \exp(my)$ for some $m > 0$.]

8. Suppose you have a decision problem with the following consequence set $C = \{c_1, c_2, c_3\}$:

$c_1 \equiv$ "receive, free, the dinner of your choice at the restaurant of your choice";
$c_2 \equiv$ "receive ten dollars"; and
$c_3 \equiv$ "wash my car five times."

Assess your utility function $u: C \rightarrow R^1$.

9. By using Axion 2, reduce the (three-stage) compound lottery in Figure 15.17 to a (one-stage) simple lottery. (*Hint*: Use Axiom 2 to reduce stages 2 and 3 to a single stage, and then use Axiom 2 again to reduce the resulting two-stage lottery to a one-stage lottery.)

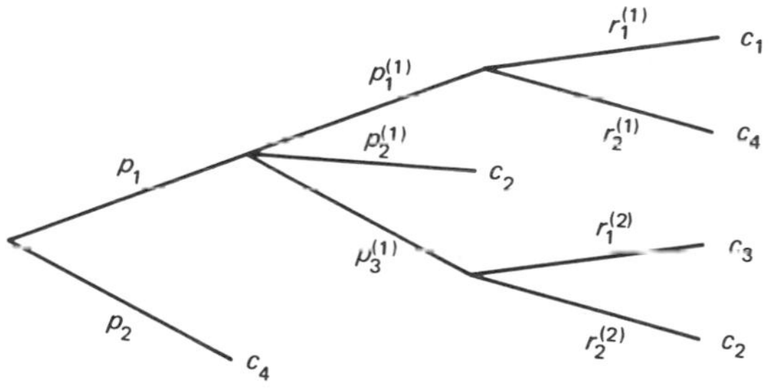

Figure 15.17

10. Utility functions can be defined on C when $C = [c_0, c_1]$ but c is not a monetary consequence. Suppose $C = [0, 1] = \{c: c$ is the proportion of coffee and $1 - c$ is the proportion of cocoa in a cup$\}$. For this example you will probably wish to reject assumption (a) of Section 15.4; namely, that $c > c'$ implies $u(c) > u(c')$.

(a) Assess *your* utility function $u: C \rightarrow R^1$.

(b) Some students have mistakenly confused the *mixture* consisting of proportions c of coffee and 1-c of cocoa with the *lottery* in Figure 15.18.

Why is this a mistake? Does it matter for utility purposes? [*Hint*: Are *you* indifferent between: (1) a cup filled with equal parts of coffee and cocoa, and (2) playing the lottery which yields as prizes a cup of coffee and a cup of cocoa, each with probability $\frac{1}{2}$?]

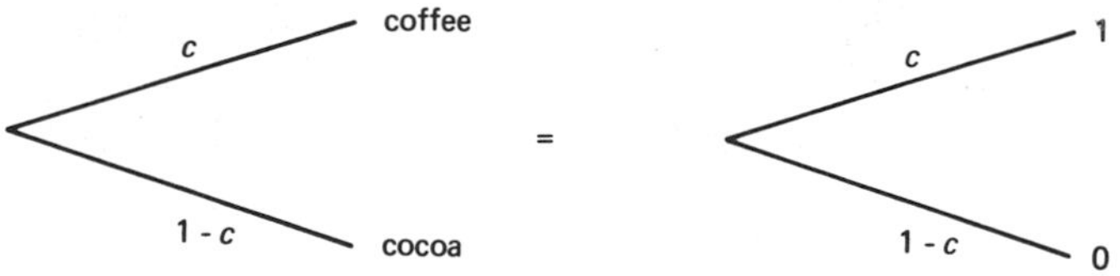

Figure 15.18

16

ELEMENTS OF BAYESIAN DECISION THEORY (II): SUBJECTIVE PROBABILITY

16.1 INTRODUCTION

Chapter 15 showed that six axioms imply the existence of a utility function which usefully mirrors the decision-maker's preferences between consequences and between lotteries with consequences as ultimate prizes. This chapter shows how two additional axioms, together with the results of Chapter 15, imply the existence of subjective probabilities $P(\theta_1), \ldots, P(\theta_s)$ of the states $\theta_1, \ldots, \theta_s$ in a decision problem.

This demonstration was first made by Anscombe and Aumann [**3**] and constitutes Section 16.2. It is restricted to the case in which θ is a finite set. However, it remains valid when Θ is an infinite set, $\{\Theta_1, \ldots, \Theta_s\}$ is any finite partition of Θ, and it is necessary to prove the existence of $P(\Theta_1), \ldots, P(\Theta_s)$.

Section 16.3 discusses how a decision-maker may actually assess subjection probabilities $P(\theta_1), \ldots, P(\theta_s)$ when Θ is a finite set, while Section 16.4 concerns the assessment problem when Θ is infinite and the decision-maker wishes to select that member of a conjugate family which adequately expresses his prior judgments.

Section 16.5 concerns other assessment procedures and other axiom systems for utility and subjective probability. Thus the basic outline of this chapter parallels that of Chapter 15.

As in Chapter 15, we shall need some prefatory comments on notation. The symbol L, with or without a subscript or subscripts, will denote any simple or compound lottery which is objective in the sense that its prizes are conditional upon events with *objective* probabilities.

We shall also denote the objective compound lottery (as shown in

Figure 16.1) which yields objective lottery L_{ij} as prize with objective probability p_j, for $j = 1, \ldots, k$, by $(p_1L_{i1}, \ldots, p_kL_{ik})$. (The justification for the subscript i will be clear later.) This notation is obviously a space saver.

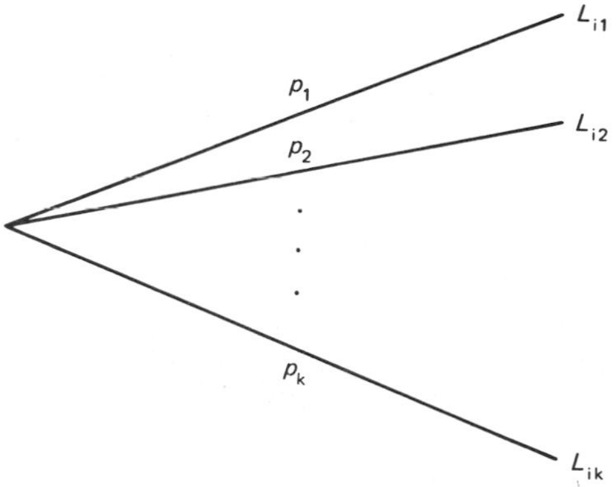

Figure 16.1

We denote by $\mathcal{L}$ the set of *all* simple or compound objective lotteries with ultimate prizes in C.

The discerning reader will have suspected that our present use of the qualifier *objective* in referring to lotteries portends the introduction of subjective lotteries. If $L_1, \ldots, L_s$ are any (objective) lotteries in $\mathcal{L}$, we use the symbol $[L_1, \ldots, L_s]$ to denote the subjective lottery in which the decision-maker receives L_j as prize if θ_j occurs, for $j = 1, \ldots, s$. Thus we have Figure 16.2.

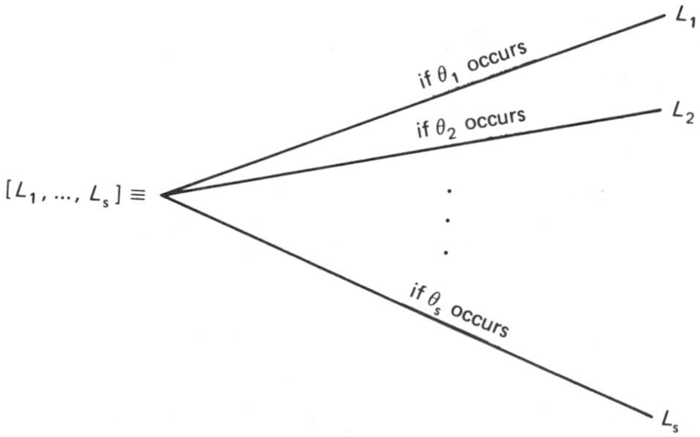

Figure 16.2

We denote by $\mathcal{S}$ the set of all subjective lotteries of the form $[L_1, \ldots, L_s]$.

Finally, we shall need to consider compound lotteries in which the first stage is objective and the second stage is subjective; that is, lotteries of the form shown in Figure 16.3, where k is arbitrary and $p_1, \ldots, p_k$ are any nonnegative numbers that sum to one. We shall denote such a lottery by $(p_1[L_{11}, \ldots, L_{1s}], \ldots, p_k[L_{k1}, \ldots, L_{ks}])$, and the set of all such lotteries by $\mathcal{L}^*$. Roughly speaking, $\mathcal{L}$ is to C as $\mathcal{L}^*$ is to $\mathcal{S}$.

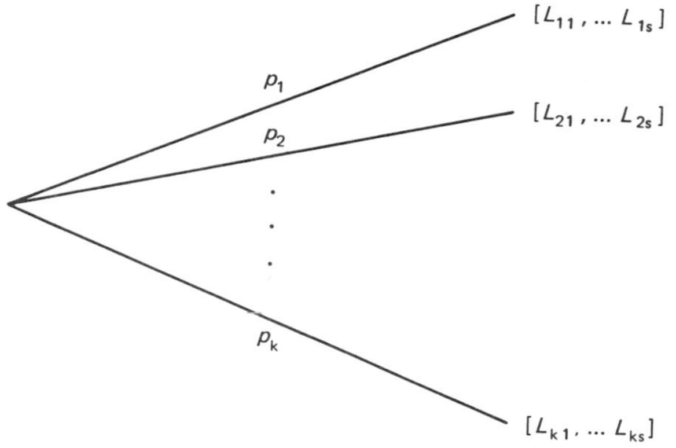

Figure 16.3

16.2 AN AXIOMATIC DEVELOPMENT OF SUBJECTIVE PROBABILITY

We shall assume, as in Chapter 15, that $c_b \succ c_w$, since otherwise the entire decision problem is trivial. We shall further assume that the decision-maker accepts Axioms 1 through 6 of Chapter 15 and hence has a utility function $u: C \to R^1$ which satisfies (15.2.2) and, in view of the $u(L)$-notation, can be thought of as having domain $\mathcal{L}$ rather than C.

We shall also assume that the decision-maker has preferences between elements of $\mathcal{L}^*$; that is, between compound lotteries of the form $(p_1[L_{11}, \ldots, L_{1s}], \ldots, p_k[L_{k1}, \ldots, L_{ks}])$. Preference-or-indifference, preference, and indifference between elements of $\mathcal{L}^*$ will be denoted by $\succsim^*$, $\succ^*$, and $\sim^*$, respectively to prevent confusion with $\succsim$, $\succ$, and $\sim$ between elements of $\mathcal{L}$.

Our last informal assumption is that the decision-maker wishes $\succsim^*$, $\succ^*$, and $\sim^*$ to obey, *in principle*, Axioms 1 through 6. We stress "in principle," because this section emphasizes that if the decision-maker accepts Axioms 7 and 8, then he will never have to assess his utility function u^*, on $\mathcal{L}^*$ [see (16.2.6)].

If the decision-maker has a utility function u^* on L*, then it follows immediately from (b) of (15.2.2) that

$$u^*((p_1[L_{11}, \ldots, L_{1s}], \ldots, p_k[L_{k1}, \ldots, L_{ks}])) = \sum_{i=1}^{k} u^*([L_{i1}, \ldots, L_{is}])p_i, \tag{16.2.1}$$

for every $(p_1[L_{11}, \ldots, L_{1s}], \ldots, p_k[L_{k1}, \ldots, L_{ks}]) \in \mathcal{L}^*$, and hence our primary attention will be focused on $u^*\colon \mathcal{S} \to R^1$, where $\mathcal{S}$ is the set of all subjective lotteries of the form $[L_1, \ldots, L_s]$.

We shall now state the first of two axioms which relate the decision-maker's preferences ($\gtrsim^*$) in $\mathcal{L}^*$ with those ($\gtrsim$) in $\mathcal{L}$. Axiom 7 roughly says that if one modifies any $[L_1, \ldots, L_s] \in \mathcal{S}$ by replacing some L_i with a "no-more-desirable" L'_i and keep the other L's fixed, then the modified lottery is not more desirable than $[L_1, \ldots, L_s]$.

Axiom 7

For any $[L_1, \ldots, L_s] \in \mathcal{S}$ and any $i \in \{1, \ldots, s\}$, if $L_i \gtrsim L'_i$, then

$$[L_1, \ldots, L_i, \ldots, L_s] \gtrsim^* [L_1, \ldots, L'_i, \ldots, L_s].$$

In other words, the desirability ($\gtrsim^*$) of a lottery in $\mathcal{S}$ is not decreased by replacing one of its prizes with a prize which is at least as desirable ($\gtrsim$) as the one replaced.

Axiom 7 expresses two idealizations. The first involves the joy of gambling and is illustrated by cases in which $L_i > L'_i$ but $[L_1, \ldots, L'_i, \ldots, L_s] > [L_1, \ldots, L_i, \ldots, L_s]$. For example, suppose $L_i \equiv$ (0 "fall off mountain," 1 "receive \$10"), while $L'_i \equiv$ (.01 "fall off mountain," .99 "receive \$10"); a decision-maker might strictly prefer L_i to L'_i, but at the same time hold that $[L_1, \ldots, L'_i, \ldots, L_s]$ furnishes a preferable amount of excitement vis-à-vis $[L_1, \ldots, L_i, \ldots, L_s]$.

The second idealization is really an independence assumption, which requires that the outcomes $\theta_1, \ldots, \theta_s$ are not affected by the prizes. For example, suppose $L_1, \ldots, L_{i-1}, L_{i+1}, \ldots, L_s$ are all the consequence (1 "receive \$10"), while $L'_i \equiv$ (1 "receive \$15") and $L_i \equiv$ (1 "receive \$1,000,000"). Obviously $L_i > L'_i$; but if the subjective lottery is "rigged," the decision-maker may feel sure of receiving \$10 from $[L_1, \ldots, L_i, \ldots, L_s]$ and of receiving at least \$10 from $[L_1, \ldots, L'_i, \ldots, L_s]$. Hence he would have $[L_1, \ldots, L'_i, \ldots, L_s] \gtrsim^* [L_1, \ldots, L_i, \ldots, L_s]$ despite $L_i > L'_i$.

From Axiom 7 we may draw some conclusions about $u^*\colon \mathcal{S} \to R^1$. First, recall that every $c \in C$ can be considered an element of $\mathcal{L}$; namely, $L(c) \equiv (1c) \in \mathcal{L}$. Hence $[c, \ldots, c] \equiv [(1c), \ldots, (1c)] \in \mathcal{S}$, for every $c \in C$.

But clearly $[c, \ldots, c]$ is the subjective lottery which yields c as its prize regardless of which θ_i occurs. Hence every $c \in C$ determines a $[c, \ldots, c] \in \mathcal{S}$. (16.2.2)

(a) For any $[L_1, \ldots, L_s] \in \mathcal{S}$. $[c_b, \ldots, c_b] \gtrsim^* [L_1, \ldots, L_s]$ and $[L_1, \ldots, L_s] \gtrsim^* [c_w, \ldots, c_w]$;
(b) $[c_b, \ldots, c_b] \succ^* [c_w, \ldots, c_w]$;
(c) u^* can be chosen such that $u^*([c_b. \ldots, c_b]) = 1$ and $u^*([c_w, \ldots, c_w]) = 0$; and
(d) if $i \in \{1, \ldots, s\}$ and $L_i \sim L'_i$, then $[L_1, \ldots, L_i, \ldots, L_s] \sim^* [L_1, \ldots, L'_i, \ldots, L_s]$.

Proof

For (a), note that $c_b \equiv (1c_b) \gtrsim L_i$ and apply Axiom 7 s times to obtain

$$[c_b, L_2, \ldots, L_s] \gtrsim^* [L_1, \ldots, L_s],$$
$$[c_b, c_b, L_3, \ldots, L_s] \gtrsim^* [c_b, L_2, \ldots, L_s],$$
$$\vdots$$
$$[c_b, c_b, \ldots, c_b] \gtrsim^* [c_b, c_b, \ldots, c_b, L_s],$$

which implies $[c_b, \ldots, c_b] \gtrsim^* [L_1, \ldots, L_s]$ by transitivity. Similarly for $L_1, \ldots, L_s] \gtrsim^* [c_w, \ldots, c_w]$. Part (b) is obvious from (a) and the assumption that $c_b \succ c_w$. For (c), note that (a) says that $[c_b, \ldots, c_b]$ is the most desirable element of $\mathcal{S}$, and $[c_w, \ldots, c_w]$ is the least desirable. In view of (b), the origin and scale of u^* can be chosen so that $u^*([c_b, \ldots, c_b]) = 1$ and $u^*([c_w, \ldots, c_w]) = 0$. You may prove (d) as Exercise 1. QED

In order to simplify the ensuing development, we shall make the following assumption:

The values u and u^* have been chosen such that

$$u(c_b) = 1 = u^*([c_b, \ldots, c_b]) \text{ and}$$
$$u(c_w) = 0 = u^*([c_w, \ldots, c_w]). \tag{16.2.3}$$

Next, we shall show that the (u^*—) utility of a lottery $[L_1, \ldots, L_s]$ is completely determined by the utilities $u(L_1), \ldots, u(L_s)$.

If $u(L_i) = u(L'_i)$, for every $i \in \{1, \ldots, s\}$, then

$$u^*([L_1, \ldots, L_s]) = u^*([L'_1, \ldots, L'_s]). \tag{16.2.4}$$

Proof

By (b) of (15.2.2), $u(L_i) = u(L'_i)$ if and only if $L_i \sim L'_i$. Hence the hypothesis of (16.2.4) is tantamount to assuming that $L_i \sim L'_i$, for every i. Now apply (d) of (16.2.2) s times to conclude that $[L_1, \ldots, L_s] \sim^* [L'_1, \ldots, L'_s]$. Hence these subjective lotteries must have the same (u^*-) utility. QED

In view of (16.2.3) and (16.2.4), every $[L_1, \ldots, L_s]$ can be represented by the s-tuple $[u_1 \ldots, u_s]$, where $u_i \equiv u(L_i)$, for $i = 1, \ldots, s$; and we shall occasionally consider u^* as if it were defined on the set of all s-tuples $[u_1, \ldots, u_s]$ such that $u_i \in [0, 1]$, for every $i \in \{1, \ldots, s\}$. Hence we define $u^*([u_i, \ldots, u_s])$ as equal to $u^*([L_1, \ldots, L_s])$, where L_i is any lottery in $\mathfrak{L}$ such that $u(L_i) = u_i$.

Axiom 8 is rather similar to, but milder than, Axiom 2. It implies, for example, that if one's prize L_{ij} is determined both by the spin of a roulette wheel and by a horserace, then one should not care whether the race is run before the wheel is spun or vice versa.

Axiom 8

For every $(p_1(L_{11}, \ldots, L_{1s}], \ldots, p_k[L_{k1}, \ldots, L_{ks}]) \in \mathfrak{L}^*$,

$$(p_1[L_{11}, \ldots, L_{1s}], \ldots, p_k[L_{k1}, \ldots, L_{ks}]) \sim^* [(p_1L_{11}, \ldots, p_kL_{k1}), \ldots, (p_1L_{1s}, \ldots, p_kL_{ks})].$$

In tree form, Axiom 8 reads as shown in Figure 16.4 and is equally as desirable as the tree shown in Figure 16.5.

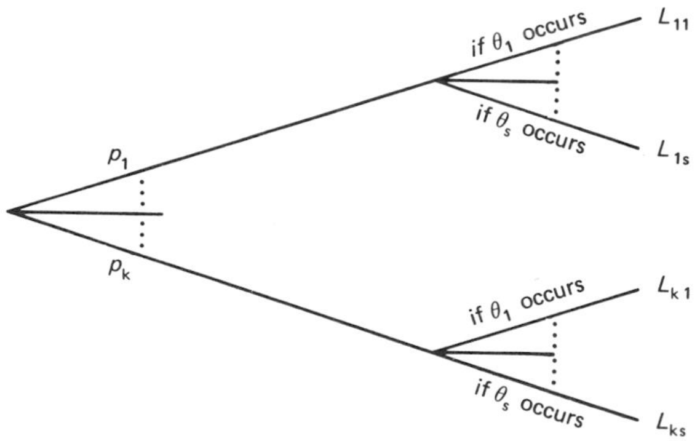

Figure 16.4

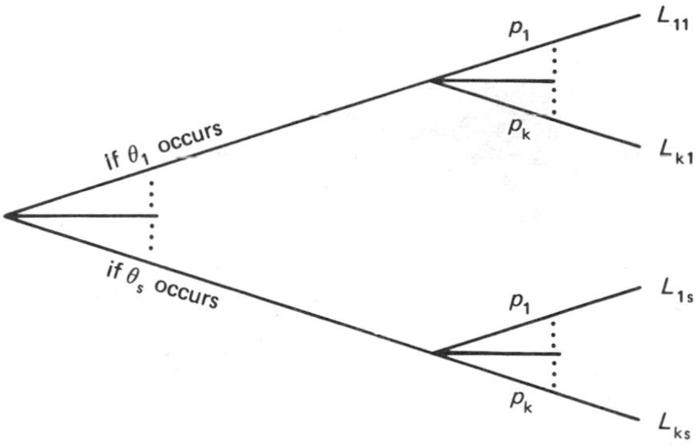

Figure 16.5

Next, we state a result which is of only passing interest in its own right but which will be useful in proving the existence of subjective probabilities.

Let x be a given positive number, and suppose that $u_i \in [0, 1]$ and $xu_i \in [0, 1]$, for every $i \in \{1, \ldots, s\}$. Then (16.2.5)

$$u^*([xu_1, \ldots, xu_s]) = xu^*([u_1, \ldots, u_s]).$$

Proof

First suppose that $x \leqq 1$. Then $xu_i = xu_i + (1 - x)0$, and hence if $u_i = u(L_i)$, then xu_i is the utility of a compound (objective) lottery of the form $(xL_i, (1 - x)c_w) \in \mathfrak{L}$. This is true, for every $i \in \{1, \ldots, s\}$. Hence $[xu_1, \ldots, xu_s]$ represents a subjective lottery of the form shown in Figure 16.6, which has utility $u^*([xu_1, \ldots, xu_s])$. But by Axiom 8, this lottery is equally as desirable as that shown in Figure 16.7, which has utility

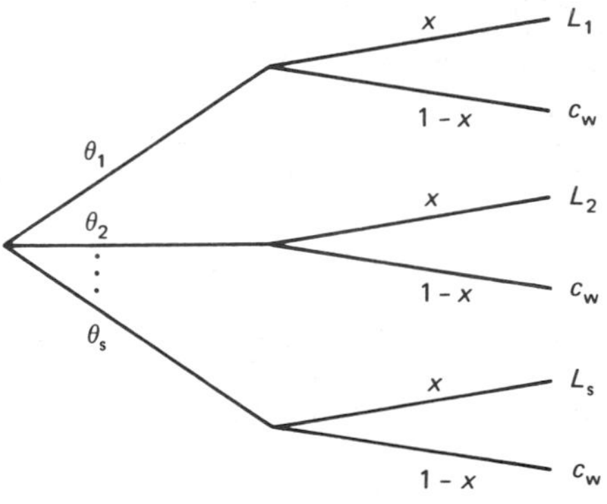

Figure 16.6

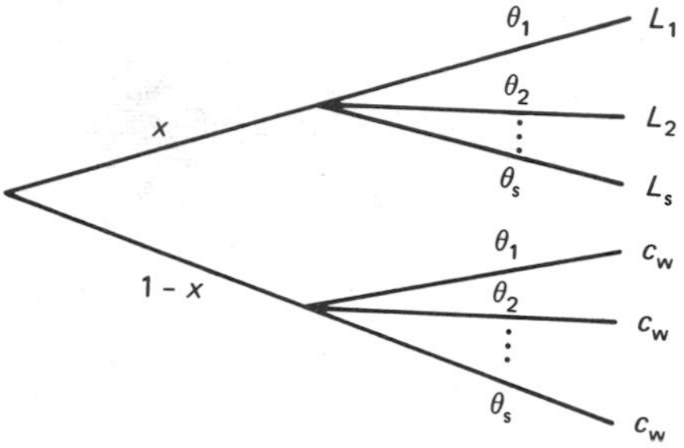

Figure 16.7

$u^*((x[L_1, \ldots, L_s], (1 - x)[c_w, \ldots, c_w]))$. Hence $u^*([xu_1, \ldots, xu_s]) = u^*((x[L_1, \ldots, L_s], (1 - x)[c_w, \ldots, c_w]))$. But by (16.2.1) and (16.2.3),

$$\begin{aligned} &u^*((x[L_1, \ldots, L_s], (1 - x)[c_w, \ldots, c_w])) \\ &\quad = xu^*([L_1, \ldots, L_s]) + (1 - x)u^*([c_w, \ldots, c_w]) \\ &\quad = xu^*([u_1, \ldots, u_s]) + (1 - x)0, \end{aligned}$$

concluding the proof for $x \leqq 1$. If $x > 1$, then $1/x < 1$, and we have

$$\begin{aligned} xu^*([u_1, \ldots, u_s]) &= xu^*\left(\left[\left(\frac{1}{x}\right)xu_1, \ldots, \left(\frac{1}{x}\right)xu_s\right]\right) \\ &= \left(\frac{1}{x}\right)xu^*([xu_1, \ldots, xu_s]) \\ &= u^*([xu_1, \ldots, xu_s]), \end{aligned}$$

where the second equality follows from the proof of the assertion for $1/x < 1$. QED

We may now state and prove the desired result.

There is a unique set of probabilities $P(\theta_1), \ldots, P(\theta_s)$ such that

$$u^*([L_1, \ldots, L_s]) = \sum_{i=1}^{s} u(L_i)P(\theta_i), \tag{16.2.6}$$

for every $[L_1, \ldots, L_s] \in \mathcal{S}$.

Proof

Let $U \equiv \sum_{i=1}^{s} u(L_i)$. If $U = 0$ then, all $u(L_i)$'s are zero and $[L_1, \ldots, L_s] \sim^* [c_w, \ldots, c_w]$. Hence both sides of the asserted equality are zero. If $U \neq 0$, then by (16.2.3) $U > 0$. In this case we have

$$\begin{aligned} u^*([L_1, \ldots, L_s]) &= u^*([u_1, \ldots, u_s]) \\ &= u^*([Uu_1/U, \ldots, Uu_s/U]) \\ &= Uu^*([u_1/U, \ldots, u_s/U]) \quad \text{[by (16.2.5)]}. \end{aligned}$$

Next, note that $[u_1/U, \ldots, u_s/U]$ represents the lottery shown in Figure 16.8, which by Axiom 8 is equally as desirable as the lottery in Figure 16.9, the utility of which is

$$\left(\frac{u_1}{U}\right)u^*([c_b, c_w, \ldots, c_w]) + \ldots + \left(\frac{u_s}{U}\right)u^*([c_w, \ldots, c_w, c_b]) \quad \text{[by (16.2.1)]}.$$

Hence

$$Uu^*\left(\left[\frac{u_1}{U}, \ldots, \frac{u_s}{U}\right]\right)$$
$$= u_1u^*([c_b, c_w, \ldots, c_w]) + \ldots + u_su^*([c_w, \ldots, c_w, c_b]).$$

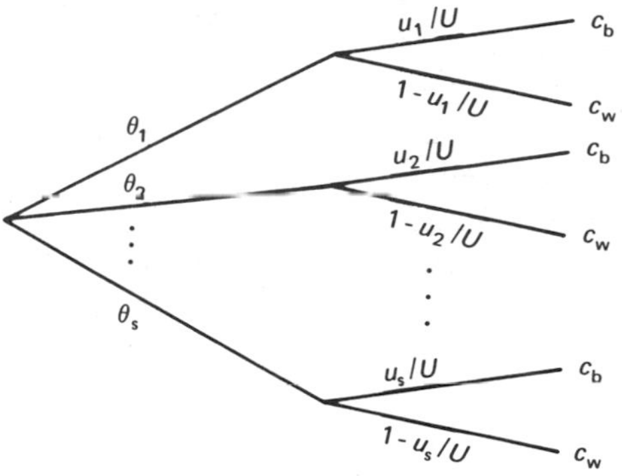

Figure 16.8

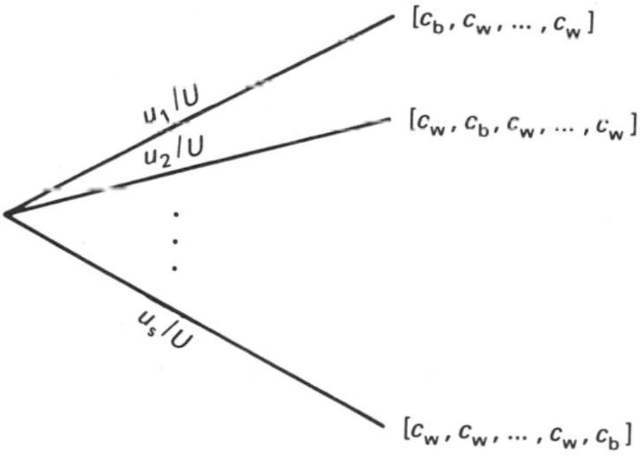

Figure 16.9

Now define

$$P(\theta_i) \equiv u^*([c_w, \ldots, c_w, c_b, c_w, \ldots, c_w]),$$

where on the right-hand side c_b is the prize given θ_i and c_w is the prize otherwise. Combining equalities and using the definition of $P(\theta_i)$ yields the asserted equality. By (16.2.3), $P(\theta_i) \in [0, 1]$, for every $i \in \{1, \ldots, s\}$; and to show that $\sum_{i=1}^{s} P(\theta_i) = 1$, it suffices to put $[L_1, \ldots, L_s] = [c_b, \ldots, c_b]$ in order to obtain, from (16.2.3),

$$\begin{aligned} 1 &= u^*([c_b, \ldots, c_b]) \\ &= \sum_{i=1}^{s} u(c_b)P(\theta_i) \\ &= \sum_{i=1}^{s} (1)P(\theta_i). \end{aligned}$$

Hence $P(\cdot)$: $\{\theta_1, \ldots, \theta_s\} \to R^1$ is a valid probability function. QED

The definition of $P(\theta_i)$ as the utility of $[c_w, \ldots, c_b, \ldots, c_w]$, where c_b is the prize only if θ_i occurs, has important implications for the interpretation and assessment of subjective probabilities. The reader will recall, by reviewing Axiom 3 in Chapter 15 if necessary, that $u^*([c_w, \ldots, c_b, \ldots, c_w])$ is that *objective* probability such that the decision-maker is indifferent between: (1) playing $[c_w, \ldots, c_b, \ldots, c_w]$; and (2) playing the lottery shown in Figure 16.10.

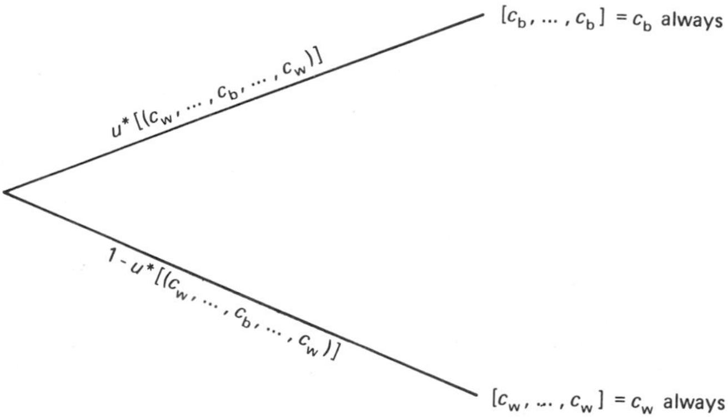

Figure 16.10

If we rewrite $[c_w, \ldots, c_b, \ldots, c_w]$ suggestively as shown in Figure 16.11, then the definition

$$P(\theta_i) \equiv u^*([c_w, \ldots, c_b, \ldots, c_w]) \tag{16.2.7}$$

says that the decision-maker's *subjective* probability of θ_i is that objective probability p such that he is indifferent between: (1) the objective lottery yielding c_b with probability p and c_w with probability $1 - p$; and (2) the

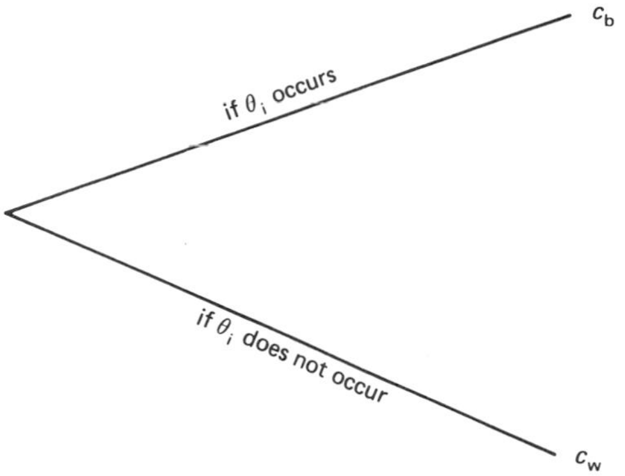

Figure 16.11

subjective lottery yielding c_b if θ_i occurs and c_w otherwise. This fact is of extreme importance in suggesting a simple procedure for assessing $P(\theta_i)$, as we shall see in the next section (also see Section 5.4 for example).

By now the reader might have begun to worry about cases in which the θ_i's have objective probabilities, say, $p(\theta_1), \ldots, p(\theta_s)$. Can it happen that $P(\theta_i) \neq p(\theta_i)$ for some i in such cases? The answer is yes if the decision-maker does not know $p(\theta_1), \ldots, p(\theta_s)$, since then the probabilities of $\theta_1, \ldots, \theta_s$ are unknown *to him*.

On the other hand, if the decision-maker *does* know $p(\theta_1), \ldots, p(\theta_k)$, then $[L_1, \ldots, L_s] \sim (p(\theta_1)L_1, \ldots, p(\theta_s)L_s)$ for every $L_1, \ldots L_s$; and so in particular

$$[c_w, \ldots, c_b, \ldots, c_w] \sim (p(\theta_i)c_b, (1 - p(\theta_i))c_w), \tag{16.2.8}$$

for every $i \in \{1, \ldots, s\}$. By (16.2.6), the utility of the left-hand side of (16.2.8) is $P(\theta_i)$, while the right-hand side of (16.2.8) is in $\mathcal{L}$ and so by (b) of (15.2.2) has utility $p(\theta_i)\cdot 1 + [1 - p(\theta_i)]\cdot(0) = p(\theta_i)$. Since the two sides of (16.2.8) are equally desirable to the decision-maker, they have the same utility and hence $P(\theta_i) = p(\theta_i)$. Since this argument holds for every $i \in \{1, \ldots, s\}$, it follows that there can be no conflict between subjective and objective probabilities, since the former reduce to the latter when the latter are known.

Finally, it is useful to note that while $P(\theta_i)$ was defined as that probability p for which $[c_w, \ldots, c_b, \ldots, c_w]$ is equally as desirable as $(pc_b, (1 - p)c_w)$, for assessment purposes c_b and c_w may be replaced by any two consequences c and c' respectively, provided only that c and c' are not equally desirable. To see this, it suffices to show that the lottery in Figure 16.12, in $\mathcal{S}$, is at least as desirable as $(pc, (1 - p)c') \in \mathcal{L}$, if and only if $P(\theta_i) \geqq p$, given that $c \succ c'$. This is left to the reader as Exercise 2.

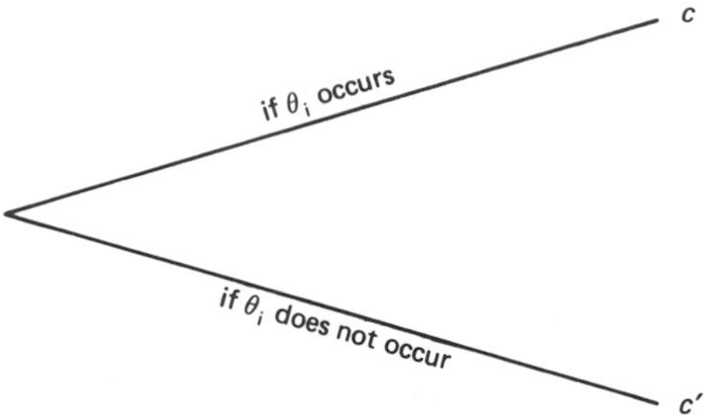

Figure 16.12

Thus the decision-maker may assess his subjective probabilities $P(\theta_1)$, . . . , $P(\theta_s)$ by using any two unequally desirable consequences c and c', even consequences which do not appear in the given decision problem. For convenience, we may take $c \equiv$ "receive \$100 with no strings attached" and $c' \equiv$ "receive \$0 with no strings attached."

16.3 ASSESSING $P: \Theta \rightarrow R^1$ WHEN Θ IS FINITE

Suppose in a given decision problem that Θ is a finite set, say $\Theta = \{\theta_1, \ldots, \theta_s\}$. We shall describe the procedure suggested in the preceding section first for $\Theta = \{\theta_1, \theta_2\}$, and second, for cases in which $s > 2$.

If $\Theta = \{\theta_1, \theta_2\}$, and if c and c' are any prizes such that c and c' are not equally desirable, the decision-maker might consider the subjective lottery SL, defined in Figure 16.13. For definiteness, we shall suppose that $c =$ \$500 and $c' =$ \$0. As suggested in Section 16.2, the decision-maker may compare SL for desirability with an objective lottery $L(p)$ of the form shown in Figure 16.14. One may take $L(p)$ as being a physical lottery which pays $c =$ \$500 if a red ball is drawn from an urn in which there is proportion p of red balls, and which pays $c' =$ \$0 otherwise.

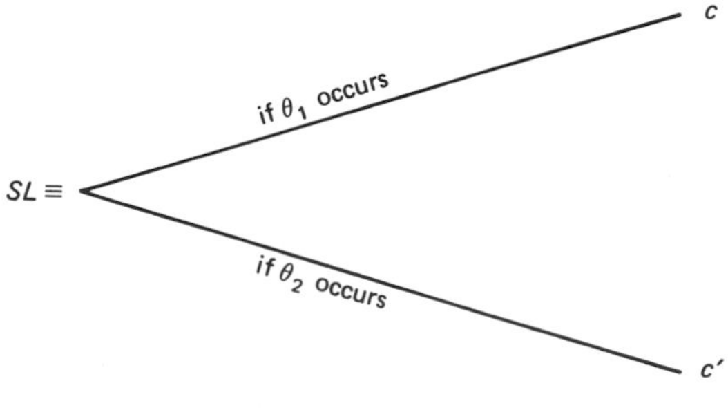

Figure 16.13

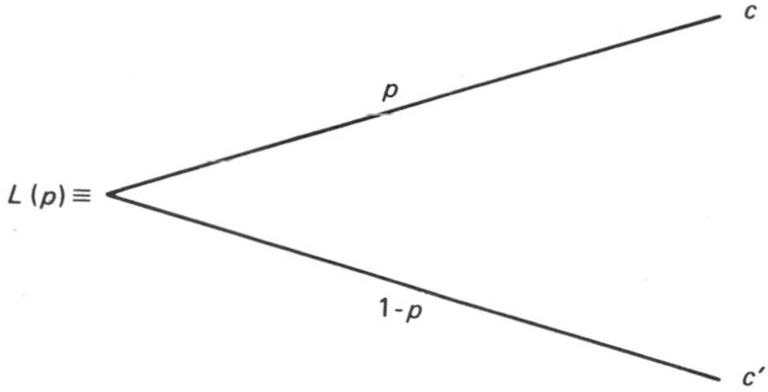

Figure 16.14

The decision-maker then asks himself questions of the following sort: "If $p = p'$, do I prefer $L(p')$ to SL (which pays c if and only if θ_1 occurs), or vice versa?" Presumably, for very low values of p he will prefer SL to $L(p)$, but he will prefer $L(p)$ to SL for very high values of p. The essence of (16.2.6) is that there is a *unique* value $p_0 \equiv P(\theta_1)$ such that

(1) he prefers $L(p)$ to SL whenever $p > p_0$;
(2) he prefers SL to $L(p)$ whenever $p < p_0$; and
(3) he is indifferent between SL and $L(p)$ when $p = p_0$.

A glance at step two in Section 15.3 indicates how much the assessment of subjective probabilities has in common with the assessment of utilities. Here, as there, it is apparent that the decision-maker is called upon to perform some very difficult judgments when p is close to $P(\theta_1)$. He might be much more comfortable naming two numbers, $p_1 \in [0, 1]$ and $p_2 \in [p_1, 1]$, such that

(1′) he prefers $L(p)$ to SL whenever $p > p_2$;
(2′) he prefers SL to $L(p)$ whenever $p < p_1$; and
(3′) he is indifferent between SL and $L(p)$ whenever $p \in [p_1, p_2]$.

In this case, all that we can say is that $P(\theta_1) \in [p_1, p_2]$. Fishburn's work [**23**], referred to in Chapter 15, covers the case of incompletely assessed probabilities as well as incompletely assessed utilities.

Example 16.3.1

Let $\theta_1 \equiv$ "my Empress camellia will yield at least six good blooms this winter," and $\theta_2 \equiv$ "θ_1 won't occur." I shall fix in my mind the subjective lottery SL in which I win \$500 if θ_1 occurs and nothing if θ_2 occurs. Then I imagine an urn containing 100 balls, exactly x of which are red, and I consider $L(x/100)$, in which I win \$500 if a red ball is drawn and nothing otherwise. If $x = 100$, I obviously prefer the sure-\$500 win of $L(100/100)$

to *SL*. Conversely, since my Empress camellia appears to be in good health, I prefer *SL* to the sure-\$0 win of $L(0/100)$. I vary x (and hence $x/100$) and obtain the following table:

$x/100$	*Preference*
1	$L(1)$
0	*SL*
.5	*SL*
.9	*SL*
.95	$L(.95)$
.92	indifferent

Now suppose that $\Theta = \{\theta_1, \theta_2, \theta_3\}$. The decision-maker could assess $P(\theta_1)$ by the preceding procedure if *SL* is defined as in Figure 16.15 and could then assess $P(\theta_2)$ by the preceding procedure, but now defining *SL* as in Figure 16.16. He might then set $P(\theta_3) = 1 - P(\theta_1) - P(\theta_2)$, or he might assess $P(\theta_3)$ directly, defining *SL* as shown in Figure 16.17. If he follows the latter approach, he may find that $P(\theta_1) + P(\theta_2) + P(\theta_3) \neq 1$, which cannot happen for probabilities. Experimentally, it *does* frequently happen; an excellent account and analysis of such situations may be found

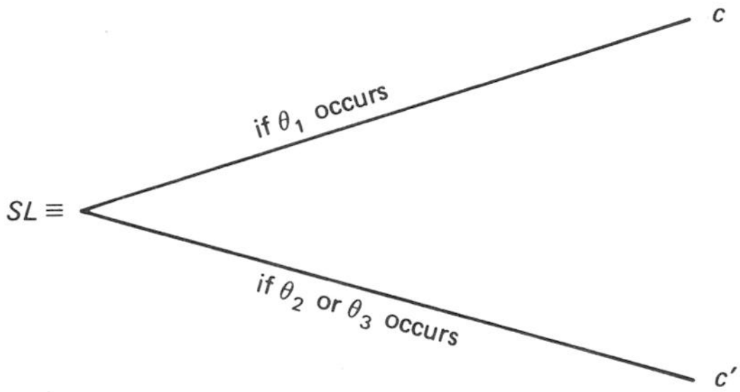

Figure 16.15

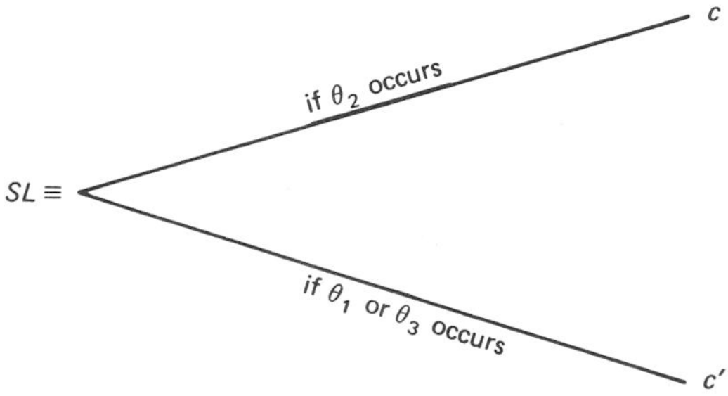

Figure 16.16

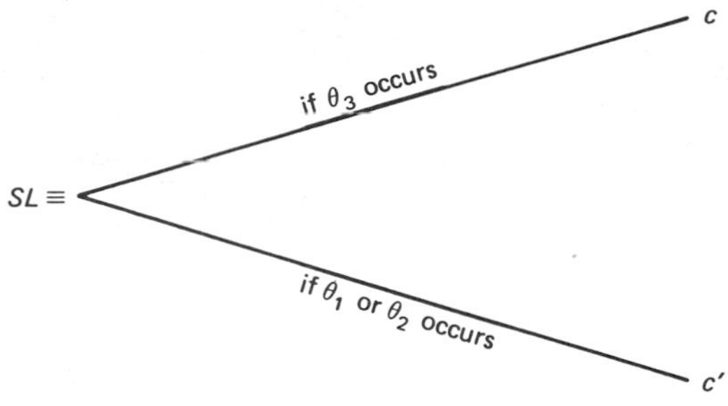

Figure 16.17

in Fellner [**21**]. One "correction" that can be applied when Axiom (*P*1) fails is to divide each $P(\theta_1)$ by $P(\theta_1) + P(\theta_2) + P(\theta_3)$ and thus obtain probabilities that do sum to one. However, the decision-maker should then check his indifferences between each *SL* and the related *L*(corrected $P(\theta_i)$).

It should be clear that the preceding procedure can be extended to the case where $\Theta \equiv \{\theta_1, \ldots, \theta_s\}$ for any positive integer *s*: the reader can assess $P(\theta_i)$ by comparing $L(p)$ with Figure 16.18, then $P(\theta_i)$ satisfies $L(P(\theta_i)) \sim SL_i$; and correct the $P(\theta_i)$'s if necessary. The reader should be cautioned, however, that as *s* increases, $\sum_{i=1}^{s} P(\theta_i)$ may differ significantly from one due to the compounding of assessment difficulties and "introspection errors."

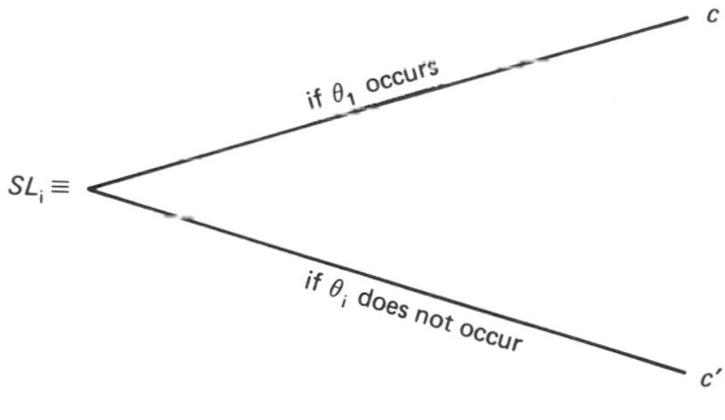

Figure 16.18

Another procedure for use when $s > 2$ can be illustrated quite easily for $s = 3$; that is, $\Theta = \{\theta_1, \theta_2, \theta_3\}$. Suppose that the decision-maker has assessed $P(\theta_1)$ and puts $P(\{\theta_2, \theta_3\}) = 1 - P(\theta_1)$, using the simple procedure for $s = 2$. He may then consider "called-off lotteries" (Pratt, Raiffa, and Schlaifer [**64**]) of the form shown in Figures 16.19 and 16.20.

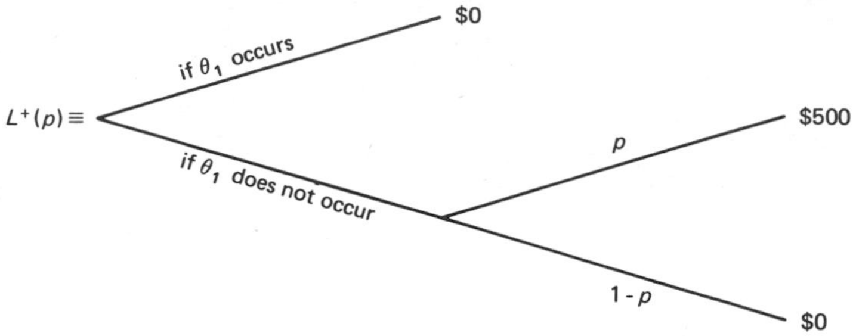

Figure 16.19

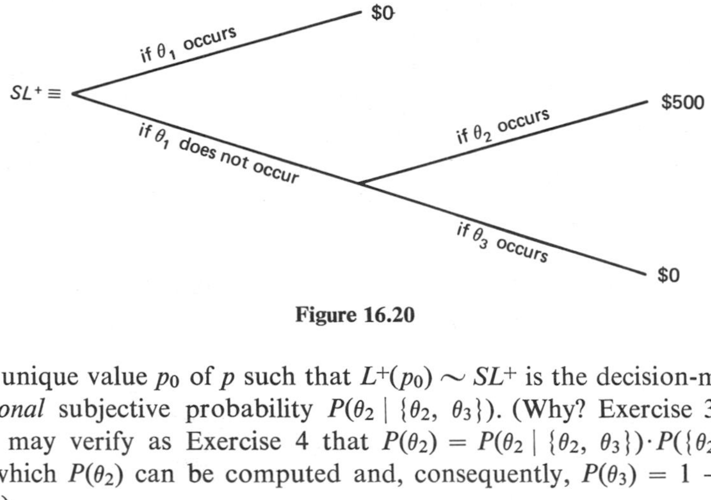

Figure 16.20

The unique value p_0 of p such that $L^+(p_0) \sim SL^+$ is the decision-maker's *conditional* subjective probability $P(\theta_2 \mid \{\theta_2, \theta_3\})$. (Why? Exercise 3.) The reader may verify as Exercise 4 that $P(\theta_2) = P(\theta_2 \mid \{\theta_2, \theta_3\}) \cdot P(\{\theta_2, \theta_3\})$, from which $P(\theta_2)$ can be computed and, consequently, $P(\theta_3) = 1 - P(\theta_1) - P(\theta_2)$.

The "called-off lottery" terminology stems from the observation that $L^+(p)$ can be considered to be a lottery that will not be played if θ_1 occurs (that is, will be called off if θ_1 occurs); and similarly for SL^+.

We have discussed called-off lotteries as assessment devices when $s = 3$, but they are applicable also when $s > 3$. To see this, suppose that in the preceding discussion θ_3 is not an elementary event but rather the union of two elementary events, say θ_{31} and θ_{32}. By comparing Figure 16.21 with 16.22, the decision-maker may find the unique $p_0 = P(\theta_{31} \mid \{\theta_{31}, \theta_{32}\})$ for which $L^{++}(p_0) \sim SL^{++}$. Since $P(\theta_3) = P(\{\theta_{31}, \theta_{32}\})$ has already been found, he may then compute $P(\theta_{31}) = P(\theta_{31} \mid \{\theta_{31}, \theta_{32}\})\ P(\{\theta_{31},\theta_{32}\})$ and $P(\theta_{32}) = 1 - P(\theta_1) - P(\theta_2) - P(\theta_{31})$. Hence the called-off-lottery procedure works for $\Theta = \{\theta_1, \theta_2, \theta_{31}, \theta_{32}\}$; that is, $s = 4$. It extends in a similar fashion to any $s > 4$.

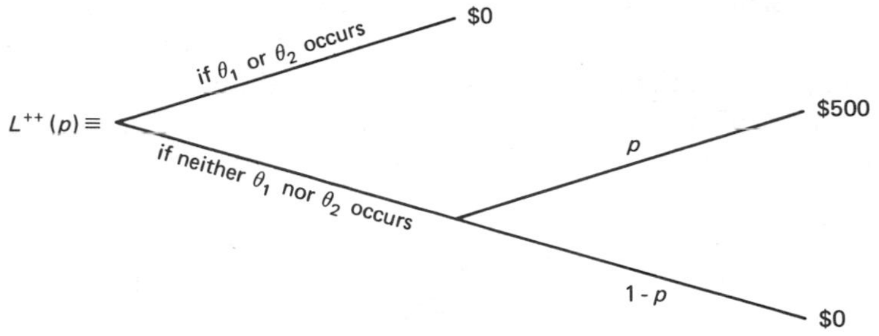

Figure 16.21

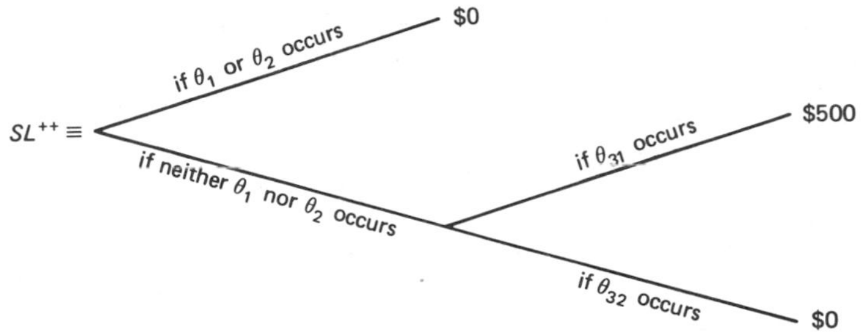

Figure 16.22

16.4 ASSESSING F: $\Theta \rightarrow R^1$ WHEN $\Theta \subset R^m$

In this section we consider the problem of assessing subjective probabilities when Θ is an infinite set, and hence when a θ_i-by-θ_{i+1}, step-by-step assessment of $\{P(\theta): \theta \in \Theta\}$ is futile.

More particularly, we shall suppose *first*, that Θ is a subset of a finite-dimensional real space R^m; and *second*, that the decision-maker wishes to use a conjugate prior df to express his judgments. Since each of the conjugate families in Chapter 14 consists of df's which depend upon a finite number of parameters, it will follow that a member of a given family will be determined uniquely be a finite number of assessments on the part of the decision-maker.

We consider specifically how the decision-maker may "fit" his prior judgments to beta, gamma, univariate Normal, k-variate beta, k-variate Normal, and inverted Wishart distributions, in Subsections 16.4.2 through 16.4.7, respectively. Subsection 16.4.1 describes the general procedure used when Θ is a subset of R^1; that is, when $\tilde{\theta}$ is a random variable.

16.4.1 The Fractile Assessment Procedure

Suppose $\tilde{\theta}$ is a random variable with an unknown dgf $f(\theta)$. Such random variables appear in a natural manner as the parameters of data-generating processes; for example, $\tilde{\theta} \equiv \tilde{p}$ for the Bernoulli process, and $\tilde{\theta} \equiv \tilde{\lambda}$ for the Poisson process.

Also suppose that the decision-maker wishes to express his prior judgments about $\tilde{\theta}$ by choosing a member of the natural conjugate family $\{f(\theta \mid \alpha_1, \ldots, \alpha_j)$: all $\alpha_1, \ldots, \alpha_j\}$ of df's of $\tilde{\theta}$. He will want to choose a conjugate rather than an arbitrary df of $\tilde{\theta}$ for two reasons: first, because the mathematical computations required to solve the general decision problem are much easier with a conjugate prior df; and second, because "fitting" a conjugate prior df is generally less onerous than assessing an arbitrary dgf. Nevertheless, it occasionally happens that *no* conjugate prior df adequately expresses the decision-maker's prior judgments, and in such cases there is little to do but resign oneself to using a nonconjugate prior dgf.

If a conjugate prior df $f(\theta \mid \alpha_1, \ldots, \alpha_j)$ depends upon j parameters, it follows in all cases considered in ensuing subsections that $\alpha_1, \ldots, \alpha_j$ are uniquely specified by the assessment of j different fractiles ${}_{p_1}\theta, \ldots, {}_{p_j}\theta$. That is, each α_i is a function of $({}_{p_1}\theta, \ldots, {}_{p_j}\theta)$.

Now, the α_i's are functions of other summary measures such as moments, and hence it is natural to ask why it is desirable to assess fractiles rather than moments. The reason is that to assess fractiles the decision-maker may use a lottery-comparison procedure quite closely related to those of Sections 15.3 and 16.3.

For example, suppose the decision-maker wishes to express his judgments about the parameter $\tilde{p}(= \tilde{\theta})$ of a Bernoulli process by a special beta df of the form

$$f_\beta(p \mid \rho, \rho + 1) = \begin{cases} \rho p^{\rho-1}, & p \in (0, 1) \\ 0, & p \notin (0, 1). \end{cases} \tag{16.4.1}$$

By comparing an objective lottery of the form in Figure 16.23, (in which he wins \$500 with objective probability $\frac{1}{2}$) with subjective lotteries of the form in Figure 16.24 (in which he wins \$500 if $\tilde{p} \leqq x$) he finds a number $x_0 \in (0, 1)$ such that $SL(x_0) \sim L(\frac{1}{2})$. Now, it is clear that $P(\tilde{p} \leqq x_0) = \frac{1}{2}$; that is, $x_0 = {}_{1/2}p =$ the median of his prior df of $\tilde{p}$. But, by (16.4.1),

$$F_\beta(p \mid \rho, 1) = p^\rho, \; p \in (0, 1); \tag{16.4.2}$$

and hence from Section 7.3 we have $\frac{1}{2} = ({}_{1/2}p)^\rho$, which implies

$$\rho = \log\left(\tfrac{1}{2}\right)/\log\left({}_{1/2}p\right). \tag{16.4.3}$$

Thus the decision-maker can use (16.4.3) to solve for ρ in terms of his assessed median ${}_{1/2}p$ of $\tilde{p}$.

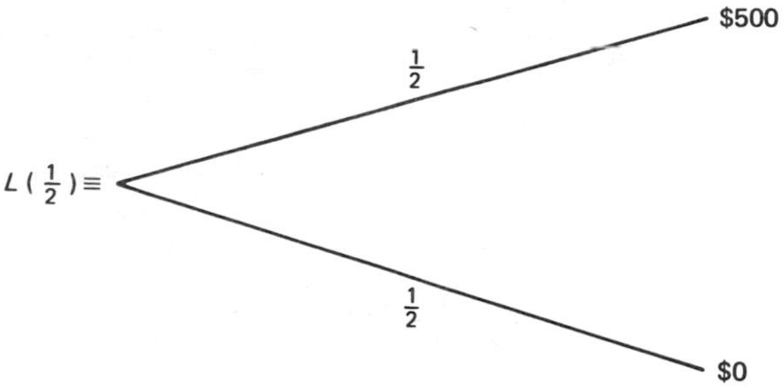

Figure 16.23

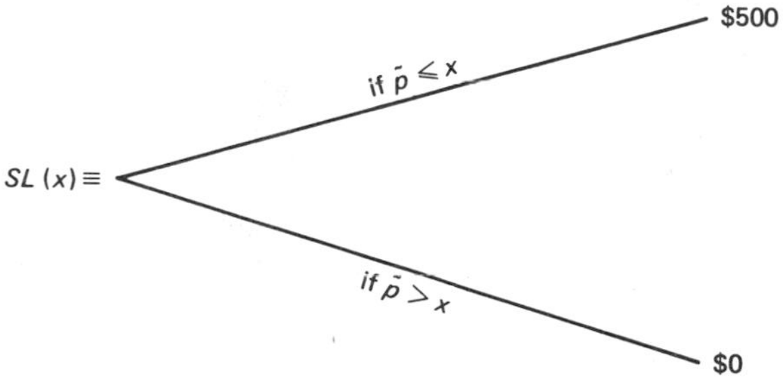

Figure 16.24

The essential point in this example is not the derivation of (16.4.3) (because each such example requires a different derivation) but lies in the procedure for assessing a fractile.

In general, let $\pi \in (0, 1)$ and suppose that you wish to assess the πth fractile of $\tilde{\theta}$. You may compare an objective lottery $L(\pi)$ of the form in Figure 16.25 with subjective lotteries of the form in Figure 16.26. Now,

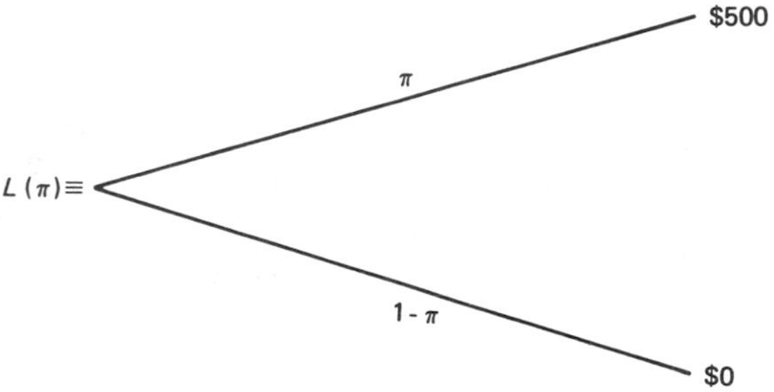

Figure 16.25

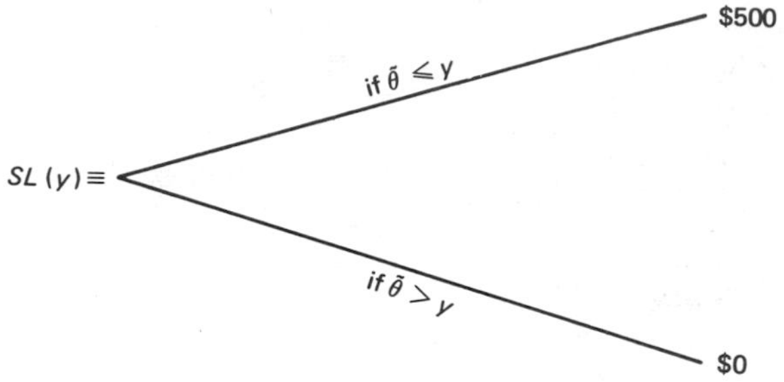

Figure 16.26

if y is "too small," then $L(\pi) \succ SL(y)$; whereas if y is "too large," then $SL(y) \succ L(\pi)$. In all cases of interest, there should be a unique y_0 such that $L(\pi) \sim SL(y_0)$; and from $\pi = P\tilde{\theta} \leqq (y_0) = F(y_0)$ you may conclude that $y_0 = {}_{\pi}\theta$.

If only one parameter characterizes the conjugate prior df, then only one fractile must be assessed, and it is usually easiest to assess the median ($\pi = \frac{1}{2}$). If (as in the univariate beta and Normal cases) two parameters must be specified, it is usually most convenient to first assess the median, ${}_{1/2}\theta$, and to secondly assess either the first quartile ${}_{0.25}\theta$ or the third quartile ${}_{0.75}\theta$ of $\tilde{\theta}$. There are two ways of making this second assessment: either use the $L(.25)$ or $L(.75)$ comparison with $SL(y)$, or use a called-off lottery procedure. For assessing the third quartile ${}_{0.75}\theta$ of $\tilde{\theta}$, the called-off lottery procedure requires the decision-maker to compare $L^+(.75)$ with $SL^+(y)$, as defined by Figures 16.27 and 16.28, respectively. If y is "too small," $L^+(.75) \succ SL^+(y)$, and if y is "too large," $SL^+(y) \succ L^+(.75)$. You may easily verify that if $L^+(.75) \sim SL^+(y_0)$, then $y_0 = {}_{0.75}\theta$. You may modify this procedure, (so that it will yield $y_0 = {}_{0.25}\theta$) as Exercise 5. Note that, since ${}_{1/2}\theta$ has already been assessed, each of $L^+(.75)$ and $SL^+(y)$ is completely specified; and the same will be true of the $SL^+(y)$ and $L(.25)$ needed for assessing ${}_{0.25}\theta$.

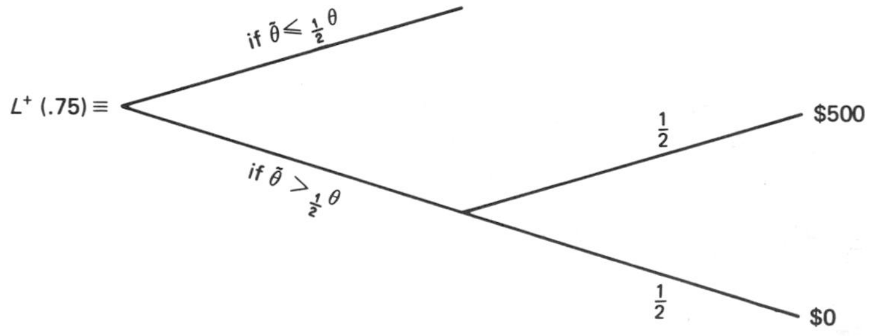

Figure 16.27

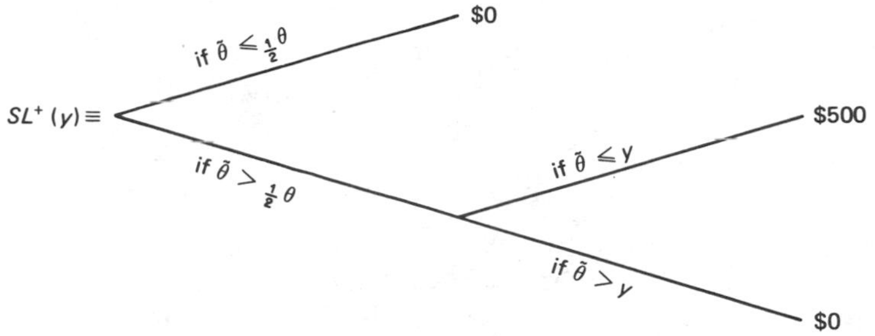

Figure 16.28

If $\tilde{\boldsymbol{\theta}}$ is a random vector and you must assess a k-variate prior df of $\tilde{\boldsymbol{\theta}}$, the preceding procedures continue to obtain with some modifications. For example, if $k = 2$ and you wish to assess a conjugate prior df $f_N^{(2)}(\boldsymbol{\mu} \mid \boldsymbol{\mu}_0, \boldsymbol{\sigma}_0^2)$ for $\tilde{\mu}$, you could use the preceding procedures twice to assess $f_N^{(1)}(\mu_i \mid \mu_{0i}, \sigma_{0ii}^2)$ for $i = 1, 2$, and then assess one fractile of the *conditional* distribution of $\tilde{\mu}_2$ given μ_1 in order to specify the covariance σ_{012}^2 of $\tilde{\mu}_1$ and μ_2 (see Subsection 16.4.5).

More generally, one tries to specify as many parameters of $f^{(k)}(\boldsymbol{\theta})$ as possible by assessing the *marginal* df's of the components θ_i, and then uses *conditional* df's of some θ_1's given others to specify the remaining parameters.

16.4.2 Assessing Standardized Beta Distributions

The introductory example of the preceding subsection showed how a standardized beta distribution with ν fixed at $\rho + 1$ could be assessed. This short subsection shows how a decision-maker can express his prior judgments about a Bernoulli parameter $\tilde{p}$ by assessing two fractiles of $\tilde{p}$ and then finding that (ρ, ν) pair for which the corresponding fractiles agree most closely with his assessed fractiles.

We shall suppose that the methods of Subsection 16.4.1 have been used to determine the median, ${}_{1/2}p$, and either the first or third quartile, ${}_{0.25}p$ or ${}_{0.75}p$. The decision-maker may then refer to Table III (in the Appendix) and examine the tabulated fractiles for various values of ρ and ν.

A methodical procedure for making this comparison is to start by finding all (ρ, ν) pairs for which, in the notation of Table III, ${}_{.50}\beta_{(\rho,\nu)} \doteq {}_{1/2}p$. In so doing, use the prefatory note to Table III to the effect that if $\rho > \nu/2$, then ${}_{.50}\beta_{(\rho,\nu)} = 1 - {}_{.50}\beta_{(\nu-\rho,\nu)}$. After listing these pairs, find the 0.75 fractile for each—if you assessed ${}_{0.75}p$; otherwise, find the 0.25 fractile of each. Recall that ${}_{.75}\beta_{(\rho,\nu)} = 1 - {}_{.25}\beta_{(\nu-\rho,\nu)}$ and ${}_{.25}\beta_{(\rho,\nu)} = 1 - {}_{.75}\beta_{(\nu-\rho,\nu)}$. Find that pair for which ${}_{.75}\beta_{(\rho,\nu)} \doteq {}_{.75}p$ (or ${}_{.25}\beta_{(\rho,\nu)} \doteq {}_{.25}p$). This yields the desired distribution, but you will want to check other fractiles of that distribution to see if it really does represent your prior judgments adequately.

Example 16.4.1

A decision-maker wishes to express his prior judgments about the proportion $\tilde{p}$ of housewives who will purchase his new kitchen utensil. By comparing $L(\frac{1}{2})$ with $SL(x)$ (as defined in 16.4.1), he finds that $SL(.40) \sim L(\frac{1}{2})$; and by comparing Figure 16.29 with Figure 16.30, he finds that $SL^{+}(.55) \sim L^{+}(\frac{1}{2})$. Hence ${}_{1/2}p = .40$ and ${}_{.75}p = .55$. By an interpolation in Table III, he concludes that $(\rho, \nu) \doteq (2.5, 6)$, since we have:

(ν,ρ)	${}_{.50}\beta$	${}_{.75}\beta$	
(6,2)	.3138	.4572	
(6,3)	.5000	.6406	
(6,2.5) =	.4069	.5489	= averages

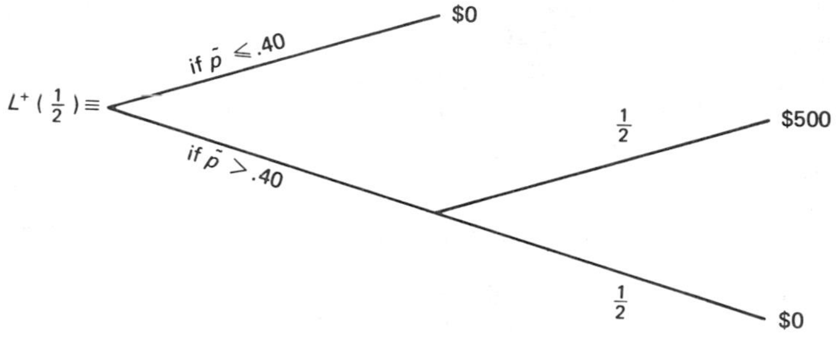

Figure 16.29

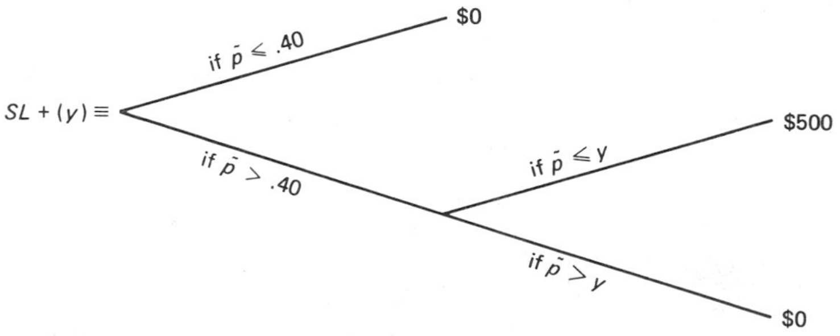

Figure 16.30

He then checks the reasonableness of this distribution by noting that ${}_{.25}\beta_{(2.5,6)} \doteq \frac{1}{2}(.1938 + .3594) = .2766$, which is roughly in line with his prior judgments that "the odds are three to one in favor of at least 25 percent of the housewives buying the utensil." Now, the sophisticated reader will worry about the accuracy of the linear interpolation in the table and also about the advisability of approximating, for example, .40 by .4069. We shall not be concerned with such matters because if a subsequent sample is taken in which n is at all substantial, the posterior df $f_\beta(p \mid \rho + r, \nu + n)$ will be virtually unaffected by *minor* changes in the parameters ρ and ν of the prior df $f_\beta(p \mid \rho, \nu)$. For more discussion of this point, see Edwards, Lindman and Savage [**16**].

Example 16.4.2

A production manager has assessed ${}_{1/2}p = .05$ and ${}_{.75}p = .10$, where $\tilde{p}$ is the proportion of defective items in a given production run. Since ${}_{.50}\beta_{(1,14)} = .0519 \doteq .05$ and ${}_{.75}\beta_{(1,14)} = .1011 \doteq .10$, he decides to use $f_\beta(p \mid 1, 14)$ as his prior df of $\tilde{p}$.

Example 16.4.3

An engineer wishes to express his prior judgments regarding the reliability (probability of successful operation) p of a new component. He uses the procedure of Subsection 16.4.1 to determine that ${}_{.50}p = .995$ and ${}_{.25}p = .990$. A brief check of Table III indicates that ν must exceed 29, since ${}_{.50}\beta_{(28,29)} = 1 - {}_{.50}\beta_{(1,29)} = .9755$. We use Note (3) to Table III to compute

$$a_{.25} = \{[.995(1 - .990)]^{1/2} - [.990(1 - .995)]^{1/2}\}^2$$
$$\doteq .000864;$$
$$c = \frac{.112}{.000864} \doteq 129.5;$$
$$\nu \doteq 129.5 + .67 \doteq 130; \text{ and}$$
$$\rho \doteq (129.5)(.995) + .33 = 129.$$

Hence his prior judgments are expressed (approximately) by $f_\beta(p \mid 129, 130)$. However, this df is on the form $f_\beta(p \mid \rho, \rho + 1)$, and we may therefore solve for ${}_q\beta_{(\rho,\rho+1)}$ in terms of q. From (16.4.2), we deduce that

$${}_qp = q^{1/\rho},$$

and hence

$${}_{.50}p = (.5)^{1/129} = .9946,$$

and

$$_{.25}p = (.25)^{1/129} \doteq .9893,$$

which are close enough to the assessed values, .995 and .990, respectively.

16.4.3 Assessing Gamma Distributions

If the decision-maker wishes to express his prior judgments about the intensity parameter $\tilde{\lambda}$ of a Poisson process in the form of a natural conjugate gamma df $f_\gamma(\lambda \mid \rho_0, \tau_0)$, it again suffices for him to make two fractile assessments, $_{.50}\lambda$ and either $_{.25}\lambda$ or $_{.75}\lambda$, from which ρ_0 and τ_0 may be determined via the following procedure:

As *step one*, ρ_0 is determined by comparing either $_{.75}\lambda/_{.50}\lambda$ with $_{.75}\chi^2{}_{(2\rho)}/{}_{.50}\chi^2{}_{(2\rho)}$ or $_{.50}\lambda/_{.25}\lambda$ with $_{.50}\chi^2{}_{(2\rho)}/{}_{.25}\chi^2{}_{(2\rho)}$ for various values of ρ until the best fit of the ratios is obtained, at $\rho = \rho_0$. The chi-square fractiles in Table II (of the Appendix) may be used for the computation of the ratios $_{p}\chi^2{}_{(2\rho)}/{}_{q}\chi^2{}_{(2\rho)}$.

To verify that step one does approximate ρ_0, it suffices to show that if $p \in (0, 1)$, $q \in (0, 1)$, and $\rho > 0$, then

$$_{p}\gamma_{(\rho,\tau)}/{}_{q}\gamma_{(\rho,\tau)} = {}_{p}\chi^2{}_{(2\rho)}/{}_{q}\chi^2{}_{(2\rho)}. \tag{16.4.4}$$

But (16.4.4) is immediate from (12.2.13).

As *step two*, τ_0 is determined from $_{.50}\chi^2{}_{(2\rho_0)}$ and $_{.50}\lambda$ via

$$\tau_0 = \tfrac{1}{2}[{}_{.50}\chi^2{}_{(2\rho_0)}/{}_{.50}\lambda], \tag{16.4.5}$$

another immediate consequence of (12.2.13) [in which λ there is τ_0 here, r there is ρ_0 here, and $_{.50}\lambda = {}_{.50}\gamma_{(\rho_0,\tau_0)}$].

Example 16.4.4

The owner of a filling station now under construction wishes to express his judgments about the average number $\tilde{\lambda}$ of customer cars per hour, in the form of a gamma df $f_\gamma(\lambda \mid \rho_0, \tau_0)$. He assesses, by the lottery-comparison method of 16.4.1, $_{.50}\lambda = 15$ and $_{.75}\lambda = 30$. Hence $_{.75}\lambda/_{.50}\lambda = 2.00$. From Table II, the ratio $_{.75}\chi^2{}_{(2.1)}/{}_{.50}\chi^2{}_{(2.1)} = 1.99$ is closest to two, and hence he sets $\rho_0 = 2$. (The ratio of chi-square fractiles here is exactly two; the value 1.99 reflects the roundings-off in Table II.) Finally, he uses $_{.50}\lambda = 15$, $_{.50}\chi^2{}_{(2)} = 1.39$, and (16.4.5) to obtain $\tau_0 = \frac{1}{2}[1.39/15] \doteq .046$.

In Section 12.2 we defined a number of useful relatives of the gamma distribution, including the gamma-2, inverted gamma, and inverted gamma-2. In Chapter 14 we showed that the inverted gamma is natural conjugate for the Normal-process variance $\tilde{\sigma}^2$, and therefore the inverted gamma-2 is natural conjugate for the standard deviation $\tilde{\sigma}$ and the gamma-2 is natural conjugate for the "precision" $\tilde{\sigma}^{-2}$.

Natural conjugacy for $\tilde{\sigma}$ and $\tilde{\sigma}^{-2}$ follow from the facts in Section 12.2 that $\tilde{\sigma}^2$ has df $f_{i\gamma}(\sigma^2 \mid \psi_0, \nu_0)$ if and only if $\tilde{\sigma}$ has df $f_{i\gamma 2}(\sigma \mid \psi_0, \nu_0)$ if and only if $\tilde{\sigma}^{-2}$ has df $f_{i\gamma}(\sigma^{-2} \mid \psi_0\ \nu_0)$. Hence the decision-maker can assess a natural conjugate prior df for $\tilde{\sigma}$ or $\tilde{\sigma}^{-2}$ and readily obtain the natural conjugate prior df of $\tilde{\sigma}^2$. In this regard it is instructive to note that Figures 12.1 through 12.3 are comparable, in that $f_\gamma(\cdot \mid r, r) = f_{\gamma 2}(\cdot \mid 1, \frac{1}{2}r)$, so that $r = \frac{1}{2}, 1, 3, 10$ correspond to $\nu = 1, 2, 6, 20$, respectively. Thus, for example, the inverted gamma-2 df $f_{i\gamma 2}(\cdot \mid 1, 6)$ in Figure 12.3 of a Normal standard deviation $\tilde{\sigma}$ implies that $\tilde{\sigma}^2$ has the df $f_{i\gamma}(\cdot \mid 1, 6)$ in Figure 12.2 and that $\tilde{\sigma}^{-2}$ has the df $f_\gamma(\cdot \mid 3, 3)$ in Figure 12.1.

It is easier for most people to formalize their judgments about a standard deviation $\tilde{\sigma}$ rather than about the corresponding variance $\tilde{\sigma}^2$ or precision $\tilde{\sigma}^{-2}$; after all, the standard deviation of a random variable $\tilde{x}$ is expressed in x-units, rather than x-units-squared, or inverse-x-units-squared. Hence we shall show how the lottery-comparison method of Subsection 16.4.1 can be used to fit the parameters ψ_0 and ν_0 of $f_{i\gamma 2}(\cdot \mid \psi_0, \nu_0)$ to reflect prior judgments about $\tilde{\sigma}$.

The decision-maker again assesses two fractiles of $\tilde{\sigma}$, one being ${}_{.50}\sigma$ and the other either ${}_{.75}\sigma$ or ${}_{.25}\sigma$. The relation analogous to (16.4.4) for the inverted gamma-2 case is

$$[{}_p(i\gamma 2)_{(\psi,\nu)}/{}_q(i\gamma 2)_{(\psi,\nu)}]^2 = {}_{(1-q)}\chi^2{}_{(\nu)}/{}_{(1-p)}\chi^2{}_{(\nu)}, \tag{16.4.6}$$

which follows readily from (12.2.16).

Hence *step one* is to find that value ν_0 of ν for which

$$[{}_{.75}\sigma/{}_{.50}\sigma]^2 = {}_{.50}\chi^2{}_{(\nu_0)}/{}_{.25}\chi^2{}_{(\nu_0)}$$

or

$$[{}_{.50}\sigma/{}_{.25}\sigma]^2 = {}_{.75}\chi^2{}_{(\nu_0)}/{}_{.50}\chi^2{}_{(\nu_0)},$$

by approximation using the chi-square fractiles in Table II.

Step two consists in using ${}_{.50}\chi^2{}_{(\nu_0)}$, ${}_{.50}\sigma$, and

$$\psi_0 = {}_{.50}\chi^2{}_{(\nu_0)}[{}_{.50}\sigma]^2/\nu_0, \tag{16.4.7}$$

which is also immediate from (12.2.16), to solve for ψ_0.

Example 16.4.5

A foreman wishes to express his judgments about the reliability of a machine by assessing an inverted gamma-2 distribution for the standard deviation $\tilde{\sigma}$ of the length of the bolts it cuts. By lottery comparisons, he finds ${}_{.50}\sigma = .003$ and ${}_{.75}\sigma = .004$. Hence $[{}_{.75}\sigma/{}_{.50}\sigma]^2 = [1.33]^2 = 1.78$, which is approximately equal to ${}_{.50}\chi^2{}_{(4)}/{}_{.25}\chi^2{}_{(4)} = 1.75$. Hence $\nu_0 = 4$. By (16.4.7),

$$\psi_0 = 3.36(.003)^2/4 = 7.56 \times 10^{-6}.$$

Thus his df of $\tilde{\sigma}$ is $f_{i\gamma 2}(\cdot \mid 7.56 \times 10^{-6}, 4)$, implying that $\tilde{\sigma}^2$ has df $f_{i\gamma}(\cdot \mid 7.56 \times 10^{-6}, 4)$, and so forth.

Somewhat more accuracy can be obtained by using the graphs in Raiffa and Schlaifer [**66**], 230, instead of computations from Table II.

There are individuals who prefer expressing their judgments directly about $\tilde{\sigma}^2$. For them, the appropriate analogues of (16.4.6) and (16.4.7) are

$$_p(i\gamma)_{(\psi,\nu)}/_q(i\gamma)_{(\psi,\nu)} = {}_{(1-q)}\chi^2{}_{(\nu)}/{}_{(1-p)}\chi^2{}_{(\nu)} \tag{16.4.8}$$

and

$$\psi_0 = {}_{.50}\chi^2{}_{(\nu_0)}[{}_{.50}\sigma^2]/\nu_0, \tag{16.4.9}$$

respectively.

16.4.4 Assessing Univariate Normal Distributions

Perhaps the easiest assessment task of all is the univariate Normal df. Suppose a decision-maker wishes to express his judgments about a quantity $\tilde{\mu}$ in the form of a Normal df $f_N(\mu \mid \mu_0, \sigma_0{}^2)$.

As before, he may assess two fractiles of $\tilde{\mu}$ by the lottery-comparison method of Subsection 16.4.1. Suppose these fractiles are ${}_{.50}\mu$ and ${}_p\mu$ for some $p \neq .50$. Since $f_N(\cdot \mid \mu_0, \sigma_0{}^2)$ is symmetric about μ_0, it follows that

$$\mu_0 = {}_{.50}\mu. \tag{16.4.10}$$

But by virtue of Equation (12.4.6),

$$\sigma_0{}^2 = [({}_p\mu - \mu_0)/{}_pN^*]^2. \tag{16.4.11}$$

Example 16.4.6

A teacher wishes to express her judgments about the true average IQ $\tilde{\mu}$ in her class by a Normal df $f_N(\cdot \mid \mu_0, \sigma_0{}^2)$. By the methods of Subsection 16.4.1 she determines that ${}_{.5}\mu = 112$ and ${}_{.75}\mu = 119$. From Table I we find ${}_{.75}N^* = .674$. Hence (16.4.11) yields $\sigma_0{}^2 = [(119\text{-}112)/.674]^2 \doteq 108$ (and $\sigma_0 \doteq 10.4$). Hence $f_N(\mu \mid \mu_0, \sigma_0{}^2) = f_N(\mu \mid 112, 108)$.

Example 16.4.7

I wish to express my judgments regarding the average number $\tilde{\mu}$ of pages per book for the books in my library. I realize that some books are very short, with fewer than 80 pages, while others are very long, with over

700 pages. Most, however, are between 180 and 350 pages long. I have used the lottery-comparison method of Subsection 16.4.1 to determine that ${}_{.5}\mu = 230$ and ${}_{.75}\mu = 270$. Hence $\mu_0 = 230$, $\sigma_0 = (270–230)/.674 \doteq 59.3$, and $\sigma_0 \doteq 3{,}520$. Hence I determine that $f_N(\mu \mid \mu_0, \sigma_0{}^2) = f_N(\mu \mid 230, 3520)$. But I check by noting that $f_N(\mu \mid 230, 3520)$ assigns probability .25 to the event that $\tilde{\mu} \leqq 190$. This is in line with my prior judgments and, after a few similar checks, I am content to use $f_N(\mu \mid 230, 3520)$.

16.4.5 Assessing k-variate Beta Distributions

This subsection is the first of three concerned with assessing joint distributions of several random variables. The k-variate beta distribution is, in a sense, the easiest multivariate distribution to assess, since it has only $k + 1$ parameters.

You will recall from (12.6.3) that if $\tilde{p} \equiv (\tilde{p}_1, \ldots, \tilde{p}_k)$ has df $f_\beta{}^{(k)}(\mathbf{p} \mid \boldsymbol{\rho}, \nu)$, then every $\tilde{p}_i$ has the marginal df $f_\beta(p_i \mid \rho_i, \nu)$. This suggests that the decision-maker assess the k marginal df's $f_\beta(p_i \mid \rho_i, \nu)$ according to the guidelines of Subsection 16.4.2.

However, the catch to this procedure is that assessing the k marginal df's will lead in general to k values of ν, which is inadmissible. Nevertheless, the k independent assessments can be made, and if the resulting values of ν are not too different and if $\sum_{i=1}^{k} \rho_i < \nu$ for som: of these assessed ν's, the decision-maker may take a typical ν and then reassess his ρ_i's, being careful that in so doing he preserves the inequality $\sum_{i=1}^{k} \rho_i < \nu$. This procedure is not too difficult when k is small and when the decision-maker does not simultaneously hold pronounced opinions about some $\tilde{p}_i$'s (implying large ν) and vague opinions about others (implying small ν).

Example 16.4.8

Suppose a politician classifies the voters in his constituency as being (a) his supporters; (b) supporters of his opponent; and (c) undecided. Let $\tilde{p}_i$ denote the proportion in category (i), for $i = 1, 2, 3$. Suppose that he wishes to assess a bivariate beta df $f_\beta{}^{(2)}(\mathbf{p} \mid \boldsymbol{\rho}, \nu)$ to $(\tilde{p}_1, \tilde{p}_3) \equiv \tilde{\mathbf{p}}$. [Recall that $\tilde{p}_2$ is functionally dependent upon $(\tilde{p}_1, \tilde{p}_3)$ in that $\tilde{p}_2 = 1 - \tilde{p}_1 - \tilde{p}_3$.] By the methods of Subsection 16.4.2 he assesses the df $f_\beta(p_1 \mid 11, 22)$ for $\tilde{p}_1$ (because ${}_{.5}p_1 = .50$ and ${}_{.75}p_1 = .57$ by his lottery comparisons). Similarly, he assesses the df $f_\beta(p_3 \mid 2, 24)$ for $\tilde{p}_3$ (because ${}_{.5}p_3 = .07$ and ${}_{.75}p_3 = .11$ by his lottery comparisons). He decides to use $\nu = 23$, $\rho_1 = 11.5$, and $\rho_3 = 1.96$, as the adjusted values of ρ_1 and ρ_3 leave $E(\tilde{p}_1)$ and $E(\tilde{p}_3)$ unchanged. As a check, he observes that the proportion $\tilde{p}_2$ of supporters of his opponent must necessarily have df $f_\beta(p_2 \mid \nu - \rho_1 - \rho_3, \nu) =$

$f_\beta(p_2 \mid 9.54,\ 23)$, and that ${}_{.5}\beta_{(9.54,23)} \doteq .413$ and ${}_{.75}\beta_{(9.54,23)} \doteq .483$. These fractiles are in line with his judgments, and hence he accepts $f_\beta^{(2)}(\mathbf{p} \mid (11.5,\ 1.96)',\ 23)$ as an adequate embodiment of his prior judgments about $\tilde{\mathbf{p}} \equiv (\tilde{p}_1, \tilde{p}_3)$.

An *alternative* assessment procedure is based upon (12.6.4). As *step one*, the decision-maker assesses the marginal df $f_\beta(p_1 \mid \rho_1,\ \nu)$, determining ρ_1 and ν from Table III and two fractiles of $\tilde{p}_1$ obtained via lottery comparisons.

Now, given that $\tilde{p}_1 = p_1$, the df of $\tilde{p}_2$ is $f_{\beta-}(p_2 \mid \rho_2,\ \nu - \rho_1,\ 0,\ 1 - p_1)$ by (12.6.4) and (12.3.8). Hence as *step two* the decision-maker assumes a value of p_1 of $\tilde{p}_1$ and assesses one conditional fractile ${}_q p_{2|1}(p_1)$ of $\tilde{p}_2$ given $\tilde{p}_1 = p_1$; from (12.3.12) it follows that

$$ {}_q\beta_{(\rho_2, \nu - \rho_1)} = (1 - p_1)\,{}_q p_{2|1}(p_1), \tag{16.4.12} $$

from which ρ_2 can be determined via Table III.

If $k > 2$, this procedure must be continued. Given $\tilde{p}_1 = p_1$ and $\tilde{p}_2 = p_2$, the df of $\tilde{p}_3$ is $f_{\beta-}(p_3 \mid \rho_3,\ \nu - \rho_1 - \rho_2,\ 0,\ 1 - p_1 - p_2)$ by (12.6.4) and (12.3.8), so as *step three* the decision-maker assesses one conditional fractile ${}_q p_{3|12}(p_1,\ p_2)$ of $\tilde{p}_3$ given $(\tilde{p}_1,\ \tilde{p}_2) = (p_1,\ p_2)$; again, (12.3.12) implies

$$ {}_q\beta_{(\rho_3, \nu - \rho_1 - \rho_2)} = (1 - p_1 - p_2)\,{}_q p_{3|12}(p_1,\ p_2), \tag{16.4.13} $$

so that another scanning of Table III will yield ρ_3.

It should be clear how the preceding procedure continues when $k > 3$.

16.4.6 Assessing k-variate Normal Distributions

Suppose the decision-maker wishes to express his prior judgments about a k-dimensional vector $\tilde{\boldsymbol{\mu}}$ in the form of a k-variate Normal df $f_N^{(k)}(\boldsymbol{\mu} \mid \boldsymbol{\mu}_0,\ \boldsymbol{\sigma}_0^2)$.

The procedure of Subsection 16.4.5, assessment of the marginal df's of $\tilde{\mu}_1, \ldots,$ and $\tilde{\mu}_k$, is the first step. In view of (12.7.6), df $\tilde{\boldsymbol{\mu}}$ has df $f_N^{(k)}(\boldsymbol{\mu} \mid \boldsymbol{\mu}_0,\ \boldsymbol{\sigma}_0^2)$, then each $\tilde{\mu}_1$ has df $f_N(\mu_i \mid \mu_{oi},\ \sigma_{0ii}^2)$, where μ_{0i} is the ith component of $\boldsymbol{\mu}_0$ and σ_{0ii}^2 is the $(i,\ i)$th element of $\boldsymbol{\sigma}_0^2$. Hence the first step is for the decision-maker to assess the k marginal df's $f_N(\mu_i \mid \mu_{oi},\ \sigma_{0ii}^2)$, according to the procedure suggested in Subsection 16.4.4.

But more remains to be done, as the off-diagonal elements of $\boldsymbol{\sigma}_0^2$ are as yet unspecified. One convenient procedure for specifying the $k(k-1)/2$ covariances σ_{oij}, for $i \neq j$ is to exploit the fact that

$$ E(\tilde{\mu}_j \mid \mu_i) = \mu_{0j} + \frac{\sigma_{0ij}^2}{\sigma_{0ii}^2}(\mu_i - \mu_{0i}), \tag{16.4.14} $$

which is apparent from (12.7.9′). We use (16.4.14) for each of the $k(k-1)/2$ covariances as follows: *first*, choose a value μ_i of $\tilde{\mu}_i$; and *second*, assess the conditional median $_{.5}\mu_{j|i}(\mu_i)$ of $\tilde{\mu}_j$ assuming that $\tilde{\mu}_i = \mu_i$. By the symmetry of the conditional normal df of $\tilde{\mu}_j$ given $\tilde{\mu}_i = \mu_i$, we have that $_{.5}\mu_{j|i}(\mu_i) = E(\tilde{\mu}_j \mid \mu_i)$ and hence can solve for σ^2_{0ij} from Equation (16.4.14):

$$\sigma^2_{0ij} = \sigma^2_{0ii}[_{.5}\mu_{j|i}(\mu_i) - \mu_{oj}]/(\mu_i - \mu_{0i}). \qquad (16.4.15)$$

These $k(k-1)/2$ conditional-median assessments and subsequent computations complete the specification of parameters of $f_{N^{(k)}}(\boldsymbol{\mu} \mid \boldsymbol{\mu}_0, \boldsymbol{\sigma}_0{}^2)$, since $\boldsymbol{\sigma}_0{}^2$ is symmetric.

However, the decision-maker is not yet finished. The covariance matrix $\boldsymbol{\sigma}_0{}^2$ may fail to be positive-definite, and in that case reassessment is necessary. A means of avoiding this difficulty is suggested in Pratt, Raiffa, and Schlaifer [**64**], Chapter 22, and it involves assuming that $\tilde{\mu}_1, \ldots, \tilde{\mu}_k$ are conditionally independent given another (smaller) set $\tilde{w}_1, \ldots, \tilde{w}_j$ of "predictor" variables.

But the fact remains that at this time there is no simple way of avoiding the difficulties apparently inherent in assessing joint distributions of a large number (k) of random variables. The next subsection, on the inverted Wishart distribution, takes a very informal approach to perhaps the most difficult assessment task of all; namely, the distribution of a covariance matrix.

Example 16.4.9

Let $\tilde{\mu}_i$ denote the year-end price of one share of the common stock of Company i, for $i = 1, 2, 3$; and suppose that an investor wishes to express his judgments about $\tilde{\boldsymbol{\mu}} \equiv (\tilde{\mu}_1, \tilde{\mu}_2, \tilde{\mu}_3)$ in the form of a trivariate Normal df $f_{N^{(3)}}(\boldsymbol{\mu} \mid \boldsymbol{\mu}_0, \boldsymbol{\sigma}_0{}^2)$. By the procedure suggested in Subsection 16.4.4, he determines that

$$\mu_{01} = 29, \quad \sigma^2_{011} = 9;$$
$$\mu_{02} = 78, \quad \sigma^2_{022} = 16; \text{ and}$$
$$\mu_{03} = 47, \quad \sigma^2_{033} = 12.25.$$

Next, he asks himself, "Suppose I knew that $\tilde{\mu}_1 = 32$; what then would my medians $_{.5}\mu_{2|1}(32)$ and $_{.5}\mu_{3|1}(32)$ be?" Let us suppose that by the lottery comparison method of Subsection 16.4.1 he determines that $_{.5}\mu_{2|1}(32) = 79$ and $_{.5}\mu_{3|1}(32) = 46$. Then, by (16.4.8), we may compute

$$\sigma^2_{012} = \frac{9[79 - 29]}{(32 - 29)} = 3$$

and

$$\sigma^2_{013} = \frac{9[46 - 47]}{(32 - 29)} = -3.$$

Finally, suppose that the investor assesses ${}_{.5}\mu_{3|2}(82) = 45\frac{1}{2}$. Then, by (16.4.8),

$$\sigma_{023}{}^2 = \frac{16[45\frac{1}{2} - 47]}{(82 - 78)} = -6.$$

Hence $\boldsymbol{\mu}_0 = (29, 78, 47)$ and

$$\boldsymbol{\sigma}_0{}^2 = \begin{bmatrix} 9 & 3 & -3 \\ 3 & 16 & -6 \\ -3 & -6 & 12.25 \end{bmatrix}.$$

The reader may check that $\boldsymbol{\sigma}_0{}^2$ is positive-definite by verifying that all leading minors are positive. (*Note*: In actual fact, stock prices are much more highly correlated than this example implies.)

16.4.7 Assessing Inverted Wishart Distributions

Our final major assessment problem is the inverted Wishart distribution $f_{iW}{}^{(k)}(\boldsymbol{\sigma}^2 \mid \boldsymbol{\psi}_0, \nu_0)$, natural conjugate for the covariance matrix $\tilde{\boldsymbol{\sigma}}^2$ of a k-variate Normal process. No really satisfactory assessment procedure is known for this case; and the thoughtful reader—having reflected upon the operational difficulties inherent in assessing, for example, $k \geqq 4$-variate Normal df's in the preceding subsection—might well be appalled at a confrontation with the effectively $k(k + 1)/2$-variate inverted Wishart distribution.

It *is* an appalling confrontation when $k > 2$, and hence we shall suggest a procedure which differs quite radically from those of the preceding subsections (which were based upon lottery comparisons). Our procedure here will be to apply judgments to fit the *mode* of the inverted Wishart df, as given by (12.11.32), which we record here for convenience: if $\tilde{\boldsymbol{\sigma}}^2$ has df $f_{iW}{}^{(k)}(\cdot \mid \boldsymbol{\psi}_0, \nu_0)$, then

$$\hat{\boldsymbol{\sigma}}^2 = \left(\frac{\nu_0}{\nu_0 + 2k}\right)\boldsymbol{\psi}_0. \tag{16.4.16}$$

As *step one*, the decision-maker records his "best guess" $\hat{\sigma}_{ii}{}^2$ as to each marginal variance $\sigma_{ii}{}^2$, and also his "best guess" $\hat{\rho}_{ij}$ as to each correlation ρ_{ij} for $i \neq j$, finally setting

$$\hat{\sigma}_{ij}{}^2 = \hat{\rho}_{ij}(\hat{\sigma}_{ii}{}^2\hat{\sigma}_{jj}{}^2)^{1/2} \tag{16.4.17}$$

in conformity with (10.3.17). Thus in step one the decision-maker obtains $\hat{\boldsymbol{\sigma}}^2$. (We advise "best-guessing" correlations rather than covariances since most people seem to find them easier.) *The result must be checked for positive definiteness.*

As *step two*, we wish to obtain $\boldsymbol{\psi}_0$ and ν_0 from $\hat{\boldsymbol{\sigma}}^2$ via (16.4.16), but it is clear that the $k(k + 1)/2$ functionally independent elements of $\hat{\boldsymbol{\sigma}}^2$ do not suffice to determine the same number of functionally independent elements of $\boldsymbol{\psi}_0$, plus ν_0 as well. A rather simple approach is to pick a small value ν_0, keeping in mind the following facts: (1) the smaller the ν_0, the more heavily the posterior df of $\hat{\boldsymbol{\sigma}}^2$ depends upon the outcome of the experiment; (2) the $E(\tilde{\boldsymbol{\sigma}}^2)$ does not exist if $\nu_0 \leqq 2$; and (3) the prior mode $\tilde{\boldsymbol{\sigma}}^{-2}$ of $\tilde{\boldsymbol{\sigma}}^{-2}$ [which has df $f_{W^{(k)}}(\cdot \mid \boldsymbol{\psi}_0, \nu_0)$ if and only if $\tilde{\boldsymbol{\sigma}}^2$ has df $f_{iW^{(k)}}(\cdot \mid \boldsymbol{\psi}_0, \nu_0)$] does not exist if $\nu_0 \leqq 2$. These facts are immediate from (14.6.4), (12.9.11), and (12.11.31), respectively.

Example 16.4.10

Suppose $\tilde{x}_1$ = length, $\tilde{x}_2$ = breadth, and $\tilde{x}_3$ = height of supposedly cubic blocks produced by some erratic machine, and let $\tilde{\mathbf{x}} = (\tilde{x}_1, \tilde{x}_2, \tilde{x}_3)'$ have df $f_{N^{(3)}}(\mathbf{x} \mid \boldsymbol{\mu}, \boldsymbol{\sigma}^2)$. Suppose a decision-maker wishes to express his judgments about $\tilde{\boldsymbol{\sigma}}^2$ in the form of $f_{iW^{(3)}}(\cdot \mid \boldsymbol{\psi}_0, \nu_0)$, and that he has assessed $\hat{\sigma}_{11}{}^2 = 4$, $\hat{\sigma}_{22}{}^2 = 4$, $\hat{\sigma}_{33}{}^2 = 9$, $\hat{\rho}_{12} = .7$, $\hat{\rho}_{13} = 0$, and $\hat{\rho}_{23} = .5$. By (16.4.17), $\hat{\sigma}_{12}{}^2 = (.7)(4 \times 4)^{\frac{1}{2}} = 2.8$, $\hat{\sigma}_{13}{}^2 = 0$, and $\hat{\sigma}_{23}{}^2 = (.5)(4 \times 9)^{\frac{1}{2}} = 3.0$. Hence

$$\hat{\boldsymbol{\sigma}}^2 = \begin{bmatrix} 4.0 & 2.8 & 0 \\ 2.8 & 4.0 & 3.0 \\ 0 & 3.0 & 9.0 \end{bmatrix},$$

which is positive definite. He chooses $\nu_0 = 1$, so that

$$\boldsymbol{\psi}_0 = \left(\frac{\nu_0 + 2k}{\nu_0}\right)\hat{\boldsymbol{\sigma}}^2 = \left(\frac{7}{6}\right)\hat{\boldsymbol{\sigma}}^2$$

for $\hat{\boldsymbol{\sigma}}^2$ as previously given.

As an alternative to simply picking a small value of ν_0, the decision-maker may select an i such that his judgment can best be brought to bear upon the marginal variance $\sigma_{ii}{}^2$ of $\tilde{x}_i$, assess its [by (12.9.15)] marginal df $f_{i\gamma}(\sigma_{ii}{}^2 \mid \psi_{ii}, \nu_0)$, and thus determine ν_0.

16.4.8 Assessing Normal-Inverted Gamma and Normal-Inverted Wishart Distributions

It is not difficult to apply the methods of the preceding subsections to the assessment of Normal-inverted gamma and Normal-inverted Wishart distributions.

For the univariate Normal inverted gamma df $f_N(\mu \mid \mu_0, \sigma^2/n_0)$ $f_{i\gamma}(\sigma^2 \mid \psi_0, \nu_0)$, the decision-maker first assesses the marginal df $f_{i\gamma}(\sigma^2 \mid \psi_0, \nu_0)$, perhaps by applying judgments directly to assessing an inverted gamma-2 df for $\tilde{\sigma}$. Second, he chooses a positive number y, assumes that $\tilde{\sigma}^2 = y$, and assesses a univariate Normal df $f_N(\mu \mid \mu_0, \sigma_0^2)$ for $\tilde{\mu}$ conditional upon $\tilde{\sigma}^2 = y$. Finally, by equating σ_0^2 and $\sigma^2/n_0 = y/n_0$ it follows that

$$n_0 = \frac{y}{\sigma_0^2}. \tag{16.4.18}$$

Example 16.4.11

Suppose that the amount $\tilde{x}$ of money per year that any given family spends for frozen vegetables is normally distributed about some mean μ with some variance σ^2. An advertising manager wishes to assess a Normal-inverted gamma prior df to $(\tilde{\mu}, \tilde{\sigma}^2)$. First, he uses lottery comparisons to determine that ${}_{.50}\sigma = 110$ and ${}_{.75}\sigma = 132$. Hence $[{}_{.75}\sigma/{}_{.50}\sigma]^2 = 1.75$, so that from (16.4.6) and Table II we obtain $\nu_0 = 4$, which then yields $\psi_0 = 3.36/4 = .84$ via (16.4.7). Hence his prior df of $\tilde{\sigma}$ is $f_{i\gamma 2}(\sigma \mid .84, 4)$ and his prior df of $\tilde{\sigma}^2$ is $f_{i\gamma}(\sigma^2 \mid .84, 4)$. To find μ_0 and n_0, he then assumes that $\tilde{\sigma} = 100$, implying $\tilde{\sigma}^2 = 10{,}000 = y$ and assesses ${}_{.50}\mu = 400$, ${}_{.75}\mu = 600$ conditional upon $\tilde{\sigma}^2 = y$. From (16.4.10), $\mu_0 = 400$, and from (16.4.11), $\sigma_0^2 = (200/.674)^2 \doteq 87{,}600$, so that by (16.4.18), $n_0 = 10{,}000/87{,}600 \doteq .114$. Hence the prior parameters are $\mu_0 = 400$, $\psi_0 = .84$, $n_0 = .114$, and $\nu_0 = 4$.

For the k-variate Normal-inverted gamma df $f_N^{(k)}(\boldsymbol{\mu} \mid \boldsymbol{\mu}_0, \sigma_*^2 \boldsymbol{\sigma}_{0*}^2) f_{i\gamma}(\sigma_*^2 \mid \psi_0, \nu_0)$ the procedure is similar: *first*, assess the unconditional df $f_{i\gamma}(\sigma_*^2 \mid \psi_0, \nu_0)$ of $\tilde{\sigma}_*^2$; *second*, assume a given value y of $\tilde{\sigma}_*^2$; and *third*, assess a k-variate Normal df $f_N^{(k)}(\boldsymbol{\mu} \mid \boldsymbol{\mu}_0, \boldsymbol{\sigma}_0^2)$ for $\tilde{\boldsymbol{\mu}}$ conditional upon $\tilde{\sigma}_*^2 = y$. It follows readily that $\boldsymbol{\sigma}_0^2 = y\boldsymbol{\sigma}_{0*}^2$, so that

$$\boldsymbol{\sigma}_{0*}^2 = y^{-1}\boldsymbol{\sigma}_0^2. \tag{16.4.19}$$

Obtaining the unconditional df of $\tilde{\sigma}_*^2$ directly would require the decision-maker to ponder $|\boldsymbol{\sigma}^2|^{1/k}$, which is usually hard to do. An easier procedure is for him to choose an $i \in \{1, \ldots, k\}$ such that he has some feelings about $\tilde{\sigma}_{ii}^2$, and to assess a df $f_{i\gamma}(\sigma_{ii}^2 \mid a_{ii}, b_{ii})$ for $\tilde{\sigma}_{ii}^2$. Since $\sigma_{ii}^2 = \sigma_*^2 \sigma_{*ii}^2$, then $\tilde{\sigma}_*^2 = \tilde{\sigma}_{ii}^2/\sigma_{*ii}^2$, so that by (12.9.13) for $k = 1$ the df of $\tilde{\sigma}_*^2$ is

$$f_{i\gamma}(\sigma_*^2 \mid a_{ii}/\sigma_{*ii}^2, b_{ii}).$$

Hence the df of $\tilde{\sigma}_*^2$ is $f_{i\gamma}(\sigma_*^2 \mid \psi_0, \nu_0)$ with $\psi_0 = a_{ii}/\sigma_{*ii}^2$ and $\nu_0 = b_{ii}$.

Finally, for the Normal-inverted Wishart df

$$f_N^{(k)}(\boldsymbol{\mu} \mid \boldsymbol{\mu}_0, n_0^{-1}\boldsymbol{\sigma}^2) f_{iW}^{(k)}(\boldsymbol{\sigma}^2 \mid \boldsymbol{\psi}_0, \nu_0),$$

a similar three-step procedure is available: (1) assess $f_{iW}^{(k)}(\boldsymbol{\sigma}^2 \mid \boldsymbol{\psi}_0, \nu_0)$; (2) assume that $\tilde{\boldsymbol{\sigma}}^2 = \mathbf{Y}$ for a chosen covariance matrix $\mathbf{Y}$; and (3) assess a Normal df $f_N^{(k)}(\boldsymbol{\mu} \mid \boldsymbol{\mu}_0, n_0^{-1}\mathbf{Y})$ for $\tilde{\boldsymbol{\mu}}$ given $\tilde{\boldsymbol{\sigma}}^2 = \mathbf{Y}$. Note the lack of flexibility in the third step due to all relative covariances being assumed given. Lack of flexibility makes for ease in assessment but also for discomfort with the result. Because of it, step (3) requires the decision-maker to specify all medians ${}_{.50}\mu_i = \mu_{0i}$, which then constitute $\boldsymbol{\mu}_0$; and he must specify *one* other marginal fractile, from which n_0 can be determined. Suppose he assesses ${}_p\mu_i$ for some i and $p \neq .50$. Then he can use (16.4.11) to solve for $\sigma_{0ii}^2 = [({}_p\mu_i - \mu_{0i})/{}_pN^*]^2$, from which it follows (Exercise 6) that

$$n_0 = Y_{ii}/\sigma_{0ii}^2. \tag{16.4.20}$$

16.4.9 A Caveat

The examples in the preceding subsections may convey the erroneous impression that the various subassessments are easy and that the only difficulties lie in using the appropriate formulae and tables. Nothing could be further from the truth, since insofar as applications are concerned, the formulae and tables could (with training) be handled by anyone.

The reader will find, upon a little practice, that the *real* difficulty lies in making the lottery-comparison choices in Subsection 16.4.1. The reader will wish to name an interval, $[{}_p\theta_L, {}_p\theta_L]$, within which the pth fractile lies, but it is necessary to name a single point.

Some actual practice in making assessments is afforded by the exercises. Such practice would dispel the over-enthusiastic claims of those in decision theory who claim that the analysis eliminates the need for making value judgments (preferences, utility, Chapter 15) and fact judgments (probability, this chapter).

What Bayesian decision theory *does* accomplish is to change the *form* of the decision-maker's judgments *from* rather global judgments culminating in the choice of an act or of an act for each experimental outcome of a chosen experiment *to* a set of many more, very small choices between much simpler entities than his actual decision problem. In our assessment procedures of this and the preceding chapter, the decision-maker's choice at each step has been between two very simple lotteries involving the same prizes.

Now, once the decision-maker has made all these small judgments, the formal analysis in the *next* chapter can be brought to bear to solve his actual problem. This formal analysis ensures that the chosen act (or chosen experiment and act for each outcome of that experiment) is *consistent* with the decision-maker's preferences and judgments.

16.5 ALTERNATIVE AXIOMATIZATIONS AND ASSESSMENT PROCEDURES

Several alternative axiomatizations of subjective probability and/or utility have been discovered. The interested reader should consult Pratt, Raiffa and Schlaifer [64] and Savage [69] for joint axiomatizations of utility and subjective probability. Our rationale for using the Anscombe-Aumann approach to subjective probability is that it enables us to treat utility first, and then only subsequently treat subjective probability. Anscombe and Aumann [**3**] and Fishburn [**24**] cite additional references.

Assessment procedures other than the lottery-comparison method used here have been developed. The interested reader should consult Winkler, [**84**] and [**85**]. In this author's opinion and experience, none of the alternatives to some form of the lottery-comparison method possesses its directness and immediacy.

There are alternative ways of *using* the lottery-comparison method to fit the parameters of a conjugate family; for example, assess an approximating mf, compute moments according to the approximating mf, and equate these moments to the moments of a df in the conjugate family. Since the conjugate family moments are given by formulae involving the parameters, this equating of approximation to conjugate family moments allows one to solve for the parameters. See Pratt, Raiffa and Schlaifer [**64**], Chapter 11, for an application of this method to the assessment of beta distributions.

Finally, when one's prior opinions are weak, in the sense that they can be drastically changed by even a little additional, objective evidence, it may not be worth the effort to worry about a very precise specification of the prior dgf if one intends to obtain a fairly substantial amount of additional evidence.

Natural conjugate prior df's which result in posterior df's reflecting almost solely the likelihood function include $f_\beta(p \mid \rho, \nu)$ for small ρ and ν, $f_\gamma(\lambda \mid \rho_0, \tau_0)$ for small ρ_0 and τ_0, and $f_N(\mu \mid \mu_0, \sigma_0^2)$ for *large* σ_0^2. Passing to appropriate limits in each of these cases produces an *improper* df: $f_\beta(p \mid 0, 0)$ is not a df because it is proportional to $p^{-1}(1 - p)^{-1}$, which has infinite area under it and hence cannot be scaled to have unit area; and $f_N(\mu_0 \mid \mu_0, \infty) = 0$ for all μ and any μ_0, and hence cannot be scaled.

Such df's that are easy to influence often yield results by Bayesian arguments (based on posterior dgf's) which coincide numerically with the corresponding results obtained by orthodox arguments. Some examples of such coincidences are developed in Chapter 20 of this book. The interested reader may consult Edwards, Lindman, and Savage [**16**], Chapter 3 of Raiffa and Schlaifer [**66**], and Jeffreys [**40**], which offer various justifications for using such prior df's.

EXERCISES

1. Prove part (d) of (16.2.2).
2. Show that if $c \succ c'$, then $P(\theta_i) \geqq p$ if and only if the preference in Figure 16.31 holds true.

Figure 16.31

3. Verify for Section 16.3 that $L^+(p_0) \sim SL^+$ if and only if $P(\theta_2 \mid \{\theta_2, \theta_3\}) = p_0$.

4. Show that $P(\theta_2) = P(\theta_2 \mid \{\theta_2, \theta_3\}) \cdot P(\{\theta_2, \theta_3\})$.

5. Show how to define called-off lotteries in Subsection 16.4.1 so as to obtain $y_0 = {}_{0.25}\theta$.

6. Derive (16.4.20).

7. *Football probabilities*: Let $\theta_1 \equiv$ "Harvard will defeat Yale next year"; $\theta_2 \equiv$ "Yale will defeat Harvard next year"; and $\theta_3 \equiv$ "the Harvard-Yale game will be tied or called off next year." Assess your probabilities $P(\theta_1)$, $P(\theta_2)$, and $P(\theta_3)$. (You may substitute in this exercise another pair of teams—for example, Army-Navy, Harvard-Princeton, Princeton-Yale, Michigan-Wisconsin Tulane-L.S.U, Trinity-Wesleyan, and so forth.)

8. *Election probabilities*: Consider the next presidential election; and let $\tilde{p}_1 \equiv$ proportion of votes the Democratic candidate will receive; $\tilde{p}_2 \equiv$ proportion of votes the Republican candidate will receive, and $\tilde{p}_3 \equiv$ proportion of votes that other candidates will receive.

   (a) Express your judgments about $\tilde{\mathbf{p}} \equiv (\tilde{p}_1, \tilde{p}_2)'$ in the form of a bivariate beta df.

   (b) Let $\theta_1 \equiv$ "Democratic candidate has plurality," $\theta_2 \equiv$ "Republican candidate has plurality," and $\theta_3 \equiv$ "another candidate has plurality." Show how to obtain $P(\theta_i)$ for $i = 1, 2, 3$ from the bivariate beta df of $\tilde{p}$, and argue that $P(\Theta \setminus \{\theta_1, \theta_2, \theta_3\}) = 0$.

9. Think of some 10 to 20 person group to which you belong; for example, a class, a large committee, a small firm, or a board. Let $\tilde{\boldsymbol{\mu}} \equiv (\tilde{\mu}_1, \tilde{\mu}_2)'$, where $\tilde{\mu}_1$ and $\tilde{\mu}_2$ denote respectively the average weight and average height of the members (including yourself). Express your judgments about $\tilde{\boldsymbol{\mu}}$ in the form of a bivariate Normal df.

10. Let $\tilde{\mu}_1 \equiv$ year end value of the Dow-Jones Industrial average; $\tilde{\mu}_2 \equiv$ year-end price of IBM common stock; and $\tilde{\mu}_3 \equiv$ year-end price of GM common stock. Express your judgments regarding $\tilde{\boldsymbol{\mu}} \equiv (\tilde{\mu}_1, \tilde{\mu}_2, \tilde{\mu}_3)$ in the form of a trivariate normal df.

11. Think of a dangerous intersection near your home, and suppose that the interarrival times between accidents there are independent and identically distributed with common df $f_\gamma(t \mid r, \lambda)$. Assess r and λ.

17

SOLVING THE GENERAL DECISION PROBLEM AND EVALUATING INFORMATION

17.1 INTRODUCTION

The definition of the general decision problem was given in Section 4.3, which the reader is advised to review at this point. Subsequent chapters have fulfilled the promises, made there, to develop the concepts of probability for branches emanating from "chance" forks in the tree, and utility for the ultimate consequences c.

The gist of Chapters 15 and 16 is that the decision-maker chooses between two lotteries with prizes in the set C of all consequences c; that lottery offers the larger expectation of utility, the expectations being taken with respect to the decision-maker's probability function, regardless of whether it is objective or subjective. In Section 17.2 we apply this principle to show how the solution to the general decision problem may be obtained. Our analysis proceeds from finite trees—in which E, all $Z(e)$, all $A(e, z)$, and Θ are finite sets—to general trees in which $\tilde{\theta}$ and $\tilde{z}$ are random variables and E and $A(e, z)$ are fairly general sets.

Section 17.3 introduces an alternative to the utility function, called the (opportunity) loss function. The loss function is desirable from the utility function and, in many applied problems, much easier to assess. We show that the loss function produces exactly the same solutions to the general decision problem (provided that the appropriate operations are performed). All of the analysis of specific problems in Chapter 18 is conducted in terms of loss functions.

In Section 17.4 we impose the restriction that the consequences $c(e, z, a, \theta)$ of the decision problem be monetary; and, under some other restrictions, provide monetary measures of the value $I(e)$ of each experiment e to the decision-maker. The important property of this value is that the decision-maker should be willing to spend up to, but not more than $I(e)$ for the privilege of choosing his act after having observed $z \in Z(e)$ rather than after having observed only the "null outcome" z_0 of the null experiment e_0.

Section 17.5 concerns an approach to the general decision problem which is alternative to that given in Sections 17.2 and 17.3. The general approach in Sections 17.2 and 17.3 is often called the "extensive form," while the approach in 17.5 is called the "normal form." For Bayesian purpose the normal form is often more cumbersome than the extensive form. However, in non-Bayesian decision theory the extensive form cannot be applied at all. Conversely, in Bayesian decision theory the normal form furnishes the natural vehicle for analyzing decision rules of a specified form and for testing the sensitivity of an optimal decision rule to changes in the decision-maker's prior probability or dgf.

The discussion in 17.5 of the normal form constitutes a prelude to statistical inference and is used as such in the Introduction to Part IV.

17.2 SOLVING THE GENERAL DECISION PROBLEM

In Chapter 4 we defined the general decision problem and showed that it could be represented by a decision tree, reproduced in abbreviated form in Figure 17.1.

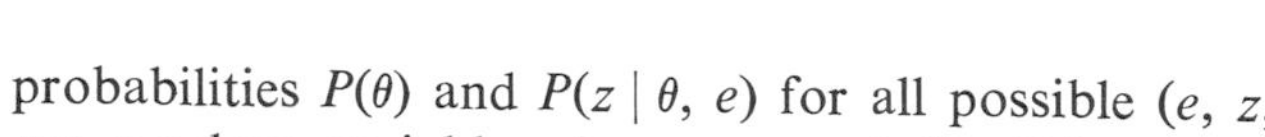

Figure 17.1

Subsequent chapters have added to this structure by showing the decision-maker can:

(1) assess probabilities $P(\theta)$ and $P(z \mid \theta, e)$ for all possible (e, z, θ)'s [if $\tilde{z}$ and $\tilde{\theta}$ are random variables, he assesses dgf's $f(\theta)$ and $f(z \mid \theta, e)$];
(2) use Bayes' Theorem to convert the assessments in (1) to probabilities $P(z \mid e)$ and $P(\theta \mid z, e)$ for all possible (e, z, θ)'s [if $\tilde{z}$ an $\tilde{\theta}$ are random variables, he derives $f(z \mid e)$ and $f(\theta \mid z, e)$]; and
(3) assess his utility function $u(e, z, a, \theta) \equiv u[c(e, z, a, \theta)]$ for the consequences of the decision.

Assume now that the decision tree is finite; that is, E, every $Z(e)$, every $A(e, z)$, and Θ are all finite sets. Let E have η elements, $Z(e)$ have $\zeta(e)$ elements, $A(e, z)$ have $\alpha(e, z)$ elements, and Θ have τ elements; and suppose that all assessments and derivations required by steps (1) through (3) above have been performed. The decision-maker is now prepared to solve his problem.

A *solution* to the general decision problem will consist of:

(a) selection of an experiment $e \in E$; and
(b) for every $z \in Z(e)$, selection of an act $a_{z,e} \in A(e, z)$, where e is the chosen experiment.

We shall seek an *optimal* solution to the general decision problem by proceeding from (b) and (a):

(b*) for every $e \in E$ and every $z \in Z(e)$, select an *optimal* act $a^0_{z,e} \in A(e, z)$; and then
(a*) select an *optimal* experiment $e^0 \in E$.

The reason for proceeding in reverse will be clear in a moment, as we "collapse" the decision tree by applying the basic utility and probability results (15.2.2) and (16.2.6).

Step One

Suppose the decision-maker has chosen e, observed $z \in Z(e)$, and chosen $a \in A(e, z)$. At this point the remainder of the tree appears as in Figure 17.2, which is a simple lottery—the decision-maker's utility $u(e, z, a, *)$ of which is found by applying (15.2.2):

$$u(e, z, a, *) = \sum_{i=1}^{T} u(e, z, a, \theta_i)P(\theta_i \mid z, e). \tag{17.2.1}$$

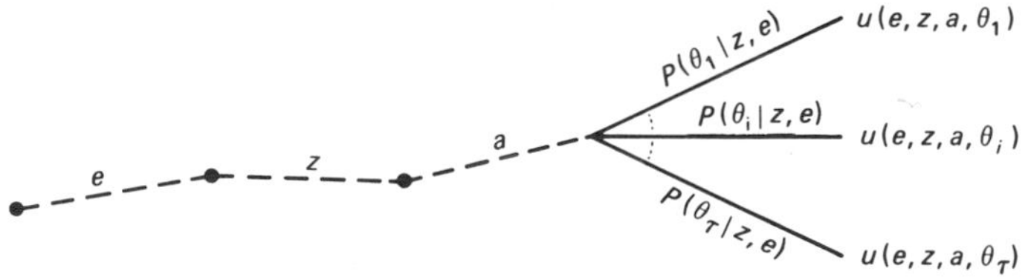

Figure 17.2

As step one, (17.2.1) is applied for every (e, z, a) in the tree, and the final θ-branches are replaced for each (e, z, a) by the number $u(e, z, a, *)$, which expresses the decision-maker's utility of being at that point in the tree. The result is a reduced tree, a typical path of which is given in Figure 17.3.

Figure 17.3

Step Two

Suppose the decision-maker has completed step one and now wishes to choose an optimal act $a^0_{z,e} \in A(e, z)$, for each (e, z). Note that for each fixed (e, z), the utility $u(e, z, \cdot, *)\colon A(e, z) \to R^1$ measures the relative desirability of the elements $a \in A(e, z)$, and that he should choose that a with the largest utility. Hence an optimal $a^0_{z,e}$ must satisfy

$$u(e, z, a^0_{z,e}, *) = \max \{u(e, z, a, *)\colon a \in A(e, z)\}. \qquad (17.2.2)$$

Note that $a^0_{z,e}$ need not be unique due to the fact that two or more elements of $A(e, z)$ may satisfy (17.2.2). Furthermore, when $A(e, z)$ is infinite, $a^0_{z,e}$ may not even exist (see Exercise 1 for an example). This pathology cannot occur when every $A(e, z)$ is finite. As step two, the decision-maker determines $u(e, z, a^0_{z,e}, *)$ and an $a^0_{z,e}$, for every (e, z). Let $u(e, z, {}^0, *) \equiv u(e, z, a^0_{z,e}, *)$. Then the tree may be further reduced from that represented in Figure 17.3 to Figure 17.4, since every (e, z) path is now evaluated by the utility $u(e, z, {}^0, *)$.

Figure 17.4

Step two has completed the process (b*) of selecting an optimal act $a^0_{z,e}$, for every (e, z) pair. Two operations were involved in (b*): first (step one), evaluating all "terminal lotteries" of the form given in Figure 17.2 by their utilities $u(e, z, a, *)$ (that is, the expectations of their utilities); and second (step two), choosing that lottery with the largest utility and evaluating the choice by the highest obtainable utility $u(e, z, {}^0, *)$.

Now suppose the decision-maker has chosen e. What is his utility? Given the choice e, he now has the lottery given in Figure 17.5, in which with probability $P(z_j \mid e)$ he receives (choices and subsequent lotteries with)

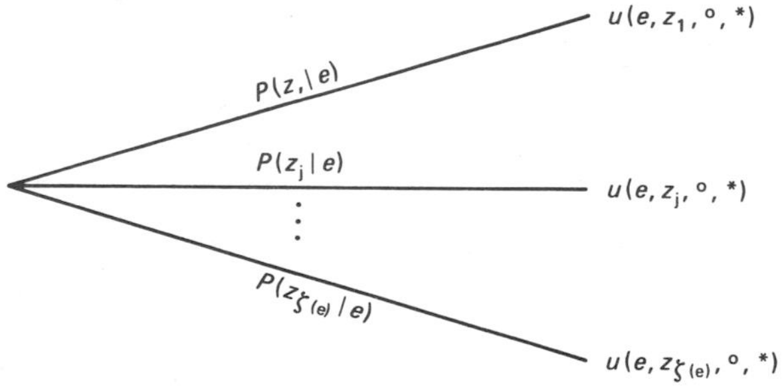

Figure 17.5

utility $u(e, z_j, {}^0, *)$, for $j = 1, \ldots, \zeta(e)$. The lottery in Figure 17.5 clearly has utility $u(e, *, {}^0, *)$, given by

$$u(e, *, {}^0, *) = \sum_{j=1}^{\zeta} u(e, z_j, {}^0, *)P(z_j \mid e). \qquad (17.2.3)$$

Step Three

The decision-maker computes $u(e, *, {}^0, *)$ via (17.2.3) for every $e \in E$.

The tree in Figure 17.4 can now be further simplified, to that in Figure 17.6.

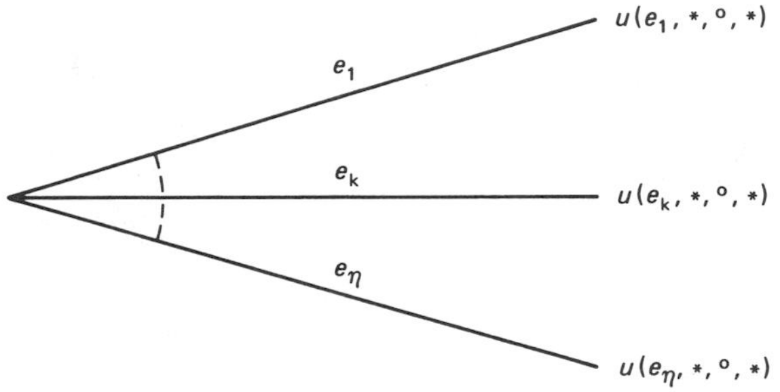

Figure 17.6

Step Four

This step should be apparent. The decision-maker chooses that experiment (or *an* experiment) which maximizes $u(e, *, {}^0, *)$; that is, an optimal experiment e^0 must satisfy

$$\begin{aligned} u(e^0, *, {}^0, *) &= \max \{u(e, *, {}^0, *): e \in E\} \\ &\equiv \max_e u(e, *, {}^0, *). \end{aligned} \qquad (17.2.4)$$

Thus process (a*) is now complete, and (17.2.4) gives the decision-maker's utility of the general decision problem.

To express the solution, it is now desirable to make the statements (a) and (b) from the previous computations. That is, first give e^0 and then read off a^0_{z,e^0}, for every $z \in Z(e^0)$.

The preceding may appear somewhat recondite in the abstract, but the basic procedure is really very simple. An example will clarify it.

Example 17.2.1

Consider the general decision problem given by the tree in Figure 17.7. The numbers at the endpoints of this tree are utilities of the consequences.

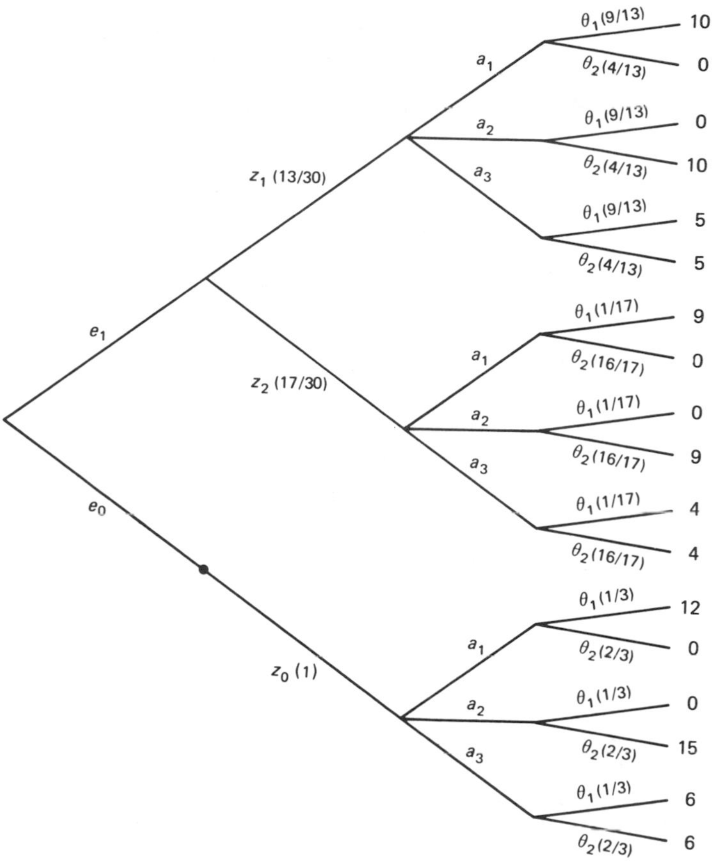

Figure 17.7

Suppose the decision-maker has assessed $P(\theta_1) = \frac{1}{3}$ [hence $P(\theta_2) = \frac{2}{3}$], $P(z_1 \mid \theta_1, e_1) = 0.9$, and $P(z_1 \mid \theta_2, e_1) = 0.2$ [and hence $P(z_2 \mid \theta_1, e_1) = 0.1$ and $P(z_2 \mid \theta_2, e_1) = 0.8$]. As an easy application of Bayes' Theorem, the reader may verify that

$$P(\theta_1 \mid z_1, e_1) = \frac{9}{13}, \; P(\theta_2 \mid z_1, e_1) = \frac{4}{13}; \text{ and}$$

$$P(\theta_1 \mid z_2, e_1) = \frac{1}{17}, \; P(\theta_2 \mid z_2, e_1) = \frac{16}{17};$$

while $P(z_1) = 13/30$ and $P(z_2) = 17/30$. These probabilities have been inserted in Figure 17.7 on the appropriate branches. Clearly, given e_0,

$P(z_0 \mid e_0) = 1$ and $P(\theta_i \mid z_0, e_0) = P(\theta_i)$, for $i = 1, 2$. As *step one*, we compute all $u(e, z, a, *)$'s:

$$u(e_0, z_0, a_1, *) = 12\left(\frac{1}{3}\right) + 0\left(\frac{2}{3}\right) = 4$$

$$u(e_0, z_0, a_2, *) = 0\left(\frac{1}{3}\right) + 15\left(\frac{2}{3}\right) = 10$$

$$u(e_0, z_0, a_3, *) = 6\left(\frac{1}{3}\right) + 6\left(\frac{2}{3}\right) = 6$$

$$u(e_1, z_1, a_1, *) = 10\left(\frac{9}{13}\right) + 0\left(\frac{4}{13}\right) = \frac{90}{13}$$

$$u(e_1, z_1, a_2, *) = 0\left(\frac{9}{13}\right) + 10\left(\frac{4}{13}\right) = \frac{40}{13}$$

$$u(e_1, z_1, a_3, *) = 5\left(\frac{9}{13}\right) + 5\left(\frac{4}{13}\right) = \frac{65}{13}$$

$$u(e_1, z_2, a_1, *) = 9\left(\frac{1}{17}\right) + 0\left(\frac{16}{17}\right) = \frac{9}{17}$$

$$u(e_1, z_2, a_2, *) = 0\left(\frac{1}{17}\right) + 9\left(\frac{16}{17}\right) = \frac{144}{17}$$

$$u(e_1, z_2, a_3, *) = 4\left(\frac{1}{17}\right) + 4\left(\frac{16}{17}\right) = \frac{68}{17}.$$

Using these figures, we obtain the tree in Figure 17.8.

Next, as *step two*, it is clear that:

$$a^0_{z_0,e_0} = a_2,\ u(e_0, z_0, {}^0, *) = 10;$$

$$a^0_{z_1,e_1} = a_1,\ u(e_1, z_1, {}^0, *) = \frac{90}{13}; \text{ and}$$

$$a^0_{z_2,e_1} = a_2,\ u(e_1, z_2, {}^0, *) = \frac{144}{17}.$$

We thus obtain Figure 17.9.

As *step three*, we compute

$$u(e_1, *, {}^0, *) = \left(\frac{90}{13}\right)\left(\frac{13}{30}\right) + \left(\frac{144}{17}\right)\left(\frac{17}{30}\right)$$
$$= \frac{234}{30}$$

and

$$u(e_0, *, {}^0, *) = 10 = \frac{300}{30}.$$

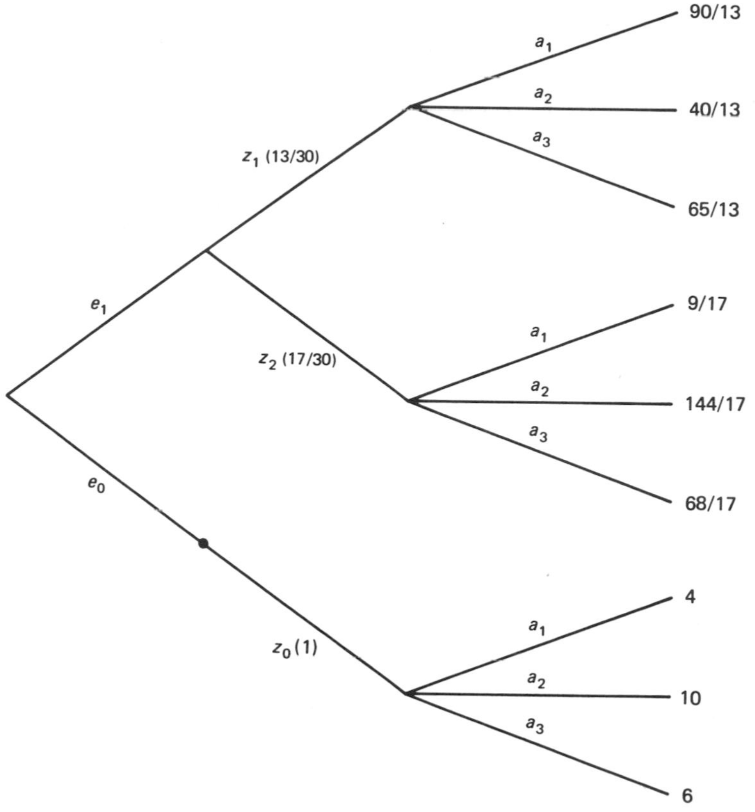

Figure 17.8

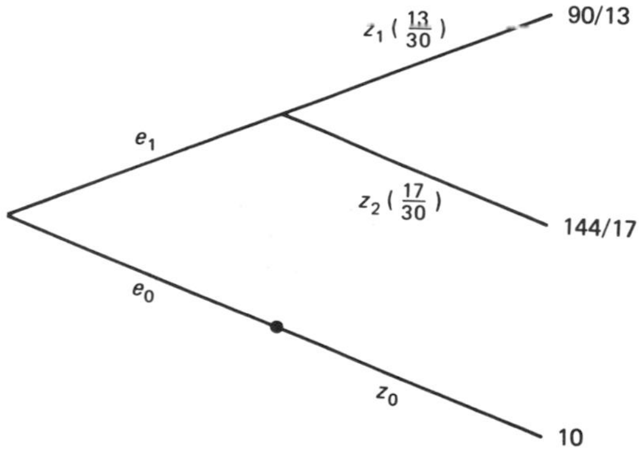

Figure 17.9

Thus we obtain Figure 17.10, from which it is clear (step four) that $e^0 = e_0$ and $u(e^0, *, {}^0, *) = 10$. This demonstrates that occasionally information is too expensive. (Examine the utilities in Figure 17.7.) Now, much redrawing of the tree can be avoided by putting all information and calculations on the original tree. Figure 17.11 contains all the data of Figures 17.7 through 17.10; a branch marked with a "//" has to be "pruned"; that is, that act (or experiment) is *non*optimal. From Figure 17.11 it is clear that the solution is $e^0 = e_0$, and $a^0_{z,e^0} = a^0_{z_0,e_0} = a_2$; that is, do not obtain information, and choose act a_2.

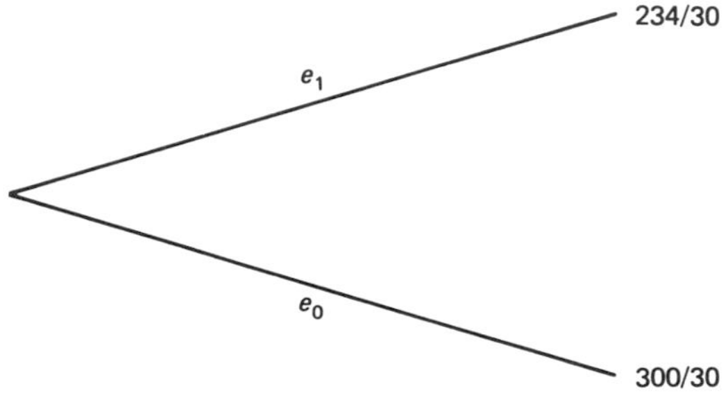

Figure 17.10

We shall now define some notation which will lead immediately to the general solution when $\tilde{z}$ and $\tilde{\theta}$ are random variables. We define

$$E_{\theta|z,e}u(e, z, a, \tilde{\theta})$$

by

$$\begin{aligned} E_{\theta|z,e}u(e, z, a, \tilde{\theta}) &\equiv u(e, z, a, *) \\ &\equiv E(u(e, z, a, \tilde{\theta}) \mid z, e). \end{aligned} \tag{17.2.5}$$

The subscripted E is simply a compact way of denoting that the expectation is with respect to $f(\theta \mid z, e)$. Next, we define $\max_a E_{\theta|z,e}u(e, z, a, \tilde{\theta})$ by

$$\begin{aligned} \max_a E_{\theta|z,e}u(e, z, a, \tilde{\theta}) &\equiv u(e, z, {}^0, *) \\ &\equiv \max\{E_{\theta|z,e}u(e, z, a, \tilde{\theta}) : a \in A(e, z)\}; \end{aligned} \tag{17.2.6}$$

and, finally, define $E_{z|e} \max_a E_{\theta|z,e}u(e, \tilde{z}, a, \tilde{\theta})$ by

$$\begin{aligned} E_{z|e} \max_a E_{\theta|z,e}u(e, \tilde{z}, a, \tilde{\theta}) &\equiv u(e, *, {}^0, *) \\ &\equiv E(u(e, \tilde{z}, {}^0, *)). \end{aligned} \tag{17.2.7}$$

When z and $\tilde{\theta}$ are random variables, (17.2.5) and (17.2.7) indicate which dgf's are relevant to which expectation. When z and θ (now untilded) are

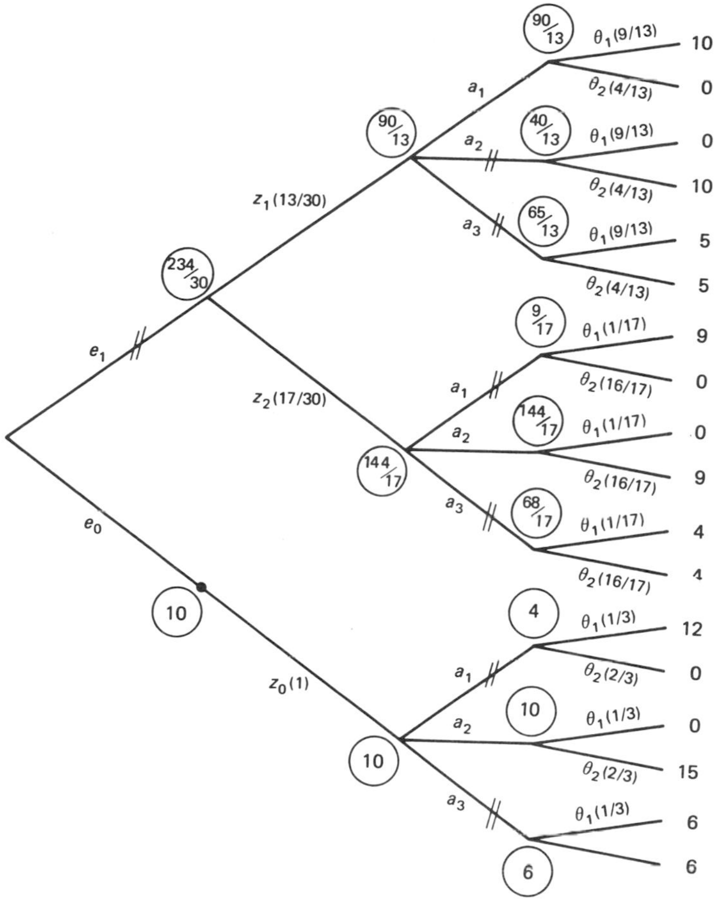

Figure 17.11

abstract elementary events, (17.2.5) and (17.2.7) indicate how the expectations are to be determined from (17.2.1) and (17.2.3). [For z and θ abstract elementary events, (17.2.5) and (17.2.7) are manifestations of a more general concept of expectation, based upon the Lebesgue integral. See Chapter 8 of Pratt, Raiffa and Schlaifer [**64**] for a brief introduction, or Halmos [**34**] or Loéve [**5**] for a much more complete treatment.] The reader should recall that when $\tilde{z}$ and/or $\tilde{\theta}$ are random vectors, they appear in boldface type. The requisite mathematical operations are apparent.

In view of (17.2.5) through (17.2.7), we deduce that

$$u(e^0, *, {}^0, *) = \max_e E_{z|e} \max_a E_{\theta|z,e} u(e, \tilde{z}, a, \tilde{\theta}), \qquad (17.2.8)$$

which is to be read as follows: *first* (step 1), "$E_{\theta|z,e}$" operates on $u(e, \tilde{z}, a, \tilde{\theta})$, producing a function $E_{\theta|z,e}u(e, \tilde{z}, a, \tilde{\theta}) \equiv u(e, z, a, *)$ depending only upon (e, z, a). *Second* (step 2), "max" operates on $E_{\theta|z,e}u(e, \tilde{z}, a, \tilde{\theta})$, producing a function $\max_a E_{\theta|z,e}u(e, \tilde{z}, a, \tilde{\theta}) \equiv u(e, z, {}^0, *)$ depending only upon (e, z). *Third* (step 3), "$E_{z|e}$" operates on $\max_a E_{\theta|z,e}u(e, \tilde{z}, a, \tilde{\theta})$, producing a function $E_{z|e} \max_a E_{\theta|z,e}u(e, \tilde{z}, a, \tilde{\theta}) \equiv u(e, *, {}^0, *)$ depending only upon e. And *fourth* (step 4), "max" operates on $E_{z|e} \max_a E_{\theta|z,e}u(e, \tilde{z}, a, \tilde{\theta})$.
_e

Equation (17.2.8) could be (but is not) called the "fundamental theorem of Bayesian decision theory," as it expresses in a single line all operations required to obtain an optimal solution of the general decision problem. It is indeed a theorem, as it is a consequence of the more basic utility theory developed in Chapter 15, of the definition of the general decision problem, and of (16.2.6) on the admissibility of subjective probability.

Example 17.2.2

Suppose $\Theta \subset (-\infty, \infty)$, $E = \{e_0, e_1\}$, $Z(e_0) = \{z_0\}$, $Z(e_1) \subset (-\infty, \infty)$, and $A = A(e_0, z_0) = A(e_1, z) = (-\infty, \infty)$, for every $z \in (-\infty, \infty)$. Then $f(\theta \mid z_0, e_0) = f(\theta)$ (naturally), while $f(\theta \mid z, e_1)$ may depend upon z. Suppose also that

$$u(e_1, z, a, \theta) = 25 - 20(\theta - a)^2 - 5z,$$

while

$$u(e_0, z_0, a, \theta) = 50 - 20(\theta - a)^2.$$

Then

$$\begin{aligned} u(e_0, z_0, a, *) & \\ &= E_{\theta|z_0,e_0}u(e_0, z_0, a, \tilde{\theta}) \\ &= \int_{-\infty}^{\infty} \{50 - 20(\theta - a)^2\} f(\theta)d\theta \\ &= 50 - 20 \int_{-\infty}^{\infty} ([\theta - E(\tilde{\theta})] + [E(\tilde{\theta}) - a])^2 f(\theta)d\theta \\ &= 50 - 20[E(\tilde{\theta}) - a]^2 - 20V(\tilde{\theta}), \end{aligned}$$

from which it follows readily that

$$a^0_{z_0,e_0} = E(\tilde{\theta})$$

and

$$u(e_0, z_0, {}^0, *) = 50 - 20V(\tilde{\theta}).$$

Similarly,

$$
\begin{aligned}
&u(e_1, z, a, *) \\
&= E_{\theta|z,e}u(e_1, z, a, \tilde{\theta}) \\
&= \int_{-\infty}^{\infty} \{25 - 20(\theta - a)^2 - 5z\} f(\theta \mid z, e_1) d\theta \\
&= 25 - 5z - 20 \int_{-\infty}^{\infty} ([\theta - E(\tilde{\theta} \mid z, e_1)] + [E(\tilde{\theta} \mid z, e_1) - a])^2 f(\theta \mid z, e_1) d\theta \\
&= 25 - 5z - 20[E(\tilde{\theta} \mid z, e_1) - a]^2 - 20V(\tilde{\theta} \mid z, e_1),
\end{aligned}
$$

which implies that

$$a^0_{z,e_1} = E(\tilde{\theta} \mid z, e_1)$$

and

$$u(e_1, z, {}^0, *) = 25 - 5z - 20V(\tilde{\theta} \mid z, e_1).$$

Now, it is clear that

$$u(e_0, *, {}^0, *) = u(e_0, z_0, {}^0, *),$$

while

$$
\begin{aligned}
u(e_1, *, {}^0, *)& \\
&= E_{z|e_1}u(e, \tilde{z}, {}^0, *) \\
&= 25 - 5E(\tilde{z} \mid e_1) - 20E_{z|e_1}V(\tilde{\theta} \mid \tilde{z}, e_1).
\end{aligned}
$$

Finally, e_1 is at least as desirable as e_0 to the decision-maker if and only if

$$u(e_1, *, {}^0, *) \geqq u(e_0, *, {}^0, *),$$

which with a little algebra is found to hold if and only if

$$25 + 5E(\tilde{z} \mid e_1) \leqq 20\{V(\tilde{\theta}) - E_{z|e_1}V(\tilde{\theta} \mid \tilde{z}, e_1)\}.$$

This inequality can be tested for any set of dgf's $f(z \mid \theta, e_1)$ and $f(\theta)$.

17.3 OPPORTUNITY LOSS

Suppose that the decision-maker has chosen e_0, "observed" z_0, and is just about to choose $a^0_{z_0,e_0} \in A(e_0, z_0)$ when an omniscient and scrupulously truthful agent informs him of the true $\theta \in \Theta$ at no cost whatsoever. Given this state of affairs, and given that he still must choose an act from $A(e_0, z_0)$, it is clear that he should choose an act $a_\theta{}^0$ such that

$$
\begin{aligned}
u(e_0, z_0, a_\theta{}^0, \theta) &\equiv \max \{u(e_0, z_0, a, \theta): a \in A(e_0, z_0)\} \\
&\equiv \max_a u(e_0, z_0, a, \theta). \qquad (17.3.1)
\end{aligned}
$$

Now, the information that the decision-maker receives gratis is called *perfect information*, as it completely eliminates his uncertainty about θ once it has been received. (See Chapter 6, Exercise 6, in this regard.) Since, as suggested by Example 17.2.1, experiments $e \neq e_0$ usually impose some cost to the decision-maker in the form of lower utilities, and since we have assumed that the perfect information is gratis, it often happens that

$$u(e_0, z_0, a_\theta{}^0, \theta) \geqq u(e, z, a, \theta), \text{ all } \theta \in \Theta, \tag{17.3.2}$$

for every (e, z, a).

If (17.3.2) obtains, it is natural to call $u(e_0, z_0, a_\theta{}^0, \theta) - u(e, z, a, \theta)$ the opportunity loss: if θ is true, the decision-maker loses the opportunity of increasing his utility by $u(e_0, z_0, a_\theta{}^0, \theta) - u(e, z, a, \theta)$ given (e, z, a).

We define the *opportunity loss* function $\ell(e, z, a, \theta)$ by

$$\ell(e, z, a, \theta) = u(e_0, z_0, a_\theta{}^0, \theta) - u(e, z, a, \theta). \tag{17.3.3}$$

Observe that $\ell(e, z, a, \theta) \geqq 0$, for every (e, z, a, θ) if and only if (17.3.2) holds. Also note that $u(e_0, z_0, a_\theta{}^0, \theta)$ in (17.3.3) depends solely upon θ, since e_0 and z_0 are fixed and θ determines $a_\theta{}^0$ sufficiently precisely to fix the value of $u(e_0, z_0, a_\theta{}^0, \theta)$.

We shall now show that, roughly speaking, minimizing (the expectation of) opportunity loss is just as valid a means of determining e^0 and $a^0_{z,e}$ as is maximizing (the expectation of) utility. More precisely, we have the following:

Let (e, z) be fixed but arbitrary, and suppose that $a^\dagger \in A(e, z)$. Then

$$E_{\theta|z,e}\ell(e, z, a^\dagger, \tilde{\theta}) = \min_a E_{\theta|z,e}\ell(e, z, a, \tilde{\theta})$$

if and only if (17.3.4)

$$E_{\theta|z,e}u(e, z, a^\dagger, \tilde{\theta}) = \max_a E_{\theta|z,e}u(e, z, a, \tilde{\theta}).$$

Proof

By definition of minimum,

(1) $$E_{\theta|z,e}\ell(e, z, a^\dagger, \tilde{\theta}) = \min_a E_{\theta|z,e}\ell(e, z, a, \tilde{\theta}),$$

if and only if

(2) $$E_{\theta|z,e}\ell(e, z, a^\dagger, \tilde{\theta}) \leqq E_{\theta|z,e}\ell(e, z, a, \tilde{\theta}),$$

for every $a \in A(e, z)$. From (17.3.3), (2) is true if and only if

$$E_{\theta|z,e}\{u(e_0, z_0, a_{\tilde{\theta}}^0, \tilde{\theta}) - u(e, z, a^\dagger, \tilde{\theta})\} \leqq E_{\theta|z,e}\{u(e_0, z_0, a_{\tilde{\theta}}^0, \tilde{\theta}) - u(e, z, a, \tilde{\theta})\},$$

for every $a \in A(e, z)$, that is, if and only if

(3) $$E_{\theta|z,e}u(e_0, z_0, a_{\tilde{\theta}}^0, \tilde{\theta}) - E_{\theta|z,e}u(e, z, a^\dagger, \tilde{\theta}) \leqq E_{\theta|z,e}u(e_0, z_0, a_{\tilde{\theta}}^0, \tilde{\theta}) - E_{\theta|z,e}u(e, z, a, \tilde{\theta}),$$

for every $a \in A(e, z)$. But, by cancellation of $E_{\theta|z,e}u(e_0, z_0, a_{\tilde{\theta}}^0, \tilde{\theta})$ and rearrangement, (3) holds if and only if

(4) $$E_{\theta|z,e}u(e, z, a^\dagger, \tilde{\theta}) \geqq E_{\theta|z,e}u(e, z, a, \tilde{\theta}),$$

for every $a \in A(e, z)$. But (4) holds if and only if

(5) $$E_{\theta|z,e}u(e, z, a^\dagger, \tilde{\theta}) = \max_a E_{\theta|z,e}u(e, z, a, \tilde{\theta}). \quad \text{QED}$$

Thus $a_{z,e}^0$ can be found by *either* minimizing $E_{\theta|z,e}\ell(e, z, a, \tilde{\theta})$ *or* maximizing $E_{\theta|z,e}u(e, z, a, \tilde{\theta})$. A similar result obtains for determining e^0.

$$E_{z|e^\dagger} \max_a E_{\theta|z,e^\dagger}u(e^\dagger, \tilde{z}, a, \tilde{\theta})$$

$$= \max_e E_{z|e} \max_a E_{\theta|z,e}u(e, \tilde{z}, a, \tilde{\theta})$$

if and only if (17.3.5)

$$E_{z|e^\dagger} \min_a E_{\theta|z,e^\dagger}\ell(e^\dagger, \tilde{z}, a, \tilde{\theta})$$

$$= \min_e E_{z|e} \min_a E_{\theta|z,e}\ell(e, \tilde{z}, a, \tilde{\theta}).$$

Proof

The second equality of (17.3.5) is equivalent to

(1) $E_{z|e^\dagger}E_{\theta|z,e^\dagger}\ell(e^\dagger, \tilde{z}, a_{\tilde{z},e^\dagger}^0, \tilde{\theta}) \leqq E_{z|e}E_{\theta|z,e}\ell(e, \tilde{z}, a_{\tilde{z},e}^0, \tilde{\theta})$, for every $e \in E$, which, by (17.3.3), is true if and only if

(2) $E_{z|e^\dagger}E_{\theta|z,e^\dagger}u(e_0, z_0, a_{\tilde{\theta}}^0, \tilde{\theta}) - E_{z|e^\dagger}E_{\theta|z,e^\dagger}u(e^\dagger, \tilde{z}, a_{\tilde{z}}^0, e^\dagger, \tilde{\theta}) \leqq E_{z|e}E_{\theta|z,e}u(e_0, z_0, a_{\tilde{\theta}}^0, \tilde{\theta}) - E_{z|e}E_{\theta|z,e}u(e, \tilde{z}, a_{\tilde{z},e}^0, \tilde{\theta})$, for every $e \in E$. But for any random variable $\tilde{x}$, it is clear that $E_{z|e}E_{\theta|z,e}\tilde{x} = E_\theta\tilde{x}$ and is independent of e. Hence the first terms on each side of (2) can be canceled, and so (2) is true if and only if

(3) $E_{z|e^\dagger}E_{\theta|z,e^\dagger}u(e^\dagger, \tilde{z}, a_{\tilde{z},e^\dagger}^0, \tilde{\theta}) \geqq E_{z|e}E_{\theta|z,e}u(e, \tilde{z}, a_{\tilde{z},e}^0, \tilde{\theta})$, for every $e \in E$. But (3) is equivalent to the first equality of (17.3.5). QED

Example 17.3.1

The loss function for the decision problem of Example 17.2.1 can be found by first noting that $a_{\theta_1}{}^0 = a_1$ and $a_{\theta_2}{}^0 = a_2$ given (e_0, z_0), and hence that $u(e_0, z_0, a_{\theta_1}{}^0, \theta_1) = 12$ and $u(e_0, z_0, a_{\theta_2}{}^0, \theta_2) = 15$. Hence:

$$\begin{aligned}
\ell(e_1, z_1, a_1, \theta_1) &= 12 - 10 = 2,\\
\ell(e_1, z_1, a_1, \theta_2) &= 15 - 0 = 15,\\
\ell(e_1, z_1, a_2, \theta_1) &= 12 - 0 = 12,\\
\ell(e_1, z_1, a_2, \theta_2) &= 15 - 10 = 5,\\
\ell(e_1, z_1, a_3, \theta_1) &= 12 - 5 = 7,\\
\ell(e_1, z_1, a_3, \theta_2) &= 15 - 5 = 10,\\
\ell(e_1, z_2, a_1, \theta_1) &= 12 - 9 = 3,\\
\ell(e_1, z_2, a_1, \theta_2) &= 15 - 0 = 15,\\
\ell(e_1, z_2, a_2, \theta_1) &= 12 - 0 = 12,\\
\ell(e_1, z_2, a_2, \theta_2) &= 15 - 9 = 6,\\
\ell(e_1, z_2, a_3, \theta_1) &= 12 - 4 = 8,\\
\ell(e_1, z_2, a_3, \theta_2) &= 15 - 4 = 11,\\
\ell(e_0, z_0, a_1, \theta_1) &= 12 - 12 = 0,\\
\ell(e_0, z_0, a_1, \theta_2) &= 15 - 0 = 15,\\
\ell(e_0, z_0, a_2, \theta_1) &= 12 - 0 = 12,\\
\ell(e_0, z_0, a_2, \theta_2) &= 15 - 15 = 0,\\
\ell(e_0, z_0, a_3, \theta_1) &= 12 - 6 = 6,\\
\ell(e_0, z_0, a_3, \theta_2) &= 15 - 6 = 9.
\end{aligned}$$

Example 17.3.2

For the decision problem of Example 17.2.2, it is clear that $a_\theta{}^0 = \theta$, and hence that $u(e_0, z_0, a_\theta{}^0, \theta) = 50$, for every θ. Hence

$$\begin{aligned}
\ell(e_0, z_0, a, \theta) &= 50 - [50 - 20(\theta - a)^2]\\
&= 20(\theta - a)^2, \text{ and}
\end{aligned}$$

$$\begin{aligned}
\ell(e_1, z, a, \theta) &= 50 - [25 - 20(\theta - a)^2 - 5z]\\
&= 20(\theta - a)^2 + 25 + 5z.
\end{aligned}$$

The essence of (17.3.4) and (17.3.5) is that the loss function is equally as useful for solving the general decision problem as the utility function. In fact, the relevant analogue of (17.2.8) is

$$\ell(e^0, *, {}^0, *) = \min_e E_{z|e} \min_a E_{\theta|z,e}\ell(e, \tilde{z}, a, \tilde{\theta}). \tag{17.3.6}$$

The loss function is no *more* useful than the utility function, and hence it is reasonable to ask why it is introduced at all. The basic justification for the loss function is that it is often easier for the decision-maker to determine his loss function rather than his utility function. This is particularly true in the estimation problems discussed in Chapter 18, in which $A \equiv A(e, z)$ is independent of (e, z), $\Theta \subset A \subset R^1$, and the decision-maker suffers a penalty depending upon the magnitude $a - \theta$ of his error.

17.4 THE MONETARY EVALUATION OF INFORMATION

In this section we restrict attention to those decision problems in which the consequences $c(e, z, a, \theta)$ are adequately expressed by a single real number, representing some attribute relevant to the decision-maker. For definiteness, we shall suppose that this attribute is *monetary profit.*

Hence $c(e, z, a, \theta)$ is assumed to be a real number for every path (e, z, a, θ) through the tree. Furthermore, we shall assume that $c(e, z, a, \theta)$ is composed of two parts: a *terminal return* $v_t(a, \theta)$ from choosing act a when θ occurs, and a cost $k(e, z)$ of choosing experiment e and observing outcome z, so that $c(e, z, a, \theta) = v_t(a, \theta) - k(e, z)$, for every (e, z, a, θ). The phrase *terminal return* is intended to connote that $v_t(a, \theta)$ arises from a no-information decision problem which various experiments and their outcomes may precede. Thus *terminal* refers to the decision problem following experimentation. We assume that "no information" costs nothing; that is, $k(e_0, z_0) = 0$.

These assumptions can be combined in the following form:

There exist real-valued functions $v_t(a, \theta)$ and $k(e, z)$ such that

$$c(e, z, a, \theta) = v_t(a, \theta) - k(e, z) \in R^1, \tag{17.4.1}$$

for every (e, z, a, θ), where $k(e_0, z_0) = 0$.

It is clear that the utilities will have to be obtained from the decision-maker's utility function for money. In subsequent developments it will be most convenient for us to use the decision-maker's utility function u^* for *asset positions*, as defined in Chapter 15, Exercise 6. If the decision-maker has initial assets y, chooses e, observes z, chooses a, and θ is true, his final asset position is $y + v_t(a, \theta) - k(e, z)$ in view of (17.4.1). Thus the relevant utilities for the endpoints of the tree are given by

$$u(e, z, a, \theta) = u^*[y + v_t(a, \theta) - k(e, z)], \text{ all } (e, z, a, \theta). \tag{17.4.2}$$

Now, since the decision-maker's utility function $u(x)$ for returns x incremental to his initial asset position y (as discussed in Section 15.4) is assumed to be a strictly increasing and continuous function, and since u^* is related to u by

$$u^*[y + x] = u(x)$$

(see Chapter 15, Exercise 6), it follows that u^* is strictly an increasing and continuous function. This fact will be of importance in what follows.

Finally, we assume

$$A \equiv A(e, z) \text{ is independent of } (e, z). \tag{17.4.3}$$

It is worth noting that (17.4.3) can always be forced to hold, by defining $A \equiv \bigcup\{A(e, z)\text{: all } (e, z)\}$, and by requiring that the consequences of any $a \in A \backslash A(e, z)$ be c_w. Then no $a \in A\backslash A(e, z)$ can be $a^0_{z|e}$ for any (e, z).

In this section we shall consider the following fundamental question:

> What is the maximum amount $I(e; y)$ of money that the decision-maker should be willing to pay in order to be able to choose $a \in A$ after observing the outcome z of e rather than after "observing" the null outcome z_0 of e_0, given initial assets y?

Subsection 17.4.1 considers this question directly and without making further assumptions, while Subsection 17.4.2 specializes to the important class of problems for which utility is linear in money; that is,

$$u^*[y + v_t(a, \theta) - k(e, z)] = y + v_t(a, \theta) - k(e, z). \tag{17.4.2$'$}$$

A much fuller treatment of the subject matter of Subsection 17.4.1 can be found in LaValle [**42**] through [**44**]. The subject matter of Subsection 17.4.2 is considered more completely in Raiffa and Schlaifer [**66**] and Pratt, Raiffa, and Schlaifer [**64**].

17.4.1 The General Case

The fundamental question of this subsection is equivalent to asking for the maximum *nonrandom* cost $k(e)$ of experiment e such that e is at least as desirable as e_0.

Suppose the decision-maker commits himself to obtain experiment e at cost $k(e)$, but has not yet observed the outcome z. If his initial assets are y, then his final assets will be

$$y + v_t(a, \theta) - k(e),$$

given θ and his choice of act a posterior to observing z. Thus his ultimate utility will be

$$u^*[y + v_t(a, \theta) - k(e)],$$

and he will choose $a = a^0_{z,e}$ such that

$$\begin{aligned} E_{\theta|z,e}u^*[y + v_t(a^0_{z,e}, \tilde{\theta}) - k(e)] \\ &= \max_a E_{\theta|z,e}u^*[y + v_t(a, \tilde{\theta}) - k(e)] \\ &\equiv U(e, y - k(e) \mid z), \end{aligned} \tag{17.4.4}$$

the second, definitional equality introducing an obviously useful shorthand.

Before observing z, the decison-maker's utility is the expectation of (17.4.4) by virtue of (17.2.8); and we define

$$U(e, y - k(e)) \equiv E_{t|e}U(e, y - k(e) \mid \tilde{z}). \tag{17.4.5}$$

Now, $U(e, y - k(e))$ is the decision-maker's *prior* utility of: (1) committing himself to pay $k(e)$ for e-information (2) observing the outcome z of e, (3) choosing $a^0_{z,e}$, and (4) receiving final assets $y + v_t(a^0_{z,e}, \theta) - k(e)$. As such, it fully expresses the relative desirability of paying $k(e)$ for e vis-à-vis other opportunities, to obtain e' at cost $k(e')$.

Suppose that the decision-maker decides not to obtain e at cost $k(e)$, but rather stay with e_0 at no cost and choose an act without further information. His prior utility of so doing is, from (17.2.8),

$$U(e_0, y) = \max_a E_\theta u^*[y + v_t(a, \tilde{\theta})]. \tag{17.4.6}$$

Should the decision-maker be willing to pay $k(e)$ for e? The answer is yes, if and only if

$$U(e, y - k(e)) \geqq U(e_0, y). \tag{17.4.7}$$

Now, it can be shown that under certain not too restrictive conditions (for example, $u^*\colon R^1 \to R^1$ is bounded or $v\colon A \times \Theta \to R^1$ is bounded),

for every $e \in E$, the function $U(e, \cdot)\colon R^1 \to R^1$ is continuous and strictly increasing, with range coinciding with the range of u^*. (17.4.8)

Hence there exists a unique value $I(e, y)$ of $k(e)$ for which equality holds in (17.4.7). We call $I(e, y)$ the *prior value* of experiment e. Formally, the *prior value* $I(e, y)$ *of experiment* e is defined implicitly by

$$U(e, y - I(e, y)) = U(e_0, y). \tag{17.4.9}$$

From (17.4.7) through (17.4.9) it is clear that $I(e, y)$ is the maximum nonrandom cost of experiment e for which e is at least as desirable as e_0 when initial assets are y. In general, $I(e, y)$ does depend upon the initial assets y.

We shall now show that $I(e, y) \geqq 0$, for every $e \in E$ and every $y \in R^1$. For this it suffices that

$$U(e, y) \geqq U(e_0, y), \text{ every } y \in R^1, e \in E,$$

since $U(e, y - x)$ is strictly decreasing in x. By definition of $U(e, y)$ and $a^0_{z,e}$, we have

$$U(e, y) = E_{z|e}E_{\theta|z,e}u^*[y + v_t(a^0_{\tilde{z},e}, \tilde{\theta})],$$

but by definition of $a^0_{z,e}$ it follows that

$$E_{\theta|z,e}u^*[y + v_t(a^0_{\tilde{z},e}, \tilde{\theta})] \geqq E_{\theta|z,e}u^*[y + v_t(a^0_{z_0,e_0}, \tilde{\theta})],$$

for every $z \in Z(e)$, and hence

$$\begin{aligned} U(e, y) &\geqq E_{z|e}E_{\theta|z,e}u^*[y + v_t(a^0_{z_0,e_0}, \tilde{\theta})] \\ &= E_\theta u^*[y + v_t(a^0_{z_0,e_0}, \tilde{\theta})] \\ &= U(e_0, y). \end{aligned}$$

Hence we conclude that

$$I(e, y) \geqq 0,\ e \in E,\ y \in R^1. \tag{17.4.10}$$

At the other end of the spectrum, consider a (real or hypothetical) experiment e^* which yields *perfect information*, in the sense that for every $z \in Z(e^*)$ there is a unique $\theta_z \in \Theta$ such that $P(\tilde{\theta} = \theta_z \mid z, e) = 1$. In other words, the outcomes of e^* completely eliminate the decision-maker's uncertainty about $\tilde{\theta}$.

If we define $a_\theta{}^0$ as in Section 17.3, it follows that, for every $e \in E$,

$$\begin{aligned} U(e^*, y) &= E_\theta u^*[y + v_t(a^0_{\tilde{\theta}}, \tilde{\theta})] \\ &= E_{z|e}E_{\theta|z,e}u^*[y + v_t(a^0_{\tilde{\theta}}, \tilde{\theta})] \\ &\geqq E_{z|e}E_{\theta|z,e}u^*[y + v_t(a^0_{\tilde{z},e}, \tilde{\theta})] \\ &= U(e, y). \end{aligned} \tag{17.4.11}$$

Hence

$$U(e^*, y) \geqq U(e, y),\ y \in R^1,\ e \in E. \tag{17.4.12}$$

From (17.4.12) and (17.4.9) we obtain

$$\begin{aligned} U(e^*, y - I(e, y)) &\geqq U(e, y - I(e, y)) \\ &= U(e_0, y) \\ &= U(e^*, y - I(e^*, y)), \end{aligned} \tag{17.4.13}$$

implying that

$$U(e^*, y - I(e, y)) \geqq U(e^*, y - I(e^*, y)). \tag{17.4.14}$$

Applying (17.4.8) to (17.4.14) thus implies the desired result:

$$I(e, y) \leqq I(e^*, y), \text{ for every } e \in E \text{ and every } y \in R^1. \tag{17.4.15}$$

Now, $I(e^*, y)$ is called the *prior value of perfect information.* It establishes an upper bound for the prior value of any experiment e, and this upper bound is often useful in practice. If $I(e^*, y)$ were known to equal ten and if the decision-maker was contemplating paying fifteen for experiment e, it is clear that the purchase of e can be ruled out without further ado, since $I(e, y) \leqq 10 < 15 = k(e)$.

The preceding results can be collected in the following summary:

For every $y \in R^1$ and every $e \in E$, we have:

$$0 \leqq I(e,y) \leqq I(e^*,y);\ \text{and} \tag{17.4.16}$$

e at nonrandom cost $k(e)$ is at least as desirable as (respectively: less desirable than) e_0 if and only if

$$U(e, y - k(e)) \geqq \text{(respectively: } <)U(e_0, y), \tag{17.4.17}$$

which obtains if and only if

$$k(e) \leqq \text{(respectively: } >) \ I(e, y).$$

The preceding results can be extended to the case of *random* costs $k(e, \tilde{z})$ fairly easily. Define the *certainty equivalent* $\#(k(e, \tilde{z}) \mid e, y)$ of $k(e, \tilde{z})$ by

$$U(e, y - k(e, \tilde{z})) = U(e, y - \#(k(e, \tilde{z}) \mid e, y)), \tag{17.4.18}$$

for every e, y, and random cost $k(e, \tilde{z})$ of e, where the left-hand side of (17.4.18) is the natural extension of (17.4.4) and (17.4.5):

$$U(e, y - k(e, \tilde{z})) \equiv E_{z|e} \max_a E_{\theta|z,e} u^*[y + v_t(a, \tilde{\theta}) - k(e, \tilde{z})]. \tag{17.4.19}$$

Existence and uniqueness of $\#(k(e, \tilde{z}) \mid e, y)$ follow from (17.4.8); there is one and only one $\# \in R^1$ such that

$$U(e, y - \#) = U(e, y - k(e, \tilde{z})).$$

The reader may prove as Exercise 4 the following:

Experiment e at cost $k(e, \tilde{z})$ is at least as desirable as (respectively: less desirable than) e_0 if and only if

$$\#(k(e, \tilde{z}) \mid e, y) \leqq \text{(respectively: } >) \ I(e, y). \tag{17.4.20}$$

A comparison of (17.4.17) and (17.4.20) shows that the nonrandom $\#(k(e, \tilde{z}) \mid e, y)$ in (17.4.20) plays the role of $k(e)$ in (17.4.17). The reader may show, as Exercise 5, that when $k(e, \tilde{z})$ is nonrandom [that is, $k(e, z) = k(e) =$ const. for every z] then $k(e) = \#(k(e) \mid e, y)$. Hence (17.4.20) generalizes (17.4.17).

The reader is cautioned against supposing that

$$\#(k(e, \tilde{z}) \mid e, y) = E_{z|e} k(e, \tilde{z}). \tag{17.4.21}$$

In fact, (17.4.12) is true in general only when u^* is linear in money, which is the assumption of the next subsection. However, $\#(\cdot \mid e, y)$ does possess many of the properties of expectation. See LaValle, [**42**] and [**44**].

The theory developed so far enables the decision-maker to compare experiments with cost—$(e, k(e, \tilde{z}))$ pairs—individually with $(e_0, 0)$. It does *not* enable the decision-maker to choose the *most* desirable e unless $(e_0, 0)$ happens to be at least as desirable as every available $(e, k(e, \tilde{z}))$. Reference [**44**] discusses several possible definitions of net *gain* functions $G(e, k(e, \tilde{z}), y)$, each of which has the property that e^0 satisfies

$$G(e^0, k(e^0, \tilde{z}), y) = \max_e G(e, k(e, \tilde{z}), y).$$

We shall omit this discussion here because the definitions of net gain are not mutually equivalent unless u^* has one of the forms cited in Chapter 15, Exercise 7 (e). In general, however, (17.2.8) implies that

$$U(e^0, y - k(e^0, \tilde{z})) \geqq U(e, y - k(e, \tilde{z})), \text{ for every } e \in E. \quad (17.4.22)$$

We remark that (17.4.22) rarely implies that perfect information is optimal; that is, that $e^* = e^0$, since in many applied problems the e's are random samples of various sizes n from $f(x \mid \theta)$, perfect information corresponds to an infinitely large sample, and there is a fixed positive cost of each observation. These assumptions imply that $k(e^*, z) = \infty$ for each z. Chapter 18 considers several related classes of such problems in detail.

Example 17.4.1

Suppose $v_t(a, \theta)$ is as given by the following table, which also tacitly specifies A and Θ.

	$v_t(a, \theta)$	
	θ_1	θ_2
a_1	10	0
a_2	0	10

Suppose that $P(\theta_1) = P(\theta_2) = \frac{1}{2}$; and, for $y < 100$, let

$$u^*(y) = 10^4 - (100 - y)^2.$$

(The reader may verify that u^* is continuous and strictly increasing for $-\infty < y < 100$.) Hence the table of utilities for this decision is as follows:

	$u^*[v_t(a, \theta) + y - x]$	
	θ_1	θ_2
a_1	$10^4 - (90 + x - y)^2$	$10^4 - (100 + x - y)^2$
a_2	$10^4 - (100 + x - y)^2$	$10^4 - (90 + x - y)^2$

where we have included a subtrahend x for subsequent convenience. We shall henceforth suppose that $y < 90$. It is easy to verify that $a^0_{z_0, e_0} = a_1$ or a_2, and that

(1) $$U(e_0, y) = \{10^4 - (100 - y)^2\}(\tfrac{1}{2}) + \{10^4 - (90 - y)^2\}(\tfrac{1}{2}) = 10^4 - [25 + (95 - y)^2] \text{ (by a little algebra).}$$

Now consider perfect information e^*. It is clear that $a_{\theta_i}{}^0 = a_i$, for $i = 1, 2$, and that

(2) $$U(e^*, y - I(e^*, y)) = 2\{10^4 - (90 + I(e^*, y) - y)^2\} = 10^4 - (90 + I(e^*, y) - y)^2.$$

We may now apply (17.4.9) and solve for $I(e^*, y)$; by algebra we obtain

(3) $$I(e^*, y) = [25 + (95 - y)^2]^{1/2} - (90 - y),\ y < 90;$$

for example,

$$I(e^*, 83) = 6.$$

Finally, suppose that $Z(e^*) = \{\theta_1, \theta_2\}$, that $k(e^*, \theta_i) \equiv k_i \geqq 0$, for $i = 1, 2$, and that $\tilde{k} \equiv k(e^*, \tilde{\theta})$. We shall now derive

$$\# \equiv \#(k(e^*, \tilde{\theta}) \mid e^*, y).$$

The reader may verify that $a_{\theta_i}{}^0 = a_i$, as before, but that

(4) $$\begin{aligned} &U(e^*, y - \tilde{k}) \\ &\quad = \{10^4 - (90 + k_1 - y)^2\}(\tfrac{1}{2}) + \{10^4 - (90 + k_2 - y)^2\}(\tfrac{1}{2}) \\ &\quad = 10^4 - \{[(90 - y) + E(\tilde{k})]^2 + [(k_1 - k_2)/2]^2\}, \end{aligned}$$

where, naturally, $E(\tilde{k}) = k_1(\tfrac{1}{2}) + k_2(\tfrac{1}{2}) = (k_1 + k_2)/2$. Equating $U(e^*, y - \tilde{k})$ to $U(e^*, y - \#)$ as given by the obvious analogue of (2) yields, after algebra,

(5) $\#(\tilde{k} \mid e^*, y) = \{[(90 - y) + E(\tilde{k})]^2 + [(k_1 - k_2)/2]^2\}^{1/2} - (90 - y)$, which exceeds $E(\tilde{k})$, whenever $k_1 \neq k_2$. It is worth noting parenthetically that $[(k_1 - k_2)/2]^2 = V(\tilde{k})$.

17.4.2 Utility Linear in Money

When utility is linear in money, we may write

$$u^*[y + v_t(a, \theta) - k(e, z)] = y + v_t(a, \theta) - k(e, z).$$

Hence

$$\begin{aligned} &\max_a E_{\theta|z,e} u^*[y + v_t(a, \tilde{\theta}) - k(e, z)] \\ &\qquad = \max_a \{y + E_{\theta|z,e} v_t(a, \tilde{\theta}) - k(e, z)\} \\ &\qquad = y + \max_a E_{\theta|z,e} v_t(a, \tilde{\theta}) - k(e, z), \end{aligned}$$

the $E_{z|e}$-expectation of which is $U(e, y - k(e, \tilde{z}))$; that is,

$$\begin{aligned} &U(e, y - k(e, \tilde{z})) \\ &\qquad = y + E_{z|e} \max_a E_{\theta|z,e} v_t(a, \tilde{\theta}) - E_{z|e} k(e, \tilde{z}). \end{aligned} \quad (17.4.23)$$

Now, (17.4.23) readily yields the prior value $I(e)$ of experiment e, which is independent of y:

$$I(e) = E_{z|e} \max_a E_{\theta|z,e} v_t(a, \tilde{\theta}) - \max_a E_\theta v_t(a, \tilde{\theta}). \quad (17.4.24)$$

Proof

$$\begin{aligned} y + \max_a E_\theta v_t(a, \tilde{\theta}) & \\ &= U(e_0, y) \qquad [\text{by (17.4.23) and } k(e_0, z_0) = 0] \\ &= U(e, y - I(e)) \qquad [\text{by (17.4.9)}] \\ &= y - I(e) + E_{z|e} \max_a E_{\theta|z,e} v_t(a, \tilde{\theta}) \qquad [\text{by (17.4.23)}]. \end{aligned}$$

Equating the first and last terms of this equality chain, canceling y, and rearranging results in (17.4.24). QED

For the case of perfect information e^*, we have

$$I(e^*) = E_\theta \max_a v_t(a, \tilde{\theta}) - \max_a E_\theta v_t(a, \tilde{\theta}). \tag{17.4.25}$$

Proof

For $e = e^*$, knowledge of z removes all uncertainty about $\tilde{\theta}$, and hence

$$E_{\theta|z,e^*} v_t(a, \tilde{\theta}) = v_t(a, \tilde{\theta}).$$

Hence

$$E_{z|e^*} \max_a E_{\theta|z,e^*} v_t(a, \tilde{\theta}) = E_{z|e^*} \max_a v_t(a, \tilde{\theta}).$$

But by (10.2.5) it follows that

$$\begin{aligned} E_{z|e^*} \max_a v_t(a, \tilde{\theta}) &= E_\theta \{E_{z|\theta,e^*} \max_a v_t(a, \tilde{\theta})\}, \\ &= E_\theta \max_a v_t(a, \tilde{\theta}), \end{aligned}$$

the second equality obtaining because $\max_a v_t(a, \tilde{\theta})$ is constant with respect to $\tilde{z}$, implying that

$$E_{z|\theta,e^*} \max_a v_t(a, \tilde{\theta}) = \max_a v_t(a, \tilde{\theta}).$$

Thus

$$E_{z|e^*} \max_a E_{\theta|z,e^*} v_t(a, \tilde{\theta}) = E_\theta \max_a v_t(a, \tilde{\theta}),$$

from which (17.4.25) follows via (17.4.24). QED

Next, we show in the following that (17.4.21) obtains for linear utility.

If u^* is linear in money, then $\#(k(e, \tilde{z}) \mid e)$ is independent of y and (17.4.26)

$$\#(k(e, \tilde{z}) \mid e) = E_{z|e} k(e, \tilde{z}). \tag{17.4.21}$$

Proof

Equation (17.4.23) implies that $\#(k(e, \tilde{z}) \mid e)$ is independent of y, and hence it suffices to prove (17.4.21).

$$
\begin{aligned}
y + E_{z|e} \max_a E_{\theta|z,e} v_t(a, \tilde{\theta}) - E_{z|e} k(e, \tilde{z}) & \\
= U(e, y - k(e, \tilde{z})) \qquad & [\text{by } (17.4.23)] \\
= U(e, y - \#(k(e, \tilde{z}) \mid e)) \qquad & [\text{by } (17.4.18)] \\
= y + E_{z|e} \max_a E_{\theta|z,e} v_t(a, \tilde{\theta}) - \#(k(e, \tilde{z}) \mid e) \qquad & [\text{by } (17.4.23)].
\end{aligned}
$$

Equating the first and last terms of this equality chain and canceling the common summands yields (17.4.21). QED

From (17.4.24) and (17.4.26), we obtain the following important corollary of (17.4.20):

If u^* is linear in money, then e at cost $k(e, \tilde{z})$ is at least as desirable as (respectively: less desirable than) e_0 if and only if (17.4.27)

$$E_{z|e} k(e, \tilde{z}) \leqq (\text{respectively: } >) \quad E_{z|e} \max_a E_{\theta|z,e} v_t(a, \tilde{\theta}) - \max_a E_\theta v_t(a, \tilde{\theta}).$$

When utility is linear in money, however, we may define the *prior net gain* $G(e, k(e, \tilde{z}))$ accruing to the decision-maker from purchasing experiment e at cost $k(e, \tilde{z})$ by

$$G(e, k(e, \tilde{z})) = I(e) - E_{z|e} k(e, \tilde{z}). \tag{17.4.28}$$

The reader may verify as Exercise 7 that the inequalities in (17.4.27) are equivalent to

$$G(e, k(e, \tilde{z})) \geqq (\text{respectively: } <) 0. \tag{17.4.29}$$

Hence paying $k(e, \tilde{z})$ for e is rationally justifiable only if the prior net gain of $(e, k(e, \tilde{z}))$ is nonnegative.

We may go further, however, by showing that e^0 maximizes G. To show this, it clearly suffices to show

$$U(e_1, y - k_1(e_1, \tilde{z})) \geqq U(e_2, y - k_2(e_2, \tilde{z})) \tag{17.4.30}$$

if and only if

$$G(e_1, k_1(e_1, \tilde{z}_1)) \geqq G(e_2, k_2(e_2, \tilde{z})),$$

where e_1 and e_2 are any two experiments in E and $k_1(e_1, \tilde{z})$ and $k_2(e_2, \tilde{z})$ are their respective associated costs.

Proof

$$U(e_1, y - k_1(e_1, \tilde{z}_1)) \geqq U(e_2, y - k_2(e_2, \tilde{z}_2))$$

if and only if

$$y + E_{z|e_1} \max_a E_{\theta|z,e_1} v_t(a, \tilde{\theta}) - E_{z|e_1} k_1(e_1, \tilde{z})$$
$$\geqq y + E_{z|e_2} \max_a E_{\theta|z,e_2} v_t(a, \tilde{\theta}) - E_{z|e_2} k_2(e_2, \tilde{z}) \qquad \text{[by (17.4.23)]}$$

if and only if

$$\{E_{z|e_1} \max_a E_{\theta|z,e_1} v_t(a, \tilde{\theta}) - \max_a E_\theta v_t(a, \tilde{\theta})\} - E_{z|e_1} k_1(e_1, \tilde{z})$$
$$\geqq \{E_{z|e_2} \max_a E_{\theta|z,e_2} v_t(a, \tilde{\theta}) - \max_a E_\theta v_t(a, \tilde{\theta})\} - E_{z|e_2} k_2(e_2, \tilde{z})$$

[by canceling y's and subtracting $\max_a E_\theta v_t(a, \tilde{\theta})$ from both sides]

if and only if

$$I(e_1) - E_{z|e_1} k_1(e_1, \tilde{z}) \geqq I(e_2) - E_{z|e_2} k_2(e_2, \tilde{z}) \qquad \text{[by (17.4.24)]}$$

if and only if

$$G(e_1, k_1(e_1, \tilde{z})) \geqq G(e_2, k_2(e_2, \tilde{z})). \quad \text{QED}$$

Thus the decision-maker may select e^0 by maximizing his prior net gain function; that is, (17.4.30) implies

$$G(e^0, k(e^0, \tilde{z})) = \max_e G(e, k(e, \tilde{z})). \qquad (17.4.31)$$

We shall use (17.4.31) extensively in the next chapter.

When u^* is linear in money, there is a useful alternative development of information evaluation in terms of opportunity loss. However, the pertinent loss function differs from that defined by (17.3.3).

We define the *terminal* (opportunity) *loss* function $\ell_t(a, \theta)$ by

$$\ell_t(a, \theta) \equiv \max_a v_t(a, \theta) - v_t(a, \theta), \qquad (17.4.32)$$

for every $a \in A, \theta \in \Theta$. Thus $\ell_t(a, \theta)$ expresses the decision-maker's foregone profit from choosing a rather than a_θ^0 when θ is true.

Terminal opportunity loss is particularly useful in the estimation problems alluded to in Section 17.3. In fact, almost the entire analysis of Chapter 18 uses terminal opportunity loss rather than utility (or profit).

All of the results in this subsection are expressible in terms of ℓ_t rather than v. The relevant facts are

$$I(e) = \min_a E_\theta \ell_t(a, \tilde{\theta}) - E_{z|e} \min_a E_{\theta|z,e} \ell_t(a, \tilde{\theta}) \qquad (17.4.33)$$

and

$$I(e^*) = \min_a E_\theta \ell_t(a, \tilde{\theta}). \qquad (17.4.34)$$

Proof

Equations (17.4.33) and (17.4.34) are immediate from (17.4.24) and (17.4.25) respectively, provided that we prove

(1) $$\min_a E_\theta \ell_t(a, \tilde{\theta}) = E_\theta \max_a v_t(a, \tilde{\theta}) - \max_a E_\theta v_t(a, \tilde{\theta}),$$

and

(2) $$E_{z|e} \min_a E_{\theta|z,e} \ell_t(a, \tilde{\theta})$$

$$= E_\theta \max_a v_t(a, \tilde{\theta}) - E_{z|e} \max_a E_{\theta|z,e} v_t(a, \tilde{\theta}).$$

(See Exercise 8.) For (1), we have

$$\begin{aligned}
\min_a E_\theta \ell_t(a, \tilde{\theta}) &= \min_a E_\theta \{\max_a v_t(a, \tilde{\theta}) - v_t(a, \tilde{\theta})\} \\
&= \min_a \{E_\theta \max_a v_t(a, \tilde{\theta}) - E_\theta v_t(\text{a}, \tilde{\theta})\} \\
&= E_\theta \max_a v_t(a, \tilde{\theta}) + \min_a \{-E_\theta v_t(a, \tilde{\theta})\} \\
&= E_\theta \max_a v_t(a, \tilde{\theta}) - \max_a E_\theta v_t(a, \tilde{\theta}),
\end{aligned}$$

while (2) follows similarly from

$$\begin{aligned}
E_{z|e} \min_a E_{\theta|z,e} \ell_t(a, \tilde{\theta}) &= E_{z|e} \min_a E_{\theta|z,e} \{\max_a v_t(a, \tilde{\theta}) - v_t(a, \tilde{\theta})\} \\
&= E_{z|e} \min_a \{E_{\theta|z,e} \max_a v_t(a, \tilde{\theta}) - E_{\theta|z,e} v_t(a, \tilde{\theta})\} \\
&= E_{z|e} \{E_{\theta|z,e} \max_a v_t(a, \tilde{\theta}) - \max_a E_{\theta|z,e} v_t(\text{a}, \tilde{\theta})\} \\
&= E_{z|e} E_{\theta|z,e} \max_a v_t(a, \tilde{\theta}) - E_{z|e} \operatorname{maz}_a E_{\theta|z,e} v_t(a, \tilde{\theta}) \\
&= E_\theta \max_a v_t(a, \tilde{\theta}) - E_{z|e} \max_a E_{\theta|z,e} v_t(a, \tilde{\theta}). \quad \text{QED}
\end{aligned}$$

Moreover, the reader may prove as Exercise 9 that $a^0_{z,e}$ can be found by minimizing $E_{\theta|z,e}\ell_t(a, \tilde{\theta})$ instead of maximizing $E_{\theta|z,e}v_t(a, \tilde{\theta})$:

$$E_{\theta|z,e}v_t(a^\dagger, \tilde{\theta}) = \max_a E_{\theta|z,e}v_t(a, \tilde{\theta}) \tag{17.4.35}$$

if and only if

$$E_{\theta|z,e}\ell_t(a^\dagger, \tilde{\theta}) = \min_a E_{\theta|z,e}\ell_t(a, \tilde{\theta}).$$

Prior net gain continues to be defined by (17.4.28), where now $I(e)$ is given by (17.4.33).

Example 17.4.2

Let $v_t(a, \theta)$ be as given in Example 17.4.1, but now assume that u^* is linear in money. From the $v_t(a, \theta)$-table we readily determine that $a_{\theta_i}{}^0 = a_i$, as before, and hence $\max_a v_t(a, \theta_i) = 10$, for $i = 1, 2$. Therefore (17.4.32) implies the following loss table:

	$\ell_t(a, \theta)$	
	θ_1	θ_2
a_1	0	10
a_2	10	0

We find $E_\theta\ell_t(a_i, \tilde{\theta}) = (10)(\frac{1}{2}) + (0)(\frac{1}{2}) = 5$ for $i = 1, 2$, and hence $a^0_{z_0,e_0} = a_1$ or a_2, and

$$\min_a E_\theta\ell_t(a, \tilde{\theta}) = 5 = I(e^*).$$

Note that $I(e^*) \neq I(e^*, 83) = 6$ in Example 17.4.1. In the linear utility case, we also have $\#(\tilde{k} \mid e, y) = E(\tilde{k})$, independent of y and $V(\tilde{k})$.

Example 17.4.3

Let $v_t(a, \theta)$ be given by the following table, in which $a_{\theta_1}{}^0 = a_2$, $a_{\theta_2}{}^0 = a_1$, and $a_{\theta_3}{}^0 = a_4$.

		$V_t(a, \theta)$	
	θ_1	θ_2	θ_3
a_1	6	10	0
a_2	15	0	4
a_3	0	8	9
a_4	0	0	30
$v_t(a^0,\theta) = \equiv \max_a V_t(a, \theta)$	15	10	30

From the preceding table we apply (17.4.32) to obtain the following corresponding table of $\ell_t(a, \theta)$'s:

	$\ell_t(a, \theta)$		
	θ_1	θ_2	θ_3
a_1	9	0	30
a_2	0	10	26
a_3	15	2	21
a_4	15	10	0

Suppose now that $P(\theta_1) = .5$, $P(\theta_2) = .2$, and $P(\theta_3) = .3$. Then

$$E_\theta \max_a v_t(a, \tilde{\theta}) = (15)(.5) + 10(.2) + 30(.3) = 18.50,$$

while

$$\begin{aligned} E_\theta v_t(a_1, \tilde{\theta}) &= (\ 6)(.5) + (10)(.2) + 0(.3) = 5.00, \\ E_\theta v_t(a_2, \tilde{\theta}) &= (15)(.5) + (\ 0)(.2) + 4(.3) = 8.70, \\ E_\theta v_t(a_3, \tilde{\theta}) &= (\ 0)(.5) + (\ 8)(.2) + 9(.3) = 4.30, \end{aligned}$$

and

$$E_\theta v_t(a_4, \tilde{\theta}) = (0)(.5) + (0)(.2) + 30(.3) = 9.00.$$

Hence $\max_a E_\theta v_t(a, \tilde{\theta}) = 9.0$ and $a^0_{z_0, e_0} = a_4$. By (17.4.25), $I(e^*) = 18.5 - 9.0 = 9.5$. If perfect information is available at (say, nonrandom) cost 8.0, the decision-maker's prior net gain $G(e^*, 8.0)$ is $9.5 - 8.0 = 1.5$. Using the loss function table and (17.4.34), we compute

$$\begin{aligned} E_\theta \ell_t(a_1, \tilde{\theta}) &= (\ 9)(.5) + (\ 0)(.2) + (30)(.3) = 13.5, \\ E_\theta \ell_t(a_2, \tilde{\theta}) &= (\ 0)(.5) + (10)(.2) + (26)(.3) = \ \ 9.8, \\ E_\theta \ell_t(a_3, \tilde{\theta}) &= (15)(.5) + (\ 2)(.2) + (21)(.3) = 14.2, \end{aligned}$$

and

$$E_\theta \ell_t(a_4, \tilde{\theta}) = (15)(.5) + (10)(.2) + (\ 0)(.3) = 9.5,$$

and hence we corroborate (17.4.35) by finding that $a^0_{z_0, e_0} = a_4$. Clearly, $\min_a E_\theta \ell_t(a, \tilde{\theta}) = 9.5$, which equals $I(e^*)$ as derived in the preceding and also as given by (17.4.34).

Example 17.4.4

Same problem as in Example 17.4.3, except that now we consider an experiment e with $Z(e) = \{z_1, z_2\}$ and the following table of conditional probabilities $P(z \mid \theta, e)$:

	$P(z \mid \theta, e)$	
	z_1	z_2
θ_1	0.9	0.1
θ_2	0.5	0.5
θ_3	0.2	0.8

As an easy exercise on elementary probability and Bayes' theorem, the reader may verify the following table of posterior probabilities:

		$P(\theta \mid z,e)$	
	θ_1	θ_2	θ_3
z_1	$\frac{45}{61}$	$\frac{10}{61}$	$\frac{6}{61}$
z_2	$\frac{5}{39}$	$\frac{10}{39}$	$\frac{24}{39}$

and also that $P(z_1 \mid e) = .61$, $P(z_2 \mid e) = .39$. We now derive $E_{\theta|z,e}\ell_t(a, \tilde{\theta})$ for each a and z and hence determine $a^0_{z,e}$ and $\min_a E_{\theta|z,e}\ell_t(a, \tilde{\theta})$. For z_1, we have:

$$E_{\theta|z_1,e}\ell_t(a_1, \tilde{\theta}) = (\ 9)\left(\frac{45}{61}\right) + (\ 0)\left(\frac{10}{61}\right) + (30)\left(\frac{6}{61}\right) = \frac{585}{61},$$

$$E_{\theta|z_1,e}\ell_t(a_2, \tilde{\theta}) = (\ 0)\left(\frac{45}{61}\right) + (10)\left(\frac{10}{61}\right) + (26)\left(\frac{6}{61}\right) = \frac{256}{61},$$

$$E_{\theta|z_1,e}\ell_t(a_3, \tilde{\theta}) = (15)\left(\frac{45}{61}\right) + (\ 2)\left(\frac{10}{61}\right) + (21)\left(\frac{6}{61}\right) = \frac{821}{61},$$

and

$$E_{\theta|z_1,e}\ell_t(a_4, \tilde{\theta}) = (15)\left(\frac{45}{61}\right) + (10)\left(\frac{10}{61}\right) + (\ 0)\left(\frac{6}{61}\right) = \frac{775}{61}.$$

Hence $a_{z_1,e} = a_2$ and $\min_a E_{\theta|z_1,e}\ell_t(a, \tilde{\theta}) = \frac{256}{61}$. For z_2, we have

$$E_{\theta|z_2,e}\ell_t(a_1, \tilde{\theta}) = (\ 9)\left(\frac{5}{39}\right) + (\ 0)\left(\frac{10}{39}\right) + (30)\left(\frac{24}{39}\right) = \frac{765}{39},$$

$$E_{\theta|z_2,e}\ell_t(a_2, \tilde{\theta}) = (\ 0)\left(\frac{5}{39}\right) + (10)\left(\frac{10}{39}\right) + (26)\left(\frac{24}{39}\right) = \frac{724}{39},$$

$$E_{\theta|z_2,e}\ell_t(a_3, \tilde{\theta}) = (15)\left(\frac{5}{39}\right) + (\ 2)\left(\frac{10}{39}\right) + (21)\left(\frac{24}{39}\right) = \frac{599}{39},$$

and

$$E_{\theta|z_2,e}\ell_t(a_4, \tilde{\theta}) = (15)\left(\frac{5}{39}\right) + (10)\left(\frac{10}{39}\right) + (\ 0)\left(\frac{24}{39}\right) = \frac{175}{39}.$$

Hence

$$a^0_{z_2,e} = a_4 \text{ and } \min_a E_{\theta|z_2,e}\ell_t(a, \tilde{\theta}) = \frac{175}{39}.$$

Therefore,

$$E_{z|e} \min_a E_{\theta|z,e}\ell_t(a, \tilde{\theta}) = \left(\frac{256}{61}\right)\left(\frac{61}{100}\right) + \left(\frac{175}{39}\right)\left(\frac{39}{100}\right)$$
$$= \frac{431}{100} = 4.31.$$

Hence by (17.4.33),

$$I(e) = 9.5 - 4.31 = 5.19.$$

Since $G(e, k(e, \tilde{z})) = 5.19 - E_{z|e}k(e, \tilde{z})$, it follows that $e^0 = e^*$ if and only if

$$G(e^*, 8.0) \geqq G(e, k(e, \tilde{z}));$$

that is, if and only if

$$E_{z|e}k(e, \tilde{z}) \geqq 5.19 - 1.50 = 3.69.$$

Thus if $k(e, z_1) = 0$ and $k(e, z_2) = 5$, then

$$E_{z|e}k(e, \tilde{z}) = (0)(.61) + (5)(.39) = 1.95 < 3.69$$

and $e^0 = e$.

17.5 NORMAL-FORM ANALYSIS OF DECISIONS

In Sections 17.2, 17.3, and 17.4, our basic *modus operandi* was to proceed from (b*) to (a*), as defined in 17.2. That is, we first determined an optimal act $a^0_{z|e}$ for each $z \in Z(e)$, and then an optimal experiment.

If we examine (b*), it is clear that it requires us to determine a^0 *as a function of* z. This observation suggests that we try solving the general decision problem by selecting, all at once, an optimal function of z, rather than by defining an optimal function at each point z of its domain $Z(e)$.

In order to implement this idea, it is necessary first to define the permissible functions in a precise manner.

A function d: $Z(e) \rightarrow \bigcup \{A(e, z): z \in Z(e)\}$ is called a *decision function* (for e) if and only if $d(z) \in A(e, z)$, for every $z \in Z(e)$. (17.5.1)

Thus a decision function is a prescription of an *available* act $d(z) \in A(e, z)$, for every $z \in Z(e)$. Let $D(e)$ denote the set of all decision functions for experiment $e \in E$, and let H denote the set consisting of all (e, d) pairs in which $d \in D(e)$. That is,

$$D(e) \equiv \{d\colon d \text{ is a decision function for } e\}, \tag{17.5.2}$$

and

$$H \equiv \{(e, d)\colon d \in D(e), e \in E\}. \tag{17.5.3}$$

Suppose now that e is fixed and the decision-maker wishes to determine an optimal $d \in D(e)$. Now, his ultimate utility is $u(e, z, d(z), \theta)$. Therefore his utility of choosing $d \in D(e)$ is

$$E_{\theta|z,e}u(e, z, d(z), \tilde{\theta}) \tag{17.5.4}$$

after he has observed z, and

$$E_{z|e}E_{\theta|z,e}u(e, \tilde{z}, d(\tilde{z}), \tilde{\theta}) \tag{17.5.5}$$

$$= E_{\theta,z|e}u(e, \tilde{z}, d(\tilde{z}), \tilde{\theta}) \tag{17.5.6}$$

$$= E_{\theta}E_{z|\theta,e}u(e, \tilde{z}, d(\tilde{z}), \tilde{\theta}) \tag{17.5.7}$$

before he observes z. [Equations (17.5.6) and (17.5.7) are immediate consequences of (10.2.5).]

From (17.5.4) through (17.5.6) it is clear that if we construct a decision function d^0 by setting $d^0(z) = a^0_{z,e}$, for every $z \in Z(e)$, then d^0 satisfies

$$E_{\theta|z,e}u(e, \tilde{z}, d^0(\tilde{z}), \tilde{\theta}) = \max_{d \in D(e)} E_{\theta,z|e}u(e, \tilde{z}, d(\tilde{z}), \tilde{\theta}). \tag{17.5.8}$$

The converse is, to all practical intents and purposes, true as well: if d^0 satisfies (17.5.8), then $d^0(z)$ maximizes (17.5.4) for every $z \in z(e)$. [The "impractical" intents and purposes are that, rigorously, d^0 satisfies (17.5.8) only if $P(\{z\colon d^0(z) \text{ does not maximize (17.5.4)}\} \mid e) = 0$.]

Finally, we now obtain an optimal solution of the general decision problem by varying e, and observing from (17.5.6) that an optimal experiment e^0 must satisfy

$$E_{\theta,z|e^0}u(e^0, \tilde{z}, d^0(\tilde{z}), \tilde{\theta})$$

$$= \max_{e \in E} \{ \max_{d \in D(e)} E_{\theta,\theta|e}u(e, \tilde{z}, d(\tilde{z}), \tilde{\theta})\}$$

$$= \max_{(e,d) \in H} E_{\theta,z|e}u(e, \tilde{z}, d(\tilde{z}), \tilde{\theta}), \tag{17.5.9}$$

(because iterated maximizations can be aggregated).

This procedure suggests thinking of the general decision problem as comprised of two stages rather than four: at stage one, the decision-maker chooses (e, d), and at stage two "chance" determines $(\tilde{\theta}, \tilde{z})$ according to

$P(\theta, z \mid e)$ [or $f(\theta, z \mid e)$ when $\tilde{\theta}$ and $\tilde{z}$ are random variables]. This suggests a decision tree of which Figure 17.12 depicts a typical $((e, d), (\theta, z))$ path.

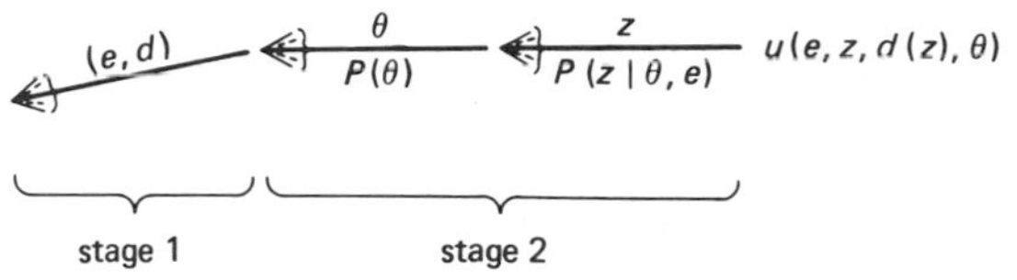

Figure 17.12

If the general decision problem is stated in terms of a loss function $\ell(e, z, a, \tilde{\theta})$, then the appropriate substitutes for (17.5.4) through (17.5.9) are

$$E_{\theta \mid z,e}\ell(e, z, d(z), \tilde{\theta}), \tag{17.5.10}$$

$$E_{z \mid e}E_{\theta \mid z,e}\ell(e, \tilde{z}, d(\tilde{z}), \tilde{\theta}) \tag{17.5.11}$$

$$= E_{\theta,z \mid e}\ell(e, \tilde{z}, d(\tilde{z}), \tilde{\theta}) \tag{17.5.12}$$

$$= E_{\theta}E_{z \mid \theta,e}\ell(e, \tilde{z}, d(\tilde{z}), \tilde{\theta}), \tag{17.5.13}$$

$$E_{\theta,z \mid e}\ell(e, \tilde{z}, d(\tilde{z}), \tilde{\theta}) = \min_{d \in D(e)} E_{\theta,z \mid e}\ell(e, \tilde{z}, d(\tilde{z}), \tilde{\theta}), \tag{17.5.14}$$

and

$$E_{\theta,z \mid e^0}\ell(e^0, \tilde{z}, d^0(\tilde{z}), \tilde{\theta}) = \min_{(e,d) \in H} E_{\theta,z \mid e}\ell(e, \tilde{z}, d(\tilde{z}), \tilde{\theta}), \tag{17.5.15}$$

respectively.

Example 17.5.1

Consider the decision problem of Example 17.2.1. In order to formulate this problem according to the procedures of this section, we must first derive $D(e_0)$ and $D(e_1)$. For $D(e_0)$, we have an easy task: $D(e_0) = \{d_1, d_2, d_3\}$, where $d_i(z_0) \equiv a_i$ for $i = 1, 2, 3$. But $D(e_1)$ consists of nine decision functions, defined in tabular form as follows:

	z_1	z_2		z_1	z_2
$d_4(z)$	a_1	a_1	$d_9(z)$	a_2	a_3
$d_5(z)$	a_1	a_2	$d_{10}(z)$	a_3	a_1
$d_6(z)$	a_1	a_3	$d_{11}(z)$	a_3	a_2
$d_7(z)$	a_2	a_1	$d_{12}(z)$	a_3	a_3
$d_8(z)$	a_2	a_2			

From these specifications and a painstaking tracing through of Figure 17.7, we obtain:

$$
\begin{array}{ll}
u(e_0, z_0, d_1(z_0), \theta_1) = 12 & u(e_0, z_0, d_2(z_0), \theta_2) = 15 \\
u(e_0, z_0, d_1(z_0), \theta_2) = 0 & u(e_0, z_0, d_3(z_0), \theta_1) = 6 \\
u(e_0, z_0, d_2(z_0), \theta_1) = 0 & u(e_0, z_0, d_3(z_0), \theta_2) = 6
\end{array}
$$

$$
\begin{array}{ll}
u(e_1, z_1, d_4(z_1), \theta_1) = 10 & u(e_1, z_1, d_5(z_1), \theta_1) = 10 \\
u(e_1, z_1, d_4(z_1), \theta_2) = 0 & u(e_1, z_1, d_5(z_1), \theta_2) = 0 \\
u(e_1, z_2, d_4(z_2), \theta_1) = 9 & u(e_1, z_2, d_5(z_2), \theta_1) = 0 \\
u(e_1, z_2, d_4(z_2), \theta_2) = 0 & u(e_1, z_2, d_5(z_2), \theta_2) = 9
\end{array}
$$

$$
\begin{array}{ll}
u(e_1, z_1, d_6(z_1), \theta_1) = 10 & u(e_1, z_1, d_7(z_1), \theta_1) = 0 \\
u(e_1, z_1, d_6(z_1), \theta_2) = 0 & u(e_1, z_1, d_7(z_1), \theta_2) = 10 \\
u(e_1, z_2, d_6(z_2), \theta_1) = 4 & u(e_1, z_2, d_7(z_2), \theta_1) = 9 \\
u(e_1, z_2, d_6(z_2), \theta_2) = 4 & u(e_1, z_2, d_7(z_2), \theta_2) = 0
\end{array}
$$

$$
\begin{array}{ll}
u(e_1, z_1, d_8(z_1), \theta_1) = 0 & u(e_1, z_1, d_9(z_1), \theta_1) = 0 \\
u(e_1, z_1, d_8(z_1), \theta_2) = 10 & u(e_1, z_1, d_9(z_1), \theta_2) = 10 \\
u(e_1, z_2, d_8(z_2), \theta_1) = 0 & u(e_1, z_2, d_9(z_2), \theta_1) = 4 \\
u(e_1, z_2, d_8(z_2), \theta_2) = 9 & u(e_1, z_2, d_9(z_2), \theta_2) = 4
\end{array}
$$

$$
\begin{array}{ll}
u(e_1, z_1, d_{10}(z_1), \theta_1) = 5 & u(e_1, z_1, d_{11}(z_1), \theta_1) = 5 \\
u(e_1, z_1, d_{10}(z_1), \theta_2) = 9 & u(e_1, z_1, d_{11}(z_1), \theta_2) = 5 \\
u(e_1, z_2, d_{10}(z_2), \theta_1) = 9 & u(e_1, z_2, d_{11}(z_2), \theta_1) = 0 \\
u(e_1, z_2, d_{10}(z_2), \theta_2) = 0 & u(e_1, z_2, d_{11}(z_2), \theta_2) = 9
\end{array}
$$

$$
\begin{array}{l}
u(e_1, z_1, d_{12}(z_1), \theta_1) = 5 \\
u(e_1, z_1, d_{12}(z_1), \theta_2) = 5 \\
u(e_1, z_2, d_{12}(z_2), \theta_1) = 4 \\
u(e_1, z_2, d_{12}(z_2), \theta_2) = 4.
\end{array}
$$

To find the decision-maker's utility of each (e, d) pair, we shall use (17.5.7), first computing $E_{z|\theta,e}u(e, \tilde{z}, d(\tilde{z}), \theta)$:

$$
\begin{array}{l}
E_{z|\theta,e}u(e_0, z_0, d_1(z_0), \theta_1) = 12 \\
E_{z|\theta,e}u(e_0, z_0, d_1(z_0), \theta_2) = 0 \\
E_{z|\theta,e}u(e_0, z_0, d_2(z_0), \theta_1) = 0 \\
E_{z|\theta,e}u(e_0, z_0, d_2(z_0), \theta_2) = 15 \\
E_{z|\theta,e}u(e_0, z_0, d_3(z_0), \theta_1) = 6 \\
E_{z|\theta,e}u(e_0, z_0, d_3(z_0), \theta_2) = 6
\end{array}
$$

$$
\begin{array}{l}
E_{z|\theta,e}u(e_1, \tilde{z}, d_4(\tilde{z}), \theta_1) = (10)(.9) + (9)(.1) = 9.9 \\
E_{z|\theta,e}u(e_1, \tilde{z}, d_4(\tilde{z}), \theta_2) = (\ 0)(.2) + (0)(.8) = 0
\end{array}
$$

$$
\begin{array}{l}
E_{z|\theta,e}u(e_1, \tilde{z}, d_5(\tilde{z}), \theta_1) = (10)(.9) + (0)(.1) = 9.0 \\
E_{z|\theta,e}u(e_1, \tilde{z}, d_5(\tilde{z}), \theta_2) = (\ 0)(.2) + (9)(.8) = 7.2
\end{array}
$$

$$
\begin{array}{l}
E_{z|\theta,e}u(e_1, \tilde{z}, d_6(\tilde{z}), \theta_1) = (10)(\ 9) + (4)(.1) = 9.4 \\
E_{z|\theta,e}u(\mathrm{e}_1, \tilde{z}, d_6(\tilde{z}), \theta_2) = (\ 0)(.2) + (4)(.8) = 3.2
\end{array}
$$

$$E_{z|\theta,e}u(e_1, \tilde{z}, d_7(\tilde{z}), \theta_1) = (\ 0)(.9) + (9)(.1) = \ .9$$
$$E_{z|\theta,e}u(e_1, \tilde{z}, d_7(\tilde{z}), \theta_2) = (10)(.2) + (0)(.8) = 2.0$$

$$E_{z|\theta,e}u(e_1, \tilde{z}, d_8(\tilde{z}), \theta_1) = (\ 0)(.9) + (0)(.1) = 0$$
$$E_{z|\theta,e}u(e_1, \tilde{z}, d_8(\tilde{z}), \theta_2) = (10)(.2) + (9)(.8) = 9.2$$

$$E_{z|\theta,e}u(e_1, \tilde{z}, d_9(\tilde{z}), \theta_1) = (\ 0)(.9) + (4)(.1) = \ .4$$
$$E_{z|\theta,e}u(e_1, \tilde{z}, d_9(\tilde{z}), \theta_2) = (10)(.2) + (4)(.8) = 5.2$$

$$E_{z|\theta,e}u(e_1, \tilde{z}, d_{10}(\tilde{z}), \theta_1) = (\ 5)(.9) + (9)(.1) = 5.4$$
$$E_{z|\theta,e}u(e_1, \tilde{z}, d_{10}(\tilde{z}), \theta_2) = (\ 5)(.2) + (0)(.8) = 1.0$$

$$E_{z|\theta,e}u(e_1, \tilde{z}, d_{11}(\tilde{z}), \theta_1) = (\ 5)(.9) + (0)(.1) = 4.5$$
$$E_{z|\theta,e}u(e_1, \tilde{z}, d_{11}(\tilde{z}), \theta_2) = (\ 5)(.2) + (9)(.8) = 8.2$$

$$E_{z|\theta,e}u(e_1, \tilde{z}, d_{12}(\tilde{z}), \theta_1) = (\ 5)(.9) + (4)(.1) = 4.9$$
$$E_{z|\theta,e}u(e_1, \tilde{z}, d_{12}(\tilde{z}), \theta_2) = (\ 5)(.2) + (4)(.8) = 4.2,$$

where $.9 = P(z_1 \mid \theta_1, e_1)$ and $.2 = P(z_1 \mid \theta_2, e_1)$, from Example 17.2.1. Finally, from $P(\theta_1) = \frac{1}{3}$ and (17.5.7), we compute the utilities

$$E_\theta\{E_{z|\theta,e}u(e, \tilde{z}, d(\tilde{z}), \tilde{\theta})\} = E_{\theta,z|e}u(e, \tilde{z}, d(\tilde{z}), \tilde{\theta}):$$

$$E_{\theta,z|e}u(e_0, z_0, d_1(z_0), \tilde{\theta}) = (12.0)(\tfrac{1}{3}) + (\ 0\ )(\tfrac{2}{3}) = 120/30$$
$$E_{\theta,z|e}u(e_0, z_0, d_2(z_0), \tilde{\theta}) = (\ 0\ )(\tfrac{1}{3}) + (15.0)(\tfrac{2}{3}) = 300/30$$
$$E_{\theta,z|e}u(e_0, z_0, d_3(z_0), \tilde{\theta}) = (\ 6.0)(\tfrac{1}{3}) + (\ 6.0)(\tfrac{2}{3}) = 180/30$$

$$E_{\theta,z|e}u(e_1, \tilde{z}, d_4(\tilde{z}), \tilde{\theta}) = (9.9)(\tfrac{1}{3}) + (0\ )(\tfrac{2}{3}) = \ 99/30$$
$$E_{\theta,z|e}u(e_1, \tilde{z}, d_5(\tilde{z}), \tilde{\theta}) = (9.0)(\tfrac{1}{3}) + (7.2)(\tfrac{2}{3}) = 234/30$$
$$E_{\theta,z|e}u(e_1, \tilde{z}, d_6(\tilde{z}), \tilde{\theta}) = (9.4)(\tfrac{1}{3}) + (3.2)(\tfrac{2}{3}) = 159/30$$
$$E_{\theta,z|e}u(e_1, \tilde{z}, d_7(\tilde{z}), \tilde{\theta}) = (\ .9)(\tfrac{1}{3}) + (2.0)(\tfrac{2}{3}) = \ 49/30$$
$$E_{\theta,z|e}u(e_1, \tilde{z}, d_8(\tilde{z}), \tilde{\theta}) = (0\ )(\tfrac{1}{3}) + (9.2)(\tfrac{2}{3}) = 184/30$$
$$E_{\theta,z|e}u(e_1, \tilde{z}, d_9(\tilde{z}), \tilde{\theta}) = (\ .4)(\tfrac{1}{3}) + (5.2)(\tfrac{2}{3}) = 108/30$$
$$E_{\theta,z|e}u(e_1, \tilde{z}, d_{10}(\tilde{z}), \tilde{\theta}) = (5.4)(\tfrac{1}{3}) + (1.0)(\tfrac{2}{3}) = \ 74/30$$
$$E_{\theta,z|e}u(e_1, \tilde{z}, d_{11}(\tilde{z}), \tilde{\theta}) = (4.5)(\tfrac{1}{3}) + (8.2)(\tfrac{2}{3}) = 209/30$$
$$E_{\theta,z|e}u(e_1, \tilde{z}, d_{12}(\tilde{z}), \tilde{\theta}) = (4.9)(\tfrac{1}{3}) + (4.2)(\tfrac{2}{3}) = 133/30.$$

It is clear that $d^0(e_0) = d_2$; that is, choose act a_2 if e_0 was chosen. Moreover, if e_1 was chosen, then $d^0(e_1) = d_5$; that is, choose a_i if z_i is observed, for $i = 1, 2$. Moreover, (e_0, d_2) is preferable to (e_1, d_5). You may check from Figure 17.11 that these prescriptions are precisely those of the previous analysis.

The formulation of the decision problem in this section is called the *normal form*, as opposed to the *extensive form* in the previous sections. It should be obvious from a comparison of Examples 17.2.1 and 17.5.1 that the normal-form analysis can be much more cumbersome than the extensive-form analysis. In fact, this is usually the case. Hence, why introduce it?

There are basically two justifications for the normal-form analysis. *First,* it sometimes happens that $a^0_{z,e}$ is a very complicated function of z; the reader will note that $a^0_{z,e_0} = E(\tilde{\theta} \mid z, e_1)$ in Example 17.2.2 could be extremely difficult to evaluate. In such cases it may not be practical to instruct a subordinate to undertake the evaluation of $a^0_{z,e}$ after observing z, and hence the decision-maker may wish consideration restricted to a subset $D'(e)$ of $D(e)$, even at the risk of having $d^0 \notin D'(e)$, thereby not fully maximizing utility.

Example 17.5.2

Suppose in Example 17.2.2 that d must be a *linear* function of z; that is,

$$D'(e_1) = \{d: d(z) = \beta_1 + \beta_2 z, \beta_i \in R^1, \text{ for } i = 1, 2\}.$$

Then

$$(1) \qquad u(e_1, z, d(z), \theta) = 25 - 20(\theta - \beta_1 - \beta_2 z)^2 - 5z; \text{ and}$$

$$(2) \qquad E_{\theta,z|e}u(e_1, \tilde{z}, d(\tilde{z}), \tilde{\theta}) = 25 - 20E_{\theta,z|e}(\tilde{\theta} - \beta_1 - \beta_2\tilde{z})^2 - 5E(\tilde{z} \mid e_1).$$

It is clear that (β_1, β_2) maximizes $E_{\theta,z|e}u(e_1, \tilde{z}, d(\tilde{z}), \tilde{\theta})$ if and only if (β_1, β_2) minimize $E_{\theta,z|e}(\tilde{\theta} - \beta_1 - \beta_2\tilde{z})^2$. We take the first partials $\partial/\partial\beta_i$ of $E_{\theta,z|e}(\tilde{\theta} - \beta_1 - \beta_2\tilde{z})^2$, interchange the order of expectation and differentiation, and equate to zero as the first-order necessary conditions for a minimum, finding

$$-2E_{\theta,z|e}(\tilde{\theta} - \beta_1{}^0 - \beta_2{}^0\tilde{z}) = 0$$

and

$$-2E_{\theta,z|e}(\tilde{\theta}\tilde{z} - \beta_1{}^0\tilde{z} - \beta_2{}^0\tilde{z}^2) = 0,$$

which implies

$$\beta_1{}^0 + E(\tilde{z} \mid e)\beta_2{}^0 = E(\tilde{\theta})$$

$$E(\tilde{z} \mid e)\beta_1{}^0 + E(\tilde{z}^2 \mid e)\beta_2{}^0 = E(\tilde{\theta}\tilde{z} \mid e).$$

The reader may verify as Exercise 10 that the solution of these equations is

$$\beta_1{}^0 = E(\tilde{\theta}) - \frac{V(\tilde{\theta}, \tilde{z} \mid e_1)}{V(\tilde{z} \mid e_1)}E(\tilde{z} \mid e_1)$$

and

$$\beta_2{}^0 = \frac{V(\tilde{\theta}, \tilde{z} \mid e_1)}{V(\tilde{z} \mid e_1)};$$

hence the optimal decision function d^0 in $D'(e_1)$ is

$$d^0(z) = E(\tilde{\theta}) + \frac{V(\tilde{\theta}, \tilde{z} \mid e_1)}{V(\tilde{z} \mid e_1)}[z - E(\tilde{z} \mid e_1)], \tag{3}$$

and

$$E_{\theta,z|e}(\tilde{\theta} - \beta_1{}^0 - \beta_2{}^0\tilde{z})^2 = V(\tilde{\theta}) - \frac{[V(\tilde{\theta}, \tilde{z} \mid e_1)]^2}{V(\tilde{z} \mid e_1)}, \tag{4}$$

whereas in Example 17.2.2

$$E_{\theta,z|e}(\tilde{\theta} - E(\tilde{\theta} \mid \tilde{z}, e_1))^2 = V(\tilde{\theta}) - V(E(\tilde{\theta} \mid \tilde{z}, e_1)), \tag{5}$$

which you are asked to derive as Exercise 11. In Exercise 12 you are asked to show that

$$V(\tilde{z} \mid e_1) \cdot V(E(\tilde{\theta} \mid \tilde{z}, e_1)) \geqq V(\tilde{\theta}, \tilde{z} \mid e_1), \tag{6}$$

a consequence of

$$E_{\theta,z|e}(\tilde{\theta} - \beta_1{}^0 - \beta_2{}^0\tilde{z})^2 \geqq E_{\theta,z|e}(\tilde{\theta} - E(\tilde{\theta} \mid \tilde{z}, e_1))^2, \tag{7}$$

itself a consequence of

$$\max_{d \in D(e)} E_{\theta,z|e}u(e_1, \tilde{z}, d(\tilde{z}), \tilde{\theta}) \geqq \max_{d \in D'(e)} E_{\theta,z|e}u(e_1, \tilde{z}, d(\tilde{z}), \tilde{\theta}), \tag{8}$$

as must be true because $D'(e) \subset D(e)$. Equality obtains in (8) only if $E(\tilde{\theta} \mid z, e_1)$ is expressible as a linear function of z. [However, $E(\tilde{\theta} \mid z, e_1)$ may deviate from this linear function on a set of z's of probability zero.] When $E(\tilde{\theta} \mid z, e_1)$ is a linear function of z, it can be shown that

$$E(\tilde{\theta} \mid z, e_1) = \beta_1{}^0 + \beta_2{}^0 z, \tag{9}$$

where $\beta_1{}^0$ and $\beta_2{}^0$ are as given previously.

Example 17.5.2 illustrates the concept of a *suboptimal* decision function; That is, a decision function which is optimal in a given subset $D'(e)$ of $D(e)$ but not always optimal in $D(e)$ itself. Suboptimal decision functions, especially of the linear form, are considered because they are easy to apply. Ease of application is an important attribute of decision functions when we view a given decision problem in the broader context of the many decisions confronting administrators.

There is a *second* important justification for the normal-form analysis of decisions. It stems from the controversy over the foundations of probability and statistics, introduced in Chapters 5 and 9. To summarize very briefly, the conditional dgf $f(z \mid \theta, e)$ may be objectively known and noncontroversial for every $\theta \in \Theta$, whereas the prior dgf $f(\theta)$ may be subjective and very controversial, with some statisticians claiming that it does not even exist.

Hence it behooves us to see how far statistical analysis can proceed without introducing $f(\theta)$.

The normal form furnishes the appropriate vehicle, since from (17.5.7) we can consider the function $\Upsilon[(e, d), \theta]$ of $(e, d) \in H$ and $\theta \in \Theta$ defined by

$$\Upsilon[(e, d), \theta] \equiv E_{z|\theta,e}u(e, \tilde{z}, d(\tilde{z}), \theta). \qquad (17.5.16)$$

The decision problem can now be regarded as consisting of two stages, in the first of which the decision-maker chooses a grand strategy $(e, d) \in H$ (as before), while in the second of which "chance" determines θ. When Θ and H are each finite sets, $\Upsilon[(e, d), \theta]$ can be depicted in matrix format.

Example 17.5.3

For the decision problem in Example 17.5.1, $\Upsilon[(e, d), \theta]$ is easily found to be given by the following matrix.

	$T[(e,d),\theta]$	
	θ_1	θ_2
(e_0,d_1)	12	0
(e_0,d_2)	0	15
(e_0,d_3)	6	6
(e_1,d_4)	9.9	0
(e_1,d_5)	9.0	7.2
(e_1,d_6)	9.4	3.2
(e_1,d_7)	.9	2.0
(e_1,d_8)	0.0	9.2
(e_1,d_9)	.4	5.2
(e_1,d_{10})	5.4	1.0
(e_1,d_{11})	4.5	8.2
(e_1,d_{12})	4.9	4.2

The reduction of the decision to the function Υ is highly useful, because it eliminates everything noncontroversial [for example, the role of $f(z \mid \theta, e)$] and thereby focuses attention upon the role of θ in determining the decision-maker's utility.

Now, an ideal state of affairs would be to find a universally optimal $(e^0, d^0) \in H$ such that for every $\theta \in \Theta$,

$$\Upsilon[(e^0, d^0), \theta] \geqq \Upsilon[(e, d), \theta], \text{ for every } (e, d) \in H. \qquad (17.5.17)$$

When $\Upsilon[(e, d), \theta]$ is given in tabular format, (17.5.17) requires that the (e^0, d^0)-row be $\geqq$ every other row in the vector sense.

The value of having (17.5.17) obtain for some $(e^0, d^0) \in H$ consists in its consequence that

$$E_\theta\Upsilon[(e^0, d^0), \tilde{\theta}] \geqq E_\theta\Upsilon[e, d), \tilde{\theta}], \text{ for every } (e, d) \in H, \qquad (17.5.18)$$

for every prior $f(\theta)$ with respect to which the E_θ's are taken. But

$$E_\theta\Upsilon[(e, d), \tilde{\theta}] \equiv E_{\theta,z|e}u(e, \tilde{z}, d(\tilde{z}), \tilde{\theta}), \text{ for every } (e, d),$$

and hence (17.5.18) and (17.5.9) imply that (e^0, d^0) is indeed universally optimal: optimal for any prior $f(\theta)$.

Nevertheless, such a state of affairs is only an ideal and rarely encountered in actual decision problems. The usual case is to find that different (e, d)'s are optimal for sufficiently different θ's or $f(\theta)$'s. This is clearly true in Example 17.5.3, which we now pursue further.

Example 17.5.4

In the decision problem of Examples 17.2.1, 17.5.1, and 17.5.3 we have $\Theta = \{\theta_1, \theta_2\}$ and hence the prior probability function is fully specified by $P(\theta_1)$. Clearly,

$$E_\theta \Upsilon[(e, d), \tilde{\theta}] = \Upsilon[(e, d), \theta_1]P(\theta_1) + \Upsilon[(e, d), \theta_2](1 - P(\theta_2)),$$

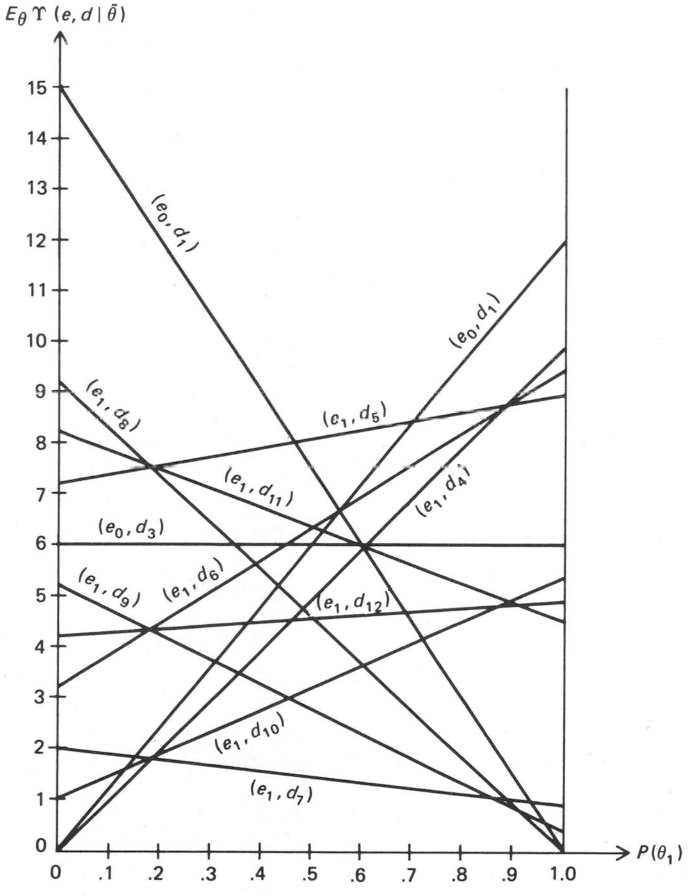

Figure 17.13

which is graphed in Figure 17.13. Note that (e^0, d^0) must belong to $\{(e_0, d_2), (e_1, d_5), (e_0, d_1)\}$ and that none of the nine other elements of H can be optimal. The reader may easily verify that

$$(e^0, d^0) = \begin{Bmatrix} (e_0, d_2) \\ (e_1, d_5) \\ (e_0, d_1) \end{Bmatrix}, \text{ if } P(\theta_1) \in \begin{cases} [0, 39/84] \\ [39/84, 36/51] \\ [36/51, 1]. \end{cases}$$

Note also the intuitively plausible fact that obtaining information (that is, choosing e_1 initially) is optimal if $P(\theta_1)$ is "close to" $\frac{1}{2}$ (that is, if the decision-maker's uncertainty about $\tilde{\theta}$ is sufficiently great). Finally, suppose that the prior $P(\theta_1)$ is controversial only in that there is disagreement as to where it falls within [0, 39/84]. The analysis shows that this controversy is irrelevant to the choice of (e, d), since all combatants should agree on the optimality of (e_0, d_2).

An exactly similar analysis of decisions in normal form can be formulated in terms of loss functions. Define

$$\Lambda[(e, d), \theta] \equiv E_{z|\theta,e}\ell(e, \tilde{z}, d(\tilde{z}), \theta), \tag{17.5.19}$$

for every $(e, d) \in H$ and $\theta \in \Theta$. Now, Λ can be easily obtained from Υ via the fact that for every $(e, d) \in H$ and $\theta \in \Theta$,

$$\Lambda[(e, d), \theta] = u(e_0, z_0, a_\theta{}^0, \theta) - \Upsilon[(e, d), \theta]. \tag{17.5.20}$$

Proof

From (17.5.19) and (17.3.1) we have

$$\begin{aligned}\Lambda[(e, d), \theta] &= E_{z|\theta,e}[u(e_0, z_0, a_\theta{}^0, \theta) - u(e, \tilde{z}, d(\tilde{z}), \theta)] \\ &= E_{z|\theta,e}u(e_0, z_0, a_\theta{}^0, \theta) - E_{z|\theta,e}u(e, \tilde{z}, d(\tilde{z}), \theta)\end{aligned}$$

by (10.2.7). But

$$E_{z|\theta,e}u(e_0, z_0, a_\theta{}^0, \theta) = u(e_0, z_0, a_\theta{}^0, \theta)$$

because $u(e_0, z_0, a_\theta{}^0, \theta)$ is a constant with respect to $E_{z|\theta,e}$-expectation; and

$$-E_{z|\theta,e}u(e, \tilde{z}, d(\tilde{z}), \tilde{\theta}) \equiv -\Upsilon[(e, d), \theta]. \quad \text{QED}$$

In view of (17.5.20) and the fact that for some $d \equiv d_\theta{}^0 \in D(e_0)$ we have

$$u(e_0, z_0, a_\theta{}^0, \theta) = \Upsilon[(e_0, d_\theta{}^0), \theta],$$

it follows that Λ can be presented in tabular format and that the appropriate table can be obtained easily from the table of Υ's.

Example 17.5.5

In example 17.5.3 it is clear that $d_1 = d_{\theta_1}{}^0$ and $d_2 = d_{\theta_2}{}^0$, and hence that

$$u(e_0, z_0, a_\theta{}^0, \theta) = \begin{cases} 12, \theta = \theta_1 \\ 15, \theta = \theta_2. \end{cases}$$

To obtain the first column of the Λ-table we subtract the first column of the Υ-table from a column of 12's, and similarly for the second column.

$\Lambda[(e,d),\theta]$

	θ_1	θ_2
(e_0,d_1)	12 − 12 = 0	15 − 0 = 15
(e_0,d_2)	12 − 0 = 12	15 − 15 = 0
(e_0,d_3)	12 − 6 = 6	15 − 6 = 9
(e_1,d_4)	12 − 9.9 = 2.1	15 − 0 = 15
(e_1,d_5)	12 − 9.0 = 3.0	15 − 7.2 = 7.8
(e_1,d_6)	12 − 9.4 = 2.6	15 − 3.2 = 11.8
(e_1,d_7)	12 − .9 = 11.1	15 − 2.0 = 13.0
(e_1,d_8)	12 − 0.0 = 12.0	15 − 9.2 = 5.8
(e_1,d_9)	12 − .4 = 11.6	15 − 5.2 = 9.8
(e_1,d_{10})	12 − 5.4 = 6.6	15 − 1.0 = 14.0
(e_1,d_{11})	12 − 4.5 = 7.5	15 − 8.2 = 6.8
(e_1,d_{12})	12 − 4.9 = 7.1	15 − 4.2 = 10.8

Now,

$$\begin{aligned} E_\theta\Lambda[(e, d), \tilde{\theta}] &= E_{\theta,z|e}\ell(e, \tilde{z}, d(\tilde{z}), \tilde{\theta}) \\ &= E_\theta u(e_0, z_0, a_\theta{}^0, \tilde{\theta}) - E_\theta\Upsilon[(e, d), \tilde{\theta}], \end{aligned} \tag{17.5.21}$$

and hence (e^0, d^0) can be obtained either by maximizing $E_\theta\Upsilon[(e, d), \tilde{\theta}]$ or by minimizing $E_\theta\Lambda[(e, d), \tilde{\theta}]$. That is,

$$E_\theta\Upsilon[(e^0, d^0), \tilde{\theta}] \geqq E_\theta\Upsilon[(e, d), \tilde{\theta}], \text{ for all } (e, d) \in H \tag{17.5.22}$$

if and only if

$$E_\theta\Lambda[(e^0, d^0), \tilde{\theta}] \leqq E_\theta\Lambda[(e, d), \tilde{\theta}].$$

Example 17.5.6

From Example 17.5.5 and $E_\theta\Lambda[(e, d), \tilde{\theta}] = \Lambda[(e, d), \theta_1]P(\theta_1) + \Lambda[(e, d), \theta_2](1 - P(\theta_1))$, we obtain Figure 17.14, which should be compared with Figure 17.13. Note that both figures indicate optimality of the same (e, d) at each $P(\theta_1)$.

We have thus shown that minimizing (the expectation of) opportunity loss is just as equivalent to maximizing (the expectation of) utility in the normal-form analysis as it is in the extensive-form analysis. This was to be expected.

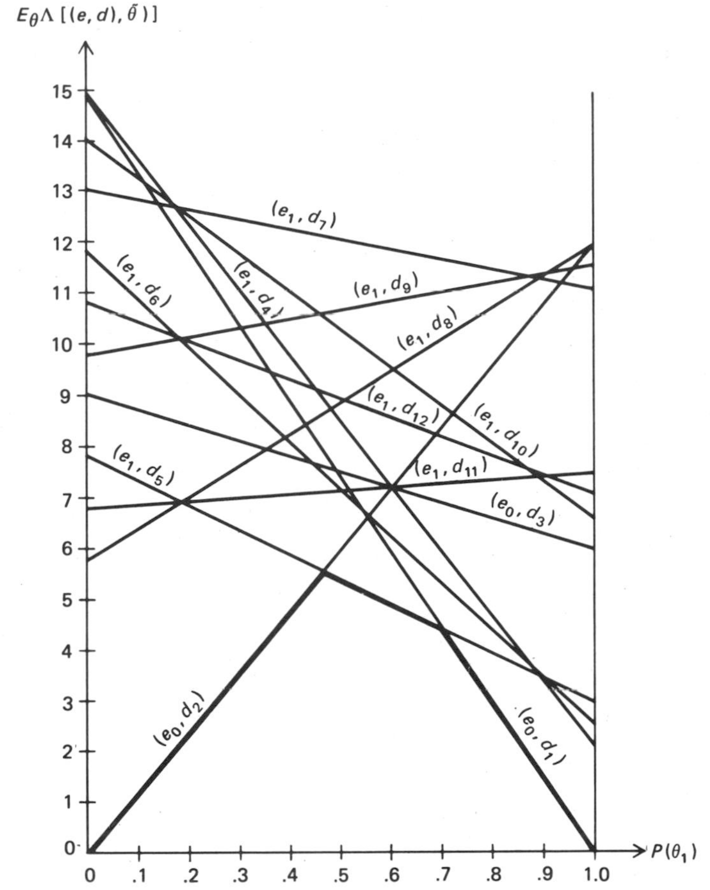

Figure 17.14

But all is not in harmony when we abandon a complete Bayesian analysis. Suppose the decision-maker is convinced that the odds are against him, in that whatever (e, d) he chooses, that θ will occur which will minimize his utility; that is, he evaluates his utility of each (e, d) by

$$\min_{\theta \in \Theta} \Upsilon[(e, d), \theta].$$

Clearly, this evaluation makes his prior probability function on (subsets of) Θ highly dependent upon his choice of (e, d), a dependence which we have ruled out for Bayesian analysis.

Now, given his assumption, it is clear that he should choose that $(e^\dagger, d^\dagger)$ for which

$$\min_{\theta \in \Theta} \Upsilon[(e^\dagger, d^\dagger), \theta] = \max_{(e,d) \in H} \{\min_{\theta \in \Theta} \Upsilon[(e, d), \theta]\}$$
$$\equiv \max_{(e,d) \in H} \min_{\theta \in \Theta} \Upsilon[(e, d), \theta]. \qquad (17.5.23)$$

Example 17.5.7

In Example 17.5.3, you may verify that $(e^\dagger, d^\dagger) = (e_1, d_5)$; that is, take e_1 and choose a_i if $\tilde{z} = z_i$ for $i = 1, 2$. We have

$$\max_{(e,d)\in H} \min_{\theta\in\Theta} \Upsilon[(e, d), \theta] = 7.2.$$

The strategy $(e^\dagger, d^\dagger)$ is called a *maximin* strategy.

If he formulates his problem in terms of opportunity loss and the fates are against him, then he will evaluate each (e, d) by

$$\max_{\theta\in\Theta} \Lambda[(e, d), \theta],$$

and he should choose that (e^*, d^*) for which

$$\begin{aligned}\max_{\theta\in\Theta} \Lambda[(e^*, d^*), \theta] &= \min_{(e,d)\in H} \{\max_{\theta\in\Theta} \Lambda[(e^*, d^*), \theta]\} \\ &\equiv \min_{(e,d)\in H} \max_{\theta\in\Theta} \Lambda[(e^*, d^*), \theta].\end{aligned} \qquad (17.5.24)$$

The strategy (e^*, d^*) is called a *minimax* strategy. Unfortunately, it does not follow that $(e^\dagger, d^\dagger) = (e^*, d^*)$.

Example 17.5.8

In Example 17.5.5, you may verify that $(e^*, d^*) = (e_1, d_{11})$ and that

$$\max_{\theta\in\Theta} \Lambda[(e^*, d^*), \theta] = 7.5.$$

Hence the maximin-utility and minimax-loss procedures are *not* equivalent; they do not furnish the same prescriptions for every problem. This means that the decision-maker who feels inclined to use one or the other of them had better think very carefully about just what his competitors have in mind. Many Bayesians (this author included) believe that the decision-maker's time would be better spent in carefully assessing his prior $f(\theta)$.

This opinion is not intended to disparage a fascinating and important branch of decision theory (namely, game theory). The subject of game theory is decisions in which there are two or more decision-maker's whose utilities depend upon the strategies chosen by all of them. Hence the decision-makers interact. The concepts of conflict and cooperation find their analytical *metier* in game theory. The interested reader is encouraged to see Luce and Raiffa [**52**] and Owen [**60**] for (much) more information about game theory.

EXERCISES

1. *Optimal acts may not exist*: Suppose $A(e, z) = (0, 1)$ and $u(e, z, a, *) = a$ for every $a \in (0, 1)$. Show that (17.2.2) does not obtain for any $a \in (0, 1)$.

2. Find $a^0_{z_0|e_0}$ and

$$E_\theta \ell(e_0, z_0, a^0_{z_0|e_0}, \tilde{\theta}) \equiv \min_a E_\theta \ell(e_0, z_0, a, \tilde{\theta})$$

for Example 17.3.2.

3. Find $a^0_{z_0|e_0}$ and $E_{z|e} \min_a E_{\theta|z,e} \ell(e_1, \tilde{z}, a, \tilde{\theta})$

for Example 17.3.2.

4. Derive (17.4.20).

5. Show that $\#(k(e) \mid e, y) = k(e)$ for any nonrandom cost $k(e)$ any e, and any y.

6. Find explicit expressions for $\#(k(e, \tilde{z}) \mid e, y)$ and $I(e, y)$ when:

   (a) $u^*[y] = -\exp(-my)$ for every y, where $m > 0$ is a parameter; and
   (b) $u^*[y] = \exp(my)$ for every y, where $m > 0$ is a parameter.
   (c) Thus show that $\#$ and I are independent of the initial asset position for these cases.

7. Show that (17.4.29) obtains if and only if the analogous inequalities in (17.4.27) obtain.

8. Show that (17.4.33) and (17.4.34) follow from (1) and (2) of the ensuing proof.

9. Prove (17.4.35).

10. Derive $\beta_1{}^0$ and $\beta_2{}^0$ in Example 17.5.2.

11. Derive (5) in Example 17.5.2. (*Hint*: See Exercise 3.)

12. Prove (8), (7), and (6), in that order, in Example 17.5.2. [*Hint*: For (8), let d' maximize $E_{\theta,z|e}u(e_1, \tilde{z}, d(\tilde{z}), \tilde{\theta})$ in $D'(e_1)$. Since $D'(e_1) \subset D(e_1)$, it follows that $d' \in D(e_1)$. Either d' maximizes $E_{\theta,z|e}u(e_1, \tilde{z}, d(\tilde{z}), \tilde{\theta})$ in $D(e_1)$ or it does not.]

13. Show that

$$\max_{(e,d)\in H} \min_{\theta\in\Theta} \Upsilon[(e, d), \theta] \leqq \max_{(e,d)\in H} E_\theta \Upsilon[e, d), \tilde{\theta}],$$

for any prior $f(\theta)$.

14. A production controller wishes to determine the amount a of a fluid to produce but does not know the amount $\tilde{\theta}$ of that fluid that will be demanded. If $\theta < a$, then he will have to discard the excess $a - \theta$ and thereby lose its direct production cost $k_0(a - \theta)$, where $k_0 > 0$ is the per-unit direct production cost. On the other hand, if $\theta > a$, then he will have to set up another production run at a fixed cost of $K_u > 0$. Utility is linear in money.

    (a) Argue that

$$\ell_t(a, \theta) = \begin{cases} k_0(a - \theta), & \theta \leqq a \\ K_u, & \theta > a. \end{cases}$$

(b) Suppose $\tilde{\theta}$ is a continuous random variable with prior density function $f(\theta)$ and prior cdf $F(\theta)$ such that $F(0) = 0$. Show that

$$E_{\theta}\ell_t(a, \tilde{\theta}) = ak_0F(a) + K_u(1 - F(a)) - k_0 \int_0^a \theta f(\theta)d\theta.$$

(c) Show that $a^0 \equiv a_{z_0^0|e_0}$ must satisfy

$$\frac{f(a^0)}{F(a^0)} = \frac{k_0}{K_u},$$

provided that

$$\frac{[(d/da)f(a^0)]}{f(a^0)} < \frac{k_0}{K_u}.$$

[*Hint*: These are first- and second-order sufficient conditions that a^0 minimize $E_{\theta}\ell_t(a, \tilde{\theta})$.]

(d) Suppose

$$f(\theta) = \begin{cases} \rho\theta^{\rho-1}, & 0 < \theta < 1 \\ 0, & \theta \notin (0, 1), \end{cases}$$

for some $\rho > 0$. Find a^0.

[*Hint*: First suppose that $K_u < k_0/\rho$, and then suppose that $K_u \geqq k_0/\rho$.]

(e) Find $I(e^*)$ for $f(\theta)$ as defined in (d).

15. In a competitive sealed auction each bidder submits one bid stipulating the amount he is willing to pay to obtain the object being auctioned; the highest bidder will obtain the object and pay the amount of his bid; and each bidder has no information about the bids to be submitted by the other bidders. Suppose the decision maker is a bidder, values the object being sold at w, has utility linear in money, and assesses a density function $f(\theta)$ for the maximum bid $\tilde{\theta}$ of his competitors. Let his bid be a. He does not have to worry about tied bids because if $\tilde{\theta}$ is a continuous random variable, then $P(\tilde{\theta} = a) = 0$ for every a.

(a) Show that

$$v_t(a, \theta) = \begin{cases} w - a, & \theta \leqq a \\ 0, & \theta > a \end{cases}$$

and that

$$\ell_t(a, \theta) = \begin{cases} a - \theta, & \theta \leqq a \leqq w \\ w - \theta, & a < \theta \leqq w \\ 0, & a \leqq w < \theta. \end{cases}$$

Why need we not bother with the cases $a > w$?

(b) Show that

$$E_\theta v_t(a, \tilde{\theta}) = (w - a)F(a)$$

and

$$E_\theta \ell_t(a, \tilde{\theta}) = aF(a) + w[F(w) - F(a)] - \int_0^w \theta f(\theta) d\theta.$$

(c) Show that $a^0 \equiv a^0_{z_0|e_0}$ must satisfy

$$\frac{f(a^0)}{F(a^0)} = \frac{1}{w - a^0},$$

provided that

$$\frac{[(d/da)f(a^0)]}{f(a^0)} < \frac{2}{w - a^0}.$$

(d) Suppose $f(\theta)$ is as given by (d) of Exercise 14. Find a^0.

(e) Suppose the decision-maker is able to obtain perfect information about $\tilde{\theta}$ before submitting his bid. Discuss any sorts of difficulties that might arise in determining $a_\theta{}^0$.

(f) Suppose the decision-maker wins whenever he is tied for highest bidder. Find $I(e^*)$ for $f(\theta)$ as given in (d).

16. Solve the general decision problem given by Figure 17.15, in extensive form. The numbers at endpoints of the tree are incremental dollar returns. Utility is linear in money; $P(\theta_1) = P(\theta_2) = \frac{1}{2}$; $P(z_1 \mid \theta_1, e_1) = \frac{1}{2}$; and $P(z_2 \mid \theta_2, e_1) = \frac{7}{8}$.

17. (a) Formulate the decision problem of Exercise 16 in normal form.
(b) Tabulate $\Upsilon[(e, d), \theta]$ and $\Lambda[(e, d), \theta]$.
(c) Graph the $E_\theta\Upsilon[(e, d), \tilde{\theta}]$'s as functions of $P(\theta_1)$, and state which (e, d)'s are optimal for each $P(\theta_1) \in [0, 1]$.
(d) Suppose $D'(e_1) \equiv \{d\colon d(z_2) \neq a_3\}$. Now which (e, d)'s are optimal for each $P(\theta_1) \in [0, 1]$?
(e) From (c) and (d), argue that the decision-maker can assign a value to having $D(e_1)$ vis-à-vis $D'(e_1)$. Express this value as a function of $P(\theta_1)$.

18. Suppose $\Theta = \{\theta_1, \ldots, \theta_r\}$; $A = \{a_1, \ldots, a_r\}$; utility is linear in money; $P(\theta_1), \ldots, P(\theta_r)$ have been given, and

$$\ell_t(a_i, \theta_j) = \begin{cases} 0, & i = j \\ 1, & i \neq j, \end{cases}$$

for $i, j = 1, \ldots, r$.

(a) Show that

$$E_\theta \ell_t(a_i, \tilde{\theta}) = 1 - P(\theta_i),$$

for $i = 1, \ldots, r$.

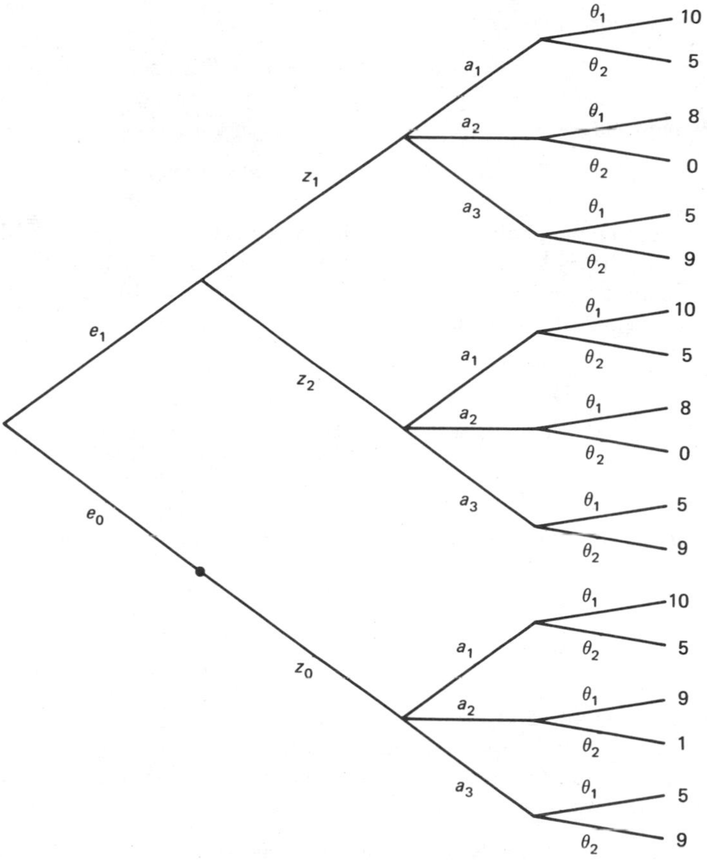

Figure 17.15

(b) Show that $a^0 \equiv a^0_{z_0|e_0} = a_i$ if

$$P(\theta_i) = \max_j P(\theta_j)$$

and hence that

$$E_\theta \ell_t(a^0, \tilde{\theta}) = 1 - \max_j P(\theta_j).$$

(c) Suppose an experiment e is available and that all probability functions $P(z \mid \theta_i, e)$ are given. Show that

$$E_{\theta|z,e}\ell_t(a_i, \tilde{\theta}) = 1 - P(\theta_i \mid z, e);$$

that $a_{z|e}{}^0 = a_i$ if

$$P(\theta_i \mid z, e) = \max_j P(\theta_j \mid z, e),$$

and hence that

$$E_{\theta|z,e}\ell_t(a_{z|e}{}^0, \tilde{\theta}) = 1 - \max_j P(\theta_j \mid z, e).$$

(d) Show that

$$I(e) = E_{z|e} \max_j P(\theta_j \mid \tilde{z}, e) - \max_j P(\theta_j).$$

(e) Interpret the choice of a_i as being a prediction that $\tilde{\theta} = \theta_i$, and argue that ℓ_t furnishes a formal rendition of "a miss is as good as a mile."

(f) Suppose $r = 3$, $P(\theta_1) = .80$, $P(\theta_2) = .10$, $P(\theta_3) = .10$, and $P(z \mid \theta_i, e)$ is given by the following table:

	z_1	z_2	z_3
θ_1	$\frac{7}{24}$	$\frac{8}{24}$	$\frac{9}{24}$
θ_2	$\frac{3}{6}$	$\frac{2}{6}$	$\frac{1}{6}$
θ_3	$\frac{3}{6}$	$\frac{2}{6}$	$\frac{1}{6}$

Show that $I(e) = 0$.

(g) Suppose in the general problem that there is an index k such that

$$\max_j P(\theta_j \mid z, e) = P(\theta_k \mid z, e),$$

for every $z \in Z(e)$. Show that $I(e) = 0$.

19. Suppose utility is linear in money and that there exists an act $a^* \in A$ such that $a^* = a^0_{z|e}$ for every $z \in Z(e)$. Show that $I(e) = 0$ and $a^* = a^0_{z_0|e_0}$. [This is intuitively plausible, since if the information cannot affect the chosen act (that is, $a^0_{z|e} = a^*$ for every $z \in Z(e)$), then the decision-maker knows what to do without bothering to obtain the information.]

20. *Randomized strategies*: Suppose the decision-maker can select (e, d) by a randomization device (for example, by drawing a ball from an urn). The resulting probability function on H is called a *randomized strategy*. Suppose $H = \{h_1, \ldots, h_m\}$ is a finite set, and let $\boldsymbol{\delta} = (\delta_1, \ldots, \delta_m)$ denote the randomized strategy which selects h_i with probability δ_i for $i = 1, \ldots, m$. Define

$$\Upsilon[\boldsymbol{\delta}, \theta] = \sum_{i=1}^{m} \Upsilon[h_i, \theta]\delta_i, \text{ for every } \theta \in \Theta$$

and

$$\Lambda[\boldsymbol{\delta}, \theta] \equiv \sum_{i=1}^{m} \Lambda[h_i, \theta]\delta_i, \text{ for every } \theta \in \Theta.$$

Let $\boldsymbol{\Delta} \equiv \{\boldsymbol{\delta}: \delta_i \geqq 0 \text{ for all } i, \sum_{i=1}^{m} \delta_i = 1\}$.

(a) Show that

$$\max_{\delta \in \Delta} \min_{\theta \in \Theta} \Upsilon[\boldsymbol{\delta}, \theta) \geqq \max_{h \in H} \min_{\theta \in \Theta} \Upsilon[h, \theta].$$

[*Hint*: What is the difference between h_i and that $\boldsymbol{\delta}$ for which $\delta_i = 1$ and $\delta_j = 0$ for $j \neq i$? Hence can H be considered a subset of $\boldsymbol{\Delta}$?]

(b) Similarly, show that

$$\min_{\delta \in \Delta} \max_{\theta \in \Theta} \Lambda[\boldsymbol{\delta}, \theta] \leqq \min_{h \in H} \max_{\theta \in \Theta} \Lambda[h, \theta].$$

(c) Under what conditions might the decision-maker wish to choose $\boldsymbol{\delta}^{\dagger}$ such that

$$\min_{\theta \in \Theta} \Upsilon[\boldsymbol{\delta}^{\dagger}. \theta] = \max_{\delta \in \Delta} \min_{\theta \in \Theta} \Upsilon[\boldsymbol{\delta}, \theta],$$

or δ^* such that

$$\max_{\theta \in \Theta} \Lambda[\boldsymbol{\delta}^*, \theta] = \min_{\delta \in \Delta} \max_{\theta \in \Theta} \Lambda[\boldsymbol{\delta}, \theta]?$$

18

BAYESIAN POINT-ESTIMATION PROBLEMS

18.1 INTRODUCTION

In a large class of general decision problems $A(e, z) \equiv A$ is independent of (e, z), and $A = \Theta \subset R^1$. Thus the state is described by a random variable $\tilde{\theta}$, and the decision-maker's choice of act a may be interpreted as an *estimate* of the as yet unobserved value θ of $\tilde{\theta}$. We shall refer to any such decision problem as a *point-estimation problem.*

In this chapter we examine several related point-estimation problems from the standpoint of Chapter 17, under the assumptions that the decision-maker's utility is linear in money and that the set E of available experiments is the set of random samples of size n from dgf $f(x \mid \theta)$. Thus each e may be identified with its n, and hence we write $E = \{0, 1, 2, \ldots\}$. However, for the Poisson process we let $E = [0, \infty)$, where $e = t$ is the experiment which consists in observing the Poisson process for a time span t.

From Chapter 17 it is evident that since $k(e, \tilde{z}) \equiv k(n, \tilde{z})$ is independent of a and θ, it follows that the optimal estimate of $\tilde{\theta}$ depends solely upon the terminal loss function $\ell_t(a, \theta)$ and the decision-maker's posterior distribution $f(\theta \mid z, n)$ of $\tilde{\theta}$ given (n, z), where z is a sufficient statistic of $x_1, \ldots, x_n$.

In Section 18.2 we characterize the optimal point-estimates under a number of distribution assumptions from Chapter 14, for the case in which terminal loss is given by

$$\ell_t(a, \theta) = k_t(\theta - a)^2, \text{ all } a \in A, \theta \in \Theta, \tag{18.1.1}$$

for some $k_t > 0$. The loss function (18.1.1) is referred to as a *quadratic loss function.* Its appropriateness in estimation problems is often taken for granted, which is dangerous. Nevertheless, quadratic loss functions are often good approximations to the true loss function, and they have the additional advantage of being very tractable mathematically.

With a quadratic loss function and experimentation cost function of the form

$$k(n) \equiv k(n, \tilde{z}) \equiv \begin{cases} 0, & n = 0 \\ K_s + k_s n, & n > 0, \end{cases} \tag{18.1.2}$$

for $k_s > 0$ and $K_s > 0$, or (for the Poisson process)

$$k(t) \equiv k(t, \tilde{z}) \equiv \begin{cases} 0, & t = 0 \\ K_s + k_s t, & t > 0, \end{cases} \tag{18.1.3}$$

we shall be able to determine in Section 18.2 not only the optimal point estimate $a^0_{z,n}$ of $\tilde{\theta}$ given the sufficient statistic z of the n observations, but also the optimal sample size n^0. This will be possible in all cases considered because we shall be able to obtain a closed-form expression for $E_{z|n}E_{\theta|z,n}$ $\ell_t(a^0_{\tilde{z},n}, \tilde{\theta})$, and hence also a closed-form expression for

$$G(n) = E_\theta \ell_t(a^0_{z_0,0}, \tilde{\theta}) - E_{z|n}E_{\theta|z,n}\ell_t(a^0_{\tilde{z},n}, \tilde{\theta}) - k(n), \tag{18.1.4}$$

which can then be maximized with respect to n.

The situation is not so good in Section 18.3, where the sampling-cost assumptions (18.1.2) and (18.1.3) continue to apply but where $\ell_t(a, \theta)$ is now defined by

$$\ell_t(a, \theta) = \begin{cases} k_u(\theta - a), & a \leqq \theta \\ k_O(a - \theta), & a \geqq \theta, \end{cases} \tag{18.1.5}$$

where $k_u > 0$ and $k_O > 0$. This so-called *linear loss function* arises in simple inventory problems, in which the decision-maker foregoes a profit of k_u per unit of any unsatisfied demand $\theta - a$ and suffers a direct loss of k_O per unit of any overstockage $a - \theta$. The only distribution assumption from Section 18.2 that leads to explicit derivation of n is the one in which $\tilde{\theta} = \tilde{\mu}$, the mean of a Normal process with known variance σ^2.

Section 18.4 concerns a multivariate generalization of point estimation with quadratic loss. Here, we assume $\mathbf{A} = \Theta \subset R^k$, and that terminal opportunity loss is defined by

$$\ell_t(\mathbf{a}, \boldsymbol{\theta}) = (\boldsymbol{\theta} - \mathbf{a})'\mathbf{k}_t(\boldsymbol{\theta} - \mathbf{a}), \tag{18.1.6}$$

where $\mathbf{k}_t$ is a positive-definite matrix. It will be apparent that the theory here parallels quite closely the results of Section 18.2.

18.2 UNIVARIATE POINT ESTIMATION WITH QUADRATIC LOSS

18.2.1 General Analysis

Figure 18.1 clarifies the nature of the quadratic loss function (18.1.1) by graphing $\ell_t(a, \cdot)$ and $\ell_t(a', \cdot)$ for two arbitrary but fixed acts $a, a' \in A$. Clearly, $l_t(a, \theta) = 0$ if and only if $a = \theta$; that is, being wrong does result in some opportunity loss. However, small errors $|\theta - a|$ are not too serious in relationship to larger ones; loss increases faster than the error. As a rough approximation, these qualitative properties are reasonably consonant with many real environments, in which censure increases more than in proportion to the error.

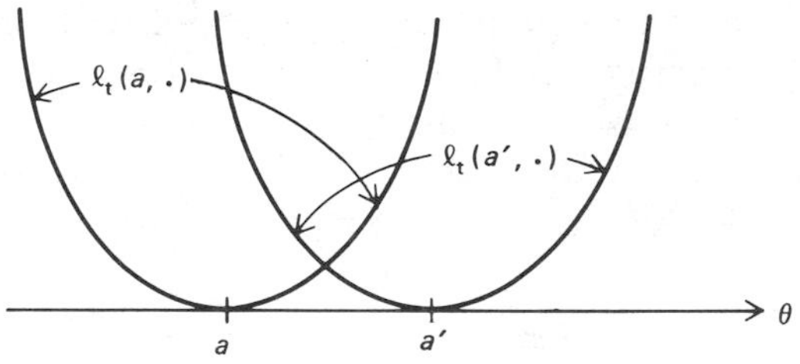

Figure 18.1

Let $f(\theta)$ be any dgf of $\tilde{\theta}$-prior, or posterior without the "$|z, n$" notation, and let E denote expectation with respect to $f(\theta)$. We shall now show that if the decision-maker's judgments about $\tilde{\theta}$ are expressed by $f(\theta)$, then his optimal point estimate a^0 of θ is $E(\tilde{\theta})$. To see this, we write

$$\begin{aligned} E(\tilde{\theta} - a)^2 &= E([\tilde{\theta} - E(\tilde{\theta})] + [E(\tilde{\theta}) - a])^2 \\ &= E([\tilde{\theta} - E(\tilde{\theta})]^2) + 2E([\tilde{\theta} - E(\tilde{\theta})][E(\tilde{\theta}) - a]) + E([E(\tilde{\theta}) - a]^2). \end{aligned}$$

Now,

$$E([\tilde{\theta} - E(\tilde{\theta})]^2) \equiv V(\tilde{\theta});$$

$$E([E(\tilde{\theta}) - a]^2) = [E(\tilde{\theta}) - a]^2$$

because $[E(\tilde{\theta}) - a]^2$ is constant for each a, and, because $[E(\tilde{\theta}) - a]$ is constant,

$$\begin{aligned} &2E([\tilde{\theta} - E(\tilde{\theta})][E(\tilde{\theta}) - a]) \\ &\qquad = 2[E(\tilde{\theta}) - a][E([\tilde{\theta} - E(\tilde{\theta})])] \\ &\qquad = 2[E(\tilde{\theta}) - a][E[E(\tilde{\theta})] - E(\tilde{\theta})] \\ &\qquad = 2[E(\tilde{\theta}) - a][E(\tilde{\theta}) - E(\tilde{\theta})] \\ &\qquad = 0. \end{aligned}$$

Hence we obtain

$$E\ell_t(a, \tilde{\theta}) = k_t E(\tilde{\theta} - a)^2$$
$$= k_t(V(\tilde{\theta}) + [E(\tilde{\theta}) - a]^2).$$

Now, $[E(\tilde{\theta}) - a]^2 \geqq 0$ and equals zero if and only if $a = E(\tilde{\theta})$. Hence $E\ell_t(a, \tilde{\theta})$ is minimized by $a^0 = E(\tilde{\theta})$, and $E\ell_t(a^0, \tilde{\theta}) = k_t V(\tilde{\theta})$.

Thus the decision-maker's optimal point estimate of $\tilde{\theta}$ is its expectation with respect to the dgf that currently expresses his judgments about $\tilde{\theta}$, be that prior or posterior.

Hence when (18.1.1) gives $\ell_t(a, \theta)$, we have

$$a^0_{z_0,0} = E(\tilde{\theta}); \tag{18.2.1}$$

$$a^0_{z,n} = E(\tilde{\theta} \mid z, n); \tag{18.2.2}$$

$$E_\theta \ell_t(a^0_{z_0,0}, \tilde{\theta}) = k_t V(\tilde{\theta}); \text{ and} \tag{18.2.3}$$

$$E_{\theta|z,n} \ell_t(a^0_{z,n}, \tilde{\theta}) = k_t V(\tilde{\theta} \mid z, n). \tag{18.2.4}$$

From these we easily obtain

$$G(n) = k_t[V(\tilde{\theta}) - E_{z|n} V(\tilde{\theta} \mid \tilde{z}, n)] - k(n), \tag{18.2.5}$$

where $k(n)$ is given by (18.1.2).

Little more can be said in the general case. In Subsection 18.2.2 we consider the Bernoulli process with a beta conjugate prior df and binomial (fixed-n) sampling. Subsection 18.2.3 concerns the Poisson process with a gamma conjugate prior df and Poisson (fixed-t) observation. Subsection 18.2.4 concerns $f(x \mid \theta) = f_N(x \mid \mu, \sigma^2)$ with σ^2 known and $\tilde{\theta} \equiv \tilde{\mu}$ having Normal prior df; and Subsection 18.2.5 generalizes this to the case of σ^2 also unknown and (μ, σ^2) having the Normal-inverted gamma prior df. The two corresponding cases of Normal sampling with $\tilde{\theta} = \tilde{\sigma}^2$ are discussed in Raiffa and Schlaifer [**66**], Chapter 6.

18.2.2 Bernoulli Process

In this subsection we assume that $\tilde{\theta} \equiv \tilde{p}$, the parameter of a Bernoulli process; the decision-maker's prior df of $\tilde{p}$ is $f_\beta(p \mid \rho, \nu)$. Then an experiment consisting of (predetermined) n observations has sufficient statistic $(\tilde{r}, n)$, with $\tilde{r}$ having the binomial mf $f_b(r \mid n, p)$ given p. Hence

$$a^0_{z_0,0} = \frac{\rho}{\nu} \quad \text{[by (18.2.1) and (12.3.6)]; and} \tag{18.2.6}$$

$$a^0_{r,n} = \frac{\rho + r}{\nu + n} \quad \text{[by (18.2.2) and (12.3.6)]} \tag{18.2.7}$$

because the posterior df of $\tilde{p}$ is $f_\beta(p \mid \rho + r, \nu + n)$ [by (14.2.3)];

$$E_p\ell_t(a^0_{z0,0}, \tilde{p}) = k_t V(\tilde{p}) \quad \text{[by (18.2.3)]}$$
$$= k_t V_\beta(\tilde{p} \mid \rho, \nu)$$
$$= \frac{k_t\rho(\nu - \rho)}{\nu^2(\nu + 1)} \quad \text{[by (12.3.7)]}; \tag{18.2.8}$$

and

$$E_{p|r,n}\ell_t(a^0_{r,n}, \tilde{p}) = k_t V_\beta(\tilde{p} \mid \rho + r, \nu + n)$$
$$= k_t \frac{(\rho + r)(\nu + n - \rho - r)}{(\nu + n)^2(\nu + n + 1)}, \tag{18.2.9}$$

by (12.3.7). Now, the marginal dgf of $\tilde{r}$ is the hyper-binomial $f_{hb}(r \mid \rho, \nu, n)$, by (14.2.4), and hence

$$E_{r|n}k_t V_\beta(\tilde{p} \mid \rho + \tilde{r}, \nu + n) = k_t E_{hb}\left\{\frac{(\rho + \tilde{r})(\nu + n - \rho - \tilde{r})}{(\nu + n)^2(\nu + n + 1)}\middle| \rho, \nu, n\right\},$$
$$= \frac{\nu}{\nu + n}\frac{k_t\rho(\nu - \rho)}{\nu^2(\nu + 1)}, \tag{18.2.10}$$
$$= \frac{\nu}{\nu + n} k_t V_\beta(\tilde{p} \mid \rho, \nu),$$

the second equality following from the first via (14.2.5), (14.2.6), (10.3.11), and elementary algebra (Exercise 1).

Hence (18.2.5), (18.2.8), and (18.2.10) imply

$$G(n) = I(n) - k(n)$$
$$= \frac{n}{\nu + n} k_t V_\beta(\tilde{p} \mid \rho, \nu) - k(n). \tag{18.2.11}$$

In Figures 18.2 through 18.5 we graph $I(n)$, $k(n)$, and $G(n)$ for typical values of K_s, under the assumption that n can be any real number rather than be restricted to integers. The purpose of these figures is to suggest that the discontinuity of $G(n)$ at $n = 0$ will lead to a slight annoyance (not really a difficulty) in characterizing n^0.

These figures suggest—and a rigorous proof establishes (Exercise 2)—that n^0 can be found by first finding the maximizer n^* of $G(n)$ for $n \in R^1$, then computing $G(n^*)$, and finally setting $n^0 = n^*$ if $n^* \geqq 0$ *and* $G(n^*) > 0$; otherwise, $n^0 = 0$.

If we continue to assume that n can be any real number, we obtain

$$n^* = \left[\frac{k_t\nu V_\beta(\tilde{p} \mid \rho, \nu)}{k_s}\right]^{1/2} - \nu \tag{18.2.12}$$

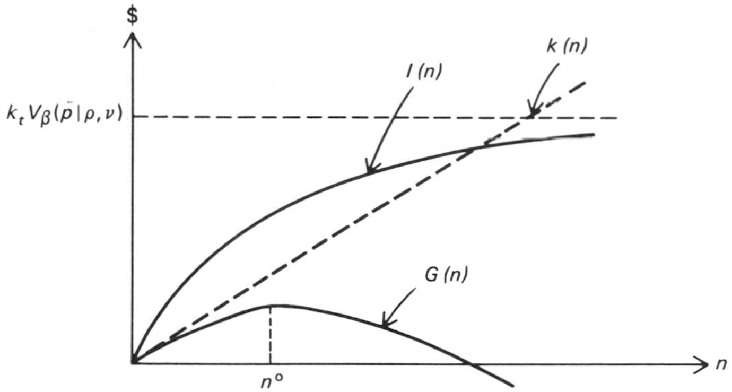

Figure 18.2 $K_s = 0, n^0 > 0$

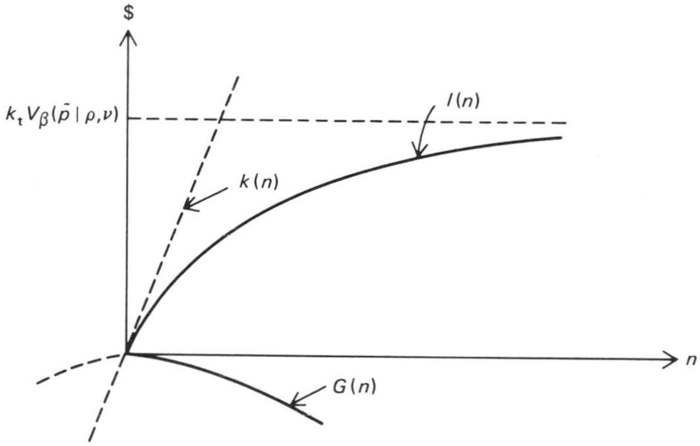

Figure 18.3 $K_s = 0, n^0 = 0$

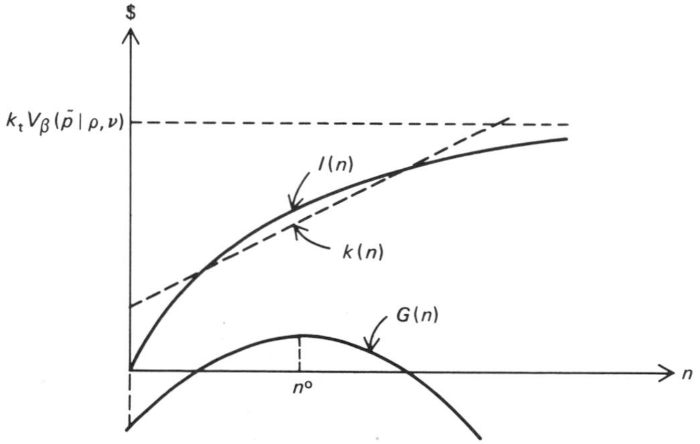

Figure 18.4 $K_s > 0, n^0 > 0$

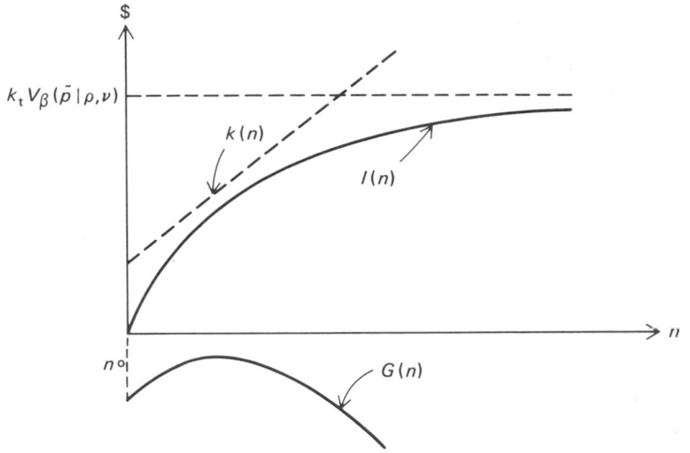

Figure 18.5 $K_s > 0$, $n^0 = 0$

from (18.2.11), (18.1.2), and elementary calculus (Exercise 2). Hence

$$n^0 = \begin{cases} n^*, & n^* \geqq 0 \text{ and } G(n^*) \geqq 0 \\ 0, & \text{otherwise.} \end{cases} \tag{18.2.13}$$

If n^0 is not an integer, there is usually a negligible reduction in net gain if it is rounded to the nearest integer. However, you may show as Exercise 3 that the integer maximizer n^* of $G(n)$, for $n \in \{1, 2, \ldots\}$, must satisfy the necessary (and sufficient) condition

$$(\nu + n^*)(\nu + n^* - 1) \leqq \frac{k_t \nu V_\beta(\tilde{p} \mid \rho, \nu)}{k_s} \leqq (\nu + n^*)(\nu + n^* + 1). \tag{18.2.14}$$

Equation (18.2.13) is just as applicable when n^* is obtained via (18.2.14) as it is when (18.2.12) is used.

We have thus shown how to determine the optimal sample size n^0; that is, the optimal number of observations to take on a Bernoulli process with a conjugate prior df of $\tilde{p}$, quadratic loss, and linear observation cost.

The qualitative dependence of n^* and hence (in a lesser sense), also of n^0, upon the costs and prior parameters can be seen by examining (18.2.12). First, n^* is an increasing function of k_t: this is reasonable, because with a large k_t it is clear that an error of given size is more serious than that same error with a small k_t; hence more information is desirable, given that other parameters do not change. Equivalently, increasing k_t and holding k_s fixed implies that sampling cost *decreases* relative to terminal opportunity loss. Second, n^* is a decreasing function of k_s: the lower the cost per observation, the more observations it is economical to take. Finally, n^* is an increasing linear function of the prior standard deviation $V_\beta^{1/2}(\tilde{p} \mid \rho, \nu)$

of $\tilde{p}$: the greater the decision-maker's uncertainty about $\tilde{p}$, the more observations he should take.

Note that these qualitative phenomena are quite in accordance with what our intuition would suggest.

Example 18.2.1

An opinion-polling firm is required to estimate the proportion p of voters supporting Candidate X. The firm's president judges that if the estimate is off by .10, then it will suffer a loss of good will, and so forth, of \$10,000. Moreover, he agrees that the quadratic loss function pertains. Hence we solve for k_t from:

$$10{,}000 = k_t(.10)^2,$$

which implies $k_t = 1{,}000{,}000$. His prior df of $\tilde{p}$ is $f_\beta(p \mid 4, 9)$. It is also known that $K_s = 500$ and $k_s = 12.50$. We compute

$$V_\beta(\tilde{p} \mid 4, 9) = \frac{4(9-4)}{(9^2)(10)} = \frac{20}{810};$$

$$\left[\frac{k_t \nu V_\beta(\tilde{p} \mid 4, 9)}{k_s}\right]^{1/2} = \left\{\frac{(10^6)(9)(20)}{[(810)(12.50)]}\right\}^{1/2}$$

$$= \frac{2000}{15}.$$

Hence

$$n^* = \frac{2000}{15} - 9 = \frac{1865}{15}$$

$$\doteq 124,$$

by (18.2.12). Now, $n^* > 0$ but we must find $G(n^*)$ to see if it is positive and hence if $n^* = n^0$. Clearly, $k(n^*) = 500 + (12.50)(124) = 2050$, while

$$I(n^*) = \left(\frac{124}{124 + 9}\right)(10^6)V_\beta(\tilde{p} \mid 4, 9)$$

$$\doteq 23{,}000.$$

Hence $G(124) \doteq 23{,}000 - 2{,}050 > 0$ and $n^* = n^0 = 124$. The reader may verify that $n^* = 124$ satisfies (18.2.14) and hence rounding produced a precisely optimal integer solution.

18.2.3 Poisson Process

In this subsection we assume that $\tilde{\theta} = \tilde{\lambda}$, the intensity parameter of a Poisson process, and that the decision-maker's prior df of $\tilde{\lambda}$ is $f_\gamma(\lambda \mid \rho, \tau)$.

If the experiment consists of observing the process for t time units, then $(\tilde{r}, t)$ is a sufficient statistic and $\tilde{r}$ has mf $f_{Po}(r \mid \lambda t)$ given λ. Hence

$$a^0_{z0,0} = \frac{\rho}{\tau} \qquad \text{[by (18.2.1) and Table 12.1];} \tag{18.2.15}$$

$$a^0_{r,t} = \frac{\rho + r}{\tau + t} \qquad \text{[by (18.2.2) and Table 12.1]} \tag{18.2.16}$$

because the posterior df of $\tilde{\lambda}$ is $f_\gamma(\lambda \mid \rho + r, \tau + t)$ [by (14.4.2)];

$$\begin{aligned} E_\lambda \ell_t(a^0_{z0,0}, \tilde{\lambda}) &= k_t V(\tilde{\lambda}) \qquad \text{[by (18.2.3)]} \\ &= k_t V_\gamma(\tilde{\lambda} \mid \rho, \tau) \\ &= \frac{k_t \rho}{\tau^2} \qquad \text{(by Table 12.1),} \end{aligned} \tag{18.2.17}$$

and

$$\begin{aligned} E_{\lambda \mid r,t} \ell_t(a^0_{r,t}, \tilde{\lambda}) &= k_t V_\gamma(\tilde{\lambda} \mid \rho + r, \tau + t) \\ &= \frac{k_t(\rho + r)}{(\tau + t)^2}. \end{aligned} \tag{18.2.18}$$

Now, the marginal mf of $\tilde{r}$ is the negative binomial $f_{nb}(r \mid \rho, \tau/[\tau + t])$ by (14.4.3), and hence

$$\begin{aligned} E_{r \mid t} k_t V_\gamma(\tilde{\lambda} \mid \rho, + \tilde{r}, \tau + t) &= k_t E_{nb}\left\{ \frac{\rho + r}{(\tau + t)^2} \middle| \rho, \frac{\tau}{\tau + t} \right\} \\ &= \frac{k_t \rho}{\tau(\tau + t)}, \end{aligned} \tag{18.2.19}$$

by (14.4.4) and using a little algebra. From (18.2.17) and (18.2.19), we obtain

$$\begin{aligned} I(t) &= \frac{k_t \rho}{\tau^2} - \frac{k_t \rho}{\tau(\tau + t)} \\ &= \frac{k_t \rho t}{\tau^2(\tau + t)} \\ &= \left(\frac{t}{\tau + t}\right) \frac{k_t \rho}{\tau^2} \\ &= \left(\frac{t}{\tau + t}\right) k_t V_\gamma(\tilde{\lambda} \mid \rho, \tau), \end{aligned} \tag{18.2.20}$$

which is of just the same form as $I(n)$ for the Bernoulli case. Hence

$$G(t) = \left(\frac{t}{\tau + t}\right) k_t V_\gamma(\tilde{\lambda} \mid \rho, \tau) - k(t), \tag{18.2.21}$$

where $k(t)$ is given by (18.1.3).

It is clear that t^0 can be derived from (18.2.21) just as n^0 was derived from (18.2.11): first compute

$$t^* = \left[\frac{k_t \tau V_\gamma(\tilde{\lambda} \mid \rho, \tau)}{k_s}\right]^{1/2} - \tau. \tag{18.2.22}$$

Then

$$t^0 = \begin{cases} t^*, & t^* \geqq 0 \text{ and } G(t^*) \geqq 0 \\ 0, & \text{otherwise.} \end{cases} \tag{18.2.23}$$

The reader will note that t^* as given by (18.2.22) is not an approximation, as was n^* in (18.2.12), because (in principle) the process may be observed for any nonnegative real time span.

Example 18.2.2

An engineering firm wishes to estimate the intensity λ of the Poisson process describing the arrival of cars at a bridge. They have determined that quadratic loss is appropriate and that $k_t = 10$. Moreover, experience with similar traffic suggests the prior df $f_\gamma(\lambda \mid 100, 2)$. There is a fixed cost $K_s = 100$ of sending an observer to the bridge, and the observer's salary plus expenses is $k_s = 5$ per time unit (say, per hour). We obtain

$$V_\gamma(\tilde{\lambda} \mid 100, 2) = \frac{100}{2^2} = 25,$$

and hence, from (18.2.22),

$$t^* = \left[\frac{(10)(2)(25)}{5}\right]^{1/2} - 2$$
$$= 8 \text{ (hours)}.$$

Now,

$$I(8) - \left(\frac{8}{8+2}\right)(10)(25) = 200,$$

while

$$k(8) = 100 + (5)(8) = 140,$$

so that

$$G(8) = 200 - 140 = 60 > 0.$$

Hence $t^0 = 8$.

18.2.4 The Univariate Normal Process: σ^2 known

In this subsection we assume that $\tilde{\theta} = \mu$, the mean of a univariate Normal process with known variance σ^2, and that the decision-maker has assessed a natural conjugate prior df $f_N(\mu \mid \mu_0, \sigma^2/n_0)$ for $\tilde{\mu}$. If an experiment consists of n independent observations $\tilde{x}_i$ with common df $f_N(x_i \mid \mu, \sigma^2)$, then $(n, \tilde{\bar{x}})$ is a sufficient statistic for $\tilde{\mu}$, and we have

$$a^0_{z_0,0} = \mu_0 \quad \text{[by (18.2.1) and (12.4.8)];} \tag{18.2.24}$$

$$a^0_{\bar{x},n} = \frac{n_0\mu_0 + n\bar{x}}{n_0 + n} \quad \text{[by (18.2.2) and (12.4.8)]} \tag{18.2.25}$$

because the posterior df of $\tilde{\mu}$ is

$$f_N\left(\mu \,\middle|\, \frac{n_0\mu_0 + n\bar{x}}{n_0 + n}, \frac{\sigma^2}{n_0 + n}\right) \quad \text{[by (14.5.1)];}$$

$$\begin{aligned} E_\mu \ell_t(a^0_{z_0,0}, \tilde{\mu}) &= k_t V_N\left(\tilde{\mu} \mid \mu_0, \frac{\sigma^2}{n_0}\right) \quad \text{[by (18.2.3)]} \\ &= \frac{k_t\sigma^2}{n_0}; \end{aligned} \tag{18.2.26}$$

and

$$\begin{aligned} E_{\mu|\bar{x},n} \ell_t(a^0_{\bar{x},n}, \tilde{\mu}) &= k_t V_N\left(\tilde{\mu} \,\middle|\, \frac{n_0\mu_0 + n\bar{x}}{n_0 + n}, \frac{\sigma^2}{n_0 + n}\right) \\ &= \frac{k_t\sigma^2}{n_0 + n}, \end{aligned} \tag{18.2.27}$$

which does not depend upon $\tilde{\bar{x}}$, and hence

$$\begin{aligned} E_{\bar{x}|n} k_t V_N\left(\tilde{\mu} \,\middle|\, \frac{n_0\mu_0 + n\tilde{\bar{x}}}{n_0 + n}, \frac{\sigma^2}{n_0 + n}\right) &= \frac{k_t\sigma^2}{n_0 + n} \\ &= \frac{n_0}{n_0 + n}\,\frac{k_t\sigma^2}{n_0} \\ &= \frac{n_0}{n_0 + n}\, k_t V_N\left(\tilde{\mu} \mid \mu_0, \frac{\sigma^2}{n_0}\right). \end{aligned} \tag{18.2.28}$$

From (18.2.26) and (18.2.28) it is clear that $I(n)$ has a familiar form:

$$I(n) = \left(\frac{n}{n_0 + n}\right) k_t V_N\left(\tilde{\mu} \mid \mu_0, \frac{\sigma^2}{n_0}\right). \tag{18.2.29}$$

Hence

$$G(n) = \left(\frac{n}{n_0 + n}\right) k_t V_N\left(\tilde{\mu} \mid \mu_0, \frac{\sigma^2}{n_0}\right) - k(n); \tag{18.2.30}$$

$$\begin{aligned} n^* &= \left[\frac{k_t n_0 V_N(\tilde{\mu} \mid \mu_0, \sigma^2/n_0)}{k_s}\right]^{1/2} - n_0 \\ &= \left[\frac{k_t\sigma^2}{k_s}\right]^{1/2} - n_0; \end{aligned} \tag{18.2.31}$$

and

$$n^0 = \begin{cases} n^*, & n^* \geqq 0 \text{ and } G(n^*) \geqq 0 \\ 0, & \text{otherwise.} \end{cases} \tag{18.2.32}$$

Example 18.2.3

A government agency wishes to estimate the average income μ per family in a certain, depressed region. It is known that individual family incomes are Normally distributed around μ with $\sigma^2 = 1{,}000{,}000$. Loss is quadratic with $k_t = 50$. A prior df of $\tilde{\mu}$ is assessed to be Normal with $\mu_0 = 2500$ and $\sigma_0{}^2 = 250{,}000$. Hence $n_0 = \sigma^2/\sigma_0{}^2 = 4$. In order to take a sample of family incomes in this region, fixed costs of $K_s = 25{,}000$ would be incurred, and in addition each observation would cost 50, representing travel time and cost, interviewer salary, and so forth. What is n^0? From (18.2.31) we obtain

$$n^* = \left[\frac{(50)(1{,}000{,}000)}{50}\right]^{1/2} - 4 = 996.$$

Now,

$$I(n^*) = \left(\frac{996}{1000}\right)(50)(250{,}000) \doteq 12{,}500{,}000,$$

while

$$k(n^*) = 25{,}000 + (50)(996) \doteq 75{,}000.$$

Hence $n^0 = 996$.

18.2.5 The Univariate Normal Process: σ^2 unknown

This subsection is the natural generalization of Subsection 18.2.4, to the case in which σ^2 is unknown but does not affect the decision-maker's terminal loss. We assume here that the decision-maker has assessed a Normal-inverted gamma prior df $f_{Ni\gamma}(\mu, \sigma^2 \mid \mu_0, \psi_0, n_0, \nu_0)$ for $(\tilde{\mu}, \tilde{\sigma}^2)$, as defined by (14.5.5). By (14.5.6), his marginal prior df of $\tilde{\mu}$ is

$$f_S(\mu \mid \mu_0, \psi_0/n_0, \nu_0),$$

and hence

$$a^0_{z_0,0} = \mu_0 \quad \text{[by (18.2.1) and (12.5.10)];} \qquad (18.2.33)$$

and

$$E_\mu \ell_t(a^0_{z_0,0}, \tilde{\mu}) = k_t V_S(\tilde{\mu} \mid \mu_0, \psi_0/n_0, \nu_0)$$

$$= k_t\left(\frac{\nu_0}{\nu_0 - 2}\right)\psi_0/n_0, \qquad (18.2.34)$$

by (18.2.3) and (12.5.11), provided that $\nu_0 > 2$.

From (13.2.6), $(n, \bar{x}, s^2)$ is sufficient for $(\tilde{\mu}, \tilde{\sigma}^2)$, where $\tilde{\bar{x}} \equiv n^{-1}\sum_{i=1}^{n} \tilde{x}_i$ of course, and $\tilde{s}^2 \equiv (n-1)^{-1}\sum_{i=1}^{n}(\tilde{x}_i - \tilde{\bar{x}})^2$. By (14.5.7), the marginal posterior df of $\tilde{\mu}$ is $f_S(\mu \mid \mu_1, \psi_1/n_1, \nu_1)$, where n_1, ν_1, μ_1, and ψ_1 are defined in (14.5.7). Hence

$$a^0_{\bar{x},s^2,n} = \mu_1 = \frac{n_0\mu_0 + n\bar{x}}{n_0 + n} \tag{18.2.35}$$

by (18.2.2) and (12.5.10), and is thus independent of s^2. From (18.2.4), (14.5.7), and (12.5.11), we obtain

$$\begin{aligned} E_{\mu\mid\bar{x},s^2,n}\ell_t(a^0_{\bar{x},s^2,n}, \tilde{\mu}) &= k_t V(\tilde{\mu} \mid \bar{x}, s^2, n) \\ &= k_t V_S\left(\tilde{\mu} \mid \mu_1, \frac{\psi_1}{n_1}, \nu_1\right) \\ &= k_t\left(\frac{\nu_1}{\nu_1 - 2}\right)\frac{\psi_1}{n_1}. \end{aligned} \tag{18.2.36}$$

Now, (14.5.9) shows that ψ_1 depends upon both $\bar{x}$ and s^2, and hence is random before the n observations are taken. Hence

$$\begin{aligned} E_{\bar{x},s^2\mid n}E_{\mu\mid\bar{x},s^2,n}\ell_t(a^0_{\bar{\tilde{x}},\tilde{s}^2,n}, \tilde{\mu}) \\ = k_t\left(\frac{\nu_1}{\nu_1 - 2}\right)\frac{E(\tilde{\psi}_1)}{n_1}. \end{aligned} \tag{18.2.37}$$

Now, by using (14.5.7), (14.5.8), (12.3.16), (12.5.11), and a some algebra, we may evaluate $E(\tilde{\psi}_1)$, simplify, and subtract (18.2.37) from (18.2.34) to obtain

$$I(n) = \left(\frac{n}{n_0 + n}\right)k_t V_S\left(\tilde{\mu} \mid \mu_0, \frac{\psi_0}{n_0}, \nu_0\right), \tag{18.2.38}$$

which is of the same form as $I(n)$ for the three previous cases. Hence

$$G(n) = \left(\frac{n}{n_0 + n}\right)k_t V_S\left(\tilde{\mu} \mid \mu_0, \frac{\psi_0}{n_0}, \nu_0\right) - k(n); \tag{18.2.39}$$

$$n^* = \left[\frac{k_t n_0 V_S(\tilde{\mu} \mid \mu_0, \psi_0/n_0, \nu_0)}{k_s}\right]^{1/2} - n_0; \tag{18.2.40}$$

and

$$n^0 = \begin{cases} n^*; & n^* \geqq 0 \text{ and } G(n^*) \geqq 0 \\ 0; & \text{otherwise.} \end{cases} \tag{18.2.41}$$

18.3 UNIVARIATE POINT ESTIMATION WITH LINEAR LOSS

18.3.1 General Analysis

Figure 18.6 graphs the linear loss function defined by (18.1.5) for two possible acts $a, a' \in A$. Such loss functions arise from a simple inventory problem in which $\theta \equiv$ demand, $a \equiv$ inventory level, retail price $\equiv p$,

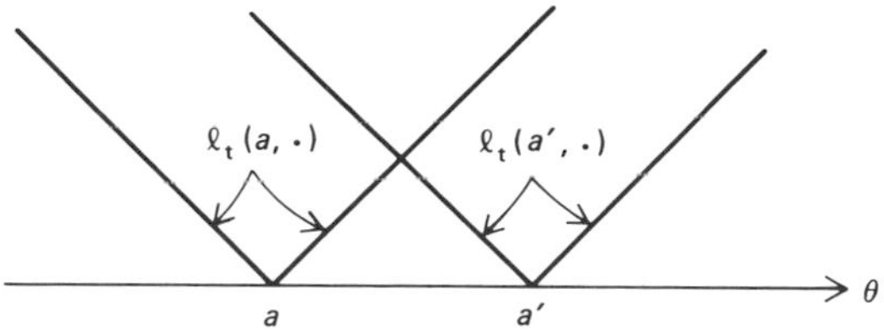

Figure 18.6

wholesale price $\equiv c$, and scrap value $\equiv r$. Assume that $r < c < p$. Assuming that any unsold merchandise will be scrapped, how much ($= a$) should the retailer stock?

It is clear that the retailer foregoes $p - c$ profit for every unit demanded that he cannot supply, and hence he suffers an opportunity loss of $(p - c)(\theta - a) \equiv k_u(\theta - a)$ if $a \leqq \theta$. On the other hand, he loses $c - r$ for every unit left over, and thus suffers an opportunity loss of $(c - r)(a - \theta) \equiv k_O(a - \theta)$ if $a \geqq \theta$. Hence (18.1.5).

Assume now that $\tilde{\theta}$ is a continuous random variable with a df $f(\tilde{\theta})$ and that $A = \Theta = R^1$. Then (18.1.5) implies

$$E_\theta \ell_t(a, \tilde{\theta}) = k_O \int_{-\infty}^{a} (a - \theta) f(\theta) d\theta + k_u \int_{a}^{\infty} (\theta - a) f(\theta) d\theta, \qquad (18.3.1)$$

for every $a \in R^1$. To find the optimal act $a \in R^1$, we differentiate (18.3.1) and equate the result to zero, obtaining

$$k_O(a^0 - a^0) f(a^0) + k_O \int_{-\infty}^{a^0} f(\theta) d\theta - k_u(a^0 - a^0) f(a^0) - k_u \int_{a^0}^{\infty} f(\theta) d\theta = 0,$$

which implies

$$0 = k_O F(a^0) - k_u[1 - F(a^0)],$$

or

$$F(a^0) = \frac{k_u}{k_u + k_O}. \qquad (18.3.2)$$

Thus *an optimal estimate is a* $k_u/(k_u + k_O)$th *fractile of* $\tilde{\theta}$. We shall henceforth assume that the $k_u/(k_u + k_O)$th fractile of $\tilde{\theta}$ is unique.

Note the dependence of a^0 upon k_u and k_O. As k_u increases, so must a^0, because an increase in k_u reflects relatively greater penalization for running out of stock. Conversely, a^0 must decrease as k_O increases, as an increase in k_O reflects a relatively greater loss on leftovers.

Now, by substituting $F(a^0) = k_u/(k_u + k_O)$ into (18.3.1), we obtain

$$E_\theta \ell_t(a^0, \tilde{\theta}) = k_O a^0 F(a^0) = k_O \int_{-\infty}^{a^0} \theta f(\theta) d\theta$$

$$+ k_u \int_{a^0}^{\infty} \theta f(\theta) d\theta - k_u a^0[1 - F(a^0)]$$

$$= k_u \int_{a^0}^{\infty} \theta f(\theta) d\theta - k_O \int_{-\infty}^{a^0} \theta f(\theta) d\theta$$

$$+ \left[k_O a^0 \left(\frac{k_u}{k_u + k_O} \right) - k_u a^0 \left(\frac{k_O}{k_u + k_O} \right) \right],$$

and the terms in brackets cancel each other. Hence

$$E_\theta \ell_t(a^0, \tilde{\theta}) = k_u \int_{a^0}^{\infty} \theta f(\theta) d\theta - k_O \int_{-\infty}^{a^0} \theta f(\theta) d\theta. \tag{18.3.3}$$

By adding and subtracting $k_u \int_{-\infty}^{a^0} \theta f(\theta) d\theta$ in (18.3.3), we obtain

$$E_\theta \ell_t(a^0, \tilde{\theta}) = \left[k_u \int_{a^0}^{\infty} \theta f(\theta) d\theta + k_u \int_{-\infty}^{a^0} \theta f(\theta) d\theta \right]$$

$$- \left[k_o \int_{-\infty}^{a^0} \theta f(\theta) d\theta + k_u \int_{-\infty}^{a^0} \theta f(\theta) d\theta \right] \tag{18.3.4}$$

$$= k_u E(\tilde{\theta}) - (k_u + k_o) \int_{-\infty}^{a^0} \theta f(\theta) d\theta.$$

Now (18.3.2) through (18.3.4) obtain for any df $f(\theta)$ expressing the decision-maker's judgments, whether prior or posterior. Hence the appropriate analogues of (18.2.1) through (18.2.5) are:

$$a^0_{z_0,0} \text{ satisfies } \int_{-\infty}^{a_{z0,0}{}^0} f(\theta) d\theta = \frac{k_u}{k_u + k_O}, \tag{18.3.5}$$

$$a^0_{z,n} \text{ satisfies } \int_{-\infty}^{a_{z,n}{}^0} f(\theta \mid z, n) d\theta = \frac{k_u}{k_u + k_O}, \tag{18.3.6}$$

$$E_\theta \ell_t(a^0_{z_0,0}, \tilde{\theta}) = k_u E(\tilde{\theta}) - (k_O + k_u) \int_{-\infty}^{a_{z0,0}{}^0} \theta f(\theta) d\theta, \tag{18.3.7}$$

$$E_{\theta \mid z,n} \ell_t(a^0_{z,n}, \tilde{\theta}) = k_u E(\tilde{\theta} \mid z, n) - (k_u + k_O) \int_{-\infty}^{a_{z,n}{}^0} \theta f(\theta \mid z, n) d\theta, \text{ and} \tag{18.3.8}$$

$$G(n) = (k_u + k_O) \left\{ E_{z|n} \left[\int_{-\infty}^{a_{\tilde{z},n}{}^0} \theta f(\theta \mid \tilde{z}, n) d\theta \right] - \int_{-\infty}^{a_{z0,0}{}^0} \theta f(\theta) d\theta \right\} - k(n). \tag{18.3.9}$$

Equations (18.3.5) through (18.3.8) are immediate from (18.3.2) and (18.3.4), while (18.3.9) is left as Exercise 4.

It would be well to continue to parallel with Section 18.2 at this point by saying that we would derive n^0 from a specialization of (18.3.9) for each case treated in Section 18.2. Unfortunately, there seems to be only one case, the Normal process with σ^2 known, which is sufficiently tractable. We consider the Normal process with known σ^2 in Subsection 18.2.2, in which n^0 is characterized in a fashion similar to Section 18.2. When σ^2 is unknown, we derive in Subsection 18.2.3 a formula for $G(n)$ which cannot be simply differentiated and equated to zero to yield n^* and thus n^0.

18.3.2 The Univariate Normal Process: σ^2 known

If $\tilde{\theta} \equiv \tilde{\mu}$, the mean of a univariate Normal process whose variance σ^2 is known, and if $\tilde{\mu}$ has the conjugate prior df $f_N(\mu \mid \mu_0, \sigma^2/n_0)$, then

$$E_\mu \ell_t(a^0_{z_0,0}, \tilde{\mu}) = n_0^{-1/2}(k_u + k_0)\sigma f_{N^*}(\alpha^0), \tag{18.3.10}$$

where $\alpha^0 \equiv (a^0_{z_0,0} - \mu_0)n_0^{1/2}/\sigma$.

Proof

Let $\sigma_0{}^2 \equiv \sigma^2/n_0$. By (18.3.4),

$$\begin{aligned} E_\mu \ell_t(a^0_{z_0,0}, \tilde{\mu}) &= k_u E_N(\tilde{\mu} \mid \mu_0, \sigma_0{}^2) \\ &\quad -(k_u + k_0)\int_{-\infty}^{a^0{}_{z_0,0}} \mu f_N(\mu \mid \mu_0, \sigma_0{}^2)d\mu \\ &= k_u\mu_0 - (k_u + k_0)\int_{-\infty}^{a^0{}_{z_0,0}} (\mu - \mu_0 + \mu_0) f_N(\mu \mid \mu_0, \sigma_0{}^2)d\mu \\ &= k_u\mu_0 - (k_u + k_0)\int_{-\infty}^{a^0{}_{z_0,0}} (\mu - \mu_0) f_N(\mu \mid \mu_0, \sigma_0{}^2)d\mu \\ &\quad -(k_u + k_0)\mu_0 F_N(a^0_{z_0,0} \mid \mu_0, \sigma_0{}^2). \end{aligned}$$

But $F_N(a^0_{z_0,0} \mid \mu_0, \sigma_0{}^2) = k_u/(k_u + k_0)$, and hence

$$-(k_u + k_0)\mu_0 F_N(a^0_{z_0,0} \mid \mu_0, \sigma_0{}^2) = -k_u\mu_0,$$

so that

$$\begin{aligned} &E_\mu \ell_t(a^0_{z_0,0}, \tilde{\mu}) \\ &= -(k_u + k_o)\int_{-\infty}^{a^0{}_{z_0,0}} (\mu - \mu_0) f_N(\mu \mid \mu_0, \sigma_0{}^2)d\mu \\ &= -(k_u + k_o)\sigma_0 \int_{-\infty}^{a^0{}_{z_0,0}} \left(\frac{\mu - \mu_0}{\sigma_0}\right)\cdot\frac{1}{\sqrt{2\pi}\sigma_0}\exp\left[-\tfrac{1}{2}\left(\frac{\mu - \mu_0}{\sigma_0}\right)^2\right]d\mu \\ &= -(k_u + k_o)\sigma_0 \int_{-\infty}^{a^0{}_{z_0,0}} \left(\frac{\mu - \mu_0}{\sigma_0}\right)\cdot\frac{1}{\sqrt{2\pi}}\exp\left[-\tfrac{1}{2}\left(\frac{\mu - \mu_0}{\sigma_0}\right)^2\right]d\left(\frac{\mu - \mu_0}{\sigma_0}\right) \\ &= -(k_u + k_o)\sigma_0 \int_{-\infty}^{\alpha^0} y\cdot\frac{1}{\sqrt{2\pi}}\exp\left[-\tfrac{1}{2}y^2\right]dy, \end{aligned}$$

where $\alpha^0 \equiv (a^0_{z_0,0} - \mu_0)/\sigma_0$. But the final integrand is simply $-df_{N*}(y)$. Hence

$$E_u \ell_t(a^0_{z_0,0}, \tilde{\mu}) = (k_u + k_0)\sigma_0 \int_{-\infty}^{\alpha_0} df_{N*}(y)$$

$$= (k_u + k_0)\sigma_0 \, f_{N*}(\alpha^0),$$

from which (18.3.10) is obvious. QED

It is very important to note in (18.3.10) that the only factor which depends upon the parameters of the prior df is $n_0^{-1/2}$. In particular, $f_{N*}(\alpha^0)$ is independent of (μ_0, σ_0^2) because

$$\frac{k_u}{(k_u + k_o)} = F_N(a^0_{z_0,0} \mid \mu_0, \sigma_0^2) = F_{N*}\left(\frac{a^0_{z_0,0} - \mu_0}{\sigma_0}\right),$$

which shows that α^0 is simply the $k_u/(k_u + k_0)$th fractile of the standardized Normal distribution. Hence α^0 can be found from Table I, and then $a^0_{z_0,0}$ computed from

$$a^0_{z_0,0} = \sigma_0 \alpha^0 + \mu_0.$$

[See (12.4.6).]

These remarks have direct application when we suppose that n observations with sufficient statistic $(n, \tilde{x})$ are made and hence the posterior df is $f_N(\mu \mid \mu_1, \sigma^2/n_1)$, where

$$\mu_1 \equiv \frac{n_0\mu_0 + n\bar{x}}{n_0 + n}$$

and

$$n_1 \equiv n_0 + n.$$

By replacing subscripts "0" with subscripts "1" in (18.3.10) and its proof, we easily obtain

$$E_{\mu|\bar{x},n} \ell_t(a^0_{\bar{x},n}, \tilde{\mu}) = (n_0 + n)^{-1/2}(k_u + k_0)\sigma f_{N*}(\alpha^0), \qquad (18.3.11)$$

where $\alpha^0 \equiv (a^0_{\bar{x},n} - \mu_1)n_1^{1/2}/\sigma$.

Now $(n_0 + n)^{-1/2}(k_u + k_0)\sigma f_{N*}(\alpha^0)$ is independent of $\tilde{x}$, and hence

$$E_{\tilde{x},n} E_{\mu|\tilde{x},n} \ell_t(a^0_{\tilde{x},n}, \tilde{\mu}) = (n_0 + n)^{-1/2}(k_u + k_0)\sigma f_{N*}(\alpha^0). \qquad (18.3.12)$$

Hence

$$I(n) = (k_u + k_0)\sigma f_{N*}(\alpha^0)[n_0^{-1/2} - (n_0 + n)^{-1/2}], \qquad (18.3.13)$$

and

$$G(n) = (k_u + k_0)\sigma f_{N*}(\alpha^0)[n_0^{-1/2} - (n_0 + n)^{-1/2}] - k(n). \quad (18.3.14)$$

Hence with $k(n)$ as given by (18.1.2), we may differentiate (18.3.14) and equate the derivative to zero, obtaining

$$n^* = \left[\frac{(k_u + k_o)\sigma f_{N^*}(\alpha^0)}{2k_s}\right]^{2/3} - n_0 \tag{18.3.15}$$

(Exercise 5). Hence

$$n^0 = \begin{cases} n^*, & n^* \geqq 0 \text{ and } G(n^*) \geqq 0 \\ 0, & \text{otherwise.} \end{cases} \tag{18.3.16}$$

Example 18.3.1

Reconsider Example 18.2.3, keeping all parameters the same, except for supposing that $k_t = 50$ had been assessed by estimating a loss of \$50,000,000 due to misestimating μ by \$1000/year. Suppose now that the loss function is linear and that $k_u = k_0 = 50{,}000$, which again implies a loss of \$50,000,000 for an error of \$1000/year. Hence $k_u/(k_u + k_0) = \frac{1}{2}$, $\alpha^0 = 0$, and (from Table VI) $f_{N^*}(0) = .3989$. With $k_s = 50$, $\sigma = 1000$, and $n_0 = 4$, we obtain

$$\begin{aligned} n^* &= \left(\frac{(100{,}000)(1000)(.3989)}{100}\right)^{2/3} - 4 \\ &= (398{,}900)^{2/3} - 4 \\ &\doteq 5400, \end{aligned}$$

by (18.3.15). Hence $n^* \geqq 0$; and direct computations show that $G(5400) \doteq \$19{,}100{,}000$. Hence $n^0 = 5400$. Compare this with the optimal sample size of 996 for Example 18.2.3. The vastly greater optimal sample size, and net gain, here reflects the fact that the more likely losses under (18.1.5) are much greater here than under (18.1.1); Figure 18.7 shows why.

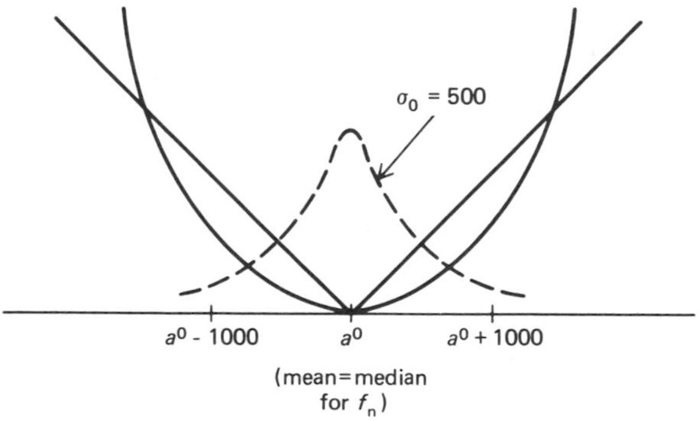

Figure 18.7

18.3.3 The Univariate Normal Process: σ^2 unknown

If n observations may be made on a univariate Normal process with unknown variance σ^2; if the loss function is as in Subsection 18.3.2 (that is, linear and independent of σ^2); and if the decision-maker's prior df of $(\tilde{\mu}, \tilde{\sigma}^2)$ is $f_{Ni\gamma}(\mu, \sigma^2 \mid \mu_0, \psi_0, n_0, \nu_0)$, as in Subsection 18.2.5, then n^0 cannot be determined, in closed form although $G(n)$ may be evaluated for each n and hence n^0 can be found by a search procedure, as shown in the following:

Let $\alpha^0(\nu)$ denote the $k_u/(k_u + k_o)$th fractile of the standardized Student distribution with ν degrees of freedom; that is,

$$F_{S^*}(\alpha^0(\nu) \mid \nu) = \frac{k_u}{k_u + k_o}. \tag{18.3.17}$$

(Table IV furnishes some standardized Student fractiles.) By Chapter 12, Exercise 23, it follows that the $k_u/(k_u + k_o)$th fractile a^0 of the Student distribution with parameters μ, ψ, and ν is given by

$$a^0 = \psi^{\frac{1}{2}}\alpha^0(\nu) + \mu. \tag{18.3.18}$$

[See (12.5.8).]

Now, if the decision-maker's prior df of $(\tilde{\mu}, \tilde{\sigma}^2)$ is Normal-inverted gamma, his marginal prior df of $\tilde{\mu}$ is $f_S(\mu_0 \mid \mu_0, \psi_0/n_0, \nu_0)$ [by (14.5.6)], and hence

$$a^0_{z_0,0} = \left(\frac{\psi_0}{n_0}\right)^{1/2}\alpha^0(\nu_0) + \mu_0. \tag{18.3.19}$$

His marginal posterior df of $\tilde{\mu}$ is $f_S(\mu \mid \mu_1, \psi_1/n_1, \nu_1)$, by (14.5.7), and hence

$$a^0_{\bar{x},s^2,n} = \left(\frac{\psi_1}{n_1}\right)^{1/2}\alpha^0(\nu_1) + \mu_1, \tag{18.3.20}$$

where n_1, ν_1, μ_1, and ψ_1 are defined in (14.5.7).

The following results can be obtained by lengthy derivations which can be found in Raiffa and Schlaifer [**66**], 198–205:

$$E_\mu \ell_t(a^0_{z_0,0}, \tilde{\mu}) = (k_u + k_o)\left(\frac{\psi_0}{n_0}\right)^{1/2}\frac{\nu_0 + [\alpha^0(\nu_0)]^2}{\nu_0 - 1}f_{S^*}(\alpha^0(\nu_0) \mid \nu_0); \tag{18.3.21}$$

$$E_{\mu|\bar{x},s^2,n}\ell_t(a^0_{\bar{x},s^2,n}, \tilde{\mu}) = (k_u + k_o)\left(\frac{\psi_1}{n_1}\right)^{1/2}\frac{\nu_1 + [\alpha^0(\nu_1)]^2}{\nu_1 - 1}f_{S^*}(\alpha^0(\nu_1) \mid \nu_1); \tag{18.3.22}$$

and

$$E_{\bar{x},s^2|n}E_{\mu|\bar{x},s^2,n}\ell_t(a^0_{\tilde{\bar{x}},\tilde{s}^2,n}, \tilde{\mu}) = (k_u + k_o)\frac{\Gamma(\frac{1}{2}\nu_1)\Gamma(\frac{1}{2}\nu_0 - \frac{1}{2})}{\Gamma(\frac{1}{2}\nu_0)\Gamma(\frac{1}{2}\nu_1 - \frac{1}{2})}\left[\frac{\nu_0\xi_0}{\nu_1 n_1}\right]^{1/2}\frac{\nu_1 + [\alpha^0(\nu_1)]^2}{\nu_1 - 1}f_{S^*}(\alpha^0(\nu_1) \mid \nu_1). \tag{18.3.23}$$

Recall that $\nu_1 = \nu_0 + n$, and hence (18.3.23) is really an exceedingly complicated function of n. Nevertheless, it can be evaluated for each fixed n, and hence

$$G(n) = E_\mu \ell_t(a^0_{z_0,0}, \tilde{\mu}) - E_{\tilde{\bar{x}},\tilde{s}^2|n} E_{\mu|\bar{x},s^2,n} \ell_t(a^0_{\tilde{\bar{x}},\tilde{s}^2,n}, \tilde{\mu}) - k(n) \qquad (18.3.24)$$

can also be evaluated for each n. So doing for every nonnegative integer n will surely enable the determination of n^0.

However, it is ridiculous to consider an infinite number of computations, especially since we can easily derive an upper bound for n^0 from the general theory of information evaluation with linear utility. Recall from (17.4.15) that the prior value $I(e^*)$ of *perfect* information must satisfy

$$I(e^*) \geqq I(n), \qquad (18.3.25)$$

for all possible experiments n; and from (17.4.34) that

$$I(e^*) = E_\mu \ell_t(a^0_{z_0,0}, \tilde{\mu}). \qquad (18.3.26)$$

Since $G(n) = I(n) - k(n)$, it follows from (18.3.25) and (18.3.26) that

$$G(n) \leqq E_\mu \ell_t(a^0_{z_0,0}, \tilde{\mu}) - k(n). \qquad (18.3.27)$$

Now, from (18.1.2) we see that $k(n)$ is increasing and unbounded, and hence there exists a smallest integer $n^\dagger$ such that

$$E_\mu \ell_t(a^0_{z_0,0}, \tilde{\mu}) - k(n) \leqq 0 \qquad (18.3.28)$$

and hence

$$G(n) \leqq 0, \qquad (18.3.29)$$

whenever $n \geqq n^\dagger$. Hence $n^0 \leqq n^\dagger$, since $G(0) = 0$. From (18.1.2), we see (Exercise 6) that

$$n^\dagger = \min \{n\colon n \geqq [E_\mu \ell_t(a^0_{z_0,0}, \tilde{\mu}) - K_s]/k_s, n \text{ a nonnegative integer}\}, \qquad (18.3.30)$$

where $E_\mu \ell_t(a^0_{z_0,0}, \tilde{\mu})$ is given by (18.3.21).

We have thus shown that

$$n^0 \in \{0, 1, \ldots, n^\dagger\}, \qquad (18.3.31)$$

and this result can be of considerable help in reducing the effort of searching for n^0.

18.4 MULTIVARIATE POINT ESTIMATION WITH QUADRATIC LOSS

18.4.1 General Theory

In this section we suppose that $\Theta = A \subset R^k$ and that

$$\ell_t(\mathbf{a}, \boldsymbol{\theta}) = (\boldsymbol{\theta} - \mathbf{a})'\mathbf{k}_t(\boldsymbol{\theta} - \mathbf{a}), \qquad (18.4.1)$$

for every $\mathbf{a}, \boldsymbol{\theta} \in R^k$, where $\mathbf{k}_t$ is a kth order, positive definite matrix. When $k = 1$, (18.4.1) reduces to (18.1.1), so that (18.4.1) represents a means of generalizing the quadratic loss function to higher dimensions. Positive definiteness of $\mathbf{k}_t$ implies, of course, that

$$\begin{aligned} \ell_t(\mathbf{a}, \boldsymbol{\theta}) &\geqq 0, \text{ for all } \mathbf{a}, \boldsymbol{\theta}; \text{ and} \\ \ell_t(\mathbf{a}, \boldsymbol{\theta}) &= 0, \text{ only if } \mathbf{a} = \boldsymbol{\theta}. \end{aligned} \tag{18.4.2}$$

Now let $f(\boldsymbol{\theta})$ by any dgf, prior or posterior, of $\tilde{\boldsymbol{\theta}}$, and let E denote expectation with respect to $f(\boldsymbol{\theta})$. Then

$$E\ell_t(\mathbf{a}, \tilde{\boldsymbol{\theta}}) = E\{[\tilde{\boldsymbol{\theta}} - E(\tilde{\boldsymbol{\theta}})]'\mathbf{k}_t[\tilde{\boldsymbol{\theta}} - E(\tilde{\boldsymbol{\theta}})]\} + [E(\tilde{\boldsymbol{\theta}}) - \mathbf{a}]'\mathbf{k}_t[E(\tilde{\boldsymbol{\theta}}) - \mathbf{a}]. \tag{18.4.3}$$

Proof

Write

$$\begin{aligned} \ell_t(\mathbf{a}, \tilde{\boldsymbol{\theta}}) &= ([\tilde{\boldsymbol{\theta}} - E(\tilde{\boldsymbol{\theta}})] + [E(\tilde{\boldsymbol{\theta}}) - \mathbf{a}])'\mathbf{k}_t([\tilde{\boldsymbol{\theta}} - E(\tilde{\boldsymbol{\theta}})] + [E(\tilde{\boldsymbol{\theta}}) - \mathbf{a}]) \\ &= [\tilde{\boldsymbol{\theta}} - E(\tilde{\boldsymbol{\theta}})]'\mathbf{k}_t(\tilde{\boldsymbol{\theta}} - E(\tilde{\boldsymbol{\theta}})] \\ &\quad + [\tilde{\boldsymbol{\theta}} - E(\tilde{\boldsymbol{\theta}})]'\mathbf{k}_t[E(\tilde{\boldsymbol{\theta}}) - \mathbf{a}] \\ &\quad + [E(\tilde{\boldsymbol{\theta}}) - \mathbf{a}]'\mathbf{k}_t[\tilde{\boldsymbol{\theta}} - E(\tilde{\boldsymbol{\theta}})] \\ &\quad + [E(\tilde{\boldsymbol{\theta}}) - \mathbf{a}]'\mathbf{k}_t[E(\tilde{\boldsymbol{\theta}}) - \mathbf{a}], \end{aligned}$$

and observe that the fourth summand is nonrandom. Hence it equals its expectation. The second and third summands are simply linear combinations of the random variables $\tilde{\theta}_i - E(\tilde{\theta}_i)$, each of which has expectation zero. Hence the expectations of the second and third summands are each zero. Now (18.4.3) is immediate. QED

We may simplify (18.4.3) by noting that the random quadratic form in braces can be written

$$\sum_{i=1}^{k}\sum_{j=1}^{k} k_{tij}[\tilde{\theta}_i - E(\tilde{\theta}_i)][\tilde{\theta}_j - E(\tilde{\theta}_j)],$$

and hence by (10.2.7) its expectation is

$$\sum_{i=1}^{k}\sum_{j=1}^{k} k_{tij}E\{[\tilde{\theta}_i - E(\tilde{\theta}_i)]\cdot[\tilde{\theta}_j - E(\tilde{\theta}_j)]\} = \sum_{i=1}^{k}\sum_{j=1}^{k} k_{tij}V_{ij},$$

where V_{ij} denotes the covariance of $\tilde{\theta}_i$ and $\tilde{\theta}_j$, and k_{tij} denotes the (i, j)th element of $\mathbf{k}_t$.

We introduce a convenient notation for the preceding double summation. Let $\mathbf{A}$ and $\mathbf{B}$ be any two kth order matrices, and define $\mathbf{A}\nabla\mathbf{B} \in R^1$ by

$$\mathbf{A}\nabla\mathbf{B} \equiv \sum_{i=1}^{k}\sum_{j=1}^{k} a_{ij}b_{ij}. \tag{18.4.4}$$

Thus $\mathbf{A}\nabla\mathbf{B}$ is simply the sum of all k^2 products of the corresponding elements of $\mathbf{A}$ and $\mathbf{B}$.

Example 18.4.1

Let

$$\mathbf{A} = \begin{bmatrix} 8 & 10 \\ 2 & -5 \end{bmatrix}$$

and

$$\mathbf{B} = \begin{bmatrix} -6 & 0 \\ 14 & 3 \end{bmatrix}.$$

Then $\mathbf{A}\nabla\mathbf{B} = (8)(-6) + (10)(0) + (2)(14) + (-5)(3) = -35$.
The following properties of the ∇-function are reasily established (Exercise 7):

$$\mathbf{A}\nabla(\mathbf{B} + \mathbf{C}) = \mathbf{A}\nabla\mathbf{B} + \mathbf{A}\nabla\mathbf{C}; \tag{18.4.5}$$

$$\mathbf{A}\nabla\mathbf{B} = \mathbf{B}\nabla\mathbf{A}; \tag{18.4.6}$$

$$E(\mathbf{A}\nabla\tilde{\mathbf{B}}) = \mathbf{A}\nabla E(\tilde{\mathbf{B}}); \text{ and} \tag{18.4.7}$$

$$\mathbf{A}\nabla(x\mathbf{B}) = (x\mathbf{A})\nabla\mathbf{B} = x\cdot(\mathbf{A}\nabla\mathbf{B}), \text{ for any real number } x. \tag{18.4.8}$$

Let $V(\tilde{\boldsymbol{\theta}})$ denote the covariance matrix of $\tilde{\boldsymbol{\theta}}$. Then (18.4.3) becomes

$$E\ell_t(\mathbf{a}, \tilde{\boldsymbol{\theta}}) = \mathbf{k}_t\nabla V(\tilde{\boldsymbol{\theta}}) + [E(\tilde{\boldsymbol{\theta}}) - \mathbf{a}]'\mathbf{k}_t[E(\tilde{\boldsymbol{\theta}}) - \mathbf{a}]. \tag{18.4.9}$$

It is clear from positive-definiteness of $\mathbf{k}_t$ that (18.4.9) is minimized for $\mathbf{a} = E(\tilde{\boldsymbol{\theta}})$. Hence

$$\mathbf{a}^0 = E(\tilde{\boldsymbol{\theta}}) \tag{18.4.10}$$

and

$$E\ell_t(\mathbf{a}^0, \tilde{\boldsymbol{\theta}}) = \mathbf{k}_t\nabla V(\tilde{\boldsymbol{\theta}}). \tag{18.4.11}$$

From (18.4.10) and (18.4.11), the parallel with the general theory of Section 18.2 should be striking. Now suppose the decision-maker can acquire information about $\tilde{\boldsymbol{\theta}}$ in the form of a sufficient statistic $(n, \tilde{z})$ of n observations on the process with parameter $\tilde{\boldsymbol{\theta}}$. Then (18.4.10) and (18.4.11) imply

$$\mathbf{a}^0_{\mathbf{z}_0,0} = E(\tilde{\boldsymbol{\theta}}); \tag{18.4.12}$$

$$\mathbf{a}^0_{\mathbf{z},n} = E(\tilde{\boldsymbol{\theta}} \mid z, n); \tag{18.4.13}$$

$$E_{\theta}\ell_t(\mathbf{a}^0_{\mathbf{z}_0,0}, \tilde{\boldsymbol{\theta}}) = \mathbf{k}_t\Delta V(\tilde{\boldsymbol{\theta}}); \text{ and} \tag{18.4.14}$$

$$E_{\theta\mid\mathbf{z},n}\ell_t(\mathbf{a}^0_{\mathbf{z},n}, \tilde{\boldsymbol{\theta}}) = \mathbf{k}_t\nabla V(\tilde{\boldsymbol{\theta}} \mid z, n). \tag{18.4.15}$$

[Compare (18.2.1) through (18.2.4).] It follows from (18.4.14), (18.4.15), (18.4.5), and (18.4.7) that

$$I(n) = \mathbf{k}_t \nabla [V(\tilde{\boldsymbol{\theta}}) - E_{\mathbf{z}|n} V(\tilde{\boldsymbol{\theta}} \mid \tilde{\mathbf{z}}, n)]. \tag{18.4.16}$$

Naturally, $G(n) = I(n) - k(n)$.

In Subsection 18.4.2 we let $\tilde{\boldsymbol{\theta}} = \tilde{\mathbf{p}}$, the parameter of a k-variate Bernoulli process. Subsection 18.4.3 is the first of three in which $\tilde{\boldsymbol{\theta}} = \tilde{\boldsymbol{\mu}}$, the mean-vector parameter of a k-variate Normal process, with $\boldsymbol{\sigma}^2$ known in 18.4.3, $\boldsymbol{\sigma}_*{}^2$ known in Subsection 18.4.4, and $\boldsymbol{\sigma}^2$ unknown in Subsection 18.4.5. We derive explicit characterizations of n^0 in Subsections 18.4.2 and 18.4.5. (Not too) surprisingly, the complicated Normal-inverted Wishart case is the only one of the Normal cases in which n^0 can be characterized explicitly.

Before proceeding to the more detailed analysis, a word on the assessment of $\mathbf{k}_t$ is in order. In many cases it is reasonable to suppose that the decision-maker's loss is the simple sum $\sum_{i=1}^{k} k_{tii}(\theta_i - a_i)^2$ of quadratic losses due to errors in estimating the θ_i's individually. Clearly, in such cases $\mathbf{k}_t$ is a diagonal matrix and $I(n)$ is given by

$$I(n) = \sum_{i=1}^{k} I_i(n), \tag{18.4.17}$$

where

$$I_i(n) \equiv k_{tii}[V(\tilde{\theta}_i) - E_{\mathbf{z}|n} V(\tilde{\theta}_i \mid \tilde{\mathbf{z}}, n)]. \tag{18.4.18}$$

When $\mathbf{k}_t$ is not diagonal, however, it reflects additional penalties (or partial reprieves) for simultaneously over- or underestimating the θ_i's, since nonzero off-diagonals k_{tij} of $\mathbf{k}_t$ produce summands of the form $k_{tij}(\theta_i - a_i)(\theta_j - a_j)$ in (18.4.1). Thus $k_{tij} > 0$ implies an additional penalty for making errors of the same sign in both a_i and a_j, while $k_{tij} < 0$ implies a partial reprieve in this case.

As originally assessed, $\mathbf{k}_t$ may not be symmetric. Asymmetry of $\mathbf{k}_t$ actually causes no problems in the analysis, but it is always possible to replace $\mathbf{k}_t$ as originally assessed by $\frac{1}{2}(\mathbf{k}_t + \mathbf{k}'_t)$—which is symmetric—without altering $\ell_t(\mathbf{a}, \boldsymbol{\theta})$ for any $\mathbf{a}, \boldsymbol{\theta}$. (See Exercise 8.)

18.4.2 The k-Variate Bernoulli Process

In this subsection we suppose that $\tilde{\boldsymbol{\theta}} = \tilde{\mathbf{p}}$, the parameter of a k-variate Bernoulli process, and that the decision-maker has assessed a k-variate beta prior df $f_\beta{}^{(k)}(\mathbf{p} \mid \boldsymbol{\rho}, \nu)$ for $\tilde{\mathbf{p}}$. Then an experiment consisting of n observations has sufficient statistic ($\tilde{\mathbf{r}}$, n) with $\tilde{\mathbf{r}}$ having the multinomial mf $f_{mu}{}^{(k)}(\mathbf{r} \mid n, \mathbf{p})$ given $\mathbf{p}$. Hence

$$\mathbf{a}^0_{z_0,0} = \nu^{-1}\boldsymbol{\varrho} \qquad \text{[by (18.4.12) and (12.6.6)];} \qquad (18.4.19)$$

$$\mathbf{a}^0_{\mathbf{r},n} = (\nu + n)^{-1}(\boldsymbol{\varrho} + \mathbf{r}) \qquad \text{[by (18.4.13) and (12.6.6)];} \quad (18.4.20)$$

because the posterior df of $\tilde{\mathbf{p}}$ is $f_\beta{}^{(k)}(\mathbf{p} \mid \boldsymbol{\varrho} + \mathbf{r}, \nu + n)$ [by (14.3.2)];

$$\begin{aligned} E_{\mathbf{p}}\ell_t(\mathbf{a}^0_{z_0,0}, \tilde{\mathbf{p}}) &= \mathbf{k}_t \nabla V(\tilde{\mathbf{p}}) \qquad \text{[by (18.4.14)]} \\ &= \mathbf{k}_t \nabla V_\beta{}^{(k)}(\tilde{\mathbf{p}} \mid \boldsymbol{\varrho}, \nu), \end{aligned} \qquad (18.4.21)$$

where $V_\beta{}^{(k)}(\tilde{\mathbf{p}} \mid \boldsymbol{\varrho}, \nu)$ is given by (12.6.7) and (12.6.8); and

$$\begin{aligned} E_{\mathbf{p}|\mathbf{r},n}\ell_t(\mathbf{a}^0_{\mathbf{r},n}, \tilde{\mathbf{p}}) &= \mathbf{k}_t \nabla V(\tilde{\mathbf{p}} \mid \mathbf{r}, n) \\ &= \mathbf{k}_t \nabla V_\beta{}^{(k)}(\tilde{\mathbf{p}} \mid \boldsymbol{\varrho} + \mathbf{r}, \nu + n). \end{aligned} \qquad (18.4.22)$$

Hence

$$\begin{aligned} I(n) &= \mathbf{k}_t \nabla [V_\beta{}^{(k)}(\tilde{\mathbf{p}} \mid \boldsymbol{\varrho}, \nu) - E_{\mathbf{r}|n} V_\beta{}^{(k)}(\tilde{\mathbf{p}} \mid \boldsymbol{\varrho} + \mathbf{r}, \nu + n)] \\ &= \frac{n}{\nu + n} \cdot \mathbf{k}_t \nabla V_\beta{}^{(k)}(\tilde{\mathbf{p}} \mid \boldsymbol{\varrho}, \nu). \end{aligned} \qquad (18.4.23)$$

Proof

The reader may show as Exercise 9 that the second equality of (18.4.23) follows from

(*) $$E_{\mathbf{r}|n} V_\beta{}^{(k)}(\tilde{\mathbf{p}} \mid \boldsymbol{\varrho} + \tilde{\mathbf{r}}, \nu + n) = \frac{\nu}{\nu + n} V_\beta{}^{(k)}(\tilde{\mathbf{p}} \mid \boldsymbol{\varrho}, \nu).$$

We shall prove (*). Consider the diagonal element

$$\begin{aligned} &E_{\mathbf{r}|n} V(\tilde{p}_i \mid \boldsymbol{\varrho} + \tilde{\mathbf{r}}, \nu + n) \\ &\qquad = E_{\mathbf{r}|n}\left\{\frac{(\rho_i + \tilde{r}_i)(\nu + n - \rho_i - \tilde{r}_i)}{(\nu + n)^2(\nu + n + 1)}\right\} \qquad \text{[by (12.6.6)].} \end{aligned}$$

Now $E_{\mathbf{r}|n}$ is expectation with respect to the marginal mf of $\tilde{\mathbf{r}}$, which is hypermultinomial. By some algebra, we obtain

$$\begin{aligned} &E_{\mathbf{r}|n} V(\tilde{p}_i \mid \boldsymbol{\varrho} + \tilde{\mathbf{r}}, \nu + n) \\ &\quad = [(\nu + n)^2(\nu + n + 1)]^{-1}\{\rho_i(\nu + n - \rho_i) + (\nu + n - 2\rho_i)E(\tilde{r}_i) - E(\tilde{r}_i)^2\}, \end{aligned}$$

where $E(\tilde{r}_i) = E^{(k)}_{hmu}(\tilde{r}_i \mid \boldsymbol{\varrho}, \nu, n) = n\rho_i/\nu$ [by (14.3.4)], and

$$\begin{aligned} E(\tilde{r}_i{}^2) &= E^{(k)}_{hmu}(\tilde{r}_i{}^2 \mid \boldsymbol{\varrho}, \nu, n) \\ &= n(n + \nu)\frac{\rho_i(\nu - \rho_i)}{\nu^2(\nu + 1)} + \frac{n^2\rho_i{}^2}{\nu^2} \end{aligned}$$

[by (14.3.5) and (10.3.11)]. By straightforward but laborious algebra, it follows that

(1) $$E_{\tilde{\mathbf{r}}|n}V(\tilde{p}_i \mid \boldsymbol{\rho} + \tilde{\mathbf{r}}, \nu + n) = \left(\frac{\nu}{\nu + n}\right)\frac{\rho_i(\nu - \rho_i)}{\nu^2(\nu + 1)},$$

which is simply the ith diagonal element of the right-hand side of (*). Now consider an off-diagonal element

$$\begin{aligned}
&E_{\tilde{\mathbf{r}}|n}V(\tilde{p}_i, \tilde{p}_j \mid \boldsymbol{\rho} + \tilde{\mathbf{r}}, \nu + n)\\
&\quad = E_{\tilde{\mathbf{r}}|n}\left\{-\frac{(\rho_i + \tilde{r}_i)(\rho_j + \tilde{r}_j)}{(\nu + n)^2(\nu + n + 1)}\right\}\\
&\quad = -[(\nu + n)^2(\nu + n + 1)]^{-1}\{\rho_i\rho_j + \rho_j E(\tilde{r}_i) + \rho_i E(\tilde{r}_j) + E(\tilde{r}_i\tilde{r}_j)\}\\
&\quad = -[(\nu + n)^2(\nu + n + 1)]^{-1}\{\rho_i\rho_j + n\rho_i\rho_j/\nu + n\rho_i\rho_j/\nu\\
&\qquad\qquad + n^2\rho_i\rho_j/\nu^2 - n(n + \nu)\rho_i\rho_j/[\nu^2(\nu + 1)]\}
\end{aligned}$$

[by (14.3.4) and (14.3.6)]. Laborious but otherwise elementary algebra then yields

(2) $$E_{\tilde{\mathbf{r}}|n}V(\tilde{p}_i, \tilde{p}_j \mid \boldsymbol{\rho} + \tilde{\mathbf{r}}, \nu + n) = \left(\frac{\nu}{\nu + n}\right)\left(-\frac{\rho_i\rho_j}{\nu^2(\nu + 1)}\right),$$

which is simply the (i, j)th off-diagonal of the right-hand side of (*). Thus (1) and (2) suffice for (*). QED

From (18.4.23) it is clear how we may proceed to find n^0 when $k(n)$ is given by (18.1.2). The reader may show as Exercise 10 that

$$n^* = \left[\frac{\nu(\mathbf{k}_t \nabla \mathbf{V}_\beta(\tilde{\mathbf{p}} \mid \boldsymbol{\rho}, \nu))}{k_s}\right]^{1/2} - \nu \tag{18.4.24}$$

and

$$n^0 = \begin{cases} n^*, & n^* \geqq 0 \text{ and } G(n^*) \geqq 0\\ 0, & \text{otherwise.}\end{cases} \tag{18.4.25}$$

Thus the k-variate Bernoulli process is just as tractable as its univariate counterpart.

18.4.3 The k-Variate Normal Process: σ^2 known

When $\tilde{\boldsymbol{\theta}} = \tilde{\boldsymbol{\mu}}$, the mean vector of a k-variate Normal process with known covariance matrix $\boldsymbol{\sigma}^2$, and when the decision-maker has assessed a prior df $f_{N^{(k)}}(\boldsymbol{\mu} \mid \boldsymbol{\mu}_0, \boldsymbol{\sigma}_0^2)$ for $\tilde{\mu}$, then

$$\mathbf{a}^0_{z_0,0} = \boldsymbol{\mu}_0, \tag{18.4.26}$$

$$\mathbf{a}^0_{\bar{\mathbf{x}},n} = \boldsymbol{\mu}_1 \equiv (\boldsymbol{\sigma}_0^{-2} + n\boldsymbol{\sigma}^{-2})^{-1}(\boldsymbol{\sigma}_0^{-2}\boldsymbol{\mu}_0 + n\boldsymbol{\sigma}^{-2}\bar{\mathbf{x}}), \tag{18.4.27}$$

$$E_\mu \ell_t(\mathbf{a}^0_{z_0,0}, \tilde{\boldsymbol{\mu}}) = \mathbf{k}_t \nabla \boldsymbol{\sigma}_0^2, \text{ and} \tag{18.4.28}$$

$$\begin{aligned} E_{\mu|\bar{\mathbf{x}},n}\ell_t(\mathbf{a}^0_{\bar{\mathbf{x}},n}, \tilde{\boldsymbol{\mu}}) &= \mathbf{k}_t \nabla(\boldsymbol{\sigma}_0^{-2} + n\boldsymbol{\sigma}^{-2})^{-1}\\ &\equiv \mathbf{k}_t \nabla \boldsymbol{\sigma}_1^2, \end{aligned} \tag{18.4.29}$$

where all symbols are as used in (14.6.1). Note that the dependence of the right-hand side of (18.4.29) upon n is not simple. Nevertheless, (18.4.29) is independent of $\bar{\tilde{\mathbf{x}}}$, and hence equals its $E_{\bar{\mathbf{x}}|n}$-expectation. Therefore

$$I(n) = \mathbf{k}_t \nabla[\boldsymbol{\sigma}_0{}^2 - \boldsymbol{\sigma}_1{}^2(n)], \tag{18.4.30}$$

where we have indicated the dependence of $\boldsymbol{\sigma}_1{}^2$ upon n for convenience. Hence

$$G(n) = \mathbf{k}_t \nabla[\boldsymbol{\sigma}_0{}^2 - \boldsymbol{\sigma}_1{}^2(n)] - k(n). \tag{18.4.31}$$

Now, we cannot find n^0 in a closed form, and hence can only state that $n^0 \in \{0, 1, \ldots, n^\dagger\}$, where

$$n^\dagger = \min\{n: n \geqq ([\mathbf{k}_t \nabla \boldsymbol{\sigma}_0{}^2] - K_s)/k_s,\ n \text{ a nonnegative integer}\}. \tag{18.4.32}$$

(Why? See Exercise 11.)

Example 18.4.1

Let

$$\mathbf{k}_t = \begin{bmatrix} 10 & 2 \\ 2 & 10 \end{bmatrix}, \quad \boldsymbol{\sigma}^2 = \begin{bmatrix} 13 & -2 \\ -2 & 8 \end{bmatrix}, \text{ and } \boldsymbol{\sigma}_0{}^2 = \begin{bmatrix} 20 & 10 \\ 10 & 55 \end{bmatrix}.$$

The reader may verify as Exercise 1 that

$$\boldsymbol{\sigma}^{-2} = \begin{bmatrix} .055 & -.010 \\ -.010 & .020 \end{bmatrix},$$

$$n\boldsymbol{\sigma}^{-2} = \begin{bmatrix} .08n & .02n \\ .02n & .13n \end{bmatrix},$$

$$\boldsymbol{\sigma}_1{}^2 = \{(.055 + .08n)(.020 + .13n) - (-.01 + .02n)^2\}^{-1} \times \begin{bmatrix} .020 + .13n & .010 - .02n \\ .010 - .02n & .055 + .08n \end{bmatrix},$$

and that

$$G(n) = 790 - K_s - (79 + 202n)/(n^2 + .914n + .100) - k_s n.$$

Try to characterize n^0 by differentiation!

18.4.4 The k-Variate Normal Process: $\sigma_*{}^2$ known

When $\tilde{\boldsymbol{\theta}} = \tilde{\boldsymbol{\mu}}$, the mean of a k-variate Normal process with covariance matrix $\boldsymbol{\sigma}^2 = \sigma_*{}^2\boldsymbol{\sigma}_*{}^2$ and $\boldsymbol{\sigma}_*{}^2$ is known, then (14.6.14) is applicable. If the decision-maker assigns $(\tilde{\boldsymbol{\mu}}, \tilde{\sigma}_*^2)$ the Normal-inverted gamma, natural conjugate prior df, then

$$\mathbf{a}^0_{z^0,0} = \boldsymbol{\mu}_0; \tag{18.4.33}$$

$$\mathbf{a}^0_{\bar{\mathbf{x}},s_*^2,n} = \boldsymbol{\mu}_1 \equiv \check{\sigma}_{1*}^2(n)[\check{\sigma}_{0*}^{-2}\boldsymbol{\mu}_0 + n\check{\sigma}_*^{-2}\bar{\mathbf{x}}]; \tag{18.4.34}$$

$$E_\mu \ell_t(\mathbf{a}^0_{z^0,0}, \tilde{\boldsymbol{\mu}}) = \frac{\nu_0}{\nu_0 - 2}\,\psi_0 \mathbf{k}_t \nabla \check{\sigma}_{0*}^2; \text{ and} \tag{18.4.35}$$

$$E_{\mu|\bar{\mathbf{x}},s_*^2,n}\ell_t(\mathbf{a}^0_{\bar{\mathbf{x}},s_*^2,n}, \tilde{\boldsymbol{\mu}}) = \frac{\nu_1}{\nu_1 - 2}\,\psi_1 \mathbf{k}_t \nabla \check{\sigma}_{1*}^2(n). \tag{18.4.36}$$

Proof

Exercise 13.

Observe in (18.4.36) that ν_1 depends upon n and ψ_1 depends upon n, $\bar{\mathbf{x}}$, and s_*^2. Hence finding the $E_{\bar{\mathbf{x}},s_*^2|n}$-expectation of (18.4.36) is a nontrivial task. We skip the details; the result is

$$E\ell_t(\mathbf{a}^0_{\tilde{\bar{\mathbf{x}}},\tilde{s}_*^2,n}, \tilde{\boldsymbol{\mu}}) = \frac{\nu_0}{\nu_0 - 2}\,\psi_0 \mathbf{k}_t \nabla \check{\sigma}_{1*}^2(n). \tag{18.4.37}$$

Hence

$$I(n) = \frac{\nu_0}{\nu_0 - 2}\,\psi_0 \mathbf{k}_t \nabla [\check{\sigma}_{0*}^2 - \check{\sigma}_{1*}^2(n)]. \tag{18.4.38}$$

Since $G(n) = I(n) - k(n)$, it follows that $n^0 \in \{0, 1, \ldots, n^\dagger\}$, where

$$n^\dagger = \min\left\{n\colon n \geqq \frac{\left[\dfrac{\nu_0}{\nu_0 - 2}\,\psi_0 \mathbf{k}_t \nabla \check{\sigma}_{0*}^2\right] - K_s}{k_s},\ n \text{ is a nonnegative integer}\right\}. \tag{18.4.39}$$

(Exercise 14.)

18.4.5 The k-Variate Normal Process: σ^2 unknown

For the Normal-inverted Wishart case [Chapter 14, Exercise 6(c)], we have

$$\mathbf{a}^0_{z^0,0} = \boldsymbol{\mu}_0; \tag{18.4.40}$$

$$\mathbf{a}^0_{\bar{\mathbf{x}},s^2,n} = \boldsymbol{\mu}_1 \equiv (n_0 + n)^{-1}(n_0\boldsymbol{\mu}_0 + n\bar{\mathbf{x}}); \tag{18.4.41}$$

$$E_\mu \ell_t(\mathbf{a}^0_{z^0,0}, \tilde{\boldsymbol{\mu}}) = \frac{\nu_0}{\nu_0 - 2}\,\mathbf{k}_t \nabla (n_0^{-1}\boldsymbol{\psi}_0); \text{ and} \tag{18.4.42}$$

$$E_{\mu|\bar{\mathbf{x}},s^2,n}\ell(\mathbf{a}^0_{\bar{\mathbf{x}},s^2,n}, \tilde{\boldsymbol{\mu}}) = \frac{\nu_1}{\nu_1 - 2}\,\mathbf{k}_t \nabla [n_1^{-1}\boldsymbol{\psi}_1(n)], \tag{18.4.43}$$

where $\boldsymbol{\psi}_1(n)$ depends upon $\bar{\mathbf{x}}$ and s^2 as well as upon n.

Proof

Exercise 15.

Taking the $E_{\bar{\mathbf{x}},\mathbf{s}^2|n}$-expectation of (18.4.43) is a nontrivial task if performed directly. However, it can be shown that the result is

$$E\ell_t(\overset{0}{\mathbf{a}}_{\tilde{\bar{\mathbf{x}}},\,\tilde{s}^2,n},\,\tilde{\boldsymbol{\mu}}) = \left(\frac{n_0}{n_0+n}\right)\cdot\frac{\nu_0}{\nu_0-2}\,\mathbf{k}_t\nabla[n_0{}^{-1}\boldsymbol{\psi}_0], \tag{18.4.44}$$

and hence (18.4.42) and (18.4.44) imply

$$I(n) = \left(\frac{n_0}{n_0+n}\right)\left[\frac{\nu_0}{\nu_0-2}\,\mathbf{k}_t\nabla(n_0{}^{-1}\boldsymbol{\psi}_0)\right]. \tag{18.4.45}$$

Thus

$$n^* = \left[\left\{\frac{\nu_0}{\nu_0-2}\right\}\frac{\mathbf{k}_t\nabla\boldsymbol{\psi}_0}{k_s}\right]^{1/2} - n_0, \tag{18.4.46}$$

and

$$n^0 = \begin{cases} n^*, & n^* \geqq 0 \text{ and } G(n^*) \geqq 0 \\ 0, & \text{otherwise.} \end{cases} \tag{18.4.47}$$

EXERCISES

1. Show in (18.2.10) that

$$k_t E_{hb}\left\{\frac{(\rho+\tilde{r})(\nu+n-\rho-\tilde{r})}{(\nu+n)^2(\nu+n+1)}\,\Big|\,\rho,\nu,n\right\}$$
$$= \frac{\nu}{\nu+n}k_t\rho(\nu-\rho)/[\nu^2(\nu+1)].$$

2. For Subsection 18.2.2. show that $I(n) - k(n)$ is a concave function and hence has a unique global maximizer n^*, provided that we redefine $k(n)$ so that $k(0) = K_s$—instead of $k(0) = 0$, as per (18.1.2). Then show that n^* and n^0 are given by (18.2.12) and (18.2.13), respectively.

   [*Hint*: Define $I(n)$ as (18.2.8) through (18.2.10) for all $n > -n_0$.]

3. Derive (18.2.14). [*Hint*: Re-express and simplify the inequalities $G(n^*) \geqq G(n^*-1)$ and $G(n^*) \geqq G(n^*+1)$.]

4. Derive (18.3.9).

5. Derive (18.3.15).

6. Derive (18.3.30).

7. Derive (18.4.5) through (18.4.8).

8. Define $h_1: R^k \to R^1$ by

$$h_1(\mathbf{x}) = \mathbf{x}'\mathbf{K}\mathbf{x},$$

where $\mathbf{K}$ is a positive-definite matrix; and define $h_2: R^k \to R^1$ by

$$h_2(\mathbf{x}) = \mathbf{x}'\mathbf{K}'\mathbf{x}.$$

(a) Show that $h_1(\mathbf{x}) = h_2(\mathbf{x})$, for every $\mathbf{x} \in R^k$. [*Hint*: $(\mathbf{x}'\mathbf{K}\mathbf{x})' = \mathbf{x}'\mathbf{K}\mathbf{x}$.]
(b) Define $h_3(\mathbf{x}) = \frac{1}{2}h_1(\mathbf{x}) + \frac{1}{2}h_2(\mathbf{x})$, for every $\mathbf{x} \in R^k$. By (a), $h_3(\mathbf{x}) = h_1(\mathbf{x})$ for every $\mathbf{x}$. Show that $h_3(\mathbf{x}) = \mathbf{x}'[\frac{1}{2}\mathbf{K} + \frac{1}{2}\mathbf{K}']\mathbf{x}$.
(c) Show that if $\mathbf{K}$ is any kth order matrix, then $\frac{1}{2}\mathbf{K} + \frac{1}{2}\mathbf{K}'$ is a symmetric matrix.
(d) Use these results to show that $\mathbf{k}_t$ in (18.4.1) may be regarded as symmetric, without loss of generality.

9. Show that (18.4.23) follows from the equality (*) in the ensuing proof.

10. Derive (18.4.24) and (18.4.25).

11. Derive (18.4.32).

12. Verify all computations in Example 18.4.1.

13. Derive (18.4.33) through (18.4.36).

14. Derive (18.4.39).

15. Derive (18.4.40) through (18.4.43).

16. *Modal Estimation*: Recall from Chapter 12 that a *mode* $\hat{\theta}$ of a random variable $\tilde{\theta}$ is a maximizer of its dgf $f(\theta)$.

(a) Interpret Exercise 17.18 as an estimation problem.
(b) Let $\tilde{\theta}$ be any random variable, $A = \Theta \subset R^1$, and $\epsilon > 0$. Define a terminal loss function by

$$\ell_t^\epsilon(a, \theta) \equiv \begin{cases} 0, & \theta \in [a - \epsilon, a + \epsilon] \\ K_t, & \theta \notin [a - \epsilon, a + \epsilon], \end{cases} \tag{18.5.1}$$

where $K_t > 0$. Show that

$$E_\theta \ell_t^\epsilon(a, \tilde{\theta}) = K_t[1 + F([a - \epsilon]^-) - F(a + \epsilon)], \tag{18.5.2}$$

for every $a \in A$, where $F(\cdot)$ is the cdf of $\tilde{\theta}$ and

$$F([a - \epsilon]^-) \equiv \lim_{\theta \to (a-\epsilon)^-} F(\theta).$$

(c) Now suppose that $\tilde{\theta}$ is a continuous random variable with df $f(\theta)$ such that:

(1) $\{\theta: f(\theta) > 0\}$ is an open interval (θ_*, θ^*), where θ_* may be $-\infty$ and/or θ^* may be $+\infty$;

(2) $f(\theta)$ is differentiable on (θ_*, θ^*); and
(3) there exists a number $\hat{\theta} \in (\theta_*, \theta^*)$ such that $(d/d\theta)f(\theta) > 0$ on $(\theta_*, \hat{\theta})$, $(d/d\theta)f(\theta) < 0$ on $(\hat{\theta}, \theta^*)$, and $(d/d\theta)f(\hat{\theta}) = 0$.

Obviously, $\tilde{\theta}$ is unimodal, with internal mode $\hat{\theta}$. Assumptions (1) through (3) are satisfied by gamma random variables with $\rho > 1$ and by all Normal and Student random variables. Show that these assumptions imply that the optimal act a_ϵ^0 given ϵ is unique and satisfies

$$f(a_\epsilon^0 - \epsilon) = f(a_\epsilon^0 + \epsilon) \tag{18.5.3}$$

and

$$\frac{d}{da}f(a_\epsilon^0 - \epsilon) > \frac{d}{da}f(a_\epsilon^0 + \epsilon). \tag{18.5.4}$$

(d) Thus verify the graphic procedure of passing horizontal lines through the graph of $f(\theta)$ until a (unique) segment of length 2_ϵ is obtained; then a^0 is the abscissa of the midpoint of that segment (Figure 18.8).

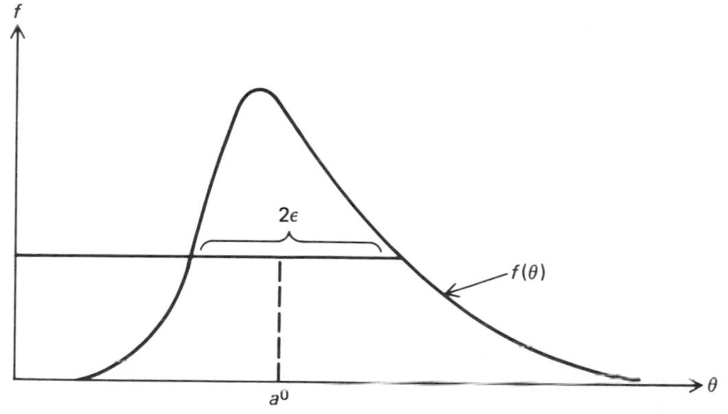

Figure 18.8

(e) Show that

$$|a_\epsilon^0 - \hat{\theta}| < \epsilon, \tag{18.5.5}$$

for all $\epsilon > 0$, and hence

$$\lim_{\epsilon \to 0} a_\epsilon^0 = \hat{\theta}. \tag{18.5.6}$$

(f) Show that if $f(\theta)$ is *symmetric* about a point x [that is, $f(x - \theta) = f(x + \theta)$, for all real θ], then $x = \hat{\theta}$ and $a_\epsilon^0 = \hat{\theta}$, for every $\epsilon > 0$.
(g) Interpret (18.5.6) and

$$\lim_{\epsilon \to 0} \ell_t^\epsilon(a, \theta) = \begin{cases} 0, & a = \theta \\ K_t, & a \neq \theta \end{cases} \tag{18.5.7}$$

as constituting the decision rule, "If a miss is as good as a mile when you are wrong and zero when you are right, maximize the likelihood of being right." [Compare part (a), and Chapter 17, Exercise 18. This exercise extends the basic ideas there to a class of continuous random variables.]

17. *Consistency*: All of the applied results in this chapter have been based upon experiments e consisting of random samples of size n, and all optimal acts $a^0_{z,n}$ have been of qualitative forms which are independent of n; for example, $a^0_{z,n} = E(\tilde{\theta} \mid z, n)$; $F(a^0_{z\ n} \mid z, n) = k_u/(k_u + k_s)$. This suggests considering a family of optimal decision functions d_n: $Z(n) \to A = \Theta \subset R^1$, one for each n, defined by

$$d_n(z) = a^0_{z,n}, \text{ all } z \in Z(n),$$

for $n = 1, 2, \ldots$. Such a family is called an *estimator* of θ. [The term *estimator* is also applied to any one of the decision functions, say $d_5(z)$.] An estimator $\{d_n\colon n = 1, 2, \ldots\}$ is said to be *consistent* (in probability) if for every $\epsilon > 0$ we have

$$\lim_{n \to \infty} P(\theta - \epsilon < d_n(\tilde{z}) < \theta + \epsilon \mid \theta) = 1, \text{ for every } \theta \in \Theta. \qquad (18.5.9)$$

Equivalently, for every $\epsilon > 0$, $\delta > 0$, and $\theta \in \Theta$ there is an integer $n(\epsilon, \delta, \theta) > 0$ such that

$$P(\theta - \epsilon < d_n(\tilde{z}) < \theta + \epsilon \mid \theta) > 1 - \delta, \qquad (18.5.10)$$

whenever $n \geqq n(\epsilon, \delta, \theta)$. Note that these probabilities are derivable from the $f(z \mid \theta, n)$'s and are *not* posterior probabilities. Show that the following are consistent estimators:

(a) $$d_n(\tilde{r}) = \frac{\rho + \tilde{r}}{\nu + n}, \text{ for } p \quad \text{(Subsection 18.2.2);}$$

(b) $$d_t(\tilde{r}) = \frac{\rho + \tilde{r}}{\tau + t}, \text{ for } \lambda \quad \text{(Subsection 18.2.3);}$$

(d) $$d_n(\tilde{\bar{x}}) = \left(\frac{n_0}{n_0 + n}\right)\mu_0 + \left(\frac{n}{n_0 + n}\right)\tilde{\bar{x}}, \text{ for } \mu \quad \text{(Subsection 18.2.4)}$$

(d) $d_n(\tilde{\bar{x}})$ as in (c), for μ (Subsection 18.2.5);

(e) $$d_n(\tilde{\bar{x}}) = \left(\frac{n_0}{n_0 + n}\right)\mu_0 + \frac{n}{n_0 + n}\tilde{\bar{x}} + {}_qN^*\frac{\sigma}{\sqrt{n_0 + n}}, \text{ for } \mu$$

(Subsection 18.3.2), where $q \in (0, 1)$, and ${}_qN^*$ is the qth fractile of a standardized Normal random variable; and

(f) $$d_n(\tilde{\bar{x}}, \tilde{s}^2) = \left(\frac{n_0}{n_0 + n}\right)\mu_0 + \left(\frac{n}{n_0 + n}\right)\tilde{\bar{x}} + {}_qS^*_{(\nu_0 + n)}\frac{\tilde{\psi}_1^{1/2}}{\sqrt{n_0 + n}},$$

for μ (Subsection 18.3.3), where $q \in (0, 1)$ and ${}_qS^*_{(\nu_0+n)}$ is the qth fractile of a standardized Student random variable with $\nu_1 \equiv \nu_0 + n$ degrees of freedom.

[*Hints*: (1) Use Chebychev's inequality to show that (18.5.9) holds if

$$\lim_{n\to\infty} |E(d_n(\tilde{z})|\theta) - \theta| = 0 \tag{18.5.11}$$

and

$$\lim_{n\to\infty} V(d_n(\tilde{z}) \mid \theta) = 0 \tag{18.5.12}$$

hold for every $\theta \in \Theta$.

(2) Verify (18.5.11) and (18.5.12) from properties of:
(a) $f_b(r \mid n, p)$
(b) $f_{Po}(r \mid \lambda t)$
(c) $f_N(\bar{x} \mid \mu, \sigma^2/n)$
(d) same as (c)
(e) same as (c)
(f) $f_N(\bar{x} \mid \mu, \sigma^2/n), f_{\gamma 2}(s^2 \mid \sigma^{-2}, n - 1)$.]

(*Note*: The importance of these results will become clear in Chapter 20. Essentially, they show that as $n \to \infty$ the discrepancy between these Bayesian estimators and their orthodox counterparts, introduced in Chapter 20, disappears provided that the prior dgf is proper and positive at every $\theta \in \Theta$.)

18. *Estimating the Area of a Blob* (*II*): Chapter 14, Exercise 19 introduced a Bayesian approach to estimating the area of a blob via Monte Carlo simulation. We extend that analysis here by assuming that if a denotes an estimate of the area $\tilde{A}$, then

$$\ell_t(a, A) = k_t(a - A)^2$$

and the cost $k(n)$ of n Monte Carlo observations is given by (18.1.2).

(a) Show that

$$a^0_{z_0,0} = \frac{R\rho}{\nu}, \tag{18.5.13}$$

$$a^0_{r,n} = \frac{R(\rho + r)}{\nu + n}, \tag{18.5.14}$$

$$E_A\ell_t(a^0_{z_0,0}, \tilde{A}) = k_t R^2 V_\beta(\tilde{p} \mid \rho, \nu), \tag{18.5.15}$$

$$E_{A|r,n}\ell_t(a^0_{r,n}, \tilde{A}) = k_t R^2 V_\beta(\tilde{p} \mid \rho + r, \nu + n), \tag{18.5.16}$$

$$I(n) = \left[\frac{n}{\nu + n}\right] k_t R^2 V_\beta(\tilde{p} \mid \rho, \nu), \tag{18.5.17}$$

$$n^* = R\left[\frac{k_t \nu\, V_\beta(\tilde{p} \mid \rho, \nu)}{k_s}\right]^{1/2} - \nu, \tag{18.5.18}$$

and

$$n^0 = \begin{cases} n^*, & n^* \geqq 0,\ I(n^*) - K_s - k_s n^* > 0 \\ 0, & \text{otherwise.} \end{cases} \tag{18.5.19}$$

(b) It is interesting to note that $n^* + \nu$ is proportional to the ratio R of the area of the original square to one. Explain in intuitive terms why this is reasonable.

19. Suppose observations $\tilde{\mathbf{x}}_1, \tilde{\mathbf{x}}_2, \ldots$ with common df $f_N^{(k)}(\mathbf{x} \mid \boldsymbol{\mu}, \boldsymbol{\sigma}^2)$ are to be obtained in order to improve judgments about $\boldsymbol{\mu}$, as in Subsections 18.4.3 through 18.4.5, except that now $a \in R^1$ is an estimate of $\mathbf{d}'\tilde{\boldsymbol{\mu}}$ and

$$\ell_t(a, \boldsymbol{\mu}) \equiv k_t(a - \mathbf{d}'\boldsymbol{\mu})^2.$$

(a) Suppose $\boldsymbol{\sigma}^2$ is known and $\tilde{\boldsymbol{\mu}}$ has the natural conjugate prior df $f_N^{(k)}(\boldsymbol{\mu} \mid \boldsymbol{\mu}_0, \boldsymbol{\sigma}_0^2)$. Show that

$$a^0_{z_0,0} = \mathbf{d}'\boldsymbol{\mu}_0, \tag{18.5.20}$$

$$E\,\ell_t(a^0_{z_0,0}, \tilde{\boldsymbol{\mu}}) = k_t\mathbf{d}'\boldsymbol{\sigma}_0^2\mathbf{d}, \tag{18.5.21}$$

$$\overset{0}{a}_{\bar{\mathbf{x}},n} = \mathbf{d}'\boldsymbol{\mu}_1, \tag{18.5.22}$$

$$E_{\mu|\bar{\mathbf{x}},n}\ell_t(\overset{0}{a}_{\bar{\mathbf{x}},n}, \tilde{\boldsymbol{\mu}}) = k_t\mathbf{d}'\boldsymbol{\sigma}_0^2\mathbf{d}, \text{ and} \tag{18.5.23}$$

$$I(n) = k_t\mathbf{d}'(\boldsymbol{\sigma}_0^2 - \boldsymbol{\sigma}_1^2)\mathbf{d}. \tag{18.5.24}$$

(b) Suppose $\boldsymbol{\sigma}^2$ is also unknown and $(\tilde{\boldsymbol{\mu}}, \tilde{\boldsymbol{\sigma}}^2)$ has the natural conjugate prior df $f_{NiW}^{(k)}(\boldsymbol{\mu}, \boldsymbol{\sigma}^2 \mid \boldsymbol{\mu}_0, \boldsymbol{\psi}_0, n_0, \nu_0)$. Show that (18.5.20) and (18.5.22) continue to obtain, but

$$E_{\mu|\sigma^2}\ell_t(a^0_{z_0,0}, \tilde{\boldsymbol{\mu}}) = \left[\frac{\nu_0}{\nu_0 - 2}\right]k_t\mathbf{d}'(n_0^{-1}\boldsymbol{\psi}_0)\mathbf{d}, \tag{18.5.25}$$

$$E_{\mu|\sigma^2|n,\bar{\mathbf{x}},\mathbf{s}^2}\ell_t(\overset{0}{a}_{\bar{\mathbf{x}},\mathbf{s}^2,n}, \tilde{\boldsymbol{\mu}}) = \left[\frac{\nu_1}{\nu_1 - 2}\right]k_t\mathbf{d}'(n_1^{-1}\boldsymbol{\psi}_1)\mathbf{d}, \tag{18.5.26}$$

$$E_{\bar{\mathbf{x}},\mathbf{s}^2|n}E_{\mu|\sigma^2|n,\bar{\mathbf{x}},\mathbf{s}^2}\ell_t(\overset{0}{a}_{\bar{\mathbf{x}},\mathbf{s}^2,n}, \tilde{\boldsymbol{\mu}}) = \left[\frac{\nu_0}{\nu_0 - 2}\right]k_t\mathbf{d}'(n_0^{-1}\boldsymbol{\psi}_0)\mathbf{d}, \tag{18.5.27}$$

$$\begin{aligned} I(n) &= \left[\frac{\nu_0}{\nu_0 - 2}\right]k_t\mathbf{d}'([n_0^{-1} - n_1^{-1})\boldsymbol{\psi}_1)\mathbf{d} \\ &= \left[\frac{n}{n_0 + n}\right]\left[\frac{\nu_0}{\nu_0 - 2}\right]k_t\mathbf{d}'(n_0^{-1}\boldsymbol{\psi}_0)\mathbf{d}, \end{aligned} \tag{18.5.28}$$

and, if $k(n)$ is given by (18.1.2),

$$n^* = \left[\left\{\frac{\nu_0}{\nu_0 - 2}\right\}\frac{k_t\mathbf{d}'(n_0^{-1}\boldsymbol{\psi}_0)\mathbf{d}}{k_s}\right]^{1/2} - n_0 \tag{18.5.29}$$

and

$$n^0 = \begin{cases} n^*, & n^* \geqq 0 \text{ and } I(n^*) - K_s - k_s n^* > 0 \\ 0, & \text{otherwise.} \end{cases} \tag{18.5.30}$$

[*Hints*: (1) To obtain (18.5.27) from (18.5.26), it is necessary to find $E_{\bar{\mathbf{x}},\mathbf{s}^2|n}(\tilde{\psi}_1)$; the appropriate definitions and distribution theory are in Chapter 14, with reference to Chapter 12 as necessary. (2) Do not confuse $\boldsymbol{\mu}_1 = n_1^{-1}(n_0\boldsymbol{\mu}_0 + n\bar{\mathbf{x}})$, in part (b), with $\boldsymbol{\mu}_1 = \boldsymbol{\sigma}_1^2(\boldsymbol{\sigma}_0^{-2}\boldsymbol{\mu}_0 + n\boldsymbol{\sigma}^{-2}\bar{\mathbf{x}})$, in part (a).]

(c) Suppose $\boldsymbol{\sigma}_*^2$ is known but $\tilde{\boldsymbol{\mu}}$ and $\tilde{\sigma}_*^2$ are unknown and have the natural conjugate prior df $f_{Ni\gamma}^{(k)}(\boldsymbol{\mu}, \boldsymbol{\sigma}_*^2 \mid \boldsymbol{\mu}_0, \psi_0, \boldsymbol{\sigma}_{0*}^2, \nu_0)$. Show that (18.5.20) and (18.5.22) continue to obtain, but

$$E_{\mu,\sigma^2}\ell_t(a^0_{z_0,0}, \tilde{\boldsymbol{\mu}}) = \left[\frac{\nu_0}{\nu_0 - 2}\right] k_t \mathbf{d}'(\psi_0 \boldsymbol{\sigma}_{0*}^2)\mathbf{d}, \tag{18.5.31}$$

$$E_{\mu,\sigma_*^2|n,\bar{\mathbf{x}},s^{*2}}\ell_t(a^0_{\bar{\mathbf{x}}s_*^2,n}, \tilde{\boldsymbol{\mu}}) = \left[\frac{\nu_1}{\nu_1 - 2}\right] k_t \mathbf{d}'(\psi_1 \boldsymbol{\sigma}_*^2)\mathbf{d}, \tag{18.5.32}$$

$$E_{\bar{\mathbf{x}},s_*^2|n}E_{\mu,\sigma^{*2}|n,\bar{\mathbf{x}},s_*^2}\ell_t(a_{\bar{\mathbf{x}}s_*^2,n}, \tilde{\boldsymbol{\mu}}) = \left[\frac{\nu_0}{\nu_0 - 2}\right] k_t \mathbf{d}'(\psi_0 \boldsymbol{\sigma}_{1*}^2)\mathbf{d}, \text{ and} \tag{18.5.33}$$

$$I(n) = \left[\frac{\nu_0}{\nu_0 - 2}\right] k_t \mathbf{d}'[\psi_0(\boldsymbol{\sigma}_{0*}^2 - \boldsymbol{\sigma}_{1*}^2)]\mathbf{d}. \tag{18.5.34}$$

The dependence of $\boldsymbol{\sigma}_{0*}^2 - \boldsymbol{\sigma}_{1*}^2$ upon n is too involved for analytic characterization of n^0.

[*Hint*: To obtain (18.5.33) from (18.5.32) it is necessary to find $E_{\bar{\mathbf{x}},s^{*2}|n}(\tilde{\psi}_1)$ for $\tilde{\psi}_1$ as defined by (14.6.14); use of (13.6.12) facilitates matters.]

20. *Stratified Sampling (III)*: Refer to Chapter 12 Exercise 54 and Chapter 14, Exercise 26 for an introduction to stratified sampling. The definitions and notation of those exercises will apply here. Suppose that an estimate of the over-all population mean $\tilde{\mu}$ is to be made, and the $\ell_t(a, \mu) = k_t(a - \mu)^2$ for all a and μ.

(a) Suppose that all stratum variances σ_i^2 are known beforehand. Show that

$$a^0_{z_0,0} = \boldsymbol{\pi}'\boldsymbol{\mu}_0, \tag{18.5.35}$$

$$a^0_{\bar{\mathbf{x}},\mathbf{n}} = \boldsymbol{\pi}'\boldsymbol{\mu}_1, \tag{18.5.36}$$

$$E_\mu \ell_t(a^0_{z_0,0}, \tilde{\boldsymbol{\mu}}) = k_t \boldsymbol{\pi}'\boldsymbol{\sigma}_0^2\boldsymbol{\pi} \tag{18.5.37}$$

$$E_{\mu|\bar{\mathbf{x}},\mathbf{n}}\ell_t(a^0_{\bar{\mathbf{x}},\mathbf{n}}, \tilde{\boldsymbol{\mu}}) = k_t \boldsymbol{\pi}'\boldsymbol{\sigma}_1^2(\mathbf{n})\boldsymbol{\pi}, \tag{18.5.38}$$

$$I(\mathbf{n}) = k_t \boldsymbol{\pi}'[\boldsymbol{\sigma}_0^2 - \boldsymbol{\sigma}_1^2(\mathbf{n})\boldsymbol{\pi}. \tag{18.5.39}$$

[Note that the prior value of information is a function of all r n_i's.]

(b) Suppose that the ratios σ_i^2/σ_j^2 of all pairs of stratum variances are known beforehand. Show that (18.5.35) and (18.5.36) remain true, while

$$E_{\mu} \ell_t(a^0_{z_0,0}, \tilde{\boldsymbol{\mu}}) = k_t\left[\frac{\nu_0}{\nu_0 - 2}\right]\boldsymbol{\pi}'[\psi_0 \boldsymbol{\sigma}_{0*}^2]\boldsymbol{\pi}, \tag{18.5.40}$$

$$E_{\mu|\bar{\mathbf{x}}, s_*^2, n} \ell_t(a^0_{\bar{\mathbf{x}}, s_*^2, n}, \tilde{\boldsymbol{\mu}}) = k_t\left[\frac{\nu_1}{\nu_1 - 2}\right]\boldsymbol{\pi}'[\psi_1 \boldsymbol{\sigma}_{1*}^2(\mathbf{n})]\boldsymbol{\pi}, \tag{18.5.41}$$

$$E_{\bar{\mathbf{x}}, s_*^2|n} E_{\mu|\bar{\mathbf{x}}, s_*^2, n} \ell_t(a^0_{\bar{\mathbf{x}}, s_*^2, n}, \tilde{\boldsymbol{\mu}}) = k_t\left[\frac{\nu_0}{\nu_0 - 2}\right]\boldsymbol{\pi}'[\psi_0 \boldsymbol{\sigma}_{1*}^2(\mathbf{n})]\boldsymbol{\pi}, \tag{18.5.42}$$

and

$$I(\mathbf{n}) = k_t\left[\frac{\nu_0}{\nu_0 - 2}\right]\boldsymbol{\pi}'(\psi_0[\boldsymbol{\sigma}_{0*}^2 - \boldsymbol{\sigma}_{1*}^2(\mathbf{n})])\boldsymbol{\pi}. \tag{18.5.43}$$

(c) Show that $I(\mathbf{n})$ can be expressed additively, as the sum $\sum_{i=1}^{r} I_i(n_i)$ of functions of the individual n_i's, if and only if $\boldsymbol{\sigma}_0^2$ [for part (a), or $\boldsymbol{\sigma}_{0*}^2$ for part (b)] is a diagonal matrix.

Stratified sampling is discussed in detail by Ericson ([**17**] through [**19**]) and introduced in greater depth than here by Pratt, Raiffa, and Schlaifer [**64**].

19

SOME OTHER BAYESIAN STATISTICAL DECISION PROBLEMS

19.1 INTRODUCTION

In Chapter 18 we applied the general theory in Chapter 17 to problems of point estimation. This chapter applies the general theory to some other decision problems which are important, both because they crop up frequently in real life and also because they furnish Bayesian analogues of problems in orthodox statistical inference.

In Section 19.2 we present a model of interval estimation which should be strongly reminiscent of point estimation with a linear loss function. A principal difference is that in this section the decision-maker is required to furnish two numbers, a and $b > a$, with the interpretation being that he asserts "$\theta \in (a, b)$." His opportunity loss depends not only upon whether or not θ does belong to (a, b), but also upon the width $b - a$ of the interval; he is penalized for vagueness.

Section 19.3 concerns problems in which there are precisely r acts $a_1, \ldots, a_r$, and in which the loss function $\ell_t(a, \theta)$ is derived from the assumption that the return $v(a_i, \theta)$ is a linear function of θ for each i. Such problems occur frequently in capital budgeting decisions (for example, whether a manufacturing process should be automated), and they constitute a Bayesian version of certain problems in the orthodox subject of hypothesis testing.

Section 19.4 applies the distribution theory in Section 14.7 to the "choice of the best of r processes" problem, in which $A = \{a_1, \ldots, a_r\}$ and $v(a_i, \mu_1, \ldots, \mu_r\} = K_i + k_i\mu_i$ for every i. As usual, we discuss optimal acts and value and net gain of information, although here we will of course replace $I(n)$ and $G(n)$ with $I(n_1, \ldots, n_r)$ and $G(n_1, \ldots, n_r)$, respectively.

Further information on the topics of Sections 19.3 and 19.4 can be found in Raiffa and Schlaifer [**66**] and in Pratt, Raiffa, and Schlaifer [**64**]. The Bracken and Schleifer tables [**13**] are of great value to readers who plan to make extensive application of the results in this chapter.

19.2 A BAYESIAN INTERVAL-ESTIMATION PROBLEM

19.2.1 General Theory

Sections 18.2 through 18.4 have dealt with situations in which the decision-maker was called upon to furnish a *single* number (18.2, 18.3) or vector (18.4) representing his best prediction of θ (18.2, 18.3) or $\boldsymbol{\theta}$ (18.4). Such problems are properly referred to as *point*-estimation problems.

In Section 9.7 we introduced point estimation as an important subject in statistical inference. Another important class of problems in inference requires the decision-maker to specify an interval $[a, b] \subset R^1$ to which it is reasonable to suppose that θ belongs. Such an interval is called a *confidence interval.* When $\Theta \subset R^k$ then, the decision-maker may name a more arbitrary subset $\Theta_I \subset R^k$ in place of an interval.

This section furnishes a Bayesian analogue of the interval-estimation problem for $\Theta \subset R^1$. The *acts* will be pairs (a, b) of real numbers such that $b \geqq a$ and the closed interval $[a, b] \subset \Theta$. Suppose also that $\tilde{\theta}$ is always a continuous random variable with a df $f(\theta)$.

The loss function $\ell_t((a, b), \theta) \equiv \ell_t(a, b; \theta)$ will be defined by

$$\ell_t(a, b; \theta) \equiv k_w(b - a) + \begin{cases} k_u(\theta - b), & \theta > b \\ 0, & a \leqq \theta \leqq b \\ k_O(a - \theta), & \theta < a, \end{cases} \tag{19.2.1}$$

where k_u, k_O, and k_w are positive real numbers and

$$k_w \leqq \frac{k_u k_o}{k_u + k_o} \equiv k_t. \tag{19.2.2}$$

The reason for (19.2.2) will be made clear shortly. Note that k_t is half the harmonic mean of k_u and k_O. The reader may show as Exercise 1 that

$$k_w \leqq k_t < \min \{k_u, k_O\}, \tag{19.2.3}$$

whenever $k_u > 0$, $k_O > 0$, $k_t \equiv k_u k_O/(k_u + k_O)$, and (19.2.2) obtains.

Figure 19.1 graphs $\ell_t(a, b; \theta)$ as a function of θ for two assumptions about (a, b). In the first, a_1 and $b_1 > a_1$ are arbitrary, while $(a_2 + b_2)/2 = (a_1 + b_1)/2$ but $b_2 - a_2 > b_1 - a_1$. Call the corresponding functions of θ ℓ_1 and ℓ_2, respectively. Thus ℓ_2 is the loss function of θ produced by "spreading (a_1, b_1) out to (a_2, b_2)."

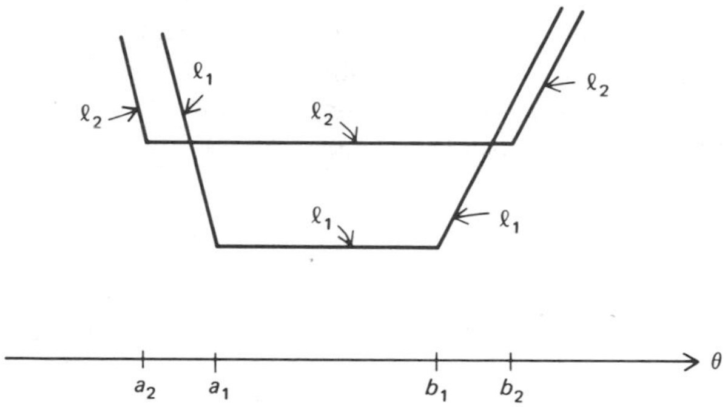

Figure 19.1

Note that (18.1.5) is the special case of (19.2.1) in which $b = a$. Thus (19.2.1) is a generalization of the linear loss function for point estimation. The symbols k_u and k_O have the same interpretations here as in Section 18.3, while k_w is a penalty imposed for "width." Clearly, stating that $\theta \in [-50, 50]$ is less informative than stating that $\theta \in [-.005, .005]$; and hence it is reasonable to suppose that the decision-maker's loss is an increasing function of $(b - a)$ for any fixed θ.

From (19.2.1) we easily obtain

$$\begin{aligned} E_\theta \ell_t(a, b; \tilde{\theta}) &= k_w(b - a) + k_u \int_b^\infty (\theta - b) f(\theta) d\theta \\ &\quad + k_O \int_{-\infty}^a (a - \theta) f(\theta) d\theta \qquad (19.2.4) \\ &= k_w(b - a) - k_u b[1 - F(b)] + k_O a F(a) \\ &\quad + k_u \int_b^\infty \theta f(\theta) d\theta - k_O \int_{-\infty}^a \theta f(\theta) d\theta. \end{aligned}$$

From (19.2.4) we obtain the first partial derivatives of $E_\theta \ell_t(a, b; \tilde{\theta})$ with respect to a and with respect to b, equate each to zero, and solve for a^0 and b^0 (Exercise 2), obtaining

$$F(a^0) = \frac{k_w}{k_O} \equiv L \qquad (19.2.5)$$

and

$$F(b^0) = \frac{k_u - k_w}{k_u} \equiv U. \qquad (19.2.6)$$

(*Note*: L and U are mnemonics for *lower* and *upper*, respectively.) Hence

$$[a^0, b^0] = [{}_L\theta, {}_U\theta], \qquad (19.2.7)$$

where ${}_x\theta \equiv x$th fractile of $\tilde{\theta}$. The reader may show as Exercise 3 that ${}_L\theta \leqq {}_U\theta$ because of (19.2.2); and as Exercise 4 that (a^0, b^0) passes the second-partials' test for a strict relative minimum.

If we substitute (19.2.7) in (19.2.4), we obtain

$$E_\theta \ell_t(a^0, b^0; \tilde{\theta}) = k_u \int_{{}_U\theta}^{\infty} \theta f(\theta) d\theta - k_O \int_{-\infty}^{{}_L\theta} \theta f(\theta) d\theta \qquad (19.2.8)$$

(Exercise 5), because the first three summands of the second equality of (19.2.4) mutually cancel at (a^0, b^0).

Two general features of the analysis to date are noteworthy. First, the width of the optimal interval estimate is the distance from the Lth to the Uth fractile of $\tilde{\theta}$, where L and U depends upon the three loss parameters k_w, k_u, and k_O.

Second, increasing k_u leaves L and hence a^0 unchanged but increases U and hence b^0. This makes sense, since the more serious the loss due to the interval falling below θ, the larger its right endpoint should be. Similarly, increasing k_O *decreases* L and hence a^0, thereby decreasing the left endpoint of the interval estimate. The intuitive interpretation is clear. Finally, increasing k_w [subject to (19.2.2)] increases L (and a^0) and decreases U (and b^0), thus producing a shorter interval estimate because of the increased penalty for width or vagueness.

Moreover, the optimal interval estimate $[a^0, b^0]$ is such that the decision-maker assesses probability

$$U - L = 1 - \frac{k_w}{k_t} \qquad (19.2.9)$$

to the event "$\tilde{\theta} \in [a^0, b^0]$." Note the this probability depends solely upon k_w, k_u, and k_O, and *not* upon $f(\theta)$. The probability $U - L$ of the prediction "$\theta \in [a^0, b^0]$" being correct is called a *Bayesian confidence coefficient*.

So far, $f(\theta)$ has been an arbitrary df for $\tilde{\theta}$. It represented either a prior or a posterior df. Accordingly, the analogues of (18.3.5) through (18.3.9) are:

$$a^0_{z_0,0} \text{ satisfies } \int_{-\infty}^{a^0_{z_0,0}} f(\theta) d\theta = \frac{k_w}{k_O}, \qquad (19.2.10a)$$

$$b^0_{z_0,0} \text{ satisfies } \int_{-\infty}^{b^0_{z_0,0}} f(\theta) d\theta = \frac{k_u - k_w}{k_u}, \qquad (19.2.10b)$$

$$a^0_{z,n} \text{ satisfies } \int_{-\infty}^{a^0_{z,n}} f(\theta \mid z, n) = \frac{k_w}{k_O}, \qquad (19.2.11a)$$

$$b^0_{z,n} \text{ satisfies } \int_{-\infty}^{b^0_{z,n}} f(\theta \mid z, n) = \frac{k_u - k_w}{k_u}. \tag{19.2.11b}$$

$$E_\theta \ell_t(a^0_{z_0,0}, b^0_{z_0,0}, \tilde{\theta}) = k_u \int_{b^0_{z_0,0}}^{\infty} \theta f(\theta) d\theta$$

$$-k_0 \int_{-\infty}^{a^0_{z_0,0}} \theta f(\theta) d\theta, \tag{19.2.12}$$

$$E_{\theta \mid z,n} \ell_t(a^0_{z,n}, b^0_{z,n}, \tilde{\theta}) = k_u \int_{b^0_{z,n}}^{\infty} \theta f(\theta \mid z, n) d\theta$$

$$- k_0 \int_{-\infty}^{a^0_{z,n}} \theta f(\theta \mid z, n) d\theta \tag{19.2.13}$$

and

$$G(n) = k_u \int_{b^0_{z_0,0}}^{\infty} \theta f(\theta) d\theta - k_0 \int_{-\infty}^{a^0_{z_0,0}} \theta f(\theta) d\theta$$

$$-E_{z \mid n} \left\{ k_u \int_{b^0_{\tilde{z},n}}^{\infty} \theta f(\theta \mid \tilde{z}, n) d\theta - k_0 \int_{-\infty}^{a^0_{\tilde{z},n}} \theta f(\theta \mid \tilde{z}, n) d\theta \right\} - k(n).$$

The perceptive reader may suspect that solving for n^0 will be subject to the same sorts of difficulties as those found in 18.3; that is correct. Hence we shall discuss only the two univariate Normal cases: σ^2 known in Subsection 19.2.2, and σ^2 unknown in Subsection 19.2.3.

19.2.2 The Univariate Normal Process: σ^2 known

In this and the next subsection we assume that $\tilde{\theta} = \tilde{\mu}$, the mean of a univariate Normal process. When σ^2 is known, $(n, \tilde{\bar{x}})$ is sufficient for $\tilde{\mu}$, which has the natural conjugate prior df $f_N(\mu \mid \mu_0, \sigma^2/n_0)$.

Before stating the principal results for this case, we cite two easily proved (Exercise 6) facts about certain Normal integrals:

Let $\tilde{x}$ have the Normal df $f_N(x \mid m, D^2)$. (The symbols m and D^2 are intended as neutral notations for mean and variance.) Then for any real numbers a and b (19.2.15)

$$\int_{-\infty}^{a} x f_N(x \mid m, D^2) dx = m F_{N*}\left(\frac{a - m}{D}\right) - D f_{N*}\left(\frac{a - m}{D}\right)$$

and

$$\int_{b}^{\infty} x f_N(x \mid m, D^2) dx = m\left[1 - F_{N*}\left(\frac{b - m}{D}\right)\right] + D f_{N*}\left(\frac{b - m}{D}\right).$$

These facts enable us to derive n^0 by almost exactly the same procedure as in Subsection 18.3.2. We have

$$a^0_{z_0,0} = \mu_0 + {}_L\alpha\frac{\sigma}{\sqrt{n_0}}, \tag{19.2.16a}$$

$$b^0_{z_0,0} = \mu_0 + {}_U\alpha\frac{\sigma}{\sqrt{n_0}}, \tag{19.2.16b}$$

$$a^0_{\bar{x},n} = \mu_1 + {}_L\alpha\frac{\sigma}{\sqrt{n_0 + n}}, \text{ and} \tag{19.2.17a}$$

$$b^0_{\bar{x},n} = \mu_1 + {}_U\alpha\frac{\sigma}{\sqrt{n_0 + n}}, \tag{19.2.17b}$$

where ${}_L\alpha$ and ${}_U\alpha$ are the Lth and Uth fractiles, respectively, of the standardized Normal random variable (see Table I) and μ_1 is the mean of the posterior distribution of $\tilde{\mu}$. The reader may prove (19.2.16a) through (19.2.17b) as Exercise 7.

From (19.2.14), (19.2.15), and $k(n)$ given by (18.1.2), it follows (Exercise 8) that

$$G(n) = K\sigma[n_0^{-1/2} - (n_0 + n)^{-1/2}] - k(n), \tag{19.2.18}$$

where

$$K \equiv k_u f_{N^*}({}_U\alpha) + k_0 f_{N^*}({}_L\alpha), \tag{19.2.19}$$

and hence

$$n^* = \left[\frac{2K\sigma}{k_s}\right]^{2/3} - n_0, \tag{19.2.20}$$

thus implying that

$$n^0 = \begin{cases} n^*, & n^* \geqq 0 \text{ and } G(n^*) \geqq 0 \\ 0, & \text{otherwise.} \end{cases} \tag{19.2.21}$$

It is instructive to compare (19.2.18) with (18.3.14), and (19.2.20) with (18.3.15). If we had defined K in (18.3.14) and (18.3.15) as $(k_u + k_0)f_{N^*}(\alpha^0)$, then (18.3.14) and (19.2.18) would be formally identical, and so would (18.3.15) and (19.2.20).

Note that the width of the posterior interval estimate for any n depends only upon n and not upon $\tilde{\bar{x}}$:

$$b^0_{\tilde{\bar{x}},n} - a^0_{\tilde{\bar{x}},n} = \frac{({}_U\alpha - {}_L\alpha)\sigma}{\sqrt{n_0 + n}}. \tag{19.2.22}$$

Furthermore, as n tends to infinity the width (19.2.22) of the interval estimate tends to zero.

19.2.3 The Univariate Normal Process: σ^2 unknown

When all assumptions are as in Subsection 19.2.2 except that σ^2 is also unknown and $(\tilde{\mu}, \tilde{\sigma}^2)$ is given a normal-inverted gamma prior df as defined by (14.5.5), the problem is more difficult than in Subsection 19.2.2. We shall be able to evaluate $G(n)$ for any n, but not n^* or n^0. Recall from (14.5.7) that the marginal prior and posterior df's of $\tilde{\mu}$ are $f_S(\mu \mid \mu_0, \psi_0/n_0, \nu_0)$ and $f_S(\mu \mid \mu_1, \psi_1/n_1, \nu_1)$, respectively, where n_1, ν_1, μ_1, and ψ_1 are defined in (14.5.7).

First, the Student analogue of (19.2.15) follows:

Let $\tilde{x}$ have the Student df $f_S(x \mid m, D^2, \nu)$. Then for any real numbers a and b, (19.2.23)

$$\int_{-\infty}^{a} x f_S(x \mid m, D^2, \nu)\ dx$$

$$= mF_{S*}\left(\frac{a-m}{D}\Big|\nu\right) - D\cdot\frac{\nu + \left(\frac{a-m}{D}\right)^2}{\nu - 1} f_{S*}\left(\frac{a-m}{D}\Big|\nu\right)$$

and

$$\int_{b}^{\infty} x f_S(x \mid m, D^2, \nu)\ dx$$

$$= m\left[1 - F_{S*}\left(\frac{b-m}{D}\Big|\nu\right)\right] + D\cdot\frac{\nu + \left(\frac{b-m}{D}\right)^2}{\nu - 1} f_{S*}\left(\frac{b-m}{D}\Big|\nu\right).$$

Proof

Omitted—see Raiffa and Schlaifer [**66**], 235.

Then

$$a^0_{z_0,0} = \mu_0 + {}_L\alpha(\nu_0)\frac{\psi_0^{1/2}}{\sqrt{n_0}}, \tag{19.2.24a}$$

$$b^0_{z_0,0} = \mu_0 + {}_U\alpha(\nu_0)\frac{\psi_0^{1/2}}{\sqrt{n_0}}, \tag{19.2.24b}$$

$$a^0_{\bar{x},s^2,n} = \mu_1 + {}_L\alpha(\nu_1)\frac{\psi_1^{1/2}}{\sqrt{n_1}}, \text{ and} \tag{19.2.25a}$$

$$b^0_{\bar{x},s^2,n} = \mu_1 + {}_U\alpha(\nu_1)\frac{\psi_1^{1/2}}{\sqrt{n_1}}, \tag{19.2.25b}$$

where ${}_L\alpha(\nu)$ and ${}_U\alpha(\nu)$ are the Lth and Uth fractiles, respectively of the standardized Student random variable with ν degrees of freedom. Recall that ψ_1 depends upon all of n, $\bar{x}$, and s^2. [Proofs of (19.2.24a) through (19.2.25b) constitute Exercise 9.] Define a function $K(\nu)$ by

$$K(\nu) \equiv k_u \frac{\nu + [{}_U\alpha(\nu)]^2}{\nu - 1} f_{S^*}({}_U\alpha(\nu) \mid \nu) + k_0 \frac{\nu + [{}_L\alpha(\nu)]^2}{\nu - 1} f_{S^*}({}_L\alpha(\nu) \mid \nu). \quad (19.2.26)$$

Then (19.2.8) and (19.2.24) through (19.2.26) imply

$$E_\mu \ell_t(a^0_{z_0,0}, b^0_{z_0,0}, \tilde{\mu}) = K(\nu_0)\left(\frac{\psi_0}{n_0}\right)^{1/2} \quad (19.2.27)$$

and

$$E_{\mu|\bar{x},s^2,n}\ell_t(a^0_{\bar{x},s^2,n}, b^0_{\bar{x},s^2,n}, \tilde{\mu}) = K(\nu_1)\left(\frac{\psi_1}{n_1}\right)^{1/2} \quad (19.2.28)$$

The reader may verify (19.2.27) and (19.2.28) as Exercise 10. Now, it can be shown that

$$E\left[\left(\frac{\tilde{\psi}_1}{n_1}\right)^{1/2}\right] = \frac{\Gamma(\frac{1}{2}\nu_1)\Gamma(\frac{1}{2}\nu_0 - \frac{1}{2})}{\Gamma(\frac{1}{2}\nu_0)\Gamma(\frac{1}{2}\nu_1 - \frac{1}{2})}\left[\frac{n_0\nu_0}{n_1\nu_1}\right]^{1/2}\left(\frac{\psi_0}{n_0}\right)^{1/2}; \quad (19.2.29)$$

this fact was used to obtain (18.3.23) from (18.3.22). From (19.2.28) and (19.2.29), we obtain

$$E_{\bar{x},s^2|n}E_{\mu|\bar{x},s^2,n}\ell_t(a^0_{\tilde{\bar{x}},\tilde{s}^2,n}, b^0_{\tilde{\bar{x}},\tilde{s}^2,n}, \tilde{\mu})$$
$$= \frac{\Gamma(\frac{1}{2}\nu_1)\Gamma(\frac{1}{2}\nu^0 - \frac{1}{2})}{\Gamma(\frac{1}{2}\nu_0)\Gamma(\frac{1}{2}\nu_1 - \frac{1}{2})}\left[\frac{n_0\nu_0}{n_1\nu_1}\right]^{1/2} K(\nu_1)\left(\frac{\psi_0}{n_0}\right)^{1/2}, \quad (19.2.30)$$

which is a very complicated function of n. From (19.2.27) and (19.2.30), we obtain

$$G(n) = \left[K(\nu_0) - \frac{\Gamma(\frac{1}{2}\nu_1)\Gamma(\frac{1}{2}\nu_0 - \frac{1}{2})}{\Gamma(\frac{1}{2}\nu_0)\Gamma(\frac{1}{2}\nu_1 - \frac{1}{2})}\left[\frac{n_0\nu_0}{n_1\nu_1}\right]^{1/2} K(\nu_1)\right]\left(\frac{\psi_0}{n_0}\right)^{1/2} - k(n). \quad (19.2.31)$$

If $k(n)$ is given by (18.1.2), then we may conclude that $n^0 \in \{0, 1, \ldots, n^\dagger\}$, where

$$n^\dagger = \min\left\{n: n \geqq \frac{K(\nu_0)(\psi_0/n_0)^{1/2} - K_s}{k_s}, n \text{ is a nonnegative integer}\right\}; \quad (19.2.32)$$

the argument parallels that in Subsection 18.3.3 precisely.

19.3 r-ACT PROBLEMS WITH LINEAR RETURN

19.3.1 General Theory

In this section we consider problems in which there are only r available acts, $A = \{a_1, \ldots, a_r\}$, regardless of experimentation, and in which the decision-maker's terminal return $v_t(a, \theta)$ is given by

$$v_t(a_i, \theta) = K_i + k_i\theta, \text{ for every } \theta \in \Theta \subset R^1 \text{ and } i \in \{1, \ldots, r\}, \quad (19.3.1)$$

for real numbers $K_1, \ldots, K_r$ and $k_1, \ldots, k_r$.

We shall now make two assumptions which can always be forced to obtain without any loss of essential generality. *First*, we shall assume that each a_i is optimal for some $\theta \in \Theta$. Otherwise, if some a_j is optimal for no θ, then a_j is optimal for no dgf $f(\theta)$ of $\tilde{\theta}$; hence a_j can be deleted from A without any conceivable reduction in the prior utility of the decision problem.

Notice that if $k_i = k_j$ for two acts a_i and a_j, then a_i should be deleted if $K_i < K_j$ and a_j should be deleted if $K_j < K_i$. If $K_i = K_j$ as well as $k_i = k_j$, then a_i and a_j yield exactly the same returns for every $\theta \in \Theta$; hence one of them can be deleted to simplify the problem; in this case, a_i and a_j are indistinguishable as far as return is concerned. Thus we can assume that no two k's are equal.

Second, we shall assume that the acts have been numbered so that $k_1 < k_2 < \cdots < k_r$. This assumption obviously can always be forced to obtain simply by properly labeling the acts.

The gist of these two assumptions is that the optimal act $a_\theta{}^0$ given θ satisfies:

$$a_\theta{}^0 = a_i, \qquad \text{if and only if } \theta_{i-1}^b \leqq \theta \leqq \theta_i{}^b,$$

where

$$\theta_o{}^b \equiv -\infty,$$

$$\theta_i{}^b = \frac{K_i - K_{i+1}}{k_{i+1} - k_i}, \qquad \text{for } i = 1, \ldots, r-1, \text{ and} \qquad (19.3.2)$$

$$\theta_r{}^b = \equiv +\infty.$$

Proof

Since $k_1 < k_2$, it follows from (19.3.1) that $v_t(a_1, \theta) \geqq v_t(a_2, \theta)$ if and only if $\theta \leqq (K_1 - K_2)/(k_2 - k_1) \equiv \theta_1{}^b$. Similarly, $v_t(a_2, \theta) \geqq v_t(a_3, \theta)$ if and only if $\theta \leqq \theta_2{}^b, \ldots,$ and $v_t(a_{r-1}, \theta) \geqq v_t(a_r, \theta)$ if and only if $\theta \leqq \theta_{r-1}^b$.

It follows from the preceding two assumptions that $\theta_1{}^b \leqq \theta_2{}^b \leqq \cdots \leqq \theta^b_{r-1}$, and hence

$$v_t(a_1, \theta) = \max_i v_t(a_i, \theta), \text{ for } \theta \leqq \theta_1{}^b,$$

$$v_t(a_2, \theta) = \max_i v_t(a_i, \theta), \text{ for } \theta_1{}^b \leqq \theta \leqq \theta_2{}^b, \ldots,$$

$$v_t(a_{r-1}, \theta) = \max_i v_t(a_i, \theta), \text{ for } \theta^b_{r-2} \leqq \theta \leqq \theta^b_{r-1}, \text{ and}$$

$$v_t(a_r, \theta) = \max_i v_t(a_i, \theta), \text{ for } \theta \geqq \theta^b_{r-1}.$$

Statement (19.3.2) now follows from the definitions of $\theta_0{}^b$ and $\theta_r{}^b$. QED

Figure 19.2 depicts all $v_t(a_i, \theta)$ for $r = 4$ and typical values of the K_i's and k_i's satisfying our two assumptions. Note that the $\theta_i{}^b$'s are *break-even* values of θ, since a_i and a_{i+1} are equally desirable when $\theta = \theta_i{}^b$.

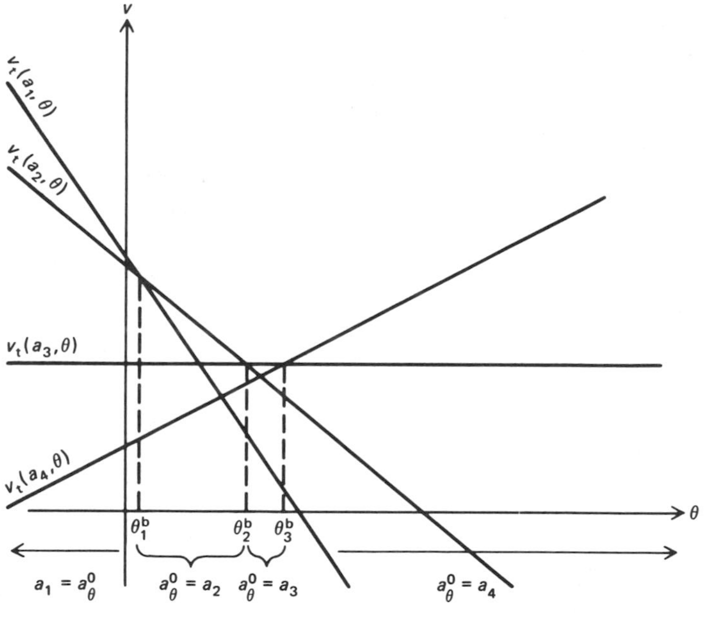

Figure 19.2

Now let $f(\theta)$ denote any dgf of $\tilde{\theta}$, and let E denote expectation with respect to $f(\theta)$. From (19.3.1), we obtain

$$Ev_t(a_i, \tilde{\theta}) = K_i + k_i E(\tilde{\theta}), \; i \in \{1, \ldots, r\}, \tag{19.3.3}$$

and hence the optimal act a^0 under uncertainty satisfies

$$a^0 = a_i, \text{ if and only if } \theta_{i-1}^b \leqq E(\tilde{\theta}) \leqq \theta_i{}^b, \tag{19.3.4}$$

where $\theta_0{}^b, \ldots, \theta_r{}^0$ are as defined in (19.3.2).

Proof

Imitate exactly the proof of (19.3.2), using (19.3.3) instead of (19.3.1). QED

Thus for $f(\theta)$ denoting a prior dgf, and $f(\theta \mid z, n)$ denoting a posterior dgf, we have

$$a^0_{z_0,0} = a_i, \text{ if and only if } \theta_{i-1}^b \leqq E(\tilde{\theta}) \leqq \theta_i{}^b \tag{19.3.5}$$

and

$$a^0_{z,n} = a_i, \text{ if and only if } \theta_{i-1}^b \leqq E(\tilde{\theta} \mid z, n) \leqq \theta_i{}^b. \tag{19.3.6}$$

We shall now derive the terminal opportunity loss function $\ell_t(a, \theta)$, for purposes of information evaluation.

$$\ell_t(a_i, \theta) = \sum_{j=1}^{i-1} (k_{j+1} - k_j) \max\{0, \theta_j{}^b - \theta\} + \sum_{j=i}^{r-1} (k_{j+1} - k_j) \max\{0, \theta - \theta_j{}^b\}, \tag{19.3.7}$$

or $i = 1, \ldots, r$.

Proof

Suppose $\theta_{i-1}^b \leqq \theta \leqq \theta_i{}^b$. Then $a^0 = a_i$ and

$$\ell_t(a_i, \theta) \equiv v_t(a_\theta{}^0, \theta) - v_t(a_i, \theta) = 0;$$

and it is straightforward to verify that (19.3.7) is also zero. Hence (19.3.7) is true for $\theta \in [\theta_{i-1}^b, \theta_i{}^b]$. Now suppose that $\theta \in [\theta_{h-1}^b, \theta_h{}^b]$ for $h < i$. Then $a_\theta{}^0 = a_h$, and

$$\ell_t(a_i, \theta) = v_t(a_h, \theta) - v_t(a_i, \theta).$$

Now,

$$\begin{aligned} v_t(a_h, \theta) - v_t(a_i, \theta) = v_t(a_h, \theta) &- v_t(a_{h+1}, \theta) \\ &+ v_t(a_{h+1}, \theta) - v_t(a_{h+2}, \theta) \\ &+ v_t(a_{h+2}, \theta) - v_t(a_{h+3}, \theta) \\ &+ \\ &\;\;\vdots \\ &+ v_t(a_{i-1}, \theta) - v_t(a_i, \theta); \end{aligned}$$

that is, both add and subtract all $v_t(a_j, \theta)$'s for all $j \in \{h + 1, \ldots, i - 1\}$— if there are any. It is easy to verify that

$$\begin{aligned} v_t(a_j, \theta) - v_t(a_{j+1}, \theta) &= (K_j + k_j\theta) - (K_{j+1} + k_{j+1}\theta) \\ &= (K_j - K_{j+1}) + (k_j - k_{j+1})\theta \\ &= (k_{j+1} - k_j)(\theta_j{}^b - \theta), \end{aligned}$$

and hence

$$\begin{aligned} \ell_t(a_i, \theta) &= \sum_{j=h}^{i-1} (k_{j+1} - k_j)(\theta_j{}^b - \theta) \\ &= \sum_{j=h}^{i-1} (k_{j+1} - k_j)(\theta_j{}^b - \theta) + \sum_{j=1}^{h-1} (k_{j+1} - k_j) \cdot 0 \\ &= \sum_{j=1}^{i-1} (k_{j+1} - k_j) \max \{0, \theta_j{}^b - \theta\}. \end{aligned}$$

Since for $\theta \in [\theta^b_{h-1}, \theta_h{}^b]$ and $h < i$ we have $\theta < \theta_j{}^b$ for all $j \in \{\text{i}, i + 1, \ldots, r - 1\}$, it follows that $\max \{0, \theta - \theta_j{}^b\} = 0$ and hence

$$\sum_{j=i}^{r-1} (k_{j+1} - k_j) \max \{0, \theta - \theta_j{}^b\} = 0$$

for this case. Hence (19.3.7) holds for $\theta \in [\theta^b_{h-1}, \theta_h{}^b]$ and $h < i$. The final case is that in which $\theta \in [\theta^b_{h-1}, \theta_h{}^b]$ and $h > i$. Then the first summation in (19.3.7) is zero and we must show that

$$\begin{aligned} \ell_t(a_i, \theta) &= \sum_{j=i}^{r-1} (k_{j+1} - k_j) \max \{0, \theta - \theta_j{}^b\} \\ &= \sum_{j=i}^{h-1} (k_{j+1} - k_j)(\theta - \theta_j{}^b). \end{aligned}$$

For this, we use the same trick of adding and subtracting $v_t(a_j, \theta)$'s to obtain

$$\begin{aligned} \ell_t(a_i, \theta) &= v_t(a_h, \theta) - v_t(a_i, \theta) \\ &= v_t(a_h, \theta) - v_t(a_{h-1}, \theta) \\ &\quad + v_t(a_{h-1}, \theta) - v_t(a_{h-2}, \theta) \\ &\quad + \\ &\quad \cdot \\ &\quad \cdot \\ &\quad \cdot \\ &\quad + v_t(a_{i+1}, \theta) - v_t(a_i, \theta). \end{aligned}$$

It is easy to verify that

$$v_t(a_{j+1}, \theta) - v_t(a_j, \theta) = (k_{j+1} - k_j)(\theta - \theta_j{}^b),$$

for any j, and hence

$$\ell_t(a_i, \theta) = \sum_{j=i}^{h-1} (k_{j+1} - k_j)(\theta - \theta_j{}^b),$$

thus concluding the verification of (19.3.7), for $\theta \in [\theta_{h-1}^b, \theta_h{}^b]$ and $h > i$. QED

From (19.3.7), it follows that

$$E\ell_t(a_i, \tilde{\theta}) = \sum_{j=1}^{i-1} (k_{j+1} - k_j) \int_{-\infty}^{\theta_j{}^b} (\theta_j{}^b - \theta) f(\theta) d\theta$$
$$+ \sum_{j=i}^{r-1} (k_{j+1} - k_j) \int_{\theta_j{}^b}^{\infty} (\theta - \theta_j{}^b) f(\theta) d\theta, \qquad (19.3.8)$$

which the reader may derive as Exercise 11. Now, (19.3.8) enables one to find the prior value of perfect information, which is simply $E\ell_t(a_{i_0}, \tilde{\theta})$, where $a_{i_0} \equiv a^0_{z_0,0}$ as determined by (19.3.5).

From the apparent complexity of the terminal opportunity loss function, one might suspect that any closed-form expression for $I(n)$ would be truly formidable. Fortunately, $I(n)$ is expressible in a form very similar to $E\ell_t(a^0_{z_0,0}, \tilde{\theta})$, the value of perfect information. Specifically:

Let $\omega \equiv E(\tilde{\theta} \mid z, n)$ and $\tilde{\omega} \equiv E(\tilde{\theta} \mid \tilde{z}, n)$; that is, $\tilde{\omega}$ is the expectation of $\tilde{\theta}$ posterior to observing $(\tilde{z}, n)$ and is hence a function of $(n, \tilde{z})$ and therefore a random variable. Denote the dgf of $\tilde{\omega}$ by $f_n(\omega)$. Then (19.3.9)

$$I(n) = \sum_{j=1}^{i_0} (k_{j+1} - k_j) \int_{-\infty}^{\theta_j{}^b} (\theta_j{}^b - \omega) f_n(\omega) d\omega$$
$$+ \sum_{j=i_0}^{r-1} (k_{j+1} - k_j) \int_{\theta_j{}^b}^{\infty} (\omega - \theta_j{}^b) f_n(\omega) d\omega,$$

where $a_{i_0} \equiv a^0_{z_0,0}$.

Proof

Since $E_\theta = E_{z|n} E_{\theta|z,n}$, it follows that

$$\begin{aligned} I(n) &= E_{z|n} E_{\theta|z,n} v_t(a^0_{\tilde{z},n}, \tilde{\theta}) - E_\theta v_t(a^0_{z_0,0}, \tilde{\theta}) \\ &= E_{z|n} E_{\theta|z,n} v_t(a^0_{\tilde{z},n}, \tilde{\theta}) - E_{z|n} E_{\theta|z,n} v_t(a^0_{z_0,0}, \tilde{\theta}) \\ &= E_{z|n} \{ E_{\theta|z,n} v_t(a^0_{\tilde{z},n}, \tilde{\theta}) - E_{\theta|z,n} v_t(a^0_{z_0,0}, \tilde{\theta}). \end{aligned}$$

Now, from (19.3.1) it follows that

$$E_{\theta|z,n} v_t(a_i, \tilde{\theta}) = v_t(a_i, E(\tilde{\theta} \mid z, n) \equiv v_t(a_i, \omega),$$

for every $a_i \in A$, as the reader may verify for Exercise 12. Hence

$$\begin{aligned} I(n) &= E_{z|n}\{v_t(a^0_{\tilde{z},n}, E(\tilde{\theta} \mid \tilde{z}, n)) - v_t(a^0_{z_0,0}, E(\tilde{\theta} \mid \tilde{z}, n))\}. \\ &= E_{z|n}\{\max_i v_t(a_i, E(\tilde{\theta} \mid \tilde{z}, n)) - v_t(a^0_{z_0,0}, E(\tilde{\theta} \mid \tilde{z}, n))\}. \end{aligned}$$

Now define $\tilde{\omega} \equiv E(\tilde{\theta} \mid \tilde{z}, n)$ and note that the $E_{z|n}$-expectation may be replaced by the $E_{\omega|n}$ expectation, so that

$$\begin{aligned} I(n) &= E_{\omega|n}\{\max_i v_t(a_i, \tilde{\omega}) - v_t(a^0_{z_0,0}, \tilde{\omega})\}. \\ &\equiv E_{\omega|n}\{\max_i v_t(a_i, \tilde{\omega}) - v_t(a_{i_0}, \tilde{\omega})\}. \end{aligned}$$

But

$$\max_i v_t(a_i, \tilde{\omega}) - v_t(a_{i_0}, \tilde{\omega}) = \ell_t(a_{i_0}, \tilde{\omega}), \text{ where } \ell_t(a_{i_0}, \omega)$$

is as given by (19.3.7) with ω replacing θ everywhere. Hence $I(n)$ is expressible by (19.3.9), which is simply (19.3.8) with ω, $f_n(\omega)$, and $d\omega$ replacing θ, $f(\theta)$, and $d\theta$, respectively. QED

The random variables $\tilde{\omega} \equiv E(\tilde{\theta} \mid \tilde{z}, n)$ play an important role in decision theory, since they are relevant in any decision problem in which $a^0_{z,n}$ depends solely upon ω and upon no other feature of the posterior dgf $f(\theta \mid z, n)$. Such problems include the quadratic loss estimation problems (Sections 18.2 and 18.4) and the r-processes problems (Section 19.5) as well as the r-act problems of this section.

The distribution of $\tilde{\omega}$ given n is, technically, the *prior* distribution of the mean of the *posterior* distribution, or the *prior* distribution of the *posterior* mean. Raiffa and Schlaifer [**66**] abbreviated this to *preposterior distribution*, while students refer to it apocryphally as the "preposterous" distribution.

Exercise 13 requires the derivation of the following facts about $\tilde{\omega}$.

If the prior expectation $E(\tilde{\theta})$ exists, then $E(\tilde{\omega} \mid n) = E(\tilde{\theta})$, for every n, (19.3.10)

and if the prior variance $V(\tilde{\theta})$ exists, then (19.3.11)

$$V(\tilde{\omega} \mid n) = V(\tilde{\theta}) - E_{z|n}[V(\tilde{\theta} \mid \tilde{z}, n)].$$

Other important properties of $\tilde{\omega}$ can be found in Raiffa and Schlaifer [**66**], Section 5.4.

In Subsection 19.3.2 we specialize the preceding analysis to the case of $\tilde{\theta}$ = the mean $\tilde{\mu}$ of a univariate Normal process with known variance σ^2. The same case but with σ^2 unknown is considereed in Subsection 19.3.3.

19.3.2 The Univariate Normal Process: σ^2 known

We shall require the following result about

$$\tilde{\omega} \equiv E_N(\tilde{\mu} \mid \tilde{\mu}_1, \sigma^2/n_1).$$

If n observations are to be obtained from $f_N(x \mid \mu, \sigma^2)$, if σ^2 is known, and if $\tilde{\mu}$ has the natural conjugate prior df $f_N(\mu_0, \sigma^2/n_0)$, then (19.3.12)

$$\tilde{\omega} \equiv \left(\frac{n_0}{n_0 + n}\right)\mu_0 + \left(\frac{n}{n_0 + n}\right)\tilde{\bar{x}}$$

has df $f_N\left(\omega \mid \mu_0, \left(\frac{n}{n_0 + n}\right)\sigma^2/n_0\right)$.

Proof

By (14.5.2), $\tilde{\bar{x}}$ has unconditional df $f_N(\bar{x} \mid \mu_0, (n_0^{-1} + n^{-1})\sigma^2)$. By definition, $\tilde{\omega}$ is a linear function of $\tilde{\bar{x}}$, and hence (19.3.12) follows by very little algebra from (12.4.10). QED

We shall also require some specific information about the integrals which appear in (19.3.8) and (19.3.9) for the Normal case.

Let

$$Ł_{N*}(z) \equiv \int_z^\infty (y - z) f_{N*}(y) dy. \tag{19.3.13}$$

Then

$$Ł_{N*}(z) = f_{N*}(z) - z[1 - F_{N*}(z)]; \tag{19.3.14}$$

$$Ł_{N*}(-z) = Ł_{N*}(z) + z; \tag{19.3.15}$$

$$\int_a^\infty (x - a) f_N(x \mid m, D^2) dx = DŁ_{N*}\left(\frac{a - m}{D}\right); \tag{19.3.16}$$

and

$$\int_{-\infty}^a (a - x) f_N(a \mid m, D^2) dx = DŁ_{N*}\left(-\frac{a - m}{D}\right). \tag{19.3.17}$$

Proof

From (19.3.13),

$$\begin{aligned} Ł_{N*}(z) &= \int_z^\infty y f_{N*}(y) dy - z \int_z^\infty f_{N*}(y) dy \\ &= \int_z^\infty y f_{N*}(y) dy - z[1 - F_{N*}(z)] \\ &= f_{N*}(z) - z[1 - F_{N*}(z)], \end{aligned}$$

by (19.2.15), thus establishing (19.3.14). For (19.3.15), (19.3.14) implies

$$\begin{aligned} Ł_{N*}(-z) &= f_{N*}(-z) + z[1 - F_{N*}(-z)] \\ &= f_{N*}(z) + zF_{N*}(z) \end{aligned}$$

because $f_{N*}(z) = f_{N*}(-z)$ and $F_{N*}(z) = 1 - F_{N*}(-z)$. Equation (19.3.15) now follows by subtraction from (19.3.14). Equation (19.3.16) follows from (19.3.13) by the change of variable from x to $(x - m)/D$. For (19.3.17), observe that for any df $f(x)$ and related expectation E, we have

$$\begin{aligned} E(\tilde{x}) - a &= \int_{-\infty}^{\infty} (x - a) f(x) dx \\ &= \int_{a}^{\infty} (x - a) f(x) dx + \int_{-\infty}^{a} (x - a) f(x) dx \\ &= \int_{a}^{\infty} (x - a) f(x) dx - \int_{-\infty}^{a} (a - x) f(x) dx, \end{aligned}$$

which implies

$$\int_{-\infty}^{a} (a - x) f(x) dx = \int_{a}^{\infty} (x - a) f(x) dx + a - E(\tilde{x}).$$

Hence by (19.3.16),

$$\begin{aligned} \int_{-\infty}^{a} (a - x) f_N(x \mid m, D^2) dx &= D Ł_{N*}\left(\frac{a - m}{D}\right) + a - m \\ &= D\left[Ł_{N*}\left(\frac{a - m}{D}\right) + \frac{a - m}{D}\right] = D Ł_{N*}\left(-\frac{a - m}{D}\right), \end{aligned}$$

by (19.3.15). QED

The function L_{N*} is tabulated in Pratt, Raiffa, and Schlaifer [**64**], and also in Bracken and Schleifer [**13**] as the special case of the Student distribution with $\nu = \infty$ degrees of freedom. Equation (19.3.14) shows how its values can be determined from Tables I and VI, while (19.3.15) shows how to find its values for untabulated, negative arguments.

We are now in a position to state the main results for the univariate Normal processes with known variance σ^2, where $\tilde{\mu}$ is given the conjugate prior dgf $f_N(\mu \mid \mu_0, \sigma^2/n_0)$ and hence has the conjugate posterior df $f_N(\mu \mid \mu_1, \sigma^2/n_1)$:

$$a^0_{0z,0} = a_i, \text{ if and only if } \mu^b_{i-1} \leqq \mu_0 \leqq \mu_i{}^b; \tag{19.3.18}$$

$$a^0_{\tilde{x},n} = a_i, \text{ if and only if } \mu^b_{i-1} \leqq \mu_1 \leqq \mu_i{}^b; \tag{19.3.19}$$

$$E_\mu \ell_t(a^0_{z_0,0}, \tilde{\mu}) = \sigma_0 \sum_{j=1}^{r-1} (k_{j+1} - k_j) Ł_{N*} \frac{d_j}{\sigma_0}; \tag{19.3.20}$$

and

$$I(n) = \left[\frac{n}{n_0 + n}\right]^{1/2} \sigma_0 \sum_{j=1}^{r-1} (k_{j+1} - k_j) Ł_{N*}\left(d_j \Big/ \left\{\left[\frac{n}{n_0 + n}\right]^{1/2} \sigma_0\right\}\right), \tag{19.3.21}$$

where

$$\sigma_0 \equiv \left(\frac{\sigma^2}{n_0}\right)^{1/2} \tag{19.3.22}$$

and

$$d_j \equiv |\mu_j{}^b - \mu_0|. \tag{19.3.23}$$

Proof

Equations (19.3.18) and (19.3.19) are immediate from (19.3.5) and (19.3.6) and the fact that the prior and posterior means of $\tilde{\mu}$ are μ_0 and μ_1, respectively. For (19.3.2), we apply (19.3.16) and (19.3.17) with $m = \mu_0$ and $D^2 = \sigma_0{}^2$ to (19.3.8), obtaining

$$E\ell_t(a_{i_0}, \tilde{\mu}) = \sum_{j=1}^{i_0-1} (k_{j+1} - k_j)\sigma_0 L_{N^*}\left(-\frac{u_j{}^b - \mu_0}{\sigma_0}\right) + \sum_{j=i_0}^{r-1} (k_{j+1} - k_j)\sigma_0 L_{N^*}\left(\frac{u_j{}^b - \mu_0}{\sigma_0}\right)$$

and since $\mu_{i_0-1}^b \leqq \mu_0 \leqq \mu_{i_0}{}^b$, it follows that $\mu_j{}^b \leqq \mu_0$ for $j \leqq i_0 - 1$ and $\mu_j{}^b \geqq \mu_0$ for $j \geqq i_0$. Hence all arguments of L_{N^*} are positive; therefore, with $a_{i_0} = a^0_{z_0,0}$,

$$E\ell_t(a_{i_0}, \tilde{\mu}) = \sum_{j=1}^{i_0-1} (k_{j+1} - k_j)\sigma_0 L_{N^*}\left(\frac{|\mu_j{}^b - \mu_0|}{\sigma_0}\right) + \sum_{j=i_0}^{r-1} (k_{j+1} - k_j)\sigma_0 L_{N^*}\left(\frac{|\mu_j{}^b - \mu_0|}{\sigma_0}\right),$$

from which (19.3.20) is immediate. The same argument suffices for (19.3.21) with $f_N\left(\omega \mid \mu_0, \left(\frac{n}{n_0 + n}\right)\sigma_0{}^2\right)$ replacing $f_N(\mu \mid \mu_0, \sigma_0{}^2)$. QED

Note that $I(n)$ can be determined via (19.3.21), and hence n^0 can always be found by a search procedure given the cost $k(n)$ of experimentation. Our previous reasoning establishes that $n^0 \in \{0, 1, \ldots, n^\dagger\}$, where

$$n^\dagger = \min\left\{n\colon n \geqq \frac{[E_\mu l_t(a^0_{z_0,0}, \tilde{\mu}) - K_s]}{k_s}, n \text{ is a nonnegative integer}\right\}. \tag{19.3.24}$$

The quantity n^0 can be determined analytically when $r = 2$. The details can be found in Pratt, Raiffa, and Schlaifer [**64**] and are summarized in Bracken and Schleifer [**13**]. These references also furnish analytic procedures for finding n^0 when $r = 2$ and σ^2 is unknown.

Example 19.3.1

A manufacturer currently uses hand labor to produce widgets, at a cost in labor and material of \$0.30 per widget. In six months a new, automated plant will be in operation; but in the meantime the manufacturer has the opportunity to lease a stamping machine for \$10,000. This machine will produce widgets in any amount desired at a cost in labor and materials of \$0.20 per widget. Widgets spoil rapidly, and hence the desirability of leasing the machine depends entirely upon demand $\tilde{\mu}$ for widgets over the next six months. Let $a_1 \equiv$ do not lease the machine and $a_2 \equiv$ lease the machine; hence $r = 2$. Then $v(a_1, \mu) = (p - .30)\mu$, where p is the selling price per widget, while $v(a_2, \mu) = (p - .20)\mu - 10{,}000$. Hence $K_1 = 0$, $K_2 = -10{,}000$, $k_1 = p - .30$, $k_2 = p - .20$, $K_1 - K_2 = 10{,}000$, and $k_2 - k_1 = .10$. Hence $\mu_1{}^b = 100{,}000$. Therefore, foregoing the machine is optimal if $\mu \leqq 100{,}000$ and leasing it is optimal if $\mu \geqq 100{,}000$. Suppose the decision-maker's prior df of $\tilde{\mu}$ is $f_N(\mu \mid 9 \times 10^4, 10^8)$. Then

$$a^0_{z_0,0} = a_1;$$

$$\sigma_0 = 10^4;$$

$$d_1 = |100{,}000 - 90{,}000| = 10{,}000;$$

$$\begin{aligned} Ł_{N*}\left(\frac{d_1}{\sigma_0}\right) &= Ł_{N*}\left(\frac{10{,}000}{10{,}000}\right) \\ &= Ł_{N*}(1.00) \\ &= f_{N*}(1.00) - (1.00)[1 - F_{N*}(1.00)] \qquad \text{[by (19.3.14)]} \\ &= .0833; \end{aligned}$$

and, from (19.3.20),

$$\begin{aligned} E_\mu \ell_t(a^0_{z_0,0}, \tilde{\mu}) &= 10^4(k_2 - k_1)Ł_{N*}(1.00) \\ &= (10^4)(.10)(.0833) \\ &= 83.30. \end{aligned}$$

Hence the prior value of perfect information is small in relation to the total labor and materials costs involved in either process. Now suppose that observations $\tilde{x}_1, \tilde{x}_2, \ldots$ on μ are available and are iid given μ with common df $f_N(x \mid \mu, 10^8)$. Then $\sigma = 10^4$ and $n_0 = \sigma^2/\sigma_0{}^2 = 10^8/10^8 = 1$. Tables I, VI, and (19.3.21) imply the following values of $I(n)$:

n	$I(n)$	n	$I(n)$
1	25.10	7	68.26
2	43.66	8	69.90
3	53.28	9	71.46
4	59.18	10	72.16
5	63.27	20	77.59
6	66.23	50	82.62

These values show that even only ten observations yield the major portion of the 83.30 prior value of perfect information.

Suppose now that $k(n)$ is given by (18.1.2) for $K_s = 20$ and $k_s = 6$. Then

$$\frac{E_\mu \ell_t(a^0_{z_0,0}, \tilde{\mu}) - K_s}{k_s} = \frac{83.30 - 20}{6} = 10.55,$$

and hence $n^\dagger = 10$, by (19.3.21). The reader may check from the preceding table and $k(n) = 20 + 6n$ that $n^0 = 3$ and $G(n^0) = 15.28$.

19.3.3 The Univariate Normal Process: σ^2 unknown

When σ^2 is unknown, the posterior expectation of $\tilde{\mu}$ is

$$\tilde{\omega} \equiv E_S(\tilde{\mu} \mid \tilde{\mu}_1, \tilde{\psi}_1/n_1, \nu_1).$$

In place of (19.3.12) we need the following:

If n observations are to be obtained from $f_N(x \mid \mu, \sigma^2)$, if σ^2 is unknown, and if $(\tilde{\mu}, \tilde{\sigma}^2)$ has the natural conjugate prior df $f_{Ni\gamma}(\mu, \sigma^2 \mid \mu_0, \psi_0, n_0, \nu_0)$, then (19.3.25)

$$\tilde{\omega} \equiv \left(\frac{n_0}{n_0 + n}\right)\mu_0 + \left(\frac{n}{n_0 + n}\right)\tilde{\bar{x}}$$

has df $f_S\left(\omega \mid \mu_0, \left(\frac{n}{n_0 + n}\right)\psi_0/n_0, \nu_0\right)$.

Proof

By (14.5.8b), $\tilde{\bar{x}}$ has unconditional df $f_S(\bar{x} \mid \mu_0, (n_0^{-1} + n^{-1})\psi_0, \nu_0)$. Since $\tilde{\omega}$ is a linear function of $\tilde{\bar{x}}$, (12.5.12) is applicable and readily yields (19.3.25).

To evaluate the integrals appearing in (19.3.8) and (19.3.9), we shall use the following results:

Let

$$Ł_{S^*}(z \mid \nu) \equiv \int_z^\infty (y - z) f_{S^*}(y \mid \nu)\,dy. \qquad (19.3.26)$$

Then

$$Ł_{S^*}(z \mid \nu) = \frac{\nu + z^2}{\nu - 1} f_{S^*}(z \mid \nu) - z[1 - F_{S^*}(z \mid \nu)]; \qquad (19.3.27)$$

$$Ł_{S^*}(-z \mid \nu) = Ł_{S^*}(z \mid \nu) + z; \qquad (19.3.28)$$

$$\int_a^\infty (x - a) f_S(x \mid m, D^2, \nu)\,dx = D Ł_{S^*}\left(\frac{a - m}{D} \,\middle|\, \nu\right); \qquad (19.3.29)$$

and

$$\int_{-\infty}^{a} (a - x) f_S(x \mid m, D^2, \nu) dx = D Ł_{S^*}\left(-\frac{a - m}{D} \,\Big|\, \nu\right). \quad (19.3.30)$$

Proof

Exercise 14.

The essential results for the univariate Normal process with unknown variance σ^2 and $(\tilde{\mu}, \tilde{\sigma}^2)$ having the conjugate prior df $f_{Ni\gamma}(\mu, \sigma^2 \mid \mu_0, \psi_0, n_0, \nu_0)$ and hence the conjugate posterior df $f_{Ni\gamma}(\mu, \sigma^2 \mid \mu_0, \psi_0, n_0, \nu_0)$, are:

$$a^0_{z_0,0} = a_i, \text{ if and only if } \mu^b_{i-1} \leqq \mu_0 \leqq \mu_i{}^b; \quad (19.3.31)$$

$$a^0_{\bar{x},n} = a_i, \text{ if and only if } \mu^b_{i-1} \leqq \mu_1 \leqq \mu_i{}^b; \quad (19.3.32)$$

$$E_\mu \ell_t(a^0_{z_0},0, \tilde{\mu}) = \sigma_0 \sum_{j=1}^{r-1} (k_{j+1} - k_j) Ł_{S^*}\left(\frac{d_j}{\sigma_0} \,\Big|\, \nu_0\right); \quad (19.3.33)$$

and

$$I(n) = \left[\frac{n}{n_0 + n}\right]^{1/2} \sigma_0 \sum_{j=1}^{r-1} (k_{j+1} - k_j) Ł_{S^*}\left(d_j \Big/ \left\{\left[\frac{n}{n_0 + n}\right]^{1/2} \sigma_0\right\} \,\Big|\, \nu_0\right), \quad (19.3.34)$$

where

$$\sigma_0 \equiv \left(\frac{\psi_0}{n_0}\right)^{1/2} \quad (19.3.35)$$

and

$$d_j \equiv |\mu_j{}^b - \mu_0|. \quad (19.3.36)$$

Proof

Exercise 15.

Search procedures for n^0 continue to apply, and, as mentioned in Subsection 19.3.2, analytic procedures are available for $r = 2$.

Example 19.3.2

Reconsider the problem in Example 19.3.1, now assuming that σ^2 is unknown and that $(\tilde{\mu}, \tilde{\sigma}^2)$ has the conjugate prior df $f_{Ni\gamma}(\mu, \sigma^2 \mid 9 \times 10^4, 5 \times 10^4, 1, 4)$. Then (14.5.7) implies that the marginal prior df of $\tilde{\mu}$ is $f_S(\mu \mid 9 \times 10^4, 5 \times 10^7, 4)$, which has variance $(\psi_0/n_0)\nu_0/(\nu_0 - 2) = 10^8$, same as in Example 19.3.1 when $\tilde{\mu}$ had prior df $f_N(\mu \mid 9 \times 10^4, 10^8)$. Clearly,

$a^0_{z_0,0} = a_1$, as before; and for the prior value of perfect information we compute

$$d_1 = 10^4, \text{ (as before)},$$

$$\frac{d_1}{\sigma_0} = \frac{10^4}{7070} \doteq 1.414;$$

$$Ł_{S^*}(1.414 \mid 4) = \frac{4 + (1.414)^2}{4 - 1} f_{S^*}(1.414 \mid 4) - (1.414)[1 - F_{S^*}(1.414 \mid 4)].$$

Since $f_{S^*}(1.414 \mid 4) \doteq .1362$ and $F_{S^*}(1.414 \mid 4) \doteq .8849$, it follows that

$$Ł_{S^*}(1.414 \mid 4) \doteq .1096.$$

Hence (19.3.33) implies

$$E_\mu \ell_t(a^0_{z_0,0}, \tilde{\mu}) = (7070)(.10)(.1096) = 77.49,$$

which is less than the 83.30 of Example 19.3.1. Now suppose that $k(n)$ is given by (18.1.2) for $K_s = 20$ and $k_s = 6$, as in Example (19.3.1). It is easily verified that $n^\dagger = 9$ and $n^0 \in \{0, 1, \ldots, 9\}$. The following table furnishes some values of $I(n)$:

n	$I(n)$	n	$I(n)$
1	30.33	7	65.54
2	45.61	8	66.65
3	53.58	9	67.67
4	58.21	10	68.68
5	61.32	20	72.73
6	63.70	50	75.35

These values were determined from (19.3.34) using $Ł_{S^*}(\cdot \mid 4)$ as tabulated in Bracken and Schleifer [**13**]. It is now easy to verify that $n^0 = 3$ and $G(n^0) = 15.58$.

19.4 CHOOSING THE BEST OF r PROCESSES

19.4.1 General Theory

In Section 19.3 we considered a class of problems in which $A = \{a_1, \ldots, a_r\}$ and the return from act a_i is a linear function of a random variable $\tilde{\theta}$, given by (19.3.1). In this section we consider a class of problems in which again $A = \{a_1, \ldots, a_r\}$, but now the state is an r-dimensional random vector $\tilde{\boldsymbol{\theta}} \equiv (\tilde{\theta}_1, \ldots, \tilde{\theta}_r)$ and

$$v_t(a_i, \boldsymbol{\theta}) = K_i + k_i\theta_i, \text{ for every } \boldsymbol{\theta} \in \boldsymbol{\Theta} \subset R^r \text{ and } i \in \{1, \ldots, r\}. \quad (19.4.1)$$

Now, if $\tilde{\boldsymbol{\theta}}$ is extremely degenerate in the sense that $P(\tilde{\theta}_1 = \tilde{\theta}_2 = \cdots = \tilde{\theta}_r) = 1$, then we can define a random variable $\tilde{\theta}$ as any one of the $\tilde{\theta}_i$'s and obtain (19.3.1) as a special case of (19.4.1). Hence the topic of this section includes that of Section 19.3 as a special case. However, there are two reasons for the separate treatment accorded in Section 19.3.

First, the r-act problems with linear return considered in Section 19.3 are much more tractable analytically than the problems of this subsection. The ensuing analysis here differs substantially from that in Section 19.3 and provides less determinate results. In particular, we shall not be able to characterize the prior value of perfect or of sample information in closed form, even for the two Normal-process assumptions.

Second, the intended applications differ here from those in Section 19.3. There, the state $\tilde{\theta}$ was unrelated to the choice of act; for example, $\tilde{\theta}$ denotes demand for widgets and a_i denotes installation of the ith type of widget-making machine. Here, the state components $\tilde{\theta}_i$ are by definition related to the choice of act. In two prototypical cases, $\tilde{\theta}_i$ denotes:

(1) yield of a given crop that has been treated with fertilizer i (the choice of which constitutes act a_i); and
(2) demand for a given product as influenced by advertising campaign i (the choice of which constitutes act a_i).

We shall assume that information can be obtained about each of the $\tilde{\theta}_i$'s separately, in the form of iid observations on a data-generating process with unknown parameter θ_i, and hence the distribution-theoretic material of Section 14.7 will be needed.

It is convenient for the following analysis to define

$$\tilde{w}_i \equiv v_t(a_i, \tilde{\boldsymbol{\theta}}) = K_i + k_i\tilde{\theta}_i \tag{19.4.2}$$

and to change variable from $\tilde{\boldsymbol{\theta}}$ to $\tilde{\mathbf{w}} \equiv (\tilde{w}_1, \ldots, \tilde{w}_r)$ via the affine transformation

$$\mathbf{w} = \mathbf{K} + \mathbf{k}\boldsymbol{\theta}, \tag{19.4.3}$$

where $\mathbf{K} \equiv (K_1, \ldots, K_r)$ and

$$\mathbf{k} \equiv \begin{bmatrix} k_1 & & & & \\ & k_2 & & 0 & \\ & & \cdot & & \\ & & & \cdot & \\ & 0 & & & \cdot \\ & & & & k_r \end{bmatrix}$$

Then if $f_{\boldsymbol{\theta}}(\boldsymbol{\theta})$ is any df of $\tilde{\boldsymbol{\theta}}$ and no $k_i = 0$, the related df $f_{\mathbf{w}}(\mathbf{w})$ of $\mathbf{w}$ is given by

$$f_{\mathbf{w}}(\mathbf{w}) = f_{\boldsymbol{\theta}}(\mathbf{k}^{-1}[\mathbf{w} - \mathbf{K}])|\det(\mathbf{k}^{-1})|$$
$$= f_{\boldsymbol{\theta}}(\mathbf{k}^{-1}[\mathbf{w} - \mathbf{K}]) \prod_{i=1}^{r} |k_i|^{-1}.$$

We shall write $v_t(a_i, \mathbf{w})$, $\ell_t(a_i, \mathbf{w})$, and so forth, in place of $v_t(a_i, \boldsymbol{\theta})$, $\ell_t(a_i, \boldsymbol{\theta})$, and so forth.

We shall make two assumptions which simplify matters slightly but without loss of essential generality. *First*, we assume that for every i there is some $\boldsymbol{\theta} \in \boldsymbol{\Theta}$ and hence some $\mathbf{w}$ such that $v_t(a_i, \mathbf{w}) = \max_j v_t(a_j, \mathbf{w})$; that is, such that $w_i = \max_j w_j$. Otherwise, a_i could be dropped from consideration.

To introduce the second assumption, note that if $f(\mathbf{w})$ is the prior df of $\tilde{w}$, then an optimal act $a^0_{z_0,0}$ without information satisfies

$$a^0_{z_0,0} = a_i \text{ if and only if } E_{\mathbf{w}}(\tilde{w}_i) = \max_j E_{\mathbf{w}}(\tilde{w}_j). \qquad (19.4.4)$$

Proof

Exercise 16.

Our *second* assumption is that the acts have been numbered so that $a_r = a^0_{z_0,0}$.

We shall now derive the terminal opportunity loss function $\ell_t(a_i, \mathbf{w})$. Since

$$\max_j v_t(a_j, \mathbf{w}) = \max_j w_j,$$

and

$$\ell_t(a_i, \mathbf{w}) = \max_j v_t(a_j, \mathbf{w}) - v_t(a_i, \mathbf{w}),$$

it follows that

$$\ell_t(a_i, \mathbf{w}) = \max_j w_j - w_i. \qquad (19.4.5)$$

Since $a_r = a^0_{z_0,0}$, the prior value of perfect information is given by

$$E_{\mathbf{w}}\ell_t(a^0_{z_0,0}, \tilde{\mathbf{w}}) = E_{\mathbf{w}}\{\max_j \tilde{w}_j - \tilde{w}_r\}$$
$$= E_{\mathbf{w}} \max_j (\tilde{w}_j - \tilde{w}_r) \qquad (19.4.6)$$
$$= E_{\mathbf{w}} \max\{\tilde{\delta}_{01}, \ldots, \tilde{\delta}_{0r-1}, 0\},$$

where $\tilde{\delta}_{0i} \equiv \tilde{w}_i - \tilde{w}_r$.

In most important cases (for example, $\tilde{\boldsymbol{\theta}}$ r-variate Normal or Student), $\max\{\tilde{\delta}_{01}, \ldots, \tilde{\delta}_{0r-1}, 0\}$ is a poorly behaved random variable from the standpoint of tractability; its expectation cannot be determined in closed form. However, we can specify more tractable upper and lower bounds for $E_{\mathbf{w}}\ell_t(a^0_{z_0,0}, \tilde{\mathbf{w}})$. A *lower* bound is given by

$$\max_j E_{\mathbf{w}} \max\{\tilde{\delta}_{0j}, 0\} \leqq E_{\mathbf{w}} \max\{\tilde{\delta}_{01}, \ldots, \tilde{\delta}_{0r-1}, 0\}, \tag{19.4.7}$$

while an *upper* bound is given by

$$E_{\mathbf{w}} \max\{\tilde{\delta}_{01}, \ldots, \tilde{\delta}_{0r-1}, 0\} \leqq \sum_{j=1}^{r-1} E_{\mathbf{w}} \max\{\tilde{\delta}_{0j}, 0\}. \tag{19.4.8}$$

Proof

First, the reader may show as Exercise 17 that

$$\max\{\delta_{01}, \ldots, \delta_{0r-1}, 0\} = \max_j\{\max\{\delta_{0j}, 0\}\},$$

for every possible set of $r - 1$ numbers $\delta_{01}, \ldots, \delta_{0r-1}$; and hence the same equality obtains when all δ_{0j}'s are tilded. Thus,

$$E_{\mathbf{w}} \max\{\tilde{\delta}_{01}, \ldots, \tilde{\delta}_{0r-1}, 0\} = E_{\mathbf{w}} \max_j\{\max\{\tilde{\delta}_{0j}, 0\}\}.$$

But

$$\max_j\{\max\{\tilde{\delta}_{0j}, 0\}\} \geqq \max\{\tilde{\delta}_{0j}, 0\},$$

every $j \in \{1, \ldots, r-1\}$, which implies

$$E_{\mathbf{w}} \max_j\{\max\{\tilde{\delta}_{0j}, 0\}\} \geqq E_{\mathbf{w}} \max\{\tilde{\delta}_{0j}, 0\},$$

for every $j \in \{1, \ldots, r\}$ and, hence

$$E_{\mathbf{w}} \max_j\{\max\{\tilde{\delta}_{0j}, 0\}\} \geqq \max_j E_{\mathbf{w}} \max\{\tilde{\delta}_{0j}, 0\},$$

establishing (19.4.7). To prove (19.4.8), note that since all $\max\{\tilde{\delta}_{0j}, 0\}$'s are nonnegative, it follows that

$$\max_j\{\max\{\tilde{\delta}_{0j}, 0\}\} \leqq \sum_{j=1}^{r-1} \max\{\tilde{\delta}_{0j}, 0\},$$

and hence

$$E_{\mathbf{w}} \max_j\{\max\{\tilde{\delta}_{0j}, 0\}\} \leqq E_{\mathbf{w}} \sum_{j=1}^{r-1} \max\{\tilde{\delta}_{0j}, 0\},$$

which readily implies (19.4.8). QED

Now suppose that n_i iid observations have been obtained on $\tilde{\theta}_i$, for $i = 1, \ldots, r$. As in Section 14.7, we shall assume that $n_i \geqq 2$, for every i, although this constraint can be relaxed at the cost of slightly complicating the analysis. If z_i is the outcome of the observations on θ_i, if $\mathbf{z} \equiv (z_1, \ldots, z_r)$, and $\mathbf{n} \equiv (n_1, \ldots, n_r)$, then an optimal act $a^0_{\mathbf{z},\mathbf{n}}$ posterior to experimentation must satisfy

$$a^0_{\mathbf{z},\mathbf{n}} = a_i, \text{ if and only if } E(\tilde{w}_i \mid \mathbf{z}, \mathbf{n}) = \max_j E(\tilde{w}_j \mid \mathbf{z}, \mathbf{n}). \quad (19.4.9)$$

Proof

Exercise 16.

Hence

$$E_{\mathbf{w}\mid\mathbf{z},\mathbf{n}} v_t(a^0_{\mathbf{z},\mathbf{n}}, \tilde{\mathbf{w}}) = \max_j E(\tilde{w}_j \mid \mathbf{z}, \mathbf{n}) \quad (19.4.10)$$

and

$$E_{\mathbf{z}\mid\mathbf{n}} E_{\mathbf{w}\mid\mathbf{z},\mathbf{n}} v_t(a^0_{\tilde{\mathbf{z}},\mathbf{n}}, \tilde{\mathbf{w}}) = E_{\mathbf{z}\mid\mathbf{n}} \max_j E(\tilde{w}_j \mid \tilde{\mathbf{z}}, \mathbf{n}). \quad (19.4.11)$$

We are now in a position to characterize the prior value $I(\mathbf{n}) \equiv I(n_1, \ldots, n_r)$ of information. Let $\tilde{\omega}_j \equiv E(\tilde{w}_j \mid \tilde{\mathbf{z}}, \mathbf{n})$, for $j = 1, \ldots, r$ and $\tilde{\boldsymbol{\omega}} \equiv (\tilde{\omega}_1, \ldots, \tilde{\omega}_r)$, and let $\tilde{\delta}_{1j} \equiv \tilde{\omega}_j - \tilde{\omega}_r$, for $j = 1, \ldots, r-1$. Then

$$I(\mathbf{n}) = E_{\omega\mid\mathbf{n}} \max \{\tilde{\delta}_{11}, \ldots, \tilde{\delta}_{1r-1}, 0\}. \quad (19.4.12)$$

Proof

By the evident chain of equalities:

$$\begin{aligned}
I(\mathbf{n}) &= E_{\mathbf{z}\mid\mathbf{n}} E_{\mathbf{w}\mid\mathbf{z},\mathbf{n}} v_t(a^0_{\tilde{\mathbf{z}},\mathbf{n}}, \tilde{\mathbf{w}}) - E_{\mathbf{w}} v_t(a^0_{z_0,0}, \tilde{\mathbf{w}}) \\
&= E_{\mathbf{z}\mid\mathbf{n}} E_{\mathbf{w}\mid\mathbf{z},\mathbf{n}} v_t(a^0_{\tilde{\mathbf{z}},\mathbf{n}}, \tilde{\mathbf{w}}) - E_{\mathbf{z}\mid\mathbf{n}} E_{\mathbf{w}\mid\mathbf{z},\mathbf{n}} v_t(a^0_{z_0,0}, \tilde{\mathbf{w}}) \\
&= E_{\mathbf{z}\mid\mathbf{n}} \{E_{\mathbf{w}\mid\mathbf{z},\mathbf{n}} v_t(a^0_{\tilde{\mathbf{z}},\mathbf{n}}, \tilde{\mathbf{w}}) - E_{\mathbf{w}\mid\mathbf{z},\mathbf{n}} v_t(a^0_{z_0,0}, \tilde{\mathbf{w}}) \\
&= E_{\mathbf{z}\mid\mathbf{n}} [\max_j E(\tilde{w}_j \mid \tilde{\mathbf{z}}, \mathbf{n}) - E(\tilde{w}_r \mid \tilde{\mathbf{z}}, \mathbf{n})] \\
&\equiv E_{\mathbf{z}\mid\mathbf{n}} [\max_j \tilde{\omega}_j - \tilde{\omega}_r] \\
&= E_{\mathbf{z}\mid\mathbf{n}} \max_j (\tilde{\omega}_j - \tilde{\omega}_r) \\
&\equiv E_{\mathbf{z}\mid\mathbf{n}} \max \{\tilde{\delta}_{11}, \ldots, \tilde{\delta}_{1r-1}, 0\} \\
&= E_{\mathbf{w}\mid\mathbf{n}} \max \{\tilde{\delta}_{11}, \ldots, \tilde{\delta}_{1r-1}, 0\}. \quad \text{QED}
\end{aligned}$$

Hence the prior value $I(\mathbf{n})$ of the sample information is of precisely the same form as the prior value of perfect information, except that the posterior $\tilde{\omega}_1, \ldots, \tilde{\omega}_r$ replace the values $\tilde{w}_1, \ldots, \tilde{w}_r$. (Be careful to distinguish the w's from the omegas.)

The evaluation of $I(\mathbf{n})$ is fraught with the same difficulties as the evaluation of $E_{\mathbf{w}}\ell_t(a^0_{z_0,0}, \tilde{\mathbf{w}})$, and the bounds (19.4.7) and (19.4.8) continue to apply, with $\tilde{\delta}_{0j}$'s replaced by $\tilde{\delta}_{1j}$'s and $E_{\mathbf{w}}$ replaced by $E_{\mathbf{w}|\mathbf{n}}$.

Pratt, Raiffa, and Schlaifer [**64**] describe a procedure for simulating $I(\mathbf{n})$ and $E_{\mathbf{w}}\ell_t(a^0_{z_0|0}, \tilde{\mathbf{w}})$ when $\tilde{\boldsymbol{\theta}}$ and hence also $\tilde{\mathbf{w}}$, $\tilde{\boldsymbol{\omega}}$, $(\tilde{\delta}_{01}, \ldots, \tilde{\delta}_{0r-1})$, and $(\tilde{\delta}_{11}, \ldots, \tilde{\delta}_{1r-1})$ are multivariate Normal. We shall briefly sketch this procedure before giving tractable results for the case $r = 2$. The reader should refer to Chapter 14, Exercises 18 and 19 for a brief introduction to simulation.

The basic idea behind the simulation procedure is to obtain a large number t of independent observations on the random variable max $\{\tilde{\delta}_1, \ldots, \tilde{\delta}_{r-1}, 0\}$, where we drop the first subscript because the analysis is the same in either case. These independent observations are then averaged to provide an estimate $\underset{t}{\overline{\max}}$ of $E \max \{\tilde{\delta}_1, \ldots, \tilde{\delta}_{r-1}, 0\}$ and this estimate satisfies the weak law of large numbers: for every $\epsilon > 0$,

$$\lim_{t\to\infty} P(E \max - \epsilon < \underset{t}{\overline{\widetilde{\max}}} < E \max + \epsilon) = 1,$$
$$t\to\infty$$

where $E\,max$ is an abbreviation for $E \max \{\tilde{\delta}_1, \ldots, \tilde{\delta}_{r-1}, 0\}$. Hence the average of a *large* number s of independent observations on $\max \{\tilde{\delta}_1, \ldots, \tilde{\delta}_{r-1}, 0\}$ is, with high probability, a good estimate of $E \max \{\tilde{\delta}_1, \ldots, \tilde{\delta}_{r-1}, 0\}$.

We now derive $I(\mathbf{n})$ for two Normal-process assumptions and $r = 2$.

19.4.2 Two Univariate Normal Processes with Known Variances

Let $r = 2$ and $\tilde{\theta}_i = \tilde{\mu}_i$, the mean of a univariate Normal process, for $i = 1, 2$. Assume that the variances $\sigma_1{}^2$ and $\sigma_2{}^2$ are both known. Then (14.7.7) provides the relevant distribution theory for $\tilde{\boldsymbol{\mu}} \equiv (\tilde{\mu}_1, \tilde{\mu}_2)$, and the reader may prove the following as Exercise 18:

> If $\tilde{\boldsymbol{\mu}}$ has prior df $f_N{}^{(2)}(\boldsymbol{\mu} \mid \boldsymbol{\mu}_0, \boldsymbol{\sigma}_0{}^2)$ and $\tilde{\bar{\mathbf{x}}}$ has df $f_N{}^{(2)}(\bar{\mathbf{x}} \mid \boldsymbol{\mu}, \boldsymbol{\sigma}^2(\mathbf{n}))$ given $\boldsymbol{\mu}$, then $\tilde{\boldsymbol{\mu}}_1 \equiv E(\tilde{\boldsymbol{\mu}} \mid \tilde{\bar{\mathbf{x}}}, \mathbf{n})$ has df $f_N{}^{(2)}(\boldsymbol{\mu}_1 \mid \boldsymbol{\mu}_0, \boldsymbol{\sigma}_0{}^2 - \boldsymbol{\sigma}_1{}^2(\mathbf{n}))$. (19.4.13)

We remark that (19.4.13) obtains for any r and for $\boldsymbol{\sigma}^2$ not necessarily diagonal.

By (12.7.5), if $\tilde{\boldsymbol{\mu}}$ has df $f_N{}^{(2)}(\boldsymbol{\mu} \mid \boldsymbol{\mu}_0, \boldsymbol{\sigma}_0{}^2)$, then $\tilde{\mathbf{w}} \equiv \mathbf{K} + \mathbf{k}\tilde{\boldsymbol{\mu}}$ has df $f_N{}^{(2)}(\mathbf{w} \mid \mathbf{K} + \mathbf{k}\boldsymbol{\mu}_0, \mathbf{k}\boldsymbol{\sigma}_0{}^2\mathbf{k}')$. By assumption, $a^0_{z_0,0} = a_2$, and

$$E(\tilde{\mathbf{w}}_2) \equiv K_2 + k_2\mu_{02} \geqq K_1 + k_1\mu_{01} \equiv E(\tilde{\mathbf{w}}_1),$$

and $\delta_{01} \equiv \tilde{w}_1 - \tilde{w}_2 = (1, -1)(\tilde{w}_1, \tilde{w}_2)'$. Hence the df of δ_{01} is $f_N(\delta_{01} \mid \Delta, D_0{}^2)$, where

$$\Delta \equiv (1, -1)(\mathbf{K} + \mathbf{k}\boldsymbol{\mu}_0) \tag{19.4.14}$$

and

$$D_0{}^2 \equiv (1, -1)\mathbf{k}\boldsymbol{\sigma}_0{}^2\mathbf{k}'(1, -1)'. \tag{19.4.15}$$

Now, the prior value of perfect information is given by (19.4.6), which reduces for $r = 2$ to

$$\begin{aligned} E_{\mathbf{w}}\ell_t(a^0_{z_0,0}, \tilde{\mathbf{w}}) &= E \max \{\tilde{\delta}_{01}, 0\}. \\ &= \int_0^\infty \delta_{01} f_N(\delta_{01} \mid \Delta, D_0{}^2) d\delta_{01} \\ &= D_0 L_{N*}\left(\frac{0 - \Delta}{D_0}\right) \\ &= D_0 L_{N*}\left(\frac{|\Delta|}{D_0}\right), \end{aligned} \tag{19.4.16}$$

by (19.3.13) and (19.3.16). Since finding Δ and D_0 is a simple matter of computation, it is clear that the prior value of perfect information can be obtained readily and without resort to simulation when $r = 2$.

The same is true for $I(\mathbf{n})$. We first note that

$$\tilde{\boldsymbol{\omega}} \equiv E(\tilde{\mathbf{w}} \mid \tilde{\bar{\mathbf{x}}}, \mathbf{n}) = \mathbf{K} + \mathbf{k}\tilde{\boldsymbol{\mu}}_1 \tag{19.4.17}$$

and then use (19.4.13) to deduce that $\tilde{\boldsymbol{\omega}}$ has df

$$f_N{}^{(2)}(\tilde{\boldsymbol{\omega}} \mid \mathbf{K} + \mathbf{k}\boldsymbol{\mu}_0, \mathbf{k}[\boldsymbol{\sigma}_0{}^2 - \boldsymbol{\sigma}_1{}^2(\mathbf{n})]\mathbf{k}'),$$

and hence that $\tilde{\delta}_{11} \equiv \tilde{\omega}_1 - \tilde{\omega}_2$ has df $f_N(\delta_{11} \mid \Delta, D_1{}^2(\mathbf{n}))$, where Δ is given by (19.4.14) and $D_1{}^2(\mathbf{n})$ by

$$D_1{}^2(\mathbf{n}) \equiv (1, -1)\mathbf{k}[\boldsymbol{\sigma}_0{}^2 - \boldsymbol{\sigma}_1{}^2(\mathbf{n})]\mathbf{k}'(1, -1)'. \tag{19.4.18}$$

Then, from (19.4.12) we obtain

$$\begin{aligned} I(\mathbf{n}) &= E \max \{\tilde{\delta}_{11}, 0\} \\ &= \int_0^\infty \delta_{11} f_N(\delta_{11} \mid \Delta, D_1{}^2(\mathbf{n})) d\delta_{11} \\ &= D_1(\mathbf{n}) L_{N*}\left(\frac{|\Delta|}{D_1(\mathbf{n})}\right). \end{aligned} \tag{19.4.19}$$

The reader may check that $\boldsymbol{\sigma}_1{}^2(\mathbf{n}) \to \mathbf{0}$ as $n_1 \to \infty$ and $n_2 \to \infty$, and hence also $D_1(\mathbf{n}) \to D_0$.

Example 19.4.1

A farmer wishes to determine whether to treat his crop with a fertilizer of dubious characteristics or to forgo fertilization altogether. Suppose that the cost of fertilization is \$10,000 independent of crop yield, and that the fertilizer will increase the quality (but not necessarily the quantity) of the crop sufficiently to increase the price per bushel from .35 to .45. (With calculations beforehand) we define $a_1 \equiv$ "fertilize," $K_1 = -10{,}000$, $k_1 = .45$, and $\tilde{\mu}_1 \equiv$ "yield with fertilizer," $a_2 \equiv$ "do not fertilize," $K_2 = 0$, $k_2 = .35$, and $\tilde{\mu}_2 =$ "yield with no fertilizer." Suppose the farmer expresses his judgments about $\tilde{\boldsymbol{\mu}}$ in the form of a prior df $f_{N^{(2)}}(\boldsymbol{\mu} \mid \boldsymbol{\mu}_0, \boldsymbol{\sigma}_0{}^2)$, where

$$\boldsymbol{\mu}_0 = (9 \times 10^4, 9 \times 10^4)',$$

and

$$\boldsymbol{\sigma}_0{}^2 = \begin{bmatrix} 10^8 & 6 \times 10^7 \\ 6 \times 10^7 & 10^8 \end{bmatrix} = 10^6 \begin{bmatrix} 100 & 60 \\ 60 & 100 \end{bmatrix}.$$

Then his prior df of $\tilde{\mathbf{w}}$ is $f_{N^{(2)}}(\mathbf{w} \mid \mathbf{K} + \mathbf{k}\boldsymbol{\mu}_0, \mathbf{k}\boldsymbol{\mu}_0\mathbf{k}')$, where

$$\mathbf{K} + \mathbf{k}\boldsymbol{\mu}_0 = \begin{pmatrix} -10{,}000 \\ 0 \end{pmatrix} + \begin{bmatrix} .45 & 0 \\ 0 & .35 \end{bmatrix} \begin{pmatrix} 9 \times 10^4 \\ 9 \times 10^4 \end{pmatrix}$$

$$= \begin{pmatrix} 30{,}500 \\ 31{,}500 \end{pmatrix},$$

(so that $a^0_{z_0,0} = a_2 =$ "do not fertilize"), and

$$\mathbf{k}\boldsymbol{\sigma}_0{}^2\mathbf{k}' = 10^6 \begin{bmatrix} 20.25 & 9.45 \\ 9.45 & 12.25 \end{bmatrix},$$

which the reader may easily check by matrix multiplication. Now, $\tilde{\delta}_{01} \equiv \tilde{w}_1 - \tilde{w}_2$, and hence $\tilde{\delta}_{01}$ has df $f_N(\delta_{01} \mid \Delta, D_0{}^2)$, where (19.4.14) and (19.4.15) imply

$$\Delta = 30{,}500 - 31{,}500 = -1000$$

and

$$D_0{}^2 = 13.60 \times 10^6, \text{ implying } D_0 \doteq 4330.$$

Hence the prior value of perfect information is given by (19.4.16):

$$E_{\mathbf{w}} \ell_t(a^0_{z_0,0}, \tilde{\mathbf{w}}) = (4330) Ł_{N^*}\left(\frac{1000}{4330}\right)$$

$$\doteq (4330) Ł_{N^*}(.231)$$

$$\doteq (4330)(.2940)$$

$$= \$1273.$$

Suppose now that information on each process i is available in the form of iid observations $\tilde{x}_{i1}, \tilde{x}_{i2}, \ldots$, for $i = 1, 2$, and that the $\tilde{x}_i$'s have df $f_N(x_i \mid \mu_i, 10^8)$ given μ_i, for $i = 1, 2$. Suppose $\mathbf{n} = (3, 3)'$; we shall find $I(\mathbf{n})$. We compute first

$$\boldsymbol{\sigma}^2(\mathbf{n}) = \begin{bmatrix} \frac{10^8}{3} & 0 \\ 0 & \frac{10^8}{3} \end{bmatrix}.$$

The reader may check that

$$\begin{aligned} \boldsymbol{\sigma}_1^{-2}(\mathbf{n}) &= \boldsymbol{\sigma}_0^{-2} + \boldsymbol{\sigma}^{-2}(\mathbf{n}) \\ &= 10^{-8}\begin{bmatrix} 4.56 & -.938 \\ -.938 & 4.56 \end{bmatrix}, \end{aligned}$$

and hence

$$\boldsymbol{\sigma}_1^2(\mathbf{n}) = 10^6\begin{bmatrix} 22.9 & 4.7 \\ 4.7 & 22.9 \end{bmatrix}.$$

Therefore,

$$\boldsymbol{\sigma}_0^2 - \boldsymbol{\sigma}_1^2(\mathbf{n}) = 10^6\begin{bmatrix} 77.1 & 55.3 \\ 55.3 & 77.1 \end{bmatrix}.$$

Thus $\tilde{\boldsymbol{\omega}}$ has df $f_N^{(2)}(\tilde{\boldsymbol{\omega}} \mid \mathbf{K} + \mathbf{k}\boldsymbol{\mu}_0, \mathbf{k}[\boldsymbol{\sigma}_0^2 - \boldsymbol{\sigma}_1^2(\mathbf{n})]\mathbf{k}')$ for

$$\mathbf{k}[\boldsymbol{\sigma}_0^2 - \boldsymbol{\sigma}_1^2(\mathbf{n})]\mathbf{k}' = 10^6\begin{bmatrix} 15.61 & 8.71 \\ 8.71 & 9.45 \end{bmatrix},$$

implying via (19.4.18) that

$$D_1^2(\mathbf{n}) = 7.64 \times 10^6 \text{ and thus } D_1(\mathbf{n}) \doteq 2870.$$

Hence by (19.5.19),

$$\begin{aligned} I(\mathbf{n}) &\doteq (2870)L_{N*}\left(\frac{1000}{2870}\right) \\ &\doteq (2870)L_{N*}(.349) \\ &\doteq (2870)(.2485) \\ &\doteq \$713. \end{aligned}$$

Thus even very small sample sizes, $\mathbf{n} = (3, 3)$, suffice to yield the majority of the prior value of perfect information in this case.

19.5.3 Two Univariate Normal Processes with σ_1^2/σ_2^2 known

In this subsection all remains as in Subsection 19.5.2, with the exception that the process variances σ_1^2 and σ_2^2 are known only up to a common, positive factor; that is, $\sigma_1^2, \sigma_2^2 = \sigma_{1*}^2/\sigma_{2*}^2$ is known. The Normal-inverted gamma and Student distribution theory in Section 14.7 pertains, together with the additional result (Exercise 19) as follows:

If $(\tilde{\boldsymbol{\mu}}, \tilde{\sigma}^2)$ has prior df $f^{(2)}_{Ni\gamma}(\boldsymbol{\mu},\sigma^2 \mid \boldsymbol{\mu}_0, \psi_0, \boldsymbol{\sigma}_{0*}^2, \nu_0)$ and $\tilde{\boldsymbol{\mu}}_1$ denotes $E(\tilde{\boldsymbol{\mu}} \mid \bar{\tilde{\mathbf{x}}},\tilde{s}_*^2, \mathbf{n})$, then $\tilde{\boldsymbol{\mu}}_1$ has df $f_S^{(2)}(\boldsymbol{\mu}_1 \mid \boldsymbol{\mu}_0, \psi_0[\boldsymbol{\sigma}_{0*}^2 - \boldsymbol{\sigma}_{1*}^2(\mathbf{n})], \nu_0)$. (19.4.20)

The reader may duplicate the reasoning in the preceding subsection to show that $\tilde{\delta}_{01} \equiv \tilde{w}_1 - \tilde{w}_2$ has df $f_S(\delta_{01} \mid \Delta, D_0^2, \nu_0)$, where Δ is given by (19.4.14) and

$$D_0^2 \equiv (1, -1)\mathbf{k}\psi_0\boldsymbol{\sigma}_{0*}^2\mathbf{k}'(1, -1)'. \tag{19.4.21}$$

The prior value of perfect information is again given by (19.4.6) specialized to $r = 2$:

$$\begin{aligned} E_{\mathbf{w}}\ell_t(a^0_{z_0,0}, \tilde{\mathbf{w}}) &= E \max \{\tilde{\delta}_{01},0\} \qquad (19.4.22)\\ &= \int_0^\infty \delta_{01} f_S(\delta_{01} \mid \Delta, D_0^2, \nu_0)d\delta_{01} \\ &= D_0 Ł_{S*}\left(\frac{|\Delta|}{D_0} \mid \nu_0\right), \end{aligned}$$

by (19.3.26) and (19.3.29).

As can be expected, similar results obtain for the prior value $I(\mathbf{n})$ of (n_1, n_2) observations. The reader may verify that $\tilde{\delta}_{11} \equiv \tilde{\omega}_1 - \tilde{\omega}_2$ has df $f_S(\delta_{11} \mid \Delta, D_1^2(\mathbf{n}), \nu_0)$, where Δ is given by (19.4.14) and

$$D_1^2(\mathbf{n}) \equiv (1, -1)\mathbf{k}\psi_0[\boldsymbol{\sigma}_{0*}^2 - \boldsymbol{\sigma}_{1*}^2(\mathbf{n})]\mathbf{k}'(1, -1)', \tag{19.4.23}$$

and

$$\begin{aligned} I(\mathbf{n}) &= E \max \{\tilde{\delta}_{11}, 0\} \\ &= \int_0^\infty \delta_{11} f_S(\delta_{11} \mid \Delta, D_1^2(\mathbf{n}), \nu_0)d\delta_{11} \\ &= D_1(\mathbf{n})Ł_{S*}\left(\frac{|\Delta|}{D_1(\mathbf{n})} \mid \nu_0\right). \end{aligned}$$

Example 19.4.2

Reconsider Example 19.4.1, assuming that K_1, K_2, k_1, k_2, and $\boldsymbol{\mu}_0$ are as given there, but $\psi_0 = 10^8$, $\nu_0 = 4$,

$$\boldsymbol{\sigma}_{0*}^2 = \begin{bmatrix} 1.00 & .60 \\ .60 & 1.00 \end{bmatrix},$$

and

$$\boldsymbol{\sigma}_*^2 = \begin{bmatrix} 1 & 0 \\ 0 & 1 \end{bmatrix}.$$

Note that $\psi_0\tilde{\sigma}_{0*}^2$ equals $\tilde{\sigma}_0^2$ as given in Example 19.3.1 and $\psi_0\tilde{\sigma}_*^2$ equals $\tilde{\sigma}^2$ as given there. Hence by (19.4.15) and (19.4.21), $D_0^2 = 13.60 \times 10^6$ and $D_0 \doteq 4330$. Moreover, $\psi_0\tilde{\sigma}_{1*}^2(\mathbf{n})$ equals $\tilde{\sigma}_1^2(\mathbf{n})$ in Example 19.4.1, and hence $D_1(\mathbf{n}) \doteq 2870$ for $\mathbf{n} = (3, 3)$. Therefore

$$\begin{aligned} E_{\mathbf{w}}\ell_t(a^0_{z_0,0}, \tilde{\mathbf{w}}) &\doteq 4330\, \L_{S*}(.231 \mid 4) \\ &\doteq (4330)(.3945) \\ &\doteq \$1708 \end{aligned}$$

and

$$\begin{aligned} I(\mathbf{n}) &\doteq 2870\, \L_{S*}(.349 \mid 4) \\ &\doteq (2870)(.3581) \\ &\doteq \$1028. \end{aligned}$$

EXERCISES

1. Show that (19.2.2) implies (19.2.3).
2. Verify that (19.2.4) implies (19.2.5) and (19.2.6).
3. Show that $a^0 \equiv {}_L\theta \leqq b^0 \equiv {}_U\theta$.
4. Show that (a^0, b^0) as defined by (19.2.7) yields a relative minimum of (19.2.4).
5. Prove (19.2.8).
6. Derive (19.2.15).
7. Derive (19.2.16a) through (19.2.17b).
8. Derive (19.2.18).
9. Derive (19.2.24a) through (19.2.25b).
10. Derive (19.2.27) and (19.2.28).
11. Derive (19.3.8).
12. Show that (19.3.1) implies $E_{\theta|z,n}v_t(a_i, \tilde{\theta}) = v_t(a_i, E(\tilde{\theta} \mid z, n))$.
13. Prove (19.3.10) and (19.3.11).
14. Derive (19.3.26) through (19.3.30).
15. Derive (19.3.31) through (19.3.36).
16. Derive (19.4.4) and (19.4.9).
17. Show that $\max\{\delta_{01}, \ldots, \delta_{0r-1}, 0\} = \max_j \{\max\{\delta_{0j}, 0\}\}$.
18. Prove (19.4.13).

19. Prove (19.4.20).

20. In Example 14.4.1, suppose that the policeman was required to furnish an interval estimate of $\tilde{\lambda}$ based upon his posterior df $f_\gamma(\lambda \mid 160, 8)$ and that $k_u = k_O = 10$.

    (a) Find (a^0, b^0), if $k_w = \frac{1}{2}$.
    (b) Find (a^0, b^0), if $k_w = 4$.

21. Suppose that the politician in Example 14.4.1 wished to make an interval estimate of $\tilde{p}$ with $k_u = k_O = 10$ and $k_w = \frac{1}{2}$.

    (a) What is his optimal interval estimate *prior* to obtaining the sample information?
    (b) What is his optimal interval estimate *after* obtaining the sample information?

22. Reconsider the problem in Examples 18.2.3 and 18.3.2, now assuming that $k_u = k_O = \$50{,}000$, $k_w = 10{,}000$, and an interval estimate (a^0, b^0) of mean family income $\tilde{\mu}$ is desired. Continue to assume that $\sigma^2 = 1{,}000{,}000$ and that $\tilde{\mu}$ has prior df $f_N(\mu \mid 2500, \sigma^2/4)$.

    (a) What is the optimal sample size? Discuss its relationship to $n^0 = 5400$ as calculated in Example 18.3.2.
    (b) Suppose that a sample of size 5400 had been obtained with sufficient statistic $\bar{x} = 3000$. What is the optimal interval estimate (a^0, b^0) of $\tilde{\mu}$?

23. In Example 19.3.1, suppose that the manufacturer's cost accountant refigures the cost of producing widgets with the new stamping machine to be \$0.18 and not \$0.20 per widget. Rework the example in light of this information.

24. Again consider Example 19.3.1 as originally stated, under the assumptions that $p = \$0.50$ and the manufacturer believes demand to be dependent upon the manufacturing process, in the sense that

$$v(a_1, \boldsymbol{\mu}) = .20\mu_1$$

and

$$v(a_2, \boldsymbol{\mu}) = -10{,}000 + .30\mu_2,$$

where $\mu_1 \equiv$ "demand for widgets produced by hand" and $\mu_2 \equiv$ "demand for widgets produced by stamping machine." Suppose that his prior df of $\tilde{\boldsymbol{\mu}} \equiv (\tilde{\mu}_1, \tilde{\mu}_2)'$ is $f_N^{(2)}(\boldsymbol{\mu} \mid \boldsymbol{\mu}_0, \boldsymbol{\sigma}_0^2)$, where $\boldsymbol{\mu}_0 = (9 \times 10^4, 9 \times 10^4)'$ and

$$\boldsymbol{\sigma}_0^2 = \begin{bmatrix} 10^8 & 9 \times 10^7 \\ 9 \times 10^7 & 10^8 \end{bmatrix}.$$

What is his prior value of perfect information about $\tilde{\boldsymbol{\mu}}$? How does it compare with the \$83.30 in Example 19.3.1?

part IV

Statistical Inference

The fourth part of this book consists of an introduction to statistical inference, primarily from orthodox points of view. In order to place the various approaches to inference in proper perspective, we suggest a scheme of classifying approaches according to their use or nonuse of prior probabilities or dgf's, and of utilities or losses. The result is the following table:

Approach	*Utility or Loss?*	*Prior Probability?*
1. Bayesian Decision	Yes	Yes
2. Non-Bayesian Decision	Yes	No
3. Bayesian Inference	No	Yes
4. Orthodox Inference	No	No

Part III has furnished the reader with an introduction to Bayesian decision theory, and also with a very brief introduction to non-Bayesian decision theory (in Section 17.5). In Part IV, Bayesian and orthodox inference are introduced in Chapters 20 through 22. Brief mention is also made of non-Bayesian decision-theoretic formulations of some of the inference problems to be considered, as one can move in either of two ways from these formulations (in terms of risk functions): by introducing prior probabilities or dgf's, one obtains the Bayesian decision-theoretic formulation; and by adopting special cases of loss functions, one obtains popular orthodox formulations.

In Chapter 20 we consider the first of three major problems in inference introduced in Section 9.7—the problem of point estimation. Several methods of point estimation are discussed, primarily from orthodox viewpoints. We also consider how a Bayesian statistician may interpret an orthodox point estimate.

Chapter 21 is devoted to the second problem in Section 9.7; namely, interval estimation (and, when $\boldsymbol{\theta}$ is multivariate, region estimation). The orthodox theory is developed, its Bayesian content discussed, and a Bayesian alternative presented.

Chapter 22 concerns the most venerable problem of inference, that of testing statistical hypotheses. The emphasis in this chapter is almost entirely upon orthodox theory, formalized and extended successively by K. Pearson, R. A. Fisher, and J. Neyman, and E. S. Pearson.

A case can be made for saying that Bayesian inference, with prior probabilities but no utilities or losses, historically predates orthodox inference. However, early Bayesian inferences were based upon prior probabilities obtained by the principle of insufficient reason (see Section 5.4). Ancient and modern Bayesian inference share the property that all conclusions are based entirely upon the posterior probability function or dgf $f(\boldsymbol{\theta} \mid \mathbf{x}_T)$ of $\tilde{\boldsymbol{\theta}}$ given the (sufficient statistic $\mathbf{x}_T$ of the) outcome of the experiment.

Because of the defects and ambiguity of the principle of insufficient reason, the orthodox approach was developed and based upon frequency definitions of probability. The influence of the late Sir Ronald A. Fisher in this development cannot be overestimated; there is hardly any aspect of statistical theory and methodology to which he did not contribute. Among his many published works are two books ([**27**], [**28**]) that describe his viewpoints.

Concurrently with Fisher's development of orthodox methodology, Sir Harold Jeffreys was reviving Bayesian inference. His *Theory of Probability* [**40**] is now a classic. Lindley's second volume [**49**] is largely devoted to Bayesian inference in the Jeffreys spirit.

J. Neyman and E. S. Pearson, in a series of fundamental papers (for example, [**59**]), appear to have been the first to view the problem of testing statistical hypotheses *explicitly* as a *decision* problem, but they refrained from introducing prior probabilities and/or losses.

The next step was A. Wald's formulation of inference as a decision problem, with loss functions. His last book [**82**] is the single most complete exposition of his viewpoint. Wald was the originator of non-Bayesian decision theory; he refrained from advocating the use of prior probability functions or dgf's when they were undefined from the relative frequency standpoint.

The final methodological step was the publication in 1954 of L. J. Savage's *The Foundations of Statistics* [**69**], which incorporated subjective prior probabilities and dgf's obtained via a joint axiomatization of utility and subjective probability. Part III of this book furnishes alternative axiomatizations and several applications of general Bayesian decision theory to specific classes of problems in Chapters 18 and 19, many of the results of which are due originally to H. Raiffa and R. O. Schlaifer [**66**].

The preceding capsule history is by no means complete, and in some cases it is slightly oversimplified. The reader should not obtain the impression that all statisticians agree that the more recent approaches are necessarily better. In fact, as of now many, if not most, statisticians are rather firm adherents to orthodox theory.

No serious student of statistics and decision theory should fail to become acquainted with the orthodox theory, even if only for the purpose of becoming "statistically literate." It is for this reason that Chapters 20 through 22 are devoted substantially to orthodox approaches.

General references on orthodox inference are many; Wilks [**83**], Cramér [**15**], Anderson [**1**], and Kendall and Stuart [**41**] are excellent, as is Lehmann's work [**45**] on hypothesis testing. Jeffreys [**40**] and Lindley [**49**] have already been mentioned as references on Bayesian inference. Wald [**82**] and Ferguson [**22**] are excellent references on non-Bayesian decision theory, while the classics in Bayesian decision theory are Savage [**69**] on foundations, and Raiffa and Schlaifer [**66**] on applied theory. Three additional references for readers interested in foundational questions are L. J. Savage *et al.* [**70**] (a long paper by Savage and extended discussion of it); Edwards, Lindman and Savage [**16**]; and A. Birnbaum [**6**]. Also in the foundational category is Pratt's paper and the ensuing discussion [**63**], on Bayesian interpretations of orthodox inferences.

20

INTRODUCTION TO POINT ESTIMATION

20.1 INTRODUCTION

The first problem of inference broached in Section 9.7 was to find a "close" estimate $\hat{\theta}$ of θ based upon the outcome of an experiment. It is clear that $\hat{\theta}$ must be a *statistic* [that is, a function of the outcome $(n, x_1, \ldots, x_n)$ only, and not of θ] so that it can be calculated once the experiment has been concluded.

Throughout this chapter we shall restrict attention to experiments which consist of a predetermined number n of observations which are iid with common dgf $f(x \mid \theta)$ [or $f(\mathbf{x} \mid \boldsymbol{\theta})$ where appropriate]. This restriction can be relaxed in many instances, but imposing it makes the presentation somewhat simpler.

Since $\hat{\theta}$ is a statistic, it is a function $\hat{\theta}(n, x_1, \ldots, x_n)$ of the outcome $(n, x_1, \ldots, x_n)$ of the experiment and hence unknown prior to the experiment. We denote $\tilde{\hat{\theta}} \equiv \hat{\theta}(n, \tilde{x}_1, \ldots, \tilde{x}_n)$ and call $\tilde{\hat{\theta}}$ as *estimator* of θ, reserving the term *estimate* for the number (or function value) $\hat{\theta}(n, x_1, \ldots, x_n)$ calculated after obtaining the experimental outcome. The term *estimator* is also applied to the set $\{\hat{\theta}(n, \tilde{x}_1, \ldots, \tilde{x}_n): n = 1, 2, \ldots\}$ of estimators, but we shall take care to prevent resulting confusion. Note that the terminological problems here are no worse than with *statistic* (compare Section 9.5).

In orthodox statistical inference, θ usually possesses no prior dgf $f(\theta)$ in the frequency sense and hence (see Section 9.4) all *probabilistic statements about* $\tilde{\hat{\theta}}$ *are conditional upon* θ. The dgf of $\tilde{\hat{\theta}}$ is derivable from the joint dgf

$$f(x_1, \ldots, x_n \mid n, \theta) = \prod_{i=1}^{n} f(x_i \mid \theta)$$

of the observations $\tilde{x}_1, \ldots, \tilde{x}_n$ given n and θ. In all cases of interest in this book, the dgf of $\tilde{\theta}$ given n and θ has either been already derived in Chapters 11 through 13 or is easily derivable from some dgf in those chapters.

Now, the problems described in Section 9.7 are stated in very imprecise language, in order not to prejudge potential solutions. Thus the adjective *close* for an estimate is open to various interpretations. We define in Section 20.2 some criteria which are desirable attributes of estimators.

Section 20.3 presents three methods of obtaining estimators. One of these, the maximum likelihood method, is singled out as having particularly desirable attributes, and forms the basis for Section 20.4. Moreover, Section 20.4 applies the preceding general discussion to the description of estimators for the parameters of the five data-generating processes that are prominently featured in this book; namely, the Bernoulli, k-variate Bernoulli, Poisson, Normal, and k-variate Normal processes.

Section 20.5 concludes with a discussion of orthodox estimation procedures from the standpoint of a Bayesian statistician, who has a prior dgf $f(\theta)$ and is given an estimate $\hat{\theta}$ of θ derived by some orthodox procedure. We discuss how he might interpret the estimate.

Section 20.6 discusses how that Bayesian statistician would go about estimating θ. The section is brief, as Chapter 18 is devoted entirely to the preceding problem if the statistician has a utility (or loss) function, and the statistician would probably insist on assuming a conventional utility (or loss) function otherwise.

All preceding comments apply unchanged to vector estimators $\tilde{\boldsymbol{\theta}}$ of vector parameters $\boldsymbol{\theta}$.

20.2 SOME IMPORTANT PROPERTIES OF ESTIMATORS

Let $f(\hat{\theta} \mid \theta)$ denote the conditional dgf of $\tilde{\theta}$ given θ. All expectations, variances, and so forth, in this section will be computed from $f(\hat{\theta} \mid \theta)$.

We say that $\tilde{\theta}$ is an *unbiased* estimator of θ if

$$E(\tilde{\theta} \mid \theta) = \theta, \text{ for every } \theta \in \Theta. \tag{20.2.1}$$

The adjective *unbiased* is a very unfortunate but also universal way of describing estimators which satisfy (20.2.1). The term suggests that an unbiased estimator is always preferable to a biased one, but is this always true?

Suppose that we have two estimators, $\tilde{\theta}_1$ and $\tilde{\theta}_2$, with respective df's $f(\hat{\theta}_1 \mid \theta)$ and $f(\hat{\theta}_2 \mid \theta)$ as depicted in Figure 20.1, which suggests that $\tilde{\theta}_1$ is unbiased but $V(\tilde{\theta}_1 \mid \theta)$ is large and $\tilde{\theta}_2$ is biased but $V(\tilde{\theta}_2 \mid \theta)$ is small. In many circumstances $\tilde{\theta}_2$ would be preferable to $\tilde{\theta}_1$ because of its smaller variance.

Unbiased estimators of a parameter θ do not always exist. When they do, they need not be unique.

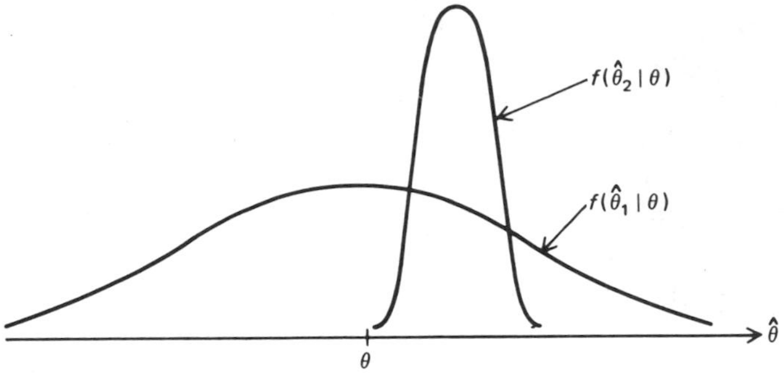

Figure 20.1

Example 20.2.1

In the univariate Normal process with σ^2 known, we seek unbiased estimators of μ. Let $\mathbf{c} \in R^n$ satisfy $\sum_{i=1}^{n} c_i = 1$, and define $\tilde{\hat{\mu}}(\mathbf{c}) \equiv \sum_{i=1}^{n} c_i \tilde{x}_i$. Then $E(\tilde{\hat{\mu}}(\mathbf{c}) \mid \mu) = E\left(\sum_{i=1}^{n} c_i \tilde{x}_i \mid \mu\right) = \sum_{i=1}^{n} c_i E(\tilde{x}_i \mid \mu) = \sum_{i=1}^{n} c_i \mu = \mu \sum_{i=1}^{n} c_i = \mu$, so that $\tilde{\hat{\mu}}(\mathbf{c})$ is unbiased for every c whose components sum to one. Thus there are uncountably many unbiased estimators of μ. This example extends verbatim to estimators of the population mean μ in non-Normal cases.

A more restrictive property than unbiasedness is minimum-variance unbiasedness. We say that $\tilde{\hat{\theta}}$ is a *minimum-variance unbiased estimator* of θ if $\tilde{\hat{\theta}}$ is unbiased and if no other unbiased estimator of θ has smaller variance for any θ.

Example 20.2.2

In Example 20.2.1 we now seek that $\mathbf{c}$ (if any) for which $V(\tilde{\hat{\mu}}(\mathbf{c}) \mid \mu)$ is minimized for every μ. It is easy to verify from (10.3.10) that $V(\tilde{\hat{\mu}}(\mathbf{c}) \mid \mu) = \sigma^2 \sum_{i=1}^{n} c_i^2$, which is independent of μ and is minimized when $\sum_{i=1}^{n} c_i^2$ is. Hence we minimize $\sum_{i=1}^{n} c_i^2$ subject to the constraint that $\sum_{i=1}^{n} c_i = 1$. Standard methods (for example, introduction of a Lagrange multiplier) suffice to show that the unique solution is $c_1 = c_2 = \cdots = c_n = 1/n$. Hence the unique minimum-variance unbiased linear $\left(\sum_{i=1}^{n} c_i \tilde{x}_i\right)$ estimator of μ is

$$\tilde{\bar{x}} = \sum_{i=1}^{n} (1/n)\tilde{x}_i = (1/n) \sum_{i=1}^{n} \tilde{x}_i.$$

Some rather sophisticated conditions and arguments have been introduced regarding the topic of minimum variance without necessarily imposing unbiasedness. See Kendall and Stuart [**41**], Chapter 17, or Wilks [**83**], Chapter 12.

Within the class of unbiased estimators, there is a result, due to Blackwell and Rao, to the effect that unbiased estimators that are functions of sufficient statistics generally have lower variance than unbiased estimators that are not functions of sufficient statistics. Thus the search for minimum variance unbiased estimators of θ can be confined to functions of sufficient statistics for θ.

Example 20.2.3

By (13.2.5), $(n, \tilde{\bar{x}})$ is sufficient for the mean μ of a univariate Normal process with known variance σ^2. Hence in Example 20.2.2 the minimum-variance unbiased estimator of μ is a function of $(n, \tilde{\bar{x}})$ and is therefore $\tilde{\bar{x}}$.

We say that an estimator $\tilde{\hat{\theta}}$ of θ is a *sufficient* estimator if it is a function of a sufficient statistic for θ. The benefits of sufficiency are the same here as for statistics in general; namely, a sufficient estimator $\tilde{\hat{\theta}}$ is based upon all the relevant information about θ contained in the outcome of the experiment. All estimators described in this book will be sufficient.

For vector estimators $\tilde{\hat{\boldsymbol{\theta}}}$ of vector parameters $\boldsymbol{\theta}$, unbiasedness is defined by (20.2.1) with the θ's in boldface type, but there are various ways of defining minimum variance, which we shall not introduce.

The next two properties of estimators pertain to sets $\{\hat{\theta}(n, \tilde{x}_1, \ldots, \tilde{x}_n): n = 1, 2, \ldots\}$ of estimators in the usual sense, and we shall write $\tilde{\hat{\theta}}_n = \hat{\theta}(n, \tilde{x}_1, \ldots, \tilde{x}_n)$ to denote dependence of the estimator upon the number of observations. This notation is convenient because the properties to be introduced are asymptotic results, of particular relevance when n is large.

We say that $\{\tilde{\hat{\theta}}_n: n = 1, 2, \ldots\}$ is a (*simply*) *consistent* estimator of θ if

$$\lim_{n\to\infty} P(\theta - \epsilon \leqq \tilde{\hat{\theta}}_n \leqq \theta + \epsilon \mid \theta) = 1, \tag{20.2.2}$$

for every $\epsilon > 0$ and every $\theta \in \Theta$. Consistency of an estimator is a very desirable property, as it means that the probability of an appreciable ($> \epsilon$) error ultimately vanishes, whatever positive ϵ is chosen.

Example 20.2.4

The *weak law of large numbers* (10.5.5) is the special case of (20.2.2) for $\theta = \mu$ and $\tilde{\hat{\theta}}_n = \tilde{\bar{x}}$. Hence the sample mean $\tilde{\bar{x}}$ is a consistent estimator of the population mean μ, provided that the latter is finite. (Recall that while our proof utilized finiteness of population variance σ^2, we stated that only μ need be finite for the assertion to obtain.)

Unfortunately, (20.2.2) masks a practical difficulty; namely, the *rate* at which the limiting probability one is approached depends not only upon ϵ, but also in general upon θ. Hence we cannot make assertions regarding lower bounds for $P(\theta - \epsilon \leqq \tilde{\tilde{\theta}}_n \leqq \theta + \epsilon \mid \theta)$ without conditioning such assertions upon θ, which we obviously do not know. In some cases this difficulty does not arise.

For vector estimators $\{\tilde{\tilde{\boldsymbol{\theta}}}_n: n = 1, 2, \ldots\}$ of vector parameters $\boldsymbol{\theta}$, (simple) consistency is defined by (20.2.2) with all θ's and ϵ's in boldface. Note that $\{\hat{\boldsymbol{\theta}}_n: \boldsymbol{\theta} - \boldsymbol{\epsilon} \leqq \hat{\boldsymbol{\theta}}_n \leqq \boldsymbol{\theta} + \boldsymbol{\epsilon}\}$ is a closed, k-dimensional (cubic) interval with side length 2ϵ, centered at $\boldsymbol{\theta}$.

An estimator $\{\hat{\theta}_n: n = 1, 2, \ldots\}$ of θ is said to be *asymptotically Normal* if there is a function $v: \Theta \rightarrow (0, \infty)$ such that $\{\tilde{\tilde{\theta}}_n: n = 1, 2, \ldots\}$ is a consistent estimator and

$$\lim_{n\rightarrow\infty} F(\hat{\theta}_n \mid \theta) = F_N\left(\hat{\theta}_n \mid \theta, \frac{v(\theta)}{n}\right). \tag{20.2.3}$$

Roughly, (20.2.3) asserts that an asymptotically Normal estimator satisfies the Central Limit Theorem, with $\tilde{\tilde{\theta}}_n$ replacing $\tilde{x}$, θ replacing μ, and $v(\theta)$ replacing σ^2. From (20.2.3) we easily conclude that

$$\lim_{n\rightarrow\infty} E(\tilde{\tilde{\theta}}_n \mid \theta) = \theta, \tag{20.2.4}$$

a condition which might be called *asymptotic unbiasedness*; and that

$$\lim_{n\rightarrow\infty} V(\tilde{\tilde{\theta}}_n \mid \theta) = \lim_{n\rightarrow\infty} \frac{v(\theta)}{n} = 0. \tag{20.2.5}$$

The limits (20.2.4) and (20.2.5) imply the consistency property (20.2.2), which suggests (but does not imply) them.

A final asymptotic property is a sharpening of asymptotic Normality. We introduce it by remarking that there may be many different asymptotically Normal (sets of) estimators of θ, each with its own *asymptotic variance function* $v(\theta)$. It clearly behooves us to choose that estimator $\tilde{\tilde{\theta}}_n{}^0$ with the smallest asymptotic variance function $v^0(\theta)$. Accordingly, we say that $\{\tilde{\tilde{\theta}}_n{}^0: n = 1, 2, \ldots\}$ is a *best asymptotically Normal* estimator of θ if it is an asymptotically Normal estimator of θ, and if no other asymptotically Normal estimator $\{\tilde{\tilde{\theta}}_n{}^*: n = 1, 2, \ldots\}$ with asymptotic variance function $v^*(\theta)$ has the property that

$$v^*(\theta) < v^0(\theta),$$

for all θ is some open interval. [Note that $v^*(\theta) < v^0(\theta)$ can obtain at isolated values of θ, and even at (say) all rational values of θ, without destroying the "bestness" of $\{\hat{\theta}_n{}^0: n = 1, 2, \ldots\}$.] We abbreviate *best asymptotically Normal* to *BAN*.

Example 20.2.5

Let $\theta = \mu$, the mean of a univariate Normal process with σ^2 known, let $\tilde{\theta}_n{}^0 = \bar{\tilde{x}} \equiv (1/n) \sum_{i=1}^{n} \tilde{x}_i$, for every n, and let

$$\tilde{\theta}_n{}^* = \begin{cases} \left(\dfrac{2}{n}\right) \sum_{i=1}^{n/2} \tilde{x}_{2i}, & \text{if } n \text{ is even} \\ \tilde{\theta}^*_{n-1}, & \text{if } n \geqq 3 \text{ is odd} \\ 0, & \text{if } n = 1. \end{cases}$$

Then $\{\tilde{\theta}_n{}^0\colon n = 1, 2, \ldots\}$ is asymptotically Normal with $\upsilon^0(\theta) = \sigma^2$ for every $\mu \in R^1$. The estimator $\{\tilde{\theta}_n{}^*\colon n = 1, 2, \ldots\}$ is also asymptotically Normal, with $\upsilon^*(\mu) = 2\sigma^2$, for every $\mu \in R^1$ because $V(\tilde{\theta}_n{}^* \mid \theta) = \sigma^2/(n/2) = 2\sigma^2/n = 2V(\tilde{\theta}_n{}^0 \mid \theta)$, for all even n. Hence $\{\tilde{\theta}_n{}^*\colon n = 1, 2, \ldots\}$ is not BAN.

It is sometimes convenient to use the following consequence of (20.2.3):

If $\tilde{\theta}_n$ satisfies (20.2.3), then $\upsilon(\theta) = \lim_{n\to\infty} nV(\tilde{\theta}_n \mid \theta)$. (20.2.6)

Proof

If $\tilde{\theta}_n$ has asymptotic variance $\upsilon(\theta)/n$ given θ, then $\sqrt{n}\tilde{\theta}_n$ has asymptotic variance $(\sqrt{n})^2\upsilon(\theta)/n = \upsilon(\theta)$, by (10.3.13). But $V(\sqrt{n}\tilde{\theta}_n \mid \theta) = nV(\tilde{\theta}_n \mid \theta)$. QED

Example 20.2.6

Let $\tilde{\theta}_n{}^* = 1/(n + 10) \sum_{i=1}^{n} \tilde{x}_i$. Then $\{\tilde{\theta}_n{}^*\colon n = 1, 2, \ldots\}$ is a BAN estimator of the mean μ of a univariate Normal process with σ^2 known, because the df of $\tilde{\theta}_n{}^*$ is

$$f_N\left(\cdot \,\Big|\, \frac{n}{n+10}\mu, \left(\frac{n}{n+10}\right)^2 \frac{\sigma^2}{n}\right), \qquad \lim_{n\to\infty} \left\{\frac{n\mu}{n+10}\right\} = \mu,$$

and

$$\begin{aligned} \upsilon^*(\mu) &= \lim_{n\to\infty} \left\{n\left(\frac{n}{n+10}\right)^2 \frac{\sigma^2}{n}\right\} \\ &= \lim_{n\to\infty} \left\{\left(\frac{n}{n+10}\right)^2 \sigma^2\right\} \\ &= \sigma^2 \lim_{n\to\infty} \left(\frac{n}{n+10}\right)^2 \\ &= \sigma^2, \end{aligned}$$

the same as $\upsilon^0(\mu)$ for $\tilde{\theta}_n{}^0 = \bar{\tilde{x}}$.

Example 20.2.6 shows that BAN estimators need not be unique.

All the preceding properties of estimators have been introduced without explicit consideration of the consequences of estimating $\hat{\theta}$ when $\theta \neq \hat{\theta}$, except insofar as all the properties reflected the philosophy that it is good to be correct and bad to be grossly incorrect. Other properties of estimators depend upon the introduction of a *loss function* $\ell(\hat{\theta}, \theta) \geqq 0$ which stipulates penalties for misestimating θ. (Compare Chapter 17, where loss functions were defined in terms of the much more fundamental concepts of preference and utility introduced in Chapter 15.)

A popular assumption is *squared-error loss*, or *quadratic loss*, defined by

$$\ell(\hat{\theta}, \theta) = K(\theta)(\hat{\theta} - \theta)^2, \tag{20.2.7}$$

where $K(\theta) > 0$ and is often assumed independent of θ as well of $\hat{\theta}$. The *risk function* $R(\tilde{\theta}, \theta)$ is defined in general by

$$R(\tilde{\theta}, \theta) \equiv E(\ell(\tilde{\theta}, \theta) \mid \theta), \text{ for every } \theta \in \Theta; \tag{20.2.8}$$

for squared-error loss,

$$\begin{aligned} R_{sq}(\tilde{\theta}, \theta) &\equiv K(\theta)E((\tilde{\theta} - \theta)^2 \mid \theta) \\ &= K(\theta)E(([(\tilde{\theta} - E(\tilde{\theta} \mid \theta)] + [E(\tilde{\theta} \mid \theta) - \theta])^2 \mid \theta) \\ &= K(\theta)[V(\tilde{\theta} \mid \theta) + (E(\tilde{\theta} \mid \theta) - \theta)^2], \end{aligned} \tag{20.2.9}$$

the last equality of which may be proved as Exercise 1.

Suppose there were an estimator $\tilde{\theta}^0$ such that

$$R(\tilde{\theta}^0, \theta) \leqq R(\tilde{\theta}, \theta), \text{ for every } \theta \in \Theta, \tag{20.2.10}$$

and for every possible estimator $\tilde{\theta}$ of θ. Then there would be no question but that $\tilde{\theta}^0$ should be used. Unfortunately, in all cases of any practical importance no $\tilde{\theta}^0$ satisfying (20.2.10) exists. For any two reasonable estimators $\tilde{\theta}^{(1)}$ and $\tilde{\theta}^{(2)}$ of θ, the corresponding risk functions behave as suggested by Figure 20.2; that is, $\tilde{\theta}^{(1)}$ is better (gives lower risk) for *some* θ's while $\tilde{\theta}^{(2)}$ is better for *other* θ's.

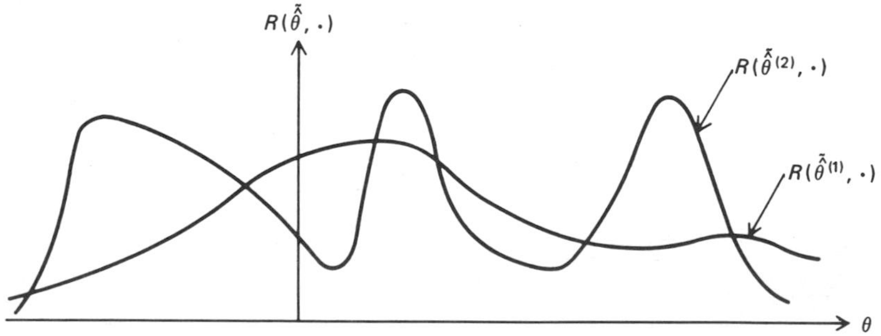

Figure 20.2

Thus in cases of practical importance a choice must be made between risk functions which cross each other. One method of choice is the *minimax risk* criterion: $\tilde{\hat{\theta}}^{(1)}$ is *defined* as preferable to $\tilde{\hat{\theta}}^{(2)}$ if and only if $\max_\theta R(\tilde{\hat{\theta}}^{(1)}, \theta) < \max_\theta R(\tilde{\hat{\theta}}^{(2)}, \theta)$; and $\tilde{\hat{\theta}}^0$ is a *minimax* estimator if

$$\max_\theta R(\tilde{\hat{\theta}}^0, \theta) = \min_{\tilde{\hat{\theta}}} \max_\theta R(\tilde{\hat{\theta}}, \theta). \tag{20.2.11}$$

Example 20.2.7

If both risk functions in Figure 20.2 vanish off the graph, then $\tilde{\hat{\theta}}^{(1)}$ is preferable to $\tilde{\hat{\theta}}^{(1)}$ according to the minimax criterion.

Another method of choice based upon risk functions is appropriate when θ is regarded as a random variable, $\tilde{\theta}$, in which case we define the *Bayes risk* $B(\tilde{\hat{\theta}})$ of any estimator $\tilde{\hat{\theta}}$ by

$$B(\tilde{\hat{\theta}}) \equiv E\{R(\tilde{\hat{\theta}}, \tilde{\theta})\}, \tag{20.2.12}$$

the expectation being with respect to the prior dgf $f(\theta)$ of $\tilde{\theta}$. An estimator $\tilde{\hat{\theta}}^0$ is called a *Bayes estimator* of θ [with respect to $f(\theta)$] if

$$B(\tilde{\hat{\theta}}^0) = \min_{\tilde{\hat{\theta}}} B(\tilde{\hat{\theta}}). \tag{20.2.13}$$

[In the notation of Chapter 18,

$$B(\tilde{\hat{\theta}}) \equiv B(a^0_{\tilde{z},n}) = E_\theta\{E_{z|n,\theta}\, \ell(a^0_{\tilde{z},n}, \tilde{\theta})\}.$$

But

$$E_\theta\{E_{z|n,\theta}\, \ell(a^0_{\tilde{z},n}, \tilde{\theta})\} = E_{z|n}\{E_{\theta|n,\tilde{z}}\, \ell(a^0_{\tilde{z},n}, \tilde{\theta})\}$$

because the order of expectation may be reversed. Hence Chapter 18 covers the subject of Bayes estimators, rather extensively in the case of squared-error loss.]

No orthodox statistician would object to using Bayes estimators when $f(\theta)$ and $\ell(\hat{\theta}, \theta)$ are known; everyone agrees on their optimality. The differences of opinion arise only when $f(\theta)$ is unknown, or when $f(\theta)$ cannot be given a frequency interpretation, and/or when $\ell(\hat{\theta}, \theta)$ is unknown.

A final, useful concept is *admissibility*. It is introduced in Exercise 19.

20.3 SOME METHODS OF OBTAINING ESTIMATORS

The preceding section introduced some desirable properties of estimators without indicating any procedure by which estimators can be found, except in the cases of Bayes and minimax estimators, where the procedures are implicit in their defintions. The purpose of this section is to introduce two additional procedures for generating estimators and to discuss the properties of the estimators so obtained.

The first, and historically earliest procedure is the *method of moments.* Suppose the common dgf $f(x \mid \boldsymbol{\theta})$ of the observations depends upon a k-dimensional vector parameter $\boldsymbol{\theta} \equiv (\theta_1, \ldots, \theta_k)$; $k = 1$ is a special case. According to the method of moments, we compute the first k *population moments* $E(\tilde{x}^r \mid \boldsymbol{\theta})$ about zero, and equate them to the corresponding *sample moments*

$$\bar{x}^r \equiv \frac{1}{n} \sum_{i=1}^{n} x_i{}^r, \; r = 1, \ldots, k, \tag{20.3.1}$$

thus obtaining a system of k equations in k unknowns $\theta_1, \ldots, \theta_k$:

$$E(\tilde{x}^r \mid \theta_1, \ldots, \theta_k) = \bar{x}^r, \; r = 1, \ldots, k. \tag{20.3.2}$$

The solution of (20.3.2), if it exists and is unique, is denoted by $\hat{\boldsymbol{\theta}}(n, x_1, \ldots, x_n)$ and constitutes the estimate of $\boldsymbol{\theta}$ obtained by the method of moments. Before the experiment is conducted, $x_1, \ldots, x_n$ are unknown and hence $\hat{\boldsymbol{\theta}}(n, \tilde{x}_1, \ldots, \tilde{x}_n)$ is an estimator with some dgf $f(\hat{\boldsymbol{\theta}} \mid \boldsymbol{\theta})$.

Example 20.3.1

For a univariate Normal process with known variance σ^2, there is only one $(= k)$ unknown parameter, μ, and $E(\tilde{x}^1 \mid \mu) = \mu$. The first sample moment $\bar{x}^1$ is the sample mean $\bar{x}$, which we equate to $E(\tilde{x}^1 \mid \mu)$ to obtain the estimator $\tilde{\hat{\mu}} = \tilde{\bar{x}}$ of μ via the method of moments.

It can be shown that estimators obtained by the method of moments are consistent and asymptotically Normal. However, they are often not best asymptotically Normal, and they need not be functions of sufficient statistics.

The second procedure to be introduced is the method of *maximum likelihood.* The estimators so obtained are called *maximum likelihood estimators* and are usually best asymptotically Normal as well as functions of sufficient statistics.

A maximum likelihood estimator $\tilde{\hat{\theta}}$ of $\boldsymbol{\theta}$ is defined at every $(n, \mathbf{x}_1, \ldots, \mathbf{x}_n)$ by

$$L(\hat{\boldsymbol{\theta}} \mid n, \mathbf{x}_1, \ldots, \mathbf{x}_n) \geqq L(\boldsymbol{\theta} \mid n, \mathbf{x}_1, \ldots, \mathbf{x}_n), \text{ for every } \boldsymbol{\theta} \in \boldsymbol{\Theta}, \tag{20.3.3}$$

where L denotes the likelihood function. Usually there is a unique $\hat{\boldsymbol{\theta}} \in \Theta$ for every $(n, \mathbf{x}_1, \ldots, \mathbf{x}_n)$ which satisfies (20.3.3), but occasionally even existence fails. In such cases, however, it is often possible to slightly modify L to obtain existence.

Example 20.3.2

Suppose the common dgf of $\tilde{x}_1, \ldots, \tilde{x}_n$ is

$$f(x \mid \theta) = \begin{cases} \theta^{-1}, & x \in (0, \theta) \\ 0, & x \notin (0, \theta). \end{cases}$$

Then

$$L(\theta \mid n, x_1, \ldots, x_n) = \prod_{i=1}^{n} f(x_i \mid \boldsymbol{\theta})$$
$$= \begin{cases} \theta^{-n}, & \text{every } x_i \in (0, \theta) \\ 0, & \text{some } x_i \notin (0, \theta). \end{cases}$$

Let $M \equiv \max_i x_i$. Then

$$L(\theta \mid n, x_1, \ldots, x_n) = \begin{cases} 0, & \theta \in (0, M] \\ \theta^{-n}, & \theta > M, \end{cases}$$

a decreasing function of θ for $\theta > M$. Hence we would like to set $\hat{\theta} = M$, but $L(M \mid n, x_1, \ldots, x_n) = 0$. In this case, the likelihood function fails to assume a maximum, and setting $\hat{\theta} = M$ is justified by redefining L at M so that $L(M \mid n, x_1, \ldots, x_n) = M^{-n}$ instead of $= 0$.

Example 20.3.2 also shows that maximum likelihood estimators cannot always be obtained by standard methods of differential calculus, which often permit $\hat{\boldsymbol{\theta}}$ to be given as the solution of

$$\left. \frac{\partial L(\boldsymbol{\theta} \mid n, \mathbf{x}_1, \ldots, \mathbf{x}_n)}{\partial \theta_i} \right|_{\boldsymbol{\theta} = \hat{\boldsymbol{\theta}}} = 0, i = 1, \ldots, k, \tag{20.3.4}$$

for each $(n, \mathbf{x}_1, \ldots, \mathbf{x}_n)$.

Suppose that $\mathbf{x}_T$ is a sufficient statistic for $\boldsymbol{\theta}$. Then

$$L(\boldsymbol{\theta} \mid \mathbf{x}_T) = g(n, \mathbf{x}_1, \ldots, \mathbf{x}_n) \cdot L(\boldsymbol{\theta} \mid n, \mathbf{x}_1, \ldots, \mathbf{x}_n)$$

by (9.5.4) and the definition of L, where $g(n, \mathbf{x}_1, \ldots, \mathbf{x}_n) > 0$, so that if $\hat{\boldsymbol{\theta}}(\mathbf{x}_T)$ maximizes $L(\boldsymbol{\theta} \mid \mathbf{x}_T)$, then it also maximizes $L(\boldsymbol{\theta} \mid n, \mathbf{x}_1, \ldots, \mathbf{x}_n)$. Hence a maximum likelihood estimator must be expressible as a function of $\mathbf{x}_T$.

Example 20.3.3

In the Bernoulli process,

$$L(p \mid n, x_1, \ldots, x_n) = p^r(1 - p)^{n-r},$$

for $r = \sum_{i=1}^{n} x_i$. If $0 < r < n$, we differentiate with respect to p to obtain

$$\frac{dL(\hat{p})}{dp} = 0 = r\hat{p}^{r-1}(1 - \hat{p})^{n-r} - (n - r)\hat{p}^r(1 - \hat{p})^{n-r-1},$$

implying $\hat{p} = r/n$. Slight modifications of the likelihood function yield $\hat{p} = 0$ if $r = 0$ and $\hat{p} = 1$ if $r = n$. Thus $\hat{p}$ is a function of the sufficient statistic (r, n).

The method of maximum likelihood satisfies the likelihood principle, introduced in Chapter 9, Exercise 7. Therefore, a maximum likelihood estimate depends only upon the actual outcome $(n, \mathbf{x}_1, \ldots, \mathbf{x}_n)$ of the experiment, and not also upon outcomes which could have occurred but did not. Many statisticians believe that this conclusion is desirable as a general principle of inference.

Another desirable property of maximum likelihood estimators is called *invariance*, and we express it as follows:

Let $\mathbf{t}: \mathbf{\Theta} \to R^k$ be an injection. If $\tilde{\hat{\boldsymbol{\theta}}}$ is the maximum likelihood estimator of $\boldsymbol{\theta}$, then $\mathbf{t}(\tilde{\hat{\boldsymbol{\theta}}})$ is the maximum likelihood estimator $\mathbf{t}(\tilde{\hat{\boldsymbol{\theta}}})$ of $\mathbf{t}(\boldsymbol{\theta})$. (20.3.5)

Proof

Let $L_{\boldsymbol{\theta}}$ denote the likelihood function of $\boldsymbol{\theta}$ and $L_{\mathbf{t}}$ denote the likelihood function of $\mathbf{t}(\boldsymbol{\theta})$. Since $\mathbf{t}$ is an injection, its inverse $\mathbf{t}^{-1}$ is a function on $\mathbf{T} \equiv \mathbf{t}(\mathbf{\Theta})$, and

$$L_{\mathbf{t}}(\mathbf{t} \mid n, \mathbf{x}_1, \ldots, \mathbf{x}_n) = L_{\boldsymbol{\theta}}(\mathbf{t}^{-1}(\mathbf{t}) \mid n, \mathbf{x}_1, \ldots, \mathbf{x}_n).$$

Thus $\hat{\mathbf{t}}$ is the maximum likelihood estimator of $\mathbf{t}$ if and only if $\mathbf{t}^{-1}(\hat{\mathbf{t}})$ is the maximum likelihood estimator $\hat{\boldsymbol{\theta}}$. Hence

$$\mathbf{t}^{-1}(\hat{\mathbf{t}}) = \hat{\boldsymbol{\theta}},$$

implying

$$\hat{\mathbf{t}}(\hat{\boldsymbol{\theta}}) \equiv \hat{\mathbf{t}} = \mathbf{t}(\hat{\boldsymbol{\theta}}). \quad \text{QED}$$

Example 20.3.4

In Example 20.3.3 we showed that the maximum likelihood estimator of the Bernoulli parameter p is $\hat{p} = r/n$. A popular alternative parameter for the Bernoulli process is the *log odds* $\omega = \log_e [p/(1 - p)]$. It is easily verified that $t: (0, 1) \to \Omega = R^1$, defined by $t(p) = \omega$, is a bijection. By (20.3.5), $\log_e [(\tilde{r}/n)/(1 - \tilde{r}/n)] = \log_e [\tilde{r}/(n - \tilde{r})]$ is the maximum likelihood estimator of the log odds ω. Note that $\log_e [r/(n - r)]$ exists only if $0 < r < n$.

Finally, we have asserted that maximum likelihood estimators are BAN. We shall not prove this fact, but rather give the Normal approximations to $f(\hat{\boldsymbol{\theta}} \mid \boldsymbol{\theta})$ for large sample sizes n. The following result is similar in nature to the multivariate Central Limit Theorem (13.9.1) and is, in fact, a consequence of (13.9.1).

Let $\tilde{\boldsymbol{\theta}}_n$ denote the maximum likelihood estimator of $\boldsymbol{\theta}$ based upon n iid observations with common dgf $f(x \mid \boldsymbol{\theta})$. Under rather general regularity conditions, (20.3.6)

$$\lim_{n\to\infty} F(\boldsymbol{\theta}_n \mid \boldsymbol{\theta}) = F_{N^{(k)}}(\boldsymbol{\theta}_n \mid \boldsymbol{\theta}, \frac{1}{n}\boldsymbol{\sigma}^2(\boldsymbol{\theta})),$$

where the (i, j)th element h_{ij} of $H \equiv [\boldsymbol{\sigma}^2(\boldsymbol{\theta})]^{-1}$ is given by

$$h_{ij} \equiv -E(\tilde{\Lambda}_{ij} \mid \boldsymbol{\theta}),$$

where

$$\tilde{\Lambda}_{ij} \equiv \frac{\partial^2[\log_e f(\tilde{\mathbf{x}} \mid \boldsymbol{\theta})]}{\partial\theta_i \partial\theta_j}.$$

The regularity conditions alluded to in (20.3.6) consist chiefly in guaranteeing that $\tilde{\Lambda}_{ij}$ is uniquely defined and has an expectation given $\boldsymbol{\theta}$. Note that, given $\boldsymbol{\theta}$, $\log_e f(\tilde{\mathbf{x}} \mid \boldsymbol{\theta})$ is a perfectly respectable function of the random variable (or, in boldface, vector) $\tilde{x}$.

20.4 SOME IMPORTANT ESTIMATORS

This section furnishes some useful estimators of the parameters of the common processes discussed in Chapter 14; namely, the Bernoulli, k-variate Bernoulli, Poisson, Normal, and k-variate Normal. Some assertions are proved, others relegated to exercises, and the remainder omitted with a reference provided.

$\tilde{\hat{p}} \equiv \tilde{r}/n$ is the maximum likelihood estimator of the parameter p of the Bernoulli process. It is unbiased provided that the number n of observations was predetermined. (20.4.1)

Proof

The first sentence is proved in Example 20.3.3. For the second, recall that if n is fixed, then $\tilde{r}$ has mf $f_b(r \mid n, p)$ given p and hence $E(\tilde{r} \mid p) = E_b(\tilde{r} \mid n, p) = np$. Therefore $E(\tilde{r}/n \mid p) = E(\tilde{r} \mid p)/n = np/n = p$. QED

In (20.4.1), the proviso that n be fixed for $\tilde{\hat{p}}$ to be unbiased is essential. If r is fixed, then $\tilde{n}$ has mf $f_{Pa}(n \mid r, p)$ and hence $E(r/\tilde{n} \mid p) = E_{Pa}(r\tilde{n}^{-1} \mid r, p) = rE_{Pa}(\tilde{n}^{-1} \mid r, p)$. Now, $E_{Pa}(\tilde{n}^{-1} \mid r, p) \neq 1/E_{Pa}(\tilde{n} \mid r, p) = p/r$, so that $\tilde{\hat{p}}$ is *not* an unbiased estimator of p if r is fixed. The appropriate unbiased estimator in this case is $(r - 1)/(\tilde{n} - 1)$, as the reader may prove in Exercise 2.

$\tilde{\hat{\mathbf{p}}} \equiv n^{-1}\tilde{\mathbf{r}}$ is the maximum likelihood estimator of the parameter $\mathbf{p}$ of the k-variate Bernoulli process. It is unbiased provided that n is predetermined. (20.4.2)

Proof

Exercise 3.

$\tilde{\hat{\lambda}} \equiv \tilde{r}/t$ is the maximum likelihood estimator of the parameter λ of the Poisson process. It is unbiased if t is predetermined. (20.4.3)

Proof

The first assertion is left to the reader as Exercise 4. To prove unbiasedness, recall that if t is predetermined the mf of $\tilde{r}$ is $f_{Po}(r \mid \lambda t)$ given λ and hence $E(\tilde{r} \mid \lambda) = E_{Po}(\tilde{r} \mid \lambda t) = \lambda t$. Therefore $E(\tilde{r}/t \mid \lambda) = E(\tilde{r} \mid \lambda)/t = \lambda t/t = \lambda$. QED

As with the Bernoulli process, the maximum likelihood estimator of λ is not unbiased if r is predetermined, since in that case $\tilde{\hat{\lambda}} = r/\tilde{t} = r\tilde{t}^{-1}$; and, because $\tilde{t}$ has df $f_\gamma(t \mid r, \lambda)$, we have $E(\tilde{\hat{\lambda}} \mid \lambda) = E_\gamma(r\tilde{t}^{-1} \mid r, \lambda) = rE_\gamma(\tilde{t}^{-1} \mid r, \lambda)$, which equals $r\lambda/(r - 1)$ provided that $r \geqq 2$. Hence the unbiased estimator of λ when $r \geqq 2$ is predetermined is $(r - 1)/\tilde{t}$.

$\tilde{\hat{\mu}} \equiv \tilde{\bar{x}}$ is the maximum likelihood estimator of the mean μ of the univariate Normal process with known variance σ^2. It is unbiased. (20.4.4)

Proof

The second sentence is obvious. For the first, (13.2.5) implies that the likelihood function of μ is a constant times $\exp\left[-\frac{1}{2}n(\bar{x} - \mu)^2/\sigma^2\right]$, which is obviously maximized at $\mu = \bar{x}$. QED

$\tilde{\xi} \equiv n^{-1} \sum_{i=1}^{n} (\tilde{x}_i - \mu)^2$ is the maximum likelihood estimator $\tilde{\hat{\sigma}}^2$ of the variance σ^2 of a univariate Normal process with known mean μ. It is unbiased. (20.4.5)

Proof

By (13.3.4), the df of $\tilde{\xi}$ is $f_{\gamma 2}(\xi \mid 1/\sigma^2, n)$, and hence by Table 12.1, $E_{\gamma 2}(\tilde{\xi} \mid 1/\sigma^2, n) = 1/(1/\sigma^2) = \sigma^2$, which proves unbiasedness. The proof of the first assertion is left as Exercise 5. QED

It is worth noting that by the invariance property of maximum likelihood estimators, the maximum likelihood estimator of the standard deviation σ is $(\tilde{\hat{\sigma}}^2)^{1/2} = \tilde{\xi}^{1/2}$, while that of the precision σ^{-2} is $\tilde{\xi}^{-1}$. However, these estimators are biased.

For the univariate Normal process with neither parameter known, the maximum likelihood estimator of μ is $\tilde{\bar{x}}$ and the maximum likelihood estimator $\tilde{\hat{\sigma}}^2$ of σ^2 is $(n-1)\tilde{\bar{s}}^2/n \equiv n^{-1}\sum_{i=1}^{n}(\tilde{x}_i - \tilde{\bar{x}})^2$. The statistic $\tilde{\bar{x}}$ is also an unbiased estimator of μ, but $(n-1)\tilde{\bar{s}}^2/n$ is biased. The statistic $\tilde{\bar{s}}^2$ is an unbiased estimator of σ^2. (20.4.6)

Proof

By (13.2.6) and (13.3.6), the likelihood function of (μ, σ^2) can be written as $f_N(\bar{x} \mid \mu, \sigma^2/n) \times f_{\gamma 2}(s^2 \mid \sigma^{-2}, n-1)$. The reader may prove the first sentence as Exercise 6. For the second, $\tilde{\bar{x}}$ is obviously an unbiased estimator of μ for every $\sigma^2 > 0$, while it follows from Table 12.1 that $E_{\gamma 2}(\tilde{\bar{s}}^2 \mid 1/\sigma^2, n-1) = 1/\sigma^{-2} = \sigma^2$. Hence $E[(n-1)\tilde{\bar{s}}^2/n] = (n-1)\sigma^2/n$, which completes the proof of the second and third sentences. QED

The result (20.4.6) and the invariance property of maximum likelihood estimators imply that the maximum likelihood estimator of σ is $\left[n^{-1}\sum_{i=1}^{n}(\tilde{x}_i - \tilde{\bar{x}})^2\right]^{1/2}$; that of the precision σ^{-2} is $n/\sum_{i=1}^{n}(\tilde{x}_i - \tilde{\bar{x}})^2$.

For the k-variate Normal process with $\boldsymbol{\sigma}^2$ known (Section 14.6, Case A), the maximum likelihood estimator of $\boldsymbol{\mu}$ is $\tilde{\bar{\mathbf{x}}}$. It is unbiased. (20.4.7)

Proof

Exercise 7.

For the k-variate Normal process with $\boldsymbol{\mu}$ known (Section 14.6, Case B), the maximum likelihood estimator of $\boldsymbol{\sigma}^2$ is $\tilde{\boldsymbol{\xi}}$. It is unbiased. (20.4.8)

Proof

Anderson [**1**], 45–48, essentially proves the first assertion. To prove the second, note that the df of $\tilde{\boldsymbol{\xi}}$ given $(\boldsymbol{\mu}, \boldsymbol{\sigma}^2)$ is

$$f_{W^{(k)}}(\boldsymbol{\xi} \mid n(n-k+1)^{-1}\boldsymbol{\sigma}^{-2}, n-k+1),$$

by (13.6.3), and hence

$$\begin{aligned} E(\tilde{\boldsymbol{\xi}} \mid \boldsymbol{\mu}, \boldsymbol{\sigma}^2) &= \frac{(n-k+1+k-1)}{(n-k+1)} \times \frac{(n-k+1)}{n}\boldsymbol{\sigma}^2 \\ &= \boldsymbol{\sigma}^2, \end{aligned}$$

by (12.9.4). QED

For the k-variate Normal process with neither $\boldsymbol{\mu}$ nor $\boldsymbol{\sigma}^2$ known (Section 14.6, Case C), the maximum likelihood estimator of $(\boldsymbol{\mu}, \boldsymbol{\sigma}^2)$ is $(\tilde{\bar{\mathbf{x}}}, [(n-1)/n]\tilde{\mathbf{s}}^2)$. It is biased; $E(\tilde{\bar{\mathbf{x}}} \mid \boldsymbol{\mu}, \boldsymbol{\sigma}^2) = \boldsymbol{\mu}$, but $E([(n-1)/n]\tilde{\mathbf{s}}^2) = [(n-1)/n]\boldsymbol{\sigma}^2$. The corresponding unbiased estimator of $(\boldsymbol{\mu}, \boldsymbol{\sigma}^2)$ is $(\tilde{\bar{\mathbf{x}}}, \tilde{\mathbf{s}}^2)$. (20.4.9)

Proof

By (20.4.8), $E(\tilde{\bar{\mathbf{x}}}) = \boldsymbol{\mu}$ regardless of $\boldsymbol{\sigma}^2$, and hence the assertions about $\tilde{\bar{\mathbf{x}}}$ are obvious. Anderson [**1**], 45–48, proves that

$$\frac{(n-1)}{n}\tilde{\mathbf{s}}^2 \equiv n^{-1}\sum_{i=1}^{n}(\tilde{\mathbf{x}}_i - \tilde{\bar{\mathbf{x}}})(\tilde{\mathbf{x}}_i - \tilde{\bar{\mathbf{x}}})'$$

is the maximum likelihood estimator of $\boldsymbol{\sigma}^2$. Since $\tilde{\mathbf{s}}^2$ has conditional df $f_{W^{(k)}}(\boldsymbol{\sigma}^2 \mid (n-1)(n-k)^{-1}\boldsymbol{\sigma}^{-2}, n-k)$ by (13.6.4), it follows from (12.9.4) that $E(\tilde{\mathbf{s}}^2 \mid \boldsymbol{\mu}, \boldsymbol{\sigma}^2) = \boldsymbol{\sigma}^2$. Hence $E([(n-1)/n]\tilde{\mathbf{s}}^2 \mid \boldsymbol{\mu}, \boldsymbol{\sigma}^2) = [(n-1)/n]\boldsymbol{\sigma}^2$. QED

For the k-variate Normal process with $\boldsymbol{\mu}$ and $\boldsymbol{\sigma}_*^2$ known (Section 14.6, Case D), the maximum likelihood estimator of σ_*^2 is $\tilde{\xi}_*$. It is unbiased. (20.4.10)

Proof

Exercise 9.

For the k-variate Normal process with $\boldsymbol{\sigma}_*^2$ known (Section 14.6, Case E), the maximum likelihood estimator of $(\boldsymbol{\mu}, \boldsymbol{\sigma}_*^2)$ is $(\tilde{\bar{\mathbf{x}}}, [k(n-1)/(kn)]\tilde{s}_*^2)$. It is biased; $E(\tilde{\bar{\mathbf{x}}} \mid \boldsymbol{\mu}, \boldsymbol{\sigma}^2) = \boldsymbol{\mu}$, but $E([k(n-1)/(kn)]\tilde{s}_*^2 \mid \boldsymbol{\mu}, \boldsymbol{\sigma}^2) = [k(n-1)/(kn)]\sigma_*^2$. The corresponding unbiased estimator of $(\boldsymbol{\mu}, \sigma_*^2)$ is $(\tilde{\bar{\mathbf{x}}}, \tilde{s}_*^2)$. (20.4.11)

Proof

From $L(\boldsymbol{\mu}, \boldsymbol{\sigma}_*^2 \mid n, \bar{\mathbf{x}}, s_*^2, \boldsymbol{\sigma}_*^2)$ as given by (13.5.13), it is clear that the likelihood is maximized at $\boldsymbol{\mu} = \bar{\mathbf{x}}$ regardless of $\boldsymbol{\sigma}_*^2$ and σ_*^2. Hence the assertions about $\tilde{\bar{\mathbf{x}}}$ and $\boldsymbol{\mu}$ are immediate. To find the maximum likelihood estimator of σ_*^2 it therefore suffices to find the σ_*^2 which maximizes

$$L(\bar{\mathbf{x}}, \sigma_*^2 \mid n, \bar{\mathbf{x}}, s_*^2, \boldsymbol{\sigma}_*^2) = K(\sigma_*^2)^{-\frac{1}{2}nk}\exp\left[-\tfrac{1}{2}k(n-1)s_*^2/\sigma_*^2\right];$$

a verification that the maximizer is as asserted constitutes Exercise 8. From (13.6.7), the conditional df of $\tilde{s}_*^2$ given $(\boldsymbol{\mu}, \boldsymbol{\sigma}^2)$ is $f_{\gamma 2}(s_*^2 \mid \sigma_*^{-2}, k(n-1))$, and hence $E(\tilde{s}_*^2 \mid \boldsymbol{\mu}, \boldsymbol{\sigma}^2) = \sigma_*^2$, by Table 12.1. QED

Note in the proof of (20.4.11) that the likelihood function was maximzed in two stages: first, with respect to $\boldsymbol{\mu}$ for any σ_*^2; and second, using $\hat{\boldsymbol{\mu}} = \bar{\mathbf{x}}$, with respect to σ_*^2.

All unbiased estimators in this section are functions of any sufficient statistics and can be shown by the Blackwell-Rao Theorem to be minimum variance unbiased.

20.5 SOME BAYESIAN INTERPRETATIONS OF ORTHODOX POINT ESTIMATES

A Bayesian statistician bases all inferences about θ upon his posterior dgf $f(\theta \mid n, x_1, \ldots, x_n)$ of $\tilde{\theta}$, which embodies all of his judgments about $\tilde{\theta}$ as revised by the outcome $(n, x_1, \ldots, x_n)$ of the experiment. Thus a Bayesian's interpretation of an orthodox inference will be in terms of what that orthodox inference implies about his posterior dgf. Moreover, the common interpretation is that the orthodox inference is an approximation to some similar numerical characteristic, such as mean or mode, of the posterior dgf.

As a first instance, we assert that an unbiased estimate is an approximation to the posterior expectation $E(\tilde{\theta} \mid n, x_1, \ldots, x_n)$, provided that the experimental outcome $(n, x_1, \ldots, x_n)$ was not surprising. To see this, we first note that both $E(\tilde{\theta} \mid n, x_1, \ldots, x_n)$ and the unbiased estimator $\hat{\theta}(n, x_1, \ldots, x_n)$ are functions of the outcome and hence are random variables before the outcome is observed. Our first result is that both $E(\tilde{\theta} \mid n, \tilde{x}_1, \ldots, \tilde{x}_n)$ and $\hat{\theta}(n, \tilde{x}_1, \ldots, \tilde{x}_n)$ have the same prior expectations if $\hat{\tilde{\theta}}$ is unbiased:

$$E(\hat{\theta}(n, \tilde{x}_1, \ldots, \tilde{x}_n)) = E(\tilde{\theta}) = E(E(\tilde{\theta} \mid n, \tilde{x}_1, \ldots, \tilde{x}_n)). \qquad (20.5.1)$$

Proof

The second equality is obvious from (10.2.8). For the first, we apply (10.2.8) to obtain

$$\begin{aligned} &E(\hat{\theta}(n, \tilde{x}_1, \ldots, \tilde{x}_n)) \\ &\qquad = E\{E(\hat{\theta}(n, \tilde{x}_1, \ldots, \tilde{x}_n) \mid \tilde{\theta})\} \\ &\qquad = E\{\tilde{\theta}\}, \end{aligned}$$

by unbiasedness of $\hat{\tilde{\theta}}$. This result is due originally to Pratt [**63**]. QED

By equating the first and last terms in (20.5.1) and using the fact that $E(\tilde{x}) = E(\tilde{y})$ if and only if $E(\tilde{x} - \tilde{y}) = 0$ [see (10.2.7)], we obtain

$$E\{\hat{\theta}(n, \tilde{x}_1, \ldots, \tilde{x}_n) - E(\tilde{\theta} \mid n, \tilde{x}_1, \ldots, \tilde{x}_n)\} = 0. \qquad (20.5.2)$$

Now, (20.5.2) expresses the important conclusion that, *before* the experiment is conducted, the unbiased estimator and the posterior expectation of θ are expected to coincide.

After the experiment is conducted, however, the unbiased estimate $\hat{\theta}(n, x_1, \ldots, x_n)$ need be close to the posterior expectation $E(\tilde{\theta} \mid n, x_1, \ldots, x_n)$ *only* if $(n, x_1, \ldots, x_n)$ is "not surprising"; for example, equals its expectation.

Example 20.5.1

By (20.4.2), $\hat{\tilde{\mathbf{p}}} \equiv n^{-1}\tilde{\mathbf{r}}$ is an unbiased estimator of the parameter $\mathbf{p}$ of a k-variate Bernoulli process. By (14.3.2), if $\tilde{\mathbf{p}}$ was given a k-variate beta prior df the posterior df is $f_\beta^{(k)}(\mathbf{p} \mid \boldsymbol{\varrho} + \mathbf{r}, \nu + n)$, and the (vector) $E(\tilde{\mathbf{p}} \mid n, x_1, \ldots, x_n) = (\nu + n)^{-1}(\boldsymbol{\varrho} + \mathbf{r})$. Now, the unconditional expectation of $\tilde{\mathbf{r}}$ is $(n/\nu)\boldsymbol{\varrho}$, by (14.3.4), so that $E(E(\tilde{\mathbf{p}} \mid n, \tilde{x}_1, \ldots, \tilde{x}_n)) = (\nu + n)^{-1}(\boldsymbol{\varrho} + (n/\nu)\boldsymbol{\varrho}) = \nu^{-1}\boldsymbol{\varrho}$, and also $E(n^{-1}\tilde{\mathbf{r}}) = n^{-1}(n/\nu)\boldsymbol{\varrho} = \nu^{-1}\boldsymbol{\varrho}$, thus corroborating (20.5.1). However, note that $\hat{\mathbf{p}} = E(\tilde{\mathbf{p}} \mid n, x_1, \ldots, x_n)$ if and only if $n^{-1}\mathbf{r} = (\nu + n)^{-1}(\boldsymbol{\varrho} + \mathbf{r})$, which obtains in two distinct instances: (1) when $\boldsymbol{\varrho} = 0$ and $\nu = 0$, for every $\mathbf{r}$; and (2) when $\boldsymbol{\varrho} > 0$ and $\nu > \sum_{i=1}^{k} \rho_i$, for $\mathbf{r} = (n/\nu)\boldsymbol{\varrho} = E(\tilde{\mathbf{r}})$. The second instance pertains whenever the prior df of $\tilde{\mathbf{p}}$ is a *proper* k-variate beta df; the first only when $\tilde{\mathbf{p}}$ is assigned the *improper* prior df $f_\beta^{(k)}(\mathbf{p} \mid \mathbf{0}, 0)$ which, for $k = 1$, is of the form $K \cdot p^{-1}(1 - p)^{-1}$, violently U-shaped and with (dgf 2) failing. Note that if $\mathbf{r}$ is "close to" $(n/\nu)\boldsymbol{\varrho}$, and hence the experimental outcome is "not too surprising," then $\hat{\mathbf{p}}$ will be "close to" $E(\tilde{\mathbf{p}} \mid n, x_1, \ldots, x_n)$ providing n is large relative to ν, since the posterior expectation can be rewritten as

$$E(\tilde{\mathbf{p}} \mid n, x_1, \ldots, x_n) = [\nu/(\nu + n)]\nu^{-1}\boldsymbol{\varrho} + [n/(\nu + n)]n^{-1}\mathbf{r},$$

and as n increases less and less relative weight $[\nu/(\nu + n)]$ is assigned to the constant $\nu^{-1}\boldsymbol{\varrho}$.

Example 20.5.2

By (20.4.3), $\hat{\tilde{\lambda}} \equiv \tilde{r}/t$ is an unbiased estimator of the intensity λ of a Poisson process. By (14.4.2), the posterior df of $\tilde{\lambda}$ is $f_\gamma(\lambda \mid \rho + r, \tau + t)$ with expectation $E(\tilde{\lambda} \mid r, t) = (\rho + r)/(\tau + t)$ if the prior df is $f_\gamma(\lambda \mid \rho, \tau)$. By (14.4.4), $E(\hat{\tilde{\lambda}}) = E(\tilde{r}/t) = t^{-1}E_{nb}(\tilde{r} \mid \rho, \tau/[\tau + t]) = t^{-1}\rho t/\tau = \rho/\tau$, and $E(E(\tilde{\lambda} \mid \tilde{r}, t)) = [\rho + E(\tilde{r})]/(\tau + t) = [\rho + (\rho t/\tau)](\tau + t) = \rho/\tau$, corroborating (20.5.1). Note that *posterior* equality of $\hat{\lambda}$ and $E(\tilde{\lambda} \mid r, t)$ obtains again in two cases: (1) for every r, if $\rho = \tau = 0$, which corresponds to an improper prior df for $\tilde{\lambda}$; and (2) for $r = \rho t/\tau = E(\tilde{r})$.

Example 20.5.3

By (20.4.4), $\hat{\tilde{\mu}} = \tilde{\bar{x}}$ is an unbiased estimator of the mean μ of a univariate Normal process with known variance σ^2. If $\tilde{\mu}$ has been assigned a prior df $f_N(\mu \mid \mu_0, \sigma^2/n_0)$, then the posterior expectation of $\tilde{\mu}$ is $E_N(\tilde{\mu} \mid \mu_1, \sigma^2/(n_0 + n)) = \mu_1 = (n_0/[n_0 + n])\mu_0 + (n/[n_0 + n])\bar{x}$, which coincides with $\bar{x}$ in two cases: (1) for every $\bar{x} \in R^1$, if $n_0 = 0$, which implies that the prior df of $\tilde{\mu}$ is improper; and (2) for $\bar{x} = E(\tilde{\bar{x}}) = \mu_0$, if $n_0 > 0$. The improper prior df in case (1) is viewed as a limit as the variance becomes infinite, and hence (compare Figure 12.6) as the df becomes broader and lower. The limit is sometimes called the "uniform df on the real line."

Example 20.5.4

By (20.4.5), $\tilde{\xi} \equiv n^{-1} \sum_{i=1}^{n} (\tilde{x}_i - \mu)^2$ is an unbiased estimator of the variance σ^2 of a univariate Normal process with known mean μ. By (14.5.3), the posterior df of $\tilde{\sigma}^2$ is $f_{i\gamma}(\sigma^2 \mid \psi_1, \nu_1)$ with expectation $E(\tilde{\sigma}^2 \mid n, x_1, \ldots, x_n) = \nu_1\psi_1/(\nu_1 - 2)$, by Table 12.1, where $\nu_1 \equiv \nu_0 + n$ and $\psi_1 \equiv (\nu_0\psi_0 + n\xi)/(\nu_0 + n)$. Hence $\hat{\sigma}^2$ and $E(\sigma^2 \mid n, x_1, \ldots, x_n)$ coincide again in two cases: (1) for any $\hat{\sigma}^2$ if $\nu_0 = 2$ and $\psi_0 = 0$, in which case the prior df of $\tilde{\sigma}^2$ is improper; and (2) for $\hat{\sigma}^2 = \nu_0\psi_0/(\nu_0 - 2)$ and $\nu_0 > 2$.

Example 20.5.5

By (20.4.6), $(\tilde{\bar{x}}, \tilde{s}^2)$ is an unbiased estimator of (μ, σ^2) in the univariate Normal process. By (14.5.7),

$$\begin{aligned} E(\tilde{\mu}, \tilde{\sigma}^2 \mid n, x_1, \ldots, x_n) & \\ &= (E_S(\tilde{\mu} \mid \mu_1, \psi_1/n_1, \nu_1), E_{i\gamma}(\tilde{\sigma}^2 \mid \psi_1, \nu_1)) \\ &= (\mu_1, \nu_1\psi_1/(\nu_1 - 2)), \end{aligned}$$

which always coincides with the unbiased estimate only if $n_0 = 0$, $\nu_0 = -1$, and $\psi_0 = 0$, given the definitions of n_1, ν_1, μ_1, and ξ_1 by (14.5.7). This joint prior df of $(\tilde{\mu}, \tilde{\sigma}^2)$ is obviously improper. Given $n_0 > 0$, $\nu_0 > 2$, and $\psi_0 > 0$, we have that $(\bar{x}, s^2)$ coincides with $(\mu_1, \nu_1\psi_1/(\nu_1 - 2))$ if $\bar{x} = \mu_0 = E(\tilde{\bar{x}})$ and $s^2 = \nu_0\psi_0/(\nu_0 - 1)$, by (14.5.7). Note that

$$E(\tilde{s}^2) = E_{i\beta 2}(\tilde{s}^2 \mid \tfrac{1}{2}(n - 1), \tfrac{1}{2}(\nu_0 + n - 1), (n - 1)^{-1}\nu_0\psi_0) = \nu_0\psi_0/(\nu_0 - 2).$$

Example 20.5.6

For the k-variate Normal process with known covariance matrix $\boldsymbol{\sigma}^2$ (Section 14.6, Case A), the unbiased estimate $\bar{\mathbf{x}}$ of $\boldsymbol{\mu}$ coincides with $\boldsymbol{\mu}_1 \equiv (\boldsymbol{\sigma}_0^{-2} + n\boldsymbol{\sigma}^{-2})^{-1}(\boldsymbol{\sigma}_0^{-2}\boldsymbol{\mu}_0 + n\boldsymbol{\sigma}^{-2}\bar{\mathbf{x}})$ in two principal cases: (1) for any $\bar{\mathbf{x}}$ if $\boldsymbol{\sigma}_0^{-2} = \mathbf{0}$, which implies an improper prior df of $\tilde{\boldsymbol{\mu}}$; and (2) for $\bar{\mathbf{x}} = \boldsymbol{\mu}_0 = E(\tilde{\bar{\mathbf{x}}})$.

Example 20.5.7

For the k-variate Normal process with known expectation vector $\boldsymbol{\mu}$ (Section 14.6, Case B), the unbiased estimate $\boldsymbol{\xi}$ of $\boldsymbol{\sigma}^2$ coincides with $E_{iW^{(k)}}(\tilde{\boldsymbol{\sigma}}^2 \mid \boldsymbol{\psi}_1, \nu_1) = (\nu_0 + n - 2)^{-1}[\nu_0\boldsymbol{\psi}_0 + n\boldsymbol{\xi}]$ in two principal cases: (1) for any $\boldsymbol{\xi}$ if $\nu_0 = 2$ and $\boldsymbol{\psi}_0 = \mathbf{0}$; and (2) if $\nu_0 > 2$ and $\boldsymbol{\psi}_0$ is positive-definite and symmetric, if $\boldsymbol{\xi} = [\nu_0/(\nu_0 - 2)]\boldsymbol{\psi}_0$. For $k = 1$, this reduces to Example 20.5.4. The prior df $f_{iW^{(k)}}(\boldsymbol{\sigma}^2 \mid \mathbf{0}, 2)$ is obviously improper.

Example 20.5.8

For the k-variate Normal process with neither $\tilde{\mathbf{\mu}}$ nor $\tilde{\mathbf{\sigma}}^2$ known (Section 14.6, Case C), the unbiased estimate $(\bar{\mathbf{x}}, \mathbf{s}^2)$ of $(\mathbf{\mu}, \mathbf{\sigma}^2)$ coincides with the pair $(\mathbf{\mu}_1, [\nu_1/(\nu_1 - 2)]\mathbf{\psi}_1)$ of posterior expectations implied by (14.6.8) in two principal cases: (1) for any $(\bar{\mathbf{x}}, s^2)$ if $n_0 = 0$, $\nu_0 = 1$, and $\mathbf{\psi}_0 = \mathbf{0}$; and (2) for any $n_0 > 0$, $\nu_0 > 2$, and positive definite and symmetric $\mathbf{\psi}_0$ if $(\bar{\mathbf{x}}, s^2) = (\mathbf{\mu}_0, [\nu_0/(\nu_0 - 1)]\mathbf{\psi}_0)$. For $k = 1$, this reduces to Example 20.5.5. The prior df $f^{(k)}_{NiW}(\mathbf{\mu}, \mathbf{\sigma}^2 \mid \mathbf{\mu}_0, 0, 0, 1)$ is obviously improper.

Example 20.5.9

For the k-variate Normal process with $\mathbf{\mu}$ and $\mathbf{\sigma}_*^2$ known (Section 14.6, Case D), the unbiased estimate ξ_* of σ_*^2 coincides with the posterior expectation $E_{i\gamma}(\tilde{\sigma}_*^2 \mid \psi_1, \nu_1) = (\nu_0 + kn - 2)^{-1}[\nu_0\psi_0 + kn\xi_*]$ in two principal cases: (1) for any ξ_* if $\nu_0 = 2$ and $\psi_0 = 0$; and (2) for any $\nu_0 > 2$ and $\psi_0 > 0$ if $\xi_* = [\nu_0/(\nu_0 - 2)]\psi_0$. This reduces to Example 20.5.4 for $k = 1$. The prior df $f_{i\gamma}(\sigma_*^2 \mid 0, 2)$ is obviously improper.

Example 20.5.10

For the k-variate Normal process with only $\mathbf{\sigma}_*^2$ known (Section 14.6, Case E), the unbiased estimate $(\bar{\mathbf{x}}, s_*^2)$ of $(\mathbf{\mu}, \sigma_*^2)$ coincides with the pair $(\mathbf{\mu}_1, [\nu_1/(\nu_1 - 2)]\psi_1)$ of posterior expectations implied by (14.6.14) in two principal cases: (1) for any $(\bar{\mathbf{x}}, s_*^2)$ if $\mathbf{\sigma}_{0*}^{-2} = \mathbf{0}$, $\psi_0 = 0$, and $\nu_0 = 2 - k$; and (2) for any positive-definite and symmetric $\mathbf{\sigma}_{0*}^2$, $\psi_0 > 0$, and $\nu_0 > 2$ if $(\bar{\mathbf{x}}, s_*^2) = (\mathbf{\mu}_0, [\nu_0/(\nu_0 + k - 2)]\psi_0$. This reduces to Example 20.5.5 for $k = 1$. The prior df $f^{(k)}_{Ni\gamma}(\mathbf{\mu}, \sigma_*^2 \mid \mathbf{\mu}_0, 0, \text{“}0^{-1}\text{”}, 2 - k)$ is obviously improper.

So far we have discussed unbiased estimators as approximations to posterior expectations. A similar interpretation obtains for *maximum likelihood* estimators, which can be considered as approximations to posterior modes. The reader will recall from Chapter 12, Exercise 53 that a *mode* $\hat{\mathbf{\theta}}$ of a continuous random vector $\tilde{\mathbf{\theta}}$ is a maximizer of its df. Thus a posterior mode $\hat{\mathbf{\theta}}(n, x_1, \ldots, x_n)$ is the natural Bayesian analogue of a maximum likelihood estimate $\hat{\mathbf{\theta}}(n, x_1, \ldots, x_n)$.

Through another series of examples we shall show that if the outcome $(n, x_1, \ldots, x_n)$ is not too surprising, then $\hat{\mathbf{\theta}}(n, x_1, \ldots, x_n)$ is approximately equal to $\hat{\mathbf{\theta}}(n, x_1, \ldots, x_n)$.

Example 20.5.11

For the parameter $\tilde{\mathbf{p}}$ of a k-variate Bernoulli process, the mode of the posterior df $f_\beta{}^{(k)}(\mathbf{p} \mid \mathbf{\varrho} + \mathbf{r}, \nu + n)$ is $\hat{\mathbf{p}}(\mathbf{r}, n) = (\nu + n - k - 1)^{-1}(\mathbf{\varrho} + \mathbf{r} - \mathbf{1})$, where $\mathbf{1} \equiv (1, 1, \ldots, 1)'$; the reader may prove this as Exercise 11. Thus

$\hat{\mathbf{p}}(\mathbf{r}, n)$ coincides with $\hat{\mathbf{p}}(\mathbf{r}, n)$ as given by (20.4.2) in two cases: (1) for any $\mathbf{r}$ if $\boldsymbol{\varrho} = \mathbf{1}$ and $\nu = k + 1$, which implies, via (12.6.1), that the prior df $f_{\beta}{}^{(k)}(\mathbf{p} \mid \mathbf{1}, k+1)$ is the uniform df on $\left\{\mathbf{p}: \mathbf{p} \in R^k, \mathbf{p} > \mathbf{0}, \sum_{i=1}^{k} p_i < 1\right\}$ rather than the improper df $f_{\beta}{}^{(k)}(\mathbf{p} \mid \mathbf{0}, 0)$ in Example 20.5.1; and (2) for any $\boldsymbol{\varrho}$, ν, provided that $\mathbf{r} = n[(\nu - k - 1)^{-1}(\boldsymbol{\varrho} - \mathbf{1})] \equiv n\hat{\mathbf{p}}$, where $\hat{\mathbf{p}}$ is the mode of the prior df. Thus by the second case, we see that an "unsurprising" outcome is one in which $\hat{\mathbf{p}} = n^{-1}\mathbf{r}$ equals the prior *mode* $\hat{\mathbf{p}}$, whereas in Example 20.5.1 the desired equality was between $\hat{\mathbf{p}}$ and the prior *expectation* $\nu^{-1}\boldsymbol{\varrho}$.

Example 20.5.12

For the parameter $\tilde{\lambda}$ of a Poisson process, the mode of the posterior df $f_{\gamma}(\lambda \mid \rho + r, \tau + t)$ is $\hat{\lambda}(r, t) = (\rho + r - 1)/(\tau + t)$, as is easily checked by differentiation, provided that $\rho + r > 1$. The mode $\hat{\lambda}(r, t)$ coincides with $\hat{\lambda}(r, t) = r/t$ as given by (20.4.3) in two cases: (1) for any r if $\rho = 1$ and $\tau = 0$, which by a limiting argument (Exercise 12) can be called the uniform df on $(0, \infty)$; and (2) for any $\rho \geqq 1, \tau > 0$ provided that $r = t[(\rho - 1)/\tau] \equiv t\hat{\lambda}$ where $\hat{\lambda}$ is the mode of the prior df. Thus by the second case, an "unsurprising" outcome is one in which $\hat{\lambda} = r/t$ equals the prior *mode* $\hat{\lambda}$, rather than the prior expectation ρ/τ, as in Example 20.5.2.

Example 20.5.13

For the mean μ of a univariate Normal process with known variance σ^2, the mode of the posterior df is μ_1, and hence all results in Example 20.5.3 apply here without change. An unsurprising outcome is one in which $\hat{\mu} \equiv \bar{x} = \mu_0 = \hat{\mu}$, the prior mode, which coincides with the prior expectation because of the symmetry of the Normal df.

Example 20.5.14

In the univariate Normal process with known mean μ, the mode of the posterior df $f_{i\gamma}(\sigma^2 \mid \psi_1, \nu_1)$ of $\tilde{\sigma}^2$ as given in (14.5.3) is $\hat{\sigma}^2(n, \xi) = \nu_1\psi_1/(\nu_1 + 2) \equiv (\nu_0\psi_0 + n\xi)/(\nu_0 + n + 2)$, which agrees with the maximum likelihood estimate ξ in two cases: (1) for any ξ if $\psi_0 = 0$ and $\nu_0 = -2$, which by a limiting argument (Exercise 13) is the improper uniform df on $(0, \infty)$; and (2) for any $\nu_0 > 0$ and $\psi_0 > 0$, provided that $\xi = \nu_0\psi_0/(\nu_0 + 2) =$ the mode $\hat{\sigma}^2$ of the prior df $f_{i\gamma}(\sigma^2 \mid \psi_0, \nu_0)$ of $\tilde{\sigma}^2$.

Example 20.5.15

In the univariate Normal process with neither parameter known beforehand, the mode $(\hat{\mu}(n, \bar{x}, s^2), \hat{\sigma}^2(n, \bar{x}, s^2))$ of the posterior df $f_{Ni\gamma}(\mu, \sigma^2 \mid \mu_1, \psi_1, n_1, \nu_1)$ of $(\tilde{\mu}, \tilde{\sigma}^2)$ is $(\mu_1, \nu_1\psi_1/(\nu_1 + 3))$, where the parameters μ_1, ψ_1, and ν_1 are as defined in (14.5.7). (The reader may prove this assertion as Exercise 14.) Hence $(\hat{\mu}(n, \bar{x}, s^2), \hat{\sigma}^2(n, \bar{x}, s^2))$ agrees with the maximum likelihood estimate $(\bar{x}, [(n-1)/n]s^2)$ in two cases: (1) for any $(\bar{x}, [(n-1)/n]s^2)$ if $n_0 = 0$, $\psi_0 = 0$, and $\nu_0 = -3$; and (2) for any $n_0 > 0$, $\nu_0 > 0$, and $\psi_0 > 0$ if $\bar{x} = \mu_0$ and $[(n-1)/n]s^2 = \nu_0\psi_0/(\nu_0 + 3)$. By a limiting argument, $f_{Ni\gamma}(\mu, \sigma^2 \mid \mu_0, 0, 0, -3)$ is the uniform df over $(-\infty, \infty) \times (0, \infty)$. In case (2), $(\mu_0, \nu_0\psi_0/(\nu_0 + 3))$ is the mode of the prior df of $(\tilde{\mu}, \tilde{\sigma}^2)$.

Example 20.5.16

In the k-variate Normal process with known covariance matrix $\boldsymbol{\sigma}^2$ (Section 14.6, Case A), all results of Example 20.5.6 continue to apply, because the expectation and mode of the k-variate Normal natural conjugate prior and posterior df's coincide, as do the maximum likelihood and unbiased estimators of $\boldsymbol{\mu}$.

Example 20.5.17

In the k-variate Normal process with $\boldsymbol{\mu}$ known (Section 14.6, Case B), it can be shown that the mode of the posterior df $f_{iW^{(k)}}(\boldsymbol{\sigma}^2 \mid \boldsymbol{\psi}_1, \nu_1)$ of $\tilde{\boldsymbol{\sigma}}^2$ is $\hat{\boldsymbol{\sigma}}^2(n, \boldsymbol{\xi}) = [\nu_1/(\nu_1 + 2k)]\boldsymbol{\psi}_1$, which coincides with the maximum likelihood estimate $\boldsymbol{\xi}$ in two cases: (1) if $\nu_0 = -2k$ and $\boldsymbol{\psi}_0 = 0$; and (2) for any $\nu_0 > 0$ and positive-definite, symmetric $\boldsymbol{\psi}_0$ if $\boldsymbol{\xi} = [\nu_0/(\nu_0 + 2k)]\boldsymbol{\psi}_0 =$ the mode of the prior df $f_{i\gamma}(\boldsymbol{\sigma}^2 \mid \boldsymbol{\psi}_0, \nu_0)$ of $\boldsymbol{\sigma}^2$.

Example 20.5.18

In the k-variate Normal process with neither $\boldsymbol{\mu}$ nor $\boldsymbol{\sigma}^2$ known (Section 14.6, Case C), the natural conjugate posterior df $f^{(k)}_{NiW}(\boldsymbol{\mu}, \boldsymbol{\sigma}^2 \mid \boldsymbol{\mu}_1, \boldsymbol{\psi}_1, n_1, \nu_1)$ has a mode the first component of which is $\boldsymbol{\mu}_1$, independent of the parameters $\boldsymbol{\psi}_1$ and ν_1. It can be shown that the posterior mode $\hat{\boldsymbol{\sigma}}^2(n, \bar{\mathbf{x}}, \mathbf{s}^2)$ of $\tilde{\boldsymbol{\sigma}}^2$ is $[\nu_1/(\nu_1 + 2k + 1)]\boldsymbol{\psi}_1$. Hence the posterior mode and maximum likelihood estimate $(\bar{\mathbf{x}}, [(n-1)/n]\mathbf{s}^2)$ coincide in two cases: (1) for any $(\bar{\mathbf{x}}, [(n-1)/n]\mathbf{s}^2)$, if $n_0 = 0$, $\boldsymbol{\psi}_0 = \mathbf{0}$, and $\nu_0 = -2k - 1$; and (2) for any $\nu_0 > 0$, positive-definite, symmetric $\boldsymbol{\psi}_0$, and $\nu_0 > 0$, if $\bar{\mathbf{x}} = \boldsymbol{\mu}_0$ and $[(n-1)/n]s^2 = [\nu_0/(\nu_0 + 2k + 1)]\boldsymbol{\psi}_0$. Again, case (1) corresponds to a suitable, improper uniform prior df, while case (2) states that the maximum likelihood estimate coincides with the mode of the prior df of $(\tilde{\boldsymbol{\mu}}, \tilde{\boldsymbol{\sigma}}^2)$.

Example 20.5.19

In the k-variate Normal process with $\boldsymbol{\mu}$ and $\boldsymbol{\sigma}_*^2$ known (Section 14.6, Case D), the natural conjugate posterior df of $\tilde{\sigma}_*^2$ is $f_{i\gamma}(\sigma_*^2 \mid \psi_1, \nu_1)$ as given by (14.6.10), and hence the posterior mode of $\tilde{\sigma}_*^2$ is $[\nu_1/(\nu_1 + 2)]\psi_1 =$

$(\nu_0\psi_0 + kn\xi_*)/(\nu_0 + kn + 2)$, which coincides with the maximum likelihood estimate ξ_* in two cases: (1) for any ξ_* if $\nu_0 = -2$ and $\psi_0 = 0$; and (2) for any $\nu_0 > 0$ and $\psi_0 > 0$ if $\xi_* = [\nu_0/(\nu_0 + 2)]\psi_0$ = the prior mode $\hat{\sigma}_*^2$ of $\tilde{\sigma}_*^2$.

Example 20.5.20

In the k-variate Normal processes with $\boldsymbol{\sigma}_*^2$ known (Section 14.6, Case E), the natural conjugate posterior df of $(\tilde{\boldsymbol{\mu}}, \tilde{\sigma}_*^2)$ is $f_{Ni\gamma}^{(k)}(\boldsymbol{\mu}, \sigma_*^2 \mid \boldsymbol{\mu}_1, \psi_1, \boldsymbol{\sigma}_{1*}^2, \nu_1)$ as given by (14.6.14). It is not hard to show that the posterior mode is therefore given by $\hat{\boldsymbol{\mu}}(n, \bar{\mathbf{x}}, s_*^2) = \boldsymbol{\mu}_1$ and $\hat{\sigma}_*^2(n, \bar{\mathbf{x}}, s_*^2) = [\nu_1/(\nu_1 + k + 2)]\psi_1$, and hence coincides with the maximum likelihood estimate $(\bar{\mathbf{x}}, [k(n-1)/(kn)]s_*^2)$ in two cases: (1) for any $(\bar{\mathbf{x}}, s_*^2)$ if $\boldsymbol{\sigma}_{0*}^{-2} = \mathbf{0}$, $\psi_0 = 0$, and $\nu_0 = -(k + 2)$; and (2) for any positive-definite and symmetric $\boldsymbol{\sigma}_{0*}^2$, $\psi_0 > 0$, and $\nu_0 > 0$ if $\bar{\mathbf{x}} = \boldsymbol{\mu}_0$ and $[k(n-1)/(kn)]s_*^2 = [\nu_0/(\nu_0 + k + 2)]\psi_0$. Case (1) is again an improper uniform df and case (2) stipulates coincidence of prior mode and maximum likelihood estimate.

Recall that with maximum likelihood estimation an unsurprising outcome is one for which the posterior mode and the maximum likelihood estimate coincide. In the preceding Examples 20.5.11 through 20.5.20, cases (2) showed that the outcome was unsurprising if the maximum likelihood estimate coincided with the *prior* mode of the parameter in question. We shall now show that coincidence of the prior mode and the maximum likelihood estimate is *sufficient* for unsurprisingness.

Let $\hat{\boldsymbol{\theta}}(n, \mathbf{x}_1, \ldots, \mathbf{x}_n)$ denote a maximum likelihood estimate of $\tilde{\boldsymbol{\theta}}$ given outcome $(n, \mathbf{x}_1, \ldots, \mathbf{x}_n)$; it need not be unique. If $\hat{\boldsymbol{\theta}}(n, \mathbf{x}_1, \ldots, \mathbf{x}_n)$ is a prior mode of $\tilde{\boldsymbol{\theta}}$, then it is a posterior mode of $\tilde{\boldsymbol{\theta}}$. (20.5.2)

Proof

Bayes' Theorem may be written

$$f(\boldsymbol{\theta} \mid n, \mathbf{x}_1, \ldots, \mathbf{x}_n) = Kf(\boldsymbol{\theta})L(\boldsymbol{\theta} \mid n, \mathbf{x}_1, \ldots, \mathbf{x}_n),$$

where $K > 0$ is a constant with respect to $\boldsymbol{\theta}$. By definition,

$$L(\hat{\boldsymbol{\theta}}(n, \mathbf{x}_1, \ldots, \mathbf{x}_n) \mid n, \mathbf{x}_1, \ldots, \mathbf{x}_n) \geqq L(\boldsymbol{\theta} \mid n, \mathbf{x}_1, \ldots, \mathbf{x}_n),$$

for all $\boldsymbol{\theta} \in \boldsymbol{\Theta}$. If $\hat{\boldsymbol{\theta}}(n, \mathbf{x}_1, \ldots, \mathbf{x}_n)$ is a prior mode of $\tilde{\boldsymbol{\theta}}$, then also

$$f(\hat{\boldsymbol{\theta}}(n, \mathbf{x}_1, \ldots, \mathbf{x}_n)) \geqq f(\boldsymbol{\theta}), \text{ for all } \boldsymbol{\theta} \in \boldsymbol{\Theta}.$$

Hence

$$\begin{aligned} f(\hat{\boldsymbol{\theta}}(n, \mathbf{x}_1, \ldots, \mathbf{x}_n) &\mid n, \mathbf{x}_1, \ldots, \mathbf{x}_n) \\ &= Kf(\hat{\boldsymbol{\theta}}(n, \mathbf{x}_1, \ldots, \mathbf{x}_n))L(\hat{\boldsymbol{\theta}}(n, \mathbf{x}_1, \ldots, \mathbf{x}_n) \mid n, \mathbf{x}_1, \ldots, \mathbf{x}_n) \\ &\geqq Kf(\boldsymbol{\theta})L(\boldsymbol{\theta} \mid n, \mathbf{x}_1, \ldots, \mathbf{x}_n) \\ &= f(\boldsymbol{\theta} \mid n, \mathbf{x}_1, \ldots, \mathbf{x}_n), \end{aligned}$$

for all $\boldsymbol{\theta} \in \boldsymbol{\Theta}$. Comparison of the extreme terms in this chain establishes the assertion. QED

Note in (20.5.2) that coincidence of $\hat{\boldsymbol{\theta}}(n, \mathbf{x}_1, \ldots, \mathbf{x}_n)$ and the prior mode $\hat{\boldsymbol{\theta}}$ is *sufficient* for "unsurprisingness"; nothing is said about *necessity*. To prove necessity requires some additional restrictions, which do not always obtain and will not be considered here.

Naturally, it would be nice to have an analogue of (20.5.2) for unbiased estimators, which are unsurprising when they coincide with posterior expectations. In most of Examples 20.5.1 through 20.5.10, the sufficient condition is that the unbiased estimate coincide with the *prior* expectation, but exceptions exist (such as, Example 20.5.5) and therefore indicate the need for additional restrictions. We have the following:

Let $\hat{\boldsymbol{\theta}}(n, \tilde{\mathbf{x}}_1, \ldots, \tilde{\mathbf{x}}_n)$ denote an unbiased estimator of $\tilde{\boldsymbol{\theta}}$, and suppose that there exists a vector $\boldsymbol{\lambda}_n$ such that the posterior expectation $E(\tilde{\boldsymbol{\theta}} \mid n, \mathbf{x}_1, \ldots, \mathbf{x}_n)$ satisfies (20.5.3)

$$E(\tilde{\boldsymbol{\theta}} \mid n, \mathbf{x}_1, \ldots, \mathbf{x}_n) = \boldsymbol{\lambda}'_n E(\tilde{\boldsymbol{\theta}}) + [\mathbf{1} - \boldsymbol{\lambda}_n]'\hat{\boldsymbol{\theta}}(n, \mathbf{x}_1, \ldots, \mathbf{x}_n), \text{ for every } (\mathbf{x}_1, \ldots, \mathbf{x}_n)$$

where $\mathbf{1} \equiv (1, 1, \ldots, 1)'$. Then $\hat{\boldsymbol{\theta}}(n, \mathbf{x}_1, \ldots, \mathbf{x}_n) = E(\tilde{\boldsymbol{\theta}} \mid n, \mathbf{x}_1, \ldots, \mathbf{x}_n)$ if $\hat{\boldsymbol{\theta}}(n, \mathbf{x}_1, \ldots, \mathbf{x}_n) = E(\tilde{\boldsymbol{\theta}})$.

Proof

The proof is obvious.

Note that Example 20.5.1 satisfies the hypothesis of (20.5.3) because $E(\tilde{p}_i \mid \boldsymbol{\rho} + \mathbf{r}, \nu + n) = (\rho_i + r_i)/(\nu + n) = [\nu(\rho_i/\nu) + n(r_i/n)]/(\nu + n) = \lambda_{n_i}E(\tilde{p}_i) + (1 - \lambda_{n_i})\hat{p}_i$, for $i = 1, \ldots, k$, where $\lambda_{n_i} \equiv \nu/(\nu + n)$. Similarly for Example 20.5.2. That hypothesis is not necessary; in fact, a slightly more general hypothesis (which covers the multivariate Normal and Student cases) is that there exist kth order matrices $\mathbf{A}_n$ and $\mathbf{B}_n$ such that $\mathbf{A}_n + \mathbf{B}_n$ is nonsingular and

$$E(\tilde{\boldsymbol{\theta}} \mid n, \mathbf{x}_1, \ldots, \mathbf{x}_n) = (\mathbf{A}_n + \mathbf{B}_n)^{-1}(\mathbf{A}_n E(\tilde{\boldsymbol{\theta}}) + \mathbf{B}_n\hat{\boldsymbol{\theta}}(n, \mathbf{x}_1, \ldots, \mathbf{x}_n)).$$

So far we have discussed Bayesian interpretations of unbiased estimators (as approximations to posterior expectations) and maximum likelihood estimators (as approximations to posterior modes). We shall now consider the Bayesian interpretation of the *consistency* property (20.2.2).

Pratt [**63**] has proved that if $\{\hat{\tilde{\boldsymbol{\theta}}}_n : n = 1, 2, \ldots\}$ is a consistent estimator of $\tilde{\boldsymbol{\theta}}$, then there is high prior probability that $\tilde{\boldsymbol{\theta}}$ will be "close to" $\hat{\boldsymbol{\theta}}_n$ as measured by the posterior distribution of $\tilde{\boldsymbol{\theta}}$. This is stated more precisely in the following:

Let (20.5.4)

$$A[n, \epsilon, \delta] \equiv \{(\mathbf{x}_1, \ldots, \mathbf{x}_n): P(|\tilde{\boldsymbol{\theta}} - \hat{\boldsymbol{\theta}}(n, \mathbf{x}_1, \ldots, \mathbf{x}_n)| \leqq \epsilon \mid n, \mathbf{x}_1, \ldots, \mathbf{x}_n) \geqq 1 - \delta\},$$

for all $n \in \{1, 2, \ldots\}$, $\delta \in (0, 1)$, and $\epsilon > 0$. Then

$$\lim_{n \to \infty} P(A[n, \epsilon, \delta]) = 1, \text{ for all } \delta, \epsilon.$$

Proof

Omitted; see Pratt [**63**].

Note in (20.5.4) that $A[n, \epsilon, \delta]$ is the set of outcomes $(\mathbf{x}_1, \ldots, \mathbf{x}_n)$ for which the posterior probability [given $(n, \mathbf{x}_1, \ldots, \mathbf{x}_n)$] of $\tilde{\boldsymbol{\theta}}$ being within ϵ of the consistent estimate $\hat{\boldsymbol{\theta}}(n, \mathbf{x}_1, \ldots, \mathbf{x}_n)$ is at least $1 - \delta$. If we plan to obtain a sufficiently large number n of observations, the assertion implies that the prior probability of the event "$(\tilde{\mathbf{x}}_1, \ldots, \tilde{\mathbf{x}}_n) \in A[n, \epsilon, \delta]$" is close to one. In fact, one might define an unsurprising outcome as one for which the posterior probability of the (k-dimensional) $\boldsymbol{\theta}$-interval $[\hat{\boldsymbol{\theta}}(n, \mathbf{x}_1, \ldots, \mathbf{x}_n) - \epsilon, \hat{\boldsymbol{\theta}}(n, \mathbf{x}_1, \ldots, \mathbf{x}_n) + \epsilon]$ is at least $1 - \delta$; with this definition, $A[n, \epsilon, \delta]$ is precisely the set of unsurprising outcomes. Note that here our definition of unsurprisingness depends upon ϵ and δ.

Another aspect of consistency is the fact that two different consistent estimators, $\{\tilde{\tilde{\boldsymbol{\theta}}}_n^{(1)}: n = 1, 2, \ldots\}$ and $\{\tilde{\tilde{\boldsymbol{\theta}}}_n^{(2)}: n = 1, 2, \ldots\}$, tend to differ very little from each other when n is large. More specifically, we have the following:

If $\{\tilde{\tilde{\boldsymbol{\theta}}}_n^{(i)}: n = 1, 2, \ldots\}$ is a consistent estimator of $\boldsymbol{\theta}$ for $i = 1, 2$; then (20.5.5)

$$\lim_{n \to \infty} P(|\tilde{\tilde{\boldsymbol{\theta}}}_n^{(1)} - \tilde{\tilde{\boldsymbol{\theta}}}_n^{(2)}| \leqq \epsilon | \boldsymbol{\theta}) = 1,$$

for every $\epsilon > 0$ and $\boldsymbol{\theta} \in \Theta$. Moreover

$$\lim_{n \to \infty} P(|\tilde{\tilde{\boldsymbol{\theta}}}^{(1)} - \tilde{\tilde{\boldsymbol{\theta}}}^{(2)}| \leqq \epsilon) = 1,$$

for every $\epsilon > 0$, provided that the prior dgf of $\tilde{\boldsymbol{\theta}}$ is positive at every $\boldsymbol{\theta} \in \Theta$.

Proof

For a given $\epsilon > 0$, $\delta > 0$, and $\boldsymbol{\theta} \in \boldsymbol{\Theta}$, choose $n(\epsilon, \delta, \boldsymbol{\theta})$ so large that

$$P(|\tilde{\tilde{\boldsymbol{\theta}}}^{(i)} - \boldsymbol{\theta}| \leqq \epsilon/2 \mid \boldsymbol{\theta}) \geqq 1 - \delta/2,$$

for $i = 1, 2$ whenever $n \geqq n(\epsilon, \delta, \boldsymbol{\theta})$; consistency of each $\tilde{\tilde{\boldsymbol{\theta}}}^{(i)}$ implies existence of the desired $n(\epsilon, \delta, \boldsymbol{\theta})$. Let $A_n{}^i(\epsilon, \delta, \boldsymbol{\theta}) \equiv \{(\mathbf{x}_1, \ldots, \mathbf{x}_n): |\hat{\boldsymbol{\theta}}^{(i)}(n, \mathbf{x}_1, \ldots, \mathbf{x}_n) - \boldsymbol{\theta}| \leqq \epsilon/2\}$, for $i = 1, 2$. Thus we have shown that for $n \geqq n(\epsilon, \delta, \boldsymbol{\theta})$,

$$P(A_n{}^i(\epsilon, \delta, \boldsymbol{\theta}) \mid \boldsymbol{\theta}) \geqq 1 - \delta/2, i = 1, 2.$$

Hence by Chapter 5, Exercise 9 it follows that

$$P(\cap_{i=1}^{2} A_n^i(\epsilon, \sigma, \boldsymbol{\theta}) \mid \boldsymbol{\theta}) \geqq 1 - \delta, \; n \geqq n(\epsilon, \delta, \boldsymbol{\theta}).$$

But if $(\mathbf{x}_1, \ldots, \mathbf{x}_n) \in \cap_{i=1}^{2} A_n^i(\epsilon, \delta, \boldsymbol{\theta})$, then

$$|\tilde{\hat{\boldsymbol{\theta}}}_n^{(i)} - \boldsymbol{\theta}| \leqq \frac{\epsilon}{2}, \quad \text{for } i = 1, 2,$$

and hence by the triangle inequality,

$$|\hat{\boldsymbol{\theta}}_n^{(1)} - \hat{\boldsymbol{\theta}}^{(2)}| \leqq |\hat{\boldsymbol{\theta}}_n^{(1)} - \boldsymbol{\theta}| + |\tilde{\hat{\theta}}_n^{(2)} - \boldsymbol{\theta}| \leqq \frac{2\epsilon}{2} = \epsilon.$$

Hence

$$P(|\tilde{\hat{\boldsymbol{\theta}}}_n^{(1)} - \tilde{\hat{\boldsymbol{\theta}}}_n^{(2)}| \leqq \epsilon \mid \boldsymbol{\theta}) \geqq P(\cap_{i=1}^{2} A_n^i(\epsilon, \delta, \boldsymbol{\theta}) \mid \boldsymbol{\theta}) \geqq 1 - \delta,$$

provided that $n \geqq n(\epsilon, \delta, \boldsymbol{\theta})$. Thus for any $\delta \in (0, 1)$, $\epsilon > 0$, and $\boldsymbol{\theta} \in \boldsymbol{\Theta}$, there exists $n(\epsilon, \delta, \boldsymbol{\theta})$ such that $n \geqq n(\epsilon, \delta, \boldsymbol{\theta})$ implies

$$P(|\tilde{\hat{\boldsymbol{\theta}}}_n^{(1)} - \tilde{\hat{\boldsymbol{\theta}}}_n^{(2)}| \leqq \epsilon \mid \boldsymbol{\theta}) \geqq 1 - \delta,$$

which proves the first assertion. A proof of the second is beyond the level of this book; it is similar to the proof of (20.5.4) in Pratt [**63**]. QED

In Chapter 18, Exercise 17 the reader was asked to show that some Bayes estimators, such as posterior expectations and fractiles, are consistent. Those results, together with (20.5.5), imply that for large n the numerical discrepancies between Bayesian point estimates and orthodox consistent estimates (for example, maximum likelihood estimates) tend with high probability to be small.

Thus for experiments consisting of many observations, the controversy between consistent Bayesian and consistent orthodox approaches to point estimation disappears as far as most practical applications are concerned.

20.6 BAYESIAN POINT ESTIMATION

If a Bayesian statistician knows the utility or loss function on the set of consequences of estimating $\boldsymbol{\theta}$ to be $\hat{\boldsymbol{\theta}}$, then he should choose his estimate so as to minimize the posterior expectation of loss, or equivalently, to maximize the posterior expectation of utility. Chapter 18 concerns Bayesian point-estimation problems as a special but important subset of Bayesian decision problems under uncertainty.

Optimal point estimates, or *Bayes estimates*, are given for a number of important cases in Chapter 18. To reconcile notation with this chapter, we remark that $\mathbf{a}_{z,n}^0$ in Chapter 18 corresponds to $\hat{\boldsymbol{\theta}}(n, \mathbf{x}_1, \ldots, \mathbf{x}_n)$ in this chapter. An estimator $\hat{\boldsymbol{\theta}}(n, \tilde{\mathbf{x}}_1, \ldots, \tilde{\mathbf{x}}_n)$, for which every $\hat{\boldsymbol{\theta}}(n, \mathbf{x}_1, \ldots, \mathbf{x}_n)$ is $\mathbf{a}^0_{z,n}$ for $\mathbf{z} \equiv (x_1, \ldots, x_n)$ is called a *Bayes estimator*. Bayes estimators are found from Chapter 18 by simply putting a tilde over $\mathbf{z}$ in $\mathbf{a}^0_{z,n}$; for example, by (18.2.7) the Bayes estimator for the Bernoulli-process parameter $\tilde{p}$ with squared-error loss, fixed-n experimentation, and a prior df $f_\beta(p \mid \rho, \nu)$ is $(\rho + \tilde{r})/(\nu + n)$, which is simply the expectation of $\tilde{p}$ with respect to the posterior df $f_\beta(p \mid \rho + r, \nu + n)$ viewed before the outcome is observed as a function of that outcome.

In Chapter 18 the derivation of Bayes estimators was only an early step in attacking the more immediate problem of determining the optimal experiment, given the economics of experimentation as expressed by $k(z, n)$. Such problems cannot be solved in a straightforward manner when no utility or loss function is introduced, but in that case neither can Bayes estimators be derived.

When no utility or loss function is introduced, a Bayesian statistician will experience grave difficulties trying to furnish a point estimate of $\tilde{\boldsymbol{\theta}}$. Recall that for a Bayesian, all information about $\tilde{\boldsymbol{\theta}}$ is embodied in his posterior dgf $f(\boldsymbol{\theta} \mid n, \mathbf{x}_1, \ldots, \mathbf{x}_n)$ once the outcome has been observed. Hence without a loss function, the Bayesian statistician will try to furnish a single point $\hat{\boldsymbol{\theta}}$ which conveys as much information about $f(\boldsymbol{\theta} \mid n, \mathbf{x}_1, \ldots, \mathbf{x}_n)$ as possible.

But no single point can shed much light on an entire distribution. *One* reasonable candidate is the posterior expectation $E(\tilde{\boldsymbol{\theta}} \mid n, \mathbf{x}_1, \ldots, \mathbf{x}_n)$ (corresponding to squared-error loss), but *other* equally reasonable candidates (for $k = 1$) are any posterior fractile [corresponding to an appropriate linear loss function (Section 18.3)], and (for any k) a posterior mode (corresponding to the zero-one loss function in Chapter 18, Exercise 16). Hence the statistician's choice of point entails an at least tacit choice of loss function.

Most Bayesian statisticians would conclude that at least *two* points (for $k = 1$) are required to convey useful information about a posterior dgf. In Chapter 21 we shall examine the subject of *credible intervals* (and, more generally, *credible regions*) which are intervals (regions) of prespecified posterior probability.

EXERCISES

1. Verify the last equality of (20.2.9).
2. Prove that if $\tilde{n}$ has mf $f_{Pa}(n \mid r, p)$, then
$$E\left[\frac{r-1}{\tilde{n}-1}\right] = p.$$
3. Prove (20.4.2).

4. Prove the first sentence of (20.4.3).

5. Prove the first sentence of (20.4.5).

6. Prove the first sentence of (20.4.6).

7. Prove (20.4.7).

8. Complete the proof of (20.4.11) as indicated there.

9. Prove (20.4.10).

10. Find the maximum likelihood estimator of σ^2 in the univariate Normal process where neither μ nor σ^2 is known, but where only σ^2 is of interest and only (n, s^2) is recorded (Chapter 14, see Exercise 12).

11. Prove the first assertion in Example 20.5.11.

12. Show that if $\lambda_i > 0$ for $i = 1, 2$, then

$$\lim_{\tau \to 0+} \left[\frac{f_\gamma(\lambda_1 \mid \rho, \tau)}{f_\gamma(\lambda_2 \mid \rho, \tau)} \right] = \left(\frac{\lambda_1}{\lambda_2} \right)^{\rho - 1},$$

and thus equals one when $\rho = 1$.

13. Show that if $\sigma_i^2 > 0$ for $i = 1, 2$, then

$$\lim_{\psi_0 \to 0+} \left[\frac{f_{i\gamma}(\sigma_1^2 \mid \psi_0, \nu_0)}{f_{i\gamma}(\sigma_2^2 \mid \psi_0, \nu_0)} \right] = (\sigma_1^2/\sigma_2^2)^{-(\frac{1}{2}\nu_0 + 1)},$$

and thus equals one when $\nu_0 = -2$.

14. Show that $(\mu_1, \nu_1 \psi_1/(\nu_1 + 3))$ is the mode of

$$f_{Ni\gamma}(\mu, \sigma^2 \mid \mu_1, \psi_1, n_1, \nu_1).$$

15. Let $\boldsymbol{\theta} \in \boldsymbol{\Theta} \subset R^k$, and define an estimator $\tilde{\hat{\boldsymbol{\theta}}}^*$ which ignores any information about $\boldsymbol{\theta}$ by $\tilde{\hat{\boldsymbol{\theta}}}^* = \boldsymbol{\theta}^*$ for some fixed k-tuple $\boldsymbol{\theta}^* \in \boldsymbol{\Theta}$.

 (a) Argue that $\tilde{\hat{\boldsymbol{\theta}}}^*$ is the best possible estimator *if* $\boldsymbol{\theta} = \boldsymbol{\theta}^*$.
 (b) Suppose $k = 1$, θ is the mean μ a univariate Normal process with known variance $\sigma^2 = 100$, and $\theta^* = \mu^* = 0$. Suppose 100 observations are to be obtained with noninformative stopping. Let $\hat{\mu}_1(\bar{x}) = \mu^* = 0$ and $\hat{\mu}_2(\bar{x}) = \bar{x}$, for every $\bar{x} \in R^1$. Suppose loss is squared-error and given by

$$\ell(\hat{\mu}, \mu) = (\hat{\mu} - \mu)^2, \text{ for all } \hat{\theta}, \theta.$$

 Graph $R(\tilde{\hat{\mu}}_1, \mu)$ and $R(\tilde{\hat{\mu}}_2, \mu)$ as function of μ. [*Hint*: $R(\tilde{\hat{\mu}}_1, \mu) = \ell(\hat{\mu}_1, \mu) = (\mu^* - \mu)^2 = \mu^2$.] Also show that $R(\tilde{\hat{\mu}}_1, \mu) \leqq R(\tilde{\hat{\mu}}_2, \mu)$ if and only if $-1 \leqq \mu \leqq 1$.

16. Suppose $\tilde{p}_1, \ldots, \tilde{p}_n$ is a random sample of size n from

$$f_\beta(p \mid \rho, \rho + 1) = \begin{cases} \rho p^{\rho - 1}, & p \in (0, 1) \\ 0, & p \notin (0, 1). \end{cases}$$

(a) Find the maximum likelihood estimator $\tilde{\hat{\rho}}$ of ρ.
(b) What is the approximate df of $\tilde{\hat{\rho}}$ given ρ for large n?

17. Suppose $\tilde{x}_1, \ldots, \tilde{x}_n$ is a random sample of size n from

$$(*) \qquad f(x \mid \rho, \lambda, \zeta) \equiv \begin{cases} \zeta\lambda \exp(-\lambda x^{\zeta}) \dfrac{(\lambda x^{\zeta})^{\rho-1} x^{\zeta-1}}{\Gamma(\rho)}, & x > 0 \\ 0, & x \leqq 0, \end{cases}$$

where ρ and ζ are known, positive real numbers. [Note that $\tilde{x}$ has df (*) if and only if $\tilde{x}^{\zeta} \equiv \tilde{y}$ has df $f_{\gamma}(y \mid \rho, \lambda)$.]

(a) Show that the maximum likelihood estimator $\tilde{\hat{\lambda}}$ of λ is given by

$$\tilde{\hat{\lambda}} = \frac{n\rho}{\sum_{i=1}^{n} \tilde{x}_i^{\zeta}}.$$

(b) Show that the df of $\tilde{\hat{\lambda}}$ is $f_{i\gamma}(\hat{\lambda} \mid \lambda, 2n\rho)$ given λ, ρ, and n.

[*Hint*: $\sum_{i=1}^{n} \tilde{x}_i^{\zeta} \equiv \tilde{y}$ has df $f_{\gamma}(y \mid n\rho, \lambda)$.]

(c) Show that $E(\tilde{\hat{\lambda}} \mid \lambda) = n\rho\lambda/(n\rho - 1)$, so that $\tilde{\hat{\lambda}}$ is biased; and that $V(\tilde{\hat{\lambda}} \mid \lambda) = (n\rho)^2\lambda^2/[(n\rho - 1)^2(n\rho - 2)]$.
(d) Suppose that λ is considered a random variable, $\tilde{\lambda}$, and given the prior df $f_{\gamma}(\lambda \mid \rho_0, \tau_0)$. Show that the posterior df of $\tilde{\lambda}$ is

$$f_{\gamma}(\lambda \mid \rho_0 + n\rho, \tau_0 + n\rho/\hat{\lambda}),$$

and hence that the posterior expectation of $\tilde{\lambda}$ is

$$E(\tilde{\lambda} \mid n, x_1, \ldots, x_n) = (\rho_0 + n\rho)/(\tau_0 + n\rho/\hat{\lambda})$$

and the posterior mode of $\tilde{\lambda}$ is

$$\check{\lambda}(n, x_1, \ldots, x_n) = (\rho_0 + n\rho - 1)/(\tau_0 + n\rho/\hat{\lambda}).$$

Show that

$$\check{\lambda} = \hat{\lambda}\ (n, x_1, \ldots, x_n)$$

in two cases: (1) for all $\hat{\lambda}$ if $\rho_0 = 1$ and $\tau_0 = 0$; and (2) for all $\rho_0 > 1$ and $\tau_0 > 0$ if $\hat{\lambda} = (\rho_0 - 1)/\tau_0$.

18. Suppose that the number n of observations on a Bernoulli process with parameter p is not predetermined but rather has mf $f_{Po}(n \mid \lambda p)$ for some, known $\lambda > 0$. (For example, the number n of arrivals at a sales counter is likely to be large if the probability p of a sale is also large, both being influenced by, say, an attractive display.) Thus the stopping process is informative.

(a) Show that the likelihood function of p is given by

$$L(p \mid r, n, \lambda) = Kp^{n+r}(1 - p)^{n-r} \exp(-\lambda p),$$

where $K \equiv \lambda^n/[r!(n - r)!]$.

[*Hint*: $f(r, n \mid p, \lambda) = f(r \mid n, p) \cdot f(n \mid p, \lambda)$.]

(b) Show that the maximum likelihood estimator of p is given by

$$\tilde{\hat{p}} = \frac{2\tilde{n} + \lambda - (4\tilde{n}^2 - 4\tilde{r}\lambda + \lambda^2)^{1/2}}{2\lambda}.$$

(c) From (b), if $n \geqq 0$ and $r = 0$, then

$$\hat{p} = \left[\frac{2n + \lambda - (4n^2 + \lambda^2)^{1/2}}{2\lambda}\right],$$

which is $(2\lambda)^{-1}$ times the excess of the sum of the side lengths $2n$ and λ of a right triangle over the length $(4n^2 + \lambda^2)^{1/2}$ of its hypotenuse. Hence $\hat{p} > 0$ if $n > 0$ even though $r = 0$. Why is this reasonable? Show that $\hat{p} = 1$ if $r = n > 0$ and $\hat{p} = 0$ if $r = n = 0$.

19. *Admissibility and Relative Optimality*: Suppose that a loss function $\ell(\hat{\boldsymbol{\theta}}, \boldsymbol{\theta})$ is given, and that $R(\tilde{\hat{\boldsymbol{\theta}}}, \boldsymbol{\theta})$ is defined by the appropriate generalization of (20.2.8); namely,

$$R(\tilde{\hat{\boldsymbol{\theta}}}, \boldsymbol{\theta}) \equiv E(\ell(\tilde{\hat{\boldsymbol{\theta}}}, \boldsymbol{\theta}) \mid \boldsymbol{\theta})$$
$$= \int \cdots \int \ell(\hat{\boldsymbol{\theta}}(n, \mathbf{x}_1, \ldots, \mathbf{x}_n), \boldsymbol{\theta}) f(n, \mathbf{x}_1, \ldots, \mathbf{x}_n \mid \boldsymbol{\theta}) dn\, d\mathbf{x}_1 \ldots d\mathbf{x}_n.$$

Let $\{\tilde{\hat{\boldsymbol{\theta}}}^{(i)}: i \in I\} \equiv \hat{\boldsymbol{\Theta}}_I$ denote a nonempty set of estimators. We have shown that it is generally futile to seek an estimator $\tilde{\hat{\boldsymbol{\theta}}}^*$ such that

$$R(\tilde{\hat{\boldsymbol{\theta}}}^*, \boldsymbol{\theta}) \leqq R(\tilde{\hat{\boldsymbol{\theta}}}^{(i)}, \boldsymbol{\theta}), \tag{20.7.1}$$

for all $\boldsymbol{\theta} \in \boldsymbol{\Theta}$ and all $\tilde{\hat{\boldsymbol{\theta}}}^{(i)}$ in the set $\hat{\boldsymbol{\Theta}}$ of all conceivable estimators of $\boldsymbol{\theta}$. If however, an estimator $\tilde{\hat{\boldsymbol{\theta}}}^*$ satisfies (20.7.1), for all $\boldsymbol{\theta} \in \boldsymbol{\Theta}$ and all $\tilde{\hat{\boldsymbol{\theta}}}^{(i)}$ belonging to a proper subset $\hat{\boldsymbol{\Theta}}_I$ of $\hat{\boldsymbol{\Theta}}$ we say that $\tilde{\hat{\boldsymbol{\theta}}}^*$ is *optimal relative to* $\hat{\boldsymbol{\Theta}}_I$.

(a) Let $k = 1$, $\theta = E(\tilde{x}) \equiv \mu =$ the common expectation of $\tilde{x}_1, \ldots, \tilde{x}_n$,

$$\hat{\boldsymbol{\Theta}}_I \equiv \left\{\sum_{i=1}^{n} c_i \tilde{x}_i \colon \sum_{i=1}^{n} c_i = 1\right\},$$ and $\ell_t(\hat{\theta}, \theta) \equiv (\hat{\theta} - \theta)^2$. Show that $\tilde{\hat{\boldsymbol{\theta}}}^* \equiv \sum_{i=1}^{n} (1/n)\tilde{x}_i$ is optimal relative to $\hat{\boldsymbol{\Theta}}_I$.

(b) If $\tilde{\hat{\boldsymbol{\theta}}}^* \in \hat{\boldsymbol{\Theta}}_I$ is such that no estimator $\tilde{\hat{\boldsymbol{\theta}}}^{(i)} \in \hat{\boldsymbol{\Theta}}_I$ satisfies

$$R(\tilde{\hat{\boldsymbol{\theta}}}^{(i)}, \boldsymbol{\theta}) \begin{cases} \leqq R(\tilde{\hat{\boldsymbol{\theta}}}^*, \boldsymbol{\theta}), \text{ for every } \boldsymbol{\theta} \in \boldsymbol{\Theta} \\ < R(\tilde{\hat{\boldsymbol{\theta}}}^*, \boldsymbol{\theta}), \text{ for some } \boldsymbol{\theta} \in \boldsymbol{\Theta}, \end{cases} \tag{20.7.2}$$

then $\tilde{\hat{\boldsymbol{\theta}}}^*$ is said to be *admissible relative* to $\hat{\boldsymbol{\Theta}}_I$. If $\hat{\boldsymbol{\Theta}}_I = \hat{\boldsymbol{\Theta}}$, then $\tilde{\hat{\boldsymbol{\theta}}}^*$ is simply called admissible. Thus $\tilde{\hat{\boldsymbol{\theta}}}^*$ is admissible relative to $\hat{\boldsymbol{\Theta}}_I$ if no other estimator $\tilde{\hat{\boldsymbol{\theta}}}^{(i)}$ in $\hat{\boldsymbol{\Theta}}$ furnishes equal or less risk at *every* $\boldsymbol{\theta} \in \boldsymbol{\Theta}$. Show that the estimator $\tilde{\hat{\theta}}^*$ defined in Exercise 15 is admissible when $\ell_t(\hat{\theta}, \theta)$ is any loss function such that $\ell_t(\hat{\theta}, \theta) \geqq 0$, with equality obtaining if and only if $\hat{\theta} = \theta$.

[*Hint*: Derive $R(\tilde{\hat{\theta}}^*, \theta^*)$.]

(c) Let $\hat{\Theta}_I \equiv \{\tilde{\tilde{\theta}}^{(1)}, \tilde{\tilde{\theta}}^{(2)}, \tilde{\tilde{\theta}}^{(3)}\}$ and $R(\tilde{\tilde{\theta}}^{(i)}, \theta)$ be as in the Figure 20.3. Show that:

(i) $\tilde{\tilde{\theta}}^{(2)}$ is inadmissible relative to $\hat{\Theta}_I$;
(ii) both $\tilde{\tilde{\theta}}^{(1)}$ and $\tilde{\tilde{\theta}}^{(3)}$ are admissible relative to $\hat{\Theta}_I$;
(iii) $\tilde{\tilde{\theta}}^{(1)}$ is minimax (relative to $\hat{\Theta}_I$); and
(iv) there is no optimal estimator relative to $\hat{\Theta}_I$.

(d) Show that if $\tilde{\tilde{\boldsymbol{\theta}}}$ is minimax (relative to some $\hat{\boldsymbol{\Theta}}_I$), then $\tilde{\tilde{\boldsymbol{\theta}}}$ is admissible (relative to that same $\hat{\boldsymbol{\Theta}}_I$).

(e) Show that if $\tilde{\tilde{\boldsymbol{\theta}}}^0$ is a Bayes estimator (relative to some $\hat{\boldsymbol{\Theta}}_I$) with respect to $f(\boldsymbol{\theta})$, then there is an admissible estimator $\tilde{\tilde{\boldsymbol{\theta}}}^*$ (relative to that same $\hat{\boldsymbol{\Theta}}_I$) such that $P[\tilde{\boldsymbol{\theta}} \in \{\boldsymbol{\theta}\colon R(\tilde{\tilde{\boldsymbol{\theta}}}^*, \boldsymbol{\theta}) < R(\tilde{\tilde{\boldsymbol{\theta}}}^0, \boldsymbol{\theta})\}] = 0$, where $P[\cdot]$ is computed with respect to $f(\boldsymbol{\theta})$, and that $B(\tilde{\tilde{\boldsymbol{\theta}}}^0) = B(\tilde{\tilde{\boldsymbol{\theta}}}^*)$. Hence $\tilde{\tilde{\boldsymbol{\theta}}}^*$ is also a Bayes estimator with respect to $f(\boldsymbol{\theta})$, and the search for a Bayes estimator (relative to $\hat{\boldsymbol{\Theta}}_I$) may be confined to the subset $\hat{\boldsymbol{\Theta}}_I{}^a$ of admissible estimators (relative to $\hat{\boldsymbol{\Theta}}_I$).

(f) Show that if $\tilde{\tilde{\boldsymbol{\theta}}}^*$ is optimal (relative to $\hat{\boldsymbol{\Theta}}_I$), then $\tilde{\tilde{\boldsymbol{\theta}}}^*$ is a minimax estimator and also a Bayes estimator with respect to any dgf $f(\cdot)$ of $\tilde{\boldsymbol{\theta}}$.

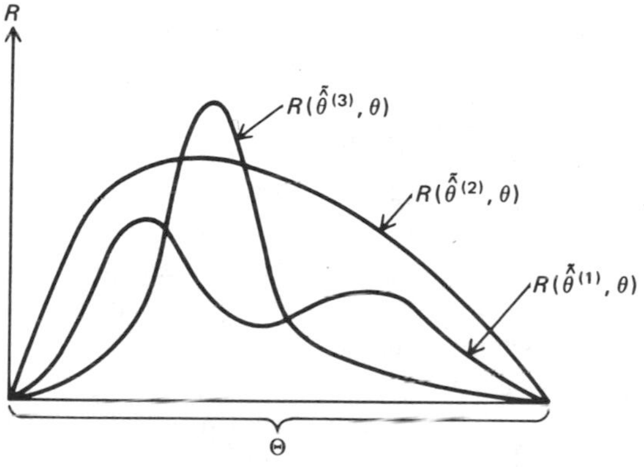

Figure 20.3

21

INTRODUCTION TO INTERVAL AND REGION ESTIMATION

21.1 INTRODUCTION

The most disquieting feature of the analysis in Chapter 20 is the failure of a point estimate to convey any information about its accuracy. For example, in the univariate Normal process with known variance σ^2, the two estimators $\tilde{\mu} \equiv \tilde{\bar{x}}$ and $\tilde{\mu}_1 \equiv n_0\mu_0/(n_0 + n) + n\tilde{\bar{x}}/(n_0 + n)$ of μ discussed in Chapters 20 are continuous random variables and hence have probability *zero* of precisely equaling μ. However, these estimators were chosen with due regard for their distributions about μ.

This chapter is concerned with making more modest estimates of a parameter θ, in the form of assertions that it is reasonable to believe that θ belongs to some subinterval of Θ. More generally, if $\boldsymbol{\theta}$ is k-dimensional, our assertions are that it is reasonable to believe that $\boldsymbol{\theta}$ belongs to some subset of $\boldsymbol{\Theta}$. The chosen subinterval or subset is often based upon an estimate $\hat{\boldsymbol{\theta}}$ of $\boldsymbol{\theta}$, and its width (or area, or volume) conveys information about the accuracy of $\hat{\boldsymbol{\theta}}$. Thus this chapter is addressed to Problem 2 of Section 9.7.

We shall consider two basic approaches to interval estimation: *confidence intervals* or *confidence regions*, and *credible intervals* or *credible regions*. A third, Fisher's *fiducial intervals*, will not be introduced. The interested reader may consult Kendall and Stuart [**41**], Chapter 21.

Section 21.2 begins with examples and continues with definitions of confidence intervals and regions, followed by a brief presentation of some of their properties. Section 21.3 then presents confidence intervals and regions for the univariate and k-variate Normal process parameters.

Section 21.4 presents two general methods for obtaining confidence intervals. The first is applicable and exact regardless of the number of observations, while the second is approximate and is based upon the approximate distribution of the maximum likelihood estimator for large n.

Section 21.5 discusses credible intervals and regions. A synonym for *credible interval* is *Bayesian confidence interval*, and similarly for credible regions. The reader can probably guess that credible intervals are determined by the posterior dgf of $\tilde{\theta}$.

Section 21.6 discusses the Bayesian interpretation of (orthodox) confidence intervals and regions.

21.2 CONFIDENCE INTERVALS AND REGIONS: DEFINITIONS AND BASIC PROPERTIES

The vast majority of consumers of statistical reports (and some producers as well) *misinterpret* confidence intervals and regions, and hence we shall take special pains to make the idea clear from the outset. We begin with three examples which should be studied carefully.

Example 21.2.1

In the univariate Normal process with known variance σ^2, we know that the sample mean $\tilde{\bar{x}}$ of 15 observations has df $f_N(\bar{x} \mid \mu, \sigma^2/15)$, so that $\tilde{z} \equiv (\tilde{\bar{x}} - \mu)/[\sigma^2/15]^{1/2}$ has df $f_{N*}(z)$. Now, a $\tilde{z}$-interval of .95 probability can be obtained by chopping off .025 probability in each tail of the standardized Normal distribution, and so doing yields

$$\begin{aligned} .95 &= P(-{}_{.975}z \leqq \tilde{z} \leqq {}_{.975}z) \\ &= P(-1.96 \leqq \tilde{z} \leqq 1.96) \\ &= P\left(-1.96 \leqq \frac{\tilde{\bar{x}} - \mu}{\sigma/\sqrt{15}} \leqq 1.96\right) \\ &= P\left(\frac{-1.96\sigma}{\sqrt{15}} \leqq \tilde{\bar{x}} - \mu \leqq \frac{1.96\sigma}{\sqrt{15}} \mid \mu\right) \\ &= P\left(\frac{-1.96\sigma}{\sqrt{15}} \leqq \mu - \tilde{\bar{x}} \leqq \frac{1.96\sigma}{\sqrt{15}} \mid \mu\right) \\ &= P\left(\tilde{\bar{x}} - \frac{1.96\sigma}{\sqrt{15}} \leqq \mu \leqq \tilde{\bar{x}} + \frac{1.96\sigma}{\sqrt{15}} \mid \mu\right). \end{aligned}$$

Note how the fourth, fifth and sixth equalities convert the argument of P from an interval of $[(\tilde{\bar{x}} - \mu)/(\sigma/\sqrt{15})]$-values with *fixed* endpoints to an interval of μ-values with *random* endpoints. Let

$$I(15, \bar{x}) \equiv [\bar{x} - 1.96\sigma/\sqrt{15}, \bar{x} + 1.96\sigma/\sqrt{15}].$$

What we have shown is that

$$P(I(15, \tilde{x}) \text{ covers } \mu \mid \mu) = .95;$$

and we say that $I(15, \bar{x})$ is a 95 percent *confidence interval for* μ. Note that the *uncertainty pertains to the interval, and not to* μ. *Before* the experiment, $\tilde{\bar{x}}$ is a random variable and hence $I(15, \tilde{\bar{x}})$ is a *random interval.* Our final probability statement is that, given μ, the probability that the as yet unknown interval $I(15, \tilde{\bar{x}})$ will cover (that is, bracket) μ is .95, regardless of μ. *After* the experiment, we can calculate $I(15, \bar{x})$ from its definition, and report "a 95 percent confidence interval for μ is $[\bar{x} - 1.96\sigma/\sqrt{15}, \bar{x} + 1.96\sigma/\sqrt{15}]$." Most users misinterpret the quoted statement as meaning, "the posterior probability that $\tilde{\mu}$ belongs to $[\bar{x} - 1.96\sigma/\sqrt{15}, \bar{x} + 1.96\sigma/\sqrt{15}]$ is .95." The second quoted statement is considered meaningless in orthodox statistics, although in Section 21.6 we shall show that in many circumstances the posterior probability that $\tilde{\mu}$ belongs to $I(n, \bar{x})$ is approximately .95 for large n.

Example 21.2.2

Notice in Example 21.2.1 that we have referred to $[\bar{x} - 1.96\sigma/\sqrt{15}, \bar{x} + 1.96\sigma/\sqrt{15}]$ as *a*, rather than *the*, 95 percent confidence interval for μ. Another 95 percent confidence interval for μ is obtained by noting that

$$\begin{aligned} .95 &= P(-\infty < \tilde{z} \leqq 1.64) \\ &= P\left(-\infty < \frac{\tilde{\bar{x}} - \mu}{\sigma/\sqrt{15}} \leqq 1.64\right) \\ &= P\left(\tilde{\bar{x}} - \mu \leqq \frac{1.64\sigma}{\sqrt{15}} \mid \mu\right) \\ &= P\left(\mu \geqq \tilde{\bar{x}} - \frac{1.64\sigma}{\sqrt{15}} \mid \mu\right), \end{aligned}$$

so that $I(15, \bar{x}) \equiv [\bar{x} - 1.64\sigma/\sqrt{15}, \infty)$ is another 95 percent confidence interval for μ. The reader may show as Exercise 1 that yet another 95 percent confidence interval for μ is $(-\infty, \bar{x} + 1.64\sigma/\sqrt{15}]$. The lengths of the two confidence intervals introduced in this example are infinite, while the length of the confidence interval in Example 21.2.1 is $3.92\sigma/\sqrt{15}$, which decreases to zero as σ tends to zero. Hence the confidence interval of Example 21.2.1 is preferable to the preceding alternatives if one's purpose is solely to be informative about μ. It is not hard to show that $[\bar{x} - 1.96\sigma/\sqrt{15}, \bar{x} + 1.96\sigma/\sqrt{15}]$ is the *shortest* 95 percent confidence interval for μ.

Example 21.2.3

Suppose that we want a 95 percent confidence interval for the mean μ of a univariate Normal process with *unknown* variance σ^2 based upon 15 observations. We use the result (13.3.12) that $\tilde{t} \equiv (\tilde{\bar{x}} - \mu)/\sqrt{\tilde{s}^2/15}$ has df $f_{S^*}(t \mid 14)$ to write

$$P\left(-2.145 \leqq \frac{\tilde{\bar{x}} - \mu}{\sqrt{\tilde{s}^2/15}} \leqq 2.145 \mid \mu\right) = .95$$

and rearrange the argument of P as in Example 21.2.1 so obtain

$$P\left(\tilde{\bar{x}} - \frac{2.145\tilde{s}}{\sqrt{15}} \leqq \mu \leqq \tilde{\bar{x}} + \frac{2.145\tilde{s}}{\sqrt{15}} \mid \mu\right) = .95,$$

where $\tilde{s} \equiv \sqrt{\tilde{s}^2}$. Thus we obtain the 95 percent confidence interval

$$I(15, \bar{x}, s^2) \equiv \left[\bar{x} - \frac{2.145s}{\sqrt{15}}, \bar{x} + \frac{2.145s}{\sqrt{15}}\right],$$

for μ. Notice that σ^2 does not enter into consideration directly here [or in (13.3.8)]. The generalization (21.3.2) of this example was a landmark contribution by Student. The length of the preceding confidence interval is random; namely, $4.29s/\sqrt{15}$. The two infinite-length intervals analogous to those in Example 21.2.2 are $[\bar{x} - 1.761s/\sqrt{15}, \infty)$ and $(-\infty, \bar{x} + 1.761s/\sqrt{15}]$, respectively as the reader may show in Exercise 2.

Before proceeding further, the reader should make sure that he understands what was done in Examples 21.2.1 through 21.2.3.

Let $\gamma \in (0, 1)$. We say that the random interval $I(n, \tilde{x}_1, \ldots, \tilde{x}_n)$ is a *confidence interval* (estimator) for θ at exact level γ if

$$P(I(n, \tilde{x}_1, \ldots, \tilde{x}_n) \text{ covers } \theta \mid \theta) = \gamma, \text{ for every } \theta \in \Theta. \qquad (21.2.1)$$

Occasionally there exists no random interval $I(n, \tilde{x}_1, \ldots, \tilde{x}_n)$ satisfying (21.2.1) for *every* $\theta \in \Theta$. For such cases, we require a somewhat less stringent concept. We say that $I(n, \tilde{x}_1, \ldots, \tilde{x}_n)$ is a confidence interval (estimator) for θ at *conservative* level γ if

$$P(I(n, \tilde{x}_1, \ldots, \tilde{x}_n) \text{ covers } \theta \mid \theta) \geqq \gamma, \text{ for every } \theta \in \Theta. \qquad (21.2.2)$$

A confidence interval for θ at conservative level γ exists for any $\gamma \in (0, 1)$ because $I(n, x_1, \ldots, x_n) \equiv (-\infty, \infty)$ for every $(n, x_1, \ldots, x_n)$ satisfies (21.2.2). A confidence interval for θ at either exact or conservative level γ is also called a 100γ percent confidence interval for θ.

More generally, suppose that $\boldsymbol{\theta}$ is k-dimensional and that $\tilde{\mathbf{x}}_1, \tilde{\mathbf{x}}_2, \ldots$ are m-dimensional. Let $\gamma \in (0, 1)$ and suppose that $I(n, \tilde{\mathbf{x}}_1, \ldots, \tilde{\mathbf{x}}_n)$ is a random subset of $\boldsymbol{\Theta} \subset R^k$; that is, each $(n, \mathbf{x}_1, \ldots, \mathbf{x}_n)$ determines a subset $I(n, \mathbf{x}_1, \ldots, \mathbf{x}_n)$ of $\boldsymbol{\Theta}$. We say that $I(n, \tilde{\mathbf{x}}_1, \ldots, \tilde{\mathbf{x}}_n)$ is a *confidence region for* $\boldsymbol{\Theta}$ *at exact level* γ if

$$P(I(n, \tilde{\mathbf{x}}_1, \ldots, \tilde{\mathbf{x}}_n) \text{ covers } \boldsymbol{\theta} \mid \boldsymbol{\theta}) = \gamma, \text{ for every } \boldsymbol{\theta} \in \boldsymbol{\Theta}; \qquad (21.2.3)$$

$I(n, \tilde{\mathbf{x}}_1, \ldots, \tilde{\mathbf{x}}_n)$ is a confidence region for $\boldsymbol{\theta}$ at *conservative* level γ if "=" in (21.2.3) is replaced by "$\geqq$." In either case, $I(n, \tilde{\mathbf{x}}_1, \ldots, \tilde{\mathbf{x}}_n)$ is also called a 100γ percent confidence region for θ.

All confidence regions in this chapter will be functions of sufficient statistics, since such regions generally have smaller volume (or area, or length) than regions not based upon sufficient statistics. Kendall and Stuart [**41**], Chapter 20, discuss the desirability of shortest-length intervals in some detail. For our purposes it suffices to note that volume is a natural measure of informativeness of confidence regions, and hence a minimum-volume confidence region at confidence level γ maximizes informativeness given γ.

21.3 CONFIDENCE INTERVALS AND REGIONS FOR NORMAL PROCESS PARAMETERS

This section consists of two subsections. In the first we derive confidence regions for the univariate and k-variate Normal process parameters μ, σ^2, (μ, σ^2), and $\boldsymbol{\mu}$, while in the second we derive confidence intervals useful in *comparing* parameters of two Normal processes. The reader is advised that heavy use will be made of Sections 13.3 and 13.5. All notation in Chapter 13 will apply without modification, while $\gamma \in (0, 1)$.

21.3.1 Confidence Regions for Normal Process Parameters

The shortest confidence interval at exact level γ for the mean μ of a univariate Normal process with known variance σ^2 is (21.3.1)

$$\left[\bar{x} - {}_{\frac{1}{2}(1+\gamma)}N^* \frac{\sigma}{\sqrt{n}}, \bar{x} + {}_{\frac{1}{2}(1+\gamma)}N^* \frac{\sigma}{\sqrt{n}}\right],$$

where ${}_{\frac{1}{2}(1+\gamma)}N^*$ is the $\frac{1}{2}(1+\gamma)$th fractile of the standardized Normal random variable $\tilde{z}$.

Proof

The proof imitates Example 21.2.1: from (13.3.3) it follows that the df of $\tilde{\bar{x}}$ is $f_N(\bar{x} \mid \mu, \sigma^2/n)$ and hence that

$$\gamma = P\left(\mu - {}_{\frac{1}{2}(1+\gamma)}N^* \frac{\sigma}{\sqrt{n}} \leqq \tilde{\bar{x}} \leqq \mu + {}_{\frac{1}{2}(1+\gamma)}N^* \frac{\sigma}{\sqrt{n}} \mid \mu\right)$$

$$= P\left(\tilde{\bar{x}} - {}_{\frac{1}{2}(1+\gamma)}N^* \frac{\sigma}{\sqrt{n}} \leqq \mu \leqq \tilde{\bar{x}} + {}_{\frac{1}{2}(1+\gamma)}N^* \frac{\sigma}{\sqrt{n}} \mid \mu\right),$$

for every $\mu \in R^1$. QED

Note in (21.3.1) that the higher the confidence level γ stipulated, the longer the confidence interval, since its length is twice ${}_{\frac{1}{2}(1+\gamma)}N^*\sigma/\sqrt{n}$, an increasing function of γ. In pathological cases this eminently plausible result may fail. Note also that the length is a decreasing function of the number n of observations: the better the information (the larger the n), the more informative the interval at the same level of confidence.

The shortest confidence interval at exact level γ for the mean μ of a univariate Normal process with unknown variance σ^2 is (21.3.2)

$$\left[\bar{x} - {}_{\frac{1}{2}(1+\gamma)}S^*{}_{(n-1)}\frac{s}{\sqrt{n}}, \bar{x} + {}_{\frac{1}{2}(1+\gamma)}S^*{}_{(n-1)}\frac{s}{\sqrt{n}}\right],$$

where $s \equiv (s^2)^{1/2}$ and ${}_{\frac{1}{2}(1+\gamma)}S^*{}_{(n-1)}$ is the $\frac{1}{2}(1+\gamma)$th fractile of the standardized Student random variable with $n-1$ degrees of freedom.

Proof

The proof imitates Example 21.2.3: from (13.3.8) it follows that

$$\gamma = P\left(-{}_{\frac{1}{2}(1+\gamma)}S^*_{(n-1)} \leqq \frac{\tilde{x}-\mu}{s/\sqrt{n}} \leqq {}_{\frac{1}{2}(1+\gamma)}S^*_{(n-1)} \mid \mu\right),$$

$$= P\{\tilde{x} - {}_{\frac{1}{2}(1+\gamma)}S^*_{(n-1)}\tilde{s}/\sqrt{n} \leqq \mu \leqq \tilde{x} + {}_{\frac{1}{2}(1+\gamma)}S^*_{(n-1)}\tilde{s}/\sqrt{n} \mid \mu),$$

for every $\mu \in R^1$. QED

A confidence interval at exact level γ for the variance σ^2 of a univariate Normal process with μ known is (21.3.3)

$$\left[\frac{n\xi}{{}_{\frac{1}{2}(1+\gamma)}\chi^2{}_{(n)}}, \frac{n\xi}{{}_{\frac{1}{2}(1-\gamma)}\chi^2{}_{(n)}}\right],$$

where ${}_{\frac{1}{2}p}\chi^2{}_{(n)}$ denotes the $\frac{1}{2}p$th fractile of the χ^2 random variables with n degrees of freedom. This confidence interval is almost the shortest if n is large.

Proof

By (13.3.4), $n\tilde{\xi}/\sigma^2$ has df $f_{\chi^2}(\cdot \mid n)$, so that by chopping off probability $\frac{1}{2}(1-\gamma)$ in each tail we obtain

$$\gamma = P\left({}_{\frac{1}{2}(1-\gamma)}\chi^2{}_{(n)} \leqq \frac{n\tilde{\xi}}{\sigma^2} \leqq {}_{\frac{1}{2}(1+\gamma)}\chi^2{}_{(n)}\right)$$

$$= P\left(\frac{1}{{}_{\frac{1}{2}(1-\gamma)}\chi^2{}_{(n)}} \geqq \frac{\sigma^2}{n\tilde{\xi}} \geqq \frac{1}{{}_{\frac{1}{2}(1+\gamma)}\chi^2{}_{(n)}}\right)$$

$$= P\left(\frac{n\tilde{\xi}}{{}_{\frac{1}{2}(1-\gamma)}\chi^2{}_{(n)}} \geqq \sigma^2 \geqq \frac{n\tilde{\xi}}{{}_{\frac{1}{2}(1+\gamma)}\chi^2{}_{(n)} \mid \sigma^2}\right),$$

from which the asserted interval is immediate. QED

The confidence interval in (21.3.3) is not the shortest due to the asymmetry of the χ^2-distribution. Without going into the details, Figure 21.1 illustrates the fact that the shortest interval is such that values *included* have higher density than points *excluded*; simply chopping equal tail probabilities off does not yield an interval with this property. Lindley [**49**], and Lindley and Miller [**50**], tabulate the equal-density endpoints of χ^2-distribution intervals for purposes of finding the *shortest* confidence intervals.

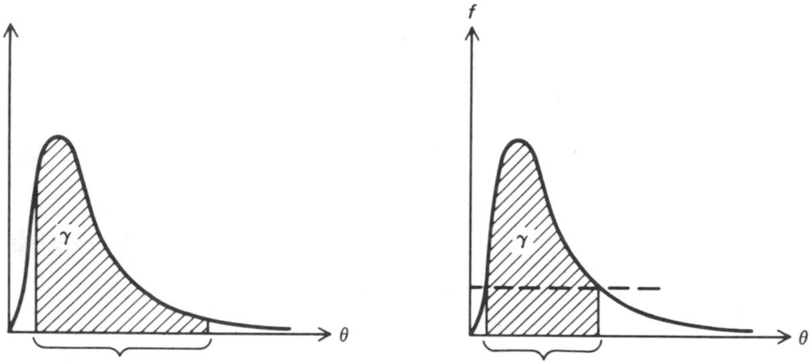

Figure 21.1 (a) Equal tail probabilities of $\frac{1}{2}(1-\gamma)$; (b) Equal density at endpoints

A confidence interval at exact level γ for the variance σ^2 of a univariate Normal process with μ unknown is (21.3.4)

$$\left[\frac{(n-1)s^2}{{}_{\frac{1}{2}(1+\gamma)}\chi^2{}_{(n-1)}}, \frac{(n-1)s^2}{{}_{\frac{1}{2}(1-\gamma)}\chi^2{}_{(n-1)}}\right];$$

this interval is almost shortest if n is large.

Proof

Exercise 3; imitate the proof of (21.3.3), using (13.3.6) in place of (13.3.4).

We shall not discuss confidence regions for (μ, σ^2) at this point. An exact but not minimum-area confidence region for (μ, σ^2) is derived in Mood and Graybill [**55**], 254–256. In Section 21.4 we derive an approximate confidence region for large n. (Compare Example 21.4.3.)

Example 21.3.1

Consider the chemical process in Example 14.5.1, in which hourly output $\tilde{x}_1, \tilde{x}_2, \ldots$ are iid with common df $f_N(x \mid \mu, 100)$. In that example, we assumed that $n = 6$ hours' outputs were observed, with $\bar{x} = 63$. By

(21.3.1), we conclude that 100γ percent confidence interval for μ is $[63 - {}_{\frac{1}{2}(1+\gamma)}N^*(4.08), 63 + {}_{\frac{1}{2}(1+\gamma)}N^*(4.08)]$, because $\sigma/\sqrt{n} = 10/\sqrt{6} \doteq 4.08$. The following table gives $I(n, \bar{x}) = I(6, 63)$ for various values of γ and thus shows how informativeness (small width) is purchased at the cost of confidence (γ):

γ	$I(6,63)$
.50	[60.25,65.75]
.90	[56.29,69.71]
.95	[55.00,71.00]
.99	[52.49,73.51]

Example 21.3.2

Reconsider Example 14.5.3, in which test scores $\tilde{x}_1, \tilde{x}_2, \ldots$ are iid with common df $f_N(x \mid \mu, \sigma^2)$; neither μ nor σ^2 is known; and $n = 99$ observations yielded $\bar{x} = 86$ and $s^2 = 25$. We may obtain a 100γ percent confidence interval from (21.3.2), which implies that

$$I(99, 86, 25) \doteq [86 - {}_{\frac{1}{2}(1+\gamma)}S^*{}_{(98)}(\tfrac{1}{2}), 86 + {}_{\frac{1}{2}(1+\gamma)}S^*{}_{(98)}(\tfrac{1}{2})]$$

because $s/\sqrt{n} \doteq \frac{1}{2}$. If $\gamma = .99$ is chosen, then by interpolation in Table IV we find ${}_{\frac{1}{2}(1.99)}S^*{}_{(.98)} \doteq 2.632$ and thus $I(99, 86, 25) \doteq [84.68, 87.32]$.

Example 21.3.3

Suppose in Example 21.3.2 that the observations had been for the purpose of obtaining information about σ^2 rather than μ, since one good feature of a test is that it separates students of differing competence. By (21.3.4), a 90 percent confidence interval for σ^2 is

$$\left[\frac{(98)(25)}{{}_{.95}\chi^2{}_{(98)}}, \frac{(98)(25)}{{}_{.05}\chi^2{}_{(98)}}\right].$$

Using the formula for ${}_p\chi^2{}_{(n)}$ for $n > 30$ in Table II, we find

$${}_{.95}\chi^2{}_{(98)} \doteq (\tfrac{1}{2})[(1.645) + (196 - 1)^{1/2}]^2 \doteq 123$$

and

$${}_{.05}\chi^2{}_{(98)} \doteq (\tfrac{1}{2})[(-1.645) + (196 - 1)^{1/2}]^2 \doteq 77,$$

from which it follows that the desired interval is [20.1, 32.2], approximately.

Example 21.3.4

Reconsider Example 14.5.2, in which readings $\tilde{x}_1, \tilde{x}_2, \ldots$ on a weight scale are iid with common df $f_N(x \mid 10, \sigma^2)$, and in which $n = 10$ readings produced the statistic $\xi = .2$. By (21.3.3), a 90 percent confidence interval for σ^2 is

$$\left[\frac{(10)(.2)}{.95\chi^2_{(10)}}, \frac{(10)(.2)}{.05\chi^2_{(10)}}\right] \doteq \left[\frac{2}{18.3}, \frac{2}{3.94}\right] \doteq [.11, .51].$$

We now give three results on confidence regions in R^k for the mean vector $\boldsymbol{\mu}$ of a k-variate Normal process, for Cases A, C, and E of Section 14.6.

A confidence region at exact level γ for the mean vector $\boldsymbol{\mu}$ of a k-variate Normal process with known covariance matrix $\boldsymbol{\sigma}^2$ is (21.3.5)

$$I(n, \bar{\mathbf{x}}) \equiv \{\boldsymbol{\mu}\colon n(\bar{\mathbf{x}} - \boldsymbol{\mu})'\boldsymbol{\sigma}^{-2}(\bar{\mathbf{x}} - \boldsymbol{\mu}) \leqq {}_{\gamma}\chi^2_{(k)}\},$$

where ${}_{\gamma}\chi^2_{(k)}$ is the γth fractile of the χ^2 random variable with k degrees of freedom.

Proof

By (13.6.8),

$$\begin{aligned}\gamma &= P(n(\bar{\tilde{\mathbf{x}}} - \boldsymbol{\mu})'\boldsymbol{\sigma}^{-2}(\bar{\tilde{\mathbf{x}}} - \boldsymbol{\mu}) \leqq {}_{\gamma}\chi^2_{(k)} \mid \boldsymbol{\mu}),\\ &= P(I(n, \bar{\tilde{\mathbf{x}}}) \text{ covers } \boldsymbol{\mu} \mid \boldsymbol{\mu}). \quad \text{QED}\end{aligned}$$

To use (21.3.5), it is instructive to note that once the observations have been made and $\bar{\mathbf{x}}$ computed, $I(n, \bar{\mathbf{x}})$ comprises the interior and boundary of an ellipsoid in R^k with center at $\bar{\mathbf{x}}$. In practice, one is often interested in determining whether a given vector $\boldsymbol{\mu}^*$ belongs to $I(n, \bar{\mathbf{x}})$; to find out, simply compute $n(\bar{\mathbf{x}} - \boldsymbol{\mu}^*)\boldsymbol{\sigma}^{-2}(\bar{\mathbf{x}} - \boldsymbol{\mu}^*)$ and see if it fails to exceed ${}_{\gamma}\chi^2_{(k)}$. Clearly, $\boldsymbol{\mu}^*$ belongs to $I(n, \bar{\mathbf{x}})$ if and only if $n(\bar{\mathbf{x}} - \boldsymbol{\mu}^*)'\boldsymbol{\sigma}^{-2}(\bar{\mathbf{x}} - \boldsymbol{\mu}^*) \leqq {}_{\gamma}\chi^2_{(k)}$.

A confidence region at exact level γ for the mean vector $\boldsymbol{\mu}$ of a k-variate Normal process with unknown covariance matrix $\boldsymbol{\sigma}^2$ is (21.3.6)

$$I(n, \bar{\mathbf{x}}, \mathbf{s}^2) \equiv \{\boldsymbol{\mu}\colon \phi(\boldsymbol{\mu}) \leqq {}_{\gamma}F_{(k,n-k)}\},$$

where

$$\phi(\boldsymbol{\mu}) \equiv \left[\frac{(n-k)n}{(n-1)k}\right](\bar{\mathbf{x}} - \boldsymbol{\mu})'\mathbf{s}^{-2}(\bar{\mathbf{x}} - \boldsymbol{\mu})$$

and ${}_{\gamma}F_{(k,n-k)}$ is the γth fractile of the F-distribution with k and $n - k$ degrees of freedom. This result obtains only when $n > k$.

Proof

Exercise 4, from (13.6.10).

In practice, one usually applies (21.3.6) by preselecting γ, obtaining the observations, computing $\bar{\mathbf{x}}$, s^2, and then determining whether a given value $\boldsymbol{\mu}^*$ of $\boldsymbol{\mu}$ belongs to $I(n, \bar{\mathbf{x}}, s^2)$ by seeing if

$$\phi(\boldsymbol{\mu}^*) \equiv \left[\frac{(n-k)n}{(n-1)k}\right](\bar{\mathbf{x}} - \boldsymbol{\mu}^*)'\mathbf{s}^{-2}(\bar{\mathbf{x}} - \boldsymbol{\mu}^*) \leqq {}_{\gamma}F_{(k,n-k)}.$$

Confidence regions in $R^{k(k+1)/2}$ for the $k(k+1)/2$ functionally independent elements of $\boldsymbol{\sigma}^2$ can be described, but their geometric interpretation is not nearly so clear as with ellipsoids for $\boldsymbol{\mu}$. Even less intelligible are confidence regions in $R^{k(k+3)/2}$ for the k elements of $\boldsymbol{\mu}$ and the $k(k+1)/2$ functionally independent elements of $\boldsymbol{\sigma}^2$. We shall not describe these regions here; the large-n methods of Section 21.4 may be used to derive approximate ellipsoidal confidence regions for $\boldsymbol{\sigma}^2$ and $(\boldsymbol{\mu}, \boldsymbol{\sigma}^2)$.

Example 21.3.5

Suppose a random sample of $(n =)$ 10 families in a given region is taken for the purpose of obtaining information on both regional average income μ_1 per capita (in thousands of dollars) and regional average savings μ_2 per capita (in hundreds of dollars). Let $\mathbf{x}_i$ denote (income of family i, savings of family i)$'$, and suppose that $\bar{\mathbf{x}} = (8.9, 8.0)'$, while

$$\mathbf{s}^2 = \begin{bmatrix} 4.00 & 2.00 \\ 2.00 & 2.00 \end{bmatrix}.$$

Since ${}_{.95}F_{(2,8)} = 4.46$ from Table V and

$$\left[\frac{(n-k)(n)}{(n-1)(k)}\right]\mathbf{s}^{-2} = \begin{bmatrix} 2.121 & -2.121 \\ -2.121 & 4.242 \end{bmatrix},$$

it follows from (21.3.6) that a 95 percent confidence region for $\boldsymbol{\mu}$ is the set of all $\boldsymbol{\mu}$ such that

$$(\bar{\mathbf{x}} - \boldsymbol{\mu})'\begin{bmatrix} 2.121 & -2.121 \\ -2.121 & 4.242 \end{bmatrix}(\bar{\mathbf{x}} - \boldsymbol{\mu}) \leq 4.46;$$

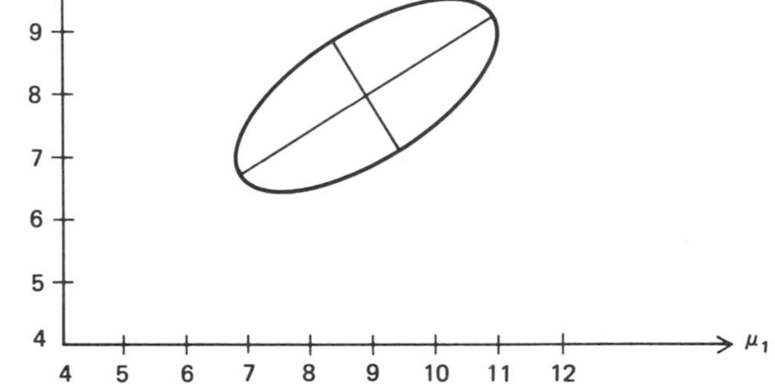

Figure 21.2

that is, the set of all points $(\mu_1, \mu_2)'$ on or inside the ellipse

$$2.121(\mu_1 - 8.9)^2 - 4.242(\mu_1 - 8.9)(\mu_2 - 8.0) + 4.242(\mu_2 - 8.0)^2 = 4.46.$$

Figure 21.2 depicts this confidence ellipsoid.

21.3.2 Confidence Regions for Comparing Two Normal Processes

The shortest confidence interval at exact level γ for the difference $\mu_1 - \mu_2$ of the mean of two univariate Normal processes with known variances σ_1^2 and σ_2^2 respectively is (21.3.7)

$$[\Delta\bar{x} - {}_{\frac{1}{2}(1+\gamma)}N^*\sigma(n_1, n_2), \Delta\bar{x} + {}_{\frac{1}{2}(1+\gamma)}N^*\sigma(n_1, n_2)],$$

where

$$\Delta x \equiv \bar{x}_1 - \bar{x}_2$$

and

$$\sigma(n_1, n_2) \equiv [\sigma_1^2/n_1 + \sigma_2^2/n_2]^{1/2}.$$

Proof

The proof follows from (13.4.9) just as (21.3.1) followed from (13.3.3); the details are left as Exercise 5.

The shortest confidence interval at exact level γ for the difference $\mu_1 - \mu_2$ of the means of two univariate Normal processes with variances σ_1^2 and σ_2^2 unknown *but assumed equal to each other* is

$$[\Delta\bar{x} - {}_{\frac{1}{2}(1+\gamma)}S^*_{(n_1+n_2-2)}s(n_1, n_2), \Delta\bar{x} + {}_{\frac{1}{2}(1+\gamma)}S^*_{(n_1+n_2-2)}s(n_1, n_2)], \tag{21.3.8}$$

where $\Delta\bar{x}$ is as defined in (21.3.7) and

$$s(n_1, n_2) \equiv [(1/n_1 + 1/n_2)\bar{s}^2]^{1/2},$$

for s^2 as defined by (13.4.4).

Proof

Exercise 6.

The hypothesis in (21.3.8) that $\sigma_1^2 = \sigma_2^2$ is restrictive but greatly simplifies the work required in obtaining a confidence interval for $\mu_1 - \mu_2$. The more general case, without the hypothesis $\sigma_1^2 = \sigma_2^2$, is known as the *Behrens-Fisher problem.*

Results (21.3.7) and (21.3.8) are commonly used to determine whether it is reasonable to believe that $\mu_1 = \mu_2$. The answer is in the affirmative if $\mu_1 - \mu_2 = 0$ is covered by the confidence interval for $\mu_1 - \mu_2$ and in the negative otherwise. The confidence level γ defines "reasonable"; γ's close to one correspond to fairly tolerant definitions of "reasonable" and make negative answers difficult, while γ's close to zero make negative answers easy. These ideas are formalized in Chapter 22, which is concerned with testing statistical hypotheses and dealing with questions of precisely the type introduced in this paragraph.

A similar question arises concerning equality of the variances σ_1^2 and σ_2^2 of two Normal processes; it can be answered by seeing if a confidence interval for the ratio σ_2^2/σ_1^2 of the process variances covers one.

A confidence interval at exact level γ for the ratio σ_2^2/σ_1^2 of the variances of two univariate Normal processes is (21.3.9)

$$\left[\left(\frac{\psi_2}{\psi_1}\right)_{\frac{1}{2}(1-\gamma)}F_{(\nu_1,\nu_2)}, \left(\frac{\psi_2}{\psi_1}\right)_{\frac{1}{2}(1-\gamma)}F_{(\nu_1,\nu_2)}\right],$$

where

$$\psi_i \equiv \begin{cases} \xi_i, & \text{if } \mu_i \text{ is known} \\ s_i^2, & \text{if } \mu_i \text{ is unknown,} \end{cases}$$

and

$$\nu_i \equiv \begin{cases} n_i, & \text{if } \mu_i \text{ is known} \\ n_i - 1, & \text{if } \mu_i \text{ is unknown,} \end{cases}$$

for $i = 1, 2$.

Proof

We shall prove (21.3.9) for μ_1 known and μ_2 unknown, in which case (21.3.9) yields the confidence interval

$$\left[\left(\frac{\tilde{s}_2^2}{\tilde{\xi}_1}\right)_{\frac{1}{2}(1-\gamma)}F_{(n_1,n_2-1)}, \left(\frac{\tilde{s}_2^2}{\tilde{\xi}_1}\right)_{\frac{1}{2}(1+\gamma)}F_{(n_1,n_2-1)}\right].$$

From (13.4.10) we know that $(\tilde{\xi}_1/\tilde{s}_2^2)(\sigma_2^2/\sigma_1^2)$ has the F distribution with n_1 and $n_2 - 1$ degrees of freedom. Hence we chop off $\frac{1}{2}(1 - \gamma)$ probability in each tail to obtain

$$\gamma = P\left({}_{\frac{1}{2}(1-\gamma)}F_{(n_1,n_2-1)} \leqq \left(\frac{\tilde{\xi}_1}{\tilde{s}_2^2}\right)\left(\frac{\sigma_2^2}{\sigma_1^2}\right) \leqq {}_{\frac{1}{2}(1+\gamma)}F_{(n_1,n_2-1)}\right),$$

from which the asserted confidence interval follows, upon multiplying each term in the argument of P by $\tilde{s}_2^2/\tilde{\sigma}_1^2$. The three other cases follow similarly from (13.4.10). QED

Confidence regions for the difference $\boldsymbol{\mu}_1 - \boldsymbol{\mu}_2$ of the mean vectors of two k-variate Normal processes are readily obtained from (13.7.12) and (13.7.13).

A confidence region at exact level γ for the difference $\boldsymbol{\mu}_1 - \boldsymbol{\mu}_2$ of mean vectors of two k-variate Normal processes with known covariance matrices $\boldsymbol{\sigma}_1^2$ and $\boldsymbol{\sigma}_2^2$ is given by (21.3.10)

$$I(n_1, n_2, \bar{\mathbf{x}}_1, \bar{\mathbf{x}}_2) \equiv \{\Delta\boldsymbol{\mu}: (\Delta\bar{\mathbf{x}} - \Delta\boldsymbol{\mu})'(n_1^{-1}\boldsymbol{\sigma}_1^2 + n_2^{-1}\boldsymbol{\sigma}_2^2)^{-1}(\Delta\bar{\mathbf{x}} - \Delta\boldsymbol{\mu}) \leqq {}_\gamma\chi^2_{(k)}\},$$

where $\Delta\bar{\mathbf{x}} \equiv \bar{\mathbf{x}}_1 - \bar{\mathbf{x}}_2$ and $\Delta\boldsymbol{\mu} \equiv \boldsymbol{\mu}_1 - \boldsymbol{\mu}_2$.

Proof

The proof follows from (13.7.12) just as (21.3.5) follows from (13.6.8). QED

A confidence region at exact level γ for the difference $\Delta\boldsymbol{\mu} \equiv \boldsymbol{\mu}_1 - \boldsymbol{\mu}_2$ of mean vectors of two k-variate Normal processes with covariance matrices $\boldsymbol{\sigma}_1^2$ and $\boldsymbol{\sigma}_2^2$ *unknown but assumed equal to each other* is given by (21.3.11)

$$I(n_1, n_2, \bar{\mathbf{x}}_1, \bar{\mathbf{x}}_2, \mathbf{s}_1^2, \mathbf{s}_2^2) \equiv \{\Delta\boldsymbol{\mu}: (\Delta\bar{\mathbf{x}} - \Delta\boldsymbol{\mu})'S^{-2}(\Delta\bar{\mathbf{x}} - \Delta\boldsymbol{\mu}) \leqq {}_\gamma F_{(k, n_1+n_2-k-1)}\},$$

where $\Delta\bar{\mathbf{x}} \equiv \bar{\mathbf{x}}_1 - \bar{\mathbf{x}}_2$ and

$$S^{-2} \equiv \left[\frac{(n_1 + n_2 - k - 1)(n_1 n_2)}{k(n_1 + n_2 - 2)(n_1 + n_2)}\right]\mathbf{s}^{-2}.$$

Proof

The proof follows from (13.7.13) just as (21.3.6) follows from (13.6.10). QED

21.4 GENERAL METHODS OF OBTAINING CONFIDENCE REGIONS

This section consists of two subsections. In the first, we present and exemplify a method of obtaining confidence regions that is always applicable; it relies upon no special properties of the statistics upon which the region is based. In the second subsection, we present and exemplify a method that yields approximate, ellipsoidal confidence regions for $\boldsymbol{\theta}$, based upon the maximum likelihood estimator $\tilde{\boldsymbol{\theta}}$, and valid for large n.

21.4.1 A Method Applicable for Every n

Suppose, for simplicity, that a fixed number n of observations $\tilde{\mathbf{x}}_1, \ldots, \tilde{\mathbf{x}}_n$ is to be acquired, and that $\tilde{\mathbf{z}}$ is a statistic (not necessarily sufficient). Let $\mathbf{Z}$ denote the set of all possible $\mathbf{z}$'s.

One can always determine, for each $\boldsymbol{\theta} \in \boldsymbol{\Theta} \subset R^k$, a subset $\mathbf{Z}(\boldsymbol{\theta}) \subset \mathbf{Z}$ such that

$$P(\tilde{\mathbf{z}} \in \mathbf{Z}(\boldsymbol{\theta}) \mid \boldsymbol{\theta}) \geqq \gamma, \tag{21.4.1}$$

for γ prespecified; at worst, one can let $\mathbf{Z}(\boldsymbol{\theta})$ be $\mathbf{Z}$ itself. Usually, however, it is desirable to make $\mathbf{Z}(\boldsymbol{\theta})$ as small in volume as possible.

Suppose now that $\mathbf{Z}(\boldsymbol{\theta})$ has been specified for every $\boldsymbol{\theta} \in \boldsymbol{\Theta}$. Then $\{(\boldsymbol{\theta}, \mathbf{z}): \mathbf{z} \in \mathbf{Z}(\boldsymbol{\theta}), \boldsymbol{\theta} \in \boldsymbol{\Theta}\} \equiv W$ is a subset of $\mathbf{Z} \times \boldsymbol{\Theta}$. If $\gamma > 0$, the reader may show as Exercise 7 that W is a *correspondence* from Θ to $\mathbf{Z}$; that is, $\mathbf{Z}(\boldsymbol{\theta}) \neq \varnothing$ for any $\boldsymbol{\theta}$. Further suppose that W^{-1} is also a correspondence, in that for every $\mathbf{z} \in \mathbf{Z}$ there exists some $\boldsymbol{\theta} \in \boldsymbol{\Theta}$ such that $\mathbf{z} \in \mathbf{Z}(\boldsymbol{\theta})$; this supposition is almost always satisfied in practical applications.

Define confidence regions pointwise by

$$I(n, \mathbf{z}) = \{\boldsymbol{\theta}: \mathbf{z} \in \mathbf{Z}(\boldsymbol{\theta})\}. \tag{21.4.2}$$

Clearly, $I(n, \mathbf{z})$ is the image of $\mathbf{z}$ under W^{-1}. From (21.4.1) and (21.4.2) it follows readily that

$$P(I(n, \tilde{\mathbf{z}}) \text{ covers } \boldsymbol{\theta} \mid \boldsymbol{\theta}) \geqq \gamma, \text{ for every } \boldsymbol{\theta} \in \boldsymbol{\Theta}. \tag{21.4.3}$$

Proof

$I(n, \tilde{\mathbf{z}})$ covers $\boldsymbol{\theta}$ if and only if $\tilde{\mathbf{z}} \in \mathbf{Z}(\boldsymbol{\theta})$, and hence both events have the same probability given $\boldsymbol{\theta}$. Hence $P(I(n, \tilde{\mathbf{z}}) \text{ covers } \boldsymbol{\theta} \mid \boldsymbol{\theta}) = P(\tilde{\mathbf{z}} \in \mathbf{Z}(\boldsymbol{\theta}) \mid \boldsymbol{\theta})$, from which (21.4.3) is immediate in view of (21.4.1). QED

To apply (21.4.1) through (21.4.3) requires the following steps (after choosing γ):

(1) Choose a statistic $\tilde{\mathbf{z}}$, the "more sufficient" the better, with known dgf $f(\mathbf{z} \mid \boldsymbol{\theta})$;
(2) For each $\boldsymbol{\theta}$, determine $\mathbf{Z}(\boldsymbol{\theta})$ so as to satisfy (21.4.1);
(3) Define—analytically or graphically—the correspondence $W \equiv \{(\boldsymbol{\theta}, \mathbf{z}): \mathbf{z} \in \mathbf{Z}(\boldsymbol{\theta}), \boldsymbol{\theta} \in \boldsymbol{\Theta}\}$;
(4) Invert the correspondence, defin*d*ing $I(n, \tilde{\mathbf{z}}) = \{\boldsymbol{\theta}: \mathbf{z} \in \mathbf{Z}(\boldsymbol{\theta})\} \equiv W^{-1}(\mathbf{z})$, for each $\mathbf{z}$.

The graphical description of W in (3) is most appropriate when θ and z are one-dimensional; it is extremely *useful* for the rather ill-behaved case of confidence intervals for the Bernoulli parameter p based upon $\tilde{r}$ with the mf $f_b(r \mid n, p)$ for fixed n. (For large n, the methods of Subsection 21.4.2 can be applied). Figure 21.3 depicts W qualitatively for n observations on a Bernoulli process, together with a "typical" resulting confidence interval.

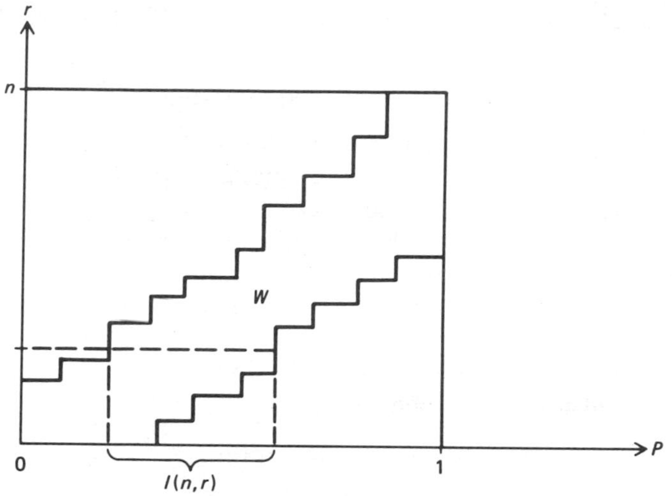

Figure 21.3

Example 21.4.1

Suppose $\Theta = \{0, 1, 2, 3, 4, 5\}$, $Z = \{5, 10, 15, 20\}$, and the $P(z \mid \theta)$'s are as given in the following table:

$\theta \backslash z$	5	10	15	20	*sum*
0	.70	.20	.10	0	1.00
1	.60	.18	.17	.05	1.00
2	.20	.40	.30	.10	1.00
3	.10	.30	.40	.20	1.00
4	.05	.17	.18	.60	1.00
5	0	.10	.20	.70	1.00

Suppose that we are required to specify a 90 percent confidence set for θ. (The term *interval* seems inappropriate in view of the discreteness of Θ.) We shall define $Z(\theta)$ so as to satisfy (21.4.1) and also so as to include as few points z as possible.

θ	$Z(\theta)$	$P(\tilde{z} \in Z(\theta) \mid \theta)$	$[\geqq .90]$
0	$\{5,10\}$	.90	
1	$\{5,10,15\}$	.95	
2	$\{5,10,15\}$	.90	
3	$\{10,15,20\}$	.90	
4	$\{10,15,20\}$	.95	
5	$\{15,20\}$	.90	

The reader may easily verify that

$$\begin{aligned} I(5) &= \{\theta: 5 \in Z(\theta)\} = \{0, 1, 2\} \\ I(10) &= \{\theta: 10 \in Z(\theta)\} = \{0, 1, 2, 3, 4\}, \\ I(15) &= \{\theta: 15 \in Z(\theta)\} = \{1, 2, 3, 4, 5\}, \text{ and} \\ I(20) &= \{\theta: 20 \in Z(\theta)\} = \{3, 4, 5\}. \end{aligned}$$

Note that the confidence level .90 is conservative for $\theta = 1$ or 4.

Example 21.4.2

Let $\tilde{x}_1, \ldots, \tilde{x}_n$ be iid with common df

$$f(x \mid \theta) = \begin{cases} \theta x^{\theta-1}, & x \in (0, 1) \\ 0, & x \notin (0, 1), \end{cases}$$

for $\theta \in \Theta = (0, \infty)$. We shall obtain confidence intervals for θ based upon the statistic $\tilde{\mathbf{z}} \equiv \min\{\tilde{x}_1, \ldots, \tilde{x}_n\}$, which (see Chapter 8, Exercise 19) has df

$$f(z \mid \theta) = \begin{cases} n\theta z^{n\theta-1}, & z \in (0, 1) \\ 0, & z \notin (0, 1). \end{cases}$$

It is easy to verify that the conditional γth fractile ${}_\gamma z(\theta)$ of $\tilde{z}$ is given by

$${}_\gamma z(\theta) = \gamma^{1/(n\theta)},$$

for any $\gamma \in (0, 1)$. Let $Z(\theta) \equiv \{z: {}_{\frac{1}{2}(1-\gamma)}z(\theta) \leqq z \leqq {}_{\frac{1}{2}(1-\gamma)}z(\theta)\}$. Clearly, $P(\tilde{z} \in Z(\theta) \mid \theta) = \gamma$, for every $\theta \in \Theta$. Figure 21.4 depicts W and a typical $I(n, z)$.

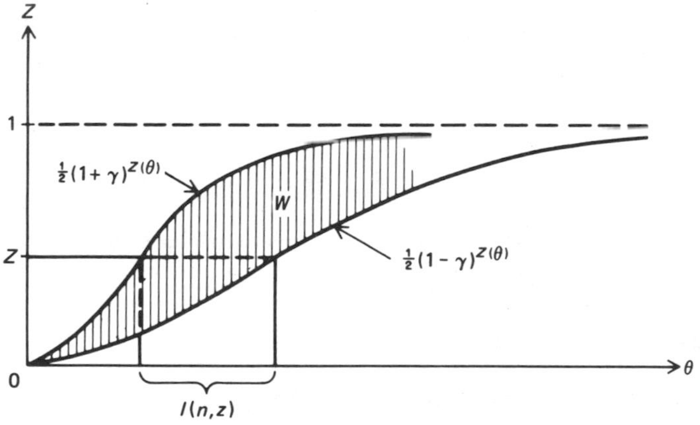

Figure 21.4

The reader may verify as Exercise 8 that

$$I(n, z) = \left[\frac{\log\{\frac{1}{2}(1+\gamma)\}}{\{n \log z\}}, \frac{\log\{\frac{1}{2}(1-\gamma)\}}{\{n \log z\}}\right].$$

We note that certain pathological situations arise in which the endpoints of $Z(\theta)$ [for $k = 1$] are not monotone functions of θ and hence $I(n, z)$ is not connected, but rather a union of intervals. Figure 21.5 furnishes an example of such a situation.

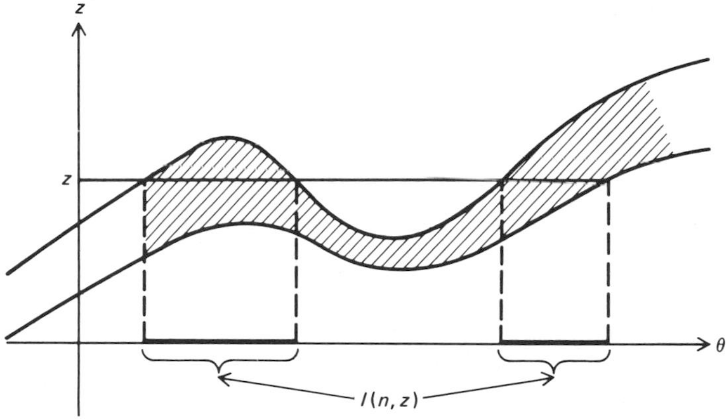

Figure 21.5

21.4.2 A Method Applicable for Large *n*

This subsection introduces a method for obtaining approximate confidence regions for $\boldsymbol{\theta}$ based upon the maximum likelihood estimator $\tilde{\hat{\boldsymbol{\theta}}}$ of $\boldsymbol{\theta}$. We shall give two results; the first is hard to apply but is a direct consequence of (20.3.6) and (13.6.8), and the second is much easier to apply but is harder to derive.

An approximate confidence region at exact level $\gamma \equiv 1 - \alpha$ for $\boldsymbol{\theta} \in \boldsymbol{\Theta} \subset R^k$ is $I(n, \tilde{\hat{\boldsymbol{\theta}}})$, defined by (21.4.4)

$$I(n, \hat{\boldsymbol{\theta}}) \equiv \{\boldsymbol{\theta}: n(\hat{\boldsymbol{\theta}} - \boldsymbol{\theta})'\hat{\boldsymbol{\sigma}}^{-2}(\boldsymbol{\theta})(\hat{\boldsymbol{\theta}} - \boldsymbol{\theta}) \leq {}_{\gamma}\chi^2{}_{(k)}\},$$

where $\hat{\boldsymbol{\theta}}$ denotes the maximum likelihood estimate of $\boldsymbol{\theta}$ based upon n observations and $\hat{\boldsymbol{\sigma}}^2(\boldsymbol{\theta})$ is as defined in (20.3.6).

Proof

By (20.3.6), the distribution of $\tilde{\hat{\boldsymbol{\theta}}}$ is approximately k-variate Normal with mean vector $\boldsymbol{\mu}$ and covariance matrix $n^{-1}\hat{\boldsymbol{\sigma}}^2(\boldsymbol{\theta})$. Hence by (13.6.8), the distribution of $n(\tilde{\hat{\boldsymbol{\theta}}} - \boldsymbol{\theta})\hat{\boldsymbol{\sigma}}^{-2}(\boldsymbol{\theta})(\tilde{\hat{\boldsymbol{\theta}}} - \boldsymbol{\theta})$ is approximately chi-square with k degrees of freedom, from which (21.4.4) follows readily. QED

Note that the confidence regions $I(n, \hat{\boldsymbol{\theta}})$ defined in (21.4.4) are *not* ellipsoidal unless $\hat{\boldsymbol{\sigma}}^2(\boldsymbol{\theta})$ is independent of $\boldsymbol{\theta}$.

Example 21.4.3

We shall derive the form of the approximate confidence region for $\boldsymbol{\theta} \equiv (\mu, \sigma^2)$ in the univariate Normal process, based upon the maximum likelihood estimator $\tilde{\hat{\boldsymbol{\theta}}} \equiv (\tilde{x}, (n-1)\tilde{s}^2/n)$. The biggest part of the job is obtaining $\boldsymbol{\sigma}^2(\boldsymbol{\theta})$. In (20.3.6), we derive first

$$\tilde{\boldsymbol{\Lambda}} = \begin{bmatrix} \dfrac{-1}{\sigma^2} & \dfrac{-(\tilde{x}-\mu)}{(\sigma^2)^2} \\ \dfrac{-(\tilde{x}-\mu)}{(\sigma^2)^2} & \dfrac{1}{2(\sigma^2)^2} - \dfrac{(\tilde{x}-\mu)^2}{(\sigma^2)^3} \end{bmatrix},$$

the matrix of second-partial derivatives of $\log_e f_N(\tilde{x} \mid \mu, \sigma^2)$ with respect to μ and σ^2. We then find

$$-E(\tilde{\boldsymbol{\Lambda}} \mid \boldsymbol{\theta}) = \begin{bmatrix} \dfrac{1}{\sigma^2} & 0 \\ 0 & \dfrac{1}{[2(\sigma^2)^2]} \end{bmatrix}$$

$$\equiv H \equiv \boldsymbol{\sigma}^{-2}(\boldsymbol{\theta}).$$

Hence the approximate confidence region is

$$I(n, \bar{x}, s^2) = \{\boldsymbol{\theta}: n(\boldsymbol{\theta} - \hat{\boldsymbol{\theta}})'\boldsymbol{\sigma}^{-2}(\boldsymbol{\theta})(\boldsymbol{\theta} - \hat{\boldsymbol{\theta}}) \leqq {}_{\gamma}\chi^2_{(2)}\}$$

$$= \left\{(\mu, \sigma^2): \frac{n(\mu - \bar{x})^2}{\sigma^2} + \frac{n(\sigma^2 - (n-1)s^2/n)^2}{2(\sigma^2)^2} \leqq {}_{\gamma}\chi^2_{(2)}\right\}.$$

It can be shown that the additional error caused by substituting $\hat{\boldsymbol{\theta}}$ for $\boldsymbol{\theta}$ in $\boldsymbol{\sigma}^2(\boldsymbol{\theta})$ is minor in relation to the error involved in the approximation (20.3.6). Hence a more convenient approximation procedure, yielding ellipsoidal confidence regions, is available.

An approximate confidence region at exact level γ, for $\boldsymbol{\theta} \in \boldsymbol{\Theta} \subset R^k$ is

$$I(n, \hat{\boldsymbol{\theta}}) \equiv \{\boldsymbol{\theta}: n(\hat{\boldsymbol{\theta}} - \boldsymbol{\theta})\boldsymbol{\sigma}^{-2}(\hat{\boldsymbol{\theta}})(\hat{\boldsymbol{\theta}} - \boldsymbol{\theta}) \leqq {}_{\gamma}\chi^2_{(k)}\}, \tag{21.4.5}$$

where $\boldsymbol{\sigma}^2(\hat{\boldsymbol{\theta}})$ is as defined in (20.3.6) but with $\hat{\boldsymbol{\theta}}$ replacing $\boldsymbol{\theta}$.

Example 21.4.4

From Example 21.4.3, we obtain

$$\boldsymbol{\sigma}^{-2}(\hat{\boldsymbol{\theta}}) = \begin{bmatrix} \dfrac{n}{(n-1)s^2} & 0 \\ 0 & \dfrac{n^2}{2\{(n-1)s^2\}^2} \end{bmatrix}$$

$$\equiv \begin{bmatrix} \dfrac{1}{\xi} & 0 \\ 0 & \dfrac{1}{(2\xi^2)} \end{bmatrix}$$

as the approximate precision matrix for the univariate Normal process, where $\xi \equiv (n - 1)s^2/n$. Hence the approximate confidence regions (ellipses) are given by

$$I(n, \bar{x}, s^2) = \left\{(\mu, \sigma^2): \frac{n(\mu - \bar{x})^2}{\xi} + \frac{n(\sigma^2 - \xi)^2}{2\xi^2} \leqq {}_{\gamma}\chi^2_{(2)}\right\}.$$

A typical confidence ellipse is depicted in Figure 21.6.

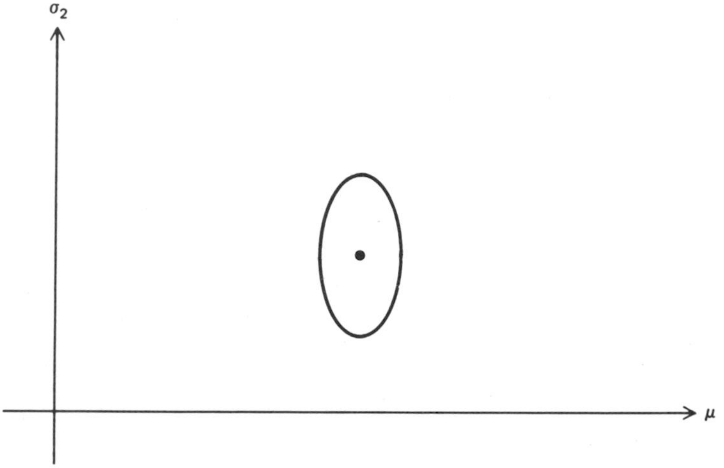

Figure 21.6

While (21.4.5) is applicable as stated when $k = 1$, an equivalent assertion is just as easy to use and appears more frequently in the literature.

An approximate confidence interval at exact level $\gamma \equiv 1 - \alpha$, for $\theta \in \Theta \subset R^1$ is (21.4.6)

$$I(n, \hat{\theta}) \equiv \left[\hat{\theta} - {}_{\frac{1}{2}(1+\gamma)}N^* \frac{\sigma(\hat{\theta})}{\sqrt{n}}, \hat{\theta} + {}_{\frac{1}{2}(1+\gamma)}N^* \frac{\sigma(\hat{\theta})}{\sqrt{n}}\right],$$

where

$$\sigma(\hat{\theta}) \equiv \left\{-E\left(\frac{\partial^2 \log_e f(\tilde{\mathbf{x}} \mid \theta)}{\partial \theta^2} \mid \theta\right)\right\}^{-1/2}\Bigg|_{\theta = \hat{\theta}}.$$

Proof

The asserted form of $\sigma(\hat{\theta})$ is immediate from (21.4.5), which asserts (for $k = 1$) that $\theta \in I(n, \hat{\tilde{\theta}})$ if and only if $n(\hat{\tilde{\theta}} - \theta)^2\sigma^{-2}(\hat{\tilde{\theta}}) \leqq {}_{\gamma}\chi^2{}_{(1)}$, which obtains if and only if $\{(\hat{\tilde{\theta}} - \theta)/[\sigma(\hat{\tilde{\theta}})/n]\}^2 \leqq {}_{\gamma}\chi^2{}_{(1)}$; that is, if and only if

$$\frac{-\{{}_{\gamma}\chi^2_{(1)}\}^{1/2}\sigma(\hat{\tilde{\theta}})}{n} \leqq \theta - \hat{\tilde{\theta}} \leqq \{{}_{\gamma}\chi^2_{(1)}\}^{1/2}\frac{\sigma(\hat{\tilde{\theta}})}{n},$$

from which (21.4.6) will follow provided that

$$_{\gamma}\chi^2_{(1)} = [_{\frac{1}{2}(1-\gamma)}N^*]^2, \; \gamma \in (0, 1).$$

This is left to the reader as Exercise 9. QED

From (21.4.5) and (21.4.6) we may obtain approximate confidence regions for the parameters of the non-Normal processes for which maximum likelihood estimates were given in Chapter 20.

An approximate confidence region at exact level $\gamma \equiv 1 - \alpha$ for the k-variate Bernoulli process parameter p is **(21.4.7)**

$$I(n, \hat{\mathbf{p}}) \equiv \{\mathbf{p}: n(\hat{\mathbf{p}} - \mathbf{p})'\boldsymbol{\sigma}^{-2}(\hat{\mathbf{p}})(\hat{\mathbf{p}} - \mathbf{p}) \leqq {}_{\gamma}\chi^2_{(k)}\},$$

where the (i, j)th element $\sigma_{ij}^{-2}(\hat{\mathbf{p}})$ of $\boldsymbol{\sigma}^{-2}(\hat{\mathbf{p}})$ is given by

$$\sigma_{ij}^{-2}(\hat{\mathbf{p}}) \equiv \begin{cases} \dfrac{1}{\hat{p}_{k+1}}, & j \neq i \\[2ex] \dfrac{1}{\hat{p}_i} + \dfrac{1}{\hat{p}_{k+1}}, & j = i, \end{cases}$$

for $\hat{\mathbf{p}} \equiv n^{-1}\mathbf{r}$ and $\hat{p}_{k+1} = 1 - \sum_{h=1}^{k} \hat{p}_h$.

Proof

By (20.4.2), $\hat{\mathbf{p}} = n^{-1}\mathbf{r}$. By (21.4.5) and (21.4.4), $\boldsymbol{\sigma}^{-2}(\hat{\mathbf{p}})$ is $\boldsymbol{\sigma}^{-2}(\mathbf{p})$ with $\hat{\mathbf{p}}$ substituted for p, and hence it suffices to show that $\sigma_{ij}^{-2}(\mathbf{p})$ has the asserted form. Since

$$f(\mathbf{x} \mid \mathbf{p}) = \prod_{h=1}^{k+1} p_h^{x_h},$$

it follows that

$$\log_e f(\mathbf{x} \mid \mathbf{p}) = \sum_{h=1}^{k+1} x_h \log_e p_h,$$

$$\frac{\partial \log_e f(\mathbf{x} \mid \mathbf{p})}{\partial p_i} = \frac{x_i}{p_i} + \frac{x_{k+1}}{p_{k+1}} \frac{\partial p_{k+1}}{\partial p_i}$$

$$= \frac{x_i}{p_i} - \frac{x_{k+1}}{p_{k+1}},$$

$$\frac{\partial^2 \log_e f(\mathbf{x} \mid \mathbf{p})}{\partial p_i^2} = \frac{-x_i}{p_i^2} + \frac{x_{k+1}}{p_{k+1}^2} \frac{\partial p_{k+1}}{\partial p_i}$$

$$= \frac{-x_i}{p_i^2} - \frac{x_{k+1}}{p_{k+1}^2}$$

$$= \Lambda_{ii},$$

and

$$\frac{\partial^2 \log_e f(\mathbf{x} \mid \mathbf{p})}{\partial p_i \partial p_j} = \frac{x_{k+1}}{p_{k+1}^2} \frac{\partial p_{k+1}}{\partial p_j}$$
$$= \frac{-x_{k+1}}{p_{k+1}^2}$$
$$= \Delta_{ij}.$$

Since $E(\tilde{x}_h \mid \mathbf{p}) = p_h$ for, $h = 1, \ldots, k+1$, it follows that every $\sigma_{ij}^{-2}(\mathbf{p}) \equiv -E(\tilde{\Lambda}_{ij} \mid \mathbf{p})$ has the asserted form. QED

An approximate confidence interval at exact level γ for the Bernoulli process parameter p is $I(n, \tilde{p})$ defined by (21.4.8)

$$I(n, \hat{p}) \equiv \left[\hat{p} - {}_{\frac{1}{2}(1+\gamma)}N^*\left[\frac{\hat{p}(1-\hat{p})}{n}\right]^{1/2}, \hat{p} + {}_{\frac{1}{2}(1+\gamma)}N^*\left[\frac{\hat{p}(1-\hat{p})}{n}\right]^{1/2}\right],$$

where $\hat{p} \equiv r/n$.

Proof

From (21.4.6) it suffices to show that $\sigma(\hat{p}) = [\hat{p}(1-\hat{p})]^{1/2}$. From (21.4.7), we have

$$\sigma^{-2}(\hat{p}) = \frac{1}{\hat{p}} + \frac{1}{1-\hat{p}}$$
$$= \frac{(1-\hat{p}+\hat{p})}{\hat{p}(1-\hat{p})}$$
$$= \frac{1}{\hat{p}([1-\hat{p})]},$$

from which the desired form of $\sigma(\hat{p})$ is obvious. QED

An approximate confidence interval at exact level γ for the intensity λ of the Poisson process is (21.4.9)

$$I(t, \hat{\lambda}) \equiv \left[\hat{\lambda} - {}_{\frac{1}{2}(1+\gamma)}N^*\left[\frac{\hat{\lambda}}{t}\right]^{1/2}, \hat{\lambda} + {}_{\frac{1}{2}(1+\gamma)}N^*\left[\frac{\hat{\lambda}}{t}\right]^{1/2}\right],$$

where $\hat{\lambda} \equiv r/t$.

Proof

For the Poisson process, t plays the role occupied by n in other processes, and (20.4.3) implies that $\hat{\lambda}$ is as asserted. From (21.4.6) and (21.4.4) it remains to be shown that $\sigma(\lambda) = \lambda^{1/2}$. For $t = 1$ we have

$$f(x \mid \lambda) = f_{Po}(x \mid \lambda) = \frac{e^{-\lambda}\lambda^x}{x!},$$

$$\log_e f(x \mid \lambda) = -\lambda + x \cdot \log_e \lambda - \log_e (x!), \text{ and}$$

$$\partial^2 \log_e f(x \mid \lambda)/\partial\lambda^2 = \frac{-x}{\lambda^2}$$

so that

$$\sigma^{-2}(\lambda) = -E\left(\frac{\partial^2 \log_e f(\tilde{x} \mid \lambda)}{\partial \lambda^2} \mid \lambda\right)$$
$$= -E\left(-\frac{\tilde{x}}{\lambda^2} \mid \lambda\right) = \frac{\lambda}{\lambda^2} = \frac{1}{\lambda},$$

from which $\sigma(\lambda) = \lambda^{1/2}$ is obvious. QED

21.5 (BAYESIAN) CREDIBLE INTERVALS AND REGIONS

As we have noted, many people misinterpret "a 95 percent confidence interval for μ is [0, 4]" as meaning, "the posterior probability that $\tilde{\mu}$ belongs to [0, 4] is .95." Pratt has observed [**63**] that people should not be blamed for this misinterpretation, since the correct interpretation ("[0, 4] is the interval which, before the observations are obtained, had probability .95 of covering μ") is simply not relevant to people concerned solely with μ and not with the observations whose only role is to furnish information about μ.

This section concerns credible intervals and regions, which are the Bayesian versions of confidence intervals and regions, and which mean what confidence intervals *appear* to mean. To start, recall that a Bayesian statistician makes all inferences on the basis of his posterior dgf

$$f(\boldsymbol{\theta} \mid n, \mathbf{x}_1, \ldots, \mathbf{x}_n)$$

of $\tilde{\boldsymbol{\theta}}$. Hence his solution to Problem 2 to Section 9.7 is to select a region $I(n, \mathbf{x}_1, \ldots, \mathbf{x}_n) \subset \boldsymbol{\Theta} \subset R^k$ such that: (1) his posterior probability $\tilde{\boldsymbol{\theta}}$ belongs to $I(n, \mathbf{x}_1, \ldots, \mathbf{x}_n)$ is at least γ; and (2) $I(n, \mathbf{x}_1, \ldots, \mathbf{x}_n)$ has minimum (or at least small) k-dimensional volume.

Formally, $I(n, \tilde{\mathbf{x}}_1, \ldots, \tilde{\mathbf{x}}_n) \subset \boldsymbol{\Theta} \subset R^k$ is a *credible region* (estimator) at exact level γ if

$$P(\tilde{\boldsymbol{\theta}} \in I(n, \mathbf{x}_1, \ldots, \mathbf{x}_n) \mid n, \mathbf{x}_1, \ldots, \mathbf{x}_n) = \gamma, \qquad (21.5.1)$$

for every possible $(n, \mathbf{x}_1, \ldots, \mathbf{x}_n)$. Credible regions at *conservative* levels γ are defined by replacing "$=$" in (21.5.1) by "$\geqq$." Credible *intervals* are connected credible regions for $k = 1$. Note that our definition does not incorporate the desideratum of minimum volume.

At this point the reader should compare (21.5.1) with (21.2.3), which can be more succinctly written as

$$P(\boldsymbol{\theta} \in I(n, \tilde{\mathbf{x}}_1, \ldots, \tilde{\mathbf{x}}_n) \mid \boldsymbol{\theta}) = \gamma, \text{ for every } \boldsymbol{\theta} \in \boldsymbol{\Theta}. \qquad (21.2.3')$$

The probabilities in (21.2.3′) and (21.5.1) are both conditional, but the former is conditional upon the unknown *parameter* $\boldsymbol{\theta}$ of interest, while the latter is conditional upon the *outcome* of the experiment. In both confidence and credible theories, the region is random before the observations are obtained. The misinterpretation of confidence regions follows from a too-rapid interpretation of "$\in$," which explains why we used the more cumbersome "covers" in (21.2.3) to define them. On the other hand, (21.5.1) means what people think it means.

Example 21.5.1

We shall re-examine Example 21.2.1 from the Bayesian viewpoint. Suppose the statistician has assessed the natural conjugate prior df $f_N(\mu \mid \mu_0, \sigma^2/n_0)$ for $\tilde{\mu}$. If 15 observations from $f_N(x \mid \mu, \sigma^2)$ result in the sufficient statistic $(n, \bar{x})$, then his posterior df of $\tilde{\mu}$ is $f_N(\mu \mid \mu_1, \sigma^2/[n_0 + 15])$, where $\mu_1 = n_0\mu_0/(n_0 + n) + n\bar{x}/(n_0 + n)$ [by (14.5.1)]. Hence

$$\tilde{z} \equiv \frac{\tilde{\mu} - \mu_1}{[\sigma^2/(n_0 + 15)]^{1/2}}$$

has df $f_{N*}(z)$, and we may chop off .025 probability in each tail of the standardized Normal distribution to obtain

$$\begin{aligned} .95 &= P(-1.96 \leqq \tilde{z} \leqq 1.96) \\ &= P\left(-1.96 \frac{\sigma}{\sqrt{n_0 + 15}} \leqq \tilde{\mu} - \mu_1 \leqq 1.96 \frac{\sigma}{\sqrt{n_0 + 15}} \,\middle|\, n, \bar{x}\right) \\ &= P\left(\mu_1 - 1.96 \frac{\sigma}{\sqrt{n_0 + 15}} \leqq \tilde{\mu} \leqq \mu_1 + 1.96 \frac{\sigma}{\sqrt{n_0 + 15}} \,\middle|\, n, \bar{x}\right), \end{aligned}$$

from which the desired 95 percent credible interval

$$I(n, \bar{x}) = \left[\frac{\mu_1 - 1.96\sigma}{\sqrt{n_0 + 15}}, \frac{\mu_1 + 1.96\sigma}{\sqrt{n_0 + 15}}\right]$$

follows readily. Before the observations are obtained, $\tilde{\bar{x}}$ is random and hence so is $\tilde{\mu}_1$. Note that this 95 percent credible interval coincides with the 95 percent confidence interval $[\bar{x} - 1.96\sigma/\sqrt{15},\ \bar{x} + 1.96\sigma/\sqrt{15}]$ only when $n_0 = 0$, which corresponds to the improper, uniform prior df $f_N(\mu \mid \mu_0, \infty)$. Also note that: (1) the length $3.92\sigma/\sqrt{n_0 + 15}$ of the credible interval is less than that of the corresponding confidence interval when $n_0 > 0$; and (2) the midpoint μ_1 of the credible interval lies on the same side of the midpoint $\bar{x}$ of the corresponding confidence interval as does the prior mean μ_0. Thus prior judgments serve to: (1) shorten the interval estimate; and (2) affect its location, in this case.

Example 21.5.2

We reconsider Example 21.4.1 under the assumption that $P(\tilde{\theta} = i) = \frac{1}{6}$, for $i = 0, 1, \ldots, 5$. The reader can easily verify that the resulting posterior mf's of $\tilde{\theta}$ are as given in the following table:

θ \ z	5	10	15	20
0	$\frac{14}{33}$	$\frac{20}{135}$	$\frac{10}{135}$	0
1	$\frac{12}{33}$	$\frac{18}{135}$	$\frac{17}{135}$	$\frac{1}{33}$
2	$\frac{4}{33}$	$\frac{40}{135}$	$\frac{30}{135}$	$\frac{2}{33}$
3	$\frac{2}{33}$	$\frac{30}{135}$	$\frac{40}{135}$	$\frac{4}{33}$
4	$\frac{1}{33}$	$\frac{17}{135}$	$\frac{18}{135}$	$\frac{12}{33}$
5	0	$\frac{10}{135}$	$\frac{20}{135}$	$\frac{14}{33}$

It follows that

$$P(\tilde{\theta} \in \{0, 1, 2\} \mid z = 5) = \frac{30}{33} > .90$$

while

$$P(\tilde{\theta} \in \{0, 1\} \mid z = 5\} = \frac{26}{33} < .90,$$

so that $\{0, 1, 2\} \equiv I(5)$ is a credible set at conservative level .90 and is the "smallest" such set in that it contains a minimal number of possible values θ. The reader may verify that

$$I(10) = \{0, 1, 2, 3, 4\},$$
$$I(15) = \{1, 2, 3, 4, 5\},$$

and

$$I(20) = \{3, 4, 5\}$$

possess similar properties. Note that the 90 percent credible sets $I(\tilde{z})$ coincide with the 90 percent confidence sets obtained in Example 21.4.1.

Example 21.5.3

Lest the reader obtains the impression from Example 21.5.2 that the practical differences between confidence and credible theories are nil, we again reconsider Example 21.4.1, this time for the prior mf $P(\tilde{\theta} \in \{0, 1, 5\})$

$= 0$, $P(\tilde{\theta} = 2) = P(\tilde{\theta} = 4) = \frac{1}{4}$, and $P(\tilde{\theta} = 3) = \frac{1}{2}$. The reader may verify the following table of posterior mf's:

θ \ z	5	10	15	20
0	0	0	0	0
1	0	0	0	0
2	.444	.342	.234	.091
3	.444	.513	.625	.364
4	.112	.145	.141	.545
5	0	0	0	0

It follows readily that

$$I(5) = I(10) = I(15) = \{2, 3, 4\}$$

and

$$I(20) = \{3, 4\}.$$

This system of 90 percent credible sets differs markedly from that obtained in Examples 21.5.2 and 21.4.1.

The remainder of this section consists of two subsections, the first of which describes credible regions for some parameters of the processes discussed in Chapter 14, and the second of which describes credible intervals based upon large-n maximum likelihood estimators.

21.5.1 Credible Regions for Some Parameters of Common Processes

Throughout this subsection we shall assume that the prior dgf of $\tilde{\boldsymbol{\theta}}$ is natural conjugate, as given in Chapter 14.

A credible interval at exact level γ for the parameter $\tilde{p}$ of a Bernoulli process is (21.5.2)

$$I(r, \text{n}) \equiv [{}_{\frac{1}{2}(1-\gamma)}\beta_{(\rho+r,\nu+n)}, {}_{\frac{1}{2}(1+\gamma)}\beta_{(\rho+r,\nu+n)}],$$

where ${}_x\beta_{(\rho+r,\nu+n)}$ is the xth fractile of a beta random variable with parameters $\rho + r$ and $\nu + n$.

Proof

By (14.2.3), the posterior df of $\tilde{p}$ is $f_\beta(p \mid \rho + r, \nu + n)$, and hence

$$\gamma = P({}_{\frac{1}{2}(1-\gamma)}\beta_{(\rho+r,\nu+n)} \leqq \tilde{p} \leqq {}_{\frac{1}{2}(1+\gamma)}\beta_{(\rho+r,\nu+n)} \mid \rho + r, \nu + n). \quad \text{QED}$$

The reader should note in (21.5.2) and its proof that the credible interval depends upon the experiment only through the sufficient statistic

(r, n), and that (21.5.2) depends for its validity only upon noninformativeness of the stopping process. Confidence intervals usually depend upon the particular stopping process of the experiment; in particular, confidence intervals for p depend upon whether n was predetermined or r was predetermined. Many people believe that the discrimination between noninformative stopping processes is to the discredit of orthodox methods in general, and confidence procedures in particular.

The logical next result is a credible region for the parameter $\tilde{\mathbf{p}}$ of the k-variate Bernoulli process, but no practical, precise credible region can be given. We will give an approximate, ellipsoidal credible region in the next subsection.

A credible interval at exact level γ for the intensity $\tilde{\lambda}$ of the Poisson process is (21.5.3)

$$I(r, t) \equiv [{}_{\frac{1}{2}(1-\gamma)}\gamma^*{}_{(\rho+r)}/(\tau + t),\ {}_{\frac{1}{2}(1+\gamma)}\gamma^*{}_{(\rho+r)}/(\tau + t)],$$

where ${}_x\gamma^*{}_{(\rho+r)}$ is the xth fractile of a standardized gamma random variable with parameter $\rho + r$.

Proof

By (14.4.2), the posterior df of $\tilde{\lambda}$ is $f_\gamma(\lambda \mid \rho + r, \tau + t)$, so that the posterior df of $\tilde{z} \equiv (\tau + t)\tilde{\lambda}$ is $f_{\gamma^*}(z \mid \rho + r)$. (See Section 12.2.) Hence

$$P({}_{\frac{1}{2}(1-\gamma)}\gamma^*{}_{(\rho+r)} \leqq (\tau + t)\tilde{\lambda} \leqq {}_{\frac{1}{2}(1+\gamma)}\gamma^*{}_{(\rho+r)} \mid r, t) = \gamma,$$

from which the assertion follows after dividing all terms in the inequalities by $(\tau + t)$. QED

A credible interval at exact level γ for the mean $\tilde{\mu}$ of the univariate Normal process with σ^2 known is (21.5.4)

$$I(n, \bar{x}) \equiv \left[\mu_1 - {}_{\frac{1}{2}(1+\gamma)}N^* \frac{\sigma}{\sqrt{n_0 + n}},\ \mu_1 + {}_{\frac{1}{2}(1+\gamma)}N^* \frac{\sigma}{\sqrt{n_0 + n}}\right],$$

where $\mu_1 = n_0\mu_0/(n_0 + n) + n\bar{x}/(n_0 + n)$ and ${}_xN^*$ is the xth fractile of the standardized Normal random variable.

Proof

Exercise 10.

A credible interval at exact level γ for the mean $\tilde{\mu}$ of the univariate Normal process with σ^2 unknown is (21.5.5)

$$I(n, \bar{x}, s) \equiv \left[\mu_1 - {}_{\frac{1}{2}(1+\gamma)}S^*{}_{(\nu_1)} \frac{\psi_1{}^{1/2}}{\sqrt{n_1}},\ \mu_1 + {}_{\frac{1}{2}(1+\gamma)}S^*{}_{(\nu_1)} \frac{\psi_1{}^{1/2}}{\sqrt{n_1}}\right],$$

where μ_1 is as defined in (21.5.4), $n_1 \equiv n_0 + n$,

$$\nu_1 \equiv \nu_0 + n, \psi_1 \equiv \frac{\nu_0\psi_0 + (n-1)s^2 + (n_0^{-1} + n^{-1})^{-1}(\bar{x} - \mu_0)^2}{\nu_1}$$

and ${}_\zeta S^*{}_{(\nu)}$ is the ζth fractile of the standardized Student random variable with ν degrees of freedom.

Proof

By (14.5.7), the posterior df of $\tilde{\mu}$ is $f_S(\mu \mid \mu_1, \psi_1/n_1, \nu_1)$, so that the posterior df of $\tilde{z} \equiv (\tilde{\mu} - \mu_1)/\sqrt{\psi_1/n_1}$ is $f_{S^*}(z \mid \nu_1)$ and

$$\gamma = P\left(-{}_{\frac{1}{2}(1+\gamma)}S^*{}_{(\nu_1)} \leqq \frac{\tilde{\mu} - \mu_1}{\sqrt{\psi_1/n_1}} \leqq {}_{\frac{1}{2}(1+\gamma)}S^*{}_{(\nu_1)}\right),$$

from which the assertion follows, via manipulations similar to those required in the proof of (21.5.4). QED

A credible interval at exact level γ for the variance $\tilde{\sigma}^2$ of the univariate Normal process with μ known is (21.5.6)

$$I(n, \xi) = \left[\frac{\nu_1\psi_1}{{}_{\frac{1}{2}(1+\gamma)}\chi^2{}_{(\nu_1)}}, \frac{\nu_1\psi_1}{{}_{\frac{1}{2}(1-\gamma)}\chi^2{}_{(\nu_1)}}\right],$$

where $\nu_1 \equiv \nu_0 + n$, $\nu_1\psi_1 \equiv \nu_0\psi_0 + n\xi$, and ${}_y\chi^2{}_{(\nu)}$ is the yth fractile of the χ^2 random variable with ν degrees of freedom.

Proof

By (14.5.3), the posterior df of $\tilde{\sigma}^2$ is $f_{i\gamma}(\sigma^2 \mid \psi_1, \nu_1)$, and hence (see Section 12.2) the posterior df of $\tilde{\chi}^2 \equiv \nu_1\psi_1/\tilde{\sigma}^2$ is $f_{\chi^2}(\chi^2 \mid \nu_1)$. Therefore

$$\gamma = P\left({}_{\frac{1}{2}(1-\gamma)}\chi^2{}_{(\nu_1)} \leqq \frac{\nu_1\psi_1}{\sigma^2} \leqq {}_{\frac{1}{2}(1+\gamma)}\chi^2{}_{(\nu_1)} \mid n, \xi\right)$$

$$= P\left(\frac{1}{{}_{\frac{1}{2}(1-\gamma)}\chi^2{}_{(\nu_1)}} \geqq \frac{\sigma^2}{\nu_1\psi_1} \geqq \frac{1}{{}_{\frac{1}{2}(1+\gamma)}\chi^2{}_{(\nu_1)}} \mid n, \xi\right),$$

from which the asserted credible interval follows by multiplying all terms in the inequalities by $\nu_1\psi_1$. QED

A credible interval at exact level γ for the variance $\tilde{\sigma}^2$ of the univariate Normal process with μ unknown is (21.5.7)

$$I(n, \bar{x}, s^2) \equiv \left[\frac{\nu_1\psi_1}{{}_{\frac{1}{2}(1+\gamma)}\chi^2{}_{(\nu_1)}}, \frac{\nu_1\psi_1}{{}_{\frac{1}{2}(1-\gamma)}\chi^2{}_{(\nu_1)}}\right],$$

where ν_1 and ψ_1 are as defined in (21.5.5) [not (21.5.6)].

Proof

By (14.5.7) the posterior df of $\tilde{\sigma}^2$ is $f_{i\gamma}(\sigma^2 \mid \psi_1, \nu_1)$ for the current definitions of ψ_1 and ν_1. The remainder of the proof duplicates that of (21.5.6). QED

A credible region at exact level γ for the mean vector $\tilde{\boldsymbol{\mu}}$ of the k-variate Normal process with $\boldsymbol{\sigma}^2$ known is (21.5.8)

$$I(n, \bar{\mathbf{x}}) \equiv \{\boldsymbol{\mu}\colon (\boldsymbol{\mu} - \boldsymbol{\mu}_1)'\boldsymbol{\sigma}_1^{-2}(\boldsymbol{\mu} - \boldsymbol{\mu}_1) \leqq {}_\gamma\chi^2_{(k)}\},$$

where $n_1 \equiv n_0 + n$, $\boldsymbol{\sigma}_1^{-2} \equiv \boldsymbol{\sigma}_0^{-2} + n\boldsymbol{\sigma}^{-2}$, and $\boldsymbol{\mu}_1 \equiv \boldsymbol{\sigma}_1^{2}(\boldsymbol{\sigma}_0^{-2}\boldsymbol{\mu}_0 + n\boldsymbol{\sigma}^{-2}\bar{\mathbf{x}})$.

Proof

By (14.6.1), the posterior df of $\tilde{\boldsymbol{\mu}}$ is $f_N{}^{(k)}(\boldsymbol{\mu} \mid \boldsymbol{\mu}_1, \boldsymbol{\sigma}_1^2)$, and hence by (13.6.5) the posterior df of $\tilde{z} \equiv (\tilde{\boldsymbol{\mu}} - \boldsymbol{\mu}_1)'\boldsymbol{\sigma}_1^{-2}(\tilde{\boldsymbol{\mu}} - \boldsymbol{\mu}_1)$ is $f_{\chi^2}(z \mid k)$. The remainder of the proof imitates that of (21.3.5). QED

A credible region at exact level γ for the mean vector $\tilde{\boldsymbol{\mu}}$ of the k-variate Normal process with $\boldsymbol{\sigma}_*^2$ known is (21.5.9)

$$I(n, \bar{\mathbf{x}}, s_*^2) \equiv \{\boldsymbol{\mu}\colon (\boldsymbol{\mu} - \boldsymbol{\mu}_1)'\psi_1^{-1}\boldsymbol{\sigma}_{1*}^{-2}(\boldsymbol{\mu} - \boldsymbol{\mu}_1) \leqq {}_\gamma(i\beta 2)_{(\frac{1}{2}k, \frac{1}{2}(\nu_1+k), \nu_1)}\},$$

where $\boldsymbol{\mu}_1$, ψ_1, ν_1, and $\boldsymbol{\sigma}_{1*}^{-2}$ are as in (14.6.14), and ${}_\gamma(i\beta 2)_{(\frac{1}{2}k, \frac{1}{2}(\nu_1+k), \nu_1)}$ is the γth fractile of the inverted beta-2 random variable with parameters $\frac{1}{2}k$, $\frac{1}{2}(\nu_1 + k)$, and ν_1.

Proof

By (14.6.14), the posterior df of $\tilde{\boldsymbol{\mu}}$ is $f_S{}^{(k)}(\boldsymbol{\mu} \mid \boldsymbol{\mu}_1, \psi_1\boldsymbol{\sigma}_{1*}^2, \nu_1)$, and hence by (13.6.12), $\tilde{z} \equiv (\tilde{\boldsymbol{\mu}} - \boldsymbol{\mu}_1)'\psi_1^{-1}\boldsymbol{\sigma}_{1*}^{-2}(\tilde{\boldsymbol{\mu}} - \boldsymbol{\mu}_1)$ has the asserted inverted beta-2 distribution. Hence

$$P((\tilde{\boldsymbol{\mu}} - \boldsymbol{\mu}_1)'\psi_1^{-1}\boldsymbol{\sigma}_{1*}^{-2}(\tilde{\boldsymbol{\mu}} - \boldsymbol{\mu}_1) \leqq {}_\gamma(i\beta 2)_{(\frac{1}{2}k, \frac{1}{2}(\nu_1+k), \nu_1)} \mid n, \bar{\mathbf{x}}, s_*^2) = \gamma. \quad \text{QED}$$

Note that the inverted beta-2 fractiles can be found from Table III via (12.3.15).

A credible region at exact level γ for the mean $\tilde{\boldsymbol{\mu}}$ of the k-variate Normal process with $\boldsymbol{\sigma}^2$ unknown is (21.5.10)

$$I(n, \bar{\mathbf{x}}, \mathbf{s}^2) \equiv \{\boldsymbol{\mu}\colon n_1(\boldsymbol{\mu} - \boldsymbol{\mu}_1)'\boldsymbol{\psi}_1^{-1}(\boldsymbol{\mu} - \boldsymbol{\mu}_1) \leqq {}_\gamma(i\beta 2)_{(\frac{1}{2}k, \frac{1}{2}(\nu_1+k), \nu_1)}\},$$

where $\boldsymbol{\mu}_1$, $\boldsymbol{\psi}_1$, n_1, and ν_1 are as in (14.6.22).

Proof

The proof follows from (14.6.8) just as (21.5.9) followed from (14.6.14). QED

21.5.2 Approximate Credible Regions for Large n

In Subsection 21.4.2 we developed approximate confidence regions for θ based upon the k-variate approximately Normal distribution of the maximum likelihood estimator $\tilde{\boldsymbol{\theta}}$ of $\boldsymbol{\theta}$. In this subsection we shall do the same for credible regions.

First, we note that (21.4.4) and the discussion leading to (21.4.5) imply that the likelihood function $L(\boldsymbol{\theta} \mid n, \mathbf{x}_1, \ldots, \mathbf{x}_n)$ is approximately $f_N^{(k)}(\hat{\boldsymbol{\theta}} \mid \boldsymbol{\theta}, (1/n)\boldsymbol{\sigma}^2(\hat{\boldsymbol{\theta}}))$, where $\boldsymbol{\sigma}^2(\hat{\boldsymbol{\theta}})$ is as defined in (20.3.6) with $\hat{\boldsymbol{\theta}}$ substituting for $\boldsymbol{\theta}$. Lindley [**49**] has noted that this definition of $\boldsymbol{\sigma}^2(\hat{\boldsymbol{\theta}})$ should be unsatisfactory to Bayesian statisticians because the expectation in (20.3.6) violates the likelihood principle, and derives an alternative, in which the (i, j)th element $n\sigma_{ij}{}^{-2}(\hat{\boldsymbol{\theta}})$ of the inverse $n\boldsymbol{\sigma}^{-2}(\hat{\boldsymbol{\theta}})$ of $(1/n)\boldsymbol{\sigma}^2(\hat{\boldsymbol{\theta}})$ is given by

$$n\sigma_{ij}{}^{-2}(\hat{\boldsymbol{\theta}}) = -\left[\frac{\partial^2}{\partial\theta_i\partial\theta_j}\sum_{h=1}^{n}\log_e f(\mathbf{x}_h \mid \boldsymbol{\theta})\right]_{\boldsymbol{\theta}=\hat{\boldsymbol{\theta}}}. \tag{21.5.11}$$

We note that the practical differences between this and the former definition of $(1/n)\boldsymbol{\sigma}^2(\hat{\boldsymbol{\theta}})$ are likely to be small when n is large enough for either to yield a satisfactory approximation to the true likelihood function.

By using $f_N^{(k)}(\hat{\boldsymbol{\theta}} \mid \boldsymbol{\theta}, (1/n)\boldsymbol{\sigma}^2(\hat{\boldsymbol{\theta}}))$ as the likelihood function, we readily obtain the following:

If $\tilde{\boldsymbol{\theta}}$ is assigned a k-variate Normal prior df $f_N^{(k)}(\boldsymbol{\theta} \mid \boldsymbol{\mu}_0, \boldsymbol{\sigma}_0^2)$ and n is large, then the posterior df of $\tilde{\boldsymbol{\theta}}$ is approximately $f_N^{(k)}(\boldsymbol{\theta} \mid \boldsymbol{\mu}_1, \boldsymbol{\sigma}_1^2)$, where (21.5.12)

$$\boldsymbol{\sigma}_1^2 \equiv [\boldsymbol{\sigma}_0^{-2} + n\boldsymbol{\sigma}^{-2}(\hat{\boldsymbol{\theta}})]^{-1},$$
$$\boldsymbol{\mu}_1 \equiv \boldsymbol{\sigma}_1^2[\boldsymbol{\sigma}_0^{-2}\boldsymbol{\mu}_0 + n\boldsymbol{\sigma}^{-2}(\hat{\boldsymbol{\theta}})\hat{\boldsymbol{\theta}}],$$

and $\hat{\boldsymbol{\theta}}$ denotes the maximum likelihood estimate of $\boldsymbol{\theta}$ based upon $(n, \mathbf{x}_1, \ldots, \mathbf{x}_n)$.

Proof

Assertion (21.5.12) is a restatement of (14.6.1) with $\boldsymbol{\theta}$ replacing $\boldsymbol{\mu}$, $\boldsymbol{\sigma}^2(\hat{\boldsymbol{\theta}})$ replacing ${}^2\boldsymbol{\sigma}$, $\hat{\boldsymbol{\theta}}$ replacing $\bar{\mathbf{x}}$, and the approximate df $f_N^{(k)}(\hat{\boldsymbol{\theta}} \mid \boldsymbol{\theta}, (1/n)\ \boldsymbol{\sigma}^2(\hat{\boldsymbol{\theta}}))$ replacing the exact df $f_N^{(k)}(\bar{\mathbf{x}} \mid \boldsymbol{\mu}, (1/n)\boldsymbol{\sigma}^2)$. QED

The reader can probably guess that (21.5.12) implies a restatement of (21.5.8), as follows:

An approximate credible region at exact level γ for $\tilde{\boldsymbol{\theta}}$ is (21.5.13)

$$I(n\ \hat{\boldsymbol{\theta}}) \equiv \{\boldsymbol{\theta}: (\boldsymbol{\theta} - \boldsymbol{\mu}_1)'\boldsymbol{\sigma}_1^{-2}(\boldsymbol{\theta} - \boldsymbol{\mu}_1) \leqq {}_\gamma\chi^2{}_{(k)}\},$$

where $\boldsymbol{\mu}_1$ and $\boldsymbol{\sigma}_1^{-2}$ are as defined in (21.5.12).

Proof

Exercise 11.

The reader may show as Exercise 12 that the approximate credible region of (21.5.13) coincides with the approximate confidence region of (21.4.5) of $\boldsymbol{\sigma}^2(\hat{\boldsymbol{\theta}})$ is computed by the procedure of (20.3.6) in each case and $\boldsymbol{\sigma}_0^{-2} = \mathbf{0}$. Recall from Chapter 20 that $\boldsymbol{\sigma}_0^{-2} = \mathbf{0}$ corresponds to the improper uniform df over R^k.

An approximate credible region at exact level γ for the k-variate Bernoulli process parameter $\tilde{\mathbf{p}}$ is (21.5.14)

$$I(n, \hat{\mathbf{p}}) \equiv \{\mathbf{p}: (\mathbf{p} - \boldsymbol{\mu}_1)'\boldsymbol{\sigma}_1^{-2}(\mathbf{p} - \boldsymbol{\mu}_1) \leqq {}_\gamma\chi^2{}_{(k)}\},$$

where $\boldsymbol{\mu}_1$ and $\boldsymbol{\sigma}_1^{-2}$ are as defined in (21.5.12).

Proof

The proof is obvious from (21.5.13).

Note in (21.5.14) that the prior df of $\tilde{\mathbf{p}}$ is assumed, through (21.5.12), to be k-variate Normal. Now, when the assessor's prior judgments about $\tilde{p}$ are rather vague, there will be no k-variate Normal distribution which approximates those judgments at all adequately. In particular, if a k-variate Normal df is "fitted" to a k-variate beta df by equating $\boldsymbol{\mu}$ with $\nu^{-1}\boldsymbol{\rho}$ and $\boldsymbol{\sigma}^2$ with the matrix of k-variate-beta covariances, the fitted Normal df will often place substantial probability on impossible $\tilde{\mathbf{p}}$'s (in which some components are negative and/or $\sum_{i=1}^{k} p_i \geqq 1$).

Nevertheless, this difficulty should not be of great concern, since when n is large the posterior df is very insensitive to even quite substantial changes in the prior df, and hence the assessor may be able to be quite cavalier about his prior df. In particular, he may decide to use the uniform df over R^k and hence *interpret* the approximate confidence region (21.4.5) as his approximate credible region. We shall apply this reasoning in the next section to some of the cases treated in Sections 20.3 and 20.5.

The insensitivity of posterior df's to changes in prior df's when n is large is a general and agreeable result in Bayesian inference. The major obstacle to complete acceptance of Bayesian methods is the fact that two individuals' prior df's may differ widely, and hence that two statisticians (or decision-makers) can be led to different conclusions (or decisions) when confronted with the same experimental outcome. This is "unscientific" and "nonobjective." But the insensitivity property means that individual differences eventually disappear as experimental evidence becomes compelling (as n increases). This property is asymptotic (increasingly valid for increasing n), but so are many of the orthodox optimum properties, such as consistency.

Complete discussions of the insensitivity property, called the *principle of precise measurement*, can be found in Edwards, Lindman, and Savage [**16**] and in Savage *et al.* [**70**].

21.6 BAYESIAN INTERPRETATION OF CONFIDENCE REGIONS

In Section 20.5 we interpreted unbiased and maximum likelihood estimates as approximations to posterior expectations and posterior modes respectively. In this section we shall interpret confidence regions as approximations to credible regions.

After presenting a result, due originally to Pratt [**63**] and analogous to (20.5.1), we compare the Normal process credible and confidence regions to obtain some idea as to the goodness of the approximation and the particular values of the prior parameters for which the approximation is exact.

Suppose $I(n, \tilde{\mathbf{x}}_1, \ldots, \tilde{\mathbf{x}}_n)$ is a 100γ percent confidence region for $\boldsymbol{\theta}$. Then, by (21.2.3′),

$$P(\boldsymbol{\theta} \in I(n, \tilde{\mathbf{x}}_1, \ldots, \tilde{\mathbf{x}}_n) \mid \boldsymbol{\theta}) = \gamma$$

if the level is exact. This does *not* imply that

$$P(\tilde{\boldsymbol{\theta}} \in I(n, \mathbf{x}_1, \ldots, \mathbf{x}_n) \mid n, \mathbf{x}_1, \ldots, \mathbf{x}_n) = \gamma$$

once the outcome $(n, \mathbf{x}_1, \ldots, \mathbf{x}_n)$ has been observed; that is, a confidence level is not a posterior probability. However, *before* the outcome is observed, the prior probability that $\tilde{\boldsymbol{\theta}}$ will belong to $I(n, \tilde{\mathbf{x}}_1, \ldots, \tilde{\mathbf{x}}_n)$ is γ. (21.6.1)

If $I(n, \tilde{\mathbf{x}}_1, \ldots, \tilde{\mathbf{x}}_n)$ is a confidence region for $\boldsymbol{\theta}$ at $\begin{Bmatrix}\text{conservative}\\ \text{exact}\end{Bmatrix}$ level $\gamma \equiv 1 - \alpha$, then

$$P(\tilde{\boldsymbol{\theta}} \in I(n, \tilde{\mathbf{x}}_1, \ldots, \tilde{\mathbf{x}}_n)) = E[P(\tilde{\boldsymbol{\theta}} \in I(n, \tilde{\mathbf{x}}_1, \ldots, \tilde{\mathbf{x}}_n) \mid n, \tilde{\mathbf{x}}_1, \ldots, \tilde{\mathbf{x}}_n)] \begin{Bmatrix}\geq\\ =\end{Bmatrix} \gamma.$$

Proof

Let $\mathbf{y} \equiv (\mathbf{x}_1, \ldots, \mathbf{x}_n)$ and $I(\mathbf{y}) \equiv I(n, \mathbf{x}_1, \ldots, \mathbf{x}_n)$, and suppress dependences upon n, as it is kept fixed in (21.6.1). The first equality follows from

$$\begin{aligned} P(\tilde{\boldsymbol{\theta}} \in I(n, \tilde{\mathbf{x}}_1, \ldots, \tilde{\mathbf{x}}_n)) &\equiv P(\tilde{\boldsymbol{\theta}} \in I(\tilde{\mathbf{y}})) \\ &= \int_{-\infty}^{\infty} \int_{I(\mathbf{y})} f(\boldsymbol{\theta}, \mathbf{y}) d\boldsymbol{\theta} d\mathbf{y} \\ &= \int_{-\infty}^{\infty} \int_{I(\mathbf{y})} f(\boldsymbol{\theta} \mid \mathbf{y}) f(\mathbf{y}) d\boldsymbol{\theta} d\mathbf{y} \\ &= \int_{-\infty}^{\infty} \left\{ \int_{I(\mathbf{y})} f(\boldsymbol{\theta} \mid \mathbf{y}) d\boldsymbol{\theta} \right\} f(\mathbf{y}) d\mathbf{y} \\ &= \int_{-\infty}^{\infty} P(\tilde{\boldsymbol{\theta}} \in I(\mathbf{y}) \mid \mathbf{y}) f(\mathbf{y}) d\mathbf{y} \\ &= E[P(\tilde{\boldsymbol{\theta}} \in I(\tilde{\mathbf{y}}) \mid \tilde{\mathbf{y}})]. \end{aligned}$$

For the second equality, we shall show that $P(\tilde{\boldsymbol{\theta}} \in I(n, \tilde{\mathbf{x}}_1, \ldots, \tilde{\mathbf{x}}_n)) = \gamma$. From the preceding equality chain, we have

$$\begin{aligned} P(\tilde{\boldsymbol{\theta}} \in I(\tilde{\mathbf{y}})) &= \int_{-\infty}^{\infty} \int_{I(\mathbf{y})} f(\boldsymbol{\theta}, \mathbf{y}) d\boldsymbol{\theta} d\mathbf{y} \\ &= \int_{-\infty}^{\infty} \int_{I(\mathbf{y})} f(\mathbf{y} \mid \boldsymbol{\theta}) f(\boldsymbol{\theta}) d\boldsymbol{\theta} d\mathbf{y}. \end{aligned}$$

We now interchange order of integration, recalling that $\boldsymbol{\theta} \in I(\mathbf{y})$ if and only if $\mathbf{y} \in Z(\boldsymbol{\theta})$:

$$\begin{aligned} \int_{-\infty}^{\infty} \int_{I(\mathbf{y})} f(\mathbf{y} \mid \boldsymbol{\theta}) f(\boldsymbol{\theta}) d\boldsymbol{\theta} d\mathbf{y} &= \int_{-\infty}^{\infty} \left\{ \int_{Z(\boldsymbol{\theta})} f(\mathbf{y} \mid \boldsymbol{\theta}) d\mathbf{y} \right\} f(\boldsymbol{\theta}) d\boldsymbol{\theta} \\ &= \int_{-\infty}^{\infty} P(\boldsymbol{\theta} \in I(\tilde{\mathbf{y}}) \mid \boldsymbol{\theta}) f(\boldsymbol{\theta}) d\boldsymbol{\theta} \\ (*) \qquad &= \int_{-\infty}^{\infty} \gamma f(\boldsymbol{\theta}) d\boldsymbol{\theta} \\ &= \gamma. \end{aligned}$$

If $I(\tilde{\mathbf{y}})$ were at conservative level γ, then equality (*) would become "$\geqq$." QED

Thus before the outcome is observed, we expect the confidence region to cover $\tilde{\boldsymbol{\theta}}$ with posterior probability γ. This means that if the outcome is not too surprising, then the posterior probability that $\tilde{\boldsymbol{\theta}} \in I(n, \mathbf{x}_1, \ldots, \mathbf{x}_n)$ will be approximately γ.

We could go on to characterize those "unsurprising" values of sufficient statistics for the Normal cases in Section 21.3 such that the posterior probability of the confidence interval is exactly γ, but the analysis is quite cumbersome; see Exercise 13. Rather, we shall simply indicate the values of the prior parameters for which the 100γ percent confidence and credible regions coincide for every possible outcome.

The confidence and credible intervals at level γ for the mean μ of a univariate Normal process with σ^2 known coincide for all $(n, x_1, \ldots, x_n)$ if and only if $n_0 = 0$. (21.6.2)

Proof

By direct comparison of (21.3.1) and (21.5.4): the respective widths $2 \cdot {}_{\frac{1}{2}(+\gamma)}N^*\sigma/\sqrt{n}$ and $2 \cdot {}_{\frac{1}{2}(1+\gamma)}N^*\sigma/\sqrt{n_0 + n}$ coincide if and only if $n_0 = 0$, in which case also $\mu_1 = \bar{x}$. QED

Note that the prior df $f_N(\mu \mid \mu_0, \sigma^2/n_0)$ for $n_0 = 0$ is the improper, uniform df over R^1, and hence $n_0 > 0$ implies that the credible interval at level γ is shorter than the corresponding confidence interval.

The confidence and credible intervals at level γ for the mean μ of a univariate Normal process with σ^2 unknown coincide for all $(n, x_1, \ldots, x_n)$ if and only if $n_0 = 0$, $\nu_0 = -1$, and $\xi_0 = 0$. (21.6.3)

Proof

Make a direct comparison of (21.3.2) and (21.5.5). The sufficiency of the specified values is clear; the necessity follows by noting that the requisite prior parameter values for equal width would otherwise depend upon n. QED

The confidence and credible intervals at level γ for the variance σ^2 of a univariate Normal process with μ known coincide for all $(n, x_1, \ldots, x_n)$ if and only if $\nu_0 = 0$. (21.6.4)

Proof

Compare (21.3.3) with (21.5.6).

The confidence and credible intervals at level γ for the variance σ^2 of a univariate Normal process with μ unknown coincide for all $(n, x_1, \ldots, x_n)$ if and only if $\nu_0 = -1$, $\xi_0 = 0$, and $n_0 = 0$. (21.6.5)

Proof

Compare (21.3.4) with (21.5.7).

The confidence and credible regions at level γ for the mean vector $\boldsymbol{\mu}$ of a k-variate Normal process with $\boldsymbol{\sigma}^2$ known coincide for all $(n, \mathbf{x}_1, \ldots, \mathbf{x}_n)$ if and only if $\boldsymbol{\sigma}_0^{-2} = 0$. (21.6.6)

Proof

Compare (21.3.5) with (21.5.8).

EXERCISES

1. Show the $I(15, \bar{x}) \equiv (-\infty, \bar{x} + 1.64\sigma/\sqrt{15}]$ is a 95 percent confidence intravel for the mean μ of a univariate Normal process with known variance σ^2, given $n = 15$ observations.

2. Show that $[\bar{x} - 1.761\bar{s}/\sqrt{15}, \infty)$ and $(-\infty, \bar{x} + 1.761\bar{s}/\sqrt{15}]$ are 95 percent confidence intervals for the mean μ of a univariate Normal process with unknown variance σ^2, given $n = 15$ observations.

3. Derive (12.3.4).

4. Derive (21.3.6).

5. Derive (21.3.7).

6. Derive (21.3.8).

7. Show that $W \equiv \{(\boldsymbol{\theta}, \mathbf{z}): \mathbf{z} \in \mathbf{Z}(\boldsymbol{\theta}), \boldsymbol{\theta} \in \boldsymbol{\Theta}\}$ is a correspondence from $\boldsymbol{\Theta}$ to $\mathbf{Z}$ if $\mathbf{Z}(\boldsymbol{\theta})$ is such that

$$P(\tilde{\mathbf{z}} \in \mathbf{Z}(\boldsymbol{\theta}) \mid \boldsymbol{\theta}) \geqq \gamma > 0, \text{ for every } \boldsymbol{\theta} \in \boldsymbol{\Theta}.$$

8. Show in Example 21.4.2 that

$$I(n, z) = \left[\frac{\log\{\frac{1}{2}(1+\gamma)\}}{n \log\{z\}}, \frac{\log\{\frac{1}{2}(1-\gamma)\}}{n \log\{z\}}\right].$$

9. Show that ${}_{\gamma}\chi^2{}_{(1)} = [{}_{\frac{1}{2}(1+\gamma)}N^*]^2$, for every $\gamma \in (0, 1)$.

10. Derive (21.5.4).

11. Derive (21.5.13).

12. Show that the approximate credible region at level γ given by (12.5.13) coincides with the approximate confidence region at level γ given by (21.4.5) if $\boldsymbol{\sigma}^2(\hat{\boldsymbol{\theta}})$ is computed by the procedure of (20.3.6) in each case and if $\boldsymbol{\sigma}_0^{-2} = \mathbf{0}$.

13. *Posterior Probabilities of Confidence Intervals*:

(a) Show that for the univariate Normal process with known variance σ^2,

$$P\left(\tilde{\mu} \in \left[\bar{x} - {}_{\frac{1}{2}(1+\gamma)}N^* \frac{\sigma}{\sqrt{n}},\ \bar{x} + {}_{\frac{1}{2}(1+\gamma)}N^* \frac{\sigma}{\sqrt{n}}\right] \mid \mu_1, \frac{\sigma^2}{n_1}\right)$$

$$= F_{N^*}\left(\Delta(\bar{x}, n) + {}_{\frac{1}{2}(1+\gamma)}N^* \sqrt{\frac{1 + n_0}{n}}\right)$$

$$- F_{N^*}\left(\Delta(\bar{x}, n) - {}_{\frac{1}{2}(1+\gamma)}N^* \sqrt{\frac{1 + n_0}{n}}\right),$$

where

$$\Delta(\bar{x}, n) \equiv \left(\frac{\bar{x} - \mu_0}{\sigma}\right) \frac{n_0}{\sqrt{n_0 + n}}.$$

(b) Denote the probability in part (a) by $\gamma(\bar{x}, n)$. What is $E[\gamma(\tilde{\bar{x}}, n)]$?

(c) Show that

$$\max_{\bar{x}} \gamma(\bar{x}, n) = \gamma(\mu_0, n)$$

and that

$$\gamma(\mu_0 + \epsilon, n) = \gamma(\mu_0 - \epsilon, n),$$

for every $\epsilon \in R^1$ and every $n > 0$. Thus show that $\gamma(\cdot, n)$ appears as in Figure 21.7.

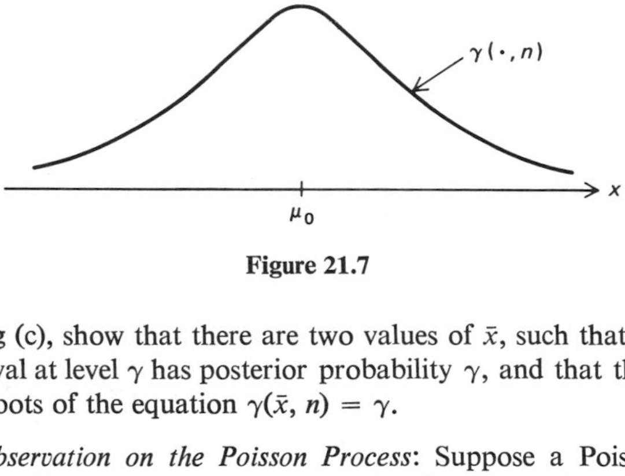

Figure 21.7

(d) Using (c), show that there are two values of $\bar{x}$, such that the confidence interval at level γ has posterior probability γ, and that these values are the roots of the equation $\gamma(\bar{x}, n) = \gamma$.

14. *Inverse Observation on the Poisson Process*: Suppose a Poisson process is observed until the rth occurrence, and that $\tilde{t} = t$ was the total observation time.

(a) Show that a confidence interval for λ at exact level γ is

$$\left[\frac{{}_{\frac{1}{2}(1-\gamma)}\gamma^*_{(r)}}{t}, \frac{{}_{\frac{1}{2}(1+\gamma)}\gamma^*_{(r)}}{t}\right],$$

where ${}_p\gamma^*_{(r)}$ is the pth fractile of the standardized gamma random variable with parameter r. [*Hint*: Recall from (11.7.12) that the df of $\tilde{t}$ given λ is $f_\gamma(t \mid r, \lambda)$.]

(b) Show that the confidence interval of (a) coincides with the credible interval at level γ if and only if $\rho = 0$ and $\tau = 0$.

15. *Pathologies in Confidence Theory (I)*: Suppose $\Theta = \{0, 1\}$, $Z = \{5, 10\}$, and the $P(z \mid \theta)$'s are as given in the following table:

θ \ z	5	10
0	.95	.05
1	.98	.02

(a) Find the smallest $Z(0)$ and $Z(1)$ such that $P(Z(\theta) \mid \theta) \geqq .90$.
(b) Find $I(5)$ and $I(10)$. Is $I(5)$ a reasonable estimate of θ? How about $I(10)$?

16. *Pathologies in Confidence Theory (II)*: Suppose the outcome $\tilde{x}$ of the cast of a die is independent of the parameter p of some Bernoulli process, and that a confidence interval $I(\tilde{x})$ for p, based upon $\tilde{x}$, is defined by

$$I(x) = \begin{cases} (1, \infty), & x = 1 \\ [0, 1], & x \in \{2, 3, 4, 5\} \\ (-\infty, 0), & x = 6. \end{cases}$$

(a) Show that $I(\tilde{x})$ is indeed a confidence interval for p at exact level $\gamma = \frac{2}{3}$.

[*Hint*: $P[I(\tilde{x})$ covers $p \mid p]$ is independent of p.]

(b) Is this confidence procedure reasonable?

17. Show that for the univariate Normal process with neither μ nor σ^2 known and μ not of interest—the case considered in Chapter 14, Exercise 12, and Chapter 20, Exercise 10—a credible interval for σ^2 at exact level γ is

$$I(n, s^2) \equiv \left[\frac{\nu_1\psi_1}{\chi^2_{\frac{1}{2}(1+\gamma)}(\nu_1)}, \frac{\nu_1\psi_1}{\chi^2_{\frac{1}{2}(1-\gamma)}(\nu_1)}\right],$$

where $\nu_1 \equiv \nu_0 + n - 1$ and $\nu_1\psi_1 \equiv \nu_0\psi_0 + (n - 1)s^2$. Show that $I(n, s^2)$ coincides with the confidence interval for σ^2 at exact level γ as given by (21.3.4) for $\nu_0 = 0$.

18. *Credible Regions for the Uniform Process*: This exercise depends heavily upon Chapter 14, Exercises 16 and 17, which should be reviewed for notation and definitions.

(a) Show the following:

A shortest credible interval at exact level γ for the parameter $\tilde{v}_2$ of the uniform process with v_1 known is (21.7.1)

$$I(n, M) = [M_1, v_1 + (M_1 - v_1)(1 - \gamma)^{-1/(\nu_1 - 1)}];$$

and

a shortest credible interval at exact level γ for the parameter $\tilde{\upsilon}_1$ of the uniform process with υ_2 known is (21.7.2)

$$I(n, m) = [\upsilon_2 - (\upsilon_2 - m_1)(1 - \gamma)^{-1/(\nu_1 - 1)}, m_1].$$

(b) From Part (c) of Chapter 14, Exercise 16, the posterior df

$$f_{ucc}(\upsilon_1, \upsilon_2 \mid m_1, M_1, \nu_1)$$

is a decreasing function of $\upsilon_2 - \upsilon_1$, and hence a minimum-area credible region for $(\tilde{\upsilon}_1, \tilde{\upsilon}_2)$ will be of the form

$\{(\upsilon_1, \upsilon_2)\colon \upsilon_1 \leqq m_1, \upsilon_2 \geqq M_1, \upsilon_2 - \upsilon_1 \leqq \text{const}\}$, as in Figure 21.8. Let $\tilde{y} \equiv \tilde{\upsilon}_2 - \tilde{\upsilon}_1$. Show that the posterior df of $\tilde{y}$ is

$$f(y) \equiv \begin{cases} \nu_1(\nu_1 + 1)(M_1 - m_1)^{\nu_1} y^{-(\nu_1+2)}(y - [M_1 - m_1]), & y > M_1 - m_1 \\ 0, & y \leqq M_1 - m_1. \end{cases}$$

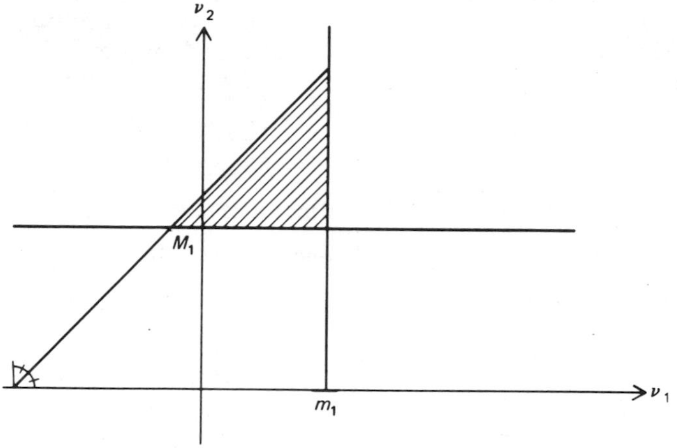

Figure 21.8

[*Hint*: Change variables in the posterior df from υ_1 and υ_2 to y and υ_2; show that

$$f(y, \upsilon_2) = \begin{cases} \nu_1(\nu_1 + 1)(M_1 - m_1)^{\nu_1} y^{-(\nu_1+2)}, & y > M_1 - m_1,\ M_1 < \upsilon_2 < y + m_1 \\ 0; & \text{elsewhere} \end{cases}$$

and integrate υ_2 out.]

(c) Show that the df of $\tilde{w} \equiv \tilde{y} - (M_1 - m_1)$ is $f_{i\beta 2}(w \mid 2, \nu_1 + 2, M_1 - m_1)$; that is (see Chapter 14, Exercise 17), the df of $\tilde{y}$ is

$$f_{\iota i\beta 2}(y \mid 2, \nu_1 + 2, M_1 - m_1; M_1 - m_1, 1).$$

(d) Show that the following is true:

A minimum-area credible region at exact level γ for the parameter (υ_1, υ_2) of the uniform process is (21.7.3)

$$I(n, m, M) \equiv \left\{(\upsilon_1, \upsilon_2)\colon \upsilon_1 \leqq m_1,\ \upsilon_2 \geqq M_1,\ \upsilon_2 - \upsilon_1 \leqq \frac{M_1 - m_1}{1 - {}_{\gamma}\beta_{(2,\ \nu_1+2)}}\right\},$$

where ${}_{\gamma}\beta_{(2,\nu_1+2)}$ is the γth fractile of the beta random variable parameters 2 and $\nu_1 + 2$.

[*Hint*: Require that

$$(w \equiv)(\upsilon_2 - \upsilon_1) - (M_1 - m_1) \leqq (M_1 - m_1)_{\gamma}(i\beta 2)_{(2,\nu_1+2,M_1-m_1)}.]$$

(e) Suppose that the *marginal* posterior credible intervals for $\tilde{\upsilon}_1$ and $\tilde{\upsilon}_2$ are desired. Show that the following obtains:

The shortest marginal credible interval at exact level γ for the parameter $\tilde{\upsilon}_2$ of the uniform process is (21.7.4)

$$I_2(n, m, M) \equiv [M_1, m_1 + (M_1 - m_1)(1 - \gamma)^{-1/(\nu_1-1)}];$$

and

the shortest marginal credible interval at exact level γ for the parameter $\tilde{\upsilon}_1$ of the uniform process is (21.7.5)

$$I_1(n, m, M) \equiv [M_1 - (M_1 - m_1)(1 - \gamma)^{-1/(\nu_1-1)}, m_1].$$

19. *Credible Regions for the Translated Exponential Process*: This exercise depends heavily upon Chapter 14, Exercises 17 and 18, which should be reviewed. Show that the following obtains:

A credible interval at exact level γ for the parameter $\tilde{\lambda}$ of the translated exponential process with Ω known is (21.7.6)

$$I(n, t) \equiv \left[\frac{{}_{\frac{1}{2}(1-\gamma)}\chi^2_{(2\rho_1)}}{2\tau_1}, \frac{{}_{\frac{1}{2}(1+\gamma)}\chi^2_{(2\rho_1)}}{2\tau_1}\right]$$

and

a shortest credible interval at exact level γ for the parameter $\tilde{\Omega}$ of the translated exponential process with λ known is (21.7.7)

$$I(n, m) \equiv [m_1 + (n_1\lambda)^{-1}\log_e(1 - \gamma), m_1].$$

[*Hint*: For (21.7.6): $\tilde{\lambda}$ has df $f_{\gamma}(\lambda \mid \rho_1, \tau_1)$ if and only if $\tilde{Q} \equiv 2\tau_1\tilde{\lambda}$ has df $f_{\chi^2}(Q \mid 2\rho_1)$.]

22

INTRODUCTION TO HYPOTHESIS TESTING

22.1 INTRODUCTION

In scientific fields of activity it has long been held that a central purpose of experimentation is to verify or refute various hypotheses, or statements of fact. There are deep philosophical arguments, dating at least as far back as Hume, concerning problems of induction, how we come to think we known certain things, and so forth; but our approach in this chapter will be considerably more concrete.

The third problem of inference broached in Section 9.7 was to use the outcome z [often expressed as $(n, \mathbf{x}_1, \ldots, \mathbf{x}_n)$] of some given experiment e as the basis for accepting or rejecting some hypothesis H_0 that $\boldsymbol{\theta}$ belongs to a given subset $\boldsymbol{\Theta}_0$ of $\boldsymbol{\Theta}$. We express this hypothesis succinctly by writing H_0: $\boldsymbol{\theta} \in \boldsymbol{\Theta}_0$. The symbol $\boldsymbol{\theta}$ as used here need not represent the value of a parameter of a family of distributions, although it will later on, when we are concerned with specific problems in hypothesis testing.

H_0: $\boldsymbol{\theta} \in \boldsymbol{\Theta}_0$ is usually referred to as the *null hypothesis*. It is natural to assume that if we reject the null hypothesis, then we must conclude only that $\boldsymbol{\theta} \notin \boldsymbol{\Theta}_0$; that is, $\boldsymbol{\theta} \in \boldsymbol{\Theta} \backslash \boldsymbol{\Theta}_0$. However, it is often useful to assume more generally that rejection of the null hypothesis leads to acceptance of the alternative hypothesis H_A: $\boldsymbol{\theta} \in \boldsymbol{\Theta}_A$, where $\boldsymbol{\Theta}_A$ is a given subset (possibly improper) of $\boldsymbol{\Theta} \backslash \boldsymbol{\Theta}_0$. We say that H_i is a *simple* hypothesis if $\boldsymbol{\Theta}_i$ consists of a single element, for $i \in \{0, A\}$. If H_i is not simple, it is called *composite*.

Example 22.1.1

Suppose that a wealth of previous information implies that a proportion defective p can either be .01 or .20, and that we wish to test the null hypothesis H_0: $p \in \{.01\}$. Rejection of H_0 should not lead to accepting the hypothesis that $p \in (0, 1)$ but $p \neq .01$; rather, because of our prior information, we should accept H_A: $p \in \{.20\}$. These hypotheses are more naturally written as H_0: $p = .01$ and H_A: $p = .20$.

In Section 22.2 we formulate the basic hypothesis testing problem as a decision and discuss some important aspects of that decision. Section 22.3 presents three different solution concepts for the testing problem, in addition to the solution concepts introduced in Section 22.2 in connection with loss and risk function.

Section 22.4 is devoted to tests of a simple null hypothesis against a simple alternative; that is, tests in which $\boldsymbol{\Theta}_0 = \{\boldsymbol{\theta}_0\}$ and $\boldsymbol{\Theta}_A = \{\boldsymbol{\theta}_A\}$.

Sections 22.5 through 22.8 concern tests of various sorts of hypotheses from the orthodox, Neyman-Pearson viewpoint presented in Section 22.3. These sections are by no means complete; there is a huge literature on hypothesis testing. Fisher [**27**], Kendall and Stuart [**41**], Wilks [**83**], and Lehmann [**45**] are excellent references; and the reader need only glance through the *Annals of Mathematical Statistics*, the *Journal of the American Statistical Association*, *Biometrika*, and the *Journal of the Royal Statistical Society* to appreciate how much theoretical work is being done in hypothesis testing. As far as applications are concerned, virtually any journal in the empirical sciences contains many papers reporting the results of experiments involving hypothesis testing.

Section 22.9 consists of general comments on some aspects of hypothesis testing.

22.2 THE COMMON STRUCTURE OF HYPOTHESIS TESTING PROBLEMS

Assume that we are required to decide between H_0: $\boldsymbol{\theta} \in \boldsymbol{\Theta}_0$ and H_A: $\boldsymbol{\theta} \in \boldsymbol{\Theta}_A$ on the basis of the outcome $z \in Z(e)$ of a given experiment e. We may *not* decide that (e, z) is so inconclusive that we will conduct another experiment; this option is considered in the subject of *sequential* testing of hypotheses, or *sequential analysis*, which will not be considered in this book. Mood and Graybill [**55**] and Wilks [**83**] furnish introductions to sequential analysis, which was introduced by Wald [**81**], [**82**].

Since a decision between H_0 and H_A will be required on the basis of the as yet unobserved outcome $z \in Z(e)$, we shall have to adopt a decision rule which tells us whether to accept or reject H_0, given outcome z, for each $z \in Z(e)$. Let a_i denote the statement "H_i: $\boldsymbol{\theta} \in \boldsymbol{\Theta}_i$ is true," for $i \in \{0, A\}$. Then a_0 amounts to accepting H_0: $\boldsymbol{\theta} \in \boldsymbol{\Theta}_0$, and a_A amounts to rejecting H_0. We must then choose a *decision function* d: $Z(e) \rightarrow \{a_0, a_A\}$. By the definition of "function," our choice of d prescribes our action unambiguously for each possible outcome $z \in Z(e)$. [There are many possible decision functions. The reader may verify that if $Z(e)$ contains n elements, then there are 2^n decision functions.]

In the hypothesis testing context, a decision function d is referred to as a *test*. Each test d induces a partition $\{Z(e)\backslash C_d, C_d\}$ of $Z(e)$, where $C_d \equiv \{z: z \in Z(e), d(z) = a_A\}$. Clearly, C_d is the set of possible outcomes of e for which the test d will reject H_0; it is called the *critical region*. The set $Z(e)\backslash C_d$ is called the *acceptance region*. [The correspondence between decision functions d and subsets C of $Z(e)$ is a bijection, as the reader may prove in Exercise 1. Thus knowledge of the critical region C_d of a test d is equivalent to knowledge of d itself.]

We have as yet said nothing about how to choose one from among the plethora of tests which are usually available in realistic problems. That subject is the concern of the next section; but all choice procedures there are based at least in part upon the probabilities $P(z \mid \boldsymbol{\theta}, e)$ [or dgf's $f(z \mid \boldsymbol{\theta}, e)$] of the outcomes, conditional upon $\boldsymbol{\theta}$ (and, of course, upon e), which will always be assumed given.

From the conditional probability distributions on $Z(e)$ given $\boldsymbol{\theta}$, it is (usually) a relatively easy matter to compute $P(C_d \mid \boldsymbol{\theta}, e)$ for any given $\boldsymbol{\theta}$ and d. The function $P(C_d \mid \cdot, e)$: $\boldsymbol{\Theta} \rightarrow [0, 1]$ is called the *power function* of the test d. Its complement, the function $[1 - P(C_d \mid \cdot, e)]$: $\boldsymbol{\Theta} \rightarrow [0, 1]$, is called the *operating characteristic* of d. Thus the value of the power function at a given $\boldsymbol{\theta}$ is simply the probability that d will reject H_0 if $\boldsymbol{\theta}$ is true, while the value of the operating characteristic is the probability that d will *accept* H_0 if $\boldsymbol{\theta}$ is true.

Naturally, it is good to accept H_0 when H_0 is true (and H_A is false) and to reject H_0 when H_0 is false (and H_A is true), while it is bad to erroneously reject or to erroneously accept H_0. An *error of the first kind*, or a *type I error*, is the erroneous rejection of H_0; an *error of the second kind*, or a *type II error*, is the erroneous acceptance of H_0. We define the *probability of type I error* by

$$\epsilon_I(\boldsymbol{\theta} \mid e, d) \equiv \begin{cases} P(C_d \mid \boldsymbol{\theta}, e), & \boldsymbol{\theta} \in \boldsymbol{\Theta}_0 \\ 0, & \boldsymbol{\theta} \in \boldsymbol{\Theta}_A, \end{cases} \tag{22.2.1}$$

and the *probability of type II error* by

$$\epsilon_{II}(\boldsymbol{\theta} \mid e, d) \equiv \begin{cases} 0, & \boldsymbol{\theta} \in \boldsymbol{\Theta}_0 \\ 1 - P(C_d \mid \boldsymbol{\theta}, e), & \boldsymbol{\theta} \in \boldsymbol{\Theta}_A. \end{cases} \tag{22.2.2}$$

Note that when $\boldsymbol{\theta} \in \boldsymbol{\Theta}_A$, a type I error is impossible because it is not erroneous to reject H_0: $\boldsymbol{\theta} \in \boldsymbol{\Theta}_0$, and hence $\epsilon_I(\boldsymbol{\theta} \mid e, d) = 0$ for $\boldsymbol{\theta} \in \boldsymbol{\Theta}_A$—similarly with $\epsilon_{II}(\boldsymbol{\theta} \mid e, d) = 0$ for $\boldsymbol{\theta} \in \boldsymbol{\Theta}_0$. Also note that the error probabilities (22.2.1) and (22.2.2) are definable in terms of the power function, or equivalently, in terms of the operating characteristic.

Before giving some examples, we define one more concept. The *error characteristic* $\epsilon(\cdot \mid e, d): \Theta_0 \cup \Theta_A \rightarrow [0, 1]$ of d is given by

$$\epsilon(\theta \mid e, d) = \epsilon_I(\theta \mid e, d) \rightarrow \epsilon_{II}(\theta \mid e, d), \tag{22.2.3}$$

for every $\theta \in \Theta_0 \cup \Theta_A$. Note that at most one of the summands on the right-hand side of (22.2.3) will be positive. The error characteristic coincides with the power function for $\theta \in \Theta_0$, and with the operating characteristic for $\theta \in \Theta_A$.

Example 22.2.1

The definitive experiment. Suppose that $Z(e) = \{z_1, z_2\}$, $\Theta = \Theta_0 \cup \Theta_A$, and

$$P(z_1 \mid \theta, e) = \begin{cases} 1, & \theta \in \Theta_0 \\ 0, & \theta \in \Theta_A. \end{cases}$$

There are four possible tests, $d_1, \ldots, d_4$, with corresponding critical regions $C_1, \ldots, C_4$ given by $C_1 \equiv \{z_1, z_2\}$, $C_2 \equiv \{z_1\}$, $C_3 \equiv \varnothing$, and $C_4 \equiv \{z_2\}$. The corresponding power functions are

$$P(C_1 \mid \theta, e) = 1, \text{ for every } \theta \in \Theta,$$

$$P(C_2 \mid \theta, e) = \begin{cases} 1, & \theta \in \Theta_0 \\ 0, & \theta \in \Theta_A, \end{cases}$$

$$P(C_3 \mid \theta, e) = 0 \text{ for every } \theta \in \Theta,$$

and

$$P(C_4 \mid \theta, e) = \begin{cases} 0, & \theta \in \Theta_0 \\ 1, & \theta \in \Theta_A. \end{cases}$$

The corresponding error characteristics are

$$\epsilon(\theta \mid e, d_1) = \begin{cases} 1, & \theta \in \Theta_0 \\ 0, & \theta \in \Theta_A \end{cases}$$

$$\epsilon(\theta \mid e, d_2) = 1, \text{ for every } \theta \in \Theta,$$

$$\epsilon(\theta \mid e, d_3) = \begin{cases} 0, & \theta \in \Theta_0 \\ 1, & \theta \in \Theta_A, \end{cases}$$

and

$$\epsilon(\theta \mid e, d_4) = 0, \text{ for every } \theta \in \Theta.$$

Note that test d_2 is sure to prescribe the wrong course of action and test d_4 is sure to prescribe the right course of action, which was to be expected: since z_1 is a sure signal of Θ_0 and z_2 is a sure signal of Θ_A, it stands to reason that H_0 should be rejected if and only if z_2 occurs. Note also that tests d_3 and d_1 work well for $\boldsymbol{\theta} \in \boldsymbol{\Theta}_0$ and $\boldsymbol{\theta} \in \boldsymbol{\Theta}_A$, respectively but are poor elsewhere.

Example 22.2.2

Unfortunately, few experiments are definitive. Suppose e consists of one observation x with df $f_e(x \mid \lambda) \equiv \lambda \exp(-\lambda x)$ for $x > 0$, and let H_0: $\lambda \geqq \lambda_0$ and H_A: $\lambda < \lambda_0$, for some $\lambda_0 > 0$. [That is, $\boldsymbol{\Theta}_0 = [\lambda_0, \infty)$ and $\boldsymbol{\Theta}_A = (0, \lambda_0)$.] Let d_δ denote the test with critical region (δ, ∞). The power function is given by $P(\tilde{x} \in [\delta, \infty) \mid \lambda) = \exp(-\lambda\delta)$ for every $\lambda > 0$ and $\delta > 0$, and the error characteristic by

$$\epsilon(\lambda \mid e, d_\delta) = \begin{cases} \exp(-\lambda\delta), & \lambda \geqq \lambda_0 \\ 1 - \exp(-\lambda\delta), & \lambda < \lambda_0. \end{cases}$$

Error characteristics for three different values of δ are graphed in Figure 22.1. Note that for each δ, we have $\epsilon_I(\lambda_0 \mid e, d_\delta) + \lim_{\lambda \to \lambda_0 -} \epsilon_{II}(\lambda \mid e, d_\delta) = 1$. (Why?)

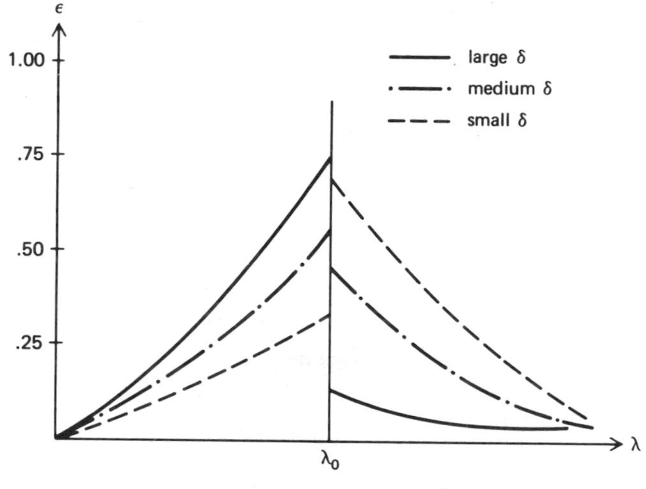

Figure 22.1

Note that setting δ at a large (positive) value drives down the probability of type I error—but at the cost of increasing the probability of type II error.

Example 22.2.3

In the preceding example it is intuitively clear that the best tests are of the form d_δ as defined there. Suppose, however, that we consider tests of the form d^*_δ with associated critical regions $(0, \delta]$. The power function of such a test d^*_δ is given by $P(\tilde{x} \in (0, \delta] \mid \lambda) = 1 - \exp(-\lambda\delta)$, for every $\lambda > 0$ and $\delta > 0$, and the error characteristic by

$$\epsilon(\lambda \mid e, d^*_\delta) = \begin{cases} 1 - \exp(-\lambda\delta), & \lambda \geqq \lambda_0 \\ \exp(-\lambda\delta), & \lambda < \lambda_0. \end{cases}$$

The error characteristic for a given $\delta > 0$ constitutes the solid line in Figure 22.2; the dotted line will be explained presently. We again have $\epsilon_I(\lambda_0 \mid e, d^*_\delta) + \lim_{\lambda \to \lambda_0 -} \epsilon_{II}(\lambda \mid e, d^*_\delta) = 1$. To show the inferiority of the tests d^*_δ, we shall find for any given d^*_δ a δ' such that, *first*, $\epsilon(\lambda \mid e, d_{\delta'}) \leqq \epsilon(\lambda \mid e, d^*_\delta)$, for every λ; and *second*, this inequality is strict for some values of λ. Hence the error probability with $d_{\delta'}$ is at least as small as (and sometimes smaller than) the error probability with d^*_δ. The dotted line in Figure 22.2 represents $\epsilon(\cdot \mid e, d_{\delta'})$. Let $\delta' = -\lambda_0^{-1} \log_e(1 - \exp(-\lambda_0\delta))$, so that $\exp(-\lambda_0\delta') = 1 - \exp(-\lambda_0\delta)$ and hence d^*_δ and $d_{\delta'}$ have the same error characteristic at λ_0. Since $\exp(-\lambda\delta')$ and $1 - \exp(-\lambda\delta)$ are respectively decreasing and increasing in λ, it follows that $\epsilon(\lambda \mid e, d_{\delta'}) < \epsilon(\lambda \mid e, d^*_\delta)$ for all $\lambda > \lambda_0$. But

$$\lim_{\lambda \to \lambda_0 -} \epsilon(\lambda \mid e, d) = 1 - \epsilon(\lambda_0 \mid e, d),$$

for both $d = d^*_\delta$ and $d = d_{\delta'}$, and hence (by definition of δ')

$$\lim_{\lambda \to \lambda_0 -} \epsilon(\lambda \mid e, d_{\delta'}) = \lim_{\lambda \to \lambda_0 -} \epsilon(\lambda \mid e, d^*_\delta).$$

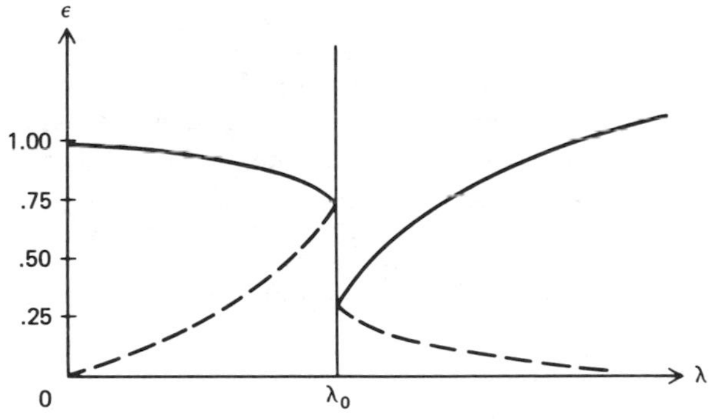

Figure 22.2

Since $\epsilon(\cdot \mid e, d_{\delta'})$ is increasing and $\epsilon(\cdot \mid e, d^*_\delta)$ is decreasing in λ for $\lambda < \lambda_0$, it follows that $\epsilon(\lambda \mid e, d_{\delta'}) < \epsilon(\lambda \mid e, d^*_\delta)$ for every $\lambda < \lambda_0$. Thus $d_{\delta'}$ has lower error probability than d^*_δ unless $\lambda = \lambda_0$, in which case the error probabilities are the same. Hence every test with critical region of the form $(0, \delta]$ is worse than some test with critical region of the form $[\delta, \infty)$.

As a preview of future developments, we remark at this point that the inferior tests with critical regions of the form $(0, \delta]$ in Example 22.2.3 are inadmissible (see Example 22.2.4). The concept of admissibility depends upon consequences of introducing a loss function, which we shall now examine.

Suppose that a loss function ℓ: $\{a_0, a_A\} \times \boldsymbol{\Theta} \to R^1$ is given or derivable from a utility function. Usually such a loss function will satisfy

$$\ell(a_i, \boldsymbol{\theta})\begin{cases} =0, & \boldsymbol{\theta} \in \boldsymbol{\Theta}_i \\ >0, & \boldsymbol{\theta} \notin \boldsymbol{\Theta}_i, \end{cases} \tag{22.2.4}$$

for $i \in \{0, A\}$, which reflects the appropriateness of a_i if and only if $\boldsymbol{\theta} \in \boldsymbol{\Theta}_i$. As in Section 20.2, a *risk function* $R(d, \boldsymbol{\theta})$ may be defined by

$$R(d, \boldsymbol{\theta}) \equiv E(\ell(d(\tilde{z}), \boldsymbol{\theta}) \mid \boldsymbol{\theta}, e), \tag{22.2.5}$$

for every $\boldsymbol{\theta} \in \boldsymbol{\Theta}$ and every test d. [We use the tilde even though $Z(e)$ may not be a numerical set, in order to emphasize that the expectation is *not* with respect to $\boldsymbol{\theta}$.]

With the definition (22.2.5) of risk function, the choice between tests (or between their corresponding critical regions) is guided by the same considerations as was the choice between estimators in Section 20.2 on the basis of $R(\tilde{\hat{\boldsymbol{\theta}}}, \boldsymbol{\theta})$.

In particular, if there exists a test d^0 such that

$$R(d^0, \boldsymbol{\theta}) \leqq R(d, \boldsymbol{\theta}), \text{ for every } \boldsymbol{\theta} \in \boldsymbol{\Theta} \tag{22.2.6}$$

and every d, then everyone would agree that d^0 is the best test (given experiment e, notation for which has been suppressed). Unfortunately, such best tests d^0 rarely exist. More usually, some tests are better, in the sense of offering lower risk, for some $\boldsymbol{\theta}$'s while other tests are better for other $\boldsymbol{\theta}$'s. Figure 20.2 remains appropriate here, but with $\tilde{\hat{\boldsymbol{\theta}}}$'s replaced by d's. A *minimax* test d^0 satisfies

$$\max_{\theta} R(d^0, \boldsymbol{\theta}) = \min_{d} \max_{\theta} R(d, \boldsymbol{\theta}); \tag{22.2.7}$$

and a *Bayes* test d^0 satisfies

$$B(d^0) = \min_{d} B(d), \tag{22.2.8}$$

where

$$B(d) \equiv E\{R(d, \tilde{\boldsymbol{\theta}})\}, \tag{22.2.9}$$

the expectation being with respect to the prior distribution of $\tilde{\boldsymbol{\theta}}$ (tilded even though possibly nonnumerical).

Finally, a test d is *inadmissible* (relative to $\Theta_0 \cup \Theta_A$) if there exists another test d^* such that

$$R(d^*, \theta)\begin{cases} \leq R(d, \theta), & \text{every } \theta \in \Theta_0 \cup \Theta_A \\ < R(d, \theta), & \text{some } \theta \in \Theta_0 \cup \Theta_A; \end{cases} \qquad (22.2.10)$$

while d is *admissible* if it is not inadmissible. The search for minimax and Bayes tests can be restricted to the set of all admissible tests.

Example 22.2.4

Let

$$\ell(a_i, \theta) = \begin{cases} 0, & \theta \in \Theta_i \\ 1, & \theta \notin \Theta_i, \end{cases} \qquad (22.2.11)$$

for $i \in \{0, A\}$. The reader may verify as Exercise 2 that

$$R(d, \theta) = \epsilon(\theta \mid e, d),$$

for every $\theta \in \Theta_0 \cup \Theta_A$ and every test d. Thus the tests introduced in Example 22.2.3 are inadmissible for the loss function of (22.2.11).

Example 22.2.5

For the loss function (22.2.11) and the problem in Example 22.2.2, it is clear from the form of $\epsilon(\lambda \mid e, d_\delta)$ that if all tests of the form d_δ are admissible (which can be proved), then a minimax test d^0 satisfies $\epsilon(\lambda_0 \mid e, d^0) = \frac{1}{2}$. Let $d^0 \equiv [\delta^0, \infty)$. Then from $\exp(-\lambda_0\delta^0) = \frac{1}{2}$ we conclude that $\delta^0 = \log_e(2)/\lambda_0$.

Example 22.2.6

Suppose in Example 22.2.2 that the loss function is given by

$$\ell(a_i, \lambda) - \begin{cases} 0, & \lambda \in \Lambda_i \\ k_i|\lambda - \lambda_0|, & \lambda \notin \Lambda_i, \end{cases} \qquad (22.2.12)$$

for $i \in \{0, A\}$, where $\Lambda_0 \equiv [\lambda_0, \infty)$ and $\Lambda_A \equiv (0, \infty)\backslash\Lambda_0$. Then

$$R(d, \lambda) = k_0(\lambda_0 - \lambda)\epsilon_{II}(\lambda \mid e, d) + k_A(\lambda - \lambda_0)\epsilon_I(\lambda \mid e, d), \qquad (22.2.13)$$

for every $\lambda > 0$ and test d. Figure 22.3 depicts $R(\cdot, d_\delta)$ for each of the three tests d_δ whose error characteristics were graphed in Figure 22.1.

The reader will note that choice between tests is no easier now than in Figure 22.2; lower risk for H_0 true can be obtained only at the expense of higher risk for H_0 false.

Equation (22.2.13) is easily generalized to yield an expression for $R(d, \theta)$ in terms of $\epsilon(\theta \mid e, d)$ for any testing problem.

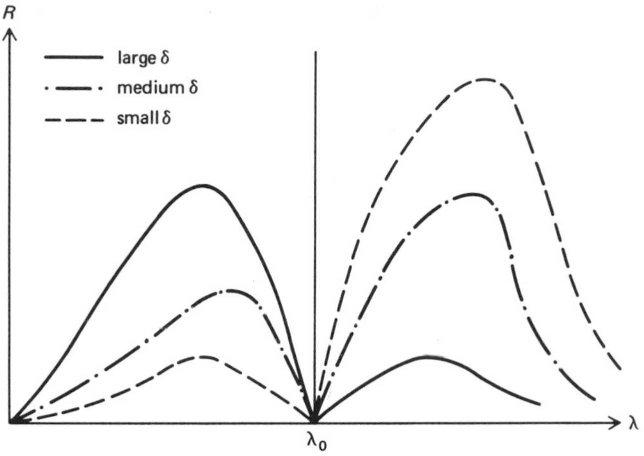

Figure 22.3

If $\ell(a, \boldsymbol{\theta})$ satisfies (22.2.4), then

$$R(d, \boldsymbol{\theta}) = \ell(a_0, \boldsymbol{\theta})\epsilon_{II}(\boldsymbol{\theta} \mid e, d) + \ell(a_A, \boldsymbol{\theta})\epsilon_I(\boldsymbol{\theta} \mid e, d)$$
$$= \begin{cases} \ell(a_A, \boldsymbol{\theta})\epsilon_I(\boldsymbol{\theta} \mid e, d), & \boldsymbol{\theta} \in \boldsymbol{\Theta}_0 \\ \ell(a_0, \boldsymbol{\theta})\epsilon_{II}(\boldsymbol{\theta} \mid e, d), & \boldsymbol{\theta} \in \boldsymbol{\Theta}_A, \end{cases} \tag{22.2.14}$$

for every test d and $\boldsymbol{\theta} \in \boldsymbol{\Theta}_0 \cup \boldsymbol{\Theta}_A$.

Proof

Exercise 3.

Generalization (22.2.14) implies the following important fact:

If $\ell(a, \boldsymbol{\theta})$ satisfies (22.2.4), then d is admissible with respect to $R(d, \boldsymbol{\theta})$ if and only if d is admissible with respect to $\epsilon(\boldsymbol{\theta} \mid e, d)$ [derivable from the special loss function (22.2.11) via (22.2.5)]. (22.2.15)

Proof

We shall show that d is inadmissible with respect to R if and only if it is inadmissible with respect to ϵ. From (22.2.14) it follows readily that $R(d^*, \boldsymbol{\theta}) \leqq R(d, \boldsymbol{\theta})$, for every $\boldsymbol{\theta} \in \boldsymbol{\Theta}_0$ if and only if $\ell(a_A, \boldsymbol{\theta})\epsilon_I(\boldsymbol{\theta} \mid e, d^*) \leqq \ell(a_A, \boldsymbol{\theta})\epsilon_I(\boldsymbol{\theta} \mid e, d)$, which [because $\ell(a_A, \boldsymbol{\theta}) > 0$ for every $\boldsymbol{\theta} \in \boldsymbol{\Theta}_0$] is equivalent to $\epsilon_I(\boldsymbol{\theta} \mid e, d^*) \leqq \epsilon_I(\boldsymbol{\theta} \mid e, d^*)$, for every $\boldsymbol{\theta} \in \boldsymbol{\Theta}_0$. Similar reasoning for $\boldsymbol{\theta} \in \boldsymbol{\Theta}_A$ establishes that

$$R(d^*, \boldsymbol{\theta}) \leqq R(d, \boldsymbol{\theta}), \text{ for every } \boldsymbol{\theta} \in \boldsymbol{\Theta}_0 \cup \boldsymbol{\Theta}_A$$

if and only if

$$\epsilon(\boldsymbol{\theta} \mid e, d^*) \leqq \epsilon(\boldsymbol{\theta} \mid e, d), \text{ for every } \boldsymbol{\theta} \in \boldsymbol{\Theta}_0 \cup \boldsymbol{\Theta}_A.$$

Now suppose there exists $\boldsymbol{\theta}^* \in \boldsymbol{\Theta}_0$ such that $R(d^*, \boldsymbol{\theta}^*) < R(d, \boldsymbol{\theta}^*)$; it follows as previously that $\epsilon_I(\boldsymbol{\theta}^* \mid e, d^*) < \epsilon_I(\boldsymbol{\theta}^* \mid e, d)$. Similarly, $\boldsymbol{\theta}^* \in \boldsymbol{\Theta}_A$ and $R(d^*, \boldsymbol{\theta}^*) < R(d, \boldsymbol{\theta}^*)$ imply $\epsilon_{II}(\boldsymbol{\theta}^* \mid e, d^*) < \epsilon_{II}(\boldsymbol{\theta}^* \mid e, d)$. Thus inadmissibility of d with respect to R implies its inadmissibility with respect to ϵ. As Exercise 4, the reader may use (22.2.3) and reverse the roles of R and ϵ in the sentence beginning "Now suppose . . ." to prove the reverse implication. QED

The importance of (22.2.15) stems from the fact that a great deal of attention has been paid to the power functions of tests, from which their error characteristics are derivable via (22.2.1) through (22.2.3), and (22.2.15) shows that all results about admissibility with respect to error characteristics also apply verbatim to admissibility with respect to any risk function derived from a loss function satisfying (22.2.4).

Under quite general conditions, the search for admissible tests can be confined to those tests d which can be expressed as functions of any sufficient statistics. By analogy with Chapter 20, we might call such tests *sufficient tests*. The argument which establishes the result is somewhat involved and is sketched in Lehmann [**45**] 16–21; a complete development is given in Wald [**82**].

Thus, if e consists in taking n iid observations with noninformative stopping from $f(x \mid \theta)$, and hence $Z(e)$ is a set of points $(x_1, \ldots, x_n)$ in R^n, and if $\mathbf{x}_T$ is sufficient for θ, then the search for admissible tests can be confined to functions of $\mathbf{x}_T$ and, the dgf $f(\mathbf{x}_T \mid \theta)$ can be used in place of $f(x_1, \ldots, x_n) \mid \theta)$ for computing power functions, and so forth. In particular, admissible tests of hypotheses about the mean μ of a univariate Normal process with known variance σ^2 are functions of $\bar{x}$, and the power functions of such tests are derivable from $f_N(\bar{x} \mid \mu, \sigma^2/n)$. The reader will appreciate the simplification possible when n is very large (say, over 100).

22.3 THREE PROCEDURES FOR CHOOSING A TEST

The preceding section furnished two procedures for choosing a test d of H_0: $\boldsymbol{\theta} \in \boldsymbol{\Theta}_0$ versus H_A: $\boldsymbol{\theta} \in \boldsymbol{\Theta}_A$ on the basis of the outcome z of experiment e. This section furnishes three other procedures, not explicitly based upon loss and risk function.

The first procedure, introduced in Subsection 22.3.1, is orthodox; the power functions or error characteristics of tests play a dominant role, and prior probabilities or dgf's are not introduced. Orthodox tests almost completely dominate current statistical practice.

The second procedure is a Bayesian modification of the orthodox approach, while the third procedure is another Bayesian approach to hypothesis testing. These two Bayesian versions are introduced in Subsections 22.3.2 and 22.3.3, respectively and rely upon prior probabilities and dgf's, but not upon loss functions. Hence they are not to be confused with Bayes tests as defined in Section 22.2; those tests might be called "fully Bayesian," as they utilize the entire machinery of Bayesian decision theory as discussed in Part III.

22.3.1 Orthodox Choice Procedures: the Neyman-Pearson Framework

The treatment of H_0 and H_A and of the probabilities of types I and II error is asymmetric in the orthodox framework. It is assumed that special care must be taken to guard against a type-I-error, and hence that $\epsilon_I(\boldsymbol{\theta} \mid e, d)$ must be controlled.

The first step in the orthodox procedure is to select an $\alpha \in (0, 1)$, called the *significance level*; α is to serve as an upper bound for $\epsilon_I(\boldsymbol{\theta} \mid e, d)$, and any test d such that $\epsilon_I(\boldsymbol{\theta} \mid e, d) > \alpha$ for some $\boldsymbol{\theta} \in \boldsymbol{\Theta}_0$ will be eliminated from consideration. One test that is never eliminated for any α is the test d_0 with empty critical region and $\epsilon_I(\boldsymbol{\theta} \mid e, d_0) = 0$ on $\boldsymbol{\Theta}_0$, and hence the imposition of the constraint α on type-I-error probability never eliminates all tests.

More usually, there are still a great many test d's satisfying $\epsilon_I(\boldsymbol{\theta} \mid e, d) \leqq \alpha$ for every $\boldsymbol{\theta} \in \boldsymbol{\Theta}_0$. To choose between them, we try to maximize the power of the test for $\boldsymbol{\theta} \in \boldsymbol{\Theta}_A$; or equivalently, to minimize $\epsilon_{II}(\boldsymbol{\theta} \mid e, d)$.

Example 22.3.1

Let $\boldsymbol{\theta} = \{\theta_1, \theta_2\}$, $\boldsymbol{\Theta}_0 = \{\theta_1\}$, $Z(e) = \{z_1, z_2\}$, and assume the following table of conditional probabilities $P(z \mid \boldsymbol{\theta}, e)$:

$\boldsymbol{\theta}$ \ z	z_1	z_2	*sum*
θ_1	.999	.001	1.000
θ_2	.002	.998	1.000

There are four possible tests: d_1 with critical region $\{z_1, z_2\} \equiv C_1$; d_2 with critical region $\{z_1\} \equiv C_2$; d_3 with critical region $\{z_2\} \equiv C_3$; and d_4 with empty critical region. The power functions are graphed in Figure 22.4. Note that d_1 is ruled out by any $\alpha < 1$; that d_2 is ruled out by any $\alpha < .999$; that d_3 is ruled out by any $\alpha < .001$; and that d_4 is never ruled out. Now suppose that due consideration of the relative seriousness of the types I and II errors has lead the decision-maker to choose $\alpha = .01$. Then the

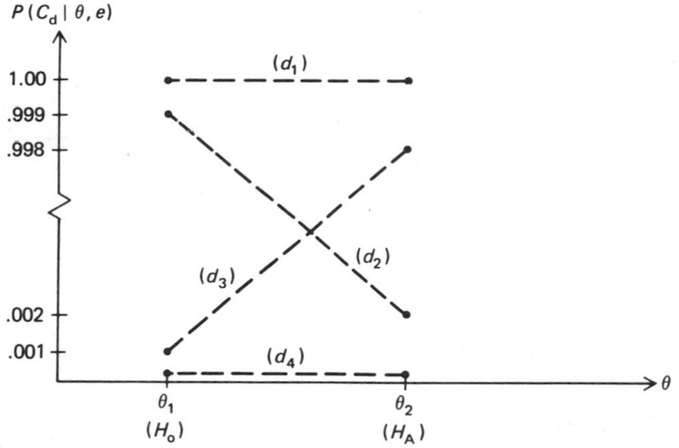

Figure 22.4

only eligible tests are d_3 and d_4, and the power of d_3 is clearly greater than d_4, so that d_3 will be chosen. This is in line with common sense, since z_i is a rather clear signal of θ_i, for $i = 1, 2$.

The orthodox procedure can be summarized as consisting of three steps:

(1) choose $\alpha \in (0, 1)$;
(2) determine $\mathfrak{D}(\alpha) \equiv \{d\colon \epsilon_I(\theta \mid e, d) \leqq \alpha$, for every $\theta \in \Theta_0 \cup \Theta_A\}$; and
(3) choose $d \in \mathfrak{D}(\alpha)$ in order to minimize $\epsilon_{II}(\theta, e, d)$, for $\theta \in \Theta_A$.

The perceptive reader will note a difficulty in step (3) when $H_A\colon \theta \in \Theta_A$ is not simple; that is, when Θ_A consists of more than one element. Step (3) is impossible to execute when $\Theta_A \equiv \{\theta_a, \theta_b\}$, $\mathfrak{D}(\alpha) = \{d', d''\}$, $\epsilon_{II}(\theta_a \mid e, d') > \epsilon_{II}(\theta_a \mid e, d'')$, and $\epsilon_{II}(\theta_b \mid e, d') < \epsilon_{II}(\theta_b \mid e, d'')$, since d'' is better than d' for $\theta_a \in \Theta_A$ while the reverse is true for $\theta_b \in \Theta_A$.

If there exists a test $d^* \in \mathfrak{D}(\alpha)$ such that $\epsilon_{II}(\theta \mid e, d^*) \leqq \epsilon_{II}(\theta \mid e, d)$, for every $\theta \in \Theta_A$ and every $d \in \mathfrak{D}(\alpha)$, then d^* is called a *uniformly most powerful test at (significance) level* α of $H_0\colon \theta \in \Theta_0$ versus $H_A\colon \theta \in \Theta_A$. The words "uniformly most powerful" are often abbreviated to UMP.

Example 22.3.2

In Example 22.3.1, d_3 is UMP at level α, for every $\alpha \in [.001, 999]$, since for such α we have $\mathfrak{D}(\alpha) = \{d_3, d_4\}$, $\Theta_A = \{\theta_2\}$, and

$$\epsilon_{II}(\theta_2 \mid e, d_3) = .002 < 1.000 = \epsilon_{II}(\theta_2 \mid e, d_4).$$

UMP tests usually (but not always) exist when H_A is simple. The exceptions arise because of slight mathematical pathologies.

Example 22.3.3

Suppose H_A: $\boldsymbol{\theta} \in \boldsymbol{\Theta}_A$ is simple with $\boldsymbol{\Theta}_A \equiv \{\boldsymbol{\theta}_A\}$, that $\epsilon_{II}(\boldsymbol{\theta}_A \mid e, d) > .2$ for every $d \in \mathfrak{D}(\alpha)$, and that for every $p < 1$ there exists a test d_p such that $\epsilon_{II}(\boldsymbol{\theta}_A \mid e, d_p) = 1 - .8p$. We would like to find a test d^* such that $\epsilon_{II}(\boldsymbol{\theta}_A \mid e, d^*) = .2$, but that is impossible (by assumption). We can, however, find tests d with $\epsilon_{II}(\boldsymbol{\theta}_A \mid e, d)$ arbitrarily close to, but strictly exceeding, .2.

The nonexistence of UMP tests has given rise to an *ad hoc* modification of step (3):

(3a) choose a specific element $\boldsymbol{\theta}_A \in \boldsymbol{\Theta}_A$; and
(3b) choose $d \in \mathfrak{D}(\alpha)$ so as to minimize $\epsilon_{II}(\boldsymbol{\theta}_A \mid e, d)$.

Steps (3a) and (3b) amount essentially to replacing a composite alternative hypothesis H_A: $\boldsymbol{\theta} \in \boldsymbol{\Theta}_A$ with a simple alternative H'_A: $\boldsymbol{\theta} = \boldsymbol{\theta}_A$, for some $\boldsymbol{\theta}_A \in \boldsymbol{\Theta}_A$. The difficulty is in choosing $\boldsymbol{\theta}_A$, and that choice does affect the final selection of a test d unless an UMP test exists.

Before introducing another property of some tests, we pause to examine just what it *means* to choose α and a test $d \in \mathfrak{D}(\alpha)$, to observe $z \in C_d$, and to reject H_0: $\boldsymbol{\theta} \in \boldsymbol{\Theta}_0$. This sequence of events is often expressed in practice by saying, "the outcome z is significant at the α level." Since $d \in \mathfrak{D}(\alpha)$, it follows that if z falls in C_d, then one of two things must be true: *either* an event (namely, C_d) of probability not exceeding α has occurred and H_0 is true, *or* H_0 is false.

Example 22.3.4

In Example 22.3.1, with $d = d_3$ and $C_d = \{z_2\}$, suppose that experiment e is conducted and the outcome is z_2. Then *either* H_0: $\boldsymbol{\theta} = \boldsymbol{\theta}_1$ is true and an event of probability .001 (a virtual miracle) has occurred, *or* H_0 is false, H_A: $\boldsymbol{\theta} = \boldsymbol{\theta}_2$ is true, and an event of probability .998 has occurred. The theory tacitly decides in favor of H_A when continued belief in H_0 would necessitate belief in the occurrence of "virtual miracles," which can be implicitly defined as events of probability not exceeding α.

Unfortunately, virtual miracles do occur. The next example shows that it is possible *both* for H_0 to be false *and* for the critical region to have probabilily not exceeding α.

Example 22.3.5

Let $\boldsymbol{\Theta} = \{\boldsymbol{\theta}_1, \boldsymbol{\theta}_2\}$, $\boldsymbol{\Theta}_0 = [\boldsymbol{\theta}_1\}$, $Z(e) = \{z_1, z_2\}$, and assume the following table of conditional probabilities $P(z \mid \boldsymbol{\theta}, e)$:

θ \ z	z_1	z_2	*sum*
θ_1	.9990	.0010	1.000
θ_2	.9995	.0005	1.000

Let $\alpha \in [.001, 999)$, and define the tests and critical regions as in Example 22.3.1. For every such α, we again have $\mathfrak{D}(\alpha) = \{d_3, d_4\}$ and d_3 has lower error probability (and higher power) than d_4, given θ_2. But $\epsilon_{II}(\theta_2 \mid e, d_3) = .9995$ and $P(C_3 \mid \theta_2, e) = .0005$. *Hence the event C_3 is even more miraculous when H_0 is false than when H_0 is true.* The power functions of d_1 through d_4 appear in Figure 22.5, which should be compared with Figure 22.4.

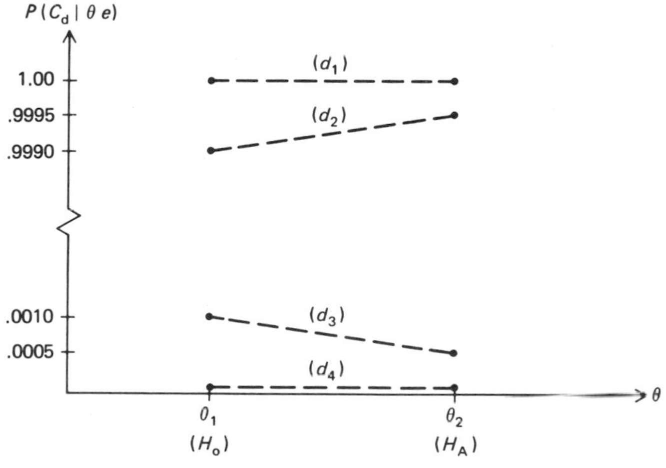

Figure 22.5

Every statistician, regardless of his philosophical persuasion, would agree that e in Example 22.3.5 is an extremely poor experiment in comparison with the almost definitive experiment in Example 22.3.1. Nevertheless, it dramatically illustrates a difficulty in the interpretation of "significant outcomes": the critical region may be even less likely when H_0 is false.

Tests which do not possess this disagreeable feature are called unbiased. Formally, a test H_0: $\theta \in \Theta_0$ versus H_A: $\theta \in \Theta_A$ at level α is *unbiased* (at level α) if (1) $P(C_d \mid \theta, e) \leqq \alpha$, for every $\theta \in \Theta_0$, and (2) $P(C_d \mid \theta, e) > \alpha$, for every $\theta \in \Theta_A$; or equivalently, if (1′) $\epsilon(\theta \mid e, d) \leqq \alpha$, for every $\theta \in \Theta_0$, and

(2′) $\epsilon(\boldsymbol{\theta} \mid e, d) < 1 - \alpha$, for every $\boldsymbol{\theta} \in \boldsymbol{\Theta}_A$. The stipulations (1) and (1′) merely state that d belongs to $\mathfrak{D}(\alpha)$, while (2) and (2′) state that the rejection of H_0 is more likely when H_0 is false.

Existence of an unbiased test at one level α does not imply existence of an unbiased test at another level. In Example 22.3.5, no unbiased test at any level $\alpha < .9990$ exists, but d_2 is an eligible, unbiased test if $\alpha \in [.9990, 1)$, as the reader may verify.

Because of the ambiguity in interpreting significant results of biased tests, it is sometimes desirable to restrict attention to the subset $\mathfrak{D}_U(\alpha)$ and $\mathfrak{D}(\alpha)$ consisting of all unbiased tests at level α. When such a restriction is made, it may happen that a uniformly most powerful test can be found while none can be found when all tests in $\mathfrak{D}(\alpha)$ are candidates. An unbiased test which is uniformly more powerful than any other unbiased test is called *uniformly most powerful unbiased* (*UMPU*).

Example 22.3.6

Let $\boldsymbol{\Theta} \equiv \{\boldsymbol{\theta}_1, \boldsymbol{\theta}_2, \boldsymbol{\theta}_3\}$, $\boldsymbol{\Theta}_0 = \{\boldsymbol{\theta}_1\}$, $Z(e) = \{z_1, z_2, z_3\}$ and assume the following table of conditional probabilities $P(z \mid \boldsymbol{\theta}, e)$:

$\boldsymbol{\theta}$ \ z	z_1	z_2	z_3	*sum*
$\boldsymbol{\theta}_1$	.001	.002	.997	1.000
$\boldsymbol{\theta}_2$	.010	.020	.970	1.000
$\boldsymbol{\theta}_3$	.970	.001	.029	1.000

Suppose that $\alpha < .997$, so that no $d \in D(\alpha)$ has $z_3 \in C_d$. Define $d_1 - d_4$ and $C_1 - C_4$ as in Example 22.3.1; their power functions are graphed in Figure 22.6. If $\alpha \in [.003, .997)$, then d_1 with critical region $\{z_1, z_2\}$ is UMP. However, d_1 is not unbiased unless $\alpha < .030$. Moreover, d_2 is biased for $\alpha \in [.010, .970)$ but unbiased for $\alpha \in [.001, .010)$; and d_3 is biased wherever available (for $\alpha \geqq .002$). In fact,

$$\mathfrak{D}_U(\alpha) = \begin{cases} \varnothing, & \alpha \in [.030, .997) \\ \{d_1\}, & \alpha \in [.010, .030) \\ \{d_1, d_2\}, & \alpha \in [.003, .010) \\ \{d_2\}, & \alpha \in [.001, .003) \\ \varnothing, & \alpha \in (0, .001). \end{cases}$$

Note that d_2 is UMPU for $\alpha \in [.002, .003)$ but is not UMP, since d_3 is eligible and $P(C_3 \mid \boldsymbol{\theta}_2, e) > P(C_2 \mid \boldsymbol{\theta}_2, e)$.

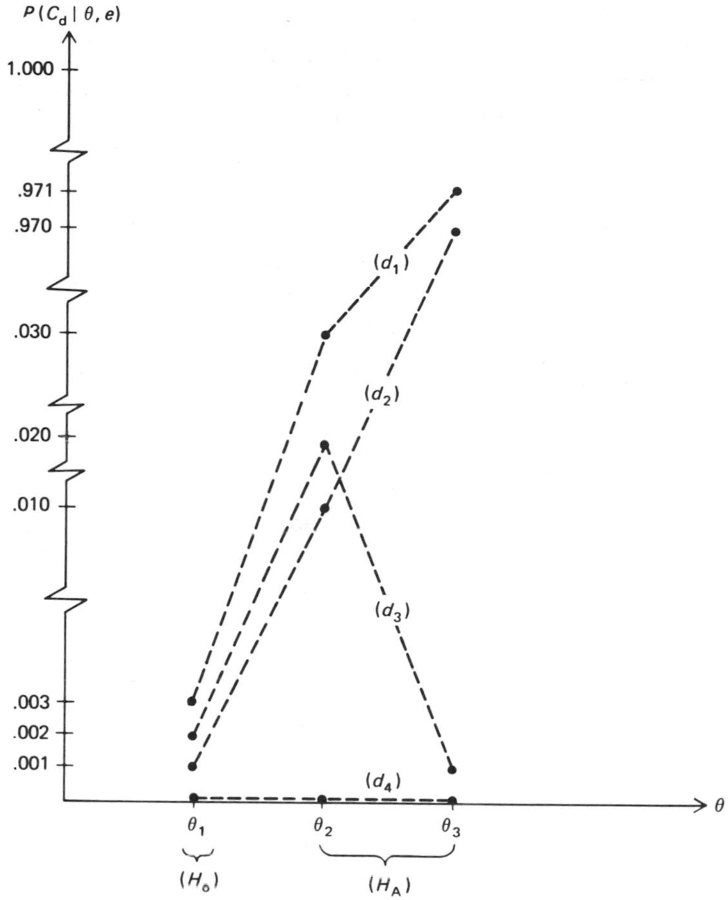

Figure 22.6

The preceding discussion has been phrased almost entirely in terms of general experiments e and outcomes z. In practice, e usually consists in taking a fixed number n of iid observations $\tilde{\mathbf{x}}_1, \ldots, \tilde{\mathbf{x}}_n$ with common dgf $f(\mathbf{x} \mid \boldsymbol{\theta})$, where $\boldsymbol{\theta} \in \boldsymbol{\Theta} \subset R^k$ for some $k > 0$. The investigator defines a convenient but not necessarily sufficient statistic $(n, \tilde{\mathbf{y}}) \equiv \mathbf{g}(n, \tilde{\mathbf{x}}_1, \ldots, \tilde{\mathbf{x}}_n)$ of fixed dimensionality m and then examines tests d based upon $(n, \tilde{\mathbf{y}})$; that is, functions d: $\{1, 2, \ldots\} \times R^{m-1} \to \{a_0, a_A\}$. Any such test gives rise to a sequence of tests d_n for $n = 1, 2, \ldots$, defined by

$$d_n(\mathbf{y}) \equiv d(n, \mathbf{y}), \text{ for every } \mathbf{y} \in R^{m-1} \text{ and } n = 1, 2, \ldots. \quad (22.3.1)$$

The advantage of the $d_n(\cdot)$ notation is that it facilitates consideration of limiting properties as $n \to \infty$. In particular, every test $d(\cdot, \cdot)$ generates a sequence $\{C_{d_n}: n = 1, 2, \ldots\}$ of critical regions in the space R^{m-1} of all possible values of the random statistic $\tilde{\mathbf{y}}$. Of course, the C_{d_n}'s yield the power functions $P(C_{d_n} \mid \boldsymbol{\theta}, n)$, error characteristics $\epsilon(\boldsymbol{\theta} \mid n, d_n)$, and so forth. We are thus led to use the term *test* in referring to $\{d_n: n = 1, 2, \ldots\}$, just as we occasionally used *estimator* in referring to $\{\tilde{\boldsymbol{\theta}}_n: n = 1, 2, \ldots\}$.

Example 22.3.7

Suppose we wish to test H_0: $\mu = \mu^*$ versus H_A: $\mu \neq \mu^*$ on the basis of n iid observations $\tilde{x}_1, \ldots, \tilde{x}_n$ with common df $f_N(x \mid \mu, \sigma^2)$, where σ^2 is known and μ_0 is a specified real number. By (13.2.5), $(n, \tilde{\bar{x}})$ is sufficient for μ when σ^2 is known, and we shall consider tests d_n of the form

$$d_n(\bar{x}) = \begin{cases} a_0, & \bar{x} \in [\mu^* - \delta, \mu^* + \delta] \\ a_A, & \bar{x} \notin [\mu^* - \delta, \mu^* + \delta], \end{cases}$$

where δ is a fixed positive number which will be determined so that $\epsilon_I(\mu^* \mid n, d_n) \leqq \alpha$ for the chosen significance level α. By (13.3.3) the df of $\tilde{\bar{x}}$ given (μ, σ^2) is $f_N(\bar{x} \mid \mu, \sigma^2/n)$, so that, when H_0: $\mu = \mu^*$ is true,

$$\begin{aligned} \epsilon_I(\mu^* \mid n, d_n) &= P_N\left[\tilde{\bar{x}} \notin [\mu^* - \delta, \mu^* + \delta] \mid \mu^*, \frac{\sigma^2}{n}\right] \\ &= 1 - F_N\left(\mu^* - \delta \mid \mu^*, \frac{\sigma^2}{n}\right) + F_N\left(\mu^* - \delta \mid \mu^*, \frac{\sigma^2}{n}\right) \\ &= 2F_N\left(\mu^* - \delta \mid \mu^*, \frac{\sigma^2}{n}\right) \\ &= 2F_{N*}\left(\frac{-\delta\sqrt{n}}{\sigma}\right). \end{aligned}$$

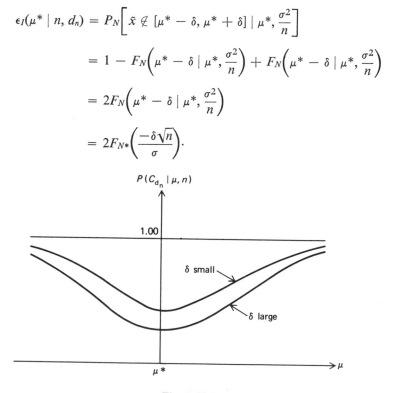

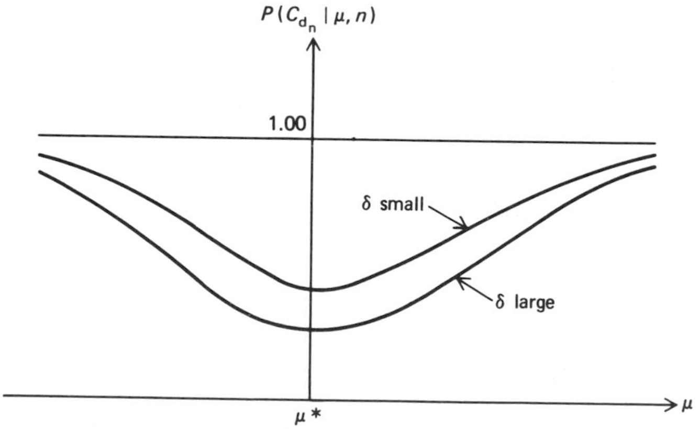

Figure 22.7

The power functions for two values of δ (and some fixed n and σ^2) are graphed in Figure 22.7; the reader may verify as Exercise 5 that

$$P(C_{d_n} \mid \mu, n) = F_{N^*}\left(\frac{-\delta - (\mu^* - \mu)}{\sigma/\sqrt{n}}\right) + F_{N^*}\left(\frac{-\delta + (\mu^* - \mu)}{\sigma/\sqrt{n}}\right), \qquad (22.3.2)$$

from which it is apparent that the power is a decreasing function of δ at every $\mu \in R^1$. Hence, to find the most powerful test d_n of the class under consideration, it suffices to equate $\epsilon_I(\mu^* \mid n, d_n) = P(C_{d_n} \mid \mu^*, n)$ to α and solve for δ. From

$$2F_{N^*}\left(\frac{-\delta\sqrt{n}}{\sigma}\right) = \alpha,$$

it follows readily that

$$\frac{-\delta\sqrt{n}}{\sigma} = {}_{\frac{1}{2}\alpha}N^* = -{}_{(1-\frac{1}{2}\alpha)}N^*,$$

which implies that

$$\delta = \frac{{}_{(1-\frac{1}{2}\alpha)}N^*\sigma}{\sqrt{n}}. \qquad (22.3.3)$$

Thus to test H_0: $\mu = \mu^*$ versus H_A: $\mu \neq \mu^*$ at level α, one rejects H_0 if $\bar{x} \notin [\mu^* - {}_{(1-\frac{1}{2}\alpha)}N^*\sigma/\sqrt{n},\ \mu^* + {}_{(1-\frac{1}{2}\alpha)}N^*\sigma/\sqrt{n}]$. It can be shown that this test is UMPU at level α; see Kendall and Stuart [**41**], Section 23.33.

An optimum property of point estimators $\{\tilde{\boldsymbol{\theta}}_n: n = 1, 2, \ldots\}$ given in Chapter 20 is consistency: the probability that $\tilde{\boldsymbol{\theta}}_n$ is arbitrarily close to the true $\boldsymbol{\theta}$ tends toward one as $n \to \infty$. A similar concept is applicable to tests. We say that a test $\{d_n: n = 1, 2, \ldots\}$ of H_0: $\boldsymbol{\theta} \in \boldsymbol{\Theta}_0$ versus H_A: $\boldsymbol{\theta} \in \boldsymbol{\Theta}_A$ at level α is *consistent* if

$$\lim_{n\to\infty} \epsilon_{II}(\boldsymbol{\theta} \mid n, d_n) = 0, \text{ for every } \boldsymbol{\theta} \in \boldsymbol{\Theta}_A; \qquad (22.3.4a)$$

or equivalently, if

$$\lim_{n\to\infty} P(C_{d_n} \mid \boldsymbol{\theta}, n) = 0, \text{ for every } \boldsymbol{\theta} \in \boldsymbol{\Theta}_A. \qquad (22.3.4b)$$

Example 22.3.8

The test $\{d_n: n = 1, 2, \ldots\}$ of H_0: $\mu = \mu^*$ versus H_A: $\mu \neq \mu^*$ derived in Example 22.3.7 is consistent, since (22.3.2) and (22.3.3) imply that

$$P(C_{d_n} \mid \mu, n) = F_{N^*}(-w - A_n) + F_{N^*}(-w + A_n),$$

where $w \equiv {}_{(1-\frac{1}{2}\alpha)}N^*$ and $A_n \equiv \sqrt{n}(\mu^* - \mu)/\sigma$. Suppose $\mu > \mu^*$. Then

$$\begin{aligned}\lim_{n\to\infty} F_{N^*}(-w - A_n) &= F_{N^*}(-w - \lim_{n\to\infty} A_n)\\ &= F_{N^*}(-\infty)\\ &= 0,\end{aligned}$$

and similarly,

$$\lim_{n\to\infty} F_{N^*}(-w + A_n) = F_{N^*}(+\infty) = 1,$$

so that

$$\lim_{n\to\infty} P(C_{d_n} \mid \mu, n) = 1.$$

The reader may derive the same result for $\mu < \mu^*$ as Exercise 6.

There is an intimate relationship between orthodox tests and confidence regions. This relationship will become apparent in Sections 22.5 and 22.6, and amounts to the following tests procedure:

(1) construct a $100(1 - \alpha)$ percent confidence region

$$I(n, \mathbf{x}_1, \ldots, \mathbf{x}_n), \text{ for } \boldsymbol{\theta}; \text{ and}$$

(2) reject H_0: $\boldsymbol{\theta} \in \boldsymbol{\Theta}_0$ in favor of H_A: $\boldsymbol{\theta} \in \boldsymbol{\Theta}_A$ if and only if

$$\boldsymbol{\Theta}_0 \cap I(n, \mathbf{x}_1, \ldots, \mathbf{x}_n) = \varnothing.$$

The specific form of $\boldsymbol{\Theta}_A$ influences the choice of region estimator $I(n, \tilde{\mathbf{x}}_1, \ldots, \tilde{\mathbf{x}}_n)$, which is not always selected on grounds of minimum length (area, or volume).

Example 22.3.9

The UMPU test of H_0: $\mu = \mu^*$ versus H_A: $\mu \neq \mu^*$ derived in Example 22.3.7 calls for rejecting H_0 if and only if $\mu^* \notin [\bar{x} - {}_{(1-\frac{1}{2}\alpha)}N^*\sigma/\sqrt{n}, \bar{x} + {}_{(1-\frac{1}{2}\alpha)}N^*\sigma/\sqrt{n}]$; but this interval is simply the shortest-length confidence interval for μ as given by (21.3.1) for $\gamma = 1 - \alpha$.

22.3.2 First Bayesian Choice Procedure

This is the first of two comparatively short subsections on Bayesian procedures which do not introduce loss and risk functions. The brevity of these subsections is due to: (1) the fact that many of the basic concepts in Section 22.2 and Subsection 22.3.1 continue to apply; (2) the predominant role of the orthodox approach in practice; and (3) the somewhat more natural interpretation of Bayesian procedures, based as they are (by definition) upon the decision-maker's posterior probabilities.

The first Bayesian procedure is closely related to credible regions as discussed in Chapter 21, just as many orthodox procedures are related to confidence regions. We proceed by:

(1) choosing $\alpha \in (0, 1)$;
(2) finding $100(1 - \alpha)$ percent credible sets $I(e, z)$ [or $I(n, x_1, \ldots, x_n)$]; and
(3) rejecting H_0: $\boldsymbol{\theta} \in \boldsymbol{\Theta}_0$ in favor of H_A: $\boldsymbol{\theta} \in \boldsymbol{\Theta}_A$ if and only if $\boldsymbol{\Theta}_0 \cap I(e, z) = \varnothing$.

This procedure implies that the posterior probability $P(\boldsymbol{\Theta}_0 \mid e, z)$ stating the null hypothesis to be true is *at most* α when it is rejected, since $\boldsymbol{\Theta}_0 \cap I(e, z) = \varnothing$ and $P(I(e, z) \mid e, z) \geqq 1 - \alpha$ imply

$$1 \geqq P(\boldsymbol{\Theta}_0 \mid e, z) + P(I(e, z) \mid e, z) \geqq P(\boldsymbol{\Theta}_0 \mid e, z) + 1 - \alpha.$$

The interpretation of this procedure is most straightforward when $I(e, z)$ consists of all elements $\boldsymbol{\theta}$ of highest posterior probability or posterior density; that is, when $\boldsymbol{\theta} \in I(e, z)$ and $\boldsymbol{\theta}' \notin I(e, z)$ imply $P(\boldsymbol{\theta} \mid e, z) > P(\boldsymbol{\theta}' \mid e, z)$ or $f(\boldsymbol{\theta} \mid e, z) > f(\boldsymbol{\theta}' \mid e, z)$. If this is the case, then the decision-maker can claim that the null hypothesis is not credible because the $100(1 - \alpha)$ percent credibility of $I(e, z)$ arises in a natural way.

Example 22.3.10

In the first Bayesian procedure for testing H_0: $\mu = \mu^*$ versus H_A: $\mu \neq \mu^*$ as in Example 22.3.7 at level α, we first note that by (21.5.4), a (shortest) credible interval for μ at level $\gamma = 1 - \alpha$ is

$$I(n, \bar{x}) = [\mu_1 - {}_{(1-\frac{1}{2}\alpha)}N^*\sigma/\sqrt{n_0 + n}, \mu_1 + {}_{(1-\frac{1}{2}\alpha)}N^*\sigma/\sqrt{n_0 + n}]$$

for

$$\mu_1 \equiv n_0\mu_0/(n_0 + n) + n\bar{x}/(n_0 + n).$$

We reject H_0: $\mu = \mu^*$ in favor of H_0: $\mu \neq \mu^*$ if and only if $\mu^* \notin I(n, \bar{x})$.

For particular applications, it is not necessarily true that one should select $I(e, z)$ according to the highest-posterior-probability criterion of the preceding example. The choice can be based upon considerations of power or error characteristic, which are as obtainable for these Bayesian tests as for orthodox tests. The previous steps (1) through (3) define a test d: $Z(e) \rightarrow \{a_0, a_A\}$ with critical region

$$C_d \equiv \{z: \boldsymbol{\Theta}_0 \cap I(e, z) = \phi\}, \tag{22.3.5}$$

power function $P(C_d \mid \cdot, e)$: $\Theta_0 \cup \Theta_A \to [0, 1]$, and error characteristic given by (22.2.1) through (22.2.3) and C_d.

22.3.3 Second Bayesian Choice Procedure

The first Bayesian choice procedure may seem somewhat abstruse to the reader because of the potential arbitrariness in selecting the system of credible regions. The second Bayesian procedure does not share this defect.

To test H_0: $\theta \in \Theta_0$ versus H_A: $\theta \in \Theta_A$, the second Bayesian procedure consists of:

(1) choosing $\alpha \in (0, 1)$;
(2) deriving the posterior probability $P(\Theta_0 \mid z, e)$ of H_0; and
(3) rejecting H_0 in favor of H_A if and only if $P(\Theta_0 \mid z, e) \leqq \alpha$.

The philosophy underlying this procedure is simply that when the posterior probability of the null hypothesis' validity is too small, then it is no longer tenable.

Example 22.3.11

In a legal application, let Θ_0 denote the event that the alleged suspect X is innocent, and Θ_A, the event that X is guilty. Then a juror may consider X "guilty beyond a reasonable doubt" if $P(\Theta_0 \mid z, e) \leqq \alpha$, where α serves to define quantitatively the "reasonableness" of doubts and, along with prior probabilities $P(\Theta_0)$, may vary between jurors. Incidentally, the preliminary questioning of jurors by the attorneys and bench can be interpreted as an attempt to exclude from the jury those individuals with prior probabilities $P(\Theta_0)$ differing substantially from $\frac{1}{2}$.

Example 22.3.12

Continuing in the legal vein, we note that the concept of circumstantial evidence is not based upon posterior probabilities $P(\Theta_0 \mid z, e)$, but rather upon $P(z \mid \Theta_0, e)$. Let z denote the event that suspect X hated the murder victim, was seen near the scene of the crime minutes after it was committed, and left his fingerprints on the murder weapon, *if* he is innocent. [The smallness of $P(z \mid \Theta_0, e)$ always encourages Lt. Tragg in the Perry Mason stories. However, it fails to discourage Mason, who appears to assess $P(\Theta_0) = 1$ and then launch an additional investigation e' so as to uncover (often dramatically) evidence z' such that $P(z \text{ and } z' \mid \Theta_0, e \text{ and } e')$ is large and X is acquitted.]

For the second Bayesian procedure we can obtain power functions and error characteristics from the critical region C_d, which is given by

$$C_d \equiv \{z: P(\Theta_0 \mid z, e) \leqq \alpha\}. \tag{22.3.6}$$

Example 22.3.13

Reconsider Example 22.3.1, assuming $P(\Theta_1) = P(\Theta_2) = \frac{1}{2}$. The table of posterior probabilities $P(\theta \mid z, e)$ is found by a simple application of Bayes' Theorem, as follows:

$\theta \backslash z$	z_1	z_2
θ_1	.998	.001
θ_2	.002	.999
sum	1.000	1.000

The second Bayesian procedure yields critical regions $C_1 \equiv \{z_1, z_2\}$ if $\alpha \geqq .998$, $C_3 \equiv \{z_2\}$ if $\alpha \in [.001, .998)$, and $C_4 \equiv \varnothing$ if $\alpha < .001$. The power functions of the associated tests d_1, d_3, and d_4 were graphed in Figure 22.4. Note that this procedure eliminates d_2 and $C_2 \equiv \{z_1\}$ from consideration. Note also that the orthodox reasoning in Example 22.3.1 leads to the same choice (d_3) as here for all $\alpha \in [.001, .998)$.

Example 22.3.14

Again consider Example 22.3.1, this time with $P(\Theta_1) = 1$ and $P(\Theta_2) = 0$. No amount of evidence against θ_1 will sway the assessor, and $P(\theta_1 \mid z, e) = 1$ for $z \in \{z_1, z_2\}$. Hence $C_d = \{z: P(\theta_1 \mid z, e) \leqq \alpha\} = \varnothing$ for every $\alpha < 1$. It should be clear why attorneys object to prejudiced jurors who cannot be swayed by evidence and rhetoric.

Example 22.3.15

Reconsider Example 22.3.7, assuming that $\tilde{\mu}$ is given a natural conjugate prior df $f_N(\mu \mid \mu_0, \sigma^2/n_0)$. Then the posterior df of $\tilde{\mu}$ is $f_N(\mu \mid \mu_1, \sigma^2/n_1)$ and $P(\Theta_0 \mid \bar{x}, n) = P_N[\tilde{\mu} = \mu^* \mid \mu_1, \sigma^2/n_1] = 0$, so that H_0: $\mu = \mu^*$ is always rejected in favor of H_A: $\mu \neq \mu^*$ by the second Bayesian procedure for any $\alpha > 0$. Some consider this and similar results an argument against the second Bayesian procedure, but the tables can be turned. We shall argue in Subsection 22.9.2 that most people do not seriously want to test whether μ *exactly* equals μ^*, but rather whether μ is *sufficiently close to* μ^*. If pressed, they would argue to the (Bayesian) effect that there is very little chance that $\mu = \mu^*$ precisely. Thus people use the orthodox test of H_0: $\mu = \mu^*$ versus H_A: $\mu \neq \mu^*$ to determine whether setting $\mu = \mu^*$ is reasonable in view of the experimental outcome $(n, \bar{x})$. Now, if the real null hypothesis is thus (say) H_0: $\mu \in [\mu^* - \Delta, \mu^* + \Delta] \equiv \mathbf{\Theta}_0$ and the alternative is H_A: $\mu \notin \mathbf{\Theta}_0$, then the posterior probability of $\mathbf{\Theta}_0$ will be positive, and there is a straightforward generalization of the orthodox UMPU test in Example 22.3.7 which takes Δ into account (Exercise 7).

22.4 TESTS OF A SIMPLE NULL HYPOTHESIS AGAINST A SIMPLE ALTERNATIVE; LIKELIHOOD RATIO TESTS

The most straightforward hypothesis testing problems are those in which both H_0 and H_A are simple, with $\mathbf{\Theta}_0 \equiv \{\boldsymbol{\theta}_0\}$ and $\mathbf{\Theta}_A \equiv \{\boldsymbol{\theta}_A\}$ for $\boldsymbol{\theta}_0 \neq \boldsymbol{\theta}_A$, and in which e consists in obtaining a fixed number n of iid observations $\tilde{\mathbf{x}}_1, \ldots, \tilde{\mathbf{x}}_n$ with common df $f(\mathbf{x} \mid \boldsymbol{\theta})$.

Under these assumptions we first develop a convenient geometric representation of the error characteristic $\epsilon(\boldsymbol{\theta} \mid e, d)$ and, more generally, the risk function $R(d, \boldsymbol{\theta})$, for any test d.

Since $\mathbf{\Theta}_0 \cup \mathbf{\Theta}_A = \{\boldsymbol{\theta}_0, \boldsymbol{\theta}_A\}$, we can represent the error characteristic $\epsilon(\cdot \mid e, d)$ of a given test d by the pair of numbers $[\epsilon(\boldsymbol{\theta}_0 \mid e, d), \epsilon(\boldsymbol{\theta}_A \mid e, d)]$, and hence by the point in the plane with Cartesian coordinates that same pair of numbers. We shall (arbitrarily) let the abscissa and ordinate denote the probabilities ϵ_I and ϵ_{II} of types I and II error, respectively. Thus each test d corresponds to a point in the $(\epsilon_I, \epsilon_{II})$-plane.

Example 22.4.1

For Example 22.3.1, with $\boldsymbol{\theta}_1 \equiv \boldsymbol{\theta}_0$ and $\boldsymbol{\theta}_2 \equiv \boldsymbol{\theta}_A$, we easily obtain Figure 22.8. The desirability of minimizing the error probabilities implies that d_3 is better than d_2, for both $\epsilon(\boldsymbol{\theta}_0 \mid e, d_3) < \epsilon(\boldsymbol{\theta}_0 \mid e, d_2)$ and $\epsilon(\boldsymbol{\theta}_A \mid e, d_3) < \epsilon(\boldsymbol{\theta}_A \mid e, d_2)$, and hence d_2 is inadmissible. The only tests worth considering are d_1, d_3, and d_4; the reader may verify that none of these are inadmissible.

The dotted lines in Figure 22.8 are intended as guides to the eye, but we shall now endow them with meaning by introducing the concept of a randomized test.

Suppose the decision-maker is having difficulty choosing between d_3 and d_4. He may decide to resort to a chance mechanism, such as a coin toss, with outcome independent of $\boldsymbol{\theta}$, which will result in selecting d_4 with probability p and d_3 with probability $1 - p$: We denote the decision to use d_4 with probability p and d_3 with probability $1 - p$ by $[p: d_4, (1 - p): d_3]$ and call it a randomized test. More generally, a *randomized test* is a probability distribution over the set $\mathfrak{D}(1)$ of all possible tests. In case $\mathfrak{D}(1)$ is a finite set $\{d_j: j = 1, \ldots, m\}$, every randomized test corresponds to an

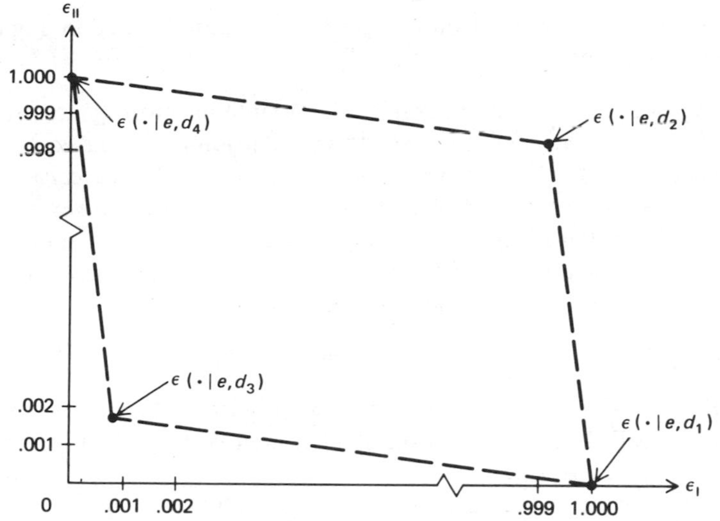

Figure 22.8

m-dimensional vector $(p_1, \ldots, p_m)$, of which the jth component p_j is the probability that the chance mechanism will select d_j.

The error probabilities of a randomized test are easily obtainable. Let $d^* \equiv [p_1: d_1, \ldots, p_m: d_m]$. Then

$$\epsilon(\boldsymbol{\theta}_i \mid e, d^*) = \sum_{j=1}^{m} \epsilon(\boldsymbol{\theta}_i \mid e, d_j)p_j, \text{ for } i \in \{0, A\}. \qquad (22.4.1a)$$

More generally, if d_t is a test for every $t \in R^1, f(t)$ is a df, and $d^* \equiv [f(t): d_t$, for every $t \in R^1]$, then

$$\epsilon(\boldsymbol{\theta}_i \mid e, d^*) = \int_{-\infty}^{\infty} \epsilon(\boldsymbol{\theta}_i \mid e, d_t)f(t)dt, \text{ for } i \in \{0, A\}. \qquad (22.4.1b)$$

The reader may prove (22.4.1) as Exercise 8. [It is evident that the definition of randomized test and (22.4.1) remain meaningful when $\boldsymbol{\Theta}_0$ and/or $\boldsymbol{\Theta}_A$ is composite.]

From (22.4.1a) it is clear that when $d^* = [p: d, (1-p): d']$, the $(\epsilon_I, \epsilon_{II})$-point for d^* belongs to the line segment joining the $(\epsilon_I, \epsilon_{II})$-points of d and d', and that as p goes from zero to one, the $(\epsilon_I, \epsilon_{II})$-point of d^* traverses the entire segment in the d'-to-d direction.

Example 22.4.2

Every point on the line joining $\epsilon(\cdot \mid e, d_4)$ and $\epsilon(\cdot \mid e, d_3)$ is $\epsilon(\cdot \mid e, d^*)$ for some d^* of the form $[p: d_4, (1-p): d_3]$. By considering more general randomized tests of the form $[p_1: d_1, \ldots, p_4: d_4]$ with $p_j \geqq 0$ and $\sum_{i=1}^{4} p_i$

= 1, we obtain every point on or inside the diamond shape in Figure 22.8 as an $(\epsilon_I, \epsilon_{II})$-point.

The second sentence of Example 22.4.2 obtains more generally. A set $A \subset R^h$ is said to be *convex* if the line segment joining $\mathbf{x}$ and $\mathbf{x}'$ is a subset of A whenever $\{\mathbf{x}, \mathbf{x}'\} \subset A$. Given any subset $B \subset R^h$, the *convex hull* of B is the smallest convex set $A \subset R^h$ such that $B \subset A$. Convex hulls exist and are unique. It then follows readily that the set of $(\epsilon_I, \epsilon_{II})$-points corresponding to the set $\mathcal{RD}(1)$ of all randomized tests is the convex hull of the set corresponding to the nonrandomized tests $\mathcal{D}(1)$.

Example 22.4.3

Figure 22.9 depicts the $(\epsilon_I, \epsilon_{II})$-graph in Figure 22.8, filled in to denote the convex hull. Every point in the shaded area is $\epsilon(\cdot \mid e, d^*)$ for some test $d^* \in [p_1: d_1, \ldots, p_4: d_4]$.

The "southwest" boundary of the convex hull has been heavily drawn to indicate that the only admissible tests d^* are those for which $\epsilon(\cdot \mid e, d^*)$ belongs to this boundary.

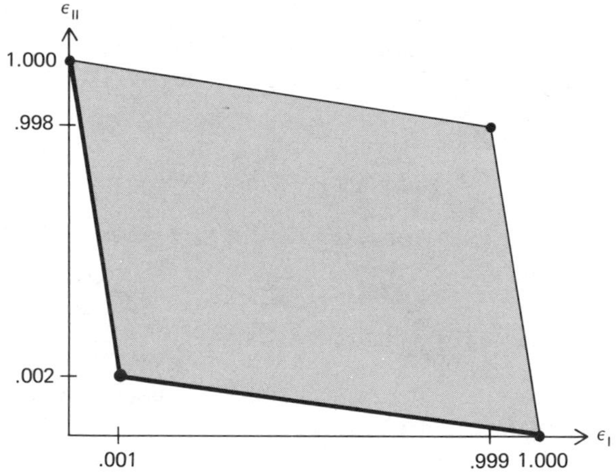

Figure 22.9

A qualitatively similar geometric representation obtains when a loss function $\ell(a, \theta)$ satisfying (22.2.4) is introduced. Instead of graphing error probabilities, we graph $(R(d, \theta_0), R(d, \theta_A))$-points. But from (22.2.5) it follows readily that

$$R(d, \theta_0) = \ell(a_A, \theta_0)\epsilon_I(\theta_0 \mid e, d) \tag{22.4.2a}$$

and

$$R(d, \theta_A) = \ell(a_0, \theta_A)\epsilon_{II}(\theta_A \mid e, d), \tag{22.4.2b}$$

so that obtaining the graph of the risk function from the graph of the error characteristic is accomplished by rescaling the abscissa values from ϵ_I to $\ell(a_A, \theta_0)\epsilon_I$ and independently rescaling the ordinate values from ϵ_{II} to $\ell(a_0, \theta_A)\epsilon_{II}$. By (22.2.15), the same tests are admissible in each case—provided that (22.2.4) obtains.

Example 22.4.4

In Example 22.4.3, suppose that $\ell(a_i, \theta_i) = 0$, for $i \in \{0, A\}$, $\ell(a_0, \theta_A) = 10$, and $\ell(a_A, \theta_0) = 1000$. Then, instead of Figure 22.9, we would use Figure 22.10.

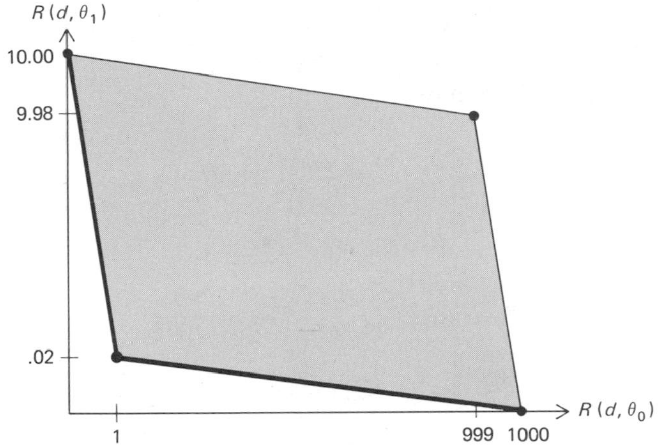

Figure 22.10

In Section 22.2 we remarked that the search for Bayes tests can be restricted to the set of all admissible tests. When both H_0 and H_A are simple, there is a strong converse, as follows:

Every admissible test d is a Bayes test for some prior probabilities $P(\theta_0) \equiv \pi_0$ and $P(\theta_A) \equiv \pi_A = 1 - \pi_0$. (22.4.3)

Proof

From (22.2.9) we have

$$B(d) = \pi_0 R(d, \theta_0) + \pi_A R(d, \theta_A),$$

for any test d. We wish to minimize $B(d)$ by optimal choice of d. For any $c \in R^1$, consider

$$S(c) \equiv \{(R_0, R_A): \pi_0 R_0 + \pi_A R_A = c\} R^2.$$

Clearly, $S(c)$ is a line with nonpositive slope (because $\pi_0 \in [0, 1]$ and $\pi_A = 1 - \pi_0$). If $S(c)$ intersects the "risk region" $\{(R(d, \boldsymbol{\theta}_0), R(d, \boldsymbol{\theta}_A)): d$ a randomized or nonrandomized test$\}$, then there is a test d for which $B(d) = c$. Now, reducing c translates $S(c)$ parallel to itself and down toward the origin, and hence a Bayes test corresponds to a risk point intersection $S(c^\circ)$, where c° is the smallest c for which $S(c)$ intersects the risk region. Clearly, this point is admissible. Furthermore, by changing π_0 one changes the slopes of the lines $S(c)$ and hence changes the indicated admissible test: as π_0 moves from zero to one, the entire sourthwest boundary of the risk region is traversed from southeasternmost to northwesternmost. QED

Example 22.4.5

For Example 22.4.4, we have

$$d^0 = \begin{cases} d_1, & 0 \leqq \pi_0 < \dfrac{2}{99{,}902} \\[2ex] d_{13}^*, & \dfrac{2}{99{,}902} = \pi_0 \\[2ex] d_3, & \dfrac{2}{99{,}902} < \pi_0 < \dfrac{998}{1098} \\[2ex] d_{34}^*, & \dfrac{998}{1098} = \pi_0 \\[2ex] d_4, & \dfrac{998}{1098} < \pi_0 \leqq 1, \end{cases}$$

where $d_{13}{}^*$ is any test of the form $[p: d_1, (1 - p): d_3]$ and $d_{34}{}^*$ is any test of the form $[p: d_3, (1 - p): d_4]$. Thus the set of Bayes tests obtained by varying π_0 coincides with the set of admissible tests. (Note the optimality of the "common-sense" test d_3 over a very wide range of π_0's.)

Next, we shall prove two results of importance, taking for convenience e to consist of n iid observations $\tilde{\mathbf{x}}_1, \ldots, \tilde{\mathbf{x}}_n$ with common dgf $f(\mathbf{x} \mid \boldsymbol{\theta})$, where $\boldsymbol{\theta} \in \{\boldsymbol{\theta}_0, \boldsymbol{\theta}_A\}$. First, a definition: d is said to be a *likelihood ratio test* of H_0: $\boldsymbol{\theta} = \boldsymbol{\theta}_0$ versus H_A: $\boldsymbol{\theta} = \boldsymbol{\theta}_A$ if there exists $w \in (0, \infty)$ such that

$$d(\lambda) = \begin{cases} a_0, & \lambda \geqq w \\ a_A, & \lambda < w, \end{cases} \tag{22.4.4}$$

where

$$\lambda \equiv \lambda(\mathbf{x}_1, \ldots, \mathbf{x}_n) \equiv \frac{L(\boldsymbol{\theta}_0 \mid n, \mathbf{x}_1, \ldots, \mathbf{x}_n)}{L(\boldsymbol{\theta}_A \mid n, \mathbf{x}_1, \ldots, \mathbf{x}_n)}. \tag{22.4.5}$$

[Recall that $L(\boldsymbol{\theta} \mid n, \mathbf{x}_1, \ldots, \mathbf{x}_n) = \prod_{i=1}^{n} f(\mathbf{x}_i \mid \boldsymbol{\theta})$ for iid observations with common dgf $f(\mathbf{x} \mid \boldsymbol{\theta})$.]

The intuitive meaning of a likelihood ratio test is straightforward. When λ is very small, considerable doubt is cast upon H_0 and hence H_0 is rejected when λ is "too small": $\lambda < w$. The converse is true for large λ's.

We have shown that every admissible test is a Bayes test. Next, we show that every Bayes test [of H_0: $\boldsymbol{\theta} = \boldsymbol{\theta}_0$ versus H_A: $\boldsymbol{\theta} = \boldsymbol{\theta}_A$] is a likelihood ratio test. These results imply that every admissible test is a likelihood ratio test.

Every Bayes test of H_0: $\boldsymbol{\theta} = \boldsymbol{\theta}_0$ versus H_A: $\boldsymbol{\theta} = \boldsymbol{\theta}_A$, based upon an experiment consisting of n iid observations $\tilde{\mathbf{x}}_1, \ldots, \tilde{\mathbf{x}}_n$ with common dgf $f(\mathbf{x} \mid \boldsymbol{\theta})$, is a likelihood ratio test. (22.4.6)

Proof

Since

$$\epsilon_I(\boldsymbol{\theta}_0 \mid e, d) = P(C_d \mid \boldsymbol{\theta}_0, e) = \int_{C_d} \prod_{i=1}^{n} f(\mathbf{x}_i \mid \boldsymbol{\theta}_0) d\mathbf{x}_1, \ldots d\mathbf{x}_n$$

and similarly,

$$\epsilon_{II}(\boldsymbol{\theta}_A \mid e, d) = 1 - P(C_d \mid \boldsymbol{\theta}_A, e) = 1 - \int_{C_d} \prod_{i=1}^{n} f(\mathbf{x}_1 \mid \boldsymbol{\theta}_A) d\mathbf{x}_1, \ldots, d\mathbf{x}_n,$$

it follows from (22.2.9) that

$$B(d) = (1 - \pi_0)\ell(a_0, \boldsymbol{\theta}_A) + \{\pi_0 \ell(a_A, \boldsymbol{\theta}_0) P(C_d \mid \boldsymbol{\theta}_0, e) - (1 - \pi_0)\ell(a_0, \boldsymbol{\theta}_A) P(C_d \mid \boldsymbol{\theta}_A, e)\},$$

and $B(d)$ is minimized by minimizing the term in braces, which can now be re-expressed as

$$\int_{C_d} \left\{\pi_0 \ell(a_A, \boldsymbol{\theta}_0) \prod_{i=1}^{n} f(\mathbf{x}_i \mid \boldsymbol{\theta}_0) - (1 - \pi_0)\ell(a_0, \boldsymbol{\theta}_A) \prod_{i=1}^{n} f(\mathbf{x}_1 \mid \boldsymbol{\theta}_A)\right\} d\mathbf{x}_1 \ldots d\mathbf{x}_n.$$

It is clear that the minimum of this integral with respect to choice of C_d is attained by assigning to C_d every $(\mathbf{x}_1, \ldots, \mathbf{x}_n)$ for which the integrand $\{\ \}$ is negative; that is, $d°$ satisfies $d°(\mathbf{x}_1, \ldots, \mathbf{x}_n) = a_A$ whenever

$$\pi_0 \ell(a_A, \boldsymbol{\theta}_0) \prod_{i=1}^{n} f(\mathbf{x}_i \mid \boldsymbol{\theta}_0) < (1 - \pi_0)\ell(a_0, \boldsymbol{\theta}_A) \prod_{i=1}^{n} f(\mathbf{x}_i \mid \boldsymbol{\theta}_A),$$

which is equivalent to rejection of H_0 whenever

$$\lambda = \frac{L(\boldsymbol{\theta}_0 \mid n, \mathbf{x}_1, \ldots, \mathbf{x}_n)}{L(\boldsymbol{\theta}_A \mid n, \mathbf{x}_1, \ldots, \mathbf{x}_n)} < w,$$

for

$$w \equiv \frac{(1 - \pi_0)\ell(a_0, \boldsymbol{\theta}_A)}{\pi_0 \ell(a_A, \boldsymbol{\theta}_0)},$$

provided that ℓ satisfies (22.2.4) and $\pi_0 \in (0, 1)$. (The other cases are trivial.) QED

This result sheds light on how losses and prior probabilities affect the critical region. Since w is as explicitly characterized in the proof of (22.4.6), we have:

(1) the lower the prior probability π_0 of H_0, the easier it is (and should be) to reject H_0;
(2) the greater the relative seriousness of type-II error vis-á-vis type-I error, the easier it is (and should be) to reject H_0.

By interchanging "lower" with "greater" and replacing "easier" with "harder," we obtain the equally valid converses of (1) and (2). We have already discussed (2) and its converse; (1) is of interest as well, even though it is often overlooked in strictly orthodox treatments.

Both points hold considerably more generally than in the present context. However, it suffices to note, apropos of (1), that with tests in general, one should not adopt a very low significance level α when testing a hypothesis H_0: $\boldsymbol{\theta} \in \boldsymbol{\Theta}_0$ that everyone strongly believes is true.

Example 22.4.6

A production foreman has purchased two rolls of wire with resistances μ_1 and μ_2 ohms, respectively. It is very important that the μ_i-wire not be used for μ_j-purposes when $i \neq j$. Unfortunately, a new stock clerk has discarded the labels, but readings on an ohmmeter can be taken on each roll. Readings on the μ_i-wire are iid random variables $\tilde{x}_j$ with common df $f_N(x \mid \mu_i, \sigma^2)$, where σ^2 is known. Wire for a μ_1-purpose is required, and hence n readings on one of the rolls are obtained. A likelihood ratio test calls for rejecting H_0: $\mu = \mu_1$ in favor of H_A: $\mu = \mu_2$ if λ is too small (that is, $\lambda < w$), where

$$\begin{aligned}\lambda &= \prod_{j=1}^{n} \left[\frac{f_N(x_j \mid \mu_1, \sigma^2)}{f_N(x_j \mid \mu_2, \sigma^2)}\right] \\ &= \prod_{j=1}^{n} \exp\,[-\tfrac{1}{2}[(x_j - \mu_1)^2 - (x_j - \mu_2)^2]/\sigma^2] \\ &= \exp\left[-\tfrac{1}{2}\left\{\sum_{j=1}^{n} (x_j - \mu_1)^2 - \sum_{j=1}^{n} (x_j - \mu_2)^2\right\}\Big/\sigma^2\right],\end{aligned}$$

and the last term in braces reduces to $n[\mu_1^2 - \mu_2^2 - 2\bar{x}(\mu_1 - \mu_2)]$. Thus H_0 is rejected whenever

$$\exp\left[-\tfrac{1}{2}n[\mu_1^2 - \mu_2^2 - 2\bar{x}(\mu_1 - \mu_2)]/\sigma^2\right] < w.$$

By taking logs and rearranging, we deduce that H_0 is rejected whenever

$$\bar{x} \begin{Bmatrix} < \\ > \end{Bmatrix} \frac{\log_e(w)\left(\dfrac{\sigma^2}{n}\right)}{(\mu_1 - \mu_2)} + \tfrac{1}{2}(\mu_1 - \mu_2)$$

if $\mu_1 \begin{Bmatrix} > \\ < \end{Bmatrix} \mu_2$; the reader may verify this fact as Exercise 9. Thus the likelihood ratio test is equivalent to a test based upon the sufficient statistic $(n, \bar{x})$. Note that if the rolls are really indistinguishable, then $\pi_0 = \pi_A = \frac{1}{2}$, and if the errors are equally serious, then $w = 1$, $\log_e(w) = 0$, and H_0 is rejected whenever $\bar{x}$ is more than halfway from μ_1 to μ_2. The power function is easily obtained from the df's $f_N(\bar{x} \mid \mu_1, \sigma^2/n)$ of $\tilde{\bar{x}}$ given μ_i.

This example suggests the readily verified fact:

If $\tilde{\mathbf{x}}_T$ is sufficient for $\boldsymbol{\theta}$, then the likelihood ratio λ defined by (22.4.5) is a function of $\mathbf{x}_T$. (22.4.7)

Proof

Exercise 10.

A test d of H_0: $\boldsymbol{\theta} \in \boldsymbol{\Theta}_0$ versus H_A: $\boldsymbol{\theta} \in \boldsymbol{\Theta}_A$ is said to be *of size* α (rather than at level α) if $P(C_d \mid \boldsymbol{\theta}, e) = \alpha$, for every $\boldsymbol{\theta} \in \boldsymbol{\Theta}_0$.

Our final result regarding likelihood ratio tests of simple hypotheses H_0 *versus* simple alternatives H_A is the celebrated *Neyman-Pearson Lemma* [**59**]:

If an UMP test of H_0: $\boldsymbol{\theta} = \boldsymbol{\theta}_0$ versus H_A: $\boldsymbol{\theta} = \boldsymbol{\theta}_A$ of size α exists, then it is a likelihood ratio test. (22.4.8)

Proof

Let $w > 0$ be fixed and define

(1) $$C \equiv \{(\mathbf{x}_1, \ldots, \mathbf{x}_n): L(\boldsymbol{\theta}_0 \mid n, \mathbf{x}_1, \ldots, \mathbf{x}_n) < wL(\boldsymbol{\theta}_A \mid n, \mathbf{x}_1, \ldots, \mathbf{x}_n)\}.$$

Then C is the critical region of the likelihood ratio test of H_0 versus H_A of size $\alpha \equiv \alpha(w) \equiv P(C \mid \boldsymbol{\theta}_0, e)$. Let C' denote the critical region of any test of H_0 versus H_A of size $P(C' \mid \boldsymbol{\theta}_0, e) \equiv \alpha'$ for $\alpha' \leqq \alpha(w)$. We shall show that

(2) $$P(C \mid \boldsymbol{\theta}_A, e) \geqq P(C' \mid \boldsymbol{\theta}_A, e)$$

and hence that the likelihood ratio test of size $\alpha(w)$ is at least as powerful as any test of size not exceeding $\alpha(w)$. Since $P(C \mid \boldsymbol{\theta}_0, e) \geqq P(C' \mid \boldsymbol{\theta}_0, e)$, it readily follows that

$$(3) \qquad P(C \backslash (C \cap C') \mid \boldsymbol{\theta}_0, e) \geqq P(C' \backslash (C \cap C') \mid \boldsymbol{\theta}_0, e).$$

Since $L(\boldsymbol{\theta}_0 \mid n, \mathbf{x}_1, \ldots, \mathbf{x}_n) < wL(\boldsymbol{\theta}_A \mid n, \mathbf{x}_1, \ldots, \mathbf{x}_n)$, for every $(\mathbf{x}_1, \ldots, \mathbf{x}_n) \in C$, and hence in particular for every $(\mathbf{x}_1, \ldots, \mathbf{x}_n) \in C \backslash (C \cap C')$, it follows that

$$(4) \qquad P(C \backslash (C \cap C') \mid \boldsymbol{\theta}_0, e) \leqq wP(C \backslash (C \cap C') \mid \boldsymbol{\theta}_A, e),$$

with equality obtaining only when both sides vanish. Since C and $C' \backslash (C \cap C')$ are disjoint and $L(\boldsymbol{\theta}_0 \mid n, \mathbf{x}_1, \ldots, \mathbf{x}_n) \geqq wL(\boldsymbol{\theta}_A \mid n, \mathbf{x}_1, \ldots, \mathbf{x}_n)$ for outcomes not in C, it follows that

$$(5) \qquad P(C \backslash (C \cap C') \mid \boldsymbol{\theta}_0, e) \geqq wP(C' \backslash (C \cap C') \mid \boldsymbol{\theta}_A, e).$$

(3), (4), and (5) imply the following equality chain:

$$\begin{aligned} wP(C' \backslash (C \cap C') \mid \boldsymbol{\theta}_A, e) & \\ & \leqq P(C' \backslash (C \cap C') \mid \boldsymbol{\theta}_0, e) \text{ [by (5)]} \\ & \leqq P(C \backslash (C \cap C') \mid \boldsymbol{\theta}_0, e) \text{ [by (3)]} \\ & \leqq wP(C \backslash (C \cap C') \mid \boldsymbol{\theta}_A, e) \text{ [by (4)], so that} \end{aligned}$$

$$(6) \qquad P(C' \backslash (C \cap C') \mid \boldsymbol{\theta}_A, e) \leqq P(C \backslash (C \cap C') \mid \boldsymbol{\theta}_A, e)$$

follows by canceling w. Adding $P(C \cap C' \mid \boldsymbol{\theta}_A, e)$ to both sides of (6) yields (2), via the additivity postulate of probability. QED

Another good feature of likelihood ratio tests is the following:

The likelihood ratio test of H_0: $\boldsymbol{\theta} = \boldsymbol{\theta}_0$ versus H_A: $\boldsymbol{\theta} = \boldsymbol{\theta}_A$ of size $\alpha(w)$ is unbiased. (22.4.9)

Proof

First assume that $w \leqq 1$. Then

$$\begin{aligned} \alpha(w) &= P(C \mid \boldsymbol{\theta}_0, e) \\ &= \int_C \prod_{i=1}^n f(\mathbf{x}_i \mid \boldsymbol{\theta}_0) d\mathbf{x}_1 \ldots d\mathbf{x}_n \\ &< w \int_C \prod_{i=1}^n f(\mathbf{x}_i \mid \boldsymbol{\theta}_A) d\mathbf{x}_1 \ldots d\mathbf{x}_n \\ &= wP(C \mid \boldsymbol{\theta}_A, e), \end{aligned}$$

so that

$$P(C \mid \boldsymbol{\theta}_A, e) > \frac{\alpha(w)}{w} \geqq \alpha(w),$$

which proves unbiasedness. Next, assume that $w > 1$. The proof for this case consists of two steps.

Step A: We prove existence of a nonempty subset $C^* \subset C$ such that

$$L(\boldsymbol{\theta}_A \mid n, \mathbf{x}_1, \ldots, \mathbf{x}_n) \geqq L(\boldsymbol{\theta}_0 \mid n, \mathbf{x}_1, \ldots, \mathbf{x}_n), \tag{1}$$

whenever $(\mathbf{x}_1, \ldots, \mathbf{x}_n) \in C^*$. Suppose to the contrary that

$$L(\boldsymbol{\theta}_A \mid n, \mathbf{x}_1, \ldots, \mathbf{x}_n) < L(\boldsymbol{\theta}_0 \mid n, \mathbf{x}_1, \ldots, \mathbf{x}_n),$$

for all $(\mathbf{x}_1, \ldots, \mathbf{x}_n) \in C$. Since

$$L(\boldsymbol{\theta}_A \mid n, \mathbf{x}_1, \ldots, \mathbf{x}_n) < wL(\boldsymbol{\theta}_A \mid n, \mathbf{x}_1, \ldots, \mathbf{x}_n) \leqq L(\boldsymbol{\theta}_0 \mid n, \mathbf{x}_1, \ldots, \mathbf{x}_n),$$

for all $(\mathbf{x}_1, \ldots, \mathbf{x}_n) \in Z(e)\ C$, it would follow that $P(C \mid \boldsymbol{\theta}_0, e) \leqq P(C \mid \boldsymbol{\theta}_A, e)$, and

$$P(Z(e)\backslash C \mid \boldsymbol{\theta}_0, e) \geqq P(Z(e)\backslash C \mid \boldsymbol{\theta}_A, e);$$

and at least one of these inequalities would be strict, which contradicts the fact that $P(Z(e) \mid \boldsymbol{\theta}, e) = 1$ for each $\boldsymbol{\theta}$. Thus C^* with the asserted property is nonempty.

Step B:
$$\begin{aligned} &P(C^* \mid \boldsymbol{\theta}_A, e) - P(C^* \mid \boldsymbol{\theta}_0, e)\\ &= [1 - P(Z(e)\backslash C^* \mid \boldsymbol{\theta}_A, e)] - [1 - P(Z(e)\backslash C^* \mid \boldsymbol{\theta}_0, e)]\\ &= P(Z(e)\backslash C^* \mid \boldsymbol{\theta}_0, e) - P(Z(e)\backslash C^* \mid \boldsymbol{\theta}_A, e)\\ &= P([Z(e)\backslash C] \cup [C\backslash C^*] \mid \boldsymbol{\theta}_0, e)\\ &\qquad - P([Z(e)\backslash C] \cup [C\backslash C^*] \mid \boldsymbol{\theta}_A, e)\\ &= \{P(Z(e)\backslash C \mid \boldsymbol{\theta}_0, e) - P(Z(e)\backslash C \mid \boldsymbol{\theta}_A, e)\}\\ &\qquad + [P(C\backslash C^* \mid \boldsymbol{\theta}_0, e) - P(C\backslash C^* \mid \boldsymbol{\theta}_A, e)]. \end{aligned}$$

But the difference within the braces is positive because

$$L(\boldsymbol{\theta}_0 \mid n, \mathbf{x}_1, \ldots, \mathbf{x}_n) > L(\boldsymbol{\theta}_A \mid n, \mathbf{x}_1, \ldots, \mathbf{x}_n) \text{ on } Z(e)\backslash C.$$

Hence

$$\begin{aligned} &P(C^* \mid \boldsymbol{\theta}_A, e) - P(C^* \mid \boldsymbol{\theta}_0, e)\\ &\qquad > P(C\backslash C^* \mid \boldsymbol{\theta}_0, e) - P(C\backslash C^* \mid \boldsymbol{\theta}_A, e), \end{aligned}$$

which yields

$$\begin{aligned} &P(C \mid \boldsymbol{\theta}_A, e)\\ &\qquad = P(C^* \mid \boldsymbol{\theta}_A, e) + P(C\backslash C^* \mid \boldsymbol{\theta}_A, e)\\ &\qquad > P(C^* \mid \boldsymbol{\theta}_0, e) + P(C\backslash C^* \mid \boldsymbol{\theta}_0, e)\\ &\qquad = P(C \mid \boldsymbol{\theta}_0, e), \end{aligned}$$

thus proving unbiasedness for $w > 1$ as well. QED

22.5 GENERALIZED LIKELIHOOD RATIO TESTS

22.5.1 Introduction

The very desirable properties of likelihood ratio tests of simple hypotheses against simple alternatives lead one to hope that similar tests might be applicable when H_0 and/or H_A is composite. This section is devoted to a relative of the likelihood ratio discussed in Section 22.4; namely, the generalized likelihood ratio $\lambda^* \equiv \lambda^*(n, \tilde{\mathbf{x}}_1, \ldots, \tilde{\mathbf{x}}_n)$, defined by

$$\lambda^* \equiv \frac{\max\limits_{\theta \in \Theta_0} L(\boldsymbol{\theta} \mid n, \mathbf{x}_1, \ldots, \mathbf{x}_n)}{\max\limits_{\theta \in \Theta} L(\boldsymbol{\theta} \mid n, \mathbf{x}_1, \ldots, \mathbf{x}_n)}, \tag{22.5.1}$$

for every $(\mathbf{x}_1, \ldots, \mathbf{x}_n) \in Z(e)$, where $\boldsymbol{\Theta} \equiv \boldsymbol{\Theta}_0 \cup \boldsymbol{\Theta}_A$.

Note that the denominator of λ^* is simply $L(\hat{\boldsymbol{\theta}} \mid n, \mathbf{x}_1, \ldots, \mathbf{x}_n)$ for $\hat{\boldsymbol{\theta}} = \hat{\boldsymbol{\theta}}(n, \mathbf{x}_1, \ldots, \mathbf{x}_n)$ the maximum likelihood estimate of $\boldsymbol{\theta}$. Finding the numerator of λ^* requires maximizing the likelihood function subject to the constraint that $\boldsymbol{\theta} \in \boldsymbol{\Theta}_0$. The reader may verify as Exercise 11 that

$$\max_{\theta \in \Theta_0} L(\boldsymbol{\theta} \mid n, \mathbf{x}_1, \ldots, \mathbf{x}_n) \leqq \max_{\theta \in \Theta} L(\boldsymbol{\theta} \mid n, \mathbf{x}_1, \ldots, \mathbf{x}_n), \tag{22.5.2}$$

so that

$$0 \leqq \lambda^* \leqq 1, \tag{22.5.3}$$

with nonnegativity following from that of the likelihood function. It is easy to show that when H_0 and H_A are simple, λ^* is related to the likelihood ratio λ in Section 22.4 by

$$\lambda^* = \frac{\lambda}{\max\{\lambda, 1\}} = \min\{\lambda, 1\}. \tag{22.5.4}$$

Proof

(Exercise 12.)

In general, a value of λ^* close to one indicates that one cannot appreciably improve upon the maximum of L in $\boldsymbol{\Theta}_0$ by relaxing the constraint to $\boldsymbol{\Theta}_0 \cup \boldsymbol{\Theta}_A$, and hence H_0: $\boldsymbol{\theta} \in \boldsymbol{\Theta}_0$ is rather plausible. Conversely, a value of λ^* close to zero casts doubt upon H_0: $\boldsymbol{\theta} \in \boldsymbol{\Theta}_0$. The critical region C of a test based upon λ^* is therefore of the form

$$C \equiv \{(\mathbf{x}_1, \ldots, \mathbf{x}_n)\colon \lambda^*(\mathbf{x}_1, \ldots, \mathbf{x}_n) < w\},$$

for some $w \in (0, 1)$, where small w's are required for low significance levels α.

Actually, we shall work for the most part with

$$\Xi \equiv -2 \log_e (\lambda^*). \tag{22.5.5}$$

Since $\lambda^* < w$ if and only if $-2 \log_e (\lambda^*) > -2 \log_e(w) \equiv W$, it follows that the general form of the critical region of tests based upon Ξ is

$$C = \{(\mathbf{x}_1, \ldots, \mathbf{x}_n): \Xi(\mathbf{x}_1, \ldots, \mathbf{x}_n) > W\}.$$

Subsection 22.5.2 consists of some general results about Ξ and tests based upon it. Proofs are advanced and omitted; the interested reader may consult Wilks [**83**], which also contains further references.

Subsections 22.5.3 through 22.5.6 apply the generalized likelihood ratio test procedure in a number of important contexts.

22.5.2 Properties of Generalized Likelihood Ratio Tests

We say that d is a *generalized likelihood ratio test* of H_0: $\boldsymbol{\theta} \in \boldsymbol{\Theta}_0$ versus H_A: $\boldsymbol{\theta} \in \boldsymbol{\Theta}_A$ at level α if there exists $w \in [0, 1]$ such that

$$d(n, \mathbf{x}_1, \ldots, \mathbf{x}_n) = \begin{cases} a_0, & \lambda^*(\mathbf{x}_1, \ldots, \mathbf{x}_n) \geqq w \\ a_A, & \text{otherwise,} \end{cases} \tag{22.5.6a}$$

for λ^* as defined by (22.5.1), and

$$P(\{(\mathbf{x}_1, \ldots, \mathbf{x}_n): \lambda^* < w\} \mid \boldsymbol{\theta}, e) \leqq \alpha, \text{ for every } \boldsymbol{\theta} \in \boldsymbol{\Theta}_0. \tag{22.5.7a}$$

Define $W \equiv -2\log_e(w)$ and Ξ by (22.5.5). Then (22.5.6a) and (22.5.7a) are equivalent to

$$d(n, \mathbf{x}_1, \ldots, \mathbf{x}_n) = \begin{cases} a_0, & \Xi \leqq W \\ a_A, & \Xi > W \end{cases} \tag{22.5.6b}$$

and

$$P(\{(\mathbf{x}_1, \ldots, \mathbf{x}_n): \Xi > W\} \mid \boldsymbol{\theta}, e) \leq \alpha, \text{ for every } \boldsymbol{\theta} \in \boldsymbol{\Theta}_0, \tag{22.5.7b}$$

respectively.

We shall consider only tests based upon n iid observations $\tilde{\mathbf{x}}_1, \ldots, \tilde{\mathbf{x}}_n$ with common dgf $f(\mathbf{x} \mid \boldsymbol{\theta})$, in which $\boldsymbol{\theta} \in \boldsymbol{\Theta} \subset R^k$ for some $k \in \{1, 2, \ldots\}$, The basic result about the approximate distribution of $\tilde{\Xi}$ for large n, due to Wilks, requires a preliminary discussion of the structure of $\boldsymbol{\Theta}$, $\boldsymbol{\Theta}_0$, and $\boldsymbol{\Theta}_0 \cup \boldsymbol{\Theta}_A$.

Suppose $\boldsymbol{\Theta} \subset R^k$ is a set of k-tuples $(\theta_1, \ldots, \theta_k)'$ of which precisely t are functionally independent. That is, if any t of the θ_i's are fixed, then the remaining $k - t$ can be solved for via some functional relationship. Then we shall say that $\boldsymbol{\Theta}$ contains t *degrees of freedom*. If A is a proper subset of $\boldsymbol{\Theta}$, it cannot contain more degrees of freedom than does $\boldsymbol{\Theta}$, and it will contain fewer if it is defined by imposing additional functional constraints upon the components of $\boldsymbol{\theta}$. Let u denote the number of degrees of freedom in $\boldsymbol{\Theta}_0 \cup \boldsymbol{\Theta}_A$ and $v < u$ the number of degrees of freedom in $\boldsymbol{\Theta}_0$. Then, under fairly general conditions, the following can be shown:

For large n, the cdf of $\tilde{\Xi}$ is approximately $F_{\chi^2}(\cdot \mid u - v)$. (22.5.8)

Proof

Wilks [**83**], 420–421.

The "fairly general conditions" are assumptions about the common df or mf $f(\mathbf{x} \mid \boldsymbol{\theta})$ of the observations; they are;

$$E\left\{\frac{\partial \log_e f(\tilde{\mathbf{x}} \mid \boldsymbol{\theta})}{\partial \theta_i} \mid \boldsymbol{\theta}\right\} = 0, \text{ for every } i \in \{1, \ldots, k\} \quad (22.5.9a)$$

and

$$E\left\{\frac{\partial^2 \log_e f(\tilde{\mathbf{x}} \mid \boldsymbol{\theta})}{\partial \theta_i \partial \theta_j} + \frac{\partial \log_e f(\tilde{\mathbf{x}} \mid \boldsymbol{\theta})}{\partial \theta_i} \times \frac{\partial \log_e f(\tilde{\mathbf{x}} \mid \boldsymbol{\theta})}{\partial \theta_j} \middle| \boldsymbol{\theta}\right\} = 0, \quad (22.5.9b)$$

for every i and j.

From (22.5.8) it is easy to construct generalized likelihood ratio tests having $\epsilon_I(\cdot \mid d, e) \doteq \alpha$, for large n.

Suppose $\boldsymbol{\Theta}_0 \cup \boldsymbol{\Theta}_A$ contains u degrees of freedom and $\boldsymbol{\Theta}_0$ contains v degrees of freedom. A test of H_0: $\boldsymbol{\theta} \in \boldsymbol{\Theta}_0$ versus H_A: $\boldsymbol{\theta} \in \boldsymbol{\Theta}_A$ at approximate level α consists in rejecting H_0 if $\Xi > {}_{(1-\alpha)}\chi^2{}_{(u-v)}$. (22.5.10)

Proof

Exercise 13.

It can be shown that the generalized likelihood ratio test given by (22.5.10) is consistent. It can be biased, but it is *asymptotically unbiased* in the sense that

$$\lim_{n \to \infty} P(C_n \mid \boldsymbol{\theta}, n) \geqq \alpha, \text{ for every } \boldsymbol{\theta} \in \boldsymbol{\Theta}_A,$$

where C_n is the critical region of the test given n observations. Finally, it need not be admissible (Lehmann [**45**], 256). Despite its occasionally less than optimal behavior, the generalized likelihood ratio test is rather convenient and widely applicable.

Example 22.5.1

Suppose that we wish to test H_0: $\mathbf{p} = \mathbf{p}^*$ versus H_A: $\mathbf{p} \neq \mathbf{p}^*$, where $\mathbf{p} \in R^m$ is the parameter of $f_{mu}{}^{(m)}(\mathbf{x} \mid 1, \mathbf{p})$, the common mf of n iid observations $\tilde{\mathbf{x}}_1, \ldots, \tilde{\mathbf{x}}_n$. Then by (11.4.3) the likelihood function is $L(\mathbf{p} \mid \mathbf{r}, n) = \prod_{j=1}^{m+1} p_j^{r_j}$, where $\mathbf{r} \equiv \sum_{i=1}^{n} \mathbf{x}_i$, $r_{m+1} \equiv n - \sum_{j=1}^{m} r_j$, and $p_{m+1} \equiv 1 - \sum_{j=1}^{m} p_j$. Then: $\boldsymbol{\Theta}_0 = \{\mathbf{p}^*\}$ has zero degrees of freedom; $\boldsymbol{\Theta}_0 \cup \boldsymbol{\Theta}_A = \{\mathbf{p}: \mathbf{p} \in R^m$, every $p_j > 0, \sum_{j=1}^{m} p_j < 1\}$ has m degrees of freedom; and a test of H_0: $\mathbf{p} = \mathbf{p}^*$ versus H_A: $\mathbf{p} \neq \mathbf{p}^*$ at approximate level α for large n rejects H_0 if

$$\Xi > {}_{(1-\alpha)}\chi^2{}_{(m)},$$

where $\Xi \equiv -2 \log_e(\lambda^*)$. Since

$$\max_{\mathbf{p} \in \Theta_0} L(\mathbf{p} \mid \mathbf{r}, n) = L(\mathbf{p}^* \mid \mathbf{r}, n) = \prod_{j=1}^{m+1} (p_j^*)^{r_j}$$

and

$$\max_{all\ p} L(\mathbf{p} \mid \mathbf{r}, n) = L(\hat{\mathbf{p}} \mid \mathbf{r}, n) = \prod_{j=1}^{m+1} \left(\frac{r_j}{n}\right)$$

with the convention that $(0/n)^0 = 1$, it follows that

$$\begin{aligned}
\Xi &= -2 \log_e \left\{ \frac{\prod_{j=1}^{m+1} (p_j^*)^{r_j}}{\prod_{j=1}^{m+1} \left(\frac{r_j}{n}\right)^{r_j}} \right\} \\
&= -2 \sum_{j=1}^{m+1} r_j \log_e \left(\frac{n p_j^*}{r_j}\right) \\
&= -2 \sum_{j=1}^{m+1} r_j [\log_e (E(\tilde{r}_j \mid H_0 \text{ true})) - \log_e (r_j)],
\end{aligned}$$

with the convention that $0 \cdot \log_e(0) = 0$.

22.5.3 Goodness-of-Fit Tests

Heretofore in this book we have always assumed that the functional form of the common dgf $f(\mathbf{x} \mid \boldsymbol{\theta})$ of the iid observations $\tilde{\mathbf{x}}_1, \tilde{\mathbf{x}}_2, \ldots$ was known. In reality this is not always or even usually the case, and situations arise in which one wishes to test a null hypothesis H_0: $P = P^*$ that the common distribution P of $\tilde{\mathbf{x}}_1, \tilde{\mathbf{x}}_2, \ldots$ is P^*, for some given P^* versus the alternative H_A: $P \neq P^*$.

Now, the set of all probability distributions P on the range R^r of $\tilde{\mathbf{x}}_1, \tilde{\mathbf{x}}_2, \ldots$ cannot be represented by a functional form with a finite number of parameters θ_i. Hypothesis testing problems in which the parametric forms of some distributions in H_0 or H_A are not specified are called *nonparametric*, although a more appropriate term might be *excessively parametric*.

The nonparametric problem of testing H_0: $P = P^*$ versus H_A: $P \neq P^*$ has been handled in a number of different ways. The interested reader is referred to Kendall and Stuart [**41**], Chapter 30, for a more complete introduction to this subject.

We convert the preceding problem to a parametric one by grouping observations. Fix a partition of the $\tilde{\mathbf{x}}$-range space R^r into $m + 1$ nonempty subsets $S_1, \ldots, S_{m+1}$. Then to every probability distribution P on R^r there corresponds a $\mathbf{p} \equiv (p_1, \ldots, p_m)' \in R^m$ such that $p_i \equiv P(\tilde{\mathbf{x}} \in S_i)$, for every i. Of course, many distributions P assign identical probabilities p_i to each S_i and hence possess the same corresponding vector $\mathbf{p}$. Finally, given n iid observations $\tilde{\mathbf{x}}_1, \ldots, \tilde{\mathbf{x}}_n$ with common distribution P to which $\mathbf{p}$ corresponds, it is clear that the m-dimensional random vector $\tilde{r} \equiv (\tilde{r}_1, \ldots, \tilde{r}_m)'$ has the multinomial mf $f_{mu}{}^{(m)}(\mathbf{r} \mid n, \mathbf{p})$, where $\tilde{r}_j$ denotes the number of $\tilde{\mathbf{x}}_i$'s belonging to S_j, for $j = 1, \ldots, m$.

It follows that H_0: $P = P^*$ gives rise to H_0': $\mathbf{p} = \mathbf{p}^*$, where $p_j{}^* \equiv P^*(\tilde{\mathbf{x}} \in S_j)$, for $j = 1, \ldots, m$. A test of H_0': $\mathbf{p} = \mathbf{p}^*$ versus H_A': $\mathbf{p} \neq \mathbf{p}^*$ actually corresponds to a test of H_0'': $P \in \{P\colon P(\tilde{\mathbf{x}} \in S_j) = p_j{}^*, j = 1, \ldots, m\}$ versus H_A'': $P \notin \{P\colon P(\tilde{\mathbf{x}} \in S_j) = p_j{}^*, j = 1, \ldots, m\}$. Note that H_0'' is a composite nonparametric hypothesis which includes the simple nonparametric hypothesis H_0.

We shall test H_0 by testing H_0'' or its equivalent H_0''. Nevertheless, the following procedure is called a test of H_0 versus H_A:

> A test of H_0: $P = P^*$ versus H_A: $P \neq P^*$ at approximate level α consists in rejecting H_0 if $\Xi > {}_{(1-\alpha)}\chi^2{}_{(m)}$, where Ξ is as defined in Example 22.5.1, r_j denotes the number of $\mathbf{x}_i$'s in S_j, and $p^*{}_j \equiv P^*(\tilde{\mathbf{x}} \in S_j)$, for $j = 1, , \ldots, m + 1$. (22.5.11)

Proof

The proof duplicates the reasoning in Example 22.5.1.

Intuition is bolstered by the case $r = 1$, for which a convenient rule-of-thumb is to select m so that $n/(m + 1) \geqq 6$ and partition R^1 into $m + 1$ disjoint intervals, each with probability $p_j{}^* = 1/(m + 1)$ given H_0.

Example 22.5.2

Suppose a strong theoretical case has been made for believing that the interrarrival times $\tilde{t}_1, \tilde{t}_2, \ldots$ of some arrival process are iid with common df $f_e(t \mid 1) = \exp(-t)$, for $t > 0$, but the sufficient doubt exists to warrant testing H_0: $F(\cdot) = F_e(\cdot \mid 1)$ versus H_A: $F(\cdot) \neq F_e(\cdot \mid 1)$. Partition $(0, \infty)$ into four intervals, $S_1 \equiv (0, .2877]$, $S_2 \equiv (.2877, .6932]$, $S_3 \equiv (.6932, 1.3863]$, and $S_4 \equiv (1.3863, \infty)$. Each S_j has probability $p_j{}^* = \frac{1}{4}$. Suppose $n = 50$, with

$r_1 = 3$ observations in S_1, $r_2 = 4$ in S_2, $r_3 = 8$ in S_3, and $r_4 = 35$ in S_4. Then

$$\Xi = -2 \sum_{j=1}^{4} r_j[\log_e(np_j^*) - \log_e(r_j)],$$

which comes out to be approximately 52.1. Since ${}_{.995}\chi^2_{(3)} = 12.8$, it is clear that H_0 is rejected at all levels $\alpha \geqq .005$, which one could anticipate from the following small table comparing $E(\tilde{r}_j \mid H_0)$ with the actual r_j:

j	$E(\tilde{r}_j \mid H_0)$	r_j
1	12.5	3
2	12.5	4
3	12.5	8
4	12.5	35

The relevant question which the test (22.5.11) answers is just how radical a departure of actual r_j's from expected r_j's (given H_0) must be, for H_0 to warrant rejecting at approximate level α.

Sometimes the actual null hypothesis is composite and parametric. For example, with $r = 1$ we may wish to test H_0: $F(\cdot) = F_N(\cdot \mid \mu, \sigma^2)$ for some μ and σ^2 versus H_A: $F(\cdot)$ is non-Normal. Then the denominator of λ^* is still $\prod_{j=1}^{m+1} (r_j/n)^{r_j}$ [with $(0/n)^0 \equiv 1$] and $\boldsymbol{\Theta}_0 \cup \boldsymbol{\Theta}_A$ contains m degrees of freedom, but the numerator of λ^* is now

$$\max\left\{\prod_{j=1}^{m+1} [p_j(\mu, \sigma^2)]^{r_j} \colon \sum_{i=1}^{m+1} p_j(\mu, \sigma^2) = 1, \mu \in R^1, \sigma^2 > 0\right\}, \tag{22.5.12}$$

where

$$p_j(\mu, \sigma^2) = \int_{S_j} f_N(x \mid \mu, \sigma^2)dx. \tag{22.5.13}$$

Performing the maximization in (22.5.12) is no easy computational task, but when n is fairly sizable it is usually approximated adequately by $\prod_{j=1}^{m+1} [p_j(\hat{\mu}, \hat{\sigma}^2)]^{r_j}$, where $\hat{\mu} = \bar{x}$ and $\hat{\sigma}^2 = (n-1)s^2/n$. Note that $\boldsymbol{\Theta}_0$ contains two degrees of freedom, since specification of μ and σ^2 completely determine $p_1(\mu, \sigma^2), \ldots, p_{m+1}(\mu, \sigma^2)$ functionally. (In problems of this kind, we always choose m to exceed the number of degrees of freedom in H_0.) From (22.5.10) it follows that $\tilde{\Xi} = -2 \sum_{j=1}^{m+1} \tilde{r}_j[\log_e(np_j(\tilde{\hat{\mu}}, \tilde{\hat{\sigma}}^2)) - \log_e(\tilde{r}_j)]$ has approximate cdf $F_{\chi^2}(\cdot \mid m-2)$, so that the test of H_0 versus H_A at approximate level α calls for rejection of H_0 if $\Xi > {}_{(1-\alpha)}\chi^2_{(m-2)}$.

More generally, under various conditions discussed by Chernoff and Lehmann [**14**], the following can be shown:

A test of H_0: $F(\cdot) \in \{F(\cdot \mid \boldsymbol{\theta})$: $\boldsymbol{\theta} \in \boldsymbol{\Theta}$, $\boldsymbol{\Theta}$ has r degrees of freedom$\}$ versus H_A: "H_0 is false" at approximate level α consists in rejecting H_0 if $\Xi > {}_{(1-\alpha)}\chi^2{}_{(m-r)}$, for

$$\Xi \equiv -2 \sum_{j=1}^{m+1} r_j[\log_e (np_j(\hat{\boldsymbol{\theta}})) - \log_e (r_j)], \tag{22.5.14}$$

where r_j denotes the number of observations in S_j and $p_j(\hat{\boldsymbol{\theta}}) \equiv P(S_j \mid \hat{\boldsymbol{\theta}})$ for $j = 1, \ldots, m + 1$; $0 \log_e (0) \equiv 0$; and $\hat{\boldsymbol{\theta}}$ denotes the maximum likelihood estimate of $\boldsymbol{\theta}$.

When H_0 is composite, one cannot *preselect* $S_1, \ldots, S_{m+1}$ to have equal probabilities given H_0; but Kendall and Stuart [**41**] argue that one can *postselect* $S_1, \ldots, S_{m+1}$ so as to have equal probabilities, without destroying the validity of the test. This is done by using the maximum likelihood estimate $\hat{\boldsymbol{\theta}}$(of the parameter $\boldsymbol{\theta}$ from the family comprising H_0) as a certainty equivalent for $\boldsymbol{\theta}$ and choosing $S_1, \ldots, S_{m+1}$ to have equal probabilities as computed from $F(\cdot \mid \hat{\boldsymbol{\theta}})$. Their argument essentially is that since (22.5.11) and (22.5.14) hold for *any* choice of $m + 1$ sets, then they hold in particular when $\{S_1, \ldots, S_{m+1}\}$ is chosen on the basis of the outcome of the experiment.

Example 22.5.3

Reconsider Example 22.5.2, now adopting the more circumspect null hypothesis H_0: $F(\cdot) \in \{F_e(\cdot \mid \lambda): \lambda > 0\}$ versus H_A: "H_0 is false." Suppose the maximum likelihood estimate $\tilde{\lambda}$ of λ is 0.1.Then $S_1 \equiv (0, 2.877]$, $S_2 \equiv (2.877, 6.932]$, $S_3 \equiv (6.932, 13.863]$, and $S_4 \equiv (13.863, \infty)$ have equal probabilities $\frac{1}{4}$ under $F_e(\cdot \mid .1)$. Further suppose that the original observations underlying the grouped data in Example 22.5.2 imply the following r_j's for the present $\{S_1, \ldots, S_4\}$:

j	r_j
1	15
2	20
3	10
4	5

We compute $\Xi \doteq 10.6$ and compare it with ${}_{(1-\alpha)}\chi^2{}_{(2)}$(why two degrees of freedom?), thus seeing that the data are just significant at the .005 level: ${}_{.995}\chi^2{}_{(2)} = 10.6$. Since 50 observations may not be large enough to make the chi-square approximation to the cdf of λ good enough, we are safer in saying that H_0 is rejected at the .01 level.

An alternative test for goodness of fit long preceded the generalized likelihood ratio test, and is called (Karl) *Pearson's chi-square test*. It is based upon precisely the same considerations of partitioning R^r into $\{S_1, \ldots, S_{m+1}\}$ and noting that the grouped-data statistic $\tilde{\mathbf{r}}$ has mf $f_{mu}{}^{(m)}(\mathbf{r} \mid n, \mathbf{p})$ for $\mathbf{p}$ determined by H_0. However, the test is based upon (13.10.4); that is, instead of comparing Ξ with ${}_{(1-\alpha)}\chi^2{}_{(m)}$ for a simple H_0, we compare Pearson's chi-square statistic g_n with ${}_{(1-\alpha)}\chi^2{}_{(m)}$. It can be shown that $\Xi_n(= \Xi)$ and g_n are *asymptotically equivalent*, in the sense that

$$\lim_{n \to \infty} P(-\epsilon < \tilde{\Xi}_n - \tilde{g}_n < \epsilon) = 1,$$

for every $\epsilon > 0$. Hence for large n the generalized likelihood ratio test (22.5.11) should yield the same conclusions for given data as the *Pearson's Chi-Square Test of Goodness of Fit*.

A test of H_0: $P = P^*$ versus H_A: $P \neq P^*$ at approximate level α consists in rejecting H_0 if $g_n > {}_{(1-\alpha)}\chi^2{}_{(m)}$, where (22.5.15)

$$g_n \equiv \sum_{j=1}^{m+1} \left[\frac{(r_j - np_j{}^*)^2}{np_j{}^*} \right].$$

Proof

The statistic g_n is large precisely when some r_j's depart substantially from their expectations $np_j{}^*$ given H_0, and hence H_0 is rejected for excessively large values of g_n. By (13.10.4) the approximate cdf of $\tilde{g}_n$ for large n is $F_{\chi^2}(\cdot \mid m)$, and hence rejection of H_0 if $g_n > {}_{(1-\alpha)}\chi^2{}_{(m)}$ implies type-I-error probability of approximately α. QED

Much attention has been devoted to generalizations of Pearson's chi-square test. In fact, our basic result (22.5.14) for generalized likelihood ratio tests of composite parametric H_0's is obtained from the corresponding result (22.5.16) for testing based upon Pearson's chi-square statistic, via asymptotic equivalence.

A test of H_0: $F(\cdot) \in \{F(\cdot \mid \boldsymbol{\theta})\text{: } \boldsymbol{\theta} \in \boldsymbol{\Theta}, \boldsymbol{\Theta}$ has r degrees of freedom$\}$ versus H_A: "H_0 is false" at approximate level α consists in rejecting H_0 if $g_n > {}_{(1-\alpha)}\chi^2{}_{(m-r)}$, for (22.5.16)

$$g_n \equiv \sum_{j=1}^{m+1} \left[\frac{r_j - np_j{}^*(\hat{\boldsymbol{\theta}})^2}{np_j{}^*(\hat{\boldsymbol{\theta}})} \right],$$

where $\hat{\boldsymbol{\theta}}$ is the maximum likelihood estimate of $\boldsymbol{\theta}$.

We caution the reader that the use of the ordinary maximum likelihood estimator $\hat{\boldsymbol{\theta}}$ of $\boldsymbol{\theta}$ in (22.5.14) and (22.5.16) can cause the distribution of $\tilde{\Xi}$ and $\tilde{g}_n$ to depart much more substantially from $F_{\chi^2}(\cdot \mid m - r)$ than they would if $\hat{\boldsymbol{\theta}}$ were replaced by the so-called *multinomial maximum likelihood estimate* $\hat{\boldsymbol{\theta}}$, defined as the maximizer of $\prod_{j=1}^{m+1} [p_j^*(\boldsymbol{\theta})]^{r_j}$. The reader is again referred to Chernoff and Lehmann [**14**] for details.

Neither the generalized likelihood ratio test nor the chi-square test of goodness of fit is consistent against alternatives P such that $P(\tilde{\mathbf{x}}_i \in S_j) = p_j^*$ for $j = 1, \ldots, m$ [why? Exercise (14)]. There are other tests of H_0: $P = P^*$ which do not share this defect. Interested readers may consult Kendall and Stuart [**41**], Vol II, 452–461, for an introduction to the important *Kolmogorov-Smirnov* tests, which make no use of grouping, and are consistent.

22.5.4 Tests of Independence in Contingency Tables

Experimenters are frequently interested in whether two (or more) attributes of a given population are distributed independently of each other in that population. For example, a marketer might be interested in whether sales of his product are different for listeners to his sponsored television show than for nonlisteners; or a politician might be interested in a possible dependence between party affiliation and level of educational attainment.

Assume that the population of interest is classified according to two criteria. Criterion one then results in a partition $\mathcal{P}_1 \equiv \{A_1, \ldots, A_s\}$ of the population, while criterion two yields a partition $\mathcal{P}_2 \equiv \{B_1, \ldots, B_t\}$. We are interested in a hypothesis about the proportions p_{ij} of the population belonging to subset, or *cell* $A_i \cap B_j$ (which is an element of the product partition $\mathcal{P}_1\mathcal{P}_2$).

Suppose a random sample of n individuals is taken and r_{ij} are found to belong to cell $A_i \cap B_j$. These data can be displayed in the form of a matrix, called a (two-way) *contingency table*.

	$B_1 \cdots$	$B_j \cdots$	B_t	*Row Sum*
A_1	$r_{11} \cdots$	$r_{1j} \cdots$	r_{1t}	$r_{1\cdot}$
$\cdot$	$\cdot$	$\cdot$	$\cdot$	$\cdot$
$\cdot$	$\cdot$	$\cdot$	$\cdot$	$\cdot$
$\cdot$	$\cdot$	$\cdot$	$\cdot$	$\cdot$
A_i	$r_{i1} \cdots$	$r_{ij} \cdots$	r_{it}	$r_{i\cdot}$
$\cdot$	$\cdot$	$\cdot$	$\cdot$	$\cdot$
$\cdot$	$\cdot$	$\cdot$	$\cdot$	$\cdot$
$\cdot$	$\cdot$	$\cdot$	$\cdot$	$\cdot$
A_s	$r_{s1} \cdots$	$r_{sj} \cdots$	r_{st}	$r_{s\cdot}$
column sum	$r_{\cdot 1} \cdots$	$r_{\cdot j} \cdots$	$r_{\cdot t}$	$r_{\cdot\cdot} = n$

where we introduce the common notational convention of replacing a subscript with a dot to denote that summation has been taken over that subscript. Thus

$$r_{i.} \equiv \sum_{j=1}^{t} r_{ij}, \qquad \text{for every } i, \tag{22.5.17a}$$

and

$$r_{.j} \equiv \sum_{i=1}^{s} r_{ij}, \qquad \text{for every } j. \tag{22.5.17b}$$

We similarly define

$$p_{i.} \equiv \sum_{j=1} p_{ij}, \qquad \text{for every } i \tag{22.5.18a}$$

and

$$p_{.j} \equiv \sum_{i=1}^{s} p_{ij}, \qquad \text{for every } j. \tag{22.5.18b}$$

Clearly, the $p_{i.}$'s and $p_{.j}$'s are the marginal proportions in the A_i's and in the B_j's, respectively.

Now, if the attribute corresponding to $\mathcal{P}_1$ is distributed independently of the attribute corresponding to $\mathcal{P}_2$, then $p_{ij} = p_{i.}p_{.j}$, for every i and j. Hence the null hypothesis of independence can be stated as

$$H_0\colon p_{ij} = p_{i.}p_{.j};\ i = 1, \ldots, s, j = 1, \ldots, t. \tag{22.5.19}$$

We now construct a test of H_0 based upon the likelihood function of $\tilde{\mathbf{r}} \equiv (\tilde{r}_{11}, \ldots, \tilde{r}_{1t}, \tilde{r}_{21}, \ldots, \tilde{r}_{2t}, \ldots, \tilde{r}_{s1}, \ldots, \tilde{r}_{s(t-1)})'$ given $\mathbf{p} \equiv (p_{11}, \ldots, p_{s(t-1)})'$, which by randomness of sampling (with replacement) is

$$L(\mathbf{p} \mid n, \mathbf{r}) = \prod_{(i,j)=(1,1)}^{(s,t)} p_{ij}{}^{r_{ij}}\ , \tag{22.5.20}$$

the $(st - 1)$-variate Bernoulli likelihood function. Clearly, $\Theta = \Theta_0 \cup \Theta_A$ has $st - 1$ degrees of freedom and

$$\max_{p \in \Theta} L(\mathbf{p} \mid n, \mathbf{r}) = \prod_{(i,j)=(1,1)}^{(s,t)} \left(\frac{r_{ij}}{n}\right)^{r_{ij}}, \tag{22.5.21}$$

with the convention that $(0/n)^0 = 1$.

If the null hypothesis (22.5.19) is true, then specifying $(s - 1)$ of the $p_{i.}$'s and $(t - 1)$ of the $p_{.j}$'s just suffices to determine all p_{ij}'s. Hence Θ^0

has $(s - 1) + (t - 1) = s + t - 2$ degrees of freedom. Furthermore, the likelihood function becomes

$$\prod_{(i,j)=(1,1)}^{(s,t)} (p_{i.}p_{.j})^{r_{ij}}$$

$$= \prod_{(i,j)=(1,1)}^{(s,t)} (p_{i.}{}^{r_{ij}} p_{j.j}{}^{r_{ij}})$$

$$= \left[\prod_{i=1}^{s} p_{i.}{}^{r_{i.}}\right]\left[\prod_{j=1}^{t} p_{,j}{}^{r_{.j}}\right];$$

that is, the product of two (independent) multivariate Bernoulli likelihood functions, of parameters with dimensions $s - 1$ and $t - 1$, respectively. Hence the maximum likelihood estimates of $p_{i.}$ and $p_{.j}$ are $\hat{p}_{i.} = r_{i.}/n$ and $r_{.j}/n$, respectively, so that

$$\max_{\mathbf{p}\in\Theta_0} L(\mathbf{p} \mid n, \mathbf{r}) = \left[\prod_{i=1}^{s} (r_{i.}/n)^{r_{i.}}\right]\left[\prod_{j=1}^{t} (r_{.j}/n)^{r_{.j}}\right], \tag{22.5.22}$$

again with the convention that $(0/n)^0 = 1$.

A test of (22.5.19) at approximate level α consists in rejecting (22.5.19) if $\Xi > {}_{(1-\alpha)}\chi^2{}_{([s-1][t-1])}$, where (22.5.23)

$$\Xi \equiv -2 \log_e (\lambda^*)$$

for

$$\lambda^* \equiv \frac{\left[\prod_{i=1}^{s} (r_{i.}/n)^{r_{i.}}\right]\left[\prod_{j=1}^{t} (r_{.j}/n)^{r_{.j}}\right]}{\prod_{(i,j)=(1,1)}^{(s,t)} (r_{ij}/n)^{r_{ij}}}.$$

Proof

It only remains to verify that subpostscript of χ^2: $(st - 1) - (s + t - 2) = (s - 1)(t - 1)$. The rest is evident from (22.5.10), (22.5.21), and (22.5.22). QED

As the reader might suspect from the preceding subsection, there is an alternative test of (22.5.19) based upon a Pearson's chi-square statistic, which for this case is defined by

$$g_n \equiv \sum_{i=1}^{s}\sum_{j=1}^{t}\left[\frac{(r_{ij} - r_{i.}r_{.j}/n)^2}{(r_{i.}r_{.j}/n)}\right]$$

$$= \sum_{i=1}^{s}\sum_{j=1}^{t}\left[\frac{(\hat{p}_{ij} - \hat{p}_{i.}\hat{p}_{.j})^2}{(\hat{p}_{i.}\hat{p}_{.j})}\right]. \tag{22.5.24}$$

Example 22.5.4

W. H. Gilby [*Biometrika* **8** (1921), 94 ff.] reported a two-way classification of 1725 school children according to (a) quality of clothing (a surrogate for home environment), and (b) intelligence. Let $\mathcal{P}_1 \equiv \{A_1, \ldots, A_4\}$ and $\mathcal{P}_2 \equiv \{B_1, B_2, B_3\}$ for $A_1 \equiv$ "very well clad," $A_2 \equiv$ "well clad," $A_3 \equiv$ "poor but passable," $A_4 \equiv$ "very badly clad," $B_1 \equiv$ "dull or worse," $B_2 \equiv$ "intelligent," and $B_3 \equiv$ "highly intelligent." (We have reduced Gilby's seven B-classifications to three.) The data were as follows:

	B_1	B_2	B_3	$r_{i\cdot}$
A_1	81	322	233	636
A_2	141	457	153	751
A_3	97	131	37	265
A_4	30	32	11	73
$r_{\cdot j}$	349	942	434	1725

Roughly one-half hour with a slide rule yields $g_{1725} \doteq 129$, which is to be compared with ${}_{(1-\alpha)}\chi^2{}_{(6)}$ and thereupon indicates that the hypothesis of independence can be rejected at significance levels far less than .005. As Exercise 15, the reader may compute Ξ and use (22.5.23) to test H_0.

So far we have assumed that individuals are randomly selected from the population (with replacement). Often, however, experiments consist in selecting predetermined numbers $r_{1\cdot}, \ldots, r_{s\cdot}$ of observations in $A_1, \ldots, A_s$, or predetermined numbers $r_{\cdot 1}, \ldots, r_{\cdot t}$ in $B_1, \ldots, B_t$. Occasionally, all $r_{1\cdot}, \ldots, r_{s\cdot}, r_{\cdot 1}, \ldots, r_{\cdot t}$ are predetermined. Such problems are referred to as having *one fixed margin* or *two fixed margins*. It is an important fact that fixing one or both margins results in the same test procedure (22.5.23) [or (22.5.23) with g_n substituted for Ξ] developed for the case in which neither margin is fixed.

A test of independence in contingency tables with one margin [say, $(r_{1\cdot}, \ldots, r_{s\cdot})'$] fixed is really a test of equality of $(t-1)$-variate Bernoulli parameters, since with $r_{i\cdot} > 0$ fixed, the mf of $(r_{i1}, \ldots, r_{i(t-1)})'$ given $r_{i\cdot}$ is $f_{mu}{}^{(t-1)}(\cdot \mid r_{i\cdot}, \mathbf{p}^i)$, where $\mathbf{p}^i \equiv p_{i\cdot}{}^{-1}(p_{i1}, \ldots, p_{i(t-1)})'$. (Why?) But $p_{i\cdot}$ is fixed and known beforehand to be $r_{i\cdot}/n$, and $p_{ij}/p_{i\cdot} = P(B_j \mid A_i)$. Hence H_0: $p_{ij} = p_{i\cdot}p_{\cdot j}$, for all i and j is equivalent to H_0: $P(B_j \mid A_i) = P(B_j)$, for all i and j, or H_0: $\mathbf{p}^i = \mathbf{p}^0 \equiv (p_{\cdot 1}, \ldots, p_{\cdot(t-1)})'$, for every i.

Example 22.5.5

Suppose that polls of voters were taken in each of four successive weeks. The potential responses were $B_1 \equiv$ "favor X," $B_2 \equiv$ "favor Y," and $B_3 \equiv$ "undecided."

week	B_1	B_2	B_3	$r_i =$ *Poll Total*
1	40	50	10	100
2	80	80	40	200
3	50	30	20	100
4	230	150	130	500
$r_{\cdot j}$	400	300	200	900

Suppose that X wishes to determine whether the campaign has affected voter preferences and decides to test H_0: $\mathbf{p}^i = \mathbf{p}^0$, for $i = 1, \ldots, 4$ versus H_A: "H_0 is false." We compute $g_{900} = 28.9$, and note that 28.9 exceeds ${}_{(1-\alpha)}\chi^2{}_{(6)}$, for all α in Table II, so that H_0 can be rejected at the .005 level of significance.

The two-way classification of the population can be generalized to an arbitrary finite number of "ways." For a three-way classification, let the third partition $\mathcal{P}_3 \equiv \{C_1, \ldots, C_u\}$, and denote by p_{ijk} and r_{ijk}, respectively the proportion and the observed number in cell $A_i \cap B_j \cap C_k$. Then straightforward generalization of the preceding analysis yields the following:

Test of mutual independence: A test of H_0: $p_{ijk} = p_{i..}p_{.j.}p_{..k}$ for every i, j, and k versus H_A: "H_0 is false" at approximate level α consists in rejecting H_0 if $\Xi > {}_{(1-\alpha)}\chi^2{}_{(stu-[s+t+u]+2)}$, where (22.5.25)

$$\Xi \equiv -2 \log_e (\lambda^*)$$

for

$$\lambda^* \equiv \frac{\left[\prod_{i=1}^{s} (r_{i..}/n)^{r_{i..}}\right]\left[\prod_{j=1}^{t} (r_{.j.}/n)^{r_{.j.}}\right]\left[\prod_{k=1}^{u} (r_{..k}/n)^{r_{..k}}\right]}{\prod_{(i,j,k)=(1,1,1)}^{(s,t,u)} (r_{ijk}/n)^{r_{ijk}}}.$$

Proof

Exercise 16.

It is also useful to test the hypothesis that the joint proportions p_{ijk} are equal for all layers k and all (i, j); that is, the partition $\mathcal{P}_3$ is independent of $\mathcal{P}_1\mathcal{P}_2$.

Test of independence of "layers" (of $\mathcal{P}_3$*)*: A test of H_0: $p_{ijk} = p_{ij.}p_{..k}$, for every i, j, and k versus H_A: "H_0 is false" at approximate level α consists in rejecting H_0 if $\Xi > {}_{(1-\alpha)}\chi^2{}_{([st-1][u-1])}$, where (22.5.26)

$$\Xi -2 \log_e (\lambda^*)$$

for

$$\lambda^* \equiv \frac{\left[\prod_{(i,j)=(1,1)}^{(s,t)} (r_{ij.}/n)^{r_{ij.}}\right]\left[\prod_{k=1}^{u} (r_{..k}/n)^{r_{..k}}\right]}{\prod_{(i,j,k)=(1,1,1)}^{(s,t,u)} (r_{ijk}/n)^{r_{ijk}}}.$$

Proof

Exercise 17.

Recent developments in contingency-table analysis are discussed by Goodman [**31**] and Mosteller [**56**].

22.5.5 Test for Equality of Poisson-Process Intensities

Assume that an experimenter is interested in whether the intensities $\lambda_1, \ldots \lambda_k$ of k Poisson processes are all equal. In practice, there is often only one *physical* process, and the λ_i's are its intensities during different time periods (for example, pre-11:00 A.M., 11:00 A.M.–2:00 P.M., and post-2:00 P.M. arrival rates at a gas station).

Suppose for $i = 1, \ldots, k$ that process i is observed and r_i occurrences were recorded in t_i time units. By (11.7.4), the statistic (r_i, t_i) is sufficient for process i (given noninformative stopping), and the likeilhood function for process i is

$$L(\lambda_i \mid r_i, t_i) = \lambda_i^{r_i} \exp[-\lambda_i t_i]. \tag{22.5.27}$$

Hence the likelihood function for $(\lambda_1, \ldots, \lambda_k)'$ is

$$L(\lambda_1, \ldots, \lambda_k \mid r_1, \ldots, r_k; t_1, \ldots, t_k)$$

$$= \exp\left[-\sum_{i=1}^{k} \lambda_i t_i\right] \prod_{i=1}^{k} \lambda_i^{r_i}. \tag{22.5.28}$$

The null hypothesis H_0: $\lambda_1 = \lambda_2 = \cdots = \lambda_k \equiv \lambda$ has one degree of freedom, and

$$\max_{\lambda} L(\lambda, \lambda, \ldots, \lambda \mid r_1, \ldots, r_k; t_1, \ldots, t_k)$$

$$= \max_{\lambda} \left\{ \exp\left[-\lambda \sum_{i=1}^{k} t_i \right] \lambda^{\Sigma_{i=1}^{k} r_i} \right\} \tag{22.5.29}$$

$$= \exp\left[-\sum_{i=1}^{k} r_i \right] \left(\frac{\sum_{i=1}^{k} r_i}{\sum_{i=1}^{k} t_i} \right)^{\Sigma_{i=1}^{k} r_i},$$

the last term following from the easily verified fact that setting

$$\lambda = \frac{\sum_{i=1}^{k} r_i}{\sum_{i=1}^{k} t_i}$$

maximizes the middle term.

There are k degrees of freedom in $\Theta = \{(\lambda_1, \ldots, \lambda_k)\colon \text{all } \lambda_i > 0\}$, and

$$\max_{\lambda_1, \ldots, \lambda_k} L(\lambda_1, \ldots, \lambda_k \mid r_1, \ldots, r_k; t_1, \ldots, t_k)$$

$$= \max_{\lambda_1, \ldots, \lambda_k} \prod_{i=1}^{k} \exp[-\lambda_i t_i] \lambda_i^{r_i}$$

$$= \prod_{i=1}^{k} \max_{\lambda_i} \{\exp[-\lambda_i t_i] \lambda_i^{r_i}\} \tag{22.5.30}$$

$$= \prod_{i=1}^{k} \exp[-r_i] \left(\frac{r_i}{t_i} \right)^{r_i}$$

$$= \exp\left[-\sum_{i=1}^{k} r_i \right] \prod_{i=1}^{k} \left(\frac{r_i}{t_i} \right)^{r_i}$$

since $\hat{\lambda}_i = r_i/t_i$.

It is now easy to conclude the following:

A test of H_0: $\lambda_1 = \lambda_2 = \cdots = \lambda_k \equiv \lambda$ versus H_A: "H_0 is false" at approximate level α consists in rejecting H_0 if

$$\Xi > {}_{(1-\alpha)}\chi^2(k-1),$$

where

$$\Xi \equiv -2 \log_e \left[\left(\frac{\sum_{i=1}^{k} r_i / \sum_{i=1}^{k} t_i}{\prod_{i=1}^{k} (r_i/t_i)^{r_i}} \right)^{\Sigma_{i=1}^{k} r_i} \right]. \tag{22.5.31}$$

Proof

Exercise 18.

22.5.6 Test for Equality of Univariate Normal Process Means

Our basic problem is similar to that considered in the preceding subsection. Assume that we have a measuring process which produces iid measurements of any quantity μ with common df $f_N(x \mid \mu, \sigma^2)$, and that this process will be used to test whether k quantities $\mu_1, \ldots, \mu_k$ are all equal.

The n_i observations on μ_i have sufficient statistic $(n_i, \bar{x}_i, s_i^2)$, which is more than sufficient if σ^2 is known, and the likelihood function of μ_i and σ^2 is

$$L(\mu_i, \sigma^2 \mid n_i, \bar{x}_i, s_i^2) = K_i(\sigma^2)^{-\frac{1}{2}n_i} \exp\left[-\tfrac{1}{2}[n_i(\bar{x}_i - \mu_i)^2 + (n_i - 1)s_i^2]/\sigma^2\right], \tag{22.5.32}$$

by (13.2.6). Hence the likelihood function of $(\mu_1, \ldots, \mu_k, \sigma^2)$ is

$$L(\mu_1, \ldots, \mu_k, \sigma^2 \mid n_1, \ldots, n_k; \bar{x}_1, \ldots, \bar{x}_k; s_1^2, \ldots, s_k^2)$$
$$= \left(\prod_{i=1}^{k} K_i\right)(\sigma^2)^{-\frac{1}{2}\Sigma_{i=1}^{k} n_i} \exp\left[-\tfrac{1}{2}\left(\sum_{i=1}^{k} n_i(\bar{x}_i - \mu_i)^2 + \sum_{i=1}^{k}(n_i - 1)s_i^2\right)/\sigma^2\right]. \tag{22.5.33}$$

When σ^2 is known, (22.5.33) can be simplified to

$$L(\mu_1, \ldots, \mu_k \mid n_1, \ldots, n_k; \bar{x}_1, \ldots, \bar{x}_k; \sigma^2)$$
$$= \left(\prod_{i=1}^{k} K'_i\right) \exp\left[-\tfrac{1}{2} \sum_{i=1}^{k} n_i(\bar{x}_i - \mu_i)^2/\sigma^2\right] \tag{22.5.34}$$

by appropriate definition of $K'_i = K_i(\sigma^2)^{-\frac{1}{2}n_i} \exp\left[-\tfrac{1}{2}(n_i - 1)s_i^2/\sigma^2\right]$.

Suppose that σ^2 is known. Then $\boldsymbol{\Theta} \equiv \{(\mu_1, \ldots, \mu_k)'\colon \text{all } \mu_i \in R^1\}$ has k degrees of freedom. The maximum of (22.5.34), attained for $\hat{\mu}_1 = \bar{x}_1, \ldots, \hat{\mu}_k = \bar{x}_k$, is

$$\max_{\mu_1,\ldots,\mu_k} L(\mu_1, \ldots, \mu_k \mid n_1, \ldots, n_k; \bar{x}_1, \ldots, \bar{x}_k; \sigma^2)$$
$$= \prod_{i=1}^{k} K'_i \tag{22.5.35}$$
$$= \left(\prod_{i=1}^{k} K_i\right)(\sigma^2)^{-\frac{1}{2}\Sigma_{i=1}^{k} n_i} \exp\left[-\tfrac{1}{2} \sum_{i=1}^{k} (n_i - 1)s_i^2/\sigma^2\right].$$

If, moreover, $H_0\colon \mu_1 = \mu_2 = \cdots = \mu_k \equiv \mu$ is true, then (22.5.34) implies that we should maximize

$$\left(\prod_{i=1}^{k} K_i\right) \exp\left[-\tfrac{1}{2} \sum_{i=1}^{k} n_i(\bar{x}_i - \mu)^2/\sigma^2\right]$$

with respect to μ, which is clearly equivalent to minimizing

$$\sum_{i=1}^{k} n_i(\bar{x}_i - \mu)^2. \tag{22.5.36}$$

As Exercise 19 the reader may verify that (22.5.36) is minimized by setting μ equal to

$$\bar{x} = \sum_{i=1}^{k} n_i\bar{x}_i / \sum_{i=1}^{k} n_i = n^{-1} \sum_{i=1}^{k} \sum_{j=1}^{n} x_{ij}, \tag{22.5.37}$$

for $n = \sum_{i=1}^{k} n_i$ and x_{ij} the jth observation on μ_i. Thus $\bar{x}$ is simply the average of *all* of the observations taken in the experiment.

Substituting (22.5.37) into (22.5.36) yields

$$\max_{\mu} L(\mu \mid n_1, \ldots, n_k; \bar{x}_1, \ldots, \bar{x}_k) = \left(\prod_{i=1}^{k} K'_i\right) \exp\left[-\tfrac{1}{2} \sum_{i=1}^{k} n_i(\bar{x}_i - \bar{x})^2/\sigma^2\right], \tag{22.5.38}$$

so that

$$\lambda^* = \exp\left[-\tfrac{1}{2} \sum_{i=1}^{k} n_i(\bar{x}_i - \bar{x})^2/\sigma^2\right]$$

and

$$\Xi \equiv \frac{\sum_{i=1}^{k} n_i(\bar{x}_i - \bar{x})^2}{\sigma^2}. \tag{22.5.39}$$

It can be shown that the *exact* distribution of $\tilde{\Xi}$ is $F_{\chi^2}(\cdot \mid k - 1)$, so that the generalized likelihood ratio test has exact level α.

A test of H_0: $\mu_1 = \mu_2 = \cdots = \mu_k \equiv \mu$ versus H_A: "H_0 is false" at level α consists in rejecting H_0 if $\Xi > {}_{(1-\alpha)}\chi^2{}_{(k-1)}$, where Ξ is as defined in (22.5.39). (22.5.40)

Suppose that σ^2 is unknown. Then

$$\boldsymbol{\Theta} \equiv \{(\mu_1, \ldots, \mu_k, \sigma^2)': \sigma^2 > 0, \text{ all } \mu_i \in R^1\}$$

has $k + 1$ degrees of freedom, while

$$\boldsymbol{\Theta}_0 \equiv \{(\mu, \sigma^2)': \sigma^2 > 0, \mu \in R^1\}$$

has two. From (22.5.33) it is clear that the maximizing value of μ_i is still $\bar{x}_i$, so that (22.5.33) becomes

$$\left(\prod_{i=1}^{k} K_i\right)(\sigma^2)^{-\frac{1}{2}n} \exp\left[-\tfrac{1}{2} \sum_{i=1}^{k} (n_i - 1)s_i^2/\sigma^2\right], \tag{22.5.41}$$

which still must be maximized with respect to σ^2. As the reader may verify in Exercise 20, the maximizing value of σ^2 is $\sum_{i=1}^{k} (n_i - 1)s_i^2 / \sum_{i=1}^{k} n_i$, so that

$$\max_{\mu_1,\ldots,\mu_k,\sigma^2} L(\mu_1, \ldots, \mu_k, \sigma^2 \mid n_1, \ldots, n_k; \bar{x}_1, \ldots, \bar{x}_k; s_1^2, \ldots, s_k^2)$$
$$= \left(\prod_{i=1}^{k} K_i\right)\left[\sum_{i=1}^{k} (n_i - 1)s_i^2 \Big/ \sum_{i=1}^{k} n_i\right]^{-\frac{1}{2}n} \exp\left[-\tfrac{1}{2}\sum_{i=1}^{k} n_i\right]. \tag{22.5.42}$$

If H_0 is true, then the maximizing value of μ regardless of σ^2 is still $\bar{x}$, as in the case of σ^2 known, and we must maximize

$$\left(\prod_{i=1}^{k} K_i\right)(\sigma^2)^{-\frac{1}{2}n} \exp\left[\left[\sum_{i=1}^{k} n_i(\bar{x}_i - \bar{x})^2 + \sum_{i=1}^{k} (n_i - 1)s_i^2\right]\Big/\sigma^2\right] \tag{22.5.43}$$

with respect to σ^2. In Exercise 20 the reader may verify that the maximum of (22.5.43) is attained at

$$\sigma^2 = \left[\frac{\sum_{i=1}^{k} n_i(\bar{x}_i - \bar{x})^2 + \sum_{i=1}^{k} (n_i - 1)s_i^2}{\sum_{i=1}^{k} n_i}\right],$$

so that

$$\max_{\mu,\sigma^2} L(\mu, \sigma^2 \mid n_1, \ldots, n_k; \bar{x}_1, \ldots, \bar{x}_k; s_1^2, \ldots, s_k^2)$$
$$= \left(\prod_{i=1}^{k} K_i\right)\left\{\left[\sum_{i=1}^{k} n_i(\bar{x}_i - \bar{x})^2 + \sum_{i=1}^{k} (n_i - 1)s_i^2\right]\Big/\sum_{i=1}^{k} n_i\right\}^{-\frac{1}{2}n} \times \exp\left[-\tfrac{1}{2}\sum_{i=1}^{k} n_i\right]. \tag{22.5.44}$$

(22.5.42) and (22.5.44) imply

$$\lambda^* = \left\{\frac{\sum_{i=1}^{k} (n_i - 1)s_i^2}{\left[\sum_{i=1}^{k} n_i(\bar{x}_i - \bar{x})^2 + \sum_{i=1}^{k} (n_i - 1)s_i^2\right]}\right\}^{-\frac{1}{2}n}$$
$$= \left\{\frac{1}{1 + \dfrac{\sum_{i=1}^{k} n_i(\bar{x}_i - \bar{x})^2}{\sum_{i=1}^{k} (n_i - 1)s_i^2}}\right\}^{-\frac{1}{2}n}. \tag{22.5.45}$$

At this point, we could briefly state that $-2 \log_e(\tilde{\lambda}^*)$ has approximately the chi-square distribution with $(k + 1) - 2 = k - 1$ degrees of freedom, but we can obtain an exact test of H_0 at little cost. The critical region of the generalized likelihood ratio test is defined by the condition $\lambda^* < w$, which by (22.5.27) is equivalent to

$$\frac{\sum_{i=1}^{k} n_i(\bar{x}_i - \bar{x})^2}{\sum_{i=1}^{k} (n_i - 1)s_i^2} > w' \tag{22.5.46}$$

for

$$w' = w^{-2/n} - 1.$$

We now obtain the exact distribution of a constant multiple of the left-hand side of (22.5.46).

Let $\tilde{\phi} \equiv [(n - k)/(k - 1)] \sum_{i=1}^{k} n_i(\tilde{\bar{x}}_i - \tilde{\bar{x}})^2 / \sum_{i=1}^{k} (n_i - 1)\tilde{s}_i^2$ for $n \equiv \sum_{i=1}^{k} n_i$ and $\tilde{\bar{x}} \equiv n^{-1} \sum_{i=1}^{k} n_i\tilde{\bar{x}}_i$. Then $\tilde{\phi}$ has df $f_F(\cdot \mid k - 1, n - k)$. (22.5.47)

Proof

$\sum_{i=1}^{k} n_i(\tilde{\bar{x}}_i - \tilde{\bar{x}})^2$ and $\sum_{i=1}^{k} (n_i - 1)\tilde{s}_i^2$ are independent, since the former is a function of $\tilde{\bar{x}}_1, \ldots, \tilde{\bar{x}}_k$, the latter is a function of $\tilde{s}_1^2, \ldots, \tilde{s}_k^2$, observations on different μ_i's are independent, and $\tilde{\bar{x}}_i$ and $\tilde{s}_i^2$ are independent, for every i. We have previously noted that the df of $\sum_{i=1}^{k} n_i(\tilde{\bar{x}}_i - \tilde{\bar{x}})^2/\sigma^2$ is $f_{X^2}(\cdot \mid k - 1)$. By (13.3.6) the df of $(n_i - 1)\tilde{s}_i^2/\sigma^2$ is $f_{X^2}(\cdot \mid n_i - 1)$, and hence by (12.2.17) the df of $\sum_{i=1}^{k} (n_i - 1)\tilde{s}_i^2/\sigma^2$ is $f_{X^2}(\cdot \mid n - k)$. Hence (13.3.9) implies that

$$\frac{\sum_{i=1}^{k} n_i(\tilde{\bar{x}}_i - \tilde{\bar{x}})^2/[\sigma^2(k - 1)]}{\sum_{i=1}^{k} (n_i - 1)\tilde{s}_i^2/[\sigma^2(n - k)]} \equiv \tilde{\phi}$$

has the asserted df. QED

A test of H_0: $\mu_1 = \mu_2 = \cdots = \mu_k \equiv \mu$ versus H_A: "H_0 is false" at level α consists in rejecting H_0 if $\phi > {}_{(1-\alpha)}F_{(k-1,n-k)}$. (22.5.48)

Proof

The proof is obvious from (22.5.47).

This subsection has dealt with a simple topic in the *analysis of variance*, a branch of statistics concerned with the comparison of Normal process means. Section 19.4 consists of a decision-theoretic approach to this problem.

22.6 TESTS OF HYPOTHESES ABOUT UNIVARIATE NORMAL PROCESS PARAMETERS

This section consists of tests of various hypotheses about one of the parameters μ, σ^2 of a univariate Normal process, under each of the two possible prior-knowledge assumptions regarding the other parameter.

One such hypothesis testing problem was introduced in Example 22.3.7. We found the following:

A test of H_0: $\mu = \mu^*$ versus H_A: $\mu \neq \mu^*$ of size α, with σ^2 known, consists in rejecting H_0 if (22.6.1)

$$\bar{x} \notin \left[\mu^* - {}_{(1-\alpha/2)}N^* \frac{\sigma}{\sqrt{n}},\ \mu^* + {}_{(1-\alpha/2)}N^* \frac{\sigma}{\sqrt{n}}\right].$$

This test is UMPU.

The critical region in (22.6.1) consists of the exceptionally small and exceptionally large values of $\bar{x}$ given the truth of H_0. Any test with such critical region is called a *two-tailed* test. A *one-tailed* test is a test with a critical region of the form $(-\infty, w)$ or (w, ∞). The tests d_δ, with critical region (δ, ∞), of H_0: $\lambda \geqq \lambda_0$ versus H_A: $\lambda < \lambda_0$ in Example 22.2.2 are one-tailed.

This section proceeds less formally than its predecessors, in that we shall not always indicate how the statistic $\tilde{\mathbf{x}}_T$ underlying the asserted test d was arrived at. Rather, we shall use the results in Section 13.4. Furthermore, the asserted optimum properties of tests will not all be derived.

The choice between one-tailed and two-tailed tests of a simple H_0: $\theta = \theta_0$ is made on the basis of H_A. If the alternative is H_A: $\theta \neq \theta_0$, then a two-tailed test is more appropriate; it stands a better chance of being unbiased. However, if the alternative is H_A: $\theta < \theta_0$ and the cdf $F(y \mid \theta)$ of $\tilde{x}_T$ given θ is a *decreasing* function of θ for each fixed y, then a one-tailed test of the form $(-\infty, w)$ is appropriate because the assumption about F implies that small values of $\tilde{x}_T$ are less likely given $\theta = \theta_0$ than given any $\theta < \theta_0$. Under the same assumption about $F(y \mid \theta)$, it follows similarly that if the alternative is H_A: $\theta > \theta_0$, then a one-tailed test of the form (w, ∞) is appropriate: large values of $\tilde{x}_T$ are less likely given $\theta = \theta_0$ than given any $\theta > \theta_0$. Such tests are unbiased. We call H_A: $\theta \neq \theta_0$ a *two-sided* alternative, while H_A: $\theta > \theta_0$ and H_A: $\theta < \theta_0$ are called *one-sided* alternatives.

The reader may verify the following result, which essentially asserts that the cdf's of our subsequent test statistics all satisfy the assumption in the preceding paragraph:

(a) $F_N(y \mid \mu, \sigma^2)$ is decreasing in μ, for every $y \in R^1$ and $\sigma^2 > 0$;
(b) $F_S(y \mid \mu, \sigma^2, \nu)$ is decreasing in μ, for every $y \in R^1$, $\sigma^2 > 0$, and $\nu > 0$; and
(c) $F_{\gamma 2}(y \mid 1/\sigma^2, k)$ is decreasing in σ^2, for every $y > 0$ and $k > 0$.

(22.6.2)

Proof

By means of (22.6.2) and the considerations in the preceding paragraph, the reader should have no difficulty in establishing the following results:

A test of H_0: $\mu = \mu^*$ versus H_A: $\mu < \mu^*$ of size α, with σ^2 known, consists in rejecting H_0 if (22.6.3)

$$\bar{x} < \mu^* - {}_{(1-\alpha)}N^* \sigma/\sqrt{n}.$$

Proof

Exercise 22.

A test of H_0: $\mu = \mu^*$ versus H_A: $\mu > \mu^*$ of size α, with σ^2 known, consists in rejecting H_0 if (22.6.4)

$$\bar{x} > \mu^* + {}_{(1-\alpha)}N^* \sigma/\sqrt{n}.$$

Proof

Exercise 22.

A test of H_0: $\mu = \mu^*$ versus H_A: $\mu \neq \mu^*$ of size α, with σ^2 unknown, consists in rejecting H_0 if (22.6.5)

$$\bar{x} \notin \left[\mu^* - {}_{(1-\alpha/2)}S^*_{(n-1)} \frac{s}{\sqrt{n}},\ \mu^* + {}_{(1-\alpha/2)}S^*_{(n-1)} \frac{s}{\sqrt{n}} \right],$$

for $s \equiv \sqrt{s^2}$ and ${}_pS^*_{(k)}$ the pth fractile of the standardized Student distribution with k degrees of freedom.

Proof

Exercise 23.

A test of H_0: $\mu = \mu^*$ versus H_A: $\mu < \mu^*$ of size α, with σ^2 unknown, consists in rejecting H_0 if (22.6.6)

$$\bar{x} < \mu^* - {}_{(1-\alpha)}S^*_{(n-1)} \frac{s}{\sqrt{n}}.$$

Proof

Exercise 24.

A test of H_0: $\mu = \mu^*$ versus H_A: $\mu > \mu^*$ of size α, with σ^2 unknown, consists in rejecting H_0 if (22.6.7)

$$\bar{x} > \mu^* + {}_{(1-\alpha)}S^*{}_{(n-1)}\frac{s}{\sqrt{n}}.$$

Proof

Exercise 24.

Tests (22.6.1) and (22.6.5) are UMPU. However, tests (22.6.3), (22.6.4), (22.6.6), and (22.6.7) are UMP and unbiased, which is better than UMPU. (Why?)

The reader should compare (22.6.5) with (21.3.2) for $\gamma \equiv 1 - \alpha$ to see that (22.6.5) is equivalent to rejection of H_0 if μ^* fails to belong to the confidence interval given by (21.3.2). One-sided, infinitely long confidence intervals correspond to the one-tailed tests.

A test of H_0: $\sigma^2 = \sigma^2{}_*$ versus H_A: $\sigma^2 \neq \sigma^2{}_*$ of size α, with μ known, consists in rejecting H_0 if (22.6.8)

$$\xi \notin \left[\frac{{}_{\alpha/2}\chi^2{}_{(n)}\sigma^2{}_*}{n}, \frac{{}_{(1-\alpha/2)}\chi^2{}_{(n)}\sigma^2{}_*}{n}\right].$$

Proof

Exercise 25.

A test of H_0: $\sigma^2 = \sigma^2{}_*$ versus H_A: $\sigma^2 < \sigma^2{}_*$ of size α, with μ known, consists in rejecting H_0 if (22.6.9)

$$\xi < \frac{{}_{\alpha}\chi^2{}_{(n)}\sigma^2{}_*}{n}.$$

Proof

Exercise 26.

A test of H_0: $\sigma^2 = \sigma^2{}_*$ versus H_A: $\sigma^2 > \sigma^2{}_*$ of size α, with μ known, consists in rejecting H_0 if (22.6.10)

$$\xi > \frac{{}_{(1-\alpha)}\chi^2{}_{n)}\sigma^2{}_*}{n}.$$

Proof

Exercise 26.

A test of H_0: $\sigma^2 = \sigma^2_*$ versus H_A: $\sigma^2 = \sigma^2_*$ of size α, with μ unknown, consists in rejecting H_0 if (22.6.11)

$$s^2 \notin \left[\frac{{}_{\alpha/2}\chi^2_{(n-1)}\sigma^2_*}{n-1}, \frac{{}_{(1-\alpha/2)}\chi^2_{(n-1)}\sigma^2_*}{n-1}\right].$$

Proof

Exercise 27.

A test of H_0: $\sigma^2 = \sigma^2_*$ versus H_A: $\sigma^2 < \sigma^2_*$ of size α, with μ unknown, consists in rejecting H_0 if (22.6.12)

$$s^2 < \frac{{}_{\alpha}\chi^2_{(n-1)}\sigma^2_*}{n-1}.$$

Proof

Exercise 28.

A test of H_0: $\sigma^2 = \sigma_*^2$ versus H_A: $\sigma^2 > \sigma_*^2$ of size α, with μ unknown, consists in rejecting H_0 if (22.6.13)

$$s^2 > \frac{{}_{(1-\alpha)}\chi^2_{(n-1)}\sigma^2_*}{n-1}.$$

Proof

Exercise 28.

The tests (22.6.9), (22.6.10), (22.6.12), and (22.6.13) are UMP and unbiased. However, (22.6.8) and (22.6.11) are not UMPU. To see why not, recall that (21.3.3) and (21.3.4) correspond to (22.6.8) and (22.6.11), respectively and are not shortest confidence intervals for σ^2. Thus UMPU tests are obtained by replacing (21.3.3) and (21.3.4) with the shortest confidence intervals for σ^2, and rejecting H_0 if σ^2_* fails to belong to those shortest confidence intervals.

By virtue of (22.6.2), certain composite null hypotheses can be tested on the basis of results presented for simple null hypotheses with one-sided alternatives. Each of the preceding UMP and unbiased tests with one-sided alternatives furnishes a critical region which corresponds to an UMP and unbiased test of H_0: "H_A is false." The relevant facts are collected in Table 22.1.

Table 22.1

H_0	H_A	*Knowledge*	*Use Rejection Criterion in*
$\mu \geqq \mu$	$\mu < \mu$	σ^2 known	(22.6.3)
$\mu \leqq \mu$	$\mu > \mu$	σ^2 known	(22.6.4)
$\mu \geqq \mu$	$\mu < \mu$	σ^2 unknown	(22.6.6)
$\mu \leqq \mu$	$\mu > \mu$	σ^2 unknown	(22.6.7)
$\sigma^2 \geqq \sigma^2$	$\sigma^2 < \sigma^2$	μ known	(22.6.9)
$\sigma^2 \leqq \sigma^2$	$\sigma^2 > \sigma^2$	μ known	(22.6.10)
$\sigma^2 \geqq \sigma^2$	$\sigma^2 < \sigma^2$	μ unknown	(22.6.12)
$\sigma^2 \leqq \sigma^2$	$\sigma^2 > \sigma^2$	μ unknown	(22.6.13)

22.7 SOME TESTS OF HYPOTHESES ABOUT k-VARIATE NORMAL PROCESS MEAN VECTORS

This section consists of the k-variate analogues of the univariate tests (22.6.1) and (22.6.5) of a simple null hypothesis H_0 against the alternative consisting of the negation of H_0. We furnish three tests, corresponding to the cases in which: (1) the process covariance matrix $\boldsymbol{\sigma}^2$ is known; (2) the relative covariances $\sigma_{gh}{}^2/\sigma_{ij}{}^2$ and hence $\boldsymbol{\sigma}_*{}^2$ are known; and (3) $\boldsymbol{\sigma}^2$ completely unknown, respectively.

A test of H_0: $\boldsymbol{\mu} = \boldsymbol{\mu}^*$ versus H_A: $\boldsymbol{\mu} \neq \boldsymbol{\mu}^*$ of size α with $\boldsymbol{\sigma}^2$ known consists in rejecting H_0 if (22.7.1)

$$n(\bar{\mathbf{x}} - \boldsymbol{\mu}^*)'\boldsymbol{\sigma}^{-2}(\bar{\mathbf{x}} - \boldsymbol{\mu}^*) > {}_{(1-\alpha)}\chi^2{}_{(k)}.$$

Proof

By (13.6.8), the df of $\tilde{Q} = n(\tilde{\bar{\mathbf{x}}} - \boldsymbol{\mu}^*)'\boldsymbol{\sigma}^{-2}(\tilde{\bar{\mathbf{x}}} - \boldsymbol{\mu}^*)$ is $f_{\chi^2}(Q \mid k)$ if H_0 is true, in which case $\bar{\mathbf{x}}$ should not be far from $\boldsymbol{\mu}^*$ and hence Q should be small. Therefore the optimal test of H_0 based upon Q should reject H_0 for excessively large Q. QED

A test of H_0: $\boldsymbol{\mu} = \boldsymbol{\mu}^*$ versus H_A: $\boldsymbol{\mu} \neq \boldsymbol{\mu}^*$ of size α, with $\boldsymbol{\sigma}_*{}^2$ known but $\sigma_*{}^2$ unknown consists in rejecting H_0 if (22.7.2)

$$\left(\frac{n}{k}\right)(\bar{\mathbf{x}} - \boldsymbol{\mu}^*)'(s_*{}^2)^{-1}\boldsymbol{\sigma}_*{}^{-2}(\bar{\mathbf{x}} - \boldsymbol{\mu}^*) > {}_{(1-\alpha)}F_{(k,k[n-1])}.$$

Proof

Exercise 29; use (13.6.11).

A test of H_0: $\boldsymbol{\mu} = \boldsymbol{\mu}^*$ versus H_A: $\boldsymbol{\mu} \neq \boldsymbol{\mu}^*$ of size α, with $\boldsymbol{\sigma}^2$ unknown consists in rejecting H_0 if (22.7.3)

$$\left(\frac{n}{k}\right)\left(\frac{n-k}{n-1}\right)(\bar{\mathbf{x}} - \boldsymbol{\mu}^*)'\mathbf{s}^{-2}(\bar{\mathbf{x}} - \boldsymbol{\mu}^*) > {}_{(1-\alpha)}F_{(k,n-k)}.$$

Proof

Exercise 30; use (13.6.10).

Naturally, in the majority of real applications $\tilde{\sigma}^2$ must be considered unknown, and often the only reason for assuming that the observations $\tilde{\mathbf{x}}_i$ have common df $f_{N^{(k)}}(\cdot \mid \mathbf{\mu}, \mathbf{\sigma}^2)$ for some $(\mathbf{\mu}, \mathbf{\sigma}^2)$ is that n is large and thus the Normality assumption is an approximation justifiable by Central Limit arguments. Hence (22.7.3) is the most useful of the three preceding tests. It is called *Hotelling's T^2-test*. [Hotelling defined the statistic

$$\tilde{T}^2 \equiv n(\bar{\tilde{\mathbf{x}}} - \mathbf{\mu})'\mathbf{s}^{-2}(\bar{\tilde{\mathbf{x}}} - \mathbf{\mu}).]$$

Example 22.7.1

Rao [**67**] wished to determine whether, on the average, cork was distributed evenly in thickness around the trunk of a cork tree. In each of 28 trees, four borings were made, one each on the north, east, south, and west sides of the trunk, and the respective amounts x_{Ni}, x_{Ei}, x_{Si}, and x_{Wi} were recorded. It was reasonable to suppose that the $(\tilde{x}_{Ni}, \tilde{x}_{Ei}, \tilde{x}_{Si}, \tilde{x}_{Wi})$'s were iid with common 4-variate Normal df $f_{N^{(4)}}(\cdot \mid \mathbf{\mu}, \mathbf{\sigma}^2)$. The null hypothesis was H_0: $\mu_N = \mu_E = \mu_S$, which is composite; but Rao defined new variables to make a simple and equivalent null hypothesis. Let

$$\tilde{\mathbf{y}}_i \equiv \begin{bmatrix} 1 & -1 & 1 & -1 \\ 0 & 0 & 1 & -1 \\ 1 & 0 & -1 & 0 \end{bmatrix} \begin{bmatrix} \tilde{x}_{Ni} \\ \tilde{x}_{Ei} \\ \tilde{x}_{Si} \\ \tilde{x}_{Wi} \end{bmatrix} \equiv \mathbf{A}\tilde{\mathbf{x}}_i.$$

Then $\tilde{\mathbf{y}}_1, \ldots, \tilde{\mathbf{y}}_{28}$ are iid with common df $f_{N^{(3)}}(\cdot \mid \mathbf{A\mu}, \mathbf{A\sigma}^2\mathbf{A}')$; and the null hypothesis is equivalent to H_0: $\mathbf{\mu}_\mathbf{y} = \mathbf{0}$, where $\mathbf{\mu}_\mathbf{y} \equiv \mathbf{A\mu}$. [That $\mu_N = \mu_E = \mu_S = \mu_W$ implies $\mathbf{\mu}_\mathbf{y} = \mathbf{0}$ is clear. The reverse implication follows from: $E(\tilde{y}_3) = 0$ implies $\mu_N = \mu_S$; $E(\tilde{y}_2) = 0$ implies $\mu_S = \mu_W$; and $\mu_N = \mu_S = \mu_W$ and $E(\tilde{y}_1) = 0$ imply $\mu_N = \mu_E = \mu_S = \mu_W$.] The statistics gathered in terms of the $\mathbf{y}_i$'s were:

$$\bar{\mathbf{y}} = (8.86, 4.50, 0.86)'$$

and

$$\mathbf{s_y}^2 = \begin{bmatrix} 128.72 & 61.41 & -21.02 \\ 61.41 & 56.93 & -28.30 \\ -21.02 & -28.30 & 63.53 \end{bmatrix},$$

from which it follows that

$$\left(\frac{n}{k}\right)\left(\frac{n-k}{n-1}\right)(\bar{\mathbf{y}} - \mathbf{0})'\mathbf{s_y}^{-2}(\bar{\mathbf{y}} - \mathbf{0}) = 6.402,$$

which is to be compared with ${}_{(1-\alpha)}F_{(3,25)} < {}_{(1-\alpha)}F_{(3,20)}$. Since ${}_{.995}F_{(3,20)} = 5.82$ from Table V, it follows that H_0 can be rejected at the .005 level. The data suggest that cork is more thickly distributed on the northern and southern sides of the trees than on the eastern and western sides.

22.8 SOME RELATED TESTS FOR INDEPENDENCE-RANDOMNESS, AND COMMON DISTRIBUTION

This section begins with a test of the composite null hypothesis that two random variables, $\tilde{w}_1$ and $\tilde{w}_2$, are independent. This null hypothesis is written H_0: $F_{12}(w_1, w_2) = F_1(w_1)F_2(w_2)$, for all $(w_1, w_2)' \in R^2$, and it can be tested by partitioning the ranges of $\tilde{w}_1$ and $\tilde{w}_2$ and using the result in Subsection 22.5.4. However, the test procedure we develop in Subsection 22.8.1 will be applicable, with slight modification, to two other problems: (a) testing the null hypothesis that observations $\tilde{w}_1, \ldots, \tilde{w}_n$ were iid with some common cdf $F(\cdot)$ (an "iid-ness" test); and (b) testing the null hypothesis that n_1 iid observations $\tilde{w}_{11}, \ldots, \tilde{w}_{1n_1}$ and n_2 iid observations $\tilde{w}_{21}, \ldots, \tilde{w}_{2n_2}$ had the same cdf.

22.8.1 A Test of Independence

Suppose an experiment consists in obtaining n iid bivariate observations $\tilde{\mathbf{w}}_1, \ldots, \tilde{\mathbf{w}}_n$ with common *continuous* cdf $F_{12}(\mathbf{w})$, and we wish to test H_0: $F_{12}(\mathbf{w}) = F_1(w_1)F_2(w_2)$, for all $(w_1, w_2)' \equiv \mathbf{w} \in R^2$ against H_A: "H_0 is false."

The basic observation upon which our test procedure rests is that *if* the $\tilde{w}_1$'s and $\tilde{w}_2$'s are independent, *then* the large values of $\tilde{w}_1$ should not always be associated with either the large or the small values of w_2.

To express this observation mathematically, we define new variables. First, for $h = 1, 2$, we define the *rank* ξ_{hj} of an observation $w_{hj} \in \{w_{h1}, \ldots, w_{hn}\}$ as one plus the number of w_{hk}'s which w_{hj} exceeds. The assumption that $F_{12}(\mathbf{w})$ is continuous implies probability zero of a tie, and hence the n observations $\mathbf{w}_1, \ldots, \mathbf{w}_n$ give rise to n pairs of ranks $\xi_1, \ldots, \xi_n$, with the property that the sets $\{\xi_{hj}: j = 1, \ldots, n\}$, for $h = 1, 2$ are each $\{1, 2, \ldots, n\}$.

Example 22.8.1

Let $\mathbf{w}_1, \ldots, \mathbf{w}_7$ be as in the following table:

	$j = 1$	2	3	4	5	6	7
w_{1j}	25	100	0	-10^{10}	3.623	18	53
w_{2j}	23	10	47	15	-20	28	-65

Then $\xi_1, \ldots, \xi_7$ are as in the following table:

	$j = 1$	2	3	4	5	6	7
ξ_{1j}	5	7	2	1	3	4	6
ξ_{2j}	5	3	7	4	2	6	1

(For example, w_{27} is the smallest w_{2j} and thus has rank $\xi_{27} = 1$; w_{25} is the next smallest w_{2j} and thus has rank $\xi_{25} = 2$; and w_{23} is the largest of the seven w_{2j}'s and thus has rank 7.)

Now, for $i = 1, \ldots, n$ we define $x_{2i} = i$, $x_{1i} = \{\xi_{1j}\colon \xi_{2j} = i\}$, and $\mathbf{x}_i = (x_{1i}, x_{2i})'$. The $\mathbf{x}_i$'s are merely a rearrangement of the ξ_j's in order of increasing ξ_{2j}.

Example 22.8.2

In Example 22.8.1, the table of $\mathbf{x}$'s is as follows:

	$i = 1$	2	3	4	5	6	7
x_{1i}	6	3	7	1	5	4	2
x_{2i}	1	2	3	4	5	6	7

Note that $(x_{11}, \ldots, x_{1n})$ is a permutation of $(1, \ldots, n)$, and that under the null hypothesis of independence all $n!$ such permutations are equally likely. On the other hand, if the correlation of $\tilde{w}_1$ and $\tilde{w}_2$ is close to one, we expect $(x_{11}, \ldots, x_{1n})$ to have low values close to the beginning and high values close to the end, and conversely for a correlation of $\tilde{w}_1$ and $\tilde{w}_2$ close to -1. Our test puts into the critical region extreme permutations and is based upon the statistic

$$\tilde{Q} \equiv \sum_{i=1}^{n-1} \sum_{j=i+1}^{n} \tilde{h}_{ij}, \tag{22.8.1}$$

where

$$h_{ij} \equiv \begin{cases} 1, & x_{1i} > x_{1j} \\ 0, & x_{1i} \leqq x_{1j} \end{cases}, \text{ for } i < j. \tag{22.8.2}$$

Example 22.8.3

For the data in Examples 22.8.1 and 22.8.2 we have the following table of h_{ij}'s:

	$j = 1$	2	3	4	5	6	7
$i = 1$		1	0	1	1	1	1
2			0	1	0	0	1
3				1	1	1	1
4					0	0	0
5						1	1
6							1
7							

It follows easily by summation that $Q = 14$.

The reader may prove as Exercise 31 that (22.8.3)

$0 \leqq Q \leqq \dfrac{n(n-1)}{2}$; $Q = 0$ for the permutation $(1, 2, 3, \ldots, n)$; and $Q = \dfrac{n(n-1)}{2}$ for the permutation $(n, n-1, \ldots, 2, 1)$.

To develop a test of H_0 based upon $\tilde{Q}$, it is necessary to determine the distribution of $\tilde{Q}$ given H_0. A recursion formula is given in Kendall and Stuart [**41**], Chapter 31, by means of which the mf of $\tilde{Q}$ given H_0 was calculated for $n = 2, 3, \ldots, 9$ and presented in Table VIII of the Appendix. For $n \geqq 10$, the following approximation to the cdf of $\tilde{Q}$ given H_0 is very accurate:

$$F(Q) \doteq F_N\left(Q + \frac{1}{2} \,\middle|\, \frac{n(n-1)}{4}, \frac{n(n-1)(2n+5)}{72}\right). \tag{22.8.4}$$

It is now clear how H_0 can be tested.

A test of H_0: $F_{12}(\mathbf{w}) = F_1(w_1)F_2(w_2)$, for all $\mathbf{w} \in R^2$ versus H_A: "H_0 is false" at level α consists in rejecting H_0 if $Q \in C$, where: (22.8.5)

(1) for $n \geqq 10$, $\mu \equiv n(n-1)/4$, and $\sigma \equiv [n(n-1)(2n+5)/72]^{1/2}$,

$$C \equiv R^1 \backslash [\mu - \tfrac{1}{2} - {}_{(1-\alpha/2)}N^*\sigma, \mu + \tfrac{1}{2} + {}_{(1-\alpha/2)}N^*\sigma];$$

and

(2) for $n < 10$, C satisfies

(a) $P(C \mid H_0) \leqq \alpha$;
(b) $P(C \cup \{x\} \mid H_0) > \alpha$, for every $x \in \{0, 1, \ldots, n(n-1)/2\} \backslash C$; and
(c) $P(y \mid H_0) \leqq P(x \mid H_0)$, for every $y \in C$ and $x \in \{0, \ldots, n(n-1)/2\} \backslash C$

Proof

Exercise 32.

The three conditions defining C for the exact test amount to putting into C all possible Q-values y of *lowest* mass given H_0 until the addition of another value would violate the type-I-error constraint α.

The reader should appreciate why the test (22.8.5) is not consistent against the alternatives for which the correlation $\rho = \rho(\tilde{w}_1, \tilde{w}_2)$ between $\tilde{w}_1$ and $\tilde{w}_2$ is zero. (Recall that $\tilde{w}_1$ and $\tilde{w}_2$ can have zero correlation and yet be dependent.)

Tests of H_0 against alternatives of the form H_A: $\rho(\tilde{w}_1, \tilde{w}_2) > 0$ and H_A: $\rho(\tilde{w}_1, \tilde{w}_2) < 0$ are easily constructed.

A test of H_0: $F_{12}(\mathbf{w}) = F_1(w_1)F_2(w_2)$, for all $\mathbf{w} \in R^2$ versus H_A: $\rho(\tilde{w}_1, \tilde{w}_2) > 0$ at level α consists in rejecting H_0 if $Q \in C$, where: (22.8.6)

(1) for $n \geqq 10$, and μ and σ as defined in (22.8.5), $C \equiv (-\infty, \mu - \frac{1}{2} - {}_{(1-\alpha)}N^*\sigma)$; and
(2) for $n < 10$, $C = \{0, 1, \ldots, x\}$, where
   (a) $P(C \mid H_0) \leqq \alpha$; and
   (b) $P(C \cup \{x + 1\} \mid H_0) > \alpha$.

Proof

Exercise 33.

A test of H_0: $F_{12}(\mathbf{w}) = F_1(w_1)F_2(w_2)$, for all $\mathbf{w} \in R^2$ versus H_A: $\rho(\tilde{w}_1, \tilde{w}_2) < 0$ at level α consists in rejecting H_0 if $Q \in C$, where: (22.8.7)

(1) for $n \geqq 10$, and μ and σ as defined in (22.8.5), $C \equiv (\mu + \frac{1}{2} + {}_{(1-\alpha)}N^*\sigma, \infty)$; and
(2) for $n < 10$, $C = \{x, x + 1, \ldots, n(n-1)/2\}$, where
   (a) $P(C \mid H_0) \leqq \alpha$; and
   (b) $P(C \cup \{x - 1\} \mid H_0) > \alpha$.

Proof

Exercise 33.

Example 22.8.4

We apply the test (22.8.5) to the data in Example 22.8.3, constructing C from Table VIII. The reader may easily verify that for $\alpha = .20$, $C = \{14, 15, \ldots, 20, 21, 0, 1, 2, \ldots, 6, 7\}$. Since $Q = 14$, it follows that we can reject H_0 at level .20 (but not at level .10).

22.8.2 A Test of "iid-ness"

If n random variables $\tilde{w}_1, \ldots, \tilde{w}_n$ are iid with some common cdf $F(\cdot)$, then one would expect absence of a trend in their observed values. That is, the high and low observed values should be reasonably well interspersed, just as in the preceding subsection the $\tilde{w}_1$-ranks should be well interspersed amongst the $\tilde{w}_2$-ranks.

Hence we define, for any sequence $\tilde{w}_1, \ldots, \tilde{w}_n$ of observed values,

$$x_{1i} = \text{rank of } w_i, \text{ and } x_{2i} = i, \text{ for } i = 1, \ldots, n. \qquad (22.8.8)$$

Under the null hypothesis that $\tilde{w}_1, \ldots, \tilde{w}_n$ are iid with a common cdf $F(\cdot)$, it follows that all $n!$ permutations $(x_{11}, \ldots, x_{1n})$ of $(1, 2, \ldots, n)$ are equally likely before the $\tilde{w}_i$'s are observed. Thus we again define $\tilde{h}_{ij}$ by (22.8.2) and $\tilde{Q}$ by (22.8.1), and note that the mf of $\tilde{Q}$ given the null hypothesis of iid-ness is the same as in Subsection 22.8.1.

A test of H_0: $w_1, \ldots, w_n$ were iid versus H_A: "H_0 is false" at level α consists in rejecting H_0 if $Q \in C$, where C is as defined in (22.8.5). (22.8.9)

Proof

Exercise 34.

Often we are concerned with particular alternatives to the iid-ness hypothesis, alternatives which specify a trend toward increasing or decreasing $\tilde{w}_i$'s as i increases. An *increasingness* trend is present if $F_1(w) > F_2(w) > \ldots > F_n(w)$, for all w, while a *decreasingness* trend is present if $F_1(w) < F_2(w) < \ldots < F_n(w)$, for all w, where $F_i(\cdot)$ denotes the cdf of $\tilde{w}_i$. To see that this is so, note that if $F_1(w) > F_2(w)$, for every $w \in R^1$, then the pth fractile of $\tilde{w}_1$ is *smaller* than the pth fractile of $\tilde{w}_2$, for every p. Under these circumstances, $\tilde{w}_2$ is said to be *stochastically larger* than $\tilde{w}_1$.

The reader will observe that an increasingness trend alternative to iid-ness corresponds to a positive correlation alternative to independence in the preceding subsection; the same holds true for decreasingness corresponding to negative correlation. Hence we obtain the one-tailed tests.

A test of H_0: $w_1, \ldots, w_n$ were iid versus H_A: $F_1(\cdot) > F_2(\cdot) > \cdots > F_n(\cdot)$ (increasingness) at level α consists in rejecting H_0 if $Q \in C$, where C is as defined in (22.8.6). (22.8.10)

A test of H_0: $w_1, \ldots, w_n$ were iid versus H_A: $F_1(\cdot) < F_2(\cdot) < \cdots < F_n(\cdot)$ (decreasingness) at level α consists in rejecting H_0 if $Q \in C$, where C is as defined in (22.8.7). (22.8.11)

Example 22.8.5

Twenty observations $w_1, \ldots, w_{20}$ were, in order taken: 15, 42, 9, 3, 18, 16, 30, 86, 48, 50, 63, 39, 58, 71, 75, 64, 68, 91, 93, 95. The corresponding sequence of ranks is 3, 8, 2, 1, 5, 4, 6, 17, 9, 10, 12, 7, 11, 15, 16, 13, 14, 18, 19, 20. Hence $Q = 26$. To test the iid-ness hypothesis against the increasingness alternative, we compute $\mu = (20)(19)/4 = 95$, $\sigma = [(20)(19)(45)/72]^{1/2} \doteq 4.87$,

$$C = (-\infty, 94.5 - 4.87_{(1-\alpha)}N^*),$$

and thus see that $Q \in C$ for all but infinitesimal α's. Thus, in particular, H_0 is rejected at the .001 level.

22.8.3 A Test of Common Distribution

We now consider testing the null hypothesis that n_1 iid observations $w_{11}, \ldots, w_{1n_1}$ and n_2 iid observations $w_{21}, \ldots, w_{2n_2}$ all came from the same continuous cdf $F(\cdot)$ rather than from different continuous cdf's $F_1(\cdot)$ and $F_2(\cdot)$. The test we shall develop is called the *Wilcoxon-Mann-Whitney test* and is related to the tests of iid-ness and independence introduced earlier in this section.

The test is based upon the ranks of the w_1's and w_2's *relative to the pooled set* $\{w_{11}, \ldots, w_{1n_1}, w_{21}, \ldots, w_{2n_2}\}$ of $n_1 + n_2$ observations. We define $n \equiv n_1 + n_2$ new bivariate observations $\mathbf{x}_i$ by setting $\mathbf{x}_{2i} = i$ for $i = 1, \ldots, n$ and

$$x_{1i} = \begin{cases} 2, & i\text{th smallest of the } n \text{ observations is a } w_1 \\ 1, & i\text{th smallest of the } n \text{ observations is a } w_2. \end{cases} \tag{22.8.12}$$

We continue to define $\tilde{h}_{ij}$ and $\tilde{Q}$ by (22.8.2) and (22.8.1), respectively. Note that Q is the total number of times a w_2-observation exceeds a w_1-observation.

Example 22.8.6

Suppose $n_1 = 5$ iid observations 10, 23, 16, 47, and 8 are taken from one population and $n_2 = 4$ iid observations 14, 27, 32, and 7 are taken from another population. Then $\mathbf{x}_1, \ldots, \mathbf{x}_9$ are given by

$i = x_{2i}$	1	2	3	4	5	6	7	8	9
Pooled data	7	8	10	14	16	23	27	32	47
x_{1i}	2	1	1	2	1	1	2	2	1

Then $Q = 10$, which can be verified by computing the h_{ij}'s and then Q, or simply by observing that all five w_2's exceed the smallest w_1, three exceed the next smallest, and one exceeds the two largest w_1's.

The range of Q is $\{0, 1, \ldots, n_1 n_2\}$. The reader can verify that $Q = 0$ only if every w_1 exceeds every w_2, and $Q = n_1 n_2$ only if every w_2 exceeds every w_1. If the null hypothesis H_0: $F_1(\cdot) = F_2(\cdot)$ is true, then Q should not deviate too far from $n_1 n_2/2$.

The distribution of Q under H_0 is *not* the same as in the preceding subsections. It depends upon both n_1 and n_2 and has been tabulated for small sample sizes by Mann and Whitney [**54**]. When each of n_1 and n_2 is at all large (say, greater than ten), the following approximation is adequate:

$$F(Q) \doteq F_N\left(Q + \frac{1}{2} \,\middle|\, \frac{n_1 n_2}{2}, \frac{n_1 n_2(n_1 + n_2 + 1)}{12}\right). \tag{22.8.13}$$

A test of H_0: $F_1(\cdot) = F_2(\cdot)$ versus H_A: "H_0 is false" at approximate level α consists in rejecting H_0 if (22.8.14)

$$Q \notin [\mu - \tfrac{1}{2} - {}_{(1-\alpha/2)}N^*\sigma,\ \mu + \tfrac{1}{2} + {}_{(1-\alpha/2)}N^*\sigma],$$

where $\mu \equiv n_1 n_2/2$ and $\sigma \equiv [n_1 n_2(n_1 + n_2 + 1)/12]^{1/2}$.

Proof

Exercise 35.

One-tailed tests are appropriate for the one-sided alternatives $F_1(\cdot) < F_2(\cdot)$ and $F_1(\cdot) > F_2(\cdot)$, that the w_1-observations are stochastically larger and smaller respectively than the w_2-observations.

A test of H_0: $F_1(\cdot) = F_2(\cdot)$ versus H_A: $F_1(\cdot) < F_2(\cdot)$ at approximate level α consists in rejecting H_0 if (22.8.15)

$$Q < \mu - \tfrac{1}{2} - {}_{(1-\alpha)}N^*\sigma,$$

where μ and σ are as defined in (22.8.14).

Also, we have the following:

A test of H_0: $F_1(\cdot) = F_2(\cdot)$ versus H_A: $F_1(\cdot) > F_2(\cdot)$ at approximate level α consists in rejecting H_0 if (22.8.16)

$$Q > \mu + \tfrac{1}{2} + {}_{(1-\alpha)}N^*\sigma,$$

where μ and σ are as defined in (22.8.14).

Proof

Exercise 35.

22.9 GENERAL COMMENTS ON HYPOTHESIS TESTING

22.9.1 Hypothesis Tests and Scientific Practice

We shall describe the most common orthodox approach to testing scientific hypotheses, and then comment on a logical fallacy underlying the editorial policies of some empirical journals regarding publication of test results.

Suppose a scientist wishes to "establish" a hypothesis A, say, that more than 40 percent of humans are males. Established canons of scientific practice caution against excessively hasty acceptance of conjectures, and hence our scientist will wish to guard against a rash acceptance of his hypothesis. He will seek an experiment, the outcome of which depends, probabilistically, upon the true proportion of males amongst humans (for example, he will consider a random sample of n humans with replacement), and he will try to *reject the negation* of A, say $B \equiv$ "at most 40 percent of humans are males," at a very low significance level. Thus the null hypothesis H_0: B is to be tested at a low significance level α. If he obtains an outcome significant at a sufficiently low α, he will be convinced of the truth of H_A: A (which is what he wanted to establish in the first place).

Note carefully that by playing devil's advocate with himself, the scientist's guarding against erroneously *rejecting* H_0: B by choosing α small is tantamount to guarding against erroneously *accepting* H_A: A. The interpretation is, of course, cleanest if the test is unbiased and reasonably powerful.

Example 22.9.1

Let p denote the proportion of males to all humans. Then H_0: $B =$ H_0: $p \leqq .40$ and H_A: $A = H_A$: $p > .40$ in the preceding discussion. Suppose the scientist takes a random sample of 100,000 humans and finds 50,000 males. Since the number $\tilde{r}$ of males in 100,000 humans had mf $f_b(r \mid 100{,}000, p)$, a critical region can be constructed from the Normal approximation

$$F_b(r \mid 100{,}000, p) \doteq F_N(r \mid 100{,}000p, 100{,}000p(1-p)).$$

Power is maximized with a critical region of the form $\{r\colon r > x\}$; and the reader may verify that

$$P_N(\tilde{r} > x \mid 40{,}000, 24{,}000) = \max_{0<p\leqq .40} P_N(\tilde{r} > x \mid 100{,}000p, 100{,}000p(1-p)),$$

whenever $x \geqq 40{,}000$. Hence we solve for x in terms of α:

$$\begin{aligned} P_N(\tilde{r} > x \mid 40{,}000, 24{,}000) & \\ &= 1 - F_N(x-1 \mid 40{,}000, 24{,}000) \\ &= 1 - F_{N*}\left(\frac{x - 40{,}001}{\sqrt{24{,}000}}\right) \\ &= \alpha \\ &= 1 - F_{N*}({}_{(1-\alpha)}N^*), \end{aligned}$$

so that $[x - 40{,}001]/\sqrt{24{,}000} = {}_{(1-\alpha)}N^*$, from which we obtain

$$x = 40{,}001 + (154.92)_{(1-\alpha)}N^*.$$

Now $r = 50{,}000$ is in $(x, 100{,}000]$ provided that ${}_{(1-\alpha)}N^* < 64.5$, which is true unless α is miniscule. Hence the scientist can reject H_0: $p \leqq .40$, thus establishing what he wished to; namely, H_A: $p > .40$.

Suppose the result of an experiment is *not* significant at the chosen level α. Can the scientist claim to have "proved" H_0? Most statisticians and philosophers of science answer in the negative, saying that hypotheses can be "disproved," "rejected," or "refuted," but not "proved."

Example 22.9.2

In Example 22.9.1, an outcome of $r = 40{,}100$ is not significant at the $\alpha = .05$ level, and hence H_0 cannot be rejected at that level. But obtaining $r = 40{,}100$ is eminently reasonable if H_0 is false and $p \doteq .401$. All we can say is that $r = 40{,}000$ is not extreme enough to warrant rejecting H_0; it by no means "proves" H_0. To "prove" H_0 we should have to "disprove" H_A and hence test $H'_0 = H_A$: $p > .40$ versus $H'_A = H_0$: $p \leqq .40$ at a low level of significance. The reader can verify as Exercise 36 that $r = 30{,}000$ would result in rejection of H'_0 at any reasonably positive significance level α.

Some editors of scientific journals have suggested that no empirical results be published which are not significant at the .05 level, reasoning that readers will thereby be assured that at most five percent of such results will be false. This reasoning is fallacious, as Pratt, Raiffa, and Schlaifer [**64**] point out via the following example: Suppose 1000 researchers formulate 1000 different *false* conjectures as alternative hypotheses, and test their negations (as null hypotheses) at size $\alpha = .05$. Then approximately 50 such (true) null hypotheses will be rejected; and 50 papers will appear announcing their "scientifically proven" but false conjectures.

We can shed some additional light on the difficulties inherent in the preceding editorial policy by considering prior and posterior probabilities. Let $\Theta = \{\theta_0, \theta_A\}$, $Z(e) = \{z_1, z_2\}$, $C = \{z_2\}$, $P(z_2 \mid \theta_0) = \alpha$, and $P(z_1 \mid \theta_A) = \beta$. Then $P(\theta_0 \mid z_2, e) = P(\theta_0)\alpha/[P(\theta_0)\alpha + (1 - P(\theta_0))(1 - \beta)]$, and hence $P(\theta_0 \mid z_2, e) \leqq \alpha$ if and only if

$$P(\theta_0) \leqq \frac{(1 - \beta)}{(1 - \beta) + (1 - \alpha)}.$$

Therefore the *posterior* probability of H_A: $\theta = \theta_A$ being false is no greater than α if and only if the *prior* probability of H_A's falsity is small enough, and "small enough" depends upon the relationship between α and the power $1 - \beta$ of the test. In the preceding paragraph, we assumed that all conjectures were false, which corresponds to $P(\theta_0) = 1$, implying $P(\theta_0 \mid z, e) = 1$ no matter how close α is to zero (so long as $\alpha \neq 0$).

Ex post facto censoring of experimental results can produce the same difficulty in interpreting significance. Suppose H_A = "John Doe has ESP (extrasensory perception)," and H_0 = "H_A is false." Various simple card-naming tests for ESP have been devised. Suppose one such is run 1000 times on Doe, but that only those outcomes significant at the .05 level were retained, because of Doe's claim that the nonsignificant results were attributable to his not being in the properly attuned state of mind at the time. Under such conditions, and given enough time, it is easy enough to "prove" that anyone has ESP. The problem here is that relevant information has been discarded on the basis of untestable assertions regarding states of mind.

22.9.2 A Difficulty with Simple Null Hypotheses

Researchers often test simple null hypotheses of the form H_0: $\theta = \theta_0$ or H_0: $P(\cdot) = P^*(\cdot)$ even though they really do not believe that the null hypothesis can be precisely true. Therefore any *consistent* test should reject H_0 for a sufficiently large sample size n. Since large samples are better than small ones (when cost is not considered), why bother to test H_0? This point was made persuasively by Berkson in 1938 [**5**], regarding the composite null hypothesis that the data-generating process is Normal. (Berkson's P is that α at which the observed chi-square statistic is just barely significant.)

> I believe that an observant statistician who has had any considerable experience with applying the chi-square test repeatedly will agree with my statement that, as a matter of observation, when the numbers in the data are quite large, the P's tend to come out small. Having observed this, and on reflection, I make the following dogmatic statement, referring for illustration to the normal curve: If the normal curve is fitted to a body of data representing any real observations whatever of quantities in the physical world, then if the number of observations is extremely large—for instance, on the order of 200,000—the chi-square P will be small beyond any usual limit of significance.
>
> This dogmatic statement is made on the basis of an extrapolation of the observation referred to and can also be defended as a prediction from *a priori* considerations. For we may assume that it is practically certain that any series of real observations does not actually follow a normal curve *with absolute exactitude* in all respects, and no matter how small the discrepancy between the normal curve and the true curve of observations, the chi-square P will be small if the sample has a sufficiently large number of observations in it.
>
> If this be so, then we have something here that is apt to trouble the conscience of a reflective statistician using the chi-square test. For I suppose it would be agreed by statisticians that a large sample is always better than a small sample. If, then, we know in advance the P that will result from an application of a chi-square test to a large sample, there would seem to be no use in doing it on a smaller one. But since the result of the former test is known, it is no test at all![1]

[1]J. Berkson, "Some Difficulties of Interpretation Encountered in the Application of the Chi-Square Test," *Journal of the American Statistical Association*, **33** (1938), 526–536. Reprinted with permission of author and publication.

See also Example 22.3.13, which expresses this difficulty with simple null hypotheses in Bayesian terms.

The counterargument is that such tests serve to determine whether the null hypothesis is sufficiently false to matter for purposes of predicting outcomes of small samples. Thus a test of H_0: $\theta = \theta_0$ based on a sample size n is really a test of whether θ is sufficiently far from θ_0 to make predictions based on samples of size n and $f(x \mid \theta_0)$ differ substantially from predictions based upon the same samples but assuming $f(x \mid \theta)$. Now, the catch here is *substantially*, and to resolve such issues completely it is often necessary to appeal to a loss function and prior probabilities or dgf's.

22.9.3 On Choosing α and n

It has long been argued that α should be chosen with due regard for the relative gravity of type I and type II errors. Moreover, it was clear in Section 22.4 that in tests of simple hypotheses against simple alternatives, the prior probabilities were also relevant in determining the optimal (admissible) test: *ceteris paribus*, α should decrease with $P(\theta_0)$. Finally, confining our attention to admissible tests, it is apparent that *both* error probabilities (and risks) can be decreased only by increasing the informativeness of e, which in practice usually means taking a large sample and thereby sustaining a larger cost of experimentation.

Many statisticians will argue that experimentation cost, losses, (and sometimes prior probabilities) should be taken into account *informally* when α, ϵ_{II}, and n are "balanced." The difficulty lies in thinking meaningfully about α, ϵ_{II}, and n on a purely intuitive level. A Bayesian would argue that judgment is more clearly brought to bear upon the prior probabilities, terminal losses, and experimentation costs which, once specified quantitatively, determine the optimal test (and thus also α, ϵ_{II}, and n) up to indifference if not uniquely. The interested reader may verify that this program is precisely what was carried out in Section 19.3 when $k = 2$.

It is all too common in practice for researchers to choose a conventional significance level such as .05, .01, or .001 without regard to the problem at hand.

22.9.4 On Nonparametric Tests

The three previous subsections have isolated some negative features of orthodox practice in hypothesis testing, and the implication has been that a fully Bayesian treatment can do better. But what about nonparametric situations? Even leaving aside losses and thus considering the Bayesian procedures of Subsections 22.3.2 and 22.3.3, we run into an essential difficulty immediately.

Any Bayesian procedure is based upon posterior probabilities, which are derived from prior probabilities (or dgf's) and the conditional probabilities

$P(z \mid \theta, e)$ [or dgf's $f(z \mid \theta, e)$]. But with nonparametric hypotheses, there are many poorly structured alternatives. No human decision-maker can assess a prior distribution over (appropriate subsets of) (P: $P \neq P^*$, P a probability distribution on R^1). Hence it appears that nonparametric problems leave Bayesians at the starting gate, whereas the validity of any orthodox test depends only upon the null hypothesis. That is, use of a goodness-of-fit test of H_0: $P = P^*$ *will not* erroneously reject H_0 with probability exceeding α.

But is *validity* of a test sufficient? Most statisticians do not think so, in view of the importance of power considerations. However, power is a function on the *alternatives*, and hence the sophisticated orthodox tester faces the same difficulties as the Bayesian. Concepts such as *relative efficiency* have been introduced for nonparametric (and parametric) tests, and they are frequently measured relative to standard tests under a greatly restricted alternative hypothesis; for example, H_A: $F(\cdot) = F_N(\cdot \mid \mu, \sigma^2)$ for some (μ, σ^2) instead of H_A: $F(\cdot) \neq F^*(\cdot)$.

Nevertheless, at the present time the orthodox procedures in the nonparametric cases do furnish information which does not appear available in Bayesian form. [One exception occurs when data are grouped and the grouping is predetermined. An m-variate beta prior df can be assessed for $\tilde{p} \equiv (\tilde{p}_1, \ldots, \tilde{p}_m)'$, revised in light of the sample, used to determine a $100(1 - \alpha)$ percent credible region I for $\tilde{\mathbf{p}}$, and H_0: $P = P^*$ rejected if the resulting point $\mathbf{p}^* \notin I$. The reader may carry through the details as Exercise 37.]

EXERCISES

1. Prove that the correspondence between tests d and critical regions C is a bijection.
2. Show that $R(d, \boldsymbol{\theta}) = \epsilon(\boldsymbol{\theta} \mid e, d)$for all $\boldsymbol{\theta} \in \boldsymbol{\Theta}_0 \cup \boldsymbol{\Theta}_A$ in Example 22.2.4.
3. Prove (22.2.14).
4. Complete the proof of (22.2.15).
5. Prove (22.3.2).
6. Prove that $\lim_{n \to \infty} P(C_{d_n} \mid \mu, n) = 1$ for $\mu < \mu^*$ in Example 22.3.8.
7. Show that the following is true for the univariate Normal process with σ^2 known:

A test of H_0: $\mu \in [\mu^* - \Delta, \mu^* + \Delta]$ versus H_A: "H_0 is false" at level α consists in rejecting H_0 if (22.10.1)

$$\bar{x} \notin \left[\mu^* - \Delta - \frac{{}_{(1-\alpha/2)}N^*\sigma}{\sqrt{n}}, \mu^* + \Delta + \frac{{}_{(1-\alpha/2)}N^*\sigma}{\sqrt{n}}\right].$$

8. Prove (22.4.1).

9. In Example 22.4.6, show that

$$\bar{x}\begin{Bmatrix} < \\ > \end{Bmatrix} \log_e (w)\frac{\sigma^2/n}{\mu_1 - \mu_2} + (\mu_1 + \mu_2)/2,$$

if $\mu_1 \begin{Bmatrix} > \\ < \end{Bmatrix} \mu_2$.

10. Prove (22.4.7).

11. Prove (22.5.2).

12. Prove (22.5.4).

13. Prove (22.5.10).

14. Why is neither the generalized likelihood ratio test nor the Pearson's chi-square test of goodness of fit consistent against alternatives P such that

$$P(\tilde{\mathbf{x}}_i \in S_j) = p_j^*, \text{ for } j = 1, \ldots, m?$$

15. Test independence in Example 22.5.4 by the generalized likelihood ratio test.

16. Prove (22.5.25).

17. Prove (22.5.26).

18. Prove (22.5.31).

19. Show that $\bar{x} \equiv n^{-1} \sum_{i=1}^{k} \sum_{j=1}^{n_i} x_{ij}$ minimizes (22.5.36).

20. Prove (22.5.42) and (22.5.44).

21. Prove (22.6.2).

22. Prove (22.6.3) and (22.6.4).

23. Prove (22.6.5).

24. Prove (22.6.6) and (22.6.7).

25. Prove (22.6.8).

26. Prove (22.6.9) and (22.6.10).

27. Prove (22.6.11).

28. Prove (22.6.12) and (22.6.13).

29. Prove (22.7.2).

30. Prove (22.7.3).

31. Prove (22.8.3).

32. Prove (22.8.5).

33. Prove (22.8.6) and (22.8.7).

34. Prove (22.8.9).

35. Prove (22.8.14) through (22.8.16).

36. In Example 22.9.2, show that H_0': $p > .40$ can be rejected at small levels α if $r = 30{,}000$.

37. Consider a partition of the range of observations $\tilde{\mathbf{x}}$ into $m + 1$ elements $S_1, \ldots, S_{m+1}$; let $\mathbf{p} \equiv (p_1, \ldots, p_m)'$, with $p_j \equiv P(\tilde{\mathbf{x}} \in S_j)$, for $j = 1, \ldots, m$; and suppose $\tilde{\mathbf{p}}$ is given a prior df $f_\beta^{(m)}(\mathbf{p} \mid \boldsymbol{\rho}, \nu)$. Show that the following is true:

A Bayesian test of H_0: $P = P^*$ versus H_A: $P \neq P^*$, of the sort considered in Subsection 22.3.2 and at approximate level α, consists in rejecting H_0 if (22.10.2)

$$g_{\nu+n} > {}_{(1-\alpha)}\chi^2{}_{(m)},$$

where

$$g_{\nu+n} \equiv (\nu + n) \sum_{i=1}^{m+1} \frac{(p^*{}_i - \pi_i)^2}{\pi_i},$$

for

$$\pi_i \equiv \frac{(\rho_i + r_i - 1)}{(\nu + n - k - 1)}.$$

[*Hint*: See (13.10.6).]

38. *Tests Comparing Two Univariate Normal Process Means*: Notation in this exercise is consistent with that in (21.3.7) and (21.3.8). Prove the following:

A test of H_0: $\Delta\mu = k$ versus H_A: $\Delta\mu \neq k$ of size α, when $\sigma_1{}^2$ and $\sigma_2{}^2$ are both known, consists in rejecting H_0 if (22.10.3)

$$\Delta\bar{x} \notin [k - {}_{(1-\alpha/2)}N^*\sigma(n_1, n_2),\ k + {}_{(1-\alpha/2)}N^*\sigma(n_1, n_2)];$$

A test of H_0: $\Delta\mu = k$(or of H_0: $\Delta\mu \leqq k$) versus H_A: $\Delta\mu > k$ at level α, when $\sigma_1{}^2$ and $\sigma_2{}^2$ are both known, consists in rejecting H_0 if (22.10.4)

$$\Delta\bar{x} > k + {}_{(1-\alpha)}N^*\sigma(n_1, n_2);$$

A test of H_0: $\Delta\mu = k$ (or of H_0: $\Delta\mu \geqq k$) versus H_A: $\Delta\mu < k$ at level α, when $\sigma_1{}^2$ and $\sigma_2{}^2$ are both known, consists in rejecting H_0 if (22.10.5)

$$\Delta\bar{x} < k - {}_{(1-\alpha)}N^*\sigma(n_1, n_2);$$

A test of H_0: $\Delta\mu = k$ versus H_A: $\Delta\mu \neq k$ of size α, when $\sigma_1{}^2$ and $\sigma_2{}^2$ are unknown but assumed equal, consists in rejecting H_0 if (22.10.6)

$$\Delta\bar{x} \notin [k - {}_{(1-\alpha/2)}S^*{}_{(n_1+n_2-2)}s(n_1, n_2),\ k + {}_{(1-\alpha/2)}S^*{}_{(n_1+n_2-2)}s(n_1, n_2)];$$

A test of H_0: $\Delta\mu = k$ (or of H_0: $\Delta\mu \leqq k$) versus H_A: $\Delta\mu > k$ at level α, when σ_1^2 and σ_2^2 are unknown but assumed equal, consists in rejecting H_0 if (22.10.7)

$$\Delta\bar{x} > k + {}_{(1-\alpha)}S^*{}_{(n_1+n_2-2)}s(n_1, n_2);$$

A test of H_0: $\Delta\mu = k$ (or of H_0: $\Delta\mu \geqq k$) versus H_A: $\Delta\mu < k$ at level α, when σ_1^2 and σ_2^2 are unknown but assumed equal, consists in rejecting H_0 if (22.10.8)

$$\Delta\bar{x} < k - {}_{(1-\alpha)}S^*{}_{(n_1+n_2-2)}s(n_1, n_2).$$

39. *Tests Comparing Two Univariate Normal Process Variances*: Notation in this exercise is consistent with that in (21.3.9). Show the following:

A test of H_0: $\sigma_2^2/\sigma_1^2 = k(>0)$ versus H_A: $\sigma_2^2/\sigma_1^2 \neq k$ at level α consists in rejecting H_0 if (22.10.9)

$$\frac{\psi_2}{\psi_1} \notin [{}_{\alpha/2}F_{(\nu_2,\nu_1)}k,\ {}_{(1-\alpha/2)}F_{(\nu_2,\nu_1)}k];$$

A test of H_0: $\sigma_2^2/\sigma_1^2 = k(>0)$ (or of H_0: $\sigma_2^2/\sigma_1^2 \leqq k$) versus H_A: $\sigma_2^2/\sigma_1^2 > k$ at level α consists in rejecting H_0 if (22.10.10)

$$\frac{\psi_2}{\psi_1} > {}_{(1-\alpha)}F_{(\nu_2,\nu_1)}k;$$

A test of H_0: $\sigma_2^2/\sigma_1^2 = k(>0)$ (or of H_0: $\sigma_2^2/\sigma_1^2 \geqq k$) versus H_A: $\sigma_2^2/\sigma_1^2 < k$ at level α consists in rejecting H_0 if (22.10.11)

$$\frac{\psi_2}{\psi_1} < {}_{\alpha}F_{(\nu_2,\nu_1)}k.$$

40. *Tests Comparing Two k-Variate Normal Process Means*: Notation in this exercise is consistent with that in (21.3.10) and (21.3.11). Show the following:

A test of H_0: $\Delta\boldsymbol{\mu} = \mathbf{k}$ versus H_A: $\Delta\boldsymbol{\mu} \neq \mathbf{k}$ of size α, when $\boldsymbol{\sigma}_1^2$ and $\boldsymbol{\sigma}_2^2$ are both known, consists in rejecting H_0 if (22.10.12)

$$(\mathbf{k} - \Delta\bar{\mathbf{x}})'(n_1^{-1}\boldsymbol{\sigma}_1^2 + n_2^{-1}\boldsymbol{\sigma}_2^2)^{-1}(\mathbf{k} - \Delta\bar{\mathbf{x}}) > {}_{(1-\alpha)}\chi^2{}_{(k)};$$

A test of H_0: $\Delta\boldsymbol{\mu} = \mathbf{k}$ versus H_A: $\Delta\boldsymbol{\mu} \neq \mathbf{k}$ of size α, when $\boldsymbol{\sigma}_1^2$ and $\boldsymbol{\sigma}_2^2$ are unknown but assumed equal, consists in rejecting H_0 if (22.10.13)

$$(\mathbf{k} - \Delta\bar{\mathbf{x}})'S^{-2}(\mathbf{k} - \Delta\bar{\mathbf{x}}) > {}_{(1-\alpha)}F_{(k,n_1+n_2-k-1)}.$$

41. *Large-Sample Tests Based Upon Maximum Likelihood Estimators*. In the notation of (21.4.5), show the following:

A test of H_0: $\boldsymbol{\theta} \in \boldsymbol{\Theta}_0 \subset R^k$ versus H_A: $\boldsymbol{\theta} \notin \boldsymbol{\Theta}_0$ of approximate size α consists in rejecting H_0 if (22.10.14)

$$n(\boldsymbol{\theta}^* - \hat{\boldsymbol{\theta}})'\boldsymbol{\sigma}^{-2}(\hat{\boldsymbol{\theta}})(\boldsymbol{\theta}^* - \hat{\boldsymbol{\theta}}) > {}_{(1-\alpha)}\chi^2{}_{(k)}, \text{ for every } \boldsymbol{\theta}^* \in \boldsymbol{\Theta}_0.$$

42. (Continuation of 41.) Let $\boldsymbol{\theta} \equiv (\boldsymbol{\theta}'_1, \boldsymbol{\theta}'_2)'$ with $\boldsymbol{\theta}_1 \in R^r$, for $r < k$, and let $\boldsymbol{\Theta}_0{}^r \subset R^r$. Let $\hat{\boldsymbol{\theta}} = (\hat{\boldsymbol{\theta}}'_1, \hat{\boldsymbol{\theta}}'_2)'$ denote the maximum likelihood estimate of $\boldsymbol{\theta} \in R^k$. Show the following:

A test of H_0: $\boldsymbol{\theta}_1 \in \boldsymbol{\Theta}_0{}^r$ versus H_A: $\boldsymbol{\theta}_1 \notin \boldsymbol{\Theta}_0{}^r$ of approximate size α consists in rejecting H_0 if $n(\boldsymbol{\theta}_1{}^* - \hat{\boldsymbol{\theta}}_1)'[\boldsymbol{\sigma}_{11}{}^2(\hat{\boldsymbol{\theta}})]^{-1}(\boldsymbol{\theta}_1{}^* - \hat{\boldsymbol{\theta}}_1) > {}_{(1-\alpha)}\chi^2{}_{(r)}$ for every $\boldsymbol{\theta}_1{}^* \in \boldsymbol{\Theta}_0{}^r$, where $\boldsymbol{\sigma}_{11}{}^2(\hat{\boldsymbol{\theta}})$ is the submatrix consisting of the elements in the first r rows and columns of $\boldsymbol{\sigma}^2(\hat{\boldsymbol{\theta}})$. (22.10.15)

43. Suppose observations are obtained from a univariate Normal process with known variance σ^2. Derive generalized likelihood ratio tests for the following hypotheses:

 (a) H_0: $\mu = \mu^*$ versus H_A: $\mu \neq \mu^*$;
 (b) H_0: $\mu \leqq \mu^*$ versus H_A: $\mu > \mu^*$;
 (c) H_0: $\mu \geqq \mu^*$ versus H_A: $\mu < \mu^*$;
 (d) Show that the resulting tests reject H_0 when and only when H_0 is also rejected by (22.6.1), (22.6.4), or (22.6.3), respectively.

44. *The Sign Test*: Let observations $\tilde{\mathbf{w}}_i \equiv (\tilde{w}_{1i}, \tilde{w}_{2i})'$ be iid with common continuous cdf $F_{12}(\mathbf{w})$. This exercise develops a popular test of the hypothesis that $P(\tilde{w}_1 > \tilde{w}_2) = \frac{1}{2}$; that is, that one component of $\tilde{\mathbf{w}}$ has no systematic tendency to exceed the other.

 (a) Let $p \equiv P(\tilde{w}_1 > \tilde{w}_2)$ and define, for $i = 1, \ldots, n$,

$$\tilde{x}_i \equiv \begin{cases} 1, & \tilde{w}_{1i} > \tilde{w}_{2i} \\ 0, & \tilde{w}_{1i} > \tilde{w}_{2i}. \end{cases}$$

 Show that $\tilde{r}$ has mf $f_b(r \mid n, p)$ for $\tilde{r} \equiv \sum_{i=1}^{n} \tilde{x}_i$.

 (b) Show that when n is large (say, at least 25) the following hold true:

A test of H_0: $P(\tilde{w}_1 > \tilde{w}_2) = \frac{1}{2}$ versus H_A: "H_0 is false" at approximate level α consists in rejecting H_0 if (22.10.16)

$$r \notin \left[\tfrac{1}{2}n - \tfrac{1}{2} - {}_{(1-\alpha/2)}N^* \frac{\sqrt{n}}{2},\ \tfrac{1}{2}n + \tfrac{1}{2} + {}_{(1-\alpha/2)}N^* \frac{\sqrt{n}}{2}\right].$$

A test of H_0: $P(\tilde{w}_1 > \tilde{w}_2) = \frac{1}{2}$ [or of H_0: $P(\tilde{w}_1 > \tilde{w}_2) \leqq \frac{1}{2}$] versus H_A: $P(\tilde{w}_1 > \tilde{w}_2) > \frac{1}{2}$ at approximate level α consists in rejecting H_0 if (22.10.17)

$$r > \tfrac{1}{2}n + \tfrac{1}{2} + {}_{(1-\alpha)}N^* \frac{\sqrt{n}}{2}.$$

A test of H_0: $P(\tilde{w}_1 > \tilde{w}_2) = \frac{1}{2}$ [or of H_0: $P(\tilde{w}_1 > \tilde{w}_2) \geqq \frac{1}{2}$] versus H_A: $P(\tilde{w}_1 > \tilde{w}_2) < \frac{1}{2}$ at approximate level α consists in rejecting H_0 if (22.10.18)

$$r < \tfrac{1}{2}n - \tfrac{1}{2} - {}_{(1-\alpha)}N^* \frac{\sqrt{n}}{2}.$$

(The name *sign test* derives from the common procedure of recording a plus sign whenever $w_{1i} > w_{2i}$, a minus sign whenever $w_{1i} < w_{2i}$, and discarding any observation for which equality obtains. Then r is the total number of plus signs out of n nontied observation pairs.)

45. *Fisher's Exact Test for Equality of Binomial Parameters*: Suppose a predetermined number n_i of observations is obtained from Bernoulli process i, for $i = 1, 2$ and the number r_i of ones recorded. The data are usually arranged in the form of a 2×2 contingency table with right margin fixed.

	Ones	*Zeros*	*sum*
Process 1	r_1	$n_1 - r_1$	n_1
Process 2	r_2	$n_2 - r_2$	n_2
sum	r	$n - r$	n

The null hypothesis is H_0: $p_1 = p_2 \equiv p$; that is, the process parameters are equal.

(a) Show that if H_0 is true, then

(1) the mf of $\tilde{r} \equiv \tilde{r}_1 + \tilde{r}_2$ is $f_b(r \mid n, p)$;
(2) the mf of $\tilde{r}_i$ is $f_b(r_i \mid n_i, p)$ for $i = 1, 2$; and
(3) the mf of $\tilde{r}_i$ given $\tilde{r} = r$ is $f_h(r_1 \mid n_1, r, n)$.

(b) Using (3) above, show the following:

A test of H_0: $p_1 = p_2 \equiv p$ versus H_A: $p_1 \neq p_2$ at level α consists in rejecting H_0 if $r_1 \in C$, where C is such that (22.10.19)

(1) $P_h(C \mid n_1, r, n) \leqq \alpha$;
(2) $P_h(C \cup \{x\} \mid n_1, r, n) > \alpha$, for all possible $x \notin C$; and
(3) $P_h(y \mid n_1, r, n) \leqq P_h(x \mid n_1, r, n)$, for every $y \in C$ and possible $x \notin C$.

A test of H_0: $p_1 = p_2$ (or of H_0: $p_1 \leqq p_2$) versus H_A: $p_1 > p_2$ at level α consists in rejecting H_0 if $r_1 \in C$, where C is such that (22.10.20)

(1) C is of the form $\{x, x + 1, \ldots, \min [n_1, r]\}$,
(2) $P_h(C \mid n_1, r, n) \leqq \alpha$; and
(3) $P_h(C \cup \{x - 1\} \mid n_1, r, n) > \alpha$.

A test of H_0: $p_1 = p_2$ (or of H_0: $p_1 \geqq p_2$) versus H_A: $p_1 < p_2$ at level α consists in rejecting H_0 if $r_1 \in C$, where C is such that (22.10.21)

(1) C is of the form $\{\max [0, r - (n - n_1)], \ldots, x\}$;
(2) $P_h(C \mid n_1, r, n) \leqq \alpha$; and
(3) $P_h(C \cup \{x + 1\} \mid n_1, r, n) > \alpha$.

46. *Extension of Exercise 45 to Multinomial Parameters*: Suppose n_i observations are obtained from k-variate Bernoulli process i, for $i = 1, 2$ and the data are arranged in the form of a $2 \times (k + 1)$ contingency table with right margin fixed, similar to the table in Exercise 45, with r_{ij} denoting the total number of observations on process i in class j, $j = 1, \ldots, k + 1$. The null hypothesis H_0: $\mathbf{p}_1 = \mathbf{p}_2$. Let $\mathbf{r}_i \equiv (r_{i1}, \ldots, r_{ik})'$.

(a) Show that if H_0 is true, then

(1) the mf of $\tilde{\mathbf{r}} \equiv \tilde{\mathbf{r}}_1 + \tilde{\mathbf{r}}_2$ is $f_{mu}^{(k)}(\mathbf{r} \mid n, \mathbf{p})$;
(2) the mf of $\tilde{\mathbf{r}}_i$ is $f_{mu}^{(k)}(\mathbf{r}_i \mid n_i, \mathbf{p})$, for $i = 1, 2$; and
(3) the mf of $\tilde{\mathbf{r}}_1$ given $\tilde{\mathbf{r}} = \mathbf{r}$ is $f_h^{(k)}(\mathbf{r}_1 \mid n_1, \mathbf{r}, n)$.

(b) Show the following:

A test of H_0: $\mathbf{p}_1 = \mathbf{p}_2 \equiv \mathbf{p}$ versus H_A: $\mathbf{p}_1 \neq \mathbf{p}_2$ at level α consists in rejecting H_0 if $\mathbf{r}_1 \notin \mathbf{C}$, where $\mathbf{C} \subset R^k$ is such that (22.10.22)

(1) $P_h^{(k)}(\mathbf{C} \mid n_1, \mathbf{r}, n) \leqq \alpha$;
(2) $P_h^{(k)}(\mathbf{C} \cup \{\mathbf{x}\} \mid n_1, \mathbf{r}, n) > \alpha$, for every possible $\mathbf{x} \notin \mathbf{C}$; and
(3) $P_h^{(k)}(\mathbf{y} \mid n_1, \mathbf{r}, n) \leqq P_h^{(k)}(\mathbf{x} \mid n_1, \mathbf{r}, n)$, for every $\mathbf{y} \in \mathbf{C}$ and possible $\mathbf{x} \notin \mathbf{C}$.

47. *An Exact Test for Equality of Two Poisson Process Parameters*: Suppose Poisson process i with parameters λ_i is observed for t_i time units during which r_i occurrences were recorded, for $i = 1, 2$. Assume that the t_i's were predetermined, and that if the processes describe the same physical process, then the observation periods do not overlap. We wish to test H_0: $\lambda_1 = \lambda_2 \equiv \lambda$. Let $r \equiv r_1 + r_2$ and $t \equiv t_1 + t_2$.

(a) Show that if H_0 is true, then

(1) the mf of $\tilde{r}$ is $f_{Po}(r \mid \lambda t)$;
(2) the mf of $\tilde{r}_i$ is $f_{Po}(r_i \mid \lambda t_i)$ for $i = 1, 2$; and
(3) the mf of $\tilde{r}_1$ given $\tilde{r} = r$ is $f_b(r_1 \mid r, t_1/t)$.

(b) Using (3) above, show that the following are true:

A test of H_0: $\lambda_1 = \lambda_2 \equiv \lambda$ versus H_A: $\lambda_1 \neq \lambda_2$ at level α consists in rejecting H_0 if $r_1 \in C$, where C is such that (22.10.23)

(1) $P_b(C \mid r, t_1/t) \leqq \alpha$;
(2) $P_b(C \cup \{x\} \mid r, t_1/t) > \alpha$, for every possible $x \notin C$; and
(3) $P_b(y \mid r, t_1/t) \leqq P_b(x \mid r, t_1/t)$, for every $y \in C$ and possible $x \notin C$.

A test of H_0: $\lambda_1 = \lambda_2$ (or of H_0: $\lambda_1 \leqq \lambda_2$) versus H_A: $\lambda_1 > \lambda_2$ at level α consists in rejecting H_0 if $r_1 \in C$, where C is such that (22.10.24)

(1) C is of the form $\{x, x + 1, \ldots, r\}$;
(2) $P_b(C \mid r, t_1/t) \leqq \alpha$; and
(3) $P_b(C \cup \{x - 1\} \mid r, t_1/t) > \alpha$.

A test of H_0: $\lambda_1 = \lambda_2$ (or of H_0: $\lambda_1 \geqq \lambda_2$) versus H_A: $\lambda_1 < \lambda_2$ at level α consists in rejecting H_0 if $r_1 \in C$, where C is such that (22.10.25)

(1) C is of the form $\{0, 1, \ldots, x\}$;
(2) $P_b(C \mid r, t_1/t) \leqq \alpha$; and
(3) $P_b(C \cup \{x + 1\} \mid r, t_1/t) > \alpha$.

APPENDIX

Table I Standardized Normal Cumulative Distribution Function

z	.00	.01	.02	.03	.04	.05	.06	.07	.08	.09
.0	.5000	.5040	.5080	.5120	.5160	.5199	.5239	.5279	.5319	.5359
.1	.5398	.5438	.5478	.5517	.5557	.5596	.5636	.5675	.5714	.5753
.2	.5793	.5832	.5871	.5910	.5948	.5987	.6026	.6064	.6103	.6141
.3	.6179	.6217	.6255	.6293	.6331	.6368	.6406	.6443	.6480	.6517
.4	.6554	.6591	.6628	.6664	.6700	.6736	.6772	.6808	.6844	.6879
.5	.6915	.6950	.6985	.7019	.7054	.7088	.7123	.7157	.7190	.7224
.6	.7257	.7291	.7324	.7357	.7389	.7422	.7454	.7486	.7517	.7549
.7	.7580	.7611	.7642	.7673	.7704	.7734	.7764	.7794	.7823	.7852
.8	.7881	.7910	.7939	.7967	.7995	.8023	.8051	.8078	.8106	.8133
.9	.8159	.8186	.8212	.8238	.8264	.8289	.8315	.8340	.8365	.8389
1.0	.8413	.8438	.8461	.8485	.8508	.8531	.8554	.8577	.8599	.8621
1.1	.8643	.8665	.8686	.8708	.8729	.8749	.8770	.8790	.8810	.8830
1.2	.8849	.8869	.8888	.8907	.8925	.8944	.8962	.8980	.8997	.9015
1.3	.9032	.9049	.9066	.9082	.9099	.9115	.9131	.9147	.9162	.9777
1.4	.9192	.9207	.9222	.9236	.9251	.9265	.9279	.9292	.9306	.9319
1.5	.9332	.9345	.9357	.9370	.9382	.9394	.9406	.9418	.9429	.9441
1.6	.9452	.9463	.9474	.9484	.9495	.9505	.9515	.9525	.9535	.9545
1.7	.9554	.9564	.9573	.9582	.9591	.9599	.9608	.9616	.9625	.9633
1.8	.9641	.9649	.9656	.9664	.9671	.9678	.9686	.9693	.9699	.9706
1.9	.9713	.9719	.9726	.9732	.9738	.9744	.9750	.9756	.9761	.9767
2.0	.9772	.9778	.9783	.9788	.9793	.9798	.9803	.9808	.9812	.9817
2.1	.9821	.9826	.9830	.9834	.9838	.9842	.9846	.9850	.9854	.9857
2.2	.9861	.9864	.9868	.9871	.9875	.9878	.9881	.9884	.9887	.9890
2.3	.9893	.9896	.9898	.9901	.9904	.9906	.9909	.9911	.9913	.9916
2.4	.9918	.9920	.9922	.9925	.9927	.9929	.9931	.9932	.9934	.9936
2.5	.9938	.9940	.9941	.9943	.9945	.9946	.9948	.9949	.9951	.9952
2.6	.9953	.9955	.9956	.9957	.9959	.9960	.9961	.9962	.9963	.9964
2.7	.9965	.9966	.9967	.9968	.9969	.9970	.9971	.9972	.9973	.9974
2.8	.9974	.9975	.9976	.9977	.9977	.9978	.9979	.9979	.9980	.9981
2.9	.9981	.9982	.9982	.9983	.9984	.9984	.9985	.9985	.9986	.9986
3.0	.9987	.9987	.9987	.9988	.9988	.9989	.9989	.9989	.9990	.9990
3.1	.9990	.9991	.9991	.9991	.9992	.9992	.9992	.9992	.9993	.9993
3.2	.9993	.9993	.9994	.9994	.9994	.9994	.9994	.9995	.9995	.9995
3.3	.9995	.9995	.9995	.9996	.9996	.9996	.9996	.9996	.9996	.9997
3.4	.9997	.9997	.9997	.9991	.9997	.9997	.9997	.9997	.9997	.9998

Note: $F_{N*}(-z) = 1 - F_{N*}(z)$.

Table II Fractiles of the Chi-Square Distribution

$\nu \backslash p$	$_p\chi^2(\nu)$												
	.005	.010	.025	.050	.100	.250	.500	.750	.900	.95	.975	.990	.995
1	.000	.000	.001	.004	.016	.102	.455	1.32	2.71	3.84	5.02	6.63	7.88
2	.010	.020	.051	.103	.211	.575	1.39	2.77	4.61	5.99	7.38	9.21	10.6
3	.072	.115	.216	.352	.584	1.21	2.37	4.11	6.25	7.81	9.35	11.3	12.8
4	.207	.297	.484	.711	1.06	1.92	3.36	5.39	7.78	9.49	11.1	13.3	14.9
5	.412	.554	.831	1.15	1.61	2.67	4.35	6.63	9.24	11.1	12.8	15.1	16.7
6	.676	.872	1.24	1.64	2.20	3.45	5.35	7.84	10.6	12.6	14.4	16.8	18.5
7	.989	1.24	1.69	2.17	2.83	4.25	6.35	9.04	12.0	14.1	16.0	18.5	20.3
8	1.34	1.65	2.18	2.73	3.48	5.07	7.34	10.2	13.4	15.5	17.5	20.1	22.0
9	1.73	2.09	2.70	3.33	4.17	5.90	8.34	11.4	14.7	16.9	19.0	21.7	23.6
10	2.16	2.56	3.25	3.94	4.87	6.74	9.34	12.5	16.0	18.3	20.5	23.2	25.2
11	2.60	3.05	3.82	4.57	5.58	7.58	10.3	13.7	17.3	19.7	21.9	24.7	26.8
12	3.07	3.57	4.40	5.23	6.30	8.44	11.3	14.8	18.5	21.0	23.3	26.2	28.3
13	3.57	4.11	5.01	5.89	7.04	9.30	12.3	16.0	19.8	22.4	24.7	27.7	29.8
14	4.07	4.66	5.63	6.57	7.79	10.2	13.3	17.1	21.1	23.7	26.1	29.1	31.3
15	4.60	5.23	6.26	7.26	8.55	11.0	14.3	18.2	22.3	25.0	27.5	30.6	32.8

Table II *(continued)*

ν \ p	.005	.010	.025	.050	.100	.250	.500	.750	.900	.95	.975	.990	.995
	$_p\chi^2(\nu)$												
16	5.14	5.81	6.91	7.96	9.31	11.9	15.3	19.4	23.5	26.3	28.8	32.0	34.3
17	7.50	6.41	7.56	8.67	10.1	12.8	16.3	20.5	24.8	27.6	30.2	33.4	35.7
18	6.26	7.01	8.23	9.39	10.9	13.7	17.3	21.6	26.0	28.9	31.5	34.8	37.2
19	6.84	7.63	8.91	10.1	11.7	14.6	18.3	22.7	27.2	30.1	32.9	36.2	38.6
20	7.43	8.26	9.59	10.9	12.4	15.5	19.3	23.8	28.4	31.4	34.2	37.6	40.0
21	8.03	8.90	10.3	11.6	13.2	16.3	20.3	24.9	29.6	32.7	35.5	38.9	41.4
22	8.64	9.54	11.0	12.3	14.0	17.2	21.3	26.0	30.8	33.9	36.8	40.3	42.8
23	9.26	10.2	11.7	13.1	14.8	18.1	22.3	27.1	32.0	35.2	38.1	41.6	44.2
24	9.89	10.9	12.4	13.8	15.7	19.0	23.3	28.2	33.2	36.4	39.4	43.0	45.6
25	10.5	11.5	13.1	14.6	16.5	19.9	24.3	29.3	34.4	37.7	40.6	44.3	46.9
26	11.2	12.2	13.8	15.4	17.3	20.8	25.3	30.4	35.6	38.9	41.9	45.6	48.3
27	11.8	12.9	14.6	16.2	18.1	21.7	26.3	31.5	36.7	40.1	43.2	47.0	49.6
28	12.5	13.6	15.3	16.9	18.9	22.7	27.3	32.6	37.8	41.3	44.5	48.3	51.0
29	13.1	14.3	16.0	17.7	19.8	23.6	28.3	33.7	39.1	42.6	45.7	48.6	52.3
30	13.8	15.0	16.8	18.5	20.6	24.5	29.3	34.8	40.3	43.8	47.0	50.9	53.7

SOURCE: Abridged from Catherine M. Thompson, "Tables of Percentage Points of the Incomplete Beta Function and of the Chi-square Distribution," *Biometrika* **32** (1941), with the kind permission of the editor of *Biometrika*.

Note: For $\nu > 30$, use the asympototic formula

$$_p\chi^2(\nu) \doteq \tfrac{1}{2}(_pN^* + \sqrt{2\nu - 1})^2,$$

where $_pN^*$ is the pth fractile of the standardized Normal distribution (Table I).

Table III Fractiles of the Beta Distribution

$_p\beta_{(\rho,\nu)}$

ν	ρ \ p	.01	.05	.10	.15	.20	.25	.30	.35	.40	.45	.50	.55	.60	.65	.70	.75	.80	.85	.90	.95	.99
2	1	0100	0500	1000	1500	2000	2500	3000	3500	4000	4500	5000	5500	6000	6500	7000	7500	8000	8500	9000	9500	9900
3	1	0050	0253	0513	0780	1056	1340	1633	1938	2254	2584	2929	3292	3675	4084	4523	5000	5528	6127	6838	7764	9000
4	1	0033	0170	0345	0527	0717	0914	1121	1338	1566	1807	2063	2337	2632	2953	3306	3700	4152	4687	5358	6316	7846
4	2	0589	1354	1958	2444	2871	3264	3633	3968	4329	4666	5000	5334	5671	6014	6367	6736	7129	7556	8042	6486	9411
5	1	0025	0127	0260	0398	0543	0694	0853	1021	1199	1388	1591	1810	2047	2308	2599	2929	3313	3777	4377	5271	6838
5	2	0420	0976	1426	1794	2123	2430	2724	3010	3292	3573	3857	4147	4445	4756	5084	5437	5825	6265	6795	7514	8591
6	1	0020	0102	0209	0320	0436	0559	0689	0825	0971	1127	1294	1476	1674	1894	2140	2421	2751	3157	3690	4507	6019
6	2	0327	0764	1122	1419	1686	1938	2180	2418	2656	2895	3138	3389	3650	3925	4220	4542	4902	5321	5839	6574	7779
6	3	1056	1893	2466	2899	3266	3594	3898	4186	4463	4733	5000	5267	5537	5814	6102	6406	6734	7101	7534	8107	8944
7	1	0017	0085	0174	0267	0365	0468	0577	0693	0816	0948	1091	1246	1416	1605	1818	2063	2353	2711	3187	3930	5358
7	2	0268	0628	0926	1174	1399	1612	1818	2022	2226	2433	2644	2864	3094	3339	3604	3895	4224	4613	5103	5818	7057
7	3	0847	1532	2009	2374	2686	2969	3233	3486	3731	3973	4214	4458	4708	4967	5239	5532	5854	6222	6668	7287	8269
8	1	0014	0073	0149	0229	0314	0403	0497	0597	0704	0819	0943	1078	1227	1393	1580	1797	2054	2374	2803	3482	4821
8	2	0227	0534	0788	1001	1195	1380	1559	1737	1916	2098	2285	2480	2685	2905	3143	3407	3709	4067	4526	5207	6434
8	3	0708	1288	1696	2011	2283	2531	2763	2987	3206	3423	3641	3863	4092	4331	4586	4861	5168	5523	5962	6587	7637
8	4	1423	2253	2786	3176	3501	3788	4052	4301	4539	4771	5000	5229	5461	5699	5948	6212	6499	6824	7214	7747	8577
9	1	0013	0064	0131	0201	0275	0353	0436	0524	0619	0720	0830	0950	1082	1230	1397	1591	1822	2111	2501	3123	4377
9	2	0197	0464	0686	0873	1044	1206	1365	1523	1682	1844	2011	2186	2371	2570	3786	3027	3304	3635	4062	4707	5899
9	3	0608	1111	1469	1746	1986	2206	2413	2614	2811	3007	3205	3408	3618	3839	4075	4332	4621	4959	5382	5997	7068
9	4	1210	1929	2397	2742	3032	3291	3530	3756	3975	4189	4402	4616	4835	5061	5299	5555	5837	6159	6554	7108	8018

Table III (*continued*)

ν	ρ \ p	.01	.05	.10	.15	.20	.25	.30	.35	.40	.45	.50	.55	.60	.65	.70	.75	.80	.85	.90	.95	.99
		$_p\beta_{(\rho,\nu)}$																				
10	1	0011	0057	0116	0179	0245	0315	0389	0467	0552	0643	0741	0849	0968	1101	1252	1428	1637	1901	2257	2831	4005
10	2	0174	0410	0608	0774	0926	1072	1214	1355	1498	1645	1796	1955	2123	2304	2501	2773	2978	3285	3684	4291	5440
10	3	0533	0977	1295	1542	1757	1955	2142	2324	2502	2681	2861	3048	3242	3446	3665	3905	4177	4496	4901	5496	6563
10	4	1053	1688	2104	2414	2675	2910	3127	3335	3535	3733	3931	4131	4336	4550	4776	5020	5291	5605	5994	6551	7500
10	5	1710	2514	3010	3367	3661	3920	4156	4378	4590	4796	5000	5204	5410	5622	5844	6080	6339	6633	6990	7486	8290
11	1	0010	0051	0105	0161	0221	0284	0350	0422	0498	0580	0670	0767	0876	0997	1134	1294	1487	1728	2057	2589	3690
11	2	0155	0368	0545	0695	0833	0964	1093	1221	1351	1485	1623	1767	1921	2087	2269	2474	2710	2996	3368	3942	5044
11	3	0475	0873	1158	1381	1576	1756	1926	2092	2255	2419	2586	2757	2936	3126	3330	3554	3809	4111	4496	5069	6117
11	4	0932	1500	1876	2156	2394	2609	2808	2999	3184	3367	3551	3738	3930	4131	4345	4577	4837	5139	5517	6066	7029
11	5	1504	2224	2673	2998	3268	3507	3726	3932	4131	4325	4517	4710	4907	5111	5325	5555	5809	6100	6458	6965	7817
12	1	0009	0047	0095	0147	0201	0258	0319	0384	0454	0529	0611	0700	0799	0910	1037	1184	1361	1584	1889	2384	3412
12	2	0141	0333	0495	0631	0756	0876	0994	1111	1230	1353	1480	1613	1755	1908	2077	2266	2486	2753	3102	3644	4698
12	3	0428	0788	1048	1251	1429	1593	1750	1902	2052	2204	2358	2517	2683	2860	3050	3261	3501	3786	4152	4701	5723
12	4	0837	1351	1692	1949	2167	2364	2548	2724	2896	3067	3238	3413	3593	3782	3984	4205	4452	4742	5108	5644	6604
12	5	1344	1996	2405	2704	2953	3173	3377	3570	3755	3938	4119	4302	4489	4684	4889	5111	5357	5642	5995	6502	7378
12	6	1940	2712	3177	3508	3779	4016	4232	4434	4627	4815	5000	5185	5373	5566	5768	5984	6221	6492	6823	7288	8060
13	1	0008	0043	0088	0135	0185	0238	0294	0354	0419	0488	0564	0647	0739	0843	0961	1099	1266	1477	1769	2256	3436
13	2	0128	0305	0452	0577	0693	0803	0911	1020	1130	1242	1360	1483	1615	1757	1914	2091	2296	2546	2875	3387	4395
13	3	0390	0719	0957	1143	1307	1459	1603	1744	1883	2024	2167	2315	2470	2635	2814	3012	3238	3508	3855	4381	5373
13	4	0759	1229	1542	1778	1979	2162	2332	2496	2656	2815	2976	3140	3309	3488	3679	3888	4124	4401	4753	5273	6222
13	5	1215	1810	2187	2463	2693	2898	3088	3268	3443	3614	3785	3958	4136	4322	4518	4731	4968	5245	5590	6091	6976
13	6	1746	2453	2882	3189	3441	3663	3866	4057	4240	4418	4595	4772	4953	5140	5336	5547	5779	6047	6377	6848	7651

Table III (*continued*)

ν	ρ \ p	.01	.05	.10	.15	.20	.25	.30	.35	.40	.45	.50	.55	.60	.65	.70	.75	.80	.85	.90	.95	.99
		$_p\beta_{(\rho,\nu)}$																				
14	1	0008	0039	0081	0124	0170	0219	0271	0326	0385	0449	0519	0596	0681	0776	0885	1011	1164	1358	1623	2058	2983
14	2	0118	0281	0417	0532	0639	0741	0841	0942	1044	1149	1258	1373	1496	1629	1775	1941	2133	2368	2678	3163	4128
14	3	0358	0660	0880	1053	1204	1345	1479	1610	1740	1871	2004	2143	2288	2443	2611	2798	3011	3267	3598	4101	5062
14	4	0695	1127	1416	1635	1822	1991	2150	2303	2453	2602	2753	2907	3067	3235	3416	3615	4105	3839	4443	4946	5878
14	5	1108	1657	2005	2261	2476	2668	2845	3014	3178	3340	3502	3666	3835	4011	4199	4403	4631	4899	5234	5726	6609
14	6	1588	2240	2637	2923	3160	3368	3559	3740	3913	4082	4251	4420	4593	4773	4963	5167	5394	5657	5982	6452	7271
14	7	2129	2870	3309	3618	3870	4090	4290	4477	4656	4829	5000	5171	5444	5523	5710	5910	6130	6382	6691	7130	7871
15	1	0007	0037	0075	0115	0158	0203	0252	0303	0358	0418	0483	0554	0634	0722	0824	0943	1086	1267	1517	1926	2803
15	2	0110	0160	0387	0494	0593	0688	0781	0875	0970	1068	1170	1278	1393	1517	1655	1810	1992	2213	2507	2967	3891
15	3	0331	0611	0815	0975	1117	1248	1373	1495	1617	1739	1865	1995	2131	2277	2436	2612	2814	3057	3372	3854	4783
15	4	0640	1040	1309	1513	1688	1846	1995	2138	2279	2419	2561	2706	2857	3017	3189	3377	3592	3845	4170	4657	5567
15	5	1019	1527	1851	2090	2291	2471	2638	2797	2952	3104	3258	3413	3574	3742	3912	4117	4336	4594	4920	5400	6274
15	6	1457	2061	2432	2699	2921	3117	3298	3468	3633	3794	3954	4116	4283	4455	4637	4835	5055	5311	5631	6096	6920
15	7	1947	2636	3046	3336	3574	3782	3972	4151	4321	4487	4651	4816	4984	5157	5339	5535	5751	6001	6309	6750	7512
16	1	0007	0034	0070	0108	0148	0190	0235	0283	0335	0391	0452	0518	0593	0676	0771	0883	1017	1188	1423	1810	2644
16	2	0102	0242	0360	0461	0553	0642	0729	0817	0906	0998	1094	1195	1303	1420	1550	1697	1868	2077	2356	2794	3679
16	3	0307	0568	0759	0909	1041	1163	1281	1395	1510	1625	1743	1866	1995	2133	2283	2540	2641	2872	3173	3634	4532
16	4	0594	0967	1218	1408	1572	1720	1860	1995	2128	2260	2394	2531	2675	2826	2989	3169	3373	3616	3928	4398	5285
16	5	0944	1417	1720	1944	2132	2301	2459	2609	2755	2900	3045	3193	3346	3506	3678	3865	4076	4325	4640	5108	5969
16	6	1346	1909	2256	2507	2716	2902	3072	3234	3390	3544	3697	3852	4010	4176	4352	4543	4756	5005	5317	5774	6597
16	7	1795	2437	2822	3096	3321	3518	3699	3869	4032	4191	4348	4507	4669	4836	5013	5204	5415	5660	5965	6404	7177
16	8	2287	3000	3415	3707	3944	4150	4337	4512	4679	4840	5000	5160	5321	5488	5663	5850	6056	6293	6585	7000	7713

Table III (*continued*)

ν	ρ \ p	.01	.05	.10	.15	.20	.25	.30	.35	.40	.45	.50	.55	.60	.65	.70	.75	.80	.85	.90	.95	.99
		$_p\beta_{(\rho,\nu)}$																				
17	1	0006	0032	0066	0101	0138	0178	0220	0266	0314	0367	0424	0487	0557	0635	0725	0830	0957	1118	1340	1707	2501
17	2	0095	0227	0337	0431	0518	0602	0684	0766	0850	0937	1027	1122	1224	1335	1457	1596	1758	1957	2222	2640	3488
17	3	0287	0531	0710	0850	0975	1090	1200	1308	1416	1525	1637	1752	1874	2005	2147	2306	2488	2709	2996	3438	4305
17	4	0554	0903	1138	1317	1471	1611	1743	1870	1995	2121	2247	2378	2514	2658	2813	2985	3180	3413	3712	4166	5029
17	5	0878	1321	1606	1816	1994	2154	2303	2445	2583	2721	2859	3000	3145	3299	3463	3642	3845	4085	4389	4844	5690
17	6	1251	1778	2104	2341	2539	2714	2876	3030	3178	3324	3471	3619	3771	3930	4099	4283	4489	4731	5035	5483	6299
17	7	1665	2267	2629	2888	3101	3239	3461	3623	3779	3931	4082	4235	4391	4553	4724	4099	5116	5355	5654	6090	6866
17	8	2117	2786	3178	3455	3680	3877	4056	4224	4384	4540	4694	4849	5006	5168	5339	5522	5726	5960	6250	6666	7393
18	1	0006	0030	0062	0095	0130	0168	0208	0250	0296	0346	0400	0459	0525	0599	0684	0783	0903	1056	1267	1616	2373
18	2	0090	0213	0317	0406	0488	0566	0644	0722	0801	0883	0968	1058	1154	1259	1375	1507	1661	1850	2102	2501	3316
18	3	0269	0499	0667	0799	0916	1025	1129	1232	1333	1436	1542	1652	1768	1892	2027	2178	2352	2562	2837	3262	4099
18	4	0519	0846	1068	1237	1382	1514	1639	1760	1879	1997	2118	2242	2371	2509	2657	2821	3008	3231	3519	3956	4796
18	5	0822	1238	1506	1705	1873	2024	2165	2300	2432	2562	2694	2828	2968	3114	3271	3444	3639	3869	4164	4605	5434
18	6	1168	1664	1972	2196	2383	2549	2703	2849	2991	3131	3270	3412	3558	3711	3874	4051	4251	4485	4781	5219	6025
18	7	1552	2119	2461	2707	2909	3088	3252	3406	3555	3702	3847	3994	4144	4300	4466	4646	4846	5080	5374	5803	6577
18	8	1971	2601	2973	3235	3450	3638	3809	3970	4124	4275	4423	4573	4725	4883	5049	5229	5428	5658	5945	6360	7094
18	9	2422	3108	3504	3780	4004	4199	4376	4541	4698	4850	5000	5150	5302	5459	5624	5801	5996	6220	6496	6892	7578
19	1	0006	0028	0058	0090	0123	0159	0196	0236	0280	0327	0378	0434	0496	0567	0647	0741	0855	1000	1201	1533	2257
19	2	0085	0201	0299	0383	0460	0535	0608	0682	0757	0834	0915	1001	1092	1192	1302	1427	1574	1754	1795	2377	3160
19	3	0254	0470	0629	0754	0865	0968	1066	1163	1260	1358	1458	1563	1673	1791	1920	2064	2230	2431	2694	3103	3912
19	4	0488	0797	1006	1166	1304	1429	1547	1662	1775	1888	2002	2121	2244	2375	2517	2674	2853	3067	3344	3767	4583
19	5	0772	1164	1418	1606	1765	1909	2043	2171	2297	2422	2547	2676	2809	2949	3100	3265	3453	3676	3960	4389	5199
19	6	1096	1563	1855	2067	2245	2404	2550	2690	2825	2958	3092	3228	3368	3515	3672	3843	4036	4263	4550	4978	5772

Table III (*continued*)

ν	ρ \ p	.01	.05	.10	.15	.20	.25	.30	.35	.40	.45	.50	.55	.60	.65	.70	.75	.80	.85	.90	.95	.99
		$_p\beta_{(\rho,\nu)}$																				
19	7	1454	1990	2134	2547	2740	2910	3067	3215	3357	3497	3637	3778	3923	4074	4235	4409	4604	4832	5118	5540	6309
19	8	1844	2440	2792	3042	3247	3427	3591	3746	3894	4039	4182	4327	4474	4628	4789	4964	5159	5385	5667	6078	6814
19	9	2263	2912	3288	3552	3767	3954	4124	4283	4434	4582	4727	4873	5022	5175	5336	5510	5702	5923	6198	6594	7290
20	1	0005	0027	0055	0085	0117	0150	0186	0224	0265	0310	0358	0412	0471	0538	0614	0704	0812	0950	1141	1459	2152
20	2	0080	0190	0283	0363	0436	0507	0576	0646	0717	0791	0868	0949	1036	1131	1236	1355	1495	1668	1898	2264	3018
20	3	0240	0445	0595	0714	0819	0916	1010	1102	1194	1287	1383	1482	1587	1700	1823	1961	2120	2312	2565	2958	3741
20	4	0461	0753	0951	1103	1233	1353	1465	1574	1682	1789	1899	2012	2130	2255	2391	2541	2713	2919	3186	3594	4387
20	5	0728	1099	1339	1518	1670	1806	1934	2057	2176	2295	2415	2538	2666	2801	2945	3105	3285	3500	3775	4191	4983
20	6	1032	1475	1751	1953	2123	2274	2414	2547	2676	2804	2932	3063	3197	3339	3490	3655	3841	4061	4340	4758	5538
20	7	1368	1875	2183	2405	2589	2752	2902	3043	3180	3315	3449	3585	3725	3871	4026	4195	4384	4606	4886	5300	6060
20	8	1733	2297	2633	2871	3067	3239	3397	3545	3688	3827	3966	4106	4249	4397	4554	4725	4915	5136	5413	5819	6553
20	9	2124	2739	3098	3351	3556	3736	3900	4053	4199	4342	4483	4625	4769	4919	5076	5246	5435	5653	5925	6319	7020
20	10	2540	3201	3579	3843	4056	4241	4408	4565	4713	4858	5000	5142	5287	5435	5592	5759	5944	6157	6421	6799	7460
21	1	0005	0026	0053	0081	0111	0143	0177	0213	0252	0294	0341	0391	0448	0511	0584	0670	0773	0905	1087	1391	2057
21	2	0076	0181	0269	0344	0414	0481	0547	0614	0682	0752	0825	0903	0986	1076	1176	1290	1424	1589	1810	2161	2888
21	3	0227	0422	0564	0677	0777	0870	0959	1047	1135	1224	1315	1410	1510	1618	1736	1867	2020	2205	2448	2826	3583
21	4	0436	0714	0902	1046	1170	1284	1391	1495	1598	1701	1805	1913	2026	2147	2277	2421	2586	2785	3042	3437	4207
21	5	0688	1041	1269	1440	1584	1714	1836	1953	2068	2182	2297	2414	2537	2666	2805	2959	3133	3340	3607	4010	4783
21	6	0975	1396	1659	1851	2013	2157	2291	2418	2542	2665	2788	2913	3043	3179	3325	3484	3665	3878	4149	4556	5321
21	7	1292	1773	2067	2278	2454	2610	2753	2889	3021	3150	3280	3411	3545	3686	3836	4000	4184	4400	4673	5078	5829
21	8	1634	2171	2491	2719	2906	3071	3223	3365	3503	3637	3771	3906	4044	4189	4341	4507	4692	4908	5180	5580	6309
21	9	2001	2587	2929	3171	3368	3541	3699	3846	3988	4126	4263	4400	4541	4686	4840	5006	5191	5405	5673	6064	6766
21	10	2390	3020	3382	3635	3840	4018	4180	4331	4476	4616	4754	4893	5034	5180	5333	5498	5680	5891	6152	6531	7199

Table III *(continued)*

ν	ρ \ p	.01	.05	.10	.15	.20	.25	.30	.35	.40	.45	.50	.55	.60	.65	.70	.75	.80	.85	.90	.95	.99
		$_p\beta_{(\rho,\nu)}$																				
22	1	0005	0024	0050	0077	0106	0136	0168	0203	0240	0281	0325	0373	0427	0488	0557	0639	0738	0864	1038	1329	1969
22	2	0072	0172	0256	0328	0394	0458	0521	0585	0650	0717	0786	0860	0940	1026	1122	1232	1360	1518	1729	2067	2768
22	3	0216	0401	0537	0644	0740	0828	0914	0997	1081	1166	1253	1344	1440	1543	1656	1783	1929	2107	2340	2706	3439
22	4	0414	0678	0858	0995	1114	1222	1325	1424	1522	1621	1721	1824	1933	2048	2173	2312	2471	2662	2910	3292	4041
22	5	0653	0988	1206	1369	1507	1631	1748	1860	1969	2079	2189	2302	2420	2544	2678	2826	2994	3194	3452	3844	4598
22	6	0925	1324	1575	1759	1914	2052	2180	2302	2421	2539	2657	2778	2903	3034	3175	3329	3503	3710	3973	4370	5120
22	7	1223	1682	1962	2164	2333	2482	2620	2750	2877	3001	3126	3252	3382	3519	3664	3823	4001	4211	4477	4874	5613
22	8	1546	2057	2363	2582	2762	2920	3066	3203	3335	3465	3594	3725	3859	3999	4147	4308	4489	4700	4966	5359	6082
22	9	1891	2450	2778	3010	3199	3366	3517	3660	3797	3930	4063	4196	4333	4475	4624	4787	4967	5178	5442	5828	6528
22	10	2257	2858	3205	3448	3646	3818	3974	4121	4261	4397	4531	4667	4804	4946	5097	5258	5438	5646	5905	6281	6953
22	11	2642	3281	3644	3896	4100	4277	4436	4585	4727	4865	5000	5135	5273	5415	5564	5723	5900	6104	6356	6719	7358
23	1	0005	0023	0048	0074	0101	0130	0161	0194	0230	0268	0310	0356	0408	0466	0533	0611	0705	0826	0994	1273	1889
23	2	0069	0164	0244	0313	0376	0437	0498	0558	0620	0684	0751	0822	0898	0981	1073	1178	1301	1453	1656	1981	2658
23	3	0206	0382	0512	0615	0706	0790	0872	0952	1032	1113	1197	1284	1376	1475	1584	1705	1846	2017	2242	2595	3305
23	4	0394	0646	0817	0949	1062	1166	1264	1359	1453	1548	1644	1743	1847	1958	2078	2212	2365	2550	2789	3159	3887
23	5	0621	0941	1149	1304	1436	1556	1667	1775	1880	1985	2091	2200	2313	2433	2562	2705	2867	3061	3310	3691	4426
23	6	0879	1260	1500	1676	1824	1956	2079	2197	2311	2425	2538	2655	2775	2902	3038	3187	3355	3556	3812	4198	4933
23	7	1162	1599	1867	2061	2223	2366	2498	2624	2746	2866	2986	3108	3234	3366	3506	3660	3833	4037	4297	4685	5412
23	8	1468	1956	2248	2458	2631	2783	2923	3056	3183	3309	3433	3560	3690	3825	3969	4126	4302	4508	4768	5155	5868
23	9	1793	2327	2642	2864	3047	3207	3353	3491	3623	3753	3881	4011	4143	4281	4427	4585	4762	4968	5227	5609	6304
23	10	2138	2713	3046	3280	3471	3637	3788	3930	4065	4198	4329	4460	4594	4733	4880	5039	5215	5419	5675	6048	6720
23	11	2501	3113	3462	3705	3902	4073	4228	4372	4510	4644	4776	4909	5043	5182	5328	5486	5660	5862	6112	6475	7119

Table III (*continued*)

ν	ρ \ p	.01	.05	.10	.15	.20	.25	.30	.35	.40	.45	.50	.55	.60	.65	.70	.75	.80	.85	.90	.95	.99
		$_p\beta_{(\rho,\nu)}$																				
24	1	0004	0022	0046	0070	0097	0124	0154	0186	0220	0257	0297	0341	0391	0446	0510	0585	0676	0792	0953	1221	1815
24	2	0066	0157	0234	0299	0360	0418	0476	0534	0594	0655	0719	0787	0860	0939	1028	1128	1247	1393	1588	1902	2557
24	3	0196	0365	0489	0587	0675	0756	0834	0911	0988	1066	1146	1229	1318	1413	1517	1634	1770	1935	2152	2492	3181
24	4	0376	0617	0781	0906	1015	1114	1209	1300	1390	1481	1573	1669	1769	1876	1992	2120	2268	2446	2678	3036	3745
24	5	0593	0898	1097	1246	1372	1487	1594	1697	1799	1899	2001	2106	2215	2331	2455	2593	2750	2938	3180	3549	4267
24	6	0838	1202	1432	1601	1742	1870	1988	2101	2211	2320	2430	2542	2658	2780	2912	3056	3220	3414	3663	4039	4758
24	7	1107	1525	1782	1968	2123	2260	2388	2509	2626	2742	2858	2976	3098	3225	3362	3511	3679	3878	4131	4510	5224
24	8	1397	1863	2144	2345	2512	2658	2793	2921	3044	3165	3286	3409	3535	3666	3806	3959	4130	4331	4586	4964	5669
24	9	1705	2216	2518	2732	2908	3062	3204	3337	3465	3590	3715	3840	3969	4103	4246	4400	4573	4775	5029	5405	6094
24	10	2031	2582	2903	3128	3312	3472	3619	3756	3888	4016	4143	4271	4402	4538	4681	4836	5009	5210	5462	5832	6502
24	11	2374	2961	3297	3532	3722	3887	4038	4178	4312	4443	4572	4701	4832	4969	5112	5267	5438	5637	5885	6246	6892
24	12	2733	3351	3701	3943	4139	4308	4461	4603	4739	4870	5000	5130	5261	5397	5539	5692	5861	6057	6299	6649	7267
25	1	0004	0021	0044	0067	0093	0119	0148	0178	0211	0246	0285	0327	0375	0428	0489	0561	0649	0760	0915	1173	1746
25	2	0063	0150	0224	0287	0345	0401	0456	0512	0569	0628	0690	0755	0825	0901	0986	1083	1197	1337	1526	1829	2462
25	3	0188	0350	0468	0563	0646	0724	0799	0873	0947	1022	1099	1179	1264	1356	1456	1569	1700	1859	2069	2398	3066
25	4	0360	0590	0747	0868	0972	1068	1158	1246	1333	1420	1509	1601	1697	1800	1912	2036	2179	2351	2575	2923	3612
25	5	0566	0859	1050	1192	1314	1424	1527	1626	1724	1821	1919	2020	2125	2237	2357	2490	2642	2824	3059	3418	4118
25	6	0800	1149	1369	1532	1668	1790	1904	2013	2119	2224	2330	2438	2550	2669	2796	2936	3094	3283	3525	3891	4595
25	7	1056	1457	1703	1882	2031	2164	2287	2403	2517	2628	2741	2855	2972	3096	3228	3373	3536	3730	3976	4347	5048
25	8	1332	1780	2049	2243	2403	2544	2675	2798	2917	3034	3151	3270	3392	3520	3656	3804	3971	4167	4416	4787	5481
25	9	1625	2116	2406	2612	2782	2931	3067	3196	3320	3441	3562	3684	3809	3940	4079	4229	4398	4596	4845	5214	5896
25	10	1935	2464	2772	2989	3167	3322	3464	3597	3725	3849	3973	4098	4225	4357	4497	4649	4819	5016	5264	5629	6295
25	11	2260	2824	3148	3374	3558	3718	3864	4001	4131	4258	4384	4510	4639	4772	4912	5064	5233	5429	5674	6032	6678
25	12	2599	3194	3532	3766	3955	4120	4268	4407	4539	4668	4795	4921	5050	5184	5324	5475	5642	5835	6076	6424	7047

Table III (*continued*)

ν	ρ \ p	.01	.05	.10	.15	.20	.25	.30	.35	.40	.45	.50	.55	.60	.65	.70	.75	.80	.85	.90	.95	.99
		$_p\beta_{(\rho,\nu)}$																				
26	1	0004	0020	0042	0065	0089	0114	0142	0171	0202	0236	0273	0314	0360	0411	0470	0539	0623	0731	0880	1129	1682
26	2	0060	0144	0215	0275	0331	0385	0438	0492	0546	0603	0662	0725	0793	0866	0948	1041	1151	1287	1469	1761	2375
26	3	0180	0335	0449	0540	0620	0695	0767	0838	0909	0981	1055	1133	1215	1303	1400	1509	1635	1788	1991	2310	2959
26	4	0345	0566	0717	0832	0933	1024	1111	1196	1279	1364	1449	1538	1631	1730	1838	1958	2096	2263	2480	2817	3488
26	5	0542	0823	1006	1143	1260	1366	1465	1561	1655	1749	1843	1941	2043	2150	2267	2396	2543	2719	2947	3296	3979
26	6	0765	1101	1312	1468	1599	1717	1827	1932	2034	2136	2238	2343	2451	2566	2689	2824	2978	3161	3397	3754	4443
26	7	1010	1395	1632	1804	1947	2075	2194	2306	2416	2524	2632	2743	2857	2977	3105	3246	3405	3593	3833	4195	4884
26	8	1273	1703	1962	2149	2303	2440	2566	2685	2800	2914	3027	3142	3260	3384	3517	3661	3823	4015	4258	4622	5306
26	9	1553	2024	2303	2502	2666	2810	2942	3066	3186	3304	3422	3540	3662	3789	3924	4071	4236	4429	4673	5036	5711
26	10	1848	2356	2653	2863	3034	3184	3322	3451	3575	3696	3816	3937	4062	4191	4328	4476	4642	4836	5080	5439	6100
26	11	2156	2699	3011	3230	3408	3564	3705	3838	3965	4088	4211	4334	4460	4590	4728	4877	5042	5235	5477	5832	6476
26	12	2479	3051	3377	3604	3788	3947	4092	4227	4356	4482	4605	4730	4856	4987	5124	5273	5438	5629	5867	6214	6837
26	13	2814	3414	3751	3985	4173	4335	4482	4619	4749	4876	5000	5124	5251	5381	5518	5665	5827	6015	6249	6586	7186
27	1	0004	0020	0040	0062	0085	0110	0136	0164	0195	0227	0263	0302	0346	0396	0453	0519	0600	0704	0848	1088	1623
27	2	0058	0138	0206	0264	0318	0370	0421	0473	0526	0580	0637	0698	0763	0834	0913	1002	1108	1239	1415	1698	2293
27	3	0173	0322	0432	0519	0596	0668	0738	0806	0874	0944	1015	1090	1169	1254	1348	1453	1575	1723	1920	2229	2859
27	4	0331	0543	0688	0800	0896	0985	1068	1150	1230	1312	1394	1480	1570	1666	1770	1886	2020	2181	2392	2719	3372
27	5	0520	0790	0966	1098	1211	1313	1409	1501	1591	1682	1774	1868	1966	2070	2183	2308	2450	2621	2842	3182	3849
27	6	0734	1056	1260	1410	1536	1650	1756	1857	1956	2054	2153	2254	2359	2470	2590	2721	2870	3048	3277	3626	4300
27	7	0968	1338	1566	1732	1870	1994	2108	2217	2323	2427	2532	2639	2750	2866	2991	3128	3282	3465	3700	4054	4729
27	8	1220	1633	1883	2062	2212	2344	2465	2581	2692	2802	2912	3024	3139	3259	3388	3528	3687	3873	4111	4468	5140
27	9	1487	1940	2209	2401	2559	2698	2826	2947	3063	3178	3292	3407	3525	3649	3781	3924	4085	4274	4513	4870	5535
27	10	1768	2257	2544	2746	2912	3058	3191	3316	3436	3554	3671	3789	3910	4036	4170	4315	4478	4668	4907	5262	5916

Table III (*continued*)

ν	ρ \ p	.01	.05	.10	.15	.20	.25	.30	.35	.40	.45	.50	.55	.60	.65	.70	.75	.80	.85	.90	.95	.99
		$_p\beta_{(\rho,\nu)}$																				
27	11	2062	2584	2886	3098	3271	3421	3559	3688	3811	3931	4051	4171	4294	4421	4556	4702	4865	5055	5293	5643	6284
27	12	2369	2921	3236	3456	3634	3789	3930	4061	4187	4310	4431	4552	4676	4804	4939	5085	5247	5436	5671	6016	6639
27	13	2688	3266	3593	3820	4002	4161	4304	4437	4565	4688	4810	4932	5056	5184	5319	5464	5625	5811	6043	6379	6982
28	1	0004	0019	0039	0060	0082	0106	0131	0158	0187	0219	0253	0291	0334	0381	0436	0500	0579	0679	0817	1050	1568
28	2	0056	0133	0199	0255	0306	0356	0406	0455	0506	0559	0614	0672	0735	0803	0880	0966	1069	1195	1366	1640	2217
28	3	0166	0310	0415	0499	0574	0643	0710	0776	0842	0909	0978	1050	1127	1209	1299	1401	1519	1663	1853	2153	2766
28	4	0318	0522	0662	0770	0862	0948	1029	1107	1185	1263	1343	1426	1513	1606	1706	1819	1948	2105	2309	2627	3264
28	5	0500	0759	0929	1057	1165	1264	1356	1445	1533	1620	1709	1800	1895	1996	2105	2226	2364	2530	2745	3076	3727
28	6	0705	1015	1211	1356	1478	1588	1690	1788	1883	1978	2074	2172	2274	2383	2497	2625	2770	2943	3166	3506	4166
28	7	0929	1285	1505	1665	1799	1918	2029	2134	2237	2338	2440	2544	2651	2764	2885	3018	3168	3347	3575	3921	4584
28	8	1170	1568	1809	1983	2127	2255	2372	2484	2592	2699	2806	2914	3026	3143	3268	3405	3559	3741	3974	4323	4984
28	9	1426	1862	2122	2308	2461	2596	2719	2837	2950	3061	3171	3283	3399	3519	3647	3787	3944	4129	4364	4714	5370
28	10	1695	2166	2443	2639	2800	2941	3070	3192	3309	3423	3537	3652	3770	3893	4023	4165	4324	4510	4746	5095	5742
28	11	1976	2479	2771	2976	3144	3290	3424	3549	3669	3786	3903	4020	4140	4265	4397	4540	4699	4886	5120	5466	6102
28	12	2268	2801	3106	3319	3492	3643	3780	3908	4031	4150	4268	4387	4508	4634	4767	4910	5070	5255	5488	5829	6450
28	13	2572	3131	3448	3668	3845	4000	4139	4270	4394	4515	4634	4754	4876	5001	5134	5277	5436	5620	5849	6184	6787
28	14	2887	3470	3796	4021	4203	4360	4501	4633	4759	4880	5000	5120	5241	5367	5499	5640	5797	5979	6204	6530	7113
29	1	0004	0018	0038	0058	0079	0102	0127	0153	0181	0211	0245	0281	0322	0368	0421	0483	0559	0655	0789	1015	1517
29	2	0054	0128	0192	0245	0295	0344	0391	0439	0488	0539	0592	0649	0709	0775	0849	0933	1032	1155	1319	1585	2146
29	3	0160	0298	0400	0481	0553	0620	0685	0748	0812	0877	0944	1013	1087	1167	1254	1353	1467	1606	1791	2082	3679
29	4	0306	0503	0638	0742	0831	0914	0992	1068	1143	1219	1296	1376	1460	1550	1647	1756	1882	2034	2232	2542	3162
29	5	0481	0731	0895	1018	1123	1218	1307	1393	1478	1563	1648	1737	1829	1927	2033	2150	2284	2445	2655	2977	3613
29	6	0678	0977	1166	1306	1424	1530	1629	1724	1816	1908	2001	2096	2195	2299	2412	2536	2677	2845	3062	3394	4039

Table III *(continued)*

ν	ρ \ p	.01	.05	.10	.15	.20	.25	.30	.35	.40	.45	.50	.55	.60	.65	.70	.75	.80	.85	.90	.95	.99
		$_p\beta_{(\rho,\nu)}$																				
29	7	0894	1237	1449	1604	1733	1848	1956	2058	2157	2255	2354	2454	2559	2668	2786	2915	3062	3236	3459	3797	4447
29	8	1125	1509	1741	1909	2049	2172	2286	2395	2499	2603	2707	2812	2920	3034	3156	3290	3440	3618	3845	4187	4837
29	9	1370	1791	2042	2221	2370	2500	2621	2734	2844	2952	3059	3168	3281	3398	3523	3660	3813	3994	4224	4567	5214
29	10	1627	2082	2350	2540	2696	2833	2958	3076	3190	3301	3412	3524	3639	3759	3887	4025	4181	4364	4594	4938	5578
29	11	1896	2383	2665	2864	3026	3169	3298	3420	3537	3651	3765	3879	3997	4118	4248	4388	4544	4728	4958	5300	5930
29	12	2176	2691	2987	3193	3361	3508	3641	3766	3886	4002	4118	4234	4353	4476	4606	4746	4903	5086	5316	5654	6271
29	13	2467	3007	3314	3528	3701	3851	3987	4114	4236	4354	4471	4588	4707	4831	4961	5102	5258	5440	5667	6000	6602
29	14	2767	3331	3648	3867	4044	4197	4335	4464	4587	4706	4824	4941	5061	5184	5314	5454	5609	5789	6013	6338	6922

Notes:

(1) It is to be understood that a decimal point precedes each four-digit table entry.

(2) Let $_p\beta_{(\rho,\nu)}$ denote the pth fractile of the beta distribution with parameters ρ and ν. To find fractiles in which $\rho \geq \nu/2$, use the relation

$$_p\beta_{(\rho,\nu)} = 1 - {}_{(1-p)}\beta_{(\rho,\nu)}.$$

(3) For $\nu \geq 30$, use the following approximation: Define

$$a_p \equiv \{[{}_{.5}\beta(1-{}_p\beta)]^{1/2} - [{}_p\beta(1-{}_{.5}\beta)]^{1/2}\}^2.$$

Then $\rho \doteq c \cdot {}_{.5}\beta + \frac{1}{3}$ and $\nu \doteq c + \frac{2}{3}$, where

(a) $c = .112/a_{.25}$, if $_{.25}\beta$ and $_{.5}\beta$ were assessed;
(b) $c = .112/a_{.75}$, if $_{.5}\beta$ and $_{.75}\beta$ were assessed; and
(c) $c = .056[(1/a_{.25}) + (1/a_{.75})]$, if $_{.25}\beta$ and $_{.75}\beta$ were assessed.

(For justification, see Pratt, Raiffa and Schlaifer [**64**], Section 11.5.)

Table IV Fractiles of the Standardized Student Distribution

$_pS^*{}_{(\nu)}$

ν \ p	.55	.60	.65	.70	.75	.80	.85
1	.158	.325	.510	.727	1.000	1.376	1.963
2	.142	.289	.445	.617	.816	1.061	1.386
3	.137	.277	.424	.584	.765	.978	1.250
4	.134	.271	.414	.569	.741	.941	1.190
5	.132	.267	.408	.559	.727	.920	1.156
6	.131	.265	.404	.553	.718	.906	1.134
7	.130	.263	.402	.549	.711	.896	1.119
8	.130	.262	.399	.546	.706	.889	1.108
9	.129	.261	.398	.543	.703	.883	1.100
10	.129	.260	.397	.542	.700	.879	1.093
11	.129	.260	.396	.540	.697	.876	1.088
12	.128	.259	.395	.539	.695	.873	1.083
13	.128	.259	.394	.538	.694	.870	1.079
14	.128	.258	.393	.537	.692	.868	1.076
15	.128	.258	.393	.536	.691	.866	1.074
16	.128	.258	.392	.535	.690	.865	1.071
17	.128	.257	.392	.534	.689	.863	1.069
18	.127	.257	.382	.534	.688	.862	1.067
19	.127	.257	.391	.533	.688	.861	1.066
20	.127	.257	.391	.533	.687	.860	1.064
21	.127	.257	.391	.532	.686	.859	1.063
22	.127	.256	.390	.532	.686	.858	1.061
23	.127	.256	.390	.532	.685	.858	1.060
24	.127	.256	.390	.531	.685	.857	1.059
25	.127	.256	.390	.531	.684	.856	1.058
26	.127	.256	.390	.531	.684	.856	1.058
27	.127	.256	.389	.531	.684	.855	1.057
28	.127	.256	.389	.530	.683	.855	1.056
29	.127	.256	.389	.530	.683	.854	1.055
30	.127	.256	.389	.530	.683	.854	1.055
40	.126	.255	.388	.529	.681	.851	1.050
60	.126	.254	.387	.527	.679	.848	1.046
120	.126	.254	.386	.526	.677	.845	1.041
∞	.126	.253	.385	.524	.674	.842	1.036

Table IV (*continued*)

p / ν	.90	.95	.975	.99	.995	.9995
			$_pS^*_{(\nu)}$			
1	3.078	6.314	12.706	31.821	63.657	636.619
2	1.886	2.910	4.303	6.965	9.925	31.598
3	1.638	2.353	3.182	4.541	5.841	12.941
4	1.533	2.132	2.776	3.747	4.604	8.610
5	1.476	2.015	2.571	3.365	4.032	6.859
6	1.440	1.943	2.447	3.143	3.707	5.959
7	1.415	1.895	2.365	2.998	3.499	5.405
8	1.397	1.860	2.306	2.896	3.355	5.041
9	1.383	1.833	2.262	2.821	3.250	4.781
10	1.372	1.812	2.228	2.764	3.169	4.587
11	1.363	1.796	2.201	2.718	3.106	4.437
12	1.356	1.782	2.179	2.681	3.055	4.318
13	1.350	1.771	2.160	2.650	3.012	4.221
14	1.345	1.761	2.145	2.624	2.977	4.140
15	1.341	1.753	2.131	2.602	2.947	4.073
16	1.337	1.746	2.120	2.583	2.921	4.015
17	1.333	1.740	2.110	2.567	2.898	3.965
18	1.330	1.734	2.101	2.552	2.878	3.922
19	1.328	1.729	2.093	2.539	2.861	3.883
20	1.325	1.725	2.086	2.528	2.845	3.850
21	1.323	1.721	2.080	2.518	2.831	3.819
22	1.321	1.717	2.074	2.508	2.819	3.792
23	1.319	1.714	2.069	2.500	2.807	3.767
24	1.318	1.711	2.064	2.492	2.797	3.745
25	1.316	1.708	2.060	2.485	2.787	3.725
26	1.315	1.706	2.056	2.479	2.779	3.707
27	1.314	1.703	2.052	2.473	2.771	3.690
28	1.313	1.701	2.048	2.467	2.763	3.674
29	1.311	1.699	2.045	2.462	2.756	3.659
30	1.310	1.697	2.042	2.457	2.750	3.646
40	1.303	1.684	2.021	2.423	2.704	3.551
60	1.296	1.671	2.000	2.390	2.660	3.460
120	1.289	1.658	1.980	2.358	2.617	3.373
∞	1.282	1.645	1.960	2.326	2.576	3.291

SOURCE: Table III of Fisher and Yates, *Statistical Tables for Biological, Agricultural and Medical Research*, Published by Oliver & Boyd, Ltd., Edinburgh, and by permission of the authors and publishers.

Note: $_{(1-p)}S^*_{(\nu)} = -\,_pS^*_{(\nu)}$.

Table V Fractiles of the F-Distribution

p	ν_2 $\nu_1 \longrightarrow 1$		2	3	4	5	6	7	8
.90	↓	39.9	49.5	53.6	55.8	57.2	58.2	58.9	59.4
.95	↓	161	200	216	225	230	234	237	239
.975	1	648	800	864	900	922	937	948	957
.99		4.05*	5.00*	5.40*	5.62*	5.76*	5.86*	5.93*	5.98*
.995		16.2*	20.0*	21.6*	22.5*	23.1*	23.4*	23.7*	23.9*
.90		8.53	9.00	9.16	9.24	9.29	9.33	9.35	9.37
.95		18.5	19.0	19.2	19.2	19.3	19.3	19.4	19.4
.975	2	38.5	39.0	39.2	39.2	39.3	39.3	39.4	39.4
.99		98.5	99.0	99.2	99.2	99.3	99.3	99.4	99.4
.995		199	199	199	199	199	199	199	199
.90		5.54	5.46	5.39	5.34	5.31	5.28	5.27	5.25
.95		10.1	9.55	9.28	9.12	9.01	8.94	8.89	8.85
.975	3	17.4	16.0	15.4	15.1	14.9	14.7	14.6	14.5
.99		34.1	30.8	29.5	28.7	28.2	27.9	27.7	27.5
.995		55.6	49.8	47.5	46.2	45.4	44.8	44.4	44.1
.90		4.54	4.32	4.19	4.11	4.05	4.01	3.98	3.95
.95		7.71	6.94	6.59	6.39	6.26	6.16	6.09	6.04
.975	4	12.2	10.6	9.98	9.60	9.36	9.20	9.07	8.98
.99		21.2	18.0	16.7	16.0	15.5	15.2	15.0	14.8
.995		31.3	26.3	24.3	23.2	22.5	22.0	21.6	21.4
.90		4.06	3.78	3.62	3.52	3.45	3.40	3.37	3.34
.95		6.61	5.79	5.41	5.19	5.05	4.95	4.88	4.82
.975	5	10.0	8.43	7.76	7.39	7.15	6.98	6.85	6.76
.99		16.3	13.3	12.1	11.4	11.0	10.7	10.5	10.3
.995		22.8	18.3	16.5	15.6	14.9	14.5	14.2	14.0
.90		3.78	3.46	3.29	3.18	3.11	3.05	3.01	2.98
.95		5.99	5.14	4.76	4.53	4.39	4.28	4.21	4.15
.975	6	8.81	7.26	6.60	6.23	5.99	5.82	5.70	5.60
.99		13.7	10.9	9.78	9.15	8.75	8.47	8.26	8.10
.995		18.6	14.5	12.9	12.0	11.5	11.1	10.8	10.6
.90		3.59	3.26	3.07	2.96	2.88	2.83	2.78	2.75
.95		5.59	4.47	4.35	4.12	3.97	3.87	3.79	3.73
.975	7	8.07	6.54	5.89	5.52	5.29	5.12	4.99	4.90
.99		12.2	9.55	8.45	7.85	7.46	7.19	6.99	6.84
.995		16.2	12.4	10.9	10.1	9.52	9.16	8.89	8.68

* Denotes (entry)·(1000)

Table V (*continued*)

p	ν_2	$\nu_1 \rightarrow$ 9	10	12	15	20	30	60	120	∞
.90	↓	59.	60.2	60.7	61.2	61.7	62.3	62.8	63.1	63.3
.95		241	242	244	246	248	250	252	253	254
.975	1	963	969	977	985	993	1000	1010	1010	1020
.99		6.02*	6.06*	6.11*	6.16*	6.21*	6.26*	6.31*	6.34*	6.37*
.995		24.1*	24.2*	24.4*	24.6*	24.8*	25.0*	25.2*	25.4*	25.5*
.90		9.38	9.39	9.41	9.42	9.44	9.46	9.47	9.48	9.49
.95		19.4	19.4	19.4	19.4	19.5	19.5	19.5	19.5	19.5
.975	2	39.4	39.4	39.4	39.4	39.4	39.5	39.5	39.5	39.5
.99		99.4	99.4	99.4	99.4	99.4	99.5	99.5	99.5	99.5
.995		199	199	199	199	199	199	199	199	199
.90		5.24	5.23	5.22	5.20	5.18	5.17	5.15	5.14	5.13
.95		8.81	8.79	8.74	8.70	8.66	8.62	8.57	8.55	8.53
.975	3	14.5	14.4	14.3	14.3	14.2	14.1	14.0	13.9	13.9
.99		27.3	27.2	27.1	26.9	26.7	26.5	26.3	26.2	26.1
.995		43.9	43.7	43.4	43.1	42.8	42.5	42.1	42.0	41.8
.90		3.93	3.92	3.90	3.87	3.84	3.82	3.79	3.78	3.76
.95		6.00	5.96	5.91	5.86	5.80	5.75	5.69	5.66	5.63
.975	4	8.90	8.84	8.75	8.66	8.56	8.46	8.36	8.31	8.26
.99		14.7	14.5	14.4	14.2	14.0	13.8	13.7	13.6	13.5
.995		21.1	21.0	20.7	20.4	20.2	19.9	19.6	19.5	19.3
.90		3.32	3.30	3.27	3.24	3.21	3.17	3.14	3.12	3.11
.95		4.77	4.74	4.68	4.62	4.56	4.50	4.43	4.40	4.37
.975	5	6.68	6.62	6.52	6.43	6.33	6.23	6.12	6.07	6.02
.99		10.2	10.1	9.89	9.72	9.55	9.38	9.20	9.11	9.02
.995		13.8	13.6	13.4	13.1	12.9	12.7	12.4	12.3	12.1
.90		2.96	2.94	2.90	2.87	2.84	2.80	2.76	2.74	2.72
.95		4.10	4.06	4.00	3.94	3.87	3.81	3.74	3.70	3.67
.975	6	5.52	5.46	5.37	5.27	5.17	5.07	4.96	4.90	4.85
.99		7.98	7.87	7.72	7.56	7.40	7.23	7.06	6.97	6.88
.995		10.4	10.2	10.0	9.81	9.59	9.36	9.12	9.00	8.88
.90		2.72	2.70	2.67	2.63	2.59	2.56	2.51	2.49	2.47
.95		3.68	3.64	3.57	3.51	3.44	3.38	3.30	3.27	3.23
.975	7	4.82	4.76	4.67	4.57	4.47	4.36	4.25	4.20	4.14
.99		6.72	6.62	6.47	6.31	6.16	5.99	5.82	5.74	5.65
.995		8.51	8.38	8.18	7.97	7.75	7.53	7.31	7.19	7.08

*Denotes (entry)·(1000)

Table V (*continued*)

p	ν_2 $\nu_1 \longrightarrow$	1	2	3	4	5	6	7	8
.90	↓	3.46	3.11	2.92	2.81	2.73	2.67	2.62	2.59
.95		5.32	4.46	4.07	3.84	3.69	3.58	3.50	3.44
.975	8	7.57	6.06	5.42	5.05	4.83	4.65	4.53	4.43
.99		11.3	8.65	7.59	7.01	6.63	6.37	6.18	6.03
.995		14.7	11.0	9.60	8.81	8.30	7.95	7.69	7.50
.90		3.36	3.01	2.81	2.69	2.61	2.55	2.51	2.47
.95		5.12	4.26	3.86	3.63	3.48	3.37	3.29	3.23
.975	9	7.21	5.71	5.08	4.72	4.48	4.32	4.20	4.10
.99		10.6	8.02	6.99	6.42	6.06	5.80	5.61	5.47
.995		13.6	10.1	8.72	7.96	7.47	7.13	6.88	6.69
.90		3.29	2.92	2.73	2.61	2.52	2.46	2.41	2.38
.95		4.96	4.10	3.71	3.48	3.33	3.22	3.14	3.07
.975	10	6.94	5.46	4.83	4.47	4.24	4.07	3.95	3.85
.99		10.0	7.56	6.55	5.99	5.64	5.39	5.20	5.06
.995		12.8	9.43	8.08	7.34	6.87	6.54	6.30	6.12
.90		3.18	2.81	2.61	2.48	2.39	2.33	2.28	2.24
.95		4.75	3.89	3.49	3.26	3.11	3.00	2.91	2.85
.975	12	6.55	5.10	4.47	4.12	3.89	3.73	3.61	3.51
.99		9.33	6.93	5.95	5.41	5.06	4.82	4.64	4.50
.995		11.8	8.51	7.23	6.52	6.07	5.76	5.52	5.35
.90		3.07	2.70	2.49	2.36	2.27	2.21	2.16	2.12
.95		4.54	3.68	3.29	3.06	2.90	2.79	2.71	2.64
.975	15	6.20	4.77	4.15	3.80	3.58	3.41	3.29	3.20
.99		8.68	6.36	5.42	4.89	4.56	4.32	4.14	4.00
.995		10.8	7.70	6.48	5.80	5.37	5.07	4.85	4.67
.90		2.97	2.59	2.38	2.25	2.16	2.09	2.04	2.00
.95		4.35	3.49	3.10	2.87	2.71	2.60	2.51	2.45
.975	20	5.87	4.46	3.86	3.51	3.29	3.13	3.01	2.91
.99		8.10	5.85	4.94	4.43	4.10	3.87	3.70	3.56
.995		9.94	6.99	5.82	5.17	4.76	4.47	4.26	4.09
.90		2.88	2.49	2.28	2.14	2.05	1.98	1.93	1.88
.95		4.17	3.32	2.92	2.69	2.53	2.42	2.33	2.27
.975	30	5.57	4.18	3.59	3.25	3.03	2.87	2.75	2.65
.99		7.56	5.39	4.51	4.02	3.70	3.47	3.30	3.17
.995		9.18	6.35	5.24	4.62	4.23	3.95	3.74	3.58
.90		2.79	2.39	2.18	2.04	1.95	1.87	1.82	1.77
.95		4.00	3.15	2.76	2.53	2.37	2.25	2.17	2.10
.975	60	5.29	3.93	3.34	3.01	2.79	2.63	2.51	2.41
.99		7.08	4.98	4.13	3.65	3.34	3.12	2.95	2.82
.995		8.49	5.80	4.73	4.14	3.76	3.49	3.29	3.13

Table V (*continued*)

p	ν_2 $\nu_1 \longrightarrow$	9	10	12	15	20	30	60	120	∞
.90	↓	2.56	2.54	2.50	2.46	2.42	2.38	2.34	2.31	2.29
.95		3.39	3.35	3.28	3.22	3.15	3.08	3.01	2.97	2.93
.975	8	4.36	4.30	4.20	4.10	4.00	3.89	3.78	3.73	3.67
.99		5.91	5.81	5.67	5.52	5.36	5.20	5.03	4.95	4.86
.995		7.34	7.21	7.01	6.81	6.61	6.40	6.18	6.06	5.95
.90		2.44	2.42	2.38	2.34	2.30	2.25	2.21	2.18	2.16
.95		3.18	3.14	3.07	3.01	2.94	2.86	2.79	2.75	2.71
.975	9	4.03	3.96	3.87	3.77	3.67	3.56	3.45	3.39	3.33
.99		5.35	5.26	5.11	4.96	4.81	4.65	4.48	4.40	4.31
.995		6.54	6.42	6.23	6.03	5.83	5.62	5.41	5.30	5.19
.90		2.35	2.32	2.28	2.24	2.20	2.15	2.11	2.08	2.06
.95		3.02	2.98	2.91	2.84	2.77	2.70	2.62	2.58	2.54
.975	10	3.78	3.72	3.62	3.52	3.42	3.31	3.20	3.14	3.08
.99		4.94	4.85	4.71	4.56	4.41	4.25	4.08	4.00	3.91
.995		5.97	5.85	5.66	5.47	5.27	5.07	4.86	4.75	4.64
.90		2.21	2.19	2.15	2.10	2.06	2.01	1.96	1.93	1.90
.95		2.80	2.75	2.69	2.62	2.54	2.47	2.38	2.34	2.30
.975	12	3.44	3.37	3.28	3.18	3.07	2.96	2.85	2.79	2.72
.99		4.39	4.30	4.16	4.01	3.86	3.70	3.54	3.45	3.36
.995		5.20	5.09	4.91	4.72	4.53	4.33	4.12	4.01	3.90
.90		2.09	2.06	2.02	1.97	1.92	1.87	1.82	1.79	1.76
.95		2.59	2.54	2.48	2.40	2.33	2.25	2.16	2.11	2.07
.975	15	3.12	3.06	2.96	2.86	2.76	2.64	2.52	2.46	2.40
.99		3.89	3.80	3.67	3.52	3.37	3.21	3.05	2.96	2.87
.995		4.54	4.42	4.25	4.07	3.88	3.69	3.48	3.37	3.26
.90		1.96	1.94	1.89	1.84	1.79	1.74	1.68	1.64	1.61
.95		2.39	2.35	2.28	2.20	2.12	2.04	1.95	1.90	1.84
.975	20	2.84	2.77	2.68	2.57	2.46	2.35	2.22	2.16	2.09
.99		3.46	3.37	3.23	3.09	2.94	2.78	2.61	2.52	2.42
.995		3.96	3.85	3.68	3.50	3.32	3.12	2.92	2.81	2.69
.90		1.85	1.82	1.77	1.72	1.67	1.61	1.54	1.50	1.46
.95		2.21	2.16	2.09	2.01	1.93	1.84	1.74	1.68	1.62
.975	30	2.57	2.51	2.41	2.31	2.20	2.07	1.94	1.87	1.79
.99		3.07	2.98	2.84	2.70	2.55	2.39	2.21	2.11	2.01
.995		3.45	3.34	3.18	3.01	2.82	2.63	2.42	2.30	2.18
.90		1.74	1.71	1.66	1.60	1.54	1.48	1.40	1.35	1.29
.95		2.04	1.99	1.92	1.84	1.75	1.65	1.53	1.47	1.39
.975	60	2.33	2.27	2.17	2.06	1.94	1.82	1.67	1.58	1.48
.99		2.72	2.63	2.50	2.35	2.20	2.03	1.84	1.73	1.60
.995		3.01	2.90	2.74	2.57	2.39	2.19	1.96	1.83	1.69

Table V *(continued)*

p	ν_2 $\nu_1 \longrightarrow$ 1		2	3	4	5	6	7	8
.90	↓	2.75	2.35	2.13	1.99	1.90	1.82	1.77	1.72
.95	↓	3.92	3.07	2.68	2.45	2.29	2.18	2.09	2.02
.975	120	5.15	3.80	3.23	2.89	2.67	2.52	2.39	2.30
.99		6.85	4.79	3.95	3.48	3.17	2.96	2.79	2.66
.995		8.18	5.54	4.50	3.92	3.55	3.28	3.09	2.93
.90		2.71	2.30	2.08	1.94	1.85	1.77	1.72	1.67
.95		3.84	3.00	2.60	2.37	2.21	2.10	2.01	1.94
.975	∞	5.02	3.69	3.12	2.79	2.57	2.41	2.29	2.19
.99		6.63	4.61	3.78	3.32	3.02	2.80	2.64	2.51
.995		7.88	5.30	4.28	3.72	3.35	3.09	2.90	2.74

p	ν_2 $\nu_1 \longrightarrow$ 9		12	10	15	20	30	60	120	∞
.90	↓	1.68	1.65	1.60	1.54	1.48	1.41	1.32	1.26	1.19
.95	↓	1.96	1.91	1.83	1.75	1.66	1.55	1.43	1.35	1.25
.975	↓	2.22	2.16	2.05	1.94	1.82	1.69	1.53	1.43	1.31
.99	120	2.56	2.47	2.34	2.19	2.03	1.86	1.66	1.53	1.38
.995		2.81	2.71	2.54	2.37	2.19	1.98	1.75	1.61	1.43
.90		1.63	1.60	1.55	1.49	1.42	1.34	1.24	1.17	1.00
.95		1.88	1.83	1.75	1.67	1.57	1.46	1.32	1.22	1.00
.975	∞	2.11	2.05	1.94	1.83	1.71	1.57	1.39	1.27	1.00
.99		2.41	2.32	2.18	2.04	1.88	1.70	1.47	1.32	1.00
.995		2.62	2.52	2.36	2.19	2.00	1.79	1.53	1.36	1.00

SOURCE: Abridged from "Tables of Percentage Points of the Inverted Beta Distribution," by Maxine Merrington and Catherine M. Thompson, *Biometrika* **33** (1943), with the kind permission of the Editor of *Biometrika*.

Note: $_{(1-p)}F_{(\nu_1,\nu_2)} = \dfrac{1}{_pF_{(\nu_2,\nu_1)}}$.

Table VI Standardized Normal Density Function

z	.00	.01	.02	.03	.04	.05	.06	.07	.08	.09
.0	.3989	.3989	.3989	.3988	.3986	.3984	.3982	.3980	.3977	.3973
.1	.3970	.3965	.3961	.3956	.3951	.3945	.3939	.3932	.3925	.3918
.2	.3910	.3902	.3894	.3885	.3876	.3867	.3857	.3847	.3836	.3825
.3	.3814	.3802	.3790	.3778	.3765	.3752	.3739	.3725	.3712	.3697
.4	.3683	.3668	.3653	.3637	.3621	.3605	.3589	.3572	.3555	.3538
.5	.3521	.3503	.3485	.3467	.3448	.3429	.3410	.3391	.3372	.3352
.6	.3332	.3312	.3292	.3271	.3251	.3230	.3209	.3187	.3166	.3144
.7	.3123	.3101	.3079	.3056	.3034	.3011	.2989	.2966	.2943	.2920
.8	.2897	.2874	.2850	.2827	.2803	.2780	.2756	.2732	.2709	.2685
.9	.2661	.2637	.2613	.2589	.2565	.2541	.2516	.2492	.2468	.2444
1.0	.2420	.2396	.2371	.2347	.2323	.2299	.2275	.2251	.2227	.2203
1.1	.2179	.2155	.2131	.2107	.2083	.2059	.2036	.2012	.1989	.1965
1.2	.1942	.1919	.1895	.1872	.1849	.1826	.1804	.1781	.1758	.1736
1.3	.1714	.1691	.1669	.1647	.1626	.1604	.1582	.1561	.1539	.1518
1.4	.1497	.1476	.1456	.1435	.1415	.1394	.1374	.1354	.1334	.1315
1.5	.1295	.1276	.1257	.1238	.1219	.1200	.1182	.1163	.1145	.1127
1.6	.1109	.1092	.1074	.1057	.1040	.1023	.1006	.0989	.0973	.0957
1.7	.0940	.0925	.0909	.0893	.0878	.0863	.0848	.0833	.0818	.0804
1.8	.0790	.0775	.0761	.0748	.0734	.0721	.0707	.0694	.0681	.0669
1.9	.0656	.0644	.0632	.0620	.0608	.0596	.0584	.0573	.0562	.0551
2.0	.0540	.0529	.0519	.0508	.0498	.0488	.0478	.0468	.0459	.0449
2.1	.0440	.0431	.0422	.0413	.0404	.0396	.0387	.0379	.0371	.0363
2.2	.0355	.0347	.0339	.0332	.0325	.0317	.0310	.0303	.0297	.0290
2.3	.0283	.0277	.0270	.0264	.0258	.0252	.0246	.0241	.0235	.0229
2.4	.0224	.0219	.0213	.0208	.0203	.0988	.0194	.0189	.0184	.0180
2.5	.0175	.0171	.0167	.0163	.0158	.0154	.0151	.0147	.0143	.0139
2.6	.0136	.0132	.0129	.0126	.0122	.0119	.0116	.0113	.0110	.0107
2.7	.0104	.0101	.0099	.0096	.0093	.0091	.0088	.0086	.0084	.0081
2.8	.0079	.0077	.0075	.0073	.0071	.0069	.0067	.0065	.0063	.0061
2.9	.0060	.0058	.0056	.0055	.0053	.0051	.0050	.0048	.0047	.0046
3.0	.0044	.0043	.0042	.0040	.0039	.0038	.0037	.0036	.0035	.0034
3.1	.0033	.0032	.0031	.0030	.0029	.0028	.0027	.0026	.0025	.0025
3.2	.0024	.0023	.0022	.0022	.0021	.0020	.0020	.0019	.0018	.0018
3.3	.0017	.0017	.0016	.0016	.0015	.0015	.0014	.0014	.0013	.0013
3.4	.0012	.0012	.0012	.0011	.0011	.0010	.0010	.0010	.0009	.0009

Note: $f_{N^*}(-z) = f_{N^*}(z)$.

Table VII Ordinates of the Standardized Student Distribution at Fractiles

$f_{S^*}({}_pS^*_{(\nu)} \mid \nu)$

ν \ p	.50	.55	.60	.65	.70	.75	.80	.85	.90	.95	.975	.99	.995	.9995
1	.3183	.3106	.2881	.2526	.2083	.1592	.1100	.0656	.0304	.0078	xxx	xxx	xxx	xxx
2	.3536	.3483	.3325	.3068	.2722	.2297	.1809	.1288	.0763	.0295	.0108	.0028	xxx	xxx
3	.3676	.3630	.3495	.3272	.2964	.2574	.2113	.1589	.1024	.0454	.0192	.0059	.0024	xxx
4	.3750	.3708	.3583	.3376	.3087	.2719	.2275	.1758	.1181	.0562	.0256	.0087	.0038	xxx
5	.3796	.3757	.3638	.3441	.3165	.2808	.2375	.1865	.1282	.0638	.0303	.0109	.0049	.0003
6	.3827	.3789	.3670	.3485	.3216	.2868	.2443	.1940	.1354	.0693	.0340	.0127	.0059	.0004
7	.3850	.3813	.3702	.3514	.3253	.2913	.2494	.1993	.1408	.0735	.0368	.0141	.0067	.0005
8	.3867	.3830	.3721	.3539	.3280	.2946	.2530	.2034	.1448	.0767	.0390	.0153	.0075	.0006
9	.3880	.3844	.3737	.3556	.3302	.2970	.2561	.2065	.1480	.0794	.0409	.0163	.0080	.0007
10	.3891	.3855	.3750	.3570	.3318	.2991	.2584	.2092	.1507	.0816	.0424	.0172	.0085	.0008
12	.3907	.3873	.3768	.3593	.3345	.3023	.2618	.2131	.1547	.0851	.0448	.0185	.0093	.0009
15	.3924	.3889	.3787	.3614	.3371	.3054	.2632	.2169	.1586	.0886	.0473	.0199	.0102	.0010
20	.3940	.3906	.3806	.3637	.3397	.3084	.2691	.2210	.1629	.0918	.0499	.0213	.0111	.0012
24	.3948	.3915	.3816	.3648	.3412	.3100	.2709	.2230	.1648	.0937	.0512	.0222	.0116	.0013
30	.3956	.3923	.3825	.3660	.3424	.3115	.2726	.2250	.1670	.0956	.0526	.0231	.0121	.0013
40	.3965	.3932	.3835	.3671	.3437	.3130	.2745	.2271	.1690	.0974	.0540	.0239	.0127	.0014
60	.3973	.3941	.3845	.3682	.3451	.3145	.2763	.2290	.1712	.0992	.0555	.0248	.0133	.0015
120	.3981	.3949	.3854	.3693	.3463	.3161	.2780	.2311	.1732	.1011	.0569	.0257	.0139	.0017
∞	.3989	.3958	.3864	.3704	.3478	.3179	.2798	.2333	.1754	.1032	.0584	.0267	.0145	.0018

SOURCE: From Table IV by linear interpolation in Table *f* of Bracken and Schleifer [**13**]. The denotation xxx means that ${}_pS^*_{(\nu)}$ is not in the range of Table *f* of [**13**].

Note: $f_{S^*}({}_{(1-p)}S^*_{(\nu)} \mid \nu) = f_{S^*}({}_pS^*_\nu \mid \nu)$.

Table VIII Distribution of the Statistic Q in Subsections 22.8.1 and 22.8.2

$f(Q \mid H_0, n)$

Q \ n	2	3	4	5	6	7	8	9
0	.50000	.16667	.04167	.00833	.00139	.00020	.00002	.00000
1	.50000	.33333	.12500	.03333	.00694	.00119	.00017	.00002
2	.00000	.33333	.20833	.07500	.01944	.00397	.00067	.00010
3		.16667	.25000	.12500	.04028	.00972	.00188	.00031
4		.00000	.20833	.16667	.06806	.01944	.00432	.00079
5			.12500	.18333	.09861	.03353	.00851	.00173
6			.04167	.16667	.12500	.05139	.01493	.00339
7			.00000	.12500	.14028	.07123	.02383	.00604
8				.07500	.14028	.09028	.03509	.00994
9				.03333	.12500	.10536	.04812	.01528
10				.00833	.09861	.11369	.06183	.02213
11				.00000	.06806	.11369	.07483	.03037
12					.04028	.10536	.08557	.03967
13					.01944	.09028	.09266	.04948
14					.00694	.07123	.09514	.05911
15					.00139	.05139	.09266	.06775
16					.00000	.03353	.08557	.07460
17						.01944	.07483	.07902
18						.00972	.06183	.08054
19						.00397	.04812	.07902
20						.00119	.03509	.07460
21						.00020	.02383	.06775
22						.00000	.01493	.05911
23							.00851	.04948
24							.00432	.03967
25							.00188	.03037
26							.00067	.02213
27							.00017	.01528
28							.00002	.00994
29							.00000	.00604
30								.00339
31								.00173
32								.00079
33								.00031
34								.00010
35								.00002
36								.00000

REFERENCES

1 Anderson, T. W., *An Introduction to Multivariate Statistical Analysis*. New York: John Wiley & Sons, Inc., 1958.

2 Ando, A., and G. M. Kaufman, Bayesian Analysis of the Independent Multinormal Process—Neither Mean nor Precision Known. *Journal of the American Statistical Association*, **60** (1965), 347–358.

3 Anscombe, F. J., and R. J. Aumann, A Definition of Subjective Probability. *Annals of Mathematical Statistics*, **34** (1963), 199–205.

4 Aumann, R. J., Utility Theory Without the Completeness Axiom. *Econometrica*, **30** (1962), 445–462.

5 Berkson, J., Some Difficulties of Interpretation Encountered in the Application of the Chi-Square Test. *Journal of the American Statistical Association*, **33** (1938), 526–536.

6 Birnbaum, A., On the Foundations of Statistical Inference. *Journal of the American Statistical Association*, **57** (1962), 269–326.

7 Blackwell, D., and M. A. Girshick, *Theory of Games and Statistical Decisions*. New York: John Wiley & Sons, Inc., 1954.

8 Bortkiewicz, L., *Das Gesetz der Kleinen Zahlen*. Leipzig: Teubner, 1898.

9 Box, G. E. P., and N. R. Draper, The Bayesian Estimation of Common Parameters from Several Responses. *Biometrika*, **52** (1965), 355–365.

10 Box, G. E. P., and G. C. Tiao, A Bayesian Approach to the Importance of Assumptions Applied to the Comparison of Variances. *Biometrika*, **51** (1964), 153–167.

11 Box, G. E. P., and G. C. Tiao, Multiparameter Problems from a Bayesian Point of View. *Annals of Mathematical Statistics*, **36** (1965), 1468–1482.

12 Box, G. E. P., and G. C. Tiao, Bayesian Estimation of Means for the Random-Effects Model. *Journal of the American Statistical Association*, **63** (1968), 174–181.

13 BRACKEN, J., and A. L. SCHLEIFER, JR., *Tables for Normal Sampling with Unknown Variance*. Cambridge: Division of Research, Graduate School of Business Administration, Harvard University (1964).

14 CHERNOFF, H., and E. L. LEHMANN, The Use of Maximum Likelihood Estimates in χ^2 Tests for Goodness of Fit. *Annals of Mathematical Statistics*, **25** (1954), 579 ff.

15 CRAMÉR, H., *Mathematical Methods of Statistics*. Princeton, N.J.: Princeton University Press (1946).

16 EDWARDS, W., H. LINDMAN, and L. J. SAVAGE, Bayesian Statistical Inference for Psychological Research. *Psychological Review*, **70** (1963), 193–242.

17 ERICSON, W. A., Optimum Stratified Sampling Using Prior Information. *Journal of the American Statistical Association*, **60** (1965), 750–771.

18 ERICSON, W. A., On the Economic Choice of Experiment Sizes for Decisions Regarding Certain Linear Combinations. *Journal of the Royal Statistical Society, Series B*, **29** (1967), 503–512.

19 ERICSON, W. A., Optimal Allocation in Stratified and Multistage Samples Using Prior Information. *Journal of the American Statistical Association*, **63** (1968), 964–983.

20 EVANS, I. G., Bayesian Estimation of Parameters of a Multivariate Normal Distribution. *Journal of the Royal Statistical Society, Series B*, **27** (1965), 279–283.

21 FELLNER, W. J., *Probability and Profit*. Homewood, Ill.: Richard D. Irwin, Inc., 1965.

22 FERGUSON, T. S., *Mathematical Statistics: a Decision-Theoretic Approach*. New York: Academic Press, 1967.

23 FISHBURN, P. C., *Decision and Value Theory*. New York: John Wiley & Sons, Inc., 1964.

24 FISHBURN, P. C., Preference-Based Definitions of Subjective Probability. *Annals of Mathematical Statistics*, **38** (1967), 1605–1617.

25 FISHBURN, P. C., Bounded Expected Utility. *Annals of Mathematical Statistics*, **38** (1967), 1054–1060.

26 FISHBURN, P. C., Methods of Estimating Additive Utilities. *Management Science*, **13** (1967), 435–453.

27 FISHER, R. A., *Statistical Methods for Research Workers*. 13th ed., Edinburgh: Oliver and Boyd, Ltd., 1958.

28 FISHER, R. A., *The Design of Experiments*. 7th ed., Edinburgh: Oliver and Boyd, Ltd., 1960.

29 GEISSER, S., and J. CORNFIELD, Posterior Distribution for the Multivariate Normal Parameters. *Journal of the Royal Statistical Society, Series B*, **25** (1963), 368–376.

30 GNEDENKO, B. V., and A. N. KOLMOGOROV, *Limit Distributions for Sums of Independent Random Variables*, tr. by K. L. Chung, Reading: Addison-Wesley Publishing Company, Inc., 1954.

31 GOODMAN, L. A., The Analysis of Cross-Classified Data: Independence, Quasi-Independence, and Interactions in Contingency Tables with or without Missing Entries. *Journal of the American Statistical Association*, **63** (1968), 1091–1131.

32 GRAYBILL, F. A., *An Introduction to Linear Statistical Models, Vol. I.*, New York: McGraw-Hill, Inc., 1961.

33 HADLEY, G., *Linear Algebra.* Reading: Addison-Wesley Publishing Company, Inc., 1961.

34 HALMOS, P. R., *Measure Theory.* Princeton, N.J.: D. Van Nostrand Company, Inc., 1950.

35 HAMMERSLEY, J. H., and D. C. HANDSCOMB, *Monte Carlo Methods.* New York: John Wiley & Sons, Inc., 1964.

36 HAUSNER, M., Multidimensional Utilities, in Thrall, *et al.*, *Decision Processes.* New York: John Wiley & Sons, Inc., 1954, 167–180.

37 HILL, B. M., Inference about Variance Components in the One-way Model. *Journal of the American Statistical Association*, **60** (1965), 806–825.

38 HILL, B. M., Correlated Errors in the Random Model. *Journal of the American Statistical Association*, **62** (1967), 1387–1400.

39 HILLIER, F. S., and G. J. LIEBERMAN, *Introduction to Operations Research.* San Francisco: Holden-Day, 1967.

40 JEFFREYS, H., *Theory of Probability.* 3rd ed., New York: Clarendon Press, 1961.

41 KENDALL, M. G., and A. STUART, *The Advanced Theory of Statistics, Vols. I–III.* 2nd ed., Vol. I, London: Griffin, 1963.

42 LAVALLE, I. H., A Nonlinear Analogue of Conditional Expectation. *Tulane University Management Science Working Papers, No. 1.* (1967).

43 LAVALLE, I. H., On Information Evaluation by Teams with Nonlinear Utility. *Tulane University Management Science Working Papers, No. 3.* (1968).

44 LAVALLE, I. H., On Cash Equivalents and Information Evaluation in Decisions Under Uncertainty, Part I: Basic Theory. Part II: Incremental Information Decisions. Part III: Exchanging Partition-J for Partition-K Information. *Journal of the American Statistical Association*, **63** (1968), 252–290.

45 LEHMANN, E. L., *Testing Statistical Hypotheses.* New York: John Wiley & Sons, Inc., 1959.

46 LÉVY, P., *Calcul des Probabilités.* Paris: Gauthier-Villars, 1925.

47 LINDEBERG, J. W., Eine Neue Herleitung des Exponentialgesetzes der Wahrscheinlichkeitsrechnung. *Mathematische Zeitschrift*, **15** (1922), 221 ff.

48 LINDGREN, B. W., *Statistical Theory.* New York: Crowell-Collier and Macmillan, Inc., 1960.

49 LINDLEY, D. V., *Introduction to Probability and Statistics, Vols. 1 and 2.* Cambridge: Cambridge University Press, 1965.

50 LINDLEY, D. V., and J. C. P. MILLER, *Cambridge Elementary Statistical Tables.* Cambridge: Cambridge University Press, 1961.

51 LOÉVE, M., *Probability Theory.* 3rd ed., Princeton, N.J.: D. Van Nostrand Company, Inc., 1963.

52 LUCE, R. D., and H. RAIFFA, *Games and Decisions.* New York: John Wiley & Sons, Inc., 1957.

53 LUKACS, E., *Characteristic Functions.* London: Charles Griffin & Co., Ltd., 1960.

54 MANN, H. B., and D. R. WHITNEY, On a Test of Whether One of the Two Random Variables is Stochastically Larger than the Other. *Annals of Mathematical Statistics*, **18** (1947), 50–60.

55 Mood, A. M., and F. A. Graybill, *Introduction to the Theory of Statistics.* New York: McGraw-Hill, Inc., 1963.

56 Mosteller, F., Association and Estimation in Contingency Tables. *Journal of the American Statistical Association*, **63** (1968), 1–28.

57 Naylor, T. H., J. L. Balintfy, D. S. Burdick, and K. Chu, *Computer Simulation Techniques.* New York: John Wiley & Sons, Inc., 1966.

58 von Neumann, J., and O. Morgenstern, *Theory of Games and Economic Behavior.* 3rd ed., Princeton, N.J.: Princeton University Press, 1953.

59 Neyman, J., and E. S. Pearson, On the Problem of the Most Efficient Tests of Statistical Hypotheses. *Philosophical Transactions of the Royal Society, Series A*, **231** (1933), 289–337.

60 Owen, G., *Game Theory.* Philadelphia: W. B. Saunders Company, 1968.

61 Pfanzagl, J., A General Theory of Measurement Applications to Utility. *Naval Research Logistics Quarterly*, **6** (1959), 283–294.

62 Pratt, J. W., Risk Aversion in the Small and in the Large. *Econometrica*, **32** (1964), 122–136.

63 Pratt, J. W., Bayesian Interpretation of Standard Inference Statements. *Journal of the Royal Statistical Society, Series B*, **27** (1965), 169–203.

64 Pratt, J. W., H. Raiffa, and R. O. Schlaifer, *Introduction to Statistical Decision Theory.* Preliminary ed., New York: McGraw-Hill, Inc., 1965.

65 The RAND Corp., *A Million Random Digits with 100,000 Normal Deviates.* Glencoe, Ill.: The Free Press, 1955.

66 Raiffa, H., and R. O. Schlaifer, *Applied Statistical Decision Theory.* Boston: Division of Research, Graduate School of Business Administration, Harvard University.

67 Rao, C. R., The Utilization of Multiple Measurements in Problems of Biological Classification. *Journal of the Royal Statistical Society, Series B*, **10** (1948), 159–193.

68 Roberts, H. V., Informative Stopping Rules About Population Size. *Journal of the American Statistical Association*, **62** (1967), 763–775.

69 Savage, L. J., *The Foundations of Statistics.* New York: John Wiley & Sons, Inc., 1954.

70 Savage, L. J., *et al.*, *The Foundations of Statistical Inference: a Discussion.* New York: John Wiley & Sons, Inc., 1962.

71 Scheffé, H., *The Analysis of Variance.* New York: John Wiley & Sons, Inc., 1959.

72 Thrall, R. M., Applications of Multidimensional Utility Theory in Thrall *et al Decision Processes*, New York: John Wiley & Sons, Inc., 1954, 181–186.

73 Thrall, R. M., C. H. Coombs, and R. L. Davis, *Decision Processes.* New York: John Wiley & Sons, Inc., 1954.

74 Tiao, G. C., Bayesian Comparison of Means of a Mixed Model with Applications to Regression Analysis. *Biometrika*, **53** (1966), 11–26.

75 Tiao, G. C., and G. E. P. Box, Bayesian Analysis of a Three-Component Hierarchical Design Model. *Biometrika*, **54** (1967), 109–127.

76 Tiao, G. C., and W. Y. Tan, Bayesian Analysis of Random-Effect Models in the Analysis of Variance I. Posterior Distribution of Variance Components. *Biometrika*, **52** (1965), 37–53.

77 TIAO, G. C., and W. Y. TAN, Bayesian Analysis of Random-Effect Models in the Analysis of Variance II. Effect of Autocorrelated Errors. *Biometrika*, **53** (1966), 477–496.

78 TIAO, G. C., and A. ZELLNER, Bayes' Theorem and the Use of Prior Knowledge in Regression Analysis. *Biometrika*, **51** (1964), 219–230.

79 TIAO, G. C., and A. ZELLNER, On the Bayesian Estimation of Multivariate Regression. *Journal of the Royal Statistical Society, Series B*, **26** (1964), 277–285.

80 TOCHER, D., *The Art of Simulation*. Princeton, N.J.: D. Van Nostrand Company, Inc., 1963.

81 WALD, A., *Sequential Analysis*. New York: John Wiley & Sons, Inc., 1947.

82 WALD, A., *Statistical Decision Functions*. New York: John Wiley & Sons, Inc., 1950.

83 WILKS, S. S., *Mathematical Statistics*. New York: John Wiley & Sons, Inc., 1962.

84 WINKLER, R. V., The Assessment of Prior Distributions in Bayesian Analysis. *Journal of the American Statistical Association*, **62** (1967), 776–800.

85 WINKLER, R. V., The Quantification of Judgment: Some Methodological Suggestions. *Journal of the American Statistical Association*, **62** (1967), 1105–1120.

86 ZELLNER, A., and V. K. CHETTY, Prediction and Decision Problems in Regression Models from the Bayesian Point of View. *Journal of the American Statistical Association*, **60** (1965), 608–616.

87 ZELLNER, A., and G. C. TIAO, Bayesian Analysis of the Regression Model with Autocorrelated Errors. *Journal of the American Statistical Association*, **59** (1964), 763–778.

INDEX